Introductory and Intermediate Algebra through Applications

Second Edition

Geoffrey Akst   **Sadie Bragg**

Borough of Manhattan Community College,
City University of New York

PEARSON

Addison
Wesley

Boston • San Francisco • New York • London • Toronto
Sydney • Tokyo • Singapore • Madrid • Mexico City
Munich • Paris • Cape Town • Hong Kong • Montreal

Editorial Director	Christine Hoag
Editor in Chief	Maureen O'Connor
Production Manager	Ron Hampton
Executive Project Manager	Kari Heen
Associate Editor	Joanna Doxey
Editorial Services	Elka Block, Lenore Parens
Senior Designer	Barbara T. Atkinson
Text and Cover Design	Leslie Haimes
Composition	Pre-Press PMG
Media Producer	Ashley O'Shaughnessy
Software Development	TestGen: Mary Durnwald; MathXL: Jozef Kubit
Marketing Managers	Michelle Renda and Marlana Voerster
Marketing Coordinator	Nathaniel Koven
Senior Prepress Supervisor	Caroline Fell
Manufacturing Manager	Evelyn Beaton
Senior Media Buyer	Ginny Michaud
Cover Photo	© Peter Stroumtos/Alamy

About the Cover: Well-designed sunglasses are not just cool—they also make it easy to see the world more clearly. Akst and Bragg's distinctive side-by-side example/practice exercise format encourages students to practice as they learn, while the authors' use of motivating applications connects mathematics to real life. With clear and simple explanations that make mathematics understandable, the Akst/Bragg series brings the world of developmental mathematics into focus for students.

Photo Credits: p. 27: Photo Researchers; p. 28: Digital Vision; p. 44: Photodisc; p. 121: Shutterstock; p. 133: Shutterstock; p. 138: Reuters/Corbis; p. 152: Photodisc; p. 161: NOAA; p. 216: Beth Anderson; p. 217: our Stockbyte CD EP002 Action Sports; p. 374: Francisco Cruz/SuperStock; p. 379: Emilio Segre Visual Archives (email photos@aip.org); p. 397: Photodisc; p. 403: National Archives and Records Administration (PAL AAACSKJ0); p. 440: Duomo/Corbis; p. 442: stamp: public domain; p. 450: Robert Maass/Corbis; p. 491: Corbis; p. 492: Shutterstock; p. 513: Brand X Pictures (RF); p. 536: NASA/Kennedy Space Center; p. 539: High Plains Rocket Society, Caspar, Wyoming; p. 547: Mario Tama/Getty Images; p. 551: Photodisc; p. 555: Shutterstock; p. 563: courtesy Taylor-White, LLC; p. 591: Photodisc; p. 597: Digital Vision; p. 599: Sara Anderson; p. 605: Photodisc; p. 609: Thinkstock Royalty Free; p. 628: PhotoDisc Blue; p. 629 (b): Bettmann/Corbis; p. 641: NASA headquarters (PAL AAAJYZK0); p. 643: AP Photo/J. D. Pooley; p. 677: Digital Vision; p. 697: PhotoDisc; p. 698: PhotoDisc Vol. BS14; p. 701: iStockphoto; p. 721: Shutterstock; p. 736: public domain; p. 757: Corbis RF; p. 803: iStockphoto; p. 852: Bettmann/Corbis; p. 889: PhotoEdit (PAL_ AAAGKFL0); p. 899: PhotoDisc (PP); p. 906: AP/Wideworld

Library of Congress Cataloging-in-Publication Data

Akst, Geoffrey.

  Introductory and intermediate algebra through applications / Geoffrey Akst, Sadie Bragg.—2nd ed.

    p. cm.

  Includes index.

  ISBN-13: 978-0-321-53578-8   ISBN-10: 0-321-53578-2 (student ed. : alk. paper)

    1. Mathematics—Textbooks.   I. Bragg, Sadie   II. Title.

  QA152.3.A474 2008

  512.9—dc22   2007061241

4 5 6 7 8 9 10—CRK—12 11 10

Class Time

BEFORE CLASS

- Review your notes from the previous class session.

- Read the section(s) of your textbook that will be covered in class to get familiar with the material. Read these sections carefully. Skimming may result in your not understanding some of the material and your inability to do the homework. If you do not understand something in the text, reread it more thoroughly or seek assistance.

DURING CLASS

- Pay attention and try to understand every point your instructor makes.

- Take good notes.

- Ask questions if you do not understand something. It is best to ask questions during class. Chances are that someone else has the same question, but may not be comfortable asking it. If you feel that way also, then write your question in your notebook and ask your instructor after class or see the instructor during office hours.

AFTER CLASS

- Review your notes as soon as possible after class. Insert additional steps and comments to help clarify the material.

- Reread the section(s) of your textbook. After reading through an example, cover it up and try to do it on your own. Do the practice problem that is paired with the example. (The answers to the practice problems are given in the back of the textbook.)

Homework

The best way to learn math is by *doing* it. Homework is designed to help you learn and apply concepts and to master related skills.

TIPS FOR DOING HOMEWORK

- Do your homework the *same day* that you have class. Keeping up with the class requires you to do homework regularly rather than "cramming" right before tests.

- Review the section of the textbook that corresponds to the homework.

- Review your notes.

- If you get stuck on a problem, look for a similar example in your textbook or notes.

- Write a question mark next to any problems that you just cannot figure out. Get help from your instructor or the tutoring center or call someone from your study group.

- Check your answer after each problem. The answers to the odd-numbered problems are in the back of the textbook. If you are assigned even-numbered problems, try the odd-numbered problem first and check the answer. If your answer is correct, then you should be able to do the even-numbered problem correctly.

Form a study group. A study group provides an opportunity to discuss class material and homework problems. Find at least two other people in your class who are committed to being successful. Exchange contact information and plan to meet or work together regularly throughout the semester either in person or via e-mail or phone.

Use your book's study resources. Additional resources and support materials to help you succeed are available with this book.

INCLUDED WITH YOUR TEXT

- The resources on the **Pass the Test CD** will help you prepare for your tests. After you've taken the Posttest at the end of each chapter in the textbook, check your work by watching an instructor work through the full solutions to all of the Posttest exercises on the Pass the Test CD. You will also find vocabulary flashcards and additional study tips on this CD.

ADDITIONAL RESOURCES

The following resources are also available through your school bookstore or www.mypearsonstore.com:

- *Student's Solutions Manual*
- *Video Lectures on CD*
- *Video Lectures on DVD*
- *Worksheets for Classroom or Lab Practice*
- *MathXL® Tutorials on CD*
- *MathXL®*
- *MyMathLab®*

Full descriptions are available in the preface of the textbook.

Notebook and Note Taking

Taking good notes and keeping a neat, well-organized notebook are important factors in being successful.

YOUR NOTEBOOK

Use a loose-leaf binder divided into four sections:

1. notes
2. homework
3. graded tests and quizzes
4. handouts

TAKING NOTES

- Copy all important information written on the board. Also, write down all points that are not clear to you so that you can discuss them with your instructor, a tutor, or your study group.

- Write explanations of what you are doing in your own words next to each step of a practice problem.

- Listen carefully to what your instructor emphasizes and make note of it.

Math Study Skills

Your overall success in mastering the material that this textbook covers depends on you. The key is to be *committed* to doing your best in this course. This commitment means dedicating the time needed to study math and to do your homework.

In order to succeed in math, you must know how to study it. The goal is to study math so that you understand and not just memorize it. The following tips and strategies will help you develop good study habits that lead to understanding.

General Tips

Attend every class. Be on time. If you must miss class, be sure to get a copy of the notes and any handouts. (Get notes from someone in your class who takes good, neat notes.)

Manage your time. Work, family, and other commitments place a lot of demand on your time. To be successful in college, you must be able to devote time to study math every day. Writing out a weekly schedule that lists your class schedule, work schedule, and all other commitments with times that are not flexible will help you to determine when you can study.

Do not wait to get help. If you are having difficulty, get help immediately. Since the material presented in class usually builds on previous material, it is very easy to fall behind. Ask your instructor if he or she is available during office hours, or get help at the tutoring center on campus.

Instructor Contact Information
Name:
Office Hours:
Office Location:
Phone Number:
E-mail Address:

Campus Tutoring Center
Location:
Hours:

Tests

Tests are a source of anxiety for many students. Being well prepared to take a test can help ease anxiety.

BEFORE A TEST

- Review your notes and the sections of the textbook that will be covered on the test.

- Read through the Key Concepts and Skills at the end of the chapter in the textbook and your own summary from your notes.

- Do additional practice problems. Select problems from your homework to rework. The textbook also contains Mixed Practice and Review Exercises that provide opportunities to strengthen your skills

- Use the Posttest at the end of the chapter as your practice test. While taking the practice test, do not refer to your notes or the textbook for help. Keep track of how long it takes to complete the test. Check your answers. If you cannot complete the practice test within the time you are allotted for the real test, you need additional practice in the tutoring center to speed up.

DURING A TEST

- Read through the test before starting.

- If you find yourself panicking, relax, take a few slow breaths, and try to do some of the problems that seem easy.

- Do the problems you know how to do first, and then go back to the ones that are more difficult.

- Watch your time. Do not spend too much time on any one problem. If you get stuck while working on a problem, skip it and move on to the next problem.

- Check your work, if there is time. Correct any errors you find.

AFTER A TEST

- When you get your test back, look through all of the problems.

- On a separate sheet of paper, do any problems that you missed. Use your notes and textbook, if necessary.

- Get help from your instructor or a tutor if you cannot figure out how to do a problem, or set up a meeting with your study group to go over the test together. Make sure you understand your errors.

- Attach the corrections to the test and place it in your notebook.

Contents

Preface

From the Authors

Our goal in writing *Introductory and Intermediate Algebra through Applications,* Second Edition, was to help motivate students and to establish a strong foundation for their success in subsequent mathematics and mathematics-related courses. This text is intended for a two-term course sequence that covers elementary algebra and intermediate algebra. Alternatively, it may be used in an accelerated one-term algebra course. In addition to a review of arithmetic, our text deals with the major topics in college elementary and intermediate algebra: real numbers and algebraic expressions; solving linear equations and inequalities, both algebraically and by graphing; functions; systems of linear equations and inequalities; exponents and polynomials; factoring polynomials; rational expressions and equations; radical expressions and equations; quadratic equations, functions and inequalities; exponential and logarithmic functions; and conic sections and nonlinear systems. In contrast to other texts, graphing is not relegated to a late chapter, a recognition of the value of visualizing mathematical relationships in fostering understanding.

For all topics covered in this text, we have carefully selected interdisciplinary applications that we believe are relevant, interesting, and motivating. This thoroughly integrated emphasis on applications reflects our view that college students need to master the mathematics in a developmental mathematics program not so much for its own sake but rather to be able to apply this understanding to their everyday lives and to the demands of subsequent college courses.

Our goal throughout the text has been to address many of the issues raised by the American Mathematical Association of Two-Year Colleges and the National Council of Teachers of Mathematics by writing a flexible, approachable, and readable text that reflects the following:

- an emphasis on applications that model real-world situations
- explanations that foster conceptual understanding
- exercises with connections to other disciplines
- appropriate use of technology
- the integration of geometric visualization with concepts and applications
- exercises in student writing and groupwork that encourage interactive and collaborative learning
- a desire to foster quantitative literacy, in part through the use of real data in charts, tables, and graphs

This text is part of a series that includes the following books:

- *Fundamental Mathematics through Applications,* Fourth Edition
- *Basic Mathematics through Applications,* Fourth Edition
- *Introductory Algebra through Applications,* Second Edition
- *Intermediate Algebra through Applications,* Second Edition
- *Introductory and Intermediate Algebra through Applications,* Second Edition

The following key content and key features stem from our strong belief that mathematics is logical, useful, and fun.

Geoffrey Akst and Sadie Bragg

Key Content

Applications   One of the main reasons to study mathematics is its application to a wide range of disciplines, to a variety of occupations, and to everyday situations. Each chapter begins with a real-world application to show the usefulness of the topic under discussion and to motivate student interest. These opening applications vary widely from Cell Phones and Algebra to Rational Expressions and Grapes (see pages 121 and 555). Applications, appropriate to the section content, are highlighted in section exercise sets with an Applications heading (see pages 131, 211, and 367).

Concepts   Explanations in each section foster intuition by promoting student understanding of underlying concepts. To stress these concepts, we include exercises on reasoning and pattern recognition that encourage students to be logical in their problem-solving techniques, promote self-confidence, and allow students with varying learning styles to be successful (see pages 237, 567, and 776).

Skills   Practice is necessary to reinforce and maintain skills. In addition to comprehensive chapter problem sets, chapter review exercises include mixed applications that require students to choose among and use skills covered in that chapter (see pages 191 and 323).

Writing   Writing both demonstrates and enhances students' understanding of concepts and skills. In addition to the user-friendly worktext format, open-ended questions throughout the text give students the opportunity to explain their answers in full sentences. Students can build on these questions by keeping individual journals (see pages 124, 240, and 371).

 Use of Technology   Each student should be able to use a range of computational techniques—mental, paper-and-pencil, and calculator arithmetic—depending on the problem and the student's level of mathematical preparation. This text includes optional calculator inserts (see pages 270, 345, and 346), which provide explanations for calculator techniques. These inserts also feature paired side-by-side examples and practice exercises, as well as a variety of optional calculator exercises in the section exercise sets that use the power of calculators to perform arithmetic operations (see pages 139 and 275). Both calculator inserts and calculator exercises are indicated by a calculator icon.

Key Features

Pretests and Posttests   To promote individualized learning—particularly in a self-paced or lab environment—pretests and posttests help students gauge their level of understanding of chapter topics both at the beginning and at the end of each chapter. The pretests and posttests also allow students to target topics for which they may need to do extra work to achieve mastery. All answers to pretests and posttests are given in the answer section of the student edition. Students can watch an instructor working through the full solutions to the Posttest exercises on the Pass the Test CD.

Section Objectives   At the beginning of each section, clearly stated learning objectives help students and instructors identify and organize individual competencies covered in the upcoming content.

Side-by-Side Example/Practice Format   A distinctive side-by-side format pairs each numbered example with a corresponding practice exercise, encouraging students to get actively involved in the mathematical content from the start. Examples are immediately followed by solutions so that students can have a ready guide to follow as they work (see pages 138, 202, and 423).

Tips   Throughout the text, students find helpful suggestions for understanding concepts, skills, or rules, and advice on avoiding common mistakes (see pages 206, 409, and 673).

Cultural Notes   To show how mathematics has evolved over the centuries—in many cultures and throughout the world—each chapter features a compelling Cultural Note that investigates and illustrates the origins of mathematical concepts. Cultural Notes give students further evidence that mathematics grew out of a universal need to find efficient solutions to everyday problems. Diverse topics include the evolution of digit notation, the popularization of decimals, and the role that taxation played in the development of the percent concept (see pages 133, 216, and 379).

Mindstretchers   For every appropriate section in the text, related investigation, critical thinking, mathematical reasoning, pattern recognition, and writing exercises—along with corresponding groupwork and historical connections—are incorporated into one broad-ranged problem set called Mindstretchers. Mindstretchers target different levels and types of student understanding and can be used for enrichment, homework, or extra credit (see pages 132, 215, and 378).

For Extra Help   These boxes, found at the top of the first page of every section's exercise set, direct students to helpful resources that will aid in their study of the material.

Key Concepts and Skills   At the end of each chapter, a comprehensive chart organized by section relates the key concepts and skills to a corresponding description and example, giving students a unique tool to help them review and translate the main points of the chapter.

Chapter Review Exercises   Following the Key Concepts and Skills at the end of each chapter, a variety of relevant exercises organized by section helps students test their comprehension of the chapter content. As mentioned earlier, included in these exercises are mixed applications, which give students an opportunity to practice their reasoning skills by requiring them to choose and apply an appropriate problem-solving method from several previously presented (see pages 190, 318, and 394).

Cumulative Review Exercises   At the end of Chapter 2, and for every chapter thereafter, Cumulative Review Exercises help students maintain and build on the skills learned in previous chapters.

What's New in the Second Edition?

NEW Appendixes to Smooth the Transition to Intermediate Algebra

Appendixes A.1–A.5 help students advance from introductory to intermediate algebra. Appendix A.1 reviews topics from Chapter 1 up to Chapter 7. Appendixes A.2 and A.3 continue the discussion of solving inequalities begun in Section 2.6. Appendixes A.4 and A.5 deal with solving systems of linear equations, supplementing the development in Chapter 4.

NEW Math Study Skills Foldout   This full-color foldout, opposite the inside front cover, provides students with tips on organization, test preparation, time management, and more.

NEW Mathematically Speaking Exercises   Located at the beginning of nearly every section's exercise set, these new exercises have been added to help students understand and use standard mathematical vocabulary.

NEW Mixed Practice Exercises   Mixed Practice exercises, located in nearly every section's exercise set, reinforce the student's knowledge of topics and problem-solving skills covered in the section.

Updated Exercise Sets   This revision includes over 1,000 new exercises, including the new Mathematically Speaking and Mixed Practice exercises.

NEW Pass the Test CD   Included with every new copy of the book, the Pass the Test CD includes video footage of an instructor working through the complete solutions for all Posttest exercises for each chapter, vocabulary flashcards, interactive Spanish glossary, and additional tips on improving time management.

What Supplements Are Available?

For a complete list of the supplements and study aids that accompany *Introductory and Intermediate Algebra through Applications*, Second Edition, see pp. xi through xiii.

Acknowledgments

We are grateful to everyone who has helped to shape this textbook by responding to questionnaires, participating in telephone surveys and focus groups, reviewing the manuscript, and using the text in their classes. We wish to thank all of you, especially the following diary and manuscript reviewers who provided feedback for this revision: Sheila Anderson, *Housatonic Community College;* James J. Ball, *Indiana State University;* Mike Benningfield, *Paris Junior College;* Amanda Bertagnolli-Comstock, *Bishop State Community College;* Norma Bisulca, *University of Maine, Augusta;* Scott Brown, *Auburn University, Montgomery;* Joanne Brunner, *Joliet Junior College;* Donald M. Carr, *College of the Desert;* Alison Carter, *California State University, Howard;* Karla Childs, *Pittsburgh State University;* O. Pauline Chow, *Harrisburg Area Community College;* Edith Cook, *Suffolk University;* Susi Curl, *McCook Community College;* Karena Curtis, *Labette Community College;* Lucy Edwards, *Las Positas College;* Sam Evers, *University of Alabama, Tuscaloosa;* Gene Forster, *Southeastern Illinois College;* Naomi Gibbs, *Pitt Community College;* Tim Hagopian, *Worcester State College;* Anthony Hearn, *Community College of Philadelphia;* Barbara Heim, *Ivy Tech Community College;* Lori Holdren, *Manatee Community College;* Marlene Ignacio, *Pierce College–Puyallup;* Marilyn Jacobi, *Gateway Community College;* John Jacobs, *Massachusetts Bay Community College;* Mary Ann Klicka, *Bucks County Community College;* Sandy Lanoue, *Tulsa Community College;* Tony Masci, *Notre Dame College;* Ken Mead, *Genesee Community College;* Kimberly McHale, *Columbia College;* Debbie Moran, *Greenville Technical College;* Merrel Pepper, *Southeast Technical Institute;* Sandra Peskin, *Queensborough Community College;* Carol Phillips-Bey, *Cleveland State University;* Lynn Rickabaugh, *Aiken Technical College;* Nancy Ressler, *Oakton Community College;* Richard Sturgeon, *University of Southern Maine;* Marcia Swope, *Nova Southeastern University;* Jane Tanner, *Onondaga Community College;* Sven Trenholm, *Herkimer County Community College;* Bernadette Turner,

Lincoln University; Betty Vix Weinberger, *Delgado Community College;* Cora S. West, *Florida Community College at Jacksonville;* Jackie Wing, *Angelina College;* Mary Wolyniak, *Broome Community College*

Writing a textbook requires the contributions of many individuals. Special thanks go to Greg Tobin, our publisher at Addison-Wesley, for encouraging and supporting us throughout the entire process. We are very grateful to Lenore Parens for her constant assistance and great support, and also to Laura Wheel, our applications contributor. We thank Elka Block for her contributions and professional guidance. We also thank Kari Heen for her patience and tact, Michelle Renda and Maureen O'Connor for keeping us abreast of market trends, Joanna Doxey for attending to the endless details connected with the project, Ron Hampton and Laura Houston for their support throughout the production process, Leslie Haimes for the text and cover design, and the entire Addison-Wesley developmental mathematics team for helping to make this text one of which we are very proud.

Geoffrey Akst   Sadie Bragg
gakst@nyc.rr.com   sbragg@bmcc.cuny.edu

For Alvin Bragg Sr. and family

Student Supplements

Student's Solutions Manual

By Deana Richmond

- Provides detailed solutions to the odd-numbered exercises in each exercise set and solutions to all chapter pretests and posttests, practice exercises, review exercises, and cumulative review exercises

ISBN-10: 0-321-55670-4   ISBN-13: 978-0-321-55670-7

New Video Lectures on CD or DVD

- Complete set of digitized videos on CD-ROM or DVD for students to use at home or on campus
- Includes a full lecture for each section of the text
- Optional captioning in English is available

CD: ISBN-10: 0-321-55646-1 and
    ISBN-13: 978-0-321-55646-2

DVD: ISBN-10: 0-321-55755-7 and
    ISBN-13: 978-0-321-55755-1

Instructor Supplements

Annotated Instructor's Edition

- Provides answers to all text exercises in color next to the corresponding problems
- Includes teaching tips

ISBN-10: 0-321-53644-4   ISBN-13: 978-0-321-53644-0

Instructor's Solutions Manual

By Deana Richmond

- Provides complete solutions to even-numbered section exercises
- Contains answers to all Mindstretcher problems

ISBN-10: 0-321-54580-X   ISBN-13: 978-0-321-54580-0

Instructor and Adjunct Support Manual

- Includes resources designed to help both new and adjunct faculty with course preparation and classroom management, including sample syllabi, tips for using supplements and technology, and useful external resources
- Offers helpful teaching tips correlated to the sections of the text

ISBN-10: 0-321-55827-8   ISBN-13: 978-0-321-55827-5

Student Supplements (*continued*)

New Pass the Test CD

Automatically included with the book, this CD-ROM contains

- Video footage of an instructor working through the complete solutions for all Posttest problems
- Vocabulary flashcards
- An interactive Spanish glossary
- A short video offering tips on time management

New Worksheets for Classroom or Lab Practice

By Carrie Green

- Provides one worksheet for each section of the text, organized by section objective
- Each worksheet lists the associated objectives from the text, provides fill-in-the-blank vocabulary practice, and exercises for each objective

ISBN-10: 0-321-55667-4   ISBN-13: 978-0-321-55667-7

MathXL® Tutorials on CD

- Provides algorithmically generated practice exercises that correlate at the objective level to the exercises in the textbook
- Includes an example and a guided solution for every exercise; selected exercises also include a video clip
- Provides helpful feedback for incorrect answers and generates printed summaries of students' progress

ISBN-10: 0-321-55826-X   ISBN-13: 978-0-321-55826-8

Pearson Tutor Center

- Staffed by qualified mathematics instructors
- Provides tutoring on examples and odd-numbered exercises from the textbook
- Accessible via toll-free telephone, toll-free fax, e-mail, or the Internet

www.pearsontutorservices.com

Instructor Supplements (*continued*)

Printed Test Bank

By Kay Haralson, *Austin Peay State University*
Loretta Griffy, *Austin Peay State University* and Nancy Matthews, *Montgomery Central Middle School*

- Contains three free-response and one multiple-choice test forms per chapter, and two final exams

ISBN-10: 0-321-54581-8   ISBN-13: 978-0-321-54581-7

PowerPoint® Lecture Slides (available online)

- Present key concepts and definitions from the text

New Active Learning Questions (available online)

- Prepared in PowerPoint for use with classroom response systems
- Provide several multiple-choice questions for each section of the book, allowing instructors to quickly assess mastery of material in class
- Available for download from within MyMathLab® or from Pearson Education's online catalog

TestGen® (available online)

- *New* Now includes a premade test for each chapter that has been correlated problem by problem to the chapter tests in the book
- Enables instructors to build, edit, print, and administer tests using a computerized bank of questions developed to cover all text objectives
- Algorithmically based, TestGen allows instructors to create multiple but equivalent versions of the same question or test with the click of a button
- Instructors can also modify test bank questions or add new questions
- Tests can be printed or administered online
- Software and test bank are available for download from Pearson Education's online catalog

Student Supplements *(continued)*

InterAct Math Tutorial Website
www.interactmath.com

- Get practice and tutorial help online
- Provides algorithmically generated practice exercises that correlate directly to the textbook exercises
- Retry an exercise as many times as desired with new values each time for unlimited practice and mastery
- Every exercise is accompanied by an interactive guided solution that gives the student helpful feedback when an incorrect answer is entered
- View the steps of a worked-out sample problem similar to the one that has been worked on

Instructor Supplements *(continued)*

Pearson Adjunct Support Center

The Adjunct Support Center is staffed by qualified mathematics instructors with over 50 years' combined experience at both the community college and university level. Assistance is provided for faculty in the following areas:

- Suggested syllabus consultation
- Tips on using materials packaged with the text
- Book-specific content assistance
- Teaching suggestions including advice on classroom strategies

For more information visit

www.pearsontutorservices.com: Additional Instructor Service and Support

Available for Students and Instructors

MathXL®   MathXL® is a powerful online homework, tutorial, and assessment system that accompanies Pearson Education's textbooks in mathematics and statistics. With MathXL, instructors can create, edit, and assign online homework and tests using algorithmically generated exercises correlated at the objective level to the textbook. They can also create and assign their own online exercises and import TestGen tests for added flexibility. All student work is tracked in MathXL's online gradebook. Students can take chapter tests in MathXL and receive personalized study plans based on their test results. The study plan diagnoses weaknesses and links students directly to tutorial exercises for the objectives they need to study and retest. Students can also access supplemental animations and video clips directly from selected exercises. MathXL is available to qualified adopters. For more information, visit our website at www.mathxl.com or contact your sales representative.

MyMathLab®   MyMathLab is a series of text-specific, easily customizable online courses for Pearson Education's textbooks in mathematics and statistics. Powered by CourseCompass™ (our online teaching and learning environment) and MathXL® (our online homework, tutorial, and assessment system), MyMathLab gives you the tools you need to deliver all or a portion of your course online, whether your students are in a lab setting or working from home. MyMathLab provides a rich and flexible set of course materials, featuring free-response exercises that are algorithmically generated for unlimited practice and mastery. Students can also use online tools, such as video lectures, animations, and a multimedia textbook, to independently improve their understanding and performance. Instructors can use MyMathLab's homework and test managers to select and assign online exercises correlated directly to the textbook, and they can also create and assign their own online exercises and import TestGen tests for added flexibility. MyMathLab's online gradebook—designed specifically for mathematics and statistics—automatically tracks students' homework and test results and gives the instructor control over how to calculate final grades. Instructors can also add offline (paper-and-pencil) grades to the gradebook. MyMathLab also includes access to the Pearson Tutor Center, which provides students with tutoring via toll-free phone, fax, e-mail, and interactive Web sessions. MyMathLab is available to qualified adopters. For more information, visit our website at www.mymathlab.com or contact your sales representative.

Feature Walk-Through

Cell Phones and Algebra

Today's generation takes wireless communication for granted. The cellular telephone, one of the more important wireless devices, was first introduced to the public in the mid-1980s. Since then, cell phones have shrunk both in size and in price.

Using a cell phone generally entails choosing a calling plan with a monthly service charge. As a rule, this charge consists of two parts: a specified flat monthly fee and a per-minute usage charge varying with the number of minutes of airtime used.

The choice among calling plans can be confusing, especially with the lure of promotional incentives. One nationally advertised calling plan requires a monthly charge of $34.99 for 300 minutes of free local calls and 79 cents per minute for additional local calls. A competing plan is for $29.99 a month for 200 minutes of free local calls plus 49 cents a minute for additional local calls. Under what circumstances is one deal better than the other? The key to answering this question is solving inequalities such as

$$34.99 + 0.79(x - 300) < 29.99 + 0.49(x - 200)$$

Chapter Openers The focus of this textbook is applications, and you will find them everywhere. Each chapter opener introduces students to the material that lies ahead through an interesting real-life application that grabs students' attention and helps them understand the relevance of topics they are learning.

Chapter 2 PRETEST

To see if you have already mastered the topics in this chapter, take this test.

1. Is 4 a solution of the equation $7 - 2x = 3x - 11$?

2. Solve and check: $n + 2 = -6$

3. Solve and check: $\dfrac{y}{-5} = 1$

4. Solve and check: $-n = 8$

5. Solve and check: $\dfrac{2}{3}x - 3 = -9$

6. Solve and check: $4x - 8 = -10$

7. Solve and check: $6 - y = -5$

8. Solve and check: $9x + 13 = 7x + 19$

9. Solve and check: $-2(3n - 1) = -7n$

10. Solve and check: $14x - (8x - 13) = 12x + 3$

11. Solve $v - 5u = w$ for v in terms of u and w.

12. 9 is what percent of 36?

13. 60% of what number is 12?

14. Draw the graph of $x \le 2$.

$$\begin{array}{c} \hline -3\ -2\ -1\ \ 0\ \ 1\ \ 2\ \ 3 \end{array}$$

15. Solve and graph: $x + 3 > 3$

$$\begin{array}{c} \hline -3\ -2\ -1\ \ 0\ \ 1\ \ 2\ \ 3 \end{array}$$

16. An office photocopier makes 30 copies per minute. How long will it take to copy a 360-page document?

17. A florist charges a flat fee of $100 plus $70 for each centerpiece for a wedding. If the total bill for the centerpieces was $1500, how many centerpieces did the florist make?

18. The formula for finding the amount of kinetic energy used is $E = \dfrac{1}{2}mv^2$. Solve this equation for m in terms of E and v.

19. A certain amount of money was invested in two different accounts. The amount invested at 8% simple interest was twice the amount invested at 5% simple interest. If the total interest earned on the investments was $420, how much was invested at each rate?

20. A gym offers two membership options: Option A is $55 per month for unlimited use of the gym and Option B is $10 per month plus $3 for each hour a member uses the gym. For how many hours of use per month will Option A be a better deal?

Pretests Pretests, found at the beginning of each chapter, help students gauge their understanding of the chapter ahead. Answers can be found in the back of the book.

EXAMPLE 1

Is 2 a solution of the equation $3x + 1 = 11 - 2x$?

Solution

$$3x + 1 = 11 - 2x$$
$$3(2) + 1 \stackrel{?}{=} 11 - 2(2) \qquad \text{Substitute 2 for } x.$$
$$6 + 1 \stackrel{?}{=} 11 - 4 \qquad \text{Evaluate each side of the equation.}$$
$$7 = 7 \qquad \text{True}$$

Since $7 = 7$ is a true statement, 2 is a solution of $3x + 1 = 11 - 2x$.

PRACTICE 1

Determine whether 4 is a solution of the equation $5x - 4 = 2x + 5$.

EXAMPLE 2

Determine whether -1 is a solution of the equation $2x - 4 = 6(x + 2)$.

Solution

$$2x - 4 = 6(x + 2)$$
$$2(-1) - 4 \stackrel{?}{=} 6(-1 + 2) \qquad \text{Substitute } -1 \text{ for } x.$$
$$-2 - 4 \stackrel{?}{=} 6(1) \qquad \text{Evaluate each side of the equation.}$$
$$-6 = 6 \qquad \text{False}$$

Since $-6 = 6$ is a false statement, -1 is *not* a solution of the equation.

PRACTICE 2

Is -8 a solution of the equation $5(x + 3) = 3x - 1$?

Side-by-Side Format This format pairs examples and their step-by-step solutions side by side with corresponding practice exercises, encouraging active learning from the start. Students use this format for solving skill exercises, application problems, and technology exercises throughout the text.

New **Math Study Skills Foldout** This insert, found at the very front of the book, provides students with tips on organization, test preparation, time management, and more.

Math Study Skills

Your overall success in mastering the material that this textbook covers depends on you. The key is to be *committed* to doing your best in this course. This commitment means dedicating the time needed to study math and to do your homework.

In order to succeed in math, you must know how to study it. The goal is to study math so that you understand and not just memorize it. The following tips and strategies will help you develop good study habits that lead to understanding.

General Tips

Attend every class. Be on time. If you must miss class, be sure to get a copy of the notes and any handouts. (Get notes from someone in your class who takes good, neat notes.)

Manage your time. Work, family, and other commitments place a lot of demand on your time. To be successful in college, you must be able to devote time to study math every day. Writing out a weekly schedule that lists your class schedule, work schedule, and all other commitments with times that are not flexible will help you to determine when you can study.

Do not wait to get help. If you are having difficulty, get help immediately. Since the material presented in class usually builds on previous material, it is very easy to fall behind. Ask your instructor if he or she is available during office hours, or get help at the tutoring center on campus.

Instructor Contact Information

Name:

Office Hours:

Office Location:

Phone Number:

E-mail Address:

Campus Tutoring Center

Location:

Hours:

Form a study group. A study group provides an opportunity to discuss class material and homework problems. Find at least two other people in your class who are committed to being successful. Exchange contact information and plan to meet or work together regularly throughout the semester either in person or via e-mail or phone.

Use your book's study resources. Additional resources and support materials to help you succeed are available with this book.

INCLUDED WITH YOUR TEXT

- The resources on the **Pass the Test CD** will help you prepare for your tests. After you've taken the Posttest at the end of each chapter in the textbook, check your work by watching an instructor work through the full solutions to all of the Posttest exercises on the Pass the Test CD. You will also find vocabulary flashcards and additional study tips on this CD.

ADDITIONAL RESOURCES

The following resources are also available through your school bookstore or www.mypearsonstore.com:

- *Student's Solutions Manual*
- *Video Lectures on CD*
- *Video Lectures on DVD*
- *Worksheets for Classroom or Lab Practice*
- *MathXL® Tutorials on CD*
- *MathXL®*
- *MyMathLab®*

Full descriptions are available in the preface of the textbook.

Notebook and Note Taking

Taking good notes and keeping a neat, well-organized notebook are important factors in being successful.

YOUR NOTEBOOK

Use a loose-leaf binder divided into four sections:

1. notes
2. homework
3. graded tests and quizzes
4. handouts

TAKING NOTES

- Copy all important information written on the board. Also, write down all points that are not clear to you so that you can discuss them with your instructor, a tutor, or your study group.

- Write explanations of what you are doing in your own words next to each step of a practice problem.

- Listen carefully to what your instructor emphasizes and make note of it.

Teaching Tips These tips, found only in the Annotated Instructor's Edition, help instructors with explanations, reminders of previously covered material, and tips on encouraging students to write in a journal.

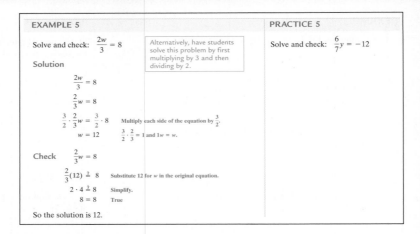

EXAMPLE 5

Solve and check: $\dfrac{2w}{3} = 8$

> Alternatively, have students solve this problem by first multiplying by 3 and then dividing by 2.

Solution

$$\frac{2w}{3} = 8$$

$$\frac{2}{3}w = 8$$

$$\frac{3}{2} \cdot \frac{2}{3}w = \frac{3}{2} \cdot 8 \qquad \text{Multiply each side of the equation by } \tfrac{3}{2}.$$

$$w = 12 \qquad\qquad \tfrac{3}{2} \cdot \tfrac{2}{3} = 1 \text{ and } 1w = w.$$

Check $\dfrac{2}{3}w = 8$

$$\frac{2}{3}(12) \stackrel{?}{=} 8 \qquad \text{Substitute 12 for } w \text{ in the original equation.}$$

$$2 \cdot 4 \stackrel{?}{=} 8 \qquad \text{Simplify.}$$

$$8 = 8 \qquad \text{True}$$

So the solution is 12.

PRACTICE 5

Solve and check: $\dfrac{6}{7}y = -12$

Tip When the same number is added to or subtracted from each side of an inequality, the *direction* of the inequality is unchanged.

The way we solve inequalities is similar to the way we solve equations.

Equation	Inequality
$x + 3 = 5$	$x + 3 < 5$
$x + 3 - 3 = 5 - 3$	$x + 3 - 3 < 5 - 3$
$x = 2$ Solution	$x < 2$ Solution
$x + 3 = 5$ is equivalent to $x = 2$	$x + 3 < 5$ is equivalent to $x < 2$.

We solve inequalities by expressing them as equivalent inequalities in which the variable term is isolated on one side.

Student Tips These insightful tips help students avoid common errors and provide other helpful suggestions to foster understanding of the material at hand.

Cultural Notes Necessity is the mother of invention, and mathematics was created out of a need to solve problems in everyday life. Cultural Notes investigate the origins of mathematical concepts, discussing and illustrating the evolution of mathematics throughout the world.

CULTURAL NOTE

Just as we solve an equation to identify an unknown number, we use a balance scale to determine an unknown weight. In this picture, an image from 3400 years ago shows an Egyptian weighing gold rings against a counterbalance in the form of a bull's head.

The balance scale is an ancient measuring device. These scales were used by Sumerians for weighing precious metals and gems at least 9000 years ago.

Source: O. A. W. Dilke, *Mathematics and Measurement* (Berkeley: University of California Press/British Museum, 1987).

 Calculators and Scientific Notation

Calculators vary as to how numbers are displayed or entered in scientific notation.

Display

In order to avoid an overflow error, many calculator models change to scientific notation an answer that is either too small or too large to fit into the calculator's display. Calculators generally use the base 10 without displaying it. Some calculators display scientific notation with either an E or e, and others show a space. For example, 3.1E−4 or 3.1■−4 can represent 3.1×10^{-4}. What other differences do you see between written scientific notation and displayed scientific notation?

EXAMPLE 13

Multiply 1,000,000,000 by 2,000,000,000.

Solution

Press	Display
1000000000 $\boxed{\times}$ 2000000000 $\boxed{\text{ENTER}}$	$1000000000 * 2000000000$ $2 \varepsilon 18$

Does your calculator display the product in scientific notation, that is, as 2E18, 2e18, or 2.■18?

Technology Inserts In order to familiarize students with a range of computational methods—mental, paper and pencil, and calculator arithmetic—the authors include optional technology inserts that instruct students on how to use the calculator to perform arithmetic or graphing operations. The side-by-side format is also used here to provide consistency across the text.

Exercise Variety   The Akst/Bragg texts provide instructors with a variety of exercise types.

MINDSTRETCHERS

Mathematical Reasoning

1. Give an example of an equation that involves combining like terms and that has

   a. no solution.

   b. an infinite number of solutions.

Patterns

2. Tables can be useful in solving equations.

   a. Complete the following table. After examining your results, identify the solution to the following equation: $3(x - 2) = 2(x + 1)$.

x	0	2	4	6	8	10	12
$3(x - 2)$							
$2(x + 1)$							

   b. Try a similar approach to solving the equation $7x = 5x + 11$. What conclusion can you draw about the solution?

x	0	2	4	6	8	10	12
$7x$							
$5x + 11$							

Groupwork

3. Working with a partner, choose a month on a calendar.

   a. Ask your partner to secretly select four days of the month that form a square, but only to tell you the sum of the four days. Determine the four days.

   b. Reverse roles with your partner and repeat part (a).

   c. Compare how you and your partner responded to part (a).

Mindstretchers   Found at the end of most sections, Mindstretchers are engaging activities that incorporate investigation, critical thinking, reasoning, pattern recognition, and writing exercises along with corresponding group work and historical connections in one comprehensive problem set. These problem sets target different levels and types of student understanding.

Definition

The **slope** m of a line passing through the points (x_1, y_1) and (x_2, y_2) is defined to be

$$m = \frac{y_2 - y_1}{x_2 - x_1}, \quad \text{where } x_1 \neq x_2.$$

Have students write their responses in a journal.

Can you explain why in the definition of slope, x_1 and x_2 must not be equal?

Use Example 1 to illustrate this concept.

Note that when using the formula for slope, it does not matter which point is chosen for (x_1, y_1) and which point for (x_2, y_2) as long as the order of subtraction of the coordinates is the same in both the numerator and denominator.

EXAMPLE 1

Find the slope of the line that passes through the points $(2, 1)$ and $(4, 2)$. Plot the points and then sketch the line.

Solution Let $(2, 1)$ stand for (x_1, y_1) and $(4, 2)$ for (x_2, y_2).

$$\underset{x_1 \; y_1}{(2, 1)} \qquad \underset{x_2 \; y_2}{(4, 2)}$$

Substituting into the formula for slope, we get:

$$m = \frac{y_2 - y_1}{x_2 - x_1} = \frac{2 - 1}{4 - 2} = \frac{1}{2}$$

Now, let's plot $(2, 1)$ and $(4, 2)$ and then sketch the line passing through them.

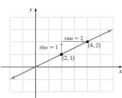

PRACTICE 1 ⊙

Find the slope of a line that contains the points $(1, 2)$ and $(4, 3)$. Plot the points and then sketch the line.

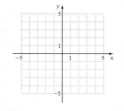

Writing Exercises   Students will understand a concept better if they have to explain it in their own words. Journal assignments (provided as instructor's edition teaching tips) allow students to work on their mathematical communication skills, thus improving their understanding of concepts.

Video Lectures on CD   The ⊙ icon indicates examples, practice exercises, and exercises that are covered in the Video Lectures on CD.

Calculator Exercises These optional exercises, denoted with a ⊞ icon, can be found in the exercise sets, giving students the opportunity to use a calculator to solve a variety of real-life applications.

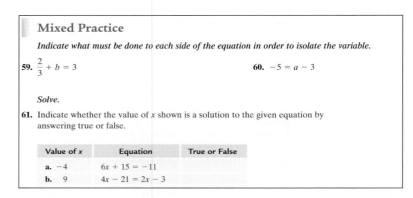

Solve. Round to the nearest hundredth.

⊞ **37.** $\dfrac{x}{-1.515} = 1.515$ ⊞ **38.** $\dfrac{n}{-2.968} = -3.85$ ⊞ **39.** $-3.14x = 21.4148$

Translate each sentence to an equation. Then solve and check.

41. The product of -4 and a number is 56.   **42.** $\dfrac{3}{4}$ of a number is eq

For Extra Help These boxes, found at the top of the first page of every section's exercise set, direct students to helpful resources that will aid in their study of the material.

New **Mathematically Speaking Exercises** Located at the begining of nearly every exercise section of the text. Mathematically Speaking exercises help students understand and master mathematical vocabulary.

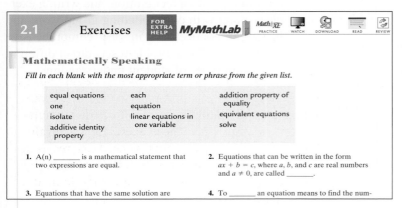

2.1 Exercises   FOR EXTRA HELP **MyMathLab** Math XL PRACTICE WATCH DOWNLOAD READ REVIEW

Mathematically Speaking

Fill in each blank with the most appropriate term or phrase from the given list.

equal equations	each	addition property of equality
one	equation	
isolate	linear equations in one variable	equivalent equations
additive identity property		solve

1. A(n) _____ is a mathematical statement that two expressions are equal.

2. Equations that can be written in the form $ax + b = c$, where a, b, and c are real numbers and $a \ne 0$, are called _____.

3. Equations that have the same solution are

4. To _____ an equation means to find the num-

New **Mixed Practice Exercises** Mixed Practice exercises reinforce skills within the section and encourage review.

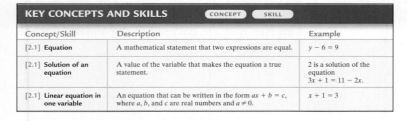

Mixed Practice

Indicate what must be done to each side of the equation in order to isolate the variable.

59. $\dfrac{2}{3} + b = 3$ **60.** $-5 = a - 3$

Solve.

61. Indicate whether the value of x shown is a solution to the given equation by answering true or false.

Value of x	Equation	True or False
a. -4	$6x + 15 = -11$	
b. 9	$4x - 21 = 2x - 3$	

End-of-Chapter Material At the end of each chapter, students will find a wealth of review- and retention-oriented material to reinforce the concepts presented in current and previous chapters.

KEY CONCEPTS AND SKILLS   CONCEPT   SKILL

Concept/Skill	Description	Example
[2.1] **Equation**	A mathematical statement that two expressions are equal.	$y - 6 = 9$
[2.1] **Solution of an equation**	A value of the variable that makes the equation a true statement.	2 is a solution of the equation $3x + 1 = 11 - 2x$.
[2.1] **Linear equation in one variable**	An equation that can be written in the form $ax + b = c$, where a, b, and c are real numbers and $a \ne 0$.	$x + 1 = 3$

Key Concepts and Skills These give students quick reminders of the chapter's most important elements and provide a one-stop quick review of the chapter material. Each concept/skill is keyed to the section in which it was introduced, and students are given a brief description and example for each.

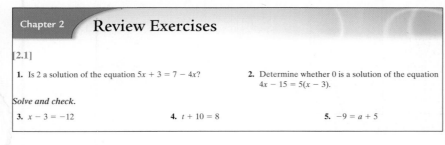

Chapter 2 Review Exercises

[2.1]

1. Is 2 a solution of the equation $5x + 3 = 7 - 4x$?

2. Determine whether 0 is a solution of the equation $4x - 15 = 5(x - 3)$.

Solve and check.

3. $x - 3 = -12$ **4.** $t + 10 = 8$ **5.** $-9 = a + 5$

Chapter Review Exercises These exercises are keyed to the corresponding sections for easy student reference. Numerous mixed application problems complete each of these exercise sets, reinforcing the applicability of what students are learning.

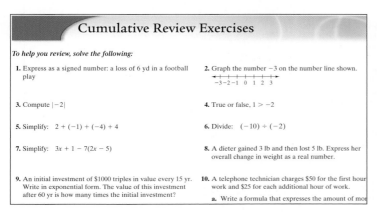

Chapter 2   POSTTEST   FOR EXTRA HELP   Pass the Test

To see if you have mastered the topics in this chapter, take this test.

1. Determine whether -2 is a solution of the equation $3x - 4 = 6(x + 2)$.

2. Solve and check: $x - 1 = -$

3. Solve and check: $\dfrac{n}{2} = -3$

4. Solve and check: $-y = -11$

5. Solve and check: $\dfrac{3y}{4} = 6$

6. Solve and check: $2x + 5 = 1$

7. Solve and check: $-s + 4 = 2$

8. Solve and check: $10x + 1 =$

9. Solve and check: $16a = -4(a - 5)$

10. Solve and check: $2(x + 5) -$

Chapter Posttest   Just as every chapter begins with a Pretest to test student understanding *before* attempting the material, every chapter ends with a Posttest to measure student understanding *after* completing the chapter material. Answers to these tests are provided in the back of the book. Students can watch an instructor working through the full solutions to the Posttest exercises on the Pass the Test CD.

Cumulative Review Exercises

To help you review, solve the following:

1. Express as a signed number: a loss of 6 yd in a football play

2. Graph the number -3 on the number line shown.
   $-3\ -2\ -1\ \ 0\ \ 1\ \ 2\ \ 3$

3. Compute $|-2|$

4. True or false, $1 > -2$

5. Simplify: $2 + (-1) + (-4) + 4$

6. Divide: $(-10) \div (-2)$

7. Simplify: $3x + 1 - 7(2x - 5)$

8. A dieter gained 3 lb and then lost 5 lb. Express her overall change in weight as a real number.

9. An initial investment of \$1000 triples in value every 15 yr. Write in exponential form. The value of this investment after 60 yr is how many times the initial investment?

10. A telephone technician charges \$50 for the first hour work and \$25 for each additional hour of work.

   a. Write a formula that expresses the amount of mor

Cumulative Review Exercises   Beginning at the end of Chapter 2, students have the opportunity to maintain their skills by completing the Cumulative Review Exercises. These exercises are invaluable, especially when students need to recall a previously learned concept or skill before beginning the next chapter, or when studying for midterm and final examinations.

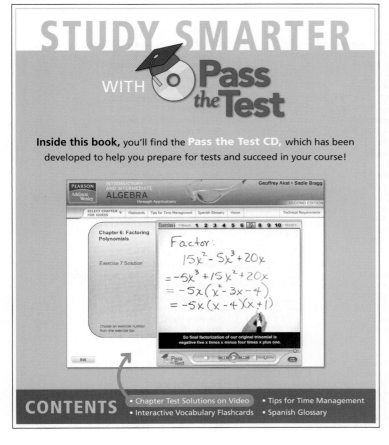

STUDY SMARTER
WITH **Pass the Test**

Inside this book, you'll find the **Pass the Test CD**, which has been developed to help you prepare for tests and succeed in your course!

CONTENTS   • Chapter Test Solutions on Video   • Tips for Time Management
• Interactive Vocabulary Flashcards   • Spanish Glossary

New **Pass the Test CD**   Included with every new copy of the book, the Pass the Test CD includes the following resources:

• video footage of an instructor working through the complete solutions for all Posttest exercises for each chapter

• vocabulary flashcards

• a video that offers an interactive Spanish glossary

• additional tips on improving time management

Index of Applications

CHAPTER R

Prealgebra Review

Why a Review of Prealgebra Topics Is Important

This chapter is a brief review of the major procedures, concepts, and vocabulary in basic mathematics that you will need to succeed in working through this text. Taking the pretest, which is given on page 2, will help you to pinpoint topics that you need to review. For these topics, study the explanations and worked-out examples and then solve the related practice problems and exercises until you have mastered the material.

OBJECTIVES
■ To review exponents and order of operations
■ To review factors, primes, and least common multiples
■ To review fractions
■ To review decimals
■ To review percents
■ To review applications of basic mathematics

Knowledge of basic mathematics is important to the study of algebra not only for calculating the value of arithmetic expressions but also for understanding algebraic procedures that are based on arithmetic procedures. For instance, familiarity with adding numerical fractions puts you one step ahead when learning how to add algebraic fractions.

In this chapter, we assume that you are proficient in whole-number arithmetic. Accordingly, we begin with a discussion of exponents and order of operations. Then we consider factors, primes, and least common multiples. These topics are needed for working with fractions—the focus of the following section. Next comes a brief review of decimals and finally percent conversions.

If you find that you need to delve more deeply into any particular arithmetic topic, take the initiative of exploring what resources—books, print material, tutoring, computer software, or videos—are available at your college to support your personal arithmetic review. You will benefit from doing this sooner rather than later.

To see if you have already mastered the topics in this chapter, take this test.

1. Rewrite $7 \cdot 7 \cdot 7$ as a power of 7.

2. Compute: $2^4 \cdot 5^2$

3. Simplify: $16 - 2^3$

4. Find the value of $20 - 2(3 - 1^2)$.

5. What are the factors of 12?

6. Rewrite 20 as a product of prime factors.

7. Write $\dfrac{16}{24}$ in simplest form.

8. Find the sum: $\dfrac{3}{10} + \dfrac{1}{10}$

9. Subtract: $10\dfrac{1}{3} - 2\dfrac{5}{6}$

10. Find the product of $\dfrac{2}{9}$ and $\dfrac{2}{3}$.

11. Calculate: $4 \div 1\dfrac{1}{2}$

12. Write 9.013 in words.

13. Round 3.072 to the nearest tenth.

14. Find the sum: $7 + 4.01 + 9.3003$

15. Find the difference: $8 - 2.34$

16. What is the product of 9.23 and 4.1?

17. Evaluate: 0.235×100

18. What is 0.045 divided by 0.25?

19. Compute: $\dfrac{3.1}{1000}$

20. What is the decimal equivalent of 7%?

21. A university launched a campaign to collect $3 million to build a new dormitory. If $1.316 million has been collected so far, how much more money, to the nearest million dollars, is needed?

22. Yesterday, an athlete swam $\dfrac{2}{3}$ of a mile. This morning, she swam $\dfrac{1}{2}$ that distance. How far did she swim this morning?

23. A doctor increases a patient's daily dosage of thyroxin from 0.05 mg to 0.1 mg. The new dosage is how many times as great as the previous dosage?

24. Snow weighs 0.1 as much as water. Express this decimal as a percent.

25. About 95% of all animal species on Earth are insects. Express this percent as a decimal.

● *Check your answers on page A-1.*

R.1   Exponents and Order of Operations

Exponents

There are many mathematical situations in which we multiply a number by itself repeatedly. Writing such expressions using *exponents* (or *powers*) provides a short-hand method for representing this repeated multiplication of the same factor:

$$\underbrace{2 \cdot 2 \cdot 2 \cdot 2}_{4 \, factors \, of \, 2} = 2^{\overset{\displaystyle\longleftarrow \text{exponent}}{4}}$$
                    base

The expression 2^4 is read "2 to the fourth power" or simply "2 to the fourth."

To evaluate 2^4, we multiply 4 factors of 2:

$$2^4 = \underbrace{2 \cdot 2}_{} \cdot 2 \cdot 2$$
$$= \underbrace{4 \cdot 2}_{} \cdot 2$$
$$= \underbrace{8 \cdot 2}_{}$$
$$= 16$$

So $2^4 = 16$.

Sometimes we prefer to shorten expressions by using exponents. For instance,

$$\underbrace{3 \cdot 3}_{2 \, factors \, of \, 3} \cdot \underbrace{4 \cdot 4 \cdot 4}_{3 \, factors \, of \, 4} = 3^2 \cdot 4^3$$

EXAMPLE 1	**PRACTICE 1**
Rewrite $6 \cdot 6 \cdot 6$ using exponents.	Write $2 \cdot 2 \cdot 2 \cdot 2 \cdot 2$ as a power of 2.
Solution $\underbrace{6 \cdot 6 \cdot 6}_{3 \, factors \, of \, 6} = 6^3$	

EXAMPLE 2	**PRACTICE 2**
Calculate: $4^3 \cdot 5^3$	Compute: $7^2 \cdot 2^4$
Solution $4^3 \cdot 5^3 = (4 \cdot 4 \cdot 4) \cdot (5 \cdot 5 \cdot 5)$ $= 64 \cdot 125$ $= 8000$	

It is especially easy to compute powers of 10:

$$10^2 = \underbrace{10 \cdot 10}_{2 \, factors} = \underbrace{100}_{2 \, zeros}$$

$$10^3 = \underbrace{10 \cdot 10 \cdot 10}_{3 \, factors} = \underbrace{1000}_{3 \, zeros}$$

and so on.

EXAMPLE 3

The distance from the Sun to the star Alpha-one Crucis is about 1,000,000,000,000,000 miles. Express this number as a power of 10.

Solution $\underbrace{1,000,000,000,000,000}_{15\ zeros} = 10^{15}$ (miles)

PRACTICE 3

In 1850, the world population was approximately 1,000,000,000. Represent this number as a power of 10.
(*Source: World Almanac and Book of Facts 2000*)

Order of Operations

Some mathematical expressions involve more than one mathematical operation. For instance, consider the expression $5 + 3 \cdot 2$. This expression seems to have two different values, depending on the order in which we perform the given operations.

Add first
$$\underbrace{5 + 3} \cdot 2$$
$$= \underbrace{8 \cdot 2}$$
$$= 16$$

Multiply first
$$5 + \underbrace{3 \cdot 2}$$
$$= 5 + \underbrace{6}$$
$$= 11$$

How do we know which operation to carry out first? By consensus we agree to follow the rule called the **order of operations** so that everyone always gets the same value as the answer.

Order of Operations Rule

To evaluate mathematical expressions, carry out the operations *in the following order:*

1. First, perform the operations within any grouping symbols, such as parentheses ( ) or brackets [ ].
2. Then raise any number to its power.
3. Next, perform all multiplications and divisions as they appear from left to right.
4. Finally, do all additions and subtractions as they appear from left to right.

Applying the order of operations rule to the previous example gives us the following result:

$$5 + \underbrace{3 \cdot 2} \quad \text{Multiply first.}$$
$$= 5 + \underbrace{6} \quad \text{Then add.}$$
$$= 11$$

So 11 is the correct answer.

EXAMPLE 4

Simplify: $20 - 18 \div 9 \cdot 8$

Solution The order of operations rule gives us

$$
\begin{aligned}
20 - 18 \div 9 \cdot 8 &= 20 - 2 \cdot 8 &&\text{Divide first.} \\
&= 20 - 16 &&\text{Then multiply.} \\
&= 4 &&\text{Finally subtract.}
\end{aligned}
$$

PRACTICE 4

Evaluate: $8 \div 2 + 4 \cdot 3$

EXAMPLE 5

Find the value of $3 + 2 \cdot (8 + 3^2)$.

Solution

$$
\begin{aligned}
3 + 2 \cdot (8 + 3^2) &= 3 + 2 \cdot (8 + 9) &&\text{Perform operations in parentheses: square the 3.} \\
&= 3 + 2 \cdot \quad 17 &&\text{Add 8 and 9.} \\
&= 3 + \quad 34 &&\text{Multiply 2 and 17.} \\
&= \quad 37 &&\text{Add 3 and 34.}
\end{aligned}
$$

PRACTICE 5

Simplify: $(4 + 1)^2 - 4 \cdot 6$

Tip When a division problem is written as a fraction, parentheses are understood to be around both the numerator and the denominator. For instance,

$$\frac{10 - 2}{3 - 1} \qquad \text{means} \qquad (10 - 2) \div (3 - 1)$$

EXAMPLE 6

A student's test scores in a class were 85, 94, 93, 86, and 92. If the average of these scores was 90 or above, she earned an A. Did the student earn an A?

Solution

$$
\begin{aligned}
\text{Average} &= \frac{\text{The sum of the scores}}{\text{The number of scores}} \\
&= \frac{85 + 94 + 93 + 86 + 92}{5} \\
&= (85 + 94 + 93 + 86 + 92) \div 5 \\
&= 450 \div 5 \\
&= 90
\end{aligned}
$$

Therefore, the student did earn an A.

PRACTICE 6

Last year, a tenant's average monthly electricity bill was $110. This year, these bills amounted to $84, $85, $88, $92, $80, $96, $150, $175, $100, $95, $75, and $80. On average, were the monthly bills higher this year?

R.2   Factors, Primes, and Least Common Multiples

Recall that in a multiplication problem involving two or more whole numbers, the whole numbers that are multiplied are called **factors**. For example, since $2 \cdot 4 = 8$, we say that 2 is a factor of 8, or that 8 is divisible by 2. The factors of 8 are 1, 2, 4, and 8.

Now let's discuss the difference between prime numbers and composite numbers. A **prime number** is a whole number that has exactly two factors, namely, itself and 1. A **composite number** is a whole number that has more than two factors. For instance, 5 is prime because its only factors are 1 and 5. However, 9 is a composite because it has more than two factors, namely, 1, 3, and 9. Note that both 0 and 1 are considered neither prime nor composite.

Every composite number can be written as the product of prime factors, called its **prime factorization**. For instance, the prime factorization of 12 is $2 \cdot 2 \cdot 3$, or $2^2 \cdot 3$. This factorization is unique.

The **multiples** of a number are the products of that number and the whole numbers. For instance some of the multiples of 5 are

$$\underbrace{0}_{0 \times 5} \quad \underbrace{5}_{1 \times 5} \quad \underbrace{10}_{2 \times 5} \quad \underbrace{15}_{3 \times 5}$$

A number that is a multiple of two or more numbers is called a **common multiple** of these numbers. Some of the common multiples of 6 and 8 are 24, 48, and 72.

The **least common multiple** (LCM) of two or more numbers is the smallest nonzero number that is a multiple of each number. For example, the LCM of 6 and 8 is 24.

A good way to find the LCM involves prime factorization.

EXAMPLE 1	PRACTICE 1

EXAMPLE 1

Find the LCM of 8 and 12.

Solution We first find the prime factorization of each number.

$$8 = 2 \cdot 2 \cdot 2 = 2^3 \qquad 12 = 2 \cdot 2 \cdot 3 = 2^2 \cdot 3$$

Since 2 appears *three* times in the factorization of 8 and *twice* in the factorization of 12, it must be included three times in forming the least common multiple. So we use 2^3 (the highest power of 2 in the factorizations). The factor of 3 must also be included. Therefore, the LCM of 8 and 12 is

$$\text{LCM} = 2^3 \cdot 3 = 8 \cdot 3 = 24$$

PRACTICE 1

What is the LCM of 10 and 25?

Example 1 suggests the following procedure.

**To Compute the Least Common Multiple (LCM)
of Two or More Numbers**

- Find the prime factorization of each number.
- Identify the prime factors that appear in each factorization.
- Multiply these prime factors, using each factor the greatest number of times
  that it occurs in any of the factorizations.

EXAMPLE 2

What is the LCM of 18, 30, and 45?

Solution We first find the prime factorization of each number.

$$18 = 2 \cdot 3 \cdot 3 = 2 \cdot 3^2 \qquad 30 = 2 \cdot 3 \cdot 5 \qquad 45 = 3 \cdot 3 \cdot 5 = 3^2 \cdot 5$$

The prime factors that appear in the prime factorizations are
2, 3, and 5. Since 2 and 5 occur at most one time and 3 occurs
at most two times in any of the factorizations, we get

$$\text{LCM} = 2 \cdot 3^2 \cdot 5 = 2 \cdot 9 \cdot 5 = 90$$

PRACTICE 2

Find the LCM of 20, 36, and 60.

EXAMPLE 3

A gym that is open every day of the week offers aerobic
classes every third day and yoga classes every fourth day.
If both classes were offered today, in how many days will the
gym again offer both classes on the same day?

Solution To answer this question, we find the LCM of 3 and 4.
As usual, we begin by finding prime factorizations.

$$3 = 3 \qquad 4 = 2 \cdot 2 = 2^2$$

To find the LCM, we multiply 3 by 2^2.

$$\text{LCM} = 2^2 \cdot 3 = 12$$

So both classes will be offered again on the same day 12 days
from today.

PRACTICE 3

A patient gets two medications while
in the hospital. One medication is
given every 6 hours and the other is
given every 8 hours. If the patient
is given both medications now, in how
many hours will both medications
again be given at the same time?

R.3   Fractions

A **fraction** can mean a part of a whole. For example, $\frac{2}{3}$ of a class means two of every three students. A fraction can also mean the quotient of two whole numbers.

A fraction has three components:

- the **denominator** (on the bottom), that stands for the number of parts into which the whole is divided,
- the **numerator** (on top) that tells us how many parts of the whole the fraction contains,
- the **fraction line** (or **fraction bar**) that separates the numerator from the denominator and stands for the phrase *out of* or *divided by*.

Note that the denominator of a fraction cannot be zero.

A fraction whose numerator is smaller than its denominator is a **proper fraction**. A **mixed number** consists of a whole number and a proper fraction. A mixed number can also be expressed as an **improper fraction**, which is a fraction whose numerator is larger than or equal to its denominator. For example, the mixed number $1\frac{1}{2}$ can be written as the improper fraction $\frac{3}{2}$.

To Change a Mixed Number to an Improper Fraction

- Multiply the denominator of the fraction by the whole-number part of the mixed number.
- Add the numerator of the fraction to this product.
- Write this sum over the original denominator to form the improper fraction.

EXAMPLE 1	PRACTICE 1
Write $12\frac{1}{4}$ as an improper fraction.	Express $3\frac{2}{9}$ as an improper fraction.

Solution
$$12\frac{1}{4} = \frac{(4 \times 12) + 1}{4}$$
$$= \frac{48 + 1}{4} = \frac{49}{4}$$

To Change an Improper Fraction to a Mixed Number

- Divide the numerator by the denominator.
- If there is a remainder, write it over the denominator.

EXAMPLE 2	PRACTICE 2

Write $\dfrac{11}{2}$ as a mixed number.

Express $\dfrac{8}{3}$ as a mixed number.

Solution

$$\dfrac{11}{2} = 2\overline{)11} \quad \overset{5}{}$$ Divide the numerator by the denominator.

$$\dfrac{10}{1} \leftarrow \text{ Remainder}$$

$$\dfrac{11}{2} = 5\dfrac{1}{2} \quad \text{Write the remainder over the denominator.}$$

Two fractions are **equivalent** if they represent the same value. To generate fractions equivalent to a given fraction, say $\dfrac{1}{3}$, multiply both its numerator and denominator by the same nonzero whole number. For instance,

$$\dfrac{1}{3} = \dfrac{1 \cdot 2}{3 \cdot 2} = \dfrac{2}{6} \qquad \dfrac{1}{3} = \dfrac{1 \cdot 3}{3 \cdot 3} = \dfrac{3}{9}$$

A fraction is said to be in **simplest form** (or **reduced to lowest terms**) when the only common factor of its numerator and its denominator is 1. To simplify a fraction, we divide its numerator and denominator by the same number, or common factor. To find these common factors, it is often helpful to express both the numerator and denominator as the product of prime factors. We can then divide out (or cancel) all common factors.

EXAMPLE 3	PRACTICE 3

Write $\dfrac{42}{28}$ in lowest terms.

Reduce $\dfrac{24}{30}$ to lowest terms.

Solution

$$\dfrac{42}{28} = \dfrac{2 \cdot 3 \cdot 7}{2 \cdot 2 \cdot 7} \qquad \text{Express the numerator and denominator as the product of primes.}$$

$$= \dfrac{\overset{1}{\cancel{2}} \cdot 3 \cdot \overset{1}{\cancel{7}}}{\underset{1}{\cancel{2}} \cdot 2 \cdot \underset{1}{\cancel{7}}} \qquad \text{Divide out common factors.}$$

$$= \dfrac{3}{2} \qquad \text{Multiply the remaining factors.}$$

Fractions with the same denominator are said to be **like**; those with different denominators are called **unlike**.

To Add (or Subtract) Like Fractions

- Add (or subtract) the numerators.
- Use the given denominator.
- Write the answer in simplest form.

EXAMPLE 4

Find the sum of $\dfrac{7}{12}$ and $\dfrac{2}{12}$.

Solution Using the rule for adding like fractions, we get

$$\dfrac{7}{12} + \dfrac{2}{12} = \dfrac{\overbrace{7+2}^{\text{Add numerators.}}}{\underbrace{12}_{\text{Keep same denominator.}}} = \dfrac{9}{12}, \text{ or } \dfrac{3}{4} \;\underset{\text{Simplest form}}{\uparrow}$$

PRACTICE 4

Add: $\dfrac{7}{15} + \dfrac{3}{15}$

EXAMPLE 5

Find the difference between $\dfrac{11}{12}$ and $\dfrac{7}{12}$.

Solution

$$\dfrac{11}{12} - \dfrac{7}{12} = \dfrac{\overbrace{11-7}^{\text{Subtract numerators.}}}{\underbrace{12}_{\text{Keep same denominator.}}} = \dfrac{4}{12}, \text{ or } \dfrac{1}{3} \;\underset{\text{Simplest form}}{\uparrow}$$

PRACTICE 5

Subtract: $\dfrac{19}{20} - \dfrac{11}{20}$

Unlike fractions are more complicated to add (or subtract) than like fractions because we must first change the unlike fractions to equivalent like fractions. Typically, we use their **least common denominator (LCD)**, that is, the least common multiple of their denominators, to find equivalent fractions.

> ### To Add (or Subtract) Unlike Fractions
> - Rewrite the fractions as equivalent fractions with a common denominator, usually the LCD.
> - Add (or subtract) the numerators, keeping the same denominator.
> - Write the answer in simplest form.

EXAMPLE 6

Add: $\dfrac{5}{12} + \dfrac{5}{16}$

Solution First find the LCD, which is 48.

$$\dfrac{5}{12} + \dfrac{5}{16} = \dfrac{20}{48} + \dfrac{15}{48} \quad \text{Find equivalent fractions.}$$

$$= \dfrac{35}{48} \quad \begin{array}{l}\text{Add the numerators, keeping}\\\text{the same denominator.}\end{array}$$

The fraction $\dfrac{35}{48}$ is already in lowest terms because 35 and 48 have no common factors other than 1.

PRACTICE 6

Add: $\dfrac{11}{12} + \dfrac{3}{4}$

EXAMPLE 7

Subtract $\frac{1}{12}$ from $\frac{1}{3}$.

Solution First find the LCD, which is 12.

$$\frac{1}{3} - \frac{1}{12} = \frac{4}{12} - \frac{1}{12}$$ Write equivalent fractions with a common denominator.

$$= \frac{3}{12}$$ Subtract the numerators, keeping the same denominator.

$$= \frac{1}{4}$$ Reduce $\frac{3}{12}$ to lowest terms.

PRACTICE 7 ⊙

Calculate: $\frac{4}{5} - \frac{1}{2}$

To Add (or Subtract) Mixed Numbers

- Rewrite the fractions as equivalent fractions with a common denominator, usually the LCD.
- When subtracting, rename (or borrow from) the whole number on top if the fraction on the bottom is larger than the fraction on top.
- Add (or subtract) the fractions.
- Add (or subtract) the whole numbers.
- Write the answer in simplest form.

EXAMPLE 8

Find the sum of $3\frac{3}{5}$ and $7\frac{2}{3}$.

Solution Since the denominators have no common factor other than 1, the least common denominator is the product of 5 and 3, which is 15.

$$3\frac{3}{5} = \quad 3\frac{9}{15}$$
$$+7\frac{2}{3} = +7\frac{10}{15}$$

Now we use the rule for adding mixed numbers with same denominator.

$$3\frac{3}{5} = \quad 3\frac{9}{15}$$ Add the fractions.
$$+7\frac{2}{3} = +7\frac{10}{15}$$ Add the whole numbers.
$$\overline{\qquad 10\frac{19}{15}}$$

Because $\frac{19}{15}$ is an improper fraction, we need to rewrite $10\frac{19}{15}$:

$$10\frac{19}{15} = 10 + \frac{19}{15} = 10 + 1\frac{4}{15} = 11\frac{4}{15}$$

$$\frac{19}{15} = 1\frac{4}{15}$$

So the sum of $3\frac{3}{5}$ and $7\frac{2}{3}$ is $11\frac{4}{15}$.

PRACTICE 8

Add $4\frac{5}{8}$ to $3\frac{1}{2}$.

EXAMPLE 9

Compute: $13\frac{2}{9} - 7\frac{8}{9}$

Solution First, write the problem, $13\frac{2}{9} - 7\frac{8}{9}$, in a vertical format.

$$13\frac{2}{9}$$

$$-7\frac{8}{9}$$

Since $\frac{8}{9}$ is larger than $\frac{2}{9}$, we need to rename $13\frac{2}{9}$:

$$13\frac{2}{9} = \underbrace{12 + 1}_{13 \,=\, 12 + 1} + \frac{2}{9} = 12 + \frac{9}{9} + \frac{2}{9} = 12\frac{11}{9}$$

The problem now becomes

$$13\frac{2}{9} = 12\frac{11}{9}$$

$$-7\frac{8}{9} = -7\frac{8}{9}$$

Finally, we subtract and then write the answer in simplest form.

$$13\frac{2}{9} = 12\frac{11}{9}$$

$$-7\frac{8}{9} = -7\frac{8}{9}$$

$$5\frac{3}{9} = 5\frac{1}{3}$$

PRACTICE 9

Find the difference between

$15\frac{1}{12}$ and $9\frac{11}{12}$.

Now let's look at how we multiply fractions.

> **To Multiply Fractions**
> - Multiply the numerators.
> - Multiply the denominators.
> - Write the answer in simplest form.

EXAMPLE 10

Multiply: $\frac{2}{3} \cdot \frac{4}{5}$

Solution $\dfrac{2}{3} \cdot \dfrac{4}{5} = \dfrac{2 \cdot 4}{3 \cdot 5} = \dfrac{8}{15}$

PRACTICE 10

Compute: $\frac{1}{2} \cdot \frac{3}{4}$

In multiplying some fractions, we can first simplify (or cancel) by dividing *any* numerator and *any* denominator by a common factor. Simplifying before multiplying allows us to work with smaller numbers and still gives us the same answer.

EXAMPLE 11	PRACTICE 11
Find the product of $\frac{4}{9}$ and $\frac{5}{8}$.	Multiply $\frac{7}{10}$ by $\frac{5}{11}$.

Solution

Divide the numerator (4) and the denominator (8) by the same number (4). Then multiply.

$$\frac{4}{9} \cdot \frac{5}{8} = \frac{\overset{1}{\cancel{4}} \cdot 5}{9 \cdot \underset{2}{\cancel{8}}} = \frac{5}{18}$$

To Multiply Mixed Numbers

- Change each mixed number to its equivalent improper fraction.
- Follow the steps for multiplying fractions.
- Write the answer in simplest form.

EXAMPLE 12	PRACTICE 12
Compute: $2\frac{1}{2} \cdot 1\frac{1}{4}$	Find the product of $3\frac{3}{4}$ and $2\frac{1}{10}$.

Solution

$$2\frac{1}{2} \cdot 1\frac{1}{4} = \frac{5}{2} \cdot \frac{5}{4} \quad \text{Change each mixed number to an improper fraction.}$$

$$= \frac{5 \cdot 5}{2 \cdot 4} \quad \text{Multiply the fractions.}$$

$$= \frac{25}{8}, \quad \text{or } 3\frac{1}{8} \quad \text{Simplify.}$$

Dividing fractions is equivalent to multiplying by the *reciprocal* of the divisor. The reciprocal is found by *inverting*—switching the position of the numerator and denominator of the divisor.

To Divide Fractions

- Change the divisor to its reciprocal.
- Multiply the resulting fractions.
- Write the answer in simplest form.

EXAMPLE 13

Divide: $\dfrac{4}{5} \div \dfrac{3}{10}$

Solution

$$\dfrac{4}{5} \div \dfrac{3}{10} = \dfrac{4}{\overset{}{\underset{1}{5}}} \times \dfrac{\overset{2}{10}}{3} = \dfrac{4 \times 2}{1 \times 3} = \dfrac{8}{3}, \quad \text{or } 2\dfrac{2}{3}$$

$\dfrac{3}{10}$ and $\dfrac{10}{3}$ are reciprocals.

PRACTICE 13

Divide: $\dfrac{3}{4} \div \dfrac{1}{8}$

To Divide Mixed Numbers

- Change each mixed number to its equivalent improper fraction.
- Follow the steps for dividing fractions.
- Write the answer in simplest form.

EXAMPLE 14

Calculate: $9 \div 2\dfrac{7}{10}$

Solution

$$9 \div 2\dfrac{7}{10} = \dfrac{9}{1} \div \dfrac{27}{10} \qquad \text{Change all mixed numbers to improper fractions.}$$

$$= \dfrac{\overset{1}{9}}{1} \times \dfrac{10}{\underset{3}{27}} \qquad \text{Invert the divisor and multiply.}$$

$$= \dfrac{10}{3}, \quad \text{or } 3\dfrac{1}{3} \qquad \text{Simplify.}$$

PRACTICE 14

Divide: $6 \div 3\dfrac{3}{4}$

EXAMPLE 15

Suppose that you spend $\dfrac{3}{8}$ of your monthly salary on rent. If that salary is $960, how much money do you have left after paying the rent?

Solution Let's break up the question into two parts:

1. First find $\dfrac{3}{8}$ of 960.

$$\dfrac{3}{\underset{1}{8}} \times \dfrac{\overset{120}{960}}{1} = 360$$

2. Then subtract that result from 960.

$$960 - 360 = 600$$

So you have $600 left after paying for rent.

PRACTICE 15

At a college with 1000 students, $\dfrac{3}{5}$ of the students take a math course. How many students in the college are not taking a math course?

R.4   Decimals

A number written as a *decimal* has

- a whole-number part, which precedes the decimal point, and
- a fractional part, which follows the decimal point.

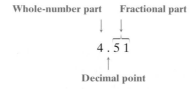

A decimal without a decimal point shown is understood to have one at the end of the last digit and is the same as a whole number. For instance, 32 and 32. are the same number.

Each digit in a decimal has a place value. The place value system for decimals is an extension of the place value system for whole numbers.

The places to the right of the decimal point are called *decimal places.* For instance, the number 64.149 is said to have three decimal places.

For a whole number, the place values are 1, 10, 100, and other powers of 10. By contrast, the place values for the fractional part of a decimal are $\frac{1}{10}, \frac{1}{100}, \frac{1}{1000}$, and the reciprocals of other powers of 10.

The first decimal place after the decimal point is the ten**ths** place. Working to the right, the next decimal places are the hundred**ths** place, the thousand**ths** place, the ten-thousand**ths**, and so forth.

The following chart shows the place values in the numbers 0.54 and 513.285.

... thousands	hundreds	tens	ones	.	tenths	hundredths	thousandths ...
			0	.	5	4	
	5	1	3	.	2	8	5

Knowing the place value system is the key to changing a decimal to its equivalent fraction and to reading the decimal. For a given decimal, the place value of the rightmost digit is the denominator of the equivalent fraction.

$$0.9 = \frac{9}{10}$$

Read "nine tenths"

$$0.21 = \frac{21}{100}$$

Read "twenty-one hundredths"

Let's look at how to rewrite any decimal as a fraction or mixed number.

To Change a Decimal to the Equivalent Fraction or Mixed Number

- Copy the nonzero whole-number part of the decimal and drop the decimal point.
- Place the fractional part of the decimal in the numerator of the equivalent fraction.
- Make the denominator of the equivalent fraction the same as the place value of the rightmost digit.

EXAMPLE 1

Write each decimal as a fraction or mixed number.

a. 0.25   **b.** 1.398

Solution

a. Write 0.25 as $\frac{25}{100}$, which simplifies to $\frac{1}{4}$. So $0.25 = \frac{1}{4}$.

b. The decimal 1.398 is equivalent to a mixed number. The whole-number part is 1. The fractional part (.398) of the decimal (without the decimal point) is the numerator of the equivalent fraction. Since the decimal has three decimal places, the denominator of the fraction is 1000. So $1.398 = 1\frac{398}{1000} = 1\frac{199}{500}$.

PRACTICE 1

Express each decimal in fractional form.

a. 0.5

b. 2.073

EXAMPLE 2

Write 2.019 in words.

Solution The fractional part of 2.019 is .019, which is equivalent to $\frac{19}{1000}$. We read the original decimal as "two and nineteen thousandths," keeping the whole number unchanged. Note that we use the word *and* to separate the whole number and the fractional part.

PRACTICE 2

Write 4.003 in words.

In computations with decimals, we sometimes *round* the decimal to a certain number of decimal places.

To Round a Decimal to a Given Decimal Place

- Underline the place to which you are rounding.
- Look at the digit to the right of the underlined digit—*the critical digit*. If this digit is 5 or more, add 1 to the underlined digit; if it is less than 5, leave the underlined digit unchanged.
- Drop all places to the right of the underlined digit.

EXAMPLE 3

Round 94.735 to the nearest tenth.

Solution First we underline the digit 7 in the tenths place: 94.735. Since the critical digit 3 is less than 5, we do not add 1 to the underlined digit. Dropping all digits to the right of the 7, we get 94.7. So 94.735 ≈ 94.7 (the symbol ≈ is read "is approximately equal to"). Note that our answer has only one decimal place because we are rounding to the nearest tenth.

PRACTICE 3

Round 748.0772 to the nearest hundredth.

To Add Decimals

- Rewrite the numbers vertically, lining up the decimal points.
- Add.
- Insert a decimal point in the answer below the other decimal points.

EXAMPLE 4

Add: 2.7 + 80.13 + 5.036

Solution Rewrite the numbers with decimal points lined up vertically so that digits with the same place value are in the same column. Then add.

$$
\begin{array}{r}
2.7 \\
80.13 \\
+\ 5.036 \\
\hline
87.866
\end{array}
$$

└── Insert the decimal point in the answer.

PRACTICE 4

Add: 5.92 + 35.872 + 0.3

To Subtract Decimals

- Rewrite the numbers vertically, lining up the decimal points.
- Subtract, inserting extra zeros if necessary for borrowing.
- Insert a decimal point in the answer below the other decimal points.

EXAMPLE 5

Subtract: 5 − 2.14

Solution

$$
\begin{array}{r}
5.00 \\
-\ 2.14 \\
\hline
2.86
\end{array}
$$

Rewrite 5 as 5.00 and line up decimal points vertically.

Subtract.

└── Insert the decimal point in the answer.

PRACTICE 5 ⊙

Find the difference: 3.8 − 2.621

> **To Multiply Decimals**
> - Multiply the factors as if they were whole numbers.
> - Find the total number of decimal places in the factors.
> - Count that many places from the right end of the product and insert a decimal point.

EXAMPLE 6

Multiply: 6.1×3.7

Solution First multiply: $61 \times 37 = 2257$

$$
\begin{array}{r}
6\,1 \\
\times\ \ 3\,7 \\
\hline
4\,2\,7 \\
1\,8\,3\ \ \\
\hline
2\,2\,5\,7
\end{array}
$$

Then count the total number of decimal places.

$$
\begin{array}{r}
6.1 \quad \longleftarrow \textbf{ One decimal place (tenths)} \\
\times\ \ 3.7 \quad \longleftarrow \textbf{ One decimal place (tenths)} \\
\hline
4\,2\,7 \quad\quad\quad\quad\quad\quad \\
1\,8\,3\ \ \quad\quad\quad\quad\quad\quad \\
\hline
2\,2.5\,7 \quad \longleftarrow \textbf{ Two decimal places (hundredths)} \\
\textbf{in the product}
\end{array}
$$

So the answer is 22.57.

PRACTICE 6

Find the product of 2.81 and 3.5.

A shortcut for multiplying a decimal by a power of 10 is to *move the decimal point to the right the same number of places as the power of 10 has zeros.*

EXAMPLE 7

Find the product: $(1000)(2.89)$

Solution We notice that 1000 is a power of 10 and has three zeros. To multiply 1000 by 2.89, we move the decimal point in 2.89 to the right three places.

$$(1000)\,(2.890) = 2\,8\,9\,0. = 2890$$

Need to insert a 0 to move three places.

The product is 2890.

PRACTICE 7

Multiply 32.7 by 10,000.

To Divide Decimals

- If the divisor is not a whole number, move the decimal point in the divisor to the right end of the number.
- Move the decimal point in the dividend the same number of places to the right as we did in the divisor.
- Insert a decimal point in the quotient directly above the decimal point in the dividend.
- Divide the new dividend by the new divisor, inserting zeros at the right end of the dividend as necessary.

EXAMPLE 8

Divide 0.035 by 0.25.

Solution

Move the decimal point to the right end, making the divisor a whole number.

$$0.25\overline{)0.035} \quad \Rightarrow \quad 0.25\overline{)0.035}$$

Move the decimal point in the dividend the same number of places.

Finally, we divide 3.5 by 25, which gives us 0.14.

```
      0.1 4
25)3.5 0
   2 5
   1 0 0
   1 0 0
```

PRACTICE 8

Divide: $2.706 \div 0.15$

A shortcut for dividing a decimal by a power of 10 is to *move the decimal point to the left the same number of places as the power of 10 has zeros.*

EXAMPLE 9

Compute: $\dfrac{7.2}{100}$

Solution Since we are dividing by the power of 10 with two zeros, we can find this quotient simply by moving the decimal point in 7.2 to the left two places.

Need to insert a 0 to move two places.

$$0\,7.2 = .072, \quad \text{or } 0.072$$

The quotient is 0.072.

PRACTICE 9

Calculate: $0.86 \div 1000$

EXAMPLE 10

The following graph shows the number of people who attended a Broadway show in selected seasons.

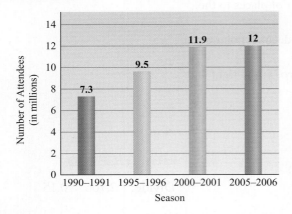

The number of Broadway attendees in the 2005–2006 season was how many times as great as the number of attendees 15 years earlier? Round to the nearest tenth. (*Source:* The League of American Theatres and Producers, Inc.)

Solution The number of attendees in 2005–2006 (in millions) was 12, and the number in 1990–1991 was 7.3. To find how many times as great 12 is as compared to 7.3, we find their quotient.

$$7.3\overline{)12} = 7.3\overline{)12.0} = 73\overline{)120}$$

$$
\begin{array}{r}
1.64 \\
73\overline{)120.00} \\
\underline{73} \\
470 \\
\underline{438} \\
320 \\
\underline{292} \\
28
\end{array}
$$

Rounding to the nearest tenth, we conclude that the number of Broadway attendees in 2005–2006 was 1.6 times the corresponding figure in 1990–1991.

PRACTICE 10

The table below gives the amount of selected foods consumed per capita in the United States in a recent year.

Food	Annual per Capita Consumption (in pounds)
Red meat	112.0
Poultry	72.7
Fish and shellfish	16.5

The amount of red meat consumed was how many times as great as the amount of poultry, rounded to the nearest tenth? (*Source:* USDA/Economic Research Service)

R.5   Percents

Percent means divided by 100. For instance, 18% means 18 divided by 100, or $\dfrac{18}{100}$, which simplifies to $\dfrac{9}{50}$. Therefore, the fraction $\dfrac{9}{50}$ is just another way of writing 18%. This result suggests the following rule.

To Change a Percent to the Equivalent Fraction
- Drop the % sign from the given percent and place the number over 100.
- Simplify the resulting fraction, if possible.

EXAMPLE 1	PRACTICE 1
Write 20% as a fraction.	Write 7% as a fraction.
Solution To change this percent to a fraction, drop the percent sign and write the 20 over 100. Then simplify.	
$$20\% = \dfrac{20}{100} = \dfrac{1}{5}$$	

A percent can also be written as a decimal, since $18\% = \dfrac{18}{100} = 0.18$. This suggests the following rule.

To Change a Percent to the Equivalent Decimal
- Move the decimal point two places *to the left* and drop the % sign.

EXAMPLE 2	PRACTICE 2
Find the decimal equivalent of 1%.	Write 5% as a decimal.
Solution The unwritten decimal point lies to the right of the 1. Moving the decimal point two places to the left and dropping the % sign, we get	
$$1\% = 0\,1\,.\% = .01, \quad \text{or } 0.01$$	

To Change a Decimal to the Equivalent Percent
- Move the decimal point two places *to the right* and insert a % sign.

EXAMPLE 3

Write 0.125 as a percent.

Solution First move the decimal point two places to the right. Then insert a % sign.

$$0.1\,2\,5 = 0\,1\,2.5\% = 12.5\%$$

PRACTICE 3

What percent is equivalent to the decimal 0.025?

To Change a Fraction to the Equivalent Percent

- Change the fraction to a decimal.
- Change the decimal to a percent.

EXAMPLE 4

Rewrite $\dfrac{1}{2}$ as a percent.

Solution To change the given fraction to a percent, we first find the equivalent decimal, which we then express as a percent.

$$\frac{1}{2} = 2\overline{)1.0}^{\,0.5} \quad \text{and} \quad 0.5\,0 = 50\%$$
$$\underline{1\,0}$$

PRACTICE 4

Rewrite $\dfrac{1}{4}$ as a percent.

EXAMPLE 5

A local sales tax rate is 0.0825 of selling prices. Express this tax rate as a percent.

Solution $0.0825 = 0\,0\,8.\,2\,5\% = 8.25\%$

PRACTICE 5

Suppose that 40% of a student's income goes to paying college expenses. Rewrite this percent as a decimal.

| R | | Exercises | | **MyMathLab** | | | | | |

To help you review this chapter, solve these problems.

[R.1] *Rewrite each product using exponents.*

1. $6 \cdot 6 \cdot 6 \cdot 6 \cdot 6$
2. $7 \cdot 7 \cdot 7 \cdot 7 \cdot 7 \cdot 7 \cdot 7 \cdot 7$
3. $2 \cdot 2 \cdot 10 \cdot 10 \cdot 10$
4. $5 \cdot 5 \cdot 5 \cdot 5 \cdot 4 \cdot 4 \cdot 4$

Calculate.

5. $5^2 \cdot 10^3$
6. $2^4 \cdot 6^2$

Simplify.

7. $6 + 3^2$
8. $20 - 4 \cdot 5$
9. $2 + 18 \div 3(9 - 7)$

10. $3 + 10(20 - 2^3)$
○ **11.** $\dfrac{4^2 + 8}{9 - 3}$
12. $\dfrac{12 + 2^2}{15 - 9 + 2}$

[R.2] *Find all the factors of each number.*

13. 150
14. 57

Indicate whether each number is prime or composite.

15. 23
16. 38
17. 51
18. 47

Write the prime factorization of each number.

19. 42
20. 54
21. 48
22. 100

Find the LCM.

23. 6 and 8
24. 15 and 20
○ **25.** 24, 36, and 72
26. 7, 28, and 42

[R.3] *Write each mixed number as an improper fraction.*

27. $3\dfrac{4}{5}$
28. $7\dfrac{3}{10}$

Write each fraction as a mixed number.

29. $\dfrac{23}{4}$
30. $\dfrac{31}{9}$

Simplify.

31. $\dfrac{14}{28}$
32. $\dfrac{30}{45}$
33. $5\dfrac{2}{4}$
34. $6\dfrac{12}{42}$

Calculate. Write answers in lowest terms.

35. $\dfrac{1}{9} + \dfrac{4}{9}$
36. $\dfrac{3}{10} + \dfrac{7}{10}$
37. $\dfrac{3}{8} - \dfrac{1}{8}$
38. $\dfrac{5}{3} - \dfrac{2}{3}$

39. $\dfrac{2}{5} + \dfrac{4}{7}$
40. $\dfrac{8}{9} + \dfrac{1}{2}$
41. $\dfrac{3}{10} - \dfrac{1}{20}$
42. $\dfrac{3}{5} - \dfrac{1}{4}$

43. $1\dfrac{1}{8} + 5\dfrac{3}{8}$
44. $3\dfrac{1}{5} + 4\dfrac{1}{5}$
45. $8\dfrac{7}{10} + 1\dfrac{9}{10}$
46. $5\dfrac{5}{6} + 2\dfrac{1}{6}$

47. $9\frac{11}{12} - 6\frac{7}{12}$ **48.** $2\frac{5}{9} - 2\frac{4}{9}$ **49.** $6\frac{1}{10} - 4\frac{3}{10}$ **50.** $5\frac{1}{4} - 2\frac{3}{4}$

51. $12 - 5\frac{1}{2}$ **52.** $3 - 1\frac{4}{5}$ **53.** $7\frac{1}{2} - 4\frac{5}{8}$ **54.** $5\frac{1}{12} + 4\frac{1}{2}$

55. $\frac{2}{3} \cdot \frac{1}{5}$ **56.** $\frac{3}{4} \cdot \frac{8}{9}$ **57.** $1\frac{2}{5} \cdot 10$ **58.** $20 \cdot 1\frac{5}{6}$

59. $3\frac{1}{4} \cdot 4\frac{2}{3}$ **60.** $2\frac{1}{2} \cdot 1\frac{1}{5}$ **61.** $2\frac{5}{6} \div \frac{1}{2}$ **62.** $1\frac{1}{3} \div \frac{4}{5}$

63. $\frac{2}{3} \div 6$ **64.** $\frac{1}{10} \div 4$ **65.** $8 \div 2\frac{1}{3}$ **66.** $4\frac{1}{2} \div 2\frac{1}{2}$

67. $\left(\frac{3}{4}\right)^2 - \frac{3}{8} \div 6$ **68.** $8\frac{2}{5} + 2 \div \left(\frac{1}{2} - \frac{1}{3}\right)$

[R.4]   *Write each decimal as a fraction or mixed number.*

69. 0.875   **70.** 2.006

Name the place that the underlined digit occupies.

71. 18.35$\underline{9}$ **72.** 8024.$\underline{5}$

Write each decimal in words.

73. 0.72   **74.** 0.05

75. 3.009   **76.** 12.235

Round as indicated.

77. 7.31 to the nearest tenth   **78.** 0.0387 to the nearest thousandth

79. 4.3868 to two decimal places   **80.** $899.09 to the nearest dollar

Calculate.

81. $8.2 + 3.91 + 6$ **82.** $8 + 3.25 + 12.88$ **83.** $3.8 - 1.927$ **84.** $2.5 - 1.6$

85. 7.28×0.4 **86.** $(0.005)(0.002)$ **87.** $2.71 \cdot 1000$ **88.** 100×5.3

89. $0.006 \div 4$ **90.** $31.9 \div 10$ **91.** $12 \div 2.4$ **92.** $\dfrac{7.11}{0.3}$

93. $7.1 + 0.5^2$ **94.** $20.8 - 7(4 - 3.1)$

[R.5]   *Change each percent to a fraction or mixed number and simplify.*

◉ **95.** 75%   **96.** 4%   **97.** 106%   **98.** 250%

Change each percent to a decimal.

99. 6%   **100.** 29%   ◉ **101.** 150%   **102.** 0.2%

Change each decimal to a percent.

103. 0.31   **104.** 0.05   ◉ **105.** 0.0145   **106.** 2

Change each fraction to a percent.

107. $\dfrac{1}{10}$

108. $\dfrac{3}{8}$

◎ **109.** $\dfrac{4}{5}$

110. $\dfrac{7}{20}$

▍Mixed Applications

Solve.

111. In a city, the daily high temperatures for a week in May were 67°F, 72°F, 78°F, 70°F, 65°F, 77°F, and 82°F. What was the average daily high temperature for the week?

112. Two airport shuttle buses arrive at a subway station at 7:30 A.M. If one airport shuttle bus arrives every 8 min and the other airport shuttle bus arrives every 10 min, when is the next time that both airport shuttles will arrive at the subway station together?

113. A trucker drives to a town $\dfrac{1}{2}$ mi away. If he then drives an additional $\dfrac{1}{4}$ mi, how far did he drive in all?

114. A sea otter eats an amount that is about $\dfrac{1}{5}$ of its body weight each day. How much will a 35-lb otter eat in a day?

115. At a florist, a shopper buys a dozen roses for her mother. The regular price of roses is $27 a dozen. Based on the store sign shown, how much does she pay for the roses on sale?

ROSES
Special Price
One-third off

116. Find the missing length in the figure shown:

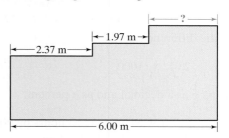

117. A supermarket sells a 4 lb package of 85% lean ground beef for $7.96 and a 3 lb package of 93% lean ground beef for $9.87. What is the difference between the costs per pound of the two types of beef?

118. Most compact discs are sold in plastic jewel cases. Suppose that in a CD collection, there are 29 jewel cases 0.4 in. thick and 3 jewel cases 0.94 in. thick. How many inches of shelf space will be needed to house this CD collection?

119. In a recent year, a single species of parasite reduced the U.S. corn crop by 15%. Express this percent as a fraction.

120. The solar wind streams off the Sun at speeds of about 1,000,000 miles per hour. Express this number as a power of 10. (*Source:* NASA)

● *Check your answers on page A-1.*

Chapter R   **POSTTEST**

 FOR EXTRA HELP

 Pass the Test   Test solutions are found on the enclosed CD.

To see if you have mastered the topics in this chapter, take this test.

1. Calculate: $8^2 \cdot 2^3$

2. Find the value of $11 \cdot 2 + 5 \cdot 3$.

3. What are the factors of 20?

4. Write $3\frac{1}{4}$ as an improper fraction.

5. Reduce $\frac{10}{36}$ to lowest terms.

6. Find the sum: $\frac{5}{8} + \frac{7}{8}$

7. Add: $7\frac{7}{8} + 4\frac{1}{6}$

8. Calculate: $\frac{4}{9} - \frac{3}{10}$

9. Subtract: $12\frac{1}{4} - 8\frac{3}{10}$

10. Find the product of $\frac{3}{4}$ and $\frac{4}{5}$.

11. Find the quotient: $\frac{2}{3} \div \frac{1}{3}$

12. Calculate: $7 \div 3\frac{1}{5}$

13. Write 2.396 in words.

14. Find the sum: $5.2 + 3 + 8.002$

15. Find the difference: $10 - 3.01$

16. What is the product of 5.02 and 8.9?

17. Evaluate: 2.07×1000

18. Compute: $\frac{0.05}{100}$

19. Express $\frac{1}{8}$ as a decimal and as a percent.

20. Write 0.7 as a percent and as a fraction.

21. A trip to a nearby island takes 3 hours by boat and half an hour by airplane. How many times as fast as the boat is the plane?

22. According to a recent survey, the cost of medical care is approximately 1.77 times what it was a decade ago. Round this decimal to the nearest tenth.

23. Find the area (in square meters) of the room pictured, rounded to one decimal place.

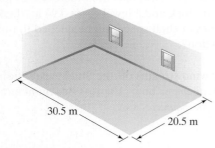

30.5 m   20.5 m

24. Typically, the heaviest organ in the human body is the skin, weighing about 9 lb. By contrast, the heart weighs about 0.6 lb. How many times the weight of the heart is the weight of the skin?

25. The following graph shows the distribution of investments for a retiree. Express as a decimal the percent of investments that are in equities.

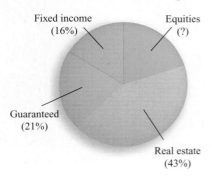

Fixed income (16%)

Equities (?)

Guaranteed (21%)

Real estate (43%)

● *Check your answers on page A-1.*

CHAPTER 1

Real Numbers and Algebraic Expressions

Osteoporosis and Real Numbers

Approximately 10 million Americans have osteoporosis, a disease in which bones become weak and are likely to break. To diagnose osteoporosis, doctors commonly employ a bone mineral density (BMD) test. A person's BMD score is compared to a norm based on the optimal density of a healthy 30-year-old adult. Scores below the norm are indicated in negative numbers. For instance, a score of -2 indicates a low bone mass and a score of -2.5 or less is diagnosed as osteoporosis. Generally, a score of -1 is equivalent to a 10% loss of bone density.

The National Osteoporosis Foundation recommends treatment if you have a result that is:

- less than -1.5 with risk factors, or
- less than -2 with no risk factors.

The results of many medical tests are reported in terms of positive and negative numbers.

Source: Dempster, D. W. et al., *Journal of Bone and Mineral Research* 1:15–21, 1986.

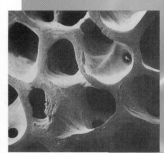

Normal bone

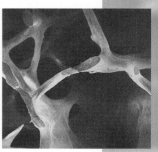

Osteoporotic bone

To see if you have already mastered the topics in this chapter, take this test.

1. Express as a real number: a profit of $2000

2. Is 0 a rational number?

3. Graph the number -0.5 on the following number line.

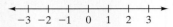

4. What is the opposite of -5?

5. Evaluate $\left| -\dfrac{2}{3} \right|$.

6. Use $>$ or $<$ to fill in the box to make a true statement:
$-31 \ \boxed{} \ -1$

7. Identify the property of real numbers illustrated in the statement: $2 \cdot (a + 6) = 2 \cdot (6 + a)$

8. Add: $9 + (-4) + 2 + (-9)$

9. Simplify: $3(-7) - 5$

10. What is the reciprocal of $\dfrac{1}{4}$?

11. Divide: $(-72) \div (-8)$

12. Translate the algebraic expression to words: $3n - 10$

13. Express in exponential form: $-6 \cdot 6 \cdot 6 \cdot 6$

14. Evaluate: $2x - 4y + 8$, if $x = -2$ and $y = 3$

15. Combine like terms: $3n - 7 + n$

16. Simplify: $-5(2 - x) + 9x$

17. Mount Kilimanjaro, Africa's highest point, is 5895 meters (m) above sea level. Lake Assal, its lowest point, is 156 m below sea level. What is the difference in the elevations of these two points? (*Source: National Geographic Family Reference Atlas of the World*, 2002)

18. A wire of length L is to be cut into three pieces of equal length. Write an expression that represents the length of each piece of wire.

19. The formula for the perimeter of a rectangle is $P = 2l + 2w$. Calculate the perimeter of the following rectangle.

12 cm

3 cm

20. In golf, scores are given in terms of par; scores above par are positive, and scores below par are negative. The table shows Phil Mickelson's scores for each round of a recent PGA Championship tournament.

Round	First	Second	Third	Fourth
Score	-3	-1	-4	$+2$

(*Source:* http://www.pga.com)

In which round did he have the lowest score?

● *Check your answers on page A-1.*

1.1   Real Numbers and the Real-Number Line

- To identify integers, rational numbers, and real numbers
- To graph rational numbers on the real-number line
- To find the opposite and the absolute value of real numbers
- To compare real numbers
- To solve applied problems involving the comparison of real numbers

What Algebra Is and Why It Is Important

Algebra is a language that allows us to express the patterns and rules of arithmetic. Consider, for instance, the rule for finding the product of two fractions: Multiply the numerators to get the numerator of the product and multiply the denominators to get the denominator of the product.

$$\frac{3}{4} \cdot \frac{1}{2} = \frac{3 \cdot 1}{4 \cdot 2} = \frac{3}{8}$$

In the language of algebra, we can write the general rule using letters to represent numbers:

$$\frac{a}{b} \cdot \frac{c}{d} = \frac{a \cdot c}{b \cdot d} = \frac{ac}{bd}$$

Can you think of another arithmetic rule that can be expressed algebraically?

This chapter reviews the basics of algebra. The focus is on two of the key concepts of algebra, namely, *real numbers* and *algebraic expressions*. We begin with a discussion of real numbers, including their operations, properties, and application.

Sets and Set Notation

A *set* is a collection of objects. Each of these objects is said to be an *element* or a *member* of the set.

A set can be represented by listing its elements separated by commas and placing braces around the listing. The set of numbers 2, 4, and 6 is written in set notation as {2, 4, 6}.

The symbol ∅ represents the set with no elements, called the *empty set*.

The Real Numbers

Real numbers are numbers that can be represented as points on a number line. They extend the numbers used in arithmetic and allow us to solve problems that we could not otherwise solve.

Let's begin our discussion by looking at different kinds of real numbers.

Integers

In arithmetic, we use *natural numbers* for counting. The set of natural numbers is {1, 2, 3, 4, 5, 6, 7, 8, 9, …}. The three dots mean that these numbers go on forever in the same pattern. The set of *whole numbers* consists of 0 and the natural numbers: {0, 1, 2, 3, 4, 5, 6, 7, 8, 9 …}.

We can represent the whole numbers on a number line. A number to the right of another on the number line is the greater number, as shown on the following number line:

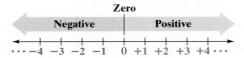

Sometimes we need to consider numbers that are to the left of 0, that is, numbers that are less than 0. For instance, temperatures that are below 0 and a company's quarterly loss are values that are less than 0.

Numbers to the left of 0 on the number line are said to be *negative*. All negative numbers are less than 0. We write $-1, -2, -3, -4$, and so on. By contrast, the whole numbers 1, 2, 3, 4, and so on are said to be *positive* and can also be written as $+1$, $+2, +3, +4$, and so on. All positive numbers are larger than 0. The numbers $-1, -2$, -3, and so on, together with the whole numbers, are called *integers*, as shown on the following number line. Note that the number line extends continuously to the right and to the left as the arrows indicate. Also note that numbers become smaller going in the negative direction and become larger going in the positive direction.

Definition

The set of **integers** is

$$\{\ldots, -4, -3, -2, -1, 0, +1, +2, +3, +4, \ldots\}.$$

These numbers continue indefinitely in both directions.

EXAMPLE 1

The Dow Jones Industrial Average on the stock market declined 4 points today. Express this situation as an integer.

Solution The number in question represents a decline (or loss), so we write a negative integer, namely, -4.

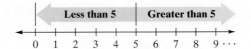

PRACTICE 1

Represent as an integer: A carnation plant freezes and dies at a temperature of 5° below 0° Fahrenheit (°F).

Rational Numbers

Suppose that we want to represent the following situation: *The New York Giants lost one-half yard on a play.* To express a loss of one-half yard, we need a kind of number other than whole numbers or integers. We need a *rational number*—a number that can be written as the quotient of two integers, where the denominator is not equal to zero.

Some examples of rational numbers are

$$\frac{2}{5}, -\frac{1}{2}, 4, 0, 7\frac{1}{4}, -0.03$$

This number represents negative one-half.
We can express a loss of one-half yard as $-\frac{1}{2}$ yd.

> **Definition**
> **Rational numbers** are numbers that can be written in the form $\frac{a}{b}$, where a and b are integers and $b \neq 0$.

A rational number can be expressed in several ways. For instance, the numbers 4 and 0 can also be written as $\frac{4}{1}$ and $\frac{0}{1}$, respectively. And we can write the mixed number $7\frac{1}{4}$ as $\frac{29}{4}$ and -0.03 as $-\frac{3}{100}$. It is important to note that all rational numbers have corresponding decimal representations that either *terminate* or *repeat*. Here are some examples.

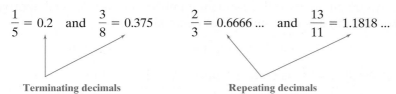

$$\frac{1}{5} = 0.2 \quad \text{and} \quad \frac{3}{8} = 0.375 \qquad \frac{2}{3} = 0.6666\ldots \quad \text{and} \quad \frac{13}{11} = 1.1818\ldots$$

Terminating decimals   Repeating decimals

Just as with integers, we can picture rational numbers as points on a number line. To *graph* a rational number, we locate the point on the number line and mark it as shown.

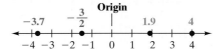

On the above number line, note that

- the point at 0 is called the *origin*,
- numbers to the right of 0 are positive, and numbers to the left of 0 are negative,
- the number 0 is neither positive nor negative.

In this text, we will work mostly with rational numbers until Chapter 8.

EXAMPLE 2

Graph each number on the same number line.

a. $-\dfrac{7}{2}$ **b.** 2.1

Solution

a. Because $-\dfrac{7}{2}$ can be expressed as $-3\frac{1}{2}$, the point $-\dfrac{7}{2}$ is graphed halfway between -3 and -4.

b. The number 2.1, or $2\frac{1}{10}$, is between 2 and 3 but is closer to 2.

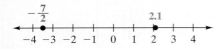

PRACTICE 2

Graph each number on the number line.

a. -1.7 **b.** $\dfrac{5}{4}$

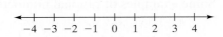

More Real Numbers

Recall that any rational number can be written as the quotient of two integers. However, there are other real numbers that cannot be written in this form. Such numbers are called *irrational numbers*, and their corresponding decimal representations continue indefinitely and have no repeating pattern. Examples of irrational numbers are

$$\sqrt{2} = 1.4142\ldots \qquad \text{The square root of 2}$$
$$-\sqrt{3} = -1.7320\ldots \qquad \text{The negative square root of 3}$$
$$\pi = 3.14159\ldots \qquad \text{Pi, the ratio of the circumference of a circle to its diameter}$$

The *square root* of a number is the number which when multiplied by itself gives the original number. So, for instance, $\sqrt{2} \cdot \sqrt{2}$ is 2.

In many computations with irrational numbers, we use decimal approximations that are rounded to a certain number of decimal places:

$$\sqrt{2} \approx 1.41 \qquad -\sqrt{3} \approx -1.73 \qquad \pi \approx 3.14$$

Recall that the symbol $\approx$ is read "is approximately equal to."

Irrational numbers, like rational numbers, can be graphed on the number line. The rational numbers and the irrational numbers together make up the *real numbers*.

Absolute Value

On the number line, the numbers $+3$ and -3 (that is, 3) are opposites of each other. Similarly, $-\dfrac{1}{2}$ and $+\dfrac{1}{2}$ are opposites. What is the opposite of 0?

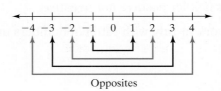

Opposites

Definition

Two real numbers that are the same distance from 0 on the number line but on opposite sides of 0 are called **opposites**. For any real number n, its opposite is $-n$.

	EXAMPLE 3		PRACTICE 3

EXAMPLE 3

Find the opposite of each number in the table.

	a.	b.	c.	d.
Number	10	$-\dfrac{1}{3}$	$10\frac{1}{2}$	-0.3
Opposite				

Solution

	a.	b.	c.	d.
Number	10	$-\dfrac{1}{3}$	$10\frac{1}{2}$	-0.3
Opposite	-10	$\dfrac{1}{3}$	$-10\frac{1}{2}$	0.3

PRACTICE 3

Find the opposite of each number.

Number	Opposite
a. -41	
b. $-\dfrac{8}{9}$	
c. 1.7	
d. $-\dfrac{2}{5}$	

Because the number -3 is negative, it lies 3 units to the left of 0. The number 3 is positive and lies 3 units to the right of 0.

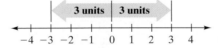

When we locate a number on the number line, the distance that number is from 0 is called its *absolute value*. Thus, the absolute value of $+3$, which we write as $|3|$, is 3. Similarly, the absolute value of -3, or $|-3|$, is 3.

Definition

The **absolute value** of a number is its distance from 0 on the number line. The absolute value of the number n is written as $|n|$.

The following properties help us to compute the absolute value of any number:

- The absolute value of a positive number is the number itself.
- The absolute value of a negative number is its opposite.
- The absolute value of 0 is 0.
- The absolute value of a number is always positive or 0.
- Opposites have the same absolute value.

EXAMPLE 4

Evaluate.

a. $|-6|$ **b.** $|4.6|$ **c.** $\left|-\dfrac{1}{3}\right|$ **d.** $-|-2|$

Solution

a. $|-6| = 6$ Because the absolute value of a negative number is its opposite, the absolute value of -6 is 6.

b. $|4.6| = 4.6$ Because the absolute value of a positive number is the number itself, the absolute value of 4.6 is 4.6.

c. $\left|-\dfrac{1}{3}\right| = \dfrac{1}{3}$ Because $-\frac{1}{3}$ is negative, its absolute value is its opposite, namely, $\frac{1}{3}$.

d. $-|-2| = -(2) = -2$ Because $|-2| = 2$, we get $-|-2| = -2$.

PRACTICE 4

Find the value.

a. $\left|\dfrac{1}{2}\right|$ **b.** $|0|$

c. $|-9|$ **d.** $-|-3|$

Comparing Real Numbers

The number line helps us to compare two real numbers, that is, to determine which number is greater.

Given two numbers on the number line, *the number to the right is greater than the number to the left*. Similarly, *the number to the left is less than the number to the right*. The *equal sign* $(=)$ and the *inequality symbols* $(\neq, <, \leq, >,$ and $\geq)$ are used to compare numbers.

$=$	means *is equal to*	$\dfrac{5}{2} = 2\frac{1}{2}$ is read "$\dfrac{5}{2}$ is equal to $2\frac{1}{2}$."
$\neq$	means *is not equal to*	$3 \neq -3$ is read "3 is not equal to -3."
$<$	means *is less than*	$-1 < 0$ is read "-1 is less than 0."
$\leq$	means *is less than or equal to*	$4 \leq 7$ is read "4 is less than or equal to 7."
$>$	means *is greater than*	$-2 > -5$ is read "-2 is greater than -5."
$\geq$	means *is greater than or equal to*	$3 \geq 1$ is read "3 is greater than or equal to 1."

The statements $3 \neq -3$, $-1 < 0$, $4 \leq 7$, $-2 > -5$, and $3 \geq 1$ are *inequalities*. The statement $4.5 = 4\frac{1}{2}$ is an *equation*.

When comparing real numbers, it is important to remember the following:

- Zero is greater than any negative number because all negative numbers on the number line lie to the left of 0. For example, $0 > -5$.

- Zero is less than any positive number because all positive numbers lie to the right of 0 on the number line. For example, $0 < 3$.

- Any positive number is greater than any negative number because on the number line all positive numbers lie to the right of all negative numbers. For example, $4 > -1$.

EXAMPLE 5

Indicate whether each inequality is true or false. Explain.

a. $0 > -1$ **b.** $0 \geq \dfrac{1}{2}$ **c.** $-3.5 < 1.5$

d. $-10 \leq -10$ **e.** $-4 > -2$

Solution

a. $0 > -1$ True, because 0 is to the right of -1.

b. $0 \geq \dfrac{1}{2}$ False, because 0 is to the left of $\frac{1}{2}$.

c. $-3.5 < 1.5$ True, because -3.5 is to the left of 1.5.

d. $-10 \leq -10$ True, because -10 is equal to -10.

e. $-4 > -2$ False, because -4 is to the left of -2.

PRACTICE 5

Determine whether each inequality is true or false. Explain.

a. $-2 < -1$

b. $0 \leq -5$

c. $\dfrac{10}{4} \geq \dfrac{5}{2}$

d. $0.3 > 0$

e. $-2.4 > 1.6$

We noted in Example 5(a) that $0 > -1$. Is $-1 < 0$? Explain why.

EXAMPLE 6

Graph the numbers $-2, \dfrac{1}{2}, -1,$ and $-\dfrac{1}{4}$. Then list them in order from least to greatest.

Solution

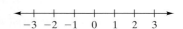

Reading the graph, we see that the numbers in order from the least to the greatest are $-2, -1, -\dfrac{1}{4},$ and $\dfrac{1}{2}$.

PRACTICE 6

Graph the numbers $-2.4, 3, -\dfrac{1}{2},$ and -1.6. Then write them in order from largest to smallest.

EXAMPLE 7

The average temperature in Fairbanks, Alaska, in the month of December is $-6°F$. The average temperature in Barrow, Alaska, in the same month is $-11°F$. Which place is warmer in December? Explain. (*Source: The National Climatic Data Center*)

Solution We need to compare -6 with -11. Because $-6 > -11$, it is warmer in Fairbanks in December.

PRACTICE 7

The table below shows the elevation of three lakes.

Lakes	Elevation (ft)
Caspian Sea (Asia-Europe)	−92
Lake Maracaibo (South America)	0
Lake Eyre (Australia)	−52

(*Source: Geological Survey, U.S. Department of the Interior*)

Which lake has the lowest elevation? Explain.

EXAMPLE 8

Historians use number lines, called *timelines*, to show dates of historical events. On a timeline, the years before the birth of Christ are denoted as B.C., and after the birth of Christ as A.D. The B.C. dates are considered to be negative numbers on the timeline.

a. Locate on the following timeline the world history events before A.D. 1600 shown in the table.

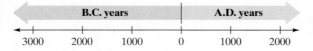

Event	Date
(A) Hieroglyphic writing developed in Egypt.	3200 B.C.
(B) Charlemagne (Charles the Great) was crowned emperor by Pope Leo III in Rome.	A.D. 800
(C) Hun invaders from Asia entered Europe.	A.D. 372
(D) Mayan civilization began to develop in Central America.	1500 B.C.
(E) Sweden seceded from the Scandinavian Union.	A.D. 1523
(F) In Greece, the Parthenon was built.	438 B.C.
(G) The city of Rome was founded, according to legend, by Romulus.	753 B.C.
(H) The evolution of England's unique political institutions began with the *Magna Carta*.	A.D. 1215

(*Source: The World Almanac Book of Facts, 2000*)

b. Order the events from the most recent to the earliest event.

Solution B.C. dates are considered to be negative numbers, and A.D. dates positive numbers. So we graph on a timeline similar to the way we graph on a number line. That is, we locate each year on the timeline and mark it as shown below.

a.

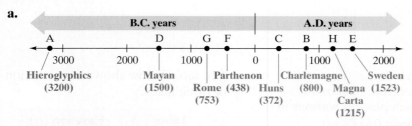

b. The events from the most recent to the earliest: E, H, B, C, F, G, D, and A.

PRACTICE 8

Consider the following table that shows the highlights in the development of algebra.

Event	Date
(A) Babylonian algebra found on cuneiform clay tablets dates back to the reign of King Hammurabi.	1700 B.C.
(B) Greek algebra (as formulated by the Pythagoreans) was geometric.	500 B.C.
(C) The Greek mathematician Diophantus introduced a style of writing equations.	A.D. 250
(D) Greek algebra (as formulated by Euclid) was geometric.	300 B.C.
(E) Bhaskara was one of the most prominent Hindu mathematicians in algebra.	A.D. 1100
(F) Mohammed ibn-Musa al-Khowarizmi wrote the book *Al-jabr* (translated as "algebra" in Latin).	A.D. 825
(G) Modern symbols and notation in algebra emerged.	A.D. 1500

(***Source:*** NCTM, *Historical Topics for the Mathematics Classroom*, 1969)

a. Locate on the timeline below the events listed in the table.

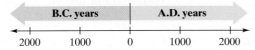

b. Then order these highlights from the earliest to the most recent event.

1.1   Exercises

Mathematically Speaking

Fill in each blank with the most appropriate term or phrase from the given list.

opposite	center	origin
rational numbers	natural numbers	negative numbers
whole numbers	zero	empty
absolute value	irrational numbers	real numbers
either terminate or	integers	neither terminate nor

1. The symbol ∅ represents the _____ set.

2. The set of _____ is {1, 2, 3, 4, ...}.

3. The _____ consist of 0 and the natural numbers.

4. Numbers to the left of 0 on the number line are _____.

5. The set of _____ is {..., −4, −3, −2, −1, 0, +1, +2, +3, +4, ...}.

6. The point at 0 on the number line is called the _____.

7. Real numbers that cannot be written as the quotient of two integers are called _____.

8. The corresponding decimal representations of irrational numbers _____ repeat.

9. The rational and the irrational numbers together make up the _____.

10. The _____ of 6.3 is −6.3.

Express each quantity as a positive or negative number.

11. 5 km below sea level

12. A profit of $1000

13. A temperature drop of 22.5°C

14. A gain of $6\frac{1}{2}$ lb

15. A withdrawal of $160 from an account

16. A debt of $1500.45

Graph each number on the number line.

17. −3

18. 2.2

19. $-\frac{1}{2}$

20. 0

21. $-\frac{3}{8}$

22. $2\frac{9}{10}$

23. −2.9

24. $-\frac{1}{4}$

Classify each number by writing a check in the appropriate boxes.

	Whole Numbers	Integers	Rational Numbers	Real Numbers
25. -7				
26. $3\frac{1}{6}$				
27. -1.9				
28. 0				
29. 10				
30. $\frac{2}{5}$				

Find the opposite of each number.

31. -3 **32.** 21 **33.** 0 **34.** $\frac{2}{3}$

35. -3.5 **36.** $1\frac{1}{2}$

Evaluate.

37. $|-4|$ **38.** $\left|-\frac{2}{3}\right|$ **39.** $|0|$ **40.** $|-15|$

41. $|-4.6|$ **42.** $|2.6|$ **43.** $-\left|\frac{1}{2}\right|$ **44.** $-\left|-\frac{6}{5}\right|$

Solve. If impossible, explain why.

45. Name all numbers that have an absolute value of 4.   **46.** Name all numbers that have an absolute value of 0.4.

47. Name a number whose absolute value is -2.   **48.** Name three different numbers that have the same absolute value.

Indicate whether each inequality is true or false.

49. $-7 < -5$ **50.** $-1 > 3$ **51.** $-1 < 2.5$

52. $-3.2 > -3$ **53.** $0 \geq -1\frac{1}{4}$ **54.** $-6 \leq -6$

Replace each ▮ with $<$, $>$, or $=$ to make a true statement.

55. 0 ▮ -1 **56.** -7 ▮ 4 **57.** -1.5 ▮ -2 **58.** -1 ▮ -1.6

59. 2.5 ▮ $2\frac{1}{2}$ **60.** $-\frac{1}{5}$ ▮ 0.2 **61.** $|-4|$ ▮ $|4|$ **62.** $-|5|$ ▮ $|-5|$

63. 6.2 ▮ $|-7.1|$ **64.** $-\left|\frac{1}{4}\right|$ ▮ $-\frac{2}{3}$

Graph the numbers in each group on the number line. Then, write the numbers from largest to smallest.

65. $3\frac{1}{2}, -1\frac{1}{2}, -\frac{1}{2}, 0$ ← ┼─┼─┼─┼─┼─┼─┼─┼─┼ →
$\qquad\qquad\qquad\quad$ −4 −3 −2 −1 0 1 2 3 4

66. $-1, 2, -2, -3, 1$ ← ┼─┼─┼─┼─┼─┼─┼─┼─┼ →
$\qquad\qquad\qquad\quad$ −4 −3 −2 −1 0 1 2 3 4

67. $-3, 3, -3.5, 3.5$ ← ┼─┼─┼─┼─┼─┼─┼─┼─┼ →
$\qquad\qquad\qquad\quad$ −4 −3 −2 −1 0 1 2 3 4

68. $2\frac{1}{2}, -4, 3, -2.5$ ← ┼─┼─┼─┼─┼─┼─┼─┼─┼ →
$\qquad\qquad\qquad\quad$ −4 −3 −2 −1 0 1 2 3 4

Mixed Practice

Solve.

69. Express the quantity as a positive or negative number: a loss of $53.

70. Graph the number $-1\frac{5}{8}$ on the number line.

$\qquad$ ← ┼─┼─┼─┼─┼─┼─┼ →
$\qquad\qquad$ −3 −2 −1 0 1 2 3

71. Classify the number by writing a check in the appropriate boxes.

	Whole Numbers	Integers	Rational Numbers	Real Numbers
2.6				

72. Find the opposite of the number $-\dfrac{5}{6}$.

73. Evaluate $-|-1.5|$

74. Name all numbers that have an absolute value of –3. If impossible, explain why.

75. Indicate whether the inequality $-5 \leq -5\frac{1}{3}$ is true or false.

Replace the ▨ with <, >, or = to make a true statement.

76. -7.8 ▨ -8.2

77. $-|3|$ ▨ $|-3|$

78. Graph the numbers $\dfrac{1}{2}, -2\frac{1}{2}, 1\frac{1}{2}, -2$ on a number line. Then write the numbers from largest to smallest.

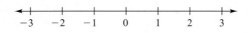

Applications

Solve.

79. Today, a person in debt owes $200. Last week, he owed $2000. Was he better off financially last week, or is he better off today?

80. Will a dieter weigh more if she loses 6 lb or gains 2 lb?

81. Three of the coldest temperature readings ever recorded on Earth were −64.8°C, −64.3°C, and −54.5°C. Of these three temperatures, which was the coldest?

82. Each liquid has its own boiling point—the temperature at which it changes to a gas. Liquid chlorine, for example, boils at −35°C, whereas liquid fluorine boils at −188°C. Which liquid has the higher boiling point? (***Source:*** *Handbook of Chemistry and Physics*)

83. Astronomers use the term *apparent magnitude* to indicate the brightness of a star as seen from Earth. The following number line shows the apparent magnitude of various stars and other objects. For historical reasons, the brighter a star or object is, the farther to the left it is graphed on the number line.

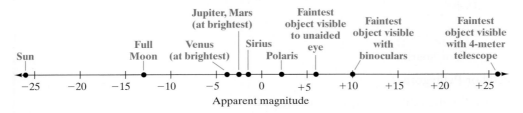

a. Which star is brighter as seen from Earth—Polaris or Sirius?

b. Use this number line to estimate the apparent magnitude of the full Moon.

c. The giant star Beta Sagittae lies hundreds of light-years from Earth, with an apparent magnitude of 4.4. Plot this star's apparent magnitude on the number line.

84. The following timeline shows the year of some major technological innovations.

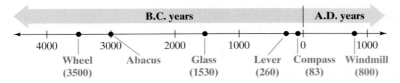

(***Source:*** Bill Yenne, *100 Inventions That Shaped World History*, 1993)

a. Which was invented earlier—the lever or the compass?

b. According to the timeline, what was invented between 2000 B.C. and A.D. 1000?

c. The mechanical clock was invented around A.D. 950. Plot a point on the timeline to represent this invention.

85. Below are the birth years for a variety of famous people throughout history.

Leonardo da Vinci (Italian painter)	A.D. 1452	Aristotle (Greek philosopher/scientist)	384 B.C.
Keith Richards (English musician)	A.D. 1943	Wolfgang Amadeus Mozart (Austrian composer)	A.D. 1756
Julius Caesar (Roman statesman)	100 B.C.	Euclid (Greek mathematician)	1300 B.C.
Charles Dickens (English writer)	A.D. 1812	Miles Davis (American jazz musician)	A.D. 1926
William Shakespeare (English playwright)	A.D. 1564	Marie Antoinette (French queen)	A.D. 1755
Pythagoras (Greek mathematician)	580 B.C.	Oliver Cromwell (English statesman)	A.D. 1599
Tiger Woods (American golfer)	A.D. 1976	Socrates (Greek philosopher)	469 B.C.
Attila (Hun king)	A.D. 406		

(*Source: Chambers Biographical Dictionary*)

a. Who was born earlier, Aristotle or Socrates?

b. Who was born later, Euclid or Pythagoras?

c. Which of the individuals listed in the table was born the earliest?

d. Which of the individuals listed in the table was born most recently?

86. The following table gives the change from the previous month in the opening stock price per share for Apple, Inc. (stock symbol: AAPL) for each month of 2006.

Jan.	Feb.	Mar.	Apr.	May	Jun.
$3.13	−$6.46	−$6.12	$6.72	−$11.00	−$2.58

Jul.	Aug.	Sept.	Oct.	Nov.	Dec.
$10.44	$0.63	$8.50	$5.98	$10.56	−$6.96

(*Source:* Yahoo! Finance)

a. What was the greatest increase in the opening stock price?

b. What was the smallest increase in the opening stock price?

c. What was the greatest decrease in the opening stock price?

d. What was the smallest decrease in the opening stock price?

• Check your answers on page A-2.

MINDSTRETCHERS

Groupwork

1. Working with a partner, develop a diagram to show the relationship among the real numbers, the rational numbers, the irrational numbers, the integers, the noninteger rational numbers, the whole numbers, the negative integers, the natural numbers, and zero.

Mathematical Reasoning

2. Is there a largest number less than 5 that is

  a. an integer?

  b. a rational number?

Research

3. Using your college library or the Web, investigate the role the Pythagoreans played in discovering irrational numbers. Write a few sentences to summarize your findings.

CULTURAL NOTE

Sources:
Lancelot Hogben, *Mathematics in the Making* (London: Galahad Books, 1960).
Henri Michel, *Scientific Instruments in Art and History* (New York: Viking Press, 1966).
Calvin C. Clawson, *The Mathematical Traveler* (New York and London: Plenum Press, 1994).

Up until the work of sixteenth-century Italian physicists, no one was able to measure temperature. Liquid-in-glass thermometers were invented around 1650, when glass blowers in Florence were able to create the intricate shapes that thermometers require. Thermometers from the seventeenth and eighteenth centuries provided a model for working with negative numbers that led to their wider acceptance in the mathematical and scientific communities. Numbers above and below 0 represented temperatures above and below the freezing point of water, just as they do on the Centigrade scale today. Before the introduction of these thermometers, a number such as −1 was difficult to interpret for those who believed that the purpose of numbers is to count or to measure.

By contrast, the early Greek mathematicians had rejected negative numbers, calling them absurd. A thousand years later in the seventeenth century A.D., the Indian mathematician Brahmagupta argued for accepting negative numbers and wrote the first comprehensive rules for computing with them.

1.2   Addition of Real Numbers

In Chapter R, computations were restricted to zero and positive numbers—whether those positive numbers happened to be whole numbers, fractions, or decimals. Now we consider computations involving negative numbers as well. Let's start with addition.

Suppose that we want to add two negative numbers: -2 and -3. It is helpful to look at this problem using a real-world situation. If you have a debt of $2 and another debt of $3, altogether you owe $5.

$$-2 + (-3) = -5$$

The number line gives us a picture of adding real numbers. To add -2 and -3, we start at the point corresponding to the first number, -2. The second number, -3, is negative, so we move 3 units to the *left*. We end at -5.

Move 3 units to the *left*.

-6 -5 -4 -3 -2 -1 0 1 2 3 4 5 6
   End   Start

Note that both -2 and -3 are negative and their sum is negative.

To Add Real Numbers, *a* and *b*, on a Number Line

● Start at *a*.
● Move according to *b* as follows:

      If *b* is positive, move to the right.

      If *b* is negative, move to the left.

      If *b* is zero, remain at *a*.

Consider these examples.

EXAMPLE 1	PRACTICE 1
Add -1 and 3 on the number line.	Add 2 and -3 on the number line.

Solution

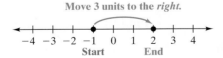

Move 3 units to the *right*.

-4 -3 -2 -1 0 1 2 3 4
      Start   End

So $-1 + 3 = 2$.

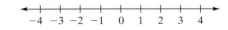

-4 -3 -2 -1 0 1 2 3 4

EXAMPLE 2	PRACTICE 2
Add 2 and -2 on the number line.	Add -5 and 5 on the number line.

Solution

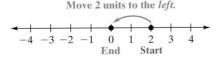

Move 2 units to the *left*.

-4 -3 -2 -1 0 1 2 3 4
         End Start

So $2 + (-2) = 0$. Note that 2 and (-2) are opposites, and their sum is 0.

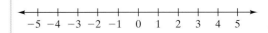

-5 -4 -3 -2 -1 0 1 2 3 4 5

Example 2 suggests that when adding two numbers that are opposites, such as 2 and -2, the sum is 0. We call such numbers **additive inverses**. Every real number has an opposite, or additive inverse, as stated in the following property.

Additive Inverse Property

For any real number a, there is exactly one real number $-a$ such that

$$a + (-a) = 0 \quad \text{and} \quad (-a) + a = 0.$$

EXAMPLE 3

Using a number line, add $-1\frac{1}{2}$ and 0.

Solution

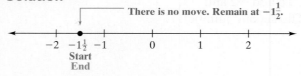

So $-1\frac{1}{2} + 0 = -1\frac{1}{2}$. Note that adding 0 to $-1\frac{1}{2}$ gives us $-1\frac{1}{2}$, the same number we started with.

PRACTICE 3

Add 0 and 0.5 on the following number line.

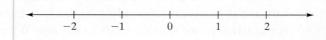

The previous examples illustrated how to add real numbers on a number line. However, this method is not very efficient, especially for adding large numbers. Instead, let's look at a *rule* for the addition of real numbers.

To Add Real Numbers

- If the numbers have the same sign, add their absolute values and keep their sign.
- If the numbers have different signs, find the difference between the larger absolute value and the smaller one and take the sign of the number with the larger absolute value.

Now let's consider some examples of applying this rule for adding real numbers.

EXAMPLE 4

Add -8 and -19.

Solution Because both numbers are negative, we find their absolute values and then add:

$$|-8| + |-19| = 8 + 19 = 27$$

The sum of two negative numbers is negative.

$$-8 + (-19) = -27$$

Negative numbers ——⌐   ⌐— Negative sum

PRACTICE 4

Find the sum: $-13 + (-18)$

Tip When adding numbers with the same sign, the sum has that sign.

EXAMPLE 5	PRACTICE 5

Find the sum of -6.7 and 5.2.

Solution Here we are adding numbers with *different* signs. First, we find the absolute values:

$$|-6.7| = 6.7 \quad \text{and} \quad |5.2| = 5.2$$

Then we find the difference between the larger absolute value and the smaller one:

$$6.7 - 5.2 = 1.5$$

Because -6.7 has the larger absolute value and its sign is negative, the sum is also negative. So

$$-6.7 + (+5.2) = -6.7 + 5.2 = -1.5$$

Negative number ⟶ │   │⟶ Positive number   │⟶ The sum takes the sign of the number with the larger absolute value.

Add: $10.1 + (-6.6)$

Can you use the number line to explain why the sum in Example 5 is negative?

EXAMPLE 6	PRACTICE 6

Combine: $-\dfrac{2}{9} + \dfrac{2}{9}$

Solution

$$-\frac{2}{9} + \frac{2}{9} = 0$$

Note that $-\dfrac{2}{9}$ and $\dfrac{2}{9}$ are additive inverses. So their sum is 0.

Combine: $-\dfrac{1}{3} + \dfrac{1}{3}$

EXAMPLE 7	PRACTICE 7

Add: $\dfrac{3}{8} + \left(-\dfrac{5}{16}\right)$

Solution

First we find the LCD of the fractions. The LCD is 16. So

$$\frac{3}{8} + \left(-\frac{5}{16}\right) = \frac{6}{16} + \frac{-5}{16} = \frac{1}{16}.$$

Add: $-\dfrac{7}{18} + \dfrac{1}{9}$

EXAMPLE 8	PRACTICE 8

In a chemistry class, a student studies the properties of atomic particles, including protons and electrons. She learns that a proton has an electric charge of $+1$, whereas an electron has an electric charge of -1. The charge of a proton cancels out that of an electron. If a charged particle of magnesium has 12 protons and 10 electrons, what is its total charge?

The price of a certain stock on Monday was $37.50 per share. On Tuesday, the price of a share went up $2; on Wednesday, it went down $1; and on Thursday, it went down another $2. What was the share price of the stock on Thursday?

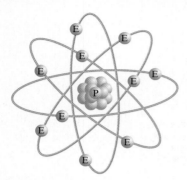

Solution We can represent the 12 protons and 10 electrons by $+12$ and -10, respectively. To find the total charge of the particle, we add $+12$ and -10.

$$+12 + (-10) = +2$$

So the total charge of the particle of magnesium is $+2$.

 Calculators and Real Numbers

To enter a negative number, we need to hit the key that indicates that the sign of the number is negative. Some calculators have a *change-of-sign key*, $\boxed{+/-}$. Others have a negative sign key, $\boxed{(-)}$. Do not confuse either of these keys with the subtraction key, $\boxed{-}$. When entering a negative number, the order in which the change-of-sign key or the negative key must be pressed will depend on the calculator being used. The $\boxed{+/-}$ key is usually pressed *after* the number is entered, whereas the $\boxed{(-)}$ key is usually pressed *before* the number is entered.

EXAMPLE 9	PRACTICE 9

Calculate: $-1.3 + (-5.8)$

Solution

Add: $6.002 + (-9.37) + (-0.22)$

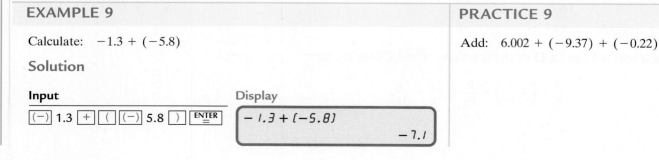

1.2   **Exercises**   FOR EXTRA HELP   *MyMathLab* Math XL PRACTICE   WATCH   DOWNLOAD   READ   REVIEW

Mathematically Speaking

Fill in each blank with the most appropriate term or phrase from the given list.

positive	opposites	subtraction
identities	smaller	property
larger	additive inverse	
negative	property	

1. To add real numbers a and b on a number line, start at a and move to the left if b is _____.

2. The _____ states that for any real number a, there is exactly one real number $-a$ such that $a + (-a) = 0$ and $(-a) + a = 0$.

3. Additive inverses are also called _____.

4. To add real numbers with different signs, find the difference between their absolute values and take the sign of the number with the _____ absolute value.

Find the sum of each pair of numbers using the number line.

5. $4 + (-3)$

6. $-8 + 7$

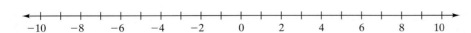

7. $8 + (-8)$

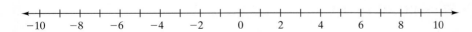

8. $-3 + (-4)$

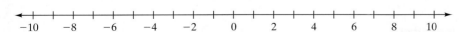

9. $-5\frac{1}{2} + 10$

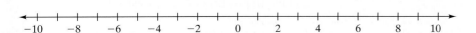

10. $-1.5 + 2$

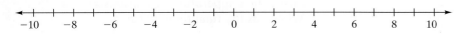

Find the additive inverse.

11. −18   **12.** 6.9   **13.** 0   **14.** $-\dfrac{12}{17}$

Find the sum.

15. 24 + (−1)   **16.** −12 + 5   **17.** −10 + 5   **18.** −6 + (−6)

19. 10 + (−6)   **20.** 60 + (−90)   **21.** −50 + (−30)   **22.** −9 + (−4)

23. −10 + 2   **24.** 2 + (−10)   **25.** −18 + 18   **26.** (−18) + (−18)

27. 5.2 + (−0.9)   **28.** −0.6 + 2   **29.** −0.2 + 0.8   **30.** (−0.1) + 0.1

31. −9.6 + 3.9   **32.** 6.1 + (−5.9)   **33.** (−9.8) + (−6.5) **34.** −0.8 + (−0.9)

35. $-\dfrac{1}{2} + \left(-\dfrac{1}{2}\right)$ **36.** $\left(-\dfrac{1}{4}\right) + \left(-\dfrac{3}{4}\right)$ **37.** $\dfrac{4}{15} + \left(-\dfrac{2}{3}\right)$ **38.** $-\dfrac{5}{6} + \dfrac{7}{12}$

39. $-1\frac{3}{5} + 2$ **40.** $-10 + 3\frac{1}{2}$ **41.** $2\frac{1}{3} + \left(-1\frac{1}{2}\right)$ **42.** $-6\frac{1}{4} + \left(-1\frac{1}{3}\right)$

43. −24 + (25) + (−89)   **44.** 36 + (−17) + (−28)   **45.** −0.4 + (−2.6) + (−4)

46. (−6.25) + (−0.4) + 3   **47.** 107 + (−97) + (−45) + 23   **48.** −64 + 7 + (−10) + (−19)

▦ **49.** −2.001 + (0.59) + (−8.1) + 10.756   ▦ **50.** −10 + 6.17 + (−10.005) + (−4.519)

Mixed Practice

Find the sum of each pair of numbers using the number line.

51. 2 and (−9)

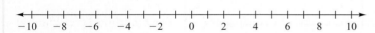

52. −4.5 and −3.5

Find the sum.

53. 53 + (−38)   **54.** (−15) + (−22)   **55.** −8.5 + 4.8

56. $2\frac{2}{3} + (-4)$ **57.** −4.1 + 2.3 + 1.5

▦ **58.** 4.81 + (−0.63) + (−10.002) + 1.05

Applications

Solve. Express your answer as a real number.

59. The temperature on the top of a mountain was 2° below 0°. If it then got 7° warmer, what was the temperature?

60. An elevator goes up 2 floors, down 3 floors, up 1 floor, and finally up 1 floor. What is the overall change in position of the elevator?

61. Last year, a corporation took in $132,000 with expenses of $148,000. How much money did the corporation make or lose?

62. In order to conduct an experiment, a chemist cooled a substance down to −5°C. In the course of this experiment, a chemical reaction took place that raised the temperature of the substance by 15°. What was the final temperature?

63. In the first quarter of Super Bowl XXIII, The San Francisco 49ers outscored the Cincinnati Bengals by three points. In the second quarter, the 49ers were outscored by three points. The Bengals scored seven more points than the 49ers in the third quarter. Finally, in the fourth quarter, the 49ers outscored the Bengals by 11 points. Who won the game and by how many points? (*Source:* ESPN Pro Football Encyclopedia)

64. In April of 2006, the average price of a gallon of regular unleaded gasoline was $2.76. In each of the next five months, May to September, the average price rose $0.19, decreased $0.03, increased $0.08, decreased $0.01, and decreased $0.04, respectively. What was the average price in September? (*Source:* Bureau of Labor Statistics)

65. With $371.25 in his checking account, an artist writes checks for $71.33 and $51.66. He also deposits $35. After the two checks clear and his deposit is credited, will he still have enough money to cover a check of $250?

66. Estimate the total annual profits for Dot.com Corporation according to the following chart. (Note that a red number written in parentheses is negative.)

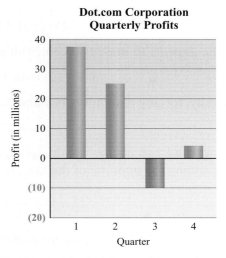

67. A plane cruising at an altitude of 32,000 ft hit an air pocket and dropped 700 ft. What was its new altitude?

68. The balance on a customer's credit card account at the start of this month was $782.48. This month, purchases of $72.99, $125.00, and $271.88 were charged to her account. There was also a credit of $250 for items returned. What was the account balance at the end of the month?

● *Check your answers on page A-2.*

1.3   Subtraction of Real Numbers

- To subtract real numbers
- To solve applied problems involving the subtraction of real numbers

The subtraction of real numbers is based on two topics that we have previously covered—adding real numbers and finding the opposite of a number.

We already know how to compute the difference between two positive numbers, for instance, $10 - 2$:

$$10 - (+2) = 8$$

In the previous section, we learned how to add a positive number and a negative number, for example, $10 + (-2)$:

$$10 + (-2) = 8$$

Note that the answers to these two computations are the same. So,

$$10 - (+2) = 10 + (-2)$$

More generally, for any real numbers a and b,

$$a - b = a + (-b)$$

In other words, we can change a subtraction problem to an equivalent addition problem by adding the *opposite* of the number being subtracted.

To Subtract Real Numbers

- Change the operation of subtraction to addition and change the number being subtracted to its opposite.
- Follow the rule for adding real numbers.

Consider these examples.

EXAMPLE 1	PRACTICE 1
Find the difference: $2 - (-5)$	Find the difference: $4 - (-1)$
Solution	

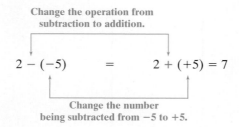

Change the operation from subtraction to addition.

$$2 - (-5) \quad = \quad 2 + (+5) = 7$$

Change the number being subtracted from -5 to $+5$.

EXAMPLE 2	**PRACTICE 2**
Subtract: $-3 - (-9)$	Subtract: $-12 - (-15)$

Solution

$$\begin{array}{c} \text{Negative 9} \quad\quad \text{Positive 9} \\ \downarrow \quad\quad\quad\quad \downarrow \\ -3 - (-9) = -3 + (+9) = 6 \\ \uparrow \quad\quad\quad\quad \uparrow \\ \text{Subtract} \quad\quad\quad \text{Add} \end{array}$$

EXAMPLE 3	**PRACTICE 3**
Find the difference: $2 - 5$	Find the difference: $8 - 12$

Solution

$$2 - 5 = 2 - (+5) = 2 + (-5) = -3$$

EXAMPLE 4	**PRACTICE 4**
Subtract: $9 - (-13.2)$	Subtract: $-8.1 - 7.6$

Solution

$$9 - (-13.2) = 9 + (+13.2) = 22.2$$

In Chapter R, we discussed the order of operations rule, which states that when a problem involves addition and subtraction, we work from left to right. Let's look at how this rule works with positive and negative numbers.

EXAMPLE 5	**PRACTICE 5**
Simplify: $-2 - (-6) - (-11)$	Simplify: $5 - (-8) - (-15)$

Solution

$$\begin{aligned} -2 - (-6) - (-11) &= \underbrace{-2 + (+6)} - (-11) && \text{Subtract } -6 \text{ from } -2. \\ &= \quad\quad 4 \quad\quad - (-11) && \text{Add } -2 \text{ and } +6. \\ &= 4 + 11 && \text{Subtract } -11 \text{ from } 4. \\ &= 15 && \text{Add } 4 \text{ and } 11. \end{aligned}$$

EXAMPLE 6

Evaluate: $-7 - (-1) + 5 + (-3)$

Solution

$-7 - (-1) + 5 + (-3)$

$= \underbrace{-7 + (+1)} + 5 + (-3)$ Subtract -1 from -7.

$= \underbrace{-6 + 5} + (-3)$ Add -7 and $+1$.

$= \underbrace{-1 + (-3)}$ Add -6 and 5.

$= -4$ Add -1 and -3.

PRACTICE 6

Simplify: $4 + (-6) - (-11) + 8$

EXAMPLE 7

Egypt emerged as a nation in about 3100 B.C. and Ethiopia in about 3000 B.C. How much older is Egypt than Ethiopia? (*Source: The World Book Encyclopedia*)

Solution Recall that a B.C. year corresponds to a negative integer. So 3100 B.C. and 3000 B.C. are represented by -3100 and -3000, respectively. Because $-3000 > -3100$, we write -3000 first:

$$-3000 - (-3100) = -3000 + (+3100)$$
$$= 100$$

So Egypt is 100 years older than Ethiopia.

PRACTICE 7

Paper was invented in China in about 100 B.C., and wood block printing in about A.D. 770. How much older is the invention of paper than that of wood block printing? (*Source: The World Book Encyclopedia*)

1.3 Exercises   FOR EXTRA HELP   *MyMathLab*

Find the difference.

1. $25 - 8$

2. $8 - 25$

3. $-24 - 7$

4. $-6 - 9$

5. $(-19) - 25$

6. $-49 - 2$

7. $52 - (-19)$

8. $24 - (-31)$

9. $60 - 95$

10. $95 - 60$

11. $-34 - (-2)$

12. $-30 - (-1)$

13. $16 - (-16)$

14. $(-16) - 16$

15. $0 - 45$

16. $45 - 0$

17. $-31 - 31$

18. $31 - 31$

19. $22 - (-22)$

20. $8 - (-19)$

21. $200 - (-800)$

22. $30 - (-10)$

23. $6 - 7.42$

24. $10.1 - 11.84$

25. $-7.3 - (0.5)$

26. $-3 - (-0.1)$

27. $(-5.6) - (-5.6)$

28. $0.4 - (-0.4)$

29. $8.6 - (-1.7)$

30. $-1.7 - 8.6$

31. $-\dfrac{1}{3} - \dfrac{5}{6}$

32. $\dfrac{4}{5} - \left(-\dfrac{7}{8}\right)$

33. $-12 - \dfrac{1}{4}$

34. $-12 - \left(-\dfrac{1}{4}\right)$

35. $4\frac{3}{5} - (-1\frac{1}{2})$

36. $-6\frac{1}{2} - (-1\frac{1}{4})$

Simplify.

37. $3 + (-6) - (-15)$

38. $-12 - (-4) + 9$

39. $8 - 10 + (-5)$

40. $12 + (-16) - (-5)$

41. $-9 + (-4) - 9 + 4$

42. $-6 - (-1) + 5 + (-8)$

43. $9 - 12 - 18$

44. $-7 - (-4) - 2$

45. $-10.722 + (-3.913) - 8.36 - 3.492$

46. $1.884 - 0.889 + (-6.12) - (-4.001)$

Mixed Practice

Find the difference.

47. $17 - (-31)$

48. $-7.6 - 5.8$

49. $(-23) - 15$

50. $-28 - (-17)$

51. $\dfrac{5}{6} - \left(-\dfrac{3}{4}\right)$

Simplify.

52. $-5 - (-3) + 2 + (-9)$

53. $18 + (-13) - (-9)$

54. $-3.712 + (-5.003) - 0.762 - (-4.73)$

Applications

Solve. Express your answer as a real number.

55. The first Olympic Games occurred in 776 B.C. Approximately how many centuries were there between the first Olympic games and the Olympic games held in A.D. 2002?

56. Going into her last round of a golf tournament, a golfer's score was 4 strokes over par. At the end of the tournament, it was 2 strokes under par. During the last round, what was her score relative to par?

57. Two airplanes take off from the same airport. One flies north and the other flies south, as pictured. How far apart are the airplanes? (*Hint:* Consider north to be positive and south to be negative.)

58. After 15 min a cheetah running due east has gone 17.5 mi while a coyote heading due west from the same point at the same time runs 10.75 mi. How far apart are the two animals at that time?

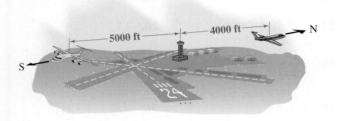

59. A company reported a loss of $281,330 last year and a loss of $5291 this year. By how much money was the company's loss reduced?

60. The value of a computer company's stock rose by $0.50 per share and then dropped by $0.75. What was the overall change in value?

61. The U.S. city with the greatest elevation is Leadville, Colorado. The U.S. city with the lowest elevation is Calipatria, California. If their respective elevations are 10,152 ft above sea level and 184 ft below sea level, what is the difference in the elevations of these cities?

62. The following chart shows the record high and low temperatures (in degrees Fahrenheit) for a number of U.S. states.

State	Record High (°F)	Record Low (°F)
Alabama	112	−27
California	134	−45
Louisiana	114	−16
Minnesota	114	−60
New York	108	−52
Virginia	110	−30

Which state had the greatest difference in record extreme temperatures?

63. Matter is liquid when its temperature is between its melting point and its boiling point. The following table shows the melting and boiling points (in degrees Celsius) of various elements.

Element	Melting Point (°C)	Boiling Point (°C)
Radon	−71	−61.8
Neon	−248.7	−246
Bromine	−7.2	58.8

(*Source: The New York Times Almanac,* 2000)

a. For each of these elements, find the difference between its boiling and melting points.

b. Which of the elements is liquid in the widest range of temperatures?

c. Which of the elements is liquid at 0°C?

64. Superconductors allow for very efficient passage of electric currents. For practical use, a superconductor must work above −196°C, which is the boiling point of nitrogen.

a. In 1986, the first high-temperature superconductor that was able to conduct electricity without resistance at a temperature of −238°C was discovered. How many degrees below the boiling point of nitrogen did this first high-temperature superconductor work?

b. In 1987, the researcher Paul Chu discovered a new class of materials that conduct electricity at −178°C. How many degrees above the boiling point of nitrogen did the new materials conduct electricity?

c. In 1990, other researchers created a miniature transistor that conducts electricity at a temperature that is 48°C above the boiling point of nitrogen. What is this temperature?
(*Sources: World Book Encyclopedia* and *Webster's New World Book of Facts,* 1999)

● *Check your answers on page A-2.*

MINDSTRETCHERS

Mathematical Reasoning

1. Consider the following two problems: $8 - (-2) = 8 + 2 = 10$

$$8 \div \frac{4}{7} = 8 \times \frac{7}{4} = 14$$

a. Explain in what way the two problems are *similar*.

b. Explain in what way the two problems are *different*.

Writing

2. Explain whether it is always true that the difference between two numbers is smaller than either of the numbers. Give an example to justify your answer.

1.4   Multiplication and Division of Real Numbers

Multiplication and division are the other two of the four basic operations on real numbers.

Multiplying Real Numbers

From arithmetic, we know that when multiplying two positive numbers, we get a positive number. Now let's look at multiplying a positive number by a negative number.

Consider, for example, finding the product of 3 and -2. Looking at multiplication as repeated addition, we know that multiplying a number by 3 means adding that number to itself three times.

$$3(-2) = -2 + (-2) + (-2)$$
$$= -6$$

Note that when multiplying a positive number by a negative number, we get a negative answer.

What about multiplying two negative numbers? Let's consider the following pattern:

This number is **decreasing** by 1 each time.   The product is **increasing** by 2 each time.

$$3\,(-2) = -6$$
$$2\,(-2) = -4$$
$$1\,(-2) = -2$$
$$0\,(-2) = 0$$
$$-1\,(-2) = +2 \quad \longleftarrow \text{ The pattern continues.}$$
$$-2\,(-2) = +4$$
$$-3\,(-2) = +6$$

Note that when we multiply two negative numbers, the product is positive. This suggests the following rule.

> **To Multiply Real Numbers**
>
> - Multiply their absolute values.
> - If the numbers have the same sign, their product is positive; if they have different signs, their product is negative.

OBJECTIVES

- To multiply real numbers
- To use the order of operations rule
- To solve applied problems involving the multiplication of real numbers
- To divide real numbers
- To solve applied problems involving the division of real numbers

---

EXAMPLE 1	PRACTICE 1
Multiply -2 by -9.	Find the product of -1 and -100.

Solution

$|-2| = 2$ and $|-9| = 9$ Find the absolute values.

$2 \cdot 9 = 18$ Multiply the absolute values.

$-2(-9) = 18$ The product of two negatives is positive.

Tip

Same Sign

Positive · Positive = Positive   Negative · Negative = Positive

Different Signs

Positive · Negative = Negative   Negative · Positive = Negative

The following property states that the product of any number and zero is zero.

Multiplication Property of Zero

For any real number a,

$$a \cdot 0 = 0 \quad \text{and} \quad 0 \cdot a = 0.$$

EXAMPLE 2

Find the product.

a. $-9\left(\dfrac{1}{3}\right)$ **b.** $-\dfrac{1}{5}(-25)$ **c.** $-4(0)$

d. $-1.5(-1.5)$ **e.** $0.4(-6)$ **f.** $-7(1)$

Solution

a. $-9\left(\dfrac{1}{3}\right) = \dfrac{\overset{3}{-\cancel{9}}}{1} \cdot \dfrac{1}{\underset{1}{\cancel{3}}} = -3$ Negative · Positive = Negative

b. $-\dfrac{1}{5}(-25) = -\dfrac{1}{\underset{1}{\cancel{5}}} \cdot \dfrac{\overset{5}{-\cancel{25}}}{1} = 5$ Negative · Negative = Positive

c. $-4(0) = 0$ Multiplication property of zero

d. $-1.5(-1.5) = 2.25$ Negative · Negative = Positive

e. $0.4(-6) = -2.4$ Positive · Negative = Negative

f. $-7(1) = -7$ Negative · Positive = Negative

PRACTICE 2

Multiply.

a. $\left(-\dfrac{2}{3}\right)(-12)$

b. $\left(-\dfrac{1}{3}\right)\left(\dfrac{5}{9}\right)$

c. $(-0.4)(-0.3)$

d. $2.5(-1.9)$

e. $0 \cdot (-2.8)$

f. $1 \cdot \dfrac{2}{3}$

EXAMPLE 3

Calculate: $-3(-2)(9)$

Solution

$$-3(-2)(9) = 6(9) \qquad \text{Multiply } -3 \text{ by } -2.$$
$$= 54 \qquad \text{Multiply 6 by 9.}$$

So $(-3)(-2)(9) = 54$.

PRACTICE 3

Multiply: $-8(4)(-2)$

According to the order of operations rule given in Chapter R, multiplication is performed before either addition or subtraction, working from left to right.

Order of Operations Rule

To evaluate mathematical expressions, carry out the operations *in the following order*:

1. First, perform the operations within any grouping symbols, such as parentheses ( ) or brackets [ ].
2. Then raise any number to its power.
3. Next, perform all multiplications and divisions as they appear from left to right.
4. Finally, do all additions and subtractions as they appear from left to right.

EXAMPLE 4

Simplify: $-2(24) - 5(-6)$

Solution We use the order of operations rule.

$$
\begin{aligned}
-2(24) - 5(-6) &= -48 - (-30) && \text{Multiply first.}\\
&= -48 + 30 && \text{Subtract } -30 \text{ from } -48.\\
&= -18 && \text{Add } -48 \text{ and } 30.
\end{aligned}
$$

So $-2(24) - 5(-6) = -18$.

PRACTICE 4

Calculate: $4(-25) - (-2)(36)$

Following the order of operation rule, we simplify mathematical expressions by first performing the operations within any grouping symbols such as parentheses ( ) or brackets [ ].

EXAMPLE 5

Simplify: $-3(6 - 10)$

Solution

$$
\begin{aligned}
-3(6 - 10) &= -3(-4) && \text{Subtract within parentheses.}\\
&= 12 && \text{Multiply.}
\end{aligned}
$$

So $-3(6 - 10) = 12$.

PRACTICE 5

Calculate: $-5(-9 + 15)$

EXAMPLE 6

Calculate: $5 - [3(2 - 4)]$

Solution

$$
\begin{aligned}
5 - [3(2 - 4)] &= 5 - [3(-2)] && \text{Subtract within parentheses.}\\
&= 5 - (-6) && \text{Multiply within brackets.}\\
&= 11 && \text{Subtract.}
\end{aligned}
$$

So $5 - [3(2 - 4)] = 11$.

PRACTICE 6

Simplify: $10 - [5(3 + 1)]$

EXAMPLE 7

Temperatures can be measured in both the Fahrenheit and Celsius systems. The normal *melting point* of the element mercury is about $-37.9°F$. To find the Celsius equivalent of this temperature, we need to compute $\frac{5}{9}(-37.9 - 32)$. Simplify this expression. (**Source:** *CRC Handbook of Chemistry and Physics*)

Solution

$$\frac{5}{9}(-37.9 - 32) = \frac{5}{9}(-69.9)$$

$$\approx -38.8$$

So $-37.9°F$ is equivalent to $-38.8°C$.

PRACTICE 7

If a rock is thrown upward on the moon with an initial velocity of 10 ft/sec, the rock's velocity after 3 sec will be $10 - (5.3)(3)$ ft/sec. Simplify this expression and interpret the result. (Note: Objects moving upward have positive velocity and objects moving downward have negative velocity.) (**Source:** NASA)

Dividing Real Numbers

We now consider division—the last of the four basic operations on real numbers. From arithmetic, we know that every division problem has a related multiplication problem.

$$8 \div 2 = 4 \qquad \text{because} \qquad 2 \cdot 4 = 8$$

Division   **Related multiplication**

Now suppose that the division problem involves a negative number, for instance, $-8 \div 2$. From the previous section we know that $(-4) \cdot 2 = -8$. So $(-8) \div 2 = -4$. This problem suggests that when we divide a negative number by a positive number, we get a negative quotient.

Similarly, suppose that we want to calculate $(-8) \div (-2)$. Since $4 \cdot (-2) = -8$, it follows that $(-8) \div (-2) = 4$. This example suggests that the quotient of two negative numbers is a positive number.

We can use the following rule for dividing real numbers.

To Divide Real Numbers

- Divide their absolute values.
- If the numbers have the same sign, their quotient is positive; if the numbers have different signs, their quotient is negative.

EXAMPLE 8

Find the quotient.

a. $(-16) \div (-2)$ **b.** $\dfrac{-24}{6}$ **c.** $\dfrac{-2}{-8}$

d. $\dfrac{9.4}{-2}$ **e.** $\dfrac{-15}{-0.3}$

Solution In each problem, first we find the absolute values. Next, we divide them. Then we attach the appropriate sign to this quotient.

a. $(-16) \div (-2)$

   $|-16| = 16$ and $|-2| = 2$

   $16 \div 2 = 8$

   Since the numbers have the *same sign*, their quotient is positive. So $(-16) \div (-2) = 8$.

b. $\dfrac{-24}{6}$ $|-24| = 24$ and $|6| = 6$

   $\dfrac{24}{6} = 4$

   Since the numbers have *different* signs, the quotient is negative. So $\dfrac{-24}{6} = -4$.

c. $\dfrac{-2}{-8}$ $|-2| = 2$ and $|-8| = 8$

   $\dfrac{2}{8} = \dfrac{1}{4}$

   Since the numbers have the same signs, the quotient is positive. So $\dfrac{-2}{-8} = \dfrac{1}{4}$.

d. $\dfrac{9.4}{-2}$ $|-9.4| = 9.4$ and $|-2| = 2$

   $\dfrac{9.4}{2} = 4.7$

   Since the numbers have different signs, the quotient is negative. So $\dfrac{9.4}{-2} = -4.7$.

e. $\dfrac{-15}{-0.3}$ $|-15| = 15$ and $|-0.3| = 0.3$

   $\dfrac{15}{0.3} = \dfrac{15.0}{0.3} = \dfrac{150}{3} = 50$

   Since the numbers have the same signs, the quotient is positive. So $\dfrac{-15}{-0.3} = 50$.

PRACTICE 8

Divide.

a. $40 \div (-5)$

b. $\dfrac{-42}{-6}$

c. $\dfrac{-5}{10}$

d. $\dfrac{-6.3}{9}$

e. $\dfrac{-24}{-0.4}$

Tip

Same Sign

$$\frac{\text{Positive}}{\text{Positive}} = \text{Positive} \qquad \frac{\text{Negative}}{\text{Negative}} = \text{Positive}$$

Different Signs

$$\frac{\text{Positive}}{\text{Negative}} = \text{Negative} \qquad \frac{\text{Negative}}{\text{Positive}} = \text{Negative}$$

Some division problems involve 0. For instance, $0 \div (-5) = 0$ because $(-5) \cdot 0 = 0$. On the other hand, $(-5) \div 0$ is *undefined* because there is no real number that when multiplied by 0 gives -5.

These two examples lead us to the following conclusion.

Division Involving Zero

For any nonzero real number a,

$$0 \div a = 0.$$

For any real number a,

$$a \div 0 \text{ is undefined.}$$

EXAMPLE 9	PRACTICE 9
Find the value of each of the following expressions, if possible.	Evaluate each expression.
a. $\dfrac{0}{5}$ **b.** $0 \div 4$ **c.** $-3 \div 0$	**a.** $\dfrac{0}{-2}$
Solution	**b.** $-2 \div 0$
a. $\dfrac{0}{5} = 0$	**c.** $0 \div 7$
b. $0 \div 4 = 0$	
c. Not possible, because $-3 \div 0$ is undefined	

Recall that in Chapter R we expressed the rule for dividing fractions in terms of the *reciprocal* of a number. In algebra, we also refer to the reciprocal of a number as its *multiplicative inverse.*

Multiplicative Inverse Property

For any nonzero real number a,

$$a \cdot \frac{1}{a} = 1 \quad \text{and} \quad \frac{1}{a} \cdot a = 1,$$

where a and $\dfrac{1}{a}$ are **multiplicative inverses** (or **reciprocals**) of each other.

This property says that the product of a number and its multiplicative inverse is 1. For example,

- $\dfrac{1}{3}$ and 3 are multiplicative inverses because $\dfrac{1}{3} \cdot 3 = 1$

- $\dfrac{5}{-6}$ and $\dfrac{-6}{5}$ are multiplicative inverses because $\dfrac{5}{-6} \cdot \dfrac{-6}{5} = 1$

> **Tip**   A number and its reciprocal have the same sign.

EXAMPLE 10

Complete the following table.

Solution

Number	Reciprocal
a. 4	$\dfrac{1}{4}$ is the reciprocal of 4 because $4 \cdot \dfrac{1}{4} = 1$
b. $-\dfrac{3}{4}$	$-\dfrac{4}{3}$ is the reciprocal of $-\dfrac{3}{4}$ because $-\dfrac{3}{4} \cdot \left(-\dfrac{4}{3}\right) = 1$
c. -10	$\dfrac{1}{-10}$ or $-\dfrac{1}{10}$ is the reciprocal of -10 because $-10\left(\dfrac{1}{-10}\right) = 1$
d. $1\dfrac{1}{2}$	$\dfrac{2}{3}$ is the reciprocal of $1\dfrac{1}{2}$ because $1\dfrac{1}{2} \cdot \dfrac{2}{3} = \dfrac{3}{2} \cdot \dfrac{2}{3} = 1$

PRACTICE 10

Fill in the following table.

Number	Reciprocal
a. -5	
b. $\dfrac{1}{-8}$	
c. $1\frac{1}{3}$	
d. $-\dfrac{8}{5}$	

> **Tip**   We can rewrite a fraction such as $\dfrac{1}{-10}$ or $\dfrac{-1}{10}$ as $-\dfrac{1}{10}$. That is,
> $$\frac{1}{-10} = \frac{-1}{10} = -\frac{1}{10}.$$

Now, let's consider division of real numbers by using reciprocals. Recall that we subtract by adding an opposite. Similarly, we can divide by multiplying by a reciprocal.

Division of Real Numbers

For any real numbers a and b, where b is nonzero,

$$a \div b = \frac{a}{b} = a \cdot \frac{1}{b}.$$

This rule states that to divide two real numbers, multiply by the reciprocal of the divisor.

EXAMPLE 11

Divide.

a. $-\dfrac{1}{3} \div \dfrac{5}{6}$

b. $-\dfrac{1}{2} \div 4$

Solution

a. $-\dfrac{1}{3} \div \dfrac{5}{6} = -\dfrac{1}{3} \cdot \dfrac{6}{5} = -\dfrac{1}{\overset{}{\underset{1}{3}}} \cdot \dfrac{\overset{2}{6}}{5} = -\dfrac{2}{5}$

b. $-\dfrac{1}{2} \div 4 = -\dfrac{1}{2} \div \dfrac{4}{1} = -\dfrac{1}{2} \cdot \dfrac{1}{4} = -\dfrac{1}{8}$

PRACTICE 11

Divide.

a. $-\dfrac{8}{9} \div \dfrac{2}{3}$

b. $-10 \div \left(-\dfrac{2}{5}\right)$

We use the order of operations rule to simplify the following expressions.

EXAMPLE 12

Simplify.

a. $-8 \div (-2)(-2)$

b. $\dfrac{-5 + (-7)}{2}$

Solution

a. $-8 \div (-2)(-2) = 4(-2)$ Perform multiplications and divisions as they occur from left to right. Divide −8 by −2.
$ = -8$ Multiply.

b. $\dfrac{-5 + (-7)}{2} = \dfrac{-12}{2}$ Parentheses are understood to be around the numerator. Add −5 and −7.
$\phantom{\dfrac{-5 + (-7)}{2}} = -6$ Divide.

PRACTICE 12

Simplify.

a. $(-3)(-4) \div (2)(-2)$

b. $\dfrac{-9 - (-3)}{2}$

EXAMPLE 13

During clinical practice, a student nurse took care of a patient with a fever. He recorded the patient's temperature at the same time every day for five days. The following table shows the change in the patient's temperature each day.

Day	Temperature Change
Monday	Up 2.5°
Tuesday	Down 2°
Wednesday	Down 1.5°
Thursday	Up 1°
Friday	Down 3°

What was the average daily change in the patient's temperature?

Solution To compute the average daily change, we add the five temperature changes and divide the sum by 5—the number of days the temperature was recorded.

$$\frac{2.5 + (-2) + (-1.5) + 1 + (-3)}{5}$$

Since parentheses are understood to be around both the numerator and the denominator, we find the sum in the numerator before dividing by the denominator.

$$\frac{2.5 + (-2) + (-1.5) + 1 + (-3)}{5} = \frac{-3}{5} = -0.6$$

Therefore, the average daily change during the five days was $-0.6°$.

PRACTICE 13

During the past four weeks, the value of a particular stock market investment changed as follows.

Week	Change
1	Down $300
2	Up $200
3	Down $500
4	Up $100

What was the average weekly change in the value of the investment?

Mathematically Speaking

Fill in each blank with the most appropriate term or phrase from the given list.

associative	absolute values	undefined
reciprocal	divisor	positive
quotient	negative	zero
one	opposite	commutative
multiplicative identity		

1. The product of two real numbers with different signs is _____.

2. The _____ property of multiplication allows us to multiply two numbers in either order.

3. The _____ property tells us that the product of any number and 1 is the original number.

4. The _____ property of multiplication allows us to regroup the product of three numbers.

5. To divide real numbers, first divide the _____.

6. The quotient of two real numbers with the same sign is _____.

7. Any real number a divided by zero is _____.

8. Any nonzero real number a has a multiplicative inverse, or _____, which is written $\frac{1}{a}$.

9. The product of any real number and its multiplicative inverse is _____.

10. To divide two real numbers, multiply the dividend by the reciprocal of the _____.

Solve.

11. What property is illustrated by $-8 \cdot \frac{1}{-8} = 1$?

12. What property is illustrated by $0 \cdot \frac{1}{2} = 0$?

13. Is a number divided by zero defined?

14. Is zero divided by a nonzero number defined?

15. Can a number and its reciprocal have different signs?

16. $-4 \div 7$ can be rewritten as -4 multiplied by what number?

Answer the questions about the order of operations rule.

17. Do we raise numbers to powers before we perform operations within grouping symbols?

18. What operations do we perform last?

Find the product.

19. $6(-2)$

20. $-10 \cdot 300$

21. $-7(-3)$

22. $-5(-5)$

23. $-12\left(\frac{1}{4}\right)$

24. $15\left(-\frac{2}{3}\right)$

25. $-\frac{1}{3} \cdot \frac{4}{9}$

26. $\left(-\frac{5}{6}\right)\left(-\frac{2}{7}\right)$

27. $\left(1\dfrac{1}{3}\right)\left(-\dfrac{4}{9}\right)$ **28.** $\left(-2\dfrac{1}{5}\right)\left(-\dfrac{2}{7}\right)$ **29.** $-1.5(-0.6)$ **30.** $0.7(0.4)$

31. $1.2(-50)$ **32.** $-1.7(-4.6)$ **33.** $3(-2)(-20)$ **34.** $-9(-12)(2)$

35. $-15(-3)(0)$ **36.** $-8.5(0)(2.6)$ **37.** $-6(1)(-2)(-3)(-4)$

38. $6(-1)(-2)(3)(-4)$ **39.** $-4(5)(-6)(1)$ **40.** $10(1)(-10)(-1)$

41. $\left(-\dfrac{1}{3}\right)\left(-\dfrac{1}{3}\right)\left(-\dfrac{1}{3}\right)$ **42.** $\left(-\dfrac{1}{2}\right)\left(-\dfrac{1}{2}\right)\left(-\dfrac{1}{2}\right)$

Multiply and round to the nearest hundredth.

43. $-6.24(0.08)(-1.97)$ **44.** $-5.42(-0.19)(-4.8)$

Simplify.

45. $-7 + 3(-2) - 10$ **46.** $4 - 2(-5) - (-3)$ **47.** $-3 - 5(-6)$ **48.** $10 - 2(-8)$

49. $\left(\dfrac{3}{5}\right)(-15) - 6$ **50.** $\left(\dfrac{3}{4}\right)(-16) + 20$ **51.** $-5 \cdot (-3 + 4)$ **52.** $(-10 + 7) \cdot (-3)$

53. $-6 - [3(5 - 9)]$ **54.** $5 - [-(4 - 10)]$

Complete each table.

55.

Input	Output
a. -2	$-4(-2) - 3 = ?$
b. -1	$-4(-1) - 3 = ?$
c. 0	$-4(0) - 3 = ?$
d. 1	$-4(1) - 3 = ?$
e. 2	$-4(2) - 3 = ?$

56.

Input	Output
a. 2	$-6(2) + 2 = ?$
b. 1	$-6(1) + 2 = ?$
c. 0	$-6(0) + 2 = ?$
d. -1	$-6(-1) + 2 = ?$
e. -2	$-6(-2) + 2 = ?$

57.

	a.	**b.**	**c.**	**d.**	**e.**
Number	$-\dfrac{1}{2}$	5	$-\dfrac{3}{4}$	$3\dfrac{1}{5}$	-1
Reciprocal					

58.

	a.	**b.**	**c.**	**d.**	**e.**
Number	-12	$\dfrac{1}{4}$	7	$-2\dfrac{1}{3}$	$-\dfrac{5}{6}$
Reciprocal					

Divide.

59. $-8 \div (-1)$ **60.** $\dfrac{-12}{3}$ **61.** $-63 \div 7$ **62.** $16 \div (-4)$

63. $\dfrac{0}{-9}$ **64.** $0 \div (-10)$ **65.** $-2500 \div 100$ **66.** $-300 \div (-10)$

67. $-200 \div (-8)$ **68.** $400 \div (-5)$ **69.** $-64 \div (-16)$ **70.** $81 \div (-9)$

71. $\dfrac{-25}{-5}$ **72.** $\dfrac{-125}{5}$ **73.** $\dfrac{-2}{16}$ **74.** $\dfrac{2}{-10}$

75. $\dfrac{10}{-20}$ **76.** $\dfrac{-35}{-40}$ **77.** $\dfrac{4}{5} \div \left(-\dfrac{2}{3}\right)$ **78.** $-\dfrac{7}{12} \div \left(-\dfrac{1}{6}\right)$

79. $8 \div \left(-\dfrac{1}{4}\right)$ **80.** $(-5) \div \left(-\dfrac{1}{6}\right)$ **81.** $2\frac{1}{2} \div (-20)$ **82.** $\dfrac{4}{5} \div \left(-1\frac{1}{15}\right)$

83. $(-3.5) \div 7$ **84.** $(-0.56) \div (-8)$ **85.** $10 \div (-0.5)$ **86.** $4.5 \div -3$

87. $\dfrac{-7.2}{0.9}$ **88.** $\dfrac{-2.5}{5}$ **89.** $\dfrac{-3}{-0.3}$ **90.** $\dfrac{1.8}{-0.6}$

▦ *Divide and round to the nearest hundredth.*

91. $(-15.5484) \div (-6.13)$ **92.** $6.4516 \div (-3.54)$ **93.** $-0.8385 \div (0.715)$ **94.** $0.3102 \div (-0.129)$

Simplify.

95. $-16 \div (-2)(-2)$ **96.** $-3(-8) \div (-2)$ **97.** $(3 - 7) \div (-4)$ **98.** $[10 - (-8)] \div (-9)$

99. $\dfrac{2 + (-6)}{-2}$ **100.** $\dfrac{-10 - (-4)}{3}$ **101.** $(4 - 6) \div (1 - 5)$ **102.** $(-15 - 3) \div (-2 - 4)$

103. $-56 \div 7 - 4 \cdot (-3)$ **104.** $32 \div (-8) + (-5) \cdot 6$ **105.** $(-4)\left(\dfrac{1}{2}\right) - 2 \div \left(-\dfrac{1}{8}\right)$

106. $(-6) \div \left(\dfrac{2}{3}\right) + (-10)\left(\dfrac{2}{5}\right)$

Mixed Practice

Find the product.

107. $-2.8(-1.3)$

108. $\dfrac{2}{5}\left(-\dfrac{3}{4}\right)$

109. $3(-5)(1)(-4)(-2)$

▦ **110.** Multiply and round to the nearest hundredth: $-3.51(-0.23)(-6.4)$

111. Complete the table.

Input	Output
a. 2	$-5(2) + 4 = ?$
b. 1	$-5(1) + 4 = ?$
c. 0	$-5(0) + 4 = ?$
d. -1	$-5(-1) + 4 = ?$
e. -2	$-5(-2) + 4 = ?$

Solve.

112. Name the property of multiplication that $1.7(-6.3) = -6.3(1.7)$ illustrates.

Simplify.

113. $-5 - 6(-2) + (-3)$

114. $\left(\dfrac{4}{9}\right)(-18) - (-3)$

Divide.

115. $\left(-4\tfrac{1}{2}\right) \div 3$

116. $\dfrac{5}{6} \div \left(-\dfrac{3}{8}\right)$

117. $(-0.72) \div (-6)$

118. Divide and round to the nearest hundredth:
$(-0.7882) \div (2.36)$

119. $-65 \div (-13)$

Solve.

120. Complete the table.

	a.	b.	c.	d.	e.
Number	8	$\dfrac{2}{3}$	$-\dfrac{1}{4}$	-6	$-1\tfrac{1}{3}$
Reciprocal					

Simplify.

121. $-12 \div (5 - 7)$

122. $\dfrac{4 - (-6)}{-2}$

Applications

Solve.

123. On a double-or-nothing wager, a gambler bets $5 and loses. Express as a signed number the amount of money he lost.

124. On January 31, the high temperature in Chicago was 40°F. The high temperature then dropped 3°F per day for 3 days. If 32°F is freezing on the Fahrenheit scale, was it below freezing on February 3?

125. In the 10 games played this season, a team won 3 games by 4 points, won 2 games by 1 point, lost 4 games by 3 points, and tied in the final game. In these games, did the team score more or fewer points than its opposing teams?

126. A start-up company lost $5000 a month for the first 3 months of business. Express this loss as a signed number.

127. During a drought, the water level in a reservoir dropped 3 in. each week for 5 straight weeks. Express the overall change in water level as a signed number.

128. The submarine shown dives to 3 times its current depth of 150 ft below sea level. What is its new depth?

150 ft

129. The following tables show the number of calories in servings of various foods and the number of calories burned by various activities.

Food	Number of Calories Per Serving
Apple	80
Banana	105
Pretzel, stick	30
Ginger ale, can	125
Donut	210

(*Source: The World Almanac and Book of Facts*, 2000)

Activity (1 hr)	Number of Calories Burned*
Swimming	−288
Bicycling	−612
Football	−498
Basketball	−450
Scrubbing floors	−440

*For a 150-lb person.
(*Source: Exercise & Weight Control*, President's Council on Physical Fitness and Sports, 1986)

Find the net calories in each situation.

a. Suppose that a weight watcher eats 3 servings of pretzel sticks and then plays basketball for $\frac{1}{2}$ hour.

b. Suppose that a dieter swims for 30 minutes and drinks 2 cans of ginger ale.

c. Suppose that an athlete eats 3 servings of apples and 2 servings of donuts. Later he goes bicycling for 2 hours and then swims for 1 hour.

130. To discourage guessing on a test, an instructor takes off for wrong answers, grading according to the following scheme:

Performance on a Test Item	Score
Correct	5
Incorrect	−2
Blank	0

What score would the instructor give to each of the following tests?

	Number of Items Correct	Number of Items Incorrect	Number of Items Blank	Test Grade
a.	17	1	2	
b.	19	1	0	
c.	12	7	1	

Solve. Express your answer as a signed number.

131. The change in a stock market index over a 5-day period was −130. What was the average daily change?

132. Two investors bought an equal number of shares of stock in a company. The value of their stock fell by $7000. How much did each of the investors lose?

133. The population of a city decreased by 47,355 people in 10 years. Find the average annual change in population.

134. The federal deficit in 1940 was about $3 billion. Five years later at the end of World War II, it was about $48 billion. How many times the deficit of 1940 was that of 1945? (*Source: Budget of the United States Government, Fiscal Year 2000*, 1999)

135. A football running back lost a total of 24 yd in 6 plays. What was the average number of yards he lost on each play?

136. A client at a weight-loss clinic lost 20 lb in 15 weeks. What was the client's average weekly change in weight?

137. A small company's business expenses for the year totaled $72,000. What were the company's average monthly expenses?

138. Over a 5-year period, the height of a cliff eroded by 4.5 ft. By how many feet did the cliff erode per year?

139. A meteorologist predicted an average daily high temperature of $-3°F$ for a five-day period. During this period, the daily high temperatures (in Fahrenheit degrees) were $2°, 0°, -7° -11°$, and $1°$. Was the meteorologist's prediction correct?

140. The value of a house decreased from $183,000 to $174,000 during a decade. What was the average amount of depreciation per year?

• *Check your answers on page A-2.*

MINDSTRETCHERS

Mathematical Reasoning

1. Explain how the number line can be used to find the product of two integers.

Historical

2. At a very early age, the eighteenth-century mathematician Carl Friedrich Gauss found the sum of the first 100 positive integers within a few minutes by using the following method. First he wrote the sum both forward and backward.

$$1 + 2 + 3 + \cdots + 98 + 99 + 100$$
$$100 + 99 + 98 + \cdots + 3 + 2 + 1$$

Then he added the 100 vertical pairs, getting a sum of 101 for each pair. He concluded that the product $100(101)$ was twice the correct answer, which turns out to be $\frac{100(101)}{2} = 50(101) = 5050$. Show how to find the sum of the first 1000 *negative* integers using Gauss's method. Explain your work.

Groupwork

3. Consider the following five integers: $-2, 6, -9, 18, -36$. Working with a partner, explain which two of the five integers you would choose

  a. to find the smallest quotient.
  b. to find the largest quotient.

1.5   Properties of Real Numbers

In this section, we focus on some of the key properties of real numbers—the commutative properties, the associative properties, the identity properties, the inverse properties, and the distributive property. These properties are very important to the study of algebra, for they underlie algebraic procedures and, in particular, they help us to simplify complicated expressions.

You may wish to review the multiplicative property of 0 and the material about division involving 0 in Section 1.4.

OBJECTIVES

- To use the commutative properties
- To use the associative properties
- To use the identity properties
- To use the inverse properties
- To use the distributive property
- To solve applied problems involving the properties of real numbers

The Commutative Properties

Let's begin by considering the two commutative properties: the *commutative property of addition* and the *commutative property of multiplication*. The commutative property of addition states that we get the same sum when we add two numbers regardless of order. For instance, the sums $3 + 5$ and $5 + 3$ give the same result, 8, even though the order in which the 3 and 5 are added differs. So we write: $3 + 5 = 5 + 3$.

Commutative Property of Addition

For any real numbers a and b,
$$a + b = b + a.$$

Note that in this statement of the commutative property, as well as in the other properties of real numbers, symbols are used to represent real numbers: The letter a represents one number and the letter b represents another number. In algebra, a *variable* is a quantity that is unknown, that is, one that can change or vary in value.

EXAMPLE 1	PRACTICE 1
Rewrite each expression using the commutative property of addition.	Use the commutative property of addition to rewrite each expression.
a. $7 + (-2)$ **b.** $2x + y$	**a.** $-3 + 5$
Solution	**b.** $b + 3a$
a. $7 + (-2) = (-2) + 7$, or $-2 + 7$	
b. $2x + y = y + 2x$	

The commutative property of multiplication says that we get the same product when we multiply two numbers in any order. For example, the products $3 \cdot 5$ and $5 \cdot 3$ are the same, 15, even though the order in which the 3 and 5 are multiplied differs. So we write: $3 \cdot 5 = 5 \cdot 3$.

Commutative Property of Multiplication

For any real numbers a and b,

$$a \cdot b = b \cdot a.$$

Note that when writing an expression involving a product, we can either omit the multiplication symbol or use parentheses. For instance, we can write $a \cdot b$ as either ab or $(a)(b)$.

EXAMPLE 2	PRACTICE 2
Rewrite each expression using the commutative property of multiplication.	Use the commutative property of multiplication to rewrite each expression.
a. $(-5)(-3)$ **b.** $x(-7)$	**a.** $(-8)(2)$
Solution	**b.** $-4n$
a. $(-5)(-3) = (-3)(-5)$ **b.** $x(-7) = (-7)x$, or $-7x$	

The Associative Properties

Next we consider the two associative properties: the *associative property of addition* and the *associative property of multiplication*. The first of these properties says that if we regroup the three numbers, we get the same sum. For instance, the result of adding 2, 3, and 6 is the same, 11, whether we add 2 and 3 first and then add 6 to this sum or whether we add 3 and 6 first and then add 2 to this sum. So we write: $(2 + 3) + 6 = 2 + (3 + 6)$.

Associative Property of Addition

For any real numbers a, b, and c,

$$(a + b) + c = a + (b + c).$$

EXAMPLE 3	PRACTICE 3
Rewrite each expression using the associative property of addition.	Use the associative property of addition to rewrite each expression.
a. $[(-4) + (-3)] + 6$ **b.** $(2p + q) + r$	**a.** $[8 + (-1)] + 2$
Solution	**b.** $(x + 3y) + z$
a. $[(-4) + (-3)] + 6 = (-4) + [(-3) + 6]$	
b. $(2p + q) + r = 2p + (q + r)$	

Similarly, the associative property of multiplication states that we get the same result if we regroup the result of three numbers.

Associative Property of Multiplication

For any real numbers a, b, and c,

$$(a \cdot b)c = a(b \cdot c).$$

EXAMPLE 4	PRACTICE 4
Rewrite each expression using the associative property of multiplication.	Use the associative property of multiplication to rewrite each expression.
a. $[(-5)(-1)](2)$ **b.** $(2)(4x)$	**a.** $[(-3)(5)](-2)$
Solution	**b.** $(3)(-6n)$
a. $[(-5)(-1)](2) = (-5)[(-1)(2)]$	
b. $(2)(4x) = (2 \cdot 4)x$	

When adding three or more numbers, their sum is the same regardless of the order or grouping because of the commutative and associative properties. Note that when adding or multiplying three or more positive numbers and negative numbers, it is usually easier to work with the positives separately from the negatives.

EXAMPLE 5	PRACTICE 5
Find the sum: $5 + (-6) + 3 + (-5)$	Find the sum: $-8 + (-4) + 3 + (-8)$

Solution Let's group the numbers by sign:

$$\underbrace{5 + 3}_{\text{Positives}} \quad + \quad \underbrace{(-6) + (-5)}_{\text{Negatives}}$$

$$
\begin{aligned}
5 + 3 &= 8 && \text{Add the positives.} \\
-6 + (-5) &= -11 && \text{Add the negatives.} \\
8 + (-11) &= -3 && \text{Find the sum of the positives and the negatives.}
\end{aligned}
$$

So $5 + (-6) + 3 + (-5) = -3$.

EXAMPLE 6

Multiply.   **a.** $5(-2)(-1)(3)$
            b. $4(-1)(-3)(-2)(2)$

Solution We group the numbers by sign.

a.
$$5(3) \qquad (-2)(-1)$$
Positives   Negatives

$5(3) = 15$ Multiply the positives.
$(-2)(-1) = 2$ Multiply the negatives.
$15 \cdot 2 = 30$ Find the product of the positives and the negatives.

So $5(-2)(-1)(3) = 30$.

b.
$$4(2) \qquad -1(-3)(-2)$$
Positives   Negatives

$4(2) = 8$ Multiply the positives.
$-1(-3)(-2) = 3(-2) = -6$ Multiply the negatives.
$8(-6) = -48$ Find the product of the positive and the negative products.

So $4(-1)(-3)(-2)(2) = -48$.

PRACTICE 6

Find the product.
a. $(-6)(-1)(4)(-5)$
b. $(3)(-4)(2)(-1)(6)$

In Example 6a, the product was positive because it had *two* negative factors. By contrast, in Example 6b, the answer was negative because it had *three* negative factors. Can you explain why a product is positive if it has an even number of negative factors, but negative if it has an odd number of negative factors?

The Identity Properties

Two other important properties of real numbers are the *additive identity property* (also called the *identity property of 0*) and the *multiplicative identity property* (the *identity property of 1*).

According to the additive identity property, the sum of a number and 0 is the number itself.

Additive Identity Property

For any real number a,
$$a + 0 = a \quad \text{and} \quad 0 + a = a.$$

The multiplicative identity property states that the product of a number and 1 is the number itself.

Multiplicative Identity Property

For any real number a,
$$a \cdot 1 = a \quad \text{and} \quad 1 \cdot a = a.$$

EXAMPLE 7	PRACTICE 7
Perform the indicated operation using an identity property.	Use an identity property to simplify.
a. $-6 + 0$ **b.** $0 + 3x$	**a.** $-5 + 0$
c. $(-4)(1)$ **d.** $(-7n)(1)$	**b.** $0 + 6y$
Solution	**c.** $(1)(-2)$
a. $-6 + 0 = -6$ **b.** $0 + 3x = 3x$	**d.** $(-5x)(1)$
c. $(-4)(1) = -4$ **d.** $(-7n)(1) = -7n$	

The Inverse Properties

Next, we review the two inverse properties. We introduced the *additive inverse property* in Section 1.2.

Additive Inverse Property

For any real number a, there is exactly one number, $-a$, such that
$$a + (-a) = 0 \quad \text{and} \quad -a + a = 0,$$
where a and $-a$ are said to be *additive inverses* of each other.

EXAMPLE 8	PRACTICE 8
Find the additive inverse of each.	Find the additive inverse of each.
a. 4.5 **b.** -3 **c.** x	**a.** -2
Solution	**b.** $-\dfrac{2}{3}$
a. The additive inverse of 4.5 is -4.5.	
b. The additive inverse of -3 is 3.	**c.** y
c. The additive inverse of x is $-x$.	

We introduced the *multiplicative inverse property* in Section 1.4.

Multiplicative Inverse Property

For any nonzero real number a, there is exactly one number $\dfrac{1}{a}$ such that

$a \cdot \dfrac{1}{a} = 1$ and $\dfrac{1}{a} \cdot a = 1$, where a and $\dfrac{1}{a}$ are said to be *multiplicative inverses* (or *reciprocals*) of each other.

EXAMPLE 9

Find the multiplicative inverse of each.

a. $\dfrac{1}{3}$ **b.** 3   **c.** -2 **d.** $-\dfrac{3}{4}$

Solution

a. The multiplicative inverse of $\dfrac{1}{3}$ is $\dfrac{3}{1}$, or 3.

b. The multiplicative inverse of 3 $\left(\text{or } \dfrac{3}{1}\right)$ is $\dfrac{1}{3}$.

c. The multiplicative inverse of -2 $\left(\text{or } \dfrac{-2}{1}\right)$ is $\dfrac{1}{-2}$, or $-\dfrac{1}{2}$.

d. The multiplicative inverse of $-\dfrac{3}{4}$ $\left(\text{or } \dfrac{-3}{4}\right)$ is $\dfrac{4}{-3}$, or $-\dfrac{4}{3}$.

PRACTICE 9

Find the multiplicative inverse of each.

a. $\dfrac{1}{5}$ **b.** 2

c. -5 **d.** $-\dfrac{2}{7}$

The Distributive Property

The *distributive property of multiplication with respect to addition*, or simply the *distributive property*, involves two operations—multiplication and addition. This property allows us to multiply a number by the sum of two other numbers.

> **The Distributive Property**
>
> For any real numbers a, b, and c,
> $$a \cdot (b + c) = a \cdot b + a \cdot c.$$

Note:

- When we rewrite $a(b + c)$ as $ab + ac$, we think of this as "removing parentheses" or "distributing out."
- Rewriting $a \cdot b + a \cdot c$ as $a \cdot (b + c)$ is sometimes referred to as applying the distributive property "in reverse."
- Another way to express the distributive property is $(b + c)a = ba + ca$.
- Substituting $-c$ for c gives still another form of the distributive property: $a(b - c) = ab - ac$.

EXAMPLE 10

Rewrite each expression using the distributive property.

a. $(-5)(2 + 6)$ **b.** $0.4(x + y)$ **c.** $(3 - x) \cdot y$

Solution

a. $(-5)(2 + 6) = (-5)(2) + (-5)(6)$

b. $0.4(x + y) = (0.4)x + (0.4)y = 0.4x + 0.4y$

c. $(3 - x) \cdot y = 3 \cdot y - x \cdot y = 3y - xy$

PRACTICE 10

Rewrite each expression using the distributive property.

a. $(-2)(9 + 4.3)$

b. $0.2(a + b)$

c. $(2 - p) \cdot q$

The properties of real numbers are typically applied not in isolation but in combination. Consider the following example.

EXAMPLE 11	PRACTICE 11

EXAMPLE 11

Show that $(x + 0) + (-x) = 0$, justifying each step.

Solution

$$(x + 0) + (-x) = (0 + x) + (-x) \quad \text{The commutative property of addition}$$
$$= 0 + [x + (-x)] \quad \text{The associative property of addition}$$
$$= 0 + 0 \quad \text{The additive inverse property}$$
$$= 0 \quad \text{The additive identity property}$$

PRACTICE 11

Show that $4 \cdot \left(\dfrac{1}{4}x\right) = x$, justifying each step.

$$4 \cdot \left(\frac{1}{4}x\right) = \left(4 \cdot \frac{1}{4}\right)x \qquad \textbf{a.} \underline{\qquad}$$
$$= 1x \qquad \textbf{b.} \underline{\qquad}$$
$$= x \qquad \textbf{c.} \underline{\qquad}$$

Finally, let's examine the basic properties of real numbers in the context of real-world applications.

EXAMPLE 12

The volume of the box shown is the product of its length, width, and height: *lwh*.

Here are two ways to compute this volume:

- First find the product of the length and the width, and then multiply this product by the height: $(lw)h$.

- First find the product of the width and the height, and then multiply the length by this product: $l(wh)$.

Must the two answers be equal? Explain.

Solution The two answers must be equal. The associative property of multiplication states that $(lw)h = l(wh)$.

PRACTICE 12

A financier receives a 100% return on an investment of *n* dollars. Express this return in dollars, justifying your answer.

1.5   Exercises

PRACTICE   WATCH   DOWNLOAD   READ   REVIEW

Mathematically Speaking

Fill in each blank with the most appropriate term or phrase from the given list.

reciprocals	additive inverse property	opposites
associative property	multiplicative identity property	positive
distributive property	additive identity property	negative
commutative property		

1. The _____ of multiplication states that we get the same product when we multiply two numbers in any order.

2. The _____ of addition states that we get the same result if we regroup the sum of three numbers.

3. A product with an odd number of negative factors is _____.

4. The _____ states that the sum of a number and 0 is the number itself.

5. The _____ states that the product of a number and 1 is the number itself.

6. The _____ states that the sum of a number and its additive inverse is the additive identity 0.

7. If $a \cdot \frac{1}{a} = 1$ and $\frac{1}{a} \cdot a = 1$, then a and $\frac{1}{a}$ are multiplicative inverses, or _____, of each other.

8. The _____ involves two operations, namely, addition and multiplication.

Rewrite each expression using the indicated property of real numbers.

9. Commutative property of addition: $3.7 + 2$

10. Commutative property of multiplication: $5(-8)$

11. Associative property of addition: $[(-1) + (-6)] + 7$

12. Associative property of multiplication: $[(-2)(-3)]\,(6)$

13. Additive identity property: $-3 + 0$

14. Multiplicative identity property: $1 \cdot (-3)$

15. Distributive property: $3(1 + 9)$

16. Distributive property in reverse: $8x + 2x$

17. Commutative property of multiplication: $(2 + 7) \cdot 5$

18. Commutative property of addition: $3 \cdot 2 + 1 \cdot (-4)$

19. Distributive property: $2a + 2b$

20. Associative property of addition:
$6(8) + [7(-3) + (-1)(2)]$

Indicate the definition, property, or number fact that justifies each statement.

21. $n(-3) = -3n$

22. $(4 - y) \cdot y = 4y - y^2$

23. $7a + b = b + 7a$

24. $(3x + y) + z = 3x + (y + z)$

25. $2s + 2t = 2(s + t)$

26. $(-8)(2n) = [(-8)(2)]n$

27. $(6x^2)(1) = 6x^2$

28. $0 = 6n + (-6n)$

29. $-(x + 1) + (x + 1) = 0$

30. $(n + 3) + 0 = n + 3$

31. $5 + (y - x) = (5 + y) - x$

32. $\left(8 \cdot \dfrac{1}{8}\right)x = 1x$

33. $-3(xy) = (-3x)y$

34. $(n + 2) \cdot 5 = 5 \cdot (n + 2)$

Calculate, if possible.

35. $8 + (-3) + (-5)$

36. $12 + 6 + (-10)$

37. $2 + (-3.8) + 9.13 + (-1)$

38. $(-2) + (-3.5) + 7.4 + (-4)$

39. $(-7)(-2)(3)$

40. $(10)(-1)(-6)$

41. $(-2)(-2)(-2)$

42. $(-1)(-1)(-1)$

43. $(-5)(-7)(-2)(10)$

44. $(7)(-1)(-5)(2)$

Find each of the following.

45. The additive inverse of 2

46. The additive inverse of 9.1

47. The additive inverse of -7

48. The additive inverse of 0

49. The multiplicative inverse of 7

50. The multiplicative inverse of $\dfrac{1}{2}$

51. The multiplicative inverse of -1

52. The multiplicative inverse of $-\dfrac{2}{3}$

Rewrite each expression without the grouping symbols.

53. $(-4)(2 + 5)$

54. $6(1 - 6)$

55. $(x + 10) \cdot 3$

56. $(y + 4) \cdot 5$

57. $-(a + 6b)$

58. $-(x - 7y)$

59. $n(n - 2)$

60. $a(a + 1)$

To prove the given statement, justify each step.

61. $(2n) \cdot 5 = 10n$

$\begin{aligned} (2n) \cdot 5 &= 5 \cdot (2n) & \textbf{a.} \quad \underline{\hspace{4cm}} \\ &= (5 \cdot 2)n & \textbf{b.} \quad \underline{\hspace{4cm}} \\ &= 10n & \textbf{c.} \quad \underline{\hspace{4cm}} \end{aligned}$

62. $t + [5 + (-t)] = 5$

$\begin{aligned} t + [5 + (-t)] &= t + [(-t) + 5] & \textbf{a.} \quad \underline{\hspace{4cm}} \\ &= [t + (-t)] + 5 & \textbf{b.} \quad \underline{\hspace{4cm}} \\ &= 0 + 5 & \textbf{c.} \quad \underline{\hspace{4cm}} \\ &= 5 & \textbf{d.} \quad \underline{\hspace{4cm}} \end{aligned}$

63. $\left(\dfrac{1}{2}\right)(2p) = p$

$\begin{aligned} \left(\dfrac{1}{2}\right)(2p) &= \left(\dfrac{1}{2} \cdot 2\right)p & \textbf{a.} \quad \underline{\hspace{4cm}} \\ &= (1)p & \textbf{b.} \quad \underline{\hspace{4cm}} \\ &= p & \textbf{c.} \quad \underline{\hspace{4cm}} \end{aligned}$

64. $3x + 2x = 5x$

$\begin{aligned} 3x + 2x &= x \cdot 3 + x \cdot 2 & \textbf{a.} \quad \underline{\hspace{4cm}} \\ &= x(3 + 2) & \textbf{b.} \quad \underline{\hspace{4cm}} \\ &= x \cdot 5 & \textbf{c.} \quad \underline{\hspace{4cm}} \\ &= 5x & \textbf{d.} \quad \underline{\hspace{4cm}} \end{aligned}$

Mixed Practice

Solve.

65. Find the additive inverse of -0.2.

66. Rewrite the expression $9y + 2y$ using the distributive property.

67. Indicate the definition, property, or number fact that justifies the statement $5(ab) = (5a)b$.

68. Calculate: $(-6)(5) + (-2)$

69. Justify the statement $\left(\dfrac{1}{4} \cdot 4\right)n = 1n$.

70. Rewrite without the grouping symbols: $t(t - 5)$

Applications

Explain each answer in terms of the appropriate property of real numbers or definition.

71. Each morning, you take a bus from home to a stop, and then you walk to your office. At the end of the day, you make the trip in reverse. Is the distance you travel going to work the same as the distance you travel returning home?

72. A hiker walks *m* mi west and then *m* mi east. How far does she wind up from her original starting point?

73. The rate of sales tax, expressed as a decimal, is *r*. A shopper buys two items, one costing *p* dollars and the other *q* dollars, before tax. One expression for the total tax paid on the two items is $r(p + q)$ and another is $rp + rq$. Will the shopper pay the same amount of taxes regardless of which expression is used?

74. A sales representative makes the trip from Company A to Company B to Company C, and then back to Company A, as shown. The length of this trip can be represented by either $(AB + BC) + CA$ or by $AB + (BC + CA)$. (The notation *AB* represents the "length of line segment *AB*.") Must the two representations be equal?

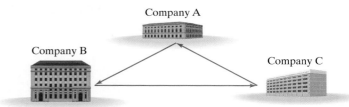

75. An athlete is trying to lose weight. At the beginning of one week, he weighs *p* lb. At the end of the week, the athlete has neither gained nor lost weight. What is his weight at the end of the week?

76. In a photograph, the pictured height of a child is *h* in. A photographer enlarges the photograph, multiplying the pictured height by a factor of *f*. By what factor must the photographer multiply the enlarged picture to shrink the pictured height of the child back to *h* in.?

77. You calculate the area of the triangle shown using the product $\frac{1}{2} \cdot 9 \cdot 12$. Your friend calculates the area of the same triangle using the product $\frac{1}{2} \cdot 12 \cdot 9$. Are both calculations correct?

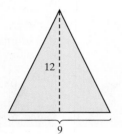

78. An expression for the surface area of the cereal box shown is given by the expression $2lh + 2wh + 2lw$. Can the surface area also be calculated using the expression $2(lh + wh + lw)$? Explain.

● *Check your answers on p. A-3.*

MINDSTRETCHERS

Groupwork

1. With a partner, determine whether the commutative and associative properties hold for subtraction and division. Give examples to support your answer.

Critical Thinking

2. Consider the following diagram:

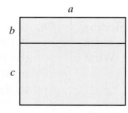

a. Determine the area of the large rectangle formed by the two smaller rectangles in two ways.

b. What property of real numbers does your response in part (a) illustrate?

Writing

3. Explain why 0 has no multiplicative inverse.

1.6 Algebraic Expressions, Translations, and Exponents

Algebraic Expressions

Suppose that each week, you spend $\frac{2}{5}$ of your time on campus working in the student activities area. If the *variable n* represents the total number of hours that you spend during a particular week on campus, then the *algebraic expression* $\frac{2}{5} \cdot n$ stands for the amount of time you worked in the student activities area that week.

In algebra, a variable is used in two ways—as an unknown quantity or as a quantity that can change in value. We can use any letter or symbol to represent a variable. By contrast, a *constant* is a known quantity whose value does not change. So n is a variable, whereas $\frac{2}{5}$ is a constant. An algebraic expression is an expression in which constants and variables are combined using standard arithmetic operations. So $\frac{2}{5} \cdot n$ is an example of an algebraic expression. When writing an algebraic expression involving a product, we usually omit any multiplication symbol. For instance, we would write $\frac{2}{5} n$ rather than $\frac{2}{5} \cdot n$.

Algebraic expressions consist of one or more *terms*, separated by addition signs. If there are subtraction signs, we can rewrite the expression in an equivalent form using addition. For instance, we can think of the algebraic expression

$$2x + \frac{y}{3} - 4 \quad \text{as} \quad 2x + \frac{y}{3} + (-4).$$

This algebraic expression is made up of three terms.

$$\overset{\text{Terms}}{2x + \frac{y}{3} + (-4)}$$

Definition

A **term** is a number, a variable, or the product or quotient of numbers and variables.

EXAMPLE 1	PRACTICE 1
Find the number of terms in each expression.	Determine how many terms are in each expression.
a. $3y + 1$ **b.** $\dfrac{a}{b}$	**a.** $2a + 3 - b$ **b.** $-4xy$
Solution	
a. The expression $3y + 1$ has two terms. **b.** The expression $\dfrac{a}{b}$ has one term.	

Translating Algebraic Expressions to Word Phrases and Word Phrases to Algebraic Expressions

In solving word problems, we may need to translate algebraic expressions to word phrases and vice versa. First, let's consider the many ways we can translate algebraic expressions to words.

$x + 5$ translates to
• x plus 5
• x increased by 5
• the sum of x and 5
• 5 more than x

$y - 4$ translates to
• y minus 4
• y decreased by 4
• the difference between y and 4
• 4 less than y

$\frac{2}{5}n$, $\frac{2}{5} \cdot n$, or $\left(\frac{2}{5}\right)(n)$ translates to
• $\frac{2}{5}$ times n
• the product of $\frac{2}{5}$ and n
• $\frac{2}{5}$ of n

$z \div 3$ or $\frac{z}{3}$ translates to
• z divided by 3
• the quotient of z and 3
• the ratio of z and 3
• z over 3

Note that there are other possible translations as well.

EXAMPLE 2	PRACTICE 2
Translate each algebraic expression to words.	Translate each algebraic expression to words.

Solution

Alegraic Expression	Translation
a. $5x$	5 times x
b. $y - (-2)$	the difference between y and -2
c. $-3 + z$	the sum of -3 and z
d. $\dfrac{m}{-4}$	m divided by -4
e. $\dfrac{3}{5}n$	$\dfrac{3}{5}$ of n

Algebraic Expression	Translation
a. $\dfrac{1}{3}p$	
b. $9 - x$	
c. $s \div (-8)$	
d. $n + (-6)$	
e. $\dfrac{3}{8}m$	

Note that in Example 2 other translations are also correct.

EXAMPLE 3

Translate each algebraic expression to words.

Solution

Algebraic Expression	Translation
a. $2m + 5$	5 more than twice m
b. $1 - 3y$	the difference between 1 and $3y$
c. $4(x + y)$	4 times the sum of x and y
d. $\dfrac{a + b}{a - b}$	the sum of a and b divided by the difference between a and b

PRACTICE 3

Translate each algebraic expression to words.

Algebraic Expression	Translation
a. $2x - 3y$	
b. $4 + 3m$	
c. $5(a - b)$	
d. $\dfrac{r - s}{r + s}$	

Note that in Example 3(c) the sum $x + y$ is considered to be a single quantity, because it is enclosed in parentheses. Similarly in Example 3(d), the numerator $a + b$ and the denominator $a - b$ are each viewed as a single quantity.

In the previous examples, we discussed translating algebraic expressions to word phrases. Now, let's look at some examples of translating word phrases to algebraic expressions.

EXAMPLE 4

Translate each word phrase to an algebraic expression.

Solution

Word Phrase	Translation
a. twice x	$2x$
b. n decreased by -7	$n - (-7)$
c. the quotient of -6 and z	$(-6) \div z$ or $\dfrac{-6}{z}$
d. $\dfrac{1}{2}$ of n	$\dfrac{1}{2}n$
e. 10 more than y	$y + 10$

PRACTICE 4

Express each word phrase as an algebraic expression.

Word Phrase	Translation
a. $\dfrac{1}{6}$ of n	
b. n increased by -5	
c. the difference between m and -4	
d. the ratio of 100 and x	
e. the product of -2 and y	

EXAMPLE 5

Translate each word phrase to an algebraic expression.

Solution

Word Phrase	Translation
a. the difference between x and the product of 3 and y	$x - 3y$
b. 6 more than 4 times z	$4z + 6$
c. 10 times the quantity r minus s	$10(r - s)$
d. twice q divided by the sum of p and q	$\dfrac{2q}{p + q}$

PRACTICE 5

Express each word phrase as an algebraic expression.

Word Phrase	Translation
a. the sum of m and $-n$	
b. 11 less than the product of 5 and y	
c. the sum of m and n divided by the product of m and n	
d. negative 6 times the sum of x and y	

EXAMPLE 6

If a polygon has n sides, the sum of the measures of its interior angles, in degrees, is 180 times the quantity n minus 2. Write this expression symbolically.

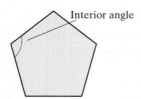
Interior angle

Solution We translate the word phrase into an algebraic expression as follows:

$$
\underset{180}{\underset{\downarrow}{180}} \quad \underset{\cdot}{\underset{\downarrow}{\text{times}}} \quad \underset{(n-2)}{\underset{\downarrow}{\text{the quantity } n \text{ minus } 2}}
$$

So the algebraic expression is $180(n - 2)$, which is in degrees.

PRACTICE 6

In a computer class, a student types 60 words per minute. How many words does he type in $(m + 1)$ minutes?

Exponents

As discussed in Chapter R, we can use *exponential notation* as a shorthand method for representing repeated multiplication of the same factor. For instance, we can write $5 \cdot 5 \cdot 5 \cdot 5$ in *exponential form* as

This expression is read "5 to the fourth power."

Definition

For any real number x and any positive integer a,

$$x^a = \underbrace{x \cdot x \cdots x \cdot x,}_{a \text{ factors}}$$

where x is called the **base** and a is called the **exponent** (or **power**).

In exponential notation, the exponent indicates how many times the base is used as a factor.

The expression x^a is read "x to the ath power," or "x to the a"; however, the exponents 2 and 3 are usually read in a special way. For instance, we generally read 5^2 as "5 *squared*" rather than "5 to the second power." Similarly, we read 5^3 as "5 *cubed*" instead of "5 to the third power."

In Chapter R, we evaluated 2^4. Now, let's consider the expression $(-2)^4$. To evaluate this expression, we multiply 4 factors of -2:

$$(-2)^4 = \underbrace{(-2)(-2)(-2)(-2)}_{4 \text{ factors of } -2}$$

Base, Exponent

$$(-2)^4 = (-2)(-2)(-2)(-2)$$
$$= 4(-2)(-2)$$
$$= (-8)(-2)$$
$$= 16$$

In short, $(-2)^4 = 16$.

Next, let's consider the expression -2^4. To evaluate this expression, we multiply 4 factors of 2. Then we take the opposite:

Base, Exponent

$$-2^4 = -\underbrace{(2)(2)(2)(2)}_{4 \text{ factors of } 2}$$
$$= -16$$

Note that in $(-2)^4$ the base is -2, whereas in -2^4 the base is 2.

EXAMPLE 7

Evaluate.

a. $(-3)^4$ **b.** $-3^4 \cdot (-2)^2$

Solution

a. $(-3)^4 = (-3)(-3)(-3)(-3) = 81$

b. $-3^4 \cdot (-2)^2 = -(3)(3)(3)(3)(-2)(-2)$
$$= -(81)(4)$$
$$= -324$$

PRACTICE 7

Evaluate.

a. -6^2

b. $(-6)^2 \cdot (-3)^2$

Sometimes we put an expression into exponential form. Such expressions may involve more than one base. For instance, the expression $-4(-4)(-4)(-3)(-3)$ can be rewritten in terms of powers of -4 and -3:

$$\underbrace{-4(-4)(-4)}_{\substack{3 \text{ factors} \\ \text{of } -4}}\underbrace{(-3)(-3)}_{\substack{2 \text{ factors} \\ \text{of } -3}} = (-4)^3(-3)^2$$

Consider the following examples.

EXAMPLE 8

Express $6(6)(-10)(-10)(-10)(-10)$ in exponential form.

Solution

$$\underbrace{6(6)}_{\substack{2 \text{ factors} \\ \text{of } 6}}\underbrace{(-10)(-10)(-10)(-10)}_{4 \text{ factors of } -10} = 6^2(-10)^4$$

PRACTICE 8

Write $2(2)(2)(2)(-5)(-5)$ using exponents.

EXAMPLE 9

Rewrite each expression using exponents.

a. $-2n \cdot n$ **b.** $-3x \cdot x \cdot y \cdot y \cdot y \cdot y$

Solution

a. $-2n \cdot n = -2n^2$

b. $-3x \cdot x \cdot y \cdot y \cdot y \cdot y = -3x^2 \cdot y^4$

Can you explain the difference between the expressions $2n$ and n^2?

PRACTICE 9

Express in exponential form.

a. $-x \cdot x \cdot x \cdot x \cdot x$

b. $2m \cdot m \cdot m \cdot n \cdot n \cdot n \cdot n$

EXAMPLE 10

The population of a small town doubles every 5 yr. If the town's population started with n people, what is its population after 20 yr?

Solution We know that the town's population started with n people and doubles every 5 yr. To find the population after 20 yr, consider the following table.

Time (in yr)	Number of 5-Yr Time Periods	Pattern	Population of the Town
Initial	0	$2^0 \cdot n$	$1n$
5	1	$2^1 \cdot n$	$2n$
10	2	$2^2 \cdot n$	$4n$
15	3	$2^3 \cdot n$	$8n$
20	4	$2^4 \cdot n$	$16n$

↑
Each time period, the population is doubled, that is, multiplied by 2.

So the population after 20 yr is $16n$, or $2^4 \cdot n$ in exponential form.

PRACTICE 10

A bacteriologist observes that the population of a bacteria growing in a petri dish triples in size every 2 hr. If x cells were present in the initial population, what was the population after 10 hr? Write the answer using exponents.

1.6   **Exercises**

Mathematically Speaking

Fill in each blank with the most appropriate term or phrase from the given list.

base	variable	the difference between x and 7
7 decreased by x	7 less than twice x	
constant	exponent	the quotient of x and 7
2 times the difference between x and 7	a to the xth power	x to the ath power
the ratio of 7 and x	algebraic expression	

1. In algebra, a(n) _____ can be used as an unknown quantity.

2. A(n) _____ is a known quantity whose value does not change.

3. A(n) _____ consists of one or more terms, separated by addition signs.

4. The algebraic expression $x - 7$ can be translated as _____.

5. The algebraic expression $\frac{x}{7}$ can be translated as _____.

6. The word phrase _____ can be translated as $2x - 7$.

7. In the expression x^a, x is called the _____.

8. The expression x^a can be read as _____.

Determine the number of terms in each expression.

9. $-5x$

10. $a + b$

11. $xy + \frac{x}{y} - z$

12. $\frac{m}{n}$

13. $10 + 2y$

14. $x - y + z$

Translate each algebraic expression to a word phrase.

15. $3 + t$

16. $r + 2$

17. $x - 4$

18. $y - 10$

19. $7r$

20. $-4x$

21. $\frac{a}{4}$

22. $x \div 3$

23. $\frac{4}{5}w$

24. $\frac{1}{2}y$

25. $-3 + z$

26. $m - (-5)$

27. $2n + 1$ **28.** $-3c + 4$ **29.** $4(x - y)$ **30.** $2(m + n)$

31. $1 - 3x$ **32.** $5x - 2$ **33.** $\dfrac{ab}{a + b}$ **34.** $\dfrac{x + y}{x - y}$

35. $2x - 5y$ **36.** $4a + 7b$

Translate each word phrase to an algebraic expression.

37. 5 more than x **38.** y increased by 10   **39.** d minus 4

40. 12 less than n **41.** the product of -6 and a **42.** -7 times x

43. the sum of y and -15 **44.** the difference of 2 and z **45.** $\dfrac{1}{8}$ of k

46. $\dfrac{1}{2}$ of m **47.** the quotient of m and n **48.** n divided by y

49. the difference between a and twice b **50.** 10 less than 3 times x

51. 5 more than 4 times z **52.** the sum of twice n and 3 times m

53. 12 times the quantity x minus y **54.** twice the sum of 3 times m and n

55. b divided by the difference between a and b **56.** x times y divided by the quantity x minus y

Evaluate.

57. -3^2 **58.** $(-4)^2$ **59.** $(-3)^3 \cdot (-4)^2$ **60.** $-4^2 \cdot (-3)^2$

Write using exponents.

61. $-2(-2)(-2)(4)(4)$ **62.** $-5(-5)(-5)(-5)(2)(2)$ **63.** $6(6)(-3)(-3)(-3)$

64. $-2(-2)(4)(4)(4)(4)$ **65.** $3(n)(n)(n)$ **66.** $-2x \cdot x$

67. $-4a \cdot a \cdot a \cdot b \cdot b$ **68.** $5r \cdot s \cdot s \cdot s \cdot s$ **69.** $-y \cdot y \cdot y$

70. $(-y)(-y)(-y)$ **71.** $10a \cdot a \cdot a \cdot b \cdot b \cdot c$ **72.** $-5x \cdot y \cdot y \cdot z$

73. $-x \cdot x \cdot y \cdot y \cdot y$ **74.** $(-x)(-x)(-x)(-y)(-y)$

Mixed Practice

Translate each algebraic expression into a word phrase.

75. $8(w - y)$ **76.** $-5m + 3$

77. $\dfrac{sr}{r - s}$

Write using exponents

78. $-3p \cdot p \cdot p \cdot q$

79. $a \cdot a(-b)(-b)$

Solve.

80. Determine the number of terms in the expression $a - \dfrac{b}{2}$.

81. Evaluate: $-3^2(-2)^2$

Translate each word phrase into an algebraic expression.

82. $\dfrac{2}{3}$ of n

83. The sum of x and twice y

84. The sum of x and -4

Applications

Solve.

85. In the triangle shown, write an expression for the sum of the measures of the three angles.

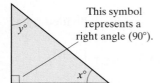

This symbol represents a right angle (90°).

86. Write an expression for the sum of the lengths of the sides in the figure shown.

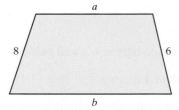

87. Suppose that p partners share equally in the profits of an e-business. What is each partner's share, if the profits were $30,000?

88. An investment of a dollars doubles in value. What is the new value of this investment?

89. A plane ticket costs t dollars before taxes. If the taxes on the plane ticket were x dollars, what was the total cost of the ticket?

90. Tonight's attendance at a concert was 100 more than double the attendance last night. If p people attended the concert last night, how many people attended the concert tonight?

91. An initial investment of 5000 dollars doubles every 10 years. Write in exponential form the value of the investment after 30 years.

92. A colony of bacteria *E. coli* doubles in size every 20 min when grown in a rich medium. If the colony started with m bacteria, how many bacteria were in the colony 2 hr later? Express the answer using exponents.

93. The area of a square can be found by squaring the length of a side. Write an expression in exponential form to represent the area of the square shown.

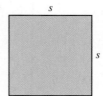

94. The volume of a cube can be found by cubing the length of an edge. Write an expression in exponential form to represent the volume of this cube.

95. A company buys a copier for $10,000. After n years, the Internal Revenue Service values the copier at $10,000 times $\dfrac{1}{20}$ times the quantity 20 minus n. Write this value as an algebraic expression.

96. In a math lab, a tutors were each assisting b students, and c tutors were each assisting d students. There were an additional e students in the lab. How many students were there in the lab altogether?

97. An area rug is placed on the wood floor shown. Find the area of the floor not covered by the rug.

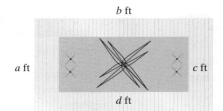

b ft

a ft   *c* ft

d ft

98. In a baseball game, a team scored 2 runs in each of *x* innings and 1 run in each of *y* innings. How many runs did the team score?

● *Check your answers on page A-3.*

MINDSTRETCHERS

Writing

1. Algebra is used in all countries of the world regardless of the language spoken. If you know how to speak a language other than English, translate each of the following algebraic expressions to that language as well as to English.

   a. $\dfrac{x}{2}$ **b.** $2 - x$ **c.** $6 + x$ **d.** $3x$

Mathematical Reasoning

2. Can there be two different numbers *a* and *b* for which $a^b = b^a$? Justify your answer.

Groupwork

3. The *algebra tiles* shown below represent the expressions $3x - 2$ and $x^2 + 1$, respectively.

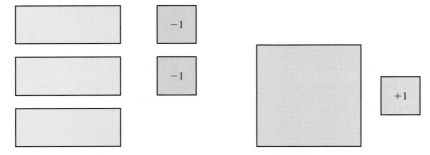

Working with a partner, represent each of the following expressions by algebra tiles.

   a. $x - 4$ **b.** $2x + 3$ **c.** $x^2 + x + 1$

1.7   Simplifying Algebraic Expressions

Each variable term of an algebraic expression is said to have a *numerical coefficient* or, simply, a *coefficient*. For instance, the term $3y$ has coefficient 3 and the term $-4y$ has coefficient -4. Note that the coefficient of y is 1, because $y = 1y$. Likewise, the coefficient of $-y$ is -1, because $-y = -1y$.

The terms $3y$, $-4y$, and y are said to be *like* terms.

OBJECTIVES

- To combine like terms
- To simplify algebraic expressions involving parentheses
- To solve applied problems involving algebraic expressions

Definition

Like terms are terms that have the same variables with the same exponents. Terms that are not like are called **unlike terms**.

EXAMPLE 1	PRACTICE 1
For each algebraic expression, identify the terms and indicate whether they are like or unlike.	Identify the terms of each algebraic expression and state whether they are like or unlike.

EXAMPLE 1

For each algebraic expression, identify the terms and indicate whether they are like or unlike.

a. $3a + 5b$ **b.** $n - 6n$

c. $4x^2 + 2x$ **d.** $2a^2b - 3a^2b$

Solution

a. $3a + 5b$ has terms $3a$ and $5b$. Unlike

different variables

b. $n - 6n$ has terms n and $-6n$. Like

same variable

c. $4x^2 + 2x$ has terms $4x^2$ and $2x$. Unlike

same variable
but different exponents

d. $2a^2b - 3a^2b$ has terms $2a^2b$ and $-3a^2b$. Like

same variables
with same exponents

PRACTICE 1

Identify the terms of each algebraic expression and state whether they are like or unlike.

a. $m - 3m$

b. $5x + 7$

c. $2x^2y - 3xy^2$

d. $m + 2m - 4m$

Combining Like Terms

We can use the distributive property to simplify algebraic expressions involving like terms. Adding or subtracting like terms using the distributive property is called *combining like terms*. Unlike terms, such as $4x^2$ and $2x$, cannot be combined.

EXAMPLE 2	PRACTICE 2

EXAMPLE 2

Simplify.

a. $2x + 6x$ **b.** $a - 8a$

c. $6y - y + 5$ **d.** $3b + 2b - 5b$

Solution

a. $2x + 6x = (2 + 6)x$ Use the distributive property.

$\qquad = 8x$ Add 2 and 6.

b. $a - 8a = (1 - 8)a$ Recall that the coefficient of a is 1.
Use the distributive property.

$\qquad = -7a$

c. $6y - y + 5 = (6 - 1)y + 5$ Recall that the coefficient of $-y$ is -1. Use the distributive property.

$\qquad = 5y + 5$

d. $3b + 2b - 5b = (3 + 2 - 5)b$ Use the distributive property.
Recall that $0 \cdot b = 0$.

$\qquad = 0 \cdot b$

$\qquad = 0$

PRACTICE 2

Simplify.

a. $5x + x$

b. $-5y - y$

c. $a - 3a + b$

d. $-9t + 3t + 6t$

EXAMPLE 3	PRACTICE 3

EXAMPLE 3

Combine like terms if possible.

a. $2x^2 + x$ **b.** $8m^2n^2 + 2m^2n^2$ **c.** $-6a^2b + 5a^2b$

Solution

a. $2x^2 + x$ This algebraic expression cannot be simplified because the terms are unlike.

Unlike terms

b. $8m^2n^2 + 2m^2n^2 = (8 + 2)m^2n^2 = 10m^2n^2$

Like terms

c. $-6a^2b + 5a^2b = (-6 + 5)a^2b = -1a^2b = -a^2b$

Like terms

PRACTICE 3

Combine like terms if possible.

a. $y^2 - 3y^2$

b. $7a^2b + 3ab^2$

c. $4xy^2 - xy^2$

Simplifying Algebraic Expressions Involving Parentheses

Some algebraic expressions involve parentheses. We can use the distributive property to remove the parentheses in order to simplify these expressions.

EXAMPLE 4	PRACTICE 4

EXAMPLE 4

Simplify: $\frac{1}{2}(x + 6) - 4$

Solution

$\frac{1}{2}(x + 6) - 4 = \frac{1}{2}x + 3 - 4$ Use the distributive property.

$\qquad = \frac{1}{2}x - 1$ Combine like terms.

PRACTICE 4

Simplify: $3(y - 4) + 2$

Let's consider simplifying algebraic expressions such as $-(4y - 6)$ in which a negative sign precedes an expression in parentheses.

EXAMPLE 5	PRACTICE 5

Simplify: $-(4y - 6)$

Solution

$$
\begin{aligned}
-(4y - 6) &= -1(4y - 6) && \text{The coefficient of } -(4y - 6) \text{ is } -1. \\
&= -1 \cdot 4y + (-1)(-6) && \text{Use the distributive property.} \\
&= -4y + 6
\end{aligned}
$$

Because the terms in the expression $-4y + 6$ are unlike, it is not possible to simplify the expression further.

Simplify: $-(2a - 3b)$

EXAMPLE 6	PRACTICE 6

Simplify: $2x - 3 - (x + 4)$

Solution

$$
\begin{aligned}
2x - 3 - (x + 4) &= 2x - 3 - 1(x + 4) && \text{The coefficient of } -(x - 4) \text{ is } -1. \\
&= 2x - 3 - x - 4 && \text{Use the distributive property.} \\
&= 2x - x - 3 - 4 \\
&= x - 7 && \text{Combine like terms.}
\end{aligned}
$$

Simplify: $5y - 6 - (y - 5)$

Tip

- When removing parentheses preceded by a *minus sign*, all the terms in parentheses change to the opposite sign.

- When removing parentheses preceded by a *plus sign*, all the terms in parentheses keep the same sign.

EXAMPLE 7	PRACTICE 7

Simplify: $(14a - 9) - 2(3a + 4)$

Solution

$$
\begin{aligned}
(14a - 9) - 2(3a + 4) &= 14a - 9 - 6a - 8 \\
&= 8a - 17
\end{aligned}
$$

Simplify: $(y + 3) - 3(y + 7)$

Some algebraic expressions contain not only parentheses but also brackets. When simplifying expressions containing both types of grouping symbols, first remove the innermost grouping symbols by using the distributive property and then continue to work outward.

EXAMPLE 8	PRACTICE 8

EXAMPLE 8

Simplify: $5 + 2[-4(x - 3) + 8x]$

Solution

$5 + 2[-4(x - 3) + 8x] = 5 + 2[-4x + 12 + 8x]$ Use the distributive property inside the brackets.

$\qquad\qquad = 5 + 2[4x + 12]$ Combine like terms inside the brackets.

$\qquad\qquad = 5 + 8x + 24$ Use the distributive property.

$\qquad\qquad = 8x + 29$ Combine like terms.

PRACTICE 8

Simplify: $10 - [4y + 3(2y - 1)]$

Now, we consider applied problems that involve simplifying expressions. To solve these problems, first we translate word phrases to algebraic expressions, and then simplify.

EXAMPLE 9

When a hospital is filled to capacity, it has p patients in private rooms and 20 more patients in semiprivate rooms than in private rooms. The daily rate for a patient in a private room is $300, and in a semiprivate room, the daily rate is $200. Write an algebraic expression to represent the total amount of money that the hospital takes in per day when all of the rooms are full. Then simplify the expression.

Solution We know that p represents the number of patients in private rooms. Since the hospital has 20 more patients in semiprivate rooms than in private rooms, $p + 20$ represents the number of patients in semiprivate rooms. A patient in a private room pays $300 per day, and a patient in a semiprivate room pays $200 per day.

Organizing the information into a table can help us to clarify the relationship between the key quantities in the problem.

Type of Room	Cost per Patient	Number of Patients in Rooms	Total Amount of Money
Private	$300	p	$300p$
Semiprivate	$200	$p + 20$	$200(p + 20)$

So when the hospital is full, the total amount of money that the hospital takes in per day is represented by the algebraic expression $300p + 200(p + 20)$.

Simplifying $300p + 200(p + 20)$, we get:

$$300p + 200(p + 20) = 300p + 200p + 4000$$
$$= 500p + 4000$$

So the total amount of money that the hospital receives per day is $(500p + 4000)$ dollars.

PRACTICE 9

For a concert, a local performing arts center sold c tickets for children and 40 fewer tickets for adults. Write an algebraic expression to represent the total income received by the center if the cost of a ticket is $12 for adults and $5 for children. Then simplify the expression.

1.7   Exercises   FOR EXTRA HELP   *MyMathLab*

Mathematically Speaking

Fill in each blank with the most appropriate term or phrase from the given list.

coefficient	unlike	negative
outermost	number	like
positive	innermost	
associative property of addition	distributive property	

1. The _____ of the term $-x$ is -1.

2. Terms that have the same variables with the same exponents are called _____ terms.

3. The _____ states that for any real numbers $a, b,$ and $c, a \cdot (b + c) = a \cdot b + a \cdot c$.

4. We cannot combine _____ terms.

5. When removing parentheses preceded by a(n) _____ sign, change all the terms in parentheses to the opposite sign.

6. To simplify expressions containing brackets and parentheses, begin by removing the _____ grouping symbols.

Name the coefficient in each term.

7. $7x$

8. $100a$

9. $-5y$

10. $-18t$

11. ab

12. m

13. $-x^2$

14. $-xy^3$

15. $-0.1n$

16. $2.5s$

17. $\frac{2}{3}a^2b$

18. $-\frac{1}{4}mn$

19. $2x - 5y$

20. $-x + 10y$

Identify the terms, and indicate whether they are like or unlike.

21. $2a - a$

22. $10r + r$

23. $5p + 3$

24. $30 - 2A$

25. $4x^2 - 6x^2$

26. $-20n - 3n$

27. $x^2 + 7x^3$

28. $-2x^2 + 3x$

Simplify, if possible, by combining like terms.

29. $3x + 7x$

30. $8p + 3p$

31. $-10n - n$

32. $-y - 5y$

33. $20a - 10a + 4a$

34. $2r - 5r - r$

35. $3y - y + 2$

36. $7y - 2y - 1$

37. $8b^3 + b^3 - 9b^3$

38. $6x^2 + 2x^2 - 14x^2$

39. $-b^2 + ab^2$

40. $-3x^2 - 5x$

41. $3r^2t^2 + r^2t^2$ **42.** $m^2n^2 + 2m^2n^2$

43. $3x^2y - 5xy^2$ **44.** $10ab^2 + a^2b$

Simplify.

45. $2(x + 3) - 4$ **46.** $3(a - 5) - 1$

47. $(7x + 1) + (2x - 1)$ **48.** $(3x - 4) + 2(2x - 18)$

49. $-(3y - 10)$ **50.** $-(2x + 5)$

51. $5x - 3 - (x + 6)$ **52.** $5x - 9 - 2(x + 4)$

53. $-4(n - 9) + 3(n + 1)$ **54.** $5(2b - 1) - 4(c + b)$

55. $x - 4 - 2(x - 1) + 3(2x + 1)$ **56.** $a + 1 - 3(a + 1) + 8(4a + 5)$

57. $7 + 3[x - 2(x - 1)]$ **58.** $2 - [n - 4(n + 5)]$

59. $10 - 3[4(a + 2) - 3a]$ **60.** $5 + 2[-(z + 7) + 4z]$

Mixed Practice

Simplify, if possible, by combining like terms.

61. $4pq^2 - 6q^2$ **62.** $12m + 2 - m$

Solve.

63. Identify the terms in the expression $3a + 3a^2$ and indicate whether they are like or unlike.

64. Are the terms $5x^2y$ and $5xy^2$ like or unlike?   **65.** Name the coefficient in the term xy.

Simplify.

66. $2n - 7 - (3n + 2)$ **67.** $y - 3 - 4(y - 2) + 2(3y + 1)$ **68.** $(5a - 6) + 3(2a + 1)$

Applications

Write an algebraic expression. Then simplify.

69. What is the sum of the angles in the triangle shown?

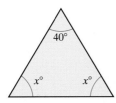

70. According to a will, an estate is to be divided among 2 children and 2 grandchildren. Each grandchild is to receive one-half the amount c that each child receives from the estate. What is the total value of the estate?

71. A baseball fan buys a ticket for a game for d dollars. Two of his friends decide to go to the game at the last minute and purchase tickets for $4 more per ticket. What is the total cost of the 3 tickets?

72. The shape of Colorado is approximately a rectangle. Its length is 100 mi more than its width, w. What is the approximate perimeter of Colorado?

73. If n is the first of 3 consecutive integers, what is the sum of the 3 consecutive integers? (*Hint:* Consecutive integers are integers that differ by 1 unit.)

74. On Tuesday, a company's stock fell by 6% from its value the day before. If v represents the value of the stock on Monday, what was its value on Tuesday?

75. Find the total annual interest on $1000 where x dollars is invested at an interest rate of 5% per year and the remainder is invested at 4% per year.

76. An apartment house contains 100 apartments. Of these, y are 3-room apartments, and the others are 4-room apartments. How many rooms are there in the apartment house?

● *Check your answers on page A-3.*

MINDSTRETCHERS

Groupwork

1. Parentheses have different meanings in different situations. Working with a partner, explain the meaning of parentheses in each context.

   a. $(-2)(-3)$

   b. $5(x + 2)$

   c.

Income	$713,014
Expenditures	$961,882
Profit	($248,868)

   d. I am studying algebra (my favorite subject!).

Research

2. By searching the Web or checking in your college's math learning center, investigate how *algebra tiles* can be used to combine like terms. Summarize your findings.

Writing

3. Explain the meaning of *combining like terms* in the following examples.

   a. 5 ft 3 in. + 7 ft 6 in.

   b. $16\frac{2}{3} - 5\frac{1}{3}$

   c. $10x - 1 + 4 + 2x$

1.8 Translating and Evaluating Algebraic Expressions

Translating and Evaluating Algebraic Expressions

In Section 1.6, we discussed translations involving algebraic expressions. For these expressions to be useful, however, we need to be able to *evaluate* them.

Consider the following example. Suppose that you had a temporary job that lasted 75 days. If you were absent from work for d days, then it follows that you were at work for $75 - d$ days. To evaluate the expression $75 - d$ for a particular value of d, replace d with that number. For instance, if you were not at work for 4 days, replace d in this expression with 4:

$$75 - d = 75 - 4 = 71$$

└ Replace ┘
d with 4.

We can conclude that you were at work 71 days.

The following method is helpful for evaluating expressions.

> **OBJECTIVES**
>
> ▪ To evaluate algebraic expressions
>
> ▪ To translate rules to formulas
>
> ▪ To evaluate formulas
>
> ▪ To solve applied word problems involving algebraic expressions and formulas

> **To Evaluate an Algebraic Expression**
> - Replace each variable with the given number.
> - Carry out the computation using the order of operations rule.

EXAMPLE 1

Evaluate each algebraic expression.

Solution

Algebraic Expression	Value
a. $9 - z$, if $z = -2$	$9 - z = 9 - (-2) = 9 + 2 = 11$
b. $-2cd$, if $c = -1$ and $d = 2$	$-2cd = -2(-1)(2) = 4$

PRACTICE 1

Find the value of each algebraic expression.

Algebraic Expression	Value
a. $25 + m$, if $m = -10$	
b. $-3xy$, if $x = -2$ and $y = 5$	

In evaluating some algebraic expressions, we need to use the order of operations rule.

EXAMPLE 2

Find the value of each expression for $x = -3$, $y = 2$, $w = -5$, and $z = 4$.

a. $2x + 5y$ **b.** $4x^2$ **c.** $3(w + z)$ **d.** $2w^2 - 3y^3$

Solution

a. $2x + 5y = 2(-3) + 5(2)$ Replace x with -3 and y with 2. Then multiply.

$= -6 + 10$ Add.

$= 4$

PRACTICE 2

Evaluate each expression for $a = 2$, $b = -3$, $c = -4$, and $d = 5$.

a. $5a - 2c$

b. $2(d - b)$

c. $3cd^2$

d. $2a^3 + 4b^2$

b. $4x^2 = 4(-3)^2$ Replace x with -3.

$\quad\quad\; = 4 \cdot 9$ Square -3.

$\quad\quad\; = 36$

c. $3(w + z) = 3(-5 + 4) = 3(-1) = -3$

d. $2w^2 - 3y^3 = 2(-5)^2 - 3(2)^3 = 2(25) - 3(8) = 50 - 24 = 26$

Consider the expression in Practice 2(c). Are $3cd^2$ and $3(cd)^2$ the same? If not, explain the difference between them.

EXAMPLE 3

Evaluate each expression if $a = 3$, $b = 4$, and $c = -2$.

a. $\dfrac{c}{5 - a}$ **b.** $\dfrac{b + c}{b - c}$ **c.** $(-b)^2$ **d.** $-b^2$

Solution

a. $\dfrac{c}{5 - a} = \dfrac{-2}{5 - 3} = \dfrac{-2}{2} = -1$

b. $\dfrac{b + c}{b - c} = \dfrac{4 + (-2)}{4 - (-2)} = \dfrac{2}{6} = \dfrac{1}{3}$

c. $(-b)^2 = (-4)^2 = (-4)(-4) = 16$

d. $-b^2 = -4^2 = -(4)(4) = -16$

PRACTICE 3

Find the value of each expression when $x = -5$, $y = -3$, and $z = 1$.

a. $\dfrac{x - 2z}{y}$ **b.** $\dfrac{x - z}{x + y}$

c. $(-y)^4$ **d.** $-y^4$

Note the difference between the answers in Examples 3(c) and 3(d). In $(-4)^2$ the exponent applies to -4, that is, the base is -4. By contrast, in -4^2 the exponent applies to 4 only, because the base is 4.

EXAMPLE 4

Physicists have shown that if an object is shot straight upward at a speed of 100 ft/sec, its location after t sec will be

$$-16t^2 + 100t$$

feet above the point of release. How far above or below the point of release will the object be after 5 sec?

100 ft/sec

PRACTICE 4

In a statistics course, a student needs to evaluate the expression

$$\dfrac{a^2 + b^2 + c^2}{3}$$

Find the value of the expression for $a = -0.5$, $b = 0.3$, and $c = 0.2$, rounded to the nearest tenth.

Solution To determine the position of the object after 5 sec, we must substitute 5 for t in the expression $-16t^2 + 100t$.

$$-16t^2 + 100t = -16(5)^2 + 100(5)$$
$$= -16(25) + 100(5)$$
$$= -400 + 500$$
$$= 100$$

So after 5 sec, the object will be 100 ft above the point of release. Note that a position above the point of release is positive, whereas a position below the point of release is negative.

Formulas

A *formula* is an equation that indicates how a number of variables are related to one another. Some formulas express geometric relationships; others express physical laws. Just as with algebraic expressions, the letters and mathematical symbols in a formula represent numbers and words.

EXAMPLE 5	PRACTICE 5
To predict the temperature T at a particular altitude a, meteorologists subtract $\frac{1}{200}$ of the altitude from the temperature g on the ground. Here, T and g are in degrees Fahrenheit and a is in feet. Translate this rule to a formula.	To convert a temperature C expressed in Celsius degrees to the temperature F expressed in Fahrenheit degrees, we multiply the Celsius temperature by $\frac{9}{5}$ and then add 32. Write a formula that expresses this relationship.

Solution Stating the rule briefly in words, the temperature at a particular altitude equals the difference between the temperature on the ground and $\frac{1}{200}$ times the altitude. Now we translate this rule to mathematical symbols.

$$T = g - \frac{1}{200}a,$$

which is the desired formula.

The method of evaluating formulas is similar to that of evaluating algebraic expressions. We substitute all the given numbers for the variables and then carry out the computations using the order of operations rule.

EXAMPLE 6	PRACTICE 6
The formula for finding simple interest is $I = Prt$, where I is the interest in dollars, P is the principal (the amount invested) in dollars, r is the annual rate of interest, and t is the time in years that the principal has been on deposit. Find the amount of interest on a principal of \$3000 that has been on deposit for 2 years at a 6% annual rate of interest.	Given the distance formula $d = rt$, find the value of d if the rate r is 50 mph and the time t is 1.6 hr.

Solution We know that $P = 3000$, $r = 6\%$, and $t = 2$. Converting 6% to its decimal form, we get 0.06. Substituting into the formula gives us

$$I = Prt$$
$$= 3000(0.06)(2)$$
$$= 360$$

So the interest earned is $360.

EXAMPLE 7

The markup m on an item is its selling price s minus its cost c.

a. Express this relationship as a formula.

b. If a digital camera cost a retailer $399.95 and was then sold for $559, how much was the markup on the camera?

Solution

a. We write the formula $m = s - c$.

b. To find the markup, we substitute for s and c.

$$m = s - c$$
$$= 559 - 399.95$$
$$= 159.05$$

So the markup was $159.05.

PRACTICE 7

Kelvin and Celsius temperature scales are commonly used in science. To convert a temperature expressed in Celsius degrees C to degrees Kelvin, K, add 273 to the Celsius temperature.

a. Write this relationship as a formula.

b. Suppose that in a chemistry experiment, C equals -6. What is the value of K?

Evaluate each algebraic expression if a = 4, b = 3, and c = −2.

1. $b - 5$ **2.** $-6 + a$ **3.** $-2ac$ **4.** $4cb$

5. $-2a^2$ **6.** $-b^2$ **7.** $2a - 15$ **8.** $5 + 3c$

9. $a + 2c$ **10.** $4b - 3a$ **11.** $2(a - c)$ **12.** $-3(a + b)$

13. $-a + b^2$ **14.** $-b + c^2$ **15.** $3a^2 - c^3$ **16.** $c^3 - 2a^2$

17. $\dfrac{a + b}{b - a}$ **18.** $\dfrac{a - c}{c + a}$ **19.** $\dfrac{3}{5}(a + b + c)^2$ **20.** $\dfrac{5}{9}(a - b - c)^2$

Evaluate each algebraic expression if w = −0.5, x = 2, y = −3, and z = 1.5.

21. $2w^2 - 3x + y - 4z$ **22.** $2x - 3w - y^2 + z$ **23.** $w - 7z - \dfrac{1}{4}(x - 6y)$

24. $5x - \dfrac{2}{5}(8w + 2y) - 4z$ **25.** $\dfrac{-10xy}{(w - z)^2}$ **26.** $\dfrac{-(2z - y)^2}{9wx}$

Complete each table.

27.

x	0	1	2	−1	−2
$2x + 5$					

28.

x	0	1	2	−1	−2
$-3x + 1$					

29.

y	0	1	2	3	4
$y - 0.5$					

30.

y	0	1	2	3	4
$y + 2.8$					

31.

x	0	2	4	−2	−4
$-\dfrac{1}{2}x$					

32.

x	0	5	10	−5	−10
$\dfrac{3}{5}x$					

33.

n	2	4	6	−2	−4
$\dfrac{n}{2}$					

34.

n	5	10	−5	−10	−15
$-\dfrac{n}{5}$					

35.

g	0	1	2	−1	−2
$-g^2$					

36.

g	0	1	2	−1	−2
g^2					

37.

a	0	1	2	−1	−2
$a^2 + 2a - 2$					

38.

a	0	1	2	−1	−2
$-a^2 - 2a + 2$					

Evaluate each formula for the quantity requested.

Formula	Given	Find
39. $C = \dfrac{5}{9}(F - 32)$	$F = -4°$	C
40. $A = P(1 + rt)$	$P = \$2000$, $r = 5\%$, and $t = 2$ yr	A
41. $P = 2l + 2w$	$l = 2\frac{1}{2}$ ft and $w = 1\frac{1}{4}$ ft	P
42. $A = \dfrac{a + b + c + d}{4}$	$a = -8$, $b = -6$, $c = 4$, and $d = -2$	A
43. $C = \pi d$	$\pi \approx 3.14$ and $d = 100$ m	C
44. $C = A \cdot \dfrac{W}{150 \text{ lb}}$	$A = 100$ mg and $W = 30$ lb	C
45. $A = 6e^2$	$e = 1.5$ cm	A
46. $V = \pi r^2 h$	$\pi \approx 3.14$, $r = 10$ in., and $h = 10$ in.	V

Mixed Practice

Solve.

47. Evaluate the algebraic expression if $w = 3$, $x = -1.5$, $y = -2$, and $z = 0.5$.

$$y^2 - 2z + \frac{1}{3}(2x - w)$$

Complete each table.

48.

x	0	8	12	−8	−12
$-\dfrac{3}{4}x$					

49.

x	0	1	2	−1	−2
$-2x + 4$					

Evaluate each algebraic expression if $a = 3$, $b = -4$, and $c = 2$.

50. $-5(c + b)$

51. $b^2 - 3a^2$

52. $\dfrac{c + a}{b - a}$

Evaluate each formula for the quantity requested.

Formula	Given	Find
53. $A = \dfrac{1}{2}bh$	$b = 7$ in. and $h = 3$ in.	A
54. $F = \dfrac{9}{5}C + 32$	$C = -10°$	F

Applications

Write each relationship as a formula.

55. The average A of three numbers a, b, and c is the sum of the numbers divided by 3.

56. For every right triangle, the sum of the squares of the two legs a and b equals the square of the hypotenuse c.

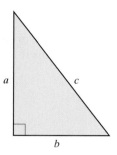

57. The perimeter P of a rectangle is twice the sum of the length l and the width w.

58. The *aspect ratio* of a hang glider is a measure of how well it can glide and soar. The aspect ratio R is the square of the glider's wingspan s divided by the wing area A.

59. The equivalent energy E of a mass equals the product of the mass m and the square of the speed of light c.

60. The weight of an object depends on the gravitational pull of the planet or moon it is on. For instance, the weight E of an astronaut on Earth is 6 times the astronaut's weight m on the moon.

61. The length l of a certain spring in centimeters is 25 more than 0.4 times the weight w in grams of the object hanging from it.

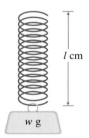

62. In electronics, when two resistors R_1 and R_2 are connected in parallel, the total resistance R between the points X and Y can be founded by dividing the product of the resistances by the sum of the resistances.

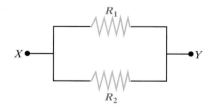

Solve.

63. The distance that a free-falling object drops, ignoring friction, is given by the formula

$$S = \frac{1}{2}gt^2,$$

where S is the distance in feet, g is the acceleration due to gravity, and t is time in seconds. Find the distance that an object falls if $g = 32$ ft/sec^2 and $t = 2$ sec.

64. The volume of a right circular cylinder is given by the formula

$$V = \pi r^2 h,$$

where r is the radius of the base of the cylinder and h is the height of the cylinder. For the cylinder shown, find the volume in cubic centimeters.

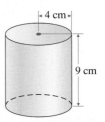

65. The percent markup m of an item based on cost is equal to 100 times the difference of the selling price s of the item and its cost c, divided by the cost.

  a. Express this relationship as a formula.

  b. If the selling price of a bottle of vitamin E is \$8.75 and the cost is \$6.25, what is the percent markup on the vitamin E?

66. To calculate the speed f of an object in feet per second, we multiply its speed m in miles per hour by $\frac{22}{15}$.

  a. Write this relationship as a formula.

  b. If an object is moving at 60 mph, what is its speed in feet per second?

                                           ● *Check your answers on page A-4.*

MINDSTRETCHERS

Mathematical Reasoning

1. Complete the first three rows of the following table. Make up your own values of a and b in the 4th and 5th rows and complete the table.

a	b	$a + b$	$\lvert a \rvert$	$\lvert b \rvert$	$\lvert a \rvert + \lvert b \rvert$	$\lvert a + b \rvert$	T or F, $\lvert a \rvert + \lvert b \rvert = \lvert a + b \rvert$	T or F, $\lvert a \rvert + \lvert b \rvert \geq \lvert a + b \rvert$
-3	4							
1	-2							
-5	-4							

What general relationship do these examples suggest? Explain why this relationship is true.

Patterns

2. Consider the following table.

Odd number, n	1	3	5	7	...	1999
Counting number, C	1	2	3	4	...	1000

Write a formula that expresses C in terms of n.

Groupwork

3. Some expressions that appear to be different are, in fact, equivalent.

   a. Confirm that the expressions $\dfrac{x^2 - 9}{x + 3}$ and $x - 3$ are equal for various values of x.

   b. Determine if there are any values of x for which the expressions are not equal.

KEY CONCEPTS AND SKILLS   (CONCEPT)   (SKILL)

Concept/Skill	Description	Example
[1.1] Set	A collection of objects called elements or members.	$\{3, 6, 9, \dots \}$
[1.1] Real Numbers	Numbers that can be represented as points on a number line.	$\frac{2}{3}, -\frac{1}{2}, 5, 7.4$, and $\sqrt{2}$
[1.1] Integers	The set of integers is $\{\dots, -4, -3, -2, -1, 0, +1, +2, +3, +4, \dots\}$.	$-20, -7, 0, +5$, and $+17$
[1.1] Rational Numbers	Numbers that can be written in the form $\frac{a}{b}$, where a and b are integers and $b \neq 0$.	$-1.6, \frac{2}{3}, -\frac{1}{2}, 5$, and 0.04
[1.1] Opposites	Two real numbers that are the same distance from 0 on a number line but on opposite sides. For any real number n, its opposite is $-n$.	3 and -3
[1.1] Absolute Value	A given number's distance from 0 on a number line. The absolute value of a number n is written as $|n|$.	$\|8\| = 8$ $\|-8\| = 8$
[1.2] To Add Real Numbers, a and b, on a Number Line	• Start at a. • Move according to b as follows: If b is positive, move to the right. If b is negative, move to the left. If b is zero, remain at a.	Add -1 and 3 on a number line: Start at -1 then move 3 units to the right. **Start** **End** $-1 + 3 = 2$
[1.2] Additive Inverse Property	For any real number a, there is exactly one real number $-a$, such that $$a + (-a) = 0 \quad \text{and} \quad -a + a = 0,$$ where a and $-a$ are **additive inverses (opposites)** of each other.	$7 + (-7) = 0$ and $-7 + 7 = 0$
[1.2] To Add Real Numbers	• If the numbers have the same sign, add the absolute values and keep the sign. • If the numbers have different signs, find the difference between the larger absolute value and the smaller one and take the sign of the number with the larger absolute value.	$-5 + (-2) = -7$ $+5 + (+2) = +7$, or 7 $-5 + (+2) = -3$ $+5 + (-2) = +3$, or 3
[1.3] To Subtract Real Numbers	• Change the operation of subtraction to addition and change the number being subtracted to its opposite. • Follow the rule for adding real numbers.	$2 - (-5)$ $= 2 + (+5) = 7$
[1.4] To Multiply Real Numbers	• Multiply their absolute values. • If the numbers have the same sign, their product is positive; if they have different signs, their product is negative.	$(-3)(-8) = 24$ $3(-8) = -24$

continued

Concept/Skill	Description	Example
[1.4] **Multiplication Property of Zero**	For any real number a, $$a \cdot 0 = 0 \quad \text{and} \quad 0 \cdot a = 0.$$	$2 \cdot 0 = 0$ and $0 \cdot 2 = 0$
[1.4] **To Divide Real Numbers**	• Divide their absolute values. • If the numbers have the same sign, their quotient is positive; if they have different signs, their quotient is negative.	$-16 \div (-2) = 8$ $-24 \div 3 = -8$
[1.4] **Division Involving Zero**	For any nonzero real number a, $$0 \div a = 0.$$ For any real number a, $$a \div 0 \text{ is undefined.}$$	$0 \div 5 = 0$ and $5 \div 0$ is undefined.
[1.4] **Multiplicative Inverse Property**	For any nonzero real number a, $$a \cdot \frac{1}{a} = 1 \text{ and } \frac{1}{a} \cdot a = 1,$$ where a and $\frac{1}{a}$ are **multiplicative inverses (reciprocals)** of each other.	$-\frac{3}{4}$ and $-\frac{4}{3}$ are multiplicative inverses because $-\frac{3}{4} \cdot \left(-\frac{4}{3}\right) = 1.$
[1.4] **Division of Real Numbers**	For any real numbers a and b where b is nonzero, $$a \div b = \frac{a}{b} = a \cdot \frac{1}{b}.$$	$-\frac{3}{5} \div 6 = -\frac{3}{5} \div \frac{6}{1}$ $= -\frac{\overset{1}{\cancel{3}}}{5} \cdot \frac{1}{\underset{2}{\cancel{6}}} = -\frac{1}{10}$
[1.5] **Commutative Property of Addition**	For any real numbers a and b, $$a + b = b + a.$$	$6 + 2 = 2 + 6$
[1.5] **Commutative Property of Multiplication**	For any real numbers a and b, $$a \cdot b = b \cdot a.$$	$(-2)(-7) = (-7)(-2)$
[1.5] **Associative Property of Addition**	For any real numbers, a, b, and c, $$(a + b) + c = a + (b + c).$$	$(8 + 4) + 1 = 8 + (4 + 1)$
[1.5] **Associative Property of Multiplication**	For any real numbers, a, b, and c, $$(a \cdot b)\, c = a\,(b \cdot c).$$	$(-1 \cdot 2)\, 3 = -1(2 \cdot 3)$
[1.5] **Additive Identity Property**	For any real number a, $$a + 0 = a \text{ and } 0 + a = a.$$	$7 + 0 = 7$ and $0 + 7 = 7$
[1.5] **Multiplicative Identity Property**	For any real number a, $$a \cdot 1 = a \quad \text{and} \quad 1 \cdot a = a.$$	$3 \cdot 1 = 3$ and $1 \cdot 3 = 3$
[1.5] **Distributive Property**	For any real numbers a, b, and c, $$a \cdot (b + c) = a \cdot b + a \cdot c$$	$2(3 + 5) = 2 \cdot 3 + 2 \cdot 5$

continued

Concept/Skill	Description	Example
[1.6] **Variable**	A letter that represents an unknown quantity or one that can change in value.	x, n
[1.6] **Constant**	A known quantity.	$5, -3.2$
[1.6] **Algebraic Expression**	An expression in which constants and variables are combined using standard arithmetic operations.	$\frac{2}{5}n + 9$
[1.6] **Term**	A number, a variable, or the product or quotient of numbers and variables.	$-11, 3x, \dfrac{n}{4}$
[1.6] **Exponential Notation**	For any real number x and any positive integer a, $$x^a = \underbrace{x \cdot x \cdots x \cdot x}_{a \text{ factors}},$$ where x is called the **base** and a is called the **exponent** (or **power**).	$\overset{\text{Exponent}}{(-2)^3} = (-2)(-2)(-2)$ $\underset{\text{Base}}{\uparrow} \qquad = -8$
[1.7] **Like Terms**	Terms that have the same variable and the same exponent.	n and $6n$ $5x^3$ and $-2x^3$
[1.7] **Unlike Terms**	Terms that are not like.	a^2 and a x and $3y$
[1.7] **To Combine Like Terms**	• Use the distributive property. • Add or subtract.	$2x + 6x = (2 + 6)x$ $\qquad\qquad = 8x$
[1.8] **To Evaluate an Algebraic Expression**	• Replace each variable with the given number. • Carry out the computation using the order of operations rule.	If $c = -1$ and $d = 2$, then $-2cd = (-2)(-1)(2) = 4.$

Chapter 1   Review Exercises

[1.1] *Express each quantity as a signed number.*

1. 3 mi above sea level

2. A withdrawal of $160 from an account

Graph each number on the number line.

3. −1

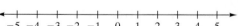

4. $2\frac{9}{10}$

5. 0.5

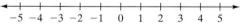

6. −3.75

Find the opposite of each number.

7. −4

8. 6.5

9. $\dfrac{2}{3}$

10. −0.7

Compute.

11. $|-4|$

12. $|0|$

13. $|2.6|$

14. $\left|-\dfrac{5}{9}\right|$

Indicate whether each inequality is true or false.

15. $-7 < -5$

16. $-1 > 3$

[1.2] *Find the sum of each pair of numbers using the number line.*

17. $-4 + (-1)$

18. $3 + (-7)$

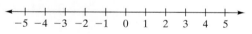

Solve.

19. Another term for opposites is _____.

20. The equation $-10 + 10 = 0$ is an example of what property?

Find the sum.

21. $9 + (-9)$

22. $4 + (-2)$

23. $-3 + 5$

24. $-3 + (-2)$

25. $-3 + 7 + (-89)$

26. $-2 + 5.3 + 12$

[1.3] *Find the difference.*

27. $12 - 3$

28. $36 - 47$

29. $-52 - 3$

30. $2 - 5$

31. $-19 - 8$

32. $24 - (-3)$

33. $8 - (-8)$

34. $0 - 5$

35. $6 - 7.42$

36. $-9 - \left(-\dfrac{3}{8}\right)$

Combine.

37. $2 + (-4) - (-7)$

38. $-3 - (-1) + 12$

[1.4]   *Name the property illustrated by each statement.*

39. $\dfrac{3}{-2} \cdot \dfrac{-2}{3} = 1$

40. $-7 \cdot 0 = 0$

Find the product.

41. $2(-5)$

42. $-3 \cdot 7$

43. $-60 \cdot 90$

44. $-8(-300)$

45. $(-2.7)(-10)$

46. $\left(\dfrac{3}{4}\right)\left(-\dfrac{1}{3}\right)$

47. $5(-4)(-300)$

48. $(-1)(-12)(3)$

Simplify.

49. $-8 + 3(-2) - 9$

50. $3 - 2(-3) - (-5)$

51. $-9 - 5(-7)$

52. $20 - 3(-6)$

53. $-4(-2 + 5)$

54. $(-12 + 6)(-1)$

Find the reciprocal.

55. $-\dfrac{2}{3}$

56. 8

Find the quotient. Simplify.

57. $-30 \div (-10)$

58. $6 \div (-1)$

59. $-\dfrac{11}{5}$

60. $\dfrac{4}{5} \div \left(-\dfrac{2}{3}\right)$

Simplify.

61. $-16 \div 2(-4)$

62. $(9 - 23) \div (-13 + 6)$

63. $\dfrac{3 + (-1)}{-2}$

64. $\dfrac{5(7 - 3)}{-8 - 2}$

65. $(-3) + 8 - 2 \cdot (-4)$

66. $10 \div (-2) + (-3) \cdot 5$

[1.5]   *Rewrite each expression using the indicated property of real numbers.*

67. Commutative property of addition: $3 + 9$

68. Distributive property: $(-3)(1 + 9)$

Identify the definition or property of real numbers that justifies each statement.

69. $(3x + y) + z = 3x + (y + z)$

70. $-(x + 1) + (x + 1) = 0$

Calculate.

71. $-10 + (-2) + (-1)$

72. $(-4)(-5)(-2)(4)$

Find.

73. The additive inverse of 4

74. The multiplicative inverse of $\dfrac{2}{3}$

Rewrite each expression using the distributive property.

75. $3(a - 4b)$

76. $-(x - 5)$

[1.6]   *Determine the number of terms in each expression.*

77. $-x + y - 3z$

78. $\dfrac{m}{n} + 4$

79. $-9t$

80. $3a - 1 + 7b + c$

Translate each algebraic expression to a word phrase.

81. $-6 + w$

82. $-\dfrac{1}{3}x$

83. $-3n + 6$

84. $5(p - q)$

Translate each word phrase to an algebraic expression.

85. 10 less than x

86. $\dfrac{1}{2}$ of s

87. the quotient of p and q

88. the difference between R and twice V

89. 6 times the quantity 2 less than 4 times n

90. the quantity negative 4 times a divided by the quantity 5 times b plus c.

Write each using exponents.

91. $-3(-3)(-3)(-3)$

92. $-5(-5)(-5)(3)(3)$

93. $4(x)(x)(x)$

94. $-5a \cdot a \cdot b \cdot b \cdot b \cdot c$

[1.7] *Solve.*

95. Indentify the coefficient in $\dfrac{3}{4}x^2y^2$.

96. Combine like terms: $4x + 10x - 2y$

97. Simplify: $3x^2 - x^2 - 4x^2$

98. Combine: $2r^2t^2 - r^2t^2$

99. Simplify: $2(a - 5) + 1$

100. Simplify: $-(3x + 2)$

101. Combine like terms: $-3x - 5 - (x + 10)$

102. Simplify: $(2a - 4) + 2(a - 5) - 3(a + 1)$

[1.8] *Evaluate each algebraic expression if a = 2, b = 5, and c = −1.*

103. $30 + c$

104. $-\dfrac{4}{9}b$

105. $-5a^2$

106. $10(b - c)$

107. $\dfrac{1 - a}{c}$

108. $4a^2 - 4ab + b^2$

Mixed Applications

Solve.

109. Express as a signed number: a gain of $700

110. A company lost $7000 last year. Express this situation as a signed number.

111. The price of a share of stock fell by $0.50 each week for 4 weeks in a row. What was the overall change in the price?

112. In exothermic chemical reactions, the surrounding temperature rises; in endothermic chemical reactions, the surrounding temperature drops. If a chemical reaction causes the surrounding temperature to change from $-7°C$ to $-4°C$, was it exothermic or endothermic?

113. A patient's temperature drops by 0.5 degrees per hour. If the initial temperature was I degrees, write an algebraic expression for the temperature after h hours.

114. Use the formula $F = \dfrac{9}{5}C + 32$ to determine the Fahrenheit temperature F that corresponds to the Celsius temperature C if $C = -10°$.

115. The following graph shows the amount of American direct investment in various other countries during a recent year. A positive amount indicates that money is flowing from the U.S. to the other country, a negative amount that money is flowing from the other country to the U.S. All amounts are rounded to the nearest billion U.S. dollars. Was more money flowing to the U.S. from the Netherlands or from Switzerland?

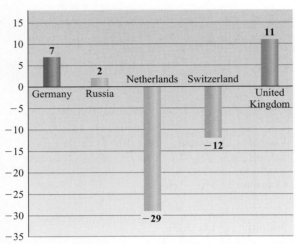

(*Source:* The U.S. Department of Commerce, Bureau of Economic Analysis, 2005)

116. The length of a rectangle is double its width w. Find the rectangle's perimeter.

117. The bar graph shows the closing price of a share of Whole Foods Market, Inc. stock for a five-day period in July of 2006. Calculate the change in the closing price per share from the previous day for Tuesday through Friday. Express each change as a signed number.

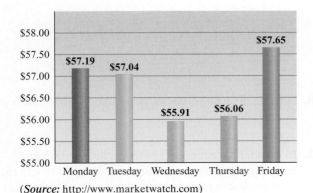

(*Source:* http://www.marketwatch.com)

118. The boiling point of the element radon is $-61.8°C$, whereas its melting point is $-71°C$. How much higher is the boiling point?

119. A colony of bacteria triples in size every 6 hr. At one point, there are 10 bacteria in the colony. Write in exponential form the number of bacteria in the colony 18 hr later.

120. Firefighters use the formula $S = 0.5N + 26$ to compute the maximum horizontal range S in feet of water from a particular hose, where N is the hose's nozzle pressure in pounds. Calculate S if $N = 90$ lb.

121. Which is warmer, a temperature of $3°$ or a temperature of $-4°$?

122. Find the annual interest earned on an investment of $600 if x dollars is invested at an interest rate of 5% per year and the remainder is invested at a rate of 7% per year.

123. The playwright Sophocles was born in 496 B.C. and died approximately 90 yr later. In what year did he die?

124. One year, a company's loss was $60,000. The next year, it was only $20,000. How many times the second loss is the first loss?

125. A homeowner needs to have sufficient current to operate the electrical appliances in the home. Electricians use the formula $I = \dfrac{P}{E}$ to compute the current I in amperes needed in terms of the power P in watts and the energy E in volts. Find I if $P = 2300$ watts and $E = 115$ volts.

126. A checking account has a balance of $410. If $900 is deposited, a check for $720 is written, and two withdrawals of $300 each are made through an ATM, by how much money is the account overdrawn?

127. In addition to a monthly flat fee of x dollars, you are charged y dollars for each hour over 20 that you surf the Web. What do you pay in a month in which you surf the Web for 32 hr?

128. The first day of the term, f students attended the computer lab. As the term went on, attendance increased by s students per day. What was the attendance on the fifth day of the term?

● *Check your answers on page A-4.*

Chapter 1   POSTTEST

FOR
EXTRA
HELP

Pass
the Test   Test solutions are found
on the enclosed CD.

To see if you have mastered the topics in this chapter, take this test.

1. Express as a real number: a loss of 10,000 jobs

2. Is the number $\frac{2}{5}$ rational?

3. Graph the number -2 on the following number line.

$$\xleftarrow{\qquad} \begin{array}{c|c|c|c|c|c|c} & & & & & & \\ -3 & -2 & -1 & 0 & 1 & 2 & 3 \end{array} \xrightarrow{\qquad}$$

4. Find the additive inverse of 7.

5. Compute: $|-3.5|$

6. True or false? $1 > -4$

7. Add: $10 + (-3)$

8. Combine: $2 + (-3) + (-1) + 5$

9. Simplify: $4 + (-1)(-6)$

10. What is the reciprocal of 12?

11. Divide: $-15 \div 0.3$

12. Translate to an algebraic expression: the sum of x and twice y

13. Write in exponential form: $-5(-5)(-5)$

14. Evaluate $3a + b - c$ if $a = -1$, $b = 0$, and $c = 2$.

15. Simplify: $4y + 3 - 7y + 10y + 1$

16. Combine: $8t + 1 - 2(3t - 1)$

17. The balance in a checking account was d dollars. Some time later, the balance was 5% higher. What was the new balance?

18. Last year, a company suffered a loss of $20,000. This year, it showed a profit of $50,000. How big an improvement was this?

● *Check your answers on page A-4.*

CHAPTER 2

Solving Linear Equations and Inequalities

Cell Phones and Algebra

Today's generation takes wireless communication for granted. The cellular telephone, one of the more important wireless devices, was first introduced to the public in the mid-1980s. Since then, cell phones have shrunk both in size and in price.

Using a cell phone generally entails choosing a calling plan with a monthly service charge. As a rule, this charge consists of two parts: a specified flat monthly fee and a per-minute usage charge varying with the number of minutes of airtime used.

The choice among calling plans can be confusing, especially with the lure of promotional incentives. One nationally advertised calling plan requires a monthly charge of $34.99 for 300 minutes of free local calls and 79 cents per minute for additional local calls. A competing plan is for $29.99 a month for 200 minutes of free local calls plus 49 cents a minute for additional local calls. Under what circumstances is one deal better than the other? The key to answering this question is solving inequalities such as

$$34.99 + 0.79(x - 300) < 29.99 + 0.49(x - 200)$$

Chapter 2   PRETEST

To see if you have already mastered the topics in this chapter, take this test.

1. Is 4 a solution of the equation $7 - 2x = 3x - 11$?

2. Solve and check: $n + 2 = -6$

3. Solve and check: $\dfrac{y}{-5} = 1$

4. Solve and check: $-n = 8$

5. Solve and check: $\dfrac{2}{3}x - 3 = -9$

6. Solve and check: $4x - 8 = -10$

7. Solve and check: $6 - y = -5$

8. Solve and check: $9x + 13 = 7x + 19$

9. Solve and check: $-2(3n - 1) = -7n$

10. Solve and check: $14x - (8x - 13) = 12x + 3$

11. Solve $v - 5u = w$ for v in terms of u and w.

12. 9 is what percent of 36?

13. 60% of what number is 12?

14. Draw the graph of $x \le 2$.

$$\xleftarrow{\quad}\overset{\displaystyle +\ +\ +\ +\ +\ +\ +}{\underset{-3\ -2\ -1\ \ 0\ \ 1\ \ 2\ \ 3}{\quad}}\xrightarrow{\quad}$$

15. Solve and graph: $x + 3 > 3$

$$\xleftarrow{\quad}\overset{\displaystyle +\ +\ +\ +\ +\ +\ +}{\underset{-3\ -2\ -1\ \ 0\ \ 1\ \ 2\ \ 3}{\quad}}\xrightarrow{\quad}$$

16. An office photocopier makes 30 copies per minute. How long will it take to copy a 360-page document?

17. A florist charges a flat fee of $100 plus $70 for each centerpiece for a wedding. If the total bill for the centerpieces was $1500, how many centerpieces did the florist make?

18. The formula for finding the amount of kinetic energy used is $E = \dfrac{1}{2}mv^2$. Solve this equation for m in terms of E and v.

19. A certain amount of money was invested in two different accounts. The amount invested at 8% simple interest was twice the amount invested at 5% simple interest. If the total interest earned on the investments was $420, how much was invested at each rate?

20. A gym offers two membership options: Option A is $55 per month for unlimited use of the gym and Option B is $10 per month plus $3 for each hour a member uses the gym. For how many hours of use per month will Option A be a better deal?

● *Check your answers on page A-4.*

2.1 Solving Linear Equations: The Addition Property

What Equations Are and Why They Are Important

In this chapter, we work with one of the most important concepts in algebra, the *equation*. Equations are important because they help us to solve a wide variety of problems.

For instance, suppose that we want to find the measure of $\angle B$ (read "angle B") in triangle ABC.

OBJECTIVES

- To determine whether a given number is a solution of a given equation

- To solve linear equations using the addition property

- To solve applied problems using the addition property

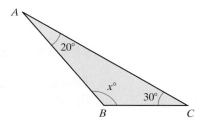

We know from geometry that for any triangle, the sum of the measures of the angles is $180°$. To solve this problem, we can write the following equation, where x represents the measure of $\angle B$.

$$20 + 30 + x = 180$$

An equation has two algebraic expressions—one on the left side of the equal sign and one on the right side of the equal sign.

Equal sign

$$\underbrace{20 + 30 + x}_{\text{Left side}} = \underbrace{180}_{\text{Right side}}$$

Definition

An **equation** is a mathematical statement that two expressions are equal.

Some examples of equations are

$$1 + 3 + 2 = 6 \qquad y - 6 = 9 \qquad y + 4 = -8 \qquad -5x = 10$$

Solving Equations

An equation may be either true or false. The equation $x + 4 = 9$ is true if 5 is substituted for the variable x.

$$x + 4 = 9$$
$$5 + 4 = 9 \qquad \text{True}$$

However, this equation is false if 6 is substituted for the variable x.

$$x + 4 = 9$$
$$6 + 4 = 9 \quad \text{False}$$

How many values of x will make $x + 4 = 9$ a true equation? How many values of x will make this equation false?

The number 5 is called a *solution* of the equation $x + 4 = 9$ because we get a true statement when we substitute 5 for x.

Definition

A **solution of an equation** is a value of the variable that makes the equation a true statement.

EXAMPLE 1

Is 2 a solution of the equation $3x + 1 = 11 - 2x$?

Solution

$$3x + 1 = 11 - 2x$$
$$3(2) + 1 \overset{?}{=} 11 - 2(2) \quad \text{Substitute 2 for } x.$$
$$6 + 1 \overset{?}{=} 11 - 4 \quad \text{Evaluate each side of the equation.}$$
$$7 = 7 \quad \text{True}$$

Since $7 = 7$ is a true statement, 2 is a solution of $3x + 1 = 11 - 2x$.

PRACTICE 1

Determine whether 4 is a solution of the equation $5x - 4 = 2x + 5$.

EXAMPLE 2

Determine whether -1 is a solution of the equation $2x - 4 = 6(x + 2)$.

Solution

$$2x - 4 = 6(x + 2)$$
$$2(-1) - 4 \overset{?}{=} 6(-1 + 2) \quad \text{Substitute } -1 \text{ for } x.$$
$$-2 - 4 \overset{?}{=} 6(1) \quad \text{Evaluate each side of the equation.}$$
$$-6 = 6 \quad \text{False}$$

Since $-6 = 6$ is a false statement, -1 is *not* a solution of the equation.

PRACTICE 2

Is -8 a solution of the equation $5(x + 3) = 3x - 1$?

In this chapter, we will work mainly with equations in one variable that have one solution. These equations are called *linear equations* or *first-degree equations* because the exponent of the variable is 1.

Definition

A **linear equation in one variable** is an equation that can be written in the form

$$ax + b = c$$

where a, b, and c are real numbers and $a \neq 0$.

Some linear equations have a special relationship. Consider the equations $x = 2$ and $x + 1 = 3$:

$x = 2$ The solution is 2.

$x + 1 = 3$ By inspection, we see the solution is 2.
$2 + 1 \stackrel{?}{=} 3$
$3 = 3$ True

The equations $x = 2$ and $x + 1 = 3$ have the same solution and so are *equivalent*.

Definition

Equivalent equations are equations that have the same solution.

To solve an equation means to find the number or constant that, when substituted for the variable, makes the equation a true statement. This solution is found by changing the equation to an equivalent equation of the form

$$x = \boxed{}$$

↑   ↑
The variable is isolated   The number or constant is
(alone on one side   isolated on the other side.
with coefficient 1).

Using the Addition Property to Solve Linear Equations

One of the properties that we use in solving equations involves adding. Suppose we have the equation $\frac{6}{3} = 2$. If we were to add 4 to each side of the equation, we would get $\frac{6}{3} + 4 = 2 + 4$. Using mental arithmetic, we see that $\frac{6}{3} + 4 = 2 + 4$ is also a true statement. This example leads us to the following property.

Addition Property of Equality

For any real numbers a, b, and c, if $a = b$, then $a + c = b + c$.

This property allows us to add any real number to each side of an equation, resulting in an equivalent equation.

EXAMPLE 3

Solve and check: $x - 5 = -11$

Solution To solve this equation, we isolate the variable by adding 5, which is the additive inverse of -5, to each side of the equation.

$$x - 5 = -11$$
$$x - 5 + 5 = -11 + 5 \qquad \text{Add 5 to each side of the equation.}$$
$$x + 0 = -6$$
$$x = -6 \qquad \text{Recall that } x + 0 = x.$$

We can check if -6 is the solution to the original equation by substituting for x.

Check $x - 5 = -11$
$$-6 - 5 \stackrel{?}{=} -11 \qquad \text{Substitute } -6 \text{ for } x \text{ in the original equation.}$$
$$-11 = -11 \qquad \text{True}$$

So the solution is -6.

PRACTICE 3

Solve and check: $y - 12 = -7$

Can you explain why checking a solution is important?

EXAMPLE 4

Solve and check: $-9 = a + 6$

Solution

$$-9 = a + 6$$
$$-9 + (-6) = a + 6 + (-6) \qquad \text{Add } -6 \text{ to each side of the equation.}$$
$$-15 = a + 0$$
$$-15 = a$$
or $\qquad a = -15$

Check $-9 = a + 6$
$$-9 \stackrel{?}{=} (-15) + 6 \qquad \text{Substitute } -15 \text{ for } a.$$
$$-9 = -9 \qquad \text{True}$$

So the solution is -15.

PRACTICE 4

Solve and check: $-2 = n + 15$

Are the solutions to the equations $-9 = a + 6$ and $a + 6 = -9$ the same? Explain.

Because subtracting a number is the same as adding its opposite, the addition property allows us to subtract the same value from each side of an equation. For instance, suppose $a = b$. Subtracting c from each side gives us $a - c = b - c$. So an alternative approach to solving Example 4 is to subtract the same number, namely 6, from each side of the equation, as shown on the next page.

$$-9 = a + 6$$
$$-9 - 6 = a + 6 - 6$$
$$-15 = a, \quad \text{or } a = -15$$

EXAMPLE 5	PRACTICE 5
Solve and check: $m - (-26.1) = 32$	Solve and check: $5 = 4.9 - (-x)$

Solution

$$m - (-26.1) = 32$$
$$m + 26.1 = 32$$
$$m + 26.1 - 26.1 = 32 - 26.1 \qquad \text{Subtract 26.1 from each side of the equation.}$$
$$m = 5.9$$

Check $m - (-26.1) = 32$
$$5.9 - (-26.1) \stackrel{?}{=} 32 \qquad \text{Substitute 5.9 for } m.$$
$$5.9 + 26.1 \stackrel{?}{=} 32$$
$$32 = 32 \qquad \text{True}$$

So the solution is 5.9.

Equations are often useful *mathematical models* that represent real-world situations. Although there is no magic formula for solving applied problems in algebra, it is a good idea to keep the following problem-solving steps in mind.

To Solve a Word Problem in Algebra

- Read the problem carefully.
- Translate the word problem to an equation.
- Solve the equation.
- Check the solution in the original equation.
- State the conclusion.

We have discussed how to translate word phrases to algebraic expressions. Now let's look at some examples of translating word sentences to equations, namely those involving addition or subtraction.

Word sentence: A number increased by 1.1 equals 8.6.

Equation: $n + 1.1 = 8.6$

Word sentence: A number minus one-half equals five.

Equation: $y - \dfrac{1}{2} = 5$

Note that in both examples we used a variable to represent the unknown number.

In solving applied problems, first we translate the given word sentences to equations, and then we solve the equations.

EXAMPLE 6

The mean distance between the planet Venus and the Sun is 31.2 million miles more than the mean distance between the planet Mercury and the Sun. (*Source: Time Almanac 2000*)

a. Using the following diagram, write an equation to find Mercury's mean distance from the Sun.

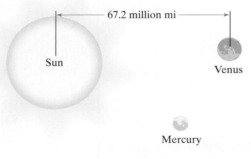

Sun

67.2 million mi

Venus

Mercury

Not to scale

b. Solve this equation.

Solution

a. Let x represent Mercury's mean (or average) distance from the Sun. Reading from the diagram, we see that the mean distance between Venus and the Sun is 67.2 million miles. This distance is 31.2 million miles more than Mercury's mean distance to the Sun. So we translate this sentence to an equation:

Word sentence: 67.2   is   31.2   plus x

Equation: 67.2   =   31.2   + x

b. Solve the equation:

$$67.2 = x + 31.2$$
$$67.2 - \mathbf{31.2} = x + 31.2 - \mathbf{31.2} \qquad \text{Subtract 31.2 from each side of the equation.}$$
$$36.0 = x$$

Check $67.2 = x + 31.2$
$$67.2 \overset{?}{=} 36.0 + 31.2 \qquad \text{Substitute 36.0 for } x \text{ in the original equation.}$$
$$67.2 = 67.2 \qquad \text{True}$$

So we conclude that Mercury's mean distance from the Sun is 36.0 million miles.

PRACTICE 6

A chemistry experiment requires you to find the mass of the solution in a plastic bottle. If the mass of the bottle with the solution is 24.56 g and the mass of the empty bottle is 9.68 g, what is the mass of the solution?

| 2.1 | Exercises | FOR EXTRA HELP |  | PRACTICE | WATCH | DOWNLOAD | READ | REVIEW |

Mathematically Speaking

Fill in each blank with the most appropriate term or phrase from the given list.

equal equations	each	addition property of equality
one	equation	
isolate	linear equations in one variable	equivalent equations
additive identity property		solve

1. A(n) _____ is a mathematical statement that two expressions are equal.

2. Equations that can be written in the form $ax + b = c$, where a, b, and c are real numbers and $a \neq 0$, are called _____.

3. Equations that have the same solution are _____.

4. To _____ an equation means to find the number or constant that, when substituted for the variable, makes the equation true.

5. The _____ tells us that for any real numbers a, b, and c, if $a = b$, then $a + c = b + c$.

6. Adding any real number to _____ side of an equation results in an equivalent equation.

Indicate whether the value of x shown is a solution to the given equation by answering true or false.

7.

Value of x	Equation	True or False
a. -8	$3x + 13 = -11$	
b. 7	$28 - x = 7 - 4x$	
c. 9	$2(x - 3) = 12$	
d. $\dfrac{2}{3}$	$12x - 2 = 6x + 2$	

8.

Value of x	Equation	True or False
a. -2	$3x - 2 = 10$	
b. 4	$2x + 3 = 5x - 9$	
c. 1	$3(5 - x) = 18$	
d. $-\dfrac{1}{2}$	$6 - 5x = 5x + 11$	

Indicate what must be done to each side of the equation in order to isolate the variable.

9. $x + 4 = -6$

10. $x - 3 = -3$

11. $-2 = -1 + x$

12. $x - (-15) = 10$

13. $z - (-3.5) = 5$

14. $x - 3.4 = 9.6$

15. $9 = x - 2\frac{1}{5}$

16. $m + \dfrac{1}{3} = 2$

Solve and check.

17. $y + 9 = -14$

18. $x + 2 = -10$

19. $t - 4 = -4$

20. $m - 6 = 1$

21. $9 + a = -3$

22. $10 + x = -4$

23. $z - 4 = -10$

24. $n - 4 = -1$

25. $12 = x + 12$

26. $-25 = y + 21$

27. $-6 = t - 12$

28. $-19 = r - 19$

29. $-4 = -10 + r$

30. $-7 = -2 + m$

31. $-15 + n = -2$

32. $-14 + c = 33$

33. $x + \dfrac{2}{3} = -\dfrac{1}{3}$

34. $z - \dfrac{1}{8} = \dfrac{3}{8}$

35. $8 + y = 4\frac{1}{2}$

36. $2 = z - 1\frac{1}{4}$

37. $m + 2.4 = 5.3$

38. $13.9 = n - 12.5$

39. $-2.3 + t = -5.9$

40. $-3.4 + r = -9.5$

41. $a - (-35) = 30$

42. $x - (-25) = 24$

43. $m - \left(-\dfrac{1}{4}\right) = -\dfrac{1}{4}$

44. $a - (-1.5) = 2$

Solve and round to the nearest hundredth.

45. $y + 2.932 = 4.811$

46. $x + 3.0245 = 9$

Translate each sentence to an equation. Then solve and check.

47. Two more than a number is 12.

48. The sum of a number and 3.2 is the same as 20.

49. If 4 is subtracted from a number, the result is 21.

50. A number minus $2\frac{1}{7}$ is the same as $1\frac{1}{2}$.

51. If -3 is added to a number, the result is -1.

52. 5.2 less than a number equals 12.

53. A number plus 7 equals 11.

54. The difference between a number and 1 is 9.

Choose the equation that best describes the situation.

55. After dieting and exercising, a featherweight boxer lost 6 lb. If the boxer now weighs 127 lb, what was his original weight?

    a. $x - 127 = -6$ **b.** $x + 6 = 127$

    c. $x + 127 = 6$ **d.** $x - 6 = 127$

56. After paying a bill of $5.25, a customer has $2.75 left. How much money did she have prior to paying the bill?

    a. $d - 5.25 = 2.75$ **b.** $d + 2.75 = 5.25$

    c. $d + 5.25 = 2.75$ **d.** $d - 2.75 = -5.25$

57. A digital picture that takes up 4.7 megabytes (Mb) of memory is saved on a computer. If 250 Mb of memory are left after the picture is saved, how many megabytes of free memory did the computer have before the picture was saved?

    a. $x - 4.7 = 250$ **b.** $x - 250 = -4.7$

    c. $x + 4.7 = 250$ **d.** $x + 250 = 4.7$

58. At a college, tuition costs students $3000 a semester. If a student received $2250 in financial aid, how much more money does the student need to pay the balance of the semester's tuition?

    a. $m - 2250 = 3000$ **b.** $m + 2250 = 3000$

    c. $m + 3000 = 2250$ **d.** $m - 3000 = 2250$

Mixed Practice

Indicate what must be done to each side of the equation in order to isolate the variable.

59. $\dfrac{2}{3} + b = 3$

60. $-5 = a - 3$

Solve.

61. Indicate whether the value of x shown is a solution to the given equation by answering true or false.

Value of x	Equation	True or False
a. -4	$6x + 15 = -11$	
b. 9	$4x - 21 = 2x - 3$	
c. 7	$-2(x - 5) = 4$	
d. $\dfrac{3}{4}$	$5 - 8x = -4x + 2$	

Translate each sentence to an equation. Then solve and check.

62. The difference between 0 and a number is 32.

63. If 2.5 is substracted from a number, the result is -3.8.

Solve and check.

64. $x - 12 = -7$

65. $-19 = m + 4$

66. $-3.8 + t = -6.6$

67. $7 + n = 2\dfrac{2}{3}$

68. $x + 7.815 = 3.298$
(Round to the nearest hundredth.)

Applications

Write an equation. Solve and check.

69. If the speed of a car is increased by 10 mph, the speed will be 44 mph. What is the speed of the car now?

70. The difference between the boiling point of gold and the boiling point of aluminum is 337°C. If aluminum boils at 2519°C, the lower temperature of the two metals, what is the boiling point of gold? (*Source: Time Almanac, 2006*)

71. A student uses about 370 calories after 1 hr of hiking. This is 190 calories less than is used after 1 hr of stretching. How many calories does the student use in an hour of stretching? (*Source:* U.S. Department of Agriculture)

72. A patient's temperature dropped by 2.4°F to 98.6°F. What had the patient's temperature been?

73. A traffic helicopter descends 170 meters to be 215 meters above the ground, as illustrated in the following diagram. What was the original height of the helicopter?

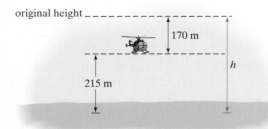

original height

170 m

h

215 m

74. In the following diagram, $\angle ABC$ and $\angle CBD$ are complementary angles, that is, the sum of their measures is 90°. Find the number of degrees in $\angle CBD$.

A

70°

C

?

B

D

75. In the diagram shown, angles x and y are supplementary angles, that is, the sum of their measures is 180°. Find the measure of $\angle x$ if the measure of $\angle y$ is 118.5°.

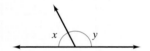

x   y

76. At the close of business today, the price per share of a stock was $21.63. This was down $0.31 from the closing price per share yesterday. What was the closing price per share of the stock yesterday?

● *Check your answers on page A-4.*

MINDSTRETCHERS

Mathematical Reasoning

1. Suppose that $x - a = b$.

    a. What happens to x if a decreases and b remains the same?

    b. What happens to x if a remains the same and b decreases?

Critical Thinking

2. In the magic square to the right, the sum of each row, column, and diagonal is the same. Calculate that sum. Then write and solve equations to find v, w, x, y, and z.

v	6	11
w	10	y
x	14	z

Research

3. In your college library or on the Web, investigate the origin of the word *algebra*. Write a few sentences to show what you have learned.

CULTURAL NOTE

Just as we solve an equation to identify an unknown number, we use a balance scale to determine an unknown weight. In this picture, an image from 3400 years ago shows an Egyptian weighing gold rings against a counterbalance in the form of a bull's head.

The balance scale is an ancient measuring device. These scales were used by Sumerians for weighing precious metals and gems at least 9000 years ago.

Source: O. A. W. Dilke, *Mathematics and Measurement* (Berkeley: University of California Press/British Museum, 1987).

2.2   Solving Linear Equations: The Multiplication Property

In the previous section, we used the addition property of equality to solve equations. Another property that is useful in solving equations involves multiplication. Consider the equation $\frac{6}{3} = 2$. If we were to multiply each side of this equation by 9, we would get $\frac{6}{3} \cdot 9 = 2 \cdot 9$. Note that each side of the equation equals 18. In general, multiplying each side of a true equation by the same number results in another true equation.

Multiplication Property of Equality

For any real numbers a, b, and c, if $a = b$, then $a \cdot c = b \cdot c$.

This property allows us to multiply each side of an equation by any real number to get an equivalent equation.

EXAMPLE 1	PRACTICE 1
Solve and check: $\dfrac{x}{4} = 11$	Solve and check: $\dfrac{y}{3} = 21$

Solution In this equation, note that $\frac{x}{4}$ is the same as $\frac{1}{4} \cdot x$. To solve this equation, we isolate the variable by multiplying each side of the equation by 4, the reciprocal of $\frac{1}{4}$.

$$\frac{x}{4} = 11$$

$$4 \cdot \frac{x}{4} = 4 \cdot 11 \qquad \text{Multiply each side of the equation by 4.}$$

$$1x = 44 \qquad 4 \cdot \frac{x}{4} = 4 \cdot \frac{1}{4}x = 1x$$

$$x = 44 \qquad 1x = x$$

Check $\dfrac{x}{4} = 11$

$$\frac{44}{4} \overset{?}{=} 11 \qquad \text{Substitute 44 for } x \text{ in the original equation.}$$

$$11 = 11 \qquad \text{True}$$

So the solution is 44.

EXAMPLE 2

Solve and check: $9x = -72$

Solution

$$9x = -72$$

$$\left(\frac{1}{9}\right)9x = \left(\frac{1}{9}\right)(-72) \qquad \text{Multiply each side of the equation by } \frac{1}{9}.$$

$$1x = -8 \qquad \frac{1}{9} \cdot 9 = 1$$

$$x = -8$$

Check $9x = -72$

$$9(-8) \stackrel{?}{=} -72 \qquad \text{Substitute } -8 \text{ for } x \text{ in the original equation.}$$

$$-72 = -72 \qquad \text{True}$$

So the solution is -8.

Because dividing by a number is the same as multiplying by its reciprocal, the multiplication property allows us to divide each side of an equation by a non-zero number. For instance, suppose $a = b$. Multiplying each side by $\frac{1}{c}$ gives us $a \cdot \frac{1}{c} = b \cdot \frac{1}{c}$, for $c \neq 0$. This equation is equivalent to $\frac{a}{c} = \frac{b}{c}$. So an alternative approach to solving Example 2 is to divide each side of the equation by the same number, namely 9, as shown.

$$9x = -72$$

$$\frac{9x}{9} = \frac{-72}{9}$$

$$x = -8$$

Note that this approach gives us the same solution, -8, that we got using the approach in Example 2.

EXAMPLE 3

Solve and check: $-y = -15$

Solution

$$-y = -15$$

$$-1y = -15 \qquad \text{The coefficient of } -y \text{ is } -1.$$

$$\frac{-1y}{-1} = \frac{-15}{-1} \qquad \text{Divide each side of the equation by } -1.$$

$$y = 15$$

Check $-y = -15$

$$-1(15) \stackrel{?}{=} -15 \qquad \text{Substitute } 15 \text{ for } y \text{ in the original equation.}$$

$$-15 = -15 \qquad \text{True}$$

So the solution is 15.

PRACTICE 2

Solve and check: $7y = 63$

PRACTICE 3

Solve and check: $-x = 10$

EXAMPLE 4

Solve and check: $46 = -4.6n$

Solution

$$46 = -4.6n$$

$$\frac{46}{-4.6} = \frac{-4.6n}{-4.6} \qquad \text{Divide each side of the equation by } -4.6.$$

$$-10 = n \quad \text{or}$$

$$n = -10$$

Check $46 = -4.6n$

$46 \overset{?}{=} -4.6\,(-10)$ Substitute -10 for n in the original equation.

$46 = 46$ True

So the solution is -10.

PRACTICE 4

Solve and check: $-11.7 = -0.9z$

EXAMPLE 5

Solve and check: $\dfrac{2w}{3} = 8$

Solution

$$\frac{2w}{3} = 8$$

$$\frac{2}{3}w = 8$$

$$\frac{3}{2} \cdot \frac{2}{3}w = \frac{3}{2} \cdot 8 \qquad \text{Multiply each side of the equation by } \frac{3}{2}.$$

$$w = 12 \qquad \frac{3}{2} \cdot \frac{2}{3} = 1 \text{ and } 1w = w.$$

Check $\dfrac{2}{3}w = 8$

$\dfrac{2}{3}(12) \overset{?}{=} 8$ Substitute 12 for w in the original equation.

$2 \cdot 4 \overset{?}{=} 8$ Simplify.

$8 = 8$ True

So the solution is 12.

PRACTICE 5

Solve and check: $\dfrac{6}{7}y = -12$

Can you show another way to solve Example 5? Explain.

Let's now consider applied problems involving the multiplication property. To solve these problems, we translate word sentences to equations involving multiplication or division as follows:

Word sentence: Three times a number x equals -12.

$$3 \quad \cdot \quad x \quad = \quad -12$$

Equation: $3x = -12$

Word sentence: A number d divided by -2 equals 8.

Equation: $d \quad \div \quad -2 \quad = \quad 8$

$$\frac{d}{-2} = 8$$

EXAMPLE 6	PRACTICE 6
A student applies for a job that pays an hourly overtime wage of \$15.90. The overtime wage is 1.5 times the regular hourly wage. What is the regular hourly wage for the job?	A mechanic billed a customer \$189.50 for labor to repair his car. If one-fourth of the bill was for labor, how much was the total bill?

Solution Letting w represent the regular hourly wage, we write the word sentence and then translate it to an equation.

Word sentence:

The overtime wage \$15.90 is 1.5 times the regular hourly wage.

Equation: $15.90 = 1.5 \quad \cdot \quad w$

$$15.90 = 1.5w$$

Next we solve the equation for w.

$$15.9 = 1.5w$$
$$\frac{15.9}{1.5} = \frac{1.5}{1.5}w$$
$$10.6 = w, \quad \text{or}$$
$$w = 10.6$$

Check $15.90 = 1.5w$

$15.90 \overset{?}{=} 1.5\,(\mathbf{10.6})$

$15.90 = 15.90$ True

So the regular wage is \$10.60.

Let's now turn to a particular kind of word problem—a problem involving motion. To solve such problems, we need to use the equation $d = rt$, where d is distance, r is the average rate or speed, and t is time.

EXAMPLE 7

One of the fastest pitchers in the history of Japanese baseball was Hideki Irabu, whose pitches were clocked at about 140 feet per second. If the distance from the pitcher's mound to home plate is 60.5 ft, approximate the time it took Irabu's pitches to reach home plate, to the nearest hundredth of a second. (*Source: The Toronto Sun*, June 9, 1998, p. 12)

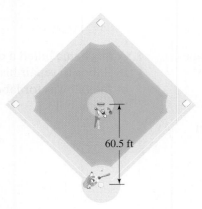

60.5 ft

Solution The distance between the pitcher's mound and home plate is 60.5 ft and the rate of the ball thrown is 140 feet per second. Using the equation $d = rt$, we can find the time.

$$d = rt$$
$$60.5 = 140t \qquad \text{Substitute 60.5 for } d \text{ and 140 for } r.$$
$$\frac{60.5}{140} = \frac{140t}{140} \qquad \text{Divide each side of the equation by 140.}$$
$$0.43 \approx t, \quad \text{or}$$
$$t \approx 0.43$$

Check Since the solution of the original equation is a rounded value, the check will not result in an exact equality. To verify the solution, check that the expressions on each side of the equation are approximately equal to one another.

$$60.5 = 140t$$
$$60.5 \stackrel{?}{\approx} 140 \cdot \mathbf{0.43} \qquad \text{Substitute 0.43 for } t \text{ in the original equation.}$$
$$60.5 \approx 60.2 \qquad \text{True}$$

So Irabu's pitches took approximately 0.43 sec.

PRACTICE 7

A driver makes a trip from Washington, D.C., the U.S. capital, to Philadelphia, Pennsylvania, the home of Independence Hall. If she averages 60 mph, how long, to the nearest tenth of an hour, will it take her to get to Philadelphia? (*Source:* http://mapquest.com)

139 mi ● Philadelphia

Washington, D.C.

| **2.2** | **Exercises** | | |

Indicate what must be done to each side of the equation in order to isolate the variable.

1. $\frac{x}{3} = -4$

2. $\frac{a}{-6} = 1$

3. $-5x = 20$

4. $-4x = 30$

5. $-2.2n = 4$

6. $1.5x = -6$

7. $\frac{3}{4}x = 12$

8. $\frac{2}{3}x = -6$

9. $-\frac{5y}{2} = 15$

10. $-\frac{8n}{5} = 4$

Solve and check.

11. $6x = -30$

12. $-8y = 8$

13. $\frac{n}{2} = 9$

14. $\frac{w}{10} = -21$

15. $\frac{a}{4} = 1.2$

16. $\frac{n}{7} = -1.3$

17. $-5x = 2.5$

18. $2y = 0.08$

19. $42 = -6c$

20. $50 = 2x$

21. $11 = -\frac{r}{2}$

22. $4 = \frac{-m}{3}$

23. $\frac{5}{6}x = 10$

24. $\frac{3}{4}d = -3$

25. $-\frac{2}{5}y = 1$

26. $\frac{2}{3}r = -8$

27. $\frac{3n}{4} = 6$

28. $\frac{5a}{6} = 5$

29. $\frac{4c}{3} = -4$

30. $-\frac{2z}{7} = 8$

31. $\frac{x}{2.4} = -1.2$

32. $-\frac{n}{0.5} = -1.3$

33. $-2.5a = 5$

34. $-2.25 = -1.5t$

35. $\frac{2}{3}y = \frac{4}{9}$

36. $\frac{5}{6}c = \frac{2}{3}$

Solve. Round to the nearest hundredth.

37. $\frac{x}{-1.515} = 1.515$

38. $\frac{n}{-2.968} = -3.85$

39. $-3.14x = 21.4148$

40. $2.54z = 6.4516$

Translate each sentence to an equation. Then solve and check.

41. The product of -4 and a number is 56.

42. $\frac{3}{4}$ of a number is equal to 12.

43. A number divided by 0.2 is 1.1.

44. Twice a number is equal to 15.

45. A number divided by -3 is 20.

46. The quotient of a number and 2.5 is 40.

47. $\frac{1}{6}$ of a number is $2\frac{4}{5}$.

48. $\frac{5}{8}$ of a number is 20.

Choose the equation that best describes the situation.

49. Suppose that a shopper used half his money to buy a backpack. If the backpack cost \$20, how much money did he have prior to this purchase?

   a. $20 = 2m$ **b.** $20m = \frac{1}{2}$

   c. $20 = \frac{m}{2}$ **d.** $\frac{m}{20} = \frac{1}{2}$

50. A child's infant brother weighs 12 lb. If this is $\frac{1}{4}$ of her weight, how much does she weigh?

   a. $4w = 12$ **b.** $\frac{w}{4} = 12$

   c. $12w = \frac{1}{4}$ **d.** $\frac{w}{12} = \frac{1}{4}$

51. A student plans to buy a DVD player 6 weeks from now. If the DVD player costs \$150, how much money must she save per week in order to buy it?

   a. $6p = 150$ **b.** $150p = 6$

   c. $\frac{p}{6} = 150$ **d.** $\frac{p}{150} = 6$

52. The student government at a college sold tickets to a play. From ticket sales, it collected \$800, which was twice the cost of the play. How much did the play cost?

   a. $\frac{c}{800} = 2$ **b.** $800c = 2$

   c. $\frac{c}{2} = 800$ **d.** $2c = 800$

Mixed Practice

Solve and check.

53. $\frac{2n}{7} = 4$

54. $8 = -\frac{a}{3}$

55. $-\frac{y}{3.8} = -0.3$

56. $\frac{n}{3.14} = -39.715$
(Round to the nearest hundredth.)

Translate each sentence to an equation. Then solve and check.

57. The quotient of a number and 5 is equal to 2.

58. $\frac{3}{8}$ of a number is 12.

Indicate what must be done to each side of the equation in order to isolate the variable.

59. $-5.2m = 4$

60. $\frac{b}{7} = 3$

Applications

Write an equation that best describes the situation. Then solve and check.

61. According to a geologist, sediment at the bottom of a local lake accumulated at the rate of 0.02 centimeter per year. How long did it take to create a layer of sediment 10.5 cm thick?

62. Because of evaporation, the water level in an aquarium drops at a rate of $\frac{1}{10}$ inch per hour. In how many hours will the level drop 2 in.?

63. The bus trip from Miami to San Francisco, a driving distance of 3348 mi, takes 70 hr. What is the average speed of the bus rounded to the nearest mile per hour? (*Source: Greyhound*)

64. A double-trailer truck is driven at an average speed of 54 mph from Atlanta to Cincinnati, a driving distance of 440 mi. To the nearest tenth of an hour, how long did the trip take? (*Source: The World Almanac and Book of Facts, 2000*)

65. A customer has only $20 to spend at a local print shop that charges $0.05 per copy. At this rate, how many copies can she afford to make?

66. Consider the two parcels of land shown—one a rectangle and the other a square. For which value of x do the two parcels have the same area?

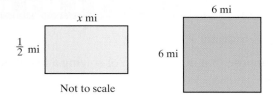

Not to scale

67. A city offers to pay a disposal company $40 per ton to bury 20,000 tons of toxic waste. If this deal represents $\frac{2}{3}$ of the disposal company's projected income, what is that income?

68. The diameter of a tree trunk increases as the tree ages and adds rings. Suppose a trunk's diameter increases by 0.2 inch per year. How many years will it take the tree to increase in diameter from 4 in. to 12 in.?

69. The maximum depth of the Caspian Sea is approximately 1000 m. If this depth is about $\frac{1}{5}$ the maximum depth of the Mediterranean Sea, what is the maximum depth of the Mediterranean Sea? (*Source: The 2006 New York Times Almanac*)

70. A top-secret plane flew 3000 mi in $1\frac{1}{2}$ hr. What was the average speed of this plane?

71. A student takes a job in the college's student center so that she can buy books that cost $187.50. How many hours must she work to make this amount if she earns $7.50/hr?

72. The coat in the ad shown is selling at $\frac{1}{4}$ the regular price. What was the regular price?

73. Last year, a young couple paid a total of $10,020 in rent for their apartment. How much money did they pay per month in rent?

74. An equilateral triangle is a triangle that has three sides equal in length. If the perimeter of an equilateral triangle is $10\frac{1}{2}$ ft, how long is each side?

● *Check your answers on page A-5.*

MINDSTRETCHERS

Mathematical Reasoning

1. Suppose that in the course of solving a linear equation, you reach the step

$$5x = 3x$$

If you then divide both sides by x, you get $5 = 3$, which is impossible. Did you make an error? Explain.

Critical Thinking

2. If you multiply each side of the equation $0.24r = -12.48$ by 100, the result is an equivalent equation. Explain why it is helpful to carry out this multiplication in solving the equation.

Writing

3. Write two different word problems that are applications of each equation.

a. $6x = 18$

b. $\frac{x}{3} = 10$

2.3 Solving Linear Equations by Combining Properties

In the previous sections, we solved simple equations involving either the addition property or the multiplication property. We now turn our attention to solving equations that involve both properties.

Solving Equations Using Both the Addition and Multiplication Properties

We need to use both the addition property and the multiplication property to solve equations such as

$$3x + 4 = 7 \quad \text{and} \quad \frac{r}{2} - 5 = 9$$

To solve these equations, we first use the addition property to get the variable term alone on one side. Then we use the multiplication property to isolate the variable.

EXAMPLE 1	PRACTICE 1
Solve and check: $3x + 4 = 7$	Solve and check: $2y + 1 = 9$

Solution

$$3x + 4 = 7$$
$$3x + 4 - 4 = 7 - 4 \qquad \text{Subtract 4 from each side of the equation.}$$
$$3x = 3$$
$$\frac{3x}{3} = \frac{3}{3} \qquad \text{Divide each side of the equation by 3.}$$
$$x = 1$$

Check $3x + 4 = 7$
$$3(1) + 4 \overset{?}{=} 7 \qquad \text{Substitute 1 for } x \text{ in the original equation.}$$
$$3 + 4 \overset{?}{=} 7$$
$$7 = 7 \qquad \text{True}$$

So the solution is 1.

In solving Example 1, would we get the same solution if we divide before subtracting? Explain.

EXAMPLE 2	PRACTICE 2

Solve and check: $\dfrac{r}{2} - 5 = 9$

Solution

$$\dfrac{r}{2} - 5 = 9$$

$$\dfrac{r}{2} - 5 + 5 = 9 + 5 \qquad \text{Add 5 to each side of the equation.}$$

$$\dfrac{r}{2} = 14$$

$$2 \cdot \dfrac{r}{2} = 2 \cdot 14 \qquad \text{Multiply each side of the equation by 2.}$$

$$r = 28$$

Check $\qquad \dfrac{r}{2} - 5 = 9$

$$\dfrac{28}{2} - 5 \overset{?}{=} 9 \qquad \text{Substitute 28 for } r \text{ in the original equation.}$$

$$14 - 5 \overset{?}{=} 9$$

$$9 = 9 \qquad \text{True}$$

So the solution is 28.

Solve and check: $\dfrac{c}{5} - 1 = 8$

EXAMPLE 3	PRACTICE 3

Solve: $-4s + 7 = 3$

Solution

$$-4s + 7 = 3$$

$$-4s + 7 - 7 = 3 - 7 \qquad \text{Subtract 7 from each side of the equation.}$$

$$-4s = -4$$

$$\dfrac{-4s}{-4} = \dfrac{-4}{-4} \qquad \text{Divide each side by } -4.$$

$$s = 1$$

So the solution is 1.

Solve: $-6b - 5 = 13$

EXAMPLE 4	PRACTICE 4

Solve: $2x - 5x = 12$

Solution

$$\underbrace{2x - 5x}_{} = 12$$

$$-3x = 12 \qquad \text{Combine like terms.}$$

$$\dfrac{-3x}{-3} = \dfrac{12}{-3} \qquad \text{Divide each side of the equation by } -3.$$

$$x = -4$$

So the solution is -4.

Solve: $8n + 10n = 24$

Solving Equations by Combining Like Terms

Now, let's consider an equation that has like terms on the same side of the equation. In order to solve this kind of equation, we combine all like terms before using the addition and multiplication properties.

EXAMPLE 5	PRACTICE 5
Solve and check: $7x - 3x - 6 = 6$	Solve and check: $5 - t - t = -1$

Solution

$$\underbrace{7x - 3x} - 6 = 6$$
$4x - 6 = 6$ Combine like terms.
$4x - 6 + 6 = 6 + 6$ Add 6 to each side of the equation.
$4x = 12$
$\dfrac{4x}{4} = \dfrac{12}{4}$ Divide each side of the equation by 4.
$x = 3$

Check $7x - 3x - 6 = 6$
$7(3) - 3(3) - 6 \overset{?}{=} 6$ Substitute 3 for x.
$21 - 9 - 6 \overset{?}{=} 6$
$6 = 6$ True

So the solution is 3.

Suppose an equation has like terms that are on opposite sides of the equation. To solve, we use the addition property to get the like terms together on the same side so that they can be combined.

EXAMPLE 6	PRACTICE 6
Solve: $13z + 5 = -z + 12$	Solve: $3f - 12 = -f - 15$

Solution

$13z + 5 = -z + 12$
$z + 13z + 5 = z + (-z) + 12$ Add z to each side of the equation.
$14z + 5 = 12$ Combine like terms.
$14z + 5 - 5 = 12 - 5$ Subtract 5 from each side of the equation.
$14z = 7$
$\dfrac{14z}{14} = \dfrac{7}{14}$ Divide each side of the equation by 14.
$z = \dfrac{1}{2}$

So the solution is $\dfrac{1}{2}$.

Solving Equations Containing Parentheses

Some equations contain parentheses. To solve these equations, we first remove the parentheses using the distributive property. Then we proceed as in previous examples.

EXAMPLE 7

Solve and check: $2c = -3(c - 5)$

Solution

$$2c = -3(c - 5)$$
$$2c = -3c + 15 \qquad \text{Use the distributive property.}$$
$$3c + 2c = 3c - 3c + 15 \qquad \text{Add } 3c \text{ to each side of the equation.}$$
$$5c = 15 \qquad \text{Combine like terms.}$$
$$\frac{5c}{5} = \frac{15}{5} \qquad \text{Divide each side of the equation by 5.}$$
$$c = 3$$

Check

$$2c = -3(c - 5)$$
$$2(3) \overset{?}{=} -3(3 - 5) \qquad \text{Substitute 3 for } c \text{ in the original equation.}$$
$$6 \overset{?}{=} -3(-2)$$
$$6 = 6 \qquad \text{True}$$

So the solution is 3.

PRACTICE 7

Solve and check: $-5(z + 6) = z$

In general, to solve linear equations, we use the following procedure.

> **To Solve Linear Equations**
> - Use the distributive property to clear the equation of parentheses, if necessary.
> - Combine like terms where appropriate.
> - Use the addition property to isolate the variable term.
> - Use the multiplication property to isolate the variable.
> - Check by substituting the solution in the original equation.

Let's apply this procedure in the next example. Recall from Section 1.7 that when removing parentheses preceded by a minus sign, all the terms in the parentheses change to the opposite sign.

EXAMPLE 8

Solve: $2(x + 5) - (x - 2) = 4x + 6$

Solution

$$2(x + 5) - (x - 2) = 4x + 6$$
$$2x + 10 - x + 2 = 4x + 6 \qquad \text{Use the distributive property.}$$
$$x + 12 = 4x + 6 \qquad \text{Combine like terms.}$$
$$x - 4x + 12 = 4x - 4x + 6 \qquad \text{Subtract } 4x \text{ from each side of the equation.}$$
$$-3x + 12 = 6 \qquad \text{Combine like terms.}$$
$$-3x + 12 - 12 = 6 - 12 \qquad \text{Subtract 12 from each side of the equation.}$$
$$-3x = -6$$
$$\frac{-3x}{-3} = \frac{-6}{-3} \qquad \text{Divide each side of the equation by } -3.$$
$$x = 2$$

So the solution is 2.

PRACTICE 8

Solve: $2(t - 3) - 3(t - 2) = t + 8$

EXAMPLE 9

Solve: $13 - [4 + 2(x - 1)] = 3(x + 2)$

Solution

$$13 - [4 + 2(x - 1)] = 3(x + 2)$$
$$13 - [4 + 2x - 2] = 3(x + 2)$$
$$13 - [2 + 2x] = 3(x + 2)$$
$$13 - 2 - 2x = 3x + 6$$
$$11 - 2x = 3x + 6$$
$$11 - 2x - 3x = 3x - 3x + 6$$
$$11 - 5x = 6$$
$$11 - 11 - 5x = 6 - 11$$
$$-5x = -5$$
$$\frac{-5x}{-5} = \frac{-5}{-5}$$
$$x = 1$$

So the solution is 1.

PRACTICE 9

Solve: $4[5y - (y - 1)] = 7(y - 2)$

Now, let's consider applications that lead to equations like those that we have discussed in this section.

EXAMPLE 10

An insurance company settles a claim by multiplying the claim by a certain factor and then subtracting the deductible. A payment of $3500 is made on a claim filed for $5000. If the company has a $500 deductible, what factor was used to settle the claim?

Solution Let x represent the factor used to settle a claim. Then write the equation:

The claim times the factor less the deductible is the payment.

$$5000 \quad \cdot \quad x \quad - \quad 500 \quad = \quad 3500$$

Next, we solve for x.

$$5000x - 500 = 3500$$
$$5000x - 500 + 500 = 3500 + 500 \qquad \text{Add 500 to each side of the equation.}$$
$$5000x = 4000$$
$$\frac{5000x}{5000} = \frac{4000}{5000} \qquad \text{Divide each side of the equation by 5000.}$$
$$x = \frac{4}{5}, \quad \text{or } 0.8$$

So the factor used to compute the $5000 claim is $\frac{4}{5}$, or 0.8.

PRACTICE 10

A car is purchased for $12,000. The car's value depreciates $1100 in value per year for each of the first 6 yr of ownership. At what point will the car have a value of $6500?

Many motion problems lead to equations of the type that we have discussed in this section, in particular, to equations containing parentheses. Some of these problems involve one or more objects traveling the same distance, at different rates and for different lengths of time. As in other motion problems, we use the formula $d = rt$.

EXAMPLE 11	PRACTICE 11

A bus leaves St. Petersburg traveling at 45 mph. An hour later, a second bus leaves the same city traveling at 55 mph in the same direction. In how many hours will the second bus overtake the first bus?

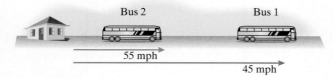

Bus 2   Bus 1

55 mph

45 mph

Solution Let t represent the number of hours traveled by the second bus. Since the first bus left an hour earlier than the second, $t + 1$ represents the number of hours traveled by the first bus.

Recall from Section 2.2 that when an object travels at a constant speed, the distance that it travels is the product of its rate of travel (speed) and the time that it has traveled. Putting these quantities in a table clarifies their relationship.

	Rate ·	Time =	Distance
First bus	45	$t + 1$	$45(t + 1)$
Second bus	55	t	$55t$

Since the two buses travel the same distance to the point where they meet, we can write the following equation and then solve:

$$45(t + 1) = 55t$$
$$45t + 45 = 55t$$
$$45t - 45t + 45 = 55t - 45t$$
$$45 = 10t$$
$$\frac{45}{10} = \frac{10}{10}t$$
$$4.5 = t, \quad \text{or}$$
$$t = 4.5$$

So the second bus will overtake the first bus in 4.5 hr, or $4\frac{1}{2}$ hr.

Two friends plan to meet in Boston. A half an hour after one friend took a local train, the other friend takes an Amtrak express train on a parallel track. If both friends leave from the same station, how long will it take the express train to catch up with the local train if their speeds are 60 mph and 50 mph, respectively?

EXAMPLE 12

A shopper walks from home to the market at a rate of 3 mph. After shopping, he returns home following the same route walking at 2 mph. If the walk back from the market takes 10 min more than the walk to the market, how far away is the market?

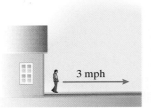 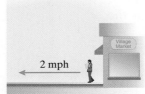

3 mph   2 mph

Solution Let t represent the time of the walk to the market. Since the rates are given in miles per hour, we express the time in hours. The return trip takes 10 min more time. We change 10 min to $\frac{10}{60}$ hr, or $\frac{1}{6}$ hr, getting $t + \frac{1}{6}$ for the time of the return trip.

	Rate · Time = Distance		
Going	3	t	$3t$
Returning	2	$t + \frac{1}{6}$	$2\left(t + \frac{1}{6}\right)$

Since the walk to and from the market followed the same route, the distances each way are equal.

$$3t = 2\left(t + \frac{1}{6}\right)$$

$$3t = 2t + \frac{1}{3}$$

$$3t - 2t = 2t - 2t + \frac{1}{3}$$

$$t = \frac{1}{3}$$

It takes $\frac{1}{3}$ hr to walk to the market. So the market is $3 \cdot \frac{1}{3}$, or 1 mi away from home.

PRACTICE 12

On a round-trip over the same roads, a car averaged 25 mph going and 30 mph returning. If the entire trip took 5 hr 30 min, what is the distance each way?

In some motion problems, an object travels at different speeds for different parts of the trip. The total distance traveled is the sum of the partial distances.

EXAMPLE 13

A car is driven for 3 hr in a rainstorm. After the weather clears, the car is driven 10 mph faster for 2 more hours, completing a 250-mile trip. How fast was the car driven during the storm?

Solution Let x represent the speed of the car during the storm. The speed of the car after the storm passes can be represented by $x + 10$. Drawing a diagram helps us to visualize the problem and to see that the total distance driven is the sum of the two partial distances.

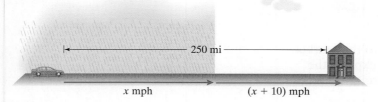

Next, let's complete a table in order to organize the relevant information. Recall that distance is the product of rate and time.

	Rate ·	Time	=	Distance
Storm	x	3		$3x$
Clear	$x + 10$	2		$2(x + 10)$

We are told that the sum of the two partial distances is 250 mi, giving us an equation to solve.

$$3x + 2(x + 10) = 250$$
$$3x + 2x + 20 = 250$$
$$5x + 20 = 250$$
$$5x = 230$$
$$x = \frac{230}{5} = 46$$

Therefore, the car was driven at a speed of 46 mph during the storm.

PRACTICE 13

A cyclist pedals uphill for 2 hr. She then continues downhill 10 mph faster for another hour. If the entire trip was 40 mi in length, what was her downhill speed?

Solve and check.

1. $3x - 1 = 8$

2. $7r - 8 = 13$

3. $9t + 17 = -1$

4. $2y + 1 = 9$

5. $20 - 5m = 45$

6. $25 - 3c = 34$

7. $\dfrac{n}{2} - 1 = 5$

8. $\dfrac{s}{3} - 2 = -4$

9. $\dfrac{x}{5} + 15 = 0$

10. $\dfrac{y}{3} + 3 = 42$

11. $3 - t = 1$

12. $2 - x = 2$

13. $-8 - b = 11$

14. $5 - y = 8$

15. $\dfrac{2}{3}x - 9 = 17$

16. $\dfrac{4}{5}d - 3 = 13$

17. $\dfrac{4}{5}r + 20 = -20$

18. $\dfrac{3y}{8} + 14 = -10$

19. $3y + y = -8$

20. $4a + 3a = -21$

21. $7z - 2z = -30$

22. $4x - x = 18$

23. $28 - a + 4a = 7$

24. $5 - 8x - 2x = -25$

25. $1 = 1 - 6t - 4t$

26. $-1 = 5 - z - z$

27. $3y + 2 = -y - 2$

28. $3n + 6 = -n - 6$

29. $5r - 4 = 2r + 6$

30. $7 - m = 5 + 3m$

31. $4(x + 7) = 7 + x$

32. $3t - 2 = 4(t - 2)$

33. $5(y - 1) = 2y + 1$

34. $5a - 4 = 7(a + 2)$

35. $3a - 2(a - 9) = 4 + 2a$

36. $5 - 2(3x - 4) = 3 - x$

37. $5(2 - t) - (1 - 3t) = 6$

38. $\dfrac{3}{5}(15y + 10) - 3(4y + 3) = 0$

39. $2y - 3(y + 1) = -(5y + 3) + y$

40. $9n + 5(n + 3) = -(n + 13) - 2$

41. $2[3z - 5(2z - 3)] = 3z - 4$

42. $5[2 - (2n - 4)] = 2(5 - 3n)$

43. $-8m - [2(11 - 2m) + 4] = 9m$

44. $7 - [4 + 2(a - 3)] = 11(a + 2)$

Solve. Round to the nearest hundredth.

 45. $\dfrac{y}{0.87} + 2.51 = 4.03$

 46. $7.02x - 3.64 = 8.29$

 47. $7.37n + 4.06 = -1.98n + 6.55$

 48. $10.13p = 3.14(p - 7.82)$

Choose the equation that best describes the situation.

49. A car leaves Seattle traveling at a rate of 45 mph. One hour later, a second car leaves from the same place, along the same road, at 54 mph. If the first car travels for *t* hr, in how many hours will the second car overtake the first car?

   a. $54(t - 1) = 45t$

   b. $45(t + 1) = 54t$

   c. $54(t + 1) = 45t$

   d. $45(t - 1) = 54t$

50. A company budgets $600,000 for an advertising campaign. It must pay $4000 for each television commercial and $1000 per radio commercial. If the company plans to air 50 fewer radio commercials than television commercials, find the number of television commercials t that will be in the advertising campaign.

a. $4000t + 1000(t + 50) = 600,000$ **b.** $4000t + 1000(t - 50) = 600,000$

c. $4000t + 1000t - 50 = 600,000$ **d.** $1000t + 4000(t - 50) = 600,000$

51. A taxi fare is $3.00 for the first mile and $1.25 for each additional mile. If a passenger's total cost was $5.50, how far did she travel in the taxi?

a. $3.00x + 1.25 = 5.50$ **b.** $3.00 + 1.25x = 5.50$

c. $3.00 + 1.25(x + 1) = 5.50$ **d.** $3.00 + 1.25(x - 1) = 5.50$

52. A family's budget allows $\frac{1}{3}$ of the family's monthly income for housing and $\frac{1}{4}$ of its monthly income for food. If a total of $1050 a month is budgeted for housing and food, what is the family's monthly income?

a. $\frac{1}{3}x = 1050 + \frac{1}{4}x$ **b.** $\frac{1}{3}x - \frac{1}{4}x = 1050$

c. $\frac{1}{3}x + \frac{1}{4}x = 1050$ **d.** $\frac{1}{4}x - \frac{1}{3}x = 1050$

Mixed Practice

Solve and check.

53. $5 - 5x + x = 21$ **54.** $16 - 3t = 31$

55. $4z + 3(5 - z) = -(z - 3) + 8$ **56.** $\frac{r}{7} + 3 = 11$

57. $-2.31y + 0.14 = -9.23$
(Round to the nearest hundredth.)   **58.** $7y - 6 = 3(y - 6)$

Applications

Write an equation and solve.

59. A part-time student at a college pays a student fee of $50 plus $120 per credit. How many credits is a part-time student carrying who pays $1010 in all?

60. A health club charges members $10 per month plus $5 per hour to use the facilities. If a member was charged $55 this month, how many hours did she use the facilities?

61. In a local election, a newspaper reported that one candidate received twice as many votes as the other. Altogether, they received a total of 3690 votes. How many votes did each candidate receive?

62. Calcium carbonate (chalk) consists of 10 parts of calcium for each 3 parts of carbon and 12 parts of oxygen by weight. Find the amount of calcium in 75 lb of chalk.

63. A parking garage charges $3 for the first hour and $2 for each additional hour or fraction thereof. If $9 was paid for parking, how many hours was the car parked in the garage?

64. A machinist earns $11.50 an hour for the first 35 hr and $15.30 for each hour over 35 per week. How many hours did he work this week if he earned $555.50?

65. The office manager of an election campaign office needs to print 5000 postcards. Suppose that it costs 2 cents to print a large postcard and 1 cent to print a small postcard. If $85 is allocated for printing postcards, how many of each type of postcard can be printed?

66. The owner of a small factory has 8 employees. Some of the employees make $10 per hour, whereas the others make $15 per hour. If the total payroll is $105 per hour, how many employees make the higher rate of pay?

67. Twenty minutes after a father left for work on the bus, he noticed that he had left his briefcase at home. His son left home, driving at 36 mph, to catch the bus that was traveling at 24 mph. How long did it take the son to catch the bus?

68. In the rush hour, a commuter drives to work at 30 mph. Returning home off-peak, she takes $\frac{1}{4}$ hr less time driving at 40 mph. What is the distance between the commuter's work and her home?

69. The two snails shown below crawl toward each other at rates that differ by 2 centimeter per minute. If it takes the snails 27 min to meet, how fast is each snail crawling?

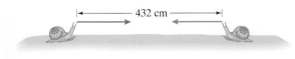

70. Suppose that a signal is sent from a station on the ground to a satellite. The signal bounces off the satellite and then is received at a second ground station. If the signal traveled 2400 mi in all, at what speed was the signal traveling?

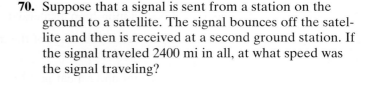
Not to scale

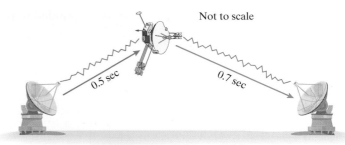

71. Two trucks leave a depot at the same time, traveling in opposite directions. One truck goes 4 mph faster than the other. After 2 hr, the trucks are 212 mi apart. What is the speed of the slower truck?

72. If a student drives from college to home at 40 mph, then he is 15 min late. However if he makes the same trip at 50 mph, he is 12 min early. What is the distance between his college and his home?

• *Check your answers on page A-5.*

MINDSTRETCHERS

Mathematical Reasoning

1. Give an example of an equation that involves combining like terms and that has

   a. no solution.

   b. an infinite number of solutions.

Patterns

2. Tables can be useful in solving equations.

   a. Complete the following table. After examining your results, identify the solution to the following equation: $3(x - 2) = 2(x + 1)$.

x	0	2	4	6	8	10	12
$3(x - 2)$							
$2(x + 1)$							

   b. Try a similar approach to solving the equation $7x = 5x + 11$. What conclusion can you draw about the solution?

x	0	2	4	6	8	10	12
$7x$							
$5x + 11$							

Groupwork

3. Working with a partner, choose a month on a calendar.

   a. Ask your partner to secretly select four days of the month that form a square, but only to tell you the sum of the four days. Determine the four days.

   b. Reverse roles with your partner and repeat part (a).

   c. Compare how you and your partner responded to part (a).

2.4   Solving Literal Equations and Formulas

In many situations, equations describe the relationship between two or more variables. Such equations are called *literal* equations and can be used to describe situations such as how the amount you pay depends on what you buy, how the distance you walk determines how long the walk takes, and how the dosage of a medicine you need to take relates to your weight.

OBJECTIVES

- To solve literal equations
- To solve and evaluate formulas
- To solve applied problems involving formulas

Definition
A **literal equation** is an equation involving two or more variables.

Consider, for instance, the literal equation

$$A = lw$$

which is the area formula for a rectangle. Since there is more than one variable, we can solve for one of the variables in terms of the others. For instance, in this equation we can solve for a in terms of b and c. Solving for a is especially useful if we want to calculate the value of a for a variety of values of b and c.

EXAMPLE 1

Solve $A = lw$ for l in terms of A and w.

Solution
$$A = lw$$
$$\frac{A}{w} = \frac{lw}{w} \quad \text{Divide each side of the equation by } w.$$
$$\frac{A}{w} = l$$

So $l = \dfrac{A}{w}$.

PRACTICE 1

Solve $q = 1 - p$ for p in terms of q.

We see from Example 1 that the solution to a literal equation is not a number, but an algebraic expression.

EXAMPLE 2

Solve $2a + b = c$ for a in terms of b and c.

Solution
$$2a + b = c$$
$$2a + b - b = c - b \quad \text{Subtract } b \text{ from each side of the equation.}$$
$$2a = c - b \quad \text{Simplify.}$$
$$\frac{2a}{2} = \frac{c - b}{2} \quad \text{Divide each side of the equation by 2.}$$
$$a = \frac{c - b}{2}$$

So $a = \dfrac{c - b}{2}$.

PRACTICE 2

Solve $3r - s = t$ for r in terms of s and t.

Note that we can solve literal equations by using the addition and multiplication properties that we have already used to solve other equations. Can you explain how to check your solution to a literal equation?

EXAMPLE 3	PRACTICE 3

EXAMPLE 3

Solve $\dfrac{2K}{3m} = n$ for K.

Solution

$$\dfrac{2K}{3m} = n$$

$3m \cdot \dfrac{2K}{3m} = 3m \cdot n$ Multiply each side of the equation by $3m$.

$2K = 3mn$

$\dfrac{2K}{2} = \dfrac{3mn}{2}$ Divide each side of the equation by 2.

$K = \dfrac{3mn}{2}$

So $K = \dfrac{3mn}{2}$.

PRACTICE 3

Solve $\dfrac{4x}{5a} = c$ for x.

EXAMPLE 4	PRACTICE 4

EXAMPLE 4

Consider the equation $Ax + By = C$.

a. Solve for y in terms of A, B, C, and x.

b. Using the equation found in part (a), find the value of y if $A = 4$, $B = 2$, $C = 2$, and $x = 5$.

Solution

a. $Ax + By = C$

$Ax - Ax + By = C - Ax$ Subtract Ax from each side of the equation.

$By = C - Ax$ Simplify.

$\dfrac{By}{B} = \dfrac{C - Ax}{B}$ Divide each side of the equation by B.

$y = \dfrac{C - Ax}{B}$

So $y = \dfrac{C - Ax}{B}$.

b. To find the value of y, we substitute $A = 4$, $B = 2$, $C = 2$, and $x = 5$ in the equation $y = \dfrac{C - Ax}{B}$.

$$y = \dfrac{C - Ax}{B} = \dfrac{2 - 4 \cdot 5}{2} = \dfrac{2 - 20}{2} = \dfrac{-18}{2} = -9$$

The value of y is -9.

PRACTICE 4

Consider the equation $y = mx + b$.

a. Solve for x in terms of y, m, and b.

b. Using the equation found in part (a), find the value of x if $y = 10$, $m = -3$, and $b = 7$.

Recall that in Section 1.8, we discussed formulas—a special type of literal equation. Here we focus on solving a formula for one variable in terms of the other variables. When we need to repeatedly find the value of a particular variable, the computation can be simplified by first solving for that variable in the formula.

EXAMPLE 5

$P = 2(l + w)$ is the formula for the perimeter P of a rectangle in terms of its length l and width w.

a. Solve for l in terms of P and w.

b. Using the formula found in part (a), find the length of a rectangle with a perimeter of 20 cm and a width of 6 cm.

Solution

a.
$$P = 2(l + w)$$
$$P = 2l + 2w \qquad \text{Use the distributive property.}$$
$$P - 2w = 2l + 2w - 2w \qquad \text{Subtract } 2w \text{ from each side of the equation.}$$
$$P - 2w = 2l$$
$$\frac{P - 2w}{2} = \frac{2l}{2} \qquad \text{Divide each side of the equation by 2.}$$
$$\frac{P - 2w}{2} = l,$$

or $\quad l = \dfrac{P - 2w}{2}$

So $l = \dfrac{P - 2w}{2}$.

b. $P = 20$ cm and $w = 6$ cm. We substitute in the formula $l = \dfrac{P - 2w}{2}$ to find the value of l.

$$l = \frac{P - 2w}{2} = \frac{20 - 2 \cdot 6}{2} = \frac{20 - 12}{2} = \frac{8}{2} = 4$$

So the length of the rectangle is 4 cm.

EXAMPLE 6

$V = lwh$ is the formula for finding the volume of a rectangular solid, where l represents the length of the solid, w the width, and h the height.

a. Solve the formula for h.

b. Using the equation found in part (a), find the value of h for $V = 48$ cu ft, $l = 8$ ft, and $w = 2$ ft.

Solution

a. Solving $V = lwh$ for h, we get

$$V = lwh$$
$$\frac{V}{lw} = \frac{lwh}{lw}$$
$$\frac{V}{lw} = h$$

So $h = \dfrac{V}{lw}$.

PRACTICE 5

$A = P(1 + rt)$ is the formula for computing the amount in an account earning simple interest. In the formula, A stands for the amount, P for the original principal, r for the annual rate of interest, and t for time.

a. Find a formula for r in terms of A, P, and t.

b. Using the equation found in part (a), evaluate r if $A = \$2100$, $P = \$2000$, and $t = 2$ years.

PRACTICE 6

$A = \dfrac{1}{2}bh$ is the formula for finding the area of a triangle, where b is the base and h is the height.

a. Express h in terms of A and b.

b. Using the formula found in part (a), find the value of h for $A = 63$ sq in. and $b = 9$ in.

b. To find the value of h, we substitute 48 for V, 8 for l, and 2 for w in the formula $h = \dfrac{V}{lw}$.

$$h = \frac{48}{8(2)} = 3$$

The height is 3 ft.

EXAMPLE 7	PRACTICE 7

To convert a temperature expressed in Fahrenheit degrees F to Celsius degrees C, a meteorologist multiplies $\dfrac{5}{9}$ by the difference between the Fahrenheit temperature and 32.

a. Write a formula for this relationship.

b. Solve the formula for F.

c. What Fahrenheit temperature corresponds to a Celsius temperature of 30°?

Solution

a. Stating the rule in words, we get

Celsius temperature, C, is equal to $\dfrac{5}{9}$ times the quantity Fahrenheit temperature, F, minus 32.

Then we translate this relationship to a formula.

$$C = \frac{5}{9}(F - 32)$$

b. Now, we solve the formula found in part (a) for F.

$$C = \frac{5}{9}(F - 32)$$

$$\frac{9}{5}C = \frac{9}{5} \cdot \frac{5}{9}(F - 32)$$

$$\frac{9}{5}C = F - 32$$

$$\frac{9}{5}C + 32 = F$$

So $F = \dfrac{9}{5}C + 32$.

c. To find the Fahrenheit temperature that corresponds to a Celsius temperature of 30°, we substitute 30 for C in the formula.

$$F = \frac{9}{5}C + 32 = \frac{9}{5}(30) + 32 = 54 + 32 = 86$$

The corresponding Fahrenheit temperature is, therefore, 86°.

To find the area A of a trapezoid, multiply $\dfrac{1}{2}$ its height h by the sum of the trapezoid's upper base b and lower base B.

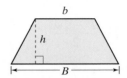

a. Translate this relationship to a formula.

b. Solve the formula for b.

c. What is the upper base of a trapezoid whose area is 32 sq cm, height is 4 cm, and lower base is 11 cm?

| 2.4 | Exercises | FOR EXTRA HELP |  | | | | |

Mathematically Speaking

Fill in each blank with the most appropriate term or phrase from the given list.

number	expression	variable
linear equation	literal equation	formula
constant	algebraic expression	

1. A(n) _____ is an equation involving two or more variables.

2. Literal equations can be solved for one _____ in terms of the others.

3. The solution to a literal equation is a(n) _____.

4. A(n) _____ is a special type of literal equation.

Solve each equation for the indicated variable.

5. $y + 10 = x$ for y

6. $b + 13 = a$ for b

7. $d - c = 4$ for d

8. $x - z = -5$ for x

9. $-3y = da$ for d

10. $ax = 5b$ for x

11. $\frac{1}{2}n = 2p$ for n

12. $\frac{3}{2}m = -4l$ for m

13. $a = \frac{1}{2}xyz$ for z

14. $w = \frac{2}{3}rst$ for r

15. $3x + y = 7$ for x

16. $x + 2y = 5$ for y

17. $3x + 4y = 12$ for y

18. $5a + 2b = 10$ for a

19. $y - 4t = 0$ for y

20. $6 = p - 4z$ for p

21. $-5b + p = r$ for b

22. $-7a + 3b = c$ for a

23. $h = 2(m - 2l)$ for l

24. $3(a - 2b) = c$ for b

Solve each formula for the indicated variable.

25. Simple interest: $I = prt$ for r

26. Electrical power: $P = iV$ for i

27. Uniform motion: $d = rt$ for r

28. Perimeter of square: $P = 4s$ for s

29. Perimeter of a triangle: $P = a + b + c$ for b

30. Perimeter of a rectangle: $P = 2l + 2w$ for w

31. Circumference of a circle: $C = \pi d$ for d

32. Aspect ratio of a hang glider: $R = \frac{s^2}{a}$ for a

33. Power: $P = I^2R$ for R

34. Centripetal force: $F = \frac{mv^2}{r}$ for m

35. Average of three numbers: $A = \dfrac{a + b + c}{3}$ for a

36. Distance of a free-falling object: $S = \dfrac{1}{2}gt^2$ for g

37. Arithmetic progression: $S = a + (n - 1)d$ for a

38. Simple interest: $A = P(1 + rt)$ for t

Solve for the variable shown in color. Then find the value of this variable for the given values of the other variables.

39. $3x + y = 6$ when $x = -12$

40. $2x - y = 5$ when $x = 4$

41. $3x - 7 = y$ when $y = 5$

42. $2x + 3y = -4$ when $y = -2$

43. $-\dfrac{1}{3}y = x$ when $x = \dfrac{1}{2}$

44. $\dfrac{1}{2}x = y$ when $y = -\dfrac{1}{4}$

45. $ax + by = c$ when $a = 1$, $b = 3$, $y = -4$, and $c = 2$

46. $dx - ey = f$ when $d = 5$, $x = -1$, $e = -2$, and $f = 7$

Mixed Practice

Solve each formula for the indicated variable.

47. Volume of a cylinder: $V = \pi r^2 h$ for h

48. Area of a rectangle: $A = lw$ for w

Solve each equation for the indicated variable.

49. $m = \dfrac{2}{5}abc$ for b

50. $-10x = yz$ for x

51. $4w + 9z = 3$ for z

52. $a + b = -3$ for a

Solve for the variable shown in color. Then find the value of this variable for the given values of the other variables.

53. $7x + 2y = -20$ when $y = 4$

54. $-\dfrac{1}{5}b = a$ when $a = \dfrac{1}{8}$

Applications

Solve.

55. In physics, Charles's Law describes the relationship between the volume of a gas and its temperature. The law states that the volume V divided by the temperature T is equal to a constant K.

    a. Write an equation to express this relationship.

    b. Solve the relationship for V.

56. In geometry, the volume V of a cylinder is the product of the area B of the circular base and the height h.

    a. Write a formula for this relationship.

    b. Solve the formula for h.

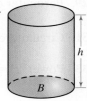

57. In nursing, Clark's Rule for medication expresses a relationship between the recommended dosages for a child and for an adult. The rule states that a child's dosage C equals the product of the weight W of the child in pounds divided by 150 and the adult's dosage A.

   a. Write a formula for this relationship.

   b. Solve the formula for A.

58. The Scholastic Aptitude Test (SAT) is a well-known test taken by high school seniors applying for college admission. Scores on such tests are often converted to standardized scores, or z-scores. In 2006, the formula for determining a z-score on the mathematics portion of the SAT was $z = \dfrac{x - 518}{115}$, where x is an individual score on the test. The value 518 is the national average score on the test whereas 115 is a measure of how spread out the scores were. (*Source:* The College Board)

   a. Solve for x in terms of z.

   b. If your z-score is 2.2, what was your score on the test?

59. During a storm, the number m of miles away a bolt of lightning strikes can be estimated by first counting the number t of seconds between the bolt of lightning and the associated clap of thunder and then dividing by 5.

   a. Translate this relationship to a formula.

   b. Solve for t in terms of m.

   c. If lightning strikes 2.5 mi away, how many seconds will elapse before the thunder is heard?

60. To estimate a man's shoe size S, triple his foot length l (expressed in inches) and subtract 22.

   a. Express this relationship as a formula.

   b. Estimate the shoe size of a man with a 12-in. foot.

   c. Estimate the foot length of a man with shoe size 12.

61. The circumference C of a circle can be found by doubling the product of the constant π and the circle's radius r.

   a. Express this relationship as a formula.

   b. Solve this formula for r in terms of C and π.

   c. Find the value of r rounded to the nearest tenth if C is 5 ft and π is approximately 3.14.

62. According to Newton's second law of motion, the force F (in newtons) applied to an object equals the product of the object's mass m (in kilograms) and its acceleration a (in m/sec^2).

   a. Translate this relationship to a formula.

   b. Solve for a in terms of m and F.

   c. Find a (in m/sec^2) if m is 3 kg and F is 10 newtons.

• Check your answers on page A-5.

MINDSTRETCHERS

Research

1. By examining books in your college library or Web sites, find several examples of literal equations that relate two or more variables. Explain what the variables represent, and write the equation that relates them.

Groupwork

2. Working with a partner, give an example of a situation that the following formula might describe.

$$y = mx + b$$

Explain what each variable represents in your example.

Patterns

3. Consider the following table.

x	0	1	2	3	4	$\cdots$	10
y	1	3	5	7	9	$\cdots$	21

Write an equation for y in terms of x.

2.5   Solving Equations Involving Percent

A percent problem typically involves three numbers—the *percent*, the *base*, and the *amount*.

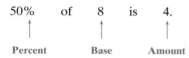

50%   of   8   is   4.

Percent   Base   Amount

We can recognize each number by noting the following:

- The base—the number that we are taking the percent of—always follows the word "of" in the statement of the problem.
- The percent typically contains the % sign.
- The remaining number is called the *amount*.

If any of these numbers is unknown, then we have a percent problem. We can translate the problem into an equation and then solve the equation.

Let's first consider a problem in which we are looking for the base.

OBJECTIVES

- To find the amount, base, or percent in a percent problem by solving an equation
- To solve percent word problems

EXAMPLE 1

20% of what number is 5?

Solution Let x represent *what number* (the base). We translate the problem into an equation as shown.

20%   of   what number   is   5?

0.2   · x =   5

Then we solve the equation for x.

$$0.2x = 5$$
$$\frac{0.2x}{0.2} = \frac{5}{0.2} \qquad \text{Divide each side of the equation by 0.2.}$$
$$x = 25$$

So 20% of 25 is 5.

PRACTICE 1

8 is 40% of what number?

EXAMPLE 2

In one year, Michael Jordan earned about $4 million, or about 10% of his total income, from salary and winnings. What was Michael Jordan's total income that year?
(*Source: 1995 Sports Almanac*)

PRACTICE 2

A company's profits amounted to 12% of its sales. If the profits were $3 million, compute the company's sales.

Solution The question is

$4,000,000 is 10% of what amount?

$$4,000,000 = 0.10 \quad \cdot \quad x$$ Translate the sentence to an equation.

$$0.1x = 4,000,000$$ Rewrite the equation and solve.

$$\frac{0.1x}{0.1} = \frac{4,000,000}{0.1}$$

$$x = 40,000,000$$

So Michael Jordan's annual income was about $40,000,000.

Next let's consider a problem in which we are looking for the amount.

EXAMPLE 3

What number is 3.5% of $200?

Solution Let x represent *what number* (the amount). Then we translate the problem into an equation.

What number   is   3.5%   of $200?

$$x \qquad = \quad 0.035 \quad \cdot \quad 200$$

Now, we solve for x.

$$x = 0.035 \cdot 200 = 7$$

So $7 is 3.5% of $200.

PRACTICE 3

What is 23% of 45 m?

EXAMPLE 4

The following graph shows the breakdown of the projected U.S. population by gender in the year 2020. If the population is expected to be 340 million people, how many more women than men will there be in 2020? (*Source:* U.S. Bureau of the Census)

Men
49%

Women
51%

Solution First, we calculate the number of women in 2020.

What amount   is   51%   of   340,000,000?

$$x \qquad = \quad 0.51 \quad \cdot \quad 340,000,000$$

$$x = 0.51 \cdot 340,000,000 = 173,400,000$$

Similarly, the number of men in 2020 is:

$$x = 0.49 \cdot 340,000,000 = 166,600,000$$

So there will be 173,400,000 − 166,600,000, or 6,800,000 more women than men.

PRACTICE 4

The number of degrees projected to be awarded in the year 2010 is shown in the following graph. How many more bachelor's degrees than associate's degrees will be awarded out of a total of 2,860,000 degrees awarded in that year? (*Source:* National Center for Education Statistics)

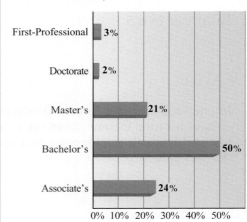

Finally, we consider a problem in which we are looking for the percent.

EXAMPLE 5	PRACTICE 5

What percent of 9 is 6?

14 is what percent of 16?

Solution Let x represent *what percent*. Then we translate the problem into an equation as shown.

$$\underbrace{\text{What percent}}_{x} \quad \text{of} \;\; 9 \;\; \text{is} \;\; 6?$$
$$x \qquad\quad \cdot \;\; 9 \;\; = \;\; 6$$

Now, we solve the equation for x.

$$x \cdot 9 = 6$$

$$9x = 6 \qquad\qquad x \cdot 9 = 6 \text{ is equivalent to } 9x = 6.$$

$$\frac{9x}{9} = \frac{6}{9} \qquad\qquad \textbf{Divide each side of the equation by 9.}$$

$$x = \frac{2}{3} = 66\tfrac{2}{3}\% \qquad \textbf{Change } \tfrac{2}{3} \textbf{ to a percent.}$$

So $66\tfrac{2}{3}\%$ of 9 is 6.

EXAMPLE 6	PRACTICE 6

Three thousand students in the fifth grade were asked the question, "What job would you most like to have when you grow up?" Of these students, 540 said that they wanted to be teachers. What percent of the fifth graders wanted to be teachers?

Of the first 42 U.S. presidents, 13 had also served as vice president. To the nearest percent, what percent of the presidents had been vice president? (*Source: The World Almanac and Book of Facts,* 1996)

Solution The question that we must answer is

$$540 \text{ is what percent of } 3000?$$

$$540 = x \cdot 3000 \qquad \textbf{Translate the sentence to an equation.}$$

$$3000x = 540 \qquad\quad \textbf{Rewrite the equation and solve.}$$

$$\frac{3000}{3000}x = \frac{540}{3000}$$

$$x = 0.18 = 18\%$$

So 18% of the 3000 fifth graders wanted to be teachers.

Now, let's consider a different type of "what percent" problem that deals with a *changing quantity*. If the quantity is increasing, we have a *percent increase*; if it is decreasing, we have a *percent decrease*.

To Find a Percent Increase or Decrease

- Compute the difference between the two given values.
- Compute what percent this difference is of the *original value*.

EXAMPLE 7

In one year, doctors performed about 10,000 kidney transplants. Four years later, the number of transplants rose to 12,000. What percent increase was this? (*Source: Time Almanac 2000*)

Solution First we compute the difference between the values.

$$12{,}000 - 10{,}000 = 2000 \quad \longleftarrow \text{ Change in value}$$

Then we answer the question

What percent of 10,000 is 2000?

$$x \cdot 10{,}000 = 2000 \qquad \text{Translate the sentence to an equation.}$$
$$10{,}000x = 2000 \qquad \text{Rewrite the equation and solve.}$$
$$\frac{10{,}000x}{10{,}000} = \frac{2000}{10{,}000}$$
$$x = 0.20 = 20\%$$

So the number of transplants increased by 20%.

PRACTICE 7

In 5 yr, the number of nursing homes in New Jersey increased from 300 to 361. To the nearest tenth of a percent, what was the percent increase in the number of nursing homes?
(*Source: Nursing Home Statistical Yearbooks*)

EXAMPLE 8

Suppose that a college will close if its student enrollment drops by more than 60%. If the enrollment falls from 4000 to 1800, will the college close?

Solution The student enrollment fell from 4000 to 1800; that is, it fell by 2200. The question is how the percent decrease of the student enrollment compares with 60%. We compute the percent decrease as follows.

What percent of 4000 is 2200?

$$x \cdot 4000 = 2200$$
$$4000x = 2200$$
$$\frac{4000}{4000}x = \frac{2200}{4000}$$
$$x = \frac{22}{40} = \frac{11}{20} = 0.55 = 55\%$$

Since the student enrollment fell by 55%, which is less than 60%, the college will not close.

PRACTICE 8

Financial crashes took place on both Friday, October 29, 1929, and Monday, October 19, 1987. On the earlier date, the stock index dropped from 300 to 230. On the later date, it dropped from 2250 to 1750. As a percent, did the stock index drop more in 1929 or in 1987?
(*Source: The Washington Post,* "Sell-Offs Rock Wall Street, World Markets," October 28, 1987)

Some percent problems involve *interest*. When you lend, deposit, or invest money, you *earn* interest. When you borrow, you *pay* interest.

The amount of interest earned (or paid) I depends on the *principal P* (the amount of money invested or borrowed), the annual rate of interest r (usually expressed as a percent), and the length of time t the money is invested or borrowed (generally expressed in years). We compute the amount of interest by multiplying the principal by the rate of interest and the number of years. This kind of interest is called *simple interest*.

EXAMPLE 9

How much simple interest was earned in 2 yr on a principal of $900 in an account with an annual interest rate of 5.25%?

Solution To compute the interest, we use the formula $I = Prt$.

$$I = \quad P \quad \cdot \quad r \quad \cdot \quad t$$
$$= (900)\,(0.0525)\,(2)$$
$$= 94.5$$

So the interest earned was $94.50.

PRACTICE 9

In 1 yr, a bank paid $130 in simple interest on an initial balance of $2000. What annual rate of interest was paid?

Some applications of percent involve *multiple investments*, as the following example illustrates.

EXAMPLE 10

A book editor invested part of her $5000 bonus in a fund that paid 4% simple interest and the rest in stocks that paid 6% simple interest. Find the amount invested at each rate if the overall interest earned in 1 yr was $260.

Solution Let's set up a table to organize the given information. Note that the interest earned for each investment is the product of each principal, the rate of interest, and the time.

	Principal ·	Rate of interest ·	Time =	Interest earned
Fund	x	0.04	1	$0.04x$
Stocks	$5000 - x$	0.06	1	$0.06(5000 - x)$
Total Interest				260

The interest earned on the total investment is the sum of the interest earned on the fund and the interest earned on the stocks.

$$0.04x + 0.06(5000 - x) = 260$$
$$0.04x + 300 - 0.06x = 260$$
$$-0.02x = -40$$
$$\frac{-0.02x}{-0.02} = \frac{-40}{-0.02}$$
$$x = 2000$$

So $2000 was invested in the fund and $5000 − $2000, or $3000, was invested in stocks.

PRACTICE 10

The amount of money a broker invested in bonds was double what she invested in a mutual fund. After 1 yr, the investment in bonds gained 10% and the investment in the mutual fund lost 10%. If the net gain was $700, how much was each of her investments?

Another type of problem involving percent is a *mixture* or *solution problem*. In this kind of problem, the amount of a particular ingredient in the solution or mixture is often expressed as a percent of the total. Chemists and pharmacists commonly need to solve mixture and solution problems.

EXAMPLE 11	PRACTICE 11
A chemist added 2 liters (L) of water to 8 L of a 48% alcohol solution. What is the alcohol concentration of the new solution, rounded to the nearest percent?	How many grams of salt must be added to 30 g of a solution that is 20% salt to make a solution that is 25% salt?

Solution By adding 2 L of water, we are diluting 8 L of 48% alcohol solution to a solution with unknown alcohol concentration. Let x represent the alcohol concentration of the new solution expressed as a decimal. It is helpful to organize the given information into a table, as follows.

Action	Substance	Amount (L)	Amount of Pure Alcohol (L)
Start with	48% alcohol solution	8	0.48(8)
Add	water	2	0(2)
Finish with	new alcohol solution	8 + 2, or 10	x(10)

The amount of pure alcohol in the final solution is equal to the sum of the amount of pure alcohol in the initial solution and the amount of pure alcohol in the water added.

$$0.48(8) + 0(2) = 10x$$

Solving the equation for x, we get

$$0.48(8) = 10x$$
$$\frac{3.84}{10} = \frac{10x}{10}$$
$$0.384 = x$$

So the alcohol concentration of the new solution is 38.4%, or 38% rounded to the nearest percent.

How would you explain each entry in the table in Example 11?

Mathematically Speaking

Fill in each blank with the most appropriate term or phrase from the given list.

plus	sum of	times
amount	base	difference between

1. In solving the percent problem "30% of what number is 7?", "what number" represents the _____.

2. In solving the percent problem "What is 10% of 4?", "What" represents the _____.

3. In translating the problem "What percent of 36 is 9?", "of" becomes _____.

4. In finding a percent increase or decrease, first compute the _____ the two given values.

Find the amount.

5. What is 75% of 8?

6. Find 50% of 48 ft.

7. Compute 100% of 23.

8. What is 200% of 6?

9. Find 41% of 7 kg.

10. Calculate 6% of 9.

11. What is 8% of $500?

12. 6% of $200 is how much money?

13. What is 12.5% of 32?

14. Compute 37.5% of 40.

Find the base.

15. 25% of what area is 8 sq in.?

16. 30% of what weight is 120 kg?

17. $12 is 10% of how much money?

18. 1% of what salary is $195?

19. $20 is 10% of how much money?

20. 2% of what amount of money is $5?

21. 3.5 is 200% of what number?

22. 150% of what number is 8.1?

23. 0.5% of what length is 23 m?

24. 0.75% of what number is 24?

Find the percent.

25. 50 cents is what percent of 80 cents?

26. What percent of 13 is 13?

27. 5 is what percent of 15?

28. What percent of 12 is 10?

29. 10 hours is what percent of 8 hours?

30. $30 is what percent of $200?

31. What percent of 5 is $\frac{1}{2}$?

32. $\frac{2}{3}$ is what percent of 6?

33. 2.5 g is what percent of 4 g?

34. 0.1 is what percent of 8?

Solve.

35. What is 35% of $400?

36. How much is 19% of $10,000?

37. What percent of 50 mi is 20 mi?

38. $15 is what percent of $20?

39. 70% of what amount is 14?

40. 6 pages is 200% of how many pages?

41. 0.1% of 35 is what number?

42. Compute 0.5% of 20.

43. $8 is what percent of $240?

44. 7 is what percent of 21?

45. 3 oz is 20% of what weight?

46. 8 is 50% of what number?

Mixed Practice

Find the missing quantity.

47. $14 is what percent of $8?

48. What is 32% of $9?

49. 80% of what amount is 96?

50. 75 is what percent of 3000?

51. 8% of 120 is what amount?

52. 350% of how much money is $77?

53. 21 is 20% of what number?

54. $\frac{3}{4}$ is what percent of 6?

Applications

Solve.

55. The following graph shows the breakdown of faculty and staff of the 160 employees at a college. How many more faculty than staff work there?

56. In 1862, the U.S. Congress enacted the nation's first income tax, at the rate of 3%. At this rate, how much in income tax would have been due on an income of $600? (**Source:** *The Seattle Times*, March 29, 1992)

Staff
25%

Faculty
75%

57. A college graduate enlisted in the army for 4 yr. So far, he has served 9 mo. What percent of his enlistment has passed?

58. A questionnaire was mailed to 5000 people. Of these recipients, 3000 responded. What was the response rate, expressed as a percent?

59. Yesterday, 8 employees in an office were absent. If these absentees constitute 25% of the employees, what is the total number of employees?

60. Payroll deductions comprise 40% of a worker's weekly gross income. If these deductions total $240, what is her gross income?

61. A house is purchased for $250,000. What percent of the purchase price is a down payment of $50,000?

62. Last year, an office worker's income was $25,000. If she donated $900 to charity, what percent of her income went to charity?

63. According to a report on a country's economic conditions, 1.5 million people, or 8% of the workforce, were unemployed. How large was the country's workforce?

64. In a heart transplant operation, the heart cost $16,000, which was about 14% of the cost of the operation. How much, to the nearest thousand dollars, did the heart transplant operation cost?

65. In a restaurant, 60% of the tables are in the no-smoking section. If the restaurant has 90 tables, how many tables are there in this section?

66. A shopper lives in a town where the sales tax is 5%. Across the river, the tax is 4%. If it costs her $6 to make the round-trip across the river, on which side of the river should she buy a $250 television set?

67. Suppose that an animal species is considered to be endangered if its population drops by more than 60%. If a species' population fell from 4000 to 1800, should we consider the species endangered?

68. The first commercial telephone exchange was set up in New Haven, Connecticut, in 1878. Between 1880 and 1890, the number of telephone systems in the United States increased from 50 to 200, in round numbers. What percent increase was this? (*Source: The World Almanac 2000*)

69. A student borrowed $2000 for 1 yr at 5% simple interest to buy a computer and printer. How much interest did she pay?

70. How much simple interest is earned on $600 in a bank account at an 8% annual interest rate for 6 mo?

71. How much simple interest is earned on a deposit of $5000 in an on-line savings account after 1 yr if the interest rate is 5%?

72. An aunt lent her nephew $2000. The nephew agreed to pay his aunt 4% simple interest. If he promised to repay her the entire amount owed at the end of 3 years, how much money must he pay her?

73. Part of $34,000 is invested at an interest rate of 8% and the rest at an interest rate of 10%. If the total interest earned in 1 yr was $3000, how much money was invested at each rate?

74. The amount of money invested at 5% simple interest was half the amount invested at 7% simple interest. If the total yearly interest earned was $380, how much money was invested at each rate?

75. An amount of $20,000 was invested in a fund with a return of 8%. How much money was invested in a fund with a 5% return if the total return on both investments was $2100?

76. A beneficiary split her $100,000 inheritance between two investments. One of the investments gained 12% and the other lost 8%. If she broke even on the two investments, how much money did she invest in each?

77. A basic lemon vinaigrette salad dressing can be made by mixing 1 cup of olive oil with $\frac{1}{3}$ cup of lemon juice along with pinches of salt and white pepper. How much olive oil should be added to make a dressing that is 20% lemon juice? (The salt and pepper contribution is negligible and can be left out of the computation.) (*Source: The Sauce Bible*)

78. A pharmacist has 10 L of a 5% drug solution. How many liters of 2% solution should be added to produce a solution that is 3%?

79. How many ounces of a 40% acetic acid solution should a photographer add to 30 oz of a 4% acetic acid solution to obtain a 10% solution?

80. A brand of antifreeze states that a radiator containing a solution that is 50% antifreeze and 50% water provides protection for a temperature as low as −34°F, whereas a solution that is 70% antifreeze provides protection down to −84°F. A car's radiator has 4 qt of a 50% solution. If the capacity of the radiator is 6 qt and the rest of the radiator is filled with pure antifreeze, what percent of the resulting solution is antifreeze? (*Source: Prestone II Antifreeze*)

● *Check your answers on page A-5.*

MINDSTRETCHERS

Writing

1. Write a word problem whose answer is 60%, in which you give the base and the amount.

Mathematical Reasoning

2. What is a% of b% of c% of 1,000,000,000?

Groupwork

3. Some properties of percent are contrary to the intuition of most people. For example, consider the following two *false* statements. Working with a partner, discuss why they are both false.

    a. If the price of an item is increased by 10% and then decreased by 10%, the final price is the same as the original price.

    b. If the price of an item is increased by 5% and then increased by another 3%, altogether it is increased by 8%.

<table>
<tr><td>

2.6

Solving Linear Inequalities
</td></tr>
</table>

In Section 1.1, we used the symbols $<$, $\leq$, $>$, $\geq$, $=$, and $\neq$ to compare two real numbers. For example, with the real numbers -5 and 4, we can write the following statements:

$$-5 < 4 \qquad 4 > -5 \qquad -5 \neq 4$$

OBJECTIVES

- To determine if a number is a solution of an inequality

- To graph the solutions of linear inequalities on the number line

- To solve linear inequalities using the addition and multiplication properties of inequalities

- To solve applied problems involving linear inequalities

Definition

An **inequality** is any mathematical statement containing $<$, $\leq$, $>$, $\geq$, or $\neq$.

Solutions of Inequalities

Now, consider an inequality that involves a variable, say $x < 2$. Let's look at the values of x that make this inequality true.

	Values for x	$x < 2$	True or False?
Values for x that are less than 2	1	$1 < 2$	True
	$\dfrac{1}{2}$	$\dfrac{1}{2} < 2$	True
	0	$0 < 2$	True
	-1	$-1 < 2$	True
Values for x that are not less than 2	2	$2 < 2$	False
	3	$3 < 2$	False
	$3\frac{1}{2}$	$3\frac{1}{2} < 2$	False
	4	$4 < 2$	False

Note that there are many values of x that make $x < 2$ true. Can you name them all? Explain.

Definition

A **solution of an inequality** is any value of the variable that makes the inequality true. To **solve an inequality** is to find all of its solutions.

EXAMPLE 1

Determine whether -3 is a solution of the inequality $2x + 5 \geq -3$.

Solution To determine if -3 is a solution of the inequality, we substitute -3 for x and simplify.

$$2x + 5 \geq -3$$
$$2(-3) + 5 \stackrel{?}{\geq} -3 \quad \text{Substitute } -3 \text{ for } x.$$
$$-6 + 5 \stackrel{?}{\geq} -3 \quad \text{Multiply.}$$
$$-1 \geq -3 \quad \text{True}$$

Because $-1 \geq -3$ is a true statement, -3 is a solution of the inequality.

PRACTICE 1

Is 4 a solution of the inequality
$$\frac{1}{2}x - 2 < -1?$$

For any inequality, we can draw on the number line a picture of its solutions—the *graph* of the inequality. Graphing the solutions of an inequality can be clearer than describing the solutions in symbols or words.

EXAMPLE 2

Draw the graph of $x < 2$.

Solution The graph of $x < 2$ includes all points on the number line to the left of 2. The open circle on the graph shows that 2 is *not* a solution.

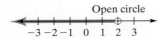
Open circle

PRACTICE 2

Draw the graph of $x > 1$.

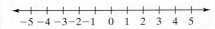

Note that solving an inequality generally results in a range of numbers rather than in a single number.

EXAMPLE 3

Draw the graph of $x \geq -\frac{1}{2}$.

Solution The graph of $x \geq -\frac{1}{2}$ includes all points on the number line to the right of $-\frac{1}{2}$ and also $-\frac{1}{2}$. The *closed circle* shows that $-\frac{1}{2}$ is a solution of $x \geq -\frac{1}{2}$.

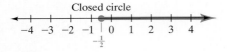
Closed circle

PRACTICE 3

Draw the graph of $x \leq -1\frac{1}{2}$.

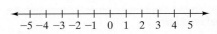

EXAMPLE 4

Draw the graph of $-1 \leq x < 2$.

Solution This inequality is read either " -1 is less than or equal to x *and* x is less than 2" or "x is greater than or equal to -1 *and* x is less than 2." The solutions of this inequality are all values of x that satisfy both $-1 \leq x$ and $x < 2$, and its graph is the overlap of the graphs of $-1 \leq x$ and $x < 2$.

Graph of $-1 \leq x$:

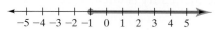

Graph of $x < 2$:

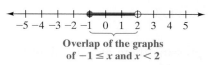

Graph of $-1 \leq x < 2$:

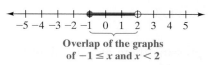

Overlap of the graphs
of $-1 \leq x$ and $x < 2$

Note that the graph includes -1 and all points *between* -1 and 2 on the number line. A closed circle at -1 means that -1 is a solution of the inequality, whereas an open circle at 2 means that 2 is not a solution of the inequality.

PRACTICE 4

Draw the graph of $-3 < x < 4$.

Solving Inequalities Using the Addition Property

Now, let's consider what happens when we perform the same operation on each side of an inequality. In the following inequality, we add 4 to each side of the inequality $5 < 12$.

$$5 < 12 \qquad \text{True}$$
$$5 + 4 \overset{?}{<} 12 + 4 \qquad \text{Add 4 to each side of the inequality.}$$
$$9 < 16 \qquad \text{True}$$

Note that $5 < 12$ and $5 + 4 < 12 + 4$ are both true, and so are equivalent.

This example suggests the addition property of inequalities.

Addition Property of Inequalities

For any real numbers a, b, and c, the following is true.
- If $a < b$, then $a + c < b + c$.
- If $a > b$, then $a + c > b + c$.

Similar statements hold for $\leq$ and $\geq$.

Alternatively, suppose we subtract 4 from each side of the original inequality:

$$5 < 12 \qquad \text{True}$$
$$5 - 4 \overset{?}{<} 12 - 4 \qquad \text{Subtract 4 from each side of the inequality.}$$
$$1 < 8 \qquad \text{True}$$

Note that $5 < 12$ and $5 - 4 < 12 - 4$ are both true, and so are equivalent. This suggests that subtracting the same number from both sides of an inequality also results in an equivalent inequality.

> **Tip** When the same number is added to or subtracted from each side of an inequality, the *direction* of the inequality is unchanged.

The way we solve inequalities is similar to the way we solve equations.

Equation	Inequality
$x + 3 = 5$	$x + 3 < 5$
$x + 3 - 3 = 5 - 3$	$x + 3 - 3 < 5 - 3$
$x = 2$ Solution	$x < 2$ Solution
$x + 3 = 5$ is equivalent to $x = 2$	$x + 3 < 5$ is equivalent to $x < 2$.

We solve inequalities by expressing them as equivalent inequalities in which the variable term is isolated on one side.

EXAMPLE 5

Solve and graph: $y + 5 > 9$

Solution $y + 5 > 9$
$y + 5 - 5 > 9 - 5$ Subtract 5 from each side of the inequality.
$y > 4$

The graph of $y > 4$ is

$$\xleftarrow{\quad\begin{array}{ccccccccc} + & + & + & + & + & \circ & + & + & + \\ -1 & 0 & 1 & 2 & 3 & 4 & 5 & 6 & 7 \end{array}\quad}\rightarrow$$

Note that an open circle is drawn at 4 to show that 4 is not a solution.

Because an inequality has many solutions, we cannot check all of the solutions as we did with an equation. However, we can do a partial check of the solutions of an inequality by substituting points on the graph in the original inequality. For instance, to check that all values for y greater than 4 are the solutions of $y + 5 > 9$, we replace y in the original inequality with some points on the graph and some points not on the graph.

Values for y	$y + 5 > 9$	True or False
6	$6 + 5 > 9$	True
5	$5 + 5 > 9$	True
4	$4 + 5 > 9$	False
3	$3 + 5 > 9$	False

The table confirms that the solution of $y + 5 > 9$ is $y > 4$. That is to say, any number greater than but not equal to 4 is a solution.

PRACTICE 5

Solve and graph: $n + 5 > 4$

EXAMPLE 6	**PRACTICE 6**

EXAMPLE 6

Solve and graph: $z - 2 \le -3\frac{1}{2}$

Solution

$$z - 2 \le -3\frac{1}{2}$$
$$z - 2 + 2 \le -3\frac{1}{2} + 2 \qquad \text{Add 2 to each side of the inequality.}$$
$$z \le -1\frac{1}{2}$$

So all numbers less than or equal to $-1\frac{1}{2}$ are solutions. The graph of $z \le -1\frac{1}{2}$ is

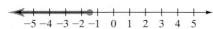

PRACTICE 6

Solve and graph: $x - 4 \le 1\frac{1}{2}$

EXAMPLE 7

Solve and graph: $6y - 3 \le 5y + 4$

Solution

$$6y - 3 \le 5y + 4$$
$$6y - 5y - 3 \le 5y - 5y + 4 \qquad \text{Subtract 5y from each side of the inequality.}$$
$$y - 3 \le 4$$
$$y - 3 + 3 \le 4 + 3 \qquad \text{Add 3 to each side of the inequality.}$$
$$y \le 7$$

So all numbers less than or equal to 7 are solutions of $6y - 3 \le 5y + 4$. The graph of $y \le 7$ is

PRACTICE 7

Solve and graph: $4x + 5 \ge 3x - 2$

Solving Inequalities Using the Multiplication Property

Consider the inequality $12 < 15$. Let's look at what happens when we multiply this inequality by a *positive* number:

$$12 < 15$$
$$12 \cdot 3 \overset{?}{<} 15 \cdot 3 \qquad \text{Multiply each side of the inequality by 3.}$$
$$36 < 45 \qquad \text{True}$$

Now, we consider multiplying each side of the original inequality by a *negative* number:

$$12 < 15$$
$$12(-3) \overset{?}{<} 15(-3) \qquad \text{Multiply each side of the inequality by –3.}$$
$$-36 \overset{?}{<} -45 \qquad \text{False, unless the direction of the inequality sign is reversed}$$
$$-36 > -45 \qquad \text{True}$$

These examples suggest the *multiplication property of inequalities.*

Multiplication Property of Inequalities

For any real numbers a, b, and c, the following is true:

- If $a < b$ and c is positive, then $ac < bc$.
- If $a < b$ and c is negative, then $ac > bc$.

Similar statements hold for $>$, $\leq$, and $\geq$.

We can demonstrate a similar property for division.

$$12 < 15$$

$$\frac{12}{3} \overset{?}{<} \frac{15}{3} \qquad \text{Divide each side of the inequality by 3.}$$

$$4 < 5 \qquad \text{True}$$

$$12 < 15$$

$$\frac{12}{-3} \overset{?}{<} \frac{15}{-3} \qquad \text{Divide each side of the inequality by } -3.$$

$$-4 \overset{?}{<} -5 \qquad \text{False, so we need to reverse the direction of the inequality sign}$$

$$-4 > -5 \qquad \text{True}$$

Note that when we multiply or divide each side of an inequality by a positive number, the direction of the inequality remains the same. But when we multiply or divide each side of an inequality by a negative number, the direction of the inequality is *reversed*.

EXAMPLE 8

Solve and graph: $\dfrac{x}{2} < 3$

Solution $\dfrac{x}{2} < 3$

$2 \cdot \dfrac{x}{2} < 2 \cdot 3$ Multiply each side of the inequality by 2.

$x < 6$

So any number less than 6 is a solution. The graph of $x < 6$ is

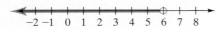

PRACTICE 8

Solve and graph: $\dfrac{x}{3} \leq 1$

EXAMPLE 9

Solve and graph: $-4z \leq 12$

Solution $-4z \leq 12$

$\dfrac{-4z}{-4} \geq \dfrac{12}{-4}$ Divide each side of the inequality by -4 and reverse the direction of the inequality.

$z \geq -3$

The solution of $-4z \leq 12$ is $z \geq -3$, so all numbers greater than or equal to -3 are solutions. The graph of $z \geq -3$ is

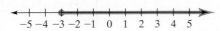

PRACTICE 9

Solve and graph: $-3x > 15$

EXAMPLE 10

Solve and graph: $-21 < 6y - 9y$

Solution

$$-21 < 6y - 9y$$

$-21 < -3y$ Combine like terms.

$\dfrac{-21}{-3} > \dfrac{-3y}{-3}$ Divide each side of the inequality by -3 and reverse the direction of the inequality.

$7 > y,$ or $y < 7$

So all numbers less than 7 are solutions. The graph is

$-2\ -1\ \ \ 0\ \ \ 1\ \ \ 2\ \ \ 3\ \ \ 4\ \ \ 5\ \ \ 6\ \ \ 7\ \ \ 8$

PRACTICE 10

Solve and graph: $10 > 5x - 7x$

As in solving equations, we may need to use more than one property of inequalities to solve some inequalities.

EXAMPLE 11

Solve: $5y - 4 - 6y \le -8$

Solution

$$5y - 4 - 6y \le -8$$

$-y - 4 \le -8$ Combine like terms.

$-y - 4 + 4 \le -8 + 4$ Add 4 to each side of the inequality.

$-y \le -4$ Simplify.

$\dfrac{-y}{-1} \ge \dfrac{-4}{-1}$ Divide each side of the inequality by -1 and reverse the direction of the inequality.

$y \ge 4$

So all numbers greater than or equal to 4 are solutions.

PRACTICE 11

Solve: $-6 \ge 3z + 4 - z$

EXAMPLE 12

Solve: $3n - 2(n + 3) < 14$

Solution

$$3n - 2(n + 3) < 14$$

$3n - 2n - 6 < 14$ Use the distributive property.

$n - 6 < 14$ Combine like terms.

$n - 6 + 6 < 14 + 6$ Add 6 to each side of the inequality.

$n < 20$

So all numbers less than 20 are solutions.

PRACTICE 12

Solve: $7x - (9x + 1) > -5$

Some common word phrases used in applied problems involving inequalities and their translations are shown in the following table.

Word Phrase	Translation
x is less than a	$x < a$
x is less than or equal to a	$x \leq a$
x is greater than a	$x > a$
x is greater than or equal to a	$x \geq a$
x is at most a	$x \leq a$
x is no more than a	$x \leq a$
x is at least a	$x \geq a$
x is no less than a	$x \geq a$

EXAMPLE 13

In geometry, the triangle inequality states that the sum of the lengths of any two sides of a triangle is greater than the length of the third side. In the isosceles triangle shown, write and solve an inequality to find the possible side lengths *a*. Graph the inequality.

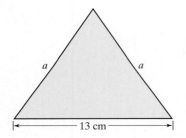

Solution The sum of the lengths of the two equal sides is greater than the length of the third side.

$$a + a > 13$$
$$2a > 13$$

Solving the inequality, we get

$$\frac{2a}{2} > \frac{13}{2} \quad \textbf{Divide each side by 2.}$$

$$a > 6.5$$

The graph is

We conclude that the length of each side *a* is any number greater than 6.5 cm.

PRACTICE 13

In the triangle shown, for which values of *x* will the perimeter be greater than or equal to 14 in.?

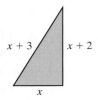

EXAMPLE 14

A factory's quality control department randomly selects a sample of 5 lightbulbs to test. In order to meet quality control standards, the lightbulbs in the sample must last an average of at least 950 hr. Four of the selected lightbulbs lasted 925 hr, 1000 hr, 950 hr, and 900 hr. How many hours must the fifth lightbulb last if the sample is to meet quality control standards?

Solution The average number of hours that the 5 lightbulbs must last is the sum of the hours each lightbulb lasts divided by the number of lightbulbs in the sample, which is 5.

$$\text{Average number of hours} = \frac{925 + 1000 + 950 + 900 + x}{5}$$

where x represents the number of hours the fifth bulb lasted. In order for the average to be *at least* 950, it must be greater than or equal to 950. So we write and solve the inequality.

$$\frac{925 + 1000 + 950 + 900 + x}{5} \geq 950$$

$$5\left(\frac{925 + 1000 + 950 + 900 + x}{5}\right) \geq 950 \cdot 5$$

$$925 + 1000 + 950 + 900 + x \geq 4750$$

$$3775 + x \geq 4750$$

$$3775 - 3775 + x \geq 4750 - 3775$$

$$x \geq 975$$

So the fifth lightbulb in the sample must last at least 975 hr for the sample to meet quality control standards.

PRACTICE 14

Suppose that a student has two part-time jobs. On the first job, she works 15 hr a week at $8.50 an hour. The second job pays only $7.50 an hour, but she can work as many hours as she wants. To make at least $300, how many hours should she work on the second job?

Mathematically Speaking

Fill in each blank with the most appropriate term or phrase from the given list.

equation	reversed	unchanged
graph	inequality	positive
negative	solution	open
closed		

1. A(n) _____ is any mathematical statement containing $<$, $\leq$, $>$, $\geq$, or $\neq$.

2. A(n) _____ of an inequality is any value of the variable that makes the inequality true.

3. In the graph of the inequality $x > -7$, the circle is _____.

4. In the graph of the inequality $x \geq -7$, the circle is _____.

5. When the same number is added to or subtracted from each side of an inequality, the direction of the inequality is _____.

6. According to the multiplication property of inequalities, if $a \geq b$ and c is _____, then $ac \geq bc$.

7. According to the multiplication property of inequalities, if $a \geq b$ and c is _____, then $ac \leq bc$.

8. When each side of an inequality is divided by a negative number, the direction of the inequality is _____.

Indicate whether the value of x shown is a solution of the given inequality by answering true or false.

9.

Value of x	Inequality	True or False
a. 1	$8 - 3x > 5$	
b. 4	$4x - 7 \leq 2x + 1$	
c. -7	$6(x + 6) < -9$	
d. $-\dfrac{3}{4}$	$8x + 10 \geq 12x + 15$	

10.

Value of x	Inequality	True or False
a. 3	$2x + 1 < -10$	
b. -5	$1 - 3x > 5x + 12$	
c. -2	$-(4x + 8) \geq 0$	
d. $\dfrac{1}{3}$	$9x - 7 \leq 6x - 11$	

Graph on the number line.

11. $x > 3$

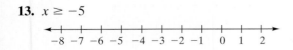

12. $x \leq -1$

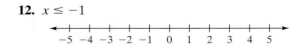

13. $x \geq -5$

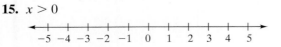

14. $x < 1$

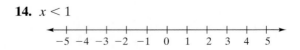

15. $x > 0$

16. $x \leq 0$

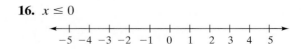

17. $x \le 0.5$

-5 -4 -3 -2 -1 0 1 2 3 4 5

18. $x < -1.5$

-6 -5 -4 -3 -2 -1 0 1 2 3 4

19. $x > -2\frac{1}{2}$

-5 -4 -3 -2 -1 0 1 2 3 4 5

20. $x \ge 4\frac{1}{2}$

-1 0 1 2 3 4 5 6 7 8 9

21. $x \ge -\frac{1}{3}$

-5 -4 -3 -2 -1 0 1 2 3 4 5

22. $x < \frac{3}{4}$

-5 -4 -3 -2 -1 0 1 2 3 4 5

23. $-3 < x < 1$

-5 -4 -3 -2 -1 0 1 2 3 4 5

24. $0 < x \le 4$

-5 -4 -3 -2 -1 0 1 2 3 4 5

25. $-\frac{1}{2} \le x < 2$

-5 -4 -3 -2 -1 0 1 2 3 4 5

26. $-4 \le x \le 1\frac{1}{2}$

-5 -4 -3 -2 -1 0 1 2 3 4 5

Solve and graph.

27. $v + 2 < -5$

28. $x - 5 > -1$

29. $y - 5 > -5$

30. $s + 1 \le 1$

31. $y + 2 \le 5.5$

32. $t - 1.5 > 1$

33. $v - 17 \le -15$

34. $n - 25 > -30$

35. $-2 \ge x - 4$

36. $4 < a - 3$

37. $\frac{1}{3}a < -1$

38. $\frac{1}{2}x \le 3$

39. $-5y > 10$

40. $-7t \le -21$

41. $2x \ge 0$

42. $-3.5m < 0$

43. $-\dfrac{3}{4}a \geq 3$

44. $\dfrac{-x}{3} > -2$

45. $6 \leq \dfrac{-2}{3}n$

46. $4 \geq \dfrac{-y}{2}$

Solve.

47. $\dfrac{n}{3} + 2 > 3$

48. $\dfrac{1}{2}y + 4 \geq -1$

49. $3x - 12 \leq 6$

50. $2v - 9 > -7$

51. $-21 - 3y > 0$

52. $7y + 4 - 2y \geq 39$

53. $5n - 11 \geq 2n + 28$

54. $24 - 9s < -13s + 8$

55. $-4m + 8 \leq -3m + 1$

56. $-6y + 13 > y + 6$

57. $-7x + 4x + 23 < 2$

58. $5t - 7 + 2t \leq 8t - 2$

59. $-3(z + 5) > -15$

60. $2(8 + w) < 22$

61. $0.5(2x + 1) \geq 3x$

62. $-0.2(10d - 5) > 9$

63. $2(x - 2) - 3x \geq -1$

64. $2(4z - 3) < 3(z + 4)$

65. $7y - (9y + 1) < 5$

66. $-(6m - 2) \leq 0$

67. $0.4(5x + 1) \geq 3x$

68. $0.2(y - 3) > 8$

69. $5x + 1 < 3x - 2(4x - 3)$

70. $-2(0.5 - 4t) > -3(4 - 3.5t)$

71. $3 + 5n \leq 6(n - 1) + n$

72. $3(4 - 2m) \geq 2(3m - 6)$

73. $-\dfrac{4}{3}x - 16 > x + \dfrac{1}{3}x$

74. $\dfrac{2}{3}z - 4 < z + \dfrac{1}{3} + \dfrac{1}{3}z$

75. $0.2y > 1500 + 2.6y$

76. $x + 1.6x \leq 52$

Choose the inequality that best describes the situation.

77. To vote in the United States, a citizen must be at least 18 years old.

a. $a < 18$ **b.** $a > 18$
c. $a \leq 18$ **d.** $a \geq 18$

78. The number of people seated in the theater is at most 650.

a. $n < 650$ **b.** $n \leq 650$
c. $n > 650$ **d.** $n \geq 650$

79. It is generally accepted that a person has a fever if the person's temperature is above 98.6°F. To convert a Celsius temperature to its Fahrenheit equivalent, we use the formula $F = \dfrac{9}{5}C + 32$. For which Celsius temperatures C does the person have a fever?

a. $\dfrac{9}{5}C + 32 \geq 98.6$ **b.** $\dfrac{9}{5}C + 32 \leq 98.6$

c. $\dfrac{9}{5}C + 32 < 98.6$ **d.** $\dfrac{9}{5}C + 32 > 98.6$

80. A teenager has a $20 gift certificate at the iTunes store. He is purchasing three episodes of his favorite television show at $1.99 each, and is also browsing the music store where songs can be downloaded at $0.99 apiece. What is the maximum number of songs he can purchase without exceeding the amount of the gift certificate?

a. $5.97 + 0.99n > 20$ **b.** $5.97 + 0.99n < 20$
c. $5.97 + 0.99n \geq 20$ **d.** $5.97 + 0.99n \leq 20$

Mixed Practice

Graph on the number line.

81. $-1 < x \leq 3\frac{1}{2}$

82. $x > -0.5$

Solve.

83. Indicate whether the value of x shown is a solution of the given inequality by answering true or false.

Value of x	Inequality	True or False
a. 2	$4 - 2x < -1$	
b. 6	$3x - 5 \geq 21 - 2x$	
c. -8	$-(2x + 4) \leq 12$	
d. $-\dfrac{1}{2}$	$8x - 5 > 4x - 8$	

Solve and graph.

84. $-3 \leq x - 7$

85. $-\dfrac{2}{3}m \geq 4$

86. $-8t > -24$

Solve.

87. $3(a + 4) \geq 2(3a - 2)$

88. $-5m + 6 \geq -4m + 4$

89. $8x + 7 - 3x < 32$

90. $0.3m - 1200 < 2.8m$

Applications

Write an inequality and solve.

91. On the last three chemistry exams a student scored 81, 85, and 91. What score must the student earn on the next exam to have an average above 85?

92. A rectangular deck that is 14 ft long is to be built onto the back of a house. What would be the area of the deck if it is at least 12 ft wide?

93. A novelty store, open Monday through Saturday, must sell at least $200 worth of merchandise per day to break even on expenses. The following table shows sales for one week.

Day	Monday	Tuesday	Wednesday	Thursday	Friday
Amount of Sales	$250	$250	$150	$130	$180

How much must the store make in sales on Saturday to at least break even for the week?

94. Molina's car rental company advertises two deals in the daily paper.

Molina's Car Rental

$29.99 per day + $0.25/mi
or
a flat rate of $39.99/day
with unlimited mileage

For which mileage is the flat rate a better deal?

95. A telemarketer claims that every call made to a customer costs at least $2. If a call costs $0.50 plus $0.10 for each minute, how long does each call last?

96. One side of a triangular garden is 2 ft longer than another side. The third side is 4 ft long. What are the maximum lengths of the other two sides if the perimeter of the garden can be no longer than 12 ft?

97. A real estate broker receives a monthly salary of $1500 plus $1200 for every house he sells during the month. His supervisor offers an alternative monthly salary: $1000 plus $1500 for each house sold. Should he accept the deal? Under what circumstances?

98. A parking garage offers two payment options: a $20 flat fee for the whole day, or $5 plus $2 per hour for each hour or part thereof that a customer parks. Which is the better option for the customer? Explain why.

99. A person weighing 200 lb volunteers for a clinical trial of a new diet pill. If he loses 2.5 lb per month using the diet pill combined with regular exercise, when will he weigh less than 180 lb?

100. A diver scored 9.2, 9.6, 9.7, 9.4, and 9.3 in the first 5 dives in a diving competition, where all scores are given in tenths. To win first place, she must beat a total score of 56.6. What is the lowest score she can get on her last dive to win the competition?

● *Check your answers on page A-6.*

MINDSTRETCHERS

Mathematical Reasoning

1. Explain for which values of x the following inequality holds: $x + 5 > x + 2$.

Groupwork

2. Working with a partner, explore whether each of the following statements is always, sometimes, or never true. Give examples.

    a. If $a < b$ and $c < d$, then $ac < bd$.

    b. $a - b \leq a + b$

Writing

3. Clearly identifying the variables, give examples of inequalities that

    a. you wish were true.

    b. you wish were false.

KEY CONCEPTS AND SKILLS (CONCEPT) (SKILL)

Concept/Skill	Description	Example
[2.1] Equation	A mathematical statement that two expressions are equal.	$y - 6 = 9$
[2.1] Solution of an equation	A value of the variable that makes the equation a true statement.	2 is a solution of the equation $3x + 1 = 11 - 2x$.
[2.1] Linear equation in one variable	An equation that can be written in the form $ax + b = c$, where a, b, and c are real numbers and $a \neq 0$.	$x + 1 = 3$
[2.1] Equivalent equations	Equations that have the same solution.	$x = 2$ and $x + 1 = 3$
[2.1] Addition property of equality	For any real numbers a, b, and c, if $a = b$, then $a + c = b + c$.	If $x - 3 = 5$, then $(x - 3) + 3 = 5 + 3$.
[2.1] To solve word problems in algebra	• Read the problem carefully. • Translate the word problem into an equation. • Solve the equation. • Check the solution in the original equation. • State the conclusion.	
[2.2] Multiplication property of equality	For any real numbers a, b, and c, if $a = b$, then $a \cdot c = b \cdot c$.	If $\dfrac{n}{2} = 8$, then $\dfrac{n}{2} \cdot 2 = 8 \cdot 2$.
[2.3] To solve linear equations	• Use the distributive property to clear the equation of parentheses, if necessary. • Combine like terms where appropriate. • Use the addition property to isolate the variable term. • Use the multiplication property to isolate the variable. • Check by substituting the solution in the original equation.	$2(x + 5) = 6$ $2x + 10 = 6$ $2x + 10 - 10 = 6 - 10$ $2x = -4$ $x = -2$ **Check** $2(x + 5) = 6$ $2(-2 + 5) \overset{?}{=} 6$ $6 = 6$ **True**
[2.4] Literal equation	An equation involving two or more variables.	$2t + b = c$
[2.5] To find a percent increase or decrease	• Compute the difference between the two given values. • Compute what percent this difference is of the *original value*.	If a quantity changes from 10 to 12, the difference is 2. Since 2 is 20% of 10, the percent increase is 20%.

continued

Concept/Skill	Description	Example
[2.6] Inequality	Any mathematical statement containing $<$, $\leq$, $>$, $\geq$, or $\neq$.	$x \geq -4$
[2.6] Solution of an inequality	Any value of the variable that makes the inequality true.	0 is a solution of $x < 2$.
[2.6] Addition property of inequalities	For any real numbers a, b, and c: • If $a < b$, then $a + c < b + c$. • If $a > b$, then $a + c > b + c$. Similar statements hold for $\leq$ and $\geq$.	If $x < 1$, then $x + 2 < 1 + 2$.
[2.6] Multiplication property of inequalities	For any real numbers a, b, and c: • If $a < b$ and c is positive, then $ac < bc$. • If $a < b$ and c is negative, then $ac > bc$. Similar statements hold for $>$, $\leq$, and $\geq$.	If $\dfrac{x}{2} > 4$, then $2 \cdot \dfrac{x}{2} > 2 \cdot 4$. If $-2x < 4$, then $\dfrac{-2x}{-2} > \dfrac{4}{-2}$.

Chapter 2   # Review Exercises

[2.1]

1. Is 2 a solution of the equation $5x + 3 = 7 - 4x$?

2. Determine whether 0 is a solution of the equation $4x - 15 = 5(x - 3)$.

Solve and check.

3. $x - 3 = -12$

4. $t + 10 = 8$

5. $-9 = a + 5$

6. $4 = n - 7$

7. $y - (-3.1) = 11$

8. $r + 4.8 = 20$

[2.2]   *Solve and check.*

9. $\dfrac{x}{3} = -2$

10. $\dfrac{z}{2} = -5$

11. $2x = -20$

12. $-5d = 15$

13. $-y = -4$

14. $-x = 3$

15. $20.5 = 0.5n$

16. $30 = -0.2r$

17. $\dfrac{2t}{3} = -6$

18. $\dfrac{5y}{6} = -10$

[2.3]   *Solve and check.*

19. $2x + 1 = 7$

20. $-t - 4 = 5$

21. $\dfrac{a}{2} - 3 = -10$

22. $\dfrac{r}{3} - 6 = 12$

23. $-y + 7 = -2$

24. $-2t + 3 = 1$

25. $4x - 2x - 5 = 7$

26. $3y - y + 12 = 6$

27. $z + 1 = -2z + 10$

28. $n - 3 = -n + 7$

29. $c = -2(c + 1)$

30. $p = -(p - 5)$

31. $2(x + 1) - (x - 8) = -x$

32. $-(x + 2) - (x - 4) = -5x$

33. $3[2n - 4(n + 1)] = 6n - 12$

34. $-4(2x - 6) = 7[x - (3x - 1)]$

35. $10 - [3 + (2x - 1)] = 3x$

36. $x - [5 + (3x - 4)] = -x$

[2.4]

37. Solve $a - 5b = 2c$ for a in terms of b and c.

38. Solve $\dfrac{2a}{b} = n$ for a in terms of b and n.

39. Consider the equation $Ax + By = C$.

   a. Solve for x in terms of A, B, C, and y.

   b. Using the equation found in part (a), find the value of x if $A = 2$, $B = -1$, $C = 0$, and $y = 5$.

40. $A = \dfrac{bh}{2}$ is the formula for the area A of a triangle in terms of its base b and height h.

   a. Solve for h in terms of A and b.

   b. Using the formula found in part (a), find the height of a triangle with an area of 12 cm^2 and a base of 4 cm.

[2.5] *Solve.*

41. 30% of what number is 12?

42. 125% of what number is 5?

43. What percent of 5 is 8?

44. What percent of 8 is 5?

45. What number is 8.5% of $300?

46. What is 3.5% of $2000?

[2.6] *Graph each inequality.*

47. $x < 2$

48. $x \geq -4.5$

49. $3\frac{1}{2} \geq x$

50. $-0.5 < x < 5$

Solve. Then graph.

51. $y + 1 > 6$

52. $-\frac{1}{2}t + 3 \leq 3$

53. $8y - 2 \leq 6y + 2$

54. $\frac{1}{2}(8 - 12x) \leq x - 10$

55. $0.5n - 0.3 < 0.2(2n + 1)$

Mixed Applications

Solve.

56. An air conditioner's energy efficiency ratio (EER) is the quotient of its British thermal unit (Btu) rating and its wattage. What is the Btu rating of a 2000-watt air conditioner if its EER is 8?

57. The sum of the measures of the angles in any triangle is 180°. If the three angles in an equilateral triangle are equal, what is the measure of each of these angles?

58. For a wedding, the reception costs will be $5000 plus $50 per guest. The bride and groom have budgeted $12,000 for the reception. How many guests can the bride and groom invite to the wedding?

59. A polygon is a closed geometric figure with straight sides. In any polygon with n sides, the sum of the measures of its angles is $180(n - 2)$ degrees. If the measures of the angles of a polygon add up to 540°, how many sides does the polygon have?

60. A newspaper reported that a candidate received 15,360 more votes than her opponent, and that 39,210 votes were cast in the election. How many votes were cast for each candidate?

61. Suppose that to send a telegram it costs $2 for the first 10 words in the telegram and y cents for each additional word.

   a. Write an equation to find the cost C of a telegram 26 words long.

   b. Solve this equation for y.

62. The road connecting two factories is 380 miles long. A truck leaves one of the factories traveling toward the other factory at 45 mph, while at the same time a second truck leaves the other factory heading at 50 mph toward the first. How long after the departure will the trucks meet?

63. Two friends leave a party at 10 P.M. driving in opposite directions. One drives at a speed of 40 mph, whereas the other drives at a speed of 32 mph. At what time are the two friends 18 mi apart?

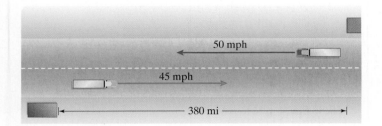

64. A plane flying between two cities at 400 mph arrives half an hour behind schedule. If the plane had flown at a speed of 500 mph, it would have been on time. Find the distance between the cities.

65. Tom Seaver received 425 out of 430 votes electing him to the Baseball Hall of Fame, surpassing the previous voting record for Ty Cobb. To the nearest whole percent, what percent of votes cast did Seaver receive? (*Source:* United Press International, Jan. 7, 1992)

66. In the decade beginning with 1901, about 8.8 million immigrants entered the United States. In the following decade, this number dropped to approximately 5.7 million people. Find the percent decrease, to the nearest whole percent. (*Source: 1999 Statistical Abstract of the United States*)

67. The rarest blood type is AB⁻, which occurs in only 0.7% of people in the United States. Of the 4000 students in your college, how many would you expect to have AB⁻ blood type? (*Source: The 1995 Guinness Book of World Records*)

68. How much interest is earned in 1 year on a principal of $500 at an annual interest rate of 6%?

69. In the presidential election of 1836, Martin Van Buren earned 170 electoral votes. Four years later, Van Buren earned 60 electoral votes. To the nearest whole percent, by what percent did Van Buren's electoral vote count drop? (*Source: Time Almanac 2000*)

70. How much pure alcohol must be mixed with 6 L of a 60% alcohol solution to make a 70% alcohol solution?

71. How many pints of a 1% solution of disinfectant must be combined with 4 pt of a 10% solution to make a 5% solution?

72. To convert a weight expressed in kilograms, k, to the equivalent number of pounds, p, multiply the number of kilograms by 2.2.

   a. Write a formula for this relationship.

   b. Solve this formula for k.

73. To print b books, it costs a publisher $(100,000 + 25b)$ dollars. The publisher receives $50b$ in revenue for selling this number of books. How many books must be printed for the revenue to equal the cost?

74. An aspiring country singer must record a demo disk of her songs to distribute to record labels. One Nashville studio charges a base rate of $350 for studio time and a small backup band. Each additional instrument desired costs $50 per instrument per song. In addition to the band, the singer would like to have a fiddle, mandolin, and dobro on each song. How many songs can the singer record if she can afford a maximum of $1000 for recording costs?

• *Check your answers on page A-6.*

| Chapter 2 | **POSTTEST** |

FOR EXTRA HELP

Pass *the***Test**   Test solutions are found on the enclosed CD.

To see if you have mastered the topics in this chapter, take this test.

1. Determine whether -2 is a solution of the equation $3x - 4 = 6(x + 2)$.

2. Solve and check: $x - 1 = -10$

3. Solve and check: $\dfrac{n}{2} = -3$

4. Solve and check: $-y = -11$

5. Solve and check: $\dfrac{3y}{4} = 6$

6. Solve and check: $2x + 5 = 11$

7. Solve and check: $-s + 4 = 2$

8. Solve and check: $10x + 1 = -x + 23$

9. Solve and check: $16a = -4(a - 5)$

10. Solve and check: $2(x + 5) - (x + 4) = 7x + 1$

11. Solve $5n + p = t$ for p in terms of n and t.

12. 40% of what number is 8?

13. What percent of 5 is 10?

14. Draw the graph of $-1 \le x < 3$.

$$\overset{\longleftarrow\;|\;\;|\;\;|\;\;|\;\;|\;\;|\;\;|\;\longrightarrow}{\;-3\;-2\;-1\;\;0\;\;1\;\;2\;\;3\;}$$

15. Solve $-2z \le 6$. Then graph.

$$\overset{\longleftarrow\;|\;\;|\;\;|\;\;|\;\;|\;\;|\;\;|\;\longrightarrow}{\;-3\;-2\;-1\;\;0\;\;1\;\;2\;\;3\;}$$

16. A taxi charges $4.00 for the first mile plus $1.25 for each additional mile. On a fare of $16.50, how long was the ride?

17. A woman's shoe size S is given by the formula $S = 3L - 21$, where L is her foot length L in inches. Solve for L in terms of S.

18. In a recent year, males experienced 37%, or 8,400,000, of the operations performed in the United States. To the nearest million, how many operations were performed in all? (*Source: The New York Times Almanac, 2000*)

19. Two friends live 33 mi apart. They cycled from their homes, riding toward one another and meeting $1\frac{1}{2}$ hr later. If one friend cycled 2 mph faster than the other, what were their two rates?

20. A cellular phone service offers two calling plans. Plan A costs $39.99 per month plus $0.79 per minute (or part thereof) for calls outside the network. Plan B costs $54.99 per month plus $0.59 per minute (or part thereof) for calls outside the network. Under what conditions will the monthly cost of Plan A exceed the monthly cost of Plan B?

• *Check your answers on page A-6.*

Cumulative Review Exercises

To help you review, solve the following:

1. Express as a signed number: a loss of 6 yd in a football play

2. Graph the number -3 on the number line shown.

$$\xleftarrow{\quad}\!\!\!+\!\!\!+\!\!\!+\!\!\!+\!\!\!+\!\!\!+\!\!\!+\!\!\!\xrightarrow{\quad}$$
$$-3\,-2\,-1\ \ 0\ \ 1\ \ 2\ \ 3$$

3. Compute $|-2|$

4. True or false, $1 > -2$

5. Simplify: $2 + (-1) + (-4) + 4$

6. Divide: $(-10) \div (-2)$

7. Simplify: $3x + 1 - 7(2x - 5)$

8. A dieter gained 3 lb and then lost 5 lb. Express her overall change in weight as a real number.

9. An initial investment of $1000 triples in value every 15 yr. The value of this investment after 60 yr is how many times the initial investment? Write your answer in exponential form.

10. A telephone technician charges $50 for the first hour of work and $25 for each additional hour of work.

   a. Write a formula that expresses the amount of money A that the technician charges in terms of the time t that the technician works.

   b. Solve this equation for t.

   c. If the repairman sent you a bill for $125, how many hours did he claim to work?

● Check your answers on page A-7.

Graphing Linear Equations and Inequalities; Functions

Graphs and Inflation: Buying Power

The U.S. Department of Labor's Bureau of Labor Statistics provides data on inflation from 1913 to the present. The bureau uses the average consumer price index for a given calendar year to calculate inflation. A good way to understand the impact of inflation on our buying power is to calculate how much $100 worth of groceries purchased in 2000 would have cost in preceding years. The following table shows a comparison of costs from 1960 to 2000. Although the table is helpful, a graph gives a much better picture of inflation.

Year	Dollars
1960	$17.19
1965	$18.29
1970	$22.53
1975	$31.24
1980	$47.85
1985	$62.49
1990	$75.90
1995	$88.50
2000	$100.00

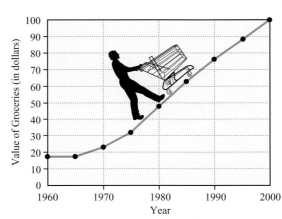

Chapter 3   PRETEST

To see if you have already mastered the topics in this chapter, take this test.

1. The following graph shows the percent of American cancer patients surviving 5 or more years, during various periods of time.

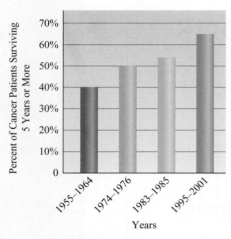

(*Source:* American Cancer Society)

In which period(s) of time shown did more than half the cancer patients survive 5 or more years?

2. The first automated teller machine (ATM) in the United States was installed in 1971 at the Citizens & Southern National Bank in Atlanta. Overall, the number of ATMs has grown rapidly. The following graph shows the number in recent years.

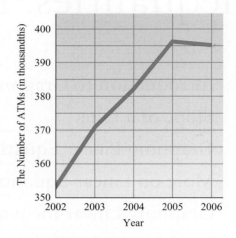

(*Source:* Insurance Information Institute)

In the year 2006, approximately how many ATMs were there in the United States?

3. **a.** On the coordinate plane below, plot the points $A(1, 4)$ and $B(-3, -6)$.

   b. In which quadrant is the point $(3, -5)$ located?

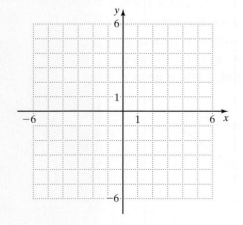

4. Given two points $C(7, 5)$ and $D(1, 2)$, compute the slope of the line that passes through the points.

5. For the points $P(-3, 3)$, $Q(1, -1)$, $R(-2, -2)$, and $S(4, 4)$, indicate whether $\overleftrightarrow{PQ}$ is perpendicular to $\overleftrightarrow{RS}$. Explain.

6. In the graph shown, find the x-intercept and the y-intercept.

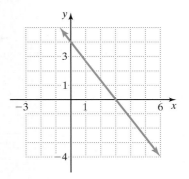

7. In the graph shown, the y-axis stands for the number of congressional representatives from a state and the x-axis stands for the state's population according to a recent U.S. census.

Draw the line that passes through the points. Is this line's slope positive, negative, or zero? How does state population relate to the number of representatives from that state?

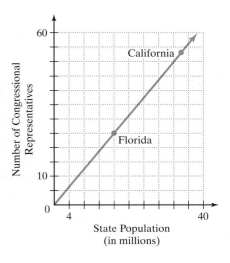

8. Agriculturalists are trying to develop varieties of wheat that grow at a faster rate. The graph shown displays the growth pattern of two new varieties of wheat. Which variety grows more quickly?

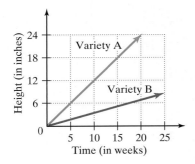

9. For the equation $y = 2x - 5$, complete the following table.

x	4	7		
y			0	-1

Graph the equation indicated in Questions 10–12.

10. $y = -3$

11. $y = -3x + 2$

12. $2x - 3y = 6$

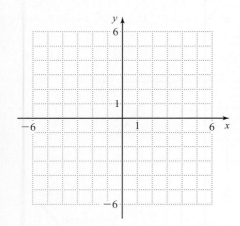

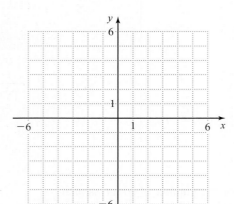

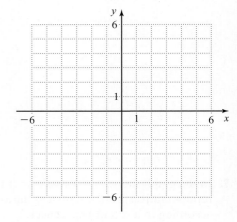

13. a. Write the equation $5x - y = 8$ in slope-intercept form.

    b. What are the slope and y-intercept of the graph of this equation?

14. Find an equation of the line with slope 2 that passes through the point $(0, 8)$.

15. Find an equation of the line that passes through the points $(4, 1)$ and $(2, -1)$.

16. Graph: $y > 3x + 1$

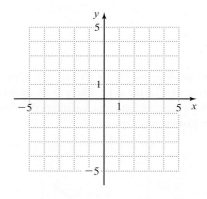

17. A local video store charges a daily rental fee of $2.50 per movie.

  a. Express the daily cost c of renting x movies.

  b. Choose an appropriate scale for the axes and graph this relationship.

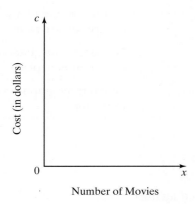

  c. What is the slope of the line in part (b)? Explain its significance in this context.

18. A textbook sales representative is driving to a college 200 mi away at a speed of 50 mph.

  a. Express the distance d (in miles) the sales representative travels in terms of the time t (in hours) he has been driving.

  b. Choose an appropriate scale for the axes and graph this relationship.

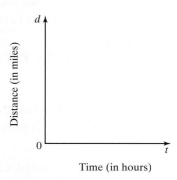

  c. What is the slope of the graph? Explain its significance in terms of the trip.

19. If $f(x) = |2x - 8|$, evaluate $f(-6)$.

20. Graph the function $f(x) = 1 - 1.5x$ for $x \geq -2$ and identify its domain and range.

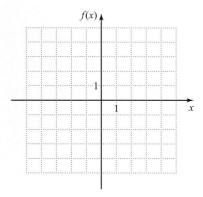

3.1   Introduction to Graphing

What Graphing Is and Why It Is Important

Many mathematical relationships can be expressed as equations, inequalities, or their graphs. Although lacking the precision of an equation, a graph can clarify at a glance patterns and trends in a relationship, helping us to understand that relationship better.

Reading Bar and Line Graphs

Bar Graphs

On a *bar graph*, quantities are represented by thin, parallel rectangles called bars. The length of each bar is proportional to the quantity that it represents.

Sometimes, bar lengths are labeled. Other times, bar lengths are read against an *axis*—a straight line parallel to the bars and similar to a number line.

Bar graphs are especially useful for making comparisons or contrasts among a few quantities, as the following example illustrates.

> **OBJECTIVES**
> - To read bar and line graphs
> - To identify and plot points on the coordinate plane
> - To identify the quadrants of the coordinate plane
> - To interpret graphs in applied problems

EXAMPLE 1	PRACTICE 1

The following graph shows the net income of U.S. Airways in recent years.

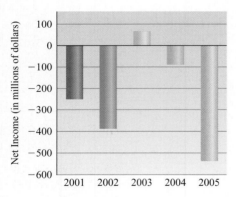

(*Source:* U.S. Airways Group)

a. What was the approximate net income of the company in the year 2005?

b. About how much greater was the net income in 2003 than in 2004?

c. Describe the graph.

Solution

a. In 2005, the net income was about −$540 million, that is, a loss of about $540 million.

The following graph shows the value of the top five agricultural commodities in the United States in a recent year.

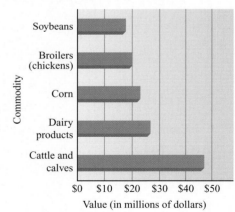

(*Source:* Department of Agriculture)

a. Which commodity had the greatest value?

b. What was the approximate value of dairy products?

c. About how much greater was the value of corn than of broilers?

b. In 2003, the net income was approximately $70 million. The next year, it was about −$90 million. So the net income for 2003 was about $160 million greater than in 2004.

c. The company operated at a profit in the year 2003. In other years, however, it operated at a loss, especially in 2005.

Line Graphs

On a *line graph*, quantities are represented as points connected by straight-line segments. The height of any point on a line is read against the vertical axis.

Line graphs are commonly used to highlight changes and trends over a period of time. Especially when we have data for many points in time, we are more likely to use a line graph than a bar graph.

EXAMPLE 2	PRACTICE 2

The following graph shows the number of Americans 65 years of age and older during the twentieth century.

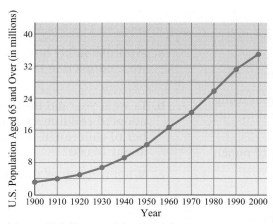

(*Source:* U.S. Bureau of the Census)

a. Approximately how big was this population in the year 2000?

b. In what year did this population number about 21 million?

c. In the year 2000, the U.S. population overall was approximately 4 times as large as it had been in the year 1900. Did the population shown in the graph grow more quickly?

Solution

a. In 2000, there were about 35 million Americans aged 65 and above.

b. There were approximately 21 million Americans aged 65 and above in the year 1970.

c. In 2000, the overall U.S. population was 4 times what it had been in 1900. But the population shown in the graph grew by a factor of about 10 and so grew more quickly.

The following graph shows the mean temperatures in Chicago over a 30-year period for each month of the year.

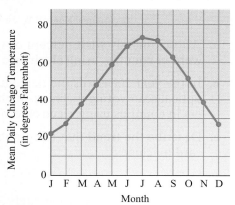

(*Source:* The U.S. National Climatic Data Center)

a. Which month in Chicago has the highest mean temperature?

b. Approximately what is the mean temperature in February?

c. What trend does the graph show?

Comparison line graphs show two or more changing quantities, as Example 3 illustrates.

EXAMPLE 3	PRACTICE 3

The following graph shows the percent of American voters, by year, in counties using various types of voting machines.

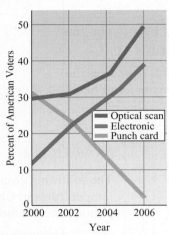

(*Source:* Kimball W. Brace, Election Data Services)

a. Approximately what percent of voters in the year 2006 were in counties using electronic voting machines?

b. In the year 2004, was the percent of voters in counties using optical scan voting machines higher than the percent of voters in counties using electronic voting machines?

c. Describe the trend that the graph shows in the use of punch card machines.

Solution

a. In the year 2006, approximately 39% of voters were in counties using electronic voting machines.

b. In 2004, about 36% of voters were in counties with optical scan machines, in contrast to approximately 29% of voters in counties with electronic machines. So the figure for optical scan machines was higher.

c. The use of punch card ballots dropped every year between 2000 and 2006. In 2006, the percent of voters in counties using punch card ballots was close to 0.

The following graph shows the number of children in New York State who are living in foster care and the number of foster children who have been adopted, for the years between 1995 and 2006.

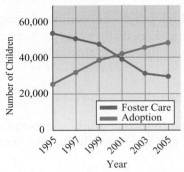

(*Source:* New York State Family Services)

a. Approximately how many children were adopted in the year 2005?

b. In what year did the number of adopted children exceed the number of children in foster care by about 14,000?

c. What trend does this graph show?

Plotting Points on the Coordinate Plane

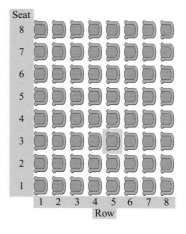

If you were to enter a theater or sports arena with a ticket for row 5, seat 3, you would know exactly where to sit. Such a system of *coordinates* in which we associate a pair of numbers in a given order with a corresponding location is commonplace.

The flat surface on which we draw graphs is called a *coordinate plane*. To create a coordinate plane, we first sketch two perpendicular number lines—one horizontal, the other vertical—that intersect at their zeros. The point where they intersect is called the *origin*. Each number line is called an *axis*. It is common practice to refer to the *horizontal* number line as the *x-axis* and the *vertical* number line as the *y-axis*.

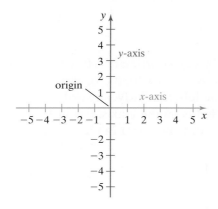

Each point in the coordinate plane is represented by a pair of numbers called an *ordered pair*. For example, the origin is the point $(0, 0)$. The first number in an ordered pair represents a horizontal distance and is called the *x-coordinate*. The second number represents a vertical distance and is called the *y-coordinate*.

To *plot* a point in the coordinate plane, we find its location represented by its ordered pair. For example, to plot the point $(3, 1)$, start at the origin and go 3 units *to the right*, then go *up* 1 unit. For this point, we say that $x = 3$ and $y = 1$.

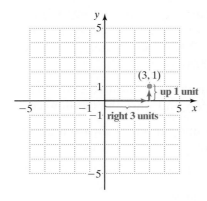

Notice that the two numbers in an ordered pair are written in parentheses, separated by a comma. Do the ordered pairs $(3, 1)$ and $(1, 3)$ correspond to different points? Why?

When an ordered pair has a negative x-coordinate, the corresponding point is to the left of the y-axis, as shown in the following coordinate plane. Similarly, when an ordered pair's y-coordinate is negative, the point is below the x-axis.

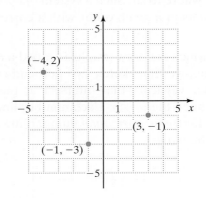

Any ordered pair whose y-coordinate is 0 corresponds to a point that is on the x-axis. For instance, the points $(-4, 0)$, $(0, 0)$, and $(3, 0)$ are on the x-axis, as shown in the graph below on the left. Similarly, any ordered pair whose x-coordinate is 0 corresponds to a point that is on the y-axis. For instance, the points $(0, 2)$, $(0, -1)$, and $(0, -4)$ are on the y-axis, as shown in the middle graph below.

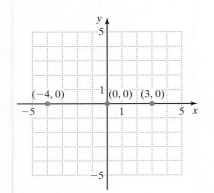 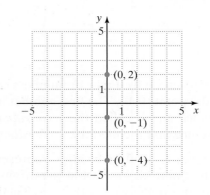 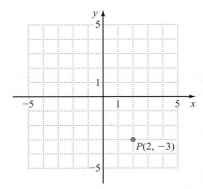

Sometimes we name points with letters. We can refer to a point as P or A or any other letter that we choose. Generally, a capital letter is used. If we want to emphasize that point P has coordinates $(2, -3)$, we can write it as $P(2, -3)$, as shown in the graph above on the right.

If there is a point whose coordinates we do not know, we can refer to it as (x, y) or $P(x, y)$, where x and y are the unknown coordinates.

Sometimes to distinguish one point from another, we use *subscripts* to name them. We may refer to a pair of points as

- P_1 and P_2 (read "P sub one" and "P sub two"),
- (x_1, y_1) and (x_2, y_2), or
- $P_1(x_1, y_1)$ and $P_2(x_2, y_2)$.

Notice that y_1 is the *y-coordinate* that corresponds to the *x-coordinate* x_1, and y_2 corresponds to x_2.

Now, let's look at some examples of plotting points.

EXAMPLE 4	PRACTICE 4

EXAMPLE 4

Plot the following points on a coordinate plane.

a. (5, 2)   **b.** (3, −4)   **c.** (−1, 1)

d. (0, 3)   **e.** (−4, −2)   **f.** (0, 0)

Solution The points are plotted on the coordinate plane as shown.

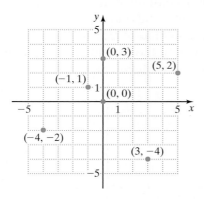

PRACTICE 4

On a coordinate plane, plot the points corresponding to each ordered pair.

a. (0, −5)   **b.** (4, 4)

c. (−2, 2)   **d.** (5, 0)

e. (−3, −2)   **f.** (2, −4)

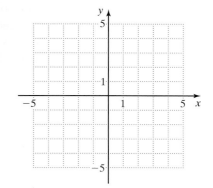

The x- and y-axes are boundaries that separate a coordinate plane into four regions called *quadrants*. These quadrants are named in counter-clockwise order starting with Quadrant I, as in the figure to the right.

The quadrant in which a point is located tells us something about its co-ordinates. For instance, in Quadrant I, any point is to the right of the y-axis and above the x-axis, so both its coordinates must be positive. For example, the point (3, 2) is located in Quadrant I. Points in this quadrant are of par-ticular interest in applications in which all quantities must be positive.

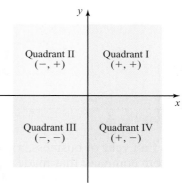

A point in Quadrant II lies to the left of the y-axis and above the x-axis, so its x-coordinate must be negative and its y-coordinate must be positive. For example, the point (−4, 4) is located in Quadrant II. In Quadrant III, points are to the left of the y-axis and below the x-axis, so both coordinates must be negative. For example, the point (−3, −2) is located in Quadrant III. And finally, in Quadrant IV, a point is to the right of the y-axis and below the x-axis, so its x-coordinate must be positive and its y-coordinate must be negative. For example, the point (2, −3) is located in Quadrant IV.

EXAMPLE 5	PRACTICE 5

EXAMPLE 5

Determine the quadrant in which each point is located.

a. (−5, 5)

b. (7, 20)

c. (1.3, −4)

d. (−4, −5)

PRACTICE 5

In which quadrant is each point located?

a. $\left(-\frac{1}{2}, 3\right)$

b. (6, −7)

c. (−1, −4)

d. (2, 9)

Solution

a. $(-5, 5)$ is in Quadrant II.

b. $(7, 20)$ is in Quadrant I.

c. $(1.3, -4)$ is in Quadrant IV.

d. $(-4, -5)$ is in Quadrant III.

Often the points that we are to plot affect how we draw the axes on a coordinate plane. For instance, in the following planes, we choose an appropriate *scale* for each axis—the length between adjacent tick marks—to conveniently plot all points in question. And depending on the location of the points to be plotted, we may choose to show only part of a coordinate plane.

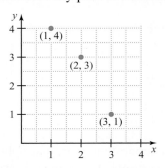
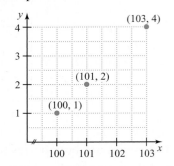

Tip   Depending on the location of the points to be plotted, we can choose to show only part of a coordinate plane.

EXAMPLE 6

In its first year of business, a young entrepreneur started a dot-com company that made a profit of $10,000. In the second year, the company's profit grew to $15,000. In the third year, however, the company lost $5000. Plot points on a coordinate plane to display this information.

Solution We let x represent the year of business and y represent the company's profit in dollars that year. The three points to be plotted are:

$$(1, 10,000), (2, 15,000), \text{ and } (3, -5000)$$

Notice that in the third year, the company's loss is represented by a negative profit. Because the y-coordinates are large, we use 5000 as the scale on the y-axis. Then we plot the points on the following coordinate plane.

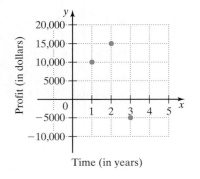

PRACTICE 6

The table shows a child's weight (in pounds) recorded at each of his annual physical examinations from birth to age 4. Plot the information on the coordinate plane.

Age, a	0	1	2	3	4
Weight, w	9	21	28	33	42

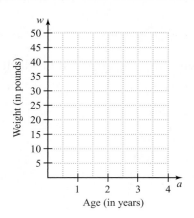

We end this discussion of plotting points with a final comment about variables. On a coordinate plane, the first coordinate of each point is a value of one quantity (or variable), and the second coordinate is a value of another quantity. For instance, in Example 6 we considered the profit that a company makes at various times. One variable represents time and the other variable the profit. Notice that the profit made by the company depends on the time rather than the other way around. So we refer to time as the *independent* variable and the profit as the *dependent* variable. It is customary when plotting points to assign the independent variable to the horizontal axis and the dependent variable to the vertical axis. However, as shown in Practice 6, letters other than *x* and *y* can be used to represent quantities.

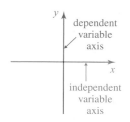

Interpreting Graphs on the Coordinate Plane

Frequently, we plot data points to make a prediction, that is, to extend the observed pattern in order to estimate the missing data. We do so by connecting the plotted points with line segments or by drawing one line through the plotted points. We can use coordinates from these lines or line segments to help us make predictions.

EXAMPLE 7

The following table indicates the size of the U.S. population according to the census in the past five decades.

Year	1960	1970	1980	1990	2000
U.S. Population (in millions)	179	203	227	249	281

(*Source:* U.S. Bureau of the Census, *Statistical Abstract of the United States*, 2002)

a. Plot this information on a coordinate plane.

b. Describe any trends that you observe.

Solution

a. After labeling the axes, we plot the points. Since all numbers are positive, we only show the first quadrant. We plot the first coordinate from the top row of the given table against the horizontal axis, and plot the second coordinate from the bottom row against the vertical axis. The points are as shown.

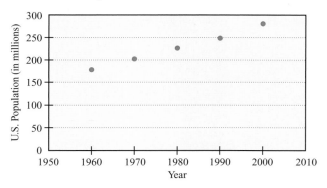

PRACTICE 7

The following table shows the amount of total federal budget outlays in three years.

Year	1980	1990	2000
Federal Budget Outlays (in billions)	$591	$1253	$1790

(*Source:* U.S. Office of Management and Budget, 2003)

a. Plot this information on the following coordinate plane.

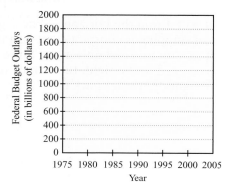

b. Describe any trends that you observe.

b. The points show the population is increasing over time.
They lie approximately in a straight line. We can use this
line to estimate the size of the population in years for
which we do not have definite information. For instance,
we might estimate the population to be about 150 million
in 1950 and about 310 million in 2010.

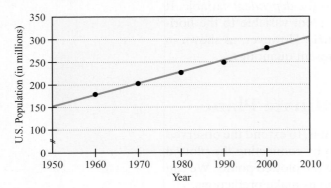

EXAMPLE 8

The following graph shows the cost that a shipping company
charges to send a package, depending on the package's weight.

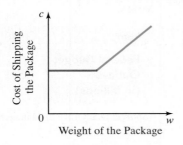

Write a brief story describing the displayed relationship. What
business practice does the horizontal line segment reflect?

Solution From the horizontal line segment, we see that the
cost of shipping is constant (that is, a flat rate) for lighter
packages up to a certain weight. Since the slanted line segment
goes upward to the right, the cost of shipping increases with
the weight of heavier packages.

PRACTICE 8

The following graph shows the number
of times per minute that a runner's
heart beats. Describe in a sentence or
two the pattern you observe, and write
a brief story based on this pattern.

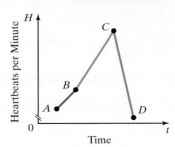

Mathematically Speaking

Fill in each blank with the most appropriate term or phrase from the given list.

above	horizontal	below
dependent	x-axis	independent
origin	coordinate	center
ordered pair	vertical	y-axis
bar graph	line graph	

1. On a(n) _____, quantities are represented by thin, parallel rectangles.

2. On a(n) _____, quantities are represented as points connected by straight-line segments.

3. A coordinate plane has two number lines that intersect at a point called the _____.

4. The horizontal number line on a coordinate plane is usually referred to as the _____.

5. Each point in a coordinate plane is represented by a pair of numbers called a(n) _____.

6. The y-coordinate of a point on a coordinate plane represents a(n) _____ distance.

7. Points in Quadrant III are to the left of the y-axis and _____ the x-axis.

8. The _____ variable is usually assigned to the horizontal axis.

On the coordinate plane to the right, plot the points with the given coordinates.

9. $A(2, 1)$
   $B(-3, -4)$
   $C(0, -1)$
   $D(-1, 0)$
   $E(3, -2)$
   $F(-2, 3)$

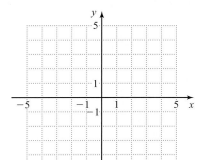

10. $A(2, 0)$
   $B(4, -1)$
   $C(-1, 4)$
   $D(0, -2)$
   $E(1, 2)$
   $F(-2, -1)$

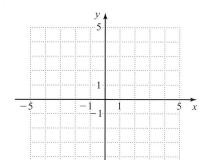

Next to each point, write its coordinates.

11.

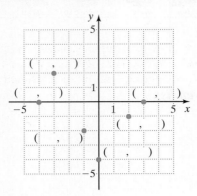

12.

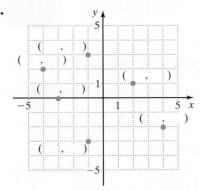

Identify the quadrant in which each point is located.

13. $(-2, -3)$　　　　**14.** $(-9, 5)$　　　　**15.** $\left(3, -\dfrac{1}{2}\right)$　　　　**16.** $\left(3\dfrac{1}{2}, -8\right)$

17. $(65, 11)$　　　　**18.** $(-13, -24)$　　　　**19.** $(-5.1, 4)$　　　　**20.** $(8, 6.2)$

Mixed Practice

Solve.

21. Plot the points with the given coordinates on the coordinate plane.

$A(3, 2)$　$B(4, 0)$　$C(-4, -1)$　$D(-2, 3)$　$E(0, -3)$

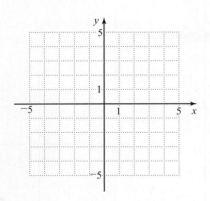

22. Write the coordinates next to each point.

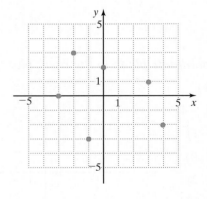

Identify the quadrant in which each point is located.

23. $(-3, 7)$　　　　**24.** $\left(6, -\dfrac{3}{4}\right)$　　　　**25.** $(27, 39)$　　　　**26.** $(-4.2, -3.8)$

Applications

Solve.

27. In chemistry, the pH scale measures how acidic or basic a solution is. A solution with a pH of 7 is considered neutral. Solutions with a pH less than 7 are acids, and solutions with a pH greater than 7 are bases. The graph below shows the pH of various solutions.

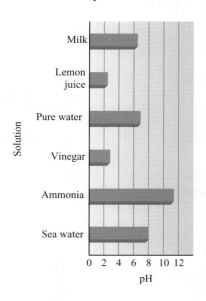

a. Which of the solutions are acids?

b. Approximate the pH of sea water.

c. Which solution is neutral?

28. The following graph shows the percent change in population between the years 2000 and 2020 projected for young Americans by age group and by race/ethnicity:

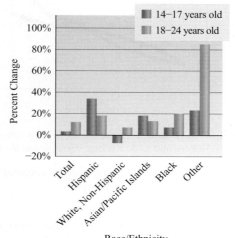

Race/Ethnicity
(*Source:* The National Center of Educational Statistics)

a. Estimate the predicted percent increase between the years 2000 and 2020 in the total population of 14–17-year-olds.

b. Among Hispanics, is a higher percent change predicted for 14–17-year-olds or for 18–24-year-olds?

c. According to the projection, which of the populations will have the greatest percent increase?

29. The graph shows the mid-year estimated number of cell phone subscribers (in millions) for the years 2000 through 2006. (*Source:* CTIA)

a. About how many subscribers were there in the year 2000?

b. In what year did the number of subscribers reach 150 million?

c. Describe the trend shown in the graph.

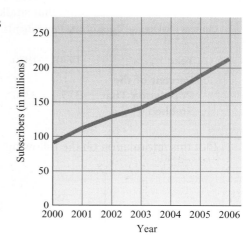

30. A human child and a chimp were raised together. Scientists graphed the number of words that the child and the chimp understood at different ages.

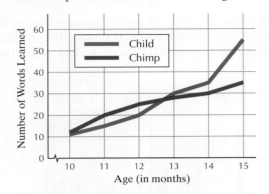

(*Source*: A. H. Kritz, *Problem solving in the Sciences*: W. N. Kellogg and L. A. Kellogg. *The Ape and the Child*)

a. At about what age was the child's vocabulary first better than that of the chimp?

b. At age 15 months, about how many more words did the child know than the chimp?

31. College students coded *A*, *B*, *C*, and *D* took placement tests in mathematics and in English. The following coordinate plane displays their scores.

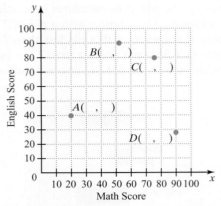

a. Estimate the coordinates of the plotted points.

b. Which students scored higher in English than in mathematics?

32. Suppose that a financier owns shares of stock in three companies—Dearborn, Inc. (*D*), Ellsworth Products (*E*), and Fairfield Publications (*F*). On the following coordinate plane, the *x*-value of a point represents the change in value of a share of the indicated stock from the previous day, and the *y*-value stands for the number of shares of that stock.

a. Name the coordinates of the plotted points.

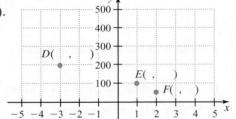

b. For each point, explain the significance of the product of the point's coordinates.

33. The following table gives the percent of the U.S. adult population in various years that smoked. (*Source*: U.S. Centers for Disease Control and Prevention)

Year	1970	1980	1990	2000
Percent of the Population That Smoked	37%	33%	26%	23%

Plot this information on the following coordinate plane.

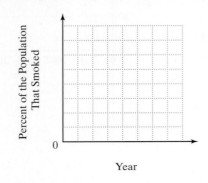

34. The number of electoral votes cast for the winning candidate in recent presidential elections is displayed in the following table. (*Source*: *The World Almanac and Book of Facts 2007*)

Year	1992	1996	2000	2004
Electoral Votes	370	379	271	286

Plot this information in the coordinate plane shown.

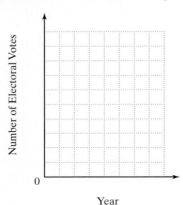

35. A chemist conducts an experiment to measure the melting and boiling points of four substances, as indicated in the following table.

Substance	Symbol	Melting Point (°C)	Boiling Point (°C)
Chlorine	Cl	−101	−35
Oxygen	O	−218	−183
Bromine	Br	−7	59
Phosphorous	P	44	280

On the following coordinate plane, an *x*-value represents a substance's melting point and a *y*-value stands for its boiling point, both in degrees Celsius.

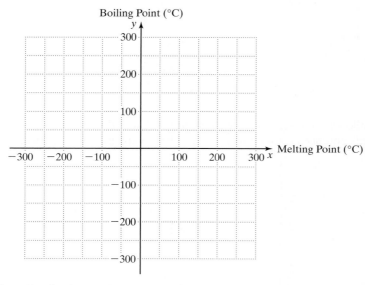

a. Plot points for the four substances. Label each point with the appropriate substance symbol.

b. For each point, which of its coordinates is larger—the *x*-value or the *y*-value? In a sentence, explain this pattern.

36. Meteorologists use the windchill index to determine the windchill temperature (how cold it feels outside) relative to the actual temperature when the wind speed is considered. The following table shows the actual temperatures in degrees Fahrenheit and the related windchill temperatures when the wind speed is 5 mph.

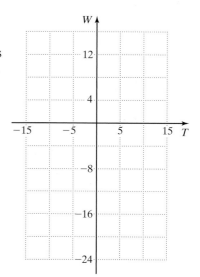

Actual Temperature, *T*	−10	−5	0	5	10	15
Windchill Temperature, *W*	−22	−16	−11	−5	1	7

a. Plot points (*T*, *W*) on the given coordinate plane.

b. For the plotted points, describe the pattern that you observe.

37. On the following coordinate plane, a *y*-coordinate stands for the number of senators from a state. The corresponding *x*-coordinate represents that state's population according to a recent U.S. census. Describe the pattern that you observe.

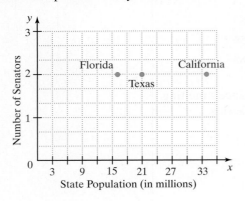

38. Last year's daily closing values (in dollars) of a share of a technology stock are plotted on the following graph. What story is this graph telling?

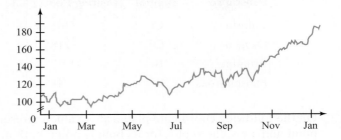

39. A child walks away from a wall, stands still, and then approaches the wall. In a couple of sentences, explain which of the following graphs could describe this motion.

a.

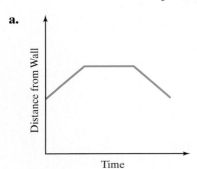

b.

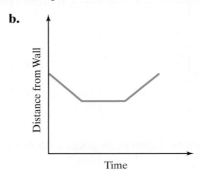

c.

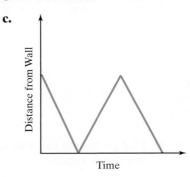

40. The following graph shows the temperature of a patient on a particular day. Describe the overall pattern you observe in the patient's temperature over the time period.

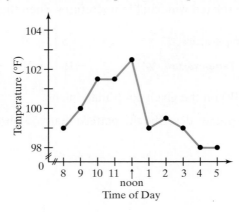

● *Check your answers on page A-8.*

MINDSTRETCHERS

Writing

1. Many situations involve using a coordinate system to identify positions. Two such situations are given below.

- a chessboard
- an atlas map

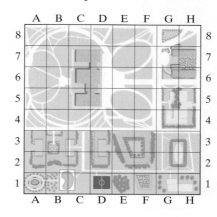

a. Explain to what extent a chessboard and an atlas map are coordinate systems.

b. Identify some other examples of coordinate systems in everyday life.

Mathematical Reasoning

2. Water is poured at a constant rate into each of the containers shown. For each container, sketch a graph that shows the height of the water in the container over time.

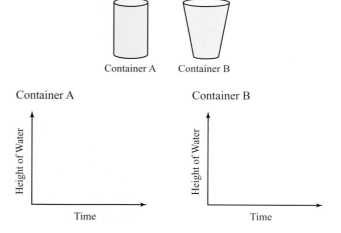

Group Work

3. Two points are plotted on a coordinate plane. Discuss with a partner what is special about a third point whose x-coordinate is the average of the first two x-coordinates, and whose y-coordinate is the average of the first two y-coordinates.

CULTURAL NOTE

It was the seventeenth-century French mathematician and philosopher René Descartes (pronounced day-KART) who developed the concepts that underlie graphing. The story goes that one morning Descartes, who liked to stay in bed and meditate, began to eye a fly crawling on his bedroom ceiling. In a flash of insight, he realized that it was possible to express mathematically the fly's position in terms of its distance to the two adjacent walls.

3.2   Slope of a Line

In the previous section, we discussed points on a coordinate plane. Now, let's look at lines connected by points. Lines have a key characteristic, namely their slopes. In this section, we focus on the slope of a line and its relationship to the corresponding equation.

Slope

On an airplane, would you rather glide downward gradually or drop like a stone? Would you rather ski down a run that drops precipitously or ski across a gently inclined snowfield? These questions relate to *slope*, the extent to which a line is slanted. In other words, slope measures a line's steepness.

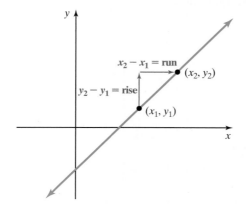

Slope, also called **rate of change**, is an important concept in the study of graphing. Examining the slope of a line can tell us if the quantity being graphed increases or decreases, as well as how fast the quantity is changing. For example, in one application, the slope of a line can represent the speed of a moving object. In another application, the slope can stand for the rate at which a share of stock is changing in value.

To understand exactly what slope means, let's suppose that a straight line on a coordinate plane passes through two arbitrary points. We can call the coordinates of the first point (x_1, y_1), read "x sub 1" and "y sub 1," and the coordinates of the second point (x_2, y_2). These coordinates are written with *subscripts* in order to distinguish them from one another. We can plot these two points on a coordinate plane, and then graph the line passing through them.

We usually represent the slope of a line by the letter m and define slope to be the ratio of the change in the y-values to the change in the x-values. Using the coordinates of the points (x_1, y_1) and (x_2, y_2) shown in the graph on the previous page, we have the formula

$$m = \frac{\text{change in } y\text{-values}}{\text{change in } x\text{-values}} = \frac{y_2 - y_1}{x_2 - x_1}, \quad \text{where } x_1 \neq x_2.$$

In this formula, the numerator of the fraction is the vertical change called the *rise* and the denominator is the horizontal change called the *run*. So another way of writing the formula for slope is $m = \dfrac{\text{rise}}{\text{run}}$.

Definition

The **slope** m of a line passing through the points (x_1, y_1) and (x_2, y_2) is defined to be

$$m = \frac{y_2 - y_1}{x_2 - x_1}, \quad \text{where } x_1 \neq x_2.$$

Can you explain why in the definition of slope, x_1 and x_2 must not be equal?

Note that when using the formula for slope, it does not matter which point is chosen for (x_1, y_1) and which point for (x_2, y_2) as long as the order of subtraction of the coordinates is the same in both the numerator and denominator.

EXAMPLE 1	PRACTICE 1
Find the slope of the line that passes through the points $(2, 1)$ and $(4, 2)$. Plot the points and then sketch the line.	Find the slope of a line that contains the points $(1, 2)$ and $(4, 3)$. Plot the points and then sketch the line.

EXAMPLE 1

Find the slope of the line that passes through the points $(2, 1)$ and $(4, 2)$. Plot the points and then sketch the line.

Solution Let $(2, 1)$ stand for (x_1, y_1) and $(4, 2)$ for (x_2, y_2).

$$\begin{matrix} (2, 1) & (4, 2) \\ \uparrow \; \uparrow & \uparrow \; \uparrow \\ x_1 \; y_1 & x_2 \; y_2 \end{matrix}$$

Substituting into the formula for slope, we get:

$$m = \frac{y_2 - y_1}{x_2 - x_1} = \frac{2 - 1}{4 - 2} = \frac{1}{2}$$

Now, let's plot $(2, 1)$ and $(4, 2)$ and then sketch the line passing through them.

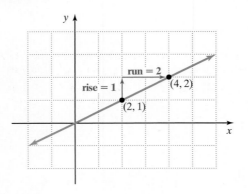

PRACTICE 1

Find the slope of a line that contains the points $(1, 2)$ and $(4, 3)$. Plot the points and then sketch the line.

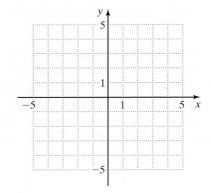

We can also find the slope of a line using its graph. From $(2, 1)$ and $(4, 2)$, we see that the change in y-values (the rise) is $2 - 1$, or 1. The change in x-values (the run) is $4 - 2$, or 2. So $m = \dfrac{\text{rise}}{\text{run}} = \dfrac{1}{2}$. Therefore, we get the same answer whether we

use the formula $m = \dfrac{y_2 - y_1}{x_2 - x_1}$ or $m = \dfrac{\text{rise}}{\text{run}}$.

A line rising to the right as shown in Example 1 has a *positive* slope. We say that such a line is *increasing* because as the x-values get larger, the corresponding y-values also get larger.

EXAMPLE 2	**PRACTICE 2**

Sketch the line passing through the points $(-3, 1)$ and $(2, -2)$. Find the slope.

Solution First, we plot the points $(-3, 1)$ and $(2, -2)$. Then we draw a line passing through them.

Sketch the line that contains the points $(-2, 1)$ and $(3, -5)$. Find the slope.

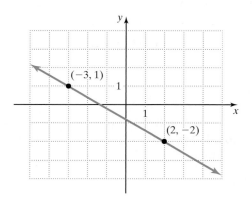
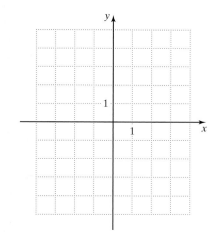

Next, find the slope.

$$
\begin{array}{cc}
(-3, 1) & (2, -2) \\
\uparrow\ \uparrow & \uparrow\quad \uparrow \\
x_1\ y_1 & x_2\quad y_2
\end{array}
$$

$$m = \frac{y_2 - y_1}{x_2 - x_1} = \frac{-2 - 1}{2 - (-3)} = \frac{-3}{5} = -\frac{3}{5}$$

A line falling to the right as shown in Example 2 has a *negative* slope. We say that such a line is *decreasing* because as the x-values get larger, the corresponding y-values get smaller.

EXAMPLE 3

Find the slope of a line that passes through the points (7, 5) and (−1, 5). Plot the points and then sketch the line.

Solution First, find the slope of the line.

$$\underset{x_1\ y_1}{(7, 5)} \qquad \underset{x_2\ y_2}{(-1, 5)}$$

$$m = \frac{y_2 - y_1}{x_2 - x_1} = \frac{5 - 5}{-1 - 7} = \frac{0}{-8} = 0$$

So the slope of this line is 0.

Next, we plot the points and then sketch the line passing through them.

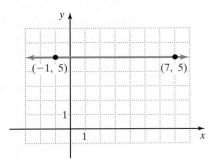

PRACTICE 3

On the following coordinate plane, plot the points (2, −1) and (6, −1). Sketch the line and then compute its slope.

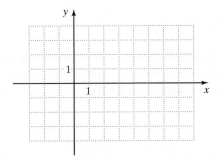

When the slope of a line is 0, its graph is a *horizontal* line as shown in Example 3. All points on a horizontal line have the same *y*-coordinate, that is, the *y*-values are constant for all *x*-values.

EXAMPLE 4

What is the slope of the line pictured on the following coordinate plane?

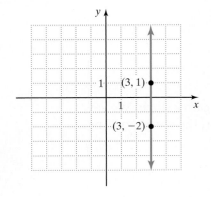

PRACTICE 4

Find the slope of the line that passes through the points (−2, 7) and (−2, 0).

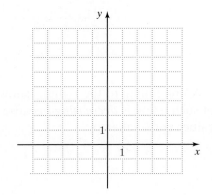

Solution

$$(3, -2) \qquad (3, 1)$$
$$\uparrow \quad \uparrow \qquad \uparrow \quad \uparrow$$
$$x_1 \quad y_1 \qquad x_2 \quad y_2$$

$$m = \frac{y_2 - y_1}{x_2 - x_1} = \frac{1 - (-2)}{3 - 3} = \frac{3}{0}$$

Since division by 0 is undefined, the slope of this line is undefined.

When the slope of a line is undefined, its graph is a *vertical* line as shown in Example 4. All points on a vertical line have the same *x*-coordinate, that is, the *x*-values are constant for all *y*-values.

As we have seen in Examples 1 through 4, the sign of the slope of a line tells us a lot about the line. As we continue graphing lines, it will be helpful to keep in mind the following graphs.

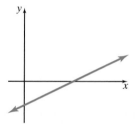

Positive *m*
The line slants upward
from left to right.

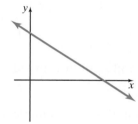

Negative *m*
The line slants downward
from left to right.

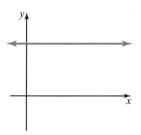

Zero *m*
The line is horizontal.

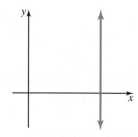

Undefined *m*
The line is vertical.

In the next example, we graph two lines on a coordinate plane.

EXAMPLE 5

Calculate the slopes for the lines shown.

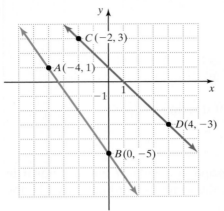

Solution For $\overleftrightarrow{AB}$ passing through $A(-4, 1)$ and $B(0, -5)$, the slope is:

$$m = \frac{y_2 - y_1}{x_2 - x_1} = \frac{1 - (-5)}{(-4) - 0} = \frac{1 + 5}{-4} = \frac{6}{-4} = -\frac{3}{2}$$

For $\overleftrightarrow{CD}$ passing through $C(-2, 3)$ and $D(4, -3)$, the slope is:

$$m = \frac{y_2 - y_1}{x_2 - x_1} = \frac{3 - (-3)}{(-2) - 4} = \frac{3 + 3}{(-2) - 4} = \frac{6}{-6} = -1$$

Note that both lines have negative slopes and slant downward from left to right.

PRACTICE 5

Compute the slopes for the lines shown.

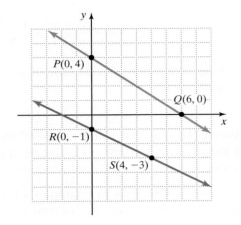

As shown in the next example, how a line slants often helps us to interpret the information given in the graph.

EXAMPLE 6

The following graph shows the amount of money that your dental insurance reimburses you, depending on the amount of your dental bill. Is the slope of the graphed line positive or negative? Explain how you know. What does this mean in terms of insurance reimbursement?

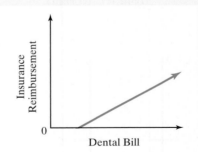

Solution Since the graphed line slants upward from left to right, its slope is positive. According to this graph, larger x-values correspond to larger y-values. So, your dental insurance reimburses you more for larger dental bills.

PRACTICE 6

A doctor is trying to help eliminate an epidemic. Explain, in terms of slope, which scenario would be the most desirable.

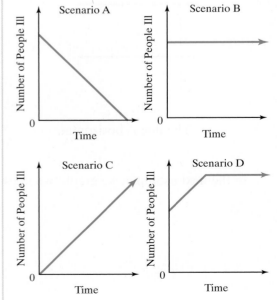

We have already graphed a line by plotting two points and drawing the line passing through them. Now, let's look at graphing a line when given the slope of the line and a point on the line.

EXAMPLE 7	PRACTICE 7

EXAMPLE 7

The slope of a line that passes through the point (2, 5) is 3. Graph the line.

Solution The line in question passes through the point (2, 5). But there are many such lines—which is the right one? We use the slope 3 to find a second point through which the line also passes. Since 3 can be written as $\frac{3}{1}$, we have

$$\text{slope} = \frac{\text{rise}}{\text{run}} = \frac{3}{1}.$$

We first plot the point (2, 5). Starting at (2, 5), we move 3 units up (for a rise of 3) and then 1 unit to the right (for a run of 1). We arrive at the point (3, 8). Finally, we sketch the line passing through the points (2, 5) and (3, 8), as shown in the graph on the left below.

Since $\frac{3}{1} = \frac{-3}{-1}$, we could have started at (2, 5) and moved down 3 units (for a rise of -3) and then 1 unit to the left (for a run of -1). In this case, we would arrive at (1, 2), which is another point on the same line, as shown in the graph below.

PRACTICE 7

Graph the line with slope 4 that passes through the point (1, -2).

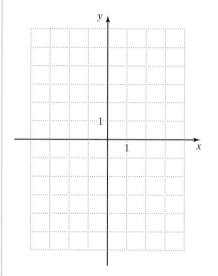

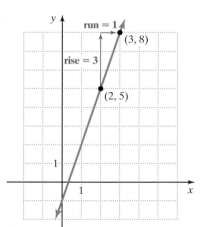

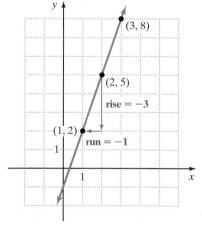

Can you find other points on this line? Explain.

| EXAMPLE 8 | PRACTICE 8 |

Suppose that a family on vacation is driving out of town at a constant speed, and that at 2 o'clock they have traveled 110 mi. By 6 o'clock, they have traveled 330 mi.

a. On a coordinate plane, label the axes and then plot the appropriate points.

b. Compute the slope of the line passing through the points.

c. Interpret the meaning of the slope in this situation.

Solution

a. Label the axes on the coordinate plane. Then plot the points (2, 110) and (6, 330).

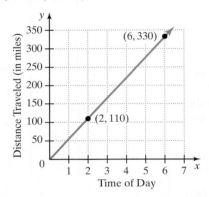

b. The slope of the line through the two points is:

$$m = \frac{y_2 - y_1}{x_2 - x_1}$$

$$= \frac{330 - 110}{6 - 2}$$

$$= \frac{220}{4}$$

$$= 55$$

c. Here the slope is the change in distance divided by the change in time. In other words, the slope is the average speed the family traveled, which is 55 mph.

An auto rental company advertises that a certain car rents for $80 with a weekly mileage of 100 mi and $90 when the weekly mileage is 200 mi.

a. On a coordinate plane, label the axes and then plot the appropriate points.

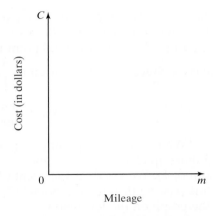

b. Compute the slope of the line that passes through the points.

c. Interpret the meaning of the slope in this situation.

Parallel and Perpendicular Lines

By examining the slopes of straight lines on a coordinate plane, we can solve problems that require us to determine if

- two given lines are parallel or
- two given lines are perpendicular.

Let's consider parallel lines first.

Since the slope of a line measures its slant, lines with equal slopes are parallel. So on the coordinate plane shown below, if we knew the coordinates of points P, Q, R, and S, we could verify that $\overleftrightarrow{PQ}$ and $\overleftrightarrow{RS}$ are parallel by computing their slopes and then checking that they are equal.

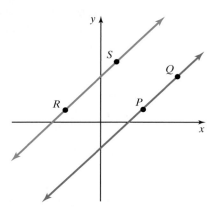

Definition

Two nonvertical lines are **parallel** if and only if their slopes are equal. That is, if the slopes are m_1 and m_2, then $m_1 = m_2$.

EXAMPLE 9

Consider points $P(0, 0)$, $Q(-2, -5)$, $R(0, 5)$, and $S(-2, 0)$. Are $\overleftrightarrow{PQ}$ and $\overleftrightarrow{RS}$ parallel?

Solution Let's check if the slopes of $\overleftrightarrow{PQ}$ and $\overleftrightarrow{RS}$ are equal. The slope of $\overleftrightarrow{PQ}$ is:

$$m = \frac{y_2 - y_1}{x_2 - x_1} = \frac{0 - (-5)}{0 - (-2)} = \frac{0 + 5}{0 + 2} = \frac{5}{2}$$

The slope of $\overleftrightarrow{RS}$ is:

$$m = \frac{y_2 - y_1}{x_2 - x_1} = \frac{5 - 0}{0 - (-2)} = \frac{5}{0 + 2} = \frac{5}{2}$$

Since the slopes of $\overleftrightarrow{PQ}$ and $\overleftrightarrow{RS}$ are equal, $\overleftrightarrow{PQ}$ and $\overleftrightarrow{RS}$ are parallel.

PRACTICE 9

Decide whether $\overleftrightarrow{EF}$ and $\overleftrightarrow{GH}$ are parallel, given points $E(0, 4)$, $F(4, -1)$, $G(0, 8)$, and $H(8, -2)$.

EXAMPLE 10

A pediatric nurse kept track of the weights of two children who are twins. Use the graph to answer the following questions.

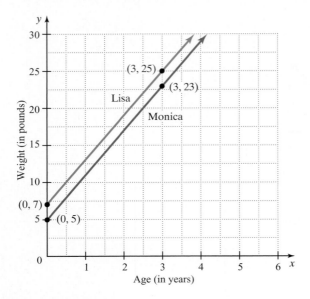

a. Which twin was heavier at birth?

b. Are the two lines parallel?

c. Which twin grew at a faster rate?

d. Could this graph be used to project the weight of the twins at age 8? Explain.

Solution

a. At birth, Lisa weighed 7 lb and Monica weighed 5 lb. So Lisa was heavier.

b. To determine if the two lines are parallel, we begin by computing the slope of the line passing through the points $(0, 7)$ and $(3, 25)$.
$$m = \frac{y_2 - y_1}{x_2 - x_1} = \frac{7 - 25}{0 - 3} = \frac{-18}{-3} = 6$$

Now, we compute the slope for the line passing through the $(0, 5)$ and $(3, 23)$.
$$m = \frac{y_2 - y_1}{x_2 - y_1} = \frac{5 - 23}{0 - 3} = \frac{-18}{-3} = 6$$

Since the slopes of the two lines are equal, the graphed lines are parallel.

c. Since the two lines are parallel, the twins grew at the same rate.

d. If we extend the x- and y-axes, we could determine the weight of each child at age 8 by reading the corresponding y-coordinate for x equal to 8.

PRACTICE 10

The graph shown gives the income of a computer lab technician and a Web site designer as related to the number of years that they have been employed. Use this graph to answer the following questions.

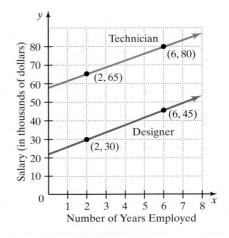

a. Are the two lines parallel?

b. Does your answer to part (a) agree with your observation of the graph? Explain.

c. Which employee's salary increased at a faster rate?

d. From the graph, estimate the starting salary of the computer lab technician.

Now, let's consider the problem of determining whether two given lines are perpendicular.

On a coordinate plane, two lines are perpendicular to one another when the product of their slopes is −1. For instance, the slope of $\overleftrightarrow{PQ}$ in the following graph is:

$$m = \frac{7 - 2}{3 - 1} = \frac{5}{2}$$

The slope of $\overleftrightarrow{QR}$ is

$$m = \frac{2 - 0}{1 - 6} = \frac{2}{-5} = -\frac{2}{5}$$

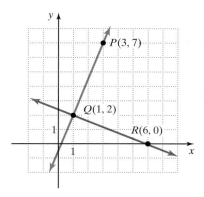

The product of these slopes is:

$$m = \left(\frac{5}{2}\right)\left(-\frac{2}{5}\right) = -1$$

We say that the slopes are *negative reciprocals* of each other. Therefore, the two lines must be perpendicular to one another.

Definition

Two nonvertical lines are **perpendicular** if and only if the product of their slopes is −1. That is, if the slopes are m_1 and m_2, then $m_1 \cdot m_2 = -1$.

EXAMPLE 11	PRACTICE 11
Determine if $\overleftrightarrow{AB}$ and $\overleftrightarrow{BC}$ shown in the graph are perpendicular to one another.	Consider points $A(1, 3)$, $B(2, 5)$, and $C(-1, 2)$. Decide if $\overleftrightarrow{AB}$ is perpendicular to $\overleftrightarrow{AC}$.

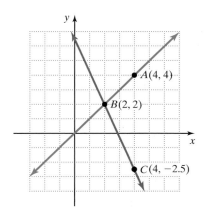

Solution First, let's compute the slopes of the two lines in question. The slope of $\overleftrightarrow{AB}$ is:

$$m = \frac{y_2 - y_1}{x_2 - x_1} = \frac{4 - 2}{4 - 2} = \frac{2}{2} = 1$$

The slope of $\overleftrightarrow{BC}$ is:

$$m = \frac{y_2 - y_1}{x_2 - x_1} = \frac{2 - (-2.5)}{2 - 4} = \frac{2 + 2.5}{2 - 4} = \frac{4.5}{-2} = -2.25$$

To check if $\overleftrightarrow{AB}$ and $\overleftrightarrow{BC}$ are perpendicular, we find the product of their slopes:

$$(-2.25)(1) = -2.25$$

Since this product is not equal to -1, the lines are not perpendicular to one another.

EXAMPLE 12	PRACTICE 12

On the coordinate plane shown, x-values represent streets and y-values represent avenues. Suppose that a road is to be constructed running straight from 4th Street and 3rd Avenue to 9th Street and 9th Avenue. A second road will run from 2nd Street and 10th Avenue to 8th Street and 5th Avenue. Will the roads meet at right angles?

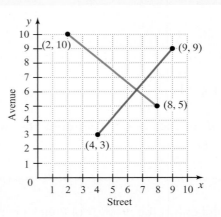

Consider the square 6 units on each side shown in the following diagram. By examining their slopes, determine whether the diagonals of the square are perpendicular to one another.

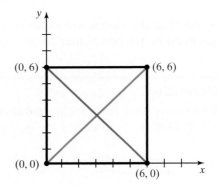

Solution Let's find the slopes of the two roads. The slope of the road from 9th Street and 9th Avenue to 4th Street and 3rd Avenue is:

$$m = \frac{y_2 - y_1}{x_2 - x_1} = \frac{9 - 3}{9 - 4} = \frac{6}{5}$$

The slope of the road from 2nd Street and 10th Avenue to 8th Street and 5th Avenue is:

$$m = \frac{y_2 - y_1}{x_2 - x_1} = \frac{10 - 5}{2 - 8} = \frac{5}{-6} = -\frac{5}{6}$$

The product of these slopes is:

$$\left(\frac{6}{5}\right)\left(-\frac{5}{6}\right) = -1$$

Therefore, the roads will be perpendicular to one another.

3.2   Exercises   FOR EXTRA HELP   *MyMathLab*   PRACTICE WATCH DOWNLOAD READ REVIEW

Mathematically Speaking

Fill in each blank with the most appropriate term or phrase from the given list.

y-coordinate	parallel	negative
positive	vertical	x-coordinate
perpendicular	rate of change	run
horizontal	rise	

1. The slope of a line is also called its _____.

2. In the slope formula, the vertical change is called the _____.

3. A line with _____ slope is decreasing.

4. If all points on a line have the same _____, then the line is vertical.

5. A line with zero slope is _____.

6. A line with undefined slope is _____.

7. Two nonvertical lines are _____ if and only if their slopes are equal.

8. Two nonvertical lines are _____ if and only if the product of their slopes is -1.

Compute the slope m of the line that passes through the given points. Plot these points on the coordinate plane, and sketch the line that passes through them.

9. $(2, 3)$ and $(-2, 0)$, $m =$

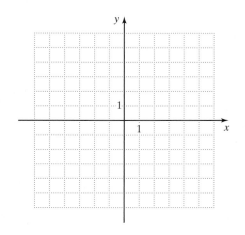

10. $(0, 0)$ and $(-2, 5)$, $m =$

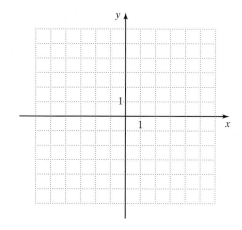

11. $(6, -4)$ and $(6, 1)$, $m =$ **12.** $(1, 1)$ and $(1, -3)$, $m =$ **13.** $(-2, 1)$ and $(3, -1)$, $m =$

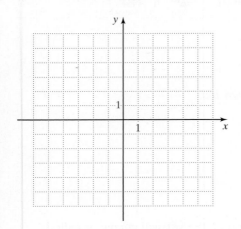 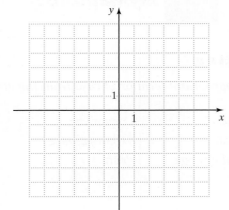 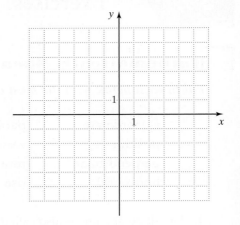

14. $(-1, 4)$ and $(0, 5)$, $m =$ **15.** $(-1, -4)$ and $(3, -4)$, $m =$ **16.** $(3, 0)$ and $(5, 0)$, $m =$

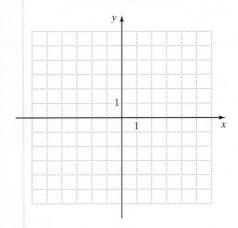 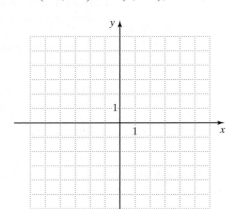 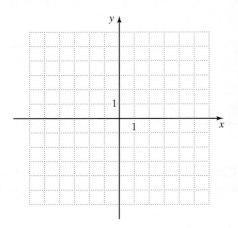

17. $(0.5, 0)$ and $(0, 3.5)$, $m =$ **18.** $(4, 4.5)$ and $(1, 2.5)$, $m =$

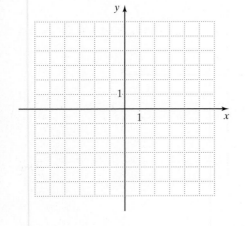 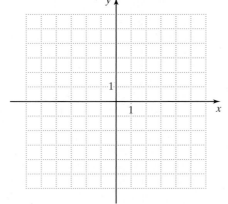

On each graph, calculate the slopes for the lines shown.

19.

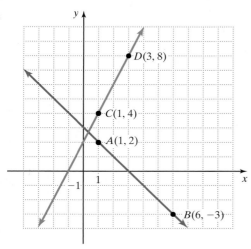

20.

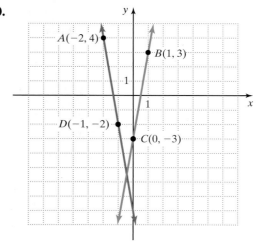

Graph the line on the coordinate plane using the given information.

21. Passes through $(-1, 1)$ and $m = \dfrac{1}{2}$

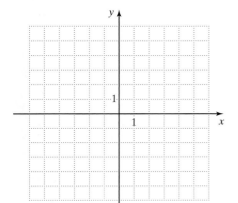

22. Passes through $(2, 5)$ and $m = -\dfrac{4}{3}$

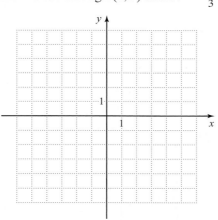

23. Passes through $(2, 5)$ and $m = 4$

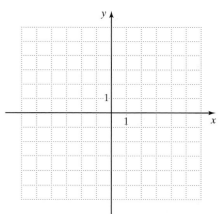

24. Passes through $(0, -6)$ and $m = 0$

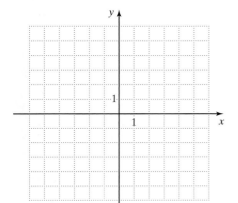

25. Passes through $(1, 5)$ and $m = -3$

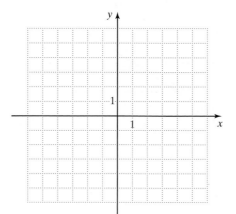

26. Passes through $(0, 0)$ and $m = 6$

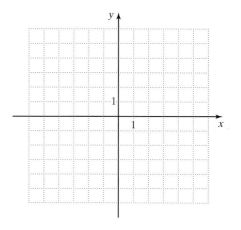

27. Passes through $(-4, 0)$ and the slope is undefined

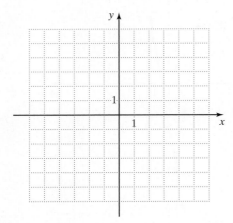

28. Passes through $(-6, 2)$ and the slope is $-\dfrac{3}{5}$

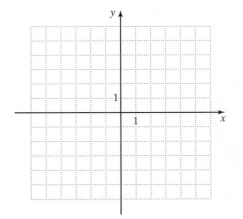

Indicate whether the slope of each graph is positive, negative, zero, or undefined. Then state whether the line is horizontal, vertical, or neither.

29.

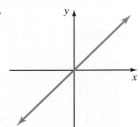

30.

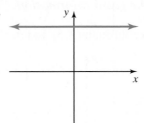

31.

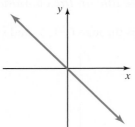

32.

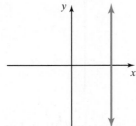

33.

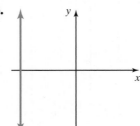

34.

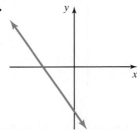

35.

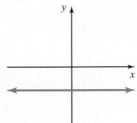

36.

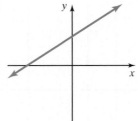

Determine whether $\overleftrightarrow{PQ}$ and $\overleftrightarrow{RS}$ are parallel or perpendicular.

37.

	P	Q	R	S
a.	$(0, -1)$	$(1, 3)$	$(5, 0)$	$(7, 8)$
b.	$(9, 1)$	$(7, 4)$	$(0, 0)$	$(6, 4)$

38.

	P	Q	R	S
a.	$(3, 3)$	$(7, 7)$	$(-5, 5)$	$(2, -2)$
b.	$(8, 0)$	$(0, 4)$	$(0, -4)$	$(-12, 2)$

Mixed Practice

Indicate whether the slope of each graph is positive, negative, zero, or undefined. Then state whether the line is horizontal, vertical, or neither.

39.

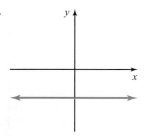

40.

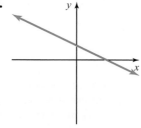

41. Graph the line on the coordinate plane if the line passes through $(-2, 3)$ and $m = 4$.

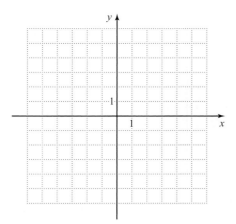

42. Calculate the slopes for the lines shown in the graph.

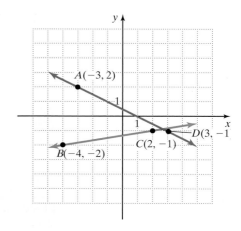

43. Determine whether $\overleftrightarrow{AB}$ and $\overleftrightarrow{CD}$ are parallel or perpendicular.

	A	B	C	D
a.	$(-3, 2)$	$(5, -2)$	$(1, -4)$	$(5, 4)$
b.	$(-1, 5)$	$(5, -3)$	$(2, 2)$	$(5, -2)$

Compute the slope m of the line that passes through the given points. Plot these points on the coordinate plane, and sketch the line that passes through them.

44. $(-4, -5)$ and $(2, -1)$, $m =$

45. $(-3, -2)$ and $(-3, 4)$, $m =$

46. $(-4, 4.5)$ and $(-2, -1.5)$, $m =$

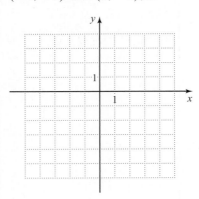
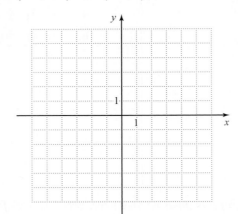
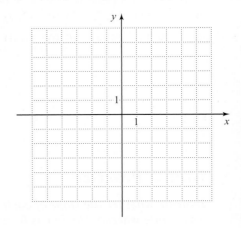

Applications

Solve.

47. A chemist conducts an experiment on the gas contained in a sealed tube. The experiment is to heat the gas and then to measure the resulting pressure in the tube. In the lab manual, points are plotted and the line is sketched to show the gas pressure for various temperatures.

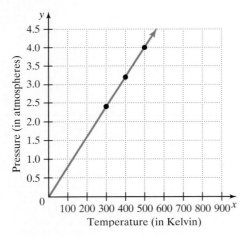

a. Is the slope of this line positive, negative, zero, or undefined?

b. In a sentence, explain the significance of the answer to part (a) in terms of temperature and pressure.

48. To reduce their taxes, many businesses use the *straight-line method of depreciation* to estimate the change in the value over time of equipment that they own. The graph shows the value of equipment owned from the time of purchase to 7 yr later.

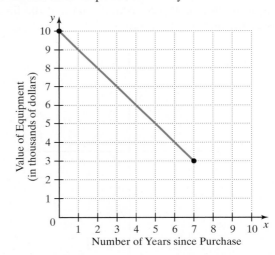

a. Is the slope of this line positive, negative, zero, or undefined?

b. In a sentence or two, explain the significance of the answer to part (a) in terms of the value of the equipment over time.

49. Two motorcyclists leave at the same time, racing down a road. Consider the lines in the graph that show the distance traveled by each motorcycle at various times.

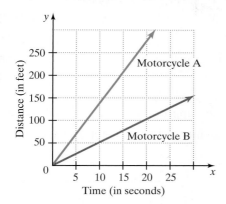

a. Which motorcycle first travels 100 ft?

b. Which motorcycle is traveling more slowly?

c. Explain what the slopes mean in this situation.

50. Most day-care centers charge parents additional fees for arriving late to pick up their children. The following graph shows the late fee for two day-care centers.

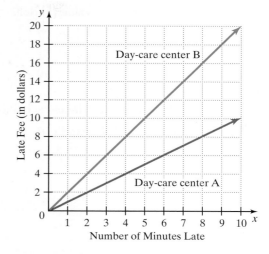

a. Which day-care center charges a higher late fee?

b. Explain what the slopes represent in this situation.

51. The following graph records the amount of garbage deposited in landfills A and B after they are opened.

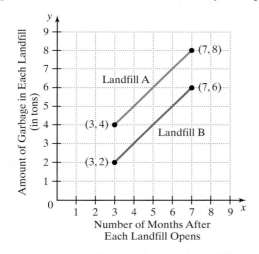

Are the garbage deposits at the two landfills growing at the same rate? Explain in a sentence or two how you know.

52. The weights of a brother and sister from age 3 yr to 7 yr are recorded in the graph. Did their weights increase at the same rate? Explain.

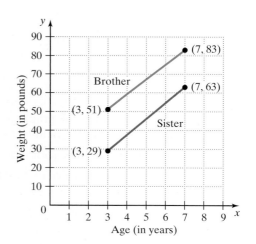

53. After a stock split, the per-share value of the stock increased, as shown in the table.

Point	Number of Days After the Split	Value (in dollars)
P	2	27
Q	4	51
R	8	79

a. Choose appropriate scales and label each axis. Plot the points and then sketch $\overleftrightarrow{PQ}$ and $\overleftrightarrow{QR}$.

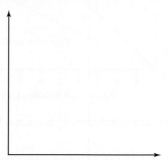

b. Determine whether the rate of increase changed over time. Explain.

54. The position of a dropped object for various times after the object is released is given in the table.

Point	Time After Release (in seconds)	Position (in feet)
A	1	−16
B	2	−64
C	3	−144

a. Choose appropriate scales and label each axis. Plot the points and then sketch $\overleftrightarrow{AB}$ and $\overleftrightarrow{BC}$.

b. Compute the slopes of $\overleftrightarrow{AB}$ and $\overleftrightarrow{BC}$.

c. Was the rate of fall for the dropped object constant throughout the experiment? Explain.

55. A hiker is at point A as shown in the graph and wants to take the shortest route through a field to reach a nearby road represented by $\overleftrightarrow{BC}$. The shortest route will be to walk perpendicular to the road. Is $\overleftrightarrow{AD}$ the shortest route? Explain.

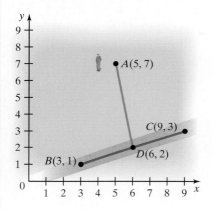

56. The coordinates of the vertices of triangle ABC are shown. Is triangle ABC a right triangle? Explain.

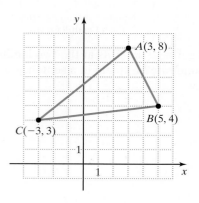

57. On a highway, a driver sets a car's cruise control for a constant speed of 55 mph.

   a. Which graph shows the distance the car travels? Using the slope of the line, explain.

   b. Which graph shows the speed of the car? Using the slope of the line, explain.

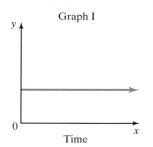

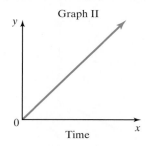

58. Suppose city leaders consider imposing an income tax on residents whose income is above a certain amount, as pictured.

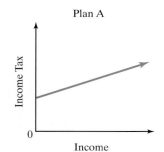

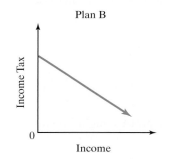

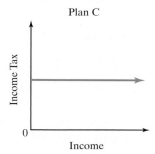

   a. Using the slopes of the lines, describe each plan.

   b. Which plan do you think is the fairest? Explain.

● *Check your answers on page A-9.*

MINDSTRETCHERS

Groupwork

1. A geoboard is a square flat surface with pegs forming a grid pattern. You can stretch rubber bands around the pegs to explore geometric questions. Pictured is a geoboard with rubber bands forming a stairway design.

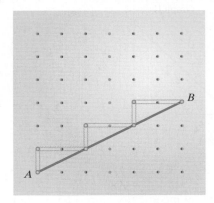

Using a geoboard or graph paper, determine whether the slope of $\overleftrightarrow{AB}$ increases or decreases if

a. the rise of each step increases by one peg.

b. the run of each step increases by one peg.

Writing

2. Describe a real-world situation that each of the following graphs might illustrate. Explain the significance of slope in the situation that you have described.

a. 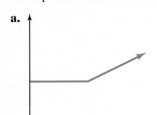   **b.** 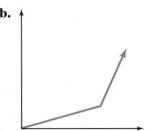   **c.**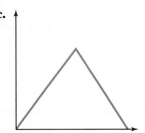

Mathematical Reasoning

3. If the slope of a line is positive, explain why any line perpendicular to it must have a negative slope.

OBJECTIVES

- To identify a given linear equation in general form
- To identify the coordinates of points that satisfy a given linear equation in two variables
- To graph a given linear equation in two variables
- To identify the *x*- and *y*-intercepts of a linear equation
- To graph horizontal and vertical lines
- To solve applied problems involving the graph of a linear equation

3.3 Graphing Linear Equations

In the first section of this chapter, we developed the idea of coordinates—an ordered pair of numbers associated with a point on the plane. In the second section, we shifted our attention from points to lines. In this section, we focus on Descartes' most startling idea—on a coordinate plane, lines (and indeed other graphs) correspond to equations. Line graphs and their associated equations, called *linear* equations, have important applications in the real world, allowing us to model many situations. In later chapters, we will discuss other types of graphs.

Solutions of a Linear Equation in Two Variables

Recall that linear equations in one variable generally have one and only one solution. For instance, the solution of $2x + 5 = 11$ is 3 because when we substitute 3 for x, the equation balances.

$$2x + 5 = 11$$
$$2 \cdot 3 + 5 \stackrel{?}{=} 11$$
$$11 = 11 \qquad \text{True}$$

So we say that $x = 3$.

We now consider linear equations in *two* variables, say $y = 2x + 5$. We can also express $y = 2x + 5$ as $-2x + y = 5$.

$$y = 2x + 5$$
$$-2x + y = 5 \qquad \text{Subtract } 2x \text{ from each side of the equation.}$$

Definition

A **linear equation in two variables**, x and y, is an equation that can be written in the *general form* $Ax + By = C$, where A, B, and C are real numbers and A and B are not both 0.

Note that in the general form of a linear equation, the two variable terms are on one side of the equation and the constant term is on the other. The equation $-2x + y = 5$, for instance, is in general form, with $A = -2$, $B = 1$, and $C = 5$.

Here are some other examples of linear equations in general form:

Equation	A	B	C
$5x + 3y = 10$	5	3	10
$x - 2y = 6 \rightarrow 1x + (-2)y = 6$	1	-2	6
$x = -4 \rightarrow 1x + 0y = -4$	1	0	-4
$y = -4 \rightarrow 0x + 1y = -4$	0	1	-4

Now, let's look at what we mean by a solution of a linear equation in two variables.

Definition

A **solution** of an equation in two variables is an ordered pair of numbers that when substituted for the variables makes the equation true.

Applying this definition to the equation $-2x + y = 5$, we observe that $x = 3$ and $y = 11$ is a solution of the equation because substituting 3 for x and 11 for y makes the equation true.

$$-2x + y = 5$$
$$-2 \cdot 3 + 11 \stackrel{?}{=} 5$$
$$5 = 5 \qquad \text{True}$$

Actually, there are many other solutions of this equation as well. For example, $x = 0$ and $y = 5$ is another solution.

$$-2x + y = 5$$
$$-2 \cdot 0 + 5 \stackrel{?}{=} 5$$
$$5 = 5 \qquad \text{True}$$

Unlike a linear equation in one variable that has at most one solution, linear equations in two variables have an infinite number of solutions. Can you explain why this is the case?

EXAMPLE 1

Determine whether each ordered pair is a solution of the equation $x + 4y = 6$.

a. $(2, 1)$ **b.** $(-3, 0)$

Solution

a. Substituting 2 for x and 1 for y, we get

$$x + 4y = 6$$
$$2 + 4(1) \stackrel{?}{=} 6$$
$$6 = 6 \qquad \text{True}$$

So $(2, 1)$ is a solution.

b. Substituting -3 for x and 0 for y, we get

$$x + 4y = 6$$
$$-3 + 4(0) \stackrel{?}{=} 6$$
$$-3 = 6 \qquad \text{False}$$

So $(-3, 0)$ is not a solution.

PRACTICE 1

Determine whether each ordered pair is a solution of the equation $2x - 3y = 5$.

a. $(3, 2)$

b. $(1, -1)$

Now, how do we *find* solutions of a linear equation in two variables? Typically, we start with the value of one of the variables and then compute the corresponding value of the other, as the following example illustrates.

EXAMPLE 2

For the equation $4x + y = -5$, find five solutions by completing the following table.

x	−2	3	0		
y			0	1	

Solution In the first row of this table, we substitute -2 for x in the given equation, and then solve for y.

$$4x + y = -5$$
$$4(-2) + y = -5$$
$$-8 + y = -5$$
$$y = 3$$

To find the corresponding y in the second row, we substitute 3 for x:

$$4x + y = -5$$
$$4(3) + y = -5$$
$$12 + y = -5$$
$$y = -17$$

And in the third row, we substitute 0 for x:

$$4x + y = -5$$
$$4(0) + y = -5$$
$$0 + y = -5$$
$$y = -5$$

The next two rows are different from the earlier ones: they give us y-values and require us to solve for x. First, we see that in the fourth row y is 0.

$$4x + y = -5$$
$$4x + 0 = -5 \qquad \text{Substitute 0 for } y.$$
$$4x = -5$$
$$x = -\frac{5}{4}$$

In the final row of the table, $y = 1$.

$$4x + y = -5$$
$$4x + 1 = -5 \qquad \text{Substitute 1 for } y.$$
$$4x = -6$$
$$x = -\frac{6}{4} = -\frac{3}{2}$$

We have found the missing values, so the table reads:

x	−2	3	0	$-\frac{5}{4}$	$-\frac{3}{2}$
y	3	−17	−5	0	1

PRACTICE 2

For the equation $-2x + y = 1$, find the missing values in the following table:

x	0	5	−3		
y				0	−3

The Graph of a Linear Equation in Two Variables

The graph of an equation, more precisely the graph of the *solutions* of that equation, is a kind of picture of the equation.

Definition

The **graph** of a linear equation in two variables consists of all points whose coordinates satisfy the equation.

Given a linear equation, how do we find its graph? A general strategy is to first isolate one of the variables, unless it is already done. Then we identify several solutions of the equation, keeping track of the x- and y-values in a table. We plot the points and then sketch the line passing through them. That line is the graph of the given equation, as the following example illustrates.

Let's graph the equation $y = 3x + 1$. The variable y is already isolated, so we find y-values by substituting arbitrary values of x. For instance, let x equal 0. To find y, we substitute 0 for x.

$$y = 3x + 1 = 3 \cdot 0 + 1 = 0 + 1 = 1$$

So $x = 0$ and $y = 1$ is a solution of this equation. We say that the ordered pair $(0, 1)$ is a solution of $y = 3x + 1$.

Let's choose three other values of x, say -1, 1, and 2. Substituting -1 for x in the equation, we get:
$$y = 3x + 1 = 3(-1) + 1 = -2$$

Substituting 1 for x, we get:

$$y = 3x + 1 = 3 \cdot 1 + 1 = 4$$

Substituting 2 for x gives:

$$y = 3x + 1 = 3 \cdot 2 + 1 = 7$$

Next, we enter these results in a table.

x	−1	0	1	2
y	−2	1	4	7

Then we plot on a coordinate plane the four points $A(-1, -2)$, $B(0, 1)$, $C(1, 4)$, and $D(2, 7)$. If we have not made a mistake, the points will all lie on the same line. We know that the line segments AB, BC, and CD are on the same line because they all have equal slopes. The graph of the equation $y = 3x + 1$ is the line passing through these points. So any point on this line satisfies the equation $y = 3x + 1$.

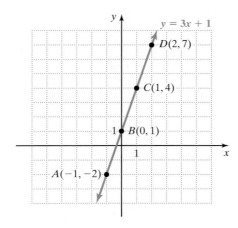

Do you think that if we had chosen three other *x*-values we would have gotten the same graph? Check to see that this is the case.

This example suggests the following procedure for graphing a linear equation.

To Graph a Linear Equation in Two Variables

- Isolate one of the variables—usually *y*—if it is not already done.
- Choose three *x*-values, entering them in a table.
- Complete the table by calculating the corresponding *y*-values.
- Plot the three points—two to draw the line and the third to serve as a *checkpoint*.
- Check that the points seem to lie on the same line.
- Draw the line passing through the points.

Notice that the point where the graph crosses the *y*-axis is $(0, 1)$. This point is called the *y-intercept*. Note that the constant term in the equation $y = 3x + 1$ is also 1.

This relationship is more than a coincidence, as we will see in Section 3.4.

EXAMPLE 3	PRACTICE 3

Graph the equation $y = -\frac{3}{2}x$ by choosing three points whose coordinates satisfy the equation.

Solution We begin by choosing *x*-values. In this case, we choose multiples of 2 for the *x*-values. Then we find the corresponding *y*-values.

x	$y = -\frac{3}{2}x$	(x, y)
0	$y = -\frac{3}{2}(0) = 0$	$(0, 0)$
2	$y = -\frac{3}{2}(2) = -3$	$(2, -3)$
4	$y = -\frac{3}{2}(4) = -6$	$(4, -6)$

Graph the equation $y = -\frac{3}{5}x$.

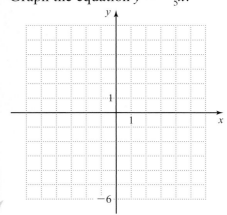

We plot the points and then draw a line passing through them to get the desired line.

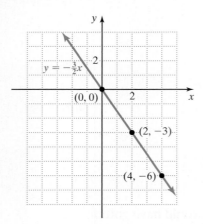

Note that in Example 3, we chose multiples of 2 for the *x*-values. Can you explain why?

EXAMPLE 4	PRACTICE 4

Consider the equation $2x + y = 3$.

a. Graph the equation.

b. Find the slope of the line.

Solution

a. We begin by solving the equation for *y*.

$$2x + y = 3$$
$$y = -2x + 3 \qquad \text{Subtract } 2x \text{ from each side.}$$

Next, we choose three values for *x*, for instance, -1, 0, and 2. Then we enter them into a table and find their corresponding *y*-values as follows.

x	y = −2x + 3	(x, y)
−1	$y = -2(-1) + 3 = 2 + 3 = 5$	$(-1, 5)$
0	$y = -2(0) + 3 = 0 + 3 = 3$	$(0, 3)$
2	$y = -2(2) + 3 = -4 + 3 = -1$	$(2, -1)$

Consider the equation $-2x + y = -5$.

a. Graph the equation.

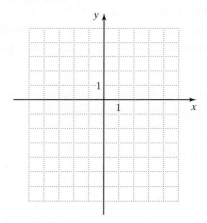

b. Find the slope of the line.

Now, we plot the points on the coordinate plane. Since the points seem to lie on the same line, we draw a line passing through the points.

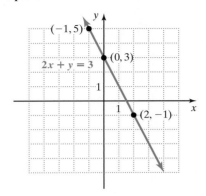

b. To find the slope of the line, we can consider the points $(0, 3)$ and $(2, -1)$.

$$m = \frac{3 - (-1)}{0 - 2} = \frac{4}{-2} = -2$$

Note that the slope is the coefficient of x in the equation $y = -2x + 3$.

We have graphed equations in general form by first isolating y and then computing y-values for arbitrary x-values. Now, we graph equations in general form with a different approach using x- and y-intercepts. Note that intercepts stand out on a graph and are easy to plot. Since an x-intercept lies on the x-axis, its y-value must be 0. Similarly, since a y-intercept lies on the y-axis, the x-value of a y-intercept must be 0.

Definition

The **_x_-intercept** of a line is the point where the graph crosses the x-axis. The **_y_-intercept** is the point where the graph crosses the y-axis.

The following graph shows a line passing through two points, $(0, 4)$ and $(3, 0)$, which are both intercepts.

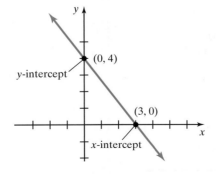

Using the x- and y-intercepts to graph equations can save work, especially when the coefficients of the two variables are factors of the constant term, as shown in the next example.

EXAMPLE 5

Consider the equation $2x + 3y = 6$. Find the x- and y-intercepts. Then graph.

Solution Since the y-intercept has x-value 0, we let $x = 0$ and then solve for y.

For $x = 0$:

$$2x + 3y = 6$$
$$2 \cdot 0 + 3y = 6 \quad \text{Substitute 0 for } x.$$
$$3y = 6$$
$$\frac{3y}{3} = \frac{6}{3}$$
$$y = 2$$

So the y-intercept is $(0, 2)$.

Similarly, the x-intercept has y-value 0. So we let $y = 0$ and then solve for x.

For $y = 0$:

$$2x + 3y = 6$$
$$2x + 3 \cdot 0 = 6 \quad \text{Substitute 0 for } y.$$
$$2x = 6$$
$$\frac{2x}{2} = \frac{6}{2}$$
$$x = 3$$

So the x-intercept is $(3, 0)$.

Before graphing, we choose a third point to be used as a checkpoint.

For $2x + 3y = 6$, let $x = 6$:

$$2x + 3y = 6$$
$$2 \cdot 6 + 3y = 6 \quad \text{Substitute 6 for } x.$$
$$12 + 3y = 6$$
$$3y = -6$$
$$y = -2$$

So the checkpoint is $(6, -2)$.

Plotting the points $(0, 2)$, $(3, 0)$, and $(6, -2)$ on a coordinate plane, we confirm that they seem to lie on the same line. Finally, we draw a line through the points, getting the desired graph.

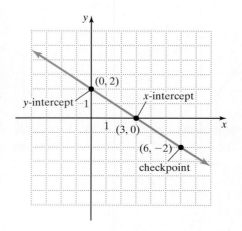

PRACTICE 5

Consider the equation $x - 2y = 4$. Find the x- and y-intercepts. Then graph.

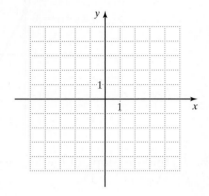

Example 5 suggests the following procedure.

To Graph a Linear Equation in Two Variables Using the *x*- and *y*-intercepts

- Let $x = 0$ and find the *y*-intercept.
- Let $y = 0$ and find the *x*-intercept.
- Find a checkpoint.
- Plot the three points.
- Check that the points seem to lie on the same line.
- Draw the line passing through the points.

We know that the general form of a linear equation in two variables is $Ax + By = C$. Sometimes in a linear equation one of the two variables is missing, that is, the coefficient of one of the two variables is zero. Consider the following equations:

$$y = 9 \quad \longrightarrow \quad 0x + y = 9$$
$$x = -5.8 \quad \longrightarrow \quad x + 0y = -5.8$$

Let's look at the graphs of these equations.

EXAMPLE 6	PRACTICE 6

EXAMPLE 6

Graph:

a. $y = 9$ **b.** $x = -5.8$

Solution

a. For the line $y = 9$, the coefficient of the *x*-term is 0. The *x*-value can be any real number and the *y*-value is always 9. Hence the graph is as follows.

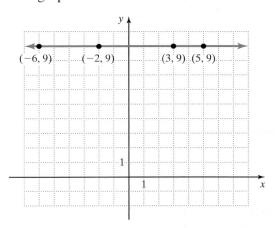

The graph of this equation is a horizontal line. Recall that the slope of a horizontal line is 0.

PRACTICE 6

On the given coordinate plane, graph:

a. $y = -1$ **b.** $x = 2.5$

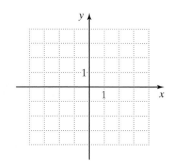

b. For the line $x = -5.8$, the coefficient of the y-term is 0. The y-value can be any real number and the x-value is always -5.8. Hence the graph is as follows.

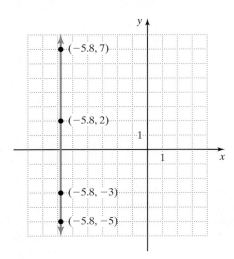

The graph of this equation is a vertical line. The slope of this line is undefined.

In general, the graph of the equation $x = a$ is a vertical line passing through the point $(a, 0)$, and the graph of the equation $y = b$ is a horizontal line passing through the point $(0, b)$.

Now, let's use our knowledge of graphing linear equations to solve applied problems.

EXAMPLE 7	PRACTICE 7

For cable television, a homeowner pays $30 per month plus $5 for each pay-per-view movie ordered.

a. Express as an equation the relationship between the monthly bill B and the number n of pay-per-view movies.

b. Draw the graph of this equation in Quadrant I of a coordinate plane.

c. Compute the slope of this graph. In terms of the cable TV bill, explain the significance of the slope.

d. In terms of the cable TV bill, explain the significance of the B-intercept of the graph.

e. From the graph in part (b), estimate what the cable bill would be if the homeowner had ordered 15 pay-per-view movies that month.

A stockbroker charges as her commission on stock transactions $40 plus 3% of the value of the sale.

a. Write as an equation the commission C in terms of the sales s.

Solution

a. The monthly bill (in dollars) amounts to the sum of 30 and 5 times the number of pay-per-view movies which the homeowner ordered, so

$$B = 5n + 30.$$

b. To draw the graph of this equation in Quadrant I, we enter several nonnegative *n*-values, say 0, 10, and 20, in a table and then compute the corresponding *B*-values.

n	*B* = 5*n* + 30	(*n*, *B*)
0	$B = 5(0) + 30 = 30$	(0, 30)
10	$B = 5(10) + 30 = 80$	(10, 80)
20	$B = 5(20) + 30 = 130$	(20, 130)

Since the monthly bill *B* depends on the number *n* of pay-per-view movies ordered each month, we label the horizontal axis with the independent variable *n* and the vertical axis with the dependent variable *B*. Now, we choose an appropriate scale for each axis and plot (0, 30), (10, 80), and (20, 130). Then we draw the line passing through these points.

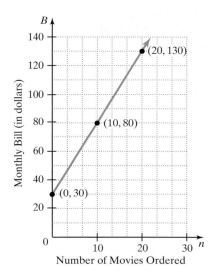

We restricted the graph to Quadrant I because the number of movies ordered *n* and the corresponding bill *B* are always positive.

c. Substituting *n* and *B* for *x* and *y*, respectively, in the slope formula, we get slope

$$m = \frac{B_2 - B_1}{n_2 - n_1} = \frac{80 - (30)}{10 - (0)} = \frac{50}{10} = 5.$$

Note that we could have predicted this answer since we know that the bill increases by $5 for every additional pay-per-view movie ordered.

d. Since the *B*-intercept is the point (0, 30), $30 would be the amount of the bill if the homeowner had watched no pay-per-view movies at all during the month.

e. From the graph, it appears that if $n = 15$, then $B = 105$.

b. Draw a graph showing this relationship on sales up to $1000. Be sure to choose an appropriate scale for each axis.

c. Compute the slope of this graph. In terms of the stock broker's commission, explain the significance of the slope.

d. If the value of a sale is $500, estimate from the graph in part (b) the broker's commission.

EXAMPLE 8

A dietician uses milk and cottage cheese as sources of calcium in her diet. One serving of milk contains 300 mg of calcium, and one serving of cottage cheese contains 100 mg of calcium. The recommended daily amount (RDA) of calcium is 1000 mg.

a. If m represents the number of servings of milk and c the number of servings of cottage cheese in a diet that contains the RDA of calcium, write an equation that relates m and c.

b. Graph this equation.

c. Explain the significance of the two intercepts in terms of the number of servings.

d. Explain how we could have predicted that the slope of this graph would be negative.

Solution

a. The amount of calcium in the milk is $300m$, and the amount of calcium in the cottage cheese is $100c$. Since the RDA of calcium is 1000 mg, the following equation holds:

$$300m + 100c = 1000, \text{ or } 3m + c = 10$$

b. To graph, we identify the m- and c-intercepts, as well as a third point, say with $m = 1$.

m	0	$3\frac{1}{3}$	1
c	10	0	7

Next, we choose an appropriate scale for each axis and plot the three points, checking that the points all lie on the same line. Then we draw the line passing through them.

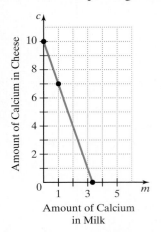

c. The m-intercept represents the number of servings she would need to meet the daily minimum requirement if she only uses milk as her source of calcium. The c-intercept represents the number of servings she would need to meet the RDA if she only uses cottage cheese as her source of calcium.

d. Even without drawing this line, we know that its slope has to be negative for the following reason: The RDA of calcium is a fixed amount (1000 mg). So larger values of m must correspond to smaller values of c. The line will therefore have to be decreasing, falling to the right and with a negative slope.

PRACTICE 8

Suppose an athlete has just signed a contract with a total value of $10 million. According to the terms of the contract, she earns $1 million in some years and $2 million in other years.

a. Let w stand for the number of years in which she earns $1 million and t the number of $2 million years. Write an equation that relates w and t.

b. Choose an appropriate scale for each axis on the coordinate plane. Graph the equation found in part (a).

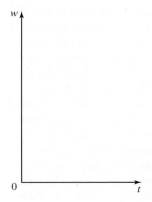

c. Describe in terms of the contract the significance of the slope of the graph.

d. Describe in terms of the contract the significance of the t- and w-intercepts of the graph.

3.3 **Exercises**

Mathematically Speaking

Fill in each blank with the most appropriate term or phrase from the given list.

three *x*-values	vertical	*x*-intercept
graph	solution	horizontal
y-intercept	three points	

1. A(n) _____ of an equation in two variables is an ordered pair of numbers that when substituted for the variables makes the equation true.

2. The _____ of a linear equation in two variables consists of all points whose coordinates satisfy the equation.

3. One way of graphing a linear equation in two variables is to plot _____.

4. The _____ of a line is the point where the graph crosses the *x*-axis.

5. One way to graph a linear equation using intercepts is to first let $x = 0$ and then find the _____.

6. The graph of the equation $y = c$ is a(n) _____ line passing through the point $(0, c)$.

Complete each table so that the ordered pairs are solutions of the given equation.

7. $y = 3x - 8$

x	4	7	
y			0

8. $y = -x + 20$

x	0		
y		3	2

9. $y = 5x$

x	3.5	6		
y			$\frac{1}{2}$	-8

10. $y = -10x$

x	$\frac{1}{5}$	2.9		
y			-6	-1

11. $3x + 4y = 12$

x	0	-4		
y			-3	0

12. $4x - y = 8$

x	5	0		
y			0	8

13. $y = \frac{1}{3}x - 1$

x	3	6	-3
y			-1

14. $y = -\frac{3}{2}x + 2$

x	$\frac{4}{3}$	6	-2	
y				2

Graph each equation by finding three points whose coordinates satisfy the equation.

15. $y = x$

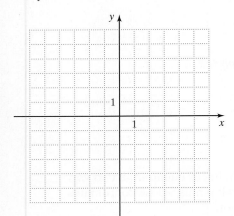

16. $y = 3x$

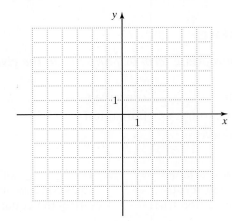

17. $y = \frac{1}{2}x$

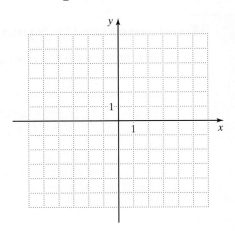

18. $y = \frac{1}{4}x$

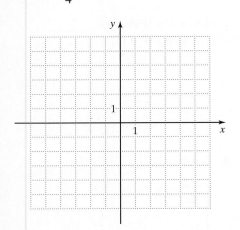

19. $y = -\frac{5}{4}x$

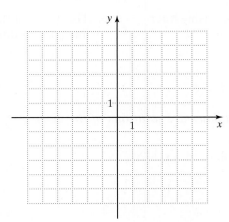

20. $y = -\frac{3}{2}x$

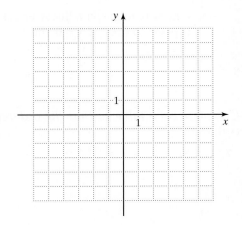

21. $y = 2x + 1$

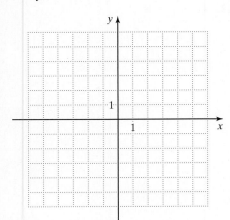

22. $y = -4x - 3$

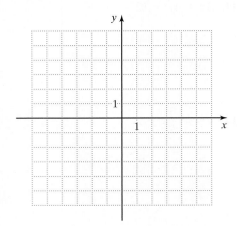

23. $y = -\frac{1}{3}x + 1$

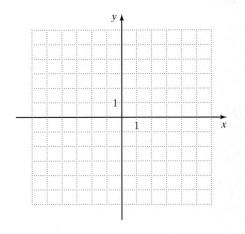

24. $y = -\dfrac{3}{4}x + 2$

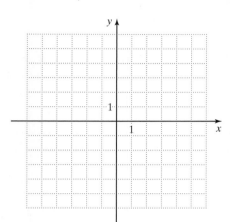

25. $y - 2x = -3$

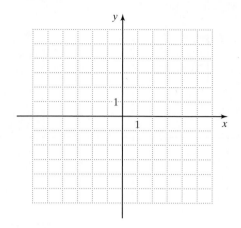

26. $y - 3x = 2$

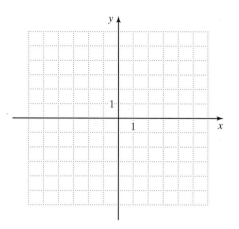

27. $x + y = 6$

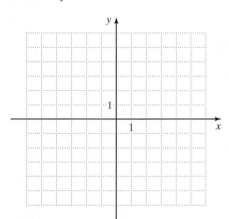

28. $3x + y = 4$

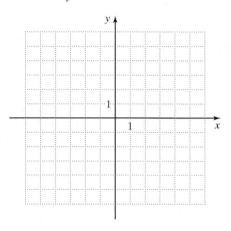

29. $x - 2y = 4$

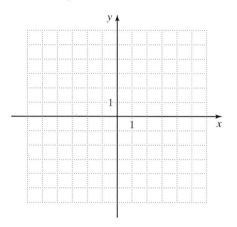

30. $x - 3y = 15$

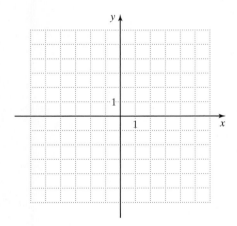

For each equation, find the x- and y-intercepts. Then use the intercepts to graph the equation.

31. $5x + 3y = 15$

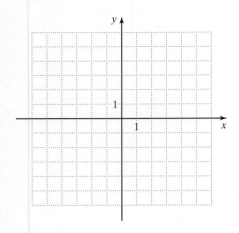

32. $4x + 5y = 20$

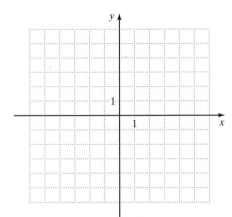

33. $3x - 6y = 18$

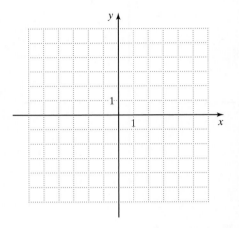

34. $3y - 2x = -6$

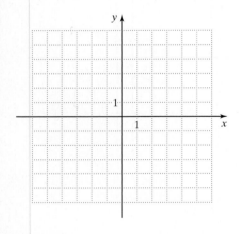

35. $4y - 5x = 10$

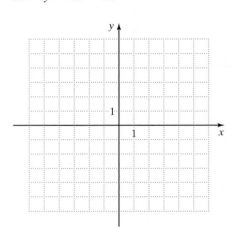

36. $7x - 2y = -7$

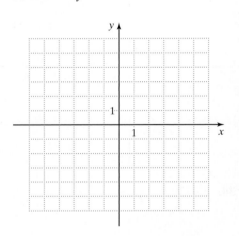

37. $9y + 6x = -9$

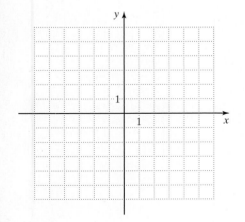

38. $4y - 8x = -4$

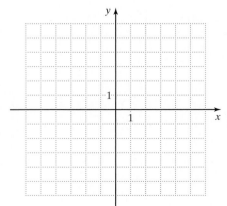

39. $y = \dfrac{1}{2}x + 2$

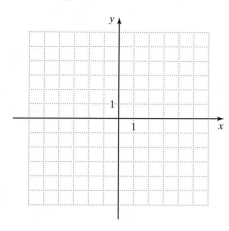

40. $y = \dfrac{5}{4}x - 5$

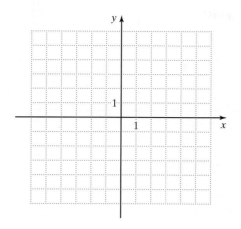

Graph.

41. $y = -2$

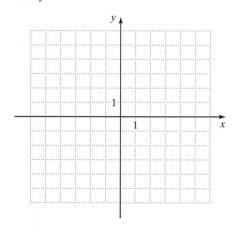

42. $x = 3$

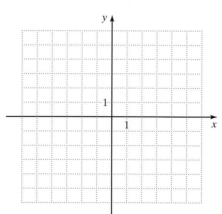

43. $x = 0$

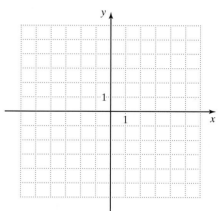

44. $y = 0$

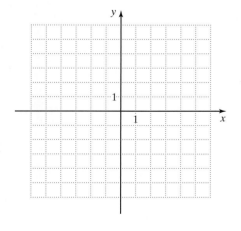

45. $x = -5.5$

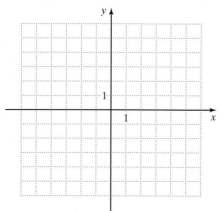

46. $y = -0.5$

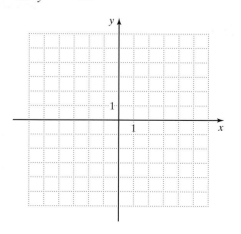

47. $y = \dfrac{1}{2}x + 3$

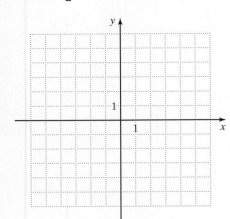

48. $y = 0.5x + 6$

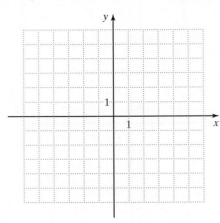

49. $3x + 5y = -15$

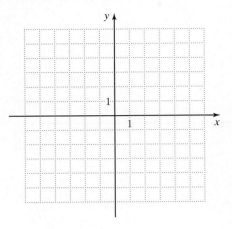

50. $3y - 5x = 15$

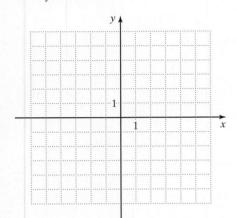

51. $y = \dfrac{3}{5}x + \dfrac{2}{5}$

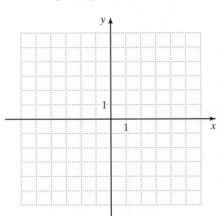

52. $y = -\dfrac{1}{4}x + \dfrac{3}{4}$

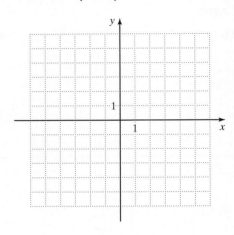

Mixed Practice

Complete each table so that the ordered pairs are solutions of the given equation.

53. $y = -2x + 6$

x	-3	$\dfrac{5}{2}$		
y			-10	4

54. $3x - 4y = 6$

x	0		-6	
y		0		3

For each equation, find the x-intercept, the y-intercept, and another point whose coordinates satisfy the equation. Then use these points to graph the equation.

55. $y = -\dfrac{1}{2}x + 2$

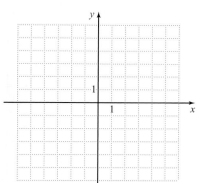

56. $2x - y = 4$

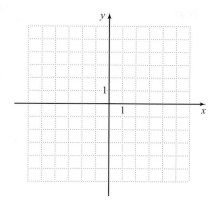

57. $6x - 2y = -12$

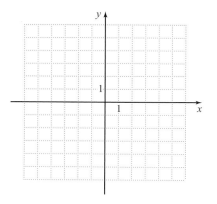

58. $y = \dfrac{3}{2}x - 3$

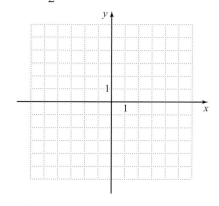

59. $y = -4x$

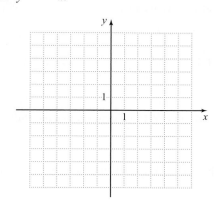

60. $y = \dfrac{3}{4}x$

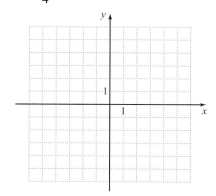

61. $3y + 6x = -3$

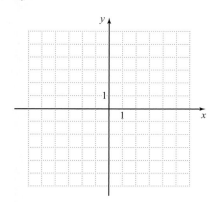

Graph.

62. $x = -3\frac{1}{2}$

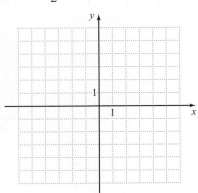

63. $y = -0.5x + 3$

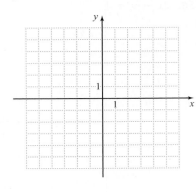

64. $2y - 4x = 8$

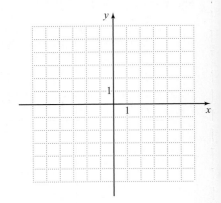

Applications

Solve.

65. In a physics lab, students study the mathematics of motion. They learn that if an object is tossed straight upward with an initial velocity of 10 feet per second, then after t sec the object will be traveling at a velocity of v feet per second, where

$$v = 10 - 32t.$$

a. Complete the following table.

t	0		1	1.5	2
v		-6			

Explain what a positive value of v means. What does a negative value of v mean?

b. Choose an appropriate scale for the each axis, and then graph this equation.

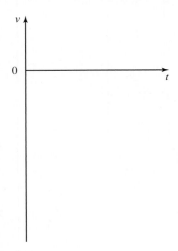

c. In terms of the object's motion, explain the significance of the v-intercept.

d. In terms of the object's motion, explain the significance of the t-intercept.

66. Each day, a local grocer varies the price p of an item in dollars and then keeps track of the number s of items sold. According to his records, the following equation describes the relationship between s and p.

$$s = -2p + 12$$

a. Complete the following table.

p	1	3	5
s			0

b. Choose appropriate scales for the axes and then graph the equation.

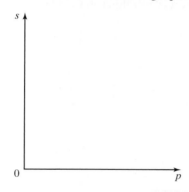

c. Explain why it makes sense to consider the graph only in Quadrant I.

d. From the graph, estimate the price required to sell 4 items.

67. A young couple buys furniture for $2000, agreeing to pay $500 down and $100 at the end of each month until the entire debt is paid off.

a. Express the amount P paid off in terms of the number m of monthly payments.

b. Complete the following table.

m	1	2	3
P			

c. Choose an appropriate scale for the axes and then graph this equation.

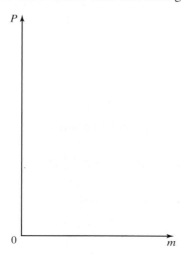

68. Students studying archaeology know that when the femur bone of an adult female is unearthed, a good estimate of her height h is 73 more than double the length l of the femur bone, where all measurements are in centimeters.

a. Express this relationship as a formula.

b. Complete the following table.

l	20	25	30
h			

c. Choose an appropriate scale for the axes, and then graph this relationship.

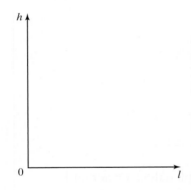

d. Use the graph to estimate the height of a woman whose femur bone was 35 cm in length.

69. The coins in a cash register, with a total value of $2, consist of n nickels and d dimes.

a. Represent this relationship as an equation.

b. Graph the equation found in part (a).

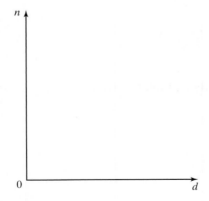

c. Explain in a sentence or two why every point on this graph in Quadrant I is *not* a reasonable solution to the problem.

70. On the first leg of a trip, a truck driver drove for x hr at a constant speed of 50 mph. On the second leg of the trip, he drove for y hr consistently at 40 mph. In all, he drove 1000 mi.

    a. Translate this information into an equation.

    b. Choose appropriate scales for the axes, and then graph the equation found in part (a).

    c. What are the x- and y-intercepts of this graph? Explain their significance in terms of the trip.

    d. Find the slope of the line. Explain whether you would have expected the slope to be positive or negative, and why.

71. At a computer rental company, the fee F for renting a laptop is $40 plus $5 for each of the d days that the computer is rented.

    a. Express this relationship as an equation.

    b. Choose appropriate scales for the axes, and then graph the equation expressed in part (a).

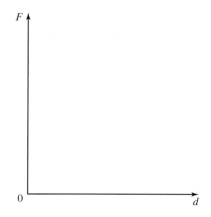

    c. Explain the significance of the F-intercept in this context.

72. At a local community center, the annual cost c to use the swimming pool includes an annual membership fee of $75 plus $5 per hour for h hr of pool time.

a. Write an equation for the annual cost of swimming at the community center in terms of the number of hours of pool time.

b. Choose appropriate scales for the axes, and then graph the equation for up to and including 150 hr.

c. Use the graph to estimate the annual cost of using the pool for 25 hr.

d. Suppose the annual cost for swimming was $500. Estimate the number of hours of pool time.

● *Check your answers on page A-11.*

MINDSTRETCHERS

Groupwork

1. Not all graphs are linear. For example, the graph of the equation $y = x^2 - 4$ is nonlinear, as the following graph illustrates.

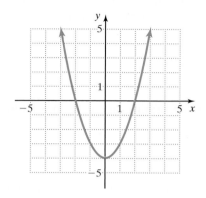

    a. Identify the x- and y-intercepts for the graph shown.

    b. Show that the x- and y-intercepts found in part (a) satisfy the equation $y = x^2 - 4$.

Writing

2. Give some advantages and disadvantages of graphing a linear equation by finding three arbitrary points versus using the intercepts.

Mathematical Reasoning

3. Recall that for a linear equation to be in general form, it must be written as $Ax + By = C$, where A, B, and C are real numbers and A and B are not both 0. What would the graph of this equation look like if A and B were both 0?

3.4   More on Linear Equations and Their Graphs

In the previous section of this chapter, we considered linear equations written in the general form and how to graph them. Now, we will discuss two other forms of linear equations: the *slope-intercept form* and the *point-slope form*. Using these two forms, we will continue to show how line graphs and their corresponding equations have real-world applications.

OBJECTIVES

■ To write a given linear equation in slope-intercept form

■ To identify the slope and *y*-intercept of an equation written in slope-intercept form

■ To graph a line using its slope and *y*-intercept

■ To write a given linear equation in point-slope form

■ To find the equation of a line, given two points on the line or its slope and one point

■ To solve applied problems relating to the graph of a linear equation

Slope-Intercept Form

Recall from Section 3.3 that one approach to graphing an equation written in general form is to isolate *y*.

$$-5x + y = 2 \qquad \text{An equation in general form}$$
$$y = 5x + 2 \qquad \text{Solve for } y.$$

The linear equation $y = 5x + 2$ is said to be in slope-intercept form. The graph of this equation has slope 5, which is equal to the coefficient of x, and y-intercept $(0, 2)$, where 2 is the constant term of the equation.

Definition

A linear equation is in **slope-intercept form** if it is written as

$$y = mx + b,$$

where m and b are constants. In this form, m is the slope and $(0, b)$ is the y-intercept of the graph of the equation.

In general, we can identify the slope and y-intercept of the graph of any linear equation in slope-intercept form without drawing the graph of the equation. The coefficient of x is the slope m, and the constant term b is the y-coordinate of the y-intercept $(0, b)$.

The following table gives additional examples of equations written in slope-intercept form:

Equation	Slope, *m*	*y*-intercept, (0, *b*)
$y = \dfrac{1}{2}x + 1$	$\dfrac{1}{2}$	$(0, 1)$
$y = 2x - 1$	2	$(0, -1)$
$y = -7x \rightarrow y = -7x + 0$	-7	$(0, 0)$
$y = 5 \rightarrow y = 0x + 5$	0	$(0, 5)$

EXAMPLE 1

Find the slope and y-intercept of the equation $y = 3x - 5$.

Solution The equation $y = 3x - 5$ or $y = 3x + (-5)$ is already in slope-intercept form, $y = mx + b$. The slope m is 3, and the y-intercept is $(0, -5)$ since the equation has constant term -5.

PRACTICE 1

Find the slope and y-intercept of the equation $y = -2x + 3$.

EXAMPLE 2

Express $3x + 5y = 6$ in slope-intercept form.

Solution Since the slope-intercept form of an equation is $y = mx + b$, we need to solve the given equation for y.

$$3x + 5y = 6$$
$$5y = -3x + 6$$
$$\frac{5y}{5} = \frac{-3}{5}x + \frac{6}{5}$$
$$y = -\frac{3}{5}x + \frac{6}{5}$$

So $y = -\frac{3}{5}x + \frac{6}{5}$ is the equation written in slope-intercept form, where m is $-\frac{3}{5}$ and b is $\frac{6}{5}$.

PRACTICE 2

Express $3x - 2y = 4$ in slope-intercept form.

EXAMPLE 3

Write $y - 1 = 5(x - 1)$ in slope-intercept form.

Solution To get the equation in the form $y = mx + b$, we must solve for y.

$$y - 1 = 5(x - 1)$$
$$y - 1 = 5x - 5$$
$$y = 5x - 5 + 1$$
$$y = 5x - 4$$

So $y = 5x - 4$ is the equation written in slope-intercept form, where m is 5 and b is -4.

PRACTICE 3

Change the equation $y - 2 = 4(x + 1)$ to slope-intercept form.

EXAMPLE 4

For the graph of $y = 3x$, find the slope and y-intercept.

Solution We can rewrite $y = 3x$ as $y = 3x + 0$. Now, the equation $y = 3x + 0$ is in slope-intercept form, with slope $m = 3$ and $b = 0$. Since $b = 0$, the y-intercept is $(0, 0)$. That is, the graph passes through the origin.

PRACTICE 4

Find the slope and y-intercept of the graph of $y = -x$.

We have already graphed equations of the form $y = mx + b$ by finding three points whose coordinates satisfy the equation. Now, let's focus on graphing such equations using the slope and the y-intercept.

Consider the equation $y = 3x - 1$, which has slope 3 and y-intercept $(0, -1)$. Since 3 is $\frac{3}{1}$, we know from the definition of slope that the *rise* is 3 and the *run* is 1. Starting at $(0, -1)$, we move up 3 units and then 1 unit to the right to find a second point $(1, 2)$ on the line. Then we draw the line through the two points $(0, -1)$ and $(1, 2)$.

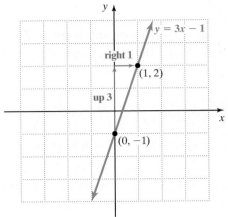

This example leads us to the following rule.

To Graph a Linear Equation in Two Variables Using the Slope and the y-intercept

- First locate the y-intercept.
- Use the slope to find a second point on the line.
- Then draw the line through the two points.

EXAMPLE 5

Graph $y = -\frac{3}{2}x + 2$ using the slope and y-intercept.

Solution Since $y = -\frac{3}{2}x + 2$ is in slope-intercept form, the slope is $-\frac{3}{2}$ and the y-intercept is $(0, 2)$. First, we locate the y-intercept $(0, 2)$. Since the slope $-\frac{3}{2}$ equals $\frac{-3}{2}$, from the point $(0, 2)$ we move *down* 3 units and then 2 units to the *right* to find the second point $(2, -1)$. Then we draw the line through the points $(0, 2)$ and $(2, -1)$ as shown in the graph.

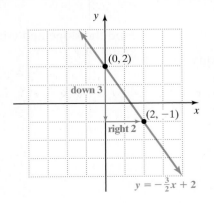

PRACTICE 5

Use the slope and y-intercept to graph $y = -\frac{1}{3}x - 4$.

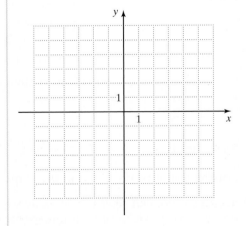

Note that since $\frac{-3}{2} = \frac{3}{-2}$, from the point $(0, 2)$ we could have moved *up* 3 units and then 2 units to the *left* to find the second point $(-2, 5)$ and then drawn the line through $(0, 2)$ and $(-2, 5)$ to obtain the same graph of the equation, as on the following coordinate plane.

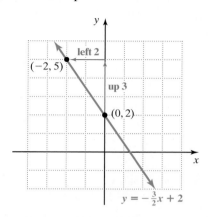

Recall that in Section 3.3 we graphed $2x + 3y = 6$ using the intercepts. We can also graph this equation using the slope and y-intercept. Which method do you prefer? Explain why.

We can use the slope-intercept form to write the equation of a line when given its slope and y-intercept.

EXAMPLE 6	PRACTICE 6
Write the equation of the line with slope $-\dfrac{4}{5}$ and y-intercept $(0, -3)$. **Solution** We are given that the slope is $-\dfrac{4}{5}$ and the y-intercept is $(0, -3)$. So $m = -\dfrac{4}{5}$ and $b = -3$. We substitute these values in the slope-intercept form: $$y = mx + b$$ $$y = -\frac{4}{5}x + (-3), \quad \text{or } y = -\frac{4}{5}x - 3$$	A line on a coordinate plane has slope 1 and intersects the y-axis 2 units above the origin. Write its equation in slope-intercept form.

We can find an equation of a line that is parallel or perpendicular to a given line.

EXAMPLE 7	PRACTICE 7
Find an equation of the line that is parallel to the graph of $y = 4x + 1$ and has y-intercept $(0, 3)$. **Solution** The line $y = 4x + 1$ is in slope-intercept form. The slope of this line is 4. Since parallel lines have the same slope, the line we want will also have slope $m = 4$. Since its y-intercept is $(0, 3)$, $b = 3$. Therefore, the desired equation is $y = 4x + 3$.	What is the equation of the line parallel to the graph of $y = -2x + 3$ with y-intercept $(0, -1)$?

EXAMPLE 8

What is the equation of the line that is perpendicular to the graph of $y = 3x - 1$ and has y-intercept $(0, 1)$?

Solution The line $y = 3x - 1$ is written in slope-intercept form. So its slope is 3. We know that the slopes of two perpendicular lines are negative reciprocals of each other. Therefore, the slope of the line we want has slope $m = -\dfrac{1}{3}$. Since its y-intercept is $(0, 1)$, $b = 1$. So the desired equation is $y = -\dfrac{1}{3}x + 1$.

PRACTICE 8

Find the equation of the line that is perpendicular to the graph of $y = 2x + 5$ and has y-intercept $(0, -2)$.

EXAMPLE 9

At her college, a student pays $75 per credit-hour plus a flat student fee of $100. Find an equation in slope-intercept form that relates the amount A that she pays to the number h of credit-hours in her program.

Solution The amount A in dollars that the student pays is the sum of 100 and 75 times the number h of credit-hours in her program. So we have:

$$A = 75h + 100$$

This equation is written in slope-intercept form.

PRACTICE 9

Suppose that a bathtub, which has a capacity of 45 gal, is filled to the top with water. The tub starts to drain at a rate of 3 gallons per minute. Write an equation in slope-intercept form that relates the amount of water w left in the tub and the time t that the tub has been draining.

Point-Slope Form

The last form of a linear equation that we will discuss is called the point-slope form.

> **Definition**
>
> The **point-slope form** of a linear equation is written as
>
> $$y - y_1 = m(x - x_1),$$
>
> where x_1, y_1, and m are constants. In this form, m is the slope and (x_1, y_1) is a point that lies on the graph of the equation.

This form, although used less frequently than other forms of a linear equation, is particularly useful for finding the equation of a line in two particular situations:

- when we know the slope of the line and a point on it, or
- when we know two points on the line.

EXAMPLE 10

A line with slope -2 passes through the point $(-1, 5)$. Find the equation of this line written in point-slope form.

Solution Since we know a point on the line and the slope of the line, we can substitute directly into the point-slope formula, where $x_1 = -1$, $y_1 = 5$, and $m = -2$.

$$y - y_1 = m(x - x_1)$$
$$y - 5 = -2[x - (-1)]$$
$$y - 5 = -2(x + 1) \quad \text{Point-slope form}$$

We can leave this equation in point-slope form, or we can simplify the equation and write it in either general form

$$y - 5 = -2x - 2$$
$$2x + y = 3 \quad \text{General form}$$

or in slope-intercept form

$$y = -2x + 3 \quad \text{Slope-intercept form}$$

PRACTICE 10

A line passing through the point $(7, 0)$ has slope 2. Find its equation in point-slope form.

EXAMPLE 11

What is the equation of the line passing through the points $(3, 5)$ and $(2, 1)$?

Solution Since we know the coordinates of two points on the line, we can find its slope.

$$m = \frac{y_2 - y_1}{x_2 - x_1} = \frac{5 - 1}{3 - 2} = \frac{4}{1} = 4$$

The line with slope $m = 4$ passing through the point $(3, 5)$ is:

$$y - y_1 = m(x - x_1)$$
$$y - 5 = 4(x - 3)$$

PRACTICE 11

Find the equation in point-slope form of the line passing through $(7, 7)$ and the origin.

The equation found in Example 11 is $y = 4x - 7$ in slope-intercept form. Had we substituted the point $(2, 1)$ rather than the point $(3, 5)$ into the point-slope formula, would the resulting equation be the same?

EXAMPLE 12

An accountant's computer decreases in value by $400 a year. The computer was worth $1600 one year after he bought it. Write an equation that gives the value V of the computer in terms of the number of years t since the purchase.

Solution Each year that passes, t increases by 1 and V decreases by 400. Therefore, the graph of the equation we are seeking has slope -400. Because the computer is worth $1600 one year after purchase, the graph passes through the

PRACTICE 12

The total weight of a box used for shipping baseballs increases by 5 oz for each baseball that is packed in the box. A box with 4 balls weighs 27 oz. Write an equation that relates the total weight w of the box to the number of baseballs b packed in the box.

point (1, 1600). Since we know a point on the line as well as its slope, we can find the point-slope form of the equation.

$$V - V_1 = m(t - t_1)$$
$$V - 1600 = -400(t - 1)$$

If we like, we can simplify the equation and write it in slope-intercept form:

$$V - 1600 = -400t + 400$$
$$V = -400t + 2000$$

Can you explain why the slope of the line in Example 12 is negative?

Using a Calculator or Computer to Graph Linear Equations

Calculators with graphing capabilities and computers with graphing software allow us to graph equations at the push of a key, even those with complicated coefficients. Although they vary somewhat in terms of features and commands, these machines all graph the equation of your choice on a coordinate plane.

To graph an equation, begin by making certain that the equation is in slope-intercept form. On many graphers, pressing the ⌐Y =⌐ key results in a screen being displayed on which you enter the equation. For instance, if you wanted to graph $2x - y = 1$, you would first solve for y, resulting in $y = 2x - 1$, and then enter $2x - 1$ to the right of \Y1 = on the screen. Pressing the ⌐GRAPH⌐ key displays a coordinate plane in which the graph of $y = 2x - 1$ is sketched, as we see on the following screen.

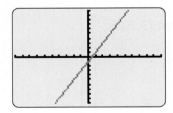

Many graphers have a **TRACE** feature that highlights a point on the graph and displays its coordinates. As you hold down an arrow key, you can see how the coordinates change as the highlighted point moves along the graph.

Tip   The viewing window allows you to set the range and scales for the axes. Before you graph an equation, be sure to set the viewing window in which you would like to display the graph.

EXAMPLE 13

Graph $y - x = 2$, and then use the **TRACE** feature to identify the y-intercept.

Solution First solve for y: $y = x + 2$. Next, press $\boxed{Y=}$ and enter $x + 2$ to the right of \Y1 =. Next, set the viewing window in which you want to display the graph. Then press $\boxed{\text{GRAPH}}$ to display the graph of the equation. If the graph of $y = x + 2$ does not appear, check your grapher's instruction manual.

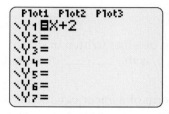   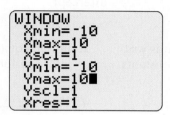

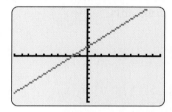

With the **TRACE** feature, run the cursor along the graph until it appears to be on the y-axis. The displayed coordinates of this y-intercept are approximately $x = 0$ and $y = 2$.

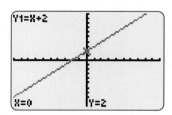

PRACTICE 13

Graph $2y + x = 3$, and then find the y-intercept with the **TRACE** feature.

3.4   Exercises

Mathematically Speaking

Fill in each blank with the most appropriate term or phrase from the given list.

standard	point-slope	slope
x-intercept	y-intercept	slope-intercept

1. A linear equation is in _____ form if it is written as $y = mx + b$, where m and b are constants.

2. For an equation of a line written in slope-intercept form, m is the _____ of the line.

3. For an equation of a line written in slope-intercept form, $(0, b)$ is the _____ of the line.

4. The _____ form of a linear equation is written as $y - y_1 = m(x - x_1)$, where x_1, y_1, and m are constants.

Complete each table.

5.

Equation	Slope, m	y-intercept, (0, b)	Which Graph Type Best Describes the Line? ╱ ╲ — ┃	x-intercept
$y = 3x - 5$				
$y = -2x$				
$y = 0.7x + 3.5$				
$y = \frac{3}{4}x - \frac{1}{2}$				
$6x + 3y = 12$				
$y = -5$				
$x = -2$				

6.

Equation	Slope, m	y-intercept, (0, b)	Which Graph Type Best Describes the Line? ╱ ╲ — ┃	x-intercept
$y = -3x + 5$				
$y = 2x$				
$y = 1.5x + 6$				
$y = \frac{2}{3}x + \frac{1}{2}$				
$4x + 6y = 24$				
$y = 0.3$				
$x = 2$				

Find the slope and y-intercept of each equation.

7. $y = -x + 2$

8. $y = 3x - 4$

9. $y = -\dfrac{1}{2}x$

10. $y = x$

Write the following equations in slope-intercept form.

11. $x - y = 10$

12. $x + y = 7$

13. $x + 10y = 10$

14. $3x - y = 15$

15. $6x + 4y = 1$

16. $4x - 8y = 12$

17. $2x - 5y = 10$

18. $3x + 5y = 15$

19. $y + 1 = 3(x + 5)$

20. $y - 1 = 3(x - 5)$

Match the equation to its graph.

21. $4x - 2y = 6$

22. $-2x + 4y = 8$

23. $2y - x = 8$

24. $6x + 3y = -9$

a.

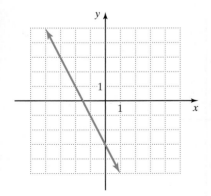

b.

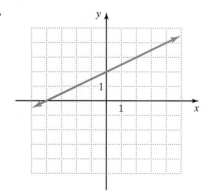

c.

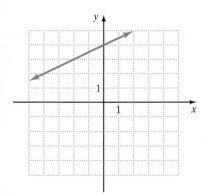

d.

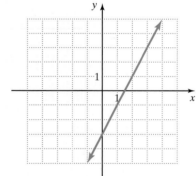

Graph the following equations using the slope and y-intercept.

25. $y = 2x + 1$

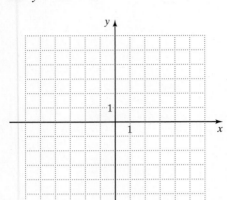

26. $y = -3x + 1$

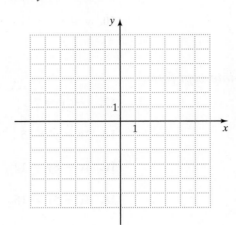

27. $y = -\dfrac{2}{3}x + 6$

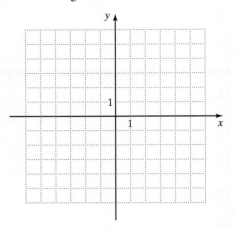

28. $y = \dfrac{3}{2}x - 6$

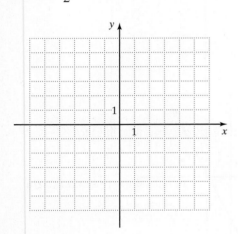

29. $x + y = 1$

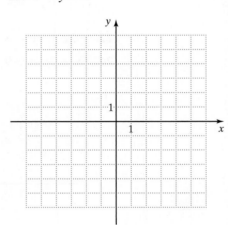

30. $x + y = -4$

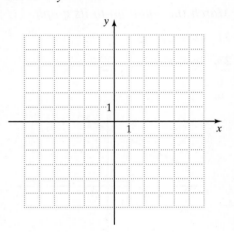

31. $y = -\dfrac{3}{4}x$

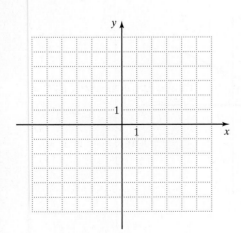

32. $y = \dfrac{1}{2}x$

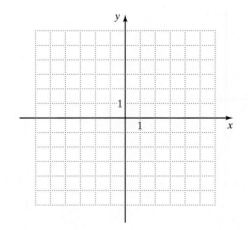

33. $x + 2y = 4$

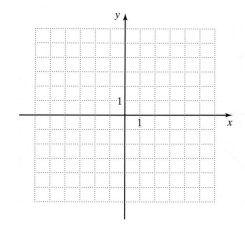

34. $-2x + 3y = 12$

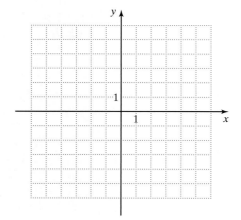

35. $y = 3.735x + 1.056$

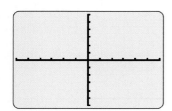

36. $y = -0.875x + 2.035$

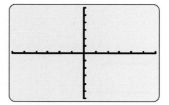

37. Find the equation of the line with slope 3 that passes through the point $(0, 7)$.

38. What is the equation of the line that has slope -1 with y-intercept $(0, -2)$?

39. Find the equation of the line that is parallel to the graph of $y = 5x - 1$ and has x-intercept $(4, 0)$.

40. What is the equation of the line that is parallel to the graph $y = \frac{1}{3}x - 1$ and has y-intercept $(0, 4)$?

41. What is the equation of the line that is perpendicular to the graph of $y = 2x$ and that passes through $(-2, 5)$?

42. Find the equation of the line that is perpendicular to the graph $y = -x$ and that passes through the point $(1, -3)$.

43. What is the equation of the line passing through the points $(2, 1)$ and $(1, 2)$?

44. The points $(5, 1)$ and $(2, -3)$ lie on a line. Find its equation.

45. Find the equation of the line passing through points $(-1, -5)$ and $(-7, -6)$.

46. What is the equation of the line passing through the origin and the point $(3, 5)$?

47. Write the equation of the vertical line that passes through the point $(-3, 5)$.

48. What is the equation of the horizontal line passing through the point $(1, -8)$?

49. What equation has the x-axis as its graph?

50. What equation has the y-axis as its graph?

Find the equation of each graph.

51.

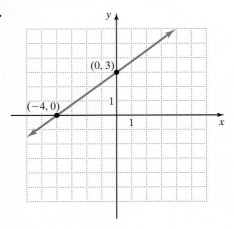

52.

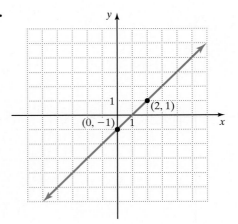

53.

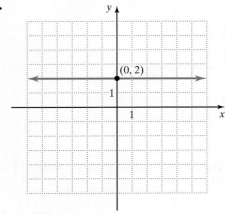

54.

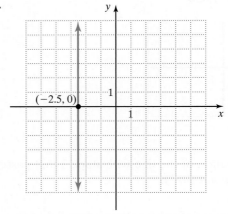

55.

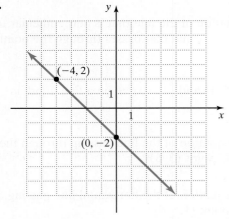

56.

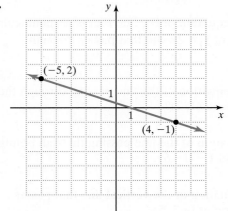

Mixed Practice

57. Complete the table.

Equation	Slope, m	y-intercept, $(0, b)$	Which Graph Type Best Describes the Line? ╱ ╲ — │	x-intercept
$y = -7x + 2$				
$y = 4x$				
$y = 2.5x + 10$				
$y = \dfrac{2}{3}x - \dfrac{1}{4}$				
$5x + 4y = 20$				
$x = 9$				
$y = -3.2$				

58. Find the slope and y-intercept of $y = -\dfrac{2}{5}x + 3$.

Write the following equations in slope-intercept form.

59. $4x - y = 5$

60. $3x - 6y = 8$

61. Which equation describes the graph?

   a. $-4x + y = 5$

   b. $4x + y = -5$

   c. $-5x + y = -4$

   d. $5x + y = -4$

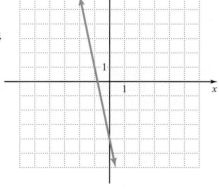

62. What is the equation of the line that is parallel to the graph $y = \dfrac{1}{2}x + 2$ and has x-intercept $(3, 0)$?

63. What is the equation of the line that is perpendicular to the graph of $y = -2x + 1$ that passes through the point $(4, 1)$?

64. What is the equation of the line that passes through the points $(2, -2)$ and $(-2, 1)$?

Graph the following equations using the slope and y-intercept.

65. $y = \dfrac{3}{5}x - 2$

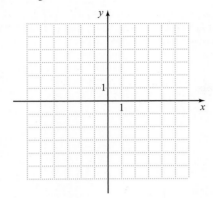

66. $3x + 2y = 4$

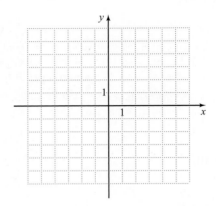

Find the equation of each graph.

67.

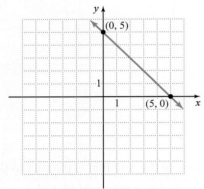

68.

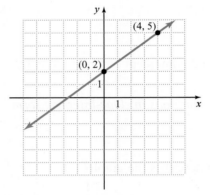

Applications

Solve.

69. The following graph describes the relationship between Fahrenheit temperature *F* and Celsius temperature *C*.

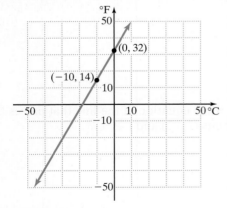

    a. Find the slope of the line.

    b. Find the equation of the line in slope-intercept form.

    c. Water boils at 212°F. Use part (b) to find the Celsius temperature at which water boils.

70. The owner of a shop buys a piece of machinery for \$1500. The value *V* of the machinery declines by \$150 per year.

    a. Write an equation for *V* after *t* years.

    b. Graph the equation found in part (a).

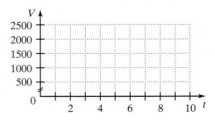

    c. Explain the significance of the two intercepts in this context.

71. A utility company charges its residential customers for electricity each month a flat fee plus 4 cents per kilowatt-hour (kWh) consumed. Last month, you used 500 kWh of electricity, and your bill amounted to $35.

a. Express as an equation in point-slope form the relationship between your monthly bill y in cents and the number x of kilowatt-hours of electricity consumed.

b. Express the equation found in part (a) in slope-intercept form.

c. What does the y-intercept represent in this situation?

72. A condo unit has been appreciating in value at $5000 per year. Three years after it was purchased, it was worth $65,000.

a. Find an equation that expresses the value y of the condo in terms of the number x of years since it was purchased.

b. Graph the equation found in part (a).

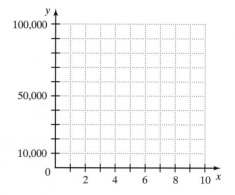

c. What is the significance of the y-intercept in this situation?

73. During a lightning storm, the relationship between the number of miles away L that lightning strikes and the number of seconds t between the flash of lightning and the sound of thunder is linear. When lightning strikes a tree 5 mi away, 1 sec passes before you hear the thunder. But when lightning strikes an old house 10 mi away, 2 sec pass between the lightning and the thunder.

a. Express the relationship between L and t as an equation in point-slope form.

b. Write the equation found in part (a) in slope-intercept form.

74. An air conditioner can reduce the temperature in a room by 8°F every 5 min. The temperature in the room was 62°F after the air conditioner had been running for 10 min.

a. Write as a linear equation the relationship between the time that the air conditioner has been running and the temperature in the room.

b. Explain how you could have predicted whether the slope of the graph of this equation is positive or negative.

75. A salesperson earns a salary of $1500 per month plus a commission of 3% of the total monthly sales.

    a. Write a linear equation giving the salesperson's total monthly income I in terms of sales S.

    b. Graph the equation found in part (a).

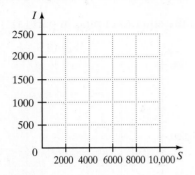

    c. Find the salesperson's income on monthly sales of $6200.

76. When the brakes on a train are applied, the speed of the train decreases by the same amount every second. Two seconds after applying the brakes, the train's speed is 88 mph. After 4 sec, its speed is 60 mph.

    a. Write an equation relating the speed s of the train and the time elapsed t seconds after applying the brakes.

    b. Graph the equation found in part (a).

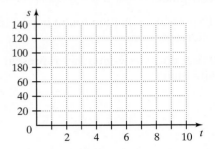

    c. What was the speed of the train when the brakes were first applied?

77. Pressure under water increases with greater depth. The pressure P on an object and the depth d below sea level are related by a linear equation. The pressure at sea level is 1 atmosphere (atm), whereas 33 ft below sea level the pressure is 2 atm. Find the equation relating P and d.

78. The length of a heated object and the temperature of the object are related by a linear equation. A rod at $0°$ Celsius is 10 m long, and at $25°$ Celsius it is 10.1 m long. Write an equation for length in terms of temperature.

79. When a force is applied to a spring, its length changes. The length L and the force F are related by a linear equation. The spring shown here was initially 10 in. long when no force was applied. What is the equation that relates length and force?

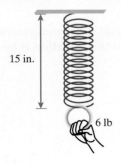

80. A company purchased a computer workstation for $8000. After 3 yr, the estimated value of the workstation was $4400. If the value V in dollars and the age a of the workstation are related by a linear equation, find an equation that relates V and a.

● *Check your answers on page A-14.*

MINDSTRETCHERS

Technology

1. Consider $2x - 7 = 0$, which is a linear equation in x.

   a. Solve the equation.

   b. On a graphing calculator or computer with graphing software, graph $y = 2x - 7$. Then find the x-intercept of the line. Explain in a sentence or two how you can use this approach to solve the equation $2x - 7 = 0$.

Critical Thinking

2. Consider the equation $y = mx + b$. Explain under what circumstances its graph lies completely in Quadrants I and II.

Mathematical Reasoning

3. What kind of line corresponds to an equation that can be written in *general form* but in neither slope-intercept form nor point-slope form?

3.5   Graphing Linear Inequalities

In Section 2.6, we showed how to graph inequalities in one variable on a number line. In such inequalities, the solutions are real numbers. For instance, the graph of $x \leq 2$ is shown below.

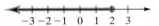

Now, we consider graphing inequalities in two variables on a coordinate plane. In this case, the solutions are ordered pairs of real numbers. For instance, if we want to graph the solutions to $x \leq 2$ in the coordinate plane, we would shade the region to the left of the vertical line $x = 2$ because every ordered pair in this region has an x-coordinate that is less than 2. Every point on the line is also a solution to $x \leq 2$ because each pair on this line has an x-coordinate equal to 2.

OBJECTIVES

- To identify the coordinates of points that satisfy a given linear inequality in two variables

- To graph a linear inequality in two variables

- To solve applied problems relating to the graph of a linear inequality

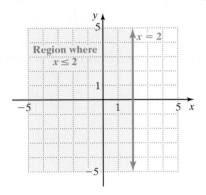

In this section, we focus our attention on graphing linear inequalities in *two* variables, such as $2x + 3y > 1$.

Definition

A **linear inequality in two variables** is an inequality that can be written in the form $Ax + By < C$, where A, B, and C are real numbers and A and B are not both 0. The inequality symbol can be $<$, $>$, $\leq$, or $\geq$.

As we will see, these inequalities have as their graphs half-planes, which are regions of the coordinate plane bounded by lines.

Consider some situations that give rise to such inequalities:

The number of men m and women w invited to a party is at least 20. $\longrightarrow$ $20 \leq m + w$

Your income i exceeds your expenses e by more than \$1000. $\longrightarrow$ $i > e + 1000$

Let's look at what we mean by a *solution* in such inequalities.

Definition

A **solution to an inequality in two variables** is an ordered pair of numbers that when substituted for the variables makes the inequality a true statement.

EXAMPLE 1

Is the ordered pair (2, 5) a solution to the inequality $y \geq x + 1$?

Solution When we substitute 2 for x and 5 for y in the given inequality, it becomes

$$y \geq x + 1$$
$$5 \overset{?}{\geq} 2 + 1$$
$$5 \geq 3 \quad \text{True}$$

Since $5 \geq 3$ is true, the ordered pair (2, 5) is a solution to $y \geq x + 1$. Is the ordered pair (6, 5) a solution to this inequality?

PRACTICE 1

Is (1, 3) a solution to the inequality $y < x - 1$?

By the *graph* of an inequality in two variables, we mean the set of all points on the plane whose coordinates satisfy the inequality. To explore what such a graph looks like, let's consider $y \geq 2x$. To find the graph of this inequality, we first graph the corresponding equation $y = 2x$. Notice that the *boundary line* $y = 2x$ is drawn as a solid line since the original inequality symbol is $\geq$. This line cuts the plane into two half-planes.

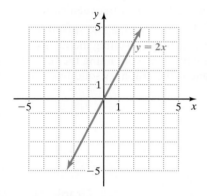

Next, we take an arbitrary point on either side of the boundary line—a *test point*. If the coordinates of the test point satisfy the inequality, then the desired graph contains the half-plane in which the test point lies. Otherwise, the desired graph contains the other half-plane.

Suppose that we take (4, 0) as our test point.

$$y \geq 2x$$
$$0 \overset{?}{\geq} 2(4)$$
$$0 \geq 8 \quad \text{False}$$

Since the inequality does not hold for the point (4, 0), the half-plane that we want is the region above the graph of $y = 2x$, so we shade this region. Therefore, the graph of $y \geq 2x$ is the boundary line and the shaded region.

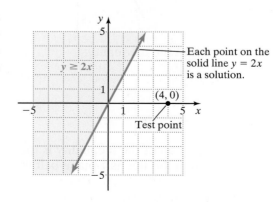

If the inequality had been $y > 2x$, the boundary line would not have been part of the graph. We would have indicated the exclusion of the boundary with a broken line, as in the following diagram.

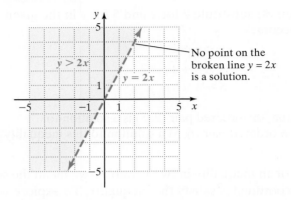

To Graph a Linear Inequality in Two Variables

- Graph the corresponding linear equation. For an inequality with the symbol ≤ or ≥, draw a solid line; for an inequality with the symbol < or >, draw a broken line. This line is the boundary between two half-planes.

- Choose a test point in either half-plane and substitute the coordinates of this point in the inequality. If the resulting inequality is true, then the graph of the inequality is the half-plane containing the test point. If it is not true, then the other half-plane is the graph. A solid line is part of the graph, and a broken line is not.

EXAMPLE 2	PRACTICE 2

EXAMPLE 2

Find the graph of $y - 2x < 4$.

Solution First, we graph the equation $y - 2x = 4$. Solving for y gives $y = 2x + 4$. Next, we graph the line. We draw a broken line since the original inequality symbol is <. Then we choose in either half-plane a test point, say $(0, 0)$.

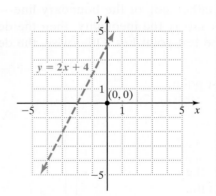

We substitute $x = 0$ and $y = 0$ into the inequality:

$$y - 2x < 4$$
$$0 - 2 \cdot 0 \overset{?}{<} 4$$
$$0 < 4 \quad \text{True}$$

Since the inequality is true, the graph of the inequality is the half-plane containing the test point. So the half-plane below

PRACTICE 2

Graph the inequality $y + 3x \geq 6$.

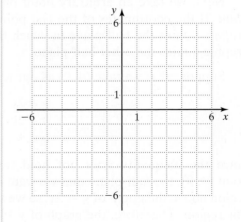

the line is our graph. Note that the graph does not include the boundary line since the original inequality symbol is $<$.

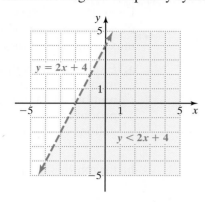

We can test any point on the graph. Why is $(0, 0)$ a good choice?

EXAMPLE 3

Graph $x < 0$.

Solution The boundary line $x = 0$ is the y-axis. It is drawn as a broken line, since the original inequality symbol is $<$. We need to select a test point on either side of the boundary line. Let's take the point, $(1, 0)$ as the test point, which is in the half-plane to the right of the line $x = 0$. Substituting into the inequality $x < 0$, we get $1 < 0$, which is not true. So the graph is the half-plane to the left of the line $x = 0$, but not including this line.

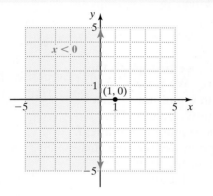

PRACTICE 3

Find the graph of $y \geq -5$.

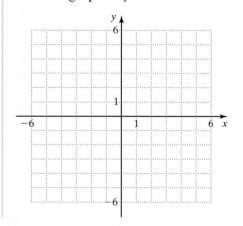

EXAMPLE 4

Graph the inequality $x - 3y \leq -9$ in the first quadrant.

Solution The corresponding equation is $x - 3y = -9$. Solving for y gives $y = \frac{1}{3}x + 3$. Next, we graph this line, drawing a solid line because the original inequality symbol is $\leq$.

Taking $(0, 0)$ as the test point, we check whether $0 - 3(0) \leq -9$. Since this inequality does not hold, the test point is not part of the graph. So the graph in Quadrant I is the region in the quadrant above the line $y = \frac{1}{3}x + 3$, and including it.

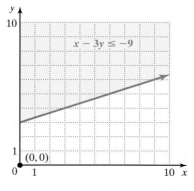

PRACTICE 4

Find the graph of $x - 2y > -6$ in Quadrant I.

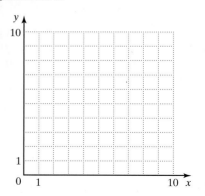

EXAMPLE 5

An online perfumery makes $50 on each bottle of perfume and $20 on each bottle of cologne. To earn a profit, the total amount of money made daily must exceed the perfumery's daily overhead of $1000.

a. Express as an inequality: The amount of money made on selling p bottles of perfume and c bottles of cologne is greater than the overhead.

b. Graph this inequality.

c. Use this graph to decide whether selling 10 bottles of perfume and 12 bottles of cologne results in a profit.

Solution

a. The factory makes $50 on each perfume bottle sold. Therefore, the factory makes $50p$ dollars on selling p bottles of perfume. Similarly, the factory makes $20c$ dollars on selling c bottles of cologne. Since the total amount of money made must be greater than the overhead of $1000, the inequality is

$$50p + 20c > 1000 \quad \text{or} \quad 5p + 2c > 100.$$

b. To graph the inequality $5p + 2c > 100$, we first graph the corresponding equation, $5p + 2c = 100$. We restrict our attention to the portion of the graph in Quadrant I, since the variables can only assume nonnegative values. The boundary line is not included since the symbol in the linear inequality is $>$. Using $(0, 0)$ as a test point, we see that the inequality $5 \cdot 0 + 2 \cdot 0 > 100$ is false. So the graph of $5p + 2c > 100$ is the region in Quadrant I above the boundary line, but not including the line.

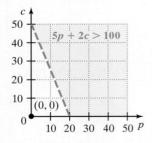

c. To decide whether there was a profit when selling 10 bottles of perfume and 12 bottles of cologne, we plot the point $(10, 12)$. Since the point lies *outside* the graph of our inequality, there was no profit.

PRACTICE 5

Each day, a refinery can produce both diesel fuel and gasoline, with a total maximum output of 3000 gal.

a. Express this relationship as an inequality, where d represents the number of gallons of diesel fuel produced and g the number of gallons of gasoline produced.

b. Graph this inequality.

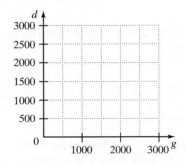

c. Explain the significance of the intercepts of this graph.

3.5 Exercises

Mathematically Speaking

Fill in each blank with the most appropriate term or phrase from the given list.

broken	solution	solid
is	graph	is not
half-plane	ray	

1. The graph of a linear inequality in two variables is a(n) _____.

2. A(n) _____ of an inequality in two variables is an ordered pair of numbers that when substituted for the variables makes the inequality a true statement.

3. The _____ of an inequality in two variables is the set of all points on the plane whose coordinates satisfy the inequality.

4. A(n) _____ boundary line is drawn when graphing a linear inequality that involves the symbol ≤ or the symbol ≥.

5. A(n) _____ boundary line is drawn when graphing a linear inequality that involves the symbol < or the symbol >.

6. If the boundary line is broken, it _____ part of the graph.

Decide if the given ordered pair is a solution to the inequality.

7. $y < 3x$ $(0, 0)$

8. $y > -5x$ $(-1, 4)$

9. $y \geq 2x - 1$ $(-\frac{1}{2}, -2)$

10. $y \leq -\frac{2}{3}x + 5$ $(6, 1)$

11. $2x - 3y > 10$ $(10, 8)$

12. $5x + 3y \geq 12$ $(0, -2)$

Each solid or broken line is the graph of $y = x$. Shade in the graph of the given inequality.

13. $y > x$

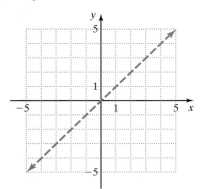

14. $y \geq x$

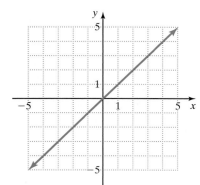

15. $x \leq y$

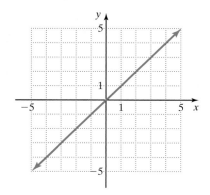

16. $x < y$

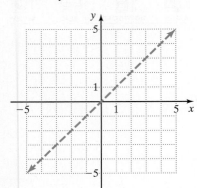

17. $y < x$

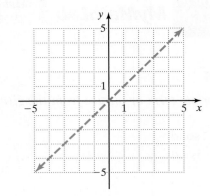

18. $y \le x$

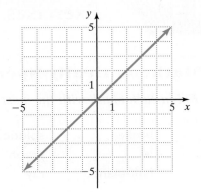

Match each inequality to its graph.

19. $y < \dfrac{1}{4}x - 1$

20. $y > -2x + 3$

21. $x - 4y \le 4$

22. $4x + 2y \ge 6$

a.

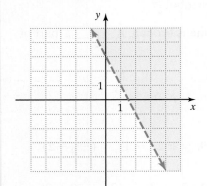

b.

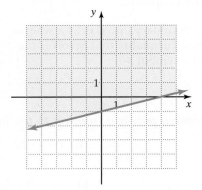

c.

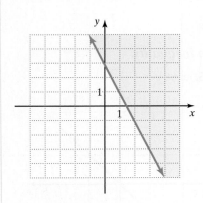

d.

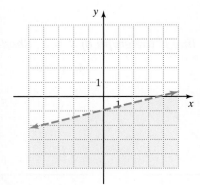

Graph the linear inequality.

23. $x > -5$

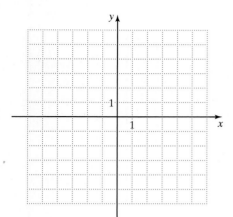

24. $x > 3$

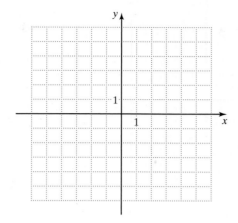

25. $y < 0$

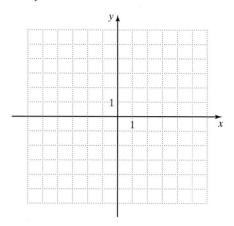

26. $y < 4$

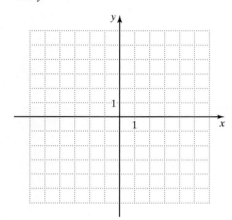

27. $y \le 3x$

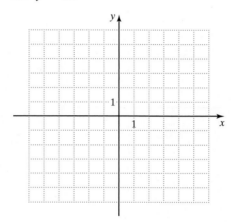

28. $y \le -x$

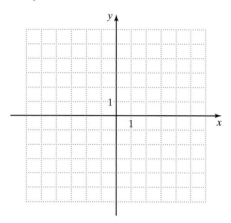

29. $y \ge -2x$

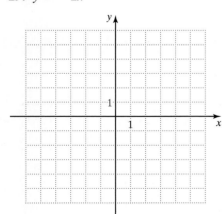

30. $y \ge 4x$

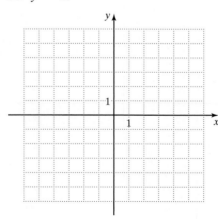

31. $y \le \dfrac{1}{2}x$

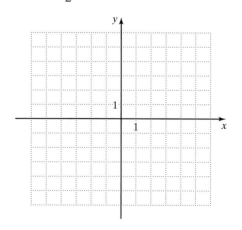

32. $y > -\dfrac{2}{3}x$

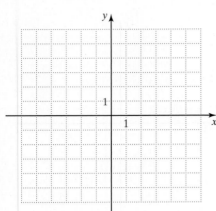

33. $y > 3x + 5$

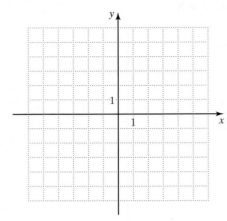

34. $y \geq -x - 1$

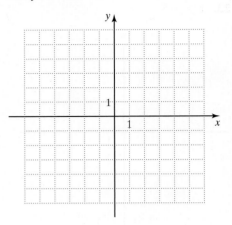

35. $5y - x > 10$

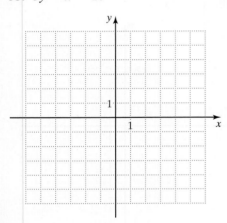

36. $4y + x < -12$

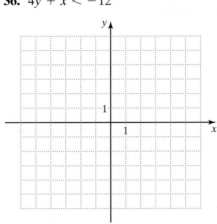

37. $2x - 3y \geq 3$

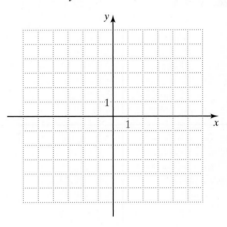

38. $6x - 4y \leq 8$

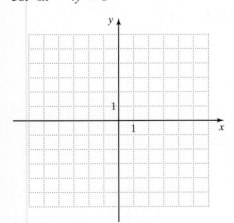

39. $4x - 5y \leq 10$

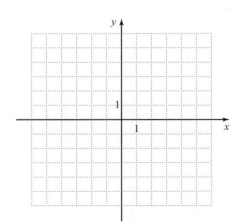

40. $-3x + 2y < 1$

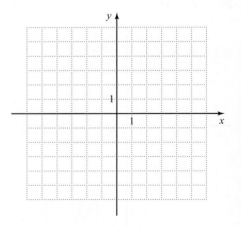

Mixed Practice

Decide if the given ordered pair is a solution to the inequality.

41. $y > -\dfrac{1}{2}x + 2$ $(4, 0)$

42. $x - 2y \leq -2$ $(8, 6)$

43. Which equation describes the graph?

    a. $y + 4x \geq 2$

    b. $y - 4x \geq 2$

    c. $y > 4x + 2$

    d. $y < 4x + 2$

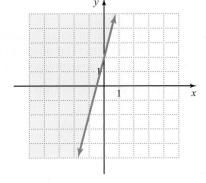

44. Shade in the graph of $x > y$.

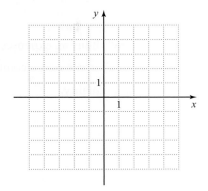

Graph the linear inequality.

45. $y < -3$

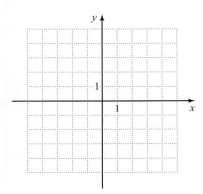

46. $y \leq \dfrac{2}{3}x$

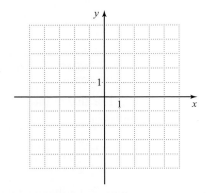

47. $y > -2x - 2$

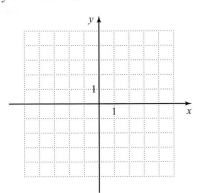

48. $2x - 3y < 6$

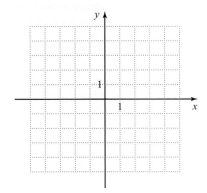

Applications

Solve.

49. According to a guideline, a family's housing expenses h should be less than $\frac{1}{4}$ of the family's combined income i.

 a. Express this guideline as an inequality.

 b. Graph this inequality.

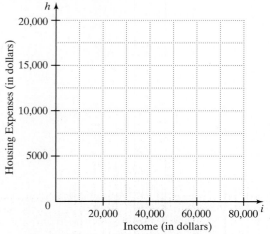

 c. Choose a point on the graph. For this point, explain why the guideline holds.

50. To purchase an apartment in a particular building, a buyer is allowed to take out a mortgage m that is at most 75% of the apartment's selling price s.

 a. Express this relationship as an inequality.

 b. Graph this inequality.

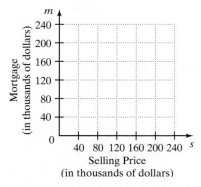

 c. Plot the point (100, 90). For this point, explain in terms of the mortgage policy of the building whether the buyer will be able to buy the apartment.

51. A printing company ships x copies of a college's telephone directory to the uptown campus and y copies to the downtown campus. The company must ship a total of at least 200 copies to these two locations.

 a. Express this relationship as an inequality.

 b. Graph this inequality.

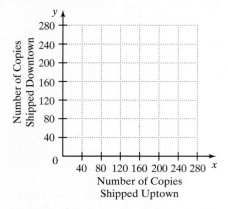

 c. Give the coordinates of a point that satisfies the inequality. Check that the coordinates satisfy the company's shipping requirement.

52. An investor has a maximum of $1000 with which to purchase stock. He purchases x shares of stock that sell for $12 apiece and y shares of stock selling for $5 apiece.

 a. Express this information as an inequality.

 b. Graph this inequality.

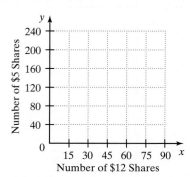

 c. Find a solution to this inequality.

53. A local gourmet coffee shop sells small and large gift baskets. A small gift basket sells for $30 and a large gift basket sells for $75. The coffee shop would like a revenue of at least $1500 per month on the sale of gift baskets.

   a. Write an inequality, where x is the number of small gift baskets sold in a month and y is the number of large gift baskets sold, to represent this situation.

   b. Graph this inequality.

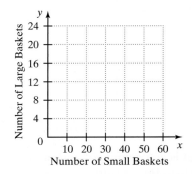

   c. Use the graph to determine if selling 20 small gift baskets and 20 large gift baskets will generate the desired revenue.

54. In moving into a new apartment, a young couple needs to borrow money for both furniture and a car. The loan for furniture has a 10% annual interest rate, whereas the car loan has a 5% annual interest rate. The couple can afford at most $2000 in interest payments for the year.

   a. Express the given information as an inequality, representing the car loan by c and the furniture loan by f.

   b. Graph this inequality.

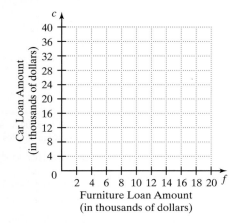

   c. What is the maximum car loan that the couple can afford?

   d. If the couple borrows $20,000 for the car loan, what is the most that they can afford to borrow for furniture?

55. While on a diet, a model wants to snack on fresh apples and bananas. An apple contains 60 calories and a banana contains 100 calories. If she wants to consume fewer than 300 calories, find the number of apples a and the number of bananas b that she can eat. Solve this problem graphically.

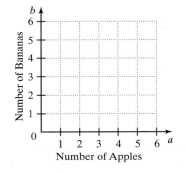

56. A student on spring break wants to drive no more than 300 mi in one day. The trip is along two highways. On the first highway, the student drives at an average speed of 60 mph for x hr and on the second highway at a speed of 50 mph for y hr. What are some possible times that the student can drive? Solve this problem graphically.

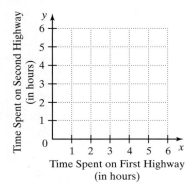

57. A plane is carrying bottled water and medicine to victims of a flood. Each bottle of water weighs 10 lb, and each container of medicine weighs 15 lb. The plane can carry a maximum of 50,000 lb of cargo.

   a. Express this weight limitation of the cargo as an inequality in terms of the number of bottles of water w and the number of medicine containers m in the plane.

   b. Graph this inequality.

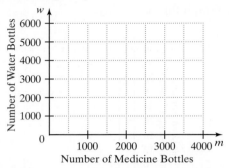

   c. Identify several quantities of water and medicine that the plane can carry.

58. An elevator has a maximum capacity of 1600 lb. Suppose that the average weight of an adult is 160 lb and the average weight of a child is 40 lb.

   a. Write an inequality that relates the number of adults a and the number of children c who can ride an elevator without overloading it.

   b. Graph this inequality.

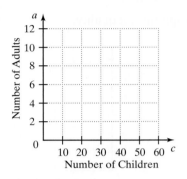

   c. What is an example of a group of people who will overload the elevator?

59. A student has two part-time jobs. One pays $8 per hour, and the other $10 per hour. Between the two jobs, the student needs to earn at least $200 per week.

   a. Write an inequality that shows the number of hours that the student can work at each job.

   b. Graph this inequality.

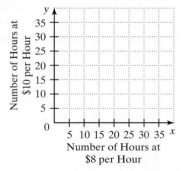

   c. Give some examples of the number of hours that the student can work at each job.

60. Scientists who study weather have developed linear models that relate a region's weather conditions to the kind of vegetation that grows in the region. One such model predicts desert conditions if $3t - 35p > 140$, where t represents the average annual temperature (in degrees Fahrenheit) and p the annual precipitation (in inches).

   a. Graph this relationship on the coordinate plane.

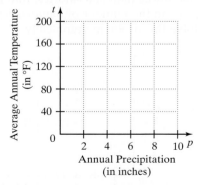

   b. Give some examples of weather conditions that this model predicts will lead to desert conditions.

• Check your answers on page A-15.

MINDSTRETCHERS

Mathematical Reasoning

1. Consider the three graphs: $y < b$, $y = b$, and $y > b$, where b is a positive number. If you were to graph $y < b$, $y = b$, and $y > b$ on the same coordinate plane, what would you get? Would you get the same answer if b were negative?

Groupwork

2. In playing a carnival game, you roll a pair of dice—a red die and a blue die—each with six faces numbered 1, 2, 3, 4, 5, and 6. The grid below shows all the possible outcomes when you roll the two dice.

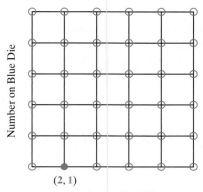

Number on Blue Die

(2, 1)

Number on Red Die

   For each point on the grid, the first coordinate represents the roll on the red die and the second coordinate represents the roll on the blue die. For instance, the point (2, 1) corresponds to rolling a 2 on the red and a 1 on the blue.

a. How many points in all are there on the grid?

b. If the number on the blue die is greater than the number on the red die, you will win a prize. Fill in the points on the grid that correspond to winning a prize. How many points did you fill in?

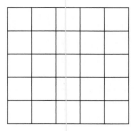

c. What fraction of the total number of points on the grid did you fill in? What does this fraction represent?

Writing

3. Compare solving a linear *equation* in two variables by graphing with solving a linear *inequality* in two variables by graphing. What are the similarities? What are the differences?

3.6   Introduction to Functions

OBJECTIVES

- To identify a function
- To determine the domain and range of a function
- To evaluate a function written in function notation
- To identify linear functions
- To recognize the graph of a function
- To solve applied problems involving functions

In this section, we introduce the concept of a *function*. This concept is one of the most important in mathematics and will be developed more fully in later courses. We begin the section by discussing the idea of a *relation*—a more general concept than function, and then explain why some relations are functions and others are not. Next we consider the domain and the range of a function, the system of notation based on functions, and linear functions. The section concludes with some applications of functions.

Functions

Let's consider a specific example that will lead us to the concept of function. Suppose that a hospital keeps a record of the birth lengths and birth weights of babies born in the maternity ward. One night, five babies were born with length and weight as shown in the following table:

Birth Length (in centimeters)	46	58	48	30	52
Birth Weight (in kilograms)	3.0	4.0	3.5	2.3	4.5

We can plot these values on a coordinate plane to get a better feel for any trends:

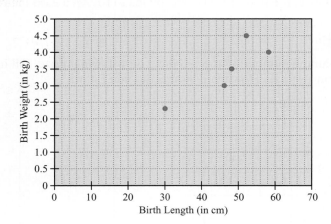

Whether expressed in a table or in a graph, the given information defines a *relation* between birth length and birth weight. We can think of this relation as a set of ordered pairs: {(46, 3.0), (58, 4.0), (48, 3.5), (30, 2.3), (52, 4.5)}. In general, we can define a relation as follows:

Definition

A **relation** is a set of ordered pairs.

The maternity ward example defines a special kind of relation called a **function**. We know that the relation is a *function* because all the first coordinates, here birth lengths, are different from one another. If two babies had been born with the same birth length but with different birth weights, which is certainly possible, the relation would not have been a function.

> **Definition**
> A **function** is a relation in which no two ordered pairs have the same first coordinates.

In any function, the first coordinates will be values of one variable (the *independent variable*) and the second coordinates will be values of another variable (the *dependent variable*). So in our example, the first coordinates are values of birth length and the second coordinates are values of birth weight. We say that the dependent variable *is a function of* the independent variable: Birth weight depends on birth length. If we write the ordered pairs in the reverse order, how will the dependent and independent variables be affected?

For any function, there is one and only one value of the dependent variable for each value of the independent variable. For instance, your income, if you worked at a fixed hourly wage, is a function of the number of hours you work and is uniquely determined by that number.

EXAMPLE 1	PRACTICE 1
Determine whether each relation represents a function.	Decide whether each relation represents a function.

EXAMPLE 1

Determine whether each relation represents a function.

a. $\{(-3, 1), (0, 5), (1, 1), (3, 2)\}$

b.

x	4	1	0	4
y	−2	−1	0	2

Solution

a. In this relation, the ordered pairs $(-3, 1)$ and $(1, 1)$ have the same second coordinate, 1. However, all the first coordinates are different from one another. So this relation represents a function.

b. From the table we see that the x-value 4 is paired with two y-values, −2 and 2. So this relation does not represent a function.

PRACTICE 1

Decide whether each relation represents a function.

a. $\{(-5, 4), (-2, 0), (0, 2), (3, 7), (0, 8)\}$

b.

x	4	2	5	−1
y	1	3	0	1

Domain and Range

Each function has a domain and a range. By the **domain** of a function, we mean the set of all first coordinates of the ordered pairs that make up the function. These first coordinates are the possible values of the independent variable. By the **range** of a function, we mean the set of all second coordinates, that is, the values of the dependent variable that the function assigns to the first coordinates.

In the maternity ward example, recall that the function is the set of ordered pairs (46, 3.0), (58, 4.0), (48, 3.5), (30, 2.3), and (52, 4.5). The following diagram shows the domain and range of this function:

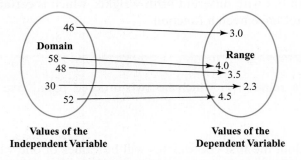

Values of the
Independent Variable

Values of the
Dependent Variable

Definition

The **domain** of a function is the set of all values of the independent variable.

The **range** of a function is the set of all values of the dependent variable.

EXAMPLE 2

Find the domain and range of each function.

a. The function defined by the set of ordered pairs:
$\{(1, 5), (8, -2), (0, 4)\}$

b. The function defined by the following graph:

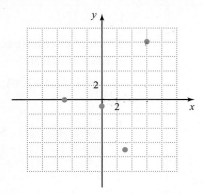

c. The function defined by the following table:

x	0	5	3
y	8	-2	-1

Solution

a. The domain is the set of first coordinates among the ordered pairs that define the function: $\{0, 1, 8\}$. The range is the set of second coordinates: $\{-2, 4, 5\}$.

b. The points plotted on the graph are $(-5, 0)$, $(0, -1)$, $(3, -7)$, and $(6, 8)$. The domain is the set of first coordinates: $\{-5, 0, 3, 6\}$. The range is the set of second coordinates: $\{-7, -1, 0, 8\}$.

c. The domain is the set of x-values in the top row: $\{0, 3, 5\}$. The range is the set of y-values in the bottom row: $\{-2, -1, 8\}$.

PRACTICE 2

Determine the range and domain of each of the following functions.

a. The function defined by the ordered pairs: $(2, 0)$, $(8, -6)$, $(5, 3)$

b. The function defined by the following graph:

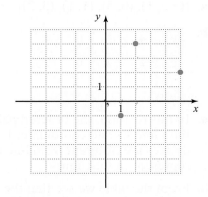

c. The function defined by the following table:

Day of the Month	1	2	3	4	5
Hours of Sleep	6.5	9	7	8	7.5

Function Notation

Another way to think about a function is as a rule (or correspondence) that assigns to each value of the independent variable a single value of the dependent variable. From this point of view, a function is a kind of input-output machine, where the independent variable is the input and the dependent variable is the output.

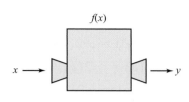

The rule that defines a function is often written as an equation. Consider, for instance, the equation $y = 4x$, which relates the length of a side of a square, x, and its perimeter, y. In this equation, the value of y depends on x and we say that y is a function of x. Using *function notation*, we can rewrite this equation as $f(x) = 4x$. The expression $f(x)$ is read "f of x." The length of a side of the square, x, is the input, and the perimeter of the square, $f(x)$, is the output. The rule of this function, $4x$, assigns to each value x one value $f(x)$. Since both y and $f(x)$ represent $4x$, we can write $y = f(x)$.

> **Tip**   In function notation, the parentheses enclose the quantity that is the independent variable. They do not indicate multiplication.

A function, such as $f(x) = 4x$, can also be represented by a graph. On the following coordinate plane, we can determine the value of $f(x)$ for a particular x-value by finding the y-coordinate of the point on the graph that has that x-value:

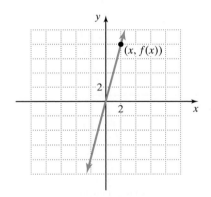

The point shown on the graph has coordinates (2, 8); therefore we can conclude that the value of the function is 8 when the value of x is 2, or that a square with side length 2 units has a perimeter of 8 units.

We can express point (2, 8) using function notation. Since $f(x) = 4x$, we write $f(2) = 8$ (read "f of 2 is 8") to indicate that when $x = 2$ the function has value $4 \cdot 2$, or 8. In general, we can evaluate a function by substituting the specific value of the independent variable given in parentheses in the rule that defines the function.

EXAMPLE 3	PRACTICE 3

EXAMPLE 3

Let $f(x) = -2x + 4$. Find:

a. $f(5)$ **b.** $f(-3)$ **c.** $f(a)$ **d.** $f(2a)$

PRACTICE 3

Given that $g(x) = 3x - 1$, evaluate each of the following.

a. $g(2)$ **b.** $g(-1)$

c. $g(n)$ **d.** $g(n + 1)$

Solution

a. By $f(5)$, we mean the value of $f(x)$ when $x = 5$.

$$f(x) = -2x + 4$$
$$f(5) = -2(5) + 4 = -10 + 4 = -6$$

b. $f(-3) = -2(-3) + 4 = 6 + 4 = 10$

c. $f(a) = -2(a) + 4 = -2a + 4$

d. $f(2a) = -2(2a) + 4 = -4a + 4$

Note that while the letters f and g are frequently used to represent functions, we can use any letter to represent either a function or its independent variable. So $f(x) = 2x$ and $g(t) = 2t$ both indicate the same function whose value is twice the value of the independent variable.

Linear Functions

A *linear* function is a function that can be defined by $f(x) = mx + b$, where m and b are real numbers. An example of a linear function is $f(x) = 3x - 1$, where $m = 3$ and $b = -1$. Let's take a look at the graph of this function.

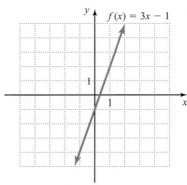

Linear Function

Recall that the graph of a line goes on indefinitely in both directions. We can see from the graph of $f(x) = 3x - 1$ that x can take on any real value, as can y. So the domain of this function is the set of all real numbers, or in interval notation, $(-\infty, \infty)$. The range of this function is also $(-\infty, \infty)$ since the graph extends as far above or below the x-axis as desired.

A *constant* function is a linear function that can be defined by $f(x) = k$, where k is a real number. For instance, consider the constant function $f(x) = 3$, whose graph is shown on the coordinate plane to the left. We see from the graph that the domain of the function is the set of all real numbers, or in interval notation, $(-\infty, \infty)$. However, the range is $\{3\}$, since the function only takes on the value 3.

Constant Function

We will investigate some nonlinear functions in later chapters.

To graph a function, we find several ordered pairs by evaluating the function. Then, we plot the ordered pairs in a coordinate plane. Finally, we connect the points. Note that since $y = f(x)$, the ordered pairs (x, y) and $(x, f(x))$ represent the same point.

EXAMPLE 4	PRACTICE 4

Graph $f(x) = x - 1$, for $x \leq 0$. What are the domain and range of this function?

Solution Let's begin by identifying some ordered pairs to plot.

x	$f(x) = x - 1$	(x, y)
-4	$f(-4) = -4 - 1 = -5$	$(-4, -5)$
-3	$f(-3) = -3 - 1 = -4$	$(-3, -4)$
-2	$f(-2) = -2 - 1 = -3$	$(-2, -3)$
-1	$f(-1) = -1 - 1 = -2$	$(-1, -2)$
0	$f(0) = 0 - 1 = -1$	$(0, -1)$

Next we plot the points and draw the line for all x-values less than or equal to 0, as shown in the following graph:

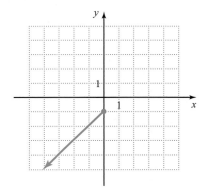

From the graph we see that the domain is $(-\infty, 0]$. The graph indicates that the dependent variable y is any negative number less than or equal to -1. So the range is $(-\infty, -1]$.

Graph $f(x) = 2x$, for $x \geq 0$, and determine the domain and range of this function.

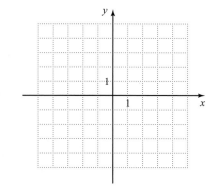

The Vertical-Line Test

One way of representing a function is as a graph. However, some graphs, while representing a relation, do not represent a function. Recall that a relation is only a function if no two distinct ordered pairs have the same first coordinate. It follows that the graph of a function cannot have two points with the same x-coordinate but with different y-coordinates. This suggests a test for determining if a graph represents a function.

The Vertical-Line Test

If any vertical line intersects a graph at more than one point, the graph does not represent a function. If no such line exists, then the graph represents a function.

EXAMPLE 5	PRACTICE 5

Consider the following graphs. Determine if each graph represents a function.

Determine whether each of the following graphs represents a function.

a.

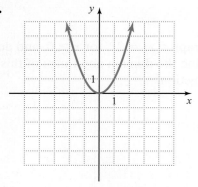

a.

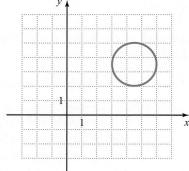

b.

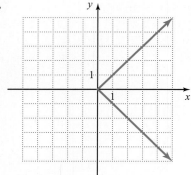

b.

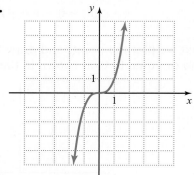

Solution

a. We see by inspecting the graph that any vertical line on the plane intersects the graph at one point. So the vertical-line test tells us that this graph represents a function.

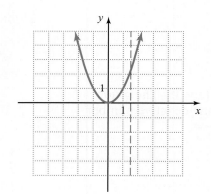

b. Since any vertical line in Quadrants I and IV intersects the graph more than once, the graph fails the vertical-line test. So this graph does not represent a function.

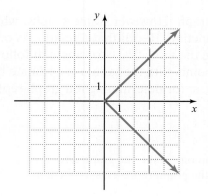

Note that in the following example, only nonnegative values of the two variables make sense, restricting the domain and the range.

EXAMPLE 6	PRACTICE 6

EXAMPLE 6

A salesperson earns $20,000 a year plus 10% commission on total sales.

a. Use function notation to express the relationship between the salesperson's yearly earnings $E(s)$ and her total sales s.

b. Express in function notation the annual earnings made on the sale of merchandise worth $30,000. Under these conditions, how much money was earned?

c. Graph the relationship in part (a).

d. Find the domain and the range of the function.

e. Explain how you could conclude that the relationship is a function by examining the graph in part (c).

Solution

a. We know that the salesperson's earnings $E(s)$ amount to $20,000 a year plus an additional 10% commission on her sales s, that is, $E(s) = 20,000 + 0.1s$. Alternatively, we can write this relation as $E(s) = 0.1s + 20,000$.

b. Using function notation, we can represent the total annual earnings on sales of $30,000 by $E(30,000)$. To evaluate, we substitute 30,000 for s in the expression $0.1s + 20,000$. So $E(30,000) = 0.1(30,000) + 20,000 = 23,000$, that is, $23,000.

c. We graph $E(s) = 0.1s + 20,000$.

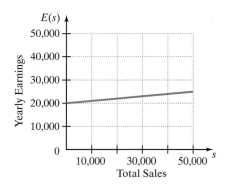

d. Inspecting the graph leads us to conclude that the domain is all real numbers greater than or equal to 0, that is, $[0, \infty)$. The range is all real numbers greater than or equal to 20,000, or in interval notation, $[20,000, \infty)$.

e. The graph in part (c) passes the vertical-line test, showing that the relation is a function.

PRACTICE 6

The cost of renting a car at a local dealer is $40 per day plus $0.10 a mile.

a. Express in function notation the relationship between the daily rental cost in dollars $C(m)$ and the number of miles driven m.

b. Express in function notation the cost of renting a car for a day and driving 200 mi. Then evaluate the function.

c. Graph the relation.

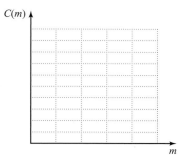

d. Find the domain and the range of the function.

e. Explain how you could conclude that the relation is a function by examining the graph in part (c).

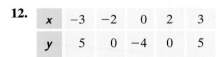

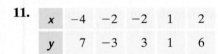

Mathematically Speaking

Fill in each blank with the most appropriate term or phrase from the given list.

output	range	represents	vertical line
does not represent	function	dependent	independent
horizontal line	domain	relation	input

1. A(n) _____ is a set of ordered pairs.

2. A function's first coordinates are values of the _____ variable.

3. For any function, there is one and only one value of the _____ variable for each value of the other variable.

4. The _____ of a function is the set of all values of the independent variable.

5. The _____ of a function is the set of all values of the dependent variable.

6. If we think of a function as an input-output machine, the independent variable is the _____.

7. The graph of a constant function is a(n) _____.

8. If any vertical line intersects a graph at more than one point, the graph _____ a function.

Determine whether each relation represents a function.

9. {(−3, 2), (0, 1), (3, 2), (4, −4)}

10. {(1, 1), (1, −1), (2, −4), (2, 4)}

11.

x	−4	−2	−2	1	2
y	7	−3	3	1	6

12.

x	−3	−2	0	2	3
y	5	0	−4	0	5

13.

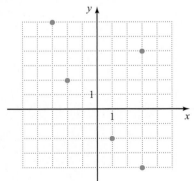

14.

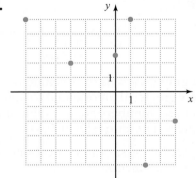

15.

Radius of a Circle, r	1	2	3	4	5
Circumference, C	2π	4π	6π	8π	10π

16.

Time, t (in seconds)	0	5	10	15
Distance, d (in feet)	0	20	40	60

Find the domain and range of the following functions.

17. $\{(-2, 6), (-1, 8), (0, 10), (1, 12), (2, 14)\}$

18. $\{(-6, 3), (-4, 2), (0, 0), (4, -2), (6, -3)\}$

19. $\{(-3, -27), (-1, -1), (0, 0), (1, 1), (3, 27)\}$

20. $\{(-4, 7), (-3, 0), (-2, -5), (-1, -8), (0, -9)\}$

21.

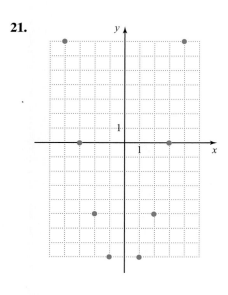

22.

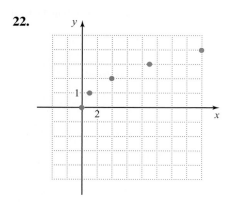

23.

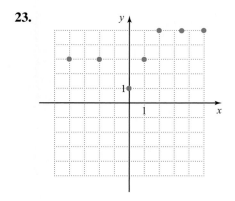

24.

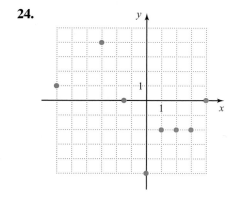

25.

x	−7	−3	−1	2	4
y	7	3	1	−2	−4

26.

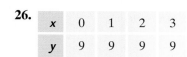

x	0	1	2	3
y	9	9	9	9

27.

Year	2000	2001	2002	2003	2004
Score of Winning Superbowl Team	23	34	20	38	32

(*Source:* National Football League)

28.

Election Year	1988	1992	1996	2000
Popular Votes Cast for Winning U.S. Presidential Candidate	47,946,000	44,908,254	45,590,703	50,456,062

(*Source:* U.S. National Archives & Records Administration, Federal Register)

Evaluate each function for the given values.

29. $f(x) = 8 - 5x$

   a. $f(2)$ **b.** $f(-1)$

   c. $f\left(\dfrac{3}{5}\right)$ **d.** $f(1.8)$

30. $g(x) = \dfrac{1}{2}x - 1$

   a. $g(0)$ **b.** $g(-6)$

   c. $g(14)$ **d.** $g(-5.2)$

31. $g(x) = 2.4x - 7$

   a. $g(5)$ **b.** $g(-2)$

   c. $g(a)$ **d.** $g(a^2)$

32. $f(x) = -3.5x + 1.5$

   a. $f(3)$ **b.** $f(-1)$

   c. $f(n)$ **d.** $f(n^3)$

33. $f(x) = \left|\dfrac{1}{2}x + 3\right|$

   a. $f(0)$ **b.** $f(-8)$

   c. $f(-4t)$ **d.** $f(t - 6)$

34. $h(x) = 1 - |2x|$

   a. $h(-5)$ **b.** $h\left(\dfrac{1}{2}\right)$

   c. $h(-a)$ **d.** $h(a + 1)$

35. $h(x) = 3x^2 - 6x - 9$

   a. $h(2)$ **b.** $h(-1)$

   c. $h(-n)$ **d.** $h(2n)$

36. $f(x) = 2x^2 + x$

   a. $f(-2)$ **b.** $f\left(-\dfrac{1}{2}\right)$

   c. $f(a)$ **d.** $f(-3a)$

37. $g(x) = 10$

   a. $g(7)$ **b.** $g(-150)$

   c. $g(t)$ **d.** $g(5 - 9t)$

38. $g(x) = -4$

   a. $g(23)$ **b.** $g(-41)$

   c. $g(5n)$ **d.** $g(2n^2)$

Graph each function. Then identify its domain and range.

39. $f(x) = 5x - 4$

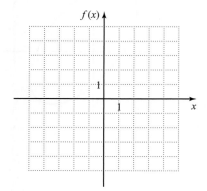

40. $f(x) = -3x + 1$

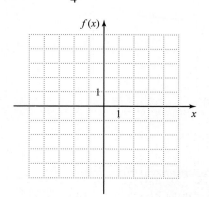

41. $f(x) = -5$

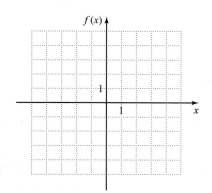

42. $h(x) = 2$

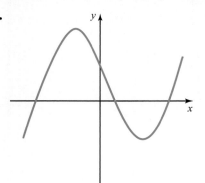

43. $f(x) = -\dfrac{1}{4}x - 1$, for $x \leq 0$

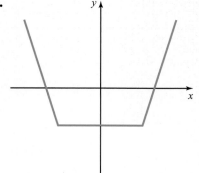

44. $f(x) = \dfrac{3}{2}x + 4$, for $x \geq -4$

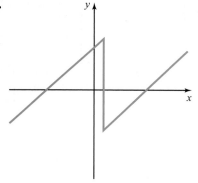

Determine whether each graph represents a function.

45.

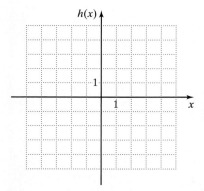

46.

47.

48.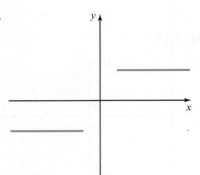

49.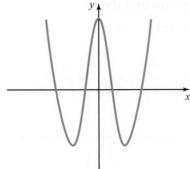

50.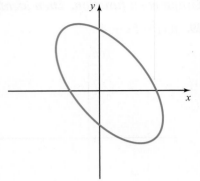

51.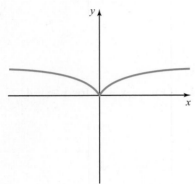

52.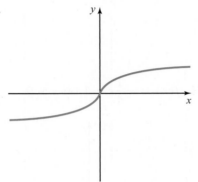

Mixed Practice

Determine whether each relation represents a function.

53. {(−2, 6), (−1, 4), (5, 8), (10, −6), (11, 4)}

54.

x	−2	0	0	1	2
y	1	3	5	−3	4

55.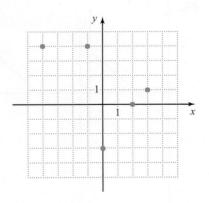

Find the domain and range of each of the following functions.

56. $\{(-5, 2), (-2, 7), (0, -6), (4, 1), (7, 33)\}$

57.

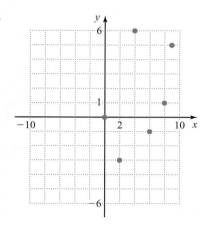

Evaluate each function for the given values.

58. $g(x) = 3.2x - 4$

   a. $g(2)$ **b.** $g(-2)$ **c.** $g(a)$ **d.** $g(b^2)$

59. $f(x) = 3x^2 + x$

   a. $f(0)$ **b.** $f(-2)$ **c.** $f(n)$ **d.** $f(-4n)$

60. Determine whether the graph to the right represents a function.

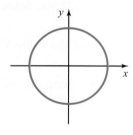

Applications

Solve.

61. A department store is running a sale that gives customers 20% off the total amount of their purchases.

  a. Use function notation to express the relationship between the total discount $d(a)$ and the amount of a purchase a before the discount.

  b. Use function notation to express the total amount of the discount on a purchase of $150. How much money was saved on the purchase?

  c. Graph the function in part (a).

  d. Identify the domain and range of the function.

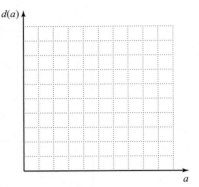

62. A paper manufacturer makes square sheets of wrapping paper.

  a. Write a function that relates the perimeter $p(s)$ and the length of a side s of a square sheet of wrapping paper.

  b. Express the perimeter of a square sheet that measures 12 in. on a side in function notation. What is the perimeter of this sheet of wrapping paper?

  c. Graph the function in part (a).

  d. Find the domain and range of the function.

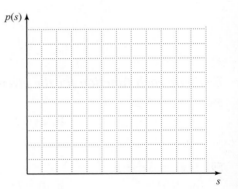

63. A car that is purchased new for $22,500 depreciates in value by $1875 per year.

  a. Use function notation to show the value $V(t)$ of the car t yr after it is purchased.

  b. What is the meaning of $V(6)$? Find this value.

  c. Graph the function in part (a).

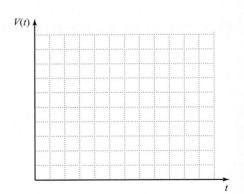

64. A cell phone company charges a $50 activation fee plus $45 per month for unlimited calling within the network.

  a. Use function notation to represent the total amount paid $A(x)$ for x mo of service.

  b. Find $A(4)$. Interpret its meaning in this situation.

  c. Graph the function in part (a).

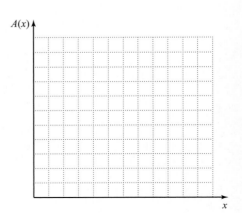

65. A patient's weekly dosage of 500 mg of a medication is reduced by 50 mg per week.

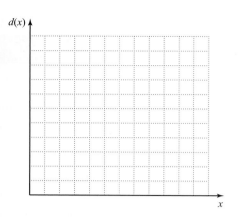

  a. Express in function notation the relationship between the patient's weekly dosage $d(x)$ and the number of weeks x.

  b. Find $d(2)$ and interpret its meaning in this situation.

  c. Graph the function in part (a).

66. A manufacturing plant has monthly fixed costs of $10,000 plus $1.20 for each unit it produces.

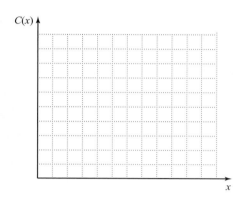

  a. Write the relationship between the number of units produced x and the manufacturing plant's monthly cost $C(x)$.

  b. What is $C(500)$? Explain its meaning in the context of the situation.

  c. Graph the function in part (a).

● *Check your answers on page A-17.*

MINDSTRETCHERS

Groupwork

1. Working with a partner, determine the domain and range of each of the following functions.

a.

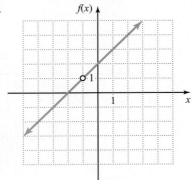

b.

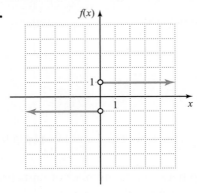

c.

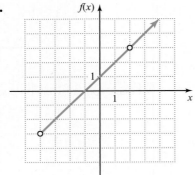

d.

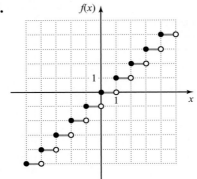

Critical Thinking

2. Some functions are described by two or more rules of correspondence that are evaluated for specific values of the domain. These functions are called *piecewise defined* functions. One such function is:

$$f(x) = \begin{cases} 2x + 5, & x \le -1 \\ -x - 2, & x > -1 \end{cases}$$

  a. Which rule is evaluated if $x = 2$? Explain.

  b. Which rule is evaluated if $x = -3$? Explain.

  c. Find the following function values.

   i. $f(-4)$ ii. $f(-1)$ iii. $f(0)$ iv. $f(3)$

Mathematical Reasoning

3. Can a relation whose domain contains fewer values than its range ever define a function? Explain.

KEY CONCEPTS AND SKILLS

CONCEPT   SKILL

Concept/Skill	Description	Example
[3.1] Bar graph	A graph in which quantities are represented by thin, parallel rectangles called bars. The length of each bar is proportional to the quantity that it represents.	
[3.1] Line graph	A graph in which quantities are represented as points connected by straight-line segments. The height of any point on a line is read against the vertical axis.	
[3.1] Coordinate Plane	A flat surface on which we draw graphs.	
[3.1] Ordered Pair	A pair of numbers that represents a point in the coordinate plane.	

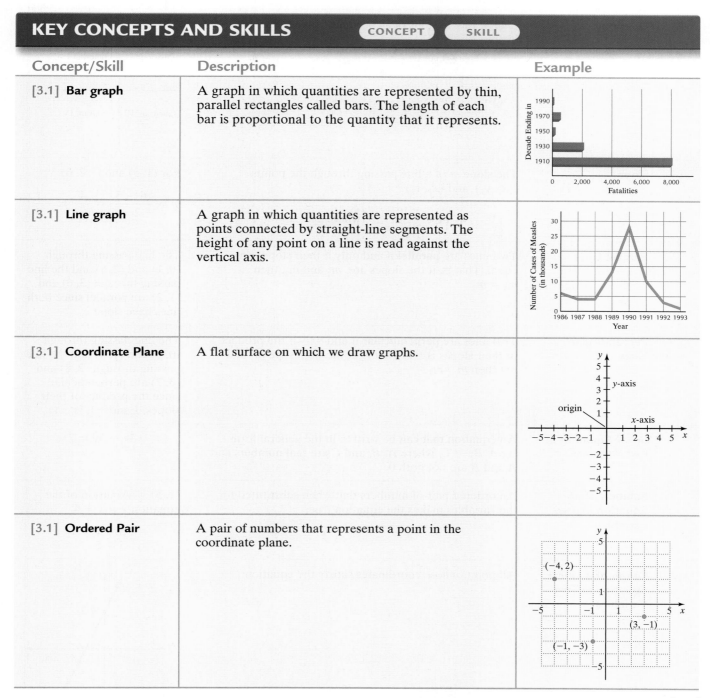

continued

Concept/Skill	Description	Example
[3.1] Quadrant	One of four regions of a coordinate plane separated by axes.	
[3.2] Slope	The **slope** m of a line passing through the points (x_1, y_1) and (x_2, y_2): $$m = \frac{y_2 - y_1}{x_2 - x_1}, \quad \text{where } x_1 \neq x_2$$	For $(1, 5)$ and $(-2, 6)$, $$m = \frac{6-5}{-2-1} = \frac{1}{-3} = -\frac{1}{3}$$
[3.2] Parallel Lines	Two lines are **parallel** if and only if their slopes are equal. That is, if the slopes are m_1 and m_2, then $m_1 = m_2$.	The line passing through $(0, 1)$ and $(2, 5)$ and the line passing through $(3, 6)$ and $(1, 2)$ are parallel since both lines have slope 2.
[3.2] Perpendicular Lines	Two lines are **perpendicular** if and only if the product of their slopes is -1. That is, if the slopes are m_1 and m_2, then $m_1 \cdot m_2 = -1$.	The line passing through $(0, 3)$ and $(1, 4)$ and the line passing through $(2, 8)$ and $(3, 7)$ are perpendicular since the product of their slopes, 1 and -1, is -1.
[3.3] Linear Equation in Two Variables	An equation that can be written in the general form $Ax + By = C$, where A, B, and C are real numbers and A and B are not both 0.	$3x + 5y = 7$
[3.3] Solution of an Equation in Two Variables	An ordered pair of numbers that when substituted for the variables makes the equation true.	$(1, 5)$ is a solution of the equation $y = x + 4$: $$5 \overset{?}{=} 1 + 4$$ $$5 = 5 \quad \text{True}$$
[3.3] Graph of a Linear Equation in Two Variables	All points whose coordinates satisfy the equation.	

continued

Concept/Skill	Description	Example
[3.3] To Graph a Linear Equation in Two Variables	• Isolate one of the variables—usually y—if it is not already done. • Choose three x-values, entering them in a table. • Complete the table by calculating the corresponding y-values. • Plot the three points—two to draw the line and the third to serve as a *checkpoint*. • Check that the points seem to lie on the same line. • Draw the line passing through the points.	To graph $y - 3x = 1$: $$y = 3x + 1$$ 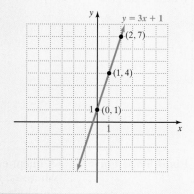
[3.3] Intercepts of a Line	The x-intercept: the point where the graph crosses the x-axis. The y-intercept: the point where the graph crosses the y-axis.	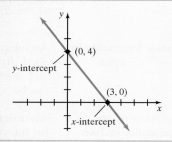
[3.3] To Graph a Linear Equation in Two Variables Using the x- and y-intercepts	• Let $x = 0$ and find the y-intercept. • Let $y = 0$ and find the x-intercept. • Find a checkpoint. • Plot the three points. • Check that the points seem to lie on the same line. • Draw the line passing through the points.	To graph $2x + 3y = 6$: 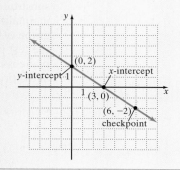

For the first example table:

x	$y = 3x + 1$	(x, y)
0	$y = 3(0) + 1$	$(0, 1)$
1	$y = 3(1) + 1$	$(1, 4)$
2	$y = 3(2) + 1$	$(2, 7)$

For the third example table:

x	$y = -\frac{2}{3}x + 2$	(x, y)
0	$y = -\frac{2}{3}(0) + 2 = 2$	$(0, 2)$
3	$y = -\frac{2}{3}(3) + 2 = 0$	$(3, 0)$
6	$y = -\frac{2}{3}(6) + 2 = -2$	$(6, -2)$

continued

Concept/Skill	Description	Example
[3.4] Slope-Intercept Form	A linear equation written as $y = mx + b$, where m and b are constants. In this form, m is the slope and $(0, b)$ is the y-intercept of the graph of the equation.	The line with slope 5 and y-intercept $(0, 2)$: $$y = mx + b$$ $$y = 5x + 2$$
[3.4] To Graph a Linear Equation in Two Variables Using the Slope and y-intercept	• First locate the y-intercept. • Use the slope to find a second point on the line. • Then draw the line through the two points.	To graph $y = 3x - 1$: The y-intercept is $(0, -1)$ and the slope is 3.
[3.4] Point-Slope Form	A linear equation written as $y - y_1 = m(x - x_1)$, where x_1, y_1, and m are constants. In this form, m is the slope and (x_1, y_1) is a point that lies on the graph of the equation.	The line with slope 5 passing through the point $(2, 1)$: $$y - y_1 = m(x - x_1)$$ $$y - 1 = 5(x - 2)$$
[3.5] Linear Inequality in Two Variables	An inequality that can be written in the form $Ax + By < C$, where A, B, and C are real numbers and A and B are not both 0. The inequality symbol can be $<$, $>$, $\geq$, or $\leq$.	$5x + 3y < 1$
[3.5] Solution of an Inequality in Two Variables	An ordered pair of numbers that when substituted for the variables makes the inequality a true statement.	$(3, 1)$ is a solution to the inequality $x < 5y$: $$3 < 5(1)$$ $$3 < 5 \quad \text{True}$$
[3.5] To Graph a Linear Inequality in Two Variables	• Graph the corresponding linear equation. For an inequality with the symbol $\leq$ or $\geq$, draw a solid line; for an inequality with the symbol $<$ or $>$, draw a broken line. This line is the boundary between two half-planes. • Choose a test point in either half-plane and substitute the coordinates of this point in the inequality. If the resulting inequality is true, then the graph of the inequality is the half-plane containing the test point. If it is not true, then the other half-plane is the graph. A solid line is part of the graph, and a broken line is not.	To graph $y > x + 1$, first graph the line $y = x + 1$. The inequality does not hold for the test point $(0, 0)$. The graph of $y > x + 1$ is the half-plane above and excluding the line $y = x + 1$.

continued

Concept/Skill	Description	Example
[3.6] **Relation**	A set of ordered pairs.	$\{(0, 1), (3, 4)\}$
[3.6] **Function**	A relation in which no two ordered pairs have the same first coordinates.	A function: $\{(1, 4), (2, 4)\}$ Not a function: $\{(4, 1), (4, 2)\}$
[3.6] **Domain of a Function; Range of a Function**	The set of all values of the independent variable. The set of all values of the dependent variable.	Function: $\{(1, 1), (2, 4), (3, 9)\}$ 1 ⟶ 1 2 ⟶ 4 3 ⟶ 9 **Domain** **Range**
[3.6] **Vertical-Line Test**	If any vertical line intersects a graph at more than one point, the graph does not represent a function. If no such line exists, then the graph represents a function.	A function: $y = x^2$ Not a function: $x^2 + y^2 = 4$

Chapter 3   ## Review Exercises

To help you review this chapter, solve these problems.

[3.1]

1. Plot the points with the given coordinates.

   $A(0, 5)$ $B(-1, -6)$ $C(3, -4)$ $D(-2, 2)$

   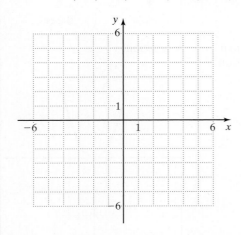

2. Fill in the coordinates of each point.

   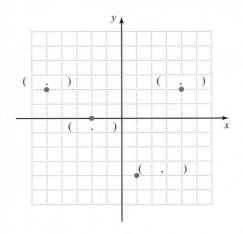

Identify the quadrant in which each point is located:

3. $(5, -1)$

4. $(-7, -2)$

[3.2] *Compute the slope m of the line that passes through the given points. Plot these points on the coordinate plane, and draw the line.*

5. $(2, 0)$ and $(3, 5)$, $m =$

   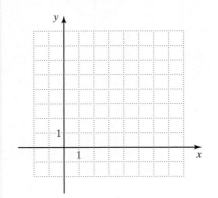

6. $(5, 7)$ and $(2, 7)$, $m =$

   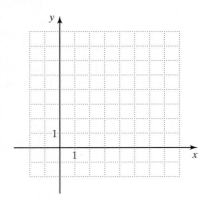

Draw the line on the coordinate plane based on the given information.

7. Passes through $(3, -1)$ and $m = 4$

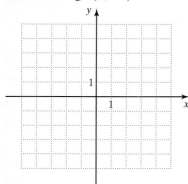

8. Passes through $(0, 0)$ and $m = -\dfrac{1}{2}$

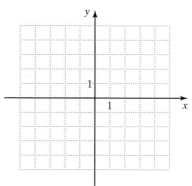

Determine whether the slope of each graph is positive, negative, zero, or undefined.

9.

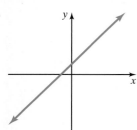

10.

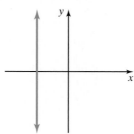

11.

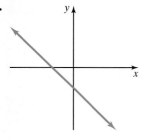

12.

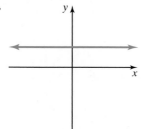

Determine whether $\overleftrightarrow{AB}$ and $\overleftrightarrow{CD}$ are parallel or perpendicular.

13. $A(5, 0)$ $B(3, 0)$ $C(-3, -2)$ $D(1, -2)$

14. $A(4, 8)$ $B(5, 9)$ $C(2, -3)$ $D(0, -1)$

For the following graphed line, find

15. the x-intercept.

16. the y-intercept.

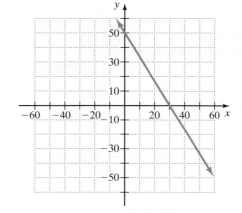

[3.3]   *Complete each table of values for the given equation.*

17. $y = 2x - 5$

x	0	1		
y			0	1

18. $y = -x + 3$

x	2	5		
y			7	-5

Graph.

19. $4x - 3y = -12$

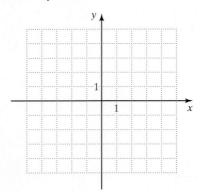

20. $x + 2y = -6$

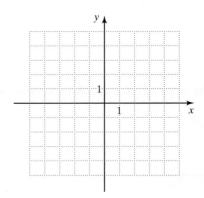

21. $y = \dfrac{1}{2}x$

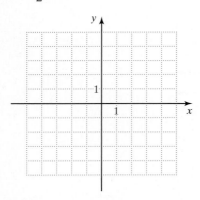

22. $y = -x + 2$

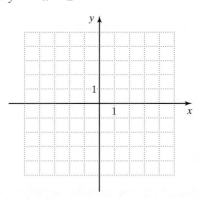

[3.4]   *Write each equation in slope-intercept form.*

23. $x - y = 10$

24. $x + 2y = -1$

Complete the following table.

	Equation	Slope, m	y-intercept (0, b)	Indicate Which Graph Type Best Describes the Line. ╱ ╲ — ▏	x-intercept
25.	$y = 4x - 16$				
26.	$y = -\dfrac{1}{3}x$				

27. Find the slope of a line perpendicular to the line $5x - 10y = 20$.

28. Find the slope of a line parallel to the line $6x - 2y = 2$.

29. Find the equation of the line with slope -1 that passes through the point $(3, 5)$.

30. Write the equation of the horizontal line that passes through the point $(3, 0)$.

31. The points $(2, 0)$ and $(1, 5)$ lie on a line. Find its equation.

32. What is an equation of the line passing through the points $(3, 1)$ and $(-2, 0)$?

Find the equation of each graph.

33.

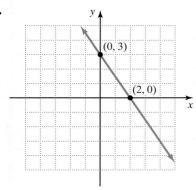

34.

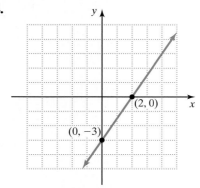

[3.5]   *Decide if the ordered pair is a solution to the inequality.*

35. $(-2, 7)$, $x + y < 1$

36. $(1, -4)$, $2x - 3y \geq 14$

Each line is the graph of $y = -x$. Shade in the graph of the given inequality.

37. $y > -x$

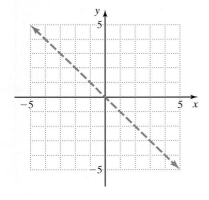

38. $y < -x$

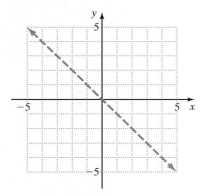

Graph the linear inequality.

39. $y \leq 2x$

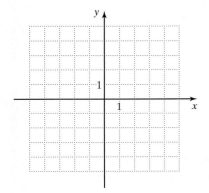

40. $y - x > -1$

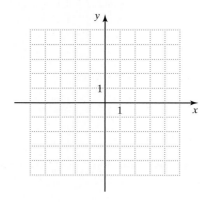

[3.6]   *Determine whether each relation represents a function.*

41. {(1, 5), (4, 9), (6, 11), (8, 5)}

42.

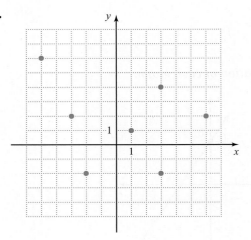

Find the domain and range of each function.

43. {(−7, 3), (−5, 3), (−3, 3), (−1, 3)}

44.

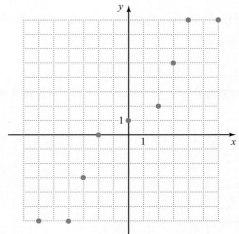

45.

x	−27	−8	−1	0	1	8	27
y	−3	−2	−1	0	1	2	3

46.

Hours Worked	32	26	38	20	24	30
Weekly Pay	$304	$247	$361	$190	$228	$285

Evaluate each function for the given values.

47. $f(x) = \dfrac{1}{3}x + 6$

    a. $f(-9)$ **b.** $f(3.6)$

    c. $f(3a)$ **d.** $f(6a - 12)$

48. $g(x) = |4x - 7|$

    a. $g(3)$ **b.** $g\left(-\dfrac{3}{4}\right)$

    c. $g(2n)$ **d.** $g\left(\dfrac{1}{4}n + 1\right)$

Graph each function. Then identify its domain and range.

49. $f(x) = 4 - \dfrac{1}{2}x$

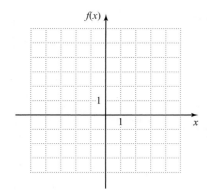

50. $g(x) = 4x - 9$, for $x \geq 1$

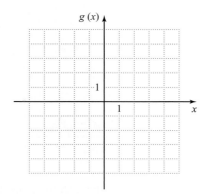

Determine whether each graph represents a function.

51.

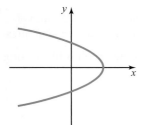

52.

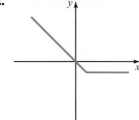

Mixed Applications

Solve.

53. The following graph shows the number of movie screens—indoor and drive-in—in the United States between 2000 and 2005. (*Source:* National Association of Theater Owners)

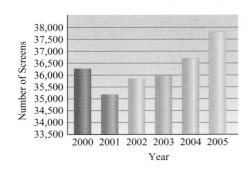

a. Approximately how many movie screens were there in 2003?

b. About how many more movie screens were there in 2005 than in 2000?

c. Describe the trend the graph illustrates.

54. The following graph displays the number of U.S. licensed drivers in a recent year, aged 20 through 84, according to their age.

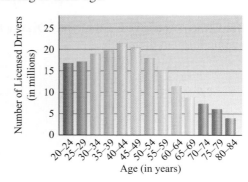

(*Source: Time Almanac 2006*)

a. About how many licensed drivers were between 35 and 49 years of age?

b. If the total number of licensed drivers was about 200 million, approximately what percent of these were between the ages of 35 and 39?

c. Describe the trend in this graph.

55. Consider the following learning curve that shows how long a rat running through a maze takes on each run:

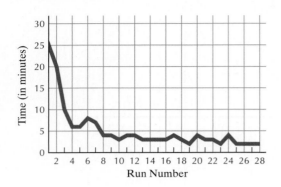

a. On which run does the rat run through the maze in 10 minutes?

b. How long does the rat take on the tenth run?

c. What general conclusion can you draw from this learning curve?

56. The graph shows the health care expenditures from out-of-pocket and insurance sources.

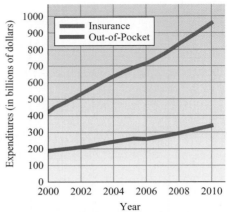

(*Source:* Centers for Medicare and Medicaid Services)

a. In what year were the insurance expenditures approximately $700 billion?

b. In the year 2000, what was the approximate ratio of out-of-pocket expenditures to insurance expenditures?

c. Express in words the trend the graph is illustrating.

57. A bed and breakfast charges R dollars for renting a room for d days.

Number of Days Stayed, d	Cost of the Rental, R
2	180
5	450

a. Graph the points given in the table. Draw a line passing through the points.

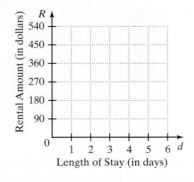

b. What is the R-intercept of this line? Explain its significance in terms of renting a room.

58. The following table shows the cost C in cents of duplicating q flyers at a print shop.

Quantity, q	Cost, C
1	4
10	40

a. Plot the points given in the table and draw the line passing through them.

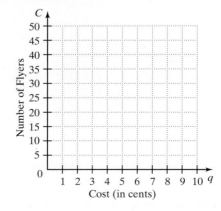

b. Calculate the slope of this line. Explain its significance in terms of the price structure at the print shop.

59. A child walks toward a wall, stands still, and then again walks toward the wall. Which of the following graphs could describe this motion? Explain your answer.

a.

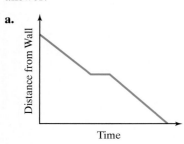

b.

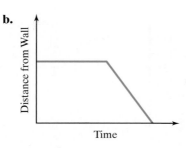

c.

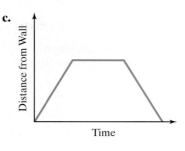

60. The graph shows the altitude of an airplane during a flight. Write a brief story describing the altitude of the plane relative to the duration of the flight.

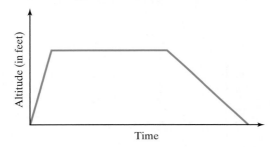

61. A novelist negotiated the following deal with her publisher: a $20,000 bonus plus 9% of book sales.

a. Express the relationship between her income *i* and book sales *s*.

b. Draw a graph of this equation for sales up to $500,000.

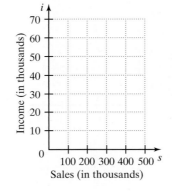

62. A bank account with an initial balance of $100 earns simple interest at an annual rate of 4%. The amount A in the account after t years is given by

$$A = 100 + 4t$$

a. Graph this equation.

b. What is the A-intercept of this graph? Explain its significance in terms of the bank account.

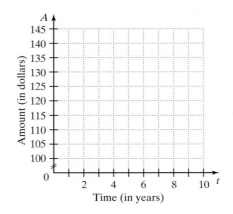

63. The width of a jewel case is $\frac{1}{4}$ of an inch for a single compact disc and $\frac{1}{2}$ of an inch for double compact discs. If there are s single jewel cases and d double jewel cases on a shelf 30 in. in length, then

$$\frac{1}{4}s + \frac{1}{2}d < 30.$$

a. Graph this inequality.

b. From this graph, identify one possible combination of single and double jewel boxes that will fit on the shelf.

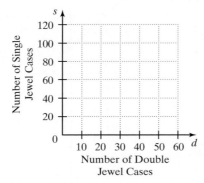

64. To be able to catch up and pass a friend driving away at a speed of 50 mph, it is necessary to cover a distance of d mi in t hr, where

$$d > 50t$$

a. Graph this inequality.

b. From this graph, choose a point in the shaded region. Explain what its coordinates mean in terms of catching up and passing the friend.

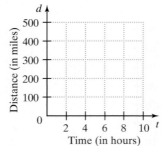

65. A cosmetics company makes a monthly profit of $300 plus $1.20 for each bottle of nail polish it sells.

   a. Write a function that relates the monthly profit $P(x)$ and the number of bottles of nail polish x the company sells.

   b. Find $P(200)$ and interpret its meaning in this situation.

   c. Graph the function in part (a).

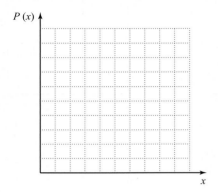

   d. Find the domain and range of the function.

66. An Internet company offers a service for publishing blogs. A customer is charged an initial subscription fee of $10 plus a monthly fee of $4.95 for the service.

   a. Express in function notation the total amount $A(m)$ that a customer pays after m mo.

   b. Graph the function in part (a).

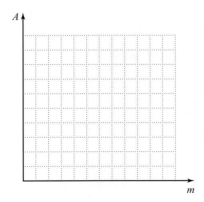

   c. From the graph, estimate the total amount a customer pays for one year of service.

● *Check your answers on page A-18.*

Chapter 3   POSTTEST

Test solutions are found on the enclosed CD.

To see whether you have mastered the topics in this chapter, take this test.

1. The following graph shows the enrollments in public and private colleges in the United States.

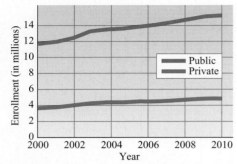

(*Source:* U.S. National Center for Education Statistics)

About how many students were enrolled in public colleges in 2004?

2. On the coordinate plane shown, plot the points $A(-2, 0)$ and $B(5, 3)$.

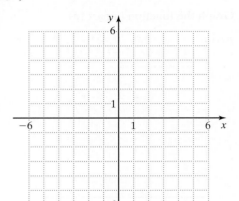

3. In which quadrant is the point $(-5, 3)$ located?

4. Given two points $C(8, 1)$ and $D(3, -4)$, compute the slope of the line that passes through the points.

5. Are the graphs of $y = 3x + 1$ and $y = 3x - 2$ parallel? Explain how you know.

6. For the points $A(0, 1)$, $B(2, 8)$, $C(0, 6)$, and $D(7, 4)$, indicate whether $\overleftrightarrow{AB}$ is perpendicular to $\overleftrightarrow{CD}$. Explain.

7. In the following graph, find the x-intercept and the y-intercept.

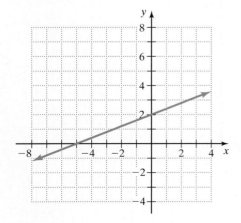

8. The graph on the following coordinate plane shows how the rental cost at a local car rental establishment relates to the number of miles that the car has been driven.

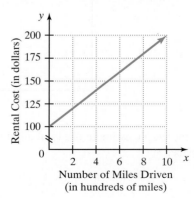

Is the slope of the graphed line positive, negative, zero, or undefined? Describe in a sentence or two the relationship between the rental cost and the number of miles driven.

9. For the equation $y = -3x + 1$, complete the following table.

x	−3	5		
y			0	−2

Graph the equation indicated in Exercises 10–12.

10. $y = 2$

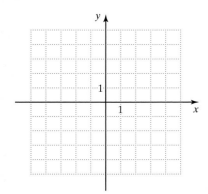

11. $y = -x - 3$

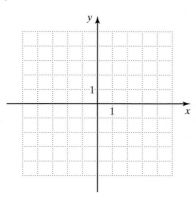

12. $3x - 2y = 6$

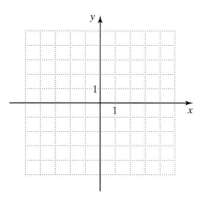

13. What are the slope and the y-intercept of the graph of $y = 3x + 1$?

14. Write the equation $2x - y = 5$ in slope-intercept form.

15. Find the equation of the line with slope -1 and that passes through the point $(0, -3)$.

16. The points $(3, 5)$ and $(-4, 2)$ lie on a line. Find its equation.

17. Graph $y \leq -\dfrac{1}{2}x + 1$.

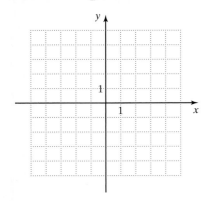

18. An entrepreneur is establishing a small business to manufacture leather bags. Her initial investment is $1000, and her unit cost to manufacture each bag is $30. Write an equation that gives the total cost C of manufacturing b bags. Plot the total cost of manufacturing 100, 200, and 300 bags.

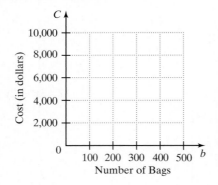

19. An investor deposits x dollars in an account that earns 4% per year and y dollars in another account earning 8% per year. His total annual earnings are at least $500. Write an inequality that describes this situation. Graph this inequality.

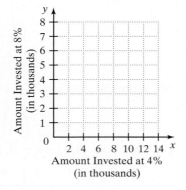

20. If $g(x) = -2x + 11$, evaluate $g(a + 3)$.

● *Check your answers on page A-20.*

Cumulative Review Exercises

To help you review, solve the following:

1. Evaluate: $2 \cdot 6 - 6^2$

2. Find the value of $f(x) = x^2 - 4x + 1$ if $x = -3$.

3. What property of real numbers is illustrated by the statement $4 \cdot a + 4 \cdot b = 4(a + b)$?

4. Solve for x: $2x - 3(5 - x) = 4x + 2$

5. Solve for x and graph: $3x + 1 > 7$

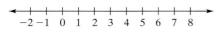

6. Graph: $2x + 5y = -12$

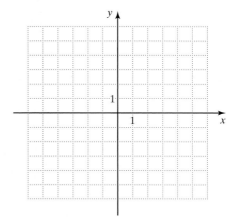

7. VA Bank and NY Bank have assets of $400 billion and $160 billion, respectively. The assets of NY Bank are what percent of those of VA Bank?

8. Lincoln's famous 1863 Gettysburg Address begins with the phrase "Four score and seven years ago." This phrase refers to 1776, the year in which the United States declared independence. How many years are there in a score of years?

9. The perimeter P of a rectangular garden is given by the formula
$$P = 2l + 2w,$$
where l is the length and w is the width of the garden. Solve for l in terms of P and w.

10. A drama club washes cars as a fund-raising activity. The club charges $6 to wash each car.

   a. Write an equation to relate the club's income y to the number x of cars they wash.

   b. Graph the equation.

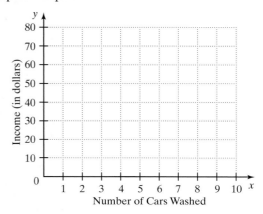

   c. What are the x- and y-intercepts of the graph?

● *Check your answers on page A-20.*

CHAPTER 4

Systems of Equations and Inequalities

Economics and Linear Curves

In a market, sellers can set the price at which their goods are offered for sale. How does this price affect the number of goods that buyers are willing to purchase? How does it affect the number of goods that producers are willing to supply the sellers?

Generally, as the *price* of an item increases, the *quantity* of items sold declines. This trend is captured in a **demand curve**—commonly approximated by a straight line with a negative slope. The coordinates of each point on this line correspond to the price at which retailers sell items and the quantity of items they sell.

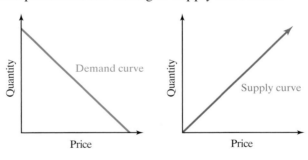

By contrast, the **supply curve** has a positive slope, meaning that as selling prices increase, wholesalers are inclined to make more goods available to retailers. The coordinates of a point on this line represent the price at which retailers sell items and the quantity of items that wholesalers supply to the retailers.

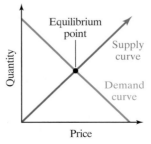

Graphing both the supply curve and the demand curve on the same coordinate plane shows the price at which the market is **at equilibrium**. At this point of intersection, all items produced are sold, and all customers are satisfied.

Chapter 4   PRETEST

To see if you have already mastered the topics in this chapter, take this test.

1. Determine which of the following ordered pairs is a solution of the system

$$x + 2y = 5$$
$$5x - y = -8$$

   a. $(5, 0)$

   b. $(-1, 3)$

   c. $(1, -3)$

2. For the system graphed, indicate the number of solutions.

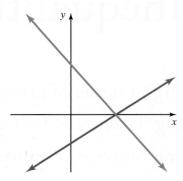

Solve each system by graphing.

3. $x + y = -2$
   $y = x + 4$

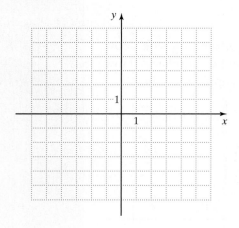

4. $x - 2y = 1$
   $y = 2$

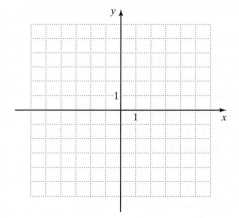

5. $9x + 3y = -6$
   $3x - 4y = 8$

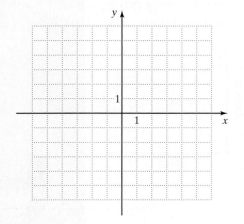

Solve each system by substitution.

6. $x - 2y = 7$
$y = -11 - x$

7. $7x - 4y = 10$
$x - 2y = 0$

8. $a + 3b = -2$
$a = 2b - 7$

Solve each system by elimination.

9. $2x + 5y = -13$
$-2x + 6y = -20$

10. $6x - 8y = 36$
$1.5x - 2y = 9$

11. $3x - 7y = -19$
$2x + 3y = -5$

Solve each system.

12. $-3n + 5m = 10$
$-4m = -2(n + 1)$

13. $x + 9 = 2y$
$8y - 13 = 4x$

14. $6x + 10y - 12 = 0$
$3x + 2.5y - 6 = 0$

Solve.

15. Ticket prices at a local movie theater are $6.50 for all shows before 5:00 P.M. and $10 for all shows after 5:00 P.M. The total revenue from ticket sales on Saturday was $17,650. If 1975 tickets were sold on Saturday, how many tickets were sold before 5:00 P.M.? After 5:00 P.M.?

16. For a student club fund-raiser, the number of $2 raffle tickets printed was three times the number of $5 tickets. If all of the tickets are sold, receipts from the $5 tickets will be $50 less than those from the $2 tickets. How many $5 tickets were printed?

17. A lottery winner invested $200,000 of her winnings in two funds earning 5% and 6% simple interest, respectively. If after one year she earned $11,200 in interest, how much did she invest in each fund?

18. On a boating trip, it took 2 hr to travel 13 mi with the current. It took the same amount of time to travel 11 mi against the current on the return trip. Find the speed of the boat and the speed of the current.

19. Graph the system of inequalities:

$y < \dfrac{1}{2}x - 1$

$y \geq -3x + 1$

$y \geq -4$

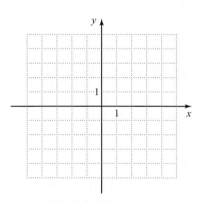

20. To save for emergencies, a young couple had at most $10,000 to invest in two accounts that earned 5% and 6% simple interest. The couple wanted the investment to earn at least $300 in the first year.

a. Express this information as a system of inequalities, letting x represent the investment in the account earning 5% simple interest and y represent the investment in the account that earned 6% simple interest.

b. Graph the system.

c. Determine one possible investment combination.

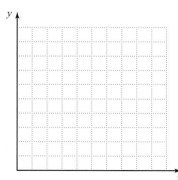

● *Check your answers on page A-21.*

4.1   Solving Systems of Linear Equations by Graphing

What Systems of Linear Equations Are and Why They Are Important

OBJECTIVES

- To decide whether an ordered pair is a solution of a system of linear equations in two variables
- To determine the number of solutions of a system of linear equations
- To solve a system of linear equations by graphing
- To solve applied problems involving systems of linear equations

Recall from previous chapters that some situations are described by a linear equation in one variable, say, $2x + 1 = 10$, whereas others are described by a single linear equation in *two* variables, for instance, $y = 3x - 5$.

Now, let's consider situations in which the relationship between two variables is described by a *pair* of linear equations, for instance:

$$x + y = 7$$
$$x - y = 3$$

Groups of equations, called *systems*, serve as a model for a wide variety of applications in fields such as business and science. A system can represent the conditions that must be satisfied in a particular situation. For instance, the system might describe how the cost of products relate to one another or the motion of a plane in various wind conditions.

In this chapter, we deal with three approaches to solving systems of linear equations, namely by *graphing, substitution,* and *elimination.*

Introduction to Systems of Linear Equations

We begin by focusing on the meaning of a system of equations.

Definition

A **system of equations** is a group of two or more equations solved simultaneously.

Systems of equations are sometimes written with large braces:

$$\begin{cases} x + y = 7 \\ x - y = 3 \end{cases} \quad \text{or} \quad \begin{cases} x + y = 7 \\ x - y = 3 \end{cases}$$

Braces are used to emphasize that any solution of a system must satisfy *all* the equations in the system. For instance, $x = 5$ and $y = 2$ is a solution of the system above, because when we substitute 5 for x and 2 for y into the equations, *both* equations are true:

$$x + y = 7 \quad \longrightarrow \quad 5 + 2 \overset{?}{=} 7 \quad \text{True}$$
$$x - y = 3 \quad \longrightarrow \quad 5 - 2 \overset{?}{=} 3 \quad \text{True}$$

A solution of a system of two equations is commonly represented as an ordered pair of numbers. For instance the solution of the system just mentioned can be written as $(5, 2)$. Can you explain why $(2, 5)$ is not a solution of this system?

Definition

A **solution** of a system of two linear equations in two variables is an ordered pair of numbers that makes both equations in the system true.

EXAMPLE 1	PRACTICE 1

EXAMPLE 1

Consider the system:

$$x - 2y = 6$$
$$2x + 5y = 3$$

a. Is $(4, -1)$ a solution of the system?

b. Is $(2, -2)$ a solution of the system?

Solution

a. To decide if the ordered pair $(4, -1)$ is a solution of this system, we substitute the x-coordinate 4 for x and the y-coordinate -1 for y in the equations and check if both equations are true.

$x - 2y = 6 \longrightarrow 4 - 2(-1) \overset{?}{=} 6 \xrightarrow{\text{Simplifies to}} 6 = 6$ **True**

$2x + 5y = 3 \longrightarrow 2(4) + 5(-1) \overset{?}{=} 3 \xrightarrow{\text{Simplifies to}} 3 = 3$ **True**

The ordered pair $(4, -1)$ satisfies both equations and so is a solution of the system.

b. To see if $(2, -2)$ is a solution, we substitute 2 for x and -2 for y in the equations and check if they are both true.

$x - 2y = 6 \longrightarrow 2 - 2(-2) \overset{?}{=} 6 \xrightarrow{\text{Simplifies to}} 6 = 6$ **True**

$2x + 5y = 3 \longrightarrow 2(2) + 5(-2) \overset{?}{=} 3 \xrightarrow{\text{Simplifies to}} -6 = 3$ **False**

The ordered pair $(2, -2)$ is not a solution of the system because it does not satisfy both equations.

PRACTICE 1

Consider the following system.

$$3x + 2y = 5$$
$$4x - 2y = -5$$

Determine whether the following ordered pairs are solutions of the system.

a. $(0, 2.5)$

b. $(1, -1)$

Number of Solutions of a System

In solving a system of linear equations, a question that immediately comes to mind is how many solutions the system has. Let's consider this question graphically. Since each equation is linear, both graphs are lines. Now suppose that we graph the two equations on the same coordinate plane. Any point at which the two graphs of the system intersect is a solution of the system, because that point must satisfy both equations.

For instance, let's again consider the system

$$x + y = 7$$
$$x - y = 3$$

and solve it by graphing both equations.

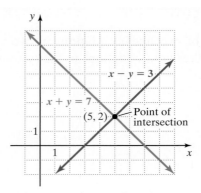

Note that the lines intersect at (5, 2)—precisely the ordered pair that we have previously determined to be a solution of this system. Since the two lines meet at a single point, the system has just one solution.

Not all systems have one solution. For instance, consider the following system in which there are *no* solutions:

$$3x - y = 2$$
$$3x - y = 4$$

Graphing this system, we get:

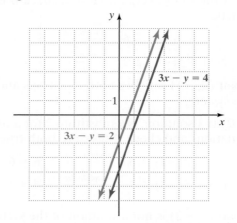

We observe that the lines are parallel and, therefore, do not intersect. So in this case, the system has no solutions.

Finally, let's examine a system that has more than one solution.

$$2x - 2y = 6$$
$$y = x - 3$$

The graph of this system is as follows:

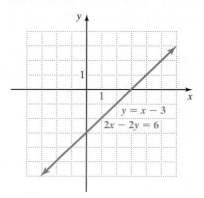

Note that only one line is shown. The reason is that both equations in the system have the same graph. Any point on this line, for instance $(3, 0)$, is a solution of both equations. A system such as this one has *infinitely many* solutions.

Every system of linear equations has one solution, no solution, or infinitely many solutions. The following table summarizes the main features of the three types of systems.

Number of Solutions	Description of the System's Graph	Possible Graph
One solution	The lines intersect at exactly one point.	
No solution	The lines are parallel.	
Infinitely many solutions	The lines coincide, that is, they are the the same line.	

EXAMPLE 2

For each system graphed, determine the number of solutions.

a.

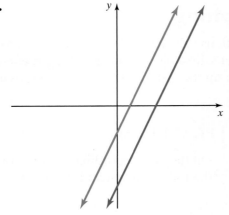

PRACTICE 2

Determine the number of solutions of each of the following systems:

a.

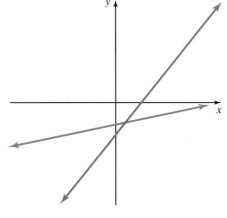

b.

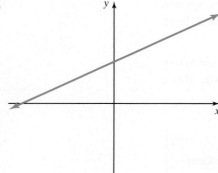

b.

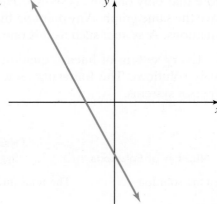

c.

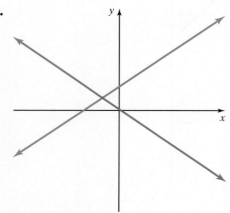

c.

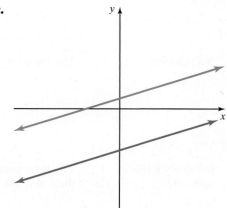

Solution

a. The graph of the system consists of two lines that appear to be parallel. Since the lines do not intersect, the system has no solution.

b. This graph is a single line. Therefore, the system has infinitely many solutions.

c. The two lines in this graph intersect at a single point. So the system has one solution.

Solving Systems by Graphing

How exactly are systems of linear equations solved? In this chapter, we consider several methods of solving systems. Let's first discuss the **graphing method** in which we graph the equations that make up the system. Any point of intersection is a solution of the system.

EXAMPLE 3	PRACTICE 3
Solve the following system by graphing. $$x + y = 6$$ $$x - y = -4$$ **Solution** Let's graph each linear equation by using the x- and y-intercept method and then sketching the line that passes through these points.	On the given coordinate plane, solve the following system by graphing. $$x + y = 2$$ $$x - y = 4$$

$x + y = 6$	
x	**y**
0	6
6	0
3	3

$x - y = -4$	
x	**y**
0	4
−4	0
2	6

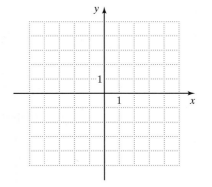

On the same coordinate plane, we plot the points and then graph both equations.

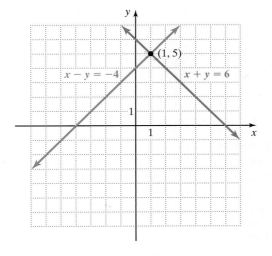

The corresponding lines appear to intersect at the point $(1, 5)$, giving $x = 1$ and $y = 5$.

Check Since solving a system of equations by graphing may result in approximate solutions, we confirm that $(1, 5)$ is the solution by substituting these values into the original equations:

$$x + y = 6 \xrightarrow{\text{Substitute 1 for } x \text{ and 5 for } y.} 1 + 5 \overset{?}{=} 6$$
$$6 = 6 \qquad \text{True}$$

$$x - y = -4 \xrightarrow{\text{Substitute 1 for } x \text{ and 5 for } y.} 1 - 5 \overset{?}{=} -4$$
$$-4 = -4 \qquad \text{True}$$

So $(1, 5)$ is the solution of the system.

To Solve a System of Linear Equations by Graphing

- Graph both equations on the same coordinate plane.
- There are three possibilities:
  a. If the lines intersect, then the solution is the ordered pair of coordinates for the point of intersection. Check that these coordinates satisfy both equations.
  b. If the lines are parallel, then there is no solution of the system.
  c. If the lines coincide, then there are infinitely many solutions, namely, all the ordered pairs of coordinates that represent points on the line.

EXAMPLE 4	PRACTICE 4

Solve by graphing.

$$y = 2x + 5$$
$$2x - y = 2$$

Solution Graphing the two equations, we get:

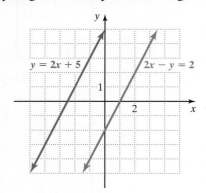

The lines appear to be parallel, which suggests that the system has no solution. To confirm that the lines are parallel, we can check that their slopes are equal. The graph of the first equation, $y = 2x + 5$, has slope 2. To find the slope of the second equation, we write $2x - y = 2$ in slope-intercept form, getting $y = 2x - 2$. The graph of this equation also has slope 2. Therefore, the lines are parallel and the system has no solution.

Solve for x and y by graphing.

$$y = x - 6$$
$$x - y = 4$$

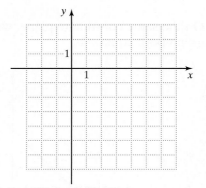

EXAMPLE 5	PRACTICE 5

Solve by graphing.

$$2y = -8x + 2$$
$$-4x - y = -1$$

Solution When we graph the two equations in this system, we get the same graph.

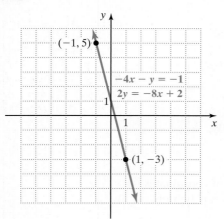

Note that when the two equations in the system are changed to slope-intercept form, the equations are identical.

Solve by graphing.

$$6x = 15 - 3y$$
$$y = 5 - 2x$$

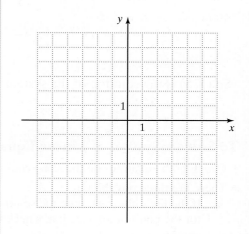

$$2y = -8x + 2 \xrightarrow{\text{Isolate } y.} y = -4x + 1$$
$$-4x - y = -1 \xrightarrow{\text{Isolate } y.} y = -4x + 1$$

We conclude that the system has infinitely many solutions. All points on the graph, some of which are indicated, are solutions. Can you identify another point on the graph and confirm that it is a solution to the system?

EXAMPLE 6	PRACTICE 6

The U.S. House of Representatives has 435 members. On a certain bill, all the representatives voted, and 15 more representatives voted for the bill than against the bill. (There were no abstentions.)

a. If s represents the number of representatives who *supported* the bill, and n the number of representatives who did *not* support the bill, express the given information as a system of equations.

b. On a coordinate plane, graph the system found in part (a).

c. Find the coordinates of the point of intersection.

d. In this problem, what is the significance of the coordinates of the point of intersection?

Solution

a. The given information can be expressed algebraically as

$$s + n = 435$$
$$s = n + 15$$

b. Let's graph s along the vertical axis and n along the horizontal axis. Since the number of representatives voting for or against a bill is between 0 and 435, we choose an appropriate scale and label the two axes accordingly. Next, we graph the two equations.

c. The lines appear to intersect approximately at the point (210, 225), that is, $n = 210$ and $s = 225$.

d. We conclude that about 210 representatives voted against the bill and 225 voted for the bill.

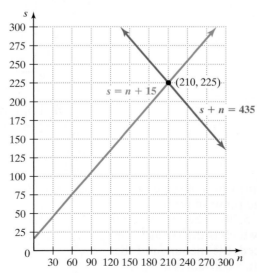

A liberal arts student transferring to a four-year institution took a test of verbal skills and a test of mathematical skills. Her total score was 1150, and the verbal score v was 100 less than the math score m.

a. Express the given information as a system of equations.

b. On a coordinate plane, graph the system found in part (a).

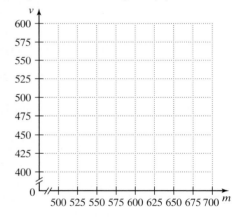

c. Name the coordinates of the point of intersection.

d. In this situation, what is the significance of the coordinates of the point of intersection?

When running a business, it is important to determine both the income that the business makes and the expenses that it takes to run the business. The business's income and its expenses depend on the number of items produced and sold. These quantities can be graphed on a coordinate plane, as shown in the next example. The point at which the income for a business equals its expenses is called the *break-even point*.

EXAMPLE 7	PRACTICE 7

EXAMPLE 7

For a start-up business, an entrepreneur determined that to produce computer-generated, silk-screen T-shirts it will cost $3.25 a shirt plus $450 in fixed overhead. Each shirt produced is sold at $5.50.

a. If x represents the number of T-shirts sold and y is the amount it costs to produce the T-shirts, write an equation that relates x and y.

b. If x represents the number of T-shirts produced and y is the amount of income from selling the T-shirts, write an equation that relates x and y.

c. On a coordinate plane, graph the lines found in parts (a) and (b).

d. Find the break-even point for producing the T-shirts. Explain its significance in terms of the x- and y-coordinates.

Solution

a. The given information can be expressed as

$$y = 3.25x + 450.$$

b. We can write the given information as

$$y = 5.50x.$$

c.

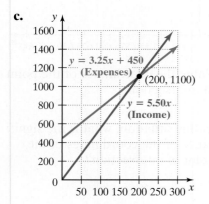

PRACTICE 7

To print a newsletter costs $450 fixed overhead plus $1.50 a copy. Every printed copy of the newsletter is sold at $3 apiece.

a. If x represents the number of copies printed and y is the amount of money it costs to print the newsletter, write an equation that relates x and y.

b. If x represents the number of copies printed and y is the amount of income from newsletter sales, write an equation that relates x and y.

c. On the coordinate plane below, graph the lines found in parts (a) and (b).

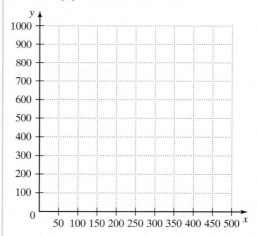

d. Find the break-even point for printing the newsletter.

d. Since the break-even point is the point where the income equals the expenses, we must find the point of intersection of the lines $y = 3.25x + 450$ and $y = 5.50x$. The intersection of the lines is the point (200, 1100), which is the break-even point. So when 200 shirts are produced, the income and the expenses will be the same, $1100. After 200 shirts are produced, the business will start making a profit.

In this section, we have examined the graphing method of solving systems of linear equations. A major advantage of this method over alternative methods is that it helps us to visualize the problem and its solution. However, a disadvantage of this approach is that our reading of a graph may be inaccurate, particularly when a coordinate of the point of intersection is not an integer or is very large. So a solution found by the graphing method may not be exact.

Solving Systems of Linear Equations on a Grapher

Both graphing calculators and computers with graphing software can help facilitate the process of solving systems of linear equations. As with the paper-and-pencil approach, a grapher displays the graphs of the equations that make up a system on the same coordinate plane. We can then use one of the special features of the grapher to read the coordinates of the point at which the graphed lines intersect, that is, the solution of the system.

The most common features of a grapher that help us to read the coordinates of the point of intersection are **TRACE**, **ZOOM**, and **INTERSECT**.

● With the **TRACE** feature, the cursor runs along either of the graphed lines until it is positioned on or near the point of intersection; the coordinates of that point are then displayed.

● The **ZOOM** feature lets us position the cursor as close as we want to the point of intersection.

● The **INTERSECT** feature automatically calculates the point of intersection.

Note that each of the features may give only an approximation for the point of intersection. However, the most accurate approximation of the point of intersection is given by the **INTERSECT** feature.

EXAMPLE 8	PRACTICE 8

EXAMPLE 8

Use either a graphing calculator or graphing software to solve.

$$4x - y = 11$$
$$x = y + 6$$

Solution Begin by solving each equation for y.

$$4x - y = 11 \xrightarrow{\text{Isolate } y.} y = 4x - 11$$
$$x = y + 6 \xrightarrow{\text{Isolate } y.} y = x - 6$$

Then press the $\boxed{Y=}$ key, and enter $4x - 11$ to the right of **Y1 =** and $x - 6$ to the right of **Y2 =**. Set the viewing window. Then press the $\boxed{\text{GRAPH}}$ key to display the coordinate plane on which the two equations are graphed. The **TRACE** feature can be used to move a cursor along one of the lines toward the intersection of the graphs by holding down an arrow key. Note that as the cursor is moved, the changing coordinates of its position will be displayed on the screen. Once the cursor reaches the point of intersection, we can read the coordinates on the screen.

PRACTICE 8

Use a grapher to solve the following system of equations.

$$8x - y = 1$$
$$y = x + 5$$

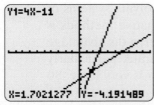

Using the TRACE feature

To get a better approximation of the solution, we can either activate the **ZOOM** feature to zoom in on the intersection point or activate the **INTERSECT** feature.

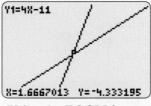
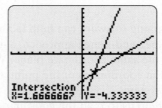

Using the ZOOM feature   **Using the INTERSECT feature**

So the approximate solution here is $(1.667, -4.333)$.

4.1   Exercises

FOR EXTRA HELP   **MyMathLab**   Math XL PRACTICE   WATCH   DOWNLOAD   READ   REVIEW

Mathematically Speaking

Fill in each blank with the most appropriate term or phrase from the given list.

| are parallel | solution | coincide |
| graph | set of equations | system of equations |

1. A(n) _____ is a group of two or more equations solved simultaneously.

2. A(n) _____ of a system of two linear equations in two variables is an ordered pair of numbers that makes both equations in the system true.

3. If a system of linear equations has no solution, the lines _____.

4. If a system has infinitely many solutions, the lines _____.

Indicate whether each ordered pair is or is not a solution to the given system.

5. $x + y = 3$
$2x - y = 6$

a. $(0, 3)$ _____

b. $(3, 3)$ _____

c. $(3, 0)$ _____

6. $x - 6y = 3$
$x - y = -7$

a. $(-2, -9)$ _____

b. $(-9, -2)$ _____

c. $(9, -2)$ _____

7. $4x + 5y = 0$
$7x - y = 0$

a. $(1, 7)$ _____

b. $(-5, 4)$ _____

c. $(0, 0)$ _____

8. $2x - 2y = 30$
$8x + 2y = -10$

a. $(1, -9)$ _____

b. $(16, 1)$ _____

c. $(2, -11)$ _____

Match each system with the appropriate graph.

9. a. A system with solution $(1, 3)$

b. A system with solution $(-1, 3)$

c. A system with infinitely many solutions

d. A system with no solution

I
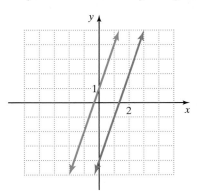

II
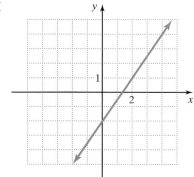

continued

III

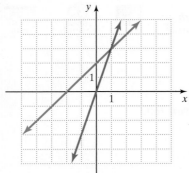

IV

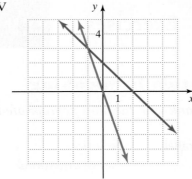

10. a. A system with solution $\left(-\frac{1}{4}, -1\frac{1}{4}\right)$ **b.** A system with solution $(-1, 3)$

    c. A system with infinitely many solutions   **d.** A system with no solution

I

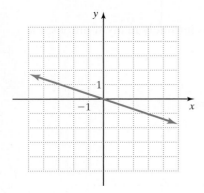

II

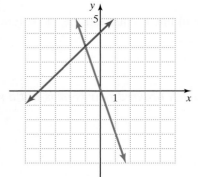

III

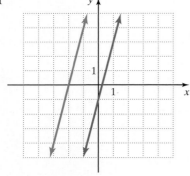

IV

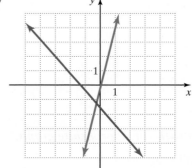

Solve by graphing.

11. $x - y = 2$
$\quad\;\; x + y = 4$

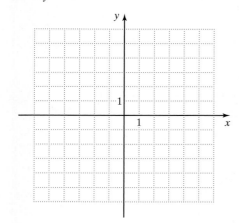

12. $x + 2y = 3$
$\quad\;\; x + \;\; y = 2$

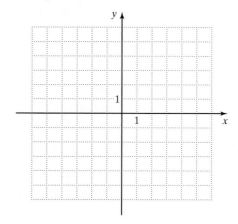

13. $\quad\;\; y = x + 4$
$\quad\; x + y = 4$

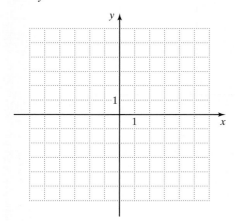

14. $-5 = 2x + y$
$\quad\;\; y = -x$

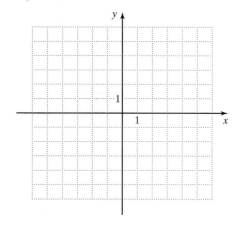

15. $y = -x + 6$
$\quad\; y = -3x + 8$

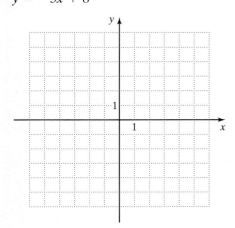

16. $y = x + 1$
$\quad\; y = -x - 3$

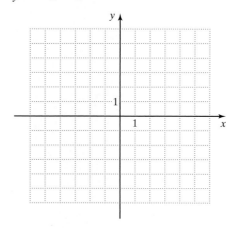

17. $y = -\frac{1}{2}x + 1$
$y = 2x + 1$

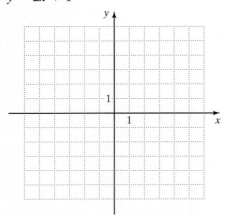

18. $y = 2x - 6$
$y = 3 - \frac{1}{4}x$

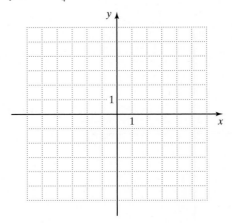

19. $y = 5 + 3x$
$x + y = -3$

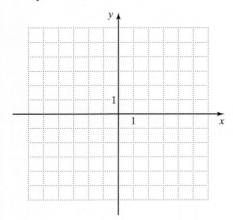

20. $2x + y = -3$
$y = -(x + 4)$

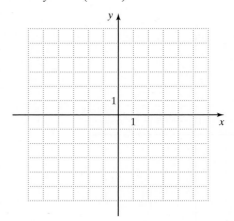

21. $2y = 6x + 2$
$3y - 9x = 3$

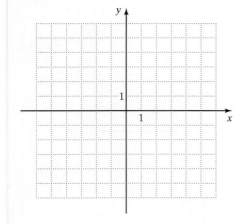

22. $4x = 8y + 4$
$5x - 10y = 5$

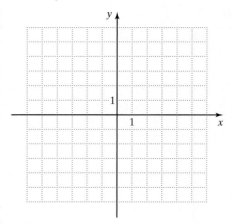

23. $2x + y = -4$
$\qquad y = -2x + 3$

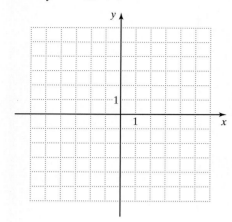

24. $3x - y = 1$
$\qquad 6x + 4 = 2y$

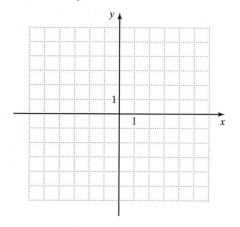

25. $\qquad x = 5 + y$
$-2x + 2y = -10$

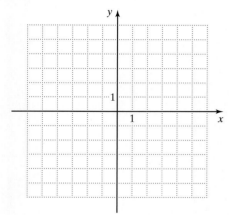

26. $\qquad x + y = 6$
$3y - 18 = -3x$

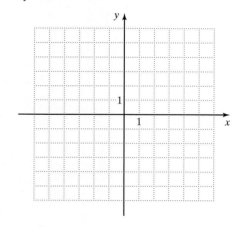

27. $x - y = -1$
$\quad x - y = 4$

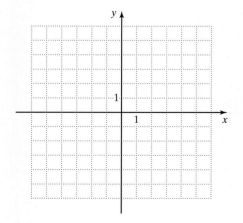

28. $x + 5y = 6$
$\quad x + 5y = 0$

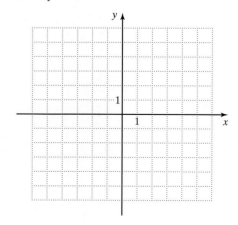

29. $3x + 2y = -10$
$\;\; 5x - y = -8$

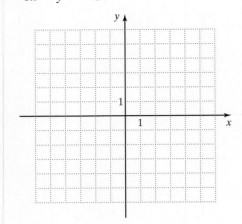

30. $-x - 2y = 8$
$\;\; -6x + 4y = 0$

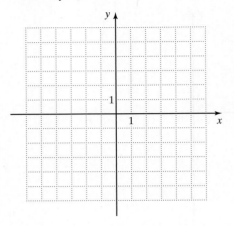

31. $-3x + y - 14 = 0$
$\;\;\; 3x - y - 11 = 0$

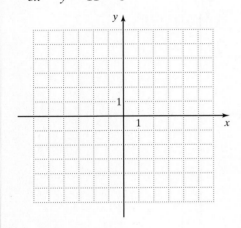

32. $7x - y - 2 = 0$
$\; 14x - 2y - 12 = 0$

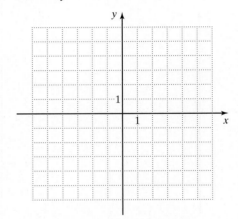

33. $10x + 2y = -6$
$\;\;\;\;\;\;\; y = 2$

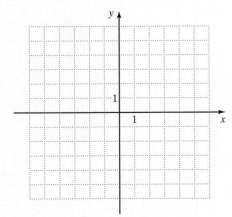

34. $x + 5y = -15$
$\;\;\; x = 5$

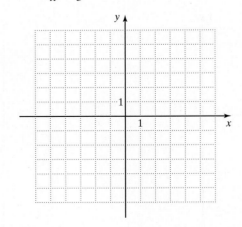

35. $x = 0$
    $x - 2y = 4$

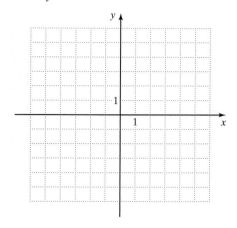

36. $y = -3$
    $y = 2x + 3$

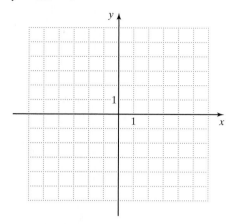

37. $4x - 4y = -8$
    $y = \dfrac{2}{3}x$

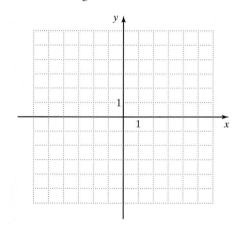

38. $3x + 2y = 6$
    $y = -\dfrac{3}{4}x$

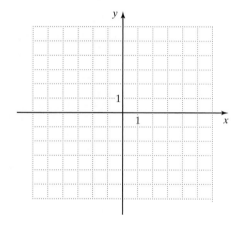

Mixed Practice

Solve by graphing.

39. $2y - 4x = -4$
    $y = -2 + 2x$

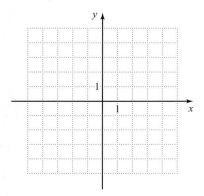

40. $4x + 2 = 2y$
    $2x - y = 3$

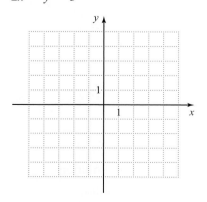

41. $y = -3 + \dfrac{1}{3}x$

$y = -2x + 4$

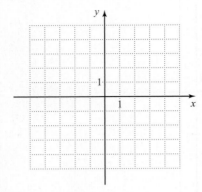

42. $-4 = 2x - y$

$x = 2y + 1$

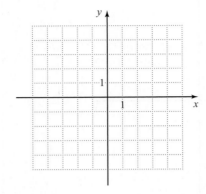

43. $3x - 2y = 2$

$4x + y = 10$

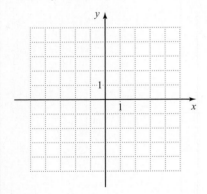

44. $x - 3y = -6$

$x = -3$

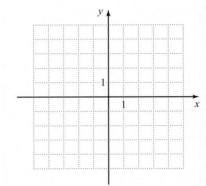

Solve.

45. Which system matches the given graph?

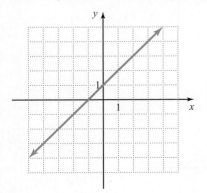

a. A system with no solution

b. A system with infinitely many solutions

c. A system with exactly one solution

46. Indicate whether each ordered pair is or is not a solution to the system.

$$x - 2y = -2$$
$$x - y = 4$$

a. $(-2, -6)$

b. $(12, 8)$

c. $(10, 6)$

Applications

Solve.

47. A young married couple had a combined annual income of $57,000.

    a. If the wife made $3000 more than the husband, write these relationships as a system of equations. Let x represents the husband's income and y represents the wife's income.

    b. Graph the equations.

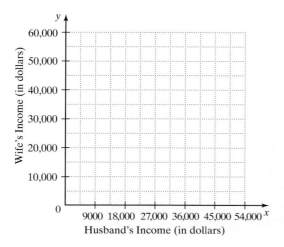

    c. Find the two incomes.

48. A plane flying with a tailwind flew at a speed of 450 mph, relative to the ground. When flying against the tailwind, it flew at a speed of 350 mph.

    a. Express these relationships as equations, where x represents the speed of the plane in calm air and y represents the speed of the wind.

    b. Graph these equations.

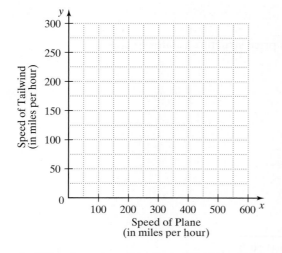

    c. Find the speed of the plane in calm air and the speed of the wind.

49. Mike the plumber charges $75 for a house call and then $40 per hour for labor. Sally the plumber charges $100 for a house call and then $30 per hour for labor.

   a. Write a cost equation for each plumber, where y is the total cost of plumbing repairs and x is the number of hours of labor.

   b. Graph the two equations.

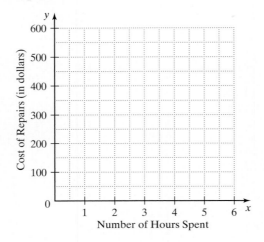

   c. Determine the number of hours of plumbing repairs that would be required for the two plumbers to charge the same amount.

   d. Determine from the graph which plumber charges less if the estimated amount of time to carry out the plumbing repairs is 5 hr.

50. To connect to the Web, you must choose between two Internet service providers (ISPs). Flat ISP charges a monthly flat fee of $20 regardless of how many hours you connect to the Web. A competing company, Variable ISP, charges $2.50 per month plus $0.50 for each hour of connection time.

   a. Express each company's price structure, p, in terms of hours connected, h.

   b. Draw a graph that shows how each company's price structure relates to connection time.

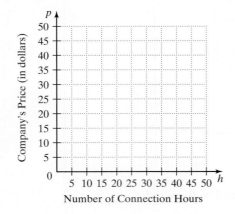

   c. For which connection time do the two companies charge the same?

51. A small company duplicates DVDs. The cost of duplicating is $30 fixed overhead plus $0.25 per DVD duplicated. The company generates revenues of $1.50 per DVD. Use a graph to determine the break-even point for duplicating DVDs.

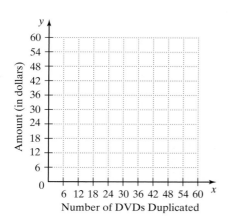

52. A clothing company sells jackets for $140 per jacket. The company's fixed costs are $9000 and the variable costs are $50 per jacket. Use a graph to determine the break-even point for production.

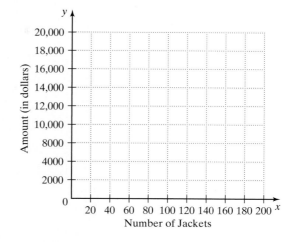

53. A movie fan rented 6 films at a local video store for one day. The daily rental charge was $2 on some films and $4 on others. If the total rental charge was $22, use a graph to determine how many $4 films were rented.

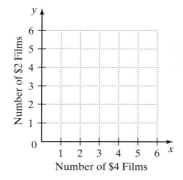

54. An appliance store sells washer-dryer combinations. If the washer costs $200 more than the dryer, use a graph to find the cost of each appliance.

Valley Appliance Center

Washer-Dryer Combos
$1500

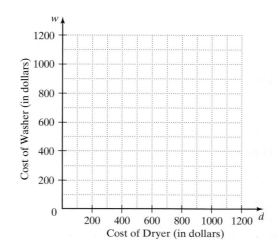

55. Silver's Gym charges a $300 initiation fee plus $30 per month. DeLuxe Fitness Center has an initial charge of $400 but only charges $25 per month. Use a graph to determine for what number of months both health clubs will charge the same amount.

56. A plant nursery is selling a 7-foot specimen of a tree that grows about 1.5 feet per year and a 6-foot specimen of a tree that grows 2 feet per year. Use a graph to determine in how many years the two trees will be the same height.

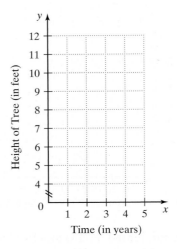

● *Check your answers on page A-22.*

MINDSTRETCHERS

Critical Thinking

1. Write a system of linear equations that has (2, 5) as its only solution.

Writing

2. Is it possible for a system of two linear equations to have exactly two solutions? If not, explain why.

Mathematical Reasoning

3. Not every system of equations is linear. For example the system

$$y = 2x$$
$$y = x^2$$

has the graph shown to the right. How many solutions does this system have? Explain how you know.

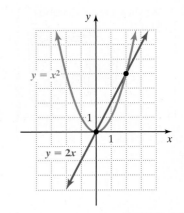

Solving Systems of Linear Equations by Substitution

This section deals with solving a system of equations by the **substitution method**. As in the previous section, we restrict the discussion to systems of two linear equations in two variables. In applying the substitution method, we solve such a system by first solving one linear equation in one variable—a much simpler problem. As compared with the graphing method, the substitution method has the advantage of being faster and also of giving us an exact solution. The exact answer can be difficult to read from a graph if the coordinates of the solution are not integral. This is of particular concern in an applied problem where precision counts. The substitution method particularly lends itself to solving systems in which a variable is isolated in one of the equations.

To see how solving by substitution works, let's look at some examples.

OBJECTIVES
■ To solve a system of linear equations by substitution
■ To solve applied problems involving systems of linear equations

---

EXAMPLE 1

Solve by substitution:
(1) $x + y = 10$
(2) $y = 2x + 4$

Solution Notice that Equation (2) is solved for y in terms of x. So we can substitute the expression $2x + 4$ from Equation (2) for y in Equation (1).

(1) $x + y = 10$
$x + (2x + 4) = 10$ Substitute $2x + 4$ for y.

We now have an equation that contains only one variable, namely x, and so is easy to solve.

$$x + (2x + 4) = 10$$
$$3x + 4 = 10$$
$$3x = 6$$
$$x = 2$$

Now that we have found the value of x, let's substitute it in either of the original equations in order to determine the corresponding value of y. Substituting 2 for x in Equation (2), we get:

(2) $y = 2x + 4$
$= 2(2) + 4$
$= 8$

So the solution to the system is $(2, 8)$, that is, $x = 2$ and $y = 8$.

Check We can check the solution by substituting 2 for x and 8 for y in the original equations.

(1) $x + y = 10$ **(2)** $y = 2x + 4$
$2 + 8 \stackrel{?}{=} 10$ $8 \stackrel{?}{=} 2 \cdot 2 + 4$
$10 = 10$ True $8 = 8$ True

The check confirms the solution $(2, 8)$.

PRACTICE 1

Use the substitution method to solve.

(1) $x - y = 7$
(2) $x = -y + 1$

In Example 1, would we have gotten the same solution if we had solved for y in the *first* equation and then replaced y in the second equation?

The preceding example suggests the following procedure for solving systems of linear equations in two variables.

To Solve a System of Linear Equations by Substitution

- In one of the equations, solve for either variable in terms of the other variable.
- In the other equation, substitute the expression equal to the variable found in the previous step. Then solve the resulting equation for the remaining variable.
- Substitute the value found in the previous step in either of the original equations and solve for the other variable.
- Check by substituting the values in both equations of the original system.

EXAMPLE 2

Use the substitution method to solve the following system.

$$\begin{array}{ll} \textbf{(1)} & d = 3q - 8 \\ \textbf{(2)} & 3d - 4q = 10 \end{array}$$

Solution Since d is isolated in Equation (1), we can substitute the expression $3q - 8$ for d in Equation (2).

$$\begin{array}{ll} \textbf{(2)} & 3d - 4q = 10 \\ & 3(3q - 8) - 4q = 10 \quad \text{Substitute } 3q - 8 \text{ for } d. \\ & 9q - 24 - 4q = 10 \\ & 5q - 24 = 10 \\ & 5q = 34 \\ & q = \dfrac{34}{5}, \quad \text{or } 6.8 \end{array}$$

Now, let's solve for d by substituting 6.8 for q in Equation (1).

$$\begin{array}{ll} \textbf{(1)} & d = 3q - 8 \\ & = 3(6.8) - 8 \\ & = 20.4 - 8 \\ & = 12.4 \end{array}$$

The solution is $q = 6.8$ and $d = 12.4$. Note that neither value is an integer. So if we solve this system by the graphing method, we have to estimate the coordinates of a point.

Check We can confirm the solution by substituting these values in both of the original equations.

$$\begin{array}{ll} \textbf{(1)} \quad d = 3q - 8 & \textbf{(2)} \quad 3d - 4q = 10 \\ 12.4 \overset{?}{=} 3(6.8) - 8 & 3(12.4) - 4(6.8) \overset{?}{=} 10 \\ 12.4 \overset{?}{=} 20.4 - 8 & 37.2 - 27.2 \overset{?}{=} 10 \\ 12.4 = 12.4 \quad \text{True} & 10 = 10 \quad \text{True} \end{array}$$

So the solution is confirmed.

PRACTICE 2

Solve for m and n by substitution.

$$\begin{array}{ll} \textbf{(1)} & m = -5n + 1 \\ \textbf{(2)} & 2m + 3n = 7 \end{array}$$

In Example 2, would we have gotten the same solution to the system if the variables had been called x and y instead of q and d? Explain.

EXAMPLE 3	**PRACTICE 3**

EXAMPLE 3

Solve the system by substitution.

$$\textbf{(1)} \quad 5x - 3y = 5$$
$$\textbf{(2)} \quad 2x - y = 1$$

Solution We first need to solve for x or y in either of the equations. Let's solve for y in Equation (2) where the coefficient of y is -1.

$$\textbf{(2)} \quad 2x - y = 1$$
$$-y = -2x + 1$$
$$y = 2x - 1 \qquad \text{Divide each side by } -1.$$

Next, we substitute the expression $2x - 1$ for y in Equation (1) and solve for x.

$$\textbf{(1)} \quad 5x - 3y = 5$$
$$5x - 3(2x - 1) = 5 \qquad \text{Substitute } 2x - 1 \text{ for } y.$$
$$5x - 6x + 3 = 5$$
$$-x + 3 = 5$$
$$-x = 2$$
$$x = -2$$

Now, we can solve for y by substituting -2 for x in Equation (2).

$$\textbf{(2)} \quad 2x - y = 1$$
$$2(-2) - y = 1$$
$$-4 - y = 1$$
$$-y = 5$$
$$y = -5$$

So the solution is $(-2, -5)$. Check this in the original system.

PRACTICE 3

Solve by substitution.

$$\textbf{(1)} \quad 2x - 7y = 7$$
$$\textbf{(2)} \quad 6x - y = 1$$

EXAMPLE 4

Solve by substitution.

$$\textbf{(1)} \quad x - 4y = 15$$
$$\textbf{(2)} \quad -2x + 8y = 5$$

Solution We first need to solve for either x or y in one of the equations. Let's solve for x in Equation (1) since the coefficient of x in this equation is 1.

$$\textbf{(1)} \quad x - 4y = 15$$
$$x = 4y + 15$$

Next, we substitute the expression $4y + 15$ for x in Equation (2).

$$\textbf{(2)} \quad -2x + 8y = 5$$
$$-2(4y + 15) + 8y = 5 \qquad \text{Substitute } 4y + 15 \text{ for } x.$$
$$-8y - 30 + 8y = 5$$
$$-30 = 5 \qquad \text{False}$$

Getting a false statement means that no value of y makes this last equation true. So the original system has no solution.

PRACTICE 4

Use the substitution method to solve the following system.

$$\textbf{(1)} \quad 3x + y = 10$$
$$\textbf{(2)} \quad -6x - 2y = 1$$

What do you think the graph of the system in Example 4 looks like?

EXAMPLE 5	PRACTICE 5

Solve by substitution.

$$(1) \quad 6x + 2y = 4$$
$$(2) \quad -y = 3x - 2$$

Solution Since the coefficient of y in Equation (2) is -1, let's solve this equation for y.

$$(2) \quad -y = 3x - 2$$
$$y = -3x + 2 \qquad \text{Divide each side by } -1.$$

Next, we substitute $-3x + 2$ for y in Equation (1).

$$(1) \qquad 6x + 2y = 4$$
$$6x + 2(-3x + 2) = 4 \qquad \text{Substitute } -3x + 2 \text{ for } y.$$
$$6x - 6x + 4 = 4$$
$$4 = 4 \qquad \text{True}$$

Getting a true statement means that every value of x makes the last equation true. Therefore, the original system has infinitely many solutions.

Solve for x and y.

$$(1) \qquad y = -2x + 4$$
$$(2) \quad 10x + 5y = 20$$

In Example 5, what do you think the graph of the system looks like? Can you identify a particular solution to the system?

> **Tip**   When solving a system of linear equations in two variables by substitution
> - if we get a false statement, then the system has no solution;
> - if we get a true statement, then the system has infinitely many solutions.

Now, let's use the substitution method to solve some applications.

EXAMPLE 6	PRACTICE 6

A car rental agency has two plans:

- In the Ambassador Plan, renting a car for one day costs $35 plus $0.25 per mile driven.
- In the Diplomat Plan, a one-day car rental costs $50 plus $0.10 per mile driven.

a. For each plan, write a linear equation that relates a day's price p for renting a car to the number of miles driven n. Express the given information as a system of equations.

b. Use the substitution method to solve the system of linear equations.

c. In the context of this problem, what is the significance of the solution?

To watch movies on premium channels, a couple decides to choose between two television cable deals:

- the TV Deal that costs $20 installation and $35 per month, and
- the Movie Deal that costs $30 installation and $25 per month.

a. Write an equation for each deal, expressing the cost of a deal c in terms of the number of months n for which the couple signs up.

Solution

a. The Ambassador Plan can be expressed as $p = 0.25n + 35$; the Diplomat Plan becomes $p = 0.10n + 50$. The system representing both plans is therefore

(1) $p = 0.25n + 35$

(2) $p = 0.10n + 50$

b. To solve the system, we can set the two expressions for p equal to each other.

$$0.25n + 35 = 0.10n + 50$$

Solving for n gives us:

$$0.25n + 35 = 0.10n + 50$$
$$0.15n = 15$$
$$n = \frac{15}{0.15}$$
$$n = 100$$

To solve for p, we can substitute 100 for n in Equation (1).

(1) $p = 0.25n + 35$
$= 0.25(\mathbf{100}) + 35$
$= 25 + 35$
$= 60$

So the solution to the system is $n = 100$ and $p = 60$.

c. With the appropriate units, the solution is $n = 100$ mi and $p = \$60$. This means that the cost of a one-day rental on the two plans is the same amount of money, namely \$60, only when the car is driven 100 mi. For other distances driven, the plans charge different amounts.

b. Solve the system of linear equations by substitution.

c. In the context of this problem, what is the significance of the solution?

Recall that we discussed mixture problems involving a single equation in Section 2.5. The following example shows how we can apply our knowledge of solving systems of linear equations to these problems.

EXAMPLE 7

How much 30% alcohol solution and 50% alcohol solution must be mixed to get 10 gal of 42% solution?

Solution We solve this problem as we did earlier mixture problems, namely by organizing the given information in a table. Let's represent the amount of 30% solution by x and the amount of 50% solution by y.

Action	Percent of Alcohol	Amount of Solution (in gallons)	Amount of Alcohol (in gallons)
Start with	30%	x	$0.3x$
Add	50%	y	$0.5y$
Finish with	42%	10	0.42 (10), or 4.2

PRACTICE 7

A chemist wishes to combine an alloy that is 20% copper with one that is 50% copper to obtain 15 oz of an alloy that is 25% copper. Find the quantities of the alloys required.

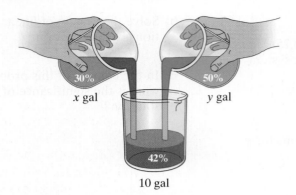

10 gal

The amount of alcohol in the 30% solution is 30% of x, or $0.3x$. The amount of alcohol in the 50% solution is 50% of y, or $0.5y$. The amount of alcohol in the 42% solution is 42% of 10, or 4.2. Since the total amount of the combined solutions is 10 gal and the total amount of alcohol is 4.2 gal, we get the following system.

$$\textbf{(1)} \qquad x + y = 10$$
$$\textbf{(2)} \quad 0.3x + 0.5y = 4.2$$

In applying the substitution method, we begin by solving for y in Equation (1).

$$\textbf{(1)} \quad x + y = 10$$
$$y = -x + 10$$

We then substitute $-x + 10$ for y in Equation (2).

$$\textbf{(2)} \qquad 0.3x + 0.5y = 4.2$$
$$0.3x + 0.5(-x + 10) = 4.2$$
$$0.3x - 0.5x + 5 = 4.2$$
$$-0.2x = -0.8$$
$$x = \frac{-0.8}{-0.2}$$
$$x = 4$$

After replacing x by 4 in Equation (1), we solve for y.

$$\textbf{(1)} \quad x + y = 10$$
$$4 + y = 10$$
$$y = 6$$

So the solution to the system is $(4, 6)$. In other words, 4 gal of 30% solution and 6 gal of 50% solution are needed to produce 10 gal of the 42% solution.

A system of linear equations can also serve as a model for investment problems, as the following example illustrates.

EXAMPLE 8

A stockbroker had $10,000 to invest for her client. The broker invested part of this amount at a low-risk, low-yield 5% rate of return per year and the rest at a high-risk, high-yield 7% rate. If the client earned a return of $550 in one year, how much money did the broker invest at each rate?

Solution The following table reflects the given information. Here, x stands for the amount of the investment at a 5% return, and y the investment at a 7% return.

Rate of Return	Amount of Investment ($)	Amount of Return ($)
5%	x	$0.05x$
7%	y	$0.07y$
TOTAL	10,000	550

The amount of return on each investment is the product of the rate of return and the amount of the investment. We add the amount of the individual investments to find the total investment, and the amount of returns on each investment to find the total amount of return. Since the total investment is $10,000 and the total return is $550, we get the following system:

(1) $\qquad x + y = 10{,}000$
(2) $\quad 0.05x + 0.07y = 550$

Now, let's solve for y in Equation (1).

(1) $\quad x + y = 10{,}000$
$\qquad\quad y = 10{,}000 - x$

We then substitute $10{,}000 - x$ for y in Equation (2).

(2) $\qquad\qquad 0.05x + 0.07y = 550$
$\qquad 0.05x + 0.07(\mathbf{10{,}000} - x) = 550$
$\qquad\quad 0.05x + 700 - 0.07x = 550$
$\qquad\qquad\quad -0.02x + 700 = 550$
$\qquad\qquad\qquad\quad -0.02x = -150$
$$x = \frac{-150}{-0.02}$$
$\qquad\qquad\qquad\qquad\quad x = 7500$

After substituting 7500 for x in Equation (1), we solve for y.

(1) $\qquad x + y = 10{,}000$
$\qquad \mathbf{7500} + y = 10{,}000$
$\qquad\qquad\quad y = 2500$

Therefore, the solution to the system is (7500, 2500). In other words, $7500 was invested at 5% and $2500 at 7%.

PRACTICE 8

The manager of a city pension fund splits $198,000 between two investments. The first investment pays 4% in simple interest per year, and the second pays 5% in simple interest per year. At the end of the first year, the two investments return the same amount of interest. How much money did the manager put into each investment?

4.2   Exercises

Solve by substitution and check.

1. $x + y = 10$
   $y = 2x + 1$

2. $x - y = 7$
   $x = 5y + 3$

3. $y = -3x - 15$
   $y = -x - 7$

4. $y = -2x - 21$
   $y = 5x$

5. $-x - y = 8$
   $x = -3y$

6. $x - y = 15$
   $y = -4x$

7. $4x + 2y = 10$
   $x = 2$

8. $2y + x = 10$
   $y = -5$

9. $-x + 20y = 0$
   $x - y = 0$

10. $5x + 3y = 0$
   $x + y = 0$

11. $6x + 4y = 2$
   $2x + y = 0$

12. $x - 3y = 0$
   $2x - 3y = 6$

13. $3x + 5y = -12$
   $x + 2y = -6$

14. $3x + 5y = -1$
   $3x + y = -5$

15. $2x + 6y = 12$
   $x + 3y = 6$

16. $x + 2y = 4$
   $3x + 6y = 3$

17. $7x - 3y = 26$
   $3x - y = 11$

18. $x + 3y = 1$
   $-3x - 5y = -2$

19. $8x + 2y = -1$
   $y = -4x + 1$

20. $4x + 2y = 4$
   $y = 5 - 2x$

21. $6x - 2y = 2$
   $y = 3x - 1$

22. $2x - 6y = -12$
   $x = 3y - 6$

23. $m = 20 - 2n$
   $2m + 4n = -22$

24. $3u + 6v = 60$
   $-2u = 4v - 40$

25. $p + 2q = 13$
   $q + 7 = 4p$

26. $a - b = -1$
   $6b = 5a$

27. $s - 3t + 5 = 0$
   $-4s + t - 9 = 0$

28. $2l - 3w + 6 = 0$
   $l - w - 10 = 0$

Mixed Practice

Solve by substitution and check.

29. $y = 2x + 12$
   $y = -3x - 3$

30. $2y - 8x = -2$
   $y = 4x - 1$

31. $6x + 2y = 2$
   $y = -3x + 2$

32. $-x - y = 12$
   $x = -5y$

33. $5x + 3y = -6$
   $-3x + y = 5$

34. $2y - 3x = 6$
   $y - 3x = 0$

Applications

Solve.

35. Two taxi companies compete in the same neighborhood. One of these companies charges $3 for the taxi drop plus $1.25 for each mile driven, while the other charges $2 for the taxi drop, plus $1.50 for each mile driven.

    a. Express these relationships as an algebraic system, where *m* represents miles driven and *c* represents cost.

    b. Solve the system and interpret the results.

36. Two electricians make house calls. One charges $75 for a visit plus $50 per hour of work. The other charges $95 per visit plus $40 per hour of work.

    a. Letting *c* represent cost and *h* represent hours worked, write a linear equation for each electrician that relates the charge for a house call in terms of the length of a visit.

    b. For how many hours of work do the two electricians charge the same?

37. On a particular airline route, a full-price coach ticket costs $310 and a discounted coach ticket costs $210. On one of these flights, there were 172 passengers in coach, which resulted in a total ticket income of $44,120. How many full-price tickets were sold?

38. During a sale, a store sells red-dot items at a 30% discount and yellow-dot items at a 20% discount. A shopper bought red- and yellow-dot items with a combined regular price of $40. If the total discount was $9.80, how much did the shopper spend on each kind of item?

39. A laboratory technician needs to make a 10-liter batch of antiseptic that is 60% alcohol. How can she combine a batch of antiseptic that is 30% alcohol with another that is 70% to get the desired concentration?

40. A bottle of fruit juice contains 20% water. How much water must be added to this bottle to produce 8 L of fruit juice that is 50% water?

41. A corporation merged two departments into one. In one department, 5% of the employees were women, whereas in the other department, 80% were women. When the departments were merged, 50% of the 150 employees were women. How many women were in each department before the merger?

42. A hospital needs 30 L of a 10% solution of disinfectant. How many liters of a 20% solution and a 4% solution should be mixed to obtain this 10% solution?

43. A student took out two loans totaling $5000. She borrowed the maximum amount she could at 6% and the remainder at 7% interest per year. At the end of the first year, she owed $310 in interest. How much was loaned at each rate?

44. A man invested three times as much money in a bond fund that earned 8% in a year as he did in a mutual fund that returned 4% in the year. How much money did he invest in each fund if the total earnings for the year were $112?

45. A $40,000 investment was split so that part was invested at a 7% annual rate of interest and the rest at 9%. If the total annual earnings were $3140, how much money was invested at each rate?

46. A financial adviser counseled a client to invest $15,000, split between two stocks. At the end of one year, the investment in one stock increased in value by 4%, and the investment in the second stock increased in value by 8%. If the total increase in value of the investment was $1120, how much money was invested in each stock?

• *Check your answers on page A-24.*

MINDSTRETCHERS

Groupwork

1. Cramer's Rule is a formula that can be used to solve a system of linear equations for x and y. Consider the following system.

$$ax + by = c$$
$$dx + ey = f$$

The formula states that $x = \dfrac{ce - bf}{ae - bd}$ and $y = \dfrac{af - cd}{ae - bd}$. Note that this mechanical approach allows machines to solve systems of equations.

a. Working with a partner, make up your own values for a, b, c, d, e, and f, and substitute these values in the system.

$$\underline{\hspace{1cm}}\ x + \underline{\hspace{1cm}}\ y = \underline{\hspace{1cm}}$$
$$\underline{\hspace{1cm}}\ x + \underline{\hspace{1cm}}\ y = \underline{\hspace{1cm}}$$

b. Use Cramer's Rule to calculate x and y.

$$x = \frac{ce - bf}{ae - bd} = \underline{\hspace{2cm}}$$
$$y = \frac{af - cd}{ae - bd} = \underline{\hspace{2cm}}$$

c. By substitution, check whether (x, y) is in fact a solution to the system.

Mathematical Reasoning

2. On the coordinate plane, consider the quadrilateral $ABCD$ shown. At what point do the diagonals $\overline{AC}$ and $\overline{BD}$ intersect? Explain how to find the answer exactly.

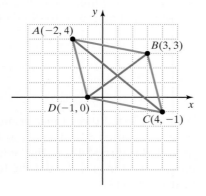

Critical Thinking

3. For what value of k will the system shown have infinitely many solutions?

$$kx - 2y = 10$$
$$4x - y = 5$$

4.3 Solving Systems of Linear Equations by Elimination

OBJECTIVES

- To solve a system of linear equations by elimination

- To solve applied problems involving systems of linear equations

The final method of solving a system of linear equations that we consider is called the **elimination** (or **addition**) **method**. As in the substitution method discussed in the previous section, the elimination method involves changing a given system of two equations in two variables to one equation in one variable. A system in which the coefficients of a variable are opposites particularly lends itself to the elimination method.

Recall that in solving linear equations in one variable, we frequently used the addition property of equality: If $a = b$, then $a + c = b + c$. That is, if we add the same number to both sides of an equation, we get an equivalent equation.

The elimination method for solving *systems* is based on a closely related property of equality: If $a = b$ and $c = d$, then $a + c = b + d$. This property allows us to "add equations."

Let's look at examples of applying this property in the elimination method.

EXAMPLE 1	PRACTICE 1
Solve the following system by the elimination method.	Solve for x and y.

EXAMPLE 1

Solve the following system by the elimination method.

$$\begin{array}{ll}(1) & x + y = 4 \\ (2) & x - y = 2\end{array}$$

Solution First we decide which variable to eliminate. Since the coefficients of the y-terms in the two equations are opposites (namely $+1$ and -1), we eliminate y if we add the equations.

$$\begin{array}{ll}(1) & x + y = 4 \\ (2) & \underline{x - y = 2} \\ & 2x + 0y = 6 \\ & \quad\;\; 2x = 6 \\ & \quad\;\;\; x = 3\end{array}$$

To find y, we substitute 3 for x in either of the original equations. Substituting in Equation (1) we get:

$$\begin{array}{ll}(1) & x + y = 4 \\ & 3 + y = 4 \\ & \quad\;\; y = 4 - 3 \\ & \quad\;\; y = 1\end{array}$$

So $x = 3$ and $y = 1$. That is, the solution is $(3, 1)$.

Check We check the solution by substituting these values for x and y in both of the original equations.

$$\begin{array}{ll}(1) \quad x + y = 4 & \qquad (2) \quad x - y = 2 \\ \quad\;\; 3 + 1 \stackrel{?}{=} 4 & \qquad \quad\;\; 3 - 1 \stackrel{?}{=} 2 \\ \quad\;\;\;\; 4 = 4 \quad \text{True} & \qquad \quad\;\;\;\; 2 = 2 \quad \text{True}\end{array}$$

So our solution $(3, 1)$ is confirmed.

PRACTICE 1

Solve for x and y.

$$\begin{array}{ll}(1) & x + y = 6 \\ (2) & x - y = -10\end{array}$$

EXAMPLE 2

Use the elimination method to solve for x and y.

$$\textbf{(1)} \quad 3x + 2y = 14$$
$$\textbf{(2)} \quad 5x + 2y = -8$$

Solution In this system, adding the two given equations does not eliminate either variable. However, we note that the two y-terms have the same coefficient, namely 2. So if we multiply Equation (1) by -1 and then add the equations, the y-terms will cancel out.

$$\textbf{(1)} \quad 3x + 2y = 14 \xrightarrow{\text{Multiply by } -1.} -3x - 2y = -14$$
$$\textbf{(2)} \quad 5x + 2y = -8 \qquad\qquad \underline{5x + 2y = -8} \qquad \text{Add the equations.}$$
$$2x \qquad\quad = -22$$
$$x = -11$$

Next, we substitute -11 for x in either of the original equations. Let's choose Equation (1), and then solve for y.

$$\textbf{(1)} \qquad\quad 3x + 2y = 14$$
$$3(-11) + 2y = 14$$
$$-33 + 2y = 14$$
$$2y = 14 + 33$$
$$2y = 47$$
$$y = \frac{47}{2} = 23.5$$

So the solution is $(-11, 23.5)$.

PRACTICE 2

Solve the following system using the elimination method.

$$\textbf{(1)} \quad 4x + 3y = -7$$
$$\textbf{(2)} \quad 5x + 3y = -5$$

EXAMPLE 3

Solve.

$$\textbf{(1)} \quad 3x - y = -2$$
$$\textbf{(2)} \quad x + 5y = 10$$

Solution Note that the coefficient of x in Equation (2) is $+1$. By multiplying this equation by -3 and then adding the two equations, we eliminate the x-terms.

$$\textbf{(1)} \quad 3x - y = -2 \qquad\qquad\qquad\quad 3x - y = -2$$
$$\textbf{(2)} \quad x + 5y = 10 \xrightarrow{\text{Multiply by } -3.} \underline{-3x - 15y = -30}$$
$$-16y = -32 \qquad \text{Add the equations.}$$
$$y = 2$$

Next, we substitute 2 for y in Equation (2) and then solve for x.

$$\textbf{(2)} \qquad x + 5y = 10$$
$$x + 5(2) = 10$$
$$x + 10 = 10$$
$$x = 0$$

So the solution is $(0, 2)$.

PRACTICE 3

Solve for x and y.

$$\textbf{(1)} \quad x - 3y = -18$$
$$\textbf{(2)} \quad 5x + 2y = 12$$

How could we have solved the system in Example 3 another way?

EXAMPLE 4	PRACTICE 4

Use the elimination method to solve the following system of linear equations.

(1) $4x + 3y = -19$
(2) $3x - 2y = -10$

Solution This system is more complicated to solve than the previous examples because there is no single integer that we can multiply either equation by that will eliminate a variable when we add the equations. Instead, we must multiply *both* equations by integers that lead to the elimination of a variable. There are a number of possible strategies to accomplish this. We can, for instance, multiply Equation (1) by 2 and Equation (2) by 3 to eliminate the y-terms when the equations are added.

(1) $4x + 3y = -19$ $\xrightarrow{\text{Multiply by 2.}}$ $8x + 6y = -38$
(2) $3x - 2y = -10$ $\xrightarrow{\text{Multiply by 3.}}$ $\underline{9x - 6y = -30}$
$17x = -68$ **Add the equations.**
$x = -4$

Now, let's substitute -4 for x in Equation (1) and then solve for y.

(1) $4x + 3y = -19$
$4(-4) + 3y = -19$
$-16 + 3y = -19$
$3y = 16 + (-19)$
$3y = -3$
$y = -1$

So the solution is $(-4, -1)$.

Solve by elimination.

(1) $5x - 7y = 24$
(2) $3x - 5y = 16$

How could we have solved the system in Example 4 by eliminating x instead of y?

To Solve a System of Linear Equations by Elimination

- Write both equations in the general form $Ax + By = C$.
- Choose the variable that you want to eliminate.
- If necessary, multiply one or both equations by appropriate numbers so that the coefficients of the variable to be eliminated are opposites.
- Add the equations. Then solve the resulting equation for the remaining variable.
- Substitute the value found in the previous step in either of the original equations and solve for the other variable.
- Check by substituting the values in both equations of the original system.

EXAMPLE 5	PRACTICE 5

Solve by elimination.

$$\begin{array}{rl} \textbf{(1)} & 5x = 3y \\ \textbf{(2)} & -3x + 2y = 9 \end{array}$$

Solution Equation (1) is not in the form $Ax + By = C$, so let's begin by rewriting it in general form.

$$\textbf{(1)} \quad 5x = 3y \quad \xrightarrow{\text{Write in general form.}} \quad 5x - 3y = 0$$
$$\textbf{(2)} \quad -3x + 2y = 9 \qquad\qquad\qquad -3x + 2y = 9$$

Now, suppose we choose to eliminate the x-terms. To do this, we can multiply Equation (1) by 3, Equation (2) by 5, and then add the equations.

$$\textbf{(1)} \quad 5x - 3y = 0 \quad \xrightarrow{\text{Multiply by 3.}} \quad 15x - 9y = 0$$
$$\textbf{(2)} \quad -3x + 2y = 9 \quad \xrightarrow{\text{Multiply by 5.}} \quad \underline{-15x + 10y = 45} \quad \text{Add the equations.}$$
$$y = 45$$

To solve for x, let's substitute 45 for y in Equation (1).

$$\begin{aligned} \textbf{(1)} \quad 5x &= 3y \\ 5x &= 3(\mathbf{45}) \\ 5x &= 135 \\ x &= 27 \end{aligned}$$

So the solution is $(27, 45)$.

Use the elimination method to solve the following system.

$$\begin{array}{rl} \textbf{(1)} & -2x + 5y = 20 \\ \textbf{(2)} & 3x = 7y - 26 \end{array}$$

EXAMPLE 6	PRACTICE 6

Solve by elimination.

$$\begin{array}{rl} \textbf{(1)} & 4x - 6y + 12 = 0 \\ \textbf{(2)} & 2x - 3y = -4 \end{array}$$

Solution We begin by writing Equation (1) in general form.

$$\textbf{(1)} \quad 4x - 6y + 12 = 0 \quad \xrightarrow{\text{Write in general form.}} \quad 4x - 6y = -12$$
$$\textbf{(2)} \qquad 2x - 3y = -4 \qquad\qquad\qquad 2x - 3y = -4$$

Now, let's eliminate the y-terms. To do this, we can multiply Equation (2) by -2, and then add the equations.

$$\textbf{(1)} \quad 4x - 6y = -12 \qquad\qquad\qquad 4x - 6y = -12$$
$$\textbf{(2)} \quad 2x - 3y = -4 \quad \xrightarrow{\text{Multiply by } -2.} \quad \underline{-4x + 6y = 8}$$
$$0 = -4 \quad \text{False}$$

Since adding the equations yields a false statement, the original system has no solution.

Solve:

$$\begin{aligned} 3x &= 4 + y \\ 9x - 3y &= 12 \end{aligned}$$

Let's use our knowledge of the elimination method to solve some applied problems, beginning with a motion problem.

EXAMPLE 7

It takes a plane 3 hr to fly between two airports, traveling with a tailwind at a ground speed of 500 mph. The plane then takes 4 hr to make the return trip against the same wind. What is the speed of the plane in still air? What is the speed of the wind?

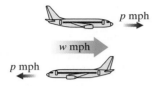

Solution Let p represent the speed of the plane in still air and w represent the speed of the wind. On the initial flight, the wind is with the plane so its speed relative to the ground is $p + w$. When returning, the wind is against the plane so its ground speed is $p - w$. We can organize the given information in the following table:

	Ground speed ·	Time =	Distance
Going	$p + w$	3	$3(p + w)$
Returning	$p - w$	4	$4(p - w)$

Note that since the distance the plane travels is the product of its ground speed and the time it travels, we can compute each entry in the distance column of the table by multiplying the corresponding entries in the ground speed and the time columns.

Since we are told that the speed going is 500 mph, we have

$$p + w = 500$$

But the distance going and the distance returning are equal, so

$$3(p + w) = 4(p - w)$$

Now, we have a system of two equations, which we must solve.

(1) $p + w = 500$
(2) $3(p + w) = 4(p - w)$

We can write Equation (2) in general form, by simplifying.

$$3(p + w) = 4(p - w)$$
$$3p + 3w = 4p - 4w$$
$$3p - 4p + 3w + 4w = 0$$
$$-p + 7w = 0$$

The system then becomes

(1) $p + w = 500$
(2) $-p + 7w = 0$

Adding the equations eliminates the p-terms.

$$8w = 500$$
$$w = 62.5$$

PRACTICE 7

A whale swimming with the current traveled 80 mi in 2 hr. Swimming against the current, the whale traveled only 40 mi in the same amount of time. Find the whale's speed in calm water and the speed of the current.

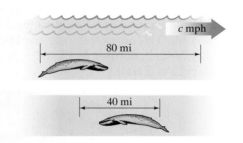

Finally, let's substitute 62.5 for *w* in Equation (1) and solve for *p*.

$$\textbf{(1)} \quad p + 62.5 = 500$$
$$p = 437.5$$

So the wind speed is 62.5 mph, and the speed of the plane in still air is 437.5 mph.

EXAMPLE 8

A student had two part-time jobs in a restaurant. One week she earned a total of $306, working 12 hr as a cashier and 10 hr as a cook. The next week, she worked 14 hr as a cashier and 22 hr cooking, earning $512. What is her hourly wage as a cashier? As a cook?

Solution Let *x* represent the student's hourly wage as a cashier and *y* represent the student's hourly wage as a cook. The first week, the student earned $306, and the second week, $512. So we must solve the following system.

$$\textbf{(1)} \quad 12x + 10y = 306$$
$$\textbf{(2)} \quad 14x + 22y = 512$$

We can divide each equation by 2 to simplify.

$$\textbf{(1)} \quad 6x + 5y = 153$$
$$\textbf{(2)} \quad 7x + 11y = 256$$

Let's eliminate the *x*-terms by multiplying Equation (1) by 7 and Equation (2) by −6. Then we add the equations.

$$\textbf{(1)} \quad 6x + 5y = 153 \quad \xrightarrow{\text{Multiply by 7.}} \quad 42x + 35y = 1071$$
$$\textbf{(2)} \quad 7x + 11y = 256 \quad \xrightarrow{\text{Multiply by } -6.} \quad -42x - 66y = -1536$$

Adding the equations eliminates the *x*-terms.

$$-31y = -465$$
$$y = \frac{-465}{-31}$$
$$y = 15$$

Finally, we substitute 15 for *y* in the original Equation (1) and solve for *x*.

PRACTICE 8

Admission prices at a football game were $10 for adults and $6 for children. The total value of the 175 tickets sold was $1450. How many adults and how many children attended the game?

$$
\begin{align}
\textbf{(1)} \qquad 12x + 10y &= 306 \\
12x + 10(\textbf{15}) &= 306 \\
12x + 150 &= 306 \\
12x &= 156 \\
x &= \frac{156}{12} \\
x &= 13
\end{align}
$$

So the student earned $13 per hour as a cashier and $15 per hour as a cook.

So far in this chapter, we have discussed three methods of solving a system of linear equations—the graphing method, the substitution method, and the elimination (or addition) method. The following table lists some advantages and disadvantages of each method, to help in deciding which method to apply in a given problem.

Method	Advantages	Disadvantages
Graphing Method	• Provides a picture that makes relationships understandable.	• Approximates solutions, particularly when they are not integers or are large. • Can be time-consuming if not using a grapher.
Substitution Method	• Gives exact solutions. • Is easy to use when a variable in one of the original equations is isolated.	• No picture. • Can result in complicated equations with parentheses and with fractions.
Elimination Method	• Gives exact solutions. • Is easy to use when the two coefficients of a variable are opposites.	• No picture.

In the next section, we will learn how to solve a system of linear inequalities.

4.3	Exercises

Solve.

1. $x + y = 3$
$x - y = 7$

2. $x - y = 10$
$x + y = -8$

3. $x + y = -4$
$-x + 3y = -6$

4. $5x - y = 8$
$2x + y = -1$

5. $10p - q = -14$
$-4p + q = -4$

6. $a + b = -4$
$-a + 2b = -8$

7. $3x + y = -3$
$4x + y = -4$

8. $x + 4y = -3$
$x - 7y = 19$

9. $3x + 5y = 10$
$3x + 5y = -5$

10. $8x + 2y = 3$
$4x + y = -9$

11. $9x + 6y = -15$
$-3x - 2y = 5$

12. $4x + y = -3$
$8x + 2y = -6$

13. $5x + 2y = -9$
$-5x + 2y = 11$

14. $7x + 4y = -6$
$-x + 4y = 10$

15. $2s + d = -2$
$5s + 3d = -6$

16. $-5x + 8y = -7$
$-6x + 9y = -9$

17. $3x - 5y = 1$
$7x - 8y = 17$

18. $3x + 2y = 9$
$-2x + 3y = -19$

19. $5x + 2y = -1$
$4x - 5y = -14$

20. $10x - 3y = 9$
$3x - 2y = -5$

21. $7p + 3q = 15$
$-5p - 7q = 16$

22. $8a + 2b = 18$
$4a - 3b = -15$

23. $6x + 5y = -8.5$
$8x + 10y = -3$

24. $6x - 6y = -3.6$
$-4x + 8y = -16$

25. $3.5x + 5y = -3$
$2x = -2y$

26. $3x - 3y = 0$
$1.5y = -6x + 30$

27. $2x - 4 = -y$
$x + 2y = 0$

28. $y = -3x + 7$
$4x + 2y = 11$

29. $8x + 10y = 1$
$-4x - 5y + 6 = 0$

30. $x - y = 6$
$3x = 3y + 10$

Mixed Practice

Solve.

31. $4a - b = -10$
$-3a + b = 7$

32. $9x - 6y = 3$
$3x - 2y = 6$

33. $3x - 5y = 4$
$-6x + 10y = -8$

34. $7p + 4q = -12$
$4p + q = -3$

35. $5x + 3y = -3$
$-7x - 5y = 4$

36. $4x + 3y = -5.5$
$5x + 6y = -3.5$

Applications

Solve.

37. A quarterback throws a pass that travels 40 yd with the wind in 2.5 sec. If he had thrown the same pass against the wind, the football would have traveled 20 yd in 2 sec. Find the speed of a pass that the quarterback would throw if there were no wind.

38. A crew team rows in a river with a current. When the team rows with the current, the boat travels 14 mi in 2 hr. Against the current, the team rows 6 mi in the same amount of time. At what speed does the team row in still water?

39. To enter a zoo, adult visitors must pay $5, whereas children and seniors pay only half price. On one day, the zoo collected a total of $765. If the zoo had 223 visitors that day, how many half-price admissions and how many full-price admissions did the zoo collect?

40. Compact discs are stored in single jewel cases, which are 0.375 in. thick, and in multiple-CD jewel cases, which have a thickness of 0.875 in. If 86 of these jewel cases exactly fit on the storage shelf shown, how many of the jewel cases are single?

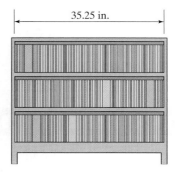

41. The annual salaries of a congressman and a senator total $227,000. If the senator makes $23,200 more than the congressman, find each of their salaries.

42. The height of the picture shown below is double its width. It took molding 180 in. long to frame the picture. Find the dimensions of the frame.

43. A particular computer takes 43 nanoseconds to carry out 5 sums and 7 products, and 42 nanoseconds to perform 2 sums and 9 products. How long does the computer take to carry out one sum? To carry out one product?

44. A wholesale novelty shop sells some embroidered scarves for $12 each and others for $15 each. A customer pays $234 for 17 scarves. How many scarves at each price did she buy?

45. One issue of a journal has 3 full-page ads and 5 half-page ads, generating $6075 in revenue. The next issue has 4 full-page ads and 4 half-page ads, resulting in advertising revenue of $6380. Determine the advertising rates in this journal for full-page and half-page ads.

46. According to a law of physics, the lever shown will balance when the products of each weight and the length of its force arm are equal.

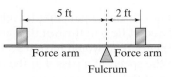

The weights shown above balance. If 10 lb are added to the left weight, which is then moved 1 ft closer to the fulcrum, the lever will again balance. Find the two original weights.

● *Check your answers on page A-24.*

MINDSTRETCHERS

Groupwork

1. Your friend performs the following magic trick: She asks you to think of two numbers but not to tell you what they are. Instead, you tell her the sum and the difference of the two numbers. She promptly tells you what the two original numbers were. Explain how your friend does the trick. Working with partners, try to perform this trick.

Writing

2. Consider the system of equations

$$\textbf{(1)} \qquad 5x - 8y = 4$$
$$\textbf{(2)} \quad 12x + 24y = 11$$

Explain how the concept of LCM relates to solving this system by elimination.

Mathematical Reasoning

3. The elimination method can be extended to three linear equations in three variables. Solve the following system:

$$4x - y - 3z = 30$$
$$3x - 2y - 6z = -5$$
$$x - z = 5$$

CULTURAL NOTE

Carl Friedrich Gauss (1777–1855) is generally considered to be one of the greatest mathematicians in history. Among his many mathematical contributions was the *Gaussian elimination method* of solving systems of linear equations, discussed in this chapter. Gauss (rhymes with "house") is credited with being the first to prove the major mathematical result known as the Fundamental Theorem of Algebra. He was also an important scientist. In astronomy, he laid the theoretical foundation for predicting a planet's orbit. To honor this German's groundbreaking work in physics, his name is given to the unit (*gauss*) used today to express the strength of a magnetic field.

Sources: Roger Cooke, *The History of Mathematics*, John Wiley, New York, 1997; D. E. Smith, *History of Mathematics*, Dover Publications, New York, 1923

4.4	Solving Systems of Linear Inequalities

In Section 3.5, we graphed on a coordinate plane a single linear inequality in two variables, such as $3x - 2y > 4$. There, we saw that a solution of such an inequality is an ordered pair of real numbers that when substituted for the variables satisfy the inequality. We also saw that the graph of this inequality (the set of all ordered pair solutions) consists of a region of points in the coordinate plane whose coordinates satisfy the inequality.

- To solve a system of linear inequalities by graphing

- To solve applied problems involving systems of linear inequalities

In this section, consider the graphs of **systems of linear inequalities**, which are two or more linear inequalities considered simultaneously, that is, together. For instance,

$$y \leq 3x - 6$$
$$y > -4x + 2$$

is a system of linear inequalities. As with single inequalities, the *solutions* of a system of linear inequalities are ordered pairs.

Definition

A **solution** of a system of linear inequalities in two variables is an ordered pair of real numbers that makes both inequalities in the system true.

Solving Systems of Linear Inequalities by Graphing

We can solve a system of linear inequalities by graphing each inequality on the same coordinate plane. Every point in the region of overlap is a solution of both inequalities and is, therefore, a solution of the system.

EXAMPLE 1	PRACTICE 1

EXAMPLE 1

Graph the solutions of the system:

$$y \leq 3x - 6$$
$$y > -4x + 2$$

Solution Let's begin by graphing each inequality. First, we graph $y \leq 3x - 6$. The boundary line is the graph of the corresponding equation $y = 3x - 6$. We draw this as a solid line since the inequality involves the $\leq$ symbol. Then, in either half-plane we choose a test point, for instance, $(3, 0)$. Since this point satisfies the inequality, the half-plane containing it is part of the graph.

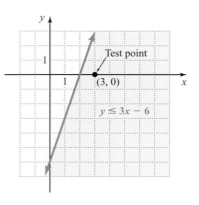

PRACTICE 1

Graph the solutions of the system:

$$y \geq x - 5$$
$$y < -3x - 2$$

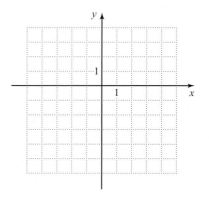

Next, we graph $y > -4x + 2$. The boundary line is the graph of the corresponding equation $y = -4x + 2$. We draw this as a broken line since the inequality involves the $>$ symbol. Then, in either half-plane we choose a test point, for instance, $(0, 4)$. Since this point satisfies the inequality, the half-plane containing it is part of the graph.

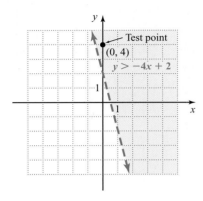

Finally, we draw each inequality on the same coordinate plane. Since a solution of the system of linear inequalities must satisfy each inequality, the solutions of the system are all the points that lie in *both* shaded regions, that is, in the overlapping region of the two graphs. Points on part of the boundary line $y = 3x - 6$ are solutions, but points on the boundary line $y = -4x + 2$ are not.

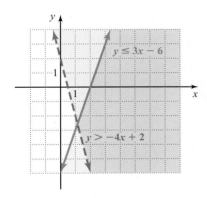

EXAMPLE 2	PRACTICE 2
Solve the following system by graphing:	Solve the following system by graphing:

EXAMPLE 2

Solve the following system by graphing:

$$-x - y < 2$$
$$y - 2x > 1$$

Solution We begin by graphing both inequalities on the same coordinate plane. To graph each boundary line, we solve each inequality for y, getting

$$\begin{array}{ll} -x - y < 2 & y - 2x > 1 \\ \quad -y < x + 2 & \quad y > 2x + 1 \\ \quad y > -x - 2 & \end{array}$$

PRACTICE 2

Solve the following system by graphing:

$$2x + y < 1$$
$$-y + 3x < 1$$

Next, we graph each boundary line. Then, for each inequality, we shade the half-plane that contains its solutions.

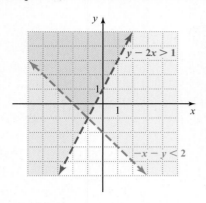

The solutions of the system are all the points that lie in both shaded regions, that is, in the overlapping region of the two graphs. Note that none of the points on the boundary lines is a solution.

Now let's look at solving a system consisting of three linear inequalities in two variables.

EXAMPLE 3

Graph the solutions of the system.

$$x + y < 8$$
$$x \geq 0$$
$$y \geq 0$$

Solution We graph all three inequalities on the same coordinate plane. First, we graph the boundary lines. Then, for each inequality we shade the half-plane that contains its solutions.

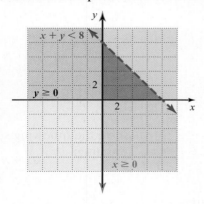

The solutions of the system are points that lie in all three shaded regions, that is, the dark purple overlapping region of all three graphs. Note that the points on parts of the lines $x = 0$ and $y = 0$ are solutions, but points on the line $x + y = 8$ are not.

PRACTICE 3

Graph the solutions of the system.

$$y + 2x \geq 4$$
$$x > -3$$
$$y \geq 1$$

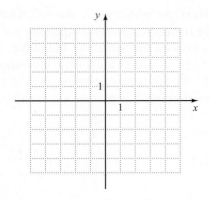

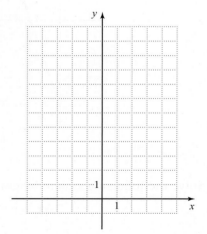

Just as with systems of linear equations, many applied problems can be modeled by systems of linear inequalities.

EXAMPLE 4

A company's employment test has two parts—a verbal subtest and a mathematics subtest. An applicant can earn a maximum combined score of 100 points. To qualify for a particular position, the applicant is required to have a math score of at least 40 and a verbal score of at least 25.

a. Write a system of inequalities to model the situation in which the applicant qualifies for a position.

b. Solve the system by graphing.

c. Describe the region in which the prospective employee qualifies for the position.

Solution

a. Let v represent the verbal score and m represent the math score. Since a prospective employee can earn a maximum combined score of 100 points, we write $v + m \leq 100$. If a successful applicant must score at least 40 on the math subtest and at least 25 on the verbal subtest, then we can write two inequalities, $m \geq 40$ and $v \geq 25$. So the system is

$$v + m \leq 100$$
$$m \geq 40$$
$$v \geq 25$$

b. Solve by graphing: $v + m \leq 100$
$$m \geq 40$$
$$v \geq 25$$

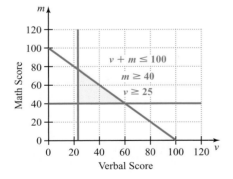

The solutions of the system are points that lie in all three shaded regions, including the highlighted parts of the boundary lines.

c. The region in which the applicant qualifies for a position is the overlapping area bounded by the lines $m = 40$, $v + m = 100$, and $v = 25$.

PRACTICE 4

Suppose the president of the student government association formed a committee to raise funds for AIDS research. The committee must have from 7 to 10 members. The number of first-time freshmen should be greater than the number of returning students.

a. Write a system of inequalities to model the problem.

b. Graph the system.

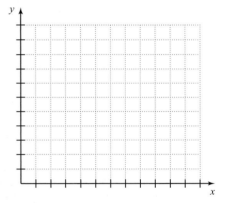

c. List the combinations of first-time freshmen and returning students that may participate in the committee. Explain why the list is finite.

4.4   Exercises   FOR EXTRA HELP    *MyMathLab*   PRACTICE   WATCH   DOWNLOAD   READ   REVIEW

Mathematically Speaking

Fill in each blank with the most appropriate term or phrase from the given list.

overlapping	triples	pairs
number line	a system of	shaded
simultaneous	coordinate plane	

1. Two or more linear inequalities considered simultaneously, that is, together, are called _____ linear inequalities.

2. The solutions of a system of linear inequalities in two variables are ordered _____.

3. A system of linear inequalities in two variables can be solved by graphing each inequality on the same _____.

4. In the graph of a system of linear inequalities, every point in the _____ region is a solution of the system.

Solve by graphing.

5. $y > 2x - 1$
   $y < -x + 3$

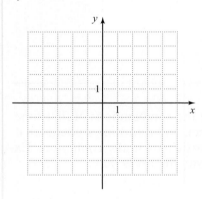

6. $y < 3x - 2$
   $y > -2x + 1$

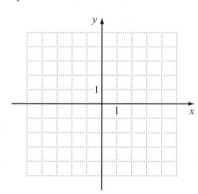

7. $y \leq \dfrac{1}{3}x + 3$

   $y < -\dfrac{1}{2}x + 1$

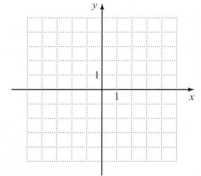

8. $y > -\dfrac{1}{4}x - 1$

   $y \geq \dfrac{3}{2}x$

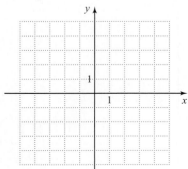

9. $4x + 2y \geq -6$
   $12x - 3y \geq -6$

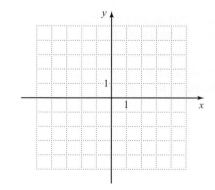

10. $3y - 9x \leq 12$
   $5y + 5x \leq 15$

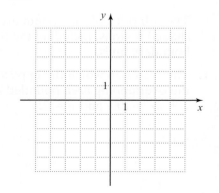

11. $3x - 2y \le 8$
   $-x - 3y > 0$

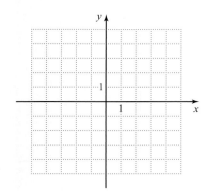

12. $-2x - 4y < 16$
   $x - 2y \ge 8$

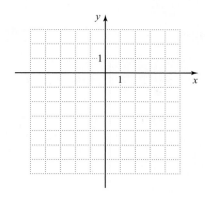

13. $4x + 2y \le 4$
   $2x + y > -3$

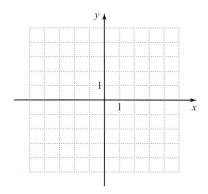

14. $x + 4y \le 20$
   $3x + 12y > -24$

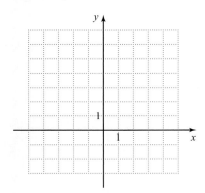

15. $3x - 9y < 18$
   $1.5x + 0.5y > 1.5$

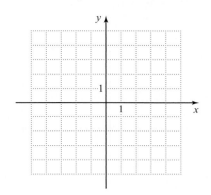

16. $1.2x + 0.4y < -0.4$
   $5x - 15y > 10$

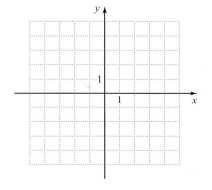

17. $y - x > 1$
   $x \le 4$
   $y > 0$

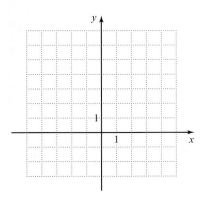

18. $x + y < 3$
   $x \le 1$
   $y \ge -3$

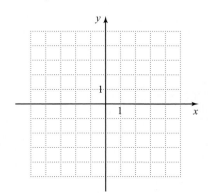

19. $-2x - y \ge 3$
   $x \ge -4$
   $y \ge -2$

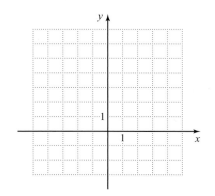

20. $y + 2x \geq -1$
     $x \leq 2$
     $y \leq 3$

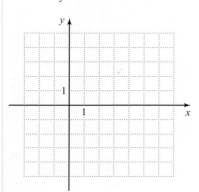

21. $3x - 2y < 2$
     $x + 3y < 12$
     $x > -2$

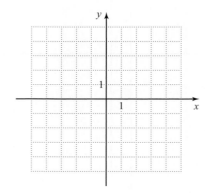

22. $4x + 8y > 8$
     $2x - y > 3$
     $x < 5$

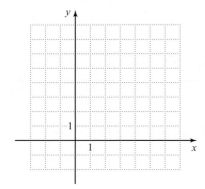

23. $12x - 4y \geq 16$
     $3x - 6y < -6$
     $y < 4$

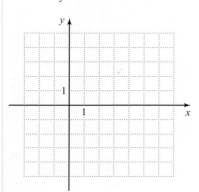

24. $10x + 5y > -5$
     $9x - 3y \geq 0$
     $y \geq -4$

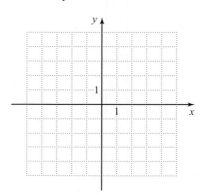

25. $-x + y > -2$
     $x + y < 2$
     $2x - y > 1$

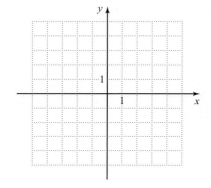

26. $y + x \leq -3$
     $y - x \leq 1$
     $2y - x > -2$

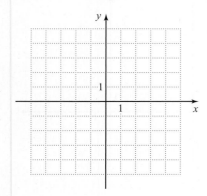

Mixed Practice

Solve by graphing.

27. $y \le -3x - 2$
       $y < x + 2$

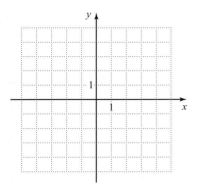

28. $-3x + 2y \le -6$
       $\frac{1}{4}x + y > 1$

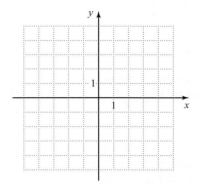

29. $6x + 3y < 9$
       $-4x - 2y < 4$

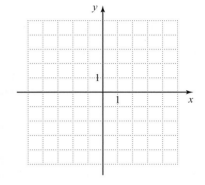

30. $x \ge 1$
       $y > -3$
       $x + y \ge 1$

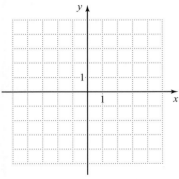

31. $x > -3$
       $y < 4$
       $-2x + y \ge 3$

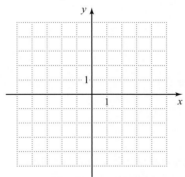

32. $y \ge -4$
       $2x + y < -1$
       $-3x + y \le 3$

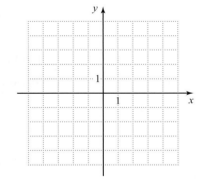

Applications

Solve.

33. A part-time student works at two jobs to support herself. She can work at her retail job no more than 20 hours per week and at her office job no more than 30 hours per week. In order to make enough money, she must work a minimum of 35 hours per week.

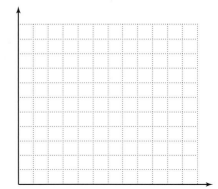

    a. Express this information as a system of inequalities.

    b. Graph the system.

    c. If she works 15 hr at her retail job, how many hours must she work at her office job?

34. An investor has a maximum of $10,000 to buy shares of stock in a pharmaceutical company and in a technology company. The pharmaceutical stock costs $40 per share, and the technology stock costs $25 per share. He wants to buy no more than 200 shares of the pharmaceutical stock and at least 50 shares of the technology stock.

   a. Write a system of inequalities to model this problem.

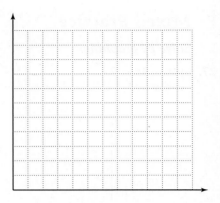

   b. Graph the system.

   c. If he buys 100 shares of the technology stock, how many shares of the pharmaceutical stock can he buy?

35. The owner of a garden shop decides to fence off part of his lot to build an outdoor nursery. The perimeter of the lot cannot exceed 400 ft. The length of the nursery is to be at least 25 ft longer than the width. The width must be at least 25 ft.

   a. Express this information as a system of inequalities.

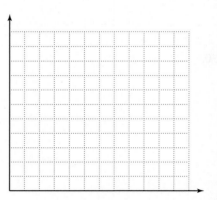

   b. Solve the system by graphing.

   c. What does the solution region represent?

36. A donut shop charges $0.40 for a donut and $0.80 for a muffin. The manager of the shop would like to have daily sales of at least $250. The shop can make at most 30 dozen donuts and 20 dozen muffins per day.

   a. Express this information as a system of inequalities.

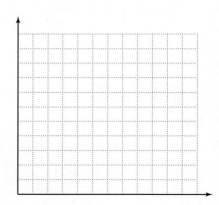

   b. Solve the system by graphing.

   c. What does the solution region represent?

37. A magazine charges advertisers $150 for a half-page ad and $225 for a full-page ad. In a particular month, the revenue from ads was less than $15,000. The magazine sold more half-page ads than full-page ads. The magazine sold at least 20 half-page ads.

   a. Write a system of inequalities to model the information.

   b. Solve the system by graphing.

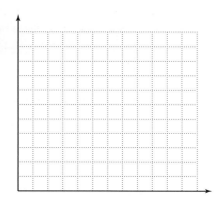

38. A game show winner decides to invest no more than $8500 of his winnings in two accounts earning 6.5% and 5% simple interest. He would like the accounts to earn at least $300 after one year. He plans to invest less than $2500 in the account earning 5% simple interest.

   a. Write a system of inequalities to model the information.

   b. Solve the system by graphing.

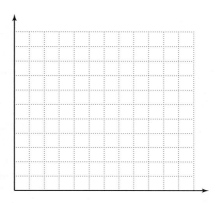

39. A furniture manufacturer produces dining tables and chairs. Each table requires 2 hr for assembly and 3 hr for finishing and painting. Each chair requires 1.5 hr for assembly and 1 hr for finishing and painting. The manufacturer allots at most 360 hours per month for assembly and at most 400 hours per month for finishing and painting.

   a. Express this information as a system of inequalities.

   b. Solve the system by graphing.

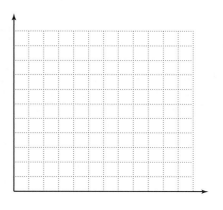

40. A factory produces two types of machine parts. Part A requires 5 hr for fabrication and $1\frac{1}{2}$ hr for testing. Part B requires 2 hr for fabrication and $\frac{1}{2}$ hr for testing. The factory manager allots no more than 150 hours per week for fabrication and no more than 40 hours per week for testing

   a. Write a system of inequalities to represent the situation.

   b. Solve the system by graphing.

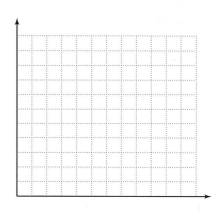

• Check your answers on page A-25.

MINDSTRETCHERS

Writing

1. Write a problem situation whose solutions lie in the region shown in the following graph.

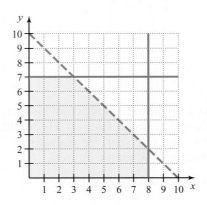

Mathematical Reasoning

2. Is it possible for a single point to be the only solution of a system of inequalities? If so, give an example of such a system. If not, explain why.

Critical Thinking

3. Write a system of inequalities whose solutions lie entirely on the line $y = 2x + 3$.

KEY CONCEPTS AND SKILLS   (CONCEPT)   (SKILL)

Concept/Skill	Description	Example
[4.1] System of equations	Two or more equations considered simultaneously, that is, together.	$x + y = 7$ $x - y = 1$
[4.1] Solution of a system of linear equations in two variables	An ordered pair of numbers that makes both equations in the system true.	Is $(4, 3)$ a solution of the following system? $x + y = 7 \rightarrow 4 + 3 \overset{?}{=} 7$ **True** $x - y = 1 \rightarrow 4 - 3 \overset{?}{=} 1$ **True** Yes, $(4, 3)$ is a solution of the system.

[4.1] To solve a system of linear equations by graphing	• Graph both equations on the same coordinate plane.
	• There are three possibilities:
	  a. If the lines intersect, then the solution is the ordered pair of coordinates for the point of intersection. Check that these coordinates satisfy both equations.
	  b. If the lines are parallel, then there is no solution of the system.
	  c. If the lines coincide, then there are infinitely many solutions, namely all the ordered pairs of coordinates that represent points on the line.
		$x + y = 7$
		$x - y = 3$
		
		$3x - y = 2$
		$3x - y = 4$
		
		$x - y = 3$
		$2x - 2y = 6$

continued

Concept/Skill	Description	Example
[4.2] To solve a system of linear equations by substitution	• In one of the equations, solve for either variable in terms of the other variable. • In the other equation, substitute the expression equal to the variable found in the previous step. Then solve the resulting equation for the remaining variable. • Substitute the value found in the previous step in either of the original equations and solve for the other variable. • Check by substituting the values in both equations of the original system.	**(1)** $2x + 3y = 2$ **(2)** $x + 4y = 6$ **(2)** $x + 4y = 6$ $x = -4y + 6$ **(1)** $2x + 3y = 2$ $2(-4y + 6) + 3y = 2$ $-8y + 12 + 3y = 2$ $-5y + 12 = 2$ $-5y = -10$ $y = 2$ **(2)** $x + 4y = 6$ $x + 4(2) = 6$ $x + 8 = 6$ $x = -2$ **Check** **(1)** $2x + 3y = 2$ $2(-2) + 3(2) \stackrel{?}{=} 2$ $-4 + 6 \stackrel{?}{=} 2$ $2 = 2 \quad$ **True** **(2)** $x + 4y = 6$ $-2 + 4(2) \stackrel{?}{=} 6$ $-2 + 8 \stackrel{?}{=} 6$ $6 = 6 \quad$ **True** The solution is $(-2, 2)$.
[4.3] To solve a system of linear equations by elimination	• Write both equations in the general form $Ax + By = C$. • Choose the variable that you want to eliminate. • If necessary, multiply one or both equations by appropriate numbers so that the coefficients of the variable to be eliminated are opposites. • Add the equations. Then solve the resulting equation for the remaining variable. • Substitute the value found in the previous step in either of the original equations and solve for the other variable. • Check by substituting the values in both equations of the original system.	**(1)** $3x - 2y = -5$ **(2)** $2x - 4y = 2$ Multiply Eq. (1) by 2. $\quad\begin{aligned}6x - 4y &= -10 \\ -2x + 4y &= -2 \\ \hline 4x \phantom{{}- 4y} &= -12 \\ x &= -3\end{aligned}$ Multiply Eq. (2) by −1. **(2)** $2x - 4y = 2$ $2(-3) - 4y = 2$ $-6 - 4y = 2$ $-4y = 8$ $y = -2$ **Check** **(1)** $3x - 2y = -5$ $3(-3) - 2(-2) \stackrel{?}{=} -5$ $-9 + 4 \stackrel{?}{=} -5$ $-5 = -5 \quad$ **True** **(2)** $2x - 4y = 2$ $2(-3) - 4(-2) \stackrel{?}{=} 2$ $-6 + 8 \stackrel{?}{=} 2$ $2 = 2 \quad$ **True** The solution is $(-3, -2)$.

Concept/Skill	Description	Example
[4.4] **Systems of Linear Inequalities**	Two or more linear inequalities considered simultaneously.	$y \geq x + 3$ $y < 3x - 2$
[4.4] **Solution of a System of Linear Inequalities in Two Variables**	An ordered pair of real numbers that makes both inequalities in the system true.	The ordered pair $(-2, 3)$ is a solution of the system of inequalities $$2x + 3y > 1$$ $$x - y < 3$$

Chapter 4   # Review Exercises

To help you review this chapter, solve these problems.

[4.1]

1. Consider the following system:

$$x + 2y = -4$$
$$3x - y = 3$$

Is $(2, -3)$ a solution of the system?

2. For each of the systems graphed, determine the number of solutions.

a.

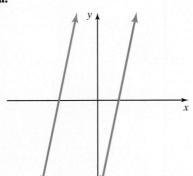

b.

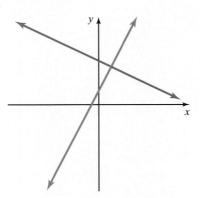

c.

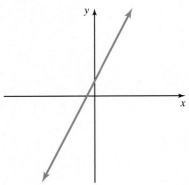

3. Match each system with the appropriate graph.

a. A system with solution $(2, 1)$ _____

b. A system with solution $(1, 2)$ _____

c. A system with infinitely many solutions _____

d. A system with no solution _____

I

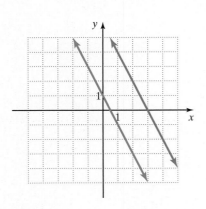

II

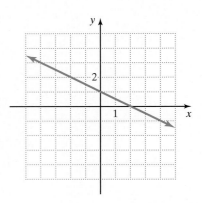

III

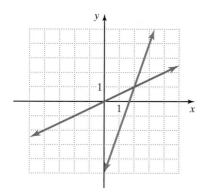

IV

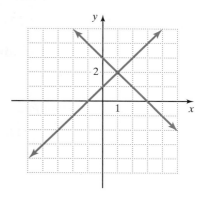

Solve each system by graphing.

4. $x + y = 6$
    $x - y = -4$

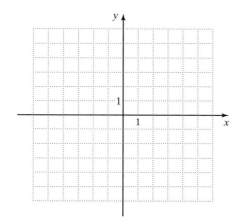

5. $y = 2x$
    $6x - 3y = 3$

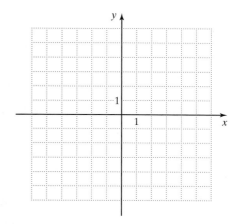

6. $2y = -8x + 2$
    $-4x - y = -1$

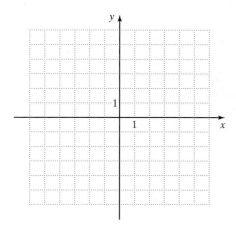

7. $x - 3y = -15$
    $y - x = 5$

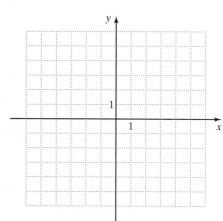

[4.2]   *Solve each system by substitution.*

8. $x + y = 3$
    $y = 2x + 6$

9. $a = 3b - 4$
    $a + 4b = 10$

10. $x - 3y = 1$
    $-2x + 6y = 7$

11. $10x + 2y = 14$
      $-y = 5x - 7$

[4.3]   *Solve each system by elimination.*

12. $x + y = 1$
    $x - y = 7$

13. $2x + 3y = 8$
    $4x + 6y = 16$

14. $4x = 9 - 5y$
    $2x + 3y = 3$

15. $3x + 2y = -4$
    $4x - 3y = 23$

[4.4]   *Graph each system of inequalities.*

16. $6x - 4y \leq -12$
    $4x + 2y > 2$

17. $y \leq -x - 2$
    $y \geq \dfrac{1}{2}x + 1$
    $y \leq 3$

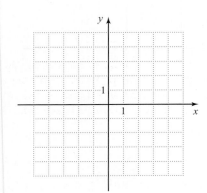

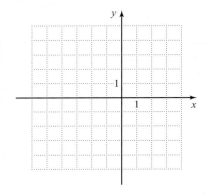

Mixed Applications

Solve.

18. A student starts a typing service. He buys a computer and a printer for $1750, and then charges $5.50 per page for typing. Expenses for ink, paper, and electricity amount to $0.50 per page.

    a. Write the given information as a system of equations.

    b. How many pages must the student type to break even?

19. A job applicant must choose between two sales positions. One position pays $10 per hour, and the other $8 per hour plus a base pay of $50 per week.

    a. Express the weekly salaries of the two positions as a system of equations.

    b. How many hours would the applicant have to work per week in order to earn the same amount of money at each position?

20. The movie screen shown has a perimeter of 332 ft. If the length is 26 ft more than the width, find the area of the screen.

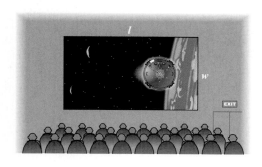

21. A doubles tennis court is 51 ft longer than it is wide. If the court's perimeter is 228 ft, find its dimensions.

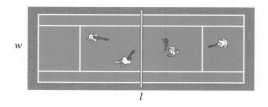

22. The coin box of a vending machine contains only nickels and dimes. There are 350 coins worth $25. How many nickels and how many dimes did the box contain?

23. Two trains start 500 mi apart and speed toward each other. The difference between the average speeds of the trains is 5 mph. If they pass one another after 4 hr, find the rate of each train.

24. During a season, a college basketball team scored 2437 points on a combination of three-point and two-point baskets. If the team made 1085 baskets, how many two-point baskets and how many three-point baskets did the team make?

25. A pharmacist has 10% and 30% alcohol solutions in stock. To prepare 200 mL of a 25% solution, how much of each should the pharmacist mix?

26. A farmer is preparing an insecticide by mixing a 50% solution with water. How much of this solution and how much water are needed to fill a 2000-liter tank with a 35% solution?

27. Last year, a financial adviser recommended that her client invest part of his $50,000 in secure municipal bonds that paid 6% and the rest in corporate stocks that paid 8%. How much money did the client put into each type of investment if the total annual return was $3200?

28. An investor split $10,000 between a high-risk mutual fund and a low-risk mutual fund. Last year, the high-risk fund paid 12% and the low-risk fund paid 2%, for a total of $900. How much money was invested in each fund?

29. Two airplanes leave an airport at the same time, one flying 100 mph faster than the other. The planes travel in opposite directions and after 2 hr they are 1800 mi apart. Determine the speed of the slower plane.

30. Flying with the wind, a bird flew 13 mi in half an hour. On the return trip against the wind, it was able to travel only 8 mi in the same amount of time. Find the speed of the bird in calm air and the speed of the wind.

31. The U.S. Senate has 100 members. After debating the merits of a treaty, all the senators voted, and 14 more voted for the treaty than against. None of the senators abstained.

   a. How many senators voted *for* the treaty?

   b. How many senators voted *against* the treaty?

32. In a chemistry lab, a piece of copper starting at 2°C is heated at the rate of 3° per minute. At the same time, a piece of iron starting at 86°C is being cooled at the rate of 4° per minute.

   a. After how much time will the two metals be at the same temperature?

   b. After how much time will the iron be 14° colder than the copper?

33. A toy manufacturer produces wooden trains and airplanes. Each train, *t*, requires 2 hr to assemble and 1.5 hr to paint. Each airplane, *a*, requires 3 hr to assemble and 2 hr to paint. The manufacturer allots no more than 500 hours per month for assembly and no more than 300 hours per month for painting.

   a. Write a system of inequalities to model this situation.

   b. Solve the system by graphing.

   c. What does the solution region represent?

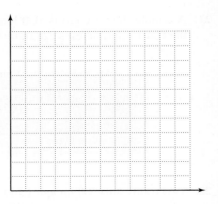

34. A family invested no more than $12,000 in two funds earning 6% and 8% simple interest. The amount of interest earned on the investment at 8% was greater than the amount of interest earned on the investment at 6%. They invested at least $4500 in the account earning 8% simple interest.

   a. Write a system of inequalities to model this problem.

   b. Solve the system by graphing.

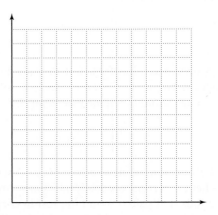

● *Check your answers on page A-26.*

Chapter 4	POSTTEST

To see if you have mastered the topics in this chapter, take this test.

1. Indicate which of the following ordered pairs is a solution of the system

$$x + y = -1$$
$$3x - y = 1$$

   a. $(0, -1)$

   b. $(-1, 0)$

   c. $(2, -3)$

2. How many solutions does the graphed system appear to have?

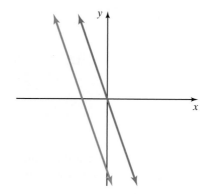

Solve each system by graphing.

3. $x - y = 3$
   $x + y = 3$

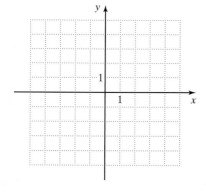

4. $5 = 2x - y$
   $y = 2x$

5. $2x + 3y = 4$
   $3x - y = -5$

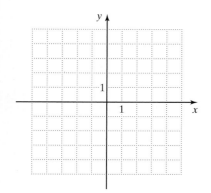

Solve each system by substitution.

6. $3x - 5y = -12$
$x + 2y = 7$

7. $u - 3v = -12$
$5u + v = 8$

Solve each system by elimination.

8. $4x + y = 3$
$7x - y = 19$

9. $x - y = 5$
$2x - 2y = 5$

10. $-5p + 2q = 1$
$4p + 3q = 1.5$

Solve each system.

11. $7x = 2y$
$-4x + y = 1$

12. $4l = -(m + 3)$
$8l + 2m = -6$

13. $5x + 2y = -1$
$x - 1 = y$

Solve.

14. In a local election, the ratio of votes for the winning candidate to the losing candidate was 2 to 1. If 6306 votes were cast in the election, how many votes did the winning candidate get?

15. The following table gives nutritional information for 3-oz servings of turkey and of salmon:

	Turkey, light meat	Salmon
Fat (in grams)	3	3
Calories	135	99

How many servings of turkey and of salmon would it take to get 9 g of fat and 333 cal?

16. A businesswoman has twice as much money invested at 7.5% as she has at 6%. The year-long income from both investments is $840. How much has she invested at each rate?

17. A 20% iodine solution is mixed with a 60% iodine solution to produce 4 gal of a 50% iodine solution. How many gallons of each solution are needed?

18. A small airplane traveled 170 mph with a tailwind and 130 mph with a headwind. Find the speed of the wind and the speed of the airplane in still air.

19. Graph the system of inequalities.

$$6x - 4y \geq -16$$
$$x + 2y > -2$$
$$x \leq 3$$

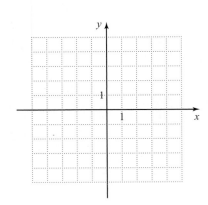

20. Two students agree to split the driving time on a cross-country road trip. One student drives at an average rate of 64 mph and the other drives at an average speed of 60 mph. They would like to drive at least 300 mi per day, and they plan to spend no more than 12 hr driving per day. The student that drives faster agrees to drive more hours.

    a. Write a system of inequalities to model the problem.

    b. Graph the system.

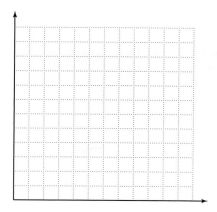

    c. Find two solutions of the system.

● *Check your answers on page A-27.*

Cumulative Review Exercises

To help you review, solve the following.

1. Calculate: $-4 \div 2 + 3\,(-1)(8)$

2. Is 5 a solution of the equation $3p + 1 = 9 - p$?

3. Solve: $5(x + 1) - (x - 2) = x - 2$

4. Solve and graph: $3x - 7 < 4(x - 2)$

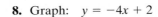

5. The maximum speed of a supersonic airplane S in miles per hour is commonly represented by its Mach number, M, where

$$M = \frac{S}{740}.$$

What is the maximum speed of an airplane flying at Mach 2.1?

6. Find the slope and y-intercept of the graph whose equation is $3x + 6y = 12$.

7. Find the slope of the line shown:

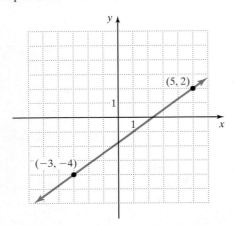

8. Graph: $y = -4x + 2$

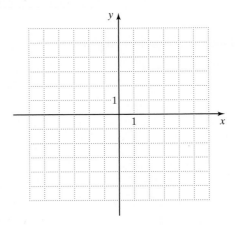

9. Does the relation represent a function?

x	−3	−2	0	1	4
y	−5	4	0	4	6

10. In renting a car for a day, you must choose between two local car agencies. One agency charges $35 per day plus $0.20 per mile and the other charges $50 per day plus $0.15 per mile. Under what circumstances do the two agencies charge the same amount for a one-day rental?

• *Check your answers on page A-27.*

CHAPTER 5

Exponents and Polynomials

Layoffs and Polynomials

Layoffs are common whenever the economy contracts. For instance, in 1933, the worst year of the Great Depression, so many American workers were laid off that one-fourth of the U.S. labor force was unemployed.

Suppose that 100 employees work in a factory. If the fraction f of these employees are laid off, the factory will still employ $100(1 - f)$, or $100 - 100f$, workers. If the factory again lays off the same fraction of employees, the remaining number of workers can be represented by $100(1 - f)(1 - f)$, an expression that can also be written as the *polynomial* $100 - 200f + 100f^2$.

(*Source:* Michael Parkin, *Economics*, Addison-Wesley, 1999)

To see if you have already mastered the topics in this chapter, take this test.

Simplify.

1. $x^5 \cdot x^4$

2. $y^7 \div y^3$

3. $-3a^0$

4. $(4x^4y^3)^2$

5. $\left(\dfrac{a}{b^5}\right)^3$

6. $(5x^{-1}y^4)^{-2}$

7. For the polynomial $6x^4 + 5x^3 + x^2 - 7x + 8$, name:

   a. the terms _____

   b. the coefficients _____

   c. the degree _____

   d. the constant term _____

8. Find the sum: $(2n^2 + 7n - 10) + (n^2 - 6n + 12)$

9. Subtract: $(8x^2 - 9) - (7x^2 - x - 5)$

10. Combine: $(6a^2b + ab - a^2) + (2a^2 + 3a^2b - 5b^2) - (7a^2 - 2ab - b^2)$

Multiply.

11. $3x^2(x^2 - 4x + 9)$

12. $(n + 3)(2n^2 + n - 6)$

13. $(4x + 9)(x - 3)$

14. $(3y - 7)(3y + 7)$

15. $(5 - 2n)^2$

Divide.

16. $\dfrac{9t^4 - 18t^3 - 45t^2}{9t^2}$

17. $(4x^2 - 3x - 10) \div (x - 2)$

Solve.

18. In chemistry, a mole (mol) of any material contains 6×10^{23} molecules. Express in scientific notation the number of molecules that 200 mol of hydrogen will contain. (*Source:* Karen Timberlake, *Chemistry: An Introduction to General Organic and Biological Chemistry*, Addison-Wesley, 1999)

19. The owner of a daycare center plans to have a rectangular sandbox built in the outdoor play area. The width (in feet) of the sandbox is given by the polynomial $2x - 5$, as shown. If the length of the sandbox is designed to be 10 ft longer than the width, write a polynomial that represents the area of the sandbox.

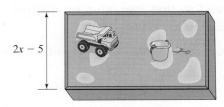

$2x - 5$

20. The average monthly cell telephone bill is approximated by the polynomial

$$-0.535t^2 + 2.64t + 45.3$$

where t represents the number of years since 2000. What was the average monthly cell telephone bill in 2006? (*Source:* Cellular Telecommunications and Internet Association)

● *Check your answers on page A-27.*

5.1 Laws of Exponents

What Exponents Are and Why They Are Important

In Section 1.8, we evaluated expressions involving exponents. In this chapter, we discuss a particular kind of expression called a *polynomial* and operations involving polynomials. Central to our discussion of these operations is a thorough understanding of the laws of exponents.

Exponents play an important role in arithmetic and algebra. Powers of 10 are key to the decimal place-value system that underlies the reading, writing, and computation of real numbers. In scientific notation, exponents are used to represent very large and very small numbers. There are many other applications of exponents in the sciences and in business.

OBJECTIVES

- To simplify expressions by using the product and quotient rules of exponents
- To simplify expressions with exponent zero
- To simplify expressions with negative exponents
- To solve applied problems involving exponents

Exponents

Recall the definition of an exponent from Chapter 1: An exponent (or power) is a number that indicates how many times another number (called the *base*) is used as a factor. For instance,

$$\overset{\text{Exponent}}{\underset{\text{Base}}{5}}{}^{3} = \underbrace{5 \cdot 5 \cdot 5}_{\text{3 factors of 5}}$$

In general, if a is a positive integer, the expression x^a means

$$x^a = \underbrace{x \cdot x \cdot x \cdot \cdots \cdot x}_{a \text{ factors}},$$

where the exponent a indicates that there are a factors of x.

Exponent of 1

For any real number x, $x^1 = x$.

Any number raised to the power 1 is the number itself. For example, $7^1 = 7$, $(-5)^1 = -5$, and $x^1 = x$.

EXAMPLE 1

Multiply.

a. 2^5 **b.** $\left(-\dfrac{1}{3}\right)^4$ **c.** $(-x)^1$ **d.** $(-x)^2$

Solution

a. $2^5 = 2 \cdot 2 \cdot 2 \cdot 2 \cdot 2 = 32$

b. $\left(-\dfrac{1}{3}\right)^4 = \left(-\dfrac{1}{3}\right)\left(-\dfrac{1}{3}\right)\left(-\dfrac{1}{3}\right)\left(-\dfrac{1}{3}\right) = \dfrac{1}{81}$

c. $(-x)^1 = -x$

d. $(-x)^2 = (-x)(-x) = x^2$

PRACTICE 1

Multiply.

a. 10^4

b. $\left(-\dfrac{1}{2}\right)^5$

c. $(-y)^6$

d. $(-y)^1$

The expression 10^0 involves raising a number to the 0 power. To understand the value of this expression, consider the following pattern:

$$10^5 = 10 \cdot 10 \cdot 10 \cdot 10 \cdot 10 = 100{,}000$$
$$10^4 = 10 \cdot 10 \cdot 10 \cdot 10 = 10{,}000$$
$$10^3 = 10 \cdot 10 \cdot 10 = 1000$$
$$10^2 = 10 \cdot 10 = 100$$
$$10^1 = 10$$
$$10^0 = ?$$

Note that to go from 10^5 to 10^4 we can divide 10^5 by 10. The pattern continues, and going from 10^2 to 10^1, we can divide 10^2 by 10. So to go from 10^1 to 10^0, it seems reasonable for us to divide 10^1 by 10, which is 1. Thus, we take 10^0 to be equal to 1. Would we have gotten the same result had we considered powers of 2 instead of powers of 10 in order to determine the value of 2^0?

Exponent of 0

For any nonzero real number x, $x^0 = 1$.

Any nonzero number raised to the power 0 is equal to 1. For example, $8^0 = 1$ and $(-100)^0 = 1$. Note that the expression 0^0 is undefined. Throughout the remainder of this text, we will assume that any variable raised to the 0 power represents a nonzero number.

EXAMPLE 2	PRACTICE 2
Simplify.	Simplify.
a. 25^0 **b.** $(-3.5)^0$ **c.** a^0	**a.** 8^0
d. $-a^0$ **e.** $4x^0$	**b.** $\left(-\dfrac{2}{3}\right)^0$
Solution	**c.** y^0
a. $25^0 = 1$ **b.** $(-3.5)^0 = 1$ **c.** $a^0 = 1$	**d.** $-y^0$
d. $-a^0 = -1 \cdot a^0 = -1 \cdot 1 = -1$	**e.** $(4x)^0$
e. $4x^0 = 4 \cdot 1 = 4$	

Laws of Exponents

The laws of exponents are rules that apply to exponents. These rules are not arbitrary but follow logically from the definition of exponent.

Let's first discuss the *product rule*. This rule applies when we multiply two powers with the same base. Consider the expression $x^4 \cdot x^6$. Using the definition of an exponent we get:

$$x^4 \cdot x^6 = \underbrace{(x \cdot x \cdot x \cdot x)}_{\text{4 factors}}\underbrace{(x \cdot x \cdot x \cdot x \cdot x \cdot x)}_{\text{6 factors}} = x^{10}$$
$$\underbrace{\qquad\qquad\qquad\qquad}_{\text{10 factors}}$$

Since 4 factors of x and 6 additional factors of x make 10 factors of x, it follows that

$$x^4 \cdot x^6 = x^{4+6} = x^{10}.$$

This result can be generalized as follows.

Product Rule of Exponents

For any nonzero real number x and for any positive integers a and b,

$$x^a \cdot x^b = x^{a+b}$$

The product rule states that when we multiply powers of the same base, we add the exponents and leave the base the same.

EXAMPLE 3

Simplify using the product rule, if possible.

a. $2^3 \cdot 2^5$ **b.** $(-5)^2 \cdot (-5)^8$ **c.** $x^8 \cdot x^{10}$

d. $m^4 \cdot m$ **e.** $x^2 \cdot y^3$

Solution When we multiply powers of the same base, the product rule tells us to add the exponents but *not to change the base.*

a. $2^3 \cdot 2^5 = 2^{3+5} = 2^8$

b. $(-5)^2 \cdot (-5)^8 = (-5)^{2+8} = (-5)^{10}$

c. $x^8 \cdot x^{10} = x^{8+10} = x^{18}$

d. $m^4 \cdot m = m^4 \cdot m^1 = m^{4+1} = m^5$

e. In the expression $x^2 \cdot y^3$, we cannot apply the product rule because the bases are not the same.

PRACTICE 3

Simplify using the product rule, if possible.

a. $10^8 \cdot 10^4$

b. $(-4)^3 \cdot (-4)^3$

c. $n^3 \cdot n^7$

d. $y^5 \cdot y^0$

e. $a \cdot b^4$

Now, we discuss another law of exponents—the *quotient rule*. This rule applies when we divide two powers of the same base. Consider the expression $\dfrac{x^6}{x^2}$. Using the definition of exponent, we get:

$$\frac{x^6}{x^2} = \frac{\overbrace{x \cdot x \cdot x \cdot x \cdot \overset{1}{\cancel{x}} \cdot \overset{1}{\cancel{x}}}^{6 \text{ factors}}}{\underset{2 \text{ factors}}{\underbrace{\underset{1}{\cancel{x}} \cdot \underset{1}{\cancel{x}}}}} = \underbrace{x \cdot x \cdot x \cdot x}_{4 \text{ factors}} = x^4$$

So

$$\frac{x^6}{x^2} = x^{6-2} = x^4,$$

which suggests the following rule.

Quotient Rule of Exponents

For any nonzero real number x and for any positive integers a and b,

$$\frac{x^a}{x^b} = x^{a-b}$$

The quotient rule states that when we divide powers of the same base, we subtract the exponent in the denominator from the exponent in the numerator and leave the base the same.

EXAMPLE 4

Simplify using the quotient rule, if possible.

a. $\dfrac{10^5}{10^2}$ **b.** $(-2)^5 \div (-2)^5$ **c.** $\dfrac{p^{12}}{p^7}$

d. $\dfrac{y^{13}}{y}$ **e.** $\dfrac{x^4}{y^2}$

Solution When we divide powers of the same base, the quotient rule tells us to subtract the exponents but *not to change the base.*

a. $\dfrac{10^5}{10^2} = 10^{5-2} = 10^3$

b. $(-2)^5 \div (-2)^5 = \dfrac{(-2)^5}{(-2)^5} = (-2)^{5-5} = (-2)^0 = 1$

c. $\dfrac{p^{12}}{p^7} = p^{12-7} = p^5$

d. $\dfrac{y^{13}}{y} = \dfrac{y^{13}}{y^1} = y^{13-1} = y^{12}$

e. In the expression $\dfrac{x^4}{y^2}$, we cannot apply the quotient rule because the bases are not the same.

PRACTICE 4

Simplify using the quotient rule, if possible.

a. $\dfrac{7^7}{7^2}$

b. $(-9)^6 \div (-9)^5$

c. $\dfrac{s^{10}}{s^{10}}$

d. $\dfrac{r^8}{r}$

e. $\dfrac{a^5}{b^3}$

Note in Example 4(b) that the quotient rule confirms the fact that any nonzero real number raised to the 0 power is 1. That is,

$$\frac{(-2)^5}{(-2)^5} = \frac{-32}{-32} = 1 \quad \text{and} \quad \frac{(-2)^5}{(-2)^5} = (-2)^{5-5} = (-2)^0 = 1$$

EXAMPLE 5

Simplify.

a. $x^3 \cdot x \cdot x^5$ **b.** $(a^2b)(ab^4)$ **c.** $\dfrac{t^3 \cdot t^5}{t^2}$

Solution

a. $x^3 \cdot x \cdot x^5 = x^9$ Use the product rule.

b. $(a^2b)(ab^4) = a^2 \cdot a^1 \cdot b^1 \cdot b^4$ Rearrange the factors.
$= a^3b^5$ Use the product rule.

c. $\dfrac{t^3 \cdot t^5}{t^2} = \dfrac{t^8}{t^2} = t^6$ Use the product rule in the numerator.
Use the quotient rule.

PRACTICE 5

Simplify.

a. $y^2 \cdot y^3 \cdot y^4$

b. $(x^3y^3)(x^2y^3)$

c. $\dfrac{a^7}{a \cdot a^4}$

Negative Exponents

Until now, we have only considered exponents that were either positive integers or 0. What meaning should we give to *negative exponents*?

The quotient rule is the key to answering this question. Consider, for instance, the quotient $\dfrac{6^4}{6^7}$. On the one hand, we can simplify this fraction by using the definition of exponent and canceling the common factors, getting:

$$\frac{6^4}{6^7} = \frac{\overset{1}{\cancel{6}} \cdot \overset{1}{\cancel{6}} \cdot \overset{1}{\cancel{6}} \cdot \overset{1}{\cancel{6}}}{\underset{1}{\cancel{6}} \cdot \underset{1}{\cancel{6}} \cdot \underset{1}{\cancel{6}} \cdot \underset{1}{\cancel{6}} \cdot 6 \cdot 6 \cdot 6} = \frac{1}{6^3}$$

On the other hand, we can simplify by using the quotient rule, getting:

$$\frac{6^4}{6^7} = 6^{4-7} = 6^{-3}$$

Since

$$\frac{6^4}{6^7} = 6^{-3} \quad \text{and} \quad \frac{6^4}{6^7} = \frac{1}{6^3},$$

we conclude that $6^{-3} = \dfrac{1}{6^3}$, which leads us to the following:

Negative Exponent

For any nonzero real number x and for any integer a,

$$x^{-a} = \frac{1}{x^a}$$

In general, an expression with exponents is considered *simplified* when it is written with only positive exponents.

EXAMPLE 6	PRACTICE 6
Simplify.	Simplify.
a. 5^{-2} **b.** p^{-8} **c.** $-(8x)^{-1}$ **d.** $(-4)^{-2}$	**a.** 9^{-2}
Solution	**b.** n^{-5}
a. $5^{-2} = \dfrac{1}{5^2} = \dfrac{1}{25}$ **b.** $p^{-8} = \dfrac{1}{p^8}$	**c.** $-(3y)^{-1}$
c. $-(8x)^{-1} = -1(8x)^{-1} = \dfrac{-1}{(8x)^1} = -\dfrac{1}{8x}$	**d.** $(5)^{-3}$
d. $(-4)^{-2} = \dfrac{1}{(-4)^2} = \dfrac{1}{16}$	

Tip A negative exponent indicates a reciprocal. For example, $5^{-2} = \dfrac{1}{25}$.

The product rule of exponents and the quotient rule of exponents, which were defined for positive-integer exponents, also hold for negative-integer exponents.

EXAMPLE 7

Simplify by writing each expression using only positive exponents.

a. $2^{-1}q$ **b.** $5x^{-2}$ **c.** $\dfrac{y^6}{y^{10}}$

d. $4^{-1} \cdot x^{-5} \cdot x^2$ **e.** $\dfrac{1}{x^{-2}}$

Solution

a. $2^{-1}q = \dfrac{1}{2} \cdot q = \dfrac{q}{2}$

b. $5x^{-2} = 5 \cdot \dfrac{1}{x^2} = \dfrac{5}{x^2}$

c. $\dfrac{y^6}{y^{10}} = y^{6-10} = y^{-4} = \dfrac{1}{y^4}$

d. $4^{-1} \cdot x^{-5} \cdot x^2 = 4^{-1} \cdot x^{-5+2} = \dfrac{1}{4}x^{-3} = \dfrac{1}{4} \cdot \dfrac{1}{x^3} = \dfrac{1}{4x^3}$

e. $\dfrac{1}{x^{-2}} = 1 \div x^{-2} = 1 \div \dfrac{1}{x^2} = 1 \cdot \dfrac{x^2}{1} = x^2$

PRACTICE 7

Write as expressions using only positive exponents.

a. $8^{-1}s$

b. $3x^{-1}$

c. $\dfrac{r^3}{r^9}$

d. $3^2 \cdot g^{-1} \cdot g^{-4}$

e. $\dfrac{1}{x^{-3}}$

The definition of a negative exponent and Example 7(e), in which we saw that $\dfrac{1}{x^{-2}} = x^2$, suggest the following.

> ### Reciprocal of x^{-a}
> For any nonzero real number x and any positive integer a,
> $$\frac{1}{x^{-a}} = x^a$$

EXAMPLE 8

Write as an expression using positive exponents.

a. $\dfrac{5}{x^{-4}}$ **b.** $\dfrac{1}{3y^{-1}}$ **c.** $\dfrac{a^2}{b^{-3}}$

Solution

a. $\dfrac{5}{x^{-4}} = 5 \cdot \dfrac{1}{x^{-4}} = 5 \cdot x^4 = 5x^4$

b. $\dfrac{1}{3y^{-1}} = \dfrac{1}{3} \cdot \dfrac{1}{y^{-1}} = \dfrac{1}{3} \cdot y^1 = \dfrac{y}{3}$

c. $\dfrac{a^2}{b^{-3}} = a^2 \cdot \dfrac{1}{b^{-3}} = a^2 \cdot b^3 = a^2b^3$

PRACTICE 8

Write as an expression using positive exponents.

a. $\dfrac{1}{a^{-3}}$

b. $\dfrac{2}{5x^{-2}}$

c. $\dfrac{r^3}{2s^{-1}}$

EXAMPLE 9

Physicists study different kinds of electromagnetic waves, including radio waves, light waves, X-rays, and gamma rays. The diagram below, called the *electromagnetic spectrum*, shows the relationship among the wavelengths of these waves.

Consider a particular X-ray whose wavelength is 10^{-10} m and a gamma ray whose wavelength is 10^{-15} m. (*Source:* Arthur Beiser, *The Mainstream of Physics*, Addison-Wesley, 1962)

a. Which of these rays has a greater length?

b. What is the ratio of the longer to the shorter wavelength?

Solution

a. The wavelengths are 10^{-10} m for the X-ray and 10^{-15} m for the gamma ray. Converting these expressions with negative exponents to equivalent expressions with positive exponents, the wavelengths can be written as $\frac{1}{10^{10}}$ m and $\frac{1}{10^{15}}$ m, respectively. Since $\frac{1}{10^{10}}$ has the smaller denominator, it is the larger number. Therefore the X-ray has the greater wavelength.

b. We use the quotient rule to compute the ratio of the greater wavelength, 10^{-10} m, to the shorter wavelength, 10^{-15} m:

$$\frac{10^{-10}}{10^{-15}} = 10^{(-10)-(-15)} = 10^{-10+15} = 10^5$$

So the ratio is 10^5.

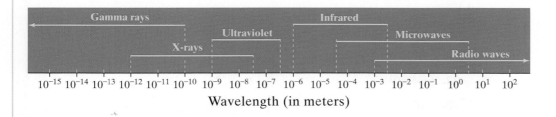

Wavelength (in meters)

PRACTICE 9

A computer's memory is often measured in *bits, bytes,* and *megabytes.* A bit (short for "binary digit") is the smallest unit of data in the memory of the computer. A byte is equal to 2^3 bits, whereas a megabyte (MB) is equal to 2^{20} bytes.

a. How many bits are in a megabyte? Write the answer as a power of 2.

b. In most computers, the hard drive's capacity is expressed in *gigabytes.* A gigabyte (GB) is 2^{10} megabytes. How many bytes are there in a gigabyte? Express the result as a power of 2.

5.1   Exercises

Mathematically Speaking

Fill in each blank with the most appropriate term or phrase from the given list.

the opposite of	added	power of 1
multiplied	0 power	subtracted
1 divided by	divided	

1. The product rule of exponents states that when powers of the same base are multiplied, the exponents are _____ and the base is left the same.

2. The quotient rule of exponents states that when powers of the same base are divided, the exponents are _____ and the base is left the same.

3. Any nonzero real number raised to the _____ is 1.

4. For any nonzero real number x and for any integer a, x^{-a} can be written as _____ x^a.

Multiply.

5. 5^3

6. 1^8

7. $\left(\dfrac{3}{4}\right)^2$

8. $\left(\dfrac{7}{8}\right)^2$

9. $(0.4)^2$

10. $(0.2)^2$

11. $-(0.5)^2$

12. $(-0.3)^2$

13. $(-2)^3$

14. -2^3

15. $(-3)^4$

16. $(-6)^3$

17. $\left(-\dfrac{1}{2}\right)^3$

18. $\left(-\dfrac{4}{5}\right)^2$

19. $(-x)^4$

20. $(-y)^3$

Simplify.

21. $(pq)^1$

22. $(-4xy)^1$

23. $(-3)^0$

24. -3^0

25. $-a^0$

26. b^0

Simplify using the product rule, if possible.

27. $10^9 \cdot 10^2$

28. $5^2 \cdot 5^3$

29. $4^3 \cdot 4^5$

30. $6^2 \cdot 6^8$

31. $a^4 \cdot a^2$

32. $x \cdot x^8$

33. $x^3 \cdot y^5$

34. $a^4 \cdot b^5$

35. $n^6 \cdot n$

36. $y \cdot y^0$

37. x^2y

38. a^4b^3

Simplify using the quotient rule, if possible.

39. $\dfrac{8^5}{8^3}$

40. $\dfrac{3^8}{3^2}$

41. $\dfrac{5^4}{5}$

42. $\dfrac{2^9}{2^3}$

43. $\dfrac{y^6}{y^5}$

44. $\dfrac{x^{12}}{x^{12}}$

45. $\dfrac{a^{10}}{a^4}$

46. $\dfrac{x^7}{x^5}$

47. $\dfrac{y^8}{x^4}$

48. $\dfrac{a^5}{b}$

49. $\dfrac{x^6}{x^6}$

50. $r^4 \div r^0$

Simplify.

51. $y^2 \cdot y^3 \cdot y$

52. $t \cdot t \cdot t^2$

53. $(p^2 q^3)(p^5 q^2)$

54. $(xy^2)(x^2 y^6)$

55. $(yx^2)(xz^2)(yz)$

56. $a(a^4 b^2)(bc^3)$

57. $\dfrac{a^2 \cdot a^3}{a^4}$

58. $\dfrac{t^4}{t^3 \cdot t}$

59. $\dfrac{x^2 \cdot x^4}{x^3 \cdot x}$

60. $\dfrac{y^5 \cdot y^5}{y^2 \cdot y^3}$

Write as an expression using only positive exponents.

61. 5^{-1}

62. 7^{-1}

63. x^{-1}

64. a^{-1}

65. $(-3a)^{-1}$

66. $-(5y)^{-1}$

67. 2^{-4}

68. 7^{-3}

69. -3^{-4}

70. -5^{-2}

71. $8n^{-3}$

72. $-2y^{-3}$

73. $(-x)^{-2}$

74. $(-a)^{-4}$

75. $-3^{-2}x$

76. $-4^{-1}y$

77. $x^{-2} y^3$

78. xy^{-3}

79. qr^{-1}

80. rs^{-1}

81. $4x^{-1}y^2$

82. $-5a^2 b^{-4}$

83. $p^{-2} \cdot p^{-3}$

84. $t^{-3} \cdot t^{-3}$

85. $p^{-1} \cdot p^4$

86. $s^4 \cdot s^{-2}$

87. $\dfrac{a^3}{a^4}$

88. $\dfrac{n}{n^5}$

89. $\dfrac{2}{n^{-4}}$

90. $\dfrac{3}{n^{-1}}$

91. $\dfrac{p^4}{q^{-1}}$

92. $\dfrac{a}{b^{-6}}$

93. $\dfrac{t^{-2}}{t^3}$

94. $\dfrac{x^{-2}}{x^5}$

95. $\dfrac{x^5}{x^{-2}}$

96. $\dfrac{n^7}{n^{-1}}$

97. $\dfrac{a^{-4}}{a^{-5}}$

98. $\dfrac{y^{-1}}{y^{-6}}$

99. $\dfrac{a^{-3}}{b^{-3}}$

100. $\dfrac{x^{-4}}{y^{-2}}$

Mixed Practice

Simplify.

101. $(s^2 t^4)(st^2)$

102. $\dfrac{a^6 \cdot a}{a^2 \cdot a^3}$

Solve.

103. Multiply $\left(-\dfrac{2}{3}\right)^3$.

104. Simplify $\dfrac{a^5}{a^4}$ using the quotient rule, if possible.

105. Simplify $x^2 y^3$ using the product rule, if possible.

106. Simplify -5^0.

Write as an expression using only positive exponents.

107. $(-y)^{-6}$

108. $\dfrac{3}{4k^{-2}}$

109. $q^{-2} \cdot q^{-3}$

110. $a^{-3}b^2$

111. $\dfrac{x^4}{y^{-3}}$

112. $\dfrac{a^{-3}}{b^{-2}}$

Applications

Solve.

113. The first day of an epidemic, 35 people got sick. Each day thereafter, the number of people who got ill doubled.

    a. How many people were ill on the sixth day of the epidemic? On the tenth day?

    b. How many times as great was the number of people ill on the tenth day as compared to the number ill on the sixth day?

114. The value of a new car t years after it is purchased is given by the expression $28{,}000(1.25)^{-t}$.

    a. What is the value of the car one year after it was purchased?

    b. Evaluate the expression for $t = 0$. Explain the significance of this value.

115. The concentration of a pollutant in a pond is 60 parts per million (ppm). The pollution level drops by 5% each month, so the amount of pollutant each month is 95% of the amount in the previous month, as shown in the following table.

Month	Pollution Level (ppm)
1	60
2	60×0.95
3	$60 \times (0.95)^2$
4	$60 \times (0.95)^3$
5	$60 \times (0.95)^4$
6	$60 \times (0.95)^5$
7	$60 \times (0.95)^6$

What will the pollution level be in the twelfth month?

116. The population of the United States in 1820 was approximately 10^7. One hundred years later, it was approximately 10^8. By what factor did the U.S. population grow during this century? (*Source:* U.S. Bureau of the Census)

117. A small cube-shaped box is packed within a larger, cube-shaped box and is surrounded by Styrofoam peanuts. The side length of the larger box is $\frac{5}{2}$ that of the smaller box.

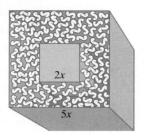

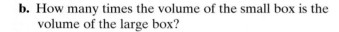

a. Write an expression for the volume of each box.

b. How many times the volume of the small box is the volume of the large box?

118. The cylindrical storage vat pictured has a base with area πr^2 and a height of r. The volume of the vat is the product of its height and the area of the base. Find the volume.

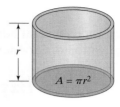

• *Check your answers on page A-28.*

MINDSTRETCHERS

Writing

1. Identify the errors that were made resulting in the following false statements:

a. $4^{-3} = -\dfrac{1}{64}$

b. $x^{-2} \cdot x^{-3} = x^6$

c. $4^3 \cdot 4^{-5} = 16^{-2}$

d. $5^4 \div 5 = 1^4$

Groupwork

2. Which is larger: x^2 or x^{-2}? Explain your answer, and give some examples.

Mathematical Reasoning

3. The expression 0^0 is undefined. Can you explain why?

5.2   More Laws of Exponents and Scientific Notation

In this section, we consider several additional laws of exponents, as well as an important application of exponents known as scientific notation.

■ To simplify expressions by using the power rule

■ To simplify expressions by raising a product to a power

■ To simplify expressions by raising a quotient to a power

■ To write a number in scientific notation

■ To solve applied problems involving laws of exponents and scientific notation

Additional Laws of Exponents

In the previous section, we discussed the product rule for the product of powers and the quotient rule for the quotient of powers. We now consider a third rule known as the *power rule*. The power rule deals with expressions in which a power is raised to a power.

Let's consider, for instance, the expression $(x^2)^3$. Using the definition of an exponent gives us:

$$(x^2)^3 = \underbrace{x^2 \cdot x^2 \cdot x^2}_{\text{3 factors of } x^2} = x^{2+2+2} = x^6$$

So $(x^2)^3 = x^6$. We can generalize this result as follows.

Power Rule of Exponents

For any nonzero real number x and any integers a and b,

$$(x^a)^b = x^{ab}$$

In words, the power rule says that when raising a power to a power, we *multiply* the exponents and *leave the base the same*.

EXAMPLE 1	PRACTICE 1
Simplify using the power rule of exponents.	Simplify using the power rule of exponents.
a. $(5^2)^2$ **b.** $(2^{-3})^2$ **c.** $-(p^4)^5$ **d.** $(q^2)^{-1}$	**a.** $(2^3)^2$
Solution We apply the power rule and then simplify.	**b.** $(7^3)^{-1}$
a. $(5^2)^2 = 5^{2 \cdot 2} = 5^4 = 625$	**c.** $(q^2)^4$
b. $(2^{-3})^2 = 2^{-3 \cdot 2}$	**d.** $-(p^3)^{-5}$

b. $(2^{-3})^2 = 2^{-3 \cdot 2}$

$= 2^{-6}$

$= \dfrac{1}{2^6}$ Use the definition of a negative exponent.

$= \dfrac{1}{64}$

c. $-(p^4)^5 = -(p^{4 \cdot 5}) = -p^{20}$

d. $(q^2)^{-1} = q^{2 \cdot (-1)} = q^{-2} = \dfrac{1}{q^2}$

Tip   Be sure to distinguish between the *product* rule and the *power* rule.

$$\textbf{Product rule: } x^a \cdot x^b = x^{a+b} \qquad \textbf{Power rule: } (x^a)^b = x^{ab}$$

Add the exponents.   Multiply the exponents.

Another law of exponents has to do with *raising a product to a power*.

For instance, consider the expression $(5x)^3$.

Rearrange the factors.

$$(5x)^3 = (5x)(5x)(5x) = (5 \cdot 5 \cdot 5)(x \cdot x \cdot x) = 5^3 \cdot x^3 = 125x^3$$

So we see that $(5x)^3$ is the same as 5^3 times x^3. We can generalize this result as follows.

Raising a Product to a Power

For any nonzero real numbers x and y and any integer a,

$$(xy)^a = x^a \cdot y^a$$

This rule states that to raise a product to a power we raise each factor to that power.

EXAMPLE 2	PRACTICE 2
Simplify using the rule for raising a product to a power.	Simplify.
a. $(2y)^4$ **b.** $(-3a)^2$ **c.** $-(3a)^2$	**a.** $(7a)^2$
Solution We apply the rule for raising a product to a power and then simplify.	**b.** $(-4x)^3$
a. $(2y)^4 = 2^4 \cdot y^4 = 16y^4$	**c.** $-(4x)^3$
b. $(-3a)^2 = (-3)^2(a)^2 = 9a^2$	
c. $-(3a)^2 = -3^2a^2 = -9a^2$	

EXAMPLE 3	PRACTICE 3
Simplify.	Simplify.
a. $(2x^4)^5$	**a.** $(-6a^9)^2$
b. $(p^3q^5)^3$	**b.** $(q^8r^{10})^2$
c. $-7(m^5n^{10})^2$	**c.** $-2(ab^7)^3$
d. $(5a^{-2}c^4)^{-2}$	**d.** $(7a^{-1}c^{-5})^2$

Solution

a. $(2x^4)^5 = 2^5(x^4)^5$ Use the rule for raising a product to a power.

$\qquad\quad = 32x^{20}$ Use the power rule.

b. $(p^3q^5)^3 = (p^3)^3(q^5)^3 = p^9q^{15}$

c. $-7(m^5n^{10})^2 = -7(m^5)^2(n^{10})^2 = -7m^{10}n^{20}$

d. $(5a^{-2}c^4)^{-2} = 5^{-2}(a^{-2})^{-2}(c^4)^{-2}$ Use the rule for raising a product to a power.

$\qquad\qquad\quad = 5^{-2}(a^4)(c^{-8})$ Use the power rule.

$\qquad\qquad\quad = \dfrac{1}{25} \cdot a^4 \cdot \dfrac{1}{c^8}$ Use the definition of a negative exponent.

$\qquad\qquad\quad = \dfrac{a^4}{25c^8}$

The final law of exponents that we discuss is *raising a quotient to a power*. For instance, consider the expression $\left(\dfrac{a}{b}\right)^4$, where a divided by b is raised to the fourth power. By the definition of exponent, we get:

$$\left(\frac{a}{b}\right)^4 = \frac{a}{b} \cdot \frac{a}{b} \cdot \frac{a}{b} \cdot \frac{a}{b} = \frac{a \cdot a \cdot a \cdot a}{b \cdot b \cdot b \cdot b} = \frac{a^4}{b^4}$$

So $\left(\dfrac{a}{b}\right)^4 = \dfrac{a^4}{b^4}$. We generalize this result as follows.

Raising a Quotient to a Power

For any nonzero real numbers x and y and any integer a,

$$\left(\frac{x}{y}\right)^a = \frac{x^a}{y^a}$$

This rule states that to raise a quotient to a power, we raise both the numerator and the denominator to that power.

EXAMPLE 4	PRACTICE 4

Simplify by using the rule for raising a quotient to a power.

a. $\left(\dfrac{x}{5}\right)^3$ **b.** $\left(\dfrac{-a}{b}\right)^4$

c. $\left(\dfrac{5}{x}\right)^{-3}$ **d.** $\left(\dfrac{-3r^2}{st^4}\right)^3$

e. $\left(\dfrac{9u}{v^{-1}}\right)^2$

Simplify.

a. $\left(\dfrac{y}{3}\right)^2$ **b.** $\left(\dfrac{-u}{v}\right)^{10}$

c. $\left(\dfrac{3}{y}\right)^{-2}$ **d.** $\left(\dfrac{-10a^5}{3b^2c}\right)^2$

e. $\left(\dfrac{5x}{y^{-2}}\right)^3$

Solution Here we use the rule for raising a quotient to a power and then simplify.

a. $\left(\dfrac{x}{5}\right)^3 = \dfrac{x^3}{5^3} = \dfrac{x^3}{125}$

b. $\left(\dfrac{-a}{b}\right)^4 = \dfrac{(-a)^4}{b^4} = \dfrac{a^4}{b^4}$

c. $\left(\dfrac{5}{x}\right)^{-3} = \dfrac{5^{-3}}{x^{-3}}$

$\qquad = 5^{-3} \cdot \dfrac{1}{x^{-3}}$

$\qquad = \dfrac{1}{5^3} \cdot x^3$

$\qquad = \dfrac{x^3}{125}$

d. $\left(\dfrac{-3r^2}{st^4}\right)^3 = \dfrac{(-3r^2)^3}{(st^4)^3} = \dfrac{(-3)^3(r^2)^3}{s^3(t^4)^3} = \dfrac{-27r^6}{s^3t^{12}}$

e. $\left(\dfrac{9u}{v^{-1}}\right)^2 = \dfrac{(9u)^2}{(v^{-1})^2}$

$\qquad = \dfrac{9^2 u^2}{v^{-2}}$

$\qquad = 81u^2 \cdot \dfrac{1}{v^{-2}}$

$\qquad = 81u^2 v^2$

Note that the simplified form of the expression in Example 4(a) is the same as the simplified form of the expression in Example 4(c). Since $\left(\dfrac{5}{x}\right)^{-3} = \dfrac{x^3}{125}$ and $\left(\dfrac{x}{5}\right)^3 = \dfrac{x^3}{125}$, we conclude that $\left(\dfrac{5}{x}\right)^{-3} = \left(\dfrac{x}{5}\right)^3$. This conclusion leads us to another law of exponents—*raising a quotient to a negative power*.

Raising a Quotient to a Negative Power

For any nonzero real numbers x and y and any integer a,

$$\left(\dfrac{x}{y}\right)^{-a} = \left(\dfrac{y}{x}\right)^a$$

EXAMPLE 5	PRACTICE 5
Simplify.	Simplify.
a. $\left(\dfrac{2}{x}\right)^{-3}$ **b.** $\left(\dfrac{3r}{10s}\right)^{-1}$	**a.** $\left(\dfrac{5}{a}\right)^{-2}$ **b.** $\left(\dfrac{4u}{v}\right)^{-1}$
c. $\left(\dfrac{x^4}{y^2}\right)^{-2}$	**c.** $\left(\dfrac{a^5}{b^3}\right)^{-2}$

Solution Use the rule for raising a quotient to a negative power.

a. $\left(\dfrac{2}{x}\right)^{-3} = \left(\dfrac{x}{2}\right)^{3} = \dfrac{x^3}{2^3} = \dfrac{x^3}{8}$

b. $\left(\dfrac{3r}{10s}\right)^{-1} = \left(\dfrac{10s}{3r}\right)^{1} = \dfrac{10s}{3r}$

c. $\left(\dfrac{x^4}{y^2}\right)^{-2} = \left(\dfrac{y^2}{x^4}\right)^{2} = \dfrac{(y^2)^2}{(x^4)^2} = \dfrac{y^4}{x^8}$

Scientific Notation

Scientific notation is an important application of exponents—whether they are positive, negative, or zero. Scientists use this notation to abbreviate very large or very small numbers. Note that scientific notation is based on powers of 10.

Example	Standard Notation	Scientific Notation
The speed of light	983,000,000 ft/sec	9.83×10^8 ft/sec
The length of a virus	0.000000000001 m	1×10^{-12} m

This way of writing numbers has several advantages over standard notation. For example, when a number contains a long string of 0's, writing it in scientific notation can take fewer digits. Also, numbers written in scientific notation can be relatively easy to multiply or divide.

Definition

A number is in **scientific notation** if it is written in the form

$$a \times 10^n$$

where n is an integer and a is greater than or equal to 1 but less than 10 ($1 \le a < 10$).

Note that any value of a that satisfies the inequality $1 \le a < 10$ must have *one nonzero digit* to the left of the decimal point. For instance, 7.3×10^5 is written in scientific notation. Do you see why the numbers 0.83×10^2, 5×3^7, and 13.8×10^{-4} are *not* written in scientific notation?

Tip When written in scientific notation, large numbers have positive powers of 10, whereas small numbers have negative powers of 10. For instance, 3×10^{23} is large, whereas 3×10^{-23} is small.

Let's now consider how to change a number from scientific notation to standard notation.

EXAMPLE 6	PRACTICE 6

Change the number 2.41×10^5 from scientific notation to standard notation.

Express 2.539×10^2 in standard notation.

Solution To express this number in standard notation, we need to multiply 2.41 by 10^5. Since $10^5 = 100,000$, multiplying 2.41 by 100,000 gives:

$$2.41 \times 10^5 = 2.41 \times 100,000 = 241,000.00 = 241,000$$

The number 241,000 is written in standard notation.

Note that the power of 10 here is *positive* and that the decimal point is moved five places *to the right*. So a shortcut for expressing 2.41×10^5 in standard notation is to move the decimal point in 2.41 five places to the right.

$$2.41 \times 10^5 = 2.41000 = 241,000$$

EXAMPLE 7	PRACTICE 7

Convert 3×10^{-5} to standard notation.

Change 4.3×10^{-9} to standard notation.

Solution Using the definition of a negative exponent, we get:

$$3 \times 10^{-5} = 3 \times \frac{1}{10^5}, \text{ or } \frac{3}{10^5}$$

Since $10^5 = 100,000$, dividing 3 by 100,000 gives us

$$\frac{3}{10^5} = \frac{3}{100,000} = 0.00003$$

Here we note that the power of 10 is *negative* and that the decimal point, which is understood to be at the right end of a whole number, is moved five places *to the left*. So a shortcut for expressing 3×10^{-5} in standard notation is to move the decimal point in 3. five places to the left.

$$3 \times 10^{-5} = 3. \times 10^{-5} = 00003. = .00003, \text{ or } 0.00003$$

Tip When converting a number from scientific notation to standard notation, move the decimal point to the *right* if the power of 10 is *positive* and to the *left* if the power of 10 is *negative*.

Now, let's consider the reverse situation, namely, changing a number in standard notation to scientific notation.

EXAMPLE 8

Express 37,000,000,000 in scientific notation.

Solution For a number to be written in scientific notation, it must be of the form

$$a \times 10^n$$

where n is an integer and $1 \le a < 10$. We know that 37,000,000,000 and 37,000,000,000. are the same. We move the decimal point *to the left* so that there is one nonzero digit to the left of the decimal point. The power of 10 we need to multiply by is the same as the number of places moved.

$$37{,}000{,}000{,}000 = 3.7\,0\,0\,0\,0\,0\,0\,0\,0\,0 \times 10^{10}$$

Move 10 places to the *left*.

$$= 3.7 \times 10^{10}$$

Since 3.7 and 3.7000000000 are equivalent, we can drop the trailing zeros. So 37,000,000,000 expressed in scientific notation is 3.7×10^{10}.

PRACTICE 8

Write 8,000,000,000,000 in scientific notation.

EXAMPLE 9

Convert 0.00000000000000002 to scientific notation.

Solution We must write the number 0.00000000000000002 in the form

$$a \times 10^n$$

where n is an integer and $1 \le a < 10$. We move the decimal point *to the right* so that there is one nonzero digit to the left of the decimal point. The number of places moved, preceded by a *negative* sign, is the power of 10 that we need.

$$0.00000000000000002 = 0\,0\,0\,0\,0\,0\,0\,0\,0\,0\,0\,0\,0\,0\,0\,0\,2. \times 10^{-17}$$

Move 17 places to the *right*.

$$= 2. \times 10^{-17} = 2 \times 10^{-17}$$

PRACTICE 9

Express 0.000000000071 in scientific notation.

Next, let's consider calculations involving numbers written in scientific notation. We focus on the operations of multiplication and division.

EXAMPLE 10

Calculate, writing the result in scientific notation.

a. $(4 \times 10^{-1})(2.1 \times 10^{6})$

b. $(1.2 \times 10^{5}) \div (2 \times 10^{-4})$

PRACTICE 10

Calculate, writing the result in scientific notation.

a. $(7 \times 10^{-2})(3.52 \times 10^{3})$

b. $(2.4 \times 10^{3}) \div (6 \times 10^{-9})$

Solution

a. $(4 \times 10^{-1})(2.1 \times 10^6)$

$= (4 \times 2.1)(10^{-1} \times 10^6)$ Regroup the factors.

$= 8.4 \times 10^{-1+6}$ Use the product rule.

$= 8.4 \times 10^5$

b. $(1.2 \times 10^5) \div (2 \times 10^{-4})$

$= \dfrac{1.2 \times 10^5}{2 \times 10^{-4}}$

$= \dfrac{1.2}{2} \times \dfrac{10^5}{10^{-4}}$ Rewrite the quotient as a product of quotients.

$= 0.6 \times 10^{5-(-4)}$

$= 0.6 \times 10^9$ Use the quotient rule.

Note that 0.6×10^9 is not in scientific notation because 0.6 is not between 1 and 10, that is, it does not have one nonzero digit to the left of the decimal point. To write 0.6×10^9 in scientific notation, we convert 0.6 to scientific notation and then simplify the product.

$0.6 \times 10^9 = (6 \times 10^{-1}) \times 10^9$ Convert 0.6 to scientific notation.

$= 6 \times (10^{-1} \times 10^9)$

$= 6 \times 10^8$ Use the product rule.

The answer is 6×10^8.

EXAMPLE 11	PRACTICE 11

EXAMPLE 11

There are about 5×10^6 red blood cells per cubic millimeter of blood. Each of these red blood cells contains about 2×10^8 hemoglobin molecules. Calculate the approximate number of hemoglobin molecules per cubic millimeter of blood, writing the result in scientific notation. (*Source:* S. Mader, *Inquiry into Life*, William C. Brown, 1991)

Solution We need to find the product of the number of red blood cells per cubic millimeter of blood and the number of hemoglobin molecules per red blood cell:

$$(5 \times 10^6)(2 \times 10^8) = \underbrace{(5 \times 2)}\underbrace{(10^6 \times 10^8)}$$

$$= \quad 10 \quad \times \quad 10^{6+8}$$

$$= \quad 10 \quad \times \quad 10^{14}$$

To write this number in scientific notation, we convert 10 to scientific notation and then simplify the product.

$$10 \times 10^{14} = (1.0 \times 10^1) \times 10^{14}$$

$$= 1.0 \times (10^1 \times 10^{14})$$

$$= 1.0 \times 10^{15}$$

So there are approximately 1×10^{15} hemoglobin molecules per cubic millimeter of blood.

PRACTICE 11

The number of hairs on the average human head is estimated to be about 1.5×10^5. If there are approximately 6×10^9 people in the world, estimate the number of human hairs in the world. (*Source: Time Almanac 2000*)

EXAMPLE 12

A certain DVD holds 9.4×10^9 bytes of information. How many files, each containing 9.4×10^6 bytes, can the DVD hold?

Solution To determine the number of files that will fit on the DVD, we divide.

$$(9.4 \times 10^9) \div (9.4 \times 10^6) = \frac{9.4 \times 10^9}{9.4 \times 10^6}$$
$$= \frac{9.4}{9.4} \times \frac{10^9}{10^6}$$
$$= 1 \times 10^{9-6}$$
$$= 1 \times 10^3$$

So the DVD can hold 1×10^3, or 1000 files.

PRACTICE 12

At the very best, a light microscope can distinguish points 2×10^{-7} m apart, whereas an electronic microscope can distinguish points that are 2×10^{-10} m apart. The second number is how many times as great as the first number? (*Source:* S. Mader, *Inquiry into Life*, William C. Brown, 1991)

 ## Calculators and Scientific Notation

Calculators vary as to how numbers are displayed or entered in scientific notation.

Display

In order to avoid an overflow error, many calculator models change to scientific notation an answer that is either too small or too large to fit into the calculator's display. Calculators generally use the base 10 without displaying it. Some calculators display scientific notation with either an E or e, and others show a space. For example, 3.1E–4 or 3.1 ▮ –4 can represent 3.1×10^{-4}. What other differences do you see between written scientific notation and displayed scientific notation?

EXAMPLE 13

Multiply 1,000,000,000 by 2,000,000,000.

Solution

Press	Display
1000000000 ⨯ 2000000000 ENTER	*1000000000 * 2000000000*
	2 ε 18

Does your calculator display the product in scientific notation, that is, as 2E18, 2e18, or 2.▮18?

PRACTICE 13

Square 0.000000005. How is the answer displayed?

Enter

Some calculators give the wrong answer to a computation if very large or very small numbers are *entered* in standard form rather than in scientific notation. To enter a number in scientific notation, many calculators have a key labeled [EE], [EXP], or [EEX]. For a negative exponent, a key labeled [+/−] or [(−)] must be pressed either before or after the exponent key, depending on the calculator.

EXAMPLE 14	PRACTICE 14
Enter the number 5,000,000,000,000 in scientific notation.	In your calculator, enter in scientific notation the number 0.00000000073.

Solution

Press | Display

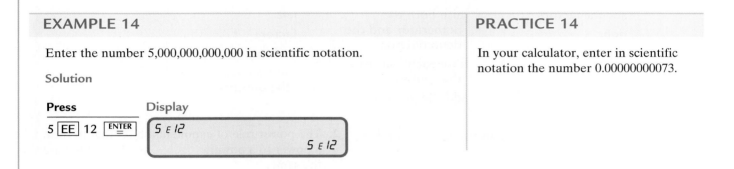

EXAMPLE 15	PRACTICE 15
Multiply 3.5×10^4 by 2.1×10^7 on a calculator.	Use a calculator to divide 9.2×10^{12} by 2×10^4.

Solution

Press | Display

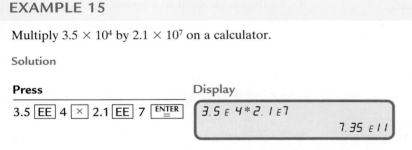

So the answer is 7.35×10^{11}. If your calculator has enough places in the display, it may give the answer to this problem in standard form: 735,000,000,000.

5.2   Exebcises

FOR EXTRA HELP

Mathematically Speaking

Fill in each blank with the most appropriate term or phrase from the given list.

add the factors	raise both the numerator and the denominator	left
right		factors
terms	raise each factor to that power	multiply the exponents
power form		raise the reciprocal of the quotient
scientific notation	add the powers	

1. The expression $(x^3)^2$ contains two _____ of x^3.

2. The power rule of exponents states that to raise a power to a power, _____ and leave the base the same.

3. To raise a product to a power, _____.

4. To raise a quotient to a power, _____ to that power.

5. To raise a quotient to a negative power, _____ to the opposite of the given power.

6. A number is written in _____ if it is in the form $a \times 10^n$, where n is an integer and a is greater than or equal to 1 but less than 10.

7. To convert a number from scientific to standard notation, move the decimal point to the _____ if the power of 10 is negative.

8. To convert a number from standard to scientific notation, move the decimal point to the _____ if the number is less than 1.

Simplify.

9. $(2^2)^4$

10. $(3^3)^2$

11. $(5^2)^2$

12. $(2^3)^3$

13. $(10^5)^2$

14. $(0^5)^3$

15. $(4^{-2})^2$

16. $(2^3)^{-4}$

17. $(x^4)^6$

18. $(p^2)^{10}$

19. $(y^4)^2$

20. $(n^3)^3$

21. $(x^{-2})^3$

22. $(y^5)^{-6}$

23. $(n^{-2})^{-2}$

24. $(a^{-5})^{-4}$

25. $(4x)^3$

26. $(2y)^5$

27. $(-8y)^2$

28. $(-7a)^2$

29. $-(4n^5)^3$

30. $-(5x^3)^3$

31. $4(-2y^2)^4$

32. $2(-3t)^3$

33. $(3a)^{-2}$

34. $-(5t)^{-3}$

35. $(pq)^{-7}$

36. $(mn)^{-6}$

37. $(r^2t)^6$

38. $(a^3b^5)^4$

39. $(-2p^5q)^2$

40. $(-3a^2b^3)^4$

41. $-2(m^4n^8)^3$

42. $4(x^2y^3)^2$

43. $(-4m^5n^{-10})^3$

44. $(3a^{-3}c^8)^2$

45. $(a^3b^2)^{-4}$ **46.** $(p^{-3}q^{-4})^2$ **47.** $(4x^{-2}y^3)^2$ **48.** $(-2x^{-2}y^3)^{-3}$

49. $\left(\dfrac{5}{b}\right)^3$ **50.** $\left(\dfrac{x}{4}\right)^2$ **51.** $\left(\dfrac{c}{b}\right)^2$ **52.** $\left(\dfrac{t}{s}\right)^5$

53. $-\left(\dfrac{a}{b}\right)^7$ **54.** $-\left(\dfrac{x}{y}\right)^3$ **55.** $\left(\dfrac{a^2}{3}\right)^3$ **56.** $\left(\dfrac{4}{y^6}\right)^3$

57. $\left(-\dfrac{p^3}{q^2}\right)^5$ **58.** $\left(\dfrac{x^2}{y^3}\right)^4$ **59.** $\left(\dfrac{a}{4}\right)^{-1}$ **60.** $\left(-\dfrac{b}{3}\right)^{-1}$

61. $\left(\dfrac{2x^5}{y^2}\right)^3$ **62.** $\left(\dfrac{n^2}{3w^5}\right)^2$ **63.** $\left(\dfrac{pq}{p^2q^2}\right)^5$ **64.** $\left(\dfrac{-s^2t^3}{st^2}\right)^2$

65. $\left(\dfrac{3x}{y^{-3}}\right)^4$ **66.** $\left(\dfrac{p^{-1}}{5q^5}\right)^2$ **67.** $\left(\dfrac{-u^2v^3}{4vu^4}\right)^2$ **68.** $\left(\dfrac{2xy^3}{xy^2}\right)^4$

69. $\left(-\dfrac{x^{-2}y}{2z^{-4}}\right)^4$ **70.** $-\left(\dfrac{4a^{-4}}{bc^{-2}}\right)^2$ **71.** $\left(\dfrac{r^5}{t^6}\right)^{-2}$ **72.** $-\left(\dfrac{y^3}{x^3}\right)^{-2}$

73. $\left(\dfrac{-2a^4}{b^2}\right)^{-3}$ **74.** $\left(\dfrac{q^5}{5p^4}\right)^{-2}$

Express in standard notation.

75. 3.17×10^8 **76.** 9.1×10^5 **77.** 1×10^{-6} **78.** 8.33×10^{-4}

79. 6.2×10^6 **80.** 7.55×10^{10} **81.** 4.025×10^{-5} **82.** 2.1×10^{-3}

Express in scientific notation.

83. 420,000,000   **84.** 100,000,000   **85.** 0.0000035   **86.** 0.00017

87. 217,000,000,000   **88.** 154,800,000,000   **89.** 0.00000000731   **90.** 0.00000005672

Complete the following tables.

91.

Standard Notation	Scientific Notation (written)	Scientific Notation (displayed on a calculator)
975,000,000		
	4.87×10^8	
		1.652E−10
0.000000067		
	1×10^{-13}	
		3.281E9

🖩 92.

Standard Notation	Scientific Notation (written)	Scientific Notation (displayed on a calculator)
975,000,000,000		
	5×10^8	
		4.988E−7
0.0000048		
	9.34×10^{-9}	
		9.772E6

🖩 *Calculate, writing the result in scientific notation.*

93. $(3 \times 10^2)(3 \times 10^5)$

94. $(5 \times 10^4)(7.1 \times 10^3)$

95. $(2.5 \times 10^{-2})(8.3 \times 10^{-3})$

96. $(2.1 \times 10^4)(8 \times 10^{-4})$

97. $(8.6 \times 10^9)(4.4 \times 10^{-12})$

98. $(9.1 \times 10^{-13})(6.3 \times 10^{-10})$

99. $(2.5 \times 10^8) \div (2 \times 10^{-2})$

100. $(3.0 \times 10^4) \div (1 \times 10^3)$

101. $(6 \times 10^5) \div (2 \times 10^3)$

102. $(4.8 \times 10^{-3}) \div (8 \times 10^2)$

103. $(9.6 \times 10^{20}) \div (3.2 \times 10^{12})$

104. $(8.4 \times 10^6) \div (4.2 \times 10^7)$

Mixed Practice

Simplify.

105. Express 3.067×10^{-4} in standard notation.

106. Express 895,600,000 in scientific notation.

107. Complete the table.

Standard Notation	Scientific Notation (written)	Scientific Notation (displayed on a calculator)
428,000,000,000		
	3.24×10^6	
		5.224E−6
0.000000057		
	6.82×10^{-7}	
		4.836E7

Simplify, using only positive exponents.

108. $(a^{-3})^5$

109. $\left(\dfrac{a^2}{3b^5}\right)^{-2}$

110. $(2r^7s^{-3})^3$

111. $-(4y)^{-3}$

112. $\left(-\dfrac{m^2}{n^3}\right)^5$

113. $-\left(\dfrac{2x^{-2}}{y^{-3}z}\right)^4$

114. $2(-3x^4)^3$

🖩 *Calculate, writing the result in scientific notation.*

115. $(6.3 \times 10^{-4}) \div (9 \times 10^3)$

116. $(4.1 \times 10^{-3})(2.7 \times 10^{-2})$

Applications

Solve.

117. Consider the two boxes shown. How many times the volume of the smaller box is the volume of the larger box?

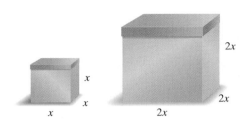

118. After a flood, the radius of a circular pond doubles. How does the area of the pond change?

119. A DVD holds between 4×10^9 and 1.7×10^{10} bytes of data. Express these quantities in standard notation.

120. The infectious part of a virus is typically between 2.5×10^{-8} m and 2×10^{-7} m in size. Express these quantities in standard notation. (*Source:* S. Mader, *Inquiry into Life*, William C. Brown, 1991)

121. The wavelength of red light is 0.0000007 m. Write this length in scientific notation.

122. In a recent year, retail sales in U.S. shopping centers totaled $1,030,000,000,000. Express this amount in scientific notation. (*Source: The 1999 Statistical Abstract of the United States*)

123. The cell is considered the basic unit of life. Each day, the body destroys and replaces more than 200 billion cells. Write this quantity in scientific notation.

124. The area of the United States is approximately 3.7×10^6 sq mi. Express this area in standard notation. (*Source: The New York Times Almanac, 2000*)

125. The mass of a proton is about 1.7×10^{-24} g. Rewrite this quantity in standard notation. (*Source:* Karen Timberlake, *Chemistry*, Addison-Wesley, 1999)

126. To measure vast distances, astronomers use a unit called a *parsec*, which is equal to about 3.086×10^{18} cm. Express this quantity in standard form. (*Source:* D. McNally, *Positional Astronomy*, 1974)

127. The population of Africa in a recent year was about 7.8×10^8. What was this population expressed in standard notation? (*Source: New York Times Almanac 2000*)

128. A supercomputer carries out 3,600,000,000 operations per second. Rewrite this capability in scientific notation. (*Source:* Apple Computer)

129. For each pound of body weight, a human body contains about 3.2×10^4 microliters (μL) of blood. In turn, a microliter of blood contains about 5×10^6 red blood cells. A person weighing 100 lb has approximately how many red blood cells?

130. On the television series *Star Trek: The Next Generation*, the android Data could carry out 60 trillion operations per second. Express this rate in scientific notation.

131. Light travels through a vacuum at a speed of 186,000 mi per sec.

a. Express this speed in scientific notation.

b. Estimate mentally how long it will take for light to travel to Earth from the star Vega, which is 1.58×10^{14} mi from Earth. (*Source: The New York Times Almanac 2000*)

132. There are 26,890,000,000,000,000,000 molecules of a gas in a cubic meter.

a. Rewrite this quantity in scientific notation.

b. What volume is required for 3.4×10^{20} molecules of the gas?

MINDSTRETCHERS

Investigation

1. On a scientific calculator, enter the number 2. Double that number. Then keep doubling the result. After how many doublings does your calculator display the number in scientific notation? Explain how you could have predicted that result.

Critical Thinking

2. What is the mathematical relationship between $(a^m)^n$ and $(a^m)^{-n}$? Justify your answer.

Research

3. In your college library or on the Web, determine the annual national debt for the United States for 5 consecutive years. Would you use scientific notation or standard notation to express these amounts? Explain why.

5.3　Basic Concepts of Polynomials

What Polynomials Are and Why They Are Important

OBJECTIVES

- To classify polynomials
- To simplify polynomials
- To evaluate polynomials
- To solve applied problems involving polynomials

In Chapter 1, we discussed algebraic expressions in general. We now consider a particular kind of algebraic expression called a *polynomial*.

Just as whole numbers are fundamental to arithmetic, polynomials play a similarly key role in algebra.

$$\text{Whole number:} \quad 3 \cdot 10^2 + 7 \cdot 10^1 + 8 \cdot 10^0 = 378$$
$$\text{Polynomial:} \quad 3 \cdot x^2 + 7 \cdot x^1 + 8 \cdot x^0 = 3x^2 + 7x + 8$$

A good deal of algebra is devoted to studying properties of polynomials and operations on polynomials. Furthermore, many phenomena in the sciences and business can be described or approximated by polynomial expressions.

Monomials

We begin by considering algebraic expressions called *monomials*.

Definition

A **monomial** is an expression that is the product of a real number and variables raised to nonnegative integer powers.

Some examples of monomials are:

$$5x \qquad -7t^4 \qquad \frac{4}{5}x^2 \qquad -2p^2q^5$$

Recall that in the expression $5x$, 5 is called the **coefficient**. Note that a constant such as -12 can be thought of as $-12x^0$. So any constant is also considered a monomial.

The expression $5x + 3$ is not a monomial since it is the sum of $5x$ and 3, which are called *terms*. A monomial consists only of factors. Can you explain whether $\frac{2}{x}$ is a monomial?

Monomials serve as building blocks (or terms) for the larger set of polynomials. Polynomials are formed by adding and subtracting monomial terms.

Definition

A **polynomial** is an algebraic expression with one or more monomials added or subtracted.

Some examples of polynomials are:

$$3x^2 - 5x + 7 \qquad 4t^2 - 3 \qquad -8x^2 \qquad 17x^4 + 5x^3 + x - 1 \qquad 20pq - p^2 - 7q^2 + 6$$

Some examples of algebraic expressions that are *not* polynomials are:

$$2x^{-1} \qquad \frac{t^2}{3} + \frac{4}{t} \qquad \frac{5x^2}{2x + 9} \qquad 2\sqrt{x} + 1$$

EXAMPLE 1	PRACTICE 1
Consider the polynomial $3x^5 + 2x^3 - 8$. **a.** Identify the terms of the polynomial. **b.** For each term, identify its coefficient. **Solution** **a.** The terms are $3x^5$, $2x^3$, and -8. **b.** The coefficients of the terms are 3, 2, and -8, respectively.	For the polynomial $-10x^2 + 4x + 20$, find (a) the terms and (b) their coefficients.

Classification of Polynomials

We can classify a polynomial according to the number of its variables, the number of its terms, or its degree. We consider each of these classifications in turn.

Number of Variables

A polynomial such as $3x^2 - 6x + 9$ is said to be *in one variable,* namely in x. The polynomial $t^5 + 11t^4 - 7t^3 - t^2 + 10t - 50$ is also in one variable, namely in t. On the other hand, the polynomial $3x^4y - 5x^2y^3 + 9$ is in *two* variables, x and y. Throughout this text, we focus on polynomials in one variable.

Number of Terms

As we have seen, a polynomial with just one term is called a *monomial*. A polynomial with *two* terms is called a **binomial**. A **trinomial** is a polynomial with *three* terms. Polynomials with four or more terms are simply called *polynomials*.

$$
\begin{array}{ll}
10x & \longleftarrow \text{Monomial} \\
3x - 2 & \longleftarrow \text{Binomial} \\
x^2 + 9x + 6 & \longleftarrow \text{Trinomial} \\
-x^3 + 8x^2 + x - 19 & \longleftarrow \text{Polynomial}
\end{array}
$$

EXAMPLE 2	PRACTICE 2
Classify each polynomial according to the number of terms. **a.** $3x - 5$ **b.** $7x^2 - 3x + 10$ **c.** $10a^2$ **Solution** **a.** Binomial **b.** Trinomial **c.** Monomial	Classify each polynomial according to the number of terms. **a.** $2x + 9$ **b.** $-4x^2$ **c.** $12p - 1$

Degree and Order of Terms

Let's consider a *monomial* in one variable. The **degree** of the monomial is the power of the variable in the monomial.

$$3x^4 \longleftarrow \text{Of the fourth degree or of degree 4}$$
$$-7y^2 \longleftarrow \text{Of the second degree or of degree 2}$$

Recall that a constant, such as 3, can be thought of as $3x^0$ and so is considered to be of degree 0.

$$8 \text{ is a monomial of degree } \mathbf{0}.$$
$$-12 \text{ is a monomial of degree } \mathbf{0}.$$

Polynomials are also classified by their degree. The degree of a polynomial is the highest degree of any of its terms. For instance, the degree of $8x^3 + 9x^2 - 7x - 1$ is **3** since 3 is the highest degree of any term.

$$\text{Degree 3 term} \longrightarrow 8x^3 + 9x^2 - 7x - 1 \longleftarrow \text{Degree 0 term or constant term}$$

Degree 2 term ↑   Degree 1 term ↑

EXAMPLE 3

Identify the degree of each polynomial.

a. $-20x^3$ **b.** $5 + x$

c. $-7x^2 + 3x + 10$ **d.** -36

Solution

a. $-20x^3$ is of degree **3**.

b. $5 + x = 5 + x^1$, which is of degree **1**.

c. $-7x^2 + 3x + 10$ is of degree **2**.

d. $-36 = -36x^0$, which is of degree **0**.

PRACTICE 3

Indicate the degree of each polynomial.

a. $7n$

b. $x - x^2$

c. $2 + 4x^2 - 10x^3$

d. -8

The **leading term** of a polynomial is the term in the polynomial with the highest degree, and the coefficient of that term is called the **leading coefficient**. The term of degree 0 is called the **constant term**. So in the polynomial $3t^2 - 8t + 1$, the leading term is $3t^2$, the leading coefficient is 3, and the constant term is 1.

EXAMPLE 4

Complete the following table.

Polynomial	Constant Term	Leading Term	Leading Coefficient
$2x^{10}$			
$4x + 25$			
$3 - 10x + 8x^2$			
$7x^3 - x - 8$			

PRACTICE 4

Complete the following table.

Polynomial	Constant Term	Leading Term	Leading Coefficient
$-3x^7 + 9$			
x^5			
$x^4 - 7x - 1$			
$3x + 5x^3 + 20$			

Solution

Polynomial	Constant Term	Leading Term	Leading Coefficient
$2x^{10}$	0	$2x^{10}$	2
$4x + 25$	25	$4x$	4
$3 - 10x + 8x^2$	3	$8x^2$	8
$7x^3 - x - 8$	-8	$7x^3$	7

The terms of a polynomial are usually arranged in *descending order of degree*. That is, we write the leading term on the left, then the term of the next highest degree, and so forth.

$$2x^3 + 5x^2 - 3x - 10 \qquad \text{Descending order of degree—the exponents get smaller from left to right.}$$

EXAMPLE 5	PRACTICE 5

EXAMPLE 5

Rearrange each polynomial in descending order.

a. $3x^4 + 9x^2 - 7x^3 + x + 10$

b. $-7x + 6x^3 + 10$

Solution

a. We rewrite the polynomial so that the term with the highest exponent of x is on the left, the next highest exponent comes second, and so on.

$$3x^4 + 9x^2 - 7x^3 + x + 10 = 3x^4 - 7x^3 + 9x^2 + x^1 + 10x^0$$
$$= 3x^4 - 7x^3 + 9x^2 + x + 10$$

b. We rewrite the polynomial so that the exponents get smaller from left to right.

$$-7x + 6x^3 + 10 = 6x^3 - 7x + 10$$

Note that a term with coefficient 0 is usually not written. The unwritten term is said to be a *missing term*. For instance, in the polynomial

$$6x^3 - 7x + 10$$

$0x^2$ is the missing term. So we can write this polynomial as $6x^3 + 0x^2 - 7x + 10$.

PRACTICE 5

Write each polynomial in descending order.

a. $-8x + 9x^5 - 7x^4 + 9x^2 - 6$

b. $x^3 + 7x^5 - 3x^2 + 8$

The concept of missing terms is important in the division of polynomials, as we will see later in this chapter.

Simplifying and Evaluating Polynomials

Recall that in Section 1.7, we discussed how to simplify algebraic expressions by combining like terms.

EXAMPLE 6	PRACTICE 6
Simplify and then put in descending order.	Combine like terms and then write in descending order.

EXAMPLE 6

Simplify and then put in descending order.

$$8 + 3x^3 + 9x + 1 - 8x + 7x^2 - 3x^2$$

Solution

$$8 + 3x^3 + 9x + 1 - 8x + 7x^2 - 3x^2$$
$$= 8 + 3x^3 + 9x + 1 - 8x + 7x^2 - 3x^2$$
$$= 9 + 3x^3 + x + 4x^2 \qquad \text{Combine like terms.}$$
$$= 3x^3 + 4x^2 + x + 9$$

PRACTICE 6

Combine like terms and then write in descending order.

$$2x^2 + 3x - x^2 + 5x^3 + 3x - 5x^3 + 20$$

Recall our discussion of evaluating algebraic expressions in Section 1.8. Polynomials, like other algebraic expressions, are evaluated by replacing each variable with the given number and then carrying out the computation.

EXAMPLE 7

Find the value of $2x^2 - 8x - 5$ when

a. $x = 3$ **b.** $x = -3$

Solution

a. $2x^2 - 8x - 5 = 2(3)^2 - 8(3) - 5$ Substitute 3 for x.
$$= 2(9) - 8(3) - 5$$
$$= 18 - 24 - 5 = -11$$

b. $2x^2 - 8x - 5 = 2(-3)^2 - 8(-3) - 5$ Substitute -3 for x.
$$= 2(9) - 8(-3) - 5$$
$$= 18 + 24 - 5$$
$$= 37$$

PRACTICE 7

Find the value of $x^2 - 5x + 5$ when

a. $x = 2$

b. $x = -2$

EXAMPLE 8

If $1000 is deposited in a savings account that pays compound interest at a rate r compounded annually, then after 2 years the balance in the account will be represented by the polynomial $(1000r^2 + 2000r + 1000)$ dollars. Find the balance if $r = 0.05$.

Solution We need to replace r by 0.05 in the polynomial.

$$
\begin{aligned}
1000r^2 + 2000r + 1000 &= 1000(0.05)^2 + 2000(0.05) + 1000 \\
&= 1000(0.0025) + 2000(0.05) + 1000 \\
&= 2.5 + 100 + 1000 \\
&= 1102.5
\end{aligned}
$$

So the account balance is $1102.50.

PRACTICE 8

If an object is dropped from a height of 500 ft above the ground, its height in feet above the ground after t sec is given by the expression $500 - 16t^2$. How high above the ground is the object after 3 sec?

5.3   Exercises

Mathematically Speaking

Fill in each blank with the most appropriate term or phrase from the given list.

ascending	constant term	descending
power	coefficient	leading term
degree	leading coefficient	
polynomial	monomial	

1. A(n) _____ is an expression that is the product of a real number and variables raised to nonnegative integer powers.

2. The _____ of the expression $-\frac{3}{4}x^3$ is $-\frac{3}{4}$.

3. The degree of a monomial is the _____ of the monomial's variable.

4. The _____ of a polynomial is the highest degree of any of its terms.

5. The _____ of a polynomial is the term in the polynomial with the highest degree.

6. The coefficient of the leading term is called the _____.

7. The term of degree 0 in a polynomial is called the _____.

8. Polynomial terms are usually written in _____ order of degree.

Indicate whether each of the following is a polynomial.

9. $7x^2$

10. $\frac{x^3}{3} - 2x^2 + 9x - 4$

11. $x - 7\sqrt{x} + 1$

12. $10p + q$

13. $4a - 3a^2 + 8$

14. $2xy - x^2$

15. $\frac{2}{x+3}$

16. $3y^{-1} + y$

Classify each polynomial according to the number of terms.

17. a. $5x - 1$ **b.** $-5a^2$

c. $-6a + 3$ **d.** $x^3 + 4x^2 + 2$

18. a. $5x + x^2$ **b.** $3x$

c. $12p - 1$ **d.** $2x^5 - x^3 + x$

Rearrange each polynomial in descending order, and then identify its degree.

19. $3x^2 - 2x + 8 - 4x^3$

20. $5x^3 + 7x + 1 - 7x^2$

21. $2 - 3y$

22. $4p^2 - p^4 + 3p^3 + 10 - p$

23. $7x - 5x^2$

24. $-8x^2 + 6x - 2$

25. $-y^3 - 2y + 2 - 4y^5$ **26.** $5x^3 + 3x^5 + 8x + 3$ **27.** $5a^2 - a$

28. $25 - y + 2y^2 + y^3 - 3y^4$ **29.** $9p + 3p^3$ **30.** $5x^2$

Complete the following table.

31.

Polynomial	Constant Term	Leading Term	Leading Coefficient
$-x^7 + 2$			
$2x - 30$			
$-5x + 1 + x^2$			
$7x^3 - 2x - 3$			

32.

Polynomial	Constant Term	Leading Term	Leading Coefficient
$5x^3 + 8$			
$-x + 10$			
$2x^2 - 3x + 4$			
$-5x + x^4 - 9$			

Simplify.

33. $9x^3 - 7x^2 + 1 + x^3 + 10x + 5$ **34.** $2y^3 - 7y^2 + 1 + 2y^2 + 3y + 8$

35. $r^3 + 2r^2 + 15 + r^2 - 8r - 1$ **36.** $n^4 - n^3 - 7n^2 + n^2 + 10n + 3$

Identify the missing terms of each polynomial.

37. $x^3 - 7x - 2$ **38.** $n^2 + 7n$ **39.** $6x^3 + 8x^2 + 1$ **40.** $x^4 - 3x$

Find the value of each polynomial for the given values of the variable.

41. $7x - 3$, for $x = 2$ and $x = -2$ **42.** $5a + 11$, for $a = 0$ and $a = 2$

43. $n^2 - 3n + 9$, for $n = 7$ and $n = -7$ **44.** $3y^2 + 2y + 1$, for $y = 2$ and $y = -1$

45. $2.1x^2 + 3.9x - 7.3$, for $x = 2.37$ and $x = -2.37$ **46.** $0.1x^3 + 4.1x - 9.1$, for $x = 3.14$ and $x = -3.14$

Combine like terms. Then write the polynomial in descending order of powers.

47. $4x^2 - 2x - x^2 - 10 - 3x + 4$ **48.** $x^3 + 5x - 7x^2 - 1 - x^3 + x$

49. $6n^3 + 20n - n^2 + 2 - 4n^3 + 15n^2 + 8$ **50.** $8y^2 + y^3 - 8y + 20 + 3y + 9y^2$

Mixed Practice

Rearrange each polynomial in descending order, and then identify its degree.

51. $-a^3 + 2a - 4 + 5a^4$ **52.** $8x + 6x^3$

53. Complete the following table.

Polynomial	Constant Term	Leading Term	Leading Coefficient
$4x - 20$			
$-7x^6 + 9$			
$3x + 2 + x^2$			
$6x - x^5 + 11$			

Solve.

54. Classify each polynomial according to the number of terms.

    a. $2p - 3p^2$ **b.** $-8x^3$ **c.** $a^2 - 4a + 5$

55. Is $x^2 + 4x^{-2} - 2$ a polynomial?

Find the value of each polynomial for the given values of the variable.

56. $5y^2 - 8y - 9$, for $y = 3$ and $y = -2$

57. $5.3x^2 - 2.7x - 6.8$, for $x = 1.25$ and $x = -3.87$

58. Simplify: $4n^3 - 8n^4 + 3 + 6n^2 + 5n^4 - 4n^2$.

59. Identify the missing terms of the polynomial $9x^4 + x^2 - 3x$.

60. Combine like terms; then write the polynomial in descending order of powers:
$3m^2 - m^3 + 9m - 11m^2 - 15 + 7m$

Applications

Solve.

61. The polynomial $1 + x^2 + x^{15} + x^{16}$ is used by computer scientists to detect errors in computer data. Classify this polynomial in terms of its variables and its degree.

62. The owner of a factory estimates that her profit (in dollars) is

$$0.003x^3 - 1.4x^2 + 300x - 1000$$

where x is the number of items that the factory produces. Describe this polynomial in terms of its variables and its degree.

63. The polynomial $x + \dfrac{x^2}{20}$ is the *stopping distance* of a car in feet after the brakes are applied, where the variable x is the speed of the car in miles per hour before braking. Find the stopping distance for a car that had been traveling at 40 mph.

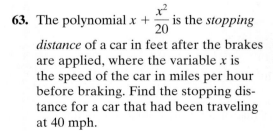

64. There are n teams that compete in a sports league, where each team plays every other team once. The following polynomial gives the total number of games that must be played:

$$0.5n^2 - 0.5n$$

If the league has 20 teams, how many games are played?

65. The polynomial $72x + 2342$ approximates the world population (in millions), where x represents the number of years since 1950. According to this model, what was the world population in 2000? (*Source:* U.S. Bureau of the Census)

66. The percent of U.S. households with Web access is approximated by $0.45x^2 + 6.75x + 26.2$ where x represents the number of years since 1998. Estimate the percent of U.S. households with Web access in 2003. (*Source:* U.S. Department of Commerce)

67. The number of U.S. radio stations with a rock music format is approximated by the polynomial $1.68x^3 + 9.95x^2 - 11.6x + 730$, where x represents the number of years since 1999. To the nearest whole number, estimate the number of U.S. radio stations with a rock music format in 2003. (*Source:* M Street Corporation)

68. The number (in thousands) of U.S. households with cable television can be approximated by the expression $24.5t^3 - 253t^2 + 2043t + 60,920$ where t represents the number of years since 1995. According to this model, how many U.S. households were there with cable television in 2004? (*Source: Statistical Abstract of the United States*)

● *Check your answers on page A-29.*

MINDSTRETCHERS

Research

1. The following prefixes are used with polynomials. Use a dictionary to fill in the table.

Prefix	Meaning of This Prefix	Three Words Beginning with This Prefix
Mono-		
Bi-		
Tri-		
Poly-		

Groupwork

2. There are some polynomials whose value is a prime number for many values of the variable. For instance, consider the second-degree polynomial $n^2 + n + 41$.

   a. Check that for n equal to a whole number between 0 and 39, the value of this polynomial is a prime number.

   b. Is the value of this polynomial a prime number for $n = 40$? Explain your answer.

Patterns

3. The degree of a monomial in more than one variable is the sum of the powers of the variables in that term. Recall that the degree of a polynomial is the highest degree of any of its terms. The following tables show polynomials that represent the area or the volume of various common geometric figures.

Area

Geometric Figure	Polynomial	Degree of the Polynomial
Square	s^2	
Triangle	$0.5bh$	
Trapezoid	$0.5hb + 0.5hB$	
Circle	$3.14r^2$	
Rectangle	lw	

Volume

Geometric Figure	Polynomial	Degree of the Polynomial
Cube	e^3	
Rectangle solid	lwh	
Sphere	$\frac{4}{3}\pi r^3$	
Cylinder	$\pi r^2 h$	

a. Complete the tables by finding the degree of each polynomial.

b. Describe the pattern you observe in the table for area. Explain your observation.

c. Describe the pattern you observe in the table for volume. Explain your observation.

CULTURAL NOTE

Muhammad ibn Musa al-Khwarizmi, a ninth-century mathematician, wrote *al-Kitab al-mukhtasar fi hisab al-jabr wa'l-muqabala* (*The Compendious Book on Calculation by Completion and Balancing*)— one of the earliest treatises on algebra and the source of the word *algebra*. This work dealt with solving equations as well as with practical applications of algebra to measurement and legacies. al-Khwarizmi, from whose name the word *algorithm* derives, also wrote influential works on astronomy and on the Hindu numeration system.

Sources:

Jan Gullberg, *Mathematics From the Birth of Numbers* (W. W. Norton, 1997).

Morris Kline, *Mathematics, A Cultural Approach* (Addison-Wesley, 1962).

5.4 Addition and Subtraction of Polynomials

In this section, we consider the addition and subtraction of polynomials and their applications to real-world situations. As our discussion proceeds, note the similarity between adding and subtracting whole numbers in arithmetic and these operations on polynomials in algebra.

OBJECTIVES

- To add polynomials
- To subtract polynomials
- To solve applied problems involving the addition or subtraction of polynomials

Adding Polynomials

Recall from Section 5.3 that we can simplify polynomials by combining like terms. We use the same approach when combining polynomials.

To Add Polynomials

- Add the like terms.

As with whole numbers, we can add polynomials using either a horizontal or vertical format. To add polynomials horizontally, we simply remove the parentheses and combine like terms.

EXAMPLE 1	**PRACTICE 1**
Find the sum: $(8x^2 + 3x + 4) + (-12x^2 + 7)$	Add: $6x - 3$ and $9x^2 - 3x - 40$

Solution Recall that when a plus sign precedes terms in parentheses, we remove the parentheses and keep the sign of each term.

$$\overbrace{(8x^2 + 3x + 4)}^{\text{First polynomial}} + \overbrace{(-12x^2 + 7)}^{\text{Second polynomial}}$$

$$= 8x^2 + 3x + 4 - 12x^2 + 7 \qquad \text{Remove parentheses.}$$

$$= -4x^2 + 3x + 11 \qquad \text{Combine like terms.}$$

EXAMPLE 2	**PRACTICE 2**
Combine: $(3st^2 - 4st + t^2) + (8s^2t - 3t^2) + (10t^2 - 5st + 7s^2)$	Combine: $(9p^2 + 4pq + 2q^2) + (-p^2 - 5q^2) + (2p^2 - 3pq - 7q^2)$

Solution

$$(3st^2 - 4st + t^2) + (8s^2t - 3t^2) + (10t^2 - 5st + 7s^2)$$

$$= 3st^2 - 4st + t^2 + 8s^2t - 3t^2 + 10t^2 - 5st + 7s^2$$

$$= 3st^2 + 8s^2t - 9st + 7s^2 + 8t^2$$

Now, let's look at how to add polynomials in a vertical format. Recall that in adding whole numbers, we position the addends so that digits with the same place value are in the same column. Similarly, when we add polynomials vertically, we position the polynomials so that like terms are in the same column. Suppose, for example, we want to add $7x^5 + x^3 - 3x^2 + 8$ and $2x^5 - 7x^4 + 9x - 9$. First we make sure that both polynomials are simplified and in descending order. Then we write the polynomials as follows:

$$7x^5 + 0x^4 + \ x^3 - 3x^2 + 0x + 8$$
$$2x^5 - 7x^4 + 0x^3 + 0x^2 + 9x - 9$$

Note that we could have just left a space for each missing term of the polynomials. Do you see that each column contains like terms? We then add the terms in each column:

$$7x^5 \qquad + x^3 - 3x^2 \qquad + 8$$
$$2x^5 - 7x^4 \qquad\qquad + 9x - 9$$
$$9x^5 - 7x^4 + x^3 - 3x^2 + 9x - 1$$

So the sum is $9x^5 - 7x^4 + x^3 - 3x^2 + 9x - 1$.

Let's consider some more examples.

EXAMPLE 3

Add vertically: $3x^2 - 5x - 6$ and $10x + 20$

Solution First we check that both polynomials are in descending powers of x. Next, we rewrite the polynomials vertically, with like terms positioned in the same column. Finally, we add within columns.

First-degree terms

Second-degree term ⌐   ⌐ Zero-degree (constant) terms

$$3x^2 - 5x - \ \ 6$$
$$10x + 20$$
$$3x^2 + 5x + 14$$

The sum is $3x^2 + 5x + 14$, which is in simplest form.

PRACTICE 3

Find the sum of $8n^2 + 2n - 1$ and $3n^2 - 2$ using a vertical format.

EXAMPLE 4

Find the sum of $7x^3 - 10x^2y + 8xy^2 + 13y^3$, $14xy^2 - 1$ and $3x^2y - 5xy^2 - y^3 + 2$ using a vertical format.

Solution

$$7x^3 - 10x^2y + 8xy^2 + 13y^3$$
$$14xy^2 \qquad\quad - 1$$
$$3x^2y - 5xy^2 \ - y^3 + 2$$
$$7x^3 - 7x^2y + 17xy^2 + 12y^3 \quad + 1$$

PRACTICE 4

Add vertically:

$7p^3 - 8p^2q - 3pq^2 + 20$,
$10p^2q + pq^2 - q^3 + 5$, and $p^3 - q^3$

Is the sum in Example 4 in simplified form? Explain.

Subtracting Polynomials

Recall from Section 1.7 that when a minus sign precedes terms in parentheses, we remove the parentheses and change the signs of each term.

EXAMPLE 5	PRACTICE 5
Remove parentheses and simplify.	Remove parentheses and simplify.
a. $2x - (3x + 4y)$	**a.** $-(4r - 3s) + 7r$
b. $(5n + 2m) - (n + m) + (3n - 4m)$	**b.** $(2p + 5q) + (p - 6q) - (3p + 2q)$

Solution

a. In the expression $2x - (3x + 4y)$, since $(3x + 4y)$ is preceded by a minus sign, we remove the parentheses and change the sign of each term in parentheses. Then we combine like terms.

$$2x - (3x + 4y) = 2x - 3x - 4y$$
$$= -x - 4y$$

b. $(5n + 2m) - (n + m) + (3n - 4m) = 5n + 2m - n - m$
$$+ 3n - 4m = 7n - 3m$$

For a polynomial preceded by a minus sign, *change* signs of terms.

For a polynomial preceded by a plus sign, *keep* signs of terms.

To subtract real numbers, we change the number being subtracted to its opposite, and then add. Subtraction of polynomials works very much in the same way.

To Subtract Polynomials

- Change the sign of each term of the polynomial being subtracted.
- Add.

For instance, suppose that we want to subtract the polynomial $2x^4 - 5x^3 + 4x^2 + x + 1$ from $3x^4 + x^3 - 4x^2 + 8x - 9$.

The polynomial from which we are subtracting

The polynomial being subtracted

$(3x^4 + x^3 - 4x^2 + 8x - 9) - (2x^4 - 5x^3 + 4x^2 + x + 1)$

$= (3x^4 + x^3 - 4x^2 + 8x - 9) + (-2x^4 + 5x^3 - 4x^2 - x - 1)$ Change the sign of each term of the polynomial being subtracted, and then add.

$= 3x^4 + x^3 - 4x^2 + 8x - 9 - 2x^4 + 5x^3 - 4x^2 - x - 1$ Remove parentheses.

$= x^4 + 6x^3 - 8x^2 + 7x - 10$ Combine like terms.

EXAMPLE 6

Subtract: $(5x^2 - 3x + 7) - (-2x^2 + 8x + 9)$

Solution

$$\begin{aligned}
(5x^2 - 3x + 7) - (-2x^2 + 8x + 9) &= (5x^2 - 3x + 7) + (2x^2 - 8x - 9) \\
&= 5x^2 - 3x + 7 + 2x^2 - 8x - 9 \\
&= 7x^2 - 11x - 2
\end{aligned}$$

PRACTICE 6

Find the difference:
$(2x - 1) - (3x^2 + 15x - 1)$

We can also subtract polynomials vertically, a skill that comes up when dividing polynomials. The key in vertical subtraction is to position the polynomials so that like terms are in the same column. Then we change the sign of each term of the polynomial being subtracted, and add.

EXAMPLE 7

Subtract using a vertical format: $(7x^2 - 3x + 7) - (x^2 + 8x - 9)$

Solution

$$\begin{array}{r}
7x^2 - 3x + 7 \\
-(x^2 + 8x - 9) \\
\end{array}$$ Position like terms in the same columns.

$$\begin{array}{r}
7x^2 - 3x + 7 \\
-x^2 - 8x + 9 \\
\end{array}$$ Change the sign of each term of the polynomial being subtracted.

$$\begin{array}{r}
7x^2 - 3x + 7 \\
-x^2 - 8x + 9 \\
\hline
6x^2 - 11x + 16 \\
\end{array}$$ Add.

PRACTICE 7

Subtract vertically:
$(20x - 13) - (5x^2 - 12x + 13)$

EXAMPLE 8

Subtract $10x^2 + 8xy + y^2$ from $4y^2 - 9x^2$, using a vertical format.

Solution

$$\begin{array}{r}
-9x^2 \qquad + 4y^2 \\
-(10x^2 + 8xy + y^2) \\
\end{array}$$ Position like terms in the same column.

$$\begin{array}{r}
-9x^2 \qquad + 4y^2 \\
-10x^2 - 8xy - y^2 \\
\end{array}$$ Change the signs of the terms in the polynomial being subtracted.

$$\begin{array}{r}
-9x^2 \qquad + 4y^2 \\
-10x^2 - 8xy - y^2 \\
\hline
-19x^2 - 8xy + 3y^2 \\
\end{array}$$ Add.

PRACTICE 8

Find the difference using a vertical format:
$(2p^2 - 7pq + 5q^2)$
$\qquad\qquad - (3p^2 + 4pq - 12q^2)$

EXAMPLE 9

The polynomial $0.1x + 25.6$ approximates the number of Americans (in millions) who voted for the Democratic candidate in the presidential election that took place x years after 1936. The corresponding polynomial for the Republican candidate is $x + 16$. Find the polynomial that approximates the number of Americans who voted for either the Democratic or the Republican candidate. (*Source: Congressional Quarterly*)

Solution To determine the number of Americans who voted for either the Democratic or the Republican candidate, we add the number who voted for the Democratic candidate and the number who voted for the Republican candidate. These numbers are approximated by the given polynomials.

$$
\begin{array}{r}
0.1x + 25.6 \\
\underline{x + 16} \\
1.1x + 41.6
\end{array}
$$

So $1.1x + 41.6$ represents the number of Americans who voted for either the Democratic or Republican candidate.

PRACTICE 9

The polynomial $0.3x + 67.2$ approximates the life expectancy (in years) at birth for males born x years after 1970. The corresponding polynomial for females is $0.3x + 74.8$. Find the polynomial that approximates how much greater the life expectancy is for females than for males.

(*Source:* U.S. National Center for Health Statistics, *Vital Statistics of the United States*)

5.4   Exercises

Add horizontally.

1. $3x^2 + 6x - 5$ and $-x^2 + 2x + 7$

2. $10x^2 + 3x + 9$ and $-x^2 - 5x + 1$

3. $2n^3 + n$ and $3n^3 + 8n$

4. $9y + 2y^2$ and $-y - 3y^2$

5. $10p + 3 + p^2$ and $p^2 - 7p - 4$

6. $x^2 + 3x - 8$ and $10 - 3x + 4x^2$

7. $8x^2 + 7xy - y^2$ and $3x^2 - 10xy + 3y^2$

8. $20p^2 + 15q^4 + 30pq$ and $-4pq + 10q^4 - p^2$

9. $2p^3 - p^2q - 5pq^2 + 1$, $3p^2q + 2pq^2 - 4q^3 + 4$,   and $p^3 + q^3$

10. $2x^3 - 4x^2y + xy^2 + y^3$, $3xy^2 - 6$,   and $2x^2y - xy^2 - 2y^3 + 3$

Add vertically.

11. $10x^2 - 3x - 8$ and $20x + 3$

12. $t^2 + 4t + 5$ and $-t + 10$

13. $5x^3 + 7x - 1$ and $x^2 + 2x + 3$

14. $2r^3 + r + 2$ and $-r^3 - 8r^2 + 5r - 6$

15. $5ab^2 - 3a^2 + a^3$ and $2ab^2 + 9a^2 - 4a^3$

16. $p^2q^3 - p^2 - q^3$ and $5p^2q^3 + p^2 + q^3$

Subtract horizontally.

17. $2x^2 + 3x - 7$ from $x^2 + x + 4$

18. $2x^3 + 7x^2 + 3x$ from $8x^3 - 10x^2 + x$

19. $3x^3 + x^2 + 5x - 8$ from $x^3 + 10x^2 - 8x + 3$

20. $5t^2 - 7t - 1$ from $8t^2 - 3t + 2$

21. $5x + 9$ from $x^2 + 3x$

22. $3x - 7$ from $x^2 - x + 4$

23. $4y^2 - 6xy - 3$ from $1 - 6xy + 5x^2 - y^2$

24. $p^4 - 7p^2q^2 + q^4$ from $8p^4 - 3q^4$

Subtract vertically.

25. $7p^2 - 10p - 1$ from $2p^2 - 3p + 5$

26. $x^2 - 5x + 2$ from $3x^2 + 10x - 2$

27. $8t^3 - 5$ from $9t^3 - 12t^2 + 3$

28. $10x + 7$ from $x^2 + 2x - 6$

29. $r^3 - 3r^2s - 5$ from $4r^3 - 20r^2s - 7$

30. $-5x^3 - 2y^2$ from $13xy^3 + 7x^3 - 10y^2$

Remove the parentheses and simplify, if possible.

31. $7x - (8x + r)$

32. $-(3x + 2y) + (4x - 5)$

33. $2p - (3q + r)$

34. $t - (4r - s)$

35. $(4y - 1) + (3y^2 - y + 5)$

36. $(2x + 6) + (x^2 - 4x + 3)$

37. $(m^3 - 6m + 7) - (-9 + 6m)$

38. $(p^2 + 3p - 5) + (p^3 + 6 - p^2)$

39. $(2x^3 - 7x + 8) - (5x^2 + 3x - 1)$

40. $(n^3 - 4n + 2) - (n^2 - 8n + 1)$

41. $(8x^2 + 3x) + (x - 2) + (x^2 + 9)$

42. $(5y^2 + y) + (9y - 1) + (y^2 + 2)$

43. $(3x - 7) + (2x + 9) - (7x - 10)$

44. $(5n + 1) - (3n + 1) + (2n + 5)$

45. $(7x^2y^2 - 10xy + 4) - (2xy + 8) + (x^2y^2 - 10)$

46. $(2m^2n^2 - mn + 7) - (mn - 3) + (m^2n^2 + 7)$

Mixed Practice

Subtract horizontally.

47. $3m^2 - 4m + 2$ from $5m^2 - 9m - 7$

48. $x^2 - 5x$ from $3x - 4$

Remove the parentheses and simplify, if possible.

49. $(5x^3 - x - 8) - (-3x^2 - 3x + 4)$

50. $3a - (4b - c)$

51. $(3m - 1) - (4m + 5) + (6m - 2)$

52. $(6t - 3) + (2t^2 - 3t + 4)$

Add horizontally.

53. $3x^2 + 5x - 4$ and $-5x^2 + x + 7$

54. $6a^2 - 8b^3 - 7ab$ and $-5ab + 3b^3 - 11a^2$

Solve.

55. Subtract $p^2 - 4p + 3$ from $6p^2 + 7p - 4$ vertically.

56. Add $t^3 - 3t^2 + t + 9$ and $-8t^3 + t^2 - 2$ vertically.

Applications

Solve.

57. The surface area of the cylindrical can shown is approximated by the polynomial expression $3.14r^2 + 3.14r^2 + 6.28rh$.

   a. Simplify this polynomial expression.

   b. Find a polynomial expression for the surface area of a can where the radius and height are equal.

58. The room shown is in the shape of a cube.

   a. Write a simplified expression for the surface area of the four walls.

   b. Find the surface area of the four walls when e is 9 ft, x is 3 ft, and y is 7 ft.

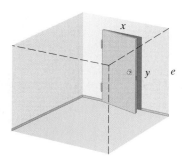

59. The total U.S. imports of petroleum (in millions of barrels per day) in a given year is approximated by the polynomial $31x^3 - 522x^2 + 2083x + 6051$, where x represents the number of years since 1975. The corresponding total of exports of petroleum is $-x^3 + 16x^2 + 22x + 189$. Write a polynomial that represents how many more barrels per day were imported than exported in a given year. (*Source:* U.S. Department of Energy's *Monthly Energy Review*)

60. The number (in thousands) of male commissioned officers in the U.S. Department of Defense in a given year is approximated by $-0.1x^2 + 5.4x + 18.3$, where x represents the number of years since 1940. The corresponding polynomial for female commissioned officers is $0.2x + 1.3$. Find the polynomial that represents, for a given year, how many more male commissioned officers than female commissioned officers there were. (*Source:* U.S. Department of Defense)

61. The number (in millions) of compact discs sold in a given year is approximated by $2x^3 - 27x^2 + 172x + 391$, where x represents the number of years since 1992. The corresponding polynomial for cassettes is $2x^3 - 22x^2 + 18x + 360$. Find the polynomial that represents how many more CDs than cassettes were sold in a given year. (*Source:* Recording Industry Association of America)

62. The total expenses (in billions of dollars) in federal U.S. hospitals can be modeled by the polynomial $0.8t + 4$, where t is the number of years since 1975. For nonfederal U.S. hospitals, the corresponding polynomial is $13t + 27$. Write a polynomial that represents the total expenses for all U.S. hospitals. (*Source:* U.S. National Center for Health Statistics)

• *Check your answers on page A-29.*

MINDSTRETCHERS

Patterns

1. A Fibonacci sequence is a list of numbers with the following property: After the first two numbers, every other number on the list is the sum of the two previous numbers. The following, for example, is a Fibonacci sequence:

$$7 + 11$$
$$\downarrow$$
$$7, \ 11, \ 18, \ 29, \ 47, \ 76, \ 123, \ldots$$
$$\uparrow$$
$$11 + 18$$

   a. What is the next number in this sequence?

   b. If the first two numbers in a Fibonacci sequence are a and b, find the third and fourth numbers. Check that the tenth number in the sequence is given by the polynomial $21a + 34b$.

Mathematical Reasoning

2. Is it possible to add two polynomials, each of degree 4, and have the sum be a polynomial of degree 2? If so, give an example. If not, explain why not.

Critical Thinking

3. Is the subtraction of polynomials a commutative operation? Give an example to support your answer.

5.5   Multiplication of Polynomials

In this section, we discuss how to multiply polynomials. We start with finding the product of two monomials.

Multiplying Monomials

When multiplying monomials such as $3x^2$ and $2x^5$, the product rule of exponents helps us to find the product.

$$\begin{aligned}(3x^2)(2x^5) &= (3 \cdot 2) \cdot (x^2 \cdot x^5) &&\text{The commutative and associative properties of multiplication}\\&= (3 \cdot 2) \cdot (x^{2+5}) &&\text{The product rule of exponents}\\&= 6x^7\end{aligned}$$

To Multiply Monomials

- Multiply the coefficients.
- Multiply the variables, using the product rule of exponents.

EXAMPLE 1	PRACTICE 1
Multiply: $-2x \cdot 8x$ **Solution** $$\begin{aligned}-2x \cdot 8x &= (-2 \cdot 8) \cdot (x \cdot x)\\&= (-2 \cdot 8)(x^{1+1})\\&= -16x^2\end{aligned}$$	Find the product of $(-10x^2)$ and $(-4x^3)$.

EXAMPLE 2	PRACTICE 2
Multiply: $(-2x^3y)(-4x^2y^4)(10xy)$ **Solution** $$\begin{aligned}(-2x^3y)(-4x^2y^4)(10xy) &= (-2 \cdot -4 \cdot 10) \cdot (x^3 \cdot x^2 \cdot x) \cdot (y \cdot y^4 \cdot y)\\&= 80(x^{3+2+1})(y^{1+4+1})\\&= 80x^6y^6\end{aligned}$$ Note that the variables in a product are generally written in alphabetical order.	Find the product: $(7ab^2)(10a^2b^3)(-5a)$

EXAMPLE 3	PRACTICE 3
Simplify: $(-3p^2r)^3$ **Solution** $\quad(-3p^2r)^3 = (-3)^3(p^2)^3(r)^3$ Use the rule for raising a product to a power. $\qquad\qquad\quad = -27p^6r^3$	Find the square of $-5xy^2$.

Multiplying a Monomial by a Polynomial

Now, let's use our knowledge of multiplying monomials to find the product of a monomial and a polynomial.

Consider, for instance, the product $(7x)(9x^2 + 5)$. We use the distributive property to find this product.

$$(7x)(9x^2 + 5) = (7x)(9x^2) + (7x)(5) = 63x^3 + 35x$$

Let's look at some more examples.

EXAMPLE 4	PRACTICE 4
Multiply: $(-8x + 9)(-3x^2)$	Find the product: $(10s^2 - 3)(7s)$
Solution	
$(-8x + 9)(-3x^2) = (-3x^2)(-8x + 9)$ $= (-3x^2)(-8x) + (-3x^2)(9)$ $= 24x^3 - 27x^2$	

EXAMPLE 5	PRACTICE 5
Multiply: $3p^2q(5p^3 - 2pq + q^3)$	Simplify: $-2m^3n^2(-6m^3n^5 + 2mn^2 + n)$
Solution	
$3p^2q(5p^3 - 2pq + q^3) = 3p^2q(5p^3) + 3p^2q(-2pq) + 3p^2q(q^3)$ $= 15p^5q - 6p^3q^2 + 3p^2q^4$	

EXAMPLE 6	PRACTICE 6
Simplify: $8x^2(3x + 1) + x^2(5x^2 - 6x + 5)$	Simplify: $7s^3(-2s^2 + 5s + 4) - s^2(s^2 + 6s - 1)$
Solution	
$8x^2(3x + 1) + x^2(5x^2 - 6x + 5)$ $= 8x^2(3x) + 8x^2(1) + x^2(5x^2) + x^2(-6x) + x^2(5)$ $= 24x^3 + 8x^2 + 5x^4 - 6x^3 + 5x^2$ $= 5x^4 + 18x^3 + 13x^2$ Combine like terms.	

Multiplying Two Binomials

Now, we extend the discussion to the multiplication of binomials. As in the case of multiplying monomials and binomials, we use the distributive property.

Consider, for example, the product $(x + 4)(7x + 2)$. To apply the distributive property, we can think of the first factor, $(x + 4)$, as a single number multiplied by the binomial $(7x + 2)$.

$$(a) \cdot (b + c) = (a) \cdot (b) + (a) \cdot (c)$$

$$(x + 4)(7x + 2) = (x + 4)(7x) + (x + 4)(2)$$ Use the distributive property.

$$= 7x(x + 4) + 2(x + 4)$$ Use the commutative property.

$$= 7x \cdot x + 7x \cdot 4 + 2 \cdot x + 2 \cdot 4$$ Use the distributive property.

$$= 7x^2 + 28x + 2x + 8$$

$$= 7x^2 + 30x + 8$$ Combine like terms.

Would we get the same answer if we multiplied $(7x + 2)$ by $(x + 4)$?

Let's consider some other examples.

EXAMPLE 7	PRACTICE 7

Find the product: $(3x + 1)(5x - 2)$

Multiply: $(a - 1)(2a + 3)$

Solution

$$\begin{aligned}(3x + 1)(5x - 2) &= (3x + 1)(5x) + (3x + 1)(-2) \\ &= 5x(3x + 1) + (-2)(3x + 1) \\ &= 5x \cdot 3x + 5x \cdot 1 + (-2) \cdot 3x + (-2) \cdot 1 \\ &= 15x^2 + 5x - 6x - 2 \\ &= 15x^2 - x - 2\end{aligned}$$

Another way to multiply two binomials is called the **FOIL method,** which is derived from the distributive property. With this method, we can memorize a formula that makes multiplying binomials quick and easy.

FOIL stands for **First, Outer, Inner,** and **Last.** Let's see how this method works, applying it to the product $(x + 4)(7x + 2)$ that we discussed above.

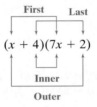

The two **first** terms in the binomials are x and $7x$, and their product is $7x^2$.

$$(x + 4)(7x + 2) \qquad x \cdot 7x = 7x^2$$

The two **outer** terms are x and 2, and their product is $2x$.

$$(x + 4)(7x + 2) \qquad x \cdot 2 = 2x$$

The two **inner** terms are 4 and $7x$, and their product is $28x$.

$$(x + 4)(7x + 2) \qquad 4 \cdot 7x = 28x$$

Next, the two **last** terms are 4 and 2, and their product is 8.

$$(x + 4)(7x + 2) \qquad 4 \cdot 2 = 8$$

Finally, we add the four products, combining any like terms.

$$7x^2 + 2x + 28x + 8 = 7x^2 + 30x + 8$$

Note that, as expected, this answer is the same as the one found using the distributive property.

To Multiply Two Binomials Using the FOIL Method

Consider $(a + b)(c + d)$.

- Multiply the two first terms in the binomials.

$$(a + b)(c + d) \qquad \text{Product is } ac.$$
F

- Multiply the two outer terms.

$$(a + b)(c + d) \qquad \text{Product is } ad.$$
O

- Multiply the two inner terms.

$$(a + b)(c + d) \qquad \text{Product is } bc.$$
I

- Multiply the two last terms.

$$(a + b)(c + d) \qquad \text{Product is } bd.$$
L

- Find the sum of these four products.

$$(a + b)(c + d) = ac + ad + bc + bd$$

With some practice, the FOIL method can be done mentally.

EXAMPLE 8	PRACTICE 8

EXAMPLE 8

Multiply: $(8x - 3)(2x - 1)$

Solution Using the FOIL method, we get:

First   Last

$$(8x - 3)(2x - 1)$$

Inner

Outer

F: $(8x)(2x) = 16x^2$
O: $(8x)(-1) = -8x$
I: $(-3)(2x) = -6x$
L: $(-3)(-1) = 3$

So $(8x - 3)(2x - 1) = 16x^2 - 8x - 6x + 3 = 16x^2 - 14x + 3$.
Note that the middle term, $-14x$, is the sum of the outer and
inner products, which are like terms.

PRACTICE 8

Find the product of $8x + 3$ and
$2x - 1$.

EXAMPLE 9

Multiply: $(3a + b)(2a - b)$

Solution

$$
\begin{array}{cccc}
\text{F} & \text{O} & \text{I} & \text{L}
\end{array}
$$

$$
\begin{aligned}
(3a + b)(2a - b) &= (3a)(2a) + (3a)(-b) + (b)(2a) + (b)(-b) \\
&= 6a^2 - 3ab + 2ab - b^2 \\
&= 6a^2 - ab - b^2
\end{aligned}
$$

PRACTICE 9

Find the product: $(7m - n)(2m + n)$

Multiplying Polynomials

Finally, let's consider multiplying polynomials in general. Here we extend the previous discussion of multiplying binomials to multiplying polynomials that can have any number of terms. Let's consider the following example in which we multiply two polynomials written in a horizontal format.

EXAMPLE 10

Multiply: $(x^2 + 3x - 1)(x + 2)$

Solution

$$
\begin{aligned}
(x^2 &+ 3x - 1)(x + 2) \\
&= (x^2 + 3x - 1)(x) + (x^2 + 3x - 1)(2) \quad \text{Use the distributive property.} \\
&= x^3 + 3x^2 - x + 2x^2 + 6x - 2 \quad \text{Use the distributive property.} \\
&= x^3 + 5x^2 + 5x - 2 \quad \text{Combine like terms.}
\end{aligned}
$$

PRACTICE 10

Multiply $(n^2 + n - 2)(n - 4)$ using a horizontal format.

Instead of multiplying horizontally, we can multiply two polynomials in a vertical format similar to that used for multiplying whole numbers.

$$
\begin{array}{r}
x^2 + 3x - 1 \\
x + 2 \\
\hline
2x^2 + 6x - 2 \\
x^3 + 3x^2 - x \\
\hline
x^3 + 5x^2 + 5x - 2
\end{array}
$$

Multiply $x^2 + 3x - 1$ by 2.
Multiply $x^2 + 3x - 1$ by x.
Add like terms.

Note that we positioned the terms of the two "partial products" in the shaded area so that each column contains like terms. Finally, to find the product of the original polynomials, we add down each column in the shaded area.

EXAMPLE 11

Find the product of $3y^2 + y + 5$ and $4y - 1$.

Solution We begin by rewriting the problem in a vertical format.

$$
\begin{array}{r}
3y^2 + y + 5 \\
4y - 1 \\
\hline
-3y^2 - y - 5 \\
12y^3 + 4y^2 + 20y \\
\hline
12y^3 + y^2 + 19y - 5
\end{array}
$$

So we conclude that

$$(3y^2 + y + 5)(4y - 1) = 12y^3 + y^2 + 19y - 5$$

PRACTICE 11

Multiply: $(8n^2 - n + 3)(n + 2)$

EXAMPLE 12

Multiply: $(4x^3 - 2x + 1)(x + 5)$

Solution Let's multiply vertically.

Use $+ 0x^2$ for the missing x^2 term.

$$
\begin{array}{r}
4x^3 + 0x^2 - 2x + 1 \\
x + 5 \\
\hline
20x^3 + 0x^2 - 10x + 5 \\
4x^4 + 0x^3 - 2x^2 + x \\
\hline
4x^4 + 20x^3 - 2x^2 - 9x + 5
\end{array}
$$

Note that we wrote $0x^2$ for the missing second-degree term in the top polynomial. (Alternatively, we could have left a blank space there.) Similarly, we write a term with a 0 coefficient for each missing term in the partial products.

EXAMPLE 13

Find the product of $a^2 + 2ab + b^2$ and $a + b$.

Solution

$$
\begin{array}{r}
a^2 + 2ab + b^2 \\
a + b \\
\hline
a^2b + 2ab^2 + b^3 \\
a^3 + 2a^2b + ab^2 \\
\hline
a^3 + 3a^2b + 3ab^2 + b^3
\end{array}
$$

EXAMPLE 14

A box factory makes an open-top box from a piece of cardboard by cutting out squares that are x ft by x ft from each corner as shown below. The area of the base of the box is given by $(4 - 2x)(3 - 2x)$ ft². Rewrite the area of the base without parentheses.

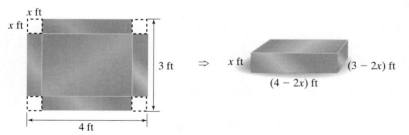

Solution To remove parentheses, we multiply the binomials using the FOIL method.

$$
\begin{array}{cccc}
\text{F} & \text{O} & \text{I} & \text{L}
\end{array}
$$
$$(4 - 2x)(3 - 2x) = (4)(3) + (4)(-2x) + (-2x)(3) + (-2x)(-2x)$$
$$= 12 - 8x - 6x + 4x^2$$
$$= 12 - 14x + 4x^2,\ \text{or}\ 4x^2 - 14x + 12$$

So the area of the base is $(4x^2 - 14x + 12)$ ft².

PRACTICE 12

Find the product of $8x^3 + 9x - 1$ and $3x + 7$.

PRACTICE 13

Multiply $p^2 - 2pq + q^2$ by $p - q$.

PRACTICE 14

At the end of 2 years, the amount of money in a savings account is given by

$$(P + Pr) + (P + Pr)r$$

where P is the initial balance and r is the annual rate of interest compounded annually. Rewrite this expression by removing parentheses and simplifying.

5.5   Exercises

Multiply.

1. $(6x)(-4x)$

2. $(-3x)(2x)$

3. $(9t^2)(-t^3)$

4. $(-6y^2)(7y^5)$

5. $(-5x^2)(-4x^4)$

6. $(-2h^2)(-3h^3)$

7. $(10x^3)(-7x^5)$

8. $(20a)(-5a^{99})$

9. $(-4pq^2)(-4p^2qr^2)$

10. $(8st^2)(6s^4t^7)$

11. $(-8x)^2$

12. $(-10p)^2$

13. $\left(\dfrac{1}{2}t^4\right)^3$

14. $\left(\dfrac{1}{3}n^3\right)^3$

15. $(7a)(10a^2)(-5a)$

16. $(-8n^3)(2n^3)(-n)$

17. $(2ab^2)(-3abc)(4a^2)$

18. $(4mn^3)(-m)(7mn^2)$

Find the product.

19. $(7x - 5)\,x$

20. $(y)(3y - 7)$

21. $(9t + t^2)\,(5t)$

22. $(5x - x^2)(4x)$

23. $6a^3(4a^2 - 7a)$

24. $2y(8y - y^3)$

25. $4x^2(3x - 2)$

26. $-5p^7(9p - 3)$

27. $x^3(x^2 - 2x + 4)$

28. $t^2(t^2 + 8t + 1)$

29. $5x(3x^2 + 5x + 6)$

30. $-4x(3x^2 - x + 2)$

31. $(5x^2 - 3x - 7)(-9x)$

32. $(10x^2 + x - 1)(2x)$

33. $6x^2(x^3 + 4x^2 - x - 1)$

34. $(x^3 + 6x^2 - 9x + 10)(-3x^4)$

35. $4p(7q - p^2)$

36. $-pq^2(3p^3 - 9q)$

37. $(v + 3w^2)(-7v)$

38. $(m^2 + n^2)\,3mn^5$

39. $2a^2b^3(3a^4b^2 + 10ab^5)$

40. $-4x^2y^2(7x^2y^3 - 4x^3y^4)$

Simplify.

41. $10x + 2x(-3x + 8)$

42. $-7x + 3x(5x - 9)$

43. $-x + 8x(x^2 - 2x + 1)$

44. $2x(8x^2 + 7x - 2) - 6x^2$

45. $9x(x^2 + 3x - 5) + 8x(-4x^2 + x)$

46. $x(x^2 + 11x) - 10x(2x^2 + 3)$

47. $-4xy(2x^2 + 4xy) + x^2y(7x^2 - 2y)$

48. $2s^3t(5s - t^2) - 7s^2t^2(9s^2 - 10t)$

49. $5a^2b^2(3ab^4 - a^3b^2) + 4a^2b^2(9ab^4 - 10a^3b^2)$

50. $(3p - 8pq)(5p^2) - (4p^3 + 1)(7q)$

Multiply.

51. $(y + 2)(y + 3)$

52. $(x + 1)(x + 4)$

53. $(x - 3)(x - 5)$

54. $(n - 4)(n - 2)$

◉ **55.** $(a - 2)(a + 2)$

56. $(x + 3)(x - 3)$

57. $(w + 3)(2w - 7)$

58. $(8x + 5)(x + 4)$

59. $(3 - 2y)(5y - 1)$

60. $(4u - 1)(3 - 2u)$

61. $(10p - 4)(2p - 1)$

62. $(7x + 1)(7x - 3)$

63. $(u + v)(u - v)$

64. $(x + y)(x - y)$

65. $(2p - q)(q - p)$

66. $(x + 4y)(x - y)$

67. $(3a - b)(a - 2b)$

68. $(5x + 4y)(x - y)$

◉ **69.** $(p - 8)(4q + 3)$

70. $(x + 7)(6y - 1)$

71. $(x - 3)(x^2 - 3x + 1)$

72. $(a + 2)(a^2 - 4a + 4)$

◉ **73.** $(2x - 1)(x^2 + 3x - 5)$

74. $(8n + 3)(2n^2 - 9n - 1)$

75. $(a - b)(a^2 + ab + b^2)$

76. $(x^2 + xy + y^2)(y - x)$

77. $(3x)(x + 5)(x - 7)$

78. $(y^2)(8 - 3y)(8 + y)$

79. $(3n)(2n - 1)(2n + 1)$

80. $(-a)(a + 2b)(a - 3b)$

Mixed Practice

Simplify.

81. $m^2n(8m^2 - 6n) - 2mn(3m^2 - 7mn)$

82. $8t^2 + 5t(3t^2 - 2t + 1)$

Multiply.

83. $(-8s^3)(-5s^4)$

84. $3p(8p - p^2)$

85. $(9y^2 + 7y - 8)(-4y)$

86. $(u - 5)(u - 7)$

87. $(3x^4)(-8x^5)(x^2)$

88. $(2w - 3)(4 - w)$

89. $-a^2b(4a - 2b^3)$

90. $(3h^2k^7)(-9hk^5l)$

91. $(x - 1)(x^2 - 3x + 2)$

92. $(a + 6)(8a - 3)$

Applications

Solve.

93. Backgammon is one of the world's oldest board games. The length and width of the distinctive board (shown below) differ by 30 mm. Find the area of the board in terms of x without using parentheses.

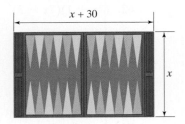

$x + 30$

x

94. To lift an object of mass m from level y_1 to level y_2 requires energy in the amount $mg(y_2 - y_1)$. Write this expression without parentheses.

95. A factory has been selling 1000 color laser printers per year for $1500 each. The company's market research indicates that for each $100 that the price is raised, sales will fall by 30 units. The expression $(1500 + 100x)(1000 - 30x)$ gives the estimated revenue that the company will take in if it adjusts the price of a printer, where x represents the number of $100 increases in the price. Rewrite this expression, multiplying out the factors.

96. A company's total revenue R is given by the equation

$$R = px$$

where p is the price of each item and x is the number of items sold. Write a polynomial for the revenue if

$$x = -\frac{1}{4}p + 100.$$

97. Investment brokers use the formula $A = P(1 + r)^t$ for the amount of money A in a client's account that earns compound interest. In this formula, P is the principal (the original amount of money that the client invested), r is the rate of return per time period (in decimal form), and t is the number of time periods that the money has been invested.

a. If the client invested $5000 for 3 periods, write this formula as a polynomial in r without parentheses.

b. If a client invested $5000, how much greater is the amount of money in the account after 3 periods as compared with 2 periods? Write your answer without parentheses.

c. If the client's rate of return on the investment is 10%, how much money is represented by the expression in part (b)?

98. The expression for the volume of a sphere is $\frac{4}{3}\pi r^3$, where r is the radius of the sphere. Assume that the shape of the Earth is approximately a sphere of radius r miles.

a. Find and simplify the expression for the volume of the sphere formed by everywhere rising 5 mi above the surface of the Earth.

b. Find and simplify the expression for the volume of the sphere formed by everywhere descending 5 mi below the surface of the Earth.

Not to scale

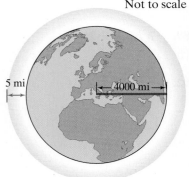

5 mi   4000 mi

c. According to some scientists, all the life discovered so far in the universe is found in the layer around the Earth that extends 5 mi above the Earth's surface to 5 mi below. What is the volume of this layer?

d. If the radius of the Earth is approximately 4000 mi, find the approximate volume of the layer described in part (c).

● *Check your answers on page A-29.*

MINDSTRETCHERS

Mathematical Reasoning

1. We can draw rectangles to visualize the product of two binomials. Consider, for example, the product $(x + 9)(x + 2)$. Using the diagram shown, explain how the product can be expressed as the sum of four areas.

	x	9
x	x^2	$9x$
2	$2x$	18

Compare your answer with the result of using the FOIL method to multiply the two binomials.

Patterns

2. Simplify the following polynomials:

$$(x + y)^0 =$$
$$(x + y)^1 =$$
$$(x + y)^2 =$$
$$(x + y)^3 =$$

In the following table, enter the coefficients from these polynomials. What pattern (known as Pascal's Triangle) do you observe in this table?

Historical

3. Lattice multiplication is a procedure that originated in India in the twelfth century. In this procedure, whole numbers are multiplied as if they were binomials. Each digit of each factor is multiplied separately. The products are recorded in little cells within a lattice, and then added along the diagonals. For instance, to find 29 · 47, the four products are placed in the small, diagonally split squares. The product of 2 and 4, shown in red, is 8.

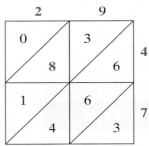

In the following diagram, the square in the upper left shows $^0/_8$, which represents 8. Since the product of 9 and 4 is 36, the square in the upper right contains $^3/_6$. The lower products are $^1/_4$ or 14 and $^6/_3$ or 63. These products are all added along the diagonal. For instance in the diagonal shaded green, the sum 6 + 6 + 4 is 16; the 6 is written below and the 1 is regrouped into the diagonal above and added into that diagonal: 1 + (3 + 8 + 1). The product 1363 appears down the left side of the lattice and across the bottom.

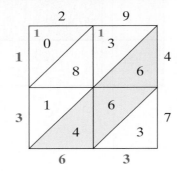

Use lattice multiplication to find the following products:

a. 53 · 89

b. 61 · 94

5.6 Special Products

Recall that in the previous section, we discussed two ways of multiplying binomials—first by directly applying the distributive property, and second by using the FOIL method. In this section, we focus on yet a third approach—using formulas. These formulas provide a shortcut for finding special products, that is, special cases of multiplying binomials.

OBJECTIVES

■ To use the formulas for squaring a binomial

■ To use the formula for multiplying the sum and the difference of two terms

■ To solve applied problems involving these formulas

The Square of a Binomial

To square a binomial means to multiply that binomial by itself. Consider, for instance, $(x + 5)^2$, in which we square the *sum* of two terms. We can rewrite this expression as a binomial multiplied by itself and then use the FOIL method to find the product.

$$(x + 5)^2 = (x + 5)(x + 5)$$
$$= x^2 + 5x + 5x + 25$$
$$= x^2 + 10x + 25$$

Note that the expression $x^2 + 10x + 25$ is equal to $(x)^2 + 2(x)(5) + (5)^2$. So the square of the sum of x and 5 equals the square of x plus twice the product of x and 5 plus the square of 5. This observation leads us to the formula for squaring the sum of two terms.

The Square of a Sum

$$(a + b)^2 = a^2 + 2ab + b^2$$

This formula states that the square of the *sum* of two terms is equal to the square of the first term plus twice the product of the two terms plus the square of the second term. Can you explain why this rule follows from the FOIL method of multiplying binomials?

Memorize the formula for the square of a sum now. With sufficient practice, you will be able to square a binomial sum mentally.

EXAMPLE 1

Simplify: $(x + 7)^2$

Solution Since we are squaring the sum of x and 7, we can use the formula for the square of a sum.

$$(x + 7)^2 = x^2 + 2(x)(7) + (7)^2$$
$$= x^2 + 14x + 49$$

PRACTICE 1

Simplify: $(p + 10)^2$

Note in Example 1 that $(x + 7)^2 = x^2 + 14x + 49$, whereas $x^2 + 7^2 = x^2 + 49$. So we see that $(x + 7)^2 \neq x^2 + (7)^2$.

Tip It is important to distinguish between the *square of the sum* of two terms and the *sum of the squares* of the terms: $(x + y)^2 \neq x^2 + y^2$

EXAMPLE 2	PRACTICE 2
Simplify: $(p + q)^2$	Simplify: $(s + t)^2$

Solution In $(p + q)^2$, the first term is p and the second term is q.

First term	Second term	The square of the first term		The square of the second term
↓	↓	↓		↓

$$(p \;\; + \;\; q)^2 \;\; = \;\; p^2 \;\; + \;\; \underbrace{2(p)(q)}_{\text{Twice the product of the two terms}} \;\; + \;\; q^2$$

$$= p^2 + 2pq + q^2$$

EXAMPLE 3	PRACTICE 3
Simplify: $(2x + 3y)^2$	Simplify: $(4p + 5q)^2$

Solution

$$(2x + 3y)^2 = (2x)^2 + 2(2x)(3y) + (3y)^2$$
$$= 4x^2 + 12xy + 9y^2$$

Now, let's examine a second and related formula, which involves the square of the *difference* of two terms rather than their sum. For instance, consider $(x - 5)^2$. Again, we rewrite the square of the binomial and then apply the FOIL method:

$$(x - 5)^2 = (x - 5)(x - 5)$$
$$= x^2 - 5x - 5x + 25$$
$$= x^2 - 10x + 25$$

Note that the expression $x^2 - 10x + 25$ is equal to $x^2 - 2(x)(5) + 25$. So the square of the difference of x and 5 equals the square of x minus twice the product of x and 5 plus the square of 5. This example leads us to the formula for squaring the difference of two terms.

The Square of a Difference

$$(a - b)^2 = a^2 - 2ab + b^2$$

This formula states that the square of the difference of two terms is equal to the square of the first term minus twice the product of the two terms plus the square of the second term.

If we compare the formula for the square of a sum with the formula for the square of a difference, we see that the signs of the middle term of the resulting trinomial differ. That is, for $(a + b)^2$, the middle term is positive, whereas for $(a - b)^2$, the middle term is negative.

EXAMPLE 4

Simplify: $(8a - 1)^2$

Solution Here we are squaring the difference of two terms, so we use the formula for the square of a difference.

First term → Second term → The square of the first term → The square of the second term

$$(8a - 1)^2 = (8a)^2 - \underbrace{2(8a)(1)}_{\substack{\text{Twice the product of} \\ \text{the two terms}}} + 1^2$$

$$= 64a^2 - 16a + 1$$

PRACTICE 4

Simplify: $(5x - 2)^2$

Another way to simplify the expression $(8a - 1)^2$ in Example 4 is to rewrite $(8a - 1)^2$ as $[8a + (-1)]^2$, and then apply the formula for squaring a sum. Can you explain how we would get the same answer with this approach?

EXAMPLE 5

Simplify: $(p - q)^2$

Solution
$$(p - q)^2 = p^2 - 2(p)(q) + q^2$$
$$= p^2 - 2pq + q^2$$

PRACTICE 5

Simplify: $(u - v)^2$

EXAMPLE 6

Simplify: $(3a - 4b)^2$

Solution
$$(3a - 4b)^2 = (3a)^2 - 2(3a)(4b) + (4b)^2$$
$$= 9a^2 - 24ab + 16b^2$$

PRACTICE 6

Simplify: $(2x - 9y)^2$

The Product of the Sum and Difference of Two Terms

The third special binomial formula relates to multiplying *the sum of two terms by the difference of the same two terms*. Explain why neither of the two previous formulas applies in this situation.

For example, consider the product $(x + 5)(x - 5)$. Using the FOIL method we get:

$$(x + 5)(x - 5) = x \cdot x + x \cdot (-5) + 5 \cdot x - 5 \cdot 5$$
$$= x^2 - \underbrace{5x + 5x}_{} - 25$$

The middle terms cancel each other out.

$$= x^2 - 25$$

> ### The Product of the Sum and Difference of Two Terms
> $$(a + b)(a - b) = a^2 - b^2$$

This formula states that the product of the sum and difference of the *same* two terms is equal to the square of the first term minus the square of the second term.

EXAMPLE 7

Multiply: $(x + 11)(x - 11)$

Solution

First term	Second term		The square of the first term	The square of the second term
↓	↓		↓	↓

$$(x \; + \; 11)(x - 11) \quad = \quad x^2 \quad - \quad (11)^2$$
$$= x^2 - 121$$

PRACTICE 7

Find the product of $(t + 10)$ and $(t - 10)$.

EXAMPLE 8

Multiply.

a. $(p - q)(p + q)$ **b.** $(3m + 2n)(3m - 2n)$

Solution

a. Since $(p - q)(p + q) = (p + q)(p - q)$, the formula for finding the product of the sum and difference of two terms applies.
$$(p - q)(p + q) = p^2 - q^2$$

b. $(3m + 2n)(3m - 2n) = (3m)^2 - (2n)^2 = 9m^2 - 4n^2$

PRACTICE 8

Find the product.

a. $(r - s)(r + s)$

b. $(8s - 3t)(8s + 3t)$

EXAMPLE 9

Find the product: $(3a^2 - 5)(3a^2 + 5)$

Solution
$$(3a^2 - 5)(3a^2 + 5) = (3a^2)^2 - (5)^2 = 9a^4 - 25$$

PRACTICE 9

Multiply: $(10 - 7k^2)(10 + 7k^2)$

Were you able to find the products in the last few examples mentally? If not, set this as a personal goal when doing similar exercises at the end of this section.

EXAMPLE 10

The nineteenth-century French physician Jean Louis Poiseuille investigated the flow of blood in the smaller blood vessels of the body. He discovered that the speed of blood varies from point to point within a blood vessel. In a blood vessel of radius r at a point b units from the center of the blood vessel, the blood flow speed is given by the expression $k(r + b)(r - b)$, where k is a constant. Write this expression without parentheses.

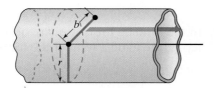

Solution We need to multiply out $k(r + b)(r - b)$.

$$k(r + b)(r - b) = k(r^2 - b^2)$$
$$= kr^2 - kb^2$$

So the expression is $kr^2 - kb^2$.

PRACTICE 10

The area of the square wooden frame shown can be represented by the polynomial $(S + s)(S - s)$, where s is the side length of the smaller square and S is the side length of the larger square. Write this expression without parentheses.

Mathematically Speaking

Fill in each blank with the most appropriate term or phrase from the given list.

minus	plus	positive
negative	times	divided by

1. The square of the sum of two terms is equal to the square of the first term _____ twice the product of the two terms plus the square of the second term.

2. In the formula for the square of the sum of two terms, the middle term of the trinomial is _____.

3. In the formula for the square of a difference of two terms, the middle term of the trinomial is _____.

4. The product of the sum and difference of the same two terms is equal to the square of the first term _____ the square of the second term.

Simplify.

5. $(y + 2)^2$

6. $(a + 3)^2$

7. $(x + 4)^2$

8. $(n + 8)^2$

9. $(x - 11)^2$

10. $(b - 10)^2$

11. $(6 - n)^2$

12. $(9 - y)^2$

13. $(x + y)^2$

14. $(s - t)^2$

15. $(3x + 1)^2$

16. $(5x + 3)^2$

17. $(4n - 5)^2$

18. $(2x - 1)^2$

19. $(9x + 2)^2$

20. $(11m + 3)^2$

21. $\left(a + \dfrac{1}{2}\right)^2$

22. $(b - 0.2)^2$

23. $(8b + c)^2$

24. $(3s + t)^2$

25. $(5x - 2y)^2$

26. $(3m - 4n)^2$

27. $(-x + 3y)^2$

28. $(-p + 2q)^2$

29. $(4x^3 + y^4)^2$

30. $(5a^2 - c^2)^2$

Multiply.

31. $(a + 1)(a - 1)$

32. $(8 + r)(8 - r)$

33. $(4x - 3)(4x + 3)$

34. $(7y - 2)(7y + 2)$

35. $(10 + 3y)(3y - 10)$

36. $(-1 + 9x)(9x + 1)$

37. $\left(m - \dfrac{1}{2}\right)\left(m + \dfrac{1}{2}\right)$

38. $(n + 0.3)(n - 0.3)$

39. $(4a + b)(4a - b)$

40. $(p - 3q)(p + 3q)$

41. $(3x - 2y)(3x + 2y)$

42. $(10t + 3s)(10t - 3s)$

43. $(1 - 5n)(5n + 1)$

44. $(2s + 3)(3 - 2s)$

45. $x(x + 5)(x - 5)$

46. $y(y - 7)(y + 7)$

47. $5n^2(n + 7)^2$

48. $-8y^3(2y - 1)^2$

49. $(n^2 - m^4)(n^2 + m^4)$

50. $(x^3 + y^5)(x^3 - y^5)$

51. $(a - b)(a + b)(a^2 + b^2)$

52. $(x + y)(x - y)(x^2 + y^2)$

Mixed Practice

Simplify each expression.

53. $(6n + 4)^2$

54. $(2h - 7k)(2h + 7k)$

55. $(4p - 9)(4p + 9)$

56. $(w + 7)^2$

57. $(8 - a)^2$

58. $(3a + 8)(8 - 3a)$

59. $-2x^2(4x - 3)^2$

60. $(3x - 5y)^2$

Applications

Solve.

61. A city laid out on a grid plans for a square-shaped park, as shown. The area of the park can be modeled by the expression $(x - 5)^2 + (y - 1)^2$. Rewrite this expression by removing parentheses and simplifying.

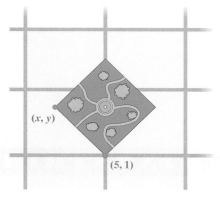

62. A student deposits $100 in an account at interest rate r (in decimal form). If the interest earned is compounded annually, the amount in the account after 2 years is given by

$$A = 100(1 + r)^2$$

Write A as a polynomial in r without parentheses.

63. An investment of A dollars increases in value by P% for each of two years. The value of the investment at the end of the two years can be represented by

$$A\left(1 + \frac{P}{100}\right)\left(1 + \frac{P}{100}\right)$$

Multiply out this expression.

64. In measuring the lengths of the sides of a square piece of wood, a carpenter can be off by length e. The carpenter measures the side of the wooden square to be x.

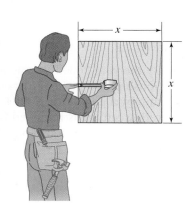

a. What are the longest possible true dimensions of the wooden square? The shortest possible true dimensions?

b. What is the difference in area between the longest possible wooden square and the shortest possible wooden square in terms of e and x?

65. To measure the spread of data, statisticians compute the sample variance of the data. For a sample of size 3, they use the following formula:

$$\text{Sample variance} = \frac{(a - m)^2 + (b - m)^2 + (c - m)^2}{2}$$

Rewrite this formula without parentheses, combining like terms.

66. As the temperature of a lightbulb's filament changes from T_1 to T_2, the energy that the filament radiates changes by the quantity

$$a(T_1 - T_2)(T_1 + T_2)(T_1^2 + T_2^2)$$

Multiply to find this change.

• *Check your answers on page A-29.*

MINDSTRETCHERS

Patterns

1. Mentally compute each product. (*Hint:* Think of these computations as "special products.")

a. $9999 \times 10{,}001$

b. $30\frac{1}{10} \times 29\frac{9}{10}$

Mathematical Reasoning

2. Suppose that you square two consecutive whole numbers and subtract the smaller square from the larger. Is it possible that the difference is an even number? Explain your answer.

Historical

3. By the year 2000 B.C., the astronomers of Mesopotamia knew the relationship $(a - b)(a + b) = a^2 - b^2$. They could demonstrate this relationship by a geometric model. Find the area of the remaining region if the yellow square is removed from the figure shown. Show that this model verifies the relationship $(a - b)(a + b) = a^2 - b^2$. (*Hint:* Find the area in two ways.)

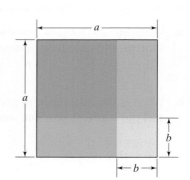

5.7 Division of Polynomials

In the final section of this chapter, we deal with dividing polynomials. We progress from dividing a monomial by another monomial to dividing a polynomial by a binomial.

OBJECTIVES

- To divide monomials
- To divide a polynomial by a monomial
- To divide a polynomial by a binomial
- To solve applied problems involving the division of polynomials

Dividing Monomials

The ability to divide monomials—the simplest of polynomial divisions—depends on our knowledge of both fractions and exponents, as the following example illustrates.

EXAMPLE 1

Simplify: $8x^6 \div 4x^2$

Solution We begin by rewriting the problem in fractional form.

$$8x^6 \div 4x^2 = \frac{8x^6}{4x^2}$$

$$\frac{8x^6}{4x^2} = \frac{8}{4} \cdot \frac{x^6}{x^2}$$

$$= 2x^{6-2} \quad \text{**Divide the coefficients and use the quotient rule of exponents.**}$$

$$= 2x^4$$

PRACTICE 1

Find the quotient: $-12n^6 \div 3n$

EXAMPLE 2

Divide: $\dfrac{-15a^2b^3c}{10ab^2}$

Solution

$$\frac{-15a^2b^3c}{10ab^2} = \frac{-15}{10} \cdot \frac{a^2b^3c}{ab^2}$$

$$= \frac{-3}{2} \cdot a^{2-1}b^{3-2}c$$

$$= \frac{-3}{2}abc$$

$$= -\frac{3abc}{2}$$

PRACTICE 2

Find the quotient:
$(20p^3q^2r^4) \div (-5p^2q^2r)$

Dividing a Polynomial by a Monomial

Now, we extend our discussion of division to finding the quotient of a polynomial divided by a monomial. We consider an example, written in fractional form.

EXAMPLE 3

Simplify the quotient: $(10x^2 + 15x) \div (5x)$

Solution Let's write the given expression in fractional form.

$$(10x^2 + 15x) \div (5x) = \dfrac{\overbrace{10x^2 + 15x}^{\text{Polynomial}}}{\underbrace{5x}_{\text{Monomial}}}$$

Since we are dividing a polynomial by a monomial, we can rewrite the given fraction as the sum of two fractions with the same denominator. Then we simplify each fraction.

$$\dfrac{10x^2 + 15x}{5x} = \dfrac{10x^2}{5x} + \dfrac{15x}{5x}$$
$$= 2x + 3$$

Example 3 suggests that when dividing a polynomial by a monomial, we can divide each term in the polynomial by the monomial and then add. Let's apply this shortcut in the following examples.

PRACTICE 3

Divide: $(21x^3 - 14x^2)$ by $7x$

EXAMPLE 4

Divide: $\dfrac{9x^4 - 6x^3 + 12x^2}{-3x^2}$

Solution Divide each term in the numerator by the denominator. Then add the quotients:

$$\dfrac{9x^4 - 6x^3 + 12x^2}{-3x^2} = \dfrac{9x^4}{-3x^2} - \dfrac{6x^3}{-3x^2} + \dfrac{12x^2}{-3x^2}$$
$$= -3x^2 + 2x - 4$$

PRACTICE 4

Simplify: $\dfrac{14x^8 + 10x^5 - 8x^3}{-2x^3}$

EXAMPLE 5

Divide: $\dfrac{10p^4q^5 + 12p^2q^4 - pq^3}{2pq^2}$

Solution

$$\dfrac{10p^4q^5 + 12p^2q^4 - pq^3}{2pq^2} = \dfrac{10p^4q^5}{2pq^2} + \dfrac{12p^2q^4}{2pq^2} - \dfrac{pq^3}{2pq^2}$$
$$= 5p^3q^3 + 6pq^2 - \dfrac{q}{2}$$

PRACTICE 5

Find the quotient:

$$\dfrac{-5a^7b^6 + a^2b^4 - 15ab^3}{5ab^3}$$

Dividing a Polynomial by a Binomial

Now, let's take a look at the general case—how to divide one polynomial by another polynomial. We will restrict our attention to dividing a polynomial by a *binomial*, but these remarks apply to dividing by polynomials with three or more terms as well.

The procedure for dividing a polynomial by a binomial is similar to that of dividing whole numbers, which is commonly called *long division*.

$$
\begin{array}{r}
\text{Divisor} \longrightarrow \quad 13\overline{)\;158}\;\; \longleftarrow \text{Dividend} \\
\end{array}
$$

$$
\begin{array}{r}
12 \longleftarrow \text{Quotient} \\
13\overline{)\;158} \longleftarrow \text{Dividend} \\
-13 \quad\;\; \\
\hline
28 \;\; \\
-26 \;\; \\
\hline
2 \longleftarrow \text{Remainder}
\end{array}
$$

We stop dividing because the remainder 2 is smaller than the divisor 13. So $158 \div 13 = 12$ with remainder 2.

Now, let's consider the following problem in dividing polynomials:

$$(x^2 - 5x - 24) \div (x + 3)$$

We set up the problem just as if we were dividing whole numbers, being careful to correctly distinguish between the dividend and the divisor. In this case, $x^2 - 5x - 24$ is the dividend and $x + 3$ is the divisor.

$$
\begin{array}{r}
x \qquad\qquad \\
x + 3\overline{)\;x^2 - 5x - 24} \\
\underline{x^2 + 3x} \qquad\quad\;\;
\end{array}
$$

Divide the first term in the dividend by the first term in the divisor: Think $x^2 \div x = x$. Place x in the quotient above the x term in the dividend. Then *multiply* x in the quotient by the divisor: $x(x + 3) = x^2 + 3x$.

$$
\begin{array}{r}
x \qquad\qquad \\
x + 3\overline{)\;x^2 - 5x - 24} \\
\underline{x^2 + 3x} \qquad\quad\;\; \\
-8x - 24
\end{array}
$$

Subtract $(x^2 + 3x)$ from $(x^2 - 5x)$ by changing the signs of each term in $(x^2 + 3x)$ and then adding to get $-8x$. Bring down the next term, **−24**. We have to divide again because the degree of $-8x - 24$ is equal to the degree of $x + 3$. Both have degree 1.

$$
\begin{array}{r}
x - 8 \qquad\;\; \\
x + 3\overline{)\;x^2 - 5x - 24} \\
\underline{x^2 + 3x} \qquad\quad\;\; \\
-8x - 24 \\
\underline{-8x - 24} \\
0
\end{array}
$$

The remainder is 0.

Divide x into $-8x$: Think $-8x \div x = -8$. Place -8 in the quotient above the constant term in the dividend. Then *multiply* -8 in the quotient by the divisor: $-8(x + 3) = -8x - 24$. Finally, *subtract* $(-8x - 24)$ from $(-8x - 24)$ by changing the signs of each term in $(-8x - 24)$ and then adding to get 0.

Note that the degree of the remainder is less than the degree of the divisor because the remainder 0 is a constant, which has degree 0, and the divisor $x + 3$ has degree 1.

So $(x^2 - 5x - 24) \div (x + 3) = x - 8$. Can you explain how you would check this quotient?

This example suggests the following general method for dividing a polynomial by a polynomial.

To Divide a Polynomial by a Polynomial

- Arrange each term of the dividend and divisor in descending order.
- Divide the first term of the dividend by the first term of the divisor. The result is the first term of the quotient.
- Multiply the first term of the quotient by the divisor and place the product under the dividend.
- Subtract the product, found in the previous step, from the dividend.
- Bring down the next term to form a new dividend.
- Repeat the process until the degree of the remainder is less than the degree of the divisor.

EXAMPLE 6

$2x - 1 \overline{)6x^2 + 9x - 6}$

Solution The dividend and divisor are already in descending order.

$$\begin{array}{r} 3x \\ 2x-1\overline{)6x^2+9x-6} \\ \underline{6x^2-3x} \\ 12x-6 \end{array}$$

Divide $6x^2$ by $2x$, getting $3x$.
Multiply $3x$ by $(2x-1)$, getting $(6x^2-3x)$.
Subtract $(6x^2-3x)$ from $(6x^2+9x)$ and bring down -6.

$$\begin{array}{r} 3x+6 \\ 2x-1\overline{)6x^2+9x-6} \\ \underline{6x^2-3x} \\ 12x-6 \\ \underline{12x-6} \\ 0 \end{array}$$

Divide $12x$ by $2x$, getting 6.
Multiply 6 by $(2x-1)$, getting $(12x-6)$.
Subtract $(12x-6)$ from $(12x-6)$.

The degree of 0 is less than the degree of $(2x-1)$, so the process stops.

So $(6x^2 + 9x - 6) \div (2x - 1) = 3x + 6$.

PRACTICE 6

Find the quotient:
$(10x^2 + 17x + 3) \div (5x + 1)$

Tip We place each term in the quotient above a term in the dividend of the same degree.

Now, we focus on problems in dividing polynomials that have *remainders*.

Dividing whole numbers:

$$\begin{array}{r} 21 \\ 45\overline{)956} \\ \underline{90} \\ 56 \\ \underline{45} \\ 11 \end{array} \leftarrow \text{Remainder}$$

Check

$45 \cdot 21 + 11 \stackrel{?}{=} 956$
$956 = 956$ **True**

So $956 \div 45 = 21\frac{11}{45}$.

Dividing polynomials:

$$\begin{array}{r} x+3 \\ x+5\overline{)x^2+8x+16} \\ \underline{x^2+5x} \\ 3x+16 \\ \underline{3x+15} \\ 1 \end{array} \leftarrow \text{Remainder}$$

Check

$(x+5)(x+3) + 1 \stackrel{?}{=} x^2 + 8x + 16$
$x^2 + 8x + 16 = x^2 + 8x + 16$ **True**

So $(x^2 + 8x + 16) \div (x + 5) = x + 3 + \frac{1}{x+5}$.

Note that we can check a problem involving division of polynomials in the same way that we check division of whole numbers:

$$\text{Divisor} \cdot \text{quotient} + \text{remainder} = \text{Dividend}$$

EXAMPLE 7	**PRACTICE 7**
Find the quotient: $(x^3 + 3x^2 - 8x + 2) \div (x + 5)$	Divide $(3x^3 + 7x^2 + 11x + 5)$ by $(3x + 1)$.

Solution

$$
\begin{array}{r}
x^2 - 2x + 2 \\
x + 5 \overline{\smash{)}x^3 + 3x^2 - 8x + 2} \\
\underline{x^3 + 5x^2} \\
-2x^2 - 8x \\
\underline{-2x^2 - 10x} \\
2x + 2 \\
\underline{2x + 10} \\
-8
\end{array}
$$

Since the degree of the remainder is less than the degree of the divisor, we stop.

Check

$$(x + 5)(x^2 - 2x + 2) + (-8) \overset{?}{=} x^3 + 3x^2 - 8x + 2$$
$$x^3 + 3x^2 - 8x + 2 = x^3 + 3x^2 - 8x + 2 \qquad \text{True}$$

So we write the answer as

$$x^2 - 2x + 2 + \frac{-8}{x + 5}$$

Some problems in dividing polynomials involve terms that are not in *descending order*.

EXAMPLE 8	**PRACTICE 8**
Divide $(6 + 8x^2 - 14x)$ by $(2x - 3)$.	Divide: $(-21s + 10 + 9s^2) \div (3s - 2)$

Solution Before dividing, we place the terms in both the divisor and the dividend in descending order. Here we need to rearrange the terms in the dividend:

$$
\begin{array}{r}
4x - 1 \\
2x - 3 \overline{\smash{)}8x^2 - 14x + 6} \\
\underline{8x^2 - 12x} \\
-2x + 6 \\
\underline{-2x + 3} \\
3 \qquad \text{The remainder is 3.}
\end{array}
$$

So $(8x^2 - 14x + 6) \div (2x - 3) = 4x - 1 + \dfrac{3}{2x - 3}$.

We leave the check to you.

In dividing polynomials, we may have *missing terms* in the dividend, as shown in the following example.

EXAMPLE 9	PRACTICE 9

EXAMPLE 9

$x + 3 \overline{)2x^3 + 7x^2 - 9}$

Solution Since there is no x-term in the dividend, we can insert $0x$ as a placeholder for the missing term.

$$
\begin{array}{r}
2x^2 + x - 3 \\
x + 3 \overline{)2x^3 + 7x^2 + 0x - 9} \\
\underline{2x^3 + 6x^2} \\
x^2 + 0x \\
\underline{x^2 + 3x} \\
-3x - 9 \\
\underline{-3x - 9} \\
0
\end{array}
$$

So $\dfrac{2x^3 + 7x^2 - 9}{x + 3} = 2x^2 + x - 3$.

PRACTICE 9

Divide: $\dfrac{4n^3 - 19n^2 - 4}{4n - 3}$

EXAMPLE 10

To find the length l of a rectangular-shaped object, we can use the formula $A = lw$, where the area A and width w are given.

a. Solve $A = lw$ for l.

b. If the area of the rectangular top of a billiard table is given by the polynomial $(5x^2 + 13x + 6)$ ft, find the length when the width is $(x + 2)$ ft.

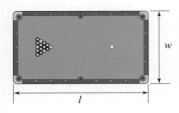

Solution

a. Since $A = lw$, $l = \dfrac{A}{w}$.

b. Using $l = \dfrac{A}{w}$, we conclude that $l = \dfrac{5x^2 + 13x + 6}{x + 2}$.

Dividing the numerator by the denominator, we find that $l = 5x + 3$. So the length is $(5x + 3)$ ft.

PRACTICE 10

If $10 is invested at an interest rate of r per year and compounded annually, the future value S in dollars at the end of the nth year is given by

$$S = 10(1 + r)^n$$

a. What is the future value of the investment after 1 yr? After 2 yr?

b. Write the answers to part (a) without parentheses and in descending order.

c. Using your answer in part (b), determine how many times as great the future value of the investment is after 2 yr as compared to the future value after 1 yr.

Mathematically Speaking

Fill in each blank with the most appropriate term or phrase from the given list.

powers	polynomial by the monomial	quotient
next degree higher	coefficients	same degree
remainder	monomial by the polynomial	

1. To divide a monomial by a monomial, divide the _____ and use the quotient rule of exponents.

2. To divide a polynomial by a monomial, divide each term in the _____ and then add.

3. In dividing one polynomial by another, repeat the process until the degree of the _____ is less than the degree of the divisor.

4. When dividing a polynomial by a polynomial, place each term in the quotient above a term in the dividend of the _____.

Simplify.

5. $\dfrac{10x^4}{5x^2}$

6. $\dfrac{6x^3}{2x^2}$

7. $\dfrac{16a^8}{-4a}$

8. $\dfrac{-35y^2}{7y}$

9. $\dfrac{-8x^5}{-6x^4}$

10. $\dfrac{-9p^5}{-12p^2}$

11. $\dfrac{12p^2q^3}{3p^2q}$

12. $\dfrac{9a^4b}{3a^2b}$

13. $\dfrac{-24u^6v^4}{-8u^4v^2}$

14. $\dfrac{-4x^5y^4}{-2x^2y}$

15. $\dfrac{-15a^2b^5}{7ab^3}$

16. $\dfrac{21x^3y^5}{-10xy^2}$

17. $\dfrac{-6u^5v^3w^3}{4u^2vw^3}$

18. $\dfrac{-10x^3yz^2}{8x^2yz}$

Divide.

19. $\dfrac{6n^2 + 10n}{2n}$

20. $\dfrac{12m^4 + 15m^3}{3m}$

21. $\dfrac{20b^4 - 10b}{10b}$

22. $\dfrac{2x^2 - 8x}{2x}$

23. $\dfrac{18a^2 + 12a}{-3a}$

24. $\dfrac{16x^2 + 10x}{-2x}$

25. $\dfrac{9x^5 - 6x^7}{3x^5}$

26. $\dfrac{6a^3 - 4a^2}{2a^2}$

27. $\dfrac{12a^4 - 18a^3 + 30a^2}{6a^2}$

28. $\dfrac{8x^5 + 4x^4 - 16x^3}{4x^2}$

29. $\dfrac{n^5 - 10n^4 - 5n^3}{-5n^3}$

30. $\dfrac{9y^6 - 3y^5 - 2y^4}{-3y^4}$

31. $\dfrac{20a^2b + 4ab^3}{8ab}$

32. $\dfrac{14xy^2 - 21x^3y^4}{7xy^2}$

33. $\dfrac{12x^2y^3 - 9xy - 3xy^2}{-3xy}$

34. $\dfrac{10ab^8 - 4ab^6 + 6ab^4}{-2ab^4}$

35. $\dfrac{8p^2q^3 - 4p^3q^3 + 6p^4q}{4p^2q}$

36. $\dfrac{6x^2y^3 - 18x^3y^4 + 9x^4y^5}{6x^2y^3}$

Find the quotient.

37. $(x^2 - 4x - 21) \div (x + 3)$

38. $(x^2 + 6x - 40) \div (x + 10)$

39. $(56x^2 - 23x + 2) \div (8x - 1)$

40. $(30x^2 + 13x - 3) \div (6x - 1)$

41. $(6x^2 + 13x - 5) \div (2x + 5)$

42. $(10x^2 - x - 2) \div (5x + 2)$

43. $(-2x + 5x^2 - 3) \div (x - 1)$

44. $(19x + 2x^2 + 35) \div (x + 7)$

45. $(4 + 20x + 21x^2) \div (2 + 3x)$

46. $(-3 + x + 2x^2) \div (3 + 2x)$

Divide.

47. $\dfrac{x^2 + 2x + 5}{x + 2}$

48. $\dfrac{x^2 + 2x + 7}{x - 2}$

49. $\dfrac{-3 - 5x + 2x^2}{x - 3}$

50. $\dfrac{-2x + 5x^2 - 3}{x - 1}$

51. $\dfrac{8x^2 - 6x - 11}{4x + 3}$

52. $\dfrac{3x^2 - x - 8}{3x - 1}$

53. $\dfrac{-x + x^3 - 5x^2 + 5}{x + 1}$

54. $\dfrac{-5 + 11x - 7x^2 + x^3}{x - 5}$

55. $\dfrac{6x^3 - 11x^2 - 5x + 19}{3x - 4}$

56. $\dfrac{2x^3 + x^2 - 4x - 8}{2x + 1}$

57. $\dfrac{5x^2 - 2}{x - 4}$

58. $\dfrac{10x^2 - 2x}{x + 3}$

59. $\dfrac{4x^3 - x + 3}{2x - 3}$

60. $\dfrac{3x^3 + x^2 - 4}{x + 1}$

61. $\dfrac{x^3 + 27}{x + 3}$

62. $\dfrac{x^3 - 1}{x - 1}$

Mixed Practice

Simplify.

63. $\dfrac{56r^2}{-8r}$

64. $\dfrac{42x^5y^2}{7xy^2}$

65. $\dfrac{-18a^2b^3c}{27ab^2c}$

Divide.

66. $\dfrac{15a^8b - 21a^4b - 24a^3b}{3a^3b}$

67. $\dfrac{18n^6 - 48n^4 - 2n^3}{-6n^3}$

68. $(36x^2 + x - 2) \div (4x + 1)$

69. $(5 + 13x + 6x^2) \div (5 + 3x)$

70. $\dfrac{10t^5 - 6t^3}{2t}$

71. $\dfrac{4x^3 + x^2 - 4x + 2}{4x + 1}$

72. $\dfrac{4x^2 - 5x}{x - 3}$

73. $\dfrac{32m^3 - 72m}{-8m}$

74. $\dfrac{2x^2 - 13x - 24}{x - 8}$

Applications

Solve.

75. A homeowner wishes to increase the length of a flower bed by twice as much as the width. The area of the new flower bed is given by $(2x^2 + 20x + 48)$ square feet.

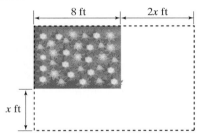

    a. Use long division to find the width of the new flower bed in terms of x.

    b. If the homeowner increases the width by 1 foot, then what are the dimensions of the new flower bed?

76. A prepaid phone card company charges a $0.79 connection fee for each call plus $0.10 for each minute. The average cost per minute of each call is given by $\dfrac{0.1x + 0.79}{x}$, where x is the number of minutes per call. Use division to rewrite this expression.

77. The formula $d = rt$ can be used to find the time t when given the distance d and the rate r.

    a. Solve the equation $d = rt$ for t.

    b. Use the answer from part (a) to find an expression for the time it takes to travel a distance of $(t^3 - 6t^2 + 7t + 14)$ mi at a rate of $(t + 1)$ mph.

78. A *geometric series* is a sum of terms where each term is formed by multiplying the previous term by a constant. For example, the series $5 + 5r + 5r^2$ has three terms where the first term is 5 and each of the other terms is r times the previous term. Use long division to show that the sum of the first three terms can be calculated from the formula $\dfrac{-5r^3 + 5}{-r + 1}$.

79. A city or county cell telephone system is divided into cells, each of which is equipped with a low-powered radio transmitter/receiver. When a cell phone moves from one cell toward another, a computer transfers the phone call to the new cell. If x is the number of years since 1984, then the number of cell systems in the United States can be modeled by the polynomial $200x - 300$ and the number of thousands of U.S. cell phone subscribers can be approximated by $600x^2 - 3700x + 4400$. Find a polynomial to model the average number of subscribers per cell system. (Ignore any remainder.) (*Source:* Cellular Telecommunications and Internet Association)

80. The United States is the world's leading producer of wheat. If x is the number of years since 1997, then the annual wheat production (in millions of bushels) of leading American states can be modeled by the polynomial $8x^3 - 150x^2 + 218x + 2520$ and the wheat acreage harvested annually (in millions of acres) by $-4x + 63$. Find a polynomial that represents the annual yield per acre of wheat harvested. (Ignore any remainder.) (*Source:* U.S. Department of Agriculture)

● *Check your answers on page A-30.*

MINDSTRETCHERS

Mathematical Reasoning

1. When you divide a polynomial by a trinomial, explain how you know if the trinomial is a factor of the polynomial.

Patterns

2. Consider the following table.

Divisor	Dividend	Quotient
$x + 1$	$x^2 - x + 1$	
$x + 1$	$x^3 - x^2 + x - 1$	
$x + 1$	$x^4 - x^3 + x^2 - x + 1$	

a. Complete the table by dividing each dividend by the divisor.

b. Predict the result of dividing $x^5 - x^4 + x^3 - x^2 + x - 1$ by $x + 1$. Verify your prediction.

Critical Thinking

3. Divide $(2y^2 + 5y + 3)$ by $(y + 1)$. For which values of y is the quotient larger in value than the divisor?

KEY CONCEPTS AND SKILLS   CONCEPT   SKILL

Concept/Skill	Description	Example
[5.1] **Exponent (or Power)**	A number that indicates how many times another number (called the *base*) is used as a factor.	$4^3 = 4 \cdot 4 \cdot 4$ Exponent, 3 factors, Base
[5.1] **Exponent of 1**	For any real number x, $x^1 = x$.	$2^1 = 2$ $(ab)^1 = ab$
[5.1] **Exponent of 0**	For any nonzero real number x, $x^0 = 1$.	$5^0 = 1$ $(ab)^0 = 1$
[5.1] **Product Rule of Exponents**	For any nonzero real number x and for any integers a and b, $$x^a \cdot x^b = x^{a+b}$$	$3^2 \cdot 3^3 = 3^{2+3} = 3^5 = 243$ $x \cdot x = x^{1+1} = x^2$
[5.1] **Quotient Rule of Exponents**	For any nonzero real number x and for any integers a and b, $$\frac{x^a}{x^b} = x^{a-b}$$	$2^5 \div 2^3 = \frac{2^5}{2^3} = 2^{5-3} = 2^2 = 4$ $\frac{x^6}{x} = x^{6-1} = x^5$
[5.1] **Negative Exponents**	For any nonzero real number x and for any integer a, $$x^{-a} = \frac{1}{x^a}$$	$6^{-2} = \frac{1}{6^2} = \frac{1}{36}$ $(2x)^{-2} = \frac{1}{(2x)^2} = \frac{1}{4x^2}$
[5.1] **Reciprocal of x^{-a}**	For any nonzero real number x and for any integer a, $$\frac{1}{x^{-a}} = x^a$$	$\frac{1}{5^{-3}} = 5^3 = 125$ $\frac{x^3}{y^{-2}} = x^3 y^2$
[5.2] **Power Rule of Exponents**	For any nonzero real number x and for any integers a and b, $$(x^a)^b = x^{ab}$$	$(2^3)^2 = 2^{3 \cdot 2} = 2^6 = 64$ $(p^7)^3 = p^{7 \cdot 3} = p^{21}$
[5.2] **Raising a Product to a Power**	For any nonzero real numbers x and y and any integer a, $$(xy)^a = x^a \cdot y^a$$	$(4y)^3 = 4^3 y^3 = 64y^3$ $(c^4 d^3)^5 = c^{4 \cdot 5} d^{3 \cdot 5}$ $= c^{20} d^{15}$
[5.2] **Raising a Quotient to a Power**	For any nonzero real numbers x and y and any integer a, $$\left(\frac{x}{y}\right)^a = \frac{x^a}{y^a}$$	$\left(\frac{2}{3}\right)^3 = \frac{2^3}{3^3} = \frac{8}{27}$ $\left(\frac{-5}{b}\right)^4 = \frac{(-5)^4}{b^4} = \frac{625}{b^4}$

continued

Concept/Skill	Description	Example
[5.2] **Raising a Quotient to a Negative Power**	For any nonzero real numbers x and y and any integer a, $$\left(\frac{x}{y}\right)^{-a} = \left(\frac{y}{x}\right)^{a}$$	$$\left(\frac{2}{3}\right)^{-1} = \left(\frac{3}{2}\right)^{1} = \frac{3}{2}$$ $$\left(\frac{p}{q}\right)^{-3} = \left(\frac{q}{p}\right)^{3} = \frac{q^3}{p^3}$$
[5.2] **Scientific Notation**	A number is in scientific notation if it is written in the form $$a \times 10^n$$ where n is an integer and a is greater than or equal to 1 but less than 10 ($1 \le a < 10$).	5.3×10^9 and 2.41×10^{-5} are in scientific notation.
[5.3] **Monomial**	An expression that is the product of a real number and variables raised to nonnegative integer powers.	$3x^3$ $-4a^2b$
[5.3] **Polynomial**	An algebraic expression with one or more monomials added or subtracted.	$-5x$ ⟵ Monomial $2x + 1$ ⟵ Binomial $x^2 - 9x + 2$ ⟵ Trinomial $-x^3 + 7x^2 - x + 19$ ⟵ Polynomial
[5.4] **To Add Polynomials**	• Add the like terms.	Find the sum horizontally: $(5x^2 + 6x - 9) + (-10x^2 + 7)$ $= 5x^2 + 6x - 9 - 10x^2 + 7$ $= -5x^2 + 6x - 2$ Add vertically: $(5x^2 + 6x - 9) + (-10x^2 + 7)$ $$\begin{array}{r}5x^2 + 6x - 9 \\ -10x^2 \qquad + 7 \\ \hline -5x^2 + 6x - 2\end{array}$$
[5.4] **To Subtract Polynomials**	• Change the signs of each term of the polynomial being subtracted. • Add.	Subtract horizontally: $(2x^2 - 6x + 1) - (-x^2 + 4x + 5)$ $= 2x^2 - 6x + 1 + x^2 - 4x - 5$ $= 3x^2 - 10x - 4$ Subtract vertically: $(2x^2 - 6x + 1) - (-x^2 + 4x + 5)$ $$\begin{array}{rr}2x^2 - 6x + 1 & 2x^2 - 6x + 1 \\ -(-x^2 + 4x + 5) & x^2 - 4x - 5 \\ \hline & 3x^2 - 10x - 4\end{array}$$
[5.5] **To Multiply Monomials**	• Multiply the coefficients. • Multiply the variables, using the product rule of exponents.	Multiply: $(-3x^2y)(5x^3y^2)$ $= (-3 \cdot 5)(x^2 \cdot x^3)(y \cdot y^2) = -15x^5y^3$

continued

Concept/Skill	Description	Example
[5.5] To Multiply Two Binomials Using the FOIL Method	Consider $(a + b)(c + d)$. • Multiply the two first terms in the binomials. $(a + b)(c + d)$ Product is ac. **F** • Multiply the two outer terms. $(a + b)(c + d)$ Product is ad. **O** • Multiply the two inner terms. $(a + b)(c + d)$ Product is bc. **I** • Multiply the two last terms. $(a + b)(c + d)$ Product is bd. **L** The product of the two binomials is the sum of these four products. $(a + b)(c + d) = ac + ad + bc + bd$	Last First $(3x - 1)(2x + 5)$ Inner Outer **F** $(3x)(2x) = 6x^2$ **O** $(3x)(5) = 15x$ **I** $(-1)(2x) = -2x$ **L** $(-1)(5) = -5$ The product $(3x - 1)(2x + 5)$ is the sum of these four products: $6x^2 + 15x - 2x - 5$ $= 6x^2 + 13x - 5$
[5.6] The Square of a Sum	$(a + b)^2 = a^2 + 2ab + b^2$	$(x + 3)^2 = x^2 + 2(x)(3) + (3)^2$ $= x^2 + 6x + 9$ $(7y + 5)^2 = (7y)^2 + 2(7y)(5) + (5)^2$ $= 49y^2 + 70y + 25$
[5.6] The Square of a Difference	$(a - b)^2 = a^2 - 2ab + b^2$	$(y - 6)^2 = y^2 - 2(y)(6) + (6)^2$ $= y^2 - 12y + 36$ $(4x - 1)^2 = (4x)^2 - 2(4x)(1) + (1)^2$ $= 16x^2 - 8x + 1$
[5.6] The Product of the Sum and Difference of Two Terms	$(a + b)(a - b) = a^2 - b^2$	$(2t + s)(2t - s) = (2t)^2 - (s)^2$ $= 4t^2 - s^2$
[5.7] To Divide a Polynomial by a Polynomial	• Arrange each term of the dividend and divisor in descending order. • Divide the first term of the dividend by the first term of the divisor. The result is the first term of the quotient. • Multiply the first term of the quotient by the divisor and place the product under the dividend. • Subtract the product, found in the previous step, from the dividend. • Bring down the next term to form a new dividend. • Repeat the process until the degree of the remainder is less than the degree of the divisor.	$(3y^2 - 4y - 7) \div (y - 2)$ Quotient $3y + 2$ $y - 2 \overline{)3y^2 - 4y - 7}$ ← Dividend $3y^2 - 6y$ Divisor $2y - 7$ $2y - 4$ Remainder → -3

Chapter 5   Review Exercises

To help you review this chapter, solve these problems.

[5.1] *Simplify.*

1. $(-x)^3$

2. -31^0

3. $n^4 \cdot n^7$

4. $x^6 \cdot x$

5. $\dfrac{n^8}{n^5}$

6. $p^{10} \div p^7$

7. $y^4 \cdot y^2 \cdot y$

8. $(a^2b)(ab^2)$

9. $x^0 y$

10. $\dfrac{n^4 \cdot n^7}{n^9}$

Write as an expression using only positive exponents.

11. $(5x)^{-1}$

12. $-3n^{-2}$

13. $8^{-2}v^4$

14. $\dfrac{1}{y^{-4}}$

15. $x^{-8} \cdot x^7$

16. $5^{-1} \cdot y^6 \cdot y^{-3}$

17. $\dfrac{a^5}{a^{-5}}$

18. $\dfrac{t^{-2}}{t^4}$

19. $\dfrac{x^{-2}}{y}$

20. $\dfrac{x^2}{y^{-1}}$

[5.2] *Simplify.*

21. $(10^2)^4$

22. $-(x^3)^3$

23. $(2x^3)^2$

24. $(-4m^5n)^3$

25. $3(x^{-2})^6$

26. $(a^3b^{-4})^{-2}$

27. $\left(\dfrac{x}{3}\right)^4$

28. $\left(\dfrac{-a}{b^3}\right)^2$

29. $\left(\dfrac{x}{y}\right)^{-6}$

30. $\left(\dfrac{x^2}{y^{-1}}\right)^5$

31. $\left(\dfrac{4a^3}{b^4c}\right)^2$

32. $\left(\dfrac{-u^{-5}v^2}{7w}\right)^2$

Express in standard notation.

33. 3.7×10^{10}

34. 1.63×10^9

35. 5.022×10^{-5}

36. 6×10^{-11}

Express in scientific notation.

37. $1,200,000,000,000$

38. $427,000,000$

39. 0.00000000000004

40. 0.00000056

▦ *Perform the indicated operation. Then write the result in scientific notation.*

41. $(1.4 \times 10^6)(4.2 \times 10^3)$

42. $(3 \times 10^{-2})(2.1 \times 10^5)$

43. $(1.8 \times 10^4) \div (3 \times 10^{-3})$

44. $(9.6 \times 10^{-4}) \div (1.6 \times 10^6)$

[5.3] *Indicate whether the expression is a polynomial.*

45. $3x^4 - 5x^3 + \dfrac{x^2}{4} - 8$

46. $-2x^2 - \dfrac{7}{x} + 1$

Classify each polynomial according to the number of terms.

47. $2x^5 + 7x^2 - 5$

48. $16 - 4t^2$

Write the polynomial in descending order. Then identify the degree, leading term, and leading coefficient of the polynomial.

49. $8y - 3y^3 + y^2 - 1$

50. $n^4 - 6n^2 - 7n^3 + n$

Simplify. Then write the polynomial in descending order of powers.

51. $10x - 8x^2 - 8x + 9x^2 - x^3 + 13$

52. $4n^3 - 7n + 9 - 3n^2 - n^3 + 7n^2 - 5 + n$

Evaluate the polynomial for the given values of the variable.

53. $2n^2 - 7n + 3$ for $n = -1$ and $n = 3$

54. $x^3 - 8$ for $x = 2$ and $x = -2$

[5.4] *Perform the indicated operations.*

55. $(4x^2 - x + 4) + (-3x^2 + 9)$

56. $(5y^4 - 2y^3 + 7y - 11) + (6 - 8y - y^2 - 5y^4)$

57. $(a^2 + 5ab + 6b^2) + (3a^2 - 9b^2) + (-7ab - 3a^2)$

58. $(5s^3t - 2st + t^2) + (s^2t - 5t^2) + (t^2 - 4st + 9s^2)$

59. $(x^2 - 5x + 2) - (-x^2 + 3x + 10)$

60. $(10n^3 + n^2 - 4n + 1) - (11n^3 - 2n^2 - 5n + 1)$

61. $\begin{aligned} 5y^4 - 4y^3 \quad\quad + y - 6 \\ -(\quad\quad y^3 - 2y^2 + 7y - 3) \end{aligned}$

62. $\begin{aligned} -9x^3 + 8x^2 - 11x - 12 \\ +(11x^3 \quad\quad - x + 15) \end{aligned}$

Simplify.

63. $14t^2 - (10t^2 - 4t)$

64. $-(5x - 6y) + (3x - 7y)$

65. $(3y^2 - 1) - (y^2 + 3y + 2) + (-2y + 5)$

66. $(1 - 4x - 6x^2) - (7x - 8) - (-11x - x^2)$

[5.5] *Multiply.*

67. $-3x^4 \cdot 2x$

68. $(3ab)(8a^2b^3)(-6b)$

69. $2xy^2(4x - 5y)$

70. $(x^2 - 3x + 1)(-5x^2)$

71. $(n + 3)(n + 7)$

72. $(3x - 9)(x + 6)$

73. $(2x - 1)(4x - 1)$

74. $(3a - b)(3a + 2b)$

75. $(2x^3 - 5x + 2)(x + 3)$

76. $(y - 2)(y^2 - 7y + 1)$

Simplify.

77. $-y + 2y(-3y + 7)$

78. $4x^2(2x - 6) - 3x(3x^2 - 10x + 2)$

[5.6] *Simplify.*

79. $(a - 1)^2$

80. $(s + 4)^2$

81. $(2x + 5)^2$

82. $(3 - 4t)^2$

83. $(5a - 2b)^2$

84. $(u^2 + v^2)^2$

Multiply.

85. $(m + 4)(m - 4)$

86. $(6 - n)(6 + n)$

87. $(7n - 1)(7n + 1)$

88. $(2x + y)(2x - y)$

89. $(4a - 3b)(4a + 3b)$

90. $x(x + 10)(x - 10)$

91. $-3t^2(4t - 5)^2$

92. $(p^2 - q^2)(p + q)(p - q)$

[5.7] *Divide.*

93. $12x^4 \div 4x^2$

94. $\dfrac{-20a^3b^5c}{10ab^2}$

95. $(18x^3 - 6x) \div (3x)$

96. $\dfrac{10x^5 + 6x^4 - 4x^3 - 2x^2}{2x^2}$

97. $(3x^2 + 8x - 35) \div (x + 5)$

98. $\dfrac{13 - 5x^2 + 2x^3}{2x - 1}$

Mixed Applications

Solve.

99. The half-life of the element thorium-232 is 13,900,000,000 yr. Express this length of time in scientific notation. (*Source:* Peter J. Nolan, *Fundamentals of College Physics*)

100. Physicists use both the joule (J) and the electron volt (eV) as units of work, where 1 J is equal to 6.24×10^{18} eV. Rewrite this quantity in standard notation. (*Source:* Peter J. Nolan, *Fundamentals of College Physics*)

101. A grain of bee pollen is about 0.00003 m in diameter. Express this length in scientific notation.

102. The diameter of an atom is about 1.1×10^{-10} m. What is this quantity in standard notation? (*Source:* Peter J. Nolan, *Fundamentals of College Physics*)

103. At a party there are n people present. If everyone shakes hands with everyone else, then the polynomial

$$\frac{n^2}{2} - \frac{n}{2}$$

gives the total number of handshakes. If 9 people are at the party, how many handshakes will there be?

104. An object falling from an altitude of 500 m will be $-4.9t^2 + 500$ m above the ground after t sec. What is the altitude of the object 2 sec into the fall?

105. The number of divorces in the United States (in thousands) in various years is approximated by the polynomial $-0.3x^2 + 36.7x + 213$, where x represents the number of years since 1950. According to this model, how many divorces to the nearest thousand were there in 1951? (*Source:* U.S. National Center for Health Statistics, *Vital Statistics of the United States*)

106. The polynomial $-0.8x^2 + 41.5x + 898.6$ approximates the number of two-year colleges in the United States in a given year, where x represents the number of years since 1970. Use this polynomial to estimate to the nearest whole number how many two-year colleges there were in the United States in the year 1970. (*Source:* U.S. National Center for Education Statistics)

107. A rectangular swimming pool is surrounded by a 6-ft-wide concrete walk, as shown in the figure.

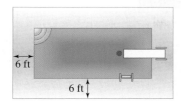

    a. If the length of the pool is 10 ft less than 3 times the width w, find a polynomial that represents the area of the swimming pool.

    b. Write a polynomial that represents the area of the concrete walk.

    c. Use your expression from part (b) to find the area of the concrete walk if the width of the swimming pool is 12 ft.

108. The number of farms (in thousands) in the United States for the years from 2000 to 2005 can be approximated by the polynomial $2164 - 13x$, where x represents the number of years since 2000. During the same time period, the total land in farms (in thousands of acres) can be approximated by the polynomial $943{,}504 - 1340x - 26x^2$. (*Source:* U.S. Department of Agriculture, National Agricultural Statistics Service, *Farms and Land in Farms*)

    a. Use division to find a polynomial that approximates the average number of acres per farm.

    b. What was the average number of acres per farm in 2002?

● *Check your answers on page A-30.*

Chapter 5   POSTTEST

FOR EXTRA HELP   Pass the Test   Test solutions are found on the enclosed CD.

To see if you have mastered the topics in this chapter, take this test.

Simplify.

1. $x^6 \cdot x$

2. $n^{10} \div n^4$

3. $7a^{-1}b^0$

4. $(-3x^2y)^3$

5. $\left(\dfrac{x^2}{y^3}\right)^4$

6. $\left(\dfrac{3x^2}{y}\right)^{-3}$

7. For the polynomial $-x^3 + 2x^2 + 9x - 1$, name:

  a. the terms _____

  b. the coefficients _____

  c. the degree _____

  d. the constant term _____

8. Find the sum: $(y^2 - 1) + (y^2 - y + 6)$

9. Subtract: $(x^2 - 7x - 4) - (2x^2 - 8x + 5)$

10. Combine:
$(4x^2y^2 - 6xy - y^2) - (3x^2 + x^2y^2 - 2y^2) - (x^2 - 6xy + y^2)$

Multiply.

11. $(2mn^2)(5m^2n - 10mn + mn^2)$

12. $(y^3 - 2y^2 + 4)(y - 1)$

13. $(3x - 1)(2x + 7)$

14. $(7 - 2n)(7 + 2n)$

15. $(2m - 3)^2$

Divide.

16. $\dfrac{12s^3 + 15s^2 - 27s}{-3s}$

17. $(3t^3 - 5t^2 - t + 6) \div (3t - 2)$

Solve.

18. Medical X-rays, with a wavelength of about 10^{-10} m, can penetrate the flesh (but not the bones) of your body. Ultraviolet rays, which cause sunburn by penetrating only the top layer of skin, have a wavelength about 1000 times as long as X-rays. Find the length of ultraviolet rays. Write the answer in scientific notation. (*Source:* Peter J. Nolan, *Fundamentals of College Physics,* 1993)

19. A real estate broker sells two houses. The first house sells for $140,000 and is expected to increase in value by $1,500 per year. The second house is purchased for $90,000 and will likely appreciate by $800 per year.

  a. Write an expression for the value of each house after x yr.

  b. Write an expression that represents the combined value of both houses after x yr.

20. A young couple is saving up to purchase a car. They deposit $1000 in an account that has an annual interest rate of r (in decimal form). At the end of 2 yr, the value of the account will be $1000(1 + r)^2$ dollars. Find the account balance at that time if the interest rate is 3%.

• *Check your answers on page A-30.*

Cumulative Review Exercises

To help you review, solve the following.

1. Solve $y = mx + b$ for m.

2. Evaluate $3a^2 - 5ab + b^2$ for $a = -3$ and $b = 2$.

3. Solve for x: $2x + 12 - 9x = 5(4 - 3x) + 6x$

4. Find the slope and y-intercept of the line $2x - 3y = 6$.

5. Graph the inequality: $y < 2x + 1$

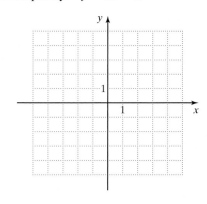

6. Solve by elimination:

$$3x - 2y = 10$$
$$2x + 3y = -2$$

7. Find the difference: $(3m^2 - 8m + 7) - (2m^2 + 8m - 9)$

8. According to Albert Einstein's famous equation $E = mc^2$, all objects, even resting ones, contain energy E. If the mass m of a raisin is 10^{-3} kg, and the speed c of light is about 3×10^8 m/sec, find the amount of energy the raisin contains. Write the answer in scientific notation.

9. The Winter Olympics occur every 4 yr.

   a. If x represents the first year that they took place, write expressions for the next three Winter Olympic years.

   b. If the Winter Olympics were held in 1972, were they also held in 1980? Explain.

10. An executive goes out to dinner and leaves a 20% tip for the service.

   a. The bill for the meal without tip is represented by b. Write an expression for the amount of the tip in terms of b.

   b. The total cost c of the meal is the original bill plus the tip. Write an equation describing this situation.

● *Check your answers on page A-30.*

Factoring Polynomials

Factoring and Cryptography

Cryptography, the science of coding and decoding messages, has been important throughout history. Sending secret messages in concealed form is particularly useful during wartime. For instance, the United States entered World War I in part because British intelligence intercepted and deciphered a message sent to the German minister in Mexico; the message called for a German–Mexican alliance against the United States. During World War II, the ability of the Allies to decode Nazi secrets allowed their commanders to eavesdrop on German plans and may have shortened the war by several years.

In peacetime, cryptographic techniques, used in electronic banking and in Web-based credit card purchases, play an increasingly important role in our lives.

Of the numerous cryptographic techniques for coding messages today, one important technique is *prime number encryption*. To crack a message coded in this way depends on finding the prime factors of a given, very large, whole number. Whereas multiplying two prime numbers is easy, reversing the process is difficult and drawn out. For instance, it has been estimated that finding the prime factorization of a five-digit whole number takes some 14 billion mathematical steps. (*Sources:* Rudolf Kippenhahn, *Code Breaking: A History and Exploration*, The Overlook Press, 1999; F. H. Hinsley and Alan Stripp (Editors), *Codebreakers: The Inside Story of Bletchley Park*, Oxford University Press, 1994)

Chapter 6   PRETEST

To see if you have already mastered the topics in this chapter, take this test.

1. Find the greatest common factor of $18ab$ and $36a^4$.

Factor.

2. $4pq + 16p$

3. $10x^2y - 5x^3y^3 + 5xy^2$

4. $3x^2 + 6x + 2x + 4$

5. $n^2 - 11n + 24$

6. $4a + a^2 - 21$

7. $9y - 12y^2 + 3y^3$

8. $5a^2 + 6ab - 8b^2$

9. $-12n^2 + 38n + 14$

10. $4x^2 - 28x + 49$

11. $25n^2 - 9$

12. $x^2y - 4y^3$

13. $y^6 - 9y^3 + 20$

Solve.

14. $n(n - 6) = 0$

15. $3x^2 + x = 2$

16. $(y + 4)(y - 2) = 7$

17. The lateral surface area of a rectangular solid is given by the formula $A = 2lw + 2lh + 2wh$. Solve this formula for h in terms of A, w, and l.

18. A baseball player hits a pop-up fly ball with an initial velocity of 63 ft/sec from a height of 4 ft above the ground. The height of the ball (in feet) t sec after it is hit is given by the expression $-16t^2 + 63t + 4$. Write this expression in factored form.

19. A homeowner wants to fence off part of her yard to build a 15-ft by 15-ft square play area for her children. Write an expression, in factored form, for the area of the yard not covered by the play area.

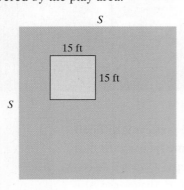

20. Find the dimensions of the 40-in. plasma television screen shown if the length is 8 in. longer than the height.

• *Check your answers on page A-30.*

What Factoring Is and Why It Is Important

In the previous chapter, we discussed how to *multiply* two factors in order to find their polynomial product.

Multiplying

$$\underbrace{(3x + 4)}_{\text{Factor}} \; \underbrace{(2x + 1)}_{\text{Factor}} = \underbrace{6x^2 + 11x + 4}_{\text{Product}}$$

In this chapter, we reverse the process beginning with a polynomial and expressing it as a product of factors. Rewriting a polynomial as a product is called *factoring* the polynomial.

Factoring

$$\underbrace{6x^2 + 11x + 4}_{\text{Polynomial}} = \underbrace{(3x + 4)}_{\text{Factor}} \underbrace{(2x + 1)}_{\text{Factor}}$$

Just as factoring integers plays a key role in arithmetic, so factoring polynomials is an important skill in algebra. Factoring polynomials helps us to simplify certain algebraic expressions and also to solve various types of equations.

Finding the Greatest Common Factor of Two or More Integers or Terms

We have already shown that every composite number can be written as the product of prime factors, called its *prime factorization*. For instance, the prime factorization of 15 is $3 \cdot 5$, and the prime factorization of 35 is $5 \cdot 7$. Since 5 appears in both factorizations, 5 is said to be a *common factor* of 15 and 35.

Definition

A **common factor** of two or more integers is an integer that is a factor of each integer.

We can use the concept of common factor to find a *greatest common factor*.

Definition

The **greatest common factor (GCF)** of two or more integers is the greatest integer that is a factor of each integer.

Let's consider an example of finding the greatest common factor of three numbers.

OBJECTIVES

- To find the greatest common factor (GCF) of two or more integers or terms
- To factor out the greatest common factor from a polynomial
- To factor by grouping
- To solve applied problems involving factoring

EXAMPLE 1

Find the GCF of 45, 63, and 81.

Solution First we find the prime factorization of each number.

$$45 = 3 \cdot 3 \cdot 5 = 3^2 \cdot 5$$
$$63 = 3 \cdot 3 \cdot 7 = 3^2 \cdot 7$$
$$81 = 3 \cdot 3 \cdot 3 \cdot 3 = 3^4$$

Then we look for the greatest common factor. We see that $3 \cdot 3 = 3^2$ is a factor of each number. Since no power of 3 higher than 3^2 and no prime number other than 3 is a factor of *all* three numbers, the GCF of 45, 63, and 81 is 3^2, or 9.

PRACTICE 1

What is the GCF of 24, 72, and 96?

We can extend the concept of greatest common factor to monomials. For instance, to find the greatest common factor of $-21x^2$ and $35xy^2$, we begin by writing the monomials in factored form.

$$-21x^2 = -1 \cdot 3 \cdot 7 \cdot x \cdot x$$

$$35xy^2 = 5 \cdot 7 \cdot x \cdot y \cdot y$$

From the factorizations we see that each monomial has a factor of 7 and a factor of x in common. Note that 7 is the greatest factor of the coefficients and x^1 is the highest power of x that is a factor of each monomial. The greatest common factor of $-21x^2$ and $35xy^2$ is the product of 7 and x, or $7x$.

EXAMPLE 2

Find the GCF of x^4, x^3, and x^2.

Solution We write each monomial in factored form.

$$x^4 = x \cdot x \cdot x \cdot x$$
$$x^3 = x \cdot x \cdot x$$
$$x^2 = x \cdot x$$

From the factored forms, we see that the monomials have at most two factors of x in common. So the GCF of x^4, x^3, and x^2 is $x \cdot x$, or x^2.

PRACTICE 2

What is the GCF of a^3, a^2, and a?

Note that we could have written the factored forms of the monomials in Example 2 as

$$x^4 = x^2 \cdot x^2$$
$$x^3 = x^2 \cdot x^1$$
$$x^2 = x^2$$

From these factored forms we see that x^2 is the highest power of the variable factor common to all three monomials.

Definition
The **greatest common factor (GCF) of two or more monomials** is the product of the greatest common factor of the coefficients and highest powers of the variable factors common to each monomial.

Next let's consider the greatest common factor of expressions involving more than one variable.

EXAMPLE 3	PRACTICE 3
Identify the GCF of $3a^3b$ and $-15a^2b^4$.	Find the GCF of $-18x^3y^4$ and $12xy^2$.

Solution First, we write each monomial in factored form.

$$3a^3b = 3 \cdot a^2 \cdot a \cdot b$$
$$-15a^2b^4 = -1 \cdot 3 \cdot 5 \cdot a^2 \cdot b \cdot b^3$$

The greatest common factor of the coefficients is 3 and the highest powers of the variable factors common to each monomial are a^2 and b. So the GCF of $3a^3b$ and $-15a^2b^4$ is $3a^2b$.

Factoring Out the Greatest Common Factor from a Polynomial

Recall from the previous chapter that we used the distributive property to multiply a monomial by a polynomial. For example, consider the product of $2x$ and $(x + 7)$.

$$2x(x + 7) = 2x \cdot x + 2x \cdot 7 = 2x^2 + 14x$$

When the terms of a polynomial have common factors, we can factor the polynomial by reversing the distributive property. So we can write $2x^2 + 14x$ in factored form by dividing out $2x$, the GCF of the terms $2x^2$ and $14x$.

$$2x^2 + 14x = 2x \cdot x + 2x \cdot 7 = 2x(x + 7)$$
$$\uparrow$$
$$\text{GCF}$$

So the factored form of $2x^2 + 14x$ is $2x(x + 7)$.

EXAMPLE 4	PRACTICE 4
Factor: $25x^3 + 10x^2$	Factor: $10y^2 + 8y^5$

Solution The GCF of $25x^3$ and $10x^2$ is $5x^2$.

$25x^3 + 10x^2 = 5x^2(5x) + 5x^2(2)$ Factor out the GCF $5x^2$ from each term.
$\qquad = 5x^2(5x + 2)$ Use the distributive property.

So the factorization of $25x^3 + 10x^2$ is $5x^2(5x + 2)$.

Now let's consider some examples of factoring polynomials in more than one variable.

EXAMPLE 5

Factor: $12c^2 - 2cd$

Solution The GCF of $12c^2$ and $-2cd$ is $2c$.

$$12c^2 - 2cd = 2c(6c) - 2c(d) \quad \text{Factor out the GCF } 2c \text{ from each term.}$$
$$= 2c(6c - d) \quad \text{Use the distributive property.}$$

PRACTICE 5

Factor: $21a^2b - 14a$

EXAMPLE 6

Factor: $3x^2y^4 + 9xy^2$

Solution

$$3x^2y^4 + 9xy^2 = 3xy^2(xy^2) + 3xy^2(3)$$
$$= 3xy^2(xy^2 + 3)$$

PRACTICE 6

Factor: $8a^2b^2 - 6ab^3$

EXAMPLE 7

Express in factored form: $20y^2 - 5y + 15$

Solution

$$20y^2 - 5y + 15 = 5(4y^2) - 5(y) + 5(3)$$
$$= 5(4y^2 - y + 3)$$

PRACTICE 7

Factor: $24a^2 - 48a + 12$

When solving some literal equations, it may be necessary to factor out common monomial factors.

EXAMPLE 8

Solve $ax + b = cx + d$, for x in terms of a, b, c, and d.

Solution To solve for x, we bring all the terms involving x to the left side of the equation and the other terms to the right side of the equation.

$$ax + b = cx + d$$
$$ax + b - b = cx + d - b \quad \text{Subtract } b \text{ from each side of the equation.}$$
$$ax = cx + d - b$$
$$ax - cx = cx - cx + d - b \quad \text{Subtract } cx \text{ from each side of the equation.}$$
$$ax - cx = d - b$$
$$x(a - c) = d - b \quad \text{Factor out } x \text{ on the left side of the equation.}$$
$$\frac{x(a - c)}{a - c} = \frac{d - b}{a - c} \quad \text{Divide each side of the equation by } (a - c), \text{ the coefficient of } x.$$
$$x = \frac{d - b}{a - c}$$

PRACTICE 8

Solve $ab = s^2 - ac$ for a in terms of b, c, and s.

Factoring by Grouping

Recall that the factorization of $2x^2 + 14x$ is $2x(x + 7)$. The factor $(x + 7)$ is called a *binomial* factor. The distributive property can be used to divide out not only a common monomial factor but also a common binomial factor, if there is one. For instance, let's consider the expression $x(x + 5) + 2(x + 5)$.

$$\underbrace{x(x + 5)}_{\text{First term}} + \underbrace{2(x + 5)}_{\text{Second term}}$$

In this polynomial, the binomial $(x + 5)$ is common to both terms. Using the distributive property, we can factor out $(x + 5)$. That is,

$$x(x + 5) + 2(x + 5) = (x + 5) \cdot x + (x + 5) \cdot 2 = (x + 5)(x + 2)$$

So the factored form of $x(x + 5) + 2(x + 5)$ is $(x + 5)(x + 2)$.

EXAMPLE 9

Factor: $x(x + 4) - 5(x + 4)$

Solution Using the distributive property, we get:

$$x(x + 4) - 5(x + 4) = (x + 4)(x - 5)$$

PRACTICE 9

Factor: $4(y - 3) + y(y - 3)$

In some algebraic expressions, such as $x(a - 7) + 3(7 - a)$, the binomial factors are opposites. In order to factor out a *common* binomial factor, we must rewrite one of the binomials by factoring out -1, as shown in the next example.

EXAMPLE 10

Factor: $x(a - 7) + 3(7 - a)$

Solution The binomial factors $(a - 7)$ and $(7 - a)$ are opposites, so we factor out -1 from the binomial $(7 - a)$ and rewrite the original expression. Note that $(7 - a) = -1(a - 7)$.

$$x(a - 7) + 3(7 - a) = x(a - 7) + 3\left[-1(a - 7)\right] \quad \text{Factor out } -1 \text{ from } (7 - a).$$

$$= x(a - 7) - 3(a - 7) \quad \text{Simplify.}$$

$$= (a - 7)(x - 3) \quad \text{Use the distributive property.}$$

PRACTICE 10

Factor: $3y(x - 1) + 2(1 - x)$

EXAMPLE 11

Factor: $5n(4n - 1) - (4n - 1)$

Solution

$$5n(4n - 1) - (4n - 1) = 5n(4n - 1) - 1(4n - 1)$$

$$= (4n - 1)(5n - 1)$$

PRACTICE 11

Factor: $(4 - 3x) + 2x(4 - 3x)$

When trying to factor a polynomial that has four terms, it may be possible to group pairs of terms in such a way that a common binomial factor can be found. This method is called **factoring by grouping**.

EXAMPLE 12	PRACTICE 12

Factor by grouping: $xy - 4x + 3y - 12$

Solution

$xy - 4x + 3y - 12 = (xy - 4x) + (3y - 12)$ Group the first two terms and the last two terms.

$= x(y - 4) + 3(y - 4)$ Factor out the GCF from each group.

$= (y - 4)(x + 3)$ Write in factored form.

Factor: $4b - 20 + ab - 5a$

EXAMPLE 13	PRACTICE 13

Express in factored form: $6h - 6k - h^2 + hk$

Solution

$6h - 6k - h^2 + hk = (6h - 6k) + (-h^2 + hk)$ Group the first two terms and the last two terms.

$= (6h - 6k) - (h^2 - hk)$ Factor out -1 in the second group.

$= 6(h - k) - h(h - k)$ Factor out the GCF from each group.

$= (h - k)(6 - h)$ Write in factored form.

Factor: $5y - 5z - y^2 + yz$

EXAMPLE 14	PRACTICE 14

Each week, a sales associate receives a salary of d dollars as well as 5% commission on the value of the sales that she makes. Last week, sales amounted to x dollars, and this week, sales rose to y dollars. How much greater was the sales associate's total income this week than last week? Express this amount in factored form.

Solution Last week, the sales associate made d dollars in salary and $0.05x$ in commission. This week, the associate made d dollars in salary and $0.05y$ in commission. So the difference in total income is:

$$(d + 0.05y) - (d + 0.05x) = d + 0.05y - d - 0.05x$$
$$= 0.05y - 0.05x$$
$$= 0.05(y - x)$$

So the sales associate made $0.05(y - x)$ dollars more this week than last week.

The distance an object under constant acceleration travels in time t is given by the expression $v_0 t + \frac{1}{2}at^2$, where v_0 is the object's initial velocity and a is its acceleration. Factor this expression.

| 6.1 | Exercises | FOR EXTRA HELP | MyMathLab | PRACTICE | WATCH | DOWNLOAD | READ | REVIEW |

Mathematically Speaking

Fill in each blank with the most appropriate term or phrase from the given list.

sum	greatest common factor (GCF)	factoring
greatest factor		product
multiplying	common factor	

1. Rewriting a polynomial as a product is called _____ the polynomial.

2. A(n) _____ of two or more integers is an integer that is a factor of each integer.

3. The _____ of two or more integers is the greatest integer that is a factor of each integer.

4. The greatest common factor of two or more monomials is the _____ of the greatest common factor of the coefficients and the highest powers of the variable factors common to each monomial.

Find the greatest common factor of each group of terms.

5. 27, 54, and 81

6. 28, 35, 63

7. x^4, x^6, x^3

8. y^2, y, y^5

9. $16b, 8b^3, 12b^2$

10. $3a, 7a^2, 5a^4$

11. $-12x^5y^7, 4y^3$

12. $9m^3, 6m^2n$

13. $18a^5b^4, -6a^4b^3, 9a^2b^2$

14. $24mn, 32mn^2, 16m^2n$

15. $x(3x - 1)$ and $8(3x - 1)$

16. $6(5n + 2)$ and $n(5n + 2)$

17. $4x(x + 7)$ and $9x(x + 7)$

18. $y(y - 4)$ and $6y(y - 4)$

Factor out the greatest common factor.

19. $3x + 6$

20. $10y + 15$

21. $24x^2 + 8$

22. $30y^2 - 6$

23. $27m - 9n$

24. $16r - 8t$

25. $2x - 7x^2$

26. $3b^2 - 18b$

27. $5b^2 - 6b^3$

28. $4z^5 + 12z^2$

29. $10x^3 - 15x$

30. $12a^2 + 18a$

31. $a^2b^2 - ab$

32. $xy + x^2y^2$

33. $6xy^2 + 7x^2y$

34. $3p^3q^2 - 5p^2q^3$

35. $27pq^2 + 18p^2q$

36. $45c^2d - 15cd^2$

37. $2x^3y - 12x^3y^4$

38. $7a^2b^3 + 9a^4b^3$

39. $3c^3 + 6c^2 + 12$

40. $5y^2 - 20y + 10$

41. $9b^4 - 3b^3 + b^2$

42. $8y^5 - y^4 - 4y^2$

43. $2m^4 + 10m^3 - 6m^2$ **44.** $3x^3 - 9x^2 - 27x$ **45.** $5b^5 - 3b^3 + 2b^2$

46. $9c^4 + c^3 + 6c^2$ **47.** $15x^4 - 10x^3 - 25x$ **48.** $12m^6 + 9m^5 + 15m^3$

49. $4a^2b + 8a^2b^2 - 12ab$ **50.** $5m^2n - 15mn^2 + 10mn$ **51.** $9c^2d^2 + 12c^3d + 3cd^3$

52. $18x^2y^4 - 24xy^3 + 30x^3y$

Factor by grouping.

53. $x(x - 1) + 3(x - 1)$ **54.** $2(n + 4) + n(n + 4)$ **55.** $5a(a - 1) - 3(a - 1)$

56. $4x(x + 3) - 7(x + 3)$ **57.** $r(s + 7) - 2(7 + s)$ **58.** $a(6 + b) - 7(b + 6)$

59. $a(x - y) - b(x - y)$ **60.** $y(a - z) - x(a - z)$ **61.** $3x(y + 2) - (y + 2)$

62. $(n - 1) - 2m(n - 1)$ **63.** $b(b - 1) + 5(1 - b)$ **64.** $x(x - 3) + 2(3 - x)$

65. $y(y - 1) - 5(1 - y)$ **66.** $n(n - 9) - 4(9 - n)$ **67.** $(t - 3) - t(3 - t)$

68. $w(w - 4) + (4 - w)$ **69.** $9a(b - 7) + 2(7 - b)$ **70.** $2y(x - 2) + 3(2 - x)$

71. $rs + 3s + rt + 3t$ **72.** $mn + 2m + np + 2p$ **73.** $xy + 6y - 4x - 24$

74. $ab - 5b - 2a + 10$ **75.** $15xy - 9yz + 20xz - 12z^2$ **76.** $6ab + 12ac - 5bc - 10c^2$

77. $2xz + 8x + 5yz + 20y$ **78.** $3ab + 9a + 4bc + 12c$

Solve for the indicated variable.

79. $TM = PC + PL$ for P **80.** $S = a + Nd - d$ for d

81. $S = 2lw + 2lh + 2wh$ for l **82.** $S = a + ar^n$ for a

Mixed Practice

Factor out the greatest common factor.

83. $16p^3 + 24p$ **84.** $4u^4 - 28u^2 + 36u$

85. $48rs^2 - 60r^2s$ **86.** $7m^3 - 4m^2$

87. $42j^2 - 6$

Solve.

88. $A = \frac{1}{2}hb_1 + \frac{1}{2}hb_2$ for h

Factor by grouping.

89. $st - 3t - 7s + 21$

90. $7b(b + 2) - 5(b + 2)$

91. $3x(y - 4) + 5(4 - y)$

92. $2bc + 8ab - 3ac - 12a^2$

Find the greatest common factor of each group of terms.

93. $16a^5b^3, -12a^2b^3, 20a^3b$

94. $14x^4, 21xy^3$

Applications

Solve.

95. When an object with mass m increases in velocity from v_1 to v_2, its momentum increases by $mv_2 - mv_1$. Factor this expression.

96. One item sells for p dollars and another for q dollars. In addition, an 8% sales tax is charged on all items sold. An expression for the total selling price is $1.08p + 1.08q$. Write this expression in factored form.

97. In a meeting of diplomats, all diplomats must shake hands with one another. If there are n diplomats, the expression $0.5n^2 - 0.5n$ represents the total number of handshakes at the meeting. Factor this expression.

98. For an investment earning simple interest, the future value of the investment is represented by the expression $P + Prt$, where P is the present value of the investment, r is the annual interest rate, and t is the time in years. Factor this expression.

99. In a polygon with n sides, the number of diagonals is given by the expression $\frac{1}{2}n^2 - \frac{3}{2}n$. Write this expression in factored form.

100. In a polygon with n sides, the interior angles (measured in degrees) add up to $180n - 360$. Find an equivalent expression by factoring out the GCF.

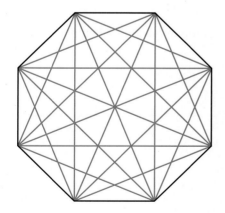

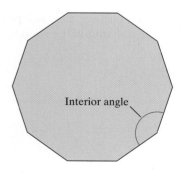

Interior angle

101. Consider the formula $P = nC + nT + D$, where P is the total price of a purchase, n is the number of items purchased, C is the cost per item, T is the tax on each item, and D is the total delivery charge. Solve for n in terms of P, C, T, and D.

102. The *harmonic mean* H of two numbers x and y is a kind of average. Solve for H in the formula $Hx + Hy = 2xy$.

● *Check your answers on page A-31.*

MINDSTRETCHERS

Mathematical Reasoning

1. A four-digit whole number can be represented by the expression

$$1000d + 100c + 10b + a$$

where a is the digit in the units place, b is the digit in the tens place, c is the digit in the hundreds place, and d is the digit in the thousands place.

a. Consider such a four-digit whole number (for instance, 8351). Then reverse the order of the digits, forming a second number $1000a + 100b + 10c + d$ (here, 1538). Subtract the smaller number from the larger ($8351 - 1538 = 6813$). Then check whether this difference is divisible by 9.

b. Show that when you reverse the order of the digits of *any* four-digit whole number and subtract the two four-digit numbers, their difference must be divisible by 9.

Critical Thinking

2. Factor the expression $a^{n+2}b^n - a^n b^{n+1}$.

Groupwork

3. Working with a partner, for each polynomial list three numbers or monomials that when placed in the ☐ will make the polynomial factorable.

a. $2xy - 7x + \boxed{} - 14$

b. $xy^2 + \boxed{} + 3y^2 - 48$

6.2   Factoring Trinomials Whose Leading Coefficient Is 1

In this section, we move on to another kind of factoring—factoring trinomials of the form $ax^2 + bx + c$, where $a = 1$. Recall that the coefficient a of the leading term, ax^2, is called the *leading* coefficient. In other words, we are examining trinomials of the form $x^2 + bx + c$, where the leading coefficient is 1. First, we factor trinomials in which the constant term is positive. For example, we factor trinomials such as

$$x^2 + 5x + 6 \qquad \text{and} \qquad x^2 - 8x + 16$$

Constant term

Next, we factor trinomials in which the constant term is negative. For example, we factor trinomials such as

$$x^2 + 3x - 10 \qquad \text{and} \qquad x^2 - 5x - 24$$

Constant term

In Section 6.6, we will see how factoring trinomials helps us to solve related equations and applied problems.

OBJECTIVES

■ To factor trinomials of the form $ax^2 + bx + c$, where $a = 1$

■ To solve applied problems involving factoring

Factoring $x^2 + bx + c$, c Positive

Recall from Section 5.5 the FOIL method of multiplying two binomials.

$$\overset{\text{F} \quad \text{O} \quad \text{I} \quad \text{L}}{(x + 2)(x + 3) = x^2 + 3x + 2x + 6}$$
$$= x^2 + 5x + 6$$

Note that when multiplying these two binomials, we get a trinomial. Also note that the leading coefficient of both binomial factors here is 1.

This suggests that factoring a trinomial of the form $x^2 + bx + c$ gives the product of two binomials of the form $(x + ?)(x + ?)$, where each question mark represents an integer. To find the binomials, we apply the FOIL method in reverse.

For instance, let's factor $x^2 + 5x + 6$. Since the leading term of the trinomial is x^2, we apply the FOIL method to two binomial factors, placing x as the first term in each of the factors.

$$\overset{\text{F} \quad \text{O} \quad \text{I} \quad \text{L}}{x^2 + 5x + 6 = (x + ?)(x + ?) = x^2 + 5x + 6}$$

We see that the product of the constant terms of the binomial factors must be 6, the constant term of the trinomial. The sum of the outer and inner products must be $5x$. So we need to find two integers whose product is 6 and whose sum is 5.

To find these integers, consider all possible factors of 6, that is, of $+6$.

Factors of 6	Possible Binomial Factors	Sum of Outer and Inner Products
1, 6	$(x + 1)(x + 6)$	$6x + x = 7x$
−1, −6	$(x − 1)(x − 6)$	$−6x − x = −7x$
2, 3	$(x + 2)(x + 3)$	$3x + 2x = 5x$ ← The correct middle term
−2, −3	$(x − 2)(x − 3)$	$−3x − 2x = −5x$

So $x^2 + 5x + 6 = (x + 2)(x + 3)$.

We check that the factors are correct by multiplying.

$$(x + 2)(x + 3) = x^2 + 3x + 2x + 6 = x^2 + 5x + 6$$

Sum of the outer and inner products   Middle term

This way of factoring a trinomial, listing all the possibilities, is sometimes called the *trial-and-error method*. Factoring a trinomial by the trial-and-error method involves recognizing patterns, looking for clues, and, once the factorization is found, multiplying to check. Note that by the commutative property of multiplication, we can also write the product $(x + 2)(x + 3)$ as $(x + 3)(x + 2)$.

EXAMPLE 1

Factor: $y^2 + 7y + 10$

Solution Applying the FOIL method in reverse, we know that the first term of each factor is y. So the factors are of the form $(y + ?)(y + ?)$. Using the trial-and-error method, we need to find two integers whose product is 10 and whose sum is 7. The following table shows how to test pairs of factors of 10, the constant term of the trinomial.

Factors of 10	Sum of Factors
1, 10	11
−1, −10	−11
2, 5	7 ← Factors 2 and 5 have a sum of 7.
−2, −5	−7

So $y^2 + 7y + 10 = (y + 2)(y + 5)$, or $(y + 5)(y + 2)$.

Same sign

PRACTICE 1

Factor: $x^2 + 5x + 4$

Can you explain why in the preceding list of possible factors we need only have tested the positive factors of the constant term?

EXAMPLE 2

Factor: $x^2 - 10x + 16$

Solution The constant term of the trinomial is positive and the coefficient of the x-term is negative. So the constant terms of the binomial factors must both be negative.

$$x^2 - 10x + 16 = (x - ?)(x - ?)$$

We need to find two negative integers whose product is 16 and whose sum is -10. In this case we need to consider only negative factors of 16.

Factors of 16	Sum of Factors
$-1, -16$	-17
$-2, -8$	-10 ← Factors -2 and -8 have a sum of -10.
$-4, -4$	-8

So $x^2 - 10x + 16 = (x - 2)(x - 8)$, or $(x - 8)(x - 2)$.

Tip When the constant term c of the trinomial $x^2 + bx + c$ is positive,

- the constant terms of the binomial factors are both positive when b, the coefficient of the x-term in the trinomial, is positive, and
- the constant terms are both negative when b is negative.

   Not every trinomial can be expressed as the product of binomial factors with coefficients that are integers. Polynomials that are not factorable are called **prime polynomials**. For instance, the trinomial $x^2 + 5x + 1$ is a prime polynomial. Can you explain why this trinomial is prime?

EXAMPLE 3

Factor: $x^2 + 4x + 6$

Solution Both the constant term 6 and the coefficient of the x-term 4 are positive. So if the trinomial is factorable, the constant terms of its binomial factors must both be positive as well.

$$x^2 + 4x + 6 = (x + ?)(x + ?)$$

We need to find two positive integers whose product is 6 and whose sum is 4.

Factors of 6	Sum of Factors
1, 6	7
2, 3	5

Since neither sum of the factors yields the correct coefficient of the x-term, we can conclude that the polynomial is not factorable. Therefore, this is a *prime polynomial*.

PRACTICE 2

Factor: $y^2 - 9y + 20$

PRACTICE 3

Factor: $x^2 + 3x + 5$

When the terms of a trinomial are not in descending order, we usually rewrite the trinomial before factoring, as shown in the next example.

EXAMPLE 4

Factor: $12 - 8x + x^2$

Solution We begin by writing the terms in descending order.

$$12 - 8x + x^2 = x^2 - 8x + 12$$
$$= (x - ?)(x - ?)$$

Since the middle term of the trinomial is negative and the constant term is positive, we are looking for two negative integers. So we need to consider only negative factors of 12.

Factors of 12	Sum of Factors
$-1, -12$	-13
$-2, -6$	-8 ← Factors -2 and -6 have a sum of -8.
$-3, -4$	-7

So $x^2 - 8x + 12 = (x - 2)(x - 6)$.

PRACTICE 4

Factor: $32 - 12y + y^2$

Note that in Example 4, we could also have factored $12 - 8x + x^2$ as $(2 - x)(6 - x)$, without rearranging the terms of the trinomial. Can you explain why our two solutions $(x - 2)(x - 6)$ and $(2 - x)(6 - x)$ are equal?

Now, let's consider a trinomial of the form $x^2 + bxy + cy^2$. We note that this trinomial contains more than one variable. When factoring these trinomials, we use the trial-and-error method where the binomial factors are of the form $(x + ?y)(x + ?y)$ and the question marks represent factors of c, the coefficient of the y^2 term, whose sum is b, the coefficient of the xy-term.

EXAMPLE 5

Factor: $x^2 + 3xy + 2y^2$

Solution $x^2 + 3xy + 2y^2 = (x + ?y)(x + ?y)$

Since the middle term and the last term are both positive, we look for only positive factors of 2, the coefficient of y^2, whose sum is 3, the coefficient of the xy-term.

Factors of 2	Sum of Factors
$1, 2$	3

So $x^2 + 3xy + 2y^2 = (x + 1y)(x + 2y) = (x + y)(x + 2y)$, or $(x + 2y)(x + y)$.

PRACTICE 5

Factor: $p^2 - 4pq + 3q^2$

Factoring $x^2 + bc + c$, c Negative

Now, let's consider how to factor trinomials in which the coefficient of the first term is 1 and the sign of the constant term is negative.

EXAMPLE 6	PRACTICE 6

Factor: $x^2 + 2x - 3$

Solution We must find factors of -3 whose sum is 2.

Factors of -3	Sum of Factors
$-1, 3$	2 ←——— Factors -1 and 3 have a sum of 2.
$1, -3$	-2

So $x^2 + 2x - 3 = (x - 1)(x + 3)$.

Check We check by multiplying.

$$(x - 1)(x + 3) = x^2 + \underbrace{3x - x}_{\substack{\text{Sum of the} \\ \text{outer and} \\ \text{inner products}}} - 3 = x^2 + \underbrace{2x}_{\substack{\text{Middle} \\ \text{term}}} - 3$$

Factor: $x^2 + x - 6$

Tip   When the constant term c of the trinomial $x^2 + bx + c$ is negative, the constant terms of the binomial factors have opposite signs.

EXAMPLE 7	PRACTICE 7

Factor: $y^2 - 3y - 10$

Solution We must find two integers whose product is -10 and whose sum is -3. Since -10 is negative, one of its factors must be positive and the other negative.

Factors of -10	Sum of Factors
$1, -10$	-9
$-1, 10$	9
$2, -5$	-3 ←——— Factors 2 and -5 have a sum of -3.
$-2, 5$	3

So $y^2 - 3y - 10 = (y + 2)(y - 5)$.

  Note that since the sum of the two factors of -10 is -3, the negative factor must have a larger absolute value than the positive factor.

Factor: $x^2 - 21x - 46$

EXAMPLE 8

Factor: $x^2 - 12 + x$

Solution

$x^2 - 12 + x = x^2 + x - 12$ Write the terms in descending order.

$= (x + ?)(x - ?)$ Since the constant term is negative, one of its factors is positive and the other factor is negative.

Now, we must find two factors of -12 whose sum is 1. So the positive factor must have a larger absolute value than the negative factor. Thus, we consider only factors of -12 for which the positive factor has the larger absolute value.

Factors of -12	Sum of Factors
$-1, 12$	11
$-2, 6$	4
$-3, 4$	1 ←——— Factors -3 and 4 have a sum of 1.

So $x^2 + x - 12 = (x - 3)(x + 4)$.

PRACTICE 8

Factor: $y^2 - 24 + 2y$

Next we consider factoring a trinomial in two variables.

EXAMPLE 9

Factor: $x^2 - 5xy - 14y^2$

Solution $x^2 - 5xy - 14y^2 = (x + ? \, y)(x - ? \, y)$

We must find two factors of -14 whose sum is -5. Since the product of these factors is a negative number, one of the factors must be positive and the other negative. Since the sum of the factors is negative, the negative factor must have a larger absolute value than the positive factor. Thus, we consider only factors of -14 for which the negative factor has the larger absolute value.

Factors of -14	Sum of Factors
$1, -14$	-13
$2, -7$	-5 ←——— Factors 2 and -7 have a sum of -5.

So $x^2 - 5xy - 14y^2 = (x + 2y)(x - 7y)$.

PRACTICE 9

Factor: $a^2 - 5ab - 24b^2$

Some trinomials have a common factor. When factoring such a trinomial, we factor out the GCF before we try factoring the trinomial into the product of two binomials.

EXAMPLE 10	PRACTICE 10

Factor: $3x^2 + 6x - 24$

Solution Each term of the trinomial has a common factor of 3, so we begin by factoring out this factor.

$$3x^2 + 6x - 24 = 3(x^2 + 2x - 8)$$
$$= 3(x + ?)(x - ?)$$

Since the product of the two missing integers is negative, we need to consider only factors of -8, one positive and the other negative, where the positive factor has a larger absolute value than the negative factor.

Factors of -8	Sum of Factors
$-1, 8$	7
$-2, 4$	2 ←——— Factors -2 and 4 have a sum of 2.

So $3x^2 + 6x - 24 = 3(x^2 + 2x - 8)$
$$= 3(x + 4)(x - 2).$$

Note that after factoring out the GCF, 3, neither of the remaining factors of the polynomial has a common factor.

Factor: $y^3 - 9y^2 - 10y$

EXAMPLE 11	PRACTICE 11

Express in factored form: $3a^4 + 27a^3 + 24a^2$

Solution

$3a^4 + 27a^3 + 24a^2 = 3a^2(a^2 + 9a + 8)$ Factor out the GCF $3a^2$ from each term.

$\qquad\qquad\qquad\quad = 3a^2(a + 1)(a + 8)$ Factor $a^2 + 9a + 8$.

So $3a^4 + 27a^3 + 24a^2 = 3a^2(a + 1)(a + 8)$.

Factor: $8x^3 - 24x^2 + 16x$

As in the previous examples, after factoring out the GCF the remaining trinomial can sometimes still be factored. A polynomial is **factored completely** when it is expressed as the product of a monomial and one or more prime polynomials. Throughout the remainder of this text, *factor* means to factor completely.

EXAMPLE 12	PRACTICE 12

Factor: $-x^2 + 9x - 14$

Solution

$$-x^2 + 9x - 14 = -1(x^2 - 9x + 14) \qquad \text{Factor out } -1 \text{ so that the leading coefficient is 1.}$$
$$= -1(x - 7)(x - 2)$$

We can also express the solution two other ways by multiplying either binomial by -1.

$$-x^2 + 9x - 14 = -1(x - 7)(x - 2)$$
$$= (-x + 7)(x - 2) \qquad \text{Multiply } (x - 7) \text{ by } -1.$$
$$= (x - 7)(-x + 2) \qquad \text{Multiply } (x - 2) \text{ by } -1.$$

Write in factored form:
$-x^2 - 10x + 11$

EXAMPLE 13	PRACTICE 13

Show that for any whole number n, the number represented by $n^2 + 7n + 12$ can be expressed as the product of two consecutive whole numbers.

Solution Since we want to represent the expression $n^2 + 7n + 12$ as a product, we factor it.

$$n^2 + 7n + 12 = (n + 3)(n + 4)$$

Note that the two factors, $n + 3$ and $n + 4$, are whole numbers that differ by 1.

$$(n + 4) - (n + 3) = n + 4 - n - 3 = 1$$

So $n^2 + 7n + 12$ can be expressed as the product of two consecutive whole numbers.

A ball is tossed upward with a velocity of 32 ft/sec from a roof 48 ft above the ground. The expression $-16t^2 + 32t + 48$ approximates the height of the ball above the ground in feet after t sec. Factor this expression.

6.2   Exercises   FOR EXTRA HELP    MyMathLab   Math XL PRACTICE   WATCH   DOWNLOAD   READ   REVIEW

Mathematically Speaking

Fill in each blank with the most appropriate term or phrase from the given list.

are both negative	not factorable	have opposite signs
a factor greater than 0	are both positive	ascending
	descending	have the same sign
factorable	a common factor	

1. Polynomials that are _____ are called prime polynomials.

2. Polynomials to be factored are generally written in _____ order.

3. If $c < 0$ in the trinomial $x^2 + bx + c$, then the constant terms of the binomial factors _____.

4. If each term in a trinomial has _____, factor out the GCF before trying to factor the trinomial into the product of two binomials.

Match each trinomial with its binomial factors.

5. $x^2 - 16x + 28$

6. $x^2 + 12x - 28$

7. $x^2 + 29x + 28$

8. $x^2 + 27x - 28$

9. $x^2 - 3x - 28$

10. $x^2 - 11x + 28$

a. $(x - 1)(x + 28)$

b. $(x - 7)(x + 4)$

c. $(x - 4)(x - 7)$

d. $(x - 2)(x + 14)$

e. $(x + 28)(x + 1)$

f. $(x - 14)(x - 2)$

Find the missing factor.

11. $x^2 - 3x - 10 = (x + 2)(\quad)$

12. $x^2 - 11x + 18 = (x - 9)(\quad)$

13. $x^2 + 5x + 4 = (x + 1)(\quad)$

14. $x^2 - x - 12 = (x - 4)(\quad)$

15. $x^2 + 5x - 6 = (x + 6)(\quad)$

16. $x^2 + 4x - 21 = (x - 3)(\quad)$

Factor, if possible.

17. $x^2 + 6x + 8$

18. $x^2 + 9x + 8$

19. $x^2 + 5x - 6$

20. $x^2 + 2x - 3$

21. $x^2 + x + 2$

22. $x^2 + 2x - 8$

23. $x^2 + 5x + 4$

24. $x^2 + 13x + 7$

25. $x^2 - 4x + 3$

26. $x^2 - 6x + 5$

27. $y^2 - 12y + 32$

28. $m^2 - 5m + 4$

29. $t^2 - 4t - 5$

30. $s^2 - 4s - 12$

31. $x^2 + 2x - 1$

32. $y^2 - 13y - 48$

33. $x^2 + 4x - 45$

34. $x^2 + 3x - 18$

35. $y^2 - 9y + 20$

36. $a^2 - 7a + 10$

37. $b^2 + 11b + 28$ **38.** $p^2 + 13p + 22$ **39.** $y^2 + 4y + 5$ **40.** $x^2 - 12x + 36$

41. $-y^2 + 5y + 50$ **42.** $-x^2 + 5x + 84$ **43.** $x^2 + 64 - 16x$ **44.** $a^2 + 20 - 12a$

45. $16 - 10x + x^2$ **46.** $20 + b^2 - 21b$ **47.** $81 - 30w + w^2$ **48.** $y^2 + 72 - 17y$

49. $p^2 - 8pq + 7q^2$ **50.** $s^2 - 10st + 25t^2$ **51.** $p^2 - 4pq - 5q^2$ **52.** $x^2 - 4xy - 12y^2$

53. $m^2 - 12mn + 35n^2$ **54.** $r^2 - rs - 30s^2$ **55.** $x^2 + 9xy + 8y^2$ **56.** $a^2 + 12ab + 27b^2$

57. $5x^2 - 5x - 30$ **58.** $4y^2 + 12y - 40$ **59.** $2x^2 + 10x - 28$ **60.** $8r^2 - 56r - 64$

61. $12 - 18t + 6t^2$ **62.** $8 - 10x + 2x^2$ **63.** $3x^2 + 24 + 18x$ **64.** $5z^2 - 15 - 10z$

65. $y^3 + 3y^2 - 10y$ **66.** $x^3 - 6x^2 - 7x$ **67.** $a^3 + 8a^2 + 15a$ **68.** $q^3 - q^2 - 42q$

69. $t^4 - 14t^3 + 24t^2$ **70.** $x^4 + 6x^3 - 27x^2$ **71.** $4a^3 - 12a^2 + 8a$ **72.** $3y^3 - 18y^2 + 24y$

73. $2x^3 + 30x + 16x^2$ **74.** $5b^3 + 10b + 15b^2$ ◉ **75.** $4x^3 + 48x - 28x^2$ **76.** $3r^3 - 42r + 15r^2$

77. $-56s + 6s^2 + 2s^3$ **78.** $-20y - 16y^2 + 4y^3$ **79.** $2c^4 + 4c^3 - 70c^2$ **80.** $4t^4 + 24t^3 - 64t^2$

81. $ax^3 - 18ax^2 + 32ax$ **82.** $b^2y^3 - 5b^2y^2 - 36b^2y$

Mixed Practice

Factor, if possible.

83. $x^2 + 6x + 3$ **84.** $x^2 - xy - 12y^2$

85. $5x^4 - 15x^3 - 50x^2$ **86.** $3z^3 + 24z^2 + 36z$

87. $-w^2 + 6w + 40$ **88.** $b^2 - 6b + 5$

89. $6m^2 - 6m - 36$ **90.** $s^4 - 17s^3 + 72s^2$

91. $t^2 + 60 - 17t$ **92.** $n^2 - 11n + 18$

Solve.

93. Choose the correct binomial factors of the trinomial $x^2 - 5x - 24$.

    a. $(x + 8)(x - 3)$ **b.** $(x + 3)(x - 8)$

    c. $(x - 6)(x + 4)$ **d.** $(x - 4)(x + 6)$

94. Find the missing factor.

    $x^2 + 5x - 36 = (x - 4)(\qquad)$

Applications

Solve.

95. Scientists who study genetics use the equation $p^2 + 2pq + q^2 = 1$, where p represents a certain dominant gene and q represents a recessive gene. Rewrite the equation so that the left side is factored.

96. Show that for any whole number n, the number represented by $n^2 + 11n + 30$ can be expressed as the product of two consecutive whole numbers.

97. A child throws a stone downward with an initial velocity of 48 ft/sec from a height of 160 ft. One step in figuring out how long it takes for the stone to reach the ground is to factor the expression $16t^2 + 48t - 160$. What is its factorization?

98. A statistician found that the cost in dollars for a company to produce x units of a certain product can be approximated by $C = x^2 - 14x + 45$. Factor the expression on the right side of this equation.

99. According to specifications, a box manufacturer makes a closed box with a length and width that are 3 in. longer than the height. Let x equal the height of the box.

    a. Find the surface area of the box.

    b. Factor this polynomial.

100. Show that for any whole number n, the number represented by $4n^2 + 20n + 24$ can be expressed as the product of two consecutive even numbers.

• *Check your answers on page A-31.*

MINDSTRETCHERS

Groupwork

1. Work with a partner. Next to each trinomial, list at least two integers that when inserted in the box will make the trinomial factorable.

    a. $x^2 + \boxed{} x + 60$

    b. $x^2 - x + \boxed{}$

Investigation

2. Copy or trace the following algebra tiles onto a piece of paper. Then cut out each tile separately. Finally position all the tiles like a jigsaw puzzle so as to form a rectangle. (*Hint:* Factor the polynomial $x^2 + 5x + 6$.)

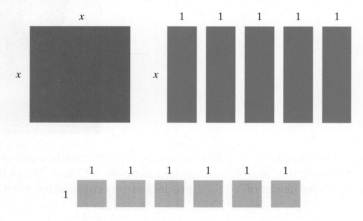

Mathematical Reasoning

3. In arithmetic, we simplify a fraction such as $\dfrac{20}{24}$ by writing the numerator and denominator in factored form and then dividing out common factors. For example, $\dfrac{20}{24} = \dfrac{\cancel{2} \cdot \cancel{2} \cdot 5}{\cancel{2} \cdot \cancel{2} \cdot 2 \cdot 3} = \dfrac{5}{6}$. Assuming $x \neq 1$ and $x \neq 3$, simplify $\dfrac{x^2 - 3x + 2}{x^2 - 4x + 3}$.

6.3 Factoring Trinomials Whose Leading Coefficient Is Not 1

OBJECTIVES

■ To factor trinomials of the form $ax^2 + bx + c$, where $a \neq 1$

■ To solve applied problems involving factoring

In the previous section we discussed factoring trinomials of the form $ax^2 + bx + c$, where $a = 1$. Here, we consider polynomials whose leading coefficient is not 1. That is, we factor trinomials such as

$$2x^2 + 5x + 3 \qquad \text{and} \qquad 5x^2 - 13x - 6$$
$$\uparrow \qquad\qquad\qquad\qquad \uparrow$$

The coefficient of the leading term is not 1.

The method of factoring that we use is, again, trial and error. However, we also discuss an alternative procedure that is based on factoring by grouping, a method that we discussed in Section 6.1.

Factoring $ax^2 + bx + c$, $a \neq 1$

Let's consider the product of two binomials.

$$
\begin{array}{cccc}
\mathbf{F} & \mathbf{O} & \mathbf{I} & \mathbf{L}
\end{array}
$$
$$(2x + 3)(x + 1) = (2x)(x) + (2x)(1) + (3)(x) + (3)(1)$$
$$= 2x^2 \quad + \quad 2x \quad + \quad 3x \quad + \quad 3$$
$$= 2x^2 + 5x + 3$$

Now, let's reverse the process: We start with the product $2x^2 + 5x + 3$, which we must factor. First, we check for common factors. Since there are no common factors other than 1 and -1, we use the process discussed in the last section for factoring trinomials, applying the FOIL method in reverse. Here, however, the leading coefficient is not 1, so we need to consider the factors of both 2 and 3. We then list and test all combinations of these factors to see if any will give us the desired middle term, $5x$. In other words, we are looking for four integers so that:

$$2x^2 + 5x + 3 = (?x + ?)(?x + ?)$$

Factors of 2	Factors of 3	Possible Binomial Factors	Sum of Outer and Inner Products
2, 1	3, 1	$(2x + 3)(x + 1)$	$2x + 3x = 5x$ ←— Correct middle term
		$(2x + 1)(x + 3)$	$6x + x = 7x$
2, 1	$-3, -1$	$(2x - 3)(x - 1)$	$-2x - 3x = -5x$
		$(2x - 1)(x - 3)$	$-6x - x = -7x$

Using this trial-and-error method, we find that $(2x + 3)(x + 1)$ is the correct factorization.

Note that this trial-and-error method for factoring a trinomial such as $2x^2 + 5x + 3$ is similar to the method for factoring trinomials with leading coefficient 1: We list and test all the possible factors of both the leading term and the constant term of the trinomial. We are looking for a combination of factors where the sum of the outer and inner products is the middle term of the trinomial. Practice and experience will shorten the process.

EXAMPLE 1	PRACTICE 1

Factor: $3x^2 + 11x + 10$

Solution The terms of $3x^2 + 11x + 10$ have no common factors. So we proceed to use the trial-and-error method to factor this trinomial. Note that both the middle term, $11x$, and the constant term, 10, are positive. So we need to consider only combinations of the positive factors of 3 and positive factors of 10 that will give us a middle term with coefficient 15.

$$3x^2 + 11x + 10 = (?x + ?)(?x + ?)$$

Factors of 3	Factors of 10	Possible Binomial Factors	Middle Term	
3, 1	2, 5	$(3x + 2)(x + 5)$	$15x + 2x = 17x$	
		$(3x + 5)(x + 2)$	$6x + 5x = \mathbf{11x}$	← Correct middle term
	10, 1	$(3x + 10)(x + 1)$	$3x + 10x = 13x$	
		$(3x + 1)(x + 10)$	$30x + x = 31x$	

So $3x^2 + 11x + 10 = (3x + 5)(x + 2)$.

Factor: $5x^2 + 14x + 8$

EXAMPLE 2	PRACTICE 2

Express in factored form: $15 - 17x + 4x^2$

Solution First, we rewrite the terms of the trinomial in descending order: $4x^2 - 17x + 15$. These terms have no common factor. To find the factorization of $4x^2 - 17x + 15$, we consider combinations of factors of 4 and factors of 15 that will result in a middle term with the coefficient -17.

Factors of 4	Factors of 15	Possible Binomial Factors	Middle Term	
2, 2	$-3, -5$	$(2x - 3)(2x - 5)$	$-10x - 6x = -16x$	
	$-15, -1$	$(2x - 15)(2x - 1)$	$-2x - 30x = -32x$	
4, 1	$-3, -5$	$(4x - 3)(x - 5)$	$-20x - 3x = -23x$	
		$(4x - 5)(x - 3)$	$-12x - 5x = \mathbf{-17x}$	← Correct middle term
	$-15, -1$	$(4x - 15)(x - 1)$	$-4x - 15x = -19x$	
		$(4x - 1)(x - 15)$	$-60x - x = -61x$	

So $4x^2 - 17x + 15 = (4x - 5)(x - 3)$.

Factor: $21 - 25x + 6x^2$

Note that in Examples 1 and 2 we could have stopped the process of testing possible factorizations after finding the correct middle term.

EXAMPLE 3	PRACTICE 3
Factor: $2y^2 + 19y - 10$	Factor: $7y^2 + 47y - 14$

Solution The terms of $2y^2 + 19y - 10$ have no common factors. So we factor the trinomial by considering combinations of the factors of 2 and factors of -10 that will give us the middle term with coefficient 19.

Factors of 2	Factors of -10	Possible Binomial Factors	Middle Term	
2, 1	2, -5	$(2y + 2)(y - 5)$	$-10y + 2y = -8y$	
		$(2y - 5)(y + 2)$	$4y - 5y = -y$	
	$-2, 5$	$(2y - 2)(y + 5)$	$10y - 2y = 8y$	
		$(2y + 5)(y - 2)$	$-4y + 5y = y$	
	10, -1	$(2y + 10)(y - 1)$	$-2y + 10y = 8y$	
		$(2y - 1)(y + 10)$	$20y - y = \mathbf{19y}$	⟵ Correct middle term
	$-10, 1$	$(2y - 10)(y + 1)$	$2y - 10y = -8y$	
		$(2y + 1)(y - 10)$	$-20y + y = -19y$	

So $2y^2 + 19y - 10 = (2y - 1)(y + 10)$.

Can you explain why $(2y + 2)$, $(2y - 2)$, $(2y + 10)$, and $(2y - 10)$ can be immediately eliminated as possible factors of $2y^2 + 19y - 10$ in Example 3?

Consider the following possible binomial factors in Example 3.

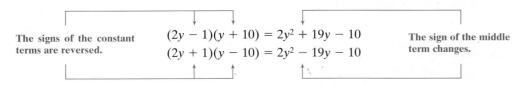

The signs of the constant terms are reversed.

$(2y - 1)(y + 10) = 2y^2 + 19y - 10$
$(2y + 1)(y - 10) = 2y^2 - 19y - 10$

The sign of the middle term changes.

Comparing these possible factors suggests the following shortcut.

Tip Reversing the signs of the constant terms in binomial factors has the effect of switching the sign of the middle term in their product.

EXAMPLE 4

Factor: $5x^2 - 13x - 6$

Solution Since the terms of $5x^2 - 13x - 6$ have no common factors, let's look at the combinations of factors of 5 and factors of -6 that will give us the middle term with coefficient -13.

Factors of 5	Factors of -6	Possible Binomial Factors	Middle Term	
5, 1	2, -3	$(5x + 2)(x - 3)$	$-15x + 2x = -13x$	← Correct middle term
		$(5x - 3)(x + 2)$	$10x - 3x = 7x$	
	$-2, 3$	$(5x - 2)(x + 3)$	$15x - 2x = 13x$	
		$(5x + 3)(x - 2)$	$-10x + 3x = -7x$	
	$-6, 1$	$(5x - 6)(x + 1)$	$5x - 6x = -x$	
		$(5x + 1)(x - 6)$	$-30x + x = -29x$	
	6, -1	$(5x + 6)(x - 1)$	$-5x + 6x = x$	
		$(5x - 1)(x + 6)$	$30x - x = 29x$	

So $5x^2 - 13x - 6 = (5x + 2)(x - 3)$. Note that it was unnecessary to examine any factors after the first trial, since we found the correct combination for the middle term, $-13x$.

EXAMPLE 5

Factor: $12y^3 + 2y^2 - 2y$

Solution Since $2y$ is the GCF of the trinomial, first we factor it out getting: $12y^3 + 2y^2 - 2y = 2y(6y^2 + y - 1)$

Next, we factor $6y^2 + y - 1$, looking for a combination of factors of 6 and factors of -1 that will give us the middle term with coefficient 1.

Factors of 6	Factors of -1	Possible Binomial Factors	Middle Term	
6, 1	1, -1	$(6y + 1)(y - 1)$	$-6y + y = -5y$	
		$(6y - 1)(y + 1)$	$6y - y = 5y$	
3, 2	1, -1	$(3y + 1)(2y - 1)$	$-3y + 2y = -y$	
		$(3y - 1)(2y + 1)$	$3y - 2y = y$	← Correct middle term

So $12y^3 + 2y^2 - 2y = 2y(6y^2 + y - 1) = 2y(3y - 1)(2y + 1)$.

PRACTICE 4

Factor: $2x^2 - x - 10$

PRACTICE 5

Factor: $18x^3 - 21x^2 - 9x$

Now, we consider factoring a trinomial of the form $ax^2 + bxy + cy^2$. This type of trinomial contains more than one variable, so we need to look for a factorization of the form $(?x + ?y)(?x + ?y)$.

EXAMPLE 6

Factor: $12x^2 + 28xy + 8y^2$

Solution Since 4 is the GCF of the trinomial, let's first factor it out.

$$12x^2 + 28xy + 8y^2 = 4(3x^2 + 7xy + 2y^2)$$

Next, we factor $3x^2 + 7xy + 2y^2$. We look for the combination of factors of 3 and factors of 2 that will give us the middle term $7xy$.

$$4(3x^2 + 7xy + 2y^2) = 4(?x + ?y)(?x + ?y)$$

Factors of 3	Factors of 2	Possible Binomial Factors	Middle Term	
3, 1	2, 1	$(3x + 2y)(x + y)$	$3xy + 2xy = 5xy$	
		$(3x + y)(x + 2y)$	$6xy + xy = 7xy$	← Correct middle term

So

$$12x^2 + 28xy + 8y^2 = 4(3x^2 + 7xy + 2y^2)$$
$$= 4(3x + y)(x + 2y).$$

Now, let's consider an alternative procedure for factoring a trinomial $ax^2 + bx + c$ based on *grouping*. This method, which the next example illustrates, is sometimes called the *ac method*.

EXAMPLE 7

Factor: $2x^2 + 5x - 3$

Solution First, we check that the terms of $2x^2 + 5x - 3$ have no common factors. Next, instead of listing the factors of 2 and the factors of -3 as in the trial-and-error method, we begin by finding their product:

$$ac = (2)(-3) = -6$$

We then look for two factors of the number ac (that is, -6) that add up to b (that is, 5). The numbers 6 and -1 satisfy these conditions, since $(6)(-1) = -6$ and $(6) + (-1) = 5$. Use these numbers to split up the middle term in the original trinomial, and then rewrite the trinomial.

$$2x^2 + 5x - 3 = 2x^2 + 6x + (-1)x - 3 \quad \text{Split up the middle term:}$$
$$5x = 6x + (-1)x.$$

$$= [2x^2 + 6x] + [(-1)x - 3] \quad \text{Group the first two terms and the last two terms.}$$

$$= 2x(x + 3) + (-1)(x + 3) \quad \text{Factor out the GCF from each group.}$$

$$= (x + 3)(2x - 1) \quad \text{Write in factored form.}$$

So $2x^2 + 5x - 3 = (x + 3)(2x - 1)$. As usual, we can check that this factorization is correct by multiplication.

PRACTICE 6

Factor: $36c^2 - 12cd - 15d^2$

PRACTICE 7

Factor: $2x^2 - 7x - 4$

Now, let's solve some applied problems involving the factoring of trinomials.

EXAMPLE 8	PRACTICE 8

Suppose a ball is thrown upward at 40 ft/sec from the top of a building 24 ft above the ground. Then the height of the ball above the ground in feet t sec after the ball is thrown is given by the expression $-16t^2 + 40t + 24$. Write this expression in factored form.

Solution We factor the expression $-16t^2 + 40t + 24$:

$$-16t^2 + 40t + 24 = -8(2t^2 - 5t - 3)$$
$$= -8(2t + 1)(t - 3)$$

So the factorization of $-16t^2 + 40t + 24$ is $-8(2t + 1)(t - 3)$, which can also be written $8(-2t - 1)(t - 3)$, or $8(2t + 1)(-t + 3)$.

A bin is made from a 7-ft by 5-ft sheet of metal by cutting out squares of equal size from each corner and then turning up the sides. The volume of the resulting bin can be represented by the expression $4x^3 - 24x^2 + 35x$. Rewrite this expression in factored form.

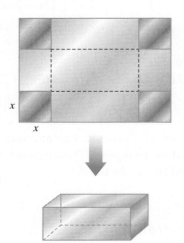

6.3 Exercises

Match each trinomial with its binomial factors.

e **1.** $2x^2 + 3x - 9$ **a.** $(2x - 9)(x + 1)$

d **2.** $2x^2 - 19x + 9$ **b.** $(2x - 9)(x - 1)$

b **3.** $2x^2 - 11x + 9$ **c.** $(x - 9)(2x + 1)$

f **4.** $2x^2 - 3x - 9$ **d.** $(x - 9)(2x - 1)$

c **5.** $2x^2 - 17x - 9$ **e.** $(2x - 3)(x + 3)$

a **6.** $2x^2 - 7x - 9$ **f.** $(x - 3)(2x + 3)$

Find the missing factor.

7. $3x^2 + 16x + 5 = (x + 5)(\quad)$ **8.** $2x^2 - 5x - 12 = (2x + 3)(\quad)$

9. $5x^2 - 13x - 6 = (5x + 2)(\quad)$ **10.** $2x^2 + x - 6 = (x + 2)(\quad)$

11. $3x^2 - 11x + 6 = (3x - 2)(\quad)$ **12.** $6x^2 - 7x + 2 = (2x - 1)(\quad)$

Factor, if possible.

13. $3x^2 + 8x + 5$ **14.** $2x^2 + 15x + 7$ **15.** $2y^2 - 11y + 5$ **16.** $3y^2 - 10y + 7$

17. $3x^2 + 14x + 8$ **18.** $2x^2 + 11x + 9$ **19.** $5x^2 + 9x - 6$ **20.** $5x^2 + 17x - 12$

21. $6y^2 - y - 5$ **22.** $4y^2 - 16y - 7$ **23.** $2y^2 - 11y + 14$ **24.** $7y^2 - 19y + 10$

25. $9a^2 - 18a - 16$ **26.** $10m^2 - m - 21$ **27.** $4x^2 - 13x + 3$ **28.** $4n^2 - 9n + 2$

29. $6 + 17y + 12y^2$ **30.** $4 + 16n + 15n^2$ **31.** $-17m + 21 + 2m^2$ **32.** $-16x + 5 + 3x^2$

33. $-6a^2 - 7a + 3$ **34.** $-5b^2 - 14b + 3$ **35.** $8y^2 + 5y - 22$ **36.** $6x^2 + 5x - 25$

37. $7y^2 + 36y - 5$ **38.** $2y^2 + 27y + 14$ **39.** $8a^2 + 65a + 8$ **40.** $8n^2 + 33n + 4$

41. $6x^2 + 25x - 9$ **42.** $10x^2 + 21x - 10$ **43.** $8y^2 - 26y + 15$ **44.** $4m^2 - 16m - 9$

45. $14y^2 - 38y + 20$ **46.** $9y^2 - 24y + 15$ **47.** $28a^2 + 24a - 4$ **48.** $6x^2 + 40x - 14$

49. $-6b^2 + 40b + 14$ **50.** $-25m^2 + 65m + 30$ **51.** $12y^3 + 50y^2 + 28y$ **52.** $6x^3 + 45x^2 + 21x$

53. $14a^4 - 38a^3 + 20a^2$ **54.** $10n^4 - 35n^3 + 15n^2$ **55.** $2x^3y + 13x^2y + 15xy$ **56.** $3xy^3 + 10xy^2 + 3xy$

57. $6ab^3 - 44ab^2 + 14ab$ **58.** $12a^3b - 34a^2b + 24ab$ **59.** $20c^2 - 9cd + d^2$ **60.** $12a^2 - 25ab + 12b^2$

61. $2x^2 - 5xy - 3y^2$ **62.** $6s^2 - st - 12t^2$ **63.** $8a^2 - 6ab + b^2$ **64.** $3m^2 - 8mn + 5n^2$

65. $18x^2 + 3xy - 6y^2$ **66.** $4s^2 + 10st - 24t^2$ **67.** $16c^2 - 44cd + 30d^2$ **68.** $16a^2 - 48ab + 36b^2$

69. $27u^2 + 18uv + 3v^2$ **70.** $4a^2 + 26ab - 48b^2$ **71.** $42x^3 + 45x^2y - 27xy^2$ **72.** $60p^3 + 28p^2q - 16pq^2$

73. $-30x^4y + 35x^3y^2 + 15x^2y^3$ **74.** $-24x^3y^2 - 6x^2y^3 + 18xy^4$

75. $5ax^2 - 28axy - 12ay^2$ **76.** $3cx^2 + 7cxy - 20cy^2$

Mixed Practice

Factor, if possible.

77. $-5m + 2 + 3m^2$ **78.** $3y^2 + 2y + 5$ **79.** $8x^2 - 2x - 3$ **80.** $8x^2 + 36x - 20$

81. $14x^3 + 44x^2 + 6x$ **82.** $30a^3b - 55a^2b + 15ab$ **83.** $8m^2 - 18mn + 9n^2$ **84.** $24a^3 + 18a^2b - 15ab^2$

85. $7r^2 - 9r + 2$ **86.** $10a^2 + 11a - 6$

Solve.

87. Choose the correct binomial factors of the trinomial $3x^2 - 22x + 7$.

    a. $(3x + 7)(x + 1)$ **b.** $(x + 7)(3x + 1)$

    c. $(3x - 7)(x - 1)$ **d.** $(x - 7)(3x - 1)$

88. Find the missing factor.

$$5x^2 - 2x - 3 = (5x + 3)(\qquad)$$

Applications

Solve.

89. An object is thrown upward so that its height in meters above the ground at time t seconds is represented by the expression $-5t^2 - 21t + 20$. Factor this expression.

90. A box with width w has a volume that can be expressed as $3w^3 - 2w^2 - w$. Rewrite this expression in factored form.

91. A homeowner decides to increase the area of his 8 ft by 10 ft deck by increasing both the length and the width, as shown in the diagram below.

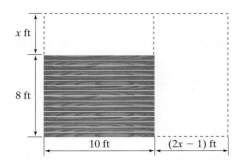

a. Find the area of the expanded deck by adding the areas of the three rectangles.

b. Express the area of the expanded deck in factored form.

93. Show that the expression $4n^2 - 12n + 5$ can be written as the product of two integers that differ by 4, no matter what integer n represents.

92. The diagram below shows a circular pad with radius r on the square top of a dining room table.

a. Find the area of the region of the table top not covered by the circular pad.

b. Express the area in part (a) in factored form.

94. The squares of the first n whole numbers add up to $\dfrac{2n^3 + 3n^2 + n}{6}$. Write this expression so that the numerator is in factored form.

● *Check your answers on page A-31.*

MINDSTRETCHERS

Groupwork

1. Work with a partner. Next to each polynomial, list at least two integers that, when inserted in the box, will make the polynomial factorable.

   a. $2x^2 + \boxed{}\ x + 5$ **b.** $\boxed{}\ x^2 - 4x + 1$ **c.** $3x^2 - x + \boxed{}$

Investigation

2. Copy or trace the following algebra tiles onto a piece of paper. Then cut out each tile separately. Finally position all the tiles like a jigsaw puzzle so as to form a rectangle. (*Hint:* Factor the polynomial $2x^2 + 7x + 3$.)

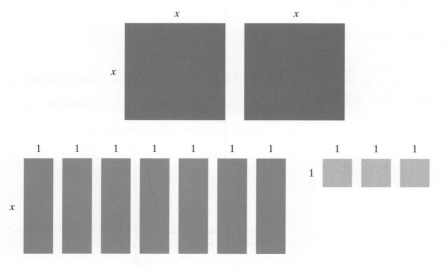

Writing

3. Do you prefer to factor trinomials using the trial-and-error method or the *ac* method? Explain why.

6.4   Special Factoring

Recall that in Section 5.6, we considered formulas for multiplying binomials in certain special cases: the square of a sum, the square of a difference, and the product of the sum and the difference of the same two terms. In this section, we show that these formulas also allow us to factor special polynomials, called *perfect square trinomials* and the *difference of squares*. Recognizing these special polynomials makes it easier to factor them.

OBJECTIVES
- To factor perfect square trinomials
- To factor the difference of squares
- To solve applied problems involving factoring

Factoring Perfect Square Trinomials

We have seen in Section 5.6 that squaring the sum or the difference of two terms gives us:

$$(a + b)^2 = a^2 + 2ab + b^2 \quad \text{and} \quad (a - b)^2 = a^2 - 2ab + b^2$$

Each of these products is called a **perfect square trinomial**. Such trinomials may be factored by reversing the multiplication process.

Factoring a Perfect Square Trinomial

$$a^2 + 2ab + b^2 = (a + b)^2$$
$$a^2 - 2ab + b^2 = (a - b)^2$$

The first formula shows us how to factor a trinomial that happens to be the sum of the squares of two terms *plus* twice their product. In this formula, the terms are a and b, the sum of their squares is $a^2 + b^2$, and $+2ab$ is twice their product. The formula says that the factorization of such a trinomial is the square of the *sum* of the two terms, namely $(a + b)^2$.

The second formula applies when we want to factor a trinomial that is the sum of the squares of two terms *minus* twice their product. In this formula, the terms are again a and b, the sum of their squares is $a^2 + b^2$, and $-2ab$ is minus twice their product. According to this formula, the factorization of such a trinomial is the square of the *difference* of the two terms, namely $(a - b)^2$.

Keep in mind that the formulas $a^2 + 2ab + b^2 = (a + b)^2$ and $a^2 - 2ab + b^2 = (a - b)^2$ only apply when a polynomial is a perfect square trinomial. Let's consider an example of recognizing these trinomials.

EXAMPLE 1

Determine whether each polynomial is a perfect square trinomial.

a. $x^2 - 8x + 16$ **b.** $-x^2 + 2x + 1$

c. $x^2 + 4x + 1$ **d.** $n^2 - 6n - 9$

e. $9x^2 + 30xy + 25y^2$

Solution

a. In the polynomial $x^2 - 8x + 16$, x^2 and 16 are perfect squares and correspond to a^2 and b^2, respectively, in the formula $a^2 - 2ab + b^2$. So a corresponds to x, and b corresponds to 4. Since the middle term, $-8x$, or $-2 \cdot x \cdot 4$, corresponds to $-2ab$, the polynomial is a perfect square trinomial.

$$x^2 - 8x + 16 = x^2 - \underbrace{2 \cdot x \cdot 4}_{} + (4)^2$$
$$ \underset{a^2}{\uparrow} \quad - \quad \underset{2ab}{\uparrow} \quad + \quad \underset{b^2}{\uparrow}$$

b. In the polynomial $-x^2 + 2x + 1$, $-x^2$ is not a perfect square since its coefficient is negative. Therefore, $-x^2 + 2x + 1$ is not a perfect square trinomial.

c. For the polynomial $x^2 + 4x + 1$, x^2 and 1 are both perfect squares with a corresponding to x, and b to 1 in the formula $a^2 + 2ab + b^2$. However, the middle term, $+4x$, is not twice the product of x and 1. So $x^2 + 4x + 1$ is not a perfect square trinomial.

d. The constant term of a perfect square trinomial, b^2, must be positive. So $n^2 - 6n - 9$ is not a perfect square, since its constant term -9 is negative.

e. In the polynomial $9x^2 + 30xy + 25y^2$, we know that $9x^2$ is equal to $(3x)^2$ and $25y^2$ equals $(5y)^2$. So a corresponds to $3x$ and b corresponds to $5y$ in the formula $a^2 + 2ab + b^2$. The middle term $+30xy$ can be expressed as $+2 \cdot 3x \cdot 5y$. So the trinomial is a perfect square.

$$9x^2 + 30xy + 25y^2 = (3x)^2 + \underbrace{2(3x)(5y)}_{} + (5y)^2$$
$$ \underset{a^2}{\uparrow} \quad + \quad \underset{2ab}{\uparrow} \quad + \quad \underset{b^2}{\uparrow}$$

Now, let's practice *factoring* perfect square trinomials.

PRACTICE 1

Indicate whether each trinomial is a perfect square.

a. $x^2 + 6x + 9$

b. $-4t^2 - 4t + 1$

c. $y^2 - 14y + 49$

d. $x^2 - 2x - 1$

e. $4p^2 - 4pq + q^2$

EXAMPLE 2

Factor: $x^2 + 12x + 36$

Solution For the trinomial $x^2 + 12x + 36$, the first term x^2 and the last term 36, or 6^2, are perfect squares. The middle term, $12x$, or $2 \cdot x \cdot 6$, is twice the product of x and 6. It follows that $x^2 + 12x + 36$ is a perfect square trinomial. Since its middle term is positive, we apply the formula for the square of a sum.

PRACTICE 2

Factor: $n^2 + 20n + 100$

$$x^2 + 12x + 36 = x^2 + \underbrace{2 \cdot x \cdot 6}_{} + (6)^2 = (x + 6)^2$$

$$a^2 \ + \quad 2ab \quad + \ b^2 \ = \ (a + b)^2$$

Check We can confirm our answer by multiplying out $(x + 6)^2$.

$$(x + 6)^2 = (x + 6)(x + 6) = x^2 + 2 \cdot 6x + 36 = x^2 + 12x + 36$$

EXAMPLE 3	PRACTICE 3

EXAMPLE 3

Write as the square of a binomial: $x^2 + 9 - 6x$

Solution Let's begin by rewriting the trinomial in descending order: $x^2 - 6x + 9$.

$$x^2 - 6x + 9 = x^2 - \underbrace{2 \cdot x \cdot 3}_{} + (3)^2 = (x - 3)^2$$

$$a^2 \ - \quad 2ab \quad + \ b^2 \ = \ (a - b)^2$$

PRACTICE 3

Factor: $t^2 + 4 - 4t$

EXAMPLE 4

Express $9x^2 - 6xy + y^2$ as the square of a binomial.

Solution Here the trinomial contains more than one variable.

$$9x^2 - 6xy + y^2 = (3x)^2 - \underbrace{2 \cdot 3x \cdot y}_{} + y^2 = (3x - y)^2$$

$$a^2 \ - \quad 2ab \quad + \ b^2 \ = \ (a - b)^2$$

PRACTICE 4

Write $25c^2 - 40cd + 16d^2$ as the square of a binomial.

EXAMPLE 5

Factor: $y^{10} + 16y^5 + 64$

Solution Since $y^{10} = (y^5)^2$ and $64 = (8)^2$, we know that y^{10} and 64 are perfect squares. Also, since $16y^5 = 2 \cdot 8 \cdot y^5$, that is, $16y^5$ is twice the product of y^5 and 8, we conclude that $y^{10} + 16y^5 + 64$ is a perfect square trinomial.

$$y^{10} + 16y^5 + 64 = (y^5)^2 + \underbrace{2 \cdot 8 \cdot y^5}_{} + (8)^2 = (y^5 + 8)^2$$

$$a^2 \ + \quad 2ab \quad + \ b^2 \ = \ (a + b)^2$$

PRACTICE 5

Express $x^4 + 8x^2 + 16$ as the square of a binomial.

   In factoring a perfect square trinomial, how do we know whether the binomial squared is the sum or the difference of two terms?

Factoring the Difference of Squares

Recall from Section 5.6 that when finding the product of the sum and the difference of the same terms, we get:

$$(a + b)(a - b) = a^2 - b^2$$

The product is a binomial that is called a **difference of squares**. We can factor such binomials by reversing the multiplication process.

Factoring the Difference of Squares

$$a^2 - b^2 = (a + b)(a - b)$$

This formula is a shortcut for factoring a binomial equal to the square of one term *minus* the square of another term. The terms are a and b, and so their squares are a^2 and b^2. The formula says that the factorization of a binomial that is the difference of the squares of two terms is the sum of the two terms times the difference of the same two terms, that is, $(a + b)(a - b)$.

EXAMPLE 6

Indicate whether each binomial is a difference of squares.

a. $x^2 - 81$ b. $x^2 + y^2$

c. $x^2 - y^3$ d. $4p^6 - q^2$

Solution

a. In the binomial $x^2 - 81$, both x^2 and 81 are perfect squares and correspond to a^2 and b^2, respectively, in our formula.

$$x^2 - 81 = x^2 - 9^2$$
$$\quad\quad a^2 \; - \; b^2$$

Here, a corresponds to x, and b corresponds to 9. So $x^2 - 81$ is a difference of squares.

b. In the expression $x^2 + y^2$, both x^2 and y^2 are perfect squares, where x corresponds to a and y to b. However, the binomial is the sum and not the difference of squares.

c. For $x^2 - y^3$, x^2 is a perfect square, but y^3 is not. So $x^2 - y^3$ is not a difference of squares.

d. The binomial $4p^6 - q^2$ can be rewritten as $(2p^3)^2 - q^2$, and so is a difference of squares.

PRACTICE 6

Determine whether each binomial is a difference of squares.

a. $x^2 - 64$

b. $x^2 + 49$

c. $x^3 - 16$

d. $r^4 - 9s^6$

Note, as the preceding example suggests, that even powers of a variable are perfect squares whereas odd powers of a variable are not. Can you explain why?

EXAMPLE 7

Factor: $x^2 - 100$

Solution Since x^2 and 100 are perfect squares, $x^2 - 100$ is a difference of squares:

$$x^2 - 100 = x^2 - 10^2 = (x + 10)(x - 10)$$
$$\quad a^2 \; - \; b^2 \; = \; (a + b) \; (a - b)$$

Check We can verify that $(x + 10)(x - 10)$ is the factorization of $x^2 - 100$ by multiplying.

$$(x + 10)(x - 10) = x^2 - 10x + 10x - 100 = x^2 - 100$$

PRACTICE 7

Factor: $y^2 - 121$

EXAMPLE 8	PRACTICE 8
Write $16x^2 - 49y^2$ in factored form.	Express $9x^2 - 25y^2$ in factored form.

Solution Since $16x^2 = (4x)^2$ and $49y^2 = (7y)^2$, we know that $16x^2$ and $49y^2$ are perfect squares. So $16x^2 - 49y^2$ is a difference of squares.

$$16x^2 - 49y^2 = (4x)^2 - (7y)^2 = \underbrace{(4x + 7y)}\,\underbrace{(4x - 7y)}$$
$$\begin{array}{ccccc} \uparrow & & \uparrow & \uparrow & \uparrow \\ a^2 & - & b^2 & = & (a+b) \quad (a-b) \end{array}$$

EXAMPLE 9	PRACTICE 9
Factor: $4x^4 - 9y^6$	Factor: $64x^8 - 81y^2$

Solution Because $4x^4 = (2x^2)^2$ and $9y^6 = (3y^3)^2$, we see that $4x^4$ and $9y^6$ are both perfect squares. So $4x^4 - 9y^6$ is a difference of squares.

$$4x^4 - 9y^6 = (2x^2)^2 - (3y^3)^2 = \underbrace{(2x^2 + 3y^3)}\,\underbrace{(2x^2 - 3y^3)}$$
$$\begin{array}{ccccc} \uparrow & & \uparrow & \uparrow & \uparrow \\ a^2 & - & b^2 & = & (a+b) \quad (a-b) \end{array}$$

EXAMPLE 10	PRACTICE 10
Find an expression for the area of the cross section of the pipe pictured. Write this expression in factored form.	A stone is dropped from a bridge 256 ft above a river. The height of the stone above the river t sec after it is dropped is given by the expression $256 - 16t^2$. Factor this expression.

Solution The cross section is a ring-shaped region between two circles with the same center. The radius of the inner circle is r and the radius of the outer circle is R. Using the formula for the area of a circle helps us to find an expression for the area of the cross section:

Area of the cross section

$$= \text{Area of the large circle} - \text{Area of the small circle}$$
$$= \pi R^2 - \pi r^2$$
$$= \pi(R^2 - r^2)$$
$$= \pi(R + r)(R - r)$$

So in factored form, the expression for the area of the cross section of the pipe is $\pi(R + r)(R - r)$.

In this section, we have discussed how to factor some special types of polynomials, namely, perfect square trinomials and the difference of squares. There are, however, other special types of polynomials. Two of these special types, the difference of cubes and the sum of cubes, are discussed in Section 6.5.

6.4   Exercises

FOR EXTRA HELP  PRACTICE WATCH DOWNLOAD READ REVIEW

Mathematically Speaking

Fill in each blank with the most appropriate term or phrase from the given list.

odd	the square	even
perfect square trinomial	half the product	twice the product
square of the sum of two terms	difference of squares	

1. A(n) _____ of the form $a^2 + 2ab + b^2$ can be factored as $(a + b)^2$.

2. The trinomial $a^2 + 5ab + 25b^2$ is not a perfect square, because the middle term is not _____ of a and $5b$.

3. The binomial $a^2 - b^2$, a(n) _____, can be factored as $(a + b)(a - b)$.

4. Powers of a variable that are _____ are not perfect squares.

Determine whether each polynomial is a perfect square trinomial, a difference of squares, or neither.

5. $x^2 + 2x + 1$

6. $y^2 + 8y + 16$

7. $-t^2 - 4t + 1$

8. $x^2 - 6x + 9$

9. $n^2 - 2n + 1$

10. $y^2 - 3y - 4$

11. $x^2 - 25$

12. $y^2 - 49$

13. $81x^2 - 36y^2$

14. $x^2 + y^2$

15. $25x^2 - 20x + 4$

16. $-y^2 + 2y + 9$

17. $y^2 + 1$

18. $4x^2 - 100y^2$

19. $25x^2 + 5xy + y^2$

20. $49x^2 + 14xy + y^2$

21. $x^3 - 1$

22. $x^4 + 9$

Factor, if possible.

23. $x^2 - 12x + 36$

24. $y^2 - 14y + 49$

25. $y^2 + 20y + 100$

26. $x^2 + 2x + 1$

27. $a^2 - 4a + 4$

28. $b^2 - 22b + 121$

29. $x^2 - 6x - 9$

30. $y^2 - 10y - 25$

31. $m^2 - 64$

32. $n^2 - 1$

33. $y^2 - 81$

34. $x^2 - 16$

35. $144 - x^2$

36. $225 - t^2$

37. $4a^2 - 36a + 81$

38. $25b^2 - 20b + 4$

39. $49x^2 + 28x + 4$

40. $9y^2 + 24y + 16$

41. $36 - 60x + 25x^2$

42. $49 - 42y + 9y^2$

43. $100m^2 - 81$ **44.** $16n^2 - 25$ **45.** $36x^2 + 121$ **46.** $64n^2 + 169$

47. $1 - 9x^2$ **48.** $81 - 4y^2$ **49.** $m^2 + 26mn + 169n^2$ **50.** $225a^2 - 30ab + b^2$

51. $4a^2 + 36ab + 81b^2$ **52.** $25s^2 - 40st + 16t^2$ **53.** $x^2 - 4y^2$ **54.** $49c^2 - d^2$

55. $100x^2 - 9y^2$ **56.** $36a^2 - 121b^2$ **57.** $y^4 + 2y^2 + 1$ **58.** $x^4 + 4x^2 + 4$

59. $6x^2 + 12x + 6$ **60.** $12y^2 + 24y + 12$ **61.** $27m^3 - 36m^2 + 12m$ **62.** $48y^3 - 24y^2 + 3y$

63. $4s^2t^3 + 80s^2t^2 + 400s^2t$ **64.** $2x^3y^2 - 52x^2y^2 + 338xy^2$ **65.** $3k^3 - 147k$ **66.** $5m^3 - 125m$

67. $4y^4 - 36y^2$ **68.** $3t^5 - 300t^3$ **69.** $27x^2y - 3x^2y^3$ **70.** $50xy - 18x^3y$

71. $2a^2b^2 - 98$ **72.** $9x^4y^2 - 81$ **73.** $256 - r^4$ **74.** $625 - t^4$

75. $5x^4 - 80y^4$ **76.** $64s^4 - 4t^4$

77. $x^2(c - d) - 4(c - d)$ **78.** $y^2(a - b) - (a - b)$

79. $16(x - y) - a^2(x - y)$ **80.** $9(y - c) - x^2(y - c)$

Mixed Practice

Factor, if possible.

81. $9c^2 + 48cd + 64d^2$ **82.** $169m^2 - 49n^2$

83. $a^2 - 225b^4$ **84.** $8t^3 - 56t^2 + 98t$

85. $54a^4b^2 - 36a^2b + 6$ **86.** $12xy - 27xy^3$

87. $81 - w^4$ **88.** $196 - r^2$

89. $36u^2 + 60u + 25$ **90.** $121s^2 + 36$

Determine whether each polynomial is a perfect square trinominal, a difference of squares, or neither.

91. $j^2 - 169$ **92.** $81a^2 - 90ab + 25b^2$

Applications

Solve.

93. If the radius of a balloon decreases from radius r_1 to radius r_2, then the drop in the balloon's surface area is given by the expression $4\pi r_1^2 - 4\pi r_2^2$. Write this expression in factored form.

94. The height (in feet) of a stone dropped from a cliff 100 ft above a river, as shown in the illustration, is given by the expression $100 - 16t^2$, where t is time in seconds. Factor this expression.

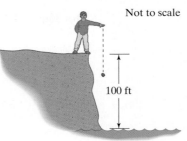

Not to scale

100 ft

95. A $16,000 investment grew by an average annual rate of return of r. After two years, the value of the investment in dollars was $16{,}000 + 32{,}000r + 16{,}000r^2$. What is the factorization of this expression?

96. Find an expression, in factored form, for the area of the wooden border of the square picture frame shown.

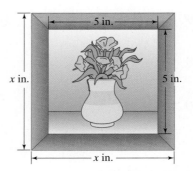

5 in.

x in.

5 in.

x in.

97. When the velocity of a rocket increases from v_1 to v_2, the force caused by air resistance increases by $kv_2^2 - kv_1^2$ for a constant k. Write this polynomial in factored form.

98. An open box is made from a 2-ft by 2-ft piece of cardboard by cutting out equal squares from each of the four corners and turning up the sides. The volume of the resulting box can be modeled by the polynomial $4x - 8x^2 + 4x^3$. Factor the expression.

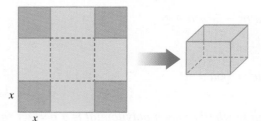

x

x

• *Check your answers on page A-32.*

MINDSTRETCHERS

Patterns

1. Find two factors of 3599 both of which are greater than 1. (*Hint:* 3599 = 3600 − 1.)

Groupwork

2. Try this trick with a partner:

    a. Take your partner's age in years.   **b.** Square it.

    c. Subtract 9.   **d.** Divide the result by 3 less than your partner's age.

    e. Subtract 53.   **f.** Add your partner's age.

    g. Divide by 2.   **h.** Add 5^2.

    i. Check that you wind up where you started—with your partner's age.

In the table, record the results for three different ages. Then in the fourth row, repeat the steps with a variable x representing your partner's age. Explain why this trick works.

(a)	(b)	(c)	(d)	(e)	(f)	(g)	(h)	(i)
x								

Writing

3. In a few sentences, explain how the following diagram shows that $(a + b)^2 = a^2 + 2ab + b^2$.

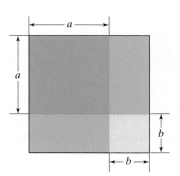

6.5   More Factoring Strategies

In this section, we discuss additional factoring strategies as well as formulas involving the *sums and differences of cubes*.

Factoring the Difference of Squares Completely

In some differences of squares, one or both of the expressions being
squared are binomials or powers of the variables. Here, we factor these
expressions completely.

EXAMPLE 1	PRACTICE 1
Factor each difference of squares. **a.** $9 - (a + b)^2$ **b.** $4x^4 - 9y^6$ **Solution** **a.** $9 - (a + b)^2 = 3^2 - (a + b)^2$ $\qquad\qquad = [3 + (a + b)][(3 - (a + b)]$ $\qquad\qquad = (3 + a + b)(3 - a - b)$ **b.** $4x^4 - 9y^6 = (2x^2)^2 - (3y^3)^2 = (2x^2 + 3y^3)(2x^2 - 3y^3)$	Write in factored form. **a.** $16 - (2x - y)^2$ **b.** $64x^8 - 25y^2$

The Sum and Difference of Cubes

Recall that the cube of a number is that number raised to the third power.

$$2^3 = 2 \cdot 2 \cdot 2 = 8$$
$$x^3 = x \cdot x \cdot x$$
$$(4y)^3 = 4 \cdot 4 \cdot 4 \cdot y \cdot y \cdot y = 64y^3$$

Note that the cube of a negative number is negative, for example,

$$(-3)^3 = (-3)(-3)(-3) = -27$$

Expressions such as 2^3, x^3, $64y^3$, and -27 are called *perfect cubes*.

An expression that is the *sum or the difference of two perfect cubes* can be
factored using the following formulas.

Factoring the Sum or Difference of Two Perfect Cubes

$$a^3 + b^3 = (a + b)(a^2 - ab + b^2)$$
$$a^3 - b^3 = (a - b)(a^2 + ab + b^2)$$

The first formula shows us how to factor a binomial that is the sum of the
cubes of two terms. In this formula, the terms are a and b and the sum of their
cubes is $a^3 + b^3$. The formula says that the factorization of a binomial of this form
is the product of the sum of the two terms, $a + b$, and the square of the first term

minus the product of the terms plus the square of the second term, that is, $a^2 - ab + b^2$.

We can apply the second formula to factor the difference of two cubes, say $a^3 - b^3$. The factorization of a binomial of this form is the product of the difference of the two terms, $a - b$, and the square of the first term *plus* the product of the terms plus the square of the second term, that is, $a^2 + ab + b^2$. Note that this factorization is identical to that in the first formula, except that the sign of the second term in each factor is reversed: $a - b$ and $a^2 + ab + b^2$.

EXAMPLE 2	PRACTICE 2

EXAMPLE 2

Indicate whether each binomial is a sum or difference of cubes.

a. $x^3 + y^3$

b. $n^6 - 1$

c. $8n^3 + 125p$

Solution

a. In the expression, x^3 and y^3 are perfect cubes, so $x^3 + y^3$ is a sum of cubes.

$$x^3 + y^3$$
$$\uparrow \quad \uparrow$$
$$a^3 + b^3$$

b. The expression $n^6 - 1$ can be written as $(n^2)^3 - (1)^3$, and so is a difference of cubes.

c. $8n^3 + 125p$ is not a sum of cubes because p is not a perfect cube.

PRACTICE 2

Determine whether each binomial is a sum or difference of cubes.

a. $p^3 - q^3$

b. $8 + x^6$

c. $x^4 + 27y^9$

Now, let's factor the sum or difference of cubes.

EXAMPLE 3	PRACTICE 3

EXAMPLE 3

Factor.

a. $8 + y^3$ **b.** $p^6 - 64q^3$ **c.** $(r + s)^3 + (r - s)^3$

Solution

a. $8 + y^3$ is a sum of cubes.

$$8 + y^3 = 2^3 + y^3 = \underbrace{(2 + y)}\underbrace{(2^2 - 2y + y^2)} = (2 + y)(4 - 2y + y^2)$$
$$\uparrow \quad \uparrow \qquad \uparrow \qquad\qquad \uparrow$$
$$a^3 + b^3 = (a + b) \quad (a^2 - ab + b^2)$$

So the factorization of $8 + y^3$ is $(2 + y)(4 - 2y + y^2)$.

b. $p^6 - 64q^3$ is a difference of cubes.

$$p^6 - 64q^3 = (p^2)^3 - (4q)^3$$
$$\qquad\qquad\uparrow \qquad\quad \uparrow$$
$$\qquad\qquad a^3 \quad - \quad b^3$$
$$= \underbrace{(p^2 - 4q)}\underbrace{[(p^2)^2 + p^2 \cdot (4q) + (4q)^2]}$$
$$\qquad\qquad\uparrow \qquad\qquad\qquad \uparrow$$
$$\qquad\quad (a - b) \qquad\quad (a^2 + ab + b^2)$$
$$= (p^2 - 4q)(p^4 + 4p^2q + 16q^2)$$

PRACTICE 3

Factor.

a. $125 - y^3$

b. $27m^3n^6 + 1$

c. $(x + 1)^3 - (x - 1)^3$

c. $(r + s)^3 + (r - s)^3$ is a sum of cubes. Here a corresponds to $(r + s)$ and b corresponds to $(r - s)$.

$(r + s)^3 + (r - s)^3$

$= [(r + s) + (r - s)][(r + s)^2 - (r + s)(r - s) + (r - s)^2]$

$= [2r][(r^2 + 2rs + s^2) - (r^2 - s^2) + (r^2 - 2rs + s^2)]$

$= [2r][r^2 + 2rs + s^2 - r^2 + s^2 + r^2 - 2rs + s^2]$

$= (2r)(r^2 + 3s^2)$

EXAMPLE 4

When the velocity of a rocket increases from v_1 to v_2, the force caused by air resistance increases by $kv_2^2 - kv_1^2$, for a constant k. Write this expression in factored form.

Solution The expression for the force is given by $kv_2^2 - kv_1^2$. Factoring this expression, we get:

$$kv_2^2 - kv_1^2 = k(v_2^2 - v_1^2) = k(v_2 + v_1)(v_2 - v_1)$$

PRACTICE 4

Suppose that a ball is dropped from the roof of a building 100 ft above the ground. The height of the ball t sec after it is dropped is given by the expression $100 - 16t^2$. Write this expression in factored form.

Factor, if possible. Remember to factor completely.

1. $(2u - v)^2 - 64$

2. $9 - (3a + b)^2$

3. $49 - 4(x + y)^2$

4. $16(p - q)^2 - 1$

5. $p^6 - 22p^3 + 121$

6. $n^4 + 8n^2 + 16$

7. $9a^8 + 48a^4b + 64b^2$

8. $81p^6 - 18p^3q^2 + q^4$

9. $4a^4 - 225$

10. $36y^6 - 169$

11. $49x^6 - 144y^4$

12. $121u^8 - 64v^4$

13. $100p^4q^2 - 9r^2$

14. $36x^2y^2 - 49z^4$

15. $(3p + q)^2 - (2p - q)^2$

16. $(a - 2b)^2 - (a - 3b)^2$

Determine whether each polynomial is a sum of cubes, a difference of cubes, or neither.

17. $x^3 - 64$

18. $125 + x^3$

19. $x^6 - 3y^3$

20. $8x^9 - 27y^6$

21. $y^{12} + 0.008x^3$

22. $1.25x^3 + 27y^3$

Factor, if possible.

23. $x^3 + 1$

24. $a^3 + 27$

25. $p^3 - 8$

26. $y^3 - 2$

27. $27t^2 - 4$

28. $8x^3 - 125$

29. $\dfrac{1}{8} - a^3$

30. $-\dfrac{1}{27} + t^3$

31. $125x^3 + y^3$

32. $u^3 + 216v^3$

33. $0.064b^3 - 0.027a^3$

34. $0.125p^3 - 0.001q^3$

35. $a^6 - 8$

36. $y^9 - 27$

37. $64x^9 + 27y^3$

38. $125a^3 - 8b^6$

39. $27 - (a + 1)^3$

40. $(x - 2)^3 + 1$

41. $(x - 2)^3 + (x + 2)^3$

42. $(p + 3)^3 - (3 - p)^3$

Mixed Practice

Solve.

43. Determine whether the polynomial $-8x^3 + 125y^6$ is a sum of cubes, a difference of cubes, or neither.

44. Determine whether the polynomial $4x^2 + 10xy + 25y^2$ is a perfect square trinomial, a difference of squares, or neither.

Factor, if possible.

45. $25(2a - b)^2 - 9$

46. $8x^3y^3 - 125$

47. $32x^3y - 18xy^3$

48. $s^9 + 27t^{12}$

Applications

Solve.

49. A furniture store sells two sizes of cube-shaped stackable storage cabinets. The dimensions of the interior storage space of the smaller cube is x in. by x in. by x in. The interior storage space of the larger cube is y in. by y in. by y in. Write an expression, in factored form, that shows the combined storage space of a large and a small cube.

50. If the radius of a spherical balloon increases from r_1 to r_2, then the increase in the volume of the balloon is given by the expression $\frac{4}{3}\pi r_2^3 - \frac{4}{3}\pi r_1^3$. Write the expression in factored form.

• *Check your answers on page A-32.*

6.6   Solving Quadratic Equations by Factoring

This section deals with a kind of equation that we have not previously considered, namely, a *quadratic equation*. Such equations come up in physics, in finance, and other fields as well.

Consider, for instance, a situation involving the movement of a rocket. If the rocket is shot straight upward from ground level with an initial velocity of 80 ft/sec, physicists approximate the rocket's height above the ground h by the expression $80t - 16t^2$, where t is the elapsed time in seconds and h is measured in feet. To find the time at which the rocket falls back and hits the ground (that is, when $h = 0$), we need to be able to solve the quadratic equation $80t - 16t^2 = 0$.

OBJECTIVES

- To solve quadratic equations by factoring

- To solve applied problems using quadratic equations

Definition

A **second-degree** or **quadratic equation** is an equation that can be written in the form $ax^2 + bx + c = 0$, where a, b, and c are real numbers and $a \neq 0$.

Some examples of quadratic equations are:

$$x^2 - x + 6 = 0 \qquad 3x^2 - 12 = 0 \qquad (x - 1)^2 = 0$$

Can you explain why these polynomials are of the second degree?

As in the case of a linear equation, a value is said to be a *solution* of a quadratic equation if substituting the value for the variable makes the equation a true statement.

Using Factoring to Solve Quadratic Equations

In this section, we consider those quadratic equations that can be solved by factoring. In Chapter 9, we will consider additional approaches to solving quadratic equations.

The key to solving quadratic equations by factoring is to apply the **zero-product property**.

> ### The Zero-Product Property
> If $ab = 0$, then $a = 0$ or $b = 0$, or both a and $b = 0$.

This property states that if the product of two factors is zero, then either one or both of the factors must be zero.

Consider these examples of the zero-product property.

- If $2x = 0$, then x must be 0 (since $2 \neq 0$).
- If $x(3x - 1) = 0$, then either $x = 0$ or $3x - 1 = 0$.
- If $(x - 3)(x + 2) = 0$, then either $x - 3 = 0$ or $x + 2 = 0$.

Let's see how to use the zero-product property to solve a quadratic equation that is already in factored form.

EXAMPLE 1	PRACTICE 1

EXAMPLE 1

Solve: $(2x - 1)(x + 6) = 0$

Solution Since $(2x - 1)(x + 6) = 0$, the zero-product property tells us that at least one of the factors must equal zero, that is either $2x - 1 = 0$ or $x + 6 = 0$.

$$(2x - 1)(x + 6) = 0$$

$2x - 1 = 0$ or $x + 6 = 0$ Set each factor equal to 0.

$\qquad 2x = 1 \qquad\qquad x = -6$ Solve each equation for x.

$$x = \frac{1}{2}$$

Check We replace x by the values $\frac{1}{2}$ and -6 in the original equation.

Substitute $\frac{1}{2}$ for x.

$$(2x - 1)(x + 6) = 0$$
$$\left[2 \cdot \left(\frac{1}{2}\right) - 1\right]\left(\frac{1}{2} + 6\right) \stackrel{?}{=} 0$$
$$(1 - 1)\left(6\tfrac{1}{2}\right) \stackrel{?}{=} 0$$
$$0 \cdot 6\tfrac{1}{2} \stackrel{?}{=} 0$$
$$0 = 0 \quad \textbf{True}$$

Substitute -6 for x.

$$(2x - 1)(x + 6) = 0$$
$$[(2)(-6) - 1](-6 + 6) \stackrel{?}{=} 0$$
$$(-13)(0) \stackrel{?}{=} 0$$
$$0 = 0 \quad \textbf{True}$$

So the solutions of the equation $(2x - 1)(x + 6) = 0$ are $\frac{1}{2}$ and -6.

PRACTICE 1

Solve: $(3x - 1)(x + 5) = 0$

In Example 1, note that we found the solutions of a quadratic equation by solving two linear equations: $2x - 1 = 0$ and $x + 6 = 0$. Explain how we know that these equations are linear.

When the quadratic expression in a second-degree equation is not given in factored form, we need to factor it before solving.

EXAMPLE 2	PRACTICE 2
Solve: $y^2 - 5y = 0$	Solve: $y^2 + 6y = 0$

Solution

$$y^2 - 5y = 0$$
$$y(y - 5) = 0 \qquad \text{Factor the left side of the equation.}$$

Next we set each factor equal to 0. Then we solve for x.

$$y = 0 \quad \text{or} \quad y - 5 = 0$$
$$y = 5$$

Check We verify our solutions in the original equation.

Substitute 0 for y.

$$y^2 - 5y = 0$$
$$(0)^2 - 5(0) \stackrel{?}{=} 0$$
$$0 - 0 \stackrel{?}{=} 0$$
$$0 = 0 \qquad \text{True}$$

Substitute 5 for y.

$$y^2 - 5y = 0$$
$$5^2 - 5(5) \stackrel{?}{=} 0$$
$$25 - 25 \stackrel{?}{=} 0$$
$$0 = 0 \qquad \text{True}$$

The solutions are 0 and 5.

In order to apply the zero-product property, the product of the factors of a quadratic must equal zero. This implies that a quadratic equation must be written in *standard form*, $ax^2 + bx + c = 0$, before it can be solved.

EXAMPLE 3	PRACTICE 3
Solve: $2x^2 + x = 1$	Solve: $4y^2 - 11y = 3$

Solution

$$2x^2 + x = 1$$
$$2x^2 + x - 1 = 0 \qquad \text{Write in standard form by adding } -1 \text{ to each side.}$$
$$(2x - 1)(x + 1) = 0 \qquad \text{Factor the left side of the equation.}$$
$$2x - 1 = 0 \quad \text{or} \quad x + 1 = 0 \qquad \text{Set each factor to equal to 0.}$$
$$2x = 1 \qquad\qquad x = -1 \qquad \text{Solve for } x.$$
$$x = \frac{1}{2}$$

Check We verify our solutions in the original equation.

Substitute $\frac{1}{2}$ for x.

$$2x^2 + x = 1$$
$$2\left(\frac{1}{2}\right)^2 + \frac{1}{2} \stackrel{?}{=} 1$$
$$2\left(\frac{1}{4}\right) + \frac{1}{2} \stackrel{?}{=} 1$$
$$1 = 1 \qquad \text{True}$$

Substitute -1 for x.

$$2x^2 + x = 1$$
$$2(-1)^2 + (-1) \stackrel{?}{=} 1$$
$$2(1) + (-1) \stackrel{?}{=} 1$$
$$1 = 1 \qquad \text{True}$$

So the solutions are $\frac{1}{2}$ and -1.

EXAMPLE 4

Solve: $2x(x - 3) = 8$

Solution We begin by writing the equation in standard form.

$$
\begin{aligned}
2x(x - 3) &= 8 \\
2x^2 - 6x &= 8 \qquad \text{Multiply.}\\
2x^2 - 6x - 8 &= 0 \qquad \text{Write in standard form by adding } -8 \text{ to each side.}\\
2(x^2 - 3x - 4) &= 0 \qquad \text{Factor out the GCF.}\\
2(x + 1)(x - 4) &= 0 \qquad \text{Write in factored form.}
\end{aligned}
$$

Next, we set factors containing variables equal to 0. Then we solve for x.

$$
\begin{aligned}
x + 1 = 0 \quad &\text{or} \quad x - 4 = 0 \\
x = -1 \qquad &\qquad x = 4
\end{aligned}
$$

Check

Substitute -1 for x.

$$
\begin{aligned}
2x(x - 3) &= 8 \\
2(-1)(-1 - 3) &\stackrel{?}{=} 8 \\
(-2)(-4) &\stackrel{?}{=} 8 \\
8 &= 8 \qquad \text{True}
\end{aligned}
$$

Substitute 4 for x.

$$
\begin{aligned}
2x(x - 3) &= 8 \\
2(4)(4 - 3) &\stackrel{?}{=} 8 \\
8(1) &\stackrel{?}{=} 8 \\
8 &= 8 \qquad \text{True}
\end{aligned}
$$

So the solutions are -1 and 4.

These examples lead us to the following strategy (or rule) for solving a quadratic equation by factoring.

> **To Solve a Quadratic Equation by Factoring**
> - If necessary, rewrite the equation in standard form with 0 on one side.
> - Factor the other side.
> - Use the zero-product property to get two simple linear equations.
> - Solve the linear equations.
> - Check by substituting the solutions in the original quadratic equation.

EXAMPLE 5

Suppose a homeowner has a square garden that she wants to make longer. If she extends one side by 5 ft and the adjacent side by 2 ft, the resulting garden would be rectangular with an area of 130 ft². How much fencing will she need to enclose the enlarged garden?

PRACTICE 4

Solve: $3t(t + 4) = 15$

PRACTICE 5

A framemaker is planning to frame a rectangular painting that has an area of 80 in².

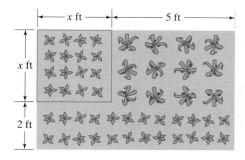

Solution Let's represent the length of each side of the square by x. The resulting rectangular garden will have dimensions $(x + 5)$ and $(x + 2)$. The area of a rectangle can be computed by multiplying its length and its width, and we are told that this area is 130.

$$(x + 5)(x + 2) = 130$$
$$x^2 + 7x + 10 = 130$$
$$x^2 + 7x - 120 = 0$$
$$(x + 15)(x - 8) = 0$$
$$x + 15 = 0 \quad \text{or} \quad x - 8 = 0$$
$$x = -15 \qquad x = 8$$

If she has 3 ft of framing to put around the picture, what should the dimensions of the frame be? (Ignore the frame's thickness.)

Check

Substitute -15 for x.

$$(x + 5)(x + 2) = 130$$
$$(-15 + 5)(-15 + 2) \stackrel{?}{=} 130$$
$$(-10)(-13) \stackrel{?}{=} 130$$
$$130 = 130 \quad \text{True}$$

Substitute 8 for x.

$$(x + 5)(x + 2) = 130$$
$$(8 + 5)(8 + 2) \stackrel{?}{=} 130$$
$$(13)(10) \stackrel{?}{=} 130$$
$$130 = 130 \quad \text{True}$$

So the two solutions of the equation are -15 and 8. Since x represents a length, we can reject the negative solution -15. We conclude that the length of each side of the square is 8 ft. To compute the perimeter of the rectangle, we can substitute into the following formula:

$$P = 2l + 2w$$
$$= 2(8 + 5) + 2(8 + 2)$$
$$= 2(13) + 2(10)$$
$$= 26 + 20$$
$$= 46$$

Therefore, 46 ft of fencing is needed to enclose the enlarged garden.

We can also apply what we know about solving quadratic equations to problems involving the Pythagorean theorem. This theorem relates to the lengths of the three sides of a right triangle, which is illustrated on the next page.

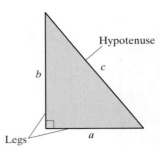

Hypotenuse

A right triangle has one 90° angle. The side opposite the 90° angle, the longest side, is called the *hypotenuse*. The other sides are called *legs*.

The Pythagorean theorem states that for every right triangle, the sum of the squares of the legs equals the square of the hypotenuse: $a^2 + b^2 = c^2$.

EXAMPLE 6	PRACTICE 6

EXAMPLE 6

How far from the base of a building should a painter place a 17-ft ladder so that it reaches 15 ft up the building?

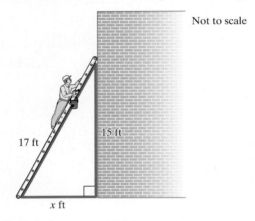

Not to scale

17 ft

15 ft

x ft

Solution Let x represent the distance from the base of the building to the bottom of the ladder. Note that the ladder, the ground, and the building form a right triangle. Using the Pythagorean theorem, we get:

$$a^2 + b^2 = c^2$$
$$x^2 + 15^2 = 17^2$$
$$x^2 + 225 = 289$$
$$x^2 - 64 = 0$$
$$(x - 8)(x + 8) = 0$$
$$x - 8 = 0 \quad \text{or} \quad x + 8 = 0$$
$$x = 8 \qquad\qquad x = -8$$

Since x represents a distance, we consider only the positive value of x, namely, 8.

Check Substitute 8 for x.

$$x^2 + 15^2 = 17^2$$
$$8^2 + 15^2 \stackrel{?}{=} 17^2$$
$$64 + 225 \stackrel{?}{=} 289$$
$$289 = 289 \quad \text{True}$$

So the painter should place the ladder 8 ft from the base of the building.

PRACTICE 6

Two scooters, traveling at constant rates, leave an intersection at the same time. One scooter travels north while the other travels east. When the scooter traveling east has gone 5 mi, the distance between the two scooters is 13 mi. How far has the scooter going north traveled?

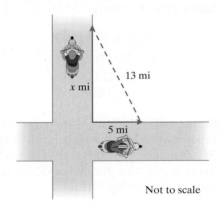

x mi

13 mi

5 mi

Not to scale

Mathematically Speaking

Fill in each blank with the most appropriate term or phrase from the given list.

sum of the squares of the legs	binomial equation	either one or both
in standard form	square of the sum of the legs	a perfect square
both	quadratic equation	

1. A second-degree or _____ is an equation that can be written in the form $ax^2 + bx + c = 0$, where a, b, and c are real numbers and $a \neq 0$.

2. The zero-product property states that if the product of two factors is zero, then _____ of the factors must be zero.

3. In using the zero-product property to solve a quadratic equation, the equation must be _____.

4. The Pythagorean theorem states that for every right triangle, the _____ equals the square of the hypotenuse.

Indicate whether each equation is linear or quadratic.

5. $x^2 - 3x + 2 = 0$

6. $x^2 = 6x$

7. $(x + 3)(x - 4) = x^2$

8. $\dfrac{x + 1}{4} = 2$

9. $2x^2 + 12x = -10$

10. $(x + 4)(x - 1) = 14$

Solve.

11. $(x + 3)(x - 4) = 0$

12. $(x - 2)(x - 1) = 0$

13. $4(x - 1) = 0$

14. $-7(2t + 3) = 0$

15. $y(3y + 5) = 0$

16. $2y(5y - 4) = 0$

17. $(2t + 1)(t - 5) = 0$

18. $(t - 3)(3t - 1) = 0$

19. $(2x + 3)(2x - 3) = 0$

20. $(5 - 4x)(5 + 4x) = 0$

21. $t(2 - 3t) = 0$

22. $t(1 + 2t) = 0$

23. $y^2 - 2y = 0$

24. $y^2 + 3y = 0$

25. $5x - 25x^2 = 0$

26. $3t + 6t^2 = 0$

27. $x^2 + 5x + 6 = 0$

28. $x^2 + x - 2 = 0$

29. $x^2 + x - 56 = 0$

30. $y^2 - y - 90 = 0$

31. $2x^2 - 5x - 3 = 0$

32. $2y^2 + 5y - 12 = 0$

33. $6x^2 - x - 2 = 0$

34. $4t^2 - 8t + 3 = 0$

35. $0 = 36x^2 - 12x + 1$

36. $0 = 25y^2 + 10y + 1$

37. $r^2 - 121 = 0$

38. $t^2 - 49 = 0$

39. $0 = (2x - 3)^2$

40. $0 = (4x + 1)^2$

41. $16x^2 - 16x + 4 = 0$

42. $9t^2 + 18t + 9 = 0$

43. $9m^2 + 15m - 6 = 0$

44. $6n^2 - 32n + 10 = 0$

45. $r^2 - r = 6$

46. $r^2 + 3r = 10$

47. $y^2 - 7y = -12$

48. $y^2 - y = 12$

49. $n^2 + 2n = 8$

50. $t^2 - 3t = 18$

51. $3y^2 + 4y = -1$

52. $3k^2 + 17k = -10$

53. $4x^2 + 6x = -2$

54. $12x^2 + 33x = 9$

55. $2n^2 = -10n$

56. $3m^2 = 6m$

57. $4x^2 = 1$

58. $9y^2 = 16$

59. $8y^2 = 2$

60. $12x^2 = 48$

61. $3r^2 + 6r = 2r^2 - 9$

62. $5n^2 + 36 = 12n + 4n^2$

63. $x(x - 1) = 12$

64. $r(r + 3) = 10$

65. $4t(t - 1) = 24$

66. $2n(n + 7) = -24$

67. $(y + 3)(y - 2) = 14$

68. $(m - 6)(m + 1) = -10$

69. $(3n - 2)(n + 5) = -14$

70. $(t - 5)(t + 2) = 18$

71. $(n + 2)(n + 4) = 12n$

72. $(3x + 5)(x - 1) = 16x$

73. $3x(2x - 5) = x^2 - 10$

74. $n^2 + 8 = 3n(n - 2)$

Mixed Practice

Solve.

75. $4r^2 + 11r - 3 = 0$

76. $0 = 9u^2 + 6u + 1$

77. $10x^2 + 25x - 15 = 0$

78. $n^2 + 3n = 28$

79. $5a^2 + 19a = 4$

80. $9b^2 = 27b$

81. $3s^2 + 64 = 16s + 2s^2$

82. $(x - 6)(x + 2) = 33$

83. $6t(3t - 5) = 0$

84. $(7 - 3m)(7 + 3m) = 0$

85. $4j + 8j^2 = 0$

Solve.

86. Is the equation $(x - 3)(x + 1) = 5$ linear or quadratic?

Applications

Solve.

87. In a certain league, the teams play each other twice in a season. It can be shown that if there are *n* teams in the league, the teams must play $n^2 - n$ games. If the league plays 210 games in a season, how many teams were in the league?

88. The number of ways to pair *n* students in a physics lab can be represented by the expression $\frac{1}{2}n(n - 1)$. If there are 325 different ways to pair students, how many students are in the lab?

89. Two cars leave an intersection, one traveling west and the other south. After some time, the faster car is 2 mi farther away from the intersection than the slower car. At that time, the two cars are 10 mi apart. How far did each car travel?

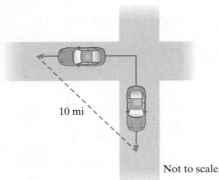

10 mi

Not to scale

90. The sail on a sailboat is a right triangle in which the hypotenuse is called the leech. A 12-ft-tall mainsail has a leech length of 13 ft. If sailcloth costs $10 per square foot, what is the cost of a new mainsail?

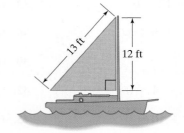

13 ft

12 ft

91. According to design specifications, the base of the Freedom Tower in New York City is a cube. The area of each side of the base is 40,000 ft². What is the height of the base of the tower? (*Source:* Lower Manhattan Development Corporation)

92. Suppose that a businessman invested $8000 in a high-risk growth fund that after two years was worth $12,500. His broker used the equation $8000(1 + r)^2 = 12,500$ to find the average annual rate of return. What is this rate?

93. A diver jumps from a diving board 24 ft above a pool. After t sec, the diver's height h above the water (in feet) is given by the expression $-16t^2 + 8t + 24$. In how many seconds will the diver hit the water ($h = 0$)?

94. The formula $h = v_0 t - 16t^2$ gives the height h in feet of an object after t sec. Here, v_0 is the initial velocity of the object expressed in feet per second. Suppose a toy rocket was launched from the ground straight up with an initial velocity of 64 ft/sec. How many seconds after launch was the rocket 48 ft above the ground?

• *Check your answers on page A-32.*

MINDSTRETCHERS

Mathematical Reasoning

1. Give an example of an equation:

   a. whose solutions are 2 and 3.

   b. whose only solution is 5.

   c. whose solutions have opposite signs.

   d. whose solutions are 2, 3, and 4.

Technology

2. Consider the quadratic equation $x^2 - 3x - 10 = 0$.

   a. Solve this equation.

   b. On a calculator or a computer, graph $y = x^2 - 3x - 10$. Use the graph to find the x-intercepts.

   c. Explain how you can use these intercepts to solve the original equation $x^2 - 3x - 10 = 0$.

Writing

3. Explain whether the zero-product property is true for more than two factors.

KEY CONCEPTS AND SKILLS   CONCEPT   SKILL

Concept/Skill	Description	Example
[6.1] **Common Factor of Two or More Integers**	An integer that is a factor of each integer.	5 is a common factor of 15 and 50.
[6.1] **Greatest Common Factor (GCF) of Two or More Integers**	The greatest integer that is a factor of each integer.	The GCF of 45, 63, and 81 is 9.
[6.1] **Greatest Common Factor (GCF) of Two or More Monomials**	The product of the greatest common factor of the coefficients and the highest powers of the variable factors common to each monomial.	The GCF of $6x^4$, $8x^3$, and $12x^2$ is $2x^2$.
[6.1] **Factoring By Grouping**	Group pairs of terms and factor out a GCF in each group, if necessary. Then factor out the common binomial factor.	$xy + x - 4y - 4$ $= (xy + x) + (-4y - 4)$ $= x(y + 1) - 4(y + 1)$ $= (y + 1)(x - 4)$
[6.2] **Factoring a Trinomial of the Form $ax^2 + bx + c$, Where $a = 1$**	List and test the factors of c to find two integers for $(x + ?)(x + ?)$ whose product is c and whose sum is b.	$x^2 - 8x + 12 = (x - 2)(x - 6)$ because **Factors of 12** $(-2) \cdot (-6) = 12$ and $(-2) + (-6) = -8$ **Sum of factors**
[6.3] **Factoring a Trinomial $ax^2 + bx + c$, Where $a \neq 1$ (Trial-and-Error Method)**	• List and test the factors of a and of c to find four integers for $(?x + ?)(?x + ?)$ so that the product of the leading coefficients of the binomial factors is a, the product of the constant terms of the binomial factors is c, and the coefficients of the inner and outer products add up to b.	$2x^2 + 15x + 7 = (2x + 1)(x + 7)$ because $2 \cdot 1 = 2,$ $1 \cdot 7 = 7,$ and $2 \cdot 7 + 1 \cdot 1 = 15.$
[6.3] **Factoring a Trinomial $ax^2 + bx + c$, Where $a \neq 1$ (ac Method)**	• Form the product ac. • Find two factors of ac that add up to b. • Use these factors to split up the middle term in the original trinomial. • Group the first two terms and the last two terms. • From each group, factor out the common factor. • Factor out the common binomial factor.	For $2x^2 + 15x + 7$, $ac = 2 \cdot 7 = 14$ $1 \cdot 14 = 14$ and $1 + 14 = 15$ $2x^2 + 15x + 7$ $= 2x^2 + x + 14x + 7$ $= (2x^2 + x) + (14x + 7)$ $= x(2x + 1) + 7(2x + 1)$ $= (2x + 1)(x + 7)$

continued

Concept/Skill	Description	Example
[6.4] Factoring a Perfect Square Trinomial	$a^2 + 2ab + b^2 = (a + b)^2$ $a^2 - 2ab + b^2 = (a - b)^2$	$x^2 + 12x + 36 = (x + 6)^2$ $x^2 - 6x + 9 = (x - 3)^2$
[6.4] Factoring the Difference of Squares	$a^2 - b^2 = (a + b)(a - b)$	$x^2 - 100 = (x + 10)(x - 10)$
[6.5] Factoring the Sum or Difference of Two Perfect Cubes	$a^3 + b^3 = (a + b)(a^2 - ab + b^2)$ $a^3 - b^3 = (a - b)(a^2 + ab + b^2)$	$x^3 + 27 = (x + 3)(x^2 - 3 \cdot x + 3^2)$ $\qquad = (x + 3)(x^2 - 3x + 9)$ $8x^3 - y^3$ $= (2x)^3 - y^3$ $= (2x - y)[(2x)^2 + 2x \cdot y + y^2]$ $= (2x - y)(4x^2 + 2xy + y^2)$
[6.6] Second-degree or Quadratic Equation	An equation that can be written in the form $ax^2 + bx + c = 0$, where a, b, and c are real numbers and $a \neq 0$.	$x^2 - x + 6 = 0$
[6.6] The Zero-Product Property	If $ab = 0$, then $a = 0$ or $b = 0$, or both a and $b = 0$.	If $x(3x - 1) = 0$, then either $x = 0$ or $3x - 1 = 0$.
[6.6] To Solve a Quadratic Equation by Factoring	• If necessary, rewrite the equation in standard form with 0 on one side. • Factor the other side. • Use the zero-product property to get two simple linear equations. • Solve the linear equations. • Check by substituting the solutions in the original quadratic equation.	$x(x - 3) = 4$ $x^2 - 3x - 4 = 0$ $(x - 4)(x + 1) = 0$ $x - 4 = 0 \quad \text{or} \quad x + 1 = 0$ $x = 4 \qquad\qquad x = -1$ **Check** Substitute 4 for x. $4(4 - 3) \stackrel{?}{=} 4$ $4(1) \stackrel{?}{=} 4$ $4 = 4 \qquad$ **True** Substitute -1 for x. $(-1)(-1 - 3) \stackrel{?}{=} 4$ $(-1)(-4) \stackrel{?}{=} 4$ $4 = 4 \qquad$ **True**

| **Chapter 6** | **Review Exercises** |

[6.1]   *To help you review this chapter, solve these problems.*

1. Find the greatest common factor of 48, 36, and 60.

2. Find the greatest common factor of $9m^3n$, $24m^4$, and $15m^2n^2$.

Factor.

3. $3x - 6y$

4. $16p^3q^2 + 18p^2q - 4pq^2$

5. $(n - 1) + n(n - 1)$

6. $xb - 5b - 2x + 10$

Solve for the indicated variable.

7. $d = rt_1 + rt_2$ for r

8. $ax + y = bx + c$ for x

[6.2]   *Factor, if possible.*

9. $x^2 + x + 1$

10. $m^2 - m + 3$

11. $y^2 + 42 + 13y$

12. $m^2 - 7mn + 10n^2$

13. $24 - 8x - 2x^2$

14. $-15xy^2 + 3x^3 - 12x^2y$

[6.3]   *Factor, if possible.*

15. $3x^2 + 5x - 2$

16. $5n^2 + 13n + 6$

17. $3n^2 - n - 1$

18. $6x^2 - x - 12$

19. $2a^2 + 3ab - 35b^2$

20. $16a - 4a^2 - 15$

21. $9y^3 - 21y + 60y^2$

22. $2p^2q - 3pq^2 - 2q^3$

[6.4]   *Factor, if possible.*

23. $b^2 - 6b + 9$

24. $64 - x^2$

25. $25y^2 - 20y + 4$

26. $9a^2 + 24ab + 16b^2$

27. $81p^2 - 100q^2$

28. $4x^8 - 28x^4 + 49$

29. $48x^4 - 3y^4$

30. $x^2(x - 1) - 9(x - 1)$

[6.5]   *Factor, if possible.*

31. $3u^3 + 81$

32. $64c^3 - 27d^3$

33. $(x + y)^2 - z^2$

34. $1 + (3a + 1)^3$

35. $32u^2 - 2v^2$

36. $x^2(x - 1) + 9(1 - x)$

[6.6]   *Solve.*

37. $(x + 2)(x - 1) = 0$

38. $t(t - 4) = 0$

39. $3x^2 + 18x = 0$

40. $4x^2 + 4x + 1 = 0$

41. $y^2 - 10y = -16$

42. $3k^2 - k = 2$

43. $4n(2n + 3) = 20$

44. $(y - 1)(y + 2) = 10$

Mixed Applications

Solve.

45. The length of an object varies with its temperature. The expression $aLt_2 - aLt_1$ represents the change in the length of an object heated to temperature t_2, where L is its length at temperature t_1 and a is the *coefficient of linear expansion*, a constant that depends on the material that the object is made of. Write this expression in factored form.

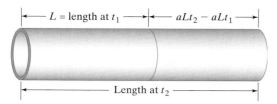

46. If a ball is thrown straight upward at a ft/sec, its height above the point of release is given by the expression $at - 16t^2$, where t is the number of seconds after release. Write an expression in factored form for the distance between the object's location at t_1 and at t_2.

47. Find the distance between the two intersections shown on the city grid pictured if each block is 500 ft long.

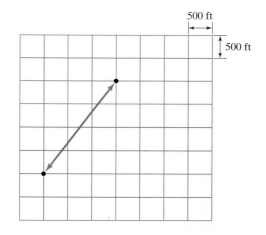

48. A kite maker designs the diamond-shaped kite shown. The diagonals of the kite cross at right angles. The vertical diagonal is 52 in. long. What is the length of the horizontal diagonal of the kite shown?

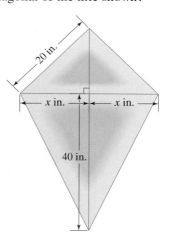

49. After t sec, the height of a rocket launched straight upward from ground level with an initial velocity of 76 ft/sec can be modeled by the polynomial $76t - 16t^2$. After how many seconds will the rocket reach a height of 18 ft above the launch (that is, equal to +18)?

50. The formula for the area of the trapezoid shown is $A = \frac{1}{2}hb + \frac{1}{2}hB$. Solve this formula for h in terms of A, b, and B.

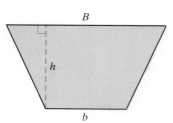

● *Check your answers on page A-32.*

**To see whether you have mastered the topics in this chapter, take this test.**

1. Find the greatest common factor of $12x^3$ and $15x^2$.

**Factor.**

2. $2xy - 14y$

3. $6pq^2 + 8p^3 - 16p^2q$

4. $ax - bx + by - ay$

5. $n^2 - 13n - 48$

6. $-8 + x^2 - 2x$

7. $15x^2 - 5x^3 + 20x$

8. $4x^2 + 13xy - 12y^2$

9. $-12x^2 + 36x - 27$

10. $9x^2 + 30xy + 25y^2$

11. $121 - 4x^2$

12. $p^2q^2 - 1$

13. $y^4 - 8y^2 + 16$

14. $64 - n^3$

**Solve.**

15. $(n + 8)(n - 1) = 0$

16. $6x^2 + 10x = 4$

17. $(2n + 1)(n - 1) = 5$

18. In the right triangle shown, find the length of the missing side.

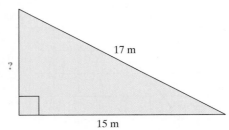

19. The energy it takes to lift an object of mass m from level y_1 to level y_2 is $mgy_2 - mgy_1$, where g is a constant. Factor this expression.

20. A rectangular garden measures 25 ft by 30 ft. The gardener wishes to surround the garden with a border of mulch x feet wide, as shown. Write an expression in factored form for the area of the mulch border in terms of x.

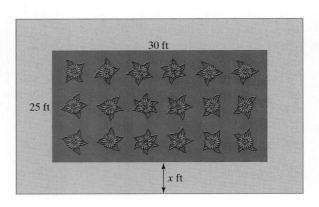

● _Check your answers on page A-32._

Cumulative Review Exercises

To help you review, solve the following:

1. Evaluate $2x + y - z$ if $x = -4$, $y = 0$, and $z = -5$.

2. Simplify: $5a + 2 - 2b + 3a + 9b$

3. Solve: $z + 4 = 5z - 2(z + 6)$

4. Solve the system:

      (1) $x + y = 6$
      (2) $y = 2x + 9$

5. Solve: $y < 1 - x$
            $y \geq \dfrac{1}{3}x - 3$

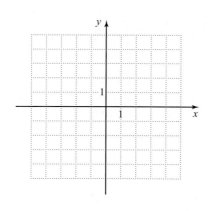

6. Simplify: $\dfrac{n^2 \cdot n^3 \cdot n}{n^5}$

7. Find the product: $(3x + 1)(7x - 2)$

8. A part-time college student pays a student fee of f dollars plus c dollars per credit. What is the charge for a student with an 8-credit schedule?

9. The owner of a shirt factory has fixed expenses of $500 per day. It costs the factory $15 to produce each shirt. If the shirts are sold wholesale at $25 apiece, how many shirts must be sold per day to break even?

10. An expression for the total surface area of a tin can is $2\pi r^2 + 2\pi rh$, where r is the radius and h is the height of the can. Rewrite this expression in factored form.

• *Check your answers on page A-33.*

CHAPTER 7

Rational Expressions and Equations

Rational Expressions and Grapes

Science plays a role in viticulture—the art of growing grapes. For instance, scientists have used DNA finger-printing—the same technique used in paternity suits and criminal trials—to trace the parentage of numerous grape varieties.

Viticulture also makes use of algebra. Since much of the flavor and color of grapes is in their skin, wine growers want to raise grapes with an increased surface-to-volume ratio. This ratio is larger for smaller grapes, which have proportionately more skin than larger grapes.

We can see why this is true by assuming that a grape is approximately a sphere of radius r. The ratio of a sphere's surface area to its volume is represented by the rational expression $\dfrac{4\pi r^2}{\frac{4}{3}\pi r^3}$, which simplifies to $\dfrac{3}{r}$. Substituting smaller values for the radius r in this expression gives larger values of the expression. Thus, the smaller the grape is, the larger the surface-to-volume ratio. (*Source:* David L. Wheeler, "Scholars Marry the Science and the Art of Winemaking," *Chronicle of Higher Education,* May 30, 1997)

Chapter 7	PRETEST

To see if you have already mastered the topics in this chapter, take this test.

1. Identify the values for which the rational expression $\dfrac{5}{x+6}$ is undefined.

2. Show that the rational expressions $\dfrac{n^2 - 2n}{3n}$ and $\dfrac{n-2}{3}$ are equivalent.

Simplify.

3. $\dfrac{24x^2y^3}{6xy^5}$

4. $\dfrac{4a^2 - 8a}{a - 2}$

5. $\dfrac{w^2 - 6w}{36 - w^2}$

6. $\dfrac{\frac{5n}{8}}{\frac{n^3}{16}}$

Perform the indicated operation.

7. $\dfrac{12y}{y+1} - \dfrac{7y+2}{y+1}$

8. $\dfrac{1}{6x^2} + \dfrac{5}{4x}$

9. $\dfrac{3}{c-3} - \dfrac{1}{c+3}$

10. $\dfrac{1}{x^2 - 2x + 1} - \dfrac{2}{1 - x^2}$

11. $\dfrac{15a^3b}{20n^4} \cdot \dfrac{16n^2}{9ab}$

12. $\dfrac{y-4}{5y^2 + 10y} \cdot \dfrac{y^2 - 2y - 8}{y^2 - 16}$

13. $\dfrac{x^2 - x - 2}{x^2 + 5x + 4} \div \dfrac{x^2 - 7x + 10}{x - 5}$

Solve and check.

14. $\dfrac{3x - 8}{x^2 - 4} + \dfrac{2}{x - 2} = \dfrac{7}{x + 2}$

15. $\dfrac{x}{x + 4} - 1 = \dfrac{2}{x - 1}$

16. $\dfrac{9}{2x} = \dfrac{x}{x - 1}$

Solve.

17. An ice cream company has fixed costs of \$1500 and variable costs of \$2 for each gallon of ice cream it produces. The cost per gallon of producing x gallons of ice cream is given by the expression $2 + \dfrac{1500}{x}$. Write this cost as a single rational expression.

18. On the first part of a 360-mi trip, a family drives 195 mi at an average speed of r mph. They drive the remainder of the trip at an average speed that is 10 mph less than their speed during the first part of the trip. If the entire trip took 6 hr, what was the average speed during each part of the trip?

19. It takes a photocopier t min to make 30 copies. If at the same rate it takes 5 min longer to make 90 copies, how long does it take the photocopier to make 30 copies?

20. Hooke's Law states that the distance, d, a spring is stretched varies directly with the force, F, applied to the spring. If a force of 12 lb stretches a spring 3 in., how far will the spring stretch when a force of 30 lb is applied?

• *Check your answers on page A-33.*

7.1   Rational Expressions and Functions

What Rational Expressions Are and Why They Are Important

In this chapter, we move beyond our previous discussion of polynomials to consider a broader type of algebraic expression, the *rational* expression. Rational expressions, sometimes called *algebraic fractions*, are useful in many disciplines, including the sciences, the social sciences, medicine, and business.

For instance, the work of some anthropologists who study the history of the human species involves rational expressions. In studying a fossil record, anthropologists examine ancient skulls and compute $\dfrac{W}{L}$, the ratio of each skull's width W to its length L. This expression is an example of a rational expression.

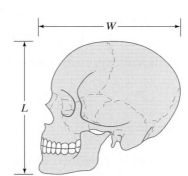

In this chapter, we discuss how to simplify rational expressions, as well as how to carry out operations on rational expressions. We also introduce graphs of rational functions and focus on how to solve equations involving rational expressions.

Introduction to Rational Expressions

Rational expressions in algebra are similar to fractions in arithmetic. In arithmetic, a fraction such as $\dfrac{3}{4}$ is the quotient of two integers, whereas in algebra, a rational expression such as $\dfrac{W}{L}$ is the quotient of two polynomials.

Definition

A **rational expression**, $\dfrac{P}{Q}$, is an algebraic expression that can be written as the quotient of two polynomials, P and Q, where $Q \neq 0$.

Other examples of rational expressions are:

$$\frac{5}{x-2} \qquad \frac{n^2 + 2n - 1}{n+1} \qquad \frac{-a^2}{7bc}$$

A rational expression can be written in terms of division. For example,

$\dfrac{5}{x-2}$ can be written as $5 \div (x - 2)$, and

$\dfrac{n^2 + 2n - 1}{n+1}$ can be written as $(n^2 + 2n - 1) \div (n + 1)$.

Since we can write a rational expression as division, we must be sure that its denominator does not equal 0. When a variable is replaced with a value that makes the denominator 0, the rational expression is undefined.

For example, consider the rational expression $\dfrac{1}{x}$. When its denominator is 0, the expression becomes $\dfrac{1}{0}$, which is undefined. The graph of the *rational function* $g(x) = \dfrac{1}{x}$, therefore, does not contain a point whose x-coordinate is equal to 0.

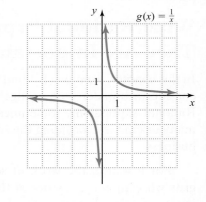

Now consider the rational expression $\dfrac{5}{x + 3}$. For what values of x is this expression undefined? Setting the denominator $x + 3$ equal to 0, we get:

$$x + 3 = 0$$
$$x = -3$$

When x is replaced with -3 in the rational expression, we get:

$$\frac{5}{x + 3} = \frac{5}{-3 + 3} = \frac{5}{0} \quad \leftarrow \text{ Undefined}$$

So $\dfrac{5}{x + 3}$ is undefined when x is equal to -3.

We can also see why this expression is undefined at this x-value by considering the graph of the rational function $f(x) = \dfrac{5}{x + 3}$. Note that the graph never intersects the line $x = -3$. This line, shown on the coordinate plane as dashed, is called an *asymptote*. No point on the graph of $f(x) = \dfrac{5}{x + 3}$ has x-value -3; that is, the graph is not defined at $x = -3$.

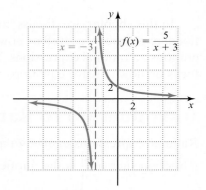

Graphically we see that the expression $\dfrac{5}{x + 3}$ is also undefined when x is equal to -3.

The graphs of $g(x) = \dfrac{1}{x}$ and $f(x) = \dfrac{5}{x + 3}$ are not straight lines. Graphs that are not straight are called *nonlinear*. We can also use the term *nonlinear* to describe equations or functions whose graphs are nonlinear.

Note that the asymptote for the graph of $g(x) = \dfrac{1}{x}$ is the line $x = 0$.

EXAMPLE 1	PRACTICE 1

EXAMPLE 1

Identify all numbers for which the following rational expressions are undefined.

a. $\dfrac{3}{x + 5}$ **b.** $\dfrac{8x}{x^2 - 3x + 2}$

Solution We find the values for which the rational expressions are undefined by setting each denominator equal to 0. Then we solve for x.

a. $\dfrac{3}{x + 5}$

$$x + 5 = 0$$
$$x = -5$$

So $\dfrac{3}{x + 5}$ is undefined when x is equal to -5.

PRACTICE 1

Indicate the values of the variable for which each rational expression is undefined.

a. $\dfrac{n + 1}{n - 3}$

b. $\dfrac{6}{n^2 - 9}$

b. $\dfrac{8x}{x^2 - 3x + 2}$

$$x^2 - 3x + 2 = 0$$

$$(x - 2)(x - 1) = 0 \qquad \text{Factor.}$$

$$x - 2 = 0 \quad \text{or} \quad x - 1 = 0 \qquad \text{Set each factor equal to 0.}$$

$$x = 2 \qquad\qquad x = 1 \qquad \text{Solve for } x.$$

So $\dfrac{8x}{x^2 - 3x + 2}$ is undefined when x is equal to 1 or 2.

Note that throughout the following discussion, we assume that all rational expressions are defined.

Equivalent Expressions

Recall from Chapter R that equivalent fractions, such as $\dfrac{1}{2}$ and $\dfrac{3}{6}$, are fractions that have the same value even though they are written differently. Similarly, equivalent rational expressions are rational expressions that have the same value, no matter what value replaces the variable.

To Find an Equivalent Rational Expression

Multiply the numerator and denominator of $\dfrac{P}{Q}$ by the same polynomial R.

$$\frac{P}{Q} = \frac{PR}{QR},$$

where $Q \neq 0$ and $R \neq 0$.

The rule for finding an equivalent rational expression allows us to multiply the numerator and denominator of a rational expression by the same nonzero polynomial.

EXAMPLE 2	PRACTICE 2

EXAMPLE 2

Indicate whether each pair of rational expressions is equivalent.

a. $\dfrac{2}{3}$ and $\dfrac{2x}{3x}$ **b.** $\dfrac{d - 5}{3}$ and $\dfrac{4d^2 - 20d}{12d}$

Solution

a. If we *multiply* both the numerator and denominator of $\dfrac{2}{3}$

by x, we see that $\dfrac{2}{3} = \dfrac{2x}{3x}$. So the expressions $\dfrac{2}{3}$ and $\dfrac{2x}{3x}$ are equivalent.

b. If we *multiply* both the numerator and denominator of $\dfrac{d - 5}{3}$ by $4d$, we get:

$$\frac{(d - 5) \cdot 4d}{3 \cdot 4d} = \frac{4d^2 - 20d}{12d}$$

So $\dfrac{d - 5}{3}$ and $\dfrac{4d^2 - 20d}{12d}$ are equivalent.

PRACTICE 2

Determine whether each pair of rational expressions is equivalent.

a. $\dfrac{b}{3}$ and $\dfrac{3b^2}{9b}$

b. $\dfrac{5}{a + 7}$ and $\dfrac{5a}{a^2 + 7a}$

Note that we can also use the rule to find an equivalent rational expression by dividing the numerator and denominator of the rational expression by the same nonzero polynomial.

$$\frac{x^2 - 4}{x^2 + 5x + 6} = \frac{(x - 2)(x + 2)}{(x + 3)(x + 2)}$$ Factor the numerator and denominator.

$$= \frac{(x - 2)(\overset{1}{\cancel{x + 2}})}{(x + 3)(\underset{1}{\cancel{x + 2}})}$$ Divide out the common factor $(x + 2)$.

$$= \frac{x - 2}{x + 3}$$

So $\dfrac{x^2 - 4}{x^2 + 5x + 6}$ is equivalent to $\dfrac{x - 2}{x + 3}$.

Tip When simplifying a rational expression, do not divide out common terms in a sum or difference in the numerator and denominator of the expression.

For instance, do not divide out the x's in $\dfrac{x - 2}{x + 3}$.

A rational expression is said to be *in simplest form* (or *reduced to lowest terms*) when its numerator and denominator have no common factor other than 1 or −1. Throughout the remainder of this book, we generally simplify any answer that is a rational expression. Can you explain the similarities and differences in simplifying a fraction in arithmetic and simplifying a rational expression in algebra?

To Simplify a Rational Expression

- Factor the numerator and denominator.
- Divide out any common factors.

EXAMPLE 3

Write in simplest form.

a. $\dfrac{10x^2}{-5xy}$ **b.** $\dfrac{-4a^3b^2}{-2ab}$

Solution

a. $\dfrac{10x^2}{-5xy} = \dfrac{5x(2x)}{5x(-y)}$ Factor the numerator and denominator.

$$= \frac{\overset{1}{\cancel{5x}}(2x)}{\underset{1}{\cancel{5x}}(-y)}$$ Divide out the common factor 5x.

$$= -\frac{2x}{y}$$ Simplify.

b. $\dfrac{-4a^3b^2}{-2ab} = \dfrac{-2ab(2a^2b)}{-2ab}$

$$= \frac{\overset{1}{\cancel{-2ab}}(2a^2b)}{\underset{1}{\cancel{-2ab}}} = 2a^2b$$

PRACTICE 3

Express in lowest terms.

a. $\dfrac{-12n^3}{-6mn}$

b. $\dfrac{-3x^4y}{2x^2y}$

EXAMPLE 4

Write in simplest form, if possible.

a. $\dfrac{3x - 6}{9x + 12}$ **b.** $\dfrac{n + 3}{2n + 6}$ **c.** $\dfrac{t + 3}{t - 1}$

Solution

a. $\dfrac{3x - 6}{9x + 12} = \dfrac{3(x - 2)}{3(3x + 4)}$ Factor the numerator and the denominator.

$= \dfrac{\overset{1}{\cancel{3}}(x - 2)}{\underset{1}{\cancel{3}}(3x + 4)}$ Divide out the common factor 3.

$= \dfrac{x - 2}{3x + 4}$

b. $\dfrac{n + 3}{2n + 6} = \dfrac{n + 3}{2(n + 3)}$ Factor the denominator.

$= \dfrac{\overset{1}{\cancel{n + 3}}}{2\underset{1}{\cancel{(n + 3)}}}$ Divide out the common factor $n + 3$.

$= \dfrac{1}{2}$

c. $\dfrac{t + 3}{t - 1}$ The numerator $t + 3$ and the denominator $t - 1$ have no common factor (other than 1). This expression cannot be simplified.

PRACTICE 4

Write in lowest terms.

a. $\dfrac{2y - 8}{4y + 6}$

b. $\dfrac{v - 4}{v + 3}$

c. $\dfrac{x + 2}{3x + 6}$

EXAMPLE 5

Simplify.

a. $\dfrac{ab - ac}{ax + 2ay}$ **b.** $\dfrac{2t^2 - 2}{t^2 + t - 2}$ **c.** $\dfrac{3x^2 + 2x - 1}{3x^2 - 4x + 1}$

Solution

a. $\dfrac{ab - ac}{ax + 2ay} = \dfrac{a(b - c)}{a(x + 2y)}$

$= \dfrac{\overset{1}{\cancel{a}}(b - c)}{\underset{1}{\cancel{a}}(x + 2y)}$

$= \dfrac{b - c}{x + 2y}$

b. $\dfrac{2t^2 - 2}{t^2 + t - 2} = \dfrac{2(t^2 - 1)}{(t - 1)(t + 2)}$

$= \dfrac{2(t - 1)(t + 1)}{(t - 1)(t + 2)}$

$= \dfrac{2\overset{1}{\cancel{(t - 1)}}(t + 1)}{\underset{1}{\cancel{(t - 1)}}(t + 2)}$

$= \dfrac{2(t + 1)}{t + 2}$

PRACTICE 5

Simplify.

a. $\dfrac{wt - wx}{wz - 3wg}$

b. $\dfrac{4n^2 - 4}{n^2 + 3n + 2}$

c. $\dfrac{2y^2 - y - 1}{2y^2 + 7y + 3}$

c. $\dfrac{3x^2 + 2x - 1}{3x^2 - 4x + 1} = \dfrac{(3x - 1)(x + 1)}{(3x - 1)(x - 1)}$

$$= \dfrac{\overset{1}{(\cancel{3x - 1})}(x + 1)}{\underset{1}{(\cancel{3x - 1})}(x - 1)} = \dfrac{x + 1}{x - 1}$$

The following examples involve factors such as $a - b$ and $b - a$ that are opposites of each other; that is, they differ only in sign. Recall that since $(-1)(a - b) = b - a$, it follows that $\dfrac{a - b}{b - a} = -1$ because

$$\dfrac{a - b}{b - a} = \dfrac{\overset{1}{(\cancel{a - b})}}{\underset{1}{-1(\cancel{a - b})}} = \dfrac{1}{-1} = -1.$$

EXAMPLE 6

Write in lowest terms.

a. $\dfrac{s - 5}{5 - s}$ **b.** $\dfrac{2p - 10}{-p + 5}$

c. $\dfrac{3x - 6}{4 - x^2}$ **d.** $\dfrac{1 - x^2}{x^2 - 3x + 2}$

Solution

a. $\dfrac{s - 5}{5 - s} = \dfrac{s - 5}{-1(s - 5)}$ Write $5 - s$ as $-1(s - 5)$.

$$= \dfrac{\overset{1}{\cancel{s - 5}}}{\underset{1}{-1(\cancel{s - 5})}}$$ Divide out the common factor $s - 5$.

$$= \dfrac{1}{-1}$$ Simplify.

$$= -1$$

b. $\dfrac{2p - 10}{-p + 5} = \dfrac{2(p - 5)}{-1(p - 5)} = \dfrac{2\overset{1}{(\cancel{p - 5})}}{-1\underset{1}{(\cancel{p - 5})}} = \dfrac{2}{-1} = -2$

c. $\dfrac{3x - 6}{4 - x^2} = \dfrac{3(x - 2)}{(2 - x)(2 + x)}$ Factor the numerator and denominator.

$$= \dfrac{3(x - 2)}{-1(x - 2)(2 + x)}$$ Write $(2 - x)$ as $-1(x - 2)$.

$$= \dfrac{3\overset{1}{(\cancel{x - 2})}}{-\underset{1}{(\cancel{x - 2})}(2 + x)}$$ Divide out the common factor $(x - 2)$.

$$= -\dfrac{3}{x + 2}$$ Simplify.

d. $\dfrac{1 - x^2}{x^2 - 3x + 2} = \dfrac{(1 + x)(1 - x)}{(x - 2)(x - 1)} = \dfrac{-(x + 1)(x - 1)}{(x - 2)(x - 1)}$

$$= \dfrac{-(x + 1)\overset{1}{(\cancel{x - 1})}}{(x - 2)\underset{1}{(\cancel{x - 1})}} = -\dfrac{x + 1}{x - 2}$$

PRACTICE 6

Simplify.

a. $\dfrac{-y - 1}{1 + y}$

b. $\dfrac{3x - 12}{-x + 4}$

c. $\dfrac{3n - 15}{25 - n^2}$

d. $\dfrac{4 - 9s^2}{3s^2 + s - 2}$

EXAMPLE 7

It costs a television manufacturer $\dfrac{95x + 10,000}{x}$ dollars per television to produce x televisions.

a. For which value of x is this rational expression undefined?

b. Explain in a sentence or two why you think that it makes sense for the cost per television set to be undefined for the value of x found in part (a).

Solution

a. A rational expression is undefined when its denominator is equal to 0. For the expression $\dfrac{95x + 10,000}{x}$, the denominator is 0 when $x = 0$.

b. When $x = 0$, the manufacturer is producing *no* television sets. So it makes no sense to speak of the cost per television set.

PRACTICE 7

For a circle of radius r, the ratio of its area to its circumference is given by the expression $\dfrac{\pi r^2}{2\pi r}$.

a. Simplify this expression.

b. For which value of r is the original rational expression undefined?

7.1 **Exercises**

Mathematically Speaking

Fill in each blank with the most appropriate term or phrase from the given list.

nonlinear	vertical	rational expression
exponential expression	undefined	equal to 0
has an asymptote at	intersects the line	
in simplest form	equivalent	

1. A(n) _____, $\dfrac{P}{Q}$, is an algebraic expression that can be written as the quotient of two polynomials, P and Q, where $Q \neq 0$.

2. A rational expression is _____ if a variable is replaced with a value that makes its denominator 0.

3. The graph of $f(x) = \dfrac{3}{x-4}$ _____ $x = 4$.

4. A graph that is not a straight line is _____.

5. Rational expressions are _____ if they have the same value no matter what value replaces the variable.

6. A rational expression is said to be _____ (or reduced to lowest terms) when its numerator and denominator have no common factor other than 1 or −1.

Identify the values for which the given rational expression is undefined.

7. $\dfrac{7}{x}$

8. $\dfrac{-2}{c}$

9. $\dfrac{8}{y-2}$

10. $\dfrac{4}{x+2}$

11. $\dfrac{x-3}{x+5}$

12. $\dfrac{y-6}{y-1}$

◉ **13.** $\dfrac{n+11}{2n-1}$

14. $\dfrac{x-4}{3x-2}$

15. $\dfrac{x^2+1}{x^2-1}$

16. $\dfrac{n+2}{n^2-16}$

◉ **17.** $\dfrac{x^2+x+1}{x^2-x-20}$

18. $\dfrac{p^2+7}{p^2-4p-21}$

Give the equation of the vertical line that the graph of the rational function does not intersect.

19.

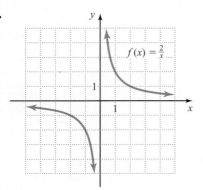

20.

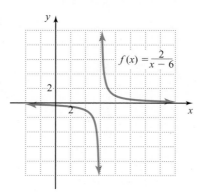

Indicate whether each pair of rational expressions is equivalent.

21. $\dfrac{p}{q}$ and $\dfrac{pr}{qr}$

22. $\dfrac{8}{3y}$ and $\dfrac{16}{6y}$

◉ 23. $\dfrac{3t+5}{t+1}$ and $\dfrac{3t^2+5t}{t^2+t}$

24. $\dfrac{2x-1}{x-4}$ and $\dfrac{14x-7}{7x-28}$

◉ 25. $\dfrac{x-1}{x+3}$ and $-\dfrac{1}{3}$

26. $\dfrac{n+4}{2n+4}$ and $\dfrac{1}{2}$

27. $\dfrac{x-2}{2-x}$ and $\dfrac{2-x}{x-2}$

28. $\dfrac{y-x}{y+x}$ and $\dfrac{y+x}{y-x}$

29. $\dfrac{x^2+4}{x+2}$ and $x+2$

30. $\dfrac{(y+5)^2}{y+5}$ and $y+5$

Simplify.

31. $\dfrac{10a^4}{12a}$

32. $\dfrac{4b}{6b^3}$

◉ 33. $\dfrac{3x^2}{12x^5}$

34. $\dfrac{-2y^6}{2y^4}$

35. $\dfrac{9s^3t^2}{6s^5t}$

36. $\dfrac{10r^3s^4}{5r^3s^4}$

37. $\dfrac{24a^4b^5}{3ab^2}$

38. $\dfrac{14yz^2}{21y^2z^2}$

39. $\dfrac{-2p^2q^3}{-10pq^4}$

40. $\dfrac{-3t^4s^5}{t^2s}$

41. $\dfrac{5x(x+8)}{4x(x+8)}$

42. $\dfrac{7n(n-1)}{2n(n-1)}$

43. $\dfrac{8x^2(5-2x)}{3x(2x-5)}$

44. $\dfrac{(a-b)ab^2}{3a^3(b-a)}$

45. $\dfrac{5x-10}{5}$

46. $\dfrac{3y+12}{3}$

47. $\dfrac{2x^2+2x}{4x^2+6x}$

48. $\dfrac{7y^3-5y^2}{3y^2+y}$

49. $\dfrac{a^2-4a}{ab-4b}$

50. $\dfrac{a^2+2b}{4b+2a^2}$

51. $\dfrac{6x-4y}{9x-6y}$

52. $\dfrac{10m-2n}{n-5m}$

53. $\dfrac{t^2-1}{t+1}$

54. $\dfrac{b^2-1}{b-1}$

55. $\dfrac{p^2-q^2}{q^2-p^2}$

56. $\dfrac{y^2-4x^2}{4x^2-y^2}$

57. $\dfrac{n-1}{2-2n}$

58. $\dfrac{x-3}{3-x}$

59. $\dfrac{(b-4)^2}{b^2-16}$

60. $\dfrac{m^2-1}{(m+1)^2}$

◉ 61. $\dfrac{2x^2+5x-3}{10x-5}$

62. $\dfrac{6p+12}{p^2-p-6}$

◉ 63. $\dfrac{a^3+9a^2+14a}{a^2-10a-24}$

64. $\dfrac{x^2+3x-4}{x^2+2x-3}$

◉ 65. $\dfrac{t^2-4t-5}{t^2-3t-10}$

66. $\dfrac{y^2+8y+15}{y^2-2y-15}$

67. $\dfrac{9-16d^2}{16d^2-24d+9}$

68. $\dfrac{6n^2+7n-10}{25-36n^2}$

69. $\dfrac{8s-2s^2}{2s^2-11s+12}$

70. $\dfrac{3x^2+2x-1}{12x^2-4x}$

71. $\dfrac{6x^2+5x+1}{6x^2-x-1}$

72. $\dfrac{2n^3+2n^2-4n}{n^3+2n^2-3n}$

73. $\dfrac{6y^2-7y+2}{6y^3+5y^2-6y}$

74. $\dfrac{3x^2+xy}{3x^2+7xy+2y^2}$

75. $\dfrac{2ab^2+4a^2b}{2b^2+5ab+2a^2}$

76. $\dfrac{p^2-4pq-12q^2}{2p^2-15pq+18q^2}$

77. $\dfrac{m^2+3mn-28n^2}{2m^2+4mn-48n^2}$

78. $\dfrac{3a^2+5ab-2b^2}{3a^2+8ab-3b^2}$

Mixed Practice

Simplify.

79. $\dfrac{12x^2y^6}{18x^3y^4}$

80. $\dfrac{-5a^7b^4}{a^3b^3}$

81. $\dfrac{3x^3 - 9x^2 + 6x}{x^3 + x^2 - 6x}$

82. $\dfrac{x^2 - 3x - 10}{x^2 + 4x + 4}$

83. $\dfrac{w^2 - 2y^2}{2y^2 - w^2}$

84. $\dfrac{h + 3k^2}{12k^2 + 4h}$

85. $\dfrac{2x^2 + x - 3}{3x^2 - 3x}$

86. $\dfrac{r^2s(r - s)}{2rs^3(s - r)}$

Identify the values for which the given rational expression is undefined.

87. $\dfrac{p - 5}{2p - 3}$

88. $\dfrac{m + 3}{m^2 - 25}$

Indicate whether each pair of rational expressions is equivalent.

89. $\dfrac{6y + 7}{y + 1}$ and $\dfrac{6y^2 + 7y}{y^2 + y}$

90. $\dfrac{2u - 9}{5u - 9}$ and $\dfrac{2}{5}$

Applications

Solve.

91. An expression important in the study of nuclear energy is $\dfrac{mu^2 - mv^2}{mu - mv}$, where m represents mass and u and v represent velocities. Write this expression in simplified form.

92. The force of gravity between two planets is given by the expression $\dfrac{kMm}{d^2}$, where m and M are the masses of the planets, d is the distance between the planets, and k is a fixed constant. Under what circumstances is this force of gravity undefined?

93. When a mathematics department had more faculty and staff members, each had an office with dimensions x ft by x ft. Now that the department has gotten smaller, its offices are being enlarged. Each faculty and each staff office will be made 2 ft wider, whereas each faculty office will also be made 5 ft longer.

   a. Write an expression for the area (in square feet) of an enlarged faculty office.

   b. Write an expression for the area (in square feet) of an enlarged staff office.

   c. Write an expression for the ratio of the area of an enlarged faculty office to the area of an enlarged staff office. Simplify.

   d. Find the value of the expression in part (c) if the length of each office had been 8 ft.

94. The baking time for bread is the ratio of its volume V to its surface area S. For each of the following types of bread, express the ratio $\dfrac{V}{S}$ in simplest form.

   a. A rectangular loaf of bread

   b. A cylindrical bread stick

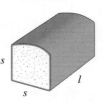

$V = s^2 l$
$S = 2s^2 + 4sl$

$V = \pi r^2 h$
$V = 2\pi r^2 + 2\pi rh$

95. An archer shoots an arrow at the target shown.

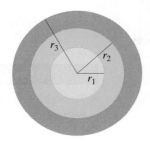

   a. Find and simplify the ratio of the area of the inner circle to the area of the adjacent ring.

   b. Find and simplify the ratio of the area of the smaller ring to the area of the larger ring.

96. A sphere with radius r is packed in a cubic box with side $2r$. The volume of the sphere is $\frac{4}{3}\pi r^3$ and the volume of the box is $(2r)^3$.

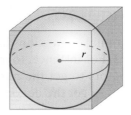

   a. The smaller volume is what fraction of the larger volume? Simplify your answer.

   b. Does the answer to part (a) depend on the value of r?

● *Check your answers on page A-33.*

MINDSTRETCHERS

Technology

1. Using a grapher, display the graphs of $y = \dfrac{x^2 - 1}{x - 1}$ and $y = x + 1$ on the same screen. Compare the two graphs. What conclusion can you draw from this comparison?

Patterns

2. Some rational expressions, when simplified, are equivalent to polynomials. Check the simplifications of the following rational expressions:

   a. $\dfrac{x^2 - 1}{x - 1} = 1 + x$ **b.** $\dfrac{x^3 - 1}{x - 1} = 1 + x + x^2$ **c.** $\dfrac{x^4 - 1}{x - 1} = 1 + x + x^2 + x^3$

   Identify this pattern and extend it.

Writing

3. Sometimes, arithmetic fractions are written with a slanted fraction line, as in 3/4, rather than a horizontal fraction line, as in $\dfrac{3}{4}$. In a sentence or two, explain the disadvantage of writing a rational expression such as $\dfrac{x + 2}{x - 3}$ with a slanted fraction line.

7.2   Multiplication and Division of Rational Expressions

Multiplying Rational Expressions

Operations on rational expressions in algebra are similar to those on fractions in arithmetic. For instance, to multiply arithmetic fractions, we multiply their numerators to get the numerator of the product and multiply their denominators to get the denominator of the product.

$$\frac{2}{3} \cdot \frac{5}{7} = \frac{2 \cdot 5}{3 \cdot 7} = \frac{10}{21}$$

In algebra, rational expressions are multiplied in the same way.

$$\frac{2}{a} \cdot \frac{b}{y} = \frac{2b}{ay}$$

Recall from arithmetic that it is often easier to first divide out any common factors and then to multiply the numerators and denominators:

$$\frac{2}{9} \cdot \frac{3}{10} = \frac{\overset{1}{\cancel{2}}}{\underset{3}{\cancel{9}}} \cdot \frac{\overset{1}{\cancel{3}}}{\underset{5}{\cancel{10}}} = \frac{1}{15}$$

Similarly, in algebra:

$$\frac{5a}{b} \cdot \frac{b}{2a^2} = \frac{\overset{1}{5\cancel{a}}}{\underset{1}{\cancel{b}}} \cdot \frac{\overset{1}{\cancel{b}}}{\underset{a}{2\cancel{a^2}}} = \frac{5}{2a}$$

EXAMPLE 1	PRACTICE 1
Multiply.	Multiply.
a. $\dfrac{3}{x} \cdot \dfrac{5}{x}$ **b.** $\dfrac{4x^2}{7y^3} \cdot \dfrac{3y}{8x}$	**a.** $\dfrac{n}{7} \cdot \dfrac{2}{m}$
Solution	**b.** $\dfrac{p^2}{6q^2} \cdot \dfrac{2q}{5p}$

Solution

a. There are no common factors in the numerators and denominators.

$$\frac{3}{x} \cdot \frac{5}{x} = \frac{3 \cdot 5}{x \cdot x} \qquad \text{Multiply the numerators and multiply the denominators.}$$

$$= \frac{15}{x^2} \qquad \text{Simplify.}$$

b. $\dfrac{4x^2}{7y^3} \cdot \dfrac{3y}{8x} = \dfrac{\overset{1}{\overset{x}{\cancel{4x^2}}}}{7\underset{y^2}{\cancel{y^3}}} \cdot \dfrac{\overset{1}{3\cancel{y}}}{\underset{2}{\cancel{8x}}\,1}$ Divide out all common factors in the numerators and denominators.

$$= \frac{x \cdot 3}{7y^2 \cdot 2} \qquad \text{Multiply the numerators and multiply the denominators.}$$

$$= \frac{3x}{14y^2} \qquad \text{Simplify.}$$

The following rule describes how we multiply rational expressions such as
$\dfrac{2n}{n^2 - 5n} \cdot \dfrac{3n - 15}{4n}$.

To Multiply Rational Expressions

- Factor the numerators and denominators.
- Divide the numerators and denominators by all common factors.
- Multiply the remaining factors in the numerators and the remaining factors in the denominators.

EXAMPLE 2

Multiply.

a. $\dfrac{2n}{n^2 - 5n} \cdot \dfrac{3n - 15}{4n}$ **b.** $\dfrac{10}{9 - x^2} \cdot \dfrac{6 + 2x}{5}$

Solution

a. $\dfrac{2n}{n^2 - 5n} \cdot \dfrac{3n - 15}{4n} = \dfrac{2n}{n(n - 5)} \cdot \dfrac{3(n - 5)}{4n}$ Factor the numerator and the denominator.

$= \dfrac{\overset{1}{\cancel{2}}\overset{1}{\cancel{n}}}{\underset{1}{\cancel{n}}(\underset{1}{\cancel{n - 5}})} \cdot \dfrac{3(\cancel{n - 5})^{1}}{\underset{2}{\cancel{4}n}}$ Divide out common factors.

$= \dfrac{3}{2n}$ Multiply the factors in the numerators and in the denominators.

b. $\dfrac{10}{9 - x^2} \cdot \dfrac{6 + 2x}{5} = \dfrac{10}{(3 - x)(3 + x)} \cdot \dfrac{2(3 + x)}{5}$

$= \dfrac{\overset{2}{\cancel{10}}}{(3 - x)(\cancel{3 + x})_{1}} \cdot \dfrac{2(\cancel{3 + x})^{1}}{\underset{1}{\cancel{5}}} = \dfrac{4}{3 - x}$

PRACTICE 2

Multiply.

a. $\dfrac{3t}{t^2 + 5t} \cdot \dfrac{3t + 15}{6t}$

b. $\dfrac{8}{36g^2 - 1} \cdot \dfrac{1 + 6g}{4}$

EXAMPLE 3

Find the product.

a. $\dfrac{x^2 - x - 20}{5x + 5} \cdot \dfrac{15x^2 - 15}{2x + 8}$

b. $\dfrac{16 + 6a - a^2}{a^2 - 10a - 24} \cdot \dfrac{a^2 - 6a - 27}{a^2 - 17a + 72}$

PRACTICE 3

Find the product.

a. $\dfrac{y^2 + 4y - 21}{3y - 9} \cdot \dfrac{6y^2 - 24}{y + 2}$

b. $\dfrac{x^2 - x - 30}{x^2 + 10x + 9} \cdot \dfrac{18 - 7x - x^2}{x^2 - 8x + 12}$

Solution

a. $\dfrac{x^2 - x - 20}{5x + 5} \cdot \dfrac{15x^2 - 15}{2x + 8}$

$= \dfrac{(x - 5)(x + 4)}{5(x + 1)} \cdot \dfrac{15(x^2 - 1)}{2(x + 4)}$

$= \dfrac{(x - 5)(x + 4)}{5(x + 1)} \cdot \dfrac{15(x - 1)(x + 1)}{2(x + 4)}$

$= \dfrac{(x - 5)\cancel{(x + 4)}^{1}}{\cancel{5}(x + 1)_{1}\,_{1}} \cdot \dfrac{\overset{3}{\cancel{15}}(x - 1)\cancel{(x + 1)}^{1}}{2\cancel{(x + 4)}_{1}}$

$= \dfrac{3(x - 5)(x - 1)}{2}$

b. $\dfrac{16 + 6a - a^2}{a^2 - 10a - 24} \cdot \dfrac{a^2 - 6a - 27}{a^2 - 17a + 72}$

$= \dfrac{(8 - a)(2 + a)}{(a - 12)(a + 2)} \cdot \dfrac{(a - 9)(a + 3)}{(a - 8)(a - 9)}$

$= \dfrac{(8 - a)\cancel{(2 + a)}^{1}}{(a - 12)\cancel{(a + 2)}_{1}} \cdot \dfrac{\cancel{(a - 9)}^{1}(a + 3)}{(a - 8)\cancel{(a - 9)}_{1}}$

$= \dfrac{-1(a - 8)}{a - 12} \cdot \dfrac{a + 3}{a - 8}$

$= \dfrac{-\cancel{(a - 8)}^{1}}{a - 12} \cdot \dfrac{a + 3}{\cancel{a - 8}_{1}}$

$= \dfrac{-(a + 3)}{a - 12}$

$= -\dfrac{a + 3}{a - 12}$

Dividing Rational Expressions

In dividing arithmetic fractions, we take the reciprocal of the divisor and change the operation to multiplication. Recall that the reciprocal of a fraction is formed by interchanging its numerator and denominator.

$$
\begin{array}{cc}
\text{The} & \text{The reciprocal} \\
\text{divisor} & \text{of the divisor} \\
\downarrow & \downarrow
\end{array}
$$

$$
\underset{\uparrow}{\dfrac{2}{3}} \div \dfrac{3}{4} = \dfrac{2}{3} \cdot \underset{\uparrow}{\dfrac{4}{3}} = \dfrac{2 \cdot 4}{3 \cdot 3} = \dfrac{8}{9}
$$

$$
\begin{array}{cc}
\text{Division} & \text{Multiplication}
\end{array}
$$

Rational expressions are divided in the same way.

$$
\begin{array}{cc}
\text{The} & \text{The reciprocal} \\
\text{divisor} & \text{of the divisor} \\
\downarrow & \downarrow
\end{array}
$$

$$
\underset{\uparrow}{\dfrac{p}{q}} \div \dfrac{r}{s} = \dfrac{p}{q} \cdot \underset{\uparrow}{\dfrac{s}{r}} = \dfrac{p \cdot s}{q \cdot r} = \dfrac{ps}{qr}
$$

$$
\begin{array}{cc}
\text{Division} & \text{Multiplication}
\end{array}
$$

In general, to divide rational expressions, we apply the following rule.

To Divide Rational Expressions

• Take the reciprocal of the divisor and change the operation to multiplication.
• Follow the rule for multiplying rational expressions.

EXAMPLE 4	PRACTICE 4

EXAMPLE 4

Divide.

a. $\dfrac{3}{x} \div \dfrac{y}{2}$ **b.** $\dfrac{6ab^2}{4a^3b} \div \dfrac{3a^4}{2b^5}$

Solution

a. $\dfrac{3}{x} \div \dfrac{y}{2} = \dfrac{3}{x} \cdot \dfrac{2}{y}$ Take the reciprocal of the divisor and change the operation to multiplication.

$\qquad = \dfrac{6}{xy}$ Multiply.

b. $\dfrac{6ab^2}{4a^3b} \div \dfrac{3a^4}{2b^5} = \dfrac{6ab^2}{4a^3b} \cdot \dfrac{2b^5}{3a^4}$

$\qquad = \dfrac{\overset{1}{\cancel{6}}\, \overset{1}{\cancel{a}}\, b^2}{\underset{2}{\cancel{4}}\, \underset{a^2}{\cancel{a^3}}\, \underset{1}{\cancel{b}}} \cdot \dfrac{\overset{1}{\cancel{2}}\, b^5}{\underset{1}{\cancel{3}}\, a^4}$

$\qquad = \dfrac{b \cdot b^5}{a^2 \cdot a^4}$

$\qquad = \dfrac{b^6}{a^6}$

PRACTICE 4

Find the quotient.

a. $\dfrac{a}{4} \div \dfrac{b}{6}$

b. $\dfrac{5pq}{7p^2q^4} \div \dfrac{p^3}{3q^2}$

EXAMPLE 5	PRACTICE 5

EXAMPLE 5

Find the quotient.

a. $\dfrac{a+1}{a-2} \div \dfrac{a+2}{a-1}$

b. $\dfrac{6x+3y}{4x-12y} \div \dfrac{10x+5y}{2x-6y}$

c. $\dfrac{x^2+3x+2}{3x+12} \div \dfrac{x^2-1}{7x}$

Solution

a. $\dfrac{a+1}{a-2} \div \dfrac{a+2}{a-1} = \dfrac{a+1}{a-2} \cdot \dfrac{a-1}{a+2}$ Take the reciprocal and change the operation to multiplication.

$\qquad = \dfrac{(a+1)(a-1)}{(a-2)(a+2)}$ Multiply.

PRACTICE 5

Find the quotient.

a. $\dfrac{x+3}{x-10} \div \dfrac{x+1}{x+3}$

b. $\dfrac{p+4q}{3p-6q} \div \dfrac{2p+8q}{2p-4q}$

c. $\dfrac{y^2+4y+3}{5y+10} \div \dfrac{y^2-1}{y}$

b. $\dfrac{6x + 3y}{4x - 12y} \div \dfrac{10x + 5y}{2x - 6y}$

$= \dfrac{6x + 3y}{4x - 12y} \cdot \dfrac{2x - 6y}{10x + 5y}$ Take the reciprocal and change the operation to multiplication.

$= \dfrac{3(2x + y)}{4(x - 3y)} \cdot \dfrac{2(x - 3y)}{5(2x + y)}$ Factor the numerators and denominators.

$= \dfrac{3\overset{1}{\cancel{(2x + y)}}}{\underset{2}{\cancel{4}}(x - 3y)} \cdot \dfrac{\overset{1}{\cancel{2}}\overset{1}{\cancel{(x - 3y)}}}{5\overset{1}{\cancel{(2x + y)}}}$ Divide out the common factors.

$= \dfrac{3}{10}$ Multiply.

c. $\dfrac{x^2 + 3x + 2}{3x + 12} \div \dfrac{x^2 - 1}{7x} = \dfrac{x^2 + 3x + 2}{3x + 12} \cdot \dfrac{7x}{x^2 - 1}$

$= \dfrac{(x + 1)(x + 2)}{3(x + 4)} \cdot \dfrac{7x}{(x + 1)(x - 1)}$

$= \dfrac{\overset{1}{\cancel{(x + 1)}}(x + 2)}{3(x + 4)} \cdot \dfrac{7x}{\underset{1}{\cancel{(x + 1)}}(x - 1)}$

$= \dfrac{7x(x + 2)}{3(x + 4)(x - 1)}$

EXAMPLE 6

According to the physicist Isaac Newton, the force of gravitation between two objects with mass m and M is given by the expression $G \cdot \dfrac{m}{d} \cdot \dfrac{M}{d}$, where d is the distance between them and G is the gravitational constant. Simplify this expression.

Solution

$$G \cdot \dfrac{m}{d} \cdot \dfrac{M}{d} = \dfrac{G}{1} \cdot \dfrac{m}{d} \cdot \dfrac{M}{d} = \dfrac{GmM}{d^2}$$

PRACTICE 6

When a body moves in a circular path, a force called the centripetal force is directed toward the center of the circle. This force has magnitude equal to the product of the object's mass m and its acceleration a. If $m = \dfrac{W}{g}$ and $a = \dfrac{v^2}{r}$, find the centripetal force in terms of W, g, v, and r, which represent the weight of the object, a constant due to gravity, the velocity of the object, and the radius of the circle, respectively.

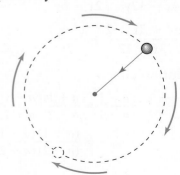

Multiply. Express the product in lowest terms.

1. $\dfrac{1}{t^2} \cdot \dfrac{t}{4}$

2. $-\dfrac{y}{8} \cdot \dfrac{10}{y^3}$

3. $\dfrac{2}{a} \cdot \dfrac{3}{b}$

4. $\dfrac{5x}{2} \cdot \dfrac{2y}{3}$

5. $\dfrac{2x^4}{3x^5} \cdot \dfrac{5}{x^8}$

6. $\dfrac{4c}{d^3} \cdot \dfrac{3d}{8c^4}$

7. $-\dfrac{7x^2y}{3} \cdot \dfrac{6}{x^3y}$

8. $-\dfrac{5p^2}{10pq^2} \cdot \left(-\dfrac{6pq}{5p^3q}\right)$

9. $\dfrac{x}{x-2} \cdot \dfrac{5x-10}{x^4}$

10. $\dfrac{t}{t+3} \cdot \dfrac{4t+12}{t}$

11. $\dfrac{8n-3}{n^2} \cdot n$

12. $\dfrac{s+1}{s} \cdot s^2$

13. $\dfrac{8x-6}{5x+20} \cdot \dfrac{2x+8}{4x-3}$

14. $\dfrac{10y-1}{3y-6} \cdot \dfrac{5y-10}{10y-1}$

15. $\dfrac{5n-1}{6n+4} \cdot \dfrac{3n+2}{1-5n}$

16. $\dfrac{-4x-2}{10-x} \cdot \dfrac{6x+1}{2x+1}$

17. $\dfrac{x^2-4y^2}{x+y} \cdot \dfrac{3x+3y}{4x-8y}$

18. $\dfrac{a^2-4b^2}{6a-6b} \cdot \dfrac{10a-10b}{3a+6b}$

19. $\dfrac{p^4-1}{p^4-16} \cdot \dfrac{p^2+4}{p^2+1}$

20. $\dfrac{x^4-81}{x^2-x-12} \cdot \dfrac{x^2-16}{x^2+9}$

21. $\dfrac{n^2-2n-24}{n^2+6n+8} \cdot \dfrac{n^2+5n+6}{n^2-5n-6}$

22. $\dfrac{t^2+4t-21}{t^2+2t-15} \cdot \dfrac{t^2+t-20}{t^2+3t-28}$

 23. $\dfrac{2y^2-y-6}{2y^2+y-3} \cdot \dfrac{2y^2-3y+1}{2y^2-9y+10}$

24. $\dfrac{2m+1}{2m^2+7m+3} \cdot \dfrac{m^2+2m-3}{2m+4m^2}$

25. $\dfrac{2}{x^3} \cdot \dfrac{4x}{5} \cdot \dfrac{10}{x^2}$

26. $\dfrac{a-3}{a^2} \cdot \dfrac{2a}{5} \cdot \dfrac{10a}{3}$

27. $\dfrac{x^2-7x+10}{2x-2} \cdot \dfrac{6x}{x^2-2x-15} \cdot \dfrac{x^2+2x-3}{x-2}$

28. $\dfrac{p^2-q^2}{p^2-pq} \cdot \dfrac{q}{2p^2-pq-q^2} \cdot \dfrac{6p+3q}{p}$

Divide. Express the quotient in lowest terms.

29. $\dfrac{7}{a} \div \dfrac{14}{a}$

30. $\dfrac{-5}{n} \div \dfrac{n}{2}$

31. $\dfrac{p^3}{10} \div \dfrac{p^3}{20}$

32. $\dfrac{-s}{v^2} \div \dfrac{s^2}{v}$

33. $\dfrac{12}{x^3} \div \dfrac{6}{5x^2}$

34. $\dfrac{1}{a^2} \div \dfrac{2}{3a}$

35. $-\dfrac{3}{t} \div t$

36. $\dfrac{5}{s} \div s$

37. $\dfrac{9xy^2}{2x^3} \div \dfrac{3x^2y}{4y}$

38. $\dfrac{10a^3}{7ab^2} \div \dfrac{-6a^2b^2}{14ab}$

39. $\dfrac{c+3}{c-5} \div \dfrac{c+9}{c-7}$

40. $\dfrac{t+4}{t-6} \div \dfrac{2t-8}{2t-4}$

41. $\dfrac{6a-12}{8a+32} \div \dfrac{9a-18}{5a+20}$

42. $\dfrac{3x+6}{6x+18} \div \dfrac{2x+4}{x+3}$

43. $\dfrac{x+1}{10} \div \dfrac{1-x^2}{5}$

44. $\dfrac{4x+8}{8} \div \dfrac{4-x^2}{4}$

45. $\dfrac{p^2 - 1}{1 - p} \div \dfrac{p + 1}{p}$

46. $\dfrac{y + 6}{3y} \div \dfrac{y^2 - 36}{6 - y}$

47. $\dfrac{x^2y + 3xy^2}{x^2 - 9y^2} \div \dfrac{5x^2y}{x^2 - 2xy - 3y^2}$

48. $\dfrac{2x + 1}{2x^2 + 7x + 3} \div \dfrac{2x + 4x^2}{x^2 + 2x - 3}$

49. $\dfrac{2t^2 - 3t - 2}{2t + 1} \div (4 - t^2)$

50. $(y - x) \div \dfrac{x^2 - y^2}{x^2 + xy}$

51. $\dfrac{x^2 - 11x + 28}{x^2 - x - 42} \div \dfrac{x^2 - 2x - 8}{x^2 + 7x + 10}$

52. $\dfrac{a^2 - a - 56}{a^2 + 8a + 7} \div \dfrac{a^2 - 13a + 40}{a^2 - 4a - 5}$

53. $\dfrac{3p^2 - 3p - 18}{p^2 + 2p - 15} \div \dfrac{2p^2 + 6p - 20}{2p^2 - 12p + 16}$

54. $\dfrac{3y^2 + 13y + 4}{16 - y^2} \div \dfrac{3y^2 - 5y - 2}{3y - 12}$

Mixed Practice

Divide. Express the quotient in lowest terms.

55. $\dfrac{r^2}{s} \div \left(-\dfrac{r^4}{s}\right)$

56. $\dfrac{x^2 + 5x - 24}{x^2 + 9x + 8} \div \dfrac{x^2 - 10x + 21}{x^2 - 6x - 7}$

57. $\dfrac{3x + 9}{12} \div \dfrac{9 - x^2}{8}$

58. $\dfrac{h^2 + 2h - 8}{h + 4} \div (4 - h^2)$

59. $-\dfrac{6q^2}{15p^2q} \div \dfrac{12p^2q^2}{5p^2q}$

Multiply. Express the product in lowest terms.

60. $\dfrac{9a}{2b^2} \cdot \dfrac{b}{3a^3}$

61. $\dfrac{3y}{4} \cdot \dfrac{y + 3}{2y^2} \cdot \dfrac{6y}{3}$

62. $\dfrac{2c + 2d}{6c - 3d} \cdot \dfrac{4c^2 - d^2}{c + d}$

63. $\dfrac{x^2 + x - 2}{x + 1} \cdot \dfrac{3x + 3x^2}{3x^2 + 7x + 2}$

64. $\dfrac{u^2}{u + 5} \cdot \dfrac{5u + 25}{u}$

Applications

Solve.

65. An investment of p dollars is growing at the annual simple interest rate r. The number of years that it will take the investment to be worth A dollars is given by the expression

$$\left(\dfrac{A - p}{p}\right) \div r$$

Simplify this expression.

66. A store is having a sale, with each item selling at a discount of $x\%$.

  a. With this discount, what percent of the normal price is a customer paying on each item?

  b. What is the sale price of 10 items that normally sell for z dollars each?

67. A company has annual expenses totaling B dollars, of which $p\%$ goes toward rents. Of the rental expenses, $q\%$ goes toward the head office. Write as a rational expression the annual cost of the head office rental.

68. Physicists studying momentum may use the expression

$$\frac{W}{g} \cdot \frac{1}{t} \cdot (v_2 - v_1),$$

where W is the weight of an object, g is a constant, t is time, v_2 is the final velocity of the object, and v_1 is the initial velocity of the object. Write as a single rational expression.

69. Electricians use the following formula when studying the resistance in a heating element:

$$P = \frac{V^2}{R + r} \cdot \frac{r}{R + r}$$

Multiply out the right-hand side of this formula.

70. Suppose that the chance that one event occurs is $\dfrac{a}{b}$ and the chance that an independent (or unrelated) event occurs is $\dfrac{c}{d}$.

   a. The chance that both events will occur is the product of their individual chances. Write this product as a rational expression.

   b. The chance that neither event will occur can be represented by $\left(1 - \dfrac{a}{b}\right)\left(1 - \dfrac{c}{d}\right)$. Write this product as a single rational expression.

71. The following diagram shows a cylinder and its inscribed sphere, where the radius of the sphere is r and the height of the cylinder is h.

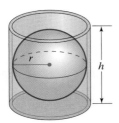

Volume of a sphere $= \dfrac{4}{3}\pi r^3$

Volume of a cylinder $= \pi r^2 h$

Surface area of a sphere $= 4\pi r^2$

Surface area of a cylinder $= 2\pi rh + 2\pi r^2$

Use these formulas to answer the following questions.

   a. Find the ratio of the volume of the sphere to the volume of the cylinder.

   b. Find the ratio of the surface area of the sphere to the surface area of the cylinder.

   c. Divide the expression in part (a) by the expression in part (b).

   d. In the diagram, note that $h = 2r$. Substitute $2r$ for h in the expression in part (c) and simplify.

72. In 1934, Harold Urey won the Nobel Prize in chemistry for discovering deuterium (also called heavy hydrogen). This discovery involved reducing the mass of an electron in an atom of deuterium.

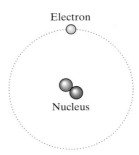

If m stands for the mass of an electron, then the reduced mass r can be represented by

$$r = \frac{mM}{M + m},$$

where M is the mass of the atom's nucleus. Find an expression for the quotient r divided by m in terms of m and M, and simplify.

● *Check your answers on page A-33.*

MINDSTRETCHERS

Critical Thinking

1. Even though the operation of division is not commutative, can you find four different polynomials P, Q, R, and S for which $\dfrac{P}{Q} \div \dfrac{R}{S} = \dfrac{R}{S} \div \dfrac{P}{Q}$? Explain.

Mathematical Reasoning

2. True or false, the expressions $\dfrac{\left(\dfrac{a}{b}\right)}{c}$ and $\dfrac{a}{\left(\dfrac{b}{c}\right)}$ are equivalent. Explain.

Groupwork

3. Working with a partner, find the following.

 a. Two different pairs of rational expressions whose quotient is $\dfrac{x+4}{x+6}$.

 b. Two different pairs of rational expressions whose product is $\dfrac{x^2 - x - 6}{2x^2 + 5x - 3}$.

7.3 Addition and Subtraction of Rational Expressions

Adding and Subtracting Rational Expressions with the Same Denominator

Recall from arithmetic that to add like fractions, we add the numerators and keep the same denominator.

$$\frac{3}{5} + \frac{1}{5} = \frac{3+1}{5} = \frac{4}{5}$$

To subtract like fractions, we subtract the numerators and keep the same denominator.

$$\frac{3}{5} - \frac{1}{5} = \frac{3-1}{5} = \frac{2}{5}$$

In algebra, adding and subtracting rational expressions with the same denominator is similar:

$$\frac{3}{x} + \frac{1}{x} = \frac{3+1}{x} = \frac{4}{x} \quad \text{and} \quad \frac{3}{x} - \frac{1}{x} = \frac{3-1}{x} = \frac{2}{x}$$

OBJECTIVES

- To add and subtract rational expressions with the same denominator
- To find the least common denominator (LCD) of two or more rational expressions
- To add and subtract rational expressions with different denominators
- To solve applied problems involving the addition or subtraction of rational expressions

To Add (or Subtract) Rational Expressions with the Same Denominator

- Add (or subtract) the numerators and keep the same denominator.
- Simplify, if possible.

EXAMPLE 1

Add.

a. $\dfrac{3}{x-1} + \dfrac{8}{x-1}$ **b.** $\dfrac{3x}{4y} + \dfrac{x}{4y}$

c. $\dfrac{3x}{x+1} + \dfrac{3}{x+1}$ **d.** $\dfrac{2t^2+5t-7}{3t-1} + \dfrac{7t^2-5t+6}{3t-1}$

Solution

a. $\dfrac{3}{x-1} + \dfrac{8}{x-1} = \dfrac{3+8}{x-1}$ Add the numerators and keep the same denominator.

$$= \frac{11}{x-1}$$ Simplify.

b. $\dfrac{3x}{4y} + \dfrac{x}{4y} = \dfrac{4x}{4y}$ Add the numerators and keep the same denominator.

$$= \frac{\overset{1}{\cancel{4}}x}{\underset{1}{\cancel{4}}y}$$ Divide out the common factor.

$$= \frac{x}{y}$$ Simplify.

PRACTICE 1

Add.

a. $\dfrac{9}{y+2} + \dfrac{1}{y+2}$

b. $\dfrac{10r}{3s} + \dfrac{5r}{3s}$

c. $\dfrac{6t-7}{t-1} + \dfrac{1}{t-1}$

d. $\dfrac{n^2+10n+1}{n+5} + \dfrac{2n^2+4n-6}{n+5}$

c. $\dfrac{3x}{x+1} + \dfrac{3}{x+1} = \dfrac{3x+3}{x+1}$ **Add the numerators and keep the same denominator.**

$= \dfrac{3\overset{1}{\cancel{(x+1)}}}{\underset{1}{\cancel{x+1}}}$ **Factor the numerator and divide out the common factor.**

$= 3$ **Simplify.**

d. $\dfrac{2t^2 + 5t - 7}{3t - 1} + \dfrac{7t^2 - 5t + 6}{3t - 1} = \dfrac{(2t^2 + 5t - 7) + (7t^2 - 5t + 6)}{3t - 1}$

$= \dfrac{9t^2 - 1}{3t - 1}$ **Combine like terms in the numerator.**

$= \dfrac{\overset{1}{\cancel{(3t-1)}}(3t+1)}{\underset{1}{\cancel{3t-1}}}$

$= 3t + 1$

EXAMPLE 2

Subtract.

a. $\dfrac{6}{z} - \dfrac{4}{z}$ **b.** $\dfrac{5y}{8} - \dfrac{3y}{8}$ **c.** $\dfrac{x}{3y} - \dfrac{2x}{3y}$

Solution

a. $\dfrac{6}{z} - \dfrac{4}{z} = \dfrac{6-4}{z}$ **Subtract the numerators and keep the same denominator.**

$= \dfrac{2}{z}$ **Simplify.**

b. $\dfrac{5y}{8} - \dfrac{3y}{8} = \dfrac{5y - 3y}{8} = \dfrac{\overset{1}{\cancel{2}}y}{\underset{4}{\cancel{8}}} = \dfrac{y}{4}$

c. $\dfrac{x}{3y} - \dfrac{2x}{3y} = \dfrac{x - 2x}{3y} = \dfrac{-x}{3y} = -\dfrac{x}{3y}$

PRACTICE 2

Find the difference.

a. $\dfrac{12}{v} - \dfrac{7}{v}$

b. $\dfrac{7t}{10} - \dfrac{t}{10}$

c. $\dfrac{7p}{3q} - \dfrac{8p}{3q}$

EXAMPLE 3

Find the difference.

a. $\dfrac{x + 5y}{2y} - \dfrac{x - 11y}{2y}$

b. $\dfrac{3ax + bx}{a + 2b} - \dfrac{2ax - bx}{a + 2b}$

c. $\dfrac{6x + 12}{x^2 - x - 6} - \dfrac{x + 2}{x^2 - x - 6}$

PRACTICE 3

Subtract.

a. $\dfrac{7a - 4b}{3a} - \dfrac{a - 4b}{3a}$

b. $\dfrac{9xy - 5xz}{4y - z} - \dfrac{xy - 3xz}{4y - z}$

c. $\dfrac{2x + 13}{x^2 - 7x + 10} - \dfrac{5x + 7}{x^2 - 7x + 10}$

Solution

a. $\dfrac{x + 5y}{2y} - \dfrac{x - 11y}{2y}$

$= \dfrac{(x + 5y) - (x - 11y)}{2y}$ Subtract the numerators and keep the same denominator.

$= \dfrac{x + 5y - x + 11y}{2y}$ Remove the parentheses.

$= \dfrac{16y}{2y}$ Combine like terms.

$= \dfrac{\overset{8}{\cancel{16}}\overset{1}{\cancel{y}}}{\underset{1}{\cancel{2}}\underset{1}{\cancel{y}}}$ Divide out the common factors.

$= 8$ Simplify.

b. $\dfrac{3ax + bx}{a + 2b} - \dfrac{2ax - bx}{a + 2b} = \dfrac{(3ax + bx) - (2ax - bx)}{a + 2b}$

$= \dfrac{3ax + bx - 2ax + bx}{a + 2b}$

$= \dfrac{ax + 2bx}{a + 2b}$

$= \dfrac{x(a + 2b)}{a + 2b}$

$= \dfrac{x(\cancel{a + 2b})}{\cancel{a + 2b}} = x$

c. $\dfrac{6x + 12}{x^2 - x - 6} - \dfrac{x + 2}{x^2 - x - 6} = \dfrac{(6x + 12) - (x + 2)}{x^2 - x - 6}$

$= \dfrac{6x + 12 - x - 2}{x^2 - x - 6}$

$= \dfrac{5x + 10}{x^2 - x - 6}$

$= \dfrac{5(x + 2)}{(x - 3)(x + 2)}$

$= \dfrac{5\cancel{(x + 2)}}{(x - 3)\cancel{(x + 2)}} = \dfrac{5}{x - 3}$

The Least Common Denominator of Rational Expressions

Recall that combining unlike arithmetic fractions involves finding the least common denominator (LCD). We write the fractions in terms of their LCD, and then add these like fractions.

$$\frac{1}{2} + \frac{4}{5} = \frac{1}{2} \cdot \frac{5}{5} + \frac{4}{5} \cdot \frac{2}{2} = \frac{5}{10} + \frac{8}{10} = \frac{13}{10}$$

10 is the LCD of $\frac{1}{2}$ and $\frac{4}{5}$.

To find the LCD of rational expressions, we begin by factoring their denominators completely. For instance, consider the rational expressions $\dfrac{8x}{x^2 + 3x + 2}$ and $\dfrac{5x - 6}{x^2 + 2x + 1}$, which we can write as

$$\underbrace{\dfrac{8x}{(x + 1)(x + 2)}} \quad \text{and} \quad \underbrace{\dfrac{5x - 6}{(x + 1)(x + 1)}}.$$

Denominators in factored form.

The LCD is the product of the different factors in the denominators, where the power of each factor is the greatest number of times that it occurs in any single denominator.

The highest power of this factor
in any denominator

$$\text{LCD} = (x + 1)(x + 1)(x + 2) = (x + 1)^2(x + 2)$$

To Find the LCD of Rational Expressions

- Factor each denominator completely.
- Multiply these factors, using for the power of each factor the greatest number of times that it occurs in any of the denominators. The product of all these factors is the LCD.

EXAMPLE 4

Find the LCD of each group of rational expressions.

a. $\dfrac{1}{6}$ and $\dfrac{2}{n}$ **b.** $\dfrac{1}{10x}$ and $\dfrac{7}{12x^2}$

c. $\dfrac{3}{8p}$, $\dfrac{7}{10p^2q}$, and $\dfrac{1}{4pqr}$

Solution

a. We begin by factoring the denominators of $\dfrac{1}{6}$ and $\dfrac{2}{n}$.

Factor 6: $2 \cdot 3$

Factor n: n

No factor is repeated more than once in any denominator. So the LCD is the product of the factors.

$$\text{LCD} = 2 \cdot 3 \cdot n = 6n$$

b. $\dfrac{1}{10x}$ and $\dfrac{7}{12x^2}$

Factor $10x$: $2 \cdot 5 \cdot x$

Factor $12x^2$: $2^2 \cdot 3 \cdot x^2$

The factors 2 and x each appear at most twice in any denominator. The factors 3 and 5 each appear at most once in any denominator. So

$$\text{LCD} = 2^2 \cdot 3 \cdot 5 \cdot x^2 = 60x^2.$$

PRACTICE 4

For each group of rational expressions, determine the least common denominator.

a. $\dfrac{8}{y}$ and $\dfrac{x}{20}$

b. $\dfrac{1}{6t}$ and $\dfrac{4}{3t^2}$

c. $\dfrac{1}{5x}$, $\dfrac{7}{15xy^3}$, and $\dfrac{x + y}{2x^2}$

c. $\dfrac{3}{8p}$, $\dfrac{7}{10p^2q}$, and $\dfrac{1}{4pqr}$

Factor $8p$: $2^3 \cdot p$
Factor $10p^2q$: $2 \cdot 5 \cdot p^2 \cdot q$
Factor $4pqr$: $2^2 \cdot p \cdot q \cdot r$
LCD $= 2^3 \cdot 5 \cdot p^2 \cdot q \cdot r = 40p^2qr$

EXAMPLE 5

Find the LCD of each group of rational expressions.

a. $\dfrac{4x+3}{3x^2-3x}$ and $\dfrac{x-6}{x^2-2x+1}$

b. $\dfrac{x+4}{x+7}$, $\dfrac{3}{x-4}$, and $\dfrac{8x-1}{x+9}$

c. $\dfrac{2y}{x^2-2xy+y^2}$, $\dfrac{x}{x^2-y^2}$, and $\dfrac{x-y}{x^2+2xy+y^2}$

Solution

a. First, we factor the denominators of $\dfrac{4x+3}{3x^2-3x}$ and $\dfrac{x-6}{x^2-2x+1}$.

Factor $3x^2-3x$: $3x(x-1)$
Factor x^2-2x+1: $(x-1)(x-1)=(x-1)^2$

The factor $3x$ appears at most once and the factor $(x-1)$ appears at most twice, so

$$\text{LCD} = 3x(x-1)^2.$$

b. $\dfrac{x+4}{x+7}$, $\dfrac{3}{x-4}$, and $\dfrac{8x-1}{x+9}$

Factor $x+7$: $x+7$
Factor $x-4$: $x-4$
Factor $x+9$: $x+9$

No factor appears more than once in any denominator, so
$$\text{LCD} = (x+7)(x-4)(x+9).$$

c. $\dfrac{2y}{x^2-2xy+y^2}$, $\dfrac{x}{x^2-y^2}$, and $\dfrac{x-y}{x^2+2xy+y^2}$

Factor $x^2-2xy+y^2$: $(x-y)(x-y)=(x-y)^2$
Factor x^2-y^2: $(x-y)(x+y)$
Factor $x^2+2xy+y^2$: $(x+y)(x+y)=(x+y)^2$
LCD $= (x-y)^2(x+y)^2$

PRACTICE 5

Determine the LCD for each group of rational expressions.

a. $\dfrac{9n+1}{2n^2+2n}$ and $\dfrac{n-5}{n^2+2n+1}$

b. $\dfrac{7p+1}{p+2}$, $\dfrac{5p}{p-1}$, and $\dfrac{3p+1}{p+5}$

c. $\dfrac{s-t}{s^2+4st+4t^2}$, $\dfrac{3t}{s^2-4t^2}$, and $\dfrac{6s}{s^2-4st+4t^2}$

Adding and Subtracting Rational Expressions with Different Denominators

In order to add or subtract rational expressions with different denominators, we will need to change each expression to an equivalent rational expression whose denominator is the LCD.

EXAMPLE 6	PRACTICE 6

EXAMPLE 6

Write the following rational expressions in terms of their LCD.

a. $\dfrac{3}{5x^2}$ and $\dfrac{x+1}{6xy}$ **b.** $\dfrac{6x+1}{x^2-4}$ and $\dfrac{x}{x^2+5x+6}$

Solution

a. The LCD of $\dfrac{3}{5x^2}$ and $\dfrac{x+1}{6xy}$ is $30x^2y$.

Since $5x^2 \cdot 6y = 30x^2y$, we multiply the numerator and denominator of $\dfrac{3}{5x^2}$ by $6y$ so that the denominator becomes the LCD.

$$\frac{3}{5x^2} = \frac{3 \cdot 6y}{5x^2 \cdot 6y} = \frac{18y}{30x^2y}$$

Note that multiplying the numerator and the denominator by $6y$ is the same as multiplying the expression by 1. So $\dfrac{3}{5x^2}$ and $\dfrac{18y}{30x^2y}$ are equivalent expressions.

Since $6xy \cdot 5x = 30x^2y$, we multiply the numerator and denominator of $\dfrac{x+1}{6xy}$ by $5x$ so that the denominator becomes the LCD.

$$\frac{x+1}{6xy} = \frac{(x+1) \cdot 5x}{6xy \cdot 5x} = \frac{5x(x+1)}{30x^2y}$$

b. To find the LCD, we begin by factoring the denominators of the expressions.

Factor $x^2 - 4$: $(x+2)(x-2)$
Factor $x^2 + 5x + 6$: $(x+2)(x+3)$

The LCD of $\dfrac{6x+1}{x^2-4}$ and $\dfrac{x}{x^2+5x+6}$ is $(x+2)(x-2)(x+3)$.

Since $(x^2-4)(x+3) = (x+2)(x-2)(x+3)$, we multiply the numerator and denominator of $\dfrac{6x+1}{x^2-4}$ by $(x+3)$ so that the denominator becomes the LCD.

$$\frac{6x+1}{x^2-4} = \frac{(6x+1)(x+3)}{(x+2)(x-2)(x+3)}$$

PRACTICE 6

Express each pair of rational expressions in terms of their LCD.

a. $\dfrac{2}{7p^3}$ and $\dfrac{p+3}{2p}$

b. $\dfrac{3y-2}{y^2-9}$ and $\dfrac{y}{y^2-6y+9}$

Since $(x^2 + 5x + 6)(x - 2) = (x + 2)(x + 3)(x - 2)$, we multiply the numerator and denominator of $\dfrac{x}{x^2 + 5x + 6}$ by $(x - 2)$ so that the denominator becomes the LCD.

$$\frac{x}{x^2 + 5x + 6} = \frac{x(x - 2)}{(x + 2)(x + 3)(x - 2)}$$

Now, let's look at how to add and subtract rational expressions with different denominators using the concept of equivalent rational expressions.

To Add (or Subtract) Rational Expressions with Different Denominators

- Find the LCD of the rational expressions.
- Write each rational expression with a common denominator, usually the LCD.
- Add (or subtract) the numerators.
- Simplify, if possible.

EXAMPLE 7

Perform the indicated operation.

a. $\dfrac{1}{2n} - \dfrac{3}{5n}$ **b.** $\dfrac{2}{3x^2} + \dfrac{5}{9x}$

Solution

a. The LCD of the rational expressions is $10n$.

$$\frac{1}{2n} - \frac{3}{5n} = \frac{1 \cdot 5}{2n \cdot 5} - \frac{3 \cdot 2}{5n \cdot 2} \quad \text{Write as equivalent rational expressions with the LCD as the denominator.}$$

$$= \frac{5}{10n} - \frac{6}{10n} \quad \text{Simplify.}$$

$$= \frac{-1}{10n} \quad \text{Subtract the numerators.}$$

$$= -\frac{1}{10n} \quad \text{Simplify.}$$

b. The LCD of the rational expressions is $9x^2$.

$$\frac{2}{3x^2} + \frac{5}{9x} = \frac{2 \cdot 3}{3x^2 \cdot 3} + \frac{5 \cdot x}{9x \cdot x} \quad \text{Write as equivalent rational expressions with the LCD as the denominator.}$$

$$= \frac{6}{9x^2} + \frac{5x}{9x^2} \quad \text{Simplify.}$$

$$= \frac{5x + 6}{9x^2} \quad \text{Add the numerators.}$$

PRACTICE 7

Combine.

a. $\dfrac{3}{4p} + \dfrac{1}{6p}$

b. $\dfrac{1}{5y} - \dfrac{2}{15y^2}$

EXAMPLE 8

Combine.

a. $\dfrac{y + 5}{y - 2} - \dfrac{y - 3}{y}$ **b.** $\dfrac{7a + 5}{a - 1} - \dfrac{a}{1 - a}$

Solution

a. The LCD of the rational expression is $y(y - 2)$.

$$\dfrac{y + 5}{y - 2} - \dfrac{y - 3}{y}$$

$$= \dfrac{(y + 5) \cdot y}{(y - 2) \cdot y} - \dfrac{(y - 3) \cdot (y - 2)}{y \cdot (y - 2)} \qquad \text{Write in terms of the LCD.}$$

$$= \dfrac{y^2 + 5y}{y(y - 2)} - \dfrac{y^2 - 5y + 6}{y(y - 2)} \qquad \text{Multiply.}$$

$$= \dfrac{(y^2 + 5y) - (y^2 - 5y + 6)}{y(y - 2)} \qquad \text{Subtract the numerators.}$$

$$= \dfrac{y^2 + 5y - y^2 + 5y - 6}{y(y - 2)} \qquad \text{Remove the parentheses.}$$

$$= \dfrac{10y - 6}{y(y - 2)} \qquad \text{Simplify.}$$

$$= \dfrac{2(5y - 3)}{y(y - 2)} \qquad \text{Factor the numerator.}$$

b. The LCD of the rational expressions can be either $a - 1$ or $1 - a$, since $1 - a = -(a - 1)$. To get the LCD $(a - 1)$, we factor out -1 in the denominator of the expression $\dfrac{a}{1 - a}$.

$$\dfrac{7a + 5}{a - 1} - \dfrac{a}{1 - a}$$

$$= \dfrac{7a + 5}{a - 1} - \dfrac{a}{-(a - 1)} \qquad \text{Write } 1 - a \text{ as } -(a - 1).$$

$$= \dfrac{7a + 5}{a - 1} - \left(-\dfrac{a}{a - 1}\right) \qquad \text{Write } \dfrac{a}{-(a - 1)} \text{ as } -\dfrac{a}{(a - 1)}.$$

$$= \dfrac{7a + 5}{a - 1} + \dfrac{a}{a - 1} \qquad \text{Simplify.}$$

$$= \dfrac{7a + 5 + a}{a - 1} \qquad \text{Add the numerators.}$$

$$= \dfrac{8a + 5}{a - 1} \qquad \text{Simplify.}$$

PRACTICE 8

Combine.

a. $\dfrac{x + 2}{x} - \dfrac{x - 4}{x + 3}$

b. $\dfrac{3x - 4}{x - 1} + \dfrac{x + 1}{1 - x}$

EXAMPLE 9

Combine.

a. $\dfrac{9}{2x + 4} + \dfrac{x}{x^2 - 4}$ **b.** $\dfrac{2}{x - 3} - \dfrac{x - 1}{9 - x^2}$

PRACTICE 9

Combine.

a. $\dfrac{1}{4x - 16} + \dfrac{2x}{x^2 - 16}$

b. $\dfrac{3x - 7}{x^2 - 1} + \dfrac{2}{1 - x}$

Solution

a. $\dfrac{9}{2x + 4} + \dfrac{x}{x^2 - 4}$

$= \dfrac{9}{2(x + 2)} + \dfrac{x}{(x + 2)(x - 2)}$ **Factor the denominators.**

$= \dfrac{9 \cdot (x - 2)}{2(x + 2) \cdot (x - 2)} + \dfrac{x \cdot 2}{(x + 2)(x - 2) \cdot 2}$ **Write in terms of the LCD $2(x + 2)(x - 2)$.**

$= \dfrac{9x - 18}{2(x + 2)(x - 2)} + \dfrac{2x}{2(x + 2)(x - 2)}$ **Multiply.**

$= \dfrac{11x - 18}{2(x + 2)(x - 2)}$ **Add the numerators and combine like terms.**

b. $\dfrac{2}{x - 3} - \dfrac{x - 1}{9 - x^2}$

$= \dfrac{2}{x - 3} - \dfrac{x - 1}{(3 - x)(3 + x)}$

$= \dfrac{2}{x - 3} - \dfrac{x - 1}{-(x - 3)(x + 3)}$ **Write $3 - x$ as $-(x - 3)$.**

$= \dfrac{2}{x - 3} - \left[-\dfrac{x - 1}{(x - 3)(x + 3)} \right]$

$= \dfrac{2}{x - 3} + \dfrac{x - 1}{(x - 3)(x + 3)}$

$= \dfrac{2 \cdot (x + 3)}{(x - 3) \cdot (x + 3)} + \dfrac{x - 1}{(x - 3)(x + 3)}$ **Write in terms of the LCD $(x - 3)(x + 3)$.**

$= \dfrac{2x + 6}{(x - 3)(x + 3)} + \dfrac{x - 1}{(x - 3)(x + 3)}$ **Add the numerators and combine like terms.**

$= \dfrac{(2x + 6) + (x - 1)}{(x - 3)(x + 3)}$

$= \dfrac{3x + 5}{(x - 3)(x + 3)}$

EXAMPLE 10	PRACTICE 10
Perform the indicated operations.	Perform the indicated operation.

EXAMPLE 10

Perform the indicated operations.

a. $\dfrac{7n}{n^2 + 4n + 3} - \dfrac{3n - 2}{n^2 + 2n + 1}$

b. $\dfrac{y}{y - 1} - \dfrac{y}{3} - \dfrac{4}{y + 2}$

PRACTICE 10

Perform the indicated operation.

a. $\dfrac{y}{y^2 + 5y + 6} - \dfrac{4y + 1}{y^2 + 3y + 2}$

b. $\dfrac{2x}{5} - \dfrac{1}{x + 1} - \dfrac{x - 1}{4x}$

Solution

a. $\dfrac{7n}{n^2 + 4n + 3} - \dfrac{3n - 2}{n^2 + 2n + 1}$

$= \dfrac{7n}{(n + 3)(n + 1)} - \dfrac{3n - 2}{(n + 1)^2}$ **Factor the denominators.**

$= \dfrac{7n \cdot (n + 1)}{(n + 3)(n + 1) \cdot (n + 1)}$

$\qquad - \dfrac{(3n - 2) \cdot (n + 3)}{(n + 1)^2 \cdot (n + 3)}$ **Write in terms of the LCD $(n + 3)(n + 1)^2$.**

$= \dfrac{7n^2 + 7n}{(n + 3)(n + 1)^2} - \dfrac{3n^2 + 7n - 6}{(n + 3)(n + 1)^2}$ **Multiply.**

$= \dfrac{(7n^2 + 7n) - (3n^2 + 7n - 6)}{(n + 3)(n + 1)^2}$ **Subtract the numerators.**

$= \dfrac{7n^2 + 7n - 3n^2 - 7n + 6}{(n + 3)(n + 1)^2}$ **Remove the parentheses.**

$= \dfrac{4n^2 + 6}{(n + 3)(n + 1)^2}$ **Combine like terms.**

$= \dfrac{2(n^2 + 3)}{(n + 3)(n + 1)^2}$ **Factor the numerator.**

b. The LCD is $3(y - 1)(y + 2)$.

$\dfrac{y}{y - 1} - \dfrac{y}{3} - \dfrac{4}{y + 2}$

$= \dfrac{y \cdot 3(y + 2)}{(y - 1) \cdot 3(y + 2)} - \dfrac{y \cdot (y - 1)(y + 2)}{3 \cdot (y - 1)(y + 2)} - \dfrac{4 \cdot 3(y - 1)}{(y + 2) \cdot 3(y - 1)}$

$= \dfrac{3y^2 + 6y}{3(y - 1)(y + 2)} - \dfrac{y^3 + y^2 - 2y}{3(y - 1)(y + 2)} - \dfrac{12y - 12}{3(y - 1)(y + 2)}$

$= \dfrac{(3y^2 + 6y) - (y^3 + y^2 - 2y) - (12y - 12)}{3(y - 1)(y + 2)}$

$= \dfrac{3y^2 + 6y - y^3 - y^2 + 2y - 12y + 12}{3(y - 1)(y + 2)}$

$= \dfrac{-y^3 + 2y^2 - 4y + 12}{3(y - 1)(y + 2)}$

EXAMPLE 11

An expression used in the design of electric motors is $\dfrac{V}{kp} - \dfrac{RT}{k^2 p^2}$. Write this difference as a single rational expression.

Solution The LCD is $k^2 p^2$.

$$\frac{V}{kp} - \frac{RT}{k^2 p^2} = \frac{V \cdot kp}{kp \cdot kp} - \frac{RT}{k^2 p^2} = \frac{Vkp}{k^2 p^2} - \frac{RT}{k^2 p^2} = \frac{Vkp - RT}{k^2 p^2}$$

PRACTICE 11

To find the percent change for the cost of an item, retailers can use the expression

$$100\left(\frac{C_1}{C_0} - 1\right),$$

where C_1 is the new cost and C_0 is the old cost. Write as a single rational expression.

Perform the indicated operation. Simplify, if possible.

1. $\dfrac{5a}{12} + \dfrac{11a}{12}$

2. $\dfrac{y}{5} + \dfrac{4y}{5}$

3. $\dfrac{5t}{3} - \dfrac{2t}{3}$

4. $\dfrac{2x}{15} - \dfrac{8x}{15}$

5. $\dfrac{10}{x} + \dfrac{1}{x}$

6. $\dfrac{4}{a} - \dfrac{3}{a}$

7. $\dfrac{6}{7y} - \dfrac{1}{7y}$

8. $\dfrac{-7}{8x} + \dfrac{3}{8x}$

9. $\dfrac{5x}{2y} + \dfrac{x}{2y}$

10. $\dfrac{4a}{3b} + \dfrac{8a}{3b}$

11. $\dfrac{2p}{5q} - \dfrac{3p}{5q}$

12. $\dfrac{x}{6y} - \dfrac{11x}{6y}$

13. $\dfrac{2}{x+1} + \dfrac{7}{x+1}$

14. $\dfrac{8}{p+q} + \dfrac{1}{q+p}$

15. $\dfrac{5}{x+2} - \dfrac{9}{2+x}$

16. $\dfrac{10}{r+s} - \dfrac{14}{s+r}$

17. $\dfrac{a}{a+3} + \dfrac{1}{a+3}$

18. $\dfrac{p}{p-1} + \dfrac{3}{p-1}$

19. $\dfrac{3x}{x-8} + \dfrac{2x+1}{x-8}$

20. $\dfrac{4p}{p-5} + \dfrac{p-2}{p-5}$

21. $\dfrac{7x+1}{5x+2} - \dfrac{3x}{5x+2}$

22. $\dfrac{2x}{2x-1} - \dfrac{x+1}{2x-1}$

23. $\dfrac{9x+17}{2x+5} - \dfrac{3x+2}{2x+5}$

24. $\dfrac{5a+1}{2a+3} - \dfrac{a-5}{2a+3}$

25. $\dfrac{-7+5n}{3n-1} + \dfrac{7n+3}{3n-1}$

26. $\dfrac{-5+8x}{5x-1} + \dfrac{4-3x}{5x-1}$

27. $\dfrac{x^2-1}{x^2-4x-2} - \dfrac{x^2-x+3}{x^2-4x-2}$

28. $\dfrac{9+3x-x^2}{x^2+x+1} + \dfrac{x^2-5}{x^2+x+1}$

29. $\dfrac{x}{x^2-3x+2} + \dfrac{2}{x^2-3x+2} + \dfrac{x^2-4x}{x^2-3x+2}$

30. $\dfrac{a}{a^2-5a+4} + \dfrac{2}{a^2-5a+4} + \dfrac{a^2-4a}{a^2-5a+4}$

31. $\dfrac{2x-1}{3x^2-x+2} + \dfrac{8}{3x^2-x+2} - \dfrac{3x}{3x^2-x+2}$

32. $\dfrac{4y-3}{y^2+7y+1} - \dfrac{y}{y^2+7y+1} + \dfrac{6-y}{y^2+7y+1}$

The following expressions represent denominators of rational expressions. Find their LCD.

33. $5(x+2)$ and $3(x+2)$

34. $9(4c-1)$ and $6(4c-1)$

35. $(p-3)(p+8)$ and $(p-3)(p-8)$

36. $(b+c)(5b)$ and $(b+c)(2b)$

37. $t, t+3,$ and $t-3$

38. $s, 4s,$ and $s+5$

39. $t^2+7t+10$ and t^2-25

40. n^2-1 and n^2+6n-7

41. $3s^2-11s+6$ and $3s^2+4s-4$

42. $2x^2+x-15$ and $-2x^2+9x-10$

Write each pair of rational expressions in terms of their LCD.

43. $\dfrac{1}{3x}$ and $\dfrac{5}{4x^2}$

44. $\dfrac{2}{5y^3}$ and $\dfrac{3}{y^2}$

45. $\dfrac{5}{2a^2}$ and $\dfrac{a-3}{7ab}$

46. $\dfrac{x+2}{4xy}$ and $\dfrac{7}{6x^2}$

47. $\dfrac{8}{n(n+1)}$ and $\dfrac{5}{(n+1)^2}$

48. $\dfrac{2}{c(c-3)^2}$ and $\dfrac{4}{(c-3)}$

49. $\dfrac{3n}{4n+4}$ and $\dfrac{2n}{n^2-1}$

50. $\dfrac{4y}{y^2-1}$ and $\dfrac{y}{2y+2}$

51. $\dfrac{2n}{n^2+6n+5}$ and $\dfrac{3n}{n^2+2n-15}$

52. $\dfrac{7t}{t^2-2t+1}$ and $\dfrac{4t}{t^2-5t+4}$

Perform the indicated operations. Simplify, if possible.

53. $\dfrac{5}{3x}+\dfrac{1}{2x}$

54. $\dfrac{3}{4a}+\dfrac{2}{5a}$

55. $\dfrac{2}{3x^2}-\dfrac{5}{6x}$

56. $\dfrac{9}{7c^2}-\dfrac{3}{5c^3}$

57. $\dfrac{-2}{3x^2y}+\dfrac{4}{3xy^2}$

58. $\dfrac{6}{p^2q}+\dfrac{8}{pq^2}$

59. $\dfrac{1}{x+1}+\dfrac{1}{x-1}$

60. $\dfrac{2}{a+b}+\dfrac{2}{a-b}$

61. $\dfrac{p+6}{3}-\dfrac{2p+1}{7}$

62. $\dfrac{b+9}{5}-\dfrac{5-7b}{2}$

63. $x-\dfrac{10-4x}{2}$

64. $\dfrac{-2t+1}{8}-3t$

65. $\dfrac{3a+1}{6a}-\dfrac{a^2-2}{2a^2}$

66. $\dfrac{y^2-2}{2y^2}-\dfrac{2y-7}{4y}$

67. $\dfrac{a^2}{a-1}-\dfrac{1}{1-a}$

68. $\dfrac{6}{x-4}-\dfrac{x}{4-x}$

69. $\dfrac{4}{c-4}+\dfrac{c}{4-c}$

70. $\dfrac{a}{a-b}+\dfrac{b}{b-a}$

71. $\dfrac{x-5}{x+1}-\dfrac{x+2}{x}$

72. $\dfrac{p+4}{p}-\dfrac{p+3}{p-1}$

73. $\dfrac{4x-5}{x-4}+\dfrac{1-3x}{4-x}$

74. $\dfrac{5n}{1-n}+\dfrac{3n-2}{n-1}$

75. $\dfrac{5x}{x^2+x-2}+\dfrac{6}{x+2}$

76. $\dfrac{p}{p^2-3p+2}+\dfrac{4}{p-1}$

77. $\dfrac{4}{3n-9}-\dfrac{n}{n^2+2n-15}$

78. $\dfrac{-2x}{x^2+7x+12}+\dfrac{5}{4x+16}$

79. $\dfrac{2}{t+5}-\dfrac{t+6}{25-t^2}$

80. $\dfrac{4}{x-2}-\dfrac{x-1}{4-x^2}$

81. $\dfrac{4x}{x^2+2x+1}-\dfrac{2x+5}{x^2+4x+3}$

82. $\dfrac{y-1}{y^2-4y+4}+\dfrac{3y}{y^2-y-2}$

83. $\dfrac{2t-1}{2t^2+t-3}+\dfrac{2}{t-1}$

84. $\dfrac{4}{y-2}+\dfrac{3y-1}{y^2-6y+8}$

85. $\dfrac{4x}{x-1}+\dfrac{2}{3x}+\dfrac{x}{x^2-1}$

86. $\dfrac{c}{c+2}+\dfrac{3}{4c}+\dfrac{2c}{c^2-4}$

87. $\dfrac{5y}{3y-1}-\dfrac{3}{y-4}+\dfrac{y+1}{3y^2-13y+4}$

88. $\dfrac{6x-1}{2x+1}+\dfrac{x}{x-1}-\dfrac{4x}{2x^2-x-1}$

89. $\dfrac{a-1}{(a+3)^2} - \dfrac{2a-3}{a+3} - \dfrac{a}{4a+12}$

90. $\dfrac{y}{8y-16} + \dfrac{3y+4}{y-2} - \dfrac{y-3}{(y-2)^2}$

Mixed Practice

Write each pair of rational expressions in terms of their LCD.

91. $\dfrac{p+3}{6p^2}$ and $\dfrac{5}{8pq}$

92. $\dfrac{3}{y(y-2)}$ and $\dfrac{8}{(y-2)^2}$

Perform the indicated operations. Simplify, if possible.

93. $\dfrac{9}{xy^2} + \dfrac{6}{x^2y}$

94. $\dfrac{7}{y-3} - \dfrac{y}{3-y}$

95. $\dfrac{c-3}{c} - \dfrac{c-2}{c+1}$

96. $\dfrac{-2a}{a^2+7a+10} + \dfrac{4}{3a+15}$

97. $\dfrac{b+3}{b^2-2b-3} - \dfrac{4}{b^2-6b+9}$

98. $\dfrac{2x}{x^2-1} + \dfrac{x}{x+1} + \dfrac{2}{3x}$

99. $\dfrac{6}{m+n} - \dfrac{11}{n+m}$

100. $\dfrac{5x-2}{3x-1} - \dfrac{2x}{3x-1}$

101. $\dfrac{r}{r^2-r-6} + \dfrac{3}{r^2-r-6} + \dfrac{r^2-5r}{r^2-r-6}$

102. If j^2-4 and j^2+2j-8 represent denominators of rational expressions, find their LCD.

Applications

Solve.

103. The distance that an object falls is given by the expression $vt + \dfrac{at^2}{2}$. Write this expression as a single rational expression.

104. The chances of an event happening can be represented by the rational expression $\dfrac{f}{t}$, whereas $\dfrac{t-f}{t}$ represents the chances of the event *not* happening. What is the sum of these two expressions?

105. A bank pays an interest rate r compounded annually on all account balances. If a customer wanted the balance in an account to be $1000 at the end of one year, she would need to have a current balance of $\dfrac{1000}{1+r}$ dollars. However, if she were willing to wait two years for the balance to reach $1000, then her current balance would need to be only $\dfrac{1000}{(1+r)^2}$ dollars. Express the difference between the quantities $\dfrac{1000}{1+r}$ dollars and $\dfrac{1000}{(1+r)^2}$ dollars as a single rational expression.

106. An expression to find the length of base b of a trapezoid is $\dfrac{2A}{h} - a$. Another expression is $\dfrac{2A-ah}{h}$. Explain whether these two expressions are equivalent.

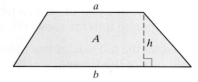

107. A trucker drove 20 mi at a speed r mph and then returned at double the speed. How long did the whole trip take? Write the answer as a single rational expression.

108. A train traveled m mi at a speed of s mph. A bus following the same route traveled 5 mph slower. How much longer did the bus take than the train to make this trip? Write the answer as a single rational expression.

109. To maintain a checking account, a bank charges a customer \$3 per month and \$0.10 per check. For x checks, the cost per check is $\left(\dfrac{3}{x} + 0.1\right)$ dollars. Write this cost as a single rational expression.

110. In baseball, an expression for *runners per inning* is $\dfrac{H}{I} + \dfrac{W}{I}$, where H represents the number of hits, W the number of walks, and I the number of innings pitched. Explain whether the expression $\dfrac{H + W}{2I}$ is equivalent to $\dfrac{H}{I} + \dfrac{W}{I}$.

● *Check your answers on page A-33.*

MINDSTRETCHERS

Groupwork

1. When we break a given rational expression into *partial fractions*, we are expressing it as the sum of two other rational expressions. For instance, we could write the rational expression $\dfrac{4x + 1}{x^2 + x}$ as $\dfrac{3}{x + 1} + \dfrac{1}{x}$.

   a. Confirm that $\dfrac{4x + 1}{x^2 + x} = \dfrac{3}{x + 1} + \dfrac{1}{x}$.

   b. Find the missing numerators so that $\dfrac{5x + 1}{x^2 - 1} = \dfrac{?}{x - 1} + \dfrac{?}{x + 1}$.

Critical Thinking

2. Find the following product.

$$\left(1 + \frac{1}{n}\right)\left(1 + \frac{1}{n + 1}\right)\left(1 + \frac{1}{n + 2}\right)\left(1 + \frac{1}{n + 3}\right)\cdots\left(1 + \frac{1}{n + 99}\right)\left(1 + \frac{1}{n + 100}\right)$$

Mathematical Reasoning

3. A formula sometimes used to add arithmetic fractions is:

$$\frac{p}{q} + \frac{r}{s} = \frac{ps + qr}{qs}$$

   a. Working with a pair of arithmetic fractions of your choice, confirm that this formula gives the correct sum by comparing it to the answer derived using an alternative method.

   b. Consider the left side of this formula as a sum of rational expressions. Explain why the right side of the formula is correct.

CULTURAL NOTE

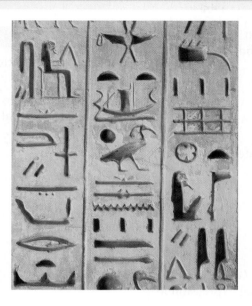

The ancient Egyptians expressed most fractions as the sum of *unit fractions* (fractions whose numerators are 1). For instance, they would write the fraction $\frac{2}{5}$ as $\frac{1}{3} + \frac{1}{15}$. More generally, the rational expression $\frac{x + y}{xy}$ can be written as the sum of the unit fractions $\frac{1}{x}$ and $\frac{1}{y}$.

(***Source:*** Jan Gullberg, *From the Birth of Numbers*, W. W. Norton & Company, New York, 1997)

7.4   Complex Rational Expressions

Recall that a *complex fraction* in arithmetic is a fraction that in turn contains one or more fractions in its numerator, denominator, or both:

$$\text{Main fraction line} \longrightarrow \cfrac{\dfrac{1}{3} \longleftarrow \text{Fraction in the numerator}}{\dfrac{4}{5} \longleftarrow \text{Fraction in the denominator}}$$

Such fractions can be put in standard form as the quotient of two integers by recalling that the main fraction line represents division:

$$\cfrac{\dfrac{1}{3}}{\dfrac{4}{5}} = \frac{1}{3} \div \frac{4}{5} = \frac{1}{3} \cdot \frac{5}{4} = \frac{5}{12}$$

A comparable expression in algebra is called *a complex rational expression* (or a *complex algebraic fraction*). Such an expression contains a rational expression in its numerator, denominator, or both. Here are some examples.

$$\cfrac{x^2 + 5x - 2}{\dfrac{3}{4}} \qquad\qquad \cfrac{7}{\dfrac{a + b}{2}} \qquad\qquad \cfrac{\dfrac{3n}{2} + \dfrac{1}{n}}{\dfrac{n - 4}{n^2}}$$

As in the case of an arithmetic complex fraction, a complex rational expression is usually simplified, that is, written as the quotient of two polynomials with no common factors.

We consider two methods of simplifying a complex rational expression: the *division method* and the *LCD method*.

In the division method, we begin by writing the numerator and denominator as single rational expressions, simplifying if necessary. Then, we write the complex rational expression as the numerator divided by the denominator.

For instance, suppose we want to simplify the complex rational expression

$$\cfrac{\dfrac{x}{2} + \dfrac{1}{2}}{\dfrac{x^2}{3} + \dfrac{x}{3}}.$$

We first write the numerator and denominator as single rational expressions.

$$\cfrac{\dfrac{x}{2} + \dfrac{1}{2}}{\dfrac{x^2}{3} + \dfrac{x}{3}} = \cfrac{\dfrac{x + 1}{2}}{\dfrac{x^2 + x}{3}} \qquad \begin{array}{l}\longleftarrow \text{Add the rational expressions} \\ \text{in the numerator.} \\ \longleftarrow \text{Add the rational expressions} \\ \text{in the denominator.}\end{array}$$

$$= \frac{x + 1}{2} \div \frac{x^2 + x}{3} \qquad \begin{array}{l}\text{Write the expression as the numerator} \\ \text{divided by the denominator.}\end{array}$$

$$= \frac{x + 1}{2} \cdot \frac{3}{x^2 + x} \qquad \begin{array}{l}\text{Take the reciprocal of the divisor and} \\ \text{change the operation to multiplication.}\end{array}$$

$$= \frac{\overset{1}{\cancel{x+1}}}{2} \cdot \frac{3}{x\cancel{(x+1)}}$$ Factor and divide out the common factor.

$$= \frac{3}{2x}$$ Simplify.

So we can conclude that the original complex rational expression

$$\frac{\dfrac{x}{2} + \dfrac{1}{2}}{\dfrac{x^2}{3} + \dfrac{x}{3}}$$ is equivalent to the rational expression $\dfrac{3}{2x}$.

To Simplify a Complex Rational Expression: Division Method

- Write the numerator and denominator as single rational expressions in simplified form.
- Write the expression as the numerator divided by the denominator.
- Divide.
- Simplify, if possible.

Let's use the division method in the following examples.

EXAMPLE 1	PRACTICE 1

EXAMPLE 1

Simplify.

a. $\dfrac{\dfrac{2}{x}}{\dfrac{6}{x^3}}$

b. $\dfrac{y}{\dfrac{y}{2} + \dfrac{y^2}{3}}$

Solution

a. Neither the numerator nor the denominator of the complex rational expression can be simplified. So we write the expression as the numerator divided by the denominator.

$$\frac{\dfrac{2}{x}}{\dfrac{6}{x^3}} = \frac{2}{x} \div \frac{6}{x^3}$$

$$= \frac{2}{x} \cdot \frac{x^3}{6}$$ Take the reciprocal of the divisor and change the operation to multiplication.

$$= \frac{\overset{1}{\cancel{2}}}{\underset{1}{\cancel{x}}} \cdot \frac{\overset{x^2}{\cancel{x^3}}}{\underset{3}{\cancel{6}}}$$ Divide out the common factors.

$$= \frac{x^2}{3}$$ Simplify.

PRACTICE 1

Simplify.

a. $\dfrac{\dfrac{3}{x^4}}{\dfrac{5}{x}}$

b. $\dfrac{2x}{\dfrac{x^2}{4} + \dfrac{x}{2}}$

b. Let's express the denominator as a single rational expression.

$$\frac{y}{\frac{y}{2} + \frac{y^2}{3}} = \frac{y}{\frac{y \cdot 3}{2 \cdot 3} + \frac{y^2 \cdot 2}{3 \cdot 2}}$$ Add the rational expressions in the denominator. The LCD is 6.

$$= \frac{y}{\frac{3y + 2y^2}{6}}$$

$$= y \div \frac{3y + 2y^2}{6}$$ Write the expression as the numerator divided by the denominator.

$$= y \cdot \frac{6}{3y + 2y^2}$$ Take the reciprocal of the divisor and change the operation to multiplication.

$$= \overset{1}{y} \cdot \frac{6}{\underset{1}{y}(3 + 2y)}$$ Factor and divide out the common factor.

$$= \frac{6}{3 + 2y}$$ Simplify.

EXAMPLE 2

Simplify: $\dfrac{1 + \dfrac{1}{x}}{1 - \dfrac{1}{x}}$

Solution We begin by simplifying the numerator and denominator so that each is a single rational expression.

$$\frac{1 + \dfrac{1}{x}}{1 - \dfrac{1}{x}} = \frac{\dfrac{x + 1}{x}}{\dfrac{x - 1}{x}}$$

$$= \frac{x + 1}{x} \div \frac{x - 1}{x}$$

$$= \frac{x + 1}{x} \cdot \frac{x}{x - 1}$$

$$= \frac{x + 1}{\underset{1}{x}} \cdot \frac{\overset{1}{x}}{x - 1}$$

$$= \frac{x + 1}{x - 1}$$

PRACTICE 2

Simplify: $\dfrac{2 - \dfrac{1}{n}}{2 + \dfrac{1}{n}}$

Now, we consider the LCD method of simplifying a complex rational expression. In this method, we multiply the numerator and denominator of the complex rational expression by the LCD of all rational expressions that appear within it.

Let's simplify the complex rational expression on the bottom of page 592 using the LCD method. First, we must find the LCD of all the rational expressions in the numerator and denominator. The denominators within the complex rational expression are 2 and 3, so the LCD is 6.

$$\frac{\dfrac{x}{2} + \dfrac{1}{2}}{\dfrac{x^2}{3} + \dfrac{x}{3}} = \frac{\left(\dfrac{x}{2} + \dfrac{1}{2}\right) \cdot 6}{\left(\dfrac{x^2}{3} + \dfrac{x}{3}\right) \cdot 6}$$ **Multiply the numerator and denominator by the LCD 6.**

$$= \frac{\dfrac{x}{2} \cdot 6 + \dfrac{1}{2} \cdot 6}{\dfrac{x^2}{3} \cdot 6 + \dfrac{x}{3} \cdot 6}$$ **Use the distributive property in the numerator and in the denominator.**

$$= \frac{3x + 3}{2x^2 + 2x}$$ **Simplify.**

$$= \frac{3(\overset{1}{\cancel{x+1}})}{2x(\underset{1}{\cancel{x+1}})}$$ **Factor and divide out the common factor.**

$$= \frac{3}{2x}$$ **Simplify.**

Note that the division method and the LCD method result in the same answer.

To Simplify a Complex Rational Expression: LCD Method

- Find the LCD of all the rational expressions *within* both the numerator and denominator.
- Multiply the numerator and denominator of the complex rational expression by this LCD.
- Simplify, if possible.

EXAMPLE 3	PRACTICE 3

Simplify.

a. $\dfrac{\dfrac{3}{y^2}}{\dfrac{1}{y}}$ **b.** $\dfrac{n^2}{\dfrac{1}{n} + \dfrac{2}{n^2}}$

Solution

a. $\dfrac{\dfrac{3}{y^2}}{\dfrac{1}{y}} = \dfrac{\dfrac{3}{y^2} \cdot y^2}{\dfrac{1}{y} \cdot y^2}$ **Multiply by the LCD y^2.**

$\qquad = \dfrac{3}{y}$ **Simplify.**

b. $\dfrac{n^2}{\dfrac{1}{n} + \dfrac{2}{n^2}} = \dfrac{n^2 \cdot n^2}{\left(\dfrac{1}{n} + \dfrac{2}{n^2}\right) \cdot n^2}$ **Multiply by the LCD n^2.**

$\qquad = \dfrac{n^2 \cdot n^2}{\dfrac{1}{n} \cdot n^2 + \dfrac{2}{n^2} \cdot n^2}$ **Use the distributive property.**

$\qquad = \dfrac{n^4}{n + 2}$ **Simplify.**

Simplify.

a. $\dfrac{\dfrac{4}{x}}{\dfrac{2}{x^3}}$

b. $\dfrac{\dfrac{1}{y} + \dfrac{3}{y^2}}{2y}$

EXAMPLE 4

Simplify.

a. $\dfrac{3 + \dfrac{1}{x^2}}{3 - \dfrac{1}{x}}$

b. $\dfrac{\dfrac{1}{x^2} - \dfrac{1}{y^2}}{\dfrac{3}{x} - \dfrac{3}{y}}$

Solution

a. Multiply the numerator and denominator by the LCD x^2.

$$\frac{3 + \dfrac{1}{x^2}}{3 - \dfrac{1}{x}} = \frac{\left(3 + \dfrac{1}{x^2}\right) \cdot x^2}{\left(3 - \dfrac{1}{x}\right) \cdot x^2}$$

$$= \frac{3 \cdot x^2 + \dfrac{1}{x^2} \cdot x^2}{3 \cdot x^2 - \dfrac{1}{x} \cdot x^2}$$

$$= \frac{3x^2 + 1}{3x^2 - x}$$

b. Multiply the numerator and denominator by the LCD x^2y^2.

$$\frac{\dfrac{1}{x^2} - \dfrac{1}{y^2}}{\dfrac{3}{x} - \dfrac{3}{y}} = \frac{\left(\dfrac{1}{x^2} - \dfrac{1}{y^2}\right) \cdot x^2 y^2}{\left(\dfrac{3}{x} - \dfrac{3}{y}\right) \cdot x^2 y^2}$$

$$= \frac{\dfrac{1}{x^2}(x^2y^2) - \dfrac{1}{y^2}(x^2y^2)}{\dfrac{3}{x}(x^2y^2) - \dfrac{3}{y}(x^2y^2)}$$

$$= \frac{y^2 - x^2}{3xy^2 - 3x^2y}$$

$$= \frac{\overset{1}{\cancel{(y - x)}}(y + x)}{3xy\underset{1}{\cancel{(y - x)}}}$$

$$= \frac{y + x}{3xy}$$

PRACTICE 4

Express in simplest terms.

a. $\dfrac{4 + \dfrac{1}{y}}{4 - \dfrac{1}{y^2}}$

b. $\dfrac{\dfrac{1}{2a^2} - \dfrac{1}{2b^2}}{\dfrac{5}{a} + \dfrac{5}{b}}$

EXAMPLE 5

A resistor is an electrical device such as a lightbulb that offers resistance to the flow of electricity. When two lightbulbs with resistance R_1 and R_2 are connected in a certain kind of circuit, the combined resistance is given by the expression

$$\frac{1}{\dfrac{1}{R_1} + \dfrac{1}{R_2}}$$

Simplify this expression for the combined resistance.

Solution

$$\frac{1}{\dfrac{1}{R_1} + \dfrac{1}{R_2}} = \frac{1 \cdot R_1 R_2}{\left(\dfrac{1}{R_1} + \dfrac{1}{R_2}\right) \cdot R_1 R_2}$$

$$= \frac{1 \cdot R_1 R_2}{\dfrac{1}{R_1} \cdot R_1 R_2 + \dfrac{1}{R_2} \cdot R_1 R_2} = \frac{R_1 R_2}{R_2 + R_1}$$

PRACTICE 5

The harmonic mean is a kind of average. The harmonic mean of the three numbers a, b, and c is given by the expression $\dfrac{3}{\dfrac{1}{a} + \dfrac{1}{b} + \dfrac{1}{c}}$. Simplify this complex rational expression.

7.4   Exercises

Simplify.

1. $\dfrac{\dfrac{x}{5}}{\dfrac{x^2}{10}}$

2. $\dfrac{\dfrac{3}{s^2}}{\dfrac{s^3}{2}}$

3. $\dfrac{\dfrac{a+1}{2}}{\dfrac{a-1}{2}}$

4. $\dfrac{\dfrac{n-1}{n}}{\dfrac{n+1}{2n}}$

5. $\dfrac{3+\dfrac{1}{x}}{3-\dfrac{1}{x^2}}$

6. $\dfrac{1+\dfrac{1}{y^2}}{8-\dfrac{1}{y}}$

7. $\dfrac{\dfrac{1}{3d}-\dfrac{1}{d^2}}{d-\dfrac{9}{d}}$

8. $\dfrac{a-\dfrac{4}{a}}{\dfrac{1}{a^2}+\dfrac{1}{2a}}$

9. $\dfrac{1-\dfrac{4y^2}{x^2}}{3+\dfrac{6y}{x}}$

10. $\dfrac{\dfrac{p}{q}+2}{\dfrac{p^2}{q^2}-4}$

11. $\dfrac{\dfrac{2}{y}-\dfrac{1}{5}}{\dfrac{5}{y}-1}$

12. $\dfrac{\dfrac{2}{t}-\dfrac{3}{t^2}}{10+\dfrac{2}{t}}$

13. $\dfrac{1+\dfrac{4}{x}+\dfrac{4}{x^2}}{1+\dfrac{5}{x}+\dfrac{6}{x^2}}$

14. $\dfrac{1-\dfrac{1}{a^2}}{4-\dfrac{5}{a}+\dfrac{1}{a^2}}$

 15. $\dfrac{3+\dfrac{1}{y+1}}{5-\dfrac{1}{y+1}}$

16. $\dfrac{2+\dfrac{1}{b-1}}{2-\dfrac{1}{b-1}}$

 17. $\dfrac{\dfrac{x}{4}-\dfrac{x}{8}}{\dfrac{2}{y^2}+\dfrac{2}{y}}$

18. $\dfrac{\dfrac{n}{2}+\dfrac{n}{3}}{\dfrac{1}{m}-\dfrac{1}{m^2}}$

19. $\dfrac{\dfrac{x}{x+1}-\dfrac{2}{x}}{\dfrac{x}{3}}$

20. $\dfrac{\dfrac{y}{5}}{\dfrac{y}{y+2}+\dfrac{3}{y}}$

Mixed Practice

Simplify.

21. $\dfrac{\dfrac{m+2}{3m}}{\dfrac{m-1}{m}}$

22. $\dfrac{3-\dfrac{1}{d-2}}{3+\dfrac{1}{d-2}}$

23. $\dfrac{\dfrac{4}{u}-\dfrac{2}{u+1}}{\dfrac{u}{2}}$

24. $\dfrac{x-\dfrac{9}{x}}{\dfrac{1}{3x}-\dfrac{1}{x^2}}$

25. $\dfrac{\dfrac{3}{y^2}+\dfrac{4}{y}}{6+\dfrac{3}{y}}$

26. $\dfrac{3-\dfrac{1}{b}-\dfrac{2}{b^2}}{1-\dfrac{1}{b^2}}$

Applications

Solve.

27. An expression from the study of electricity is

$$\frac{V}{\dfrac{1}{2R} + \dfrac{1}{2R+2}},$$

where V represents voltage and R represents resistance. Simplify this expression.

28. The expression

$$\frac{\dfrac{m}{c}}{1 - \dfrac{p^2}{c^2}}$$

is important in the design of airplanes. Write this expression in simplified form.

29. The earned run average (ERA) is a statistic used in baseball to represent the average number of earned runs that a pitcher allows. A pitcher's ERA can be calculated using the expression

$$\frac{E}{\dfrac{I}{9}},$$

where E stands for the number of earned runs a pitcher gave up after pitching I innings. Simplify this complex rational expression.

30. The ratio of the volume of a cube with side s to the volume of its inscribed sphere is

$$\frac{s^3}{\dfrac{1}{6}\pi s^3}.$$

Simplify this ratio.

31. On a trip, an airplane flew at an average speed of a mph, returning on the same route at an average speed of b mph. The plane's average speed for the round trip is given by the complex rational expression

$$\frac{2}{\dfrac{1}{a} + \dfrac{1}{b}}.$$

Simplify.

32. An object travels at speed s for a distance d and later travels at speed S for distance D. The following expression represents the average speed for the entire trip:

$$\frac{d + D}{\dfrac{d}{s} + \dfrac{D}{S}}.$$

Simplify this expression.

33. The weight of an object decreases as its distance from the Earth's surface increases. Suppose an object weighs w kg at sea level. Then at a height of h km above sea level, it will weigh (in kilograms)

$$\frac{w}{\left(1 + \dfrac{h}{6400}\right)^2}.$$

Show that $\dfrac{w}{\left(1 + \dfrac{h}{6400}\right)^2}$ and $\dfrac{6400^2 w}{(6400 + h)^2}$ are equivalent expressions.

34. The sound from a car horn as it approaches an observer seems to change its frequency. This is an example of the Doppler effect, which is studied in physics. The observer will hear a sound with frequency

$$\frac{S}{\dfrac{S - v}{f}},$$

where f is the actual frequency of the sound, S is the speed of sound, and v is the speed of the approaching object. Simplify the expression.

• *Check your answers on page A-34.*

MINDSTRETCHERS

Research

1. The harmonic mean (see Practice 5, page 597) is related to musical harmony. Either in your college library or on the Web, investigate the relationship between music and the harmonic mean. Summarize your findings in a few sentences.

Writing

2. Select a complex rational expression of your choice and simplify it using (a) the division method and (b) the LCD method. Which method do you prefer? Explain why.

Groupwork

3. Find a complex rational expression with numerator $\dfrac{4}{y}$ that simplifies to $\dfrac{3}{x + 5}$.

<table>
<tr><td>**7.5**</td><td># Solving Rational Equations</td></tr>
</table>

What Rational Equations Are and Why They Are Important

OBJECTIVES

■ To solve an equation involving rational expressions

■ To solve applied problems involving rational equations

So far in this chapter, we have discussed rational *expressions*. Now, let's consider rational or fractional *equations*, that is, equations that contain one or more rational expressions. Here are some examples:

$$\frac{n}{2} - 3 = \frac{n}{5} \qquad \frac{1}{t} + \frac{1}{3} = \frac{1}{2} \qquad \frac{3}{x+1} - \frac{1}{8} = \frac{5}{x^2 - 1}$$

Many situations involving rates of work, motion, and proportions can be modeled by rational equations.

Solving Rational Equations

In solving rational equations, it is important not to confuse a rational expression with a rational equation.

$$\frac{4x}{3} + \frac{x}{3} \qquad\qquad \frac{4x}{3} + \frac{x}{3} = 5$$

Rational *expression*   Rational *equation*

The key to solving a rational equation is to *clear the equation of rational expressions*. We do this by first determining their *least common denominator*, and then by multiplying both sides of the equation by the LCD.

EXAMPLE 1

Solve and check: $\dfrac{x}{2} + \dfrac{x}{6} = \dfrac{2}{3}$

Solution The rational expressions in this equation are $\dfrac{x}{2}, \dfrac{x}{6}$, and $\dfrac{2}{3}$. The denominators of these expressions are 2, 6, and 3, so the LCD is 6. To clear the equation of rational expressions, we multiply each side of the equation by 6.

$$\frac{x}{2} + \frac{x}{6} = \frac{2}{3}$$

$6 \cdot \left(\dfrac{x}{2} + \dfrac{x}{6}\right) = 6 \cdot \dfrac{2}{3}$ Multiply each side of the equation by the LCD.

$6 \cdot \dfrac{x}{2} + 6 \cdot \dfrac{x}{6} = 6 \cdot \dfrac{2}{3}$ Use the distributive property.

$\quad 3x + x = 4$ Simplify.

$\qquad 4x = 4$ Combine like terms.

$\qquad\quad x = 1$ Divide each side by 4.

PRACTICE 1

Solve and check: $\dfrac{y}{2} - \dfrac{y}{3} = \dfrac{1}{12}$

Check

$$\frac{x}{2} + \frac{x}{6} = \frac{2}{3}$$

$$\frac{1}{2} + \frac{1}{6} \stackrel{?}{=} \frac{2}{3} \qquad \text{Substitute 1 for } x.$$

$$\frac{3}{6} + \frac{1}{6} \stackrel{?}{=} \frac{2}{3}$$

$$\frac{4}{6} \stackrel{?}{=} \frac{2}{3}$$

$$\frac{2}{3} = \frac{2}{3} \qquad \text{True}$$

So the solution to the given equation is 1.

The following rule describes the general procedure for solving equations of this type.

To Solve a Rational Equation

- Find the LCD of all rational expressions in the equation.
- Multiply each side of the equation by the LCD.
- Solve the equation.
- Check the solution(s) in the original equation.

EXAMPLE 2

Solve and check: $\dfrac{4}{n} - \dfrac{n+1}{3} = 1$

Solution The LCD of the rational expressions in the equation is $3n$.

$$\frac{4}{n} - \frac{n+1}{3} = 1$$

$$3n \cdot \left(\frac{4}{n} - \frac{n+1}{3} \right) = 3n \cdot 1 \qquad \begin{array}{l}\text{Multiply each side of the equation}\\\text{by the LCD.}\end{array}$$

$$3n \cdot \frac{4}{n} - 3n \cdot \frac{n+1}{3} = 3n \cdot 1 \qquad \begin{array}{l}\text{Distribute } 3n \text{ on the left side}\\\text{of the equation.}\end{array}$$

$$\overset{1}{3n} \cdot \frac{4}{\underset{1}{n}} - \overset{1}{3}n \cdot \frac{n+1}{\underset{1}{3}} = 3n \cdot 1$$

$$12 - n(n+1) = 3n \qquad \text{Simplify.}$$

$$12 - n^2 - n = 3n \qquad \text{Use the distributive property.}$$

$$n^2 + n + 3n - 12 = 0$$

$$n^2 + 4n - 12 = 0 \qquad \text{Write in standard form.}$$

$$(n+6)(n-2) = 0 \qquad \text{Factor the left side of the equation.}$$

$$n+6 = 0 \quad \text{or} \quad n-2 = 0 \qquad \text{Set each factor equal to 0.}$$

$$n = -6 \qquad\qquad n = 2$$

PRACTICE 2

Solve and check: $\dfrac{x-2}{5} - 1 = -\dfrac{2}{x}$

Check

Substitute -6 for n.

$$\frac{4}{n} - \frac{n+1}{3} = 1$$

$$\frac{4}{-6} - \frac{-6+1}{3} \overset{?}{=} 1$$

$$-\frac{2}{3} - \left(-\frac{5}{3}\right) \overset{?}{=} 1$$

$$-\frac{2}{3} + \frac{5}{3} \overset{?}{=} 1$$

$$\frac{3}{3} \overset{?}{=} 1$$

$$1 = 1 \qquad \textbf{True}$$

Substitute 2 for n.

$$\frac{4}{n} - \frac{n+1}{3} = 1$$

$$\frac{4}{2} - \frac{2+1}{3} \overset{?}{=} 1$$

$$2 - 1 \overset{?}{=} 1$$

$$1 = 1 \qquad \textbf{True}$$

Our check confirms that the solutions are -6 and 2.

EXAMPLE 3

Solve and check: $\dfrac{2}{x+3} + \dfrac{1}{x-3} = -\dfrac{6}{x^2-9}$

Solution First, we find the LCD of the rational expressions in this equation, which is $x^2 - 9$, or $(x+3)(x-3)$. Then, we muliply each side of the equation by the LCD.

$$\frac{2}{x+3} + \frac{1}{x-3} = -\frac{6}{x^2-9}$$

$$\frac{2}{x+3} + \frac{1}{x-3} = -\frac{6}{(x+3)(x-3)}$$

$$(x+3)(x-3)\cdot\frac{2}{x+3} + (x+3)(x-3)\cdot\frac{1}{x-3} = (x+3)(x-3)\cdot\frac{-6}{(x+3)(x-3)}$$

$$\cancel{(x+3)}(x-3)\cdot\frac{2}{\cancel{x+3}} + (x+3)\cancel{(x-3)}\cdot\frac{1}{\cancel{x-3}} = \cancel{(x+3)}\cancel{(x-3)}\cdot\frac{-6}{\cancel{(x+3)}\cancel{(x-3)}}$$

$$2(x-3) + (x+3) = -6$$

$$2x - 6 + x + 3 = -6$$

$$3x - 3 = -6$$

$$3x = -3$$

$$x = -1$$

Check

$$\frac{2}{x+3} + \frac{1}{x-3} = -\frac{6}{x^2-9}$$

$$\frac{2}{-1+3} + \frac{1}{-1-3} \overset{?}{=} -\frac{6}{(-1)^2-9} \qquad \textbf{Substitute } -1 \textbf{ for } x.$$

$$\frac{2}{2} + \frac{1}{-4} \overset{?}{=} -\frac{6}{-8} \qquad \textbf{Simplify.}$$

$$\frac{2}{2} + \left(-\frac{1}{4}\right) \overset{?}{=} -\left(-\frac{6}{8}\right)$$

$$1 - \frac{1}{4} \overset{?}{=} \frac{3}{4}$$

$$\frac{3}{4} = \frac{3}{4} \qquad \textbf{True}$$

So the solution is -1.

PRACTICE 3

Solve and check: $\dfrac{4}{y+2} + \dfrac{2}{y-1} = \dfrac{12}{y^2+y-2}$

When multiplying each side of a rational equation by a variable expression, the resulting equation may have a solution that does not satisfy the original equation. If such a result makes a denominator in the original equation 0, then the rational expression is undefined. These *extraneous solutions* are *not* solutions of the original equation. So in solving rational equations, it is particularly important to check all possible solutions.

EXAMPLE 4

Solve and check: $\dfrac{x^2}{x-2} = \dfrac{4}{x-2}$

Solution The LCD of the rational expressions is $x-2$.

$$\frac{x^2}{x-2} = \frac{4}{x-2}$$

$$(x-2)\cdot\frac{x^2}{x-2} = (x-2)\cdot\frac{4}{x-2}$$ Multiply each side of the equation by the LCD.

$$\cancel{(x-2)}\cdot\frac{x^2}{\cancel{x-2}} = \cancel{(x-2)}\cdot\frac{4}{\cancel{x-2}}$$

$$x^2 = 4$$
$$x^2 - 4 = 0$$
$$(x+2)(x-2) = 0$$ Factor the left side of the equation.
$$x+2=0 \quad\text{or}\quad x-2=0$$ Set each factor equal to 0.
$$x=-2 \qquad\qquad x=2$$ Solve for x.

Check

Substitute -2 for x.

$$\frac{x^2}{x-2} = \frac{4}{x-2}$$
$$\frac{(-2)^2}{-2-2} \stackrel{?}{=} \frac{4}{-2-2}$$
$$-\frac{4}{4} = -\frac{4}{4} \quad\text{True}$$

Substitute 2 for x.

$$\frac{x^2}{x-2} = \frac{4}{x-2}$$
$$\frac{2^2}{2-2} \stackrel{?}{=} \frac{4}{2-2}$$
$$\frac{4}{0} = \frac{4}{0} \quad\text{Undefined}$$

Since we get undefined fractions when we substitute 2 for x in the original equation, 2 is *not* a solution. So the solution is -2. Without solving, how could we have known that 2 is *not* a solution of the original equation? Explain.

PRACTICE 4

Solve and check: $x = \dfrac{9}{x+3} + \dfrac{3x}{x+3}$

Rational equations play an important role in a kind of application known as *work problems*. In these problems, we typically want to compute how long it will take to complete a task.

The key to solving a work problem is to determine the *rate of work*, that is, the fraction of the task that is completed in one unit of time. For instance, if it takes a secretary 5 hr to type a report, then the secretary's rate of work—the fraction of the report typed in 1 hr—would be $\frac{1}{5}$. If it takes a painter 6 hr to paint a room, then $\frac{1}{6}$ of the room would be painted in an hour so that $\frac{1}{6}$ is the painter's rate of work.

Consider the following example.

EXAMPLE 5	PRACTICE 5

EXAMPLE 5

Two company employees, one senior and the other junior, are responsible for carrying out a project. If the two employees had worked alone, the junior employee would have completed the project in 6 hr and the senior employee would have completed it in 4 hr. How long would it take the two employees to carry out the project working together?

Solution Since this is a work problem, we can use the following equation to determine how long it will take the two employees to carry out the project working together.

Rate of work · Time worked = Part of the task completed

Using this equation, we can set up a table. Let t represent the time it takes the two employees to complete the project working together. Note here that the task to be completed is the project.

	Rate of Work · Time Worked =	Part of the Task Completed
Senior employee	$\frac{1}{4}$ t	$\frac{1}{4} \cdot t$
Junior employee	$\frac{1}{6}$ t	$\frac{1}{6} \cdot t$

Since the sum of the parts of the task completed must equal one complete task, we have:

$$\frac{1}{4} \cdot t + \frac{1}{6} \cdot t = 1$$

$$\frac{t}{4} + \frac{t}{6} = 1$$

To solve this equation, we multiply each side by the LCD, 12.

$$12 \cdot \frac{t}{4} + 12 \cdot \frac{t}{6} = 12 \cdot 1$$

$$3t + 2t = 12$$

$$5t = 12$$

$$t = 2\frac{2}{5} = 2.4$$

So working together, it would take the two employees 2.4 hr (or 2 hr 24 min) to complete the project. We leave the check to you.

PRACTICE 5

A town water tank has two pumps. Working alone, the less powerful pump can fill the tank in 10 hr, whereas the more powerful pump can fill it in 6 hr. How long will it take both pumps working together to fill the tank?

Another application of rational equations is motion problems. Recall that in these problems, an object moves at a constant rate r for time t and travels a distance d. These three quantities are related by the following formula:

$$r \cdot t = d$$

If we solve this equation for t, we get $t = \dfrac{d}{r}$, a variation on the formula that we apply in Example 6. In this example, note that the total time that the two objects traveled is known.

EXAMPLE 6	PRACTICE 6

EXAMPLE 6

A family on vacation drove 50 mi to a hotel and then returned home following the same route. Because of lighter traffic, the family drove at twice the speed going to the hotel as compared to returning home. If the round trip took 3 hr, at what speed did the family return home?

Solution We are looking for the family's speed returning home. Let's represent this unknown quantity by r. Since the family traveled twice as fast going to the hotel as returning home, their speed going to the hotel must have been $2r$. We use the formula $\dfrac{d}{r} = t$ to find the time traveled in each direction, and set up a table.

	Distance ÷ Rate = Time		
Going to the hotel	50	$2r$	$\dfrac{50}{2r}$
Returning home	50	r	$\dfrac{50}{r}$

Since it is given that the round trip took 3 hr in all, we write:

$$\frac{50}{2r} + \frac{50}{r} = 3$$

To solve this rational equation, we multiply each side of the equation by the LCD, $2r$.

$$2r \cdot \frac{50}{2r} + 2r \cdot \frac{50}{r} = 2r \cdot 3$$

$$\overset{1}{\cancel{2r}} \cdot \frac{50}{\cancel{2r}} + 2\overset{1}{\cancel{r}} \cdot \frac{50}{\cancel{r}} = 2r \cdot 3$$

$$50 + 100 = 6r$$
$$6r = 150$$
$$r = 25$$

So the family returned home at 25 mph. We can confirm this solution by checking.

PRACTICE 6

A business executive traveled 1800 mi by jet, continuing the trip an additional 300 mi on a propeller plane. The speed of the jet was 3 times that of the propeller plane. If the entire trip took 6 hr, what was the speed of the propeller plane?

Some rational equations are formulas that relate two or more variables. Consider the following example.

EXAMPLE 7	PRACTICE 7

EXAMPLE 7

The formula $\dfrac{1}{f} = \dfrac{1}{p} + \dfrac{1}{q}$ gives the focal length f of a lens, where p is the distance between the lens and an object and q is the distance between the image and the lens.

a. Solve this equation for f.

b. Use the formula found in part (a) to find the value of f when $p = 40$ cm and $q = 10$ cm.

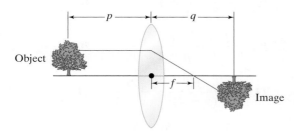

Solution

a. Our task is to solve the equation for f.

$$\frac{1}{f} = \frac{1}{p} + \frac{1}{q}$$

$$fpq \cdot \frac{1}{f} = fpq \cdot \left(\frac{1}{p} + \frac{1}{q}\right)$$

$$fpq \cdot \frac{1}{f} = fpq \cdot \frac{1}{p} + fpq \cdot \frac{1}{q}$$

$$\overset{1}{\cancel{f}}pq \cdot \frac{1}{\underset{1}{\cancel{f}}} = f\overset{1}{\cancel{p}}q \cdot \frac{1}{\underset{1}{\cancel{p}}} + fp\overset{1}{\cancel{q}} \cdot \frac{1}{\underset{1}{\cancel{q}}}$$

$$pq = fq + fp$$

Since we are solving for f, we factor it out on the right side of the equation.

$$pq = f(q + p)$$

$$f = \frac{pq}{p + q}$$

b. Substituting 40 for p and 10 for q in the formula $f = \dfrac{pq}{p + q}$

we get:

$$f = \frac{40 \cdot 10}{40 + 10}$$

$$= \frac{400}{50}$$

$$= 8$$

So the focal length of the lens is 8 cm.

PRACTICE 7

Economists are interested in how the free market works. Suppose that a commodity sells at the price x dollars per unit. A formula that might represent the number of units sold, called the demand D, is:

$$D = \frac{500}{x} + 50$$

a. Solve for x in terms of D.

b. Find the value of x when the demand is 450 units.

7.5   Exercises

FOR EXTRA HELP

Solve and check.

1. $\dfrac{y}{2} + \dfrac{7}{10} = -\dfrac{4}{5}$

2. $\dfrac{x}{6} - \dfrac{1}{10} = \dfrac{2}{5}$

3. $\dfrac{1}{t} - \dfrac{7}{3} = -\dfrac{1}{3}$

4. $\dfrac{5}{p} + \dfrac{1}{3} = \dfrac{6}{p}$

5. $x + \dfrac{1}{x} = 2$

6. $y + \dfrac{2}{y} = 3$

7. $\dfrac{t+1}{2t-1} - \dfrac{5}{7} = 0$

8. $\dfrac{b-3}{3b-2} - \dfrac{4}{5} = 0$

9. $\dfrac{t-2}{3} = 4$

10. $\dfrac{4}{x+1} = 3$

11. $\dfrac{2n}{n+1} = \dfrac{-2}{n+1} + 1$

12. $\dfrac{4}{s-3} - \dfrac{3s}{s-3} = 2$

13. $\dfrac{5x}{x+1} = \dfrac{x^2}{x+1} + 2$

14. $\dfrac{x^2}{x-2} = \dfrac{2x}{x-2} + 3$

15. $\dfrac{x}{x-3} - \dfrac{6}{x} = 1$

16. $\dfrac{2}{t} + \dfrac{t}{t+1} = 1$

17. $1 + \dfrac{4}{x^2} = \dfrac{4}{x}$

18. $\dfrac{1}{2x} + \dfrac{3}{x^2} = 1$

⊙ 19. $\dfrac{2}{p+1} - \dfrac{1}{p-1} = \dfrac{2p}{p^2-1}$

20. $\dfrac{2}{y+2} - \dfrac{5}{2-y} = \dfrac{3y}{y^2-4}$

⊙ 21. $\dfrac{3}{x} - \dfrac{1}{x+4} = \dfrac{5}{x^2+4x}$

22. $\dfrac{2}{y-2} + \dfrac{3}{y} = \dfrac{4}{-2y+y^2}$

23. $1 - \dfrac{6x}{(x-4)^2} = \dfrac{2x}{x-4}$

24. $\dfrac{4x}{x+1} - 2 = -\dfrac{3x}{(x+1)^2}$

25. $\dfrac{n+1}{n^2+2n-3} = \dfrac{n}{n+3} - \dfrac{1}{n-1}$

26. $\dfrac{7}{x-5} - \dfrac{5x+6}{x^2-3x-10} = \dfrac{x}{x+2}$

Mixed Practice

Solve and check.

27. $\dfrac{2}{b+3} = \dfrac{5b}{b^2-1} - \dfrac{3}{3-b}$

28. $\dfrac{3}{r} + \dfrac{1}{4} = \dfrac{12}{r}$

29. $\dfrac{6m}{m+1} - 3 = \dfrac{8m}{(m+1)^2}$

30. $\dfrac{a-3}{3(a-2)} - \dfrac{3}{8} = 0$

31. $2 - \dfrac{2x}{x+3} = \dfrac{9}{x+1}$

32. $\dfrac{8}{s} + \dfrac{s}{s+4} = 1$

Applications

Solve.

33. Two brothers share a house. It would take the younger brother working alone 45 min to clean the attic, whereas it would take the older brother only 30 min. If the two brothers worked together, how long would it take them to clean the attic?

34. One crew can pave a road in 24 hr and a second crew can do the same job in 16 hr. How long would it take for the two crews working together to pave the road?

35. A clerical worker takes 4 times as long to finish a job as it does an executive secretary. Working together, it takes them 3 hr to finish the job. How long would it take the clerical worker, working alone, to finish the job?

36. One pipe can fill a tank in 1 min. A second pipe takes 2 min to fill the same tank. Working together, how long will it take both pipes to fill the tank?

37. A car traveled twice as fast on a dry road as it did making the return trip on a slippery road. The trip was 60 mi each way, and the round trip took 3 hr. What was the speed of the car on the dry road?

38. A cyclist rode uphill and then turned around and made the same trip downhill. The downhill speed was 3 times the uphill speed. If the trip each way was 18 mi and the entire trip lasted 2 hr, at what speed was the cyclist going uphill?

39. Steam exerts pressure on a pipe. In the formula

$$p = \frac{P}{LD},$$

p represents the pressure per square inch in the pipe, P stands for the total pressure in the pipe, L is the pipe's length, and D is the pipe's diameter. Solve this formula for D.

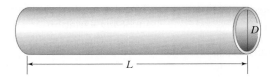

40. To approximate the appropriate dosage of medicine for a child, doctors sometimes use Young's rule:

$$C = \frac{aA}{a + 12},$$

where C represents the child's dosage, a stands for the child's age, and A is the adult dosage. Solve this formula for a.

• *Check your answers on page A-34.*

MINDSTRETCHERS

Groupwork

1. Working with a partner, show that there are no positive real numbers a and b for which the equation

$\frac{1}{a} + \frac{1}{b} = \frac{1}{a + b}$ is true.

Mathematical Reasoning

2. Use your knowledge of systems of equations to solve the following system:

$$\textbf{(1)} \quad \frac{1}{x} + \frac{3}{y} = 1$$

$$\textbf{(2)} \quad -\frac{4}{x} + \frac{3}{y} = -3$$

7.6   Ratio and Proportion; Variation

Ratio and Proportion

Recall in Section 7.5 that to solve rational equations we multiplied every term by the LCD in order to clear the denominators. However, rational equations of a particular form, where one rational expression equals another, lend themselves to an alternative method for clearing denominators. Here are some examples of equations of that form:

$$\frac{x}{12} = \frac{2}{3} \qquad \frac{5}{6} = \frac{10}{n} \qquad \frac{1}{n+1} = \frac{3}{n-1}$$

Note that in each of these equations, two rational expressions are equal to one another.

Two rational expressions are equal to one another when their *cross products* are equal. So rather than multiplying by the LCD, we can simply *set the cross products equal*. Let's apply each method to the first equation.

OBJECTIVES

- To write and solve proportions
- To solve applied problems involving proportions
- To write and solve equations expressing direct variation, inverse variation, joint variation, and combined variation
- To solve applied problems involving variation

LCD Method

$$\frac{x}{12} = \frac{2}{3}$$

$$12 \cdot \frac{x}{12} = 12 \cdot \frac{2}{3} \qquad \text{Multiply each side of the equation by the LCD, 12.}$$

$$\overset{1}{12} \cdot \frac{x}{12} = \overset{4}{12} \cdot \frac{2}{3} \qquad \text{Divide out common factors.}$$

$$1 \cdot x = 4 \cdot 2 \qquad \text{Simplify.}$$

$$x = 8$$

Cross-Product Method

$$\frac{x}{12} = \frac{2}{3}$$

$$\frac{x}{12} = \frac{2}{3} \qquad \text{Cross products}$$

$$3 \cdot x = 12 \cdot 2 \qquad \text{Cross multiply.}$$

$$3x = 24 \qquad \text{Simplify.}$$

$$\frac{3x}{3} = \frac{24}{3} \qquad \text{Divide each side of the equation by 3.}$$

$$x = 8$$

So in both methods, our solution appears to be 8, but is it correct? We can check by substituting in the original equation:

$$\frac{x}{12} = \frac{2}{3}$$

$$\frac{8}{12} \overset{?}{=} \frac{2}{3}$$

$$\frac{2}{3} = \frac{2}{3} \qquad \text{True}$$

We conclude that when two rational expressions are equal to one another, we can solve the equation by either the LCD method or the cross-product method. In this section, we solve these types of equations using the cross-product method.

EXAMPLE 1	PRACTICE 1
Solve: $\dfrac{5}{6} = \dfrac{10}{n}$	Solve: $\dfrac{4}{5} = \dfrac{p}{10}$

Solution We clear the denominators by cross multiplying.

$$\dfrac{5}{6} = \dfrac{10}{n} \qquad \text{Cross multiply.}$$

$$5n = 60$$

$$\dfrac{5n}{5} = \dfrac{60}{5} \qquad \text{Divide both sides by 5.}$$

$$n = 12$$

Check

$$\dfrac{5}{6} = \dfrac{10}{n}$$

$$\dfrac{5}{6} \overset{?}{=} \dfrac{10}{12} \qquad \text{Substitute 12 for } n.$$

$$\dfrac{5}{6} = \dfrac{5}{6} \qquad \text{True}$$

Equations of the type shown in Example 1 commonly arise from *ratio and proportion* problems.

A *ratio* is a comparison of two numbers, expressed as a quotient. For instance, if we compare the numbers 5 and 8, their ratio would be 5 to 8, written as $\dfrac{5}{8}$ or 5:8. Here are some other ratios.

- A ratio of 6 women to 4 men, written as $\dfrac{6}{4}$, or in simplest terms, $\dfrac{3}{2}$

- A ratio of 230 congressmen voting for a bill to 180 congressmen voting against the same bill, written as $\dfrac{230}{180}$, or $\dfrac{23}{18}$

A ratio of quantities with different units is called a *rate*. Here are some examples.

- A basketball team's record of 8 wins and 2 losses, written as $\dfrac{8 \text{ wins}}{2 \text{ losses}}$

- A wage of $20 per hour, written as $\dfrac{20 \text{ dollars}}{1 \text{ hour}}$

When two ratios (or rates) are equal, we say that they are *in proportion*. For instance, the ratios $\dfrac{1}{2}$ and $\dfrac{4}{8}$ are in proportion. The equation $\dfrac{1}{2} = \dfrac{4}{8}$ is called *a proportion*.

Definition

A **proportion** is a statement that two ratios $\dfrac{a}{b}$ and $\dfrac{c}{d}$ are equal, written $\dfrac{a}{b} = \dfrac{c}{d}$, where $b \neq 0$ and $d \neq 0$.

Now, let's consider an example of *solving a proportion problem*. Suppose that you saved $500 in 4 months. At this rate, how long would it take you to save for a computer that costs $750? To answer this question, we can write a proportion in which the rates compare the amount of savings in a given amount of time. We want to find the amount of time corresponding to a savings of $750. Let's call this missing value x.

Savings in dollars $\longrightarrow \quad \dfrac{500}{4} = \dfrac{750}{x} \quad \longleftarrow$ Savings in dollars
Time in months $\longrightarrow \quad \phantom{\dfrac{500}{4}} \quad \phantom{\dfrac{750}{x}} \quad \longleftarrow$ Time in months

We then set the cross products equal in order to find the missing value.

$$\frac{500}{4} = \frac{750}{x}$$

$$500 \cdot x = 4 \cdot 750 \qquad \text{Cross multiply.}$$

$$500x = 3000$$

$$x = \frac{3000}{500}$$

$$x = 6$$

So it will take you 6 months to save for the computer.

To Solve a Proportion

- Find the cross products and set them equal.
- Solve the resulting equation.
- Check the solution in the original equation.

EXAMPLE 2

The actual length of the bedroom shown in the floor plan below is 15 ft. What is the actual width of the room?

$l = 2$ in.

$w = 1.5$ in.

Solution The actual measurements and the floor plan measurements are in proportion.

The actual width in feet $\longrightarrow \quad \dfrac{w}{15} = \dfrac{1.5}{2} \quad \longleftarrow$ The floor plan width in inches
The actual length in feet $\longrightarrow \phantom{\dfrac{w}{15}} \phantom{\dfrac{1.5}{2}} \quad \longleftarrow$ The floor plan length in inches

$$2w = (15)(1.5) \qquad \text{Cross multiply.}$$

$$\frac{2w}{2} = \frac{22.5}{2} \qquad \begin{array}{l}\text{Divide each side of the} \\ \text{equation by 2.}\end{array}$$

$$w = 11.25$$

PRACTICE 2

It takes 80 lb of sodium hydroxide to neutralize 98 lb of sulfuric acid. At this rate, how many pounds of sodium hydroxide are needed to neutralize 49 lb of sulfuric acid?

Check

$$\frac{w}{15} = \frac{1.5}{2}$$

$$\frac{\mathbf{11.25}}{15} \overset{?}{=} \frac{1.5}{2}$$

$$0.75 = 0.75 \qquad \text{True}$$

Since our answer checks, we can conclude that the bedroom is actually 11.25 ft wide.

Would we get the same answer if we had set up the proportion in Example 2 as $\frac{w}{1.5} = \frac{15}{2}$? Explain.

EXAMPLE 3	PRACTICE 3

EXAMPLE 3

A 125-lb adult gets a dosage of 2 mL of a particular drug. At this rate, how much additional drug does a 175-lb adult require?

Solution We begin by writing a proportion in which we use x to represent the amount of additional drug required.

$$\frac{2}{125} = \frac{2 + x}{175} \qquad \begin{matrix} \longleftarrow \text{ Amount of drug in milliliters} \\ \longleftarrow \text{ Weight in pounds} \end{matrix}$$

$$2 \cdot 175 = 125(2 + x) \qquad \text{Cross multiply.}$$

$$350 = 250 + 125x \qquad \text{Use the distributive property.}$$

$$125x = 100 \qquad \text{Combine like terms.}$$

$$x = \frac{100}{125} \qquad \text{Divide each side of the equation by 125.}$$

$$x = 0.8$$

Check

$$\frac{2}{125} = \frac{2 + x}{175}$$

$$\frac{2}{125} \overset{?}{=} \frac{2 + \mathbf{0.8}}{175} \qquad \text{Substitute 0.8 for } x.$$

$$\frac{2}{125} \overset{?}{=} \frac{2.8}{175} \qquad \text{Simplify.}$$

$$0.016 = 0.016 \qquad \text{True}$$

So an additional 0.8 mL of the drug is required for a 175-lb adult.

PRACTICE 3

A homeowner pays $900 a month on a $100,000 mortgage. If instead she had a $75,000 mortgage, how much less money would she be paying per month at the same interest rate?

Another application of ratio and proportion is the geometric topic of *similar triangles*. These are triangles with the same shape but not necessarily the same size.

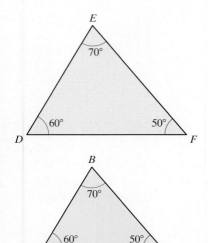

In the diagram to the right, $\triangle ABC$ and $\triangle DEF$ are similar. These triangles have corresponding angles that have equal measures, that is:

$$m\angle A = m\angle D$$
$$m\angle B = m\angle E$$
$$m\angle C = m\angle F$$

Note that in similar triangles, corresponding sides are opposite equal angles. Each side of $\triangle ABC$ corresponds to a side of $\triangle DEF$.

$\overline{AB}$ corresponds to $\overline{DE}$.
$\overline{BC}$ corresponds to $\overline{EF}$.
$\overline{AC}$ corresponds to $\overline{DF}$.

It can be shown that the lengths of the corresponding sides of similar triangles are in proportion:

$$\frac{AB}{DE} = \frac{BC}{EF} = \frac{AC}{DF} \quad \begin{matrix} \leftarrow \triangle ABC \\ \leftarrow \triangle DEF \end{matrix}$$

We can use this relationship to find missing sides of similar triangles, as the following example illustrates.

EXAMPLE 4	PRACTICE 4

In the following diagram, $\triangle PQR$ and $\triangle STU$ are similar. Find x, the length of side $\overline{SU}$.

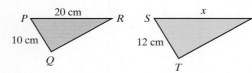

Solution Corresponding sides are in proportion. To solve for x, let's set up a proportion involving the lengths of sides $\overline{PQ}$, $\overline{ST}$, $\overline{PR}$, and $\overline{SU}$.

$$\frac{PQ}{ST} = \frac{PR}{SU}$$

$$\frac{10}{12} = \frac{20}{x} \qquad \text{Substitute the given values.}$$

$$10x = 240 \qquad \text{Cross multiply.}$$

$$x = \frac{240}{10}$$

$$x = 24$$

So the length of $\overline{SU}$ is 24 cm.

Triangle ABC and triangle DEC are similar. Find x, the length of side $\overline{DE}$.

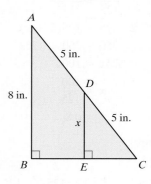

In Section 7.5, we discussed motion problems involving rational equations. Here we continue the discussion of these types of problems. In this case, we use $t = \dfrac{d}{r}$ because we must set two travel times equal to one another.

EXAMPLE 5	PRACTICE 5

EXAMPLE 5

A boat travels 60 mph in still water. Find the speed of the river's current if the boat traveled 80 mi down the river in the same time that it took to travel 70 mi up the river.

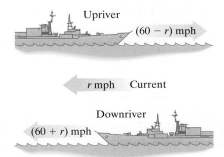

Upriver

$(60 - r)$ mph

r mph   Current

Downriver

$(60 + r)$ mph

Solution Let r represent the speed of the river's current. When traveling downriver, the speed of the current is added to that of the boat. When traveling upriver, the speed of the river is subtracted from that of the boat. Applying the formula $t = \dfrac{d}{r}$, we can set up the following table:

| | d | ÷ | r | = | t |
	Distance		**Rate**		**Time**
Downriver	80		$60 + r$		$\dfrac{80}{60 + r}$
Upriver	70		$60 - r$		$\dfrac{70}{60 - r}$

We are told that these two times are equal, so we write the following proportion and solve for r.

$$\frac{80}{60 + r} = \frac{70}{60 - r}$$
$$80(60 - r) = 70(60 + r) \qquad \text{Cross multiply.}$$
$$4800 - 80r = 4200 + 70r$$
$$600 = 150r$$
$$r = 4$$

The speed of the river's current was 4 mph. We leave the check to you.

PRACTICE 5

The speed of the jet stream was 300 mph. When flying in the direction of the jet stream, a plane flies 1000 mi in the same time that it takes to fly 250 mi against the jet stream. Find the speed of the plane in still air.

Variation

Next, we consider a type of equation or formula that relates one variable to one or more other variables by means of multiplication, division, or both operations. We call this type of equation a *variation equation*. Variation equations or formulas show how one quantity changes in relation to other quantities. Such quantities can vary *directly*, *inversely*, or *jointly*. For example, for a fixed speed, distance traveled *varies directly* as time traveled. For a fixed distance, time traveled *varies inversely* as speed. And in general, distance traveled *varies jointly* as time traveled and speed.

Now, let's look at how to write equations expressing direct variation, inverse variation, joint variation, and combined variation and how to solve applied problems involving these types of variation.

Direct Variation

Let's first discuss *direct variation*. Consider the circumference of a circle, which is given by the formula $C = \pi d$, where d is the diameter of the circle. In this formula, the circumference is always π times the diameter; that is, it is always a constant multiple of d. In such a case, we say that C *varies directly* as d, or that C is *directly proportional* to d.

Definition

If a relationship between two variables is described by an equation of the form $y = kx$, where k is a positive constant, we say that we have **direct variation**, that y **varies directly** as x, or that y is **directly proportional** to x. The number k is called the **constant of variation** or the **constant of proportionality**, and $y = kx$ is called the **direct variation equation**.

In the direct variation equation $y = kx$, the relationship between x and y is linear. Recall from Section 3.3 that the graph of this equation is a straight line with slope k that passes through the origin.

For example, the graph of the direct variation equation $C = \pi d$ is a line that passes through the origin with slope π. In the graph to the right, we see that as d increases, C increases. Similarly, as d decreases, C decreases. Note that for $C = \pi d$, the number π is the constant of proportionality.

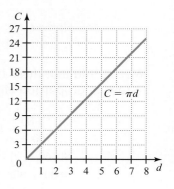

In general, for the direct variation equation $y = kx$, as x increases, y increases. Similarly, as x decreases, y decreases.

EXAMPLE 6

Suppose y varies directly as x, and $y = 25$ when $x = 15$. Find the constant of variation and the direct variation equation.

Solution We know that y varies directly as x, and so we write $y = kx$. If $y = 25$ when $x = 15$, we get:

$$y = kx$$
$$25 = k(15) \quad \text{Substitute 25 for } y \text{ and 15 for } x.$$
$$\frac{25}{15} = k$$
$$k = \frac{5}{3}$$

So the constant of variation is $\frac{5}{3}$, and the direct variation equation is $y = \frac{5}{3}x$.

Let's consider an application involving direct variation.

PRACTICE 6

Find the constant of variation and the direct variation equation in which y varies directly as x and $y = 16$ when $x = 20$.

EXAMPLE 7

In chemistry, Charles's Law deals with the properties of a sample of gas. The law states that if the pressure of the gas remains constant, then its volume, V, is directly proportional to the Kelvin temperature T. If the gas in a hot-air balloon occupies 100 cm³ at 200 K, what volume would it occupy at a temperature of 150 K while the pressure of the gas remains constant?

Solution Using Charles's Law, we know that the volume, V, of a sample of gas is directly proportional to the temperature, T. So we write
$$V = kT$$

where k is the constant of variation. We are given that the sample of gas occupies 100 cm³ at 200 K; that is, V equals 100 cu cm when T equals 200 K. Substituting this information in the variation equation, we can find the value of k.

$$V = kT$$
$$100 = k(200)$$
$$\frac{100}{200} = k$$
$$k = \frac{1}{2}$$

Substituting $\frac{1}{2}$ for k in the direct variation equation, we get:

$$V = \frac{1}{2}T$$

PRACTICE 7

In a clinical class, a student nurse was told that the amount, A, of the drug gentamicine administered to patients is directly proportional to the patient's weight, M, in kilograms. Suppose 160 mg of gentamicine is administered to a patient whose weight is 40 kg. How much of the drug should be administered to a patient whose weight is 55 kg?

Now, we find the volume when the temperature is 150 K.

$$V = \frac{1}{2}T$$

$$V = \frac{1}{2}(150)$$

$$V = 75$$

So when the temperature is 150 K, the gas will occupy 75 cm³.

Inverse Variation

In a direct variation relationship, we saw that as x increases, y increases, and as x decreases, y decreases. Here, we discuss another kind of variation called *inverse variation*. An example of inverse variation can be found by looking at the distance formula, $d = rt$. Solving the formula for t, we get:

$$t = \frac{d}{r}$$

If the distance d is constant, we say that t *varies inversely* as r or that t is *inversely proportional* to r.

Definition

If a relationship between two variables is described by an equation in the form $y = \dfrac{k}{x}$, where k is a positive constant, we say that we have **inverse variation**, that y **varies inversely** as x, or y is **inversely proportional** to x. The number k is called the **constant of variation** or the **constant of proportionality**, and $y = \dfrac{k}{x}$ is called an **inverse variation equation**.

Recall from Section 7.1 that the function $f(x) = \dfrac{1}{x}$ is nonlinear. Similarly, in the inverse variation equation $y = \dfrac{k}{x}$, the relationship between x and y is nonlinear. Likewise, in the inverse variation equation $t = \dfrac{d}{r}$, where d is a fixed distance, r is rate, and t is time traveled, the relationship between r and t is nonlinear. Suppose $d = 8$ mi. From the nonlinear graph $t = \dfrac{8}{r}$, we see that as r increases, t decreases.

r	$t = \dfrac{8}{r}$	(r, t)
1	$t = \dfrac{8}{1} = 8$	$(1, 8)$
2	$t = \dfrac{8}{2} = 4$	$(2, 4)$
4	$t = \dfrac{8}{4} = 2$	$(4, 2)$
8	$t = \dfrac{8}{8} = 1$	$(8, 1)$

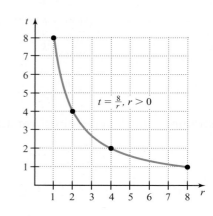

Note that for $t = \dfrac{8}{r}$, the constant of variation is 8.

In general, for the inverse variation equation $y = \dfrac{k}{x}$, as x increases, y decreases. Similarly, as y increases, x decreases.

EXAMPLE 8	PRACTICE 8

Find the constant of variation and the inverse variation equation in which y varies inversely as x if $y = 3.5$ when $x = 6$.

Solution We know that y varies inversely as x, so we write $y = \dfrac{k}{x}$. Since $y = 3.5$ when $x = 6$, we get:

$$y = \frac{k}{x}$$

$$3.5 = \frac{k}{6} \qquad \text{Substitute 3.5 for } y \text{ and 6 for } x.$$

$$k = 21.0, \text{ or } 21$$

So the constant of variation is 21, and the inverse variation equation is $y = \dfrac{21}{x}$.

Suppose y varies inversely as x and $y = 1.6$ when $x = 30$. Find the constant of variation and the inverse variation equation.

Now let's consider an applied problem that can be modeled by inverse variation.

EXAMPLE 9	PRACTICE 9

The marketing division of a large company found that the demand, d, for one of its products varies inversely with the price, p, of the product. The monthly demand is 60,000 units when the price of the product is $9. Find the monthly demand if the company reduces the price to $6.

Solution We are given that the demand, d, for a product varies inversely with the price, p, of the product. So we write

$$d = \frac{k}{p}$$

where k is the constant of variation. We can use this inverse variation equation to find k when d equals 60,000 units and p equals $9. Substituting 60,000 for d and 9 for p, we get:

$$d = \frac{k}{p}$$

$$60,000 = \frac{k}{9}$$

$$540,000 = k$$

So the value of k is 540,000. Substituting this value for k in the inverse variation equation gives us:

$$d = \frac{540,000}{p}$$

In photography, the f-stop for a particular lens is inversely proportional to the aperture (the maximum diameter of the opening of the lens). For a certain lens, an f-stop of 2 corresponds to an aperture diameter of 25 mm. Find the f-stop of this lens when the aperture diameter is 12.5 mm.

Now, let's use the inverse variation equation to find the monthly demand for the product when the price is $6.

$$d = \frac{540,000}{6} = 90,000$$

So the monthly demand is 90,000 units when the price is reduced to $6.

Joint Variation

In some relationships, one quantity may vary as the product of two or more other quantities. For example, in the interest formula, $I = prt$, if the principal p is constant, then the amount of interest earned, I, varies directly as the product of the interest rate, r, and the time, t. This kind of variation is called *joint variation*. In this case, we say the interest varies jointly as the rate and the time.

Definition
If a relationship among three variables is described by an equation of the form $y = kxz$, where k is a positive constant, we say we have **joint variation**, that y **varies jointly** as x and z, or that y is **jointly proportional** to x and z. The number k is called the **constant of variation** or the **constant of proportionality**, and $y = kxz$ is called the **joint variation equation**.

EXAMPLE 10

Suppose that y varies jointly as x and z and that $y = 240$ when $x = 8$ and $z = 12$. Find the constant of variation and the joint variation equation.

Solution We know that y varies jointly as x and z, so we write $y = kxz$. Since $y = 240$ when $x = 8$ and $z = 12$, we get:

$$y = kxz$$
$$240 = k(8)(12) \quad \text{Substitute 240 for } y, \text{ 8 for } x, \text{ and 12 for } z.$$
$$240 = 96k$$
$$\frac{240}{96} = k$$
$$k = \frac{5}{2}$$

So the constant of variation is $\frac{5}{2}$, and the joint variation equation is $y = \frac{5}{2}xz$.

PRACTICE 10

Find the constant of variation and the joint variation equation in which y varies jointly as x and z and $y = 150$ when $x = 15$ and $z = 20$.

In Example 10, we say y varies jointly as x and z. In this situation, "and" does not mean "to add." Can you explain why?

Consider the following application of joint variation.

EXAMPLE 11

The volume, V, of a rectangular structure of a given height is jointly proportional to its length, l, and its width, w. An architect plans to construct two rectangular buildings with the same height in an office complex. One building is 100 ft long and 20 ft wide and has a volume of 96,000 ft³. Find the volume of the other building with the same height that is 90 ft long and 25 ft wide.

Solution We are given that the volume, V, of a rectangular structure of a given height is jointly proportional to its length, l, and width, w. Since the height is constant, we write

$$V = klw$$

where k is the constant of variation. So we use this relationship to find k when the volume is 96,000 ft³, the length is 100 ft, and the width is 20 ft.

$$V = klw$$
$$96{,}000 = k(100)(20)$$
$$96{,}000 = 2000k$$
$$\frac{96{,}000}{2000} = k$$
$$k = 48$$

Substituting 48 for k in the joint variation equation, we get:

$$V = 48lw$$

Now we find the volume of the other building given that the length is 90 ft and the width is 25 ft.

$$V = 48lw$$
$$V = 48(90)(25)$$
$$V = 108{,}000$$

So the other building has a volume of 108,000 ft³.

PRACTICE 11

Two employees know that for a given period of time, t, the simple interest earned, I, varies jointly as the interest rate, r, and the principal, p. Suppose one employee invests $2000 at 3.5% for a given period of time earning $140 interest. For the same period of time, how much money must the other employee invest at 4% to earn $180 interest?

Combined Variation

Direct variation, inverse variation, and joint variation may be combined. A given relationship may be a combination of two or all three kinds of variation. Let's look at some examples of *combined variation*.

EXAMPLE 12

Suppose y varies directly as the square of x and inversely as z.

a. Write the variation equation.

b. If $y = 60$ when $x = 2$ and $z = 10$, find the constant of variation using the equation found in part (a).

c. Using part (b), find y for $x = 6$ and $z = 15$.

Solution

a. We are given that y varies directly as the square of x and inversely as z. So we can write the equation as

$$y = \frac{kx^2}{z}$$

b. We use the equation $y = \frac{kx^2}{z}$ to find k when $y = 60$, $x = 2$, and $z = 10$.

$$y = \frac{kx^2}{z}$$

$$60 = \frac{k(2)^2}{10} \qquad \text{Substitute 60 for } y, \text{ 2 for } x, \text{ and 10 for } z.$$

$$600 = 4k$$

$$k = 150$$

c. Now we find y for $x = 6$ and $z = 15$.

$$y = \frac{150x^2}{z}$$

$$y = \frac{150(6)^2}{15} \qquad \text{Substitute 6 for } x \text{ and 15 for } z.$$

$$y = \frac{150(36)}{15}$$

$$y = 360$$

So for this relationship, $y = 360$ for $x = 6$ and $z = 15$.

PRACTICE 12

Suppose w varies jointly as x and y and inversely as the square of z.

a. Write the equation for the variation.

b. If $w = 8$ when $x = 2$, $y = 6$, and $z = 6$, find the constant of variation using the equation found in part (a).

c. Using part (b), find w when $x = 2$, $y = 3$, and $z = 4$.

Many relationships, particularly in science, can be modeled by combined variation equations. Consider the following example.

EXAMPLE 13	PRACTICE 13

The centrifugal force, C, of a body moving in a circle varies jointly with the radius, r, of the circular path and the body's mass, m, and inversely with the square of the time, t, it takes to complete one revolution. A 9-g body moving in a circle with radius 100 cm at a rate of one revolution in 2 sec has a centrifugal force of 9000 dynes. What is the centrifugal force of a 12-g body moving in a circle with a radius of 100 cm at a rate of one revolution in 4 sec?

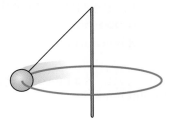

Solution We are given that the centrifugal force, C, of a body moving in a circle varies jointly with the radius, r, of the circular path and the body's mass, m, and inversely with the square of the time, t, it takes to make a complete revolution. So we write the combined variation equation.

$$C = \frac{krm}{t^2}$$

Using this equation, we can find the constant of variation k given that $C = 9000$ dynes, $r = 100$ cm, $m = 9$ g, and $t = 2$ sec.

$$C = \frac{krm}{t^2}$$

$$9000 = \frac{k(100)(9)}{(2)^2}$$

$$9000 = \frac{900k}{4}$$

$$36,000 = 900k$$

$$k = 40$$

So the equation is $C = \frac{40rm}{t^2}$.

Now, we find the centrifugal force of the 12-g body moving in a circle with radius 100 cm at a rate of one revolution in 4 sec.

$$C = \frac{40rm}{t^2}$$

$$= \frac{40(100)(12)}{4^2}$$

$$= 3000$$

So the centrifugal force is 3000 dynes.

The body mass index, or BMI, is used by physicians to determine a patient's total body fat. The BMI varies directly as a person's weight in pounds and inversely as the square of the person's height in inches. A patient who weighs 165 lb and is 70 in. tall has a BMI of approximately 23. To the nearest whole number, find the approximate BMI of a patient who weighs 120 lb with a height of 65 in.

7.6   Exercises   FOR EXTRA HELP

Mathematically Speaking

Fill in each blank with the most appropriate term or phrase from the given list.

directly	rate	joint variation
constant term	inverse variation	inversely proportional
inversely	constant of variation	proportion
directly proportional	ratio	direct variation

1. A(n) _____ is a comparison of two numbers, written as a quotient.

2. A(n) _____ is a ratio of quantities that have different units.

3. A(n) _____ is a statement that two ratios $\dfrac{a}{b}$ and $\dfrac{c}{d}$ are equal, and is written $\dfrac{a}{b} = \dfrac{c}{d}$, where $b \neq 0$ and $d \neq 0$.

The following statements concern the relationship between two variables described by an equation of the form y = kx, where k is a positive constant.

4. We say that y varies _____ as x.

5. The number k is called the _____.

The following statements concern the relationship between two variables described by an equation of the form $y = \dfrac{k}{x}$, where k is a positive constant.

6. We say that y is _____ to x.

7. The relationship is a(n) _____.

The following statement concerns the relationship among three variables described by an equation of the form y = kxz, where k is a positive constant.

8. The relationship is a(n) _____.

Solve and check.

9. $\dfrac{x}{10} = \dfrac{4}{5}$

10. $\dfrac{2}{9} = \dfrac{x}{27}$

11. $\dfrac{n}{100} = \dfrac{4}{5}$

12. $\dfrac{22}{35} = \dfrac{44}{x}$

13. $\dfrac{8}{7} = \dfrac{s}{21}$

14. $\dfrac{1}{v} = \dfrac{10}{3}$

15. $\dfrac{8+x}{12} = \dfrac{22}{36}$

16. $\dfrac{75}{20} = \dfrac{30}{p-1}$

17. $\dfrac{y+3}{14} = \dfrac{y}{7}$

18. $\dfrac{n-6}{3} = \dfrac{n}{5}$

19. $\dfrac{x-1}{8} = \dfrac{x+1}{12}$

20. $\dfrac{3}{t-2} = \dfrac{9}{t+2}$

21. $\dfrac{x}{8} = \dfrac{2}{x}$

22. $\dfrac{4}{n} = \dfrac{n}{16}$

23. $\dfrac{2}{y} = \dfrac{y-4}{16}$

24. $\dfrac{5}{x+2} = \dfrac{x}{7}$

25. $\dfrac{a}{a+3} = \dfrac{4}{5a}$

26. $\dfrac{1}{3n} = \dfrac{n}{n+2}$

27. $\dfrac{y+1}{y+6} = \dfrac{y}{y+6}$

28. $\dfrac{2x+3}{x-3} = \dfrac{3x}{x-3}$

For each pair of variables given, indicate whether the second variable increases or decreases if the first variable increases and whether the variation between the variables is direct or inverse.

29. The speed of a runner and the time it takes the runner to complete a 5-km race

30. The side of a square and its perimeter

31. The distance between two towns on a map and the actual distance between them

32. The average thickness of DVD containers and the number of containers that fit on a shelf

33. The cost of a plane ticket in dollars and its cost in euros

34. The length and width of a rectangle with area 20 sq ft

Find the constant of variation and the variation equation if y varies directly as x.

35. $y = 48$ when $x = 16$ **36.** $y = 35$ when $x = 7$ **37.** $y = 6$ when $x = 36$ **38.** $y = 15$ when $x = 40$

39. $y = 3$ when $x = \dfrac{1}{3}$ **40.** $y = 8$ when $x = \dfrac{1}{4}$ **41.** $y = 0.9$ when $x = 0.6$ **42.** $y = 0.5$ when $x = 0.8$

Find the constant of variation and the variation equation if y varies inversely as x.

43. $y = 13$ when $x = 3$ **44.** $y = 25$ when $x = 4$ **45.** $y = 1.8$ when $x = 15$ **46.** $y = 2.1$ when $x = 20$

47. $y = 0.7$ when $x = 0.4$ **48.** $y = 0.1$ when $x = 1.9$ **49.** $y = 27$ when $x = \dfrac{2}{3}$ **50.** $y = 54$ when $x = \dfrac{1}{6}$

Find the constant of variation and the variation equation if y varies jointly as x and z.

51. $y = 160$ when $x = 10$ and $z = 4$ **52.** $y = 216$ when $x = 18$ and $z = 6$

53. $y = 360$ when $x = 25$ and $z = 12$ **54.** $y = 120$ when $x = 16$ and $z = 15$

55. $y = 63$ when $x = 4.2$ and $z = 5$ **56.** $y = 90$ when $x = 2$ and $z = 0.9$

57. $y = 4.5$ when $x = 0.6$ and $z = 0.3$ **58.** $y = 5.6$ when $x = 0.5$ and $z = 0.7$

Find the constant of variation and the variation equation using the given information.

59. y varies directly as x and inversely as the square of z; $y = 20$ when $x = 4$ and $z = 5$

60. y varies directly as the cube of z and inversely as x; $y = 60$ when $x = 16$ and $z = 2$

61. y varies inversely as x and the square of z; $y = 100$ when $x = 20$ and $z = 0.5$

62. y varies inversely as the square of x and the square of z; $y = 250$ when $x = 0.2$ and $z = 10$

63. y varies jointly as x and w and inversely as the square of z; $y = 130$ when $x = 13$, $w = 16$, and $z = 0.4$

64. y varies jointly as x and the square of z and inversely as w; $y = 600$ when $x = 8$, $z = 5$, and $w = 0.5$

Mixed Practice

Solve and check.

65. $\dfrac{3}{r} = \dfrac{r-4}{7}$ **66.** $\dfrac{a}{a+3} = \dfrac{3}{2a}$ **67.** $\dfrac{4}{w-3} = \dfrac{7}{w+3}$

68. $\dfrac{7}{9} = \dfrac{v}{72}$ **69.** $\dfrac{7}{m} = \dfrac{63}{6}$ **70.** $\dfrac{45}{40} = \dfrac{18}{t+1}$

Use the given information to find the constant of variation and the variation equation.

71. y varies jointly as x and w, and inversely as the cube of z; $y = 15$ when $x = \dfrac{1}{2}$, $w = 8$, and $z = 2$.

72. y varies inversely as x; $y = 4$ when $x = 8$

73. y varies directly as x; $y = \dfrac{3}{7}$ when $x = \dfrac{9}{14}$

74. y varies jointly as x and z; $y = 9$ when $x = 0.6$ and $z = 3$

75. y varies inversely as x; $y = 36$ when $x = \dfrac{2}{3}$

76. y varies directly as x; $y = 2$ when $x = \dfrac{1}{6}$

For each pair of variables given, indicate whether the second variable increases or decreases if the first variable increases and whether the variation between the variables is direct or inverse.

77. The number of slices in a pizza and the amount of pizza in each slice

78. The number of credit hours taken and the total cost of the credit hours

Applications

Solve.

79. The owner of a small business is considering the purchase of a laser printer that can print 6 pages in 2 min. How long would it take to print a 25-page report?

80. In a mayoral election, 48 out of every 100 voters voted for a certain candidate. How many of the 200,000 voters chose the candidate?

81. A trucker drives 60 mi on 8 gal of gas. At the same rate, how many gallons of gas will it take him to drive 120 mi?

82. A chauffeur took 2 hr longer to drive 275 mi than he took to drive 165 mi. If his speed was the same on both trips, at what speed was he driving?

83. A runner traveled 3 mi in the same time that a cyclist traveled 10 mi. The speed of the cyclist was 14 mph greater than that of the runner. What was the cyclist's speed?

84. A train travels 225 mi in the same time that a bus travels 200 mi. If the speed of the train is 5 mph greater than that of the bus, find the speed of the bus.

85. A train goes 30 mph faster than a bus. The train travels 400 mi in the same time as the bus goes 250 mi. What are their speeds?

86. A pilot flies 600 mi with a tailwind of 20 mph. Against the wind, he flies only 500 mi in the same amount of time. What is the speed of the plane in still air?

87. △*ABC* and △*DEF* are similar. Find *y*, the length of side $\overline{AB}$.

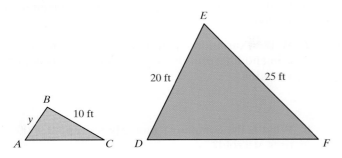

88. The two triangles shown in this map of Kansas are similar. Find the distance between Dodge City and Great Bend, rounded to the nearest 10 mi.

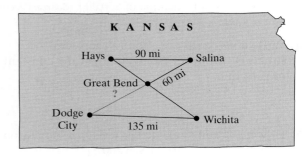

89. When a 6-ft-tall man is 8 ft away from a tree, the tip of his shadow and the tip of the tree's shadow meet 4 ft behind him. The two right triangles shown are similar. Find the height of the tree.

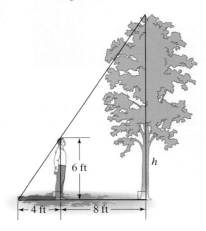

90. Similar triangles can be used to find how far a ship is from the shore. Find the value of *x* in the diagram.

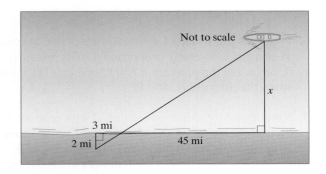

91. There are 20 guests at a party. If 4 more women were to arrive at the party, then 50% of the guests would be women. How many women are there at the party?

92. At a local bank, the simple interest rate on a savings account is 2% higher than the interest rate for a checking account. After one year, a deposit in the checking account would earn interest of $60, whereas the same deposit would earn interest of $120 in the savings account. Find the interest rate on each account.

93. The amount, A, of state income tax that taxpayers pay in Illinois in a certain year is directly proportional to their gross income, i. A person pays $1080 in income tax on a gross annual income of $36,000. (*Source:* Federation of Tax Administrators)

  a. Write a variation equation to represent the situation. Interpret the meaning of the constant of variation.

  b. How much state income tax will a person pay on a gross annual salary of $26,500?

94. A tire company's revenue, R, in dollars is directly proportional to the number of tires, n, it sells. In a particular month, the company generated revenue of $136,710 on the sale of 3255 tires.

  a. Write a variation equation to represent the situation and interpret the meaning of the constant of variation.

  b. What is the company's monthly revenue if 4000 tires are sold?

95. The length of a sound wave, w, in meters is inversely proportional to the frequency, f, of the sound in hertz (Hz). Bottlenose dolphins emit clicking sounds at different frequencies for communicating, orienting themselves to their surroundings, avoiding obstacles, and finding food. If the wavelength of a click that has a frequency of 300 Hz is 5.1 m, find the wavelength of a click that has a frequency of 500 Hz.

96. Physicists use Boyle's Law, which states that at constant temperature the volume, V, of a gas is inversely proportional to its pressure, P. If the volume of a gas is 0.5 m³ when the pressure is 850 kilopascals (kP), find the volume of the gas when the pressure is 680 kP.

97. Kinetic energy is energy associated with motion. The kinetic energy, E, of an object is jointly proportional to its mass, m, and the square of its velocity, v. A 0.142-kg baseball has 113.6 joules (J) of kinetic energy when its velocity is 40 meters per second. What is the kinetic energy of the same baseball when its velocity is 20 meters per second?

98. When an object is dropped, the distance it falls varies directly as the square of the time, t, the object has fallen. A sandbag falls 64 ft 2 sec after it is dropped from a hot-air balloon. How many feet did it fall 4 sec after it was dropped?

99. The illumination, I, from a light source varies inversely as the square of the distance, d, from the light source. If the illumination on a table from a lightbulb 2 m above the table is 21.6 lumens per square meter, find the illumination when the lightbulb is 4 m above the table.

100. The resistance of a wire varies directly as its length, L, and inversely as its cross-sectional area, A. If 2000 ft of wire with a cross-sectional area of 0.008 in² has a resistance of 1.9 ohms, what is the resistance of 5000 ft of the same wire?

● *Check your answers on page A-34.*

MINDSTRETCHERS

Research

1. Eratosthenes headed the great library at Alexandria in Egypt more than 2000 years ago. In addition to his work on prime numbers, Eratosthenes used the method of ratio and proportion to approximate the circumference of the Earth. Either in your college library or on the Web, investigate how he did this and how successful he was. Summarize your findings.

Mathematical Reasoning

2. If $\dfrac{a}{b} = \dfrac{c}{d}$, can you show that $\dfrac{a+b}{b} = \dfrac{c+d}{d}$? Explain.

Groupwork

3. Working with a partner, consider the direct variation equation $y = kx$ and the inverse variation equation $y = \dfrac{k}{x}$. Describe the effect on y if x changes as follows:

a. x is doubled
b. x is reduced by one-half
c. x is multiplied by a factor of $\dfrac{a}{b}$

CULTURAL NOTE

Hot-air balloons were developed in France in the late 1700s. The first one, unmanned but with a sheep, a rooster, and a duck for passengers, was launched from Versailles in 1783. These balloons caught the attention of two French scientists, Jacques Charles and Joseph-Louis Gay-Lussac. Working separately, they investigated the relationships between gases and temperature. Charles is credited with discovering the direct relationship between temperature and volume of a gas ($V = kT$). Gay-Lussac is credited with finding the direct relationship between temperature and pressure ($P = kT$). These two laws were later combined to show the joint variation of pressure and volume to temperature ($PV = kT$).

Source: Hugh Salzburg, *From Caveman to Chemist*, American Chemical Society, 1991.

KEY CONCEPTS AND SKILLS   CONCEPT   SKILL

Concept/Skill	Description	Example
[7.1] Rational Expression	An algebraic expression $\frac{P}{Q}$ that can be written as the quotient of two polynomials, P and Q, where $Q \neq 0$.	$\dfrac{2x}{x+5}$, where $x + 5 \neq 0$.
[7.1] To Find an Equivalent Rational Expression	Multiply the numerator and denominator of $\frac{P}{Q}$ by the same polynomial R. $$\frac{P}{Q} = \frac{PR}{QR},$$ where $Q \neq 0$ and $R \neq 0$.	$\dfrac{3}{5y+1} \cdot \dfrac{y}{y} = \dfrac{3y}{5y^2 + y}$
[7.1] To Simplify a Rational Expression	• Factor the numerator and denominator. • Divide out any common factors.	$\dfrac{3n^2 - 3}{n^2 + n - 2}$ $= \dfrac{3(n-1)(n+1)}{(n-1)(n+2)} = \dfrac{3(n+1)}{n+2}$
[7.2] To Multiply Rational Expressions	• Factor the numerators and denominators. • Divide the numerators and denominators by all common factors. • Multiply the remaining factors in the numerators and the remaining factors in the denominators.	$\dfrac{6x}{x^2-4x} \cdot \dfrac{x^2-16}{3x}$ $= \dfrac{6x}{x(x-4)} \cdot \dfrac{(x-4)(x+4)}{3x}$ $= \dfrac{6x}{x(x-4)} \cdot \dfrac{(x+4)(x-4)}{3x}$ $= \dfrac{2(x+4)}{x}$
[7.2] To Divide Rational Expressions	• Take the reciprocal of the divisor and change the operation to multiplication. • Follow the rule for multiplying rational expressions.	$\dfrac{x^2-x-6}{3x-9} \div \dfrac{4x+8}{5x}$ $= \dfrac{x^2-x-6}{3x-9} \cdot \dfrac{5x}{4x+8}$ $= \dfrac{(x+2)(x-3)}{3(x-3)} \cdot \dfrac{5x}{4(x+2)}$ $= \dfrac{5x}{12}$

continued

Concept/Skill	Description	Example
[7.3] To Add (or Subtract) Rational Expressions with the Same Denominator	• Add (or subtract) the numerators and keep the same denominator. • Simplify, if possible.	$\dfrac{3n-5}{n+1}+\dfrac{n+6}{n+1}$ $=\dfrac{(3n-5)+(n+6)}{n+1}$ $=\dfrac{3n-5+n+6}{n+1}=\dfrac{4n+1}{n+1}$ $\dfrac{2n^2+4n+6}{n-5}-\dfrac{n^2+10n+1}{n-5}$ $=\dfrac{(2n^2+4n+6)-(n^2+10n+1)}{n-5}$ $=\dfrac{2n^2+4n+6-n^2-10n-1}{n-5}$ $=\dfrac{n^2-6n+5}{n-5}$ $=\dfrac{(n-1)\overset{1}{\cancel{(n-5)}}}{\underset{1}{\cancel{(n-5)}}}=n-1$
[7.3] To Find the LCD of Rational Expressions	• Factor each denominator completely. • Multiply these factors, using for the power of each factor the greatest number of times that it occurs in any of the denominators. The product of all these factors is the LCD.	$\dfrac{x}{x^2-4x+3}$ and $\dfrac{x+5}{2x-2}$ Factor x^2-4x+3: $(x-1)(x-3)$ Factor $2x-2$: $2(x-1)$ LCD $=2(x-1)(x-3)$
[7.3] To Add (or Subtract) Rational Expressions with Different Denominators	• Find the LCD of the rational expressions. • Write each rational expression with a common denominator, usually the LCD. • Add (or subtract) the numerators. • Simplify, if possible.	$\dfrac{n}{n^2+3n+2}+\dfrac{3}{2n+2}$ $=\dfrac{n}{(n+1)(n+2)}+\dfrac{3}{2(n+1)}$ $=\dfrac{n\cdot 2}{(n+1)(n+2)\cdot 2}$ $\quad+\dfrac{3\cdot(n+2)}{2(n+1)\cdot(n+2)}$ $=\dfrac{2n}{2(n+1)(n+2)}+\dfrac{3n+6}{2(n+1)(n+2)}$ $=\dfrac{5n+6}{2(n+1)(n+2)}$ $\dfrac{5y}{y^2-9}-\dfrac{2}{y+3}$ $=\dfrac{5y}{(y+3)(y-3)}-\dfrac{2}{y+3}$ $=\dfrac{5y}{(y+3)(y-3)}-\dfrac{2\cdot(y-3)}{(y+3)\cdot(y-3)}$ $=\dfrac{5y}{(y+3)(y-3)}-\dfrac{2y-6}{(y+3)(y-3)}$ $=\dfrac{5y-(2y-6)}{(y+3)(y-3)}$ $=\dfrac{3y+6}{(y+3)(y-3)}=\dfrac{3(y+2)}{(y+3)(y-3)}$

continued

Concept/Skill	Description	Example
[7.4] To Simplify a Complex Rational Expression: Division Method	• Write the numerator and denominator as single rational expressions in simplified form. • Write the expression as the numerator divided by the denominator. • Divide. • Simplify, if possible.	$$\dfrac{\dfrac{x}{4}}{\dfrac{x^2}{3}+\dfrac{x}{2}} = \dfrac{\dfrac{x}{4}}{\dfrac{x^2}{3}\cdot\dfrac{2}{2}+\dfrac{x}{2}\cdot\dfrac{3}{3}}$$ $$= \dfrac{\dfrac{x}{4}}{\dfrac{2x^2+3x}{6}}$$ $$= \dfrac{x}{4}\div\dfrac{2x^2+3x}{6}$$ $$= \dfrac{x}{4}\cdot\dfrac{6}{2x^2+3x}$$ $$= \dfrac{\overset{1}{x}}{\underset{2}{4}}\cdot\dfrac{\overset{3}{6}}{\underset{1}{x(2x+3)}}$$ $$= \dfrac{3}{2(2x+3)}$$
[7.4] To Simplify a Complex Rational Expression: LCD Method	• Find the LCD of all the rational expressions *within* both the numerator and denominator. • Multiply the numerator and denominator of the complex rational expression by this LCD. • Simplify, if possible.	$$\dfrac{1-\dfrac{1}{y}}{1-\dfrac{1}{y^2}} = \dfrac{\left(1-\dfrac{1}{y}\right)\cdot y^2}{\left(1-\dfrac{1}{y^2}\right)\cdot y^2}$$ $$= \dfrac{1\cdot y^2-\dfrac{1}{\underset{1}{y}}\cdot\overset{y}{y^2}}{1\cdot y^2-\dfrac{1}{\underset{1}{y^2}}\cdot y^2}$$ $$= \dfrac{y^2-y}{y^2-1}$$ $$= \dfrac{y(\overset{1}{y-1})}{(y+1)\underset{1}{(y-1)}}$$ $$= \dfrac{y}{y+1}$$

continued

Concept/Skill	Description	Example
[7.5] To Solve a Rational Equation	• Find the LCD of all rational expressions in the equation. • Multiply each side of the equation by the LCD. • Solve the equation. • Check your solution(s) in the original equation.	$\dfrac{1}{x+3} + \dfrac{1}{x-3} = \dfrac{2}{x^2-9}$ $\dfrac{1}{x+3} + \dfrac{1}{x-3} = \dfrac{2}{(x+3)(x-3)}$ $(x+3)(x-3)\cdot\dfrac{1}{x+3}$ $\quad + (x+3)(x-3)\dfrac{1}{x-3}$ $\quad = (x+3)(x-3)$ $\quad\quad \cdot \dfrac{2}{(x+3)(x-3)}$ $x - 3 + x + 3 = 2$ $2x = 2$ $x = 1$ **Check** Substitute 1 for x. $\dfrac{1}{x+3} + \dfrac{1}{x-3} = \dfrac{2}{x^2-9}$ $\dfrac{1}{1+3} + \dfrac{1}{1-3} \stackrel{?}{=} \dfrac{2}{1^2-9}$ $\dfrac{1}{4} - \dfrac{1}{2} \stackrel{?}{=} -\dfrac{2}{8}$ $-\dfrac{1}{4} = -\dfrac{1}{4}$ True
[7.6] Proportion	A statement that two ratios $\dfrac{a}{b}$ and $\dfrac{c}{d}$ are equal, written $\dfrac{a}{b} = \dfrac{c}{d}$, where $b \neq 0$ and $d \neq 0$.	The equation $\dfrac{3}{4} = \dfrac{10}{x}$ is a proportion, where $\dfrac{3}{4}$ and $\dfrac{10}{x}$ are ratios.
[7.6] To Solve a Proportion	• Find the cross products and set them equal. • Solve the resulting equation. • Check the solution in the original equation.	$\dfrac{8}{x-3} = \dfrac{6}{x+4}$ $6(x-3) = 8(x+4)$ $6x - 18 = 8x + 32$ $-2x = 50$ $x = -25$ **Check** Substitute -25 for x. $\dfrac{8}{x-3} = \dfrac{6}{x+4}$ $\dfrac{8}{-25-3} \stackrel{?}{=} \dfrac{6}{-25+4}$ $\dfrac{8}{-28} \stackrel{?}{=} \dfrac{6}{-21}$ $-\dfrac{2}{7} = -\dfrac{2}{7}$ True

continued

Concept/Skill	Description	Example
[7.6] Direct Variation	If a relationship between two variables is described by an equation in the form $y = kx$, where k is a positive constant, we say that we have **direct variation**, that y **varies directly** as x, or that y is **directly proportional** to x. The number k is called the **constant of variation** or the **constant of proportionality**, and $y = kx$ is called a **direct variation equation**.	$y = 2x$ y varies directly as x; 2 is the constant of variation.
[7.6] Inverse Variation	If a relationship between two variables is described by an equation in the form $y = \dfrac{k}{x}$, where k is a positive constant, we say that we have **inverse variation**, that y **varies inversely** as x, or that y is **inversely proportional** to x. The number k is called the **constant of variation** or the **constant of proportionality**, and $y = \dfrac{k}{x}$ is called an **inverse variation equation**.	$y = \dfrac{4}{x}$ y varies inversely as x; 4 is the constant of variation.
[7.6] Joint Variation	If a relationship among three variables is described by an equation of the form $y = kxz$, where k is a positive constant, we say that we have **joint variation**, that y **varies jointly** as x and z, or that y is **jointly proportional** to x and z. The number k is called the **constant of variation** or the **constant of proportionality**, and $y = kxz$ is called the **joint variation equation**.	$y = 5xz$ y varies jointly as x and z; 5 is the constant of variation.

Chapter 7	Review Exercises

To help you review this chapter, solve these problems.

[7.1]

1. Identify the values for which the given rational expression is undefined.

    a. $\dfrac{4}{x+1}$ **b.** $\dfrac{6x+12}{x^2-x-6}$

2. Determine whether each pair of rational expressions is equivalent.

    a. $\dfrac{2x}{y} \overset{?}{=} \dfrac{10x^2y}{5xy^2}$ **b.** $\dfrac{x^2-9}{x^2+6x+9} \overset{?}{=} \dfrac{x-3}{x+3}$

Simplify.

3. $\dfrac{12m}{20m^2}$ **4.** $\dfrac{15n-18}{9n+6}$ **5.** $\dfrac{x^2+2x-8}{4-x^2}$ **6.** $\dfrac{2x^2-3x-20}{3x^2-13x+4}$

[7.2] *Perform the indicated operation.*

7. $\dfrac{10mn}{3p^2} \cdot \dfrac{9np}{5m^2}$ **8.** $\dfrac{y-5}{4y+6} \cdot \dfrac{6y+9}{3y-15}$ **9.** $\dfrac{x+6}{x^2+x-30} \cdot \dfrac{x^2-10x+25}{2x+5}$

10. $\dfrac{2a^2-2a-4}{4-a^2} \cdot \dfrac{2a^2+a-6}{4a^2-2a-6}$ **11.** $\dfrac{x^2y}{2x} \div xy^2$ **12.** $\dfrac{5m+10}{2m-20} \div \dfrac{7m+14}{14m-20}$

13. $\dfrac{5y^2}{x^2-36} \div \dfrac{25xy-25y}{x^2-7x+6}$ **14.** $\dfrac{2x^2+x-1}{x^2+8x+7} \div \dfrac{6x^2+x-2}{x^2+14x+49}$

[7.3] *Write each pair of rational expressions in terms of their LCD.*

15. $\dfrac{1}{5x}$ and $\dfrac{3}{20x^2}$ **16.** $\dfrac{4}{n-1}$ and $\dfrac{n}{n+4}$

17. $\dfrac{1}{3x+9}$ and $\dfrac{x}{x^2+4x+3}$ **18.** $\dfrac{2}{3x^2-5x-2}$ and $\dfrac{1}{4-x^2}$

Perform the indicated operation.

19. $\dfrac{3t+1}{2t} + \dfrac{t-1}{2t}$ **20.** $\dfrac{5y}{y+7} - \dfrac{y-28}{y+7}$ **21.** $\dfrac{5y+4}{4y^2-2y} - \dfrac{2}{2y-1}$

22. $\dfrac{n}{3n+15} + \dfrac{n-2}{n^2+5n}$ **23.** $\dfrac{4}{x-3} - \dfrac{4x+1}{9-x^2}$ **24.** $\dfrac{y+3}{4-y^2} + \dfrac{1}{2-y}$

25. $\dfrac{2}{m+1} + \dfrac{6m-2}{m^2-2m-3}$ **26.** $\dfrac{3x-2}{x^2-x-12} - \dfrac{x+3}{x-4}$ **27.** $\dfrac{2x}{x^2+4x+4} - \dfrac{x-1}{x^2-2x-8}$

28. $\dfrac{n+4}{2n^2-3n+1} + \dfrac{n+1}{2n^2+5n-3}$

[7.4]   *Simplify.*

29. $\dfrac{\dfrac{x}{2}}{\dfrac{3x^2}{7}}$

30. $\dfrac{1 - \dfrac{9}{y}}{1 - \dfrac{81}{y^2}}$

31. $\dfrac{\dfrac{1}{x} + \dfrac{1}{y}}{\dfrac{1}{2x} + \dfrac{1}{2y}}$

32. $\dfrac{4 - \dfrac{3}{x} - \dfrac{1}{x^2}}{2 - \dfrac{5}{x} + \dfrac{3}{x^2}}$

[7.5]   *Solve and check.*

33. $\dfrac{2x}{x-4} = 5 - \dfrac{1}{x-4}$

34. $\dfrac{y+1}{y} + \dfrac{1}{2y} = 4$

35. $\dfrac{5}{2x} + \dfrac{3}{x+1} = \dfrac{7}{x}$

36. $\dfrac{y-2}{y-4} = \dfrac{1}{y+2} + \dfrac{y+3}{y^2 - 2y - 8}$

37. $\dfrac{x}{x+2} - \dfrac{2}{2-x} = \dfrac{x+6}{x^2-4}$

38. $\dfrac{3}{n^2 - 5n + 4} - \dfrac{1}{n^2 - 4n + 3} = \dfrac{n-3}{n^2 - 7n + 12}$

[7.6]   *Solve and check.*

39. $\dfrac{8}{5} = \dfrac{72}{x}$

40. $\dfrac{28}{x+3} = \dfrac{7}{9}$

41. $\dfrac{5}{3+y} = \dfrac{3}{7y+1}$

42. $\dfrac{11}{x-2} = \dfrac{x+7}{2}$

Use the given information to find the constant of variation and the variation equation.

43. y varies directly as x; $y = 1.6$ when $x = 4$

44. y varies inversely as x; $y = \dfrac{1}{2}$ when $x = 3$

45. y varies jointly as x and z; $y = 144$ when $x = 4$ and $z = 6$

46. y varies directly as the square of x and inversely as the square of z; $y = 2$ when $x = 5$ and $z = 10$

Mixed Applications

Solve.

47. A company found that the cost per booklet for printing x booklets can be represented by $0.72 + \dfrac{200}{x}$. Write this cost as a single rational expression.

48. Four friends decide to split the cost of renting a car equally. They discover that if they let one more friend share in the rental, the cost for each of the original four friends will be reduced by $10. What is the total cost of the car rental?

49. A hiker walks a distance d miles at a speed of r mph, and then returns on the same path at a speed of s mph. The hiker's average speed for the entire trip can be represented by
$$\dfrac{2d}{\dfrac{d}{r} + \dfrac{d}{s}}.$$
Simplify this expression.

50. With the water running at full force, it takes 10 min to fill a bathtub. It then takes 15 min for the bathtub to drain. If by mistake the water is running at full force while the tub is draining, how long will it take the tub to fill?

51. A family on vacation drove 400 mi at two different speeds, 50 mph and 60 mph. The total driving time was 7 hr. How many miles did the family drive at 50 mph?

52. One student takes x hours to design a Web page. Another student takes an hour longer. What part of the job will be finished in an hour if the two students work together?

53. The director of a college cafeteria knows that it cost $31,000 to serve 15,000 meals in September. She expects to serve about 20,000 meals in October. How much should she expect to spend on the October meals?

54. Find a rational expression equal to the sum of the reciprocals of three consecutive integers, starting with n.

55. When an object is dropped, its velocity, v (in meters per second), is directly proportional to the time, t (in seconds), the object has fallen. If the velocity of an object after 2 sec is 19.6 meters per second, what is its velocity after 5 sec?

56. The accommodation (or focusing ability), A, of a person's eye is inversely proportional to the distance, d, that a person can bring an object to the eye and still see the object. If the accommodation of a person who can see an object clearly within 12.5 cm is 8 diopters, what is the accommodation of a person who can see an object clearly within 10 cm?

• *Check your answers on page A-34.*

To see if you have mastered the topics in this chapter, take this test.

1. Identify the values for which the rational expression

$\dfrac{3x}{x-8}$ is undefined.

2. Show that $\dfrac{y-3}{y}$ is equivalent to $-\dfrac{3y-y^2}{y^2}$.

Simplify.

3. $\dfrac{15a^3b}{12ab^2}$

4. $\dfrac{x^2-4x}{xy-4y}$

5. $\dfrac{3b^2-27}{b^2-4b-21}$

6. $\dfrac{\dfrac{3}{x^2}-\dfrac{1}{x}}{\dfrac{9}{x^2}-1}$

7. Write $\dfrac{4n-1}{n^2+6n-16}$, $\dfrac{2}{n+8}$, and $\dfrac{n}{4n-8}$ in terms of their LCD.

Perform the indicated operation.

8. $\dfrac{7x-10}{x+6}-\dfrac{5x-22}{x+6}$

9. $\dfrac{3}{2y-8}+\dfrac{2}{4y^2-16y}$

10. $\dfrac{5}{d-3}-\dfrac{d-4}{d^2-d-6}$

11. $\dfrac{5}{2x^2-3x-2}-\dfrac{x}{4-x^2}$

12. $\dfrac{n+1}{3n-18}\cdot\dfrac{n-6}{6n^3-6n}$

13. $\dfrac{a^2-25}{a^2-2a-24}\div\dfrac{a^2+a-30}{a^2-36}$

14. $\dfrac{x^2+6x+8}{x^2+x-2}\div\dfrac{x+4}{2x^2+12x+16}$

Solve and check.

15. $\dfrac{1}{y-5}+\dfrac{y+4}{25-y^2}=\dfrac{1}{y+5}$

16. $\dfrac{2y}{y-4}-2=\dfrac{4}{y+5}$

17. $\dfrac{x}{x+6}=\dfrac{1}{x+2}$

Solve.

18. When a circuit connected in parallel contains two resistors with resistances R_1 and R_2 ohms, electricians use the formula $\dfrac{1}{R} = \dfrac{1}{R_1} + \dfrac{1}{R_2}$ to find the total resistance R (in ohms) of the circuit. Solve the formula for R_1.

19. A company owns two electronic mail processors. The newer machine works twice as fast as the older one. Together, the two machines process 1000 pieces of mail in 20 min. How long does it take each machine, working alone, to process 1000 pieces of mail?

20. In the diagram shown, the heights and shadows of the woman and of the tree are in proportion. Find the height of the tree in meters.

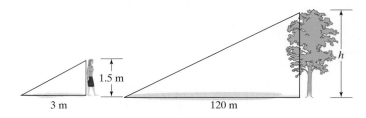

1.5 m

3 m

h

120 m

• *Check your answers on page A-35.*

Cumulative Review Exercises

To help you review, solve the following.

1. Simplify: $y - [2y - 3(y - 1)]$

2. Solve $4n - 5(n + 2) < -7$. Graph the solution on the number line.

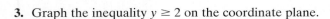

3. Graph the inequality $y \geq 2$ on the coordinate plane.

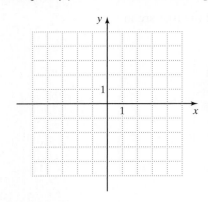

4. Solve by substitution and then check.

$$x + y = 8$$
$$y = 2x - 1$$

5. Solve: $(n + 5)(n - 2) = 0$

6. Factor: $100y^2 - 81$

7. Solve and check: $\dfrac{c}{c + 6} = \dfrac{1}{c + 2}$

8. A business woman wants to invest twice as much money in a fund that has a guaranteed return of 6% as in a fund that has a guaranteed return of 4%. What is the minimum amount that should be invested in each fund if she wants to have a return of at least $1200?

9. Two buses following the same route leave from the same station at different times. The Reston bus leaves 2 hr later than the Arlington bus. The Arlington bus is traveling at 40 mph and the Reston bus is traveling at 60 mph. How long will it take the Reston bus to overtake the Arlington bus?

10. A laser printer prints 7 pages every 3 min. At this rate, how long would it take to print a 20-page report, to the nearest minute?

● *Check your answers on page A-35.*

Radical Expressions and Equations

Radicals and Rockets

Escape velocity is the minimum initial velocity that an object must achieve to be free of the gravitational bonds of a planet. A rocket taking off from the surface of the Earth has to reach this velocity to enter orbit and deploy satellites or to fly to the Moon.

In general, the escape velocity from a planet depends on the planet's mass, m, its radius, r, and the universal gravitational constant, G. Physicists have shown that this velocity can be modeled by the radical expression:

$$\sqrt{\frac{2Gm}{r}}$$

The value of this expression for Earth turns out to be approximately 11 km (or 7 mi) per sec.

(**Sources:** Isaac Asimov, *Asimov on Astronomy*, New York: Doubleday and Company, 1974; Michael Zeilik, *Astronomy—The Evolving Universe*, Somerset: Wiley, 1997.)

To see if you have already mastered the topics in this chapter, take this test.
(Assume that all variables represent nonnegative real numbers.)

1. Evaluate:

    a. $2\sqrt{36}$ **b.** $\sqrt[3]{-64}$

2. Simplify: $\sqrt{100u^2v^4}$

3. Rewrite in radical notation, and then simplify:
$(81x^4)^{3/4}$

4. Simplify:

    a. $\dfrac{8p^{2/3}}{(36p^{4/3})^{1/2}}$ **b.** $\sqrt[6]{x^2y^4}$

5. Multiply: $\sqrt{6a} \cdot \sqrt{7b}$

6. Divide: $\dfrac{\sqrt[3]{18r^2}}{\sqrt[3]{3r}}$

7. Simplify: $\sqrt{147x^5y^4}$

Perform the indicated operation.

8. $\sqrt{\dfrac{2p}{25q^8}}$

9. $3\sqrt{12} - 5\sqrt{3} + \sqrt{108}$

10. $(\sqrt{x} - 8)(2\sqrt{x} + 1)$

11. Simplify: $\dfrac{\sqrt{2x}}{\sqrt{27y}}$

12. Rationalize the denominator:
$$\dfrac{\sqrt{n}}{\sqrt{n} + \sqrt{3}}$$

Solve.

13. $\sqrt{x + 4} - 9 = -3$

14. $\sqrt{x^2 - 2} = \sqrt{9x - 10}$

15. $\sqrt{2x} - \sqrt{x + 7} = -1$

16. Multiply: $(5 - \sqrt{-9})(4 - \sqrt{-25})$

17. Divide: $\dfrac{1 + 2i}{1 - 2i}$

18. If an object is dropped from a height of h ft above the ground, the time (in seconds) it takes for the object to be d ft above the ground is given by the following expression:
$$\sqrt{\dfrac{h - d}{16}}$$

    a. Simplify this expression.

    b. Find the time it takes for a stone dropped from a bridge 200 ft above the ground to be 125 ft above the ground. Express the answer as a radical and as a decimal rounded to the nearest tenth.

19. The length of a basketball court is 94 ft and the width is 50 ft. How long is the diagonal of the court?

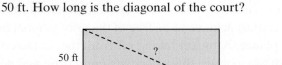

50 ft   ?

94 ft

20. When firefighters are putting out a fire, the rate at which they spray water on the fire is very important. For a hose with a nozzle 2 in. in diameter, the flow rate, f, in gallons per minute is modeled by the formula $f = 120\sqrt{p}$, where p is the nozzle pressure in pounds per square inch.

    Solve this formula for p.

● *Check your answers on page A-35.*

8.1   Radical Expressions and Functions

What Radicals Are and Why They Are Important

So far we have considered two types of algebraic expressions, namely, polynomials and rational expressions. In this chapter, we extend the discussion to a third type called *radical expressions*. Radical expressions, such as $10 - \sqrt{2}$ or $\sqrt[3]{abc}$, are algebraic expressions that contain *radicals*. These expressions are used to solve a wide variety of problems in many fields, such as geometry and the physical sciences. For instance, physicists use the radical expression $\sqrt{\dfrac{h}{16}}$ to model the number of seconds that it takes an object to fall h ft.

This chapter begins by dealing with the meaning of radical expressions. It introduces radical functions and their graphs, then shows ways of writing, simplifying, and operating on radical expressions. We cover methods of solving equations containing radical expressions. In addition, we consider a set of numbers not previously discussed—the complex numbers.

Square Roots

Let's begin our discussion of radicals with the definition of a particular type of radical, namely, the square root. Recall that we have already encountered this concept when we discussed irrational numbers in Section 1.1.

> **Definition**
>
> The number b is a **square root** of a if $b^2 = a$, for any real numbers a and b and for a nonnegative.

Every positive number has two square roots, one positive and the other negative. For example, the square roots of 4 are $+2$ and -2 because $(+2)^2 = 4$ and $(-2)^2 = 4$. The symbol $\sqrt{}$, called the **radical sign**, stands for the positive or **principal square root** but is commonly referred to as "the square root." For instance $\sqrt{9}$ is read "the square root of 9" or "radical 9" and represents $+3$, or 3. By contrast, the negative square root of 9, namely, -3, is represented by $-\sqrt{9}$. Throughout the remainder of this text, when we speak of the square root of a number we mean its principal square root.

The number under a radical sign is called the **radicand**.

Square root of a

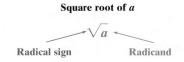

Radical sign   Radicand

Since the square of any real number is nonnegative, the square root of a negative number is not a real number. For instance, $\sqrt{-3}$ is not a real number.

A nonnegative rational number is said to be a *perfect square* if it is the square of another rational number. For instance, $\frac{1}{9}$ is a perfect square since $\frac{1}{9} = \left(\frac{1}{3}\right)^2$. On the other hand, 8 is not a perfect square since it is not the square of any rational number. *Perfect squares* play a special role in finding the value of square roots.

Now, let's look at evaluating square roots.

EXAMPLE 1	PRACTICE 1

Find the value of the following radical expressions:

a. $\sqrt{49}$ **b.** $\sqrt{\dfrac{1}{9}}$

c. $-2\sqrt{16}$ **d.** $\sqrt{-4}$

Solution

a. $\sqrt{49} = 7$ since $7^2 = 49$.

b. $\sqrt{\dfrac{1}{9}} = \dfrac{1}{3}$ since $\left(\dfrac{1}{3}\right)^2 = \dfrac{1}{9}$.

c. $-2\sqrt{16} = -2 \cdot 4 = -8$.

d. $\sqrt{-4}$ is not a real number.

Evaluate the following square roots:

a. $\sqrt{25}$

b. $\sqrt{\dfrac{9}{100}}$

c. $-4\sqrt{81}$

d. $\sqrt{-1}$

The square root of a number that is not a perfect square is an *irrational number*. Recall from Section 1.1 that irrational numbers have decimal representations that neither terminate nor repeat. For example, $\sqrt{7}$ and $\sqrt{10}$ are irrational numbers.

$$\sqrt{7} = 2.645751311\ldots, \text{ which is approximately } 2.646.$$
$$\sqrt{10} = 3.16227766\ldots, \text{ which is approximately } 3.162.$$

We can use a calculator to find the decimal approximation of these square roots rounded to a specific place value.

EXAMPLE 2	PRACTICE 2

Evaluate $\sqrt{3}$, rounded to the nearest thousandth.

Solution Since the radicand 3 is not a perfect square, its square root is an irrational number. Using a calculator, we see that $\sqrt{3} = 1.7320508\ldots$. Rounding to the nearest thousandth, we get 1.732.

Find the value of $\sqrt{6}$ rounded to the nearest thousandth.

In simplifying radicals, we use the fact that the operations of squaring and taking a square root undo each other; that is, they are inverse operations in the same way that multiplying by a number and dividing by the same number are inverse operations. The following two properties of radicals result from this relationship.

Squaring a Square Root

For any nonnegative real number a,

$$(\sqrt{a})^2 = a$$

According to this property, when we take the square root of a nonnegative number and then square the result, we get the original number. For instance, $(\sqrt{8})^2 = 8$.

Taking the Square Root of a Square

For any nonnegative real number a,

$$\sqrt{a^2} = a$$

This property says that when we square a nonnegative number and then take the square root, the result is the original number. For instance, $\sqrt{8^2} = 8$.

Some radicands contain variables. Consider, for instance, the expression $\sqrt{x}$. For negative values of x, the radicand is negative and so the radical is not a real number. For purposes of simplicity, *we will assume throughout this chapter that, unless otherwise stated, all variables in radicands are nonnegative.*

Some radicals have perfect square radicands containing variables, such as $\sqrt{x^6}$, $\sqrt{36y^8}$, and $\sqrt{49a^6b^2}$. Using the properties of exponents and the properties of square roots just stated, we can find their square roots.

$$\sqrt{x^6} = \sqrt{(x^3)^2} = x^3$$

$$\sqrt{36y^8} = \sqrt{(6y^4)^2} = 6y^4$$

$$\sqrt{49a^6b^2} = \sqrt{(7a^3b)^2} = 7a^3b$$

Note that the exponent of the variable(s) in any perfect square is always an even number. What relationship do you observe between the coefficients and the exponents of the radicands and those of the corresponding square roots?

EXAMPLE 3	PRACTICE 3
Simplify.	Simplify.
a. $\sqrt{25n^4}$ **b.** $-\sqrt{9a^8b^{10}}$	**a.** $-\sqrt{4y^2}$
Solution	**b.** $\sqrt{36x^6y^6}$
a. $\sqrt{25n^4} = \sqrt{(5n^2)^2} = 5n^2$	
b. $-\sqrt{9a^8b^{10}} = -\sqrt{(3a^4b^5)^2} = -3a^4b^5$	

Cube Roots

Not every radical is a square root. For instance, we can take the **cube root** of a number.

Definition

The number b is the **cube root** of a if $b^3 = a$, for any real numbers a and b. The cube root of a is written $\sqrt[3]{a}$, where 3 is called the **index** of the radical.

For example, the cube root of 8, written $\sqrt[3]{8}$, is 2 since $2^3 = 8$. The cube root of -8, written $\sqrt[3]{-8}$, is -2, since $(-2)^3 = -8$.

As with square roots, the operations of cubing and taking a cube root undo each other and so are inverse operations. The following two properties stem from this relationship.

Cubing a Cube Root

For any real number a,

$$(\sqrt[3]{a})^3 = a$$

The preceding property says that when we take the *cube root of a number* and then *cube the result*, we get the original number. For instance, $(\sqrt[3]{5})^3 = 5$.

Taking the Cube Root of a Cube

For any real number a,

$$\sqrt[3]{a^3} = a$$

The preceding property says that when we *cube a number* and then take the *cube root of the result*, we get the original number. For instance, $\sqrt[3]{5^3} = 5$.

A rational number is said to be a *perfect cube* if it is the cube of another rational number. For example, $\frac{1}{27}$ is a perfect cube since $\frac{1}{27} = \left(\frac{1}{3}\right)^3$. On the other hand, 16 is not a perfect cube since it is not the cube of any rational number.

EXAMPLE 4

Find the cube root.

a. $\sqrt[3]{64}$ **b.** $\sqrt[3]{-1}$ **c.** $\sqrt[3]{\dfrac{8}{27}}$ **d.** $\sqrt[3]{-27x^{12}}$

Solution

a. $\sqrt[3]{64} = \sqrt[3]{4^3} = 4$

b. $\sqrt[3]{-1} = \sqrt[3]{(-1)^3} = -1$

c. $\sqrt[3]{\dfrac{8}{27}} = \sqrt[3]{\left(\dfrac{2}{3}\right)^3} = \dfrac{2}{3}$

d. $\sqrt[3]{-27x^{12}} = \sqrt[3]{(-3x^4)^3} = -3x^4$

PRACTICE 4

Find the cube root.

a. $\sqrt[3]{216}$

b. $\sqrt[3]{-27}$

c. $\sqrt[3]{\dfrac{1}{125}}$

d. $\sqrt[3]{-64x^6}$

*n*th Roots

Just as we can raise a real number to powers other than 2 or 3, so we can find roots of a number other than the square root and the cube root.

Definition

The number b is the ***n*th root** of a if $b^n = a$, for any real number a and for any positive integer n greater than 1. The *n*th root of a is written $\sqrt[n]{a}$, where n is called the **index** of the radical.

For instance, the fourth root of 16, written $\sqrt[4]{16}$, is 2 since $2^4 = 16$.

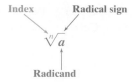

Index   Radical sign

$\sqrt[n]{a}$

Radicand

Note that for a square root, the index is understood to be 2 but is usually not written.

The *n*th root of any nonnegative real number is a real number. However, for negative radicands, the root is not a real number for every index:

- If the radicand is negative and the index is *even*, the radical expression is *not* a real number.

- If the radicand is negative and the index is *odd*, then the radical expression is a real number.

The next property of radicals involves taking the *n*th root of an *n*th power.

Taking the *n*th Root of an *n*th Power

For any real number a,

$$\sqrt[n]{a^n} = |a| \text{ for any even positive integer } n, \text{ and}$$

$$\sqrt[n]{a^n} = a \text{ for any odd positive integer } n \text{ greater than 1.}$$

This property states that the *n*th root of the *n*th power of a equals the absolute value of a when n is even and equals a when n is odd. So, for example, $\sqrt{(-5)^2} = |-5| = 5$ and $\sqrt[3]{(-5)^3} = -5$. Because we assume all variable radicands are positive, we can drop the absolute value sign even when the index is even.

EXAMPLE 5

Write without a radical sign.

a. $\sqrt[4]{81}$ **b.** $\sqrt[5]{-243}$ **c.** $\sqrt[5]{32x^{10}}$

Solution

a. $\sqrt[4]{81} = \sqrt[4]{3^4} = 3$ **b.** $\sqrt[5]{-243} = \sqrt[5]{(-3)^5} = -3$

c. $\sqrt[5]{32x^{10}} = \sqrt[5]{(2x^2)^5} = 2x^2$

PRACTICE 5

Express without a radical sign.

a. $\sqrt[6]{64}$

b. $\sqrt[5]{-32}$

c. $\sqrt[4]{256y^8}$

The following table lists some perfect powers. Committing these powers to memory can simplify working with radicals.

Number	Perfect Square	Perfect Cube	Perfect Fourth Power	Perfect Fifth Power
1	1	1	1	1
2	4	8	16	32
3	9	27	81	243
4	16	64	256	
5	25	125	625	
6	36	216		
7	49	343		
8	64	512		
9	81			
10	100			
11	121			
12	144			

Radical Functions

A *radical function* is a function in which a variable appears in a radicand. For example, the *square root function* $f(x) = \sqrt{x}$ is a radical function.

Radical functions, like other functions, can be evaluated. If a radical function's value is not an integer, it can be expressed in radical form or as a decimal approximation.

EXAMPLE 6

Given the radical functions $f(x) = \sqrt{x + 3}$ and $g(x) = 2 - \sqrt[3]{x}$, find the following function values. Round to the nearest hundredth.

a. $f(22)$ **b.** $g(27)$ **c.** $f(9)$ **d.** $g(54)$

Solution

a. $f(22) = \sqrt{22 + 3} = \sqrt{25} = 5$

b. $g(27) = 2 - \sqrt[3]{27} = 2 - 3 = -1$

c. $f(9) = \sqrt{9 + 3} = \sqrt{12} = 2\sqrt{3} \approx 3.46$

d. $g(54) = 2 - \sqrt[3]{54} = 2 - \sqrt[3]{27 \cdot 2} = 2 - 3\sqrt[3]{2} \approx -1.78$

PRACTICE 6

Given the radical function $f(x) = \sqrt{x - 4}$ and $g(x) = 3 + \sqrt{x}$, find the following function values.

a. $f(85)$

b. $g(121)$

c. $f(49)$

d. $g(48)$

Next, let's consider the graph of a radical function, such as $f(x) = \sqrt{x}$. To graph this nonlinear function, we first make a table of values that includes values of x and the corresponding values of $f(x)$. Then, we plot the points $(x, f(x))$ on a coordinate plane and draw a smooth curve through the points. The graph of this function is shown on the following coordinate plane:

x	$f(x) = \sqrt{x}$	(x, y)
0	$f(0) = \sqrt{0} = 0$	$(0, 0)$
1	$f(1) = \sqrt{1} = 1$	$(1, 1)$
4	$f(4) = \sqrt{4} = 2$	$(4, 2)$
6	$f(6) = \sqrt{6} \approx 2.45$	$(6, 2.45)$
9	$f(9) = \sqrt{9} = 3$	$(9, 3)$

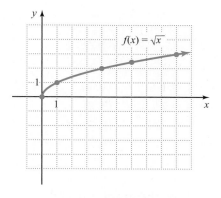

Note that no point on the graph has a negative x-coordinate. This is because the radicand under the square root, x, must be positive for the function to have a real value. Similarly, no point on the graph has a negative y-value, since all principal square roots are positive. In other words, both the range and domain of $f(x) = \sqrt{x}$ are the set of all nonnegative real numbers.

We can also graph the *cube root function, $g(x) = \sqrt[3]{x}$*, as shown below.

x	$g(x) = \sqrt[3]{x}$	(x, y)
-8	$f(-8) = \sqrt[3]{-8} = -2$	$(-8, -2)$
-1	$f(-1) = \sqrt[3]{-1} = -1$	$(-1, -1)$
0	$f(0) = \sqrt[3]{0} = 0$	$(0, 0)$
1	$f(1) = \sqrt[3]{1} = 1$	$(1, 1)$
8	$f(8) = \sqrt[3]{8} = 2$	$(8, 2)$

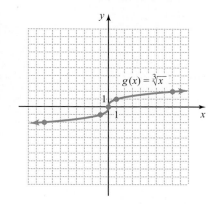

What are the domain and range of this function?

EXAMPLE 7

Graph $f(x) = \sqrt{x + 1}$.

Solution
Since the index is understood to be the even number 2, we can consider values of x that make the radicand nonnegative, that is, values of x for which $x + 1 \geq 0$.

x	$f(x) = \sqrt{x + 1}$	(x, y)
-1	$f(-1) = \sqrt{-1 + 1} = \sqrt{0} = 0$	$(-1, 0)$
0	$f(0) = \sqrt{0 + 1} = \sqrt{1} = 1$	$(0, 1)$
3	$f(3) = \sqrt{3 + 1} = \sqrt{4} = 2$	$(3, 2)$
5	$f(6) = \sqrt{6 + 1} = \sqrt{7} \approx 2.65$	$(5, 2.65)$
8	$f(8) = \sqrt{8 + 1} = \sqrt{9} = 3$	$(8, 3)$

Finally, we plot the points and draw the graph through them. We see from the graph that the domain of the function is $[-1, \infty)$, and the range is the set of all nonnegative real numbers.

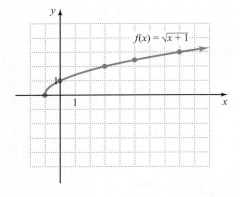

PRACTICE 7

Graph the function $h(x) = \sqrt{x - 3}$.

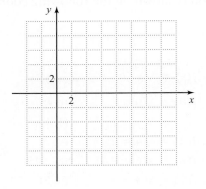

8.1 Exercises

Mathematically Speaking

Fill in each blank with the most appropriate term or phrase from the given list.

is not	square root	radicand
irrational	is	principal
square root function	cube root	rational
	index	radical function
square	cube root function	

1. The number b is a(n) _____ of a if $b^2 = a$, for any real numbers a and b and for a nonnegative.

2. The symbol $\sqrt{}$ stands for the positive or _____ square root.

3. The number under the radical sign is called the _____.

4. The square root of a negative number _____ a real number.

5. The square root of a number that is not a perfect square is a(n) _____ number.

6. The number b is the _____ of a if $b^3 = a$, for any real numbers a and b.

7. The cube root of a is written $\sqrt[3]{a}$, where 3 is called the _____ of the radical.

8. A _____ is a function in which a variable appears in a radicand.

9. The function $f(x) = \sqrt{x}$ is called the _____.

10. The function $f(x) = \sqrt[3]{x}$ is called the _____.

Evaluate, if possible.

11. $\sqrt{64}$

12. $\sqrt{49}$

13. $-\sqrt{100}$

14. $-\sqrt{144}$

15. $\sqrt{-36}$

16. $\sqrt{-9}$

17. $2\sqrt{16}$

18. $3\sqrt{25}$

19. $\sqrt[3]{27}$

20. $\sqrt[3]{125}$

21. $5\sqrt[3]{-8}$

22. $-6\sqrt[3]{-27}$

23. $\sqrt[4]{256}$

24. $\sqrt[5]{243}$

25. $8\sqrt[5]{-1}$

26. $7\sqrt[4]{81}$

27. $\sqrt{\dfrac{9}{16}}$

28. $\sqrt{\dfrac{1}{4}}$

29. $\sqrt[3]{-\dfrac{8}{125}}$

30. $-\sqrt[3]{\dfrac{27}{64}}$

31. $\sqrt{0.04}$

32. $\sqrt{1.21}$

Use a calculator to approximate the root to the nearest thousandth.

33. $\sqrt{21}$

34. $\sqrt{17}$

35. $\sqrt{46}$

36. $\sqrt{59}$

37. $\sqrt{14.25}$

38. $\sqrt{0.006}$

39. $\sqrt[3]{112}$

40. $\sqrt[3]{142}$

41. $\sqrt[5]{150}$

42. $\sqrt[4]{200}$

Simplify.

43. $\sqrt{x^8}$

44. $\sqrt{y^{10}}$

45. $\sqrt{16a^6}$

46. $\sqrt{49r^2}$

47. $9\sqrt{p^8q^4}$

48. $3\sqrt{u^8v^6}$

49. $\frac{1}{3}\sqrt{36x^{10}y^2}$

50. $-\frac{1}{4}\sqrt{64a^4b^{12}}$

51. $\sqrt[3]{-125u^9}$

52. $\sqrt[3]{-27r^6}$

53. $2\sqrt[3]{216u^3v^{12}}$

54. $7\sqrt[3]{27x^9y^9}$

55. $\sqrt[4]{16t^{12}}$

56. $\sqrt[4]{256n^{20}}$

57. $\sqrt[5]{p^5q^{15}}$

58. $\sqrt[5]{a^{10}b^{20}}$

Given $f(x) = \sqrt{2x + 5}$ and $g(x) = \sqrt[3]{x + 4}$, find the following values.
If the value is not an integer, express it in radical form.

59. $f(10)$

60. $f(2)$

61. $g(-12)$

62. $g(23)$

63. $f(0)$

64. $f(11)$

65. $g(36)$

66. $g(-28)$

Graph each function.

67. $f(x) = \sqrt{x} + 3$

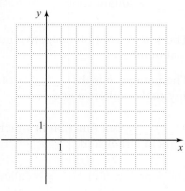

68. $g(x) = \sqrt{x} - 4$

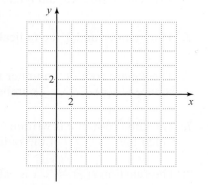

69. $h(x) = \sqrt{x + 4}$

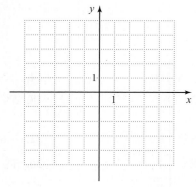

70. $g(x) = \sqrt{x - 2}$

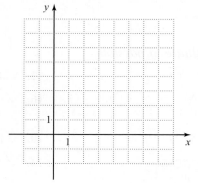

Mixed Practice

Evaluate, if possible.

71. $\sqrt{\dfrac{25}{121}}$

72. $-\sqrt[5]{32}$

73. $2\sqrt[3]{-125}$

74. $\sqrt[3]{-\dfrac{125}{216}}$

Use a calculator to approximate the root to the nearest thousandth.

75. $\sqrt{0.009}$

76. $\sqrt[3]{36}$

Simplify.

77. $\dfrac{1}{4}\sqrt{64a^6b^4}$

78. $\sqrt[3]{-8u^{12}v^3}$

If $f(x) = \sqrt[3]{2x - 10}$, find the following values. If the value is not an integer, express it in radical form.

79. $f(1)$

80. $f(32)$

Graph each function.

81. $h(x) = \sqrt{x} - 1$

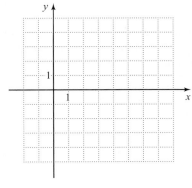

82. $g(x) = \sqrt{x + 4}$

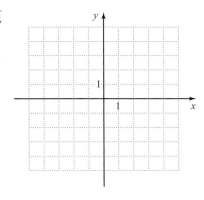

Applications

Solve.

83. If an object is dropped, the time (in seconds) it takes the object to fall s ft is given by the expression $\dfrac{1}{4}\sqrt{s}$.

Find the time it takes a stone dropped from a height of 100 ft to reach the ground.

84. The geometric mean is a statistic used in business and economics. The geometric mean of three numbers is given by the expression $\sqrt[3]{p}$, where p is the product of the three numbers. Find the geometric mean of 9, 3, and 8.

85. The length of the side of a square with area A can be computed using the expression $\sqrt{A}$. The area of the square picture, including the 1-in. wood border, is 25 in².

    a. Find x, the length of the side of the picture frame.

    b. What size photograph fits in the frame?

86. The length of one side of a cube with volume V can be computed using the expression $\sqrt[3]{V}$. A box manufacturer makes special-order cube-shaped boxes for a shipping company.

    a. If the shipping company requires a box to have a volume of 1728 in³, what is the length of each side of the box that the manufacturer must make?

    b. If the shipping company requests a box that is one-eighth the volume of the box described in part (a), by what factor does the length of each side change?

● *Check your answers on page A-35.*

MINDSTRETCHERS

Mathematical Reasoning

1. Consider the equation $y = \sqrt{x}$. Explain why only Quadrant I is used in graphing this equation.

Research

2. Before the introduction of calculators, a variety of methods were used to find the approximate value of square roots with radicands that are not perfect squares. Either in your college library or on the Web, identify two such methods and write a brief summary of your findings.

8.2   Rational Exponents

So far in this textbook, all the exponents we have discussed have been integral exponents. However, we now expand our discussion to include *rational exponents*, that is, exponents that are rational numbers. The concept of rational exponents gives us an alternative way to write radical expressions. Some examples of expressions written with rational exponents are

$$8^{1/3} \qquad -32^{4/5} \qquad x^{-3/4}$$

Let's first consider rational exponents with numerator 1, as in $8^{1/3}$. The question is, how can we attach a meaning to such an expression so that the laws of exponents hold? Cubing this expression and applying the power rule of exponents will provide an answer:

$$(8^{(1/3)})^3 = 8^{(1/3)\cdot 3} = 8^{3/3} = 8^1 = 8$$

Since the cube of $8^{1/3}$ is 8, $8^{1/3}$ must be the same as the cube root of 8, $\sqrt[3]{8}$. This suggests the following definition:

Definition of $a^{1/n}$
For any positive integer n greater than 1, and a real number a for which $\sqrt[n]{a}$ is a real number,

$$a^{1/n} = \sqrt[n]{a}$$

This definition says that a number raised to the power $\frac{1}{n}$ is the nth root of that number. Note that the denominator of the rational exponent is identical to the index of the radical.

EXAMPLE 1	PRACTICE 1
Write using radical notation. Simplify if possible.	Express as a radical expression. Simplify.
a. $49^{1/2}$ **b.** $64^{1/3}$ **c.** $-2x^{1/2}$ **d.** $(81x^8)^{1/4}$	**a.** $36^{1/2}$
Solution	**b.** $27^{1/3}$
a. $49^{1/2} = \sqrt{49} = \sqrt{7^2} = 7$	**c.** $-n^{1/4}$
b. $64^{1/3} = \sqrt[3]{64} = \sqrt[3]{4^3} = 4$	**d.** $(-125y^9)^{1/3}$
c. $-2x^{1/2} = -2(x^{1/2}) = -2\sqrt{x}$	
d. $(81x^8)^{1/4} = \sqrt[4]{81x^8} = \sqrt[4]{(3x^2)^4} = 3x^2$	

Now, let's consider the meaning that we would give to *any* positive rational exponent so that the laws of exponents hold. Consider, for instance, the expression $8^{2/3}$. Applying the laws of exponents gives us

$$8^{2/3} = (8^{1/3})^2 \qquad\qquad 8^{2/3} = (8^2)^{1/3}$$
$$= (\sqrt[3]{8})^2 \quad \text{or equivalently} \quad = \sqrt[3]{8^2}$$
$$= (2)^2 \qquad\qquad\qquad = \sqrt[3]{64}$$
$$= 4 \qquad\qquad\qquad\qquad = 4$$

This example suggests the following definition.

Definition of $a^{m/n}$

For any positive integers m and n such that n is greater than 1, where $\dfrac{m}{n}$ is in simplest form and a is a real number for which $\sqrt[n]{a}$ is a real number,

$$a^{m/n} = \sqrt[n]{a^m} \qquad \text{or equivalently} \qquad a^{m/n} = (\sqrt[n]{a})^m$$

Note that the denominator n of the rational exponent is the index of the radical. The numerator m is the power to which the base a (or $\sqrt[n]{a}$) is raised.

EXAMPLE 2	PRACTICE 2
Rewrite using radical notation. Simplify if possible.	Express in radical notation. Simplify.

EXAMPLE 2

Rewrite using radical notation. Simplify if possible.

a. $16^{5/4}$ **b.** $(-8)^{2/3}$ **c.** $\left(\dfrac{1}{4}x^2\right)^{3/2}$

Solution

a. $16^{5/4} = (\sqrt[4]{16})^5 = (\sqrt[4]{2^4})^5 = 2^5 = 32$

b. $(-8)^{2/3} = (\sqrt[3]{-8})^2 = (-2)^2 = 4$

c. $\left(\dfrac{1}{4}x^2\right)^{3/2} = \left(\sqrt{\dfrac{1}{4}x^2}\right)^3$

$\qquad = \left(\sqrt{\left(\dfrac{1}{2}x\right)^2}\right)^3$

$\qquad = \left(\dfrac{1}{2}x\right)^3$

$\qquad = \dfrac{1}{8}x^3$

PRACTICE 2

Express in radical notation. Simplify.

a. $81^{3/4}$

b. $(-64)^{2/3}$

c. $\left(\dfrac{4}{9}y^4\right)^{5/2}$

Would we have gotten the same answer to Example 2(b) if we had first squared -8 and then taken the cube root of the result?

Tip In evaluating $a^{m/n}$ for a given a, m, and n, it is usually easier to take the nth root before raising the base to the mth power, rather than the other way around.

Recall from Section 5.1 that to raise a number to a negative exponent, we raise the number to the corresponding positive exponent and then find the multiplicative inverse, that is, $a^{-n} = \dfrac{1}{a^n}$. The following definition extends the meaning of negative exponents to include negative rational exponents.

Definition of $a^{-m/n}$

For any positive integers m and n such that n is greater than 1, where $\dfrac{m}{n}$ is in simplest form and a is a real number for which $\sqrt[n]{a}$ is a nonzero real number,

$$a^{-m/n} = \frac{1}{a^{m/n}}$$

For instance,

$$8^{-2/3} = \frac{1}{8^{2/3}}$$

$$= \frac{1}{(\sqrt[3]{8})^2}$$

$$= \frac{1}{2^2} = \frac{1}{4}$$

EXAMPLE 3	PRACTICE 3

EXAMPLE 3

Simplify.

a. $16^{-3/4}$ **b.** $(-8x^3)^{-2/3}$

Solution

a. $16^{-3/4} = \dfrac{1}{16^{3/4}}$

$= \dfrac{1}{(\sqrt[4]{16})^3}$

$= \dfrac{1}{(\sqrt[4]{2^4})^3}$

$= \dfrac{1}{2^3} = \dfrac{1}{8}$

b. $(-8x^3)^{-2/3} = \dfrac{1}{(-8x^3)^{2/3}}$

$= \dfrac{1}{(\sqrt[3]{-8x^3})^2}$

$= \dfrac{1}{(-2x)^2} = \dfrac{1}{4x^2}$

PRACTICE 3

Simplify using the laws of exponents.

a. $-81^{-1/4}$

b. $(27a^6)^{-4/3}$

Expressions containing rational exponents, like those containing integral exponents, can be simplified using the laws of exponents. Review the laws in the following table.

Raising a number to a negative exponent	$x^{-a} = \dfrac{1}{x^a}$, for x nonzero
The product rule of exponents	$x^a \cdot x^b = x^{a+b}$
The quotient rule of exponents	$\dfrac{x^a}{x^b} = x^{a-b}$, for x nonzero
The power rule of exponents	$(x^a)^b = x^{ab}$
Raising a product to a power	$(xy)^a = x^a \cdot y^a$
Raising a quotient to a power	$\left(\dfrac{x}{y}\right)^a = \dfrac{x^a}{y^a}$, for y nonzero

EXAMPLE 4

Simplify, using the laws of exponents. Then, write the answer in radical notation and simplify, if possible.

a. $64^{2/3}64^{1/6}$ **b.** $\dfrac{y^{2/3}}{y^{1/3}}$ **c.** $(x^{3/8})^2$ **d.** $\left(\dfrac{8x^3}{y^6}\right)^{1/3}$

Solution

a. $64^{2/3}64^{1/6} = 64^{2/3+1/6}$ Use the product rule of exponents.

$\qquad = 64^{5/6}$ Simplify.

$\qquad = (\sqrt[6]{64})^5$ Write in radical notation.

$\qquad = 2^5$ Simplify.

$\qquad = 32$ Simplify.

b. $\dfrac{y^{2/3}}{y^{1/3}} = y^{2/3-1/3}$ Use the quotient rule of exponents.

$\qquad = y^{1/3}$ Simplify.

$\qquad = \sqrt[3]{y}$ Write in radical notation.

c. $(x^{3/8})^2 = x^{(3/8)\cdot 2}$ Use the power rule of exponents.

$\qquad = x^{3/4}$ Simplify.

$\qquad = \sqrt[4]{x^3}$ Write in radical notation.

d. $\left(\dfrac{8x^3}{y^6}\right)^{1/3} = \dfrac{(8x^3)^{1/3}}{(y^6)^{1/3}}$ Use the rule for raising a quotient to a power.

$\qquad = \dfrac{\sqrt[3]{8x^3}}{\sqrt[3]{y^6}}$ Write in radical notation.

$\qquad = \dfrac{2x}{y^2}$ Simplify.

PRACTICE 4

Use the laws of exponents to simplify. Then write the answer in radical form and simplify, if possible.

a. $81^{1/4}81^{1/2}$

b. $\dfrac{n^{3/5}}{n^{2/5}}$

c. $(r^{1/6})^3$

d. $\left(\dfrac{x^4}{4y^2}\right)^{1/2}$

EXAMPLE 5

The following formula relates the intensity of sound, I, to decibel level, D:
$$I = 10^{D/10}$$

Write this formula using radical notation.
(*Source:* Peter J. Nolan, *Fundamentals of College Physics,* 1993)

Solution

$$I = 10^{D/10} = \sqrt[10]{10^D}$$

So we can rewrite the formula as $I = \sqrt[10]{10^D}$.

PRACTICE 5

In our solar system, the *period* of a planet, that is, the time that it takes to make a complete orbit around the Sun, can be approximated by the expression $2\pi\sqrt{\dfrac{d^3}{Gm}}$, where d is the average distance of the planet from the Sun, G is a constant, and m is the mass of the planet, in appropriate units. Write this radical expression as an exponential expression in simplest form. (*Source:* Eric Chaisson and Steve McMillan, *Astronomy, A Beginner's Guide to the Universe,* 2001)

8.2 Exercises

Mathematically Speaking

Fill in each blank with the most appropriate term or phrase from the given list.

but not	can	raised to the power
multiplied by	cannot	and
power	index	

1. A number _____ $\frac{1}{n}$ is the *n*th root of that number.

2. The expression $a^{m/n}$ is equal to $\sqrt[n]{a^m}$ _____ $(\sqrt[n]{a})^m$.

3. When $a^{m/n}$ is rewritten as a radical expression, *n* is the _____.

4. Negative exponents _____ be rational.

Write using radical notation. Then simplify, if possible.

5. $16^{1/2}$
6. $81^{1/2}$
7. $-16^{1/2}$
8. $-81^{1/2}$

9. $(-64)^{1/3}$
10. $(-125)^{1/3}$
11. $6x^{1/4}$
12. $-y^{1/5}$

13. $(36a^2)^{1/2}$
14. $(25n^4)^{1/2}$
15. $(-216u^6)^{1/3}$
16. $(-32y^5)^{1/5}$

17. $27^{4/3}$
18. $4^{5/2}$
19. $-16^{3/2}$
20. $-64^{4/3}$

21. $(-27y^3)^{2/3}$
22. $(-32x^{10})^{3/5}$
23. $-81^{-3/4}$
24. $64^{-2/3}$

25. $\left(\dfrac{x^{10}}{4}\right)^{-1/2}$
26. $\left(\dfrac{8}{y^9}\right)^{-1/3}$

Write using radical notation. Simplify if possible.

27. $16 \cdot 16^{1/2}$
28. $8^{1/3} \cdot 8$
29. $\dfrac{6^{3/5}}{6^{2/5}}$
30. $\dfrac{3^{4/3}}{3^{1/3}}$

31. $\left(\dfrac{1}{2}\right)^{3/2} \cdot \left(\dfrac{1}{2}\right)^{-1/2}$
32. $\left(\dfrac{1}{16}\right)^{-1/4} \cdot \left(\dfrac{1}{16}\right)^{3/4}$
33. $(9^{3/4})^{2/3}$
34. $(81^{1/6})^{3/2}$

35. $4n^{2/5} \cdot n^{1/5}$
36. $a^{5/6} \cdot 6a^{-1/2}$
37. $(y^{-4})^{-1/8}$
38. $(x^{-4/3})^3$

39. $(4x^2)^{-1/2}$
40. $(27y^6)^{1/3}$
41. $\dfrac{5r^{3/4}}{r^{1/2}}$
42. $\dfrac{p^{5/6}}{7p^{2/3}}$

43. $\left(\dfrac{a^2}{b^6}\right)^{1/3}$
44. $\left(\dfrac{u^3}{v^4}\right)^{1/4}$
45. $3x(16x^8)^{1/2}$
46. $-2n^2(64n^9)^{2/3}$

47. $\dfrac{(2p^{1/6})^6}{16p^3}$
48. $\dfrac{12x^3}{(3x^{3/4})^4}$

Mixed Practice

Write using radical notation. Then simplify, if possible.

49. $9^{3/2}$

50. $(81x^6)^{1/2}$

51. $(-125u^3)^{2/3}$

52. $\left(\dfrac{a^8}{16}\right)^{-\frac{1}{4}}$

Write using radical notation. Simplify if possible.

53. $n^{-1/6} \cdot 3n^{2/3}$

54. $\left(\dfrac{x^2}{y^3}\right)^{1/3}$

55. $2x(64x^6)^{1/3}$

56. $\dfrac{9m^4}{(3m^{2/3})^3}$

Applications

Solve.

57. The manager of an office uses the expression $8000(0.5)^{t/3}$ to calculate the value of a piece of office equipment t years after it was purchased new for $8000.

    a. Write this expression in radical form.

    b. Find the value of the equipment 6 yr after it was purchased.

58. The number of Earth days it takes a planet in the solar system to revolve once around the Sun can be approximated by the expression $0.4(D)^{3/2}$, where D is the average distance of the planet from the Sun in millions of miles.

    a. Write this expression in radical form.

    b. The planet Mercury is an average distance of 36 million mi from the Sun. To the nearest day, how many Earth days does it take Mercury to revolve once around the Sun?

• Check your answers on page A-35.

MINDSTRETCHERS

Technology

1. Evaluate the expression $9^{1.5}$ both with and without a calculator.

Writing

2. What advantages or disadvantages do you see in writing a radical expression in terms of rational exponents?

Groupwork

3. Work with a partner to simplify each of the following radical expressions. Assume all variables and radicands are nonnegative real numbers.

    a. $\sqrt{x^2 + 2xy + y^2}$

    b. $(9a^2 + 12ab + 4b^2)^{-1/2}$

    c. $\dfrac{x + y}{(x + y)^{1/3}}$

    d. $\dfrac{(a^2 - 10a + 25)^{1/3}}{(a^2 - 10a + 25)^{1/4}}$

8.3   Simplifying Radical Expressions

Just as we rewrite rational expressions in lowest terms, so we rewrite radical expressions *in simplified form*. Simplified radical expressions are usually easier to work with and also to recognize when they are equal.

Rewriting radical expressions in terms of rational exponents and then using the laws of exponents often helps us to simplify these expressions. Once they are simplified, we can convert them back to radical notation.

EXAMPLE 1

Simplify each radical expression, if possible, by using rational exponents. Then, write in radical notation.

a. $\sqrt[8]{x^4}$ **b.** $\sqrt[4]{49}$ **c.** $\sqrt{x} \cdot \sqrt[4]{x}$ **d.** $\dfrac{\sqrt{x}}{\sqrt[3]{x}}$

Solution

a. $\sqrt[8]{x^4} = x^{4/8} = x^{1/2} = \sqrt{x}$

b. $\sqrt[4]{49} = 49^{1/4} = (7^2)^{1/4} = 7^{2/4} = 7^{1/2} = \sqrt{7}$

c. $\sqrt{x} \cdot \sqrt[4]{x} = x^{1/2} \cdot x^{1/4} = x^{1/2+1/4} = x^{3/4} = \sqrt[4]{x^3}$

d. $\dfrac{\sqrt{x}}{\sqrt[3]{x}} = \dfrac{x^{1/2}}{x^{1/3}} = x^{1/2-1/3} = x^{1/6} = \sqrt[6]{x}$

PRACTICE 1

If possible, simplify by using rational exponents. Write your answer as a radical expression.

a. $\sqrt[8]{y^2}$

b. $\sqrt[6]{25}$

c. $\sqrt{n} \cdot \sqrt[5]{n}$

d. $\dfrac{\sqrt{t}}{\sqrt[4]{t}}$

EXAMPLE 2

Simplify, if possible. Then, write using radical notation.

a. $\sqrt{\sqrt[3]{y}}$ **b.** $\sqrt[6]{r^2 s^4}$ **c.** $\sqrt[3]{3} \cdot \sqrt{2}$

Solution

a. $\sqrt{\sqrt[3]{y}} = (y^{1/3})^{1/2} = y^{1/6} = \sqrt[6]{y}$

b. $\sqrt[6]{r^2 s^4} = (r^2 s^4)^{1/6} = (r^2)^{1/6}(s^4)^{1/6} = r^{2/6} s^{4/6} = r^{1/3} s^{2/3}$
$= (rs^2)^{1/3} = \sqrt[3]{rs^2}$

c. The expression $\sqrt[3]{3} \cdot \sqrt{2} = 3^{1/3} \cdot 2^{1/2}$ cannot be simplified because the bases are not the same.

PRACTICE 2

Simplify, if possible. Then, write the answer as a radical expression.

a. $\sqrt[3]{\sqrt[4]{n}}$

b. $\sqrt[8]{a^2 b^6}$

c. $\sqrt[5]{5} \cdot \sqrt{7}$

Consider the expression $\sqrt[3]{2} \cdot \sqrt[3]{5}$. Using rational exponents and the laws of exponents, we can simplify this expression as follows:

$$\sqrt[3]{2} \cdot \sqrt[3]{5} = 2^{1/3} \cdot 5^{1/3}$$
$$= (2 \cdot 5)^{1/3}$$
$$= \sqrt[3]{2 \cdot 5}$$
$$= \sqrt[3]{10}$$

So we see that $\sqrt[3]{2} \cdot \sqrt[3]{5} = \sqrt[3]{2 \cdot 5}$. This result leads us to the *product rule of radicals*, which we use to multiply radical expressions and to simplify them.

Product Rule of Radicals

If $\sqrt[n]{a}$ and $\sqrt[n]{b}$ are real numbers, then

$$\sqrt[n]{a} \cdot \sqrt[n]{b} = \sqrt[n]{ab}$$

In words, this rule says that to multiply radicals with the *same index*, we multiply the radicands.

EXAMPLE 3	PRACTICE 3
Multiply.	Find the product.
a. $\sqrt{3} \cdot \sqrt{2}$ **b.** $\sqrt[3]{5} \cdot \sqrt[3]{x}$ **c.** $\sqrt[4]{6n^2} \cdot \sqrt[4]{2n}$ **d.** $\sqrt{\dfrac{x}{3}} \cdot \sqrt{\dfrac{y}{2}}$	**a.** $\sqrt{7} \cdot \sqrt{3}$
	b. $\sqrt[5]{2} \cdot \sqrt[5]{y^3}$
Solution	**c.** $\sqrt[3]{4p} \cdot \sqrt[3]{7p}$
a. $\sqrt{3} \cdot \sqrt{2} = \sqrt{3 \cdot 2} = \sqrt{6}$	**d.** $\sqrt{\dfrac{3}{v}} \cdot \sqrt{\dfrac{5}{u}}$
b. $\sqrt[3]{5} \cdot \sqrt[3]{x} = \sqrt[3]{5 \cdot x} = \sqrt[3]{5x}$	
c. $\sqrt[4]{6n^2} \cdot \sqrt[4]{2n} = \sqrt[4]{6n^2 \cdot 2n} = \sqrt[4]{12n^3}$	
d. $\sqrt{\dfrac{x}{3}} \cdot \sqrt{\dfrac{y}{2}} = \sqrt{\dfrac{x}{3} \cdot \dfrac{y}{2}} = \sqrt{\dfrac{xy}{6}}$	

We can use the product rule of radicals not only to multiply radicals but also to simplify them. When simplifying, we typically use the rule "in reverse," that is, in the form $\sqrt[n]{ab} = \sqrt[n]{a} \cdot \sqrt[n]{b}$.

A square root is not considered simplified if a perfect square factors the radicand. For instance, the radical $\sqrt{18}$ is not simplified, since the perfect square 9 is a factor of 18. Likewise, a cube root is not considered simplified if the radicand has a perfect cube factor. For example, the expression $\sqrt[3]{24}$ is not simplified since the perfect cube 8 is a factor of 24. In general, the expression $\sqrt[n]{a}$ is *not* simplified if the radicand a has a factor that is a perfect nth power.

EXAMPLE 4	PRACTICE 4
Simplify.	Write in simplified form.
a. $\sqrt{32}$ **b.** $-4\sqrt{75}$ **c.** $\sqrt[3]{54}$ **d.** $\sqrt[4]{80}$	**a.** $\sqrt{72}$
	b. $-7\sqrt{18}$
Solution	**c.** $\sqrt[3]{56}$
a. $\sqrt{32} = \sqrt{16 \cdot 2}$ Factor out the perfect square 16.	**d.** $\sqrt[4]{48}$
$= \sqrt{16} \cdot \sqrt{2}$ Use the product rule of radicals.	
$= 4\sqrt{2}$ Take the square root of the perfect square.	
b. $-4\sqrt{75} = -4\sqrt{25 \cdot 3}$	
$= -4\sqrt{25} \cdot \sqrt{3}$	
$= -4 \cdot 5 \cdot \sqrt{3}$	
$= -20\sqrt{3}$	
c. $\sqrt[3]{54} = \sqrt[3]{27 \cdot 2} = \sqrt[3]{27} \cdot \sqrt[3]{2} = 3\sqrt[3]{2}$	
d. $\sqrt[4]{80} = \sqrt[4]{16 \cdot 5} = \sqrt[4]{16} \cdot \sqrt[4]{5} = 2\sqrt[4]{5}$	

Some radical expressions contain radicands that have more than one factor that is a perfect nth power. For instance, the radicand in Example 4(a) has two perfect square factors, namely, 4 and 16. To simplify these radicals, we typically factor out the largest perfect nth-power factor.

EXAMPLE 5	PRACTICE 5

EXAMPLE 5

Simplify.

a. $\sqrt{x^5}$ **b.** $\sqrt{12n^3}$ **c.** $\sqrt[3]{40x^3y^4}$

Solution

a. $\sqrt{x^5} = \sqrt{x^4 \cdot x}$
$= \sqrt{x^4} \cdot \sqrt{x}$
$= x^2\sqrt{x}$

b. $\sqrt{12n^3} = \sqrt{4n^2 \cdot 3n}$
$= \sqrt{4n^2} \cdot \sqrt{3n}$
$= 2n\sqrt{3n}$

c. $\sqrt[3]{40x^3y^4} = \sqrt[3]{8x^3y^3 \cdot 5y}$
$= \sqrt[3]{8x^3y^3} \cdot \sqrt[3]{5y}$
$= 2xy\sqrt[3]{5y}$

PRACTICE 5

Simplify.

a. $\sqrt{y^7}$

b. $\sqrt{18x^5}$

c. $\sqrt[3]{81a^4b^6}$

Some radical expressions consist of the quotient of radicals. For instance, consider the expression $\dfrac{\sqrt{2}}{\sqrt{5}}$. Using rational exponents and the laws of exponents, we can write this expression as follows:

$$\frac{\sqrt{2}}{\sqrt{5}} = \frac{2^{1/2}}{5^{1/2}}$$

$$= \left(\frac{2}{5}\right)^{1/2}$$

$$= \sqrt{\frac{2}{5}}$$

So $\dfrac{\sqrt{2}}{\sqrt{5}} = \sqrt{\dfrac{2}{5}}$. This result leads us to the *quotient rule of radicals*, which we use to divide radicals and to simplify them.

Quotient Rule of Radicals

For any integer $n > 1$ and any real numbers a and b for which $\sqrt[n]{a}$ and $\sqrt[n]{b}$ are real numbers and b is nonzero,

$$\frac{\sqrt[n]{a}}{\sqrt[n]{b}} = \sqrt[n]{\frac{a}{b}}$$

In words, this rule states that to divide radicals, we divide the radicands. Note that in the quotient rule, the index is the same in all three radicals.

We can use this rule to divide two radicals.

EXAMPLE 6

Divide. Simplify, if possible.

a. $\dfrac{\sqrt{15x}}{\sqrt{3}}$
 b. $\dfrac{\sqrt{50}}{\sqrt{2}}$
 c. $\dfrac{\sqrt[3]{18a^2b}}{\sqrt[3]{3ab}}$

Solution

a. $\dfrac{\sqrt{15x}}{\sqrt{3}} = \sqrt{\dfrac{15x}{3}} = \sqrt{5x}$

b. $\dfrac{\sqrt{50}}{\sqrt{2}} = \sqrt{\dfrac{50}{2}} = \sqrt{25} = 5$

c. $\dfrac{\sqrt[3]{18a^2b}}{\sqrt[3]{3ab}} = \sqrt[3]{\dfrac{18a^2b}{3ab}} = \sqrt[3]{6a}$

PRACTICE 6

Find the quotient.

a. $\dfrac{\sqrt{42n}}{\sqrt{6}}$

b. $\dfrac{\sqrt{72}}{\sqrt{2}}$

c. $\dfrac{\sqrt[3]{10x^2y^2}}{\sqrt[3]{5x^2y}}$

A radical with a radicand in the form of a fraction is not considered to be simplified. To rewrite such radicals in simplest terms, we apply the quotient rule of radicals "in reverse": $\sqrt[n]{\dfrac{a}{b}} = \dfrac{\sqrt[n]{a}}{\sqrt[n]{b}}$.

EXAMPLE 7

Simplify.

a. $\sqrt{\dfrac{9}{49}}$
 b. $\sqrt{\dfrac{x}{4}}$
 c. $\sqrt[3]{\dfrac{8x}{27y^6}}$
 d. $\sqrt{\dfrac{5a^4b^{10}}{49c^2}}$

Solution

a. $\sqrt{\dfrac{9}{49}} = \dfrac{\sqrt{9}}{\sqrt{49}} = \dfrac{3}{7}$
    b. $\sqrt{\dfrac{x}{4}} = \dfrac{\sqrt{x}}{\sqrt{4}} = \dfrac{\sqrt{x}}{2}$

c. $\sqrt[3]{\dfrac{8x}{27y^6}} = \dfrac{\sqrt[3]{8x}}{\sqrt[3]{27y^6}} = \dfrac{2\sqrt[3]{x}}{3y^2}$

d. $\sqrt{\dfrac{5a^4b^{10}}{49c^2}} = \dfrac{\sqrt{5a^4b^{10}}}{\sqrt{49c^2}} = \dfrac{a^2b^5\sqrt{5}}{7c}$

PRACTICE 7

Express in simplest terms.

a. $\sqrt{\dfrac{16}{25}}$

b. $\sqrt{\dfrac{n}{64}}$

c. $\sqrt[3]{\dfrac{64x}{125y^9}}$

d. $\sqrt{\dfrac{7p^6q^2}{36r^4}}$

In the final examples of this section, we consider the problem of finding the distance between two points on a coordinate plane. To solve this problem, we apply our knowledge not only of radicals but also of the Pythagorean theorem from geometry.

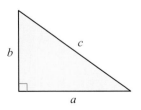

Recall that the Pythagorean theorem states that for any right triangle, the sum of the squares of the lengths of the two legs equals the square of the length of the hypotenuse: $a^2 + b^2 = c^2$.

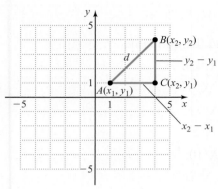

Now, consider any two points on a coordinate plane, $A(x_1, y_1)$ and $B(x_2, y_2)$. The distance between these two points, d, is the same as the length of line segment $\overline{AB}$. Note from the diagram that this line segment is the hypotenuse of a right triangle whose third vertex is $C(x_2, y_1)$. Applying the Pythagorean theorem to this triangle gives us $d^2 = (x_2 - x_1)^2 + (y_2 - y_1)^2$. This relationship can be written as

$$d = \sqrt{(x_2 - x_1)^2 + (y_2 - y_1)^2}$$

which is known as *the distance formula*. In words, this formula states that the distance between two points on the coordinate plane is the square root of the sum of the squares of the difference of the x-values and the difference of the y-values.

EXAMPLE 8

Find the distance between the points $(3, 6)$ and $(-2, 1)$ on the coordinate plane.

Solution Let $(3, 6)$ stand for (x_1, y_1) and $(-2, 1)$ for (x_2, y_2). Substituting these values in the distance formula, we get

$$d = \sqrt{(x_2 - x_1)^2 + (y_2 - y_1)^2}$$
$$= \sqrt{(-2 - 3)^2 + (1 - 6)^2}$$
$$= \sqrt{(-5)^2 + (-5)^2}$$
$$= \sqrt{25 + 25}$$
$$= \sqrt{50}$$
$$= 5\sqrt{2}$$

So the distance between $(3, 6)$ and $(-2, 1)$ on the coordinate plane is $5\sqrt{2}$ units.

PRACTICE 8

Find the distance between the points $(5, -7)$ and $(3, -1)$ on the coordinate plane.

We can use the distance formula to solve applied problems.

EXAMPLE 9

A hiker walked 3 mi north and then 3 mi east.

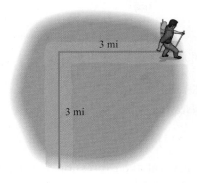

3 mi

3 mi

a. Using the hiker's starting point as the origin of a coordinate plane, find the coordinates of the starting and ending points.

b. How far is the hiker's destination from his starting point? Express the answer as a radical in simplified form.

Solution

a. The hiker's starting point is the origin of a coordinate system, with coordinates $(0, 0)$. So $x_1 = 0$ and $y_1 = 0$. The ending point for the hiker is $(3, 3)$, so $x_2 = 3$ and $y_2 = 3$.

b. To find the distance between these two points, we apply the distance formula.

$$d = \sqrt{(x_2 - x_1)^2 + (y_2 - y_1)^2}$$

$$d = \sqrt{(3 - 0)^2 + (3 - 0)^2} = \sqrt{9 + 9} = \sqrt{18} = \sqrt{9 \cdot 2} = 3\sqrt{2}$$

So the hiker was $3\sqrt{2}$ mi (or about 4.2 mi) from the starting point.

PRACTICE 9

One truck driver traveled west for 20 mi and then north for 50 mi. A second driver traveled 10 mi east and then 60 mi north. Both drivers had started from the same point.

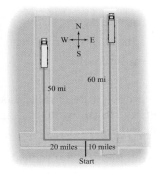

N
W ←→ E
S

60 mi

50 mi

20 miles | 10 miles

Start

a. Using their starting point as the origin of a coordinate plane, find the coordinates of the two destinations.

b. How far apart were the drivers at the end of their travels? Express the answer as a radical in simplified form.

8.3   Exercises

Mathematically Speaking

Fill in each blank with the most appropriate term or phrase from the given list.

radicands	the same index	different indices
square of the distance	perfect nth power	distance
multiple of n		

1. The product rule of radicals states that to multiply radicals with _____, multiply the radicands.

2. In general, the expression $\sqrt[n]{a}$ is not simplified if the radicand a has a factor that is a(n) _____.

3. The quotient rule of radicals states that to divide radicals with the same index, divide the _____.

4. The _____ between two points (x_1, y_1) and (x_2, y_2) on a coordinate plane is equal to $\sqrt{(x_2 - x_1)^2 + (y_2 - y_1)^2}$.

Use rational exponents to simplify the expression, if possible. Then, write the answer in radical form.

5. $\sqrt[6]{n^2}$ 6. $\sqrt[9]{a^6}$ 7. $\sqrt[8]{16}$ 8. $\sqrt[6]{64}$

9. $\sqrt[3]{x} \cdot \sqrt[6]{x}$ 10. $\sqrt[4]{t} \cdot \sqrt[8]{t}$ 11. $\sqrt[3]{p} \cdot \sqrt[4]{q}$ 12. $\sqrt{a} \cdot \sqrt[5]{b}$

13. $\dfrac{\sqrt[3]{x}}{\sqrt[4]{x}}$ 14. $\dfrac{\sqrt{n}}{\sqrt[6]{n}}$ 15. $\sqrt{\sqrt{y}}$ 16. $\sqrt[3]{\sqrt{x}}$

17. $\sqrt[4]{x^8 y^2}$ 18. $\sqrt[6]{u^4 v^{12}}$ 19. $\sqrt[6]{x^4} \cdot \sqrt[3]{x}$ 20. $\sqrt[8]{n^4} \cdot \sqrt[4]{n^2}$

21. $\dfrac{\sqrt[3]{y^2}}{\sqrt[9]{y^3}}$ 22. $\dfrac{\sqrt[4]{a^3}}{\sqrt[8]{a^6}}$ 23. $\sqrt[4]{p^2} \cdot \sqrt{q}$ 24. $\sqrt[6]{a^2} \cdot \sqrt[3]{b}$

Multiply.

25. $\sqrt{6} \cdot \sqrt{5}$ 26. $\sqrt{7} \cdot \sqrt{2}$ ⦿ 27. $\sqrt{3x} \cdot \sqrt{2y}$ 28. $\sqrt{5a} \cdot \sqrt{3b}$

29. $\sqrt[3]{4a^2} \cdot \sqrt[3]{9b}$ 30. $\sqrt[3]{10x} \cdot \sqrt[3]{2xy}$ ⦿ 31. $\sqrt[4]{3n} \cdot \sqrt[4]{7n^2}$ 32. $\sqrt[5]{6y^2} \cdot \sqrt[5]{4y^2}$

33. $\sqrt{\dfrac{x}{2}} \cdot \sqrt{\dfrac{6}{y}}$ 34. $\sqrt{\dfrac{10}{p}} \cdot \sqrt{\dfrac{q}{5}}$

Simplify.

35. $\sqrt{24}$ 36. $\sqrt{63}$ ⦿ 37. $-3\sqrt{80}$ 38. $5\sqrt{98}$

39. $\sqrt[3]{81}$ 40. $\sqrt[3]{72}$ ⦿ 41. $\sqrt[4]{96}$ 42. $\sqrt[4]{162}$

43. $\sqrt[5]{64}$

44. $\sqrt[5]{160}$

45. $6\sqrt{x^7}$

46. $-2\sqrt{n^3}$

47. $\sqrt{20y}$

48. $\sqrt{72a}$

49. $\sqrt{200r^4}$

50. $\sqrt{216t^6}$

○ 51. $\sqrt{54x^5y^7}$

52. $\sqrt{125a^7b^9}$

53. $\sqrt[3]{32n^5}$

54. $\sqrt[3]{108x^8}$

55. $\sqrt[5]{64n^8}$

56. $\sqrt[4]{243y^9}$

57. $\sqrt[3]{72x^7y^9}$

58. $\sqrt[3]{250p^2q^8}$

59. $\sqrt[4]{64x^5y^{10}}$

60. $\sqrt[4]{81p^9q^6}$

Divide.

61. $\dfrac{\sqrt{90}}{\sqrt{10}}$

62. $\dfrac{\sqrt{28}}{\sqrt{7}}$

63. $\dfrac{\sqrt{30n}}{\sqrt{6}}$

64. $\dfrac{\sqrt{65y}}{\sqrt{13}}$

○ 65. $\dfrac{\sqrt{12x^3y}}{\sqrt{3x}}$

66. $\dfrac{\sqrt{54pq}}{\sqrt{6q}}$

67. $\dfrac{\sqrt[3]{16u^2}}{\sqrt[3]{4u}}$

68. $\dfrac{\sqrt[3]{45n^2}}{\sqrt[3]{5n^2}}$

○ 69. $\dfrac{\sqrt[4]{24a^3b^2}}{\sqrt[4]{4a^2b}}$

70. $\dfrac{\sqrt[5]{36p^4q^3}}{\sqrt[5]{3p^2q^3}}$

Simplify.

71. $\sqrt{\dfrac{25}{16}}$

72. $\sqrt{\dfrac{49}{64}}$

73. $\sqrt{\dfrac{7}{81}}$

74. $\sqrt{\dfrac{3}{100}}$

75. $\sqrt{\dfrac{2}{n^6}}$

76. $\sqrt{\dfrac{3}{x^8}}$

○ 77. $\sqrt{\dfrac{7a}{121}}$

78. $\sqrt{\dfrac{5n}{144}}$

79. $\sqrt{\dfrac{a}{9b^4}}$

80. $\sqrt{\dfrac{5p}{36q^6}}$

81. $\sqrt{\dfrac{9u}{25v^2}}$

82. $\sqrt{\dfrac{4x}{81y^4}}$

83. $\sqrt[3]{\dfrac{27a^2}{64}}$

84. $\sqrt[3]{\dfrac{3x^3}{8}}$

85. $\sqrt[3]{\dfrac{9a}{8b^6c^9}}$

86. $\sqrt[3]{\dfrac{4u^2}{125v^3w^{12}}}$

87. $\sqrt[4]{\dfrac{2a^4b^3}{81c^8}}$

88. $\sqrt[5]{\dfrac{p^2}{32q^5r^{10}}}$

89. $\sqrt{\dfrac{13a^4b}{9c^6d^2}}$

90. $\sqrt{\dfrac{10pq^2}{49r^8s^4}}$

Find the distance between the two points on the coordinate plane.

91. $(10, 3)$ and $(4, -3)$

92. $(-7, 8)$ and $(2, 5)$

93. $(12, 15)$ and $(6, 12)$

94. $(9, 2)$ and $(-1, -8)$

95. $(-4, -3)$ and $(-8, 0)$

96. $(-4, 3)$ and $(-8, 0)$

Mixed Practice

Perform the indicated operation.

97. Multiply: $\sqrt[4]{5x^2} \cdot \sqrt[4]{6xy^3}$

98. Divide: $\dfrac{\sqrt[3]{63x^2}}{\sqrt[3]{7x}}$

Use rational exponents to simplify the expression. Then write the answer in radical form.

99. $\sqrt[3]{x^2} \cdot \sqrt[3]{x^4}$

100. $\dfrac{\sqrt[4]{n^3}}{\sqrt[6]{n^3}}$

Simplify.

101. $\sqrt{50r^6s^7}$

102. $\sqrt[4]{64}$

103. $\sqrt[3]{\dfrac{25x}{64y^3z^9}}$

104. $\sqrt{\dfrac{36a}{81b^2}}$

105. $\sqrt[3]{128u^4v^6}$

106. Find the distance between the points $(-1, 3)$ and $(4, -2)$ on the coordinate plane.

Applications

Solve.

107. The velocity of a fluid flowing out of a tank through a small opening d ft below the surface is given by the expression $\sqrt{64d}$. Simplify the expression.

108. The time it takes an object that is dropped to fall h ft is given by the expression $\sqrt{\dfrac{2h}{32}}$. Simplify this expression.

109. The size of a television set is commonly given by the length of its diagonal. The width, w, of the screen of a d-in. flat-panel television can be calculated using the expression $w = \sqrt{d^2 - l^2}$, where l is the length of the screen.

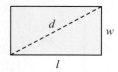

Find the width of a 42-in. flat-panel television if the length of the screen is 30 in. Express the answer as both a radical in simplest form and as a decimal rounded to the nearest inch.

110. The owner of a factory finds that the price, p (in dollars), of an item is given by the equation $p = 0.75\sqrt{x - 8}$, where x is the daily demand. Find the price of the item when the daily demand is 80 units.

111. Find the length of string, d, let out for the kite shown in the following figure. Express the answer as both a radical in simplest form and as a decimal rounded to the nearest tenth of a foot.

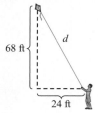

112. Two airplanes depart the same city at the same time. One flies east at a rate of 450 mph and the other flies south at a rate of 400 mph. How far apart are the two planes 1 hr after departure? Express the answer as both a radical in simplest form and as a decimal rounded to the nearest mile.

113. A sales rep drives from her home to a college 20 mi north. She then drives 48 mi east to another college.

    a. Using the origin of a coordinate plane as the sales rep's starting point, find the coordinates of the location of the second college.

    b. How far from her home is the second college?

114. To get to a friend's apartment from his apartment, a student walks 4 blocks west and 6 blocks south.

    a. Using the origin of a coordinate plane as the starting point, find the coordinates of the location of the friend's apartment.

    b. How far from the student's apartment is the friend's apartment?

• *Check your answers on page A-36.*

MINDSTRETCHERS

Writing

1. Explain what it means for the expression $\sqrt[n]{a}$ to be in simplest form.

Technology

2. Use a grapher to display the graphs of $y = \sqrt{\dfrac{1}{x^2 + 1}}$ and $y = \dfrac{1}{\sqrt{x^2 + 1}}$. Compare the two graphs. What conclusion can you draw from this comparison?

Groupwork

3. Working with a partner, simplify the following expressions.

    a. $\sqrt{x^3 + x^2 - x - 1}$

    b. $\dfrac{\sqrt{x^3 + 2x^2 - 9x - 18}}{\sqrt{x^2 + 5x + 6}}$

8.4 Addition and Subtraction of Radical Expressions

In contrast to other kinds of numbers such as fractions and decimals, sums and differences of many radicals cannot be simplified. For instance, there is no way to combine $\sqrt{5}$ and $\sqrt{2}$, and so we cannot simplify the expression $\sqrt{5} + \sqrt{2}$. Similarly, we cannot simplify the expression $\sqrt[5]{3} - 1$. However, we can approximate these expressions using a calculator.

Other radicals can be combined and simplified. In this section, we discuss how to add and subtract such radicals.

OBJECTIVES

- To add and subtract radical expressions
- To solve applied problems involving the addition or subtraction of radical expressions

Combining Like Radicals

When adding or subtracting **like radicals**, we simplify the result. Radicals are said to be *like* if they have *the same index and the same radicand*. For instance,

Same index
$7\sqrt[3]{5}$ and $4\sqrt[3]{5}$ are like radicals.
Same radicand

Different index
$\sqrt[3]{3}$ and $\sqrt[4]{3}$ are unlike radicals.
Same radicand

We use the distributive property to add or subtract like radicals, just as we do for adding or subtracting like terms.

Adding Like Terms
$5x + 4x = (5 + 4)x = 9x$
Like terms

Adding Like Radicals
$5\sqrt{3} + 4\sqrt{3} = (5 + 4)\sqrt{3} = 9\sqrt{3}$
Like radicals

Subtracting Like Terms
$9n^3 - 5n^3 = (9 - 5)n^3 = 4n^3$
Like terms

Subtracting Like Radicals
$9\sqrt[3]{2} - 5\sqrt[3]{2} = (9 - 5)\sqrt[3]{2} = 4\sqrt[3]{2}$
Like radicals

As these examples illustrate, we add and subtract like radicals by combining their coefficients and then multiply this result by the radical.

EXAMPLE 1

Combine.

a. $7\sqrt[3]{10} + 4\sqrt[3]{10}$ **b.** $8\sqrt{x} - 3\sqrt{x} - 2\sqrt{x}$
c. $5\sqrt{2y + 1} - \sqrt{2y + 1}$ **d.** $\sqrt[3]{x} + 3\sqrt{x}$

Solution

a. $7\sqrt[3]{10} + 4\sqrt[3]{10} = (7 + 4)\sqrt[3]{10}$ Combine the coefficients using the distributive property.

$= 11\sqrt[3]{10}$ Simplify.

PRACTICE 1

Add or subtract, as indicated.

a. $-\sqrt[4]{9} + 3\sqrt[4]{9}$
b. $6\sqrt{p} + \sqrt{p} - 2\sqrt{p}$
c. $-5\sqrt{t^2 - 4} + \sqrt{t^2 - 4}$
d. $5\sqrt[3]{a} - 2\sqrt[3]{b}$

b. $8\sqrt{x} - 3\sqrt{x} - 2\sqrt{x} = (8 - 3 - 2)\sqrt{x} = 3\sqrt{x}$

c. $5\sqrt{2y + 1} - \sqrt{2y + 1} = (5 - 1)\sqrt{2y + 1} = 4\sqrt{2y + 1}$

d. $\sqrt[3]{x} + 3\sqrt{x}$

The terms cannot be combined because they are not like radicals.

Tip When adding or subtracting radicals, do not combine their radicands. For example, $\sqrt{5} + \sqrt{5} \neq \sqrt{10}$.

Combining Unlike Radicals

Some *unlike* radicals become *like* when they are simplified and so can be combined.

EXAMPLE 2	**PRACTICE 2**

EXAMPLE 2

Add or subtract, as indicated.

a. $\sqrt{48} + \sqrt{12}$ **b.** $5\sqrt{72} - \sqrt{50} + \sqrt{32}$

c. $\sqrt[3]{54} + 2\sqrt[3]{16} - 7\sqrt[3]{2}$ **d.** $\sqrt{45} - \sqrt{49} + \sqrt{20}$

Solution

a. $\sqrt{48} + \sqrt{12} = \sqrt{16 \cdot 3} + \sqrt{4 \cdot 3}$ Factor out perfect squares.

$\qquad = \sqrt{16}\sqrt{3} + \sqrt{4}\sqrt{3}$ Use the product rule of radicals.

$\qquad = 4\sqrt{3} + 2\sqrt{3}$ Take the square root of a perfect square.

$\qquad = (4 + 2)\sqrt{3}$ Use the distributive property.

$\qquad = 6\sqrt{3}$ Simplify.

b. $5\sqrt{72} - \sqrt{50} + \sqrt{32}$

$\qquad = 5\sqrt{36 \cdot 2} - \sqrt{25 \cdot 2} + \sqrt{16 \cdot 2}$ Factor out perfect squares.

$\qquad = 5 \cdot 6\sqrt{2} - 5\sqrt{2} + 4\sqrt{2}$ Use the product rule and take the square root.

$\qquad = 30\sqrt{2} - 5\sqrt{2} + 4\sqrt{2}$ Simplify.

$\qquad = 29\sqrt{2}$ Combine like radicals.

c. $\sqrt[3]{54} + 2\sqrt[3]{16} - 7\sqrt[3]{2} = \sqrt[3]{27 \cdot 2} + 2\sqrt[3]{8 \cdot 2} - 7\sqrt[3]{2}$

$\qquad = 3\sqrt[3]{2} + 2 \cdot 2\sqrt[3]{2} - 7\sqrt[3]{2}$

$\qquad = (3 + 4 - 7)\sqrt[3]{2}$

$\qquad = 0$

d. $\sqrt{45} - \sqrt{49} + \sqrt{20} = \sqrt{9 \cdot 5} - \sqrt{49} + \sqrt{4 \cdot 5}$

$\qquad = 3\sqrt{5} - 7 + 2\sqrt{5}$

$\qquad = 5\sqrt{5} - 7$

PRACTICE 2

Combine.

a. $\sqrt{75} + \sqrt{27}$

b. $\sqrt{96} - 3\sqrt{54} - 7\sqrt{24}$

c. $9\sqrt[4]{2} - 4\sqrt[4]{162} - \sqrt[4]{32}$

d. $\sqrt{64} - \sqrt{28} + \sqrt{63}$

The radicands of radicals to be combined may contain one or more variables. For every radical, we assume that its value is a real number.

EXAMPLE 3

Simplify.

a. $\sqrt{9x} - 2\sqrt{x}$ **b.** $b\sqrt{a^2b} - 2a\sqrt{b}$ **c.** $\sqrt[3]{54t^4} + \sqrt[3]{-128t}$

Solution

a. $\sqrt{9x} - 2\sqrt{x} = 3\sqrt{x} - 2\sqrt{x}$

$$= (3 - 2)\sqrt{x}$$

$$= \sqrt{x}$$

b. $b\sqrt{a^2b} - 2a\sqrt{b} = b \cdot a\sqrt{b} - 2a\sqrt{b}$

$$= ab\sqrt{b} - 2a\sqrt{b}$$

$$= (ab - 2a)\sqrt{b}$$

c. $\sqrt[3]{54t^4} + \sqrt[3]{-128t} = \sqrt[3]{27 \cdot 2 \cdot t^3 \cdot t^1} + \sqrt[3]{(-64) \cdot 2 \cdot t^1}$

$$= 3t\sqrt[3]{2t} + (-4)\sqrt[3]{2t}$$

$$= (3t - 4)\sqrt[3]{2t}$$

PRACTICE 3

Combine.

a. $4\sqrt{25n} + \sqrt{n}$

b. $\sqrt[3]{27ab^3} - 11\sqrt[3]{a}$

c. $\sqrt{50x} - \sqrt{32x^3}$

EXAMPLE 4

If $f(x) = 10x\sqrt[4]{16x^3}$ and $g(x) = 2x\sqrt[4]{x^3}$, find

a. $f(x) + g(x)$ **b.** $f(x) - g(x)$

Solution

a. $f(x) + g(x) = 10x\sqrt[4]{16x^3} + 2x\sqrt[4]{x^3}$

$$= 10x \cdot 2\sqrt[4]{x^3} + 2x\sqrt[4]{x^3}$$

$$= 20x\sqrt[4]{x^3} + 2x\sqrt[4]{x^3}$$

$$= 22x\sqrt[4]{x^3}$$

b. $f(x) - g(x) = 10x\sqrt[4]{16x^3} - 2x\sqrt[4]{x^3}$

$$= 20x\sqrt[4]{x^3} - 2x\sqrt[4]{x^3}$$

$$= 18x\sqrt[4]{x^3}$$

PRACTICE 4

If $p(x) = x\sqrt[3]{81x}$ and $q(x) = \sqrt[3]{24x^4}$, find

a. $p(x) + q(x)$

b. $p(x) - q(x)$

Recall the Pythagorean theorem, which asserts that for every right triangle, the sum of the squares of the two legs, a and b, equals the square of the hypotenuse, c. This relationship is expressed as $a^2 + b^2 = c^2$. We can write this equation as:

$$c = \sqrt{a^2 + b^2}$$

EXAMPLE 5

The map to the right shows the route that a college recruiter takes in calling on colleges each week. She starts at home (*A*), visits colleges in towns *B*, *C*, and *D* in that order, and then returns home. (*ACBD* is a rectangle.)

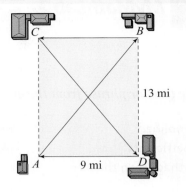
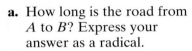

a. How long is the road from *A* to *B*? Express your answer as a radical.

b. What is the length of the recruiter's weekly trip?

Solution

a. Let *x* represent the length of $\overline{AB}$. The line segments $\overline{BD}$ and $\overline{DA}$ form the other two sides of a right triangle. The lengths of the sides are 13 mi and 9 mi. Using the Pythagorean theorem, we get:

$$x = \sqrt{(BD)^2 + (DA)^2}$$
$$x = \sqrt{13^2 + 9^2}$$
$$x = \sqrt{250}$$
$$x = \sqrt{25 \cdot 10}$$
$$x = 5\sqrt{10}$$

So the road from *A* to *B* is $5\sqrt{10}$ mi long.

b. The total length of the recruiter's trip is the sum of the lengths of the components of the trip:

$$\text{Trip length} = AB + BC + CD + DA$$
$$= 5\sqrt{10} + 9 + 5\sqrt{10} + 9$$
$$= 18 + 10\sqrt{10}$$

So the length of the recruiter's trip is $(18 + 10\sqrt{10})$ mi, or approximately 49.6 mi.

PRACTICE 5

The size of a television set is commonly given by the length of its diagonal. For the two sets pictured, find

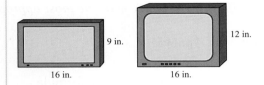

a. the lengths of their diagonals

b. the difference of the diagonal lengths

Mathematically Speaking

Fill in each blank with the most appropriate term or phrase from the given list.

unlike	simplified	reduced
indices	coefficients	like
area of a triangle formula	Pythagorean theorem	

1. Radicals that have the same index and the same radicand are called _____ radicals.

2. Like radicals can be added by combining their _____ and then multiplying this sum by the radical.

3. Some unlike radicals can become like radicals when they are _____.

4. The _____ can be expressed as $c = \sqrt{a^2 + b^2}$.

Combine, if possible.

5. $4\sqrt{3} + \sqrt{3}$

6. $2\sqrt{7} + 5\sqrt{7}$

7. $3\sqrt[3]{2} - 6\sqrt[3]{2}$

8. $11\sqrt[4]{6} - 9\sqrt[4]{6}$

9. $4\sqrt{2} - 2\sqrt{7}$

10. $10\sqrt[3]{5} + 6\sqrt{5}$

11. $-8\sqrt{y} - 12\sqrt{y}$

12. $-5\sqrt{x} - \sqrt{x}$

13. $7\sqrt{x} + 4\sqrt[3]{x}$

14. $2\sqrt[3]{r} + 10\sqrt[3]{r}$

15. $-11\sqrt{n} + 2\sqrt{n} + 9\sqrt{n}$

16. $8\sqrt{a} - 13\sqrt{a} + 3\sqrt{a}$

17. $12\sqrt[4]{p} - 5\sqrt[4]{p} + \sqrt[4]{p}$

18. $-11\sqrt[3]{x} - \sqrt[3]{x} + 15\sqrt[3]{x}$

19. $9y\sqrt{3x} + 4y\sqrt{3x}$

20. $19a\sqrt{5b} - 10a\sqrt{5b}$

21. $-7\sqrt{2r-1} + 3\sqrt{2r-1}$

22. $8\sqrt{3y-2} + 2\sqrt{3y-2}$

23. $\sqrt{x^2-9} + \sqrt{x^2-9}$

24. $6\sqrt{a^2-b^2} - 7\sqrt{a^2-b^2}$

Simplify, if possible.

25. $\sqrt{6} + \sqrt{24}$

26. $\sqrt{45} + 4\sqrt{5}$

27. $\sqrt{72} - \sqrt{8}$

28. $\sqrt{12} - \sqrt{75}$

29. $3\sqrt{18} - 4\sqrt{2} + \sqrt{50}$

30. $-6\sqrt{28} + 2\sqrt{63} - 7\sqrt{7}$

31. $-5\sqrt[3]{24} + \sqrt[3]{-81} + 9\sqrt[3]{3}$

32. $7\sqrt[4]{4} - 3\sqrt[4]{64} - \sqrt[4]{324}$

33. $\sqrt{54x^3} - x\sqrt{150x}$

34. $y\sqrt{32y} - \sqrt{98y^3}$

35. $\frac{1}{4}\sqrt{128n^5} + \sqrt{242n}$

36. $\sqrt{490t} + \frac{1}{2}\sqrt{40t^3}$

37. $-4y\sqrt{xy^3} + 7x\sqrt{x^3y}$

38. $6a\sqrt{ab^5} - 9b\sqrt{a^3b}$

39. $\sqrt[3]{-125p^9} + p\sqrt[3]{-8p^6}$

40. $\sqrt[3]{-27r^6} + 4r\sqrt[3]{64r^3}$

41. $5\sqrt[4]{48ab^3} - 2\sqrt[4]{3a^5b^3}$

42. $10\sqrt[5]{2x^2y^7} + 4\sqrt[5]{64x^2y^2}$

43. $2\sqrt{125} + \frac{1}{6}\sqrt{144} - \frac{3}{4}\sqrt{80}$ **44.** $\frac{1}{2}\sqrt{128} - \sqrt{98} + \frac{1}{3}\sqrt{225}$ **45.** $\frac{3}{5}\sqrt[3]{-125} - \frac{1}{4}\sqrt[3]{48} + \frac{1}{2}\sqrt[3]{162}$

46. $-\frac{3}{2}\sqrt[3]{192} + 6\sqrt[3]{24} + \frac{2}{3}\sqrt[3]{-216}$

Given f(x) and g(x), find f(x) + g(x) and f(x) − g(x).

47. $f(x) = 5x\sqrt{20x}$ and $g(x) = 3\sqrt{5x^3}$

48. $f(x) = -\sqrt{108x^4}$ and $g(x) = x\sqrt{147x^2}$

49. $f(x) = 2x\sqrt[4]{64x}$ and $g(x) = -\sqrt[4]{4x^5}$

50. $f(x) = 3\sqrt[3]{6x^8}$ and $g(x) = -5x^2\sqrt[3]{162x^2}$

Mixed Practice

Simplify, if possible.

51. $-8\sqrt{24} - 3\sqrt{6} + 2\sqrt{150}$

52. $\sqrt[3]{16ab^4} - \sqrt[3]{54ab}$

53. $-9\sqrt{3b} - \sqrt{3b} + 5\sqrt{3b}$

54. $-6\sqrt{r} + 5\sqrt[3]{r}$

55. $\frac{1}{3}\sqrt{405k^3} - \sqrt{180k}$

56. Given $f(x) = -2x\sqrt{63x^2}$ and $g(x) = -9\sqrt{7x^4}$, find $f(x)+g(x)$ and $f(x)-g(x)$.

Applications

Solve.

57. When an object is dropped from a given height, its velocity (in feet per second) when it has fallen h ft is given by the expression $\sqrt{64h}$. Find the difference in the velocity of an object when it has fallen 80 ft and when it has fallen 20 ft. Express the answer both as a radical in simplest form and as a decimal rounded to the nearest tenth.

58. The distance (in miles) a person can see to the horizon from an altitude of a ft is given by the expression $\sqrt{1.5a}$. How much farther can a person see from the window of an airplane flying at an altitude of 32,000 ft than from the window of an airplane flying at an altitude of 18,000 ft? Express the answer both as a radical in simplest form and as a decimal rounded to the nearest tenth.

59. The areas of two types of square floor tiles sold at a home improvement store are shown. How much longer is the side of the larger tile? Express the answer as a radical in simplest form and as a decimal to the nearest tenth.

Area = 72 in²   Area = 128 in²

60. The Great Pyramid at Giza has a square base with an area of 52,900 m². What is the perimeter of the base?

61. The length of the ramp shown in the following figure can be determined using the expression $\sqrt{b^2 + h^2}$, where b is the length of the base of the ramp and h is the height of the ramp.

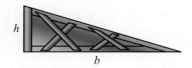

a. Find the length of the ramp if the height of the ramp is 2 ft and the base is 4 ft.

b. If the base of the ramp is increased by 2 ft, by approximately how many feet does the length of the ramp increase? (Round your answer to the nearest tenth.)

62. A guy wire of length l attached to the top of an antenna whose height is h is anchored to the ground a distance d from the center of the base of the antenna, as shown in the following diagram.

a. If the height of an antenna is 160 m, how far from the base's center will a 200-m guy wire be anchored to the ground?

b. If the same length of wire is used on an antenna that is 175 m, by approximately how many meters does the distance from the center of the antenna's base to the point where the guy wire is anchored decrease? (Round your answer to the nearest tenth.)

• *Check your answers on page A-36.*

MINDSTRETCHERS

Critical Thinking

1. Find two radical expressions whose sum is $4\sqrt{2}$ and whose difference is $7\sqrt{2}$.

Groupwork

2. Working with a partner, consider the functions $f(x) = \sqrt{9x + 18}$, $g(x) = \sqrt{4x + 8}$, and $h(x) = 5\sqrt{x + 2}$.

a. Evaluate $f(2)$, $g(2)$, and $h(2)$.

b. Find the sum of $f(2)$ and $g(2)$ and compare it to the value of $h(2)$. What do you notice?

c. Repeat parts (a) and (b) for several other values of x.

d. From the result in parts (b) and (c), what conclusion is suggested about the relationship between $f(x)$, $g(x)$, and $h(x)$?

e. Justify your answer to part (d) algebraically.

Mathematical Reasoning

3. For most values of a and b, $\sqrt{a} + \sqrt{b} \neq \sqrt{a + b}$. Are there any real numbers a and b such that $\sqrt{a} + \sqrt{b} = \sqrt{a + b}$? Justify your answer.

8.5 Multiplication and Division of Radical Expressions

In the last section, we considered the addition and subtraction of radical expressions. In this section, we discuss the multiplication and division of radical expressions.

Multiplying Radical Expressions

In considering how to find the product of radical expressions, let's begin with the simplest case—multiplying two radicals. The key is to apply the product rule of radicals discussed in Section 8.3.

$$\sqrt[n]{a} \cdot \sqrt[n]{b} = \sqrt[n]{ab}$$

EXAMPLE 1

Multiply and simplify.

a. $\sqrt{8n^3} \cdot \sqrt{3n}$ **b.** $(-5\sqrt[3]{7})(4\sqrt[3]{2})$

Solution

a. Here we use the product rule of radicals to multiply.

$$\sqrt{8n^3} \cdot \sqrt{3n} = \sqrt{8n^3 \cdot 3n}$$
$$= \sqrt{24n^4}$$
$$= \sqrt{4n^4 \cdot 6} \qquad \text{Factor out a perfect square.}$$
$$= 2n^2\sqrt{6}$$

b. $(-5\sqrt[3]{7})(4\sqrt[3]{2}) = -5 \cdot 4 \cdot \sqrt[3]{7 \cdot 2}$
$$= -20\sqrt[3]{14}$$

PRACTICE 1

Find the product.

a. $\sqrt{6y} \cdot \sqrt{12y^7}$

b. $(-2\sqrt[3]{3})(7\sqrt[3]{5})$

Next, we consider how to multiply radical expressions that may contain more than one term. Just as with the multiplication of polynomials, the key here is to apply the distributive property.

EXAMPLE 2

Find the product and simplify.

a. $\sqrt[3]{3}(5\sqrt[3]{2} - 1)$ **b.** $(5\sqrt{n} - 1)(4\sqrt{n} + 3)$

Solution

a. $\sqrt[3]{3}(5\sqrt[3]{2} - 1) = \sqrt[3]{3} \cdot 5\sqrt[3]{2} - \sqrt[3]{3} \cdot 1$ Use the distributive property.
$$= 5\sqrt[3]{6} - \sqrt[3]{3} \qquad \text{Use the product rule of radicals.}$$

PRACTICE 2

Multiply and simplify.

a. $\sqrt[4]{5}(\sqrt[4]{7} + 6\sqrt[4]{2})$

b. $(3\sqrt{x} - 1)(2\sqrt{x} + 1)$

First   Last

b. $(5\sqrt{n} - 1)(4\sqrt{n} + 3) = (5\sqrt{n} - 1)(4\sqrt{n} + 3)$

Inner

Outer

$$= (5\sqrt{n})(4\sqrt{n}) + (5\sqrt{n})3 \qquad \text{Use the FOIL}$$
$$+ (-1)(4\sqrt{n}) + (-1)(3) \qquad \text{method.}$$
$$= 5 \cdot 4 \cdot \sqrt{n^2} + 5 \cdot 3\sqrt{n} - 4\sqrt{n} - 3 \quad \text{Multiply.}$$
$$= 20n + 15\sqrt{n} - 4\sqrt{n} - 3 \qquad \text{Simplify.}$$
$$= 20n + 11\sqrt{n} - 3 \qquad \text{Combine like terms.}$$

Recall from Section 5.6 that when multiplying the sum and difference of the same two terms, we can use the following formula:

$$(a + b)(a - b) = a^2 - b^2$$

One or more of the factors in this product may contain a radical, as in the next example.

EXAMPLE 3	PRACTICE 3

Find the product.

a. $(\sqrt{7} + \sqrt{3})(\sqrt{7} - \sqrt{3})$ **b.** $(\sqrt{6n} + 1)(\sqrt{6n} - 1)$

c. $(2\sqrt{x} + 3\sqrt{y})(2\sqrt{x} - 3\sqrt{y})$

Solution

a. $(\sqrt{7} + \sqrt{3})(\sqrt{7} - \sqrt{3}) = (\sqrt{7})^2 - (\sqrt{3})^2$
$$= 7 - 3$$
$$= 4$$

b. $(\sqrt{6n} + 1)(\sqrt{6n} - 1) = (\sqrt{6n})^2 - 1^2 = 6n - 1$

c. $(2\sqrt{x} + 3\sqrt{y})(2\sqrt{x} - 3\sqrt{y}) = (2\sqrt{x})^2 - (3\sqrt{y})^2 = 4x - 9y$

Multiply.

a. $(\sqrt{2} - \sqrt{5})(\sqrt{2} + \sqrt{5})$

b. $(\sqrt{2x} + 3)(\sqrt{2x} - 3)$

c. $(\sqrt{p} - 4\sqrt{q})(\sqrt{p} + 4\sqrt{q})$

When squaring binomials that contain radical terms, we can apply the formulas for squaring a binomial discussed in Section 5.6.

$$(a + b)^2 = a^2 + 2ab + b^2$$
$$(a - b)^2 = a^2 - 2ab + b^2$$

EXAMPLE 4	PRACTICE 4

Simplify.

a. $(\sqrt{5} + 3x)^2$ **b.** $(\sqrt{x + 1} - 1)^2$

Solution

a. $(\sqrt{5} + 3x)^2 = (\sqrt{5})^2 + 2(\sqrt{5})(3x) + (3x)^2$
$$= 5 + 6\sqrt{5}x + 9x^2$$
$$= 9x^2 + 6\sqrt{5}x + 5$$

Simplify.

a. $(\sqrt{3} + 2b)^2$

b. $(\sqrt{n + 1} - 3)^2$

b. $(\sqrt{x+1} - 1)^2 = (\sqrt{x+1})^2 - 2(1)(\sqrt{x+1}) + 1$

$$= x + 1 - 2\sqrt{x+1} + 1$$

$$= x - 2\sqrt{x+1} + 2$$

EXAMPLE 5	PRACTICE 5

EXAMPLE 5

In the right triangle shown, the length of hypotenuse AB is $\dfrac{2\sqrt{3}}{3}$ times the length of side AC. Find the length of AB expressed in simplest terms.

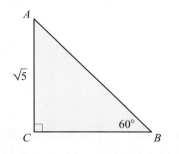

Solution

$$AB = \frac{2\sqrt{3}}{3} \cdot AC$$

$$= \frac{2\sqrt{3}}{3} \cdot \sqrt{5}$$

$$= \frac{2\sqrt{3}}{3} \cdot \frac{\sqrt{5}}{1}$$

$$= \frac{2\sqrt{3} \cdot \sqrt{5}}{3}$$

$$= \frac{2\sqrt{15}}{3}$$

So the length of AB is $\dfrac{2\sqrt{15}}{3}$.

PRACTICE 5

Find the area of the rectangle shown, expressed as a simplified radical.

$\sqrt{6}$ cm

$\sqrt{20}$ cm

Dividing Radical Expressions

Now, let's consider division. To divide radical expressions, we use the quotient rule of radicals discussed in Section 8.3:

$$\frac{\sqrt[n]{a}}{\sqrt[n]{b}} = \sqrt[n]{\frac{a}{b}}$$

When dividing nth roots, this rule allows us to bring the radicands under a single radical sign. We can then divide the radicands. This approach is particularly useful if their quotient happens to be a perfect power.

EXAMPLE 6	PRACTICE 6

EXAMPLE 6

Divide and simplify, if possible.

a. $\dfrac{3\sqrt{50}}{\sqrt{2}}$

b. $\dfrac{-x\sqrt[3]{24x^8}}{\sqrt[3]{6x^3}}$

PRACTICE 6

Find the quotient. Simplify, if possible.

a. $\dfrac{5\sqrt{12}}{\sqrt{3}}$

b. $\dfrac{\sqrt[3]{18p^7}}{-5p\sqrt[3]{2p^2}}$

Solution

a. $\dfrac{3\sqrt{50}}{\sqrt{2}} = 3\sqrt{\dfrac{50}{2}}$ Use the quotient rule of radicals.

$= 3\sqrt{25}$

$= 3 \cdot 5$

$= 15$

b. $\dfrac{-x\sqrt[3]{24x^8}}{\sqrt[3]{6x^3}} = -x\sqrt[3]{\dfrac{24x^8}{6x^3}}$

$= -x\sqrt[3]{4x^5}$

$= -x \cdot x\sqrt[3]{4x^2}$

$= -x^2\sqrt[3]{4x^2}$

A radical such as $\sqrt{\tfrac{1}{3}}$ is not considered to be simplified, since its radicand is in the form of a fraction. As we saw in Section 8.3, we can express the single radical as the quotient of two radicals by applying the quotient rule of radicals "in reverse":

$$\sqrt[n]{\dfrac{a}{b}} = \dfrac{\sqrt[n]{a}}{\sqrt[n]{b}}$$

Radicands with perfect powers in the denominator particularly lend themselves to this approach.

EXAMPLE 7	PRACTICE 7

Simplify.

a. $\sqrt{\dfrac{4p^{10}}{q^6}}$ **b.** $\sqrt[3]{\dfrac{5x^4}{8}}$

Simplify.

a. $\sqrt{\dfrac{y^2}{9x^8}}$

Solution

a. $\sqrt{\dfrac{4p^{10}}{q^6}} = \dfrac{\sqrt{4p^{10}}}{\sqrt{q^6}} = \dfrac{2p^5}{q^3}$

b. $\sqrt[3]{\dfrac{5x^4}{8}} = \dfrac{\sqrt[3]{5x^4}}{\sqrt[3]{8}} = \dfrac{x\sqrt[3]{5x}}{2}$

b. $\sqrt[3]{\dfrac{2n^5}{27}}$

Rationalizing the Denominator

Some radical expressions are written as fractions with radicals in their denominators:

$$\dfrac{1}{\sqrt{5}} \qquad \dfrac{2x}{\sqrt[3]{x}}$$

This type of expression, which can be difficult to evaluate without a calculator, is not considered to be simplified. However, we can always *rationalize the denominator* of such an expression, that is, rewrite the expression in an equivalent form that contains no radical in its denominator. To do this, we multiply the numerator and denominator by a factor that will make the radicand in the denominator a perfect power.

EXAMPLE 8	PRACTICE 8

EXAMPLE 8

Rationalize the denominator.

a. $\dfrac{2}{\sqrt{3}}$ **b.** $\dfrac{\sqrt{x}}{\sqrt{20}}$ **c.** $\dfrac{\sqrt{49y^6}}{\sqrt{12}}$ **d.** $\dfrac{\sqrt[3]{4a^2b}}{\sqrt[3]{9c}}$

Solution

a. $\dfrac{2}{\sqrt{3}} = \dfrac{2}{\sqrt{3}} \cdot \dfrac{\sqrt{3}}{\sqrt{3}}$ Multiply the numerator and denominator by $\sqrt{3}$.

$\quad = \dfrac{2\sqrt{3}}{\sqrt{3^2}}$

$\quad = \dfrac{2\sqrt{3}}{3}$

b. $\dfrac{\sqrt{x}}{\sqrt{20}} = \dfrac{\sqrt{x}}{\sqrt{20}} \cdot \dfrac{\sqrt{20}}{\sqrt{20}} = \dfrac{\sqrt{20x}}{\sqrt{20^2}} = \dfrac{2\sqrt{5x}}{20} = \dfrac{\sqrt{5x}}{10}$

c. $\dfrac{\sqrt{49y^6}}{\sqrt{12}} = \dfrac{7y^3}{\sqrt{12}}$

$\quad = \dfrac{7y^3}{\sqrt{12}} \cdot \dfrac{\sqrt{12}}{\sqrt{12}}$

$\quad = \dfrac{7y^3\sqrt{12}}{\sqrt{12^2}}$

$\quad = \dfrac{7y^3(2\sqrt{3})}{12}$

$\quad = \dfrac{7y^3\sqrt{3}}{6}$

d. $\dfrac{\sqrt[3]{4a^2b}}{\sqrt[3]{9c}} = \dfrac{\sqrt[3]{4a^2b}}{\sqrt[3]{3^2c}} \cdot \dfrac{\sqrt[3]{3c^2}}{\sqrt[3]{3c^2}} = \dfrac{\sqrt[3]{12a^2bc^2}}{\sqrt[3]{(3c)^3}} = \dfrac{\sqrt[3]{12a^2bc^2}}{3c}$

Note that we multiplied the numerator and the denominator by $\sqrt[3]{3c^2}$ to make the radicand in the denominator a perfect cube.

PRACTICE 8

Simplify.

a. $\dfrac{4}{\sqrt{7}}$

b. $\dfrac{\sqrt{y}}{\sqrt{32}}$

c. $\dfrac{\sqrt{4p^4}}{\sqrt{18}}$

d. $\dfrac{\sqrt[3]{5pq}}{\sqrt[3]{2r^2}}$

When the radicand of a radical expression is a fraction, we can simplify the radical by first applying the quotient rule and then rationalizing the denominator.

EXAMPLE 9	PRACTICE 9

EXAMPLE 9

Simplify.

a. $\sqrt{\dfrac{2a}{5b}}$ **b.** $\sqrt[3]{\dfrac{m}{2}}$

PRACTICE 9

Simplify.

a. $\sqrt{\dfrac{x}{3y}}$ **b.** $\sqrt[3]{\dfrac{5}{a}}$

Solution

a. $\sqrt{\dfrac{2a}{5b}} = \dfrac{\sqrt{2a}}{\sqrt{5b}}$

$= \dfrac{\sqrt{2a}}{\sqrt{5b}} \cdot \dfrac{\sqrt{5b}}{\sqrt{5b}}$

$= \dfrac{\sqrt{10ab}}{5b}$

b. $\sqrt[3]{\dfrac{m}{2}} = \dfrac{\sqrt[3]{m}}{\sqrt[3]{2}}$

$= \dfrac{\sqrt[3]{m}}{\sqrt[3]{2}} \cdot \dfrac{\sqrt[3]{2^2}}{\sqrt[3]{2^2}}$

$= \dfrac{\sqrt[3]{4m}}{\sqrt[3]{2^3}}$

$= \dfrac{\sqrt[3]{4m}}{2}$

Tip Remember that a radical expression is considered simplified if

- no radicand has a factor that is a perfect nth power,
- no radicand contains a fraction, and
- the expression contains no denominator with a radical.

Some radical expressions are in the form of fractions, with more than one term in the numerator. Any denominator that contains a radical can be rationalized.

EXAMPLE 10

Rationalize the denominator.

a. $\dfrac{4\sqrt{10} - \sqrt{6}}{\sqrt{2}}$ **b.** $\dfrac{\sqrt{y} - 2}{\sqrt{y}}$

Solution

a. $\dfrac{4\sqrt{10} - \sqrt{6}}{\sqrt{2}} = \dfrac{4\sqrt{10} - \sqrt{6}}{\sqrt{2}} \cdot \dfrac{\sqrt{2}}{\sqrt{2}}$

$= \dfrac{(4\sqrt{10} - \sqrt{6})\sqrt{2}}{2}$

$= \dfrac{4\sqrt{10} \cdot \sqrt{2} - \sqrt{6} \cdot \sqrt{2}}{2}$

$= \dfrac{4\sqrt{20} - \sqrt{12}}{2}$

$= \dfrac{4(2\sqrt{5}) - 2\sqrt{3}}{2}$

$= 2(2\sqrt{5}) - \sqrt{3}$

$= 4\sqrt{5} - \sqrt{3}$

b. $\dfrac{\sqrt{y} - 2}{\sqrt{y}} = \dfrac{\sqrt{y} - 2}{\sqrt{y}} \cdot \dfrac{\sqrt{y}}{\sqrt{y}}$

$= \dfrac{(\sqrt{y} - 2)\sqrt{y}}{y}$

$= \dfrac{\sqrt{y} \cdot \sqrt{y} - 2 \cdot \sqrt{y}}{y}$

$= \dfrac{y - 2\sqrt{y}}{y}$

PRACTICE 10

Express in simplified form.

a. $\dfrac{\sqrt{6} - 9}{\sqrt{3}}$

b. $\dfrac{5 + \sqrt{b}}{\sqrt{b}}$

Can you solve the problem in Example 10(a) another way?

So far, we have rationalized denominators consisting of a single radical term. But suppose that the radical expression in a denominator contains two terms. Here we consider the case in which the denominator involves one or more square roots, as in $\dfrac{1}{\sqrt{x}+5}$. The key to rationalizing such a denominator is to identify its *conjugate*. The expressions $a+b$ and $a-b$ are called conjugates of one another. When multiplying a pair of conjugates, we can apply the formula for the difference of squares:

$$(a+b)(a-b) = a^2 - b^2$$

Note that in Example 3(a), we multiplied two conjugates, $(\sqrt{7}+\sqrt{3})$ and $(\sqrt{7}-\sqrt{3})$, and found the product to be 4. Can you explain why there is no radical sign in this product?

The elimination of radical signs in the product of conjugates suggests a procedure for rationalizing a denominator with two terms, namely, *multiplying both the numerator and the denominator by the conjugate of the denominator.*

EXAMPLE 11

Rationalize the denominator.

a. $\dfrac{6}{1-\sqrt{3}}$ **b.** $\dfrac{x}{\sqrt{y}+\sqrt{z}}$

Solution

a. We multiply the numerator and the denominator by the conjugate of the denominator.

$$\frac{6}{1-\sqrt{3}} = \frac{6}{1-\sqrt{3}} \cdot \frac{1+\sqrt{3}}{1+\sqrt{3}}$$
$$= \frac{6(1+\sqrt{3})}{(1-\sqrt{3})(1+\sqrt{3})}$$
$$= \frac{6+6\sqrt{3}}{1^2 - (\sqrt{3})^2}$$
$$= \frac{6+6\sqrt{3}}{1-3}$$
$$= \frac{2(3+3\sqrt{3})}{-2}$$
$$= -3 - 3\sqrt{3}$$

b. $\dfrac{x}{\sqrt{y}+\sqrt{z}} = \dfrac{x}{\sqrt{y}+\sqrt{z}} \cdot \dfrac{\sqrt{y}-\sqrt{z}}{\sqrt{y}-\sqrt{z}}$
$$= \frac{x(\sqrt{y}-\sqrt{z})}{(\sqrt{y}+\sqrt{z})(\sqrt{y}-\sqrt{z})}$$
$$= \frac{x\sqrt{y}-x\sqrt{z}}{y-z}$$

PRACTICE 11

Express in simplified form.

a. $\dfrac{4}{2+\sqrt{2}}$

b. $\dfrac{a}{\sqrt{b}-\sqrt{c}}$

EXAMPLE 12	PRACTICE 12

EXAMPLE 12

If $f(x) = 10x\sqrt[4]{16x^3}$ and $g(x) = 2x\sqrt[4]{x^3}$, find

a. $f(x) \cdot g(x)$ **b.** $\dfrac{f(x)}{g(x)}$

Solution

a. $f(x) \cdot g(x) = 10x\sqrt[4]{16x^3} \cdot 2x\sqrt[4]{x^3}$

$= 20x^2\sqrt[4]{16x^6}$

$= 20x^2\sqrt[4]{16x^4 \cdot x^2}$ **Factor out a perfect fourth power.**

$= 20x^2 \cdot 2x\sqrt[4]{x^2}$

$= 40x^3\sqrt[4]{x^2}$

b. $\dfrac{f(x)}{g(x)} = \dfrac{10x\sqrt[4]{16x^3}}{2x\sqrt[4]{x^3}} = 5\sqrt[4]{\dfrac{16x^3}{x^3}} = 5\sqrt[4]{16} = 5 \cdot 2 = 10$

PRACTICE 12

If $p(x) = x\sqrt[3]{81x}$ and $q(x) = \sqrt[3]{24x^4}$, determine

a. $p(x) \cdot q(x)$

b. $\dfrac{p(x)}{q(x)}$

EXAMPLE 13

The expression $\sqrt{64h}$ can be used to determine the velocity of a free-falling object in feet per second when the object has fallen h ft.

a. Find the velocity of a ball after it has fallen 5 ft and after it has fallen 10 ft.

b. What is the ratio of the greater velocity to the smaller velocity?

5 ft

10 ft

Solution

a. When the ball has fallen 5 ft, its velocity is

$$\sqrt{64 \cdot 5} = 8\sqrt{5}, \text{ or } 8\sqrt{5} \text{ feet per second}$$

When the ball has fallen 10 ft, the velocity is

$$\sqrt{64 \cdot 10} = 8\sqrt{10}, \text{ or } 8\sqrt{10} \text{ feet per second}$$

b. The ratio of the greater velocity to the smaller velocity is

$$\frac{8\sqrt{10}}{8\sqrt{5}} = \frac{\sqrt{10}}{\sqrt{5}}$$

$$= \sqrt{\frac{10}{5}}$$

$$= \sqrt{2}$$

$$\approx 1.414$$

PRACTICE 13

The following diagram shows a sphere of volume V.

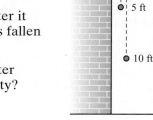

The radius of the sphere can be modeled by the expression $\dfrac{\sqrt[3]{3V}}{\sqrt[3]{4\pi}}$.

a. Rewrite this expression, rationalizing the denominator.

b. If the volume is doubled, is the radius doubled?

Mathematically Speaking

Fill in each blank with the most appropriate term or phrase from the given list.

numerator	denominator	squaring a binomial
perfect power	constant	multiply
product	rationalize	
commutative	distributive	
the sum and difference of the same two terms	quotient	

1. To multiply radical expressions containing more than one term, we use the _____ property.

2. When squaring binomials that contain radical terms, apply the formulas for _____.

3. To simplify a radical that has its radicand in the form of a fraction, we can use the _____ rule of radicals "in reverse."

4. To _____ the denominator means to rewrite an expression that has a radical in its denominator as an equivalent expression that does not have one.

5. A denominator can be rationalized by multiplying the numerator and denominator by a factor that makes the radicand in the denominator a(n) _____.

6. To rationalize a denominator with two terms, multiply both the numerator and the denominator by the conjugate of the _____.

Multiply and simplify.

7. $\sqrt{12} \cdot \sqrt{8}$

8. $\sqrt{6} \cdot \sqrt{3}$

9. $(-4\sqrt{3})(\sqrt{7})$

10. $(\sqrt{11})(-6\sqrt{2})$

11. $(5\sqrt[3]{6})(3\sqrt[3]{9})$

12. $(2\sqrt[3]{4})(4\sqrt[3]{14})$

13. $(2\sqrt{10x})(7\sqrt{5x^5})$

14. $(4\sqrt{12y^5})(3\sqrt{6y^3})$

15. $(8\sqrt{ab^3})(-2\sqrt{a^3b})$

16. $(-6\sqrt{p^3q^5})(-5\sqrt{pq})$

17. $\sqrt[3]{12x^2y} \cdot \sqrt[3]{-16xy^4}$

18. $-\sqrt[4]{8x^3y^2} \cdot \sqrt[4]{50x^2y^6}$

19. $\sqrt{2}(\sqrt{8} - 4)$

20. $\sqrt{5}(\sqrt{20} + 3)$

21. $\sqrt{6}(2\sqrt{3} + \sqrt{12})$

22. $\sqrt{3}(\sqrt{8} - 5\sqrt{2})$

23. $-2\sqrt{3}(2\sqrt{5} - 6\sqrt{3})$

24. $-3\sqrt{2}(2\sqrt{7} - 4\sqrt{2})$

25. $\sqrt[3]{4}(5\sqrt[3]{12} + 2\sqrt[3]{3})$

26. $-\sqrt[4]{8}(\sqrt[4]{8} - 7\sqrt[4]{5})$

27. $\sqrt{x}(\sqrt{x^3} + \sqrt{2x})$

28. $\sqrt{y}(\sqrt{3y} - \sqrt{y})$

29. $\sqrt[4]{a^2}(3\sqrt[4]{2a^3} - \sqrt[4]{10a^2})$

30. $\sqrt[3]{2x^2}(7\sqrt[3]{4x} - \sqrt[3]{5})$

31. $(\sqrt{2} - 3)(\sqrt{2} + 4)$

32. $(\sqrt{5} + 1)(\sqrt{5} - 6)$

33. $(2 - 4\sqrt{3})(4 + 3\sqrt{3})$

34. $(-2\sqrt{5} + 1)(2\sqrt{5} - 1)$

35. $(\sqrt{8} + \sqrt{3})(\sqrt{2} + \sqrt{12})$

36. $(\sqrt{5} + \sqrt{10})(\sqrt{10} + \sqrt{20})$

37. $(2\sqrt{r} - 4)(8\sqrt{r} + 6)$

38. $(2\sqrt{p} + 7)(3\sqrt{p} - 5)$

39. $(\sqrt{6} + \sqrt{2})(\sqrt{6} - \sqrt{2})$

40. $(\sqrt{8} + \sqrt{3})(\sqrt{8} - \sqrt{3})$

41. $(\sqrt{x} - 8)(\sqrt{x} + 8)$

42. $(\sqrt{n} + 11)(\sqrt{n} - 11)$

43. $(\sqrt{5x} + \sqrt{y})(\sqrt{5x} - \sqrt{y})$

44. $(\sqrt{7q} - \sqrt{p})(\sqrt{7q} + \sqrt{p})$

45. $(\sqrt{x - 1} - 5)(\sqrt{x - 1} + 5)$

46. $(\sqrt{y + 3} - 4)(\sqrt{y + 3} + 4)$

47. $(3\sqrt{x} - 2)^2$

48. $(3 - 2\sqrt{a})^2$

49. $(\sqrt{6a} + 4)^2$

50. $(\sqrt{8a} + 2)^2$

51. $(1 - \sqrt{n + 7})^2$

52. $(3 - \sqrt{x - 6})^2$

Divide and simplify.

53. $-\dfrac{\sqrt{32}}{4\sqrt{2}}$

54. $-\dfrac{\sqrt{108}}{2\sqrt{3}}$

55. $\dfrac{\sqrt{84x^3}}{\sqrt{7x}}$

56. $\dfrac{\sqrt{48a^4}}{\sqrt{3a^3}}$

57. $-\dfrac{n\sqrt{60n^7}}{2\sqrt{5n^3}}$

58. $-\dfrac{y\sqrt{108y^5}}{3\sqrt{6y}}$

59. $\dfrac{\sqrt[4]{128r^{10}}}{6r\sqrt[4]{4r^3}}$

60. $\dfrac{\sqrt[3]{256x^8}}{8x\sqrt[3]{2x^4}}$

61. $\sqrt{\dfrac{5a}{16}}$

62. $\sqrt{\dfrac{2x}{49}}$

63. $\sqrt{\dfrac{12x}{y^6}}$

64. $\sqrt{\dfrac{18u^2}{v^4}}$

65. $\sqrt[3]{\dfrac{54}{n^9}}$

66. $\sqrt[3]{\dfrac{r^5}{64}}$

67. $\sqrt{\dfrac{32a^5}{9b^8}}$

68. $\sqrt{\dfrac{75p^3}{81q^2}}$

Simplify.

69. $\dfrac{4}{\sqrt{5}}$

70. $\dfrac{3}{\sqrt{6}}$

71. $\dfrac{\sqrt{y}}{\sqrt{40}}$

72. $\dfrac{\sqrt{u}}{\sqrt{63}}$

73. $\dfrac{\sqrt{3a}}{\sqrt{18}}$

74. $\dfrac{\sqrt{5p}}{\sqrt{24}}$

75. $\dfrac{\sqrt{9u}}{6\sqrt{v}}$

76. $\dfrac{\sqrt{36p}}{4\sqrt{q}}$

77. $\dfrac{\sqrt{25x^4}}{\sqrt{3y}}$

78. $\dfrac{\sqrt{64a^8}}{\sqrt{5b}}$

79. $\dfrac{\sqrt{10xy}}{\sqrt{9z}}$

80. $\dfrac{\sqrt{7pq}}{\sqrt{4r}}$

81. $\dfrac{\sqrt[3]{9x}}{\sqrt[3]{2y^2}}$

82. $\dfrac{\sqrt[3]{2a}}{\sqrt[3]{3b}}$

83. $\sqrt{\dfrac{11}{x}}$

84. $\sqrt{\dfrac{3}{n}}$

85. $\sqrt{\dfrac{7x}{3y}}$

86. $\sqrt{\dfrac{5p}{2q}}$

87. $\sqrt{\dfrac{5x^3}{48y^3}}$

88. $\sqrt{\dfrac{3u^4}{98v^5}}$

89. $\sqrt[3]{\dfrac{a^2b}{32c^4}}$

90. $\sqrt[3]{\dfrac{p^3q^2}{81r^2}}$

91. $\dfrac{2 - \sqrt{3}}{\sqrt{6}}$

92. $\dfrac{\sqrt{2} - 1}{\sqrt{10}}$

93. $\dfrac{\sqrt{a} - \sqrt{b}}{\sqrt{b}}$

94. $\dfrac{\sqrt{x} + \sqrt{y}}{\sqrt{x}}$

95. $\dfrac{\sqrt{5} + 10\sqrt{t}}{\sqrt{15t}}$

96. $\dfrac{4\sqrt{3n} - \sqrt{2}}{\sqrt{6n}}$

97. $\dfrac{\sqrt[3]{x} - 4}{\sqrt[3]{x^2}}$

98. $\dfrac{3 - \sqrt[3]{a^2}}{\sqrt[3]{a}}$

Rationalize the denominator.

99. $\dfrac{1}{2 + \sqrt{2}}$

100. $\dfrac{3}{\sqrt{5} + 1}$

101. $\dfrac{6}{\sqrt{2} - \sqrt{5}}$

102. $\dfrac{12}{\sqrt{3} - \sqrt{6}}$

103. $\dfrac{8}{2 + \sqrt{2x}}$

104. $\dfrac{9}{6 - \sqrt{3y}}$

105. $\dfrac{\sqrt{x}}{\sqrt{x} + y}$

106. $\dfrac{\sqrt{b}}{a - \sqrt{b}}$

107. $\dfrac{\sqrt{a} + 3}{\sqrt{a} - \sqrt{2}}$

108. $\dfrac{\sqrt{n} - 1}{\sqrt{n} - \sqrt{5}}$

109. $\dfrac{\sqrt{x} - \sqrt{y}}{\sqrt{x} + \sqrt{y}}$

110. $\dfrac{\sqrt{p} + \sqrt{q}}{\sqrt{p} - \sqrt{q}}$

111. $\dfrac{2\sqrt{a} + 3\sqrt{b}}{3\sqrt{a} - 2\sqrt{b}}$

112. $\dfrac{4\sqrt{x} + \sqrt{y}}{\sqrt{x} - 4\sqrt{y}}$

For the given functions $f(x)$ and $g(x)$, find $f(x) \cdot g(x)$ and $\dfrac{f(x)}{g(x)}$.

113. $f(x) = 3x\sqrt{2x}$ and $g(x) = \dfrac{1}{3}\sqrt{6x}$

114. $f(x) = 4\sqrt[3]{9x^2}$ and $g(x) = x\sqrt[3]{12x^2}$

115. $f(x) = \sqrt{x} + 1$ and $g(x) = \sqrt{x} - 1$

116. $f(x) = \sqrt{x} + 2$ and $g(x) = \sqrt{x} + 4$

Mixed Practice

Simplify.

117. $(\sqrt{12} + \sqrt{3})(\sqrt{3} + \sqrt{6})$

118. $\dfrac{3\sqrt{2r} - \sqrt{5}}{\sqrt{10r}}$

119. $(-8\sqrt{ab^3})(7\sqrt{a^5b})$

120. $-\sqrt[3]{9}(4\sqrt[3]{18} - 7\sqrt[3]{15})$

121. $\sqrt{\dfrac{75y}{z^4}}$

122. $(\sqrt{x + 6} + 3)(\sqrt{x + 6} - 3)$

123. $\sqrt{\dfrac{5x}{7y}}$

124. $\dfrac{\sqrt{81m^6}}{\sqrt{5n}}$

Solve.

125. Rationalize the denominator: $\dfrac{\sqrt{p} - 4}{\sqrt{p} - \sqrt{3}}$

126. For the functions $f(x) = 2x\sqrt{7x}$ and $g(x) = \dfrac{1}{2}\sqrt{14x}$, find $f(x) \cdot g(x)$ and $\dfrac{f(x)}{g(x)}$.

Applications

Solve.

127. The altitude of an equilateral triangle with side length a bisects the base of the triangle, as shown in the diagram. (The area of a triangle equals one-half the product of its base and height.)

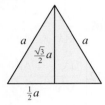

    a. Write a formula for the area, A, of an equilateral triangle.

    b. Use your formula from part (a) to find the area of a wall tile in the shape of an equilateral triangle with side length 8 in.

129. The period of a simple pendulum L feet long can be found using the expression $2\pi\sqrt{\dfrac{L}{32}}$. Simplify the expression by rationalizing the denominator.

131. The radius of a sphere with a surface area of A is given by $\sqrt{\dfrac{A}{4\pi}}$. Simplify the expression by rationalizing the denominator.

128. The length of the diagonal of a square with side length s is $s\sqrt{2}$. A circle is circumscribed about a square as shown in the following figure. (The area of a circle with radius r is πr^2.)

    a. Express the area, A, of the circle in terms of s.

    b. Use your answer from part (a) to find the area of a circle circumscribed about a square with side length 4 in.

130. The formula for the distance a person can see on any planet is given by the expression $\dfrac{\sqrt{rh}}{4\sqrt{165}}$, where r is the radius of the planet (in miles) and h is the height of the person (in feet). Simplify the expression by rationalizing the denominator.

132. The annual rate of return, r, on an initial investment of P dollars whose value is V dollars after 3 yr is given by the following expression:

$$\frac{\sqrt[3]{V} - \sqrt[3]{P}}{\sqrt[3]{P}}$$

Rationalize the denominator.

● *Check your answers on page A-36.*

MINDSTRETCHERS

Critical Thinking

1. Some equations have solutions that contain radicals. Show that $3 - 2\sqrt{2}$ is a solution of the equation $x^2 - 6x + 1 = 0$.

Groupwork

2. Working with a partner, simplify each of the following expressions:

a. $\dfrac{\dfrac{x}{\sqrt{y}}}{\dfrac{y}{\sqrt{x}}}$

b. $\dfrac{\sqrt{x} - \dfrac{1}{\sqrt{x}}}{\sqrt{x} + \dfrac{1}{\sqrt{x}}}$

c. $\dfrac{\dfrac{1}{1 - \sqrt{x}} + \dfrac{1}{1 + \sqrt{x}}}{\dfrac{1}{1 + \sqrt{x}} - \dfrac{1}{1 - \sqrt{x}}}$

Technology

3. Consider the function $f(x) = \sqrt{x + 2} \cdot \sqrt{x - 2}$.

a. Using a grapher, graph $f(x)$. To do so, press the $\boxed{Y =}$ key and enter $\sqrt{x + 2} \cdot \sqrt{x - 2}$ to the right of **\Y1 =**. Press $\boxed{\text{GRAPH}}$ to display the graph of the function.

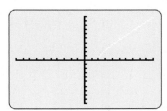

b. After examining the graph, identify the domain of this function. Justify your answer.

8.6   Solving Radical Equations

Next we consider equations containing radicals.

Definition

A **radical equation** is an equation in which a variable appears in one or more radicands.

Some examples of radical equations are

$$\sqrt{5x} = 18 \qquad \sqrt{3n} = \sqrt{2n - 1} \qquad \sqrt[3]{y + 2} - 6y = 5$$

Note that $x = \sqrt{5}$ is *not* a radical equation, since there is no variable under the radical sign.

In solving radical equations, we can use the following property of equality: If $a = b$, then $a^n = b^n$, where n is a positive integer. To use this property to solve radical equations, we first isolate the radical and then raise each side of the equation to the *same power* in order to produce a new equation that contains no radicals. When each side of the equation is raised to the same power, the solutions of the original equation are among the solutions of the new equation.

Consider, for instance, the equation $\sqrt{x} = 3$. If we square each side of the equation, that is, raise each side to the second power, we get

$$\sqrt{x} = 3$$
$$(\sqrt{x})^2 = (3)^2 \qquad \text{Square each side of the equation.}$$
$$x = 9 \qquad \text{Simplify.}$$

When raising each side of the equation to the same power, the resulting equation may have one or more solutions that do not satisfy the original equation, that is, extraneous solutions. So when solving a radical equation, it is particularly important to check all possible solutions in the *original* equation.

For instance, we can check the solution of the equation in the preceding example by substituting 9 for x in the original equation.

$$\sqrt{x} = 3$$
$$\sqrt{9} \stackrel{?}{=} 3$$
$$3 = 3 \qquad \text{True}$$

So the solution of the equation is 9.

Let's consider some examples.

EXAMPLE 1	**PRACTICE 1**
Solve and check: $\sqrt{x} - 3 = 7$	Solve and check: $\sqrt{y + 2} = 10$

Solution

$$\sqrt{x} - 3 = 7$$
$$\sqrt{x} - 3 + 3 = 7 + 3 \qquad \text{Add 3 to each side of the equation.}$$
$$\sqrt{x} = 10 \qquad \text{Simplify.}$$

$$(\sqrt{x})^2 = (10)^2 \quad \text{Square each side of the equation.}$$
$$x = 100$$

Check

$$\sqrt{x} - 3 = 7$$
$$\sqrt{100} - 3 \overset{?}{=} 7 \quad \text{Substitute 100 for } x.$$
$$10 - 3 \overset{?}{=} 7$$
$$7 = 7 \quad \text{True}$$

So 100 is the solution.

Some radical equations involve cube roots, or, more generally, radicals with an index greater than 2.

EXAMPLE 2

Solve and check: $\sqrt[3]{n - 5} = 2$

Solution To solve this equation, we cube each side of the equation, that is, raise each side of the equation to the third power.

$$\sqrt[3]{n - 5} = 2$$
$$(\sqrt[3]{n - 5})^3 = (2)^3 \quad \text{Cube each side of the equation.}$$
$$n - 5 = 8$$
$$n = 13$$

Check

$$\sqrt[3]{n - 5} = 2$$
$$\sqrt[3]{13 - 5} \overset{?}{=} 2$$
$$\sqrt[3]{8} \overset{?}{=} 2$$
$$2 = 2 \quad \text{True}$$

So 13 is the solution.

PRACTICE 2

Solve and check: $\sqrt[3]{2x + 7} = 3$

Tip When solving a radical equation, raise each side of the equation to a power that is equal to the index of the radical in the equation.

EXAMPLE 3

Solve and check: $\sqrt{x + 4} + 2 = 0$

Solution

$$\sqrt{x + 4} + 2 = 0$$
$$\sqrt{x + 4} = -2$$
$$(\sqrt{x + 4})^2 = (-2)^2$$
$$x + 4 = 4$$
$$x = 0$$

PRACTICE 3

Solve and check: $\sqrt{3t + 1} + 4 = 0$

Check

$$\sqrt{x + 4} + 2 = 0$$

$$\sqrt{0 + 4} + 2 \overset{?}{=} 0 \qquad \text{Substitute 0 for } x.$$

$$\sqrt{4} + 2 \overset{?}{=} 0$$

$$2 + 2 \overset{?}{=} 0$$

$$4 = 0 \qquad \text{False}$$

Since our check fails, 0 is *not* a solution to the original equation. Note that 0 is an extraneous solution. Therefore, the original equation has no solution.

Some radical equations contain more than one radical. We solve by first isolating either of the radicals.

EXAMPLE 4

Solve and check: $\sqrt{2x + 11} - \sqrt{4x + 1} = 0$

Solution

$$\sqrt{2x + 11} - \sqrt{4x + 1} = 0$$

$$\sqrt{2x + 11} = \sqrt{4x + 1} \qquad \text{Isolate the radical on the left.}$$

$$\left(\sqrt{2x + 11}\right)^2 = \left(\sqrt{4x + 1}\right)^2 \qquad \text{Square each side of the equation.}$$

$$2x + 11 = 4x + 1 \qquad \text{Simplify.}$$

$$2x = 10$$

$$x = 5$$

Check

$$\sqrt{2x + 11} - \sqrt{4x + 1} = 0$$

$$\sqrt{2(5) + 11} - \sqrt{4(5) + 1} \overset{?}{=} 0 \qquad \text{Substitute 5 for } x.$$

$$\sqrt{21} - \sqrt{21} \overset{?}{=} 0$$

$$0 = 0 \qquad \text{True}$$

So 5 is the solution.

PRACTICE 4

Solve and check:
$$\sqrt{4n + 5} - \sqrt{7n - 4} = 0$$

In some radical equations, we must raise each side of the equation to a power more than once.

EXAMPLE 5

Solve and check: $\sqrt{2x} - \sqrt{x-2} = 2$

Solution

$\sqrt{2x} - \sqrt{x-2} = 2$

$\sqrt{2x} = \sqrt{x-2} + 2$ Isolate one of the radicals.

$(\sqrt{2x})^2 = (\sqrt{x-2} + 2)^2$ Square each side of the equation.

$2x = (x-2) + 2 \cdot 2\sqrt{x-2} + 2 \cdot 2$ Simplify.

$2x = x + 2 + 4\sqrt{x-2}$ Combine like terms.

$2x - (x+2) = x + 2 + 4\sqrt{x-2} - (x+2)$ Subtract $(x+2)$ from each side of the equation.

$x - 2 = 4\sqrt{x-2}$ Simplify.

$(x-2)^2 = (4\sqrt{x-2})^2$ Square each side of the equation.

$x^2 - 4x + 4 = 16(x-2)$ Simplify.

$x^2 - 4x + 4 = 16x - 32$ Use the distributive property.

$x^2 - 20x + 36 = 0$ Write in standard form.

$(x-18)(x-2) = 0$ Factor.

$x - 18 = 0 \quad \text{or} \quad x - 2 = 0$

$\qquad x = 18 \qquad\qquad x = 2$

Check

Substitute 18 for x.

$\sqrt{2x} - \sqrt{x-2} = 2$

$\sqrt{2(18)} - \sqrt{18 - 2} \overset{?}{=} 2$

$\sqrt{36} - \sqrt{16} \overset{?}{=} 2$

$6 - 4 \overset{?}{=} 2$

$2 = 2$ True

Substitute 2 for x.

$\sqrt{2x} - \sqrt{x-2} = 2$

$\sqrt{2(2)} - \sqrt{2 - 2} \overset{?}{=} 2$

$\sqrt{4} - \sqrt{0} \overset{?}{=} 2$

$2 - 0 \overset{?}{=} 2$

$2 = 2$ True

So both 18 and 2 are solutions.

PRACTICE 5

Solve and check:

$\sqrt{2y+3} = \sqrt{y-2} + 2$

The preceding examples lead us to the following procedure for solving radical equations.

To Solve a Radical Equation

- Isolate a term with a radical.
- Raise each side of the equation to a power equal to the index of the radical.
- If the equation still contains a radical, repeat the preceding steps.
- Where possible, combine like terms.
- Solve the resulting equation.
- Check the possible solution(s) in the original equation.

EXAMPLE 6

Solve and check: $\sqrt[3]{x^2 - 5} + \sqrt[3]{2x - 3} = 0$

Solution

$$\sqrt[3]{x^2 - 5} + \sqrt[3]{2x - 3} = 0$$

$\sqrt[3]{x^2 - 5} = -\sqrt[3]{2x - 3}$ Isolate one of the radicals.

$(\sqrt[3]{x^2 - 5})^3 = (-\sqrt[3]{2x - 3})^3$ Cube each side of the equation.

$x^2 - 5 = -(2x - 3)$ Simplify.

$x^2 - 5 = -2x + 3$ Use the distributive property.

$x^2 + 2x - 8 = 0$ Write in standard form.

$(x - 2)(x + 4) = 0$ Factor.

$x - 2 = 0$ or $x + 4 = 0$

$x = 2$ $x = -4$

Check

Substitute 2 for x.

$$\sqrt[3]{x^2 - 5} + \sqrt[3]{2x - 3} = 0$$
$$\sqrt[3]{(2)^2 - 5} + \sqrt[3]{2(2) - 3} \overset{?}{=} 0$$
$$\sqrt[3]{4 - 5} + \sqrt[3]{4 - 3} \overset{?}{=} 0$$
$$\sqrt[3]{-1} + \sqrt[3]{1} \overset{?}{=} 0$$
$$-1 + 1 \overset{?}{=} 0$$
$$0 = 0 \quad \text{True}$$

Substitute -4 for x.

$$\sqrt[3]{x^2 - 5} + \sqrt[3]{2x - 3} = 0$$
$$\sqrt[3]{(-4)^2 - 5} + \sqrt[3]{2(-4) - 3} \overset{?}{=} 0$$
$$\sqrt[3]{16 - 5} + \sqrt[3]{-8 - 3} \overset{?}{=} 0$$
$$\sqrt[3]{11} + \sqrt[3]{-11} \overset{?}{=} 0$$
$$\sqrt[3]{11} + \sqrt[3]{(-1)11} \overset{?}{=} 0$$
$$0 = 0 \quad \text{True}$$

So both 2 and -4 are solutions.

PRACTICE 6

Solve and check:

$$\sqrt[3]{n^2 + 8} + \sqrt[3]{4 - 7n} = 0$$

EXAMPLE 7

Solve and check: $\sqrt{x + 1} + 1 = -x$

Solution

$$\sqrt{x + 1} + 1 = -x$$

$\sqrt{x + 1} = -x - 1$ Isolate the radical.

$(\sqrt{x + 1})^2 = (-x - 1)^2$ Square each side of the equation.

$x + 1 = (-x)^2 - (-2x) + (-1)^2$

$x + 1 = x^2 + 2x + 1$

$x + 1 + (-x - 1) = x^2 + 2x + 1 + (-x - 1)$ Add $(-x - 1)$ to each side of the equation.

$0 = x^2 + x$

$x(x + 1) = 0$ Factor.

$x = 0$ or $x + 1 = 0$

$x = -1$

PRACTICE 7

Solve and check: $\sqrt{y + 4} - y = 2$

Check

Substitute 0 for x.

$$\sqrt{x+1} + 1 = -x$$
$$\sqrt{0+1} + 1 \stackrel{?}{=} 0$$
$$1 + 1 \stackrel{?}{=} 0$$
$$2 = 0 \quad \text{False}$$

Substitute -1 for x.

$$\sqrt{x+1} + 1 = -x$$
$$\sqrt{-1+1} + 1 \stackrel{?}{=} -(-1)$$
$$0 + 1 \stackrel{?}{=} 1$$
$$1 = 1 \quad \text{True}$$

So the solution is -1.

Recall from Section 2.4 our discussion about solving a formula for one variable. Some of these formulas contain radicals.

EXAMPLE 8

The period of a pendulum is the time that it takes for a pendulum to swing back and forth. A pendulum L ft long will have a period of t sec, where

$$t = 2\pi\sqrt{\frac{L}{32}}$$

Rewrite this formula by solving for L.
(*Source:* Peter J. Nolan, *Fundamentals of College Physics*, 1993)

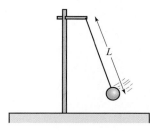

Solution

$$t = 2\pi\sqrt{\frac{L}{32}}$$

$$\frac{t}{2\pi} = \sqrt{\frac{L}{32}}$$

$$\left(\frac{t}{2\pi}\right)^2 = \left(\sqrt{\frac{L}{32}}\right)^2$$

$$\frac{t^2}{4\pi^2} = \frac{L}{32}$$

$$(32)\left(\frac{t^2}{4\pi^2}\right) = L$$

$$L = \frac{8t^2}{\pi^2}$$

PRACTICE 8

For an electrical appliance, the formula

$$I = \sqrt{\frac{P}{R}}$$

shows the relationship between its resistance, R, the amount of current, I, that it draws, and the power, P, that it consumes. Rewrite this formula by solving for P. (*Source:* Peter J. Nolan, *Fundamentals of College Physics*, 1993)

EXAMPLE 9

The formula

$$r = \sqrt{\dfrac{A}{4\pi}}$$

gives the radius of a sphere, r, in terms of the surface area, A, of the sphere.

a. Rewrite this formula by solving for A.

b. If the shape of the Earth is approximately spherical with a radius of 3964 mi, find its surface area to the nearest 100 million mi². (Use 3.14 for π.) (*Source: The World Almanac and Book of Facts 2000*)

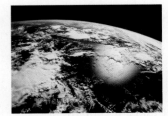

Solution

a. $r = \sqrt{\dfrac{A}{4\pi}}$

$r^2 = \dfrac{A}{4\pi}$

$A = 4\pi r^2$

b. $A = 4\pi r^2$

$= 4(3.14)(3964)^2 \approx 197{,}358{,}997.8$

We conclude that the surface area of the Earth is approximately 200,000,000 mi².

PRACTICE 9

The average annual interest rate, r, earned in a 3-year investment can be represented by

$$r = \sqrt[3]{\dfrac{A}{P}} - 1$$

where P is the initial investment and A is the value of the investment after 3 yr.

a. Rewrite this formula by solving for A.

b. If the interest rate was 0.03 (that is, 3%) on an initial investment of $10,000, find the value after 3 yr.

8.6 Exercises

Solve.

1. $\sqrt{3n} = 6$ **2.** $\sqrt{2a} = 10$ **3.** $\sqrt{x + 6} = 3$ **4.** $\sqrt{x - 7} = 4$

5. $\sqrt{5x - 6} = 2$ **6.** $\sqrt{4x + 1} = 5$ **7.** $\sqrt[3]{3y + 10} = -2$ **8.** $\sqrt[3]{12 - 5x} = 3$

9. $\sqrt{x} + 9 = 8$ **10.** $\sqrt{x} + 10 = 1$ **11.** $\sqrt{x} - 20 = -9$ **12.** $\sqrt{x} - 8 = 7$

13. $14 - \sqrt[3]{x} = 11$ **14.** $\sqrt[3]{n} + 15 = 13$ **15.** $\sqrt{6x} + 17 = 29$ **16.** $\sqrt{3n} - 11 = -2$

17. $\sqrt{2y - 1} - 8 = 5$ **18.** $\sqrt{5n - 4} + 7 = 16$ **19.** $14 - \sqrt{4a + 9} = 13$ **20.** $21 - \sqrt{3x - 15} = 24$

21. $\sqrt{5x - 1} - \sqrt{3x + 11} = 0$ **22.** $\sqrt{4n + 5} - \sqrt{2n + 7} = 0$ **23.** $2\sqrt{x - 3} = \sqrt{7x + 15}$

24. $\sqrt{6x - 11} = 3\sqrt{x - 7}$ **25.** $\sqrt[3]{3y - 19} = \sqrt[3]{6y + 26}$ **26.** $\sqrt[3]{2x + 9} = \sqrt[3]{3x + 14}$

27. $\sqrt{a^2 + 7} = \sqrt{5a + 1}$ **28.** $\sqrt{n^2 - 8} = \sqrt{2 - 3n}$ **29.** $\sqrt[3]{a^2 - 6} + \sqrt[3]{1 - 4a} = 0$

30. $\sqrt[3]{p^2 + 10} + \sqrt[3]{9p + 8} = 0$ **31.** $\sqrt{2x + 8} = -x$ **32.** $\sqrt{3y - 2} = y$

33. $2n - \sqrt{6 - 5n} = 0$ **34.** $-3a + \sqrt{6a - 1} = 0$ **35.** $x - 2 = \sqrt{4x - 11}$

36. $\sqrt{2x + 5} = x + 1$ **37.** $\sqrt{3t + 1} - 1 = 2t$ **38.** $\sqrt{5x + 9} + 3 = 2x$

39. $\sqrt{3n} + \sqrt{n - 2} = 4$ **40.** $\sqrt{x - 1} + \sqrt{2x} = 3$ **41.** $\sqrt{x - 2} + 1 = -\sqrt{x + 3}$

42. $4 - \sqrt{r + 6} = -\sqrt{r - 2}$ **43.** $\sqrt{x + 5} - 2 = \sqrt{x - 1}$ **44.** $\sqrt{14 - a} = \sqrt{a + 3} + 3$

45. $\sqrt{2y + 3} - \sqrt{3y + 7} = -1$ **46.** $\sqrt{2x - 1} + 1 = \sqrt{3x + 1}$ **47.** $\sqrt[3]{x^3 + 8} = x + 2$

48. $\sqrt[3]{8 - x^3} = 2 - x$

Mixed Practice

Solve.

49. $13 - \sqrt{4x - 7} = 18$ **50.** $\sqrt{7x + 4} - 2 = 3x$ **51.** $\sqrt{6x + 1} = 7$

52. $\sqrt[3]{p} + 13 = 9$ **53.** $\sqrt[3]{n^2 + 16} + \sqrt[3]{10n + 8} = 0$ **54.** $\sqrt{a^2 - 11} = \sqrt{7 - 3a}$

Applications

Solve.

55. The distance, d, to the horizon for an object h mi above the Earth's surface is given by the equation $d = \sqrt{8000h + h^2}$. How many miles above the Earth's surface is a satellite if the distance to the horizon is 900 mi?

56. The diagonal of a rectangular box with a square base can be calculated using the formula $d = \sqrt{2x^2 + h^2}$, where x is the length of the side of the square base. The diagonal of a box shown in the following figure is 18 in. What is the height of the box?

h

8 in.

8 in.

57. To calculate the minimum speed, S (in miles per hour), that a car was traveling before skidding to a stop, traffic accident investigators use the formula $S = \sqrt{30fL}$, where f is the drag factor of the road surface and L is the length of a skid mark (in feet).

   a. Solve the formula for L in terms of S and f.

   b. Calculate the length of the skid marks for a car traveling at a speed of 30 mph that skids to a stop on a road surface with a drag factor of 0.5.

58. If an object is dropped, the velocity, v, of the object (in feet per second) after it has fallen d ft can be calculated using the formula $v = \sqrt{64d}$.

   a. Solve the formula for d in terms of v.

   b. Find the distance an object has fallen if its velocity is 80 ft per sec.

59. How many feet from the side of a house must a painter place a 15-ft ladder in order to reach a window 12 ft above the ground?

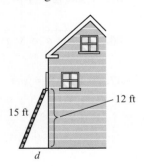

15 ft

12 ft

d

60. Two tugboats are towing a ship so that the tow ropes are perpendicular to each other. The tugboats exert forces of magnitude A and B on the ship. The formula for the magnitude of the combined force, R, is $R = \sqrt{A^2 + B^2}$. If the forces B and R have magnitude 6 tons and 10 tons, respectively, find the magnitude of A.

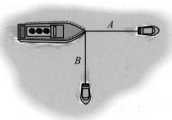

A

B

● *Check your answers on page A-37.*

MINDSTRETCHERS

Mathematical Reasoning

1. When raising each side of a radical equation to a power, is the resulting equation equivalent to the original equation? Explain.

Technology

2. Consider the equation $\sqrt{x - 1} = 3 - x$.

    a. Use a grapher to graph each side of the equation. Then find the point of intersection of the two graphs.

    b. Solve the equation algebraically.

    c. What do you notice about the x-coordinate of the point of intersection in part (a) and your answer to part (b)?

    d. Explain how a grapher can be used to solve radical equations.

Critical Thinking

3. Solve each of the following equations:

    a. $\sqrt{\sqrt{x^2 - 9}} = 2$ **b.** $\sqrt{\sqrt[3]{(2x + 1)^2}} = 2$

CULTURAL NOTE

In the 1600s, people were very interested in designing accurate clocks. The Dutch astronomer Huygens suggested using pendulums. Pendulums were chosen because they have an interesting property. The period T (the amount of time it takes for a pendulum to go back and forth once) is related to the length L of the pendulum and the force of gravity g by the following formula:

$$T = 2\pi\sqrt{\frac{L}{g}}.$$ Since the force of gravity is constant,

the only thing that influences the period of the pendulum is its length. Throughout history, this principle was used to design small wall clocks, cuckoo clocks, and grandfather clocks.

8.7   Complex Numbers

In the preceding sections of this text, our discussions have centered on operations and expressions involving real numbers. Here we extend the concept of number to include imaginary numbers and, more generally, complex numbers. The existence of these new numbers may be difficult to accept. However, working with complex numbers has several advantages—they permit us to solve equations that have no real-number solutions and also to solve applied problems in physics, engineering, and graphics.

OBJECTIVES

- To identify imaginary and complex numbers
- To add and subtract complex numbers
- To multiply complex numbers
- To identify a complex conjugate
- To divide complex numbers
- To evaluate powers of i
- To solve applied problems involving complex numbers

Imaginary and Complex Numbers

Recall that since the square of any real number is positive, there is no real number that is the square root of a negative number. Numbers represented by expressions such as $\sqrt{-1}$ and $\sqrt{-3}$ are not real numbers and so are called *imaginary numbers*. The imaginary number $\sqrt{-1}$, which is commonly represented by the letter i, will play a key role throughout this discussion.

Definition
The imaginary number i is $\sqrt{-1}$; in other words, $i^2 = -1$.

We can express any imaginary number in terms of i. For instance, consider the imaginary number $\sqrt{-9}$:

$$\sqrt{-9} = \sqrt{9(-1)}$$ Factor out -1 within the radicand.

$$= \sqrt{9} \cdot \sqrt{-1}$$ Use the product rule of radicals. It does not hold for complex numbers in general, but does hold for -1 times a positive number.

$$= 3i$$ Take the square root of the real number 9 and substitute i for $\sqrt{-1}$.

Let's see if this result is correct by checking that the square of $3i$ is the original radicand:

$$(3i)^2 = (3i)(3i) = 9i^2 = 9(-1) = -9$$

Since the square of $3i$ is -9, our result is correct. This example suggests the following general property of imaginary numbers that will allow us to rewrite any imaginary number in terms of i.

The Square Root of a Negative Number
For any positive real number a,

$$\sqrt{-a} = i\sqrt{a}$$

Note that in the product of i and a radical, i is usually written before the radical. Do you think that writing i first makes the expression easier to read?

EXAMPLE 1	PRACTICE 1

EXAMPLE 1

Write each square root in terms of i.

a. $\sqrt{-25}$ **b.** $\sqrt{-6}$ **c.** $-7\sqrt{-\dfrac{4}{9}}$

Solution

a. $\sqrt{-25} = i\sqrt{25}$
$= i \cdot 5$
$= 5i$

b. $\sqrt{-6} = i\sqrt{6}$

c. $-7\sqrt{-\dfrac{4}{9}} = -7i\sqrt{\dfrac{4}{9}}$
$= -7i\left(\dfrac{2}{3}\right)$
$= -\dfrac{14}{3}i$

PRACTICE 1

Express in terms of i.

a. $\sqrt{-36}$

b. $\sqrt{-2}$

c. $-10\sqrt{-\dfrac{1}{4}}$

We use imaginary numbers to define a broader set of numbers that are called *complex numbers*.

Definition

A **complex number** is any number that can be written in the form $a + bi$, where a and b are real numbers and $i = \sqrt{-1}$. In the complex number $a + bi$, a is called the **real part** and b is called the **imaginary part**.

Note that for any complex number, both its real part and its imaginary part are real numbers.

Every real number can be expressed as a complex number. For instance, the real number 4 can be written as $4 + 0i$. Similarly, every imaginary number can be expressed as complex, so we can write $2i$ as $0 + 2i$.

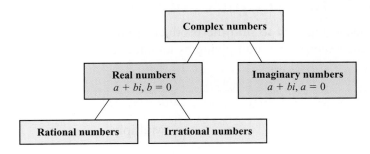

Consider the following examples of complex numbers:

Complex Number	Real Part a	Imaginary Part b
$2 + 3i$	2	3
$-1 - 5i$	-1	-5
$\dfrac{3}{4}\left(=\dfrac{3}{4} + 0i\right)$	$\dfrac{3}{4}$	0
$i\,(= 0 + 1i)$	0	1

The basic properties of real numbers apply to complex numbers as well. And just as for real numbers, we can carry out the four basic arithmetic operations on complex numbers.

Adding and Subtracting Complex Numbers

The addition and subtraction of complex numbers is similar to the addition and subtraction of algebraic binomial expressions. In other words, to add complex numbers, we add their real parts and their imaginary parts separately. Similarly, to subtract complex numbers, we subtract their real parts and their imaginary parts separately.

EXAMPLE 2	PRACTICE 2

Simplify.

a. $(2 - 5i) + (1 + 3i)$ **b.** $(-1 + 9i) - (2 - 4i)$

c. $(5 - 2\sqrt{-4}) + (-5 + 7\sqrt{-1})$

Solution

a.

$$\text{Real parts} \quad \text{Imaginary parts}$$
$$(2 - 5i) + (1 + 3i) = (2 + 1) + (-5 + 3)i = 3 - 2i$$

b. $(-1 + 9i) - (2 - 4i) = -1 + 9i - 2 + 4i$
$$= (-1 - 2) + (9 + 4)i$$
$$= -3 + 13i$$

c. $(5 - 2\sqrt{-4}) + (-5 + 7\sqrt{-1}) = (5 - 2 \cdot 2i) + (-5 + 7i)$
$$= (5 - 4i) + (-5 + 7i)$$
$$= (5 - 5) + (-4 + 7)i$$
$$= 3i$$

Combine.

a. $(6 + 5i) + (-2 - 6i)$

b. $(-4 + 3i) - (8 - i)$

c. $(1 + 5\sqrt{-9}) + (4 + \sqrt{-16})$

Multiplying Complex Numbers

Before considering how to find the product of complex numbers generally, let's look at some examples of multiplying by an imaginary number.

EXAMPLE 3	PRACTICE 3

Multiply.

a. $\sqrt{-4} \cdot \sqrt{-9}$ **b.** $\sqrt{-2} \cdot \sqrt{-7}$

c. $-3i \cdot 6i$ **d.** $5i(3 - 10i)$

Solution

a. $\sqrt{-4} \cdot \sqrt{-9} = (2i)(3i)$
$$= 6i^2$$
$$= -6$$

b. $\sqrt{-2} \cdot \sqrt{-7} = (i\sqrt{2})(i\sqrt{7})$
$$= i^2\sqrt{14}$$
$$= -1 \cdot \sqrt{14}$$
$$= -\sqrt{14}$$

Multiply.

a. $\sqrt{-36} \cdot \sqrt{-4}$

b. $\sqrt{-5} \cdot \sqrt{-3}$

c. $-4i \cdot 10i$

d. $-2i(5 + 6i)$

c. $-3i \cdot 6i = (-18)i^2$

$\qquad\qquad = (-18)(-1)$

$\qquad\qquad = 18$

d. $5i(3 - 10i) = 15i - 50i^2$

$\qquad\qquad\quad = 15i - (-50)$ Use the distributive property.

$\qquad\qquad\quad = 50 + 15i$

Note that in Example 3(a), $\sqrt{-4} \cdot \sqrt{-9} \neq \sqrt{(-4)(-9)}$. When multiplying radicals with negative radicands, we must first express them in terms of i, since the product rule for radicals does not in general hold for complex numbers.

Next, we turn to finding the product of any complex numbers. The key is to multiply the complex numbers just as if they were algebraic binomials, using the FOIL method.

EXAMPLE 4	PRACTICE 4
Multiply.	Multiply.
a. $(5 + 2i)(3 - i)$ **b.** $(2 + 3i)(4 - 6i)$	**a.** $(4 + 7i)(2 + i)$
Solution	**b.** $(3 - 2i)(1 - 3i)$

a.

$$(5 + 2i)(3 - i) = 15 - 5i + 6i - 2i^2$$

$$= 15 - 5i + 6i + 2$$

$$= 17 + i$$

b. $(2 + 3i)(4 - 6i) = 8 - 12i + 12i - 18i^2$

$$= 8 - 12i + 12i + 18$$

$$= 26$$

Complex Conjugates

The multiplication of *complex conjugates*, which play a special role in dividing complex numbers, requires special mention.

> **Definition**
> The complex numbers $a + bi$ and $a - bi$ are called **complex conjugates**.

For instance, $5 + 2i$ and $5 - 2i$ are complex conjugates. So are $-8 - 3i$ and $-8 + 3i$.

EXAMPLE 5	PRACTICE 5
Find the complex conjugate.	Find the complex conjugate.
a. $3 + 2i$ **b.** $-1 - 5i$ **c.** $6i$	**a.** $-1 - 7i$
Solution	**b.** $8 + 9i$
a. $3 - 2i$ **b.** $-1 + 5i$	**c.** $-3i$

c. The number $6i$ can be written $0 + 6i$. The conjugate is $0 - 6i$, or $-6i$.

Note that complex conjugates, such as $(3 + 6i)$ and $(3 - 6i)$, are binomials that are the sum and difference of the same two terms. Therefore, to multiply conjugates, we can use the special product formula for multiplying the sum and difference of two terms, $(a + b)(a - b) = a^2 - b^2$, as shown in the following example.

EXAMPLE 6	PRACTICE 6
Find the product of $3 + 6i$ and $3 - 6i$.	Multiply $2 - 5i$ and its complex conjugate.

Solution

$$(3 + 6i)(3 - 6i) = (3)^2 - (6i)^2$$
$$= 9 - 36i^2$$
$$= 9 + 36$$
$$= 45$$

Note that the product of these two complex numbers is a real number.

Tip   In general, the product of a complex number $a + bi$ and its complex conjugate $a - bi$ is the real number $a^2 + b^2$.

Dividing Complex Numbers

Complex numbers are best divided by expressing the division problem as a fraction. We then calculate an equivalent fraction whose denominator is a real number. To do this, we multiply the numerator and the denominator of the fraction by the complex conjugate of the denominator. Consider the following example.

EXAMPLE 7	PRACTICE 7
Divide.	Divide.

a. $\dfrac{3}{2 + i}$ **b.** $\dfrac{4}{3i}$

a. $\dfrac{7}{1 - 2i}$

b. $-\dfrac{2}{5i}$

Solution

a. To find the quotient of 3 and $2 + i$, we begin by multiplying both the numerator and the denominator by the complex conjugate of the denominator.

$$\frac{3}{2 + i} = \frac{3}{2 + i} \cdot \frac{2 - i}{2 - i}$$ Multiply both the numerator and denominator by $2 - i$, the conjugate of $2 + i$.

$$= \frac{3(2 - i)}{(2 + i)(2 - i)}$$

$$= \frac{6 - 3i}{4 + 1}$$

$$= \frac{6 - 3i}{5}$$

$$= \frac{6}{5} - \frac{3}{5}i$$ Write in $a + bi$ form.

b. $\dfrac{4}{3i} = \dfrac{4}{3i} \cdot \dfrac{-3i}{-3i}$

$= \dfrac{-12i}{-9i^2}$

$= \dfrac{-12i}{9}$

$= -\dfrac{4}{3}i$

EXAMPLE 8	PRACTICE 8

EXAMPLE 8

Divide.

a. $\dfrac{1 - 3i}{1 + 2i}$ **b.** $\dfrac{5 + \sqrt{-25}}{3 + \sqrt{-9}}$

Solution

a. $\dfrac{1 - 3i}{1 + 2i} = \dfrac{1 - 3i}{1 + 2i} \cdot \dfrac{1 - 2i}{1 - 2i}$

$= \dfrac{(1 - 3i)(1 - 2i)}{(1 + 2i)(1 - 2i)}$

$= \dfrac{1 - 5i + 6i^2}{1 + 4}$

$= \dfrac{1 - 5i - 6}{5}$

$= \dfrac{-5 - 5i}{5}$

$= \dfrac{5(-1 - i)}{5}$

$= -1 - i$

b. $\dfrac{5 + \sqrt{-25}}{3 + \sqrt{-9}} = \dfrac{5 + 5i}{3 + 3i} = \dfrac{5(1 + i)}{3(1 + i)} = \dfrac{5}{3}$

PRACTICE 8

Divide.

a. $\dfrac{2 + i}{1 - 3i}$

b. $\dfrac{1 - \sqrt{-4}}{4 - \sqrt{-64}}$

Raising i to a Power

Finally, in our discussion of operations on complex numbers, let's consider powers of the number i. When raising i to consecutive whole-number exponents, an interesting pattern emerges.

$i = \sqrt{-1} = i$ $i^5 = i^4 \cdot i = i$

$i^2 = (\sqrt{-1})^2 = -1$ $i^6 = i^4 \cdot i^2 = 1(-1) = -1$

$i^3 = i^2 \cdot i = -i$ $i^7 = i^4 \cdot i^3 = 1(-i) = -i$

$i^4 = i^2 \cdot i^2 = (-1)(-1) = 1$ $i^8 = i^4 \cdot i^4 = (1)(1) = 1$

Can you explain why this pattern continues?

Generally, an expression involving complex numbers is not considered to be simplified if it contains a power of i greater than 1. So we can use the pattern of powers just noted to simplify such expressions. Because i raised to the fourth power is 1, we can readily evaluate i raised to higher integral powers by factoring out i^4 as many times as possible.

EXAMPLE 9	PRACTICE 9

Evaluate i^{23}.

Find the value of i^{30}.

Solution $\quad i^{23} = (i^4)^5 i^3 = 1^5 i^3 = i^3 = -i$

An important application of complex numbers is the flow of alternating current in a circuit. This is the kind of current that flows through an appliance when we plug it into a wall outlet. The flow of electricity in these circuits involves three variables: the rate of flow of the current, I, measured in amperes; the voltage of the circuit, V, measured in volts; and the impedance of the circuit, Z, measured in ohms.

EXAMPLE 10	PRACTICE 10

Ohm's law for alternating-current circuits states that $V = IZ$, where V is the voltage (in volts), I is the current (in amperes), and Z is the impedance (in ohms).

a. Rewrite Ohm's law by solving for Z.

b. If V can be represented by $11 + 10i$ and I by $2 + 3i$, find Z.

c. Check your answer in part (b) by substituting into Ohm's law.

Solution

a. Ohm's law states that $V = IZ$. We solve for Z by dividing both sides of the equation by I, giving us $Z = \dfrac{V}{I}$.

b. To find Z, we use the formula found in part (a) and substitute $11 + 10i$ for V and $2 + 3i$ for I:

$$Z = \frac{V}{I} = \frac{11 + 10i}{2 + 3i} \cdot \frac{2 - 3i}{2 - 3i} = \frac{22 - 33i + 20i + 30}{4 + 9} = 4 - i$$

So the impedance, Z, can be represented by $(4 - i)$ ohms.

c. Check your answer in part (b) by substituting into Ohm's law.

$$V = IZ = (2 + 3i)(4 - i) = 8 - 2i + 12i + 3 = 11 + 10i$$

The computed voltage, $11 + 10i$, agrees with the given information, confirming our answer in part (b).

Use Ohm's law (stated in Example 10).

a. Rewrite Ohm's law by solving for I.

b. If V can be represented by $3 + 5i$ and Z by $1 - i$, find I.

c. Check your answer in part (b) by substituting into Ohm's law.

8.7   Exercises

Mathematically Speaking

Fill in each blank with the most appropriate term or phrase from the given list.

complex number	irrational number	complex part
additive inverses	imaginary number	the square root property
real part	power	
multiple	complex conjugates	FOIL method

1. The _____ i is $\sqrt{-1}$; in other words, $i^2 = -1$.

2. A(n) _____ is any number that can be written in the form $a + bi$, where a and b are real numbers and $i = \sqrt{-1}$.

3. In the complex number $a + bi$, a is called the _____ and b is called the imaginary part.

4. To multiply complex numbers, use the _____.

5. The complex numbers $a + bi$ and $a - bi$ are called _____.

6. An expression involving complex numbers is generally not considered simplified if it contains a(n) _____ of i greater than 1.

Write in terms of i.

7. $\sqrt{-4}$

8. $\sqrt{-49}$

9. $\sqrt{-\dfrac{1}{16}}$

10. $\sqrt{-\dfrac{9}{64}}$

11. $\sqrt{-3}$

12. $\sqrt{-10}$

13. $\sqrt{-18}$

14. $\sqrt{-24}$

15. $\sqrt{-500}$

16. $\sqrt{-72}$

17. $-\sqrt{-9}$

18. $-2\sqrt{-1}$

19. $6\sqrt{-\dfrac{5}{16}}$

20. $3\sqrt{-\dfrac{7}{36}}$

21. $-\dfrac{1}{4}\sqrt{-12}$

22. $-\dfrac{1}{2}\sqrt{-32}$

Perform the indicated operation.

23. $(1 + 12i) + 8i$

24. $3i + (4 + 14i)$

25. $(3 - 15i) + (2 + 9i)$

26. $(6 + 11i) + (5 - 7i)$

27. $(7 - i) - (7 + 5i)$

28. $(10 - 4i) - (9 - 4i)$

29. $(-8 - 6i) - (-1 - 3i)$

30. $(2 + 13i) - (-3 - 2i)$

31. $16 - (18 + \sqrt{-4})$

32. $20 - (7 + \sqrt{-9})$

33. $(10 - 3\sqrt{-16}) + (2 - \sqrt{-25})$

34. $(4 + 2\sqrt{-4}) - (12 - 4\sqrt{-1})$

Multiply.

35. $\sqrt{-25} \cdot \sqrt{-4}$

36. $\sqrt{-9} \cdot \sqrt{-16}$

37. $\sqrt{-3}(-\sqrt{-27})$

38. $-\sqrt{-6}(\sqrt{-24})$

39. $7i \cdot 9i$

40. $11i \cdot 5i$

41. $-2i(14i)$

42. $6i(-8i)$

43. $3i(1 - i)$

44. $2i(5 - 2i)$

45. $-i(12 + 7i)$

46. $-4i(4 + 8i)$

47. $\sqrt{-9}(7 + \sqrt{-16})$

48. $\sqrt{-1}(2 + \sqrt{-49})$

49. $-\sqrt{2}(\sqrt{8} - \sqrt{-18})$

50. $-\sqrt{3}(\sqrt{3} - \sqrt{-27})$

51. $(4 + 2i)(2 + 3i)$ **52.** $(3 + 5i)(1 + 4i)$ **53.** $(10 - i)(4 + 6i)$ **54.** $(8 + 3i)(1 - 2i)$

55. $(7i - 7)(3 - 5i)$ **56.** $(8 - i)(9i - 2)$ **57.** $(-4 - 2i)(2 - 4i)$ **58.** $(-3 - 6i)(6 - 3i)$

59. $(6 + 5i)(6 - 5i)$ **60.** $(7 + 3i)(7 - 3i)$ **61.** $(3 + 2i)^2$ **62.** $(4 + 3i)^2$

63. $(2 - 3i)^2$ **64.** $(3 - 4i)^2$

65. $(\sqrt{-1} + \sqrt{2})(\sqrt{-9} - \sqrt{8})$ **66.** $(\sqrt{3} - \sqrt{-4})(\sqrt{12} + \sqrt{-16})$

Find the complex conjugate of the complex number. Then find the product of each complex number and its complex conjugate.

67. $1 + 10i$ **68.** $7 - i$ **69.** $4 - 3i$ **70.** $5 + 2i$

71. $-9 + 6i$ **72.** $-4 + 12i$ **73.** $8i$ **74.** $5i$

75. $-11i$ **76.** $-9i$

Divide.

77. $\dfrac{7}{4 + i}$ **78.** $\dfrac{3}{3 - i}$ **79.** $\dfrac{-3}{1 - 5i}$ **80.** $\dfrac{-1}{2 + 3i}$

81. $\dfrac{5}{4i}$ **82.** $\dfrac{9}{2i}$ **83.** $-\dfrac{2}{\sqrt{-49}}$ **84.** $-\dfrac{3}{\sqrt{-25}}$

85. $\dfrac{4 - 3i}{i}$ **86.** $\dfrac{6 + 2i}{2i}$ **87.** $\dfrac{3 + 5i}{1 + i}$ **88.** $\dfrac{2 - 4i}{3 - i}$

89. $\dfrac{6 + 3i}{2 - 2i}$ **90.** $\dfrac{4 - 5i}{5 + 3i}$ **91.** $\dfrac{2 - \sqrt{-16}}{5 - \sqrt{-100}}$ **92.** $\dfrac{6 + \sqrt{-9}}{4 + \sqrt{-4}}$

93. $\dfrac{8 - \sqrt{-36}}{6 + \sqrt{-64}}$ **94.** $\dfrac{9 + \sqrt{-9}}{3 - \sqrt{-81}}$

Evaluate.

95. i^{18} **96.** i^{20} **97.** i^{35} **98.** i^{41}

99. $i^{12} \cdot i^9$ **100.** $i^{15} \cdot i^{11}$ **101.** $\dfrac{i^{38}}{i^{19}}$ **102.** $\dfrac{i^{24}}{i^7}$

Mixed Practice

Solve.

103. Find the complex conjugate of $-8 + 5i$. Then calculate the product of both numbers.

104. Evaluate: $i^{11} \cdot i^3$

Write in terms of i.

105. $4\sqrt{-\dfrac{3}{25}}$ **106.** $\sqrt{-72}$

Divide.

107. $\dfrac{2 - 7i}{1 - i}$ **108.** $\dfrac{-2}{2 - i}$

Simplify.

109. $-\sqrt{2}(\sqrt{32} - \sqrt{-50})$

110. $-6i(5 - 12i)$

111. $18 - (20 + \sqrt{-16})$

112. $(-7 + 8i)(3 - 6i)$

Applications

Solve.

113. In the study of electrical circuits, the impedance of electron flow in an alternating-current (AC) circuit is expressed as a complex number. The total impedance in an AC circuit connected in series is the sum of the individual impedances. If the impedance in one part of the circuit is $(3 + 9i)$ ohms and in another part is $(5 - 8i)$ ohms, what is the total impedance in the circuit?

114. The impedance in one part of an AC circuit connected in series is $(10 + 7i)$ ohms and in another part is $(12 - 4i)$ ohms. Find the total impedance in the circuit. (See Exercise 113.)

115. The formula $V = IZ$ expresses the relationship between the voltage, V (in volts), the current, I (in amperes), and the impedance, Z (in ohms), in an alternating-current circuit. If the current in a circuit is $(8 + 5i)$ amps and the impedance is $(9 + 3i)$ ohms, what is the voltage in the circuit?

116. Use the formula in Exercise 115 to find the current in a circuit if the voltage is $(60 + 20i)$ volts and the impedance is $(3 + i)$ ohms.

● *Check your answers on page A-37.*

MINDSTRETCHERS

Mathematical Reasoning

1. Use the properties of exponents to show that the expression $i^{4n + 3}$ is equivalent to $-i$ for any whole number n.

Research

2. Casper Wessel was credited with the geometric interpretation of a complex number as a point in the *complex plane*. Either in your college library or on the Web, investigate how to plot complex numbers. Write a summary of your findings and explain how to plot the complex number $2 - 3i$.

Critical Thinking

3. Some solutions of equations are complex numbers. Show that $1 + i$ and its conjugate are solutions of the equation $x^2 - 2x + 2 = 0$.

KEY CONCEPTS AND SKILLS   ⬭ CONCEPT   ⬭ SKILL

Concept/Skill	Description	Example
[8.1] Square Root	The number b is a square root of a if $b^2 = a$, for any real numbers a and b and for a nonnegative.	4 is a square root of 16 since $4^2 = 16$. -4 is a square root of 16 since $(-4)^2 = 16$.
[8.1] Radicand	The radicand is the number under $\sqrt{}$, the **radical sign**. This symbol stands for the positive or **principal square root**.	5 is the radicand of $\sqrt{5}$. The principal square root of 5 is $\sqrt{5}$.
[8.1] Squaring a Square Root	For any nonnegative real number a, $(\sqrt{a})^2 = a$.	$(\sqrt{6})^2 = 6$ $(\sqrt{x})^2 = x$
[8.1] Taking the Square Root of a Square	For any nonnegative real number a, $\sqrt{a^2} = a$.	$\sqrt{6^2} = 6$ $\sqrt{x^2} = x$
[8.1] Cube Root	The number b is the **cube root** of a if $b^3 = a$, for any real numbers a and b. The cube root of a is written $\sqrt[3]{a}$, where 3 is called the **index** of the radical.	2 is the cube root of 8 since $2^3 = 8$. -2 is the cube root of -8 since $(-2)^3 = -8$.
[8.1] Cubing a Cube Root	For any real number a, $(\sqrt[3]{a})^3 = a$.	$(\sqrt[3]{7})^3 = 7$ $(\sqrt[3]{n})^3 = n$
[8.1] Taking the Cube Root of a Cube	For any real number a, $\sqrt[3]{a^3} = a$.	$\sqrt[3]{7^3} = 7$ $\sqrt[3]{n^3} = n$
[8.1] nth Root	The number b is the **nth root** of a if $b^n = a$, for any real number a and for any positive integer n greater than 1. The nth root of a is written $\sqrt[n]{a}$, where n is called the **index** of the radical.	The fourth root of 81, written $\sqrt[4]{81}$, is 3 since $3^4 = 81$. The index of the radical is 4.
[8.1] Taking the nth Root of an nth Power	For any real number a, • $\sqrt[n]{a^n} = \lvert a \rvert$ for any even positive integer n, and • $\sqrt[n]{a^n} = a$ for any odd positive integer n greater than 1.	$\sqrt{(-5)^2} = \lvert -5 \rvert = 5$ $\sqrt[3]{(-5)^3} = -5$
[8.1] Radical Function	A function in which a variable appears in a radicand.	$f(x) = \sqrt{x + 3}$ $g(x) = \sqrt[4]{x}$
[8.2] $a^{1/n}$	For any positive integer n greater than 1 and a real number a for which $\sqrt[n]{a}$ is a real number, $a^{1/n} = \sqrt[n]{a}$.	$5^{1/2} = \sqrt{5}$ $x^{1/3} = \sqrt[3]{x}$

continued

Concept/Skill	Description	Example
[8.2] $a^{m/n}$	For any positive integers m and n such that n is greater than 1 where $\dfrac{m}{n}$ is in simplest form and a is a real number for which $\sqrt[n]{a}$ is a real number, $a^{m/n} = \sqrt[n]{a^m}$, or equivalently $a^{m/n} = (\sqrt[n]{a})^m$.	$5^{2/3} = (\sqrt[3]{5})^2$ or $5^{2/3} = \sqrt[3]{5^2}$ $n^{3/5} = (\sqrt[5]{n})^3$ or $n^{3/5} = \sqrt[5]{n^3}$
[8.2] $a^{-m/n}$	For any positive integers m and n such that n is greater than 1, where $\dfrac{m}{n}$ is in simplest form and a is a real number for which $\sqrt[n]{a}$ is a nonzero real number, $$a^{-m/n} = \dfrac{1}{a^{m/n}}$$	$5^{-2/3} = \dfrac{1}{5^{2/3}}$ $n^{-3/5} = \dfrac{1}{n^{3/5}}$
[8.3] **Product Rule of Radicals**	If $\sqrt[n]{a}$ and $\sqrt[n]{b}$ are real numbers, then $\sqrt[n]{a} \cdot \sqrt[n]{b} = \sqrt[n]{ab}$.	$\sqrt{2} \cdot \sqrt{3} = \sqrt{6}$ $\sqrt[4]{x} \cdot \sqrt[4]{y^2} = \sqrt[4]{xy^2}$
[8.3] **Quotient Rule of Radicals**	For any integer $n > 1$ and any real numbers a and b for which $\sqrt[n]{a}$ and $\sqrt[n]{b}$ are real numbers and b is nonzero, $$\dfrac{\sqrt[n]{a}}{\sqrt[n]{b}} = \sqrt[n]{\dfrac{a}{b}}$$	$\dfrac{\sqrt{2}}{\sqrt{3}} = \sqrt{\dfrac{2}{3}}$ $\dfrac{\sqrt[4]{x}}{\sqrt[4]{y^2}} = \sqrt[4]{\dfrac{x}{y^2}}$
[8.3] **Distance Formula**	The distance, d, between two points, (x_1, y_1) and (x_2, y_2) on a coordinate plane is given by $$d = \sqrt{(x_2 - x_1)^2 + (y_2 - y_1)^2}$$	The distance between $(3, 2)$ and $(5, 4)$ is $d = \sqrt{(x_2 - x_1)^2 + (y_2 - y_1)^2}$ $= \sqrt{(5 - 3)^2 + (4 - 2)^2}$ $= \sqrt{(2)^2 + (2)^2}$ $= \sqrt{8}$ $= 2\sqrt{2}$ units
[8.4] **Like Radicals**	Radicals that have the same radicand and the same index.	$2\sqrt[3]{5}$ and $3\sqrt[3]{5}$ are like radicals. $-\sqrt{3x}$ and $7\sqrt{3x}$ are like radicals.
[8.4] **To Add (or Subtract) Radicals**	Simplify the radicals if possible, and then add (or subtract) like radicals using the distributive property.	$4\sqrt{3} - 6\sqrt{3} = (4 - 6)\sqrt{3} = -2\sqrt{3}$ $\sqrt{8} + \sqrt{18} = 2\sqrt{2} + 3\sqrt{2}$ $= (2 + 3)\sqrt{2}$ $= 5\sqrt{2}$
[8.5] **To Multiply Radicals**	Apply the product rule of radicals. Simplify, if possible.	$(6\sqrt{2})(-2\sqrt{6}) = -12\sqrt{12}$ $= -12 \cdot 2\sqrt{3}$ $= -24\sqrt{3}$
[8.5] **To Divide Radicals**	Apply the quotient rule of radicals. Simplify, if possible.	$\dfrac{\sqrt{24x}}{\sqrt{6}} = \sqrt{\dfrac{24x}{6}}$ $= \sqrt{4x}$ $= 2\sqrt{x}$

continued

Concept/Skill	Description	Example
[8.5] To Rationalize a Denominator	Multiply the numerator and denominator by a factor that will make the radicand in the denominator a perfect power.	$\dfrac{3}{\sqrt{5}} = \dfrac{3}{\sqrt{5}} \cdot \dfrac{\sqrt{5}}{\sqrt{5}} = \dfrac{3\sqrt{5}}{\sqrt{5^2}} = \dfrac{3\sqrt{5}}{5}$ $\sqrt[3]{\dfrac{2}{x}} = \dfrac{\sqrt[3]{2}}{\sqrt[3]{x}}$ $= \dfrac{\sqrt[3]{2}}{\sqrt[3]{x}} \cdot \dfrac{\sqrt[3]{x^2}}{\sqrt[3]{x^2}}$ $= \dfrac{\sqrt[3]{2x^2}}{\sqrt[3]{x^3}}$ $= \dfrac{\sqrt[3]{2x^2}}{x}$
[8.6] Radical Equation	An equation in which a variable appears in one or more radicands.	$\sqrt{2x} = 6$ $\sqrt{x+1} = \sqrt{2x-3}$
[8.6] To Solve a Radical Equation	• Isolate a term with a radical. • Raise each side of the equation to a power equal to the index of the radicand. • If the equation still contains a radical, repeat the preceding steps. • Where possible, combine like terms. • Solve the resulting equation. • Check the possible solution(s) in the original equation.	$\sqrt{3x+4} + 7 = 12$ $\sqrt{3x+4} = 5$ $(\sqrt{3x+4})^2 = 5^2$ $3x + 4 = 25$ $3x = 21$ $x = 7$ **Check** $\sqrt{3x+4} + 7 = 12$ $\sqrt{3(7)+4} + 7 \stackrel{?}{=} 12$ $\sqrt{25} + 7 \stackrel{?}{=} 12$ $5 + 7 \stackrel{?}{=} 12$ $12 = 12$ **True**
[8.7] Imaginary Number	A number that is the square root of a negative number.	$\sqrt{-5}$
[8.7] i	The imaginary number i is $\sqrt{-1}$; in other words, $i^2 = -1$.	$\sqrt{-1}$
[8.7] Square Root of a Negative Number	For any positive real number a, $\sqrt{-a} = i\sqrt{a}$.	$\sqrt{-25} = 5i$ $\sqrt{-6} = i\sqrt{6}$
[8.7] Complex Number	Any number that can be written in the form $a + bi$, where a and b are real numbers and $i = \sqrt{-1}$. In the complex number $a + bi$, a is called the **real part** and b is called the **imaginary part**.	$6 + 2i$, 3, and $-5i$ are complex numbers.

continued

Concept/Skill	Description	Example
[8.7] **To Add (or Subtract) Complex Numbers**	Add (or subtract) the real parts and imaginary parts separately.	$(2 + 4i) + (3 - 2i)$ $= (2 + 3) + (4 - 2)i$ $= 5 + 2i$ $(4 - i) - (7 - 6i)$ $= 4 - i - 7 + 6i$ $= (4 - 7) + (-1 + 6)i$ $= -3 + 5i$
[8.7] **To Multiply Complex Numbers**	Use the distributive property or the FOIL method. Then simplify, if possible.	$(2 + 3i)(1 - 2i) = 2 - 4i + 3i - 6i^2$ $= 2 - 4i + 3i + 6$
[8.7] **Complex**	The complex numbers $a + bi$ and $a - bi$	$6 + 2i$ and $6 - 2i$
[8.7] **To Divide Complex Numbers**	Multiply the numerator and denominator by the complex conjugate of the denominator.	$\dfrac{1 + 2i}{2 - i} = \dfrac{1 + 2i}{2 - i} \cdot \dfrac{2 + i}{2 + i}$ $= \dfrac{(1 + 2i)(2 + i)}{(2 - i)(2 + i)}$ $= \dfrac{2 + 5i + 2i^2}{4 + 1}$ $= \dfrac{2 + 5i - 2}{4 + 1}$ $= \dfrac{5i}{5}$ $= i$

Chapter 8 Review Exercises

To help you review this chapter, solve these problems.

[8.1] *Evaluate.*

1. $-6\sqrt{121}$ **2.** $2\sqrt[3]{-125}$ **3.** $\sqrt{\dfrac{1}{9}}$ **4.** $\sqrt{0.36}$

Simplify.

5. $\sqrt{81y^8}$ **6.** $-\sqrt{49a^6b^2}$ **7.** $\sqrt[3]{-216x^9}$ **8.** $\sqrt[5]{243p^{15}}$

If $f(x) = \sqrt{4x}$, find the following values. If the value is not an integer, express it in radical form.

9. $f(9)$ **10.** $f(10)$

Graph each function

11. $f(x) = -\sqrt{x}$

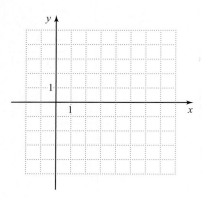

12. $g(x) = \sqrt{x} + 1$

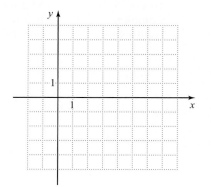

[8.2] *Write using radical notation. Then simplify, if possible.*

13. $-64^{1/2}$ **14.** $7x^{1/3}$ **15.** $-(16n^4)^{3/4}$ **16.** $8^{-2/3}$

Simplify. Then write the answer in radical notation.

17. $x^{1/4} \cdot x^{1/2}$ **18.** $\dfrac{r^{2/3}}{6r^{1/6}}$ **19.** $(25y^2)^{-1/2}$ **20.** $\dfrac{3a^{2/3}}{(6a^{1/6})^2}$

[8.3] *Use rational exponents to simplify. Then write the answer in radical notation.*

21. $\sqrt[8]{x^2}$ **22.** $\sqrt[6]{n^4} \cdot \sqrt[3]{n}$ **23.** $\sqrt{\sqrt[4]{y^2}}$

24. $\sqrt[6]{p^3q^6}$ **25.** $\dfrac{\sqrt[3]{a^2}}{\sqrt{a}}$ **26.** $\sqrt[4]{x^2} \cdot \sqrt[10]{y^5}$

Multiply.

27. $\sqrt{10r} \cdot \sqrt{3s}$ **28.** $\sqrt[3]{4p} \cdot \sqrt[3]{7pq^2}$

Simplify.

29. $\sqrt{300n^3}$ **30.** $\sqrt{45x^5y^4}$ **31.** $\sqrt[3]{128t^7}$ **32.** $\sqrt[4]{96a^5b^{10}}$

Divide.

33. $\dfrac{\sqrt{35a}}{\sqrt{7}}$

34. $\dfrac{\sqrt[3]{12p^2q^2}}{\sqrt[3]{6pq^2}}$

Simplify.

35. $\sqrt{\dfrac{n}{25}}$

36. $\sqrt{\dfrac{6}{49y^4}}$

37. $\sqrt[3]{\dfrac{64u^2}{125v^9}}$

38. $\sqrt[4]{\dfrac{4p^4q^3}{81r^4s^8}}$

[8.4] *Perform the indicated operations.*

39. $9\sqrt{x} - 5\sqrt{x}$

40. $3\sqrt[3]{q^2} + 8\sqrt[3]{q^2}$

41. $\sqrt{48} + \sqrt{27}$

42. $-\sqrt{96} - 5\sqrt{6} + 3\sqrt{54}$

43. $6\sqrt[3]{56a^4} - \sqrt[3]{189a}$

44. $\dfrac{1}{3}\sqrt[3]{27p^5q} + 2p\sqrt[3]{p^2q}$

[8.5] *Multiply and simplify.*

45. $(-2\sqrt{3a})(3\sqrt{6a})$

46. $\sqrt{5}(4\sqrt{10} - 2\sqrt{5})$

47. $\sqrt[3]{2n}(\sqrt[3]{n^2} - \sqrt[3]{4})$

48. $(4\sqrt{t} - 5)(\sqrt{t} - 3)$

49. $(\sqrt{6} - \sqrt{x})(\sqrt{6} + \sqrt{x})$

50. $(\sqrt{2y} - 1)^2$

Divide and simplify.

51. $\dfrac{\sqrt{72n^3}}{4\sqrt{6}}$

52. $\sqrt{\dfrac{32a}{9b^4}}$

Simplify.

53. $\dfrac{1}{\sqrt{8}}$

54. $\dfrac{\sqrt{16x}}{2\sqrt{y}}$

55. $\sqrt{\dfrac{14p^2}{3q}}$

56. $\sqrt[3]{\dfrac{5v}{54u^5}}$

57. $\dfrac{\sqrt{10} - 3}{\sqrt{5}}$

58. $\dfrac{2\sqrt{6a} + \sqrt{2}}{\sqrt{2a}}$

Rationalize the denominator.

59. $\dfrac{4}{\sqrt{3} - 1}$

60. $\dfrac{\sqrt{x} + 2}{\sqrt{x} - \sqrt{5}}$

[8.6] *Solve.*

61. $\sqrt{x + 8} = 4$

62. $\sqrt{n - 2} = 3$

63. $\sqrt{3n - 4} + 1 = -2$

64. $\sqrt{x^2 - 7} = \sqrt{5x + 7}$

65. $\sqrt[3]{2x - 3} = -2$

66. $\sqrt{x + 5} + 1 = \sqrt{3x + 4}$

[8.7] *Write in terms of i.*

67. $\sqrt{-36}$

68. $\sqrt{-125}$

Perform the indicated operation.

69. $(6 - 4i) + (2 + 9i)$

70. $(\sqrt{-4} - 3) - (\sqrt{-16} - 7)$

71. $\sqrt{-81} \cdot \sqrt{-1}$

72. $-2i(5i + 1)$

73. $(3 - 3i)(8 + 3i)$

74. $(5 - i)^2$

75. $\dfrac{-1}{4 - 4i}$

76. $\dfrac{3 - 4i}{6 - 2i}$

Evaluate.

77. i^{38}

78. i^{53}

Mixed Applications

Solve. Express answers in radical form.

79. When grown in a rich medium, the number of *E. coli* bacteria in a colony after *t* min can be calculated using the expression $N(2)^{t/20}$, where *N* is the number of bacteria present in the initial population.

    a. Write this expression in radical form.

    b. If 10 bacteria are present in the initial population, calculate the number of bacteria in the colony after 1 hr.

80. The velocity (in feet per second) needed for a spacecraft to escape the gravitational pull of Earth is given by the expression $\sqrt{64R}$, where *R* is the radius of the planet in feet. Simplify this expression.

81. When an object at rest is accelerated *a* meters per second², its velocity in meters per second after it has traveled *d* m is given by the expression $\sqrt{2ad}$.

    a. If a car at rest is accelerated 2 meters per second², find its velocity after it has traveled 50 m.

    b. How much faster would the object in part (a) be moving if the acceleration had been 4 meters per second²?

82. The distance associated with the illumination, *I*, of an object is given by the expression

$$\sqrt{\frac{k}{I}}$$

where *k* is a constant. Simplify this expression by rationalizing the denominator.

83. The period of oscillation for a spring if an object with mass *m* is attached is given by the expression

$$2\pi\sqrt{\frac{m}{k}}$$

where *k* is a constant associated with the spring. Simplify this expression by rationalizing the denominator.

84. To get to her office building, an administrative assistant drives 8 mi north and 4 mi east from her apartment. How far is the office building from her home?

85. A wire is to be attached to a telephone pole at a point 20 ft above the ground. If 24 ft of wire are used, how far from the pole does the wire need to be anchored to the ground?

86. The owner of a health foods store determines that the demand equation for selling a nutritional supplement is $p = 18 - 0.5\sqrt{x - 4}$, where *p* is the price in dollars and *x* is the number of bottles demanded per week. How many bottles are demanded per week if the price of the supplement is $13.50?

● *Check your answers on page A-37.*

Chapter 8   POSTTEST

 FOR EXTRA HELP

 Pass the Test   Test solutions are found on the enclosed CD.

To see if you have mastered the topics in this chapter, take this test. (Assume that all variables represent nonnegative real numbers.)

1. Evaluate:

   a. $-3\sqrt{81}$ b. $\sqrt[3]{-216}$

2. Simplify: $\sqrt{144a^6b^2}$

3. Rewrite in radical notation and simplify: $(32x^{10})^{2/5}$

4. Simplify:

   a. $\dfrac{(16p^{1/3})^{3/2}}{8p^{1/3}}$ b. $\sqrt[8]{x^6y^2}$

5. Multiply: $\sqrt[3]{5p^2} \cdot \sqrt[3]{4q}$

6. Divide: $\dfrac{\sqrt{56n}}{\sqrt{7n}}$

7. Simplify: $\sqrt{117x^3y^7}$

Perform the indicated operation.

8. $\sqrt{\dfrac{6u}{49v^6}}$

9. $-4\sqrt{24} + 2\sqrt{54} - 7\sqrt{6}$

10. $(4\sqrt{2} + 3)(2\sqrt{2} - 5)$

11. Simplify: $\dfrac{\sqrt{3a}}{\sqrt{50b}}$

12. Rationalize the denominator: $\dfrac{\sqrt{x}}{\sqrt{x} - \sqrt{y}}$

Solve.

13. $\sqrt{2x - 1} + 9 = 5$

14. $\sqrt{8 - 3x} = \sqrt{6 - x^2}$

15. $\sqrt{x + 3} + \sqrt{2x + 5} = 2$

16. Multiply: $(3 + \sqrt{-49})(1 - \sqrt{-16})$

17. Divide: $\dfrac{3 - 5i}{2 + 3i}$

18. The radius of a sphere with surface area S is given by the expression

   $$\sqrt{\dfrac{S}{4\pi}}$$

   a. Simplify the expression.

   b. Find the radius of a beach ball if its surface area is 512 in². Express the answer in radical form.

19. The distance to the horizon of a satellite h mi above Earth's surface is given by the expression $\sqrt{8000h + h^2}$. What is the distance to the horizon from a satellite that is 200 mi above Earth?

20. The demand equation for a particular item manufactured at a factory is $p = 32 - \sqrt{x - 5}$, where p is the price of the item in dollars and x is the daily demand. What is the daily demand when the price is $20?

● *Check your answers on page A-37.*

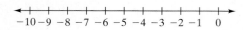

Cumulative Review Exercises

To help you review, solve these problems.

1. Solve and graph: $2(3x - 4) > 9x + 7$

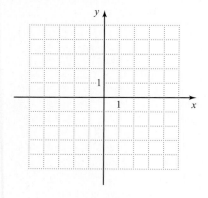

$$-10 \;\; -9 \;\; -8 \;\; -7 \;\; -6 \;\; -5 \;\; -4 \;\; -3 \;\; -2 \;\; -1 \;\; 0$$

2. Subtract $(5x^2 - x - 6)$ from $(7x^2 - 4x - 9)$.

3. Graph the equation $4x = 2y$ on the coordinate plane.

4. Solve the system of inequalities:

$$y < -\frac{1}{2}x + 3$$

$$y \geq x - 2$$

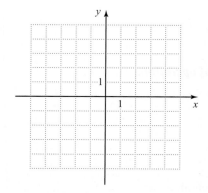

5. Factor completely: $x^3 - 2x^2 - 4x + 8$

6. Solve: $3(x - 1)^2 = 12$

7. Add: $\dfrac{3}{x^2 - 2x + 1} + \dfrac{2}{4 - 3x - x^2}$

8. A part-time employee at a grocery store has a weekly gross pay of $110 when he works 20 hr. If he works 25 hr, his gross pay is $137.50.

  a. Find a linear equation that relates the employee's gross pay, p, to the number of hours, h, he works per week.

  b. What does the slope of the line represent?

9. An object is thrown upward from a height of 250 ft above the ground with an initial velocity of 24 feet per second. The height (in feet) above the ground of the object after t sec is given by the equation $h = -16t^2 + 24t + 250$. When will the object be 115 ft above the ground?

10. Suppose that the time t that it takes to complete a job varies inversely with the number of people, n, working. If it takes 2 painters 50 hr to paint a particular size house, how long will it take 8 painters to paint the house?

• *Check your answers on page A-38.*

Quadratic Equations, Functions, and Inequalities

Dancing and Quadratic Functions

The motion of a dancer depends on a variety of physical factors. The forces on a dancer, for example, include gravity (pulling downward), support from the floor (pushing upward), and friction from the floor (exerted sideways).

For balance, the dancer's center of gravity must be directly above the area of contact with the floor. A smaller contact area with the floor makes balancing more difficult. When jumping, a dancer's center of gravity moves along a path that is parabolic, resembling the graph of a *quadratic function*. Great jumpers seem to ignore gravity as they float through the air in long, slow leaps. This illusion is created when the dancer raises or lowers her legs, changing the distance between her head and her center of gravity.

(*Sources:* Kenneth Laws and Cynthia Harvey, *Physics, Dance, and the Pas de Deux*, New York: Schirmer Books, 1994)

Chapter 9   **PRETEST**

To see if you have already mastered the topics in this chapter, take this test.

1. Solve by using the square root property:
$3x^2 + 17 = 5$

2. Solve by completing the square: $4a^2 + 12a = 7$

3. Solve by using the quadratic formula:
$2x^2 - 6x + 3 = 0$

4. Use the discriminant to determine the number and type of solutions of $3x^2 + 5x - 1 = 0$.

Solve.

5. $6(2n - 3)^2 = 48$

6. $x^2 - 4x + 12 = 0$

7. $3x^2 + 15x + 16 = x$

8. $2x^2 + 2x = 1 - 6x$

9. $5x^2 - 10x + 9 = 3$

10. $0.04x^2 - 0.12x + 0.09 = 0$

11. $x^4 - x^2 - 72 = 0$

12. Find a quadratic equation in x that has the solutions $-\dfrac{3}{2}$ and 4.

Identify the vertex, the equation of the axis of symmetry, and the x- and y-intercepts of the graph. Then graph.

13. $f(x) = x^2 - 4x + 3$

14. $f(x) = 12 - x - x^2$

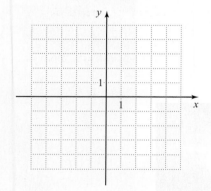

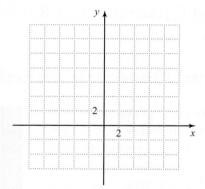

15. Graph the function $f(x) = \frac{1}{2}x^2 + x - 4$. Then identify the domain and range.

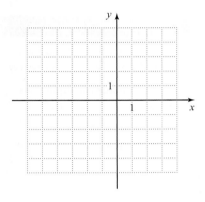

16. Solve. Then graph the solution: $x^2 - 6x + 8 > 0$

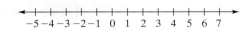

Solve.

17. A man tosses an object from the top of a tall building straight upward with an initial velocity of 20 feet per second. The height h of the object above the point of release t sec after it is tossed upward is given by the equation $h = -16t^2 + 20t$. When is the object 50 ft below the point of release?

18. It takes a student on a boating trip 30 min less time to travel 6 mi downriver than to travel the same distance upriver. If the student travels at a speed of 9 mph in still water, find the speed of the current to the nearest mile per hour.

19. A gardener wants to make a circular planting area in her yard. The planting area will be surrounded by a 1-ft-wide cement border. What radius will produce the maximum planting area if the total area including the border is to be 157 sq ft? What is the maximum planting area? Use $\pi \approx 3.14$ and round the answer to the nearest tenth.

20. A health foods company determines that its daily revenue, R, for selling x bottles of a new dietary supplement is modeled by the function $R(x) = 55x - 0.5x^2$. How many bottles must the company sell for revenue of at least \$1200?

● *Check your answers on page A-38.*

9.1 Solving Quadratic Equations: Square Root Property and Completing the Square

What Quadratic Equations Are and Why They Are Important

Recall from Section 6.6 that a **quadratic equation** (also called a *second-degree equation*) is an equation that can be written in the form $ax^2 + bx + c = 0$, where a, b, and c are real numbers and $a \neq 0$. In that section, we saw how quadratic equations can be used to model problems in physics and business. We solved such equations by factoring.

In this chapter, we consider three other methods of solving quadratic equations: the square root property of equality, completing the square, and, most importantly, the quadratic formula. Then we discuss how to graph quadratic functions, and finally how to solve quadratic and rational inequalities.

The Square Root Property of Equality

Let's consider the equation $x^2 = 64$. We already solved the equivalent equation $x^2 - 64 = 0$ in Example 6 of Section 6.6 by factoring:

$$x^2 = 64$$
$$x^2 - 64 = 0$$
$$(x - 8)(x + 8) = 0$$
$$x - 8 = 0 \quad \text{or} \quad x + 8 = 0$$
$$x = 8 \qquad\qquad x = -8$$

So the solutions of $x^2 = 64$ are 8 and -8, which can be written as ± 8 (read "plus or minus 8"). Note that ± 8 can also be expressed as $\pm\sqrt{64}$. This suggests that we could also have solved the equation $x^2 = 64$ by taking the square root of each side of the equation.

> **The Square Root Property of Equality**
>
> If b is a real number and $x^2 = b$, then $x = \pm\sqrt{b}$; that is, $x = \sqrt{b}$ or $x = -\sqrt{b}$.

We can apply the square root property of equality to solve quadratic equations of the forms $x^2 = b$ and $ax^2 = b$. Note that for some of these equations, the solutions are not real numbers.

EXAMPLE 1	PRACTICE 1
Solve by taking the square roots.	Use the square root property of equality to solve.
a. $x^2 = 36$	
b. $z^2 + 4 = 0$	**a.** $n^2 = 49$
	b. $y^2 + 9 = 0$

Solution

a. $x^2 = 36$

$x = \pm\sqrt{36}$ Use the square root property of equality.

$x = \pm 6$ Simplify.

Check

Substitute 6 for x.	Substitute -6 for x.
$x^2 = 36$	$x^2 = 36$
$(6)^2 \overset{?}{=} 36$	$(-6)^2 \overset{?}{=} 36$
$36 = 36$ True	$36 = 36$ True

So the solutions are 6 and -6.

b. $z^2 + 4 = 0$

$z^2 = -4$ Add -4 to each side of the equation.

$z = \pm\sqrt{-4}$ Use the square root property of equality.

$z = \pm 2i$ Simplify.

Check

Substitute $2i$ for z.	Substitute $-2i$ for z.
$z^2 + 4 = 0$	$z^2 + 4 = 0$
$(2i)^2 + 4 \overset{?}{=} 0$	$(-2i)^2 + 4 \overset{?}{=} 0$
$-4 + 4 \overset{?}{=} 0$	$-4 + 4 \overset{?}{=} 0$
$0 = 0$ True	$0 = 0$ True

So the solutions are $2i$ and $-2i$.

EXAMPLE 2	PRACTICE 2

EXAMPLE 2

Solve.

a. $3x^2 = 36$ **b.** $5x^2 + 40 = 0$

Solution

a. $3x^2 = 36$

$x^2 = 12$ Divide each side of the equation by 3.

$x = \pm\sqrt{12}$ Use the square root property of equality.

$x = \pm 2\sqrt{3}$ Simplify.

b. $5x^2 + 40 = 0$

$5x^2 = -40$ Subtract 40 from each side of the equation.

$x^2 = -8$ Divide each side of the equation by 5.

$x = \pm\sqrt{-8}$ Use the square root property of equality.

$x = \pm 2i\sqrt{2}$ Simplify.

PRACTICE 2

Solve.

a. $2y^2 = 16$

b. $3x^2 + 54 = 0$

We can also use the square root property of equality to solve quadratic equations of the form $(ax + b)^2 = d$ and $a(x + c)^2 = d$, that is, equations containing the square of a binomial.

EXAMPLE 3

Solve.

a. $(3x - 1)^2 = 18$ **b.** $(x + 5)^2 + 6 = 0$

c. $4(y + 1)^2 = 25$

Solution

a. $(3x - 1)^2 = 18$

$\quad\quad 3x - 1 = \pm\sqrt{18}$ Use the square root property of equality.

$\quad\quad\quad\quad 3x = 1 \pm 3\sqrt{2}$ Add 1 to each side of the equation and simplify the radical.

$\quad\quad\quad\quad\quad x = \dfrac{1 \pm 3\sqrt{2}}{3}$ Divide each side of the equation by 3.

So the solutions are $\dfrac{1 + 3\sqrt{2}}{3}$ and $\dfrac{1 - 3\sqrt{2}}{3}$.

b. $(x + 5)^2 + 6 = 0$

$\quad\quad (x + 5)^2 = -6$ Subtract 6 from each side of the equation.

$\quad\quad\quad x + 5 = \pm\sqrt{-6}$ Use the square root property of equality.

$\quad\quad\quad\quad\quad x = -5 \pm \sqrt{-6}$ Subtract 5 from each side of the equation.

$\quad\quad\quad\quad\quad x = -5 \pm i\sqrt{6}$ Simplify the radical.

So the solutions are $x = -5 + i\sqrt{6}$ and $x = -5 - i\sqrt{6}$.

c. $4(y + 1)^2 = 25$

$\quad\quad (y + 1)^2 = \dfrac{25}{4}$ Divide each side of the equation by 4.

$\quad\quad\quad y + 1 = \pm\sqrt{\dfrac{25}{4}}$ Use the square root property of equality.

$\quad\quad\quad\quad\quad y = -1 \pm \dfrac{5}{2}$ Subtract 1 from each side of the equation and simplify the radical.

$y = -1 + \dfrac{5}{2}$ or $y = -1 - \dfrac{5}{2}$

$y = \dfrac{3}{2}$ $\quad\quad\quad\quad\quad y = -\dfrac{7}{2}$

So the solutions are $\dfrac{3}{2}$ and $-\dfrac{7}{2}$.

PRACTICE 3

Solve.

a. $(2x + 1)^2 = 12$

b. $(n + 3)^2 + 2 = 0$

c. $9(y - 1)^2 = 49$

Some quadratic equations contain a perfect square trinomial. To solve these equations, we rewrite the trinomial as the square of a binomial and then apply the square root property of equality, as shown in the next example.

EXAMPLE 4	PRACTICE 4
Solve: $x^2 + 10x + 25 = 12$	Solve: $y^2 - 6y + 9 = 24$

Solution

$$x^2 + 10x + 25 = 12$$
$$(x + 5)^2 = 12$$
$$x + 5 = \pm\sqrt{12}$$
$$x = -5 \pm 2\sqrt{3}$$

So the solutions are $-5 + 2\sqrt{3}$ and $-5 - 2\sqrt{3}$.

Completing the Square

Another method of solving quadratic equations is called *completing the square.* To apply this method, we must transform one side of an equation to a perfect square trinomial. For instance, consider the equation $x^2 + 6x = 1$. Note that the expression on the left side of the equation is *not* a perfect square. Let's recall our discussion of perfect square trinomials in Section 5.6.

Perfect square		The coefficient of the quadratic term is 1.		The coefficient of the linear term		The constant term
$(x + 1)^2$	$=$	x^2	$+$	$2x$	$+$	1
$(x + 2)^2$	$=$	x^2	$+$	$4x$	$+$	4
$(x + 3)^2$	$=$	x^2	$+$	$6x$	$+$	9
$(x + 4)^2$	$=$	x^2	$+$	$8x$	$+$	16

Note that in each of these perfect square trinomials, the coefficient of the quadratic term is 1 and the constant term is the square of one-half the coefficient of the linear term. For example, in $x^2 + 8x + 16$, we see that

$$16 = \left(\frac{1}{2} \cdot 8\right)^2$$

More generally, when we square a binomial we get

$$(x + c)^2 = x^2 + 2cx + c^2$$

where

$$c^2 = \left(\frac{1}{2} \cdot 2c\right)^2$$

The constant term   The coefficient of the linear term

To complete the square of the expression $x^2 + bx$, that is, to transform it to a perfect square trinomial, we find one-half of the coefficient of the linear term x, square this value, and then add it to the expression, giving us

$$x^2 + bx + \left(\frac{1}{2}b\right)^2$$

EXAMPLE 5

Fill in the blank to make the trinomial a perfect square.

a. $y^2 - 12y + [\quad]$ **b.** $x^2 + 5x + [\quad]$

Solution

a. $y^2 - 12y + [\quad]$

The coefficient of the linear term y is -12.

$$\frac{1}{2}(-12) = -6 \qquad \text{Find one-half of the coefficient of the linear term.}$$

The constant term to be added to $y^2 - 12y$ is therefore $(-6)^2$, or 36. So we get $y^2 - 12y + 36$, which is equal to $(y - 6)^2$.

b. $x^2 + 5x + [\quad]$

First, we find one-half the coefficient of the linear term x:

$$\frac{1}{2} \cdot 5 = \frac{5}{2}$$

The constant term to be added to $x^2 + 5x$ is $\left(\dfrac{5}{2}\right)^2$, or $\dfrac{25}{4}$.

So we get $x^2 + 5x + \dfrac{25}{4}$, which is equal to $\left(x + \dfrac{5}{2}\right)^2$.

PRACTICE 5

Fill in the blank to make each trinomial a perfect square.

a. $x^2 + 6x + [\quad]$

b. $t^2 - t + [\quad]$

Now, let's apply the method of completing the square to solve a quadratic equation. Consider, for instance, the equation $2x^2 + 12x = 8$. To complete the square, the coefficient of the quadratic term x^2 must be 1, so before completing the square in this equation, we must first divide each side of the equation by 2, the coefficient of x^2.

$$2x^2 + 12x = 8$$
$$x^2 + 6x = 4 \qquad \text{Divide each side of the equation by 2.}$$

Next, we take one-half of the coefficient of the linear term x: $\dfrac{1}{2}(6) = 3$. Then we square the result: $3^2 = 9$. Now, we add 9 to the left side of the equation to complete the square. To maintain equality, we must also add 9 to the right side of the equation.

$$x^2 + 6x + 9 = 4 + 9$$

Finally, we solve the equation by rewriting $x^2 + 6x + 9$ as the square of a binomial and then applying the square root property of equality.

$$(x + 3)^2 = 13$$
$$x + 3 = \pm\sqrt{13} \qquad \text{Use the square root property of equality.}$$
$$x = -3 \pm \sqrt{13} \qquad \text{Subtract 3 from each side of the equation.}$$

So the solutions are $-3 + \sqrt{13}$ and $-3 - \sqrt{13}$. Can you check that these solutions are correct?

This example suggests the following general method for solving a quadratic equation by completing the square.

To Solve the Quadratic Equation $ax^2 + bx + c = 0$ by Completing the Square

- If $a = 1$, then proceed to the next step. If $a \neq 1$, then divide the equation by a, the coefficient of the quadratic (or second-degree) term.

- Move all terms with variables to one side of the equal sign and all constants to the other side.

- Take one-half the coefficient of the linear term, then square it and add this value to each side of the equation.

- Factor the side of the equation containing the variables, writing it as the square of a binomial.

- Use the square root property of equality and solve the resulting equation.

EXAMPLE 6

Solve $x^2 - 5x - 2 = 0$ by completing the square.

Solution

$x^2 - 5x - 2 = 0$

$x^2 - 5x = 2$ Add 2 to each side of the equation.

$x^2 - 5x + \dfrac{25}{4} = 2 + \dfrac{25}{4}$ Take one-half of the coefficient of the linear term: $\dfrac{1}{2}(-5) = -\dfrac{5}{2}$. Square it: $\left(-\dfrac{5}{2}\right)^2 = \dfrac{25}{4}$. Add $\dfrac{25}{4}$ to each side of the equation.

$x^2 - 5x + \dfrac{25}{4} = \dfrac{33}{4}$ Simplify the right side.

$\left(x - \dfrac{5}{2}\right)^2 = \dfrac{33}{4}$ Write $x^2 - 5x + \dfrac{25}{4}$ as the square of a binomial.

$x - \dfrac{5}{2} = \pm\sqrt{\dfrac{33}{4}}$ Use the square root property of equality.

$x = \dfrac{5}{2} \pm \sqrt{\dfrac{33}{4}}$ Add $\dfrac{5}{2}$ to each side.

$x = \dfrac{5}{2} + \dfrac{\sqrt{33}}{2}$ or $x = \dfrac{5}{2} - \dfrac{\sqrt{33}}{2}$

$x = \dfrac{5 + \sqrt{33}}{2}$ $x = \dfrac{5 - \sqrt{33}}{2}$

So the solutions are $\dfrac{5 + \sqrt{33}}{2}$ and $\dfrac{5 - \sqrt{33}}{2}$.

PRACTICE 6

Solve $t^2 + 3t = 1$ by completing the square.

EXAMPLE 7	PRACTICE 7

Solve: $3x^2 + 2x + 5 = 0$

Solution

$3x^2 + 2x + 5 = 0$

$x^2 + \dfrac{2}{3}x + \dfrac{5}{3} = 0$

Divide each side of the equation by 3, the coefficient of the quadratic term.

$x^2 + \dfrac{2}{3}x = -\dfrac{5}{3}$

Add $-\dfrac{5}{3}$ to each side of the equation.

$x^2 + \dfrac{2}{3}x + \dfrac{1}{9} = -\dfrac{5}{3} + \dfrac{1}{9}$

Take one-half the coefficient of the linear term: $\dfrac{1}{2}\left(\dfrac{2}{3}\right) = \dfrac{1}{3}$. Square it:

$\left(\dfrac{1}{3}\right)^2 = \dfrac{1}{9}$. Add $\dfrac{1}{9}$ to each side of the equation.

$x^2 + \dfrac{2}{3}x + \dfrac{1}{9} = -\dfrac{14}{9}$

Simplify the right side.

$\left(x + \dfrac{1}{3}\right)^2 = -\dfrac{14}{9}$

Write $x^2 + \dfrac{2}{3}x + \dfrac{1}{9}$ as the square of a binomial.

$x + \dfrac{1}{3} = \pm\sqrt{-\dfrac{14}{9}}$

Use the square root property of equality.

$x = -\dfrac{1}{3} \pm \sqrt{-\dfrac{14}{9}}$

Add $-\dfrac{1}{3}$ to each side.

$x = -\dfrac{1}{3} + i\dfrac{\sqrt{14}}{3}$ or $x = -\dfrac{1}{3} - i\dfrac{\sqrt{14}}{3}$

$x = \dfrac{-1 + i\sqrt{14}}{3}$ $\qquad$ $x = \dfrac{-1 - i\sqrt{14}}{3}$

So the solutions are $\dfrac{-1 + i\sqrt{14}}{3}$ and $\dfrac{-1 - i\sqrt{14}}{3}$.

Solve for y: $2y^2 - y + 1 = 0$

EXAMPLE 8	PRACTICE 8

If $f(x) = x^2 - 5x + 3$ and $g(x) = x + 2$, find all values of x for which $f(x) = g(x)$.

Solution

$$f(x) = g(x)$$
$$x^2 - 5x + 3 = x + 2$$
$$x^2 - 6x = -1$$
$$x^2 - 6x + 9 = -1 + 9$$
$$(x - 3)^2 = 8$$
$$x - 3 = \pm\sqrt{8}$$
$$x = 3 \pm \sqrt{8}$$
$$x = 3 \pm 2\sqrt{2}$$

So the solutions are $3 + 2\sqrt{2}$ and $3 - 2\sqrt{2}$.

If $g(x) = x^2 + 6x$ and $h(x) = 2$, find all values of x for which $g(x)$ and $h(x)$ are equal.

EXAMPLE 9	PRACTICE 9
The amount of pollution, A (in parts per million), present in the air outside a factory on a recent weekday is modeled by the equation $A = -t^2 + 16t + 6$, where t is the number of hours after 7 A.M. Find the times, to the nearest hour, when there were 42 parts per million of pollution.	Each day, a company produces x pieces of machinery. The company's daily profit, P (in dollars), depends on the value of x and can be modeled by the equation $P = -x^2 + 200x - 4000$. For which value of x, to the nearest whole number, does the company make a daily profit of $6000?

Solution Substituting 42 for A, we solve the equation $42 = -t^2 + 16t + 6$:

$$42 = -t^2 + 16t + 6$$
$$t^2 - 16t - 6 = -42$$
$$t^2 - 16t = -36$$
$$t^2 - 16t + 64 = -36 + 64$$
$$(t - 8)^2 = 28$$
$$t - 8 = \pm\sqrt{28}$$
$$t = 8 \pm \sqrt{28}$$

The solutions to the equation are $8 + \sqrt{28}$ and $8 - \sqrt{28}$. Using a calculator, we can approximate the solutions to the nearest whole number by 3 and 13, respectively. Since t represents the number of hours after 7 A.M., the times with the given amount of pollution are about 10 A.M. and 8 P.M.

In Example 9, which we solved using a calculator, why did we not express the answers as $8 \pm 2\sqrt{7}$?

Solve by using the square root property of equality.

1. $x^2 = 16$ **2.** $x^2 = 81$ **3.** $y^2 = 24$ **4.** $n^2 = 63$

5. $a^2 + 25 = 0$ **6.** $p^2 + 64 = 0$ **7.** $4n^2 - 8 = 0$ **8.** $2x^2 - 6 = 0$

9. $\frac{1}{6}y^2 = 12$ **10.** $\frac{1}{3}a^2 = 32$ **11.** $3x^2 + 6 = 21$ **12.** $5y^2 + 19 = 54$

13. $8 - 9n^2 = 14$ **14.** $10 - 4p^2 = 21$ **15.** $(x - 1)^2 = 48$ **16.** $(x + 3)^2 = 18$

17. $(2n + 5)^2 = 75$ **18.** $(3t - 2)^2 = 108$ **19.** $(4a - 3)^2 + 9 = 1$ **20.** $(2p + 1)^2 + 30 = 6$

21. $16(x + 4)^2 = 81$ **22.** $25(y - 10)^2 = 36$ **23.** $x^2 - 6x + 9 = 80$ **24.** $n^2 + 8n + 16 = 50$

25. $4p^2 + 12p + 9 = 32$ **26.** $9x^2 - 12x + 4 = 27$ **27.** $(3n - 2)(3n + 2) = -52$ **28.** $(4t + 3)(4t - 3) = -49$

29. $2x - 1 = \dfrac{18}{2x - 1}$ **30.** $3n + 1 = \dfrac{20}{3n + 1}$

Solve the formula for the indicated variable.

31. Volume of a cylinder: $V = \pi r^2 h$, for r

32. Kinetic energy: $E = \dfrac{1}{2}mv^2$, for v

33. Centripetal force: $F = \dfrac{mv^2}{r}$, for v

34. Pythagorean theorem: $a^2 + b^2 = c^2$, for b

Fill in the blank to make each trinomial a perfect square.

35. $x^2 - 12x + [\ \ ]$ **36.** $y^2 - 16y + [\ \ ]$ **37.** $n^2 + 7n + [\ \ ]$

38. $x^2 - 3x + [\ \ ]$ **39.** $t^2 - \dfrac{4}{3}t + [\ \ ]$ **40.** $a^2 + \dfrac{4}{5}a + [\ \ ]$

Solve by completing the square.

41. $x^2 - 8x = 0$ **42.** $p^2 + 12p = 0$ **43.** $n^2 - 3n = 4$ **44.** $t^2 - t = 2$

45. $x^2 + 4x - 2 = 0$ **46.** $x^2 - 6x + 6 = 0$ **47.** $a^2 + 7a = 3a - 4$ **48.** $y^2 - 4y = 4y - 16$

49. $x^2 - 9x + 4 = x - 25$ **50.** $p^2 + 2p + 35 = 10 - 6p$ **51.** $2n^2 - 8n = -24$ **52.** $3x^2 + 6x = -9$

53. $3x^2 - 12x - 84 = 0$ **54.** $5x^2 - 60x + 80 = 0$ **55.** $4a^2 + 20a - 12 = 0$ **56.** $2t^2 + 6t - 10 = 0$

57. $3y^2 - 9y + 15 = 0$ **58.** $4n^2 - 4n + 16 = 0$ **59.** $x^2 - \dfrac{4}{3}x - 4 = 0$ **60.** $x^2 - \dfrac{2}{5}x - 3 = 0$

61. $4y^2 + 11y + 6 = 0$ **62.** $2n^2 + 7n + 5 = 0$ **63.** $2p^2 + 7p = 6p - 8$ **64.** $3r^2 - r = r - 15$

Given f(x) and g(x), find all values of x for which f(x) = g(x).

65. $f(x) = x^2 - 9$ and $g(x) = 4x - 6$ **66.** $f(x) = x^2 + 3x - 5$ and $g(x) = x$

67. $f(x) = 4x^2$ and $g(x) = x^2 - 6x + 6$ **68.** $f(x) = 2x^2 - 5x$ and $g(x) = -x + 14$

Mixed Practice

Solve.

69. The formula for the volume of a cone is $V = \dfrac{1}{3}\pi r^2 h$. Solve this formula for r.

70. Fill in the blank to make the trinomial a perfect square: $n^2 - 5n + [\quad]$

Solve by completing the square.

71. Given $f(x) = x^2 - 4x$ and $g(x) = 2x - 2$, find all values of x for which $f(x) = g(x)$.

72. $(4t - 3)^2 = 32$

73. $-3x^2 - 24x + 6 = 0$

74. $x^2 + 16x + 16 = 4x - 22$

Solve by using the square root property of equality or completing the square.

75. $x^2 - 10x + 25 = 45$

76. $\dfrac{1}{2}x^2 = 24$

Applications

Solve. Use a calculator where appropriate.

77. In a search-and-rescue mission, a team maps out a circular search area from the last known location of a group of hikers. If the search region is 78.5 mi², how far from the last known location, to the nearest mile, is the team searching? Use $\pi \approx 3.14$.

78. The projection televisions made by an electronics manufacturer have screen dimensions that are in the ratio of 3:4. What are the screen dimensions of a 50-in. projection television made by the company?

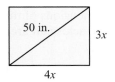

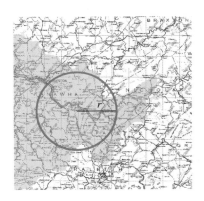

79. A student invests $1000 in an account earning r percent interest compounded annually. After two years, the amount, A, in the account is given by

$$A = 1000(1 + r)^2$$

where r is in decimal form. What is the interest rate if the account has $1102.50 after two years?

80. The parks department of a city plans to build a circular wading pool in one of its parks. The wading pool is to be surrounded by a 2-ft-wide concrete ledge for sitting. If the pool including the ledge has an area of 785 ft^2, what is the radius of the wading pool, to the nearest tenth of a foot? Use $\pi \approx 3.14$.

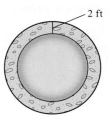

81. A sandbag is dropped from a hot-air balloon 2000 ft above the ground. The height, h, above the ground of the sandbag t sec after it is dropped is given by the equation $h = -16t^2 + 2000$.

  a. How long after it is released will the sandbag be 1000 ft above the ground?

  b. Is the time it takes the sandbag to reach the ground exactly equal to twice the length of time found in the solution to part (a)? Explain.

82. A manufacturer of athletic shoes determines that its daily revenue, R (in dollars), for selling x pairs of running shoes is modeled by the equation $R = 5x + 0.5x^2$.

  a. How many pairs of running shoes must be sold per day to generate revenue of $1500?

  b. If the answer in part (a) is doubled, does the daily revenue double? Explain.

83. A homeowner wants to build a rectangular patio using the house as one side, as shown in the figure. He decides to make the length of the patio 10 ft longer than the width.

  a. What are the dimensions of the patio if the homeowner wants to enclose an area of 144 ft^2?

  b. If the fencing costs $14.95 per foot, how much will it cost to enclose the patio?

84. The height, h, of an object t sec after it is thrown straight upward with an initial velocity of 32 feet per second relative to the point of release is given by the equation $h = 32t - 16t^2$.

  a. How long after it is thrown upward will it take the object to return to the original height at which it was released?

  b. How long after it is thrown upward will the object be 12 ft above the point of release?

85. Two trucks leave a truck stop at the same time. One truck goes north, driving at an average speed that is 16 mph slower than the truck driving east. After 15 min, the trucks are 20 mi apart. Find the average speed of each truck.

86. Two campus security officers meet at the campus center. One walks south and the other walks west. The officer walking south walks at an average rate that is 30 feet per minute faster than the officer walking west. If they are 25 ft apart after 10 sec, find the rate at which each security officer is walking.

● *Check your answers on page A-38.*

MINDSTRETCHERS

Mathematical Reasoning

1. Consider the equation $x^3 = 1$. Can this equation be solved by taking the cube root of each side? Explain.

Writing

2. Consider the two methods for solving a quadratic equation presented in this section—the square root property of equality and completing the square. Discuss when it is better to use one method over the other to solve a quadratic equation.

Groupwork

3. Working with a partner, consider the equation $x^2 + 6x + c = 0$, where c is a constant.

   a. When does this equation have a single solution?

   b. For which values of c are the solutions of the equation real?

   c. For which values of c are the solutions not real?

CULTURAL NOTE

As long ago as 2000 B.C., the Babylonians had developed techniques to solve quadratic equations. A major advance in algebra took place in the sixteenth century in connection with third-degree (cubic) and fourth-degree (quartic) equations. In 1545, Girolamo Cardano (Jerome Cardan), responding to a challenge, published solutions to these equations. This publication triggered a lengthy dispute with another Italian mathematician, Niccolò Tartaglia, over credit for the discovery—one of a number of such disputes in the history of mathematics. Cardano was a multitalented Renaissance scholar. A prominent physician, he was the first person to describe typhoid fever. An inveterate gambler, he published the earliest systematic treatise on the mathematics of probability. A talented inventor, Cardano's conviction that a code in one's mind is more secure than a physical key led to his inventing the combination lock. Many books have been written about this remarkable man's colorful life.

(**Sources:** W. Dunham, *Journey Through Genius: The Great Theorems of Mathematics*, New York: John Wiley and Sons, 1990; Peter L. Bernstein, *Against the Gods: The Remarkable Story of Risk*, New York: John Wiley and Sons, 1996; Girolamo Cardano, *The Book of My Life*, Toronto: J. M. Dent and Sons, 1931)

9.2 Solving Quadratic Equations: The Quadratic Formula

Some quadratic equations cannot be solved either by factoring or by the square root property of equality. On the other hand, any quadratic equation in standard form, $ax^2 + bx + c = 0$, *can* be solved by completing the square.

In fact, we can use the method of completing the square to derive a general formula, called the *quadratic formula*, to solve quadratic equations. In this section, we show how to use the quadratic formula to solve quadratic equations. We also discuss the *discriminant* of a quadratic equation—an expression that allows us to determine the number and type of solutions that the equation has without solving the equation.

OBJECTIVES

■ To solve quadratic equations using the quadratic formula

■ To determine the number and type of solutions to a quadratic equation using the discriminant

■ To solve applied problems using the quadratic formula

Deriving the Quadratic Formula

Consider an arbitrary quadratic equation in standard form: $ax^2 + bx + c = 0$, where $a \neq 0$. We apply the method of completing the square to this equation to solve for x.

$$ax^2 + bx + c = 0$$

$$x^2 + \frac{b}{a}x + \frac{c}{a} = 0 \qquad \text{Divide each side of the equation by } a.$$

$$x^2 + \frac{b}{a}x = -\frac{c}{a} \qquad \text{Add } -\frac{c}{a} \text{ to each side.}$$

$$x^2 + \frac{b}{a}x + \frac{b^2}{4a^2} = -\frac{c}{a} + \frac{b^2}{4a^2} \qquad \text{Complete the square: } \left(\frac{1}{2} \cdot \frac{b}{a}\right)^2 = \frac{b^2}{4a^2}; \text{ add } \frac{b^2}{4a^2} \text{ to each side of the equation.}$$

$$\left(x + \frac{b}{2a}\right)^2 = \frac{b^2 - 4ac}{4a^2} \qquad \text{Write the left side as the square of a binomial, and write the right side as a single rational expression.}$$

$$x + \frac{b}{2a} = \pm\sqrt{\frac{b^2 - 4ac}{4a^2}} \qquad \text{Use the square root property of equality.}$$

$$x = -\frac{b}{2a} \pm \frac{\sqrt{b^2 - 4ac}}{2a} \qquad \text{Add } -\frac{b}{2a} \text{ to each side of the equation and simplify the radical.}$$

$$x = \frac{-b \pm \sqrt{b^2 - 4ac}}{2a} \qquad \text{Combine the rational expressions.}$$

So the solutions are $\dfrac{-b + \sqrt{b^2 - 4ac}}{2a}$ and $\dfrac{-b - \sqrt{b^2 - 4ac}}{2a}$. This result gives us the quadratic formula.

The Quadratic Formula

If $ax^2 + bx + c = 0$, where a, b, and c are real numbers and $a \neq 0$, then

$$x = \frac{-b \pm \sqrt{b^2 - 4ac}}{2a}$$

We will be considering a number of examples of applying this formula to solve quadratic equations.

Solving Quadratic Equations Using the Quadratic Formula

The quadratic formula allows us to solve any quadratic equation of the form $ax^2 + bx + c = 0$, where $a \neq 0$. For instance, consider the equation $2x^2 - x = 3$. To solve, we first write the equation in standard form:

$$2x^2 - x = 3$$
$$2x^2 - x - 3 = 0 \qquad \text{Add } -3 \text{ to each side of the equation.}$$

In this equation, $a = 2$, $b = -1$, and $c = -3$. So we substitute these values for a, b, and c in the quadratic formula and then simplify.

$$x = \frac{-b \pm \sqrt{b^2 - 4ac}}{2a}$$

$$= \frac{-(-1) \pm \sqrt{(-1)^2 - 4(2)(-3)}}{2(2)} \qquad \text{Substitute 2 for } a, -1 \text{ for } b, \text{ and } -3 \text{ for } c.$$

$$= \frac{1 \pm \sqrt{1 + 24}}{4}$$

$$= \frac{1 \pm \sqrt{25}}{4}$$

$$= \frac{1 \pm 5}{4}$$

$$x = \frac{1 + 5}{4} \qquad \text{or} \qquad x = \frac{1 - 5}{4}$$

$$= \frac{6}{4} \qquad\qquad\qquad = \frac{-4}{4}$$

$$= \frac{3}{2} \qquad\qquad\qquad = -1$$

So the solutions are $\dfrac{3}{2}$ and -1.

Note that the equation $2x^2 - x = 3$ could also have been solved by factoring:

$$2x^2 - x - 3 = 0$$
$$(2x - 3)(x + 1) = 0$$
$$2x - 3 = 0 \quad \text{or} \quad x + 1 = 0$$
$$2x = 3 \qquad\qquad x = -1$$
$$x = \frac{3}{2}$$

We see that solving the equation by factoring yields the same solutions as those we found using the quadratic formula. Whenever the solutions to a quadratic equation are rational numbers, as in this example, the original equation can also be solved by factoring.

To Solve a Quadratic Equation Using the Quadratic Formula

- Write the equation in standard form, if necessary.
- Identify the coefficients a, b, and c.
- Substitute the values for a, b, and c in the quadratic formula: $x = \dfrac{-b \pm \sqrt{b^2 - 4ac}}{2a}$.
- Simplify.

EXAMPLE 1

Solve $x^2 + x - 3 = 0$ by using the quadratic formula.

Solution The equation $x^2 + x - 3 = 0$ is already in standard form, where $a = 1$, $b = 1$, and $c = -3$. Substitute these values in the quadratic formula.

$$x = \frac{-b \pm \sqrt{b^2 - 4ac}}{2a}$$

$$x = \frac{-(1) \pm \sqrt{(1)^2 - 4(1)(-3)}}{2(1)}$$

$$= \frac{-1 \pm \sqrt{1 + 12}}{2}$$

$$= \frac{-1 \pm \sqrt{13}}{2}$$

$$x = \frac{-1 + \sqrt{13}}{2} \quad \text{or} \quad x = \frac{-1 - \sqrt{13}}{2}$$

So the solutions are $\dfrac{-1 + \sqrt{13}}{2}$ and $\dfrac{-1 - \sqrt{13}}{2}$.

PRACTICE 1

Solve: $y^2 + 3y - 5 = 0$

Can you explain how to check that the solutions in Example 1 are correct?

EXAMPLE 2

Use the quadratic formula to solve $3x^2 + 2x = -4$.

Solution First, let's write $3x^2 + 2x = -4$ in standard form: $3x^2 + 2x + 4 = 0$. We see that $a = 3$, $b = 2$ and $c = 4$. Substitute these values in the formula.

$$x = \frac{-b \pm \sqrt{b^2 - 4ac}}{2a}$$

$$= \frac{-(2) \pm \sqrt{(2)^2 - 4(3)(4)}}{2(3)}$$

$$= \frac{-2 \pm \sqrt{4 - 48}}{6}$$

$$= \frac{-2 \pm \sqrt{-44}}{6}$$

$$= \frac{-2 \pm 2i\sqrt{11}}{6}$$

$$= \frac{2(-1 \pm i\sqrt{11})}{6_3}$$

$$= \frac{-1 \pm i\sqrt{11}}{3} \quad \text{or} \quad -\frac{1}{3} \pm \frac{i\sqrt{11}}{3}$$

So the solutions are $-\dfrac{1}{3} + \dfrac{i\sqrt{11}}{3}$ or $-\dfrac{1}{3} - i\dfrac{\sqrt{11}}{3}$.

PRACTICE 2

Solve: $6p^2 - 2p = -1$

In solving quadratic equations, we sometimes need to approximate the solutions. This is often the case when solving applied problems.

EXAMPLE 3	PRACTICE 3

Solve $\dfrac{y^2}{3} + \dfrac{y}{5} = 1$. Round the solutions to the nearest thousandth.

Solution First, we multiply each side of the equation by the LCD, 15, to clear the equation of fractions:

$$\frac{y^2}{3} + \frac{y}{5} = 1$$

$$15 \cdot \frac{y^2}{3} + 15 \cdot \frac{y}{5} = 15 \cdot 1$$

$$5y^2 + 3y = 15$$

Then, we write the equation in standard form, $5y^2 + 3y - 15 = 0$, letting $a = 5$, $b = 3$, and $c = -15$. Substituting in the quadratic formula gives us:

$$y = \frac{-b \pm \sqrt{b^2 - 4ac}}{2a}$$

$$= \frac{-3 \pm \sqrt{(3)^2 - 4(5)(-15)}}{2(5)}$$

$$= \frac{-3 \pm \sqrt{9 + 300}}{10}$$

$$= \frac{-3 \pm \sqrt{309}}{10}$$

So the solutions are

$$\frac{-3 + \sqrt{309}}{10} \approx 1.458 \text{ and } \frac{-3 - \sqrt{309}}{10} \approx -2.058$$

Solve $\dfrac{n^2}{5} - \dfrac{n}{2} = 3$. Round the solutions to the nearest thousandth.

EXAMPLE 4	PRACTICE 4

Solve: $0.04m^2 + 0.25 = 0.2m$

Solution In standard form, the equation becomes $0.04m^2 - 0.2m + 0.25 = 0$. To clear the equation of decimals, let's multiply through by 100:

$$0.04m^2 - 0.2m + 0.25 = 0$$

$$100(0.04)m^2 - 100(0.2)m + 100(0.25) = 100(0)$$

$$4m^2 - 20m + 25 = 0$$

So $a = 4$, $b = -20$, and $c = 25$. Substituting in the quadratic formula gives us:

$$m = \frac{-b \pm \sqrt{b^2 - 4ac}}{2a}$$

$$= \frac{-(-20) \pm \sqrt{(-20)^2 - 4(4)(25)}}{2(4)}$$

Use the quadratic formula to solve:
$0.01v^2 = -0.18v - 0.81$

$$= \frac{20 \pm \sqrt{400 - 400}}{8}$$

$$= \frac{20 \pm \sqrt{0}}{8}$$

$$= \frac{20}{8}$$

$$= \frac{5}{2}$$

So there is only one solution, namely, $\frac{5}{2}$.

Can you explain in terms of the quadratic formula when a quadratic equation will have a single solution?

The Discriminant

It is possible to determine the number and type of solutions to a quadratic equation without actually solving the equation. To do this, we compute the value of the radicand in the quadratic formula, that is, $b^2 - 4ac$. This quantity is called the **discriminant** of the equation.

Consider, for instance, the equation $x^2 - 5x + 3 = 0$. Here $a = 1$, $b = -5$, and $c = 3$. Using the quadratic formula, we get:

$$x = \frac{-b \pm \sqrt{b^2 - 4ac}}{2a}$$

$$= \frac{-(-5) \pm \sqrt{(-5)^2 - 4(1)(3)}}{2(1)}$$

$$= \frac{5 \pm \sqrt{13}}{2}$$

Note that the value of this discriminant, 13, is *positive* and that the quadratic equation has *two real solutions*, namely, $\frac{5 + \sqrt{13}}{2}$ and $\frac{5 - \sqrt{13}}{2}$.

Next, let's consider the equation $x^2 - 3x + 5 = 0$. Here $a = 1$, $b = -3$, and $c = 5$. Using the quadratic formula, we get

$$x = \frac{-b \pm \sqrt{b^2 - 4ac}}{2a}$$

$$= \frac{-(-3) \pm \sqrt{(-3)^2 - 4(1)(5)}}{2(1)}$$

$$= \frac{3 \pm \sqrt{-11}}{2}$$

Here the value of the discriminant, -11, is *negative* and the quadratic equation has *two complex solutions containing i*, namely, $\frac{3 + i\sqrt{11}}{2}$ and $\frac{3 - i\sqrt{11}}{2}$.

Finally, we consider the equation $x^2 - 2x + 1 = 0$. Here $a = 1$, $b = -2$, and $c = 1$. From the quadratic formula, we get

$$x = \frac{-b \pm \sqrt{b^2 - 4ac}}{2a}$$

$$= \frac{-(-2) \pm \sqrt{(-2)^2 - 4(1)(1)}}{2(1)}$$

$$= \frac{2 \pm \sqrt{0}}{2}$$

$$= 1$$

The value of the discriminant in this case is *zero* and the quadratic equation has *one real solution*, namely, 1.

The preceding examples suggest that the value of a quadratic equation's discriminant can tell us if the equation has one real solution, two real solutions, or two complex solutions. This is summarized in the following table.

Value of the Discriminant ($b^2 - 4ac$)	Number and Type of Solutions
Zero	One real solution
Positive	Two real solutions
Negative	Two complex solutions (containing i)

EXAMPLE 5

For each of the following equations, evaluate the discriminant. Then use the discriminant to determine the number and type of solutions that the equation has.

a. $6x^2 + 11x - 7 = 0$ **b.** $5y^2 - y + 3 = 0$ **c.** $4x^2 = -12x - 9$

Solution

Equation	a	b	c	Discriminant ($b^2 - 4ac$)	Number of Solutions	Type of Solution
a. $6x^2 + 11x - 7 = 0$	6	11	−7	$(11)^2 - 4(6)(-7)$ $= 121 + 168 = 289,$ which is *positive*	2	Real numbers
b. $5y^2 - y + 3 = 0$	5	−1	3	$(-1)^2 - 4(5)(3)$ $= 1 - 60 = -59,$ which is *negative*	2	Complex numbers (containing i)
c. $4x^2 = -12x - 9$	4	12	9	$(12)^2 - 4(4)(9)$ $= 144 - 144 = 0,$ which is *zero*	1	Real number

PRACTICE 5

Complete the following table.

Equation	a	b	c	Discriminant $(b^2 - 4ac)$	Number of Solutions	Type of Solution
a. $p^2 + 6p + 9 = 0$						
b. $x^2 + 6x - 9 = 0$						
c. $4n^2 = -3 - 2n$						

EXAMPLE 6

The use of cell phones has recently grown rapidly. The equation $y = 0.5x^2 + 0.2x + 1.2$ models the number of American cell phone subscribers in each of the years between 1988 and 1998, where x is the number of years after 1988 and y is the number of subscribers (in millions). Use this equation to determine the year in which there were approximately 120 million American cell phone subscribers. (*Source:* U.S. Department of Commerce, International Trade Administration)

Solution We begin by setting y equal to 120 and then solving the equation:

$$120 = 0.5x^2 + 0.2x + 1.2$$
$$0 = 0.5x^2 + 0.2x - 118.8$$
$$0.5x^2 + 0.2x - 118.8 = 0$$

Next, we substitute in the quadratic formula, letting $a = 0.5$, $b = 0.2$, and $c = -118.8$:

$$x = \frac{-b \pm \sqrt{b^2 - 4ac}}{2a}$$

$$= \frac{-0.2 \pm \sqrt{(0.2)^2 - 4(0.5)(-118.8)}}{2(0.5)}$$

$$= \frac{-0.2 \pm \sqrt{0.04 - 4(0.5)(-118.8)}}{2(0.5)}$$

$$= \frac{-0.2 \pm \sqrt{0.04 + 237.6}}{1}$$

$$= -0.2 \pm \sqrt{237.64}$$

So the solutions of the equation are $-0.2 + \sqrt{237.64} \approx 15.216$ and $-0.2 - \sqrt{237.64} \approx -15.616$. Since x cannot be negative, we conclude that x is 15.216, or approximately 15. Therefore, there were approximately 120 million American cell phone subscribers in $1988 + 15$, or 2003.

PRACTICE 6

The Standard & Poor's 500 Index is a measure of the value of 500 stocks that are traded on three major stock exchanges. The value, y, of this index between 1994 and 2000 can be approximated by the equation $y = 11.8n^2 + 83.2n + 432.9$, where n is the number of years after 1994. In what year was the value of the index approximately 1000? (*Source:* Standard and Poor's/DRI)

Mathematically Speaking

Fill in each blank with the most appropriate term or phrase from the given list.

one	quadratic equation	negative
discriminant	numerator	quadratic formula
positive	no	

1. The _____ states that if $ax^2 + bx + c = 0$, where a, b, and c are real numbers and $a \neq 0$, then $x = \dfrac{-b \pm \sqrt{b^2 - 4ac}}{2a}$.

2. The _____ is the radicand in the quadratic formula.

3. If the discriminant of a quadratic equation equals zero, the equation has _____ real solution(s).

4. If a quadratic equation has two real solutions, its discriminant is _____.

Solve.

5. $x^2 + 3x + 2 = 0$

6. $x^2 - 7x + 12 = 0$

7. $x^2 - 6x - 1 = 0$

8. $x^2 + 8x - 4 = 0$

9. $x^2 = x + 11$

10. $x^2 = 2 - 5x$

11. $x^2 - 4x + 13 = 8$

12. $x^2 + 2x + 11 = 6$

13. $3x^2 + 6x = 7$

14. $2x^2 - 10x = 1$

15. $6t^2 - 8t = 3t - 4$

16. $8n^2 + 5n = 7n + 3$

17. $2x^2 + 8x + 9 = 0$

18. $4x^2 - 10x + 7 = 0$

19. $1 - 5x^2 = 4x^2 + 6x$

20. $-3x^2 + 14x = 8x - 5$

21. $2y^2 - 9y + 10 = 1 + 3y - 2y^2$

22. $13p^2 + 8p - 2 = 4p^2 + 2p - 3$

23. $\dfrac{x^2}{4} - \dfrac{x}{2} = -3$

24. $\dfrac{x^2}{9} + \dfrac{x}{3} = -1$

25. $\dfrac{1}{2}x^2 + \dfrac{2}{3}x - \dfrac{5}{6} = 0$

26. $\dfrac{1}{8}x^2 - \dfrac{1}{4}x - \dfrac{1}{12} = 0$

27. $0.2x^2 + x + 0.8 = 0$

28. $0.5x^2 + 0.3x - 0.2 = 0$

29. $0.03x^2 - 0.12x + 0.24 = 0$

30. $0.02x^2 + 0.16x + 0.34 = 0$

31. $(x + 6)(x + 2) = 8$

32. $(x + 2)(x - 4) = 1$

33. $(2x - 3)^2 = 8(x + 1)$

34. $(3x + 1)^2 = 2(1 - 3x)$

Solve. Round the answers to the nearest thousandth.

35. $1.4x^2 - 2.7x - 0.1 = 0$ **36.** $0.6x^2 - 4.9x - 3.3 = 0$ **37.** $0.003x^2 + 0.23x + 1.124 = 0$

38. $2.04x^2 + 0.45x + 0.017 = 0$

Use the discriminant to determine the number and type of solutions for each of the following equations.

39. $x^2 + 2x + 4 = 0$ **40.** $x^2 - 8x + 16 = 0$ **41.** $4x^2 - 12x = -9$ **42.** $3x^2 + 6x = -1$

43. $6x^2 = 2 - 5x$ **44.** $10x = 16 - 2x^2$ **45.** $7x^2 - x + 3 = 0$ **46.** $8x^2 - 7x - 1 = 0$

47. $3x^2 + 10 = 0$ **48.** $2x^2 - 7 = 0$

Mixed Practice

Use the discriminant to determine the number and type of solutions for each of the following equations.

49. $2x^2 + 8 = 7x$ **50.** $5x = 2 + 3x^2$

Solve.

51. $2x^2 - 4x + 2 = -3$ **52.** $x - 2x^2 - 3 = 2x^2 - 7x - 8$ **53.** $\dfrac{x^2}{4} - \dfrac{2x}{3} - \dfrac{1}{6} = 0$

54. $0.8x^2 - 3.7x - 0.5 = 0$
(Round to the nearest thousandth.)

Applications

Solve.

55. The height h of a stone t sec after it is thrown straight downward with an initial velocity of 20 feet per second from a bridge 800 ft above the ground is given by the equation $h = -16t^2 - 20t + 800$. When is the stone 300 ft above the ground?

56. A software company's weekly profit, P (in dollars), for selling x units of a new video game can be determined by the equation $P = -0.05x^2 + 48x - 100$. What is the smaller of the two numbers of units that must be sold in order to make a profit of $9000?

57. An open box is made from a 36-in. by 20-in. rectangular piece of cardboard by cutting squares of equal area from the corners, as shown in the following figure. If the design specifications require that the base of the box have an area of 465 in², what size squares should be cut from each corner?

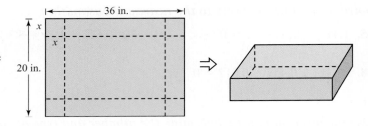

58. A homeowner decides to increase the area of her rectangular garden by increasing the length and the width, as shown in the figure. What are the new dimensions of the garden if it is to have an area of 315 ft²?

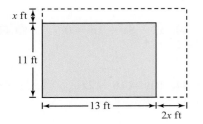

59. A travel company offers special pricing on a weekend getaway trip. For each person who buys a ticket, the price is reduced by $5. The regular price of a ticket is $300. To determine the revenue, R, for selling x tickets, the company uses the equation $R = x(300 - 5x)$. If the company wants to keep the reduced ticket price above $200, how many tickets must it sell in order to generate revenue of $3520?

60. One airplane flies east from an airport at an average speed of m mph. At the same time, a second airplane flies south from the same airport at an average speed that is 42 mph faster than the airplane traveling east. Thirty minutes after the two airplanes leave, they are 315 mi apart. Find the rate of each airplane to the nearest mile per hour.

61. Identity theft occurs when a person's name, Social Security number, or other identifying information is used without his or her knowledge to commit fraud or other crimes. The number of reports R, in thousands, of identity theft for each year from 2000 to 2003 can be approximated by the equation

$R = -\dfrac{1}{2}t^2 + 64t + 29$, where t is the number of years

after 2000. In what year were there approximately 215,000 reports of identity theft? (**Source:** Federal Trade Commission)

62. The net revenue R, in billions of dollars, for Starbucks Corporation for the years from 1992 to 2003 can be approximated by the equation $R = 0.031x^2 + 0.004x + 0.15$, where x represents the number of years after 1992. In what year was the net revenue approximately $3.3 billion? (**Source:** Starbucks Corporation, *Fiscal 2003 Annual Report*)

● *Check your answers on page A-39.*

MINDSTRETCHERS

Critical Thinking

1. Solve each of the following equations.

   a. $2x^2 + \sqrt{3}x - 3 = 0$

   b. $x^2 - ix + 6 = 0$

Mathematical Reasoning

2. Consider the equation $ax^2 + bx + c = 0$. For what values of c will the equation have two real solutions? Two complex solutions? Give examples to support your answers.

Writing

3. Do you think "discriminant" is a good name for the expression $b^2 - 4ac$? In a couple of sentences, explain why.

9.3   More on Quadratic Equations

We have discussed several methods of solving quadratic equations—factoring, the square root property, completing the square, and the quadratic formula. In this section, we use these methods to solve some additional types of problems relating to quadratic equations. The first of these problems involves equations that, while not quadratic, can be transformed to quadratic equations. In the second type of problem, we are given the solutions to an unknown quadratic equation and need to determine the equation.

OBJECTIVES

- To solve an equation that leads to a quadratic equation
- To solve an equation that is quadratic in form
- To determine a quadratic equation with given solutions
- To solve applied problems involving quadratic equations

Equations That Lead to Quadratic Equations

Recall from Section 7.5 that some equations, while not quadratic, lead to quadratic equations. For instance, in the following example, we solve a rational equation that leads to a quadratic equation.

EXAMPLE 1

Solve and check: $\dfrac{3}{x} - \dfrac{2}{2x-1} = 1$

Solution First, we multiply each side of the equation by the LCD of the denominators, $x(2x-1)$:

$$\frac{3}{x} - \frac{2}{2x-1} = 1$$

$$x(2x-1)\frac{3}{x} - x(2x-1)\frac{2}{2x-1} = x(2x-1)1$$

Multiply each side by the LCD and divide out common factors.

$$3(2x-1) - 2x = x(2x-1)$$ Simplify.
$$6x - 3 - 2x = 2x^2 - x$$ Use the distributive property.
$$4x - 3 = 2x^2 - x$$ Combine like terms.
$$2x^2 - 5x + 3 = 0$$ Write in standard form.

Next, we factor:

$$(2x-3)(x-1) = 0$$

To solve for x, we set each factor equal to 0:

$$2x - 3 = 0 \quad \text{or} \quad x - 1 = 0$$
$$2x = 3 \qquad\qquad x = 1$$
$$x = \frac{3}{2}$$

Check

Substitute $\dfrac{3}{2}$ for x.   Substitute 1 for x.

$$\frac{3}{x} - \frac{2}{2x-1} = 1 \qquad\qquad \frac{3}{x} - \frac{2}{2x-1} = 1$$

$$\frac{3}{\frac{3}{2}} - \frac{2}{2(\frac{3}{2})-1} \overset{?}{=} 1 \qquad\qquad \frac{3}{1} - \frac{2}{2(1)-1} \overset{?}{=} 1$$

PRACTICE 1

Solve and check: $1 = \dfrac{2}{x+2} + x$

$$\frac{6}{3} - \frac{2}{3-1} \overset{?}{=} 1 \qquad\qquad\qquad 3 - 2 \overset{?}{=} 1$$

$$2 - 1 \overset{?}{=} 1 \qquad\qquad\qquad\qquad 1 = 1 \quad \text{True}$$

$$1 = 1 \quad \text{True}$$

So the solutions are $\dfrac{3}{2}$ and 1.

Equations That Are Quadratic in Form

Some equations are **quadratic in form**. That is, they can be rewritten in the form $au^2 + bu + c = 0$, where $a \neq 0$ and u represents an algebraic expression. For example, the equation $x^4 - 3x^2 - 4 = 0$ is quadratic in form since it can be written as $(x^2)^2 - 3(x^2) - 4 = 0$, or $u^2 - 3u - 4 = 0$, where u stands for x^2.

To solve an equation that is quadratic in form, we transform the equation to a quadratic equation by rewriting it in terms of a new variable, as shown in the following example.

EXAMPLE 2	PRACTICE 2
Solve and check: $x^4 - 5x^2 - 6 = 0$	Solve and check: $n^4 - 9n^2 + 8 = 0$

Solution This equation is quadratic in form. If we let a new variable, say u, stand for x^2, we get

$$x^4 - 5x^2 - 6 = 0$$
$$(x^2)^2 - 5(x^2) - 6 = 0$$
$$(u)^2 - 5(u) - 6 = 0 \qquad \text{Substitute } u \text{ for } x^2.$$
$$u^2 - 5u - 6 = 0$$

Next, let's factor to solve for u:

$$(u - 6)(u + 1) = 0$$
$$u - 6 = 0 \quad \text{or} \quad u + 1 = 0$$
$$u = 6 \qquad\qquad u = -1$$

Having found the two possible values of u, we now solve for the original variable x. Since u represents x^2, we have

$$x^2 = 6 \qquad \text{or} \quad x^2 = -1$$
$$x = \pm\sqrt{6} \qquad\qquad x = \pm\sqrt{-1}$$
$$x = \pm i$$

Check We substitute these values for x in the *original equation*.

Substitute $\sqrt{6}$ for x.   Substitute $-\sqrt{6}$ for x.

$$x^4 - 5x^2 - 6 = 0 \qquad\qquad\qquad x^4 - 5x^2 - 6 = 0$$
$$(\sqrt{6})^4 - 5(\sqrt{6})^2 - 6 \overset{?}{=} 0 \qquad (-\sqrt{6})^4 - 5(-\sqrt{6})^2 - 6 \overset{?}{=} 0$$
$$6^2 - 5 \cdot 6 - 6 \overset{?}{=} 0 \qquad\qquad\qquad 6^2 - 5 \cdot 6 - 6 \overset{?}{=} 0$$
$$0 = 0 \quad \text{True} \qquad\qquad\qquad\qquad 0 = 0 \quad \text{True}$$

Substitute i for x.

$$x^4 - 5x^2 - 6 = 0$$
$$(i)^4 - 5(i)^2 - 6 \stackrel{?}{=} 0$$
$$(-1)^2 - 5(-1) - 6 \stackrel{?}{=} 0$$
$$0 = 0 \quad \text{True}$$

Substitute $-i$ for x.

$$x^4 - 5x^2 - 6 = 0$$
$$(-i)^4 - 5(-i)^2 - 6 \stackrel{?}{=} 0$$
$$(-1)^2 - 5(-1) - 6 \stackrel{?}{=} 0$$
$$0 = 0 \quad \text{True}$$

All solutions check. So the original equation has four solutions, namely, $\sqrt{6}$, $-\sqrt{6}$, i, and $-i$.

Tip When introducing a new variable to substitute for a given expression, we can choose any letter to represent the new variable.

EXAMPLE 3

Solve and check: $y - 2\sqrt{y} = 8$

Solution Recognizing that $(\sqrt{y})^2 = y$, we can write the given equation as follows:

$$(\sqrt{y})^2 - 2\sqrt{y} = 8$$

Substituting u for $\sqrt{y}$ gives us:

$$u^2 - 2u = 8$$
$$u^2 - 2u - 8 = 0$$

Now, we solve for u by factoring.

$$(u - 4)(u + 2) = 0$$
$$u - 4 = 0 \quad \text{or} \quad u + 2 = 0$$
$$u = 4 \qquad\qquad u = -2$$

Having found the possible values of u, we now solve for the original variable y. Since u represents $\sqrt{y}$, we get:

$$\sqrt{y} = 4 \qquad \text{or} \qquad \sqrt{y} = -2$$
$$(\sqrt{y})^2 = (4)^2 \qquad (\sqrt{y})^2 = (-2)^2$$
$$y = 16 \qquad\qquad y = 4$$

Check

Substitute 16 for y.

$$y - 2\sqrt{y} = 8$$
$$16 - 2\sqrt{16} \stackrel{?}{=} 8$$
$$8 = 8 \quad \text{True}$$

Substitute 4 for y.

$$y - 2\sqrt{y} = 8$$
$$4 - 2\sqrt{4} \stackrel{?}{=} 8$$
$$0 = 8 \quad \text{False}$$

So the only solution of the original equation is 16.

PRACTICE 3

Solve and check: $x - 3\sqrt{x} = 10$

EXAMPLE 4

Solve and check: $x^{1/2} - 4x^{1/4} + 3 = 0$

Solution We can rewrite this equation as follows:

$$(x^{1/4})^2 - 4(x^{1/4}) + 3 = 0$$

Substituting u for $x^{1/4}$ and solving, we get

$$u^2 - 4u + 3 = 0$$
$$(u - 3)(u - 1) = 0$$
$$u - 3 = 0 \quad \text{or} \quad u - 1 = 0$$
$$u = 3 \qquad\qquad u = 1$$

Having found the two possible values of u, we now solve for the original variable x. Since u represents $x^{1/4}$, we have

$$\begin{array}{lll}
x^{1/4} = 3 & \text{or} & x^{1/4} = 1 \\
\sqrt[4]{x} = 3 & & \sqrt[4]{x} = 1 \\
(\sqrt[4]{x})^4 = (3)^4 & & (\sqrt[4]{x})^4 = (1)^4 \\
x = 81 & & x = 1
\end{array}$$

Check

Substitute 81 for x.

$$x^{1/2} - 4x^{1/4} + 3 = 0$$
$$(81)^{1/2} - 4(81)^{1/4} + 3 \stackrel{?}{=} 0$$
$$9 - 4 \cdot 3 + 3 \stackrel{?}{=} 0$$
$$9 - 12 + 3 \stackrel{?}{=} 0$$
$$0 = 0 \quad \textbf{True}$$

Substitute 1 for x.

$$x^{1/2} - 4x^{1/4} + 3 = 0$$
$$(1)^{1/2} - 4(1)^{1/4} + 3 \stackrel{?}{=} 0$$
$$1 - 4 \cdot 1 + 3 \stackrel{?}{=} 0$$
$$1 - 4 + 3 \stackrel{?}{=} 0$$
$$0 = 0 \quad \textbf{True}$$

The solutions of the original equation are 81 and 1.

PRACTICE 4

Solve and check: $y^{2/3} - y^{1/3} - 20 = 0$

Many applications can be modeled by equations that are not quadratic, but are quadratic in form, as in the next example.

EXAMPLE 5

A young businessman cycled to and from work, which is 2 mi from his apartment. Because of traffic, he cycled 3 mph faster going to work than coming home. It took him 10 min longer coming home than going to work. To the nearest tenth of a mile per hour, at what speed did he cycle from work?

Solution Let r represent the cycling speed coming from work. Since the businessman cycled 3 mph faster going to work, we can represent this speed by $r + 3$. We know that the distance between home and work is 2 mi. Since the product of the rate and the time is the distance traveled, the cycling time is the quotient of the distance and the rate: $t = \dfrac{d}{r}$. Therefore, the time required to cycle *to* work is $\dfrac{2}{r + 3}$ and the time

PRACTICE 5

It takes a painter 1 hr longer to paint an apartment than it takes his faster partner. If they take 5 hr working together to paint the apartment, how long would it take each of them to paint the apartment working alone?

required to cycle *from* work is $\frac{2}{r}$. We can summarize this information in a table:

	Rate ·	Time =	Distance
To work	$r + 3$	$\frac{2}{r+3}$	2
From work	r	$\frac{2}{r}$	2

We also know that the cycling time from work is 10 min longer than the cycling time to work. Since the cycling speeds are expressed in mph and there are 60 min in an hour, we rewrite the 10 min in terms of hours as follows:

$$10 \text{ min} = 10 \cdot \frac{1}{60} \text{ hr} = \frac{1}{6} \text{ hr}$$

So we conclude that the cycling time from work is $\frac{1}{6}$ hr longer than the cycling time to work, leading to the following equation:

$$\frac{2}{r} = \frac{2}{r+3} + \frac{1}{6}$$

To solve this equation, we multiply by the LCD of the denominators, $6r(r + 3)$, and then divide out common factors.

$$6r\,(r+3)\frac{2}{r} = 6r(r+3)\frac{2}{r+3} + 6r(r+3)\frac{1}{6}$$
$$12(r + 3) = 12r + r(r + 3)$$
$$12r + 36 = 12r + r^2 + 3r$$
$$0 = r^2 + 3r - 36$$
$$r^2 + 3r - 36 = 0$$

Here, $a = 1$, $b = 3$, and $c = -36$. We substitute these values in the quadratic formula and then simplify.

$$r = \frac{-b \pm \sqrt{b^2 - 4ac}}{2a}$$

$$= \frac{-(3) \pm \sqrt{(3)^2 - 4(1)(-36)}}{2(1)}$$

$$= \frac{-3 \pm \sqrt{9 + 144}}{2}$$

$$= \frac{-3 \pm \sqrt{153}}{2}$$

$$= \frac{-3 \pm 3\sqrt{17}}{2}$$

So the solutions to the equation are $\dfrac{-3 + 3\sqrt{17}}{2} \approx 4.685$ and $\dfrac{-3 - 3\sqrt{17}}{2} \approx -7.685$. Since the value of r in the context of our problem must be positive, we reject the negative solution and conclude that the cycling speed from work is approximately 4.7 mph.

Finding a Quadratic Equation That Has Given Solutions

In the typical problem involving a quadratic equation, we are given the equation and are asked to find its solutions. But suppose that we know the solutions to a quadratic equation and want to find the original equation. For instance, let's say that -4 and 3 are solutions to a quadratic equation in x. Since $x = -4$, it follows that $x + 4 = 0$. Similarly, since $x = 3$, it follows that $x - 3 = 0$. Since both $x + 4$ and $x - 3$ are equal to 0, the product of these two factors must also be equal to 0.

$$(x + 4)(x - 3) = 0$$
$$x^2 + x - 12 = 0$$

We can conclude that $x^2 + x - 12 = 0$ might have been the original equation. Note that this answer is not unique, since multiplying each side of the equation by any constant results in a different quadratic equation with the same given solutions. For example, multiplying each side of the equation $x^2 + x - 12 = 0$ by 3 yields $3x^2 + 3x - 36 = 0$, which has the same solutions.

EXAMPLE 6

Find a quadratic equation that has solutions $x = 1$ and $x = -5$.

Solution

$x = 1$ $x = -5$
$x - 1 = 0$ Subtract 1 from each side. $x + 5 = 0$ Add 5 to each side.

Setting the product of the two factors equal to 0 gives us

$$(x - 1)(x + 5) = 0$$
$$x^2 + 4x - 5 = 0$$

This quadratic equation has solutions 1 and -5, as desired.

PRACTICE 6

Find a quadratic equation in y that has solutions 0 and -3.

EXAMPLE 7

Write a quadratic equation in t with integral coefficients and with solutions $\frac{1}{2}$ and $\frac{2}{3}$.

Solution

$t = \frac{1}{2}$ $t = \frac{2}{3}$

$t - \frac{1}{2} = 0$ Subtract $\frac{1}{2}$ from each side. $t - \frac{2}{3} = 0$ Subtract $\frac{2}{3}$ from each side.

$2t - 1 = 0$ Multiply each side by 2. $3t - 2 = 0$ Multiply each side by 3.

Setting the product of the two factors equal to 0 gives us

$$(2t - 1)(3t - 2) = 0$$
$$6t^2 - 7t + 2 = 0$$

This quadratic equation has the desired solutions.

PRACTICE 7

Find a quadratic equation in n with integral coefficients that has solutions $\frac{3}{4}$ and $\frac{1}{6}$.

9.3   Exercises

Mathematically Speaking

Fill in each blank with the most appropriate term or phrase from the given list.

product	in terms of a new variable	binomial
as a linear equation		quadratic
quadratic	sum	in form

1. The equation $\dfrac{4}{x-4} + \dfrac{3}{x-1} = 1$ leads to a _____ equation.

2. An equation is _____ if it can be rewritten in the form $au^2 + bu + c = 0$, where $a \neq 0$ and u represents an algebraic expression.

3. To solve an equation that is quadratic in form, rewrite it _____.

4. To find a quadratic equation whose solutions are 6 and -4, set the _____ of $x - 6$ and $x + 4$ equal to 0.

Solve and check.

5. $2 - \dfrac{3}{x} + \dfrac{1}{x^2} = 0$

6. $3 - \dfrac{7}{x} = \dfrac{6}{x^2}$

7. $\dfrac{3}{p-1} + \dfrac{4}{p-4} = 1$

8. $\dfrac{3}{t+6} - \dfrac{1}{t-2} - 1 = 0$

9. $n^4 - 8n^2 + 12 = 0$

10. $a^4 + 2a^2 - 8 = 0$

11. $3x^4 + 11x^2 - 3 = -x^4$

12. $9y^4 - 5y^2 + 2 = 6$

13. $h - 4\sqrt{h} + 4 = 0$

14. $x - \sqrt{x} - 20 = 0$

15. $x^{1/2} + 4x^{1/4} - 32 = 0$

16. $x^{1/2} - 3x^{1/4} + 2 = 0$

17. $t^{2/3} + 3t^{1/3} + 2 = 0$

18. $y^{2/3} - y^{1/3} - 12 = 0$

19. $(n-3)^2 - 5(n-3) + 6 = 0$

20. $(p+2)^2 + 4(p+2) + 3 = 0$

21. $(2x+4) + 2\sqrt{2x+4} = 24$

22. $(3x-2) - 6\sqrt{3x-2} = -8$

Find a quadratic equation that has the given solutions.

23. $x = 7, x = 2$

24. $x = 3, x = 6$

25. $y = -4, y = 9$

26. $n = 1, n = -10$

27. $t = \dfrac{2}{3}, t = -2$

28. $a = -\dfrac{1}{4}, a = -5$

29. $n = -\dfrac{3}{2}, n = -\dfrac{4}{3}$

30. $p = \dfrac{2}{5}, p = \dfrac{5}{2}$

31. $x = 4$

32. $y = -8$

33. $t = \sqrt{2}, t = -\sqrt{2}$

34. $x = \sqrt{5}, x = -\sqrt{5}$

35. $x = 3i, x = -3i$

36. $x = 6i, x = -6i$

Mixed Practice

Solve and check.

37. $y^{2/3} - y^{1/3} = 2$

38. $\dfrac{5}{x} + 2 = \dfrac{3}{x^2}$

39. $a^4 + 7a^2 - 12 = 6$

40. $b + 5\sqrt{b} - 24 = 0$

Find a quadratic equation that has the given solutions.

41. $a = -\dfrac{2}{5}$, $a = 3$

42. $x = -7$

Applications

Solve. Use a calculator where appropriate.

43. Body mass index, or BMI, is an estimate of total body fat based on a person's height, H, and weight, W. The formula for this index is BMI $= \dfrac{W}{H^2}$, where W is in kilograms and H is in meters. Find the approximate height, to the nearest tenth of a meter, of a person who weighs 52 kg and has a BMI of 20.

44. The formula $P = \dfrac{A}{(1 + r)^2}$ can be used to determine the amount, P, that must be invested in an account with an average interest rate r (in decimal form), compounded annually, in order to have A dollars at the end of two years. If an account with an initial investment of \$10,000 has \$11,290 after two years, find the interest rate to the nearest tenth of a percent.

45. Because of traffic, a commuter drives home at an average speed that is 8 mph slower than her average speed driving to work. If she lives 15 mi from work and the round-trip commute takes 1 hr, find the average speed she drives for each part of her commute to the nearest mile per hour.

46. A couple kayaks 9 mi upriver from their campsite to a waterfall. The trip downriver takes half an hour less than the trip upriver. If the couple kayaks at an average rate of 8 mph in still water, find the rate of the current. Round to the nearest tenth.

47. A swimming pool has a small and a large drain. If only the small drain is open, it takes 1.5 hr longer to empty the pool than if only the large drain is open. If both drains are open, the pool empties in 2 hr. To the nearest tenth of an hour, find the time it takes each drain to empty the pool if only one of the drains is open.

48. Working alone, it takes an inexperienced gardener 45 min longer than an experienced gardener to mow a client's lawn. If the two gardeners work together, they can mow the lawn in 30 min. Find the time it takes each gardener working alone to mow the lawn.

● *Check your answers on page A-39.*

MINDSTRETCHERS

Groupwork

1. Consider the solutions $\dfrac{-b + \sqrt{b^2 - 4ac}}{2a}$ and $\dfrac{-b - \sqrt{b^2 - 4ac}}{2a}$ of the quadratic equation $ax^2 + bx + c = 0$.
   Work with a partner to answer each of the following.

   a. Find the sum of the solutions. What do you notice?

   b. Find the product of the solutions. What do you notice?

   c. Explain how you could use the sum and product of the solutions of a quadratic equation to find its equation in standard form.

   d. Find a quadratic equation with solutions whose sum is $-\dfrac{2}{3}$ and whose product is $\dfrac{1}{2}$.

Investigation

2. Consider each of the following polynomial equations in one variable.

   a. $3x - 4 = 0$ **b.** $x^2 - 4x + 4 = 0$ **c.** $x^3 - 4x^2 + x - 4 = 0$

   d. $x^4 - 2x^2 - 3 = 0$ **e.** $x^5 - 2x^3 - x^2 + 2 = 0$ **f.** $x^6 - 7x^3 - 8 = 0$

   i. Use the techniques of this and previous sections to find *all* the solutions of each equation.

   ii. For each equation, compare the number of solutions to the degree of the polynomial. What do you notice?

Critical Thinking

3. Solve.

   a. $\dfrac{1}{(y + 1)^2} - 2\left(\dfrac{1}{y + 1}\right) = 3$ **b.** $2n^{-2} + 3n^{-1} + 1 = 0$ **c.** $x^{1/3} + 3x^{1/6} - 4 = 0$

9.4 Graphing Quadratic Functions

In this section, we focus on graphing **quadratic functions**, that is, functions of the form $f(x) = ax^2 + bx + c$. We begin with the graph of the simplest quadratic function, $f(x) = x^2$.

The graph of a quadratic function is called a **parabola**. Parabolas model many phenomena in the real world. For instance, in the water fountain below, the path of the water is parabolic.

OBJECTIVES

- To graph quadratic functions

- To identify the vertex, the axis of symmetry, and the intercepts of a parabola

- To solve applied problems relating to the graph of a quadratic function

Graphing Quadratic Functions

We can graph quadratic functions by first choosing several values of x and then computing the corresponding values of $f(x)$ to make a table of values. Then we plot the points with these coordinates on a coordinate plane and draw a smooth curve through the points. Recall that $y = f(x)$, so that the ordered pairs (x, y) and $(x, f(x))$ represent the same point.

Let's graph the function $f(x) = x^2$. We first select several values for x, say -2, $-1, 0, 1$, and 2, and then find the corresponding values for $f(x)$. We put the results in a table.

x	$f(x) = x^2$	(x, y)
-2	$f(-2) = (-2)^2 = 4$	$(-2, 4)$
-1	$f(-1) = (-1)^2 = 1$	$(-1, 1)$
0	$f(0) = (0)^2 = 0$	$(0, 0)$
1	$f(1) = (1)^2 = 1$	$(1, 1)$
2	$f(2) = (2)^2 = 4$	$(2, 4)$

Next, we plot the points on a coordinate plane and then draw a smooth curve passing through them. The result is the graph of $f(x) = x^2$, as shown to the right. Note that this nonlinear graph is U-shaped and that it opens upward. All parabolas are U-shaped.

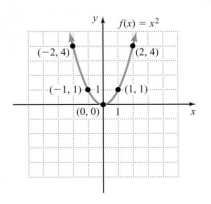

We see from the graph that the domain of the function $f(x) = x^2$ is the set of all real numbers, or, in interval notation, $(-\infty, \infty)$. Since the y-values are all nonnegative numbers, the range is $[0, \infty)$, that is, the set including all positive numbers and 0.

EXAMPLE 1

Graph $f(x) = 3x^2$.

Solution First, we select several values for x and then find the corresponding values for y.

x	$f(x) = 3x^2$	(x, y)
-2	$f(-2) = 3(-2)^2 = 3 \cdot 4 = 12$	$(-2, 12)$
-1	$f(-1) = 3(-1)^2 = 3 \cdot 1 = 3$	$(-1, 3)$
0	$f(0) = 3(0)^2 = 3 \cdot 0 = 0$	$(0, 0)$
1	$f(1) = 3(1)^2 = 3 \cdot 1 = 3$	$(1, 3)$
2	$f(2) = 3(2)^2 = 3 \cdot 4 = 12$	$(2, 12)$

Next, we plot the points on a coordinate plane and then draw a smooth curve through them. The result is a parabola, which is the graph of $f(x) = 3x^2$.

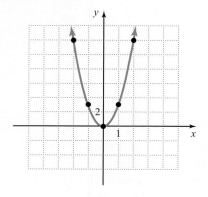

PRACTICE 1

Graph $f(x) = -3x^2$.

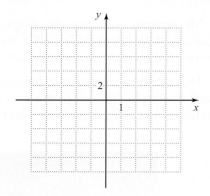

What are the differences and similarities between the graphs of $f(x) = 3x^2$ and $f(x) = -3x^2$?

The Vertex and the Axis of Symmetry

Parabolas have special characteristics that help us to sketch them. For instance, each parabola has either a highest or lowest point called a **vertex**. A graph of a quadratic equation also has an **axis of symmetry**, that is, a vertical line that passes through the vertex. If we were to fold the coordinate plane along this axis, the two parts of the parabola would coincide since each part is the mirror image of the other. That is, the parabola is symmetric with respect to its axis.

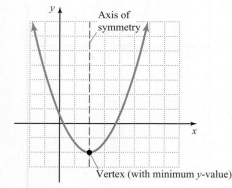

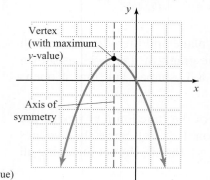

Consider the two graphs shown. The left parabola opens upward and has a lowest point with a **minimum** y-value occurring at the vertex. By contrast, the right parabola opens downward and has a highest point with a **maximum** y-value occurring at the vertex. Note that a parabola goes on indefinitely in two directions.

At the end of this section, we derive a formula for the coordinates of the vertex of a parabola given the corresponding quadratic function. The result of that derivation is as follows.

To Find the Vertex of a Parabola

If a parabola is given by the function $f(x) = ax^2 + bx + c$, its vertex is found as follows:

- The x-coordinate of the vertex is $-\dfrac{b}{2a}$.

- The y-coordinate of the vertex is found by substituting $-\dfrac{b}{2a}$ for x in $f(x)$.

Using function notation, we can express the coordinates of the vertex as

$$\left(-\frac{b}{2a},\ f\left(-\frac{b}{2a}\right)\right).$$

Knowing the coordinates of the vertex and also that the axis of symmetry is a vertical line that passes through the vertex leads us to the following conclusion.

The Equation of the Axis of Symmetry

For the graph of $f(x) = ax^2 + bx + c$, the equation of the axis of symmetry is

$$x = -\frac{b}{2a}.$$

Note that in Example 1, the vertex of the parabola is the point $(0, 0)$ and the line of symmetry is the y-axis. This is true for the graph of any quadratic function of the form $f(x) = ax^2$.

Although we can always make a table of values to graph a quadratic function, we consider two other methods of graphing quadratic equations. One method involves graphing the vertex and several points on either side of the vertex. The other method involves graphing the x- and y-intercepts as well as the vertex.

EXAMPLE 2	PRACTICE 2

EXAMPLE 2

Graph: $f(x) = 6 - 3x^2$

Solution Writing the function as $f(x) = -3x^2 + 0x + 6$, we see that $a = -3$ and $b = 0$. Next, we find the coordinates of the vertex. The x-coordinate of the vertex is

$$-\frac{b}{2a} = -\frac{0}{2(-3)} = 0$$

To find the y-coordinate of the vertex, we substitute 0 for x in $f(x)$:

$$
\begin{aligned}
f(x) &= 6 - 3x^2 \\
f(0) &= 6 - 3(0)^2 \qquad \text{\small Substitute 0 for } x \text{ in } f(x). \\
&= 6 - 0 \\
&= 6
\end{aligned}
$$

So the vertex of the parabola is the point $(0, 6)$.
Next, we choose points on each side of the vertex.

PRACTICE 2

Graph: $f(x) = -4x^2 + 1$

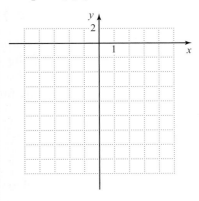

x	$f(x) = 6 - 3x^2$	(x, y)
-2	$f(-2) = 6 - 3(-2)^2 = -6$	$(-2, -6)$
-1	$f(-1) = 6 - 3(-1)^2 = 3$	$(-1, 3)$
0	$f(0) = 6 - 3(0)^2 = 6$	$(0, 6)$ ← Vertex
1	$f(1) = 6 - 3(1)^2 = 3$	$(1, 3)$
2	$f(2) = 6 - 3(2)^2 = -6$	$(2, -6)$

Now, we plot the points and draw a smooth curve through them.

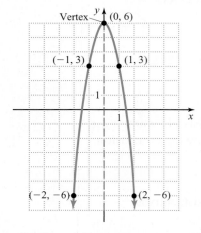

The result is a parabola, which is the graph of $f(x) = 6 - 3x^2$. Note that the y-axis, $x = 0$, is the axis of symmetry for this parabola.

In Example 1, we saw that for $f(x) = 3x^2$, a is positive and the parabola opens upward. On the other hand, for $f(x) = 6 - 3x^2$ in Example 2, a is negative and the parabola opens downward. Knowing the sign of a helps us to graph $f(x) = ax^2 + bx + c$.

> **Tip**   When a is positive, the parabola opens upward and has a minimum point. When a is negative, the parabola opens downward and has a maximum point.

The Intercepts

Recall that an x-intercept of a graph is a point where the graph crosses the x-axis, and the y-intercept is a point where the graph crosses the y-axis. The intercepts, like the vertex, are important characteristics of a parabola. We can graph a quadratic function using the vertex and the x- and y-intercepts.

EXAMPLE 3	PRACTICE 3

EXAMPLE 3

Consider $y = x^2 - 8x + 12$.

a. Find the vertex and the axis of symmetry of the graph.

b. Determine if the graph opens upward or downward.

c. Find the x- and y-intercepts.

d. Sketch the graph.

PRACTICE 3

Consider $y = -x^2 + 4x + 5$.

a. Find the vertex and the axis of symmetry of the graph.

Solution

a. First, let's find the vertex. Since $a = 1$ and $b = -8$, the x-coordinate of the vertex is

$$-\frac{b}{2a} = -\frac{(-8)}{2(1)} = \frac{8}{2} = 4$$

Substituting 4 for x, we get:

$$
\begin{aligned}
y &= x^2 - 8x + 12 \\
&= (4)^2 - 8(4) + 12 \\
&= 16 - 32 + 12 \\
&= -4
\end{aligned}
$$

So the vertex is the point $(4, -4)$. The equation of the axis of symmetry is $x = 4$, that is, the vertical line passing through the vertex.

b. The graph opens upward, since the leading coefficient, 1, is positive.

c. Next, we find the x- and y-intercepts. To find the x-intercepts of a graph, we let $y = 0$ and solve for x.

$$
\begin{aligned}
y &= x^2 - 8x + 12 \\
0 &= x^2 - 8x + 12 \\
0 &= (x - 6)(x - 2) \\
x - 6 = 0 \quad &\text{or} \quad x - 2 = 0 \\
x = 6 \qquad & \qquad x = 2
\end{aligned}
$$

So the x-intercepts are $(6, 0)$ and $(2, 0)$. Note that this parabola has two x-intercepts. To find the y-intercept, we let $x = 0$.

$$
\begin{aligned}
y &= x^2 - 8x + 12 \\
&= (0)^2 - 8(0) + 12 \\
&= 12
\end{aligned}
$$

The y-intercept is therefore $(0, 12)$.

d. To graph the parabola, let's first plot the vertex and the x- and y-intercepts on a coordinate plane. Then, we draw a smooth curve through the points. The result is the following parabola:

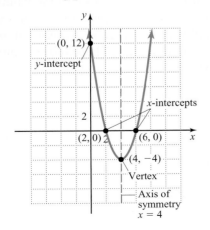

b. Determine if the graph opens upward or downward.

c. Find the x- and y-intercepts.

d. Sketch the graph.

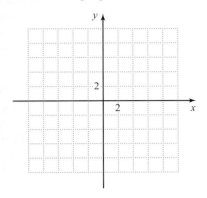

Note that in Example 3 the graph of $y = x^2 - 8x + 12$ has x-intercepts $(6, 0)$ and $(2, 0)$ and that the equation $x^2 - 8x + 12 = 0$ has solutions 6 and 2.

In general, the x-intercepts of the graph of $y = ax^2 + bx + c$ have as their x-coordinates the real solutions to the equation $ax^2 + bx + c = 0$.

The Discriminant

Recall from Section 9.2 that we can use the discriminant of the quadratic equation $ax^2 + bx + c = 0$ to determine the number and type of solutions of the equation. If the discriminant $b^2 - 4ac$ is positive, the equation will have two real solutions and the graph of $f(x) = ax^2 + bx + c$ will have two x-intercepts. If the discriminant is 0, there will be only one real solution and the graph will have only one x-intercept. And if the discriminant is negative, there will be no real solutions, and the graph will have no x-intercepts.

The following table shows the graph of $f(x) = ax^2 + bx + c$ for possible values of the leading coefficient, a, and the discriminant, $b^2 - 4ac$.

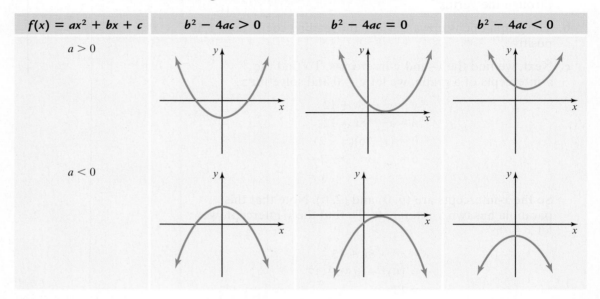

$f(x) = ax^2 + bx + c$	$b^2 - 4ac > 0$	$b^2 - 4ac = 0$	$b^2 - 4ac < 0$
$a > 0$			
$a < 0$			

Although the graph of $f(x) = ax^2 + bx + c$ may have two, one, or zero x-intercepts, it must have exactly one y-intercept. Can you explain why?

EXAMPLE 4	PRACTICE 4
For each function, determine whether its graph opens upward or downward and whether there is a minimum or a maximum point. Then find the number of x- and y-intercepts.	For each function, determine whether its graph opens upward or downward and whether there is a minimum or a maximum point. Then find the number of x- and y-intercepts.

EXAMPLE 4

For each function, determine whether its graph opens upward or downward and whether there is a minimum or a maximum point. Then find the number of x- and y-intercepts.

a. $f(x) = 2x^2 - x + 1$ **b.** $f(x) = x^2 - 6x + 9$

c. $f(x) = -2x^2 + x + 3$

Solution

a. For $f(x) = 2x^2 - x + 1$, the leading coefficient, 2, is positive, so the graph opens upward and has a minimum point.

To find the number of x-intercepts, we evaluate the discriminant, $b^2 - 4ac$, where $a = 2$, $b = -1$, and $c = 1$:

$$b^2 - 4ac = (-1)^2 - 4(2)(1) = -7$$

Since the discriminant is negative, there is no x-intercept. There is one y-intercept.

PRACTICE 4

For each function, determine whether its graph opens upward or downward and whether there is a minimum or a maximum point. Then find the number of x- and y-intercepts.

a. $f(x) = 4x^2 - 9x + 2$

b. $f(x) = -3x^2 + 5x - 7$

c. $f(x) = -4x^2 + 4x - 1$

b. For $f(x) = x^2 - 6x + 9$, the leading coefficient, 1, is positive, so the graph opens upward and has a minimum point. The discriminant is

$$b^2 - 4ac = (-6)^2 - 4(1)(9) = 0$$

Since the discriminant is 0, there is one x-intercept as well as one y-intercept.

c. For $f(x) = -2x^2 + x + 3$, the leading coefficient, -2, is negative, so the graph opens downward and has a maximum point. The discriminant is

$$b^2 - 4ac = (1)^2 - 4(-2)(3) = 25$$

Since the discriminant is positive, there are two x-intercepts and one y-intercept.

The Domain and Range of a Quadratic Function

Recall from the discussion near the beginning of this section that the domain of the function $f(x) = x^2$ is the set of all real numbers. In fact, any function of the form $f(x) = ax^2 + bx + c$ is defined for all real values of x, so the domain is the set of all real numbers, or, in interval notation, $(-\infty, \infty)$. To find the range of a quadratic function, we must identify the vertex of the parabola and determine whether it opens upward or downward.

EXAMPLE 5	PRACTICE 5

EXAMPLE 5

Consider the function $f(x) = x^2 - 2x + 4$.

a. Find the vertex of the graph.

b. Determine whether the parabola opens upward or downward.

c. Sketch the parabola.

d. Determine the domain and range of $f(x)$.

Solution

a. Here, $a = 1$, $b = -2$, and $c = 4$. For the vertex,

$x = -\dfrac{b}{2a} = -\dfrac{(-2)}{2} = 1$. To find the corresponding y-value on the graph, we evaluate the function at $x = 1$.

$$f(1) = (1)^2 - 2(1) + 4 = 3$$

So the vertex is the point $(1, 3)$.

b. Since the leading coefficient, 1, is positive, the parabola opens upward and the vertex is a minimum point.

c. The value of the discriminant is $b^2 - 4ac = (-2)^2 - 4(1)(4) = -12$. Since the discriminant is negative, we conclude that the graph has no x-intercepts. To find the y-intercept, we substitute 0 for x in $f(x)$.

$$f(0) = (0)^2 - 2(0) + 4 = 4$$

So the y-intercept is the point $(0, 4)$.

PRACTICE 5

Consider $f(x) = -x^2 - 3x - 4$.

a. Find the vertex of the graph.

b. Determine whether the parabola opens upward or downward.

c. Sketch the parabola.

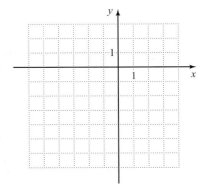

d. Determine the domain and range of $f(x)$.

To sketch the parabola, let's plot the vertex, y-intercept, and a few additional points.

x	$f(x) = x^2 - 2x + 4$	(x, y)	
-2	$f(-2) = (-2)^2 - 2(-2) + 4 = 12$	$(-2, 12)$	
-1	$f(-1) = (-1)^2 - 2(-1) + 4 = 7$	$(-1, 7)$	
0	$f(0) = (0)^2 - 2(0) + 4 = 4$	$(0, 4)$	← y-intercept
1	$f(1) = (1)^2 - 2(1) + 4 = 3$	$(1, 3)$	← Vertex
2	$f(2) = (2)^2 - 2(2) + 4 = 4$	$(2, 4)$	

The result is the following parabola:

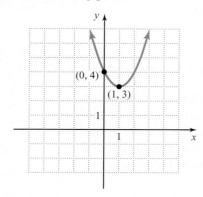

d. The domain is the set of all real numbers, that is, $(-\infty, \infty)$. From the graph, we see that the function takes on any y-value greater than or equal to 3, the y-value of the vertex. So the range of the function is the set of all real numbers greater than or equal to 3, that is, $[3, \infty)$.

In many real-life situations, we are interested in finding maximum and minimum values. For situations that can be described by a quadratic function, we can easily determine these values using our knowledge of the vertex of a parabola.

EXAMPLE 6

A boy tosses a coin into the air straight upward. The height of the coin (in feet) above the boy's hand is given by the function

$$f(t) = 32t - 16t^2$$

where t is the amount of time (in seconds) after the toss.

a. Find the vertex of the graph of this function.

b. Determine if the vertex is a minimum or a maximum point.

c. Graph the function.

d. At what time will the coin reach its maximum height? What is that height?

PRACTICE 6

A rectangular garden with width w has a perimeter of 200 ft.

a. Find the length, l, of the garden expressed in terms of w.

b. Find the area, A, of the garden expressed as a function of w.

Solution

a. We can write the function as $f(t) = -16t^2 + 32t$. Here $a = -16$, $b = 32$, and $c = 0$. To find the coordinates of the vertex, we compute

$$-\frac{b}{2a} = -\frac{32}{-32} = 1$$

Next, we compute the value of the function at $t = 1$.

$$f(1) = -16(1)^2 + 32(1) = 16$$

The vertex is the point $(1, 16)$.

b. Since the value of a, -16, is negative, the graph opens downward and has a maximum point.

c. Now, we graph the function for $t \geq 0$.

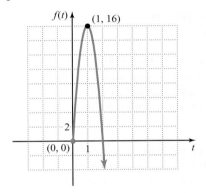

d. Since the vertex is the maximum point $(1, 16)$, the coin will reach its maximum height of 16 ft above the boy's hand 1 sec after the toss.

c. Graph the function.

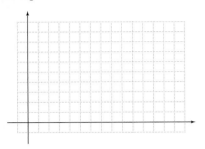

d. Determine the dimension of the garden that will maximize its area. What is that maximum area?

In Example 6(c), explain why we consider points in only the first and fourth quadrants when sketching the graph of the function.

Finding the Vertex and Intercepts of a Parabola Using a Grapher

Calculators with graphing capabilities and computers with graphing software allow us to graph quadratic functions. Recall that pressing the [Y=] key opens a window in which we enter the function. For instance, if we want to graph the function $f(x) = x^2 - 4x + 3$, using either the [^] key or the [x²] key, we enter $x^2 - 4x + 3$ to the right of \Y1 =. Pressing [GRAPH] will display the graph of the function.

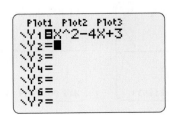

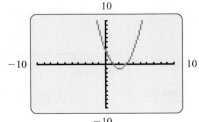

Tip When graphing a quadratic equation, set the viewing window so that it includes the vertex and the *x*- and *y*-intercepts of the parabola.

The **ZOOM** and **TRACE** features can be used to locate the coordinates of the vertex and the *x*- and *y*-intercepts of the parabola. So for the preceding graphed parabola we can use the **TRACE** feature to move the cursor to the vertex of the parabola. If we **ZOOM IN** and again use the **TRACE** feature, we will get a better approximation for the coordinates of the vertex.

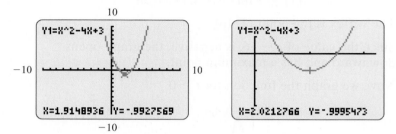

So the vertex is approximately $(2.02, -1.00)$.

Tip To get a more accurate approximation for the coordinates of the vertex, use the **ZOOM** and **TRACE** features several times.

We can also find the coordinates of the *x*- and *y*-intercepts using the **TRACE** and **ZOOM** features. To locate the *y*-intercept, use the **TRACE** feature to move the cursor until it rests on the point at which the graph crosses the *y*-axis. To locate the *x*-intercept(s), use the **TRACE** feature to move the cursor to the point at which the graph crosses the *x*-axis. As with the vertex, if we **ZOOM IN** and **TRACE** several times, we will get a better approximation of the coordinates.

To find the *y*-intercept:

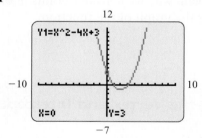

So the *y*-intercept is $(0, 3)$.

To find the *x*-intercepts:

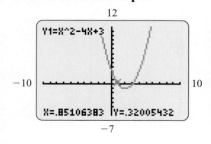

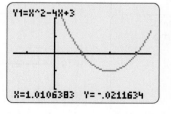

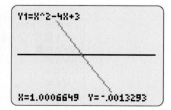

Trace to the left *x*-intercept   Zoom In and Trace   Zoom In and Trace three times

The coordinates of the left x-intercept are approximately $(1, 0)$. Carrying out the procedure for the right x-intercept, we find its coordinates to be approximately $(3, 0)$.

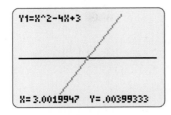

Zoom In and Trace three times

EXAMPLE 7

Consider the function $f(x) = -3x^2 - 3x + 5$.

a. Graph this equation using a grapher.

b. Use the **ZOOM** and **TRACE** features to find the coordinates of the vertex.

c. Use the **ZOOM** and **TRACE** features to find the coordinates of the x- and y-intercepts.

Solution

a. We press $\boxed{Y=}$ and enter $-3x^2 - 3x + 5$ to the right of \\Y1 =, using either the $\boxed{x^2}$ feature or $x\boxed{\wedge}2$ to enter the quadratic term. Then we press $\boxed{GRAPH}$. If necessary, we change the viewing window so that the vertex and the intercepts of the parabola are all displayed.

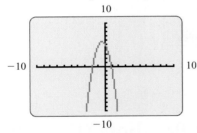

b. With the **TRACE** feature, we move the cursor until it rests on the parabola's vertex. To get a better approximation, we **ZOOM IN** and **TRACE** a few times.

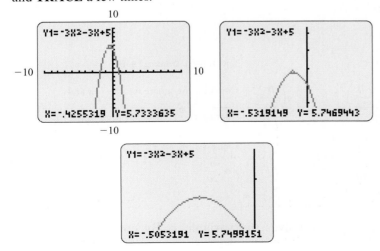

So the coordinates of the vertex are approximately $(-0.51, 5.75)$.

PRACTICE 7

Consider the equation $f(x) = 2x^2 + x - 4$.

a. Graph this equation using a graphing calculator or a computer.

b. Use the **TRACE** or **ZOOM** features to identify the coordinates of the vertex.

c. Use the **TRACE** and **ZOOM** features to find the x- and y-intercepts of the graph.

c. Using the **TRACE** feature, we locate the coordinates of the *y*-intercept.

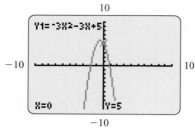

The *y*-intercept is (0, 5).

Using the **ZOOM** and **TRACE** features, we find the *x*-intercepts of the parabola:

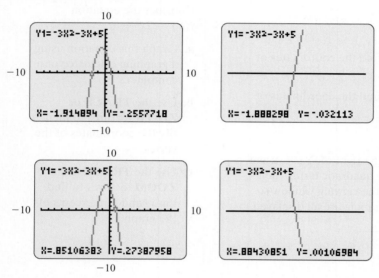

So the *x*-intercepts are approximately (−1.9, 0) and (0.9, 0).

More on the Vertex of a Parabola (Optional)

Consider the general quadratic function $f(x) = ax^2 + bx + c$, and assume that a is positive. Let's rewrite the right-hand side of this equation by completing the square for the variable terms, $ax^2 + bx$:

$$f(x) = (ax^2 + bx) + c$$

$$= a\left(x^2 + \frac{b}{a}x\right) + c \qquad \text{Factor out } a \text{ from } ax^2 + bx.$$

$$= a\left[x^2 + \frac{b}{a}x + \left(\frac{b}{2a}\right)^2\right] + c - a\left(\frac{b}{2a}\right)^2 \qquad \text{Complete the square for } x^2 + \frac{b}{a}x \text{ and}$$

$$\text{subtract } a\left(\frac{b}{2a}\right)^2 \text{ from the same side.}$$

$$= a\left(x + \frac{b}{2a}\right)^2 + \left(c - \frac{b^2}{4a}\right) \qquad \text{Express } x^2 + \frac{b}{a}x + \left(\frac{b}{2a}\right)^2 \text{ as the square of}$$

$$= a\left(x + \frac{b}{2a}\right)^2 + \frac{4ac - b^2}{4a} \qquad x + \frac{b}{2a}.$$

Note that the term $a\left(x + \dfrac{b}{2a}\right)^2$ involves the variable x. By contrast, the term $\dfrac{4ac - b^2}{4a}$ contains only constants and so does not depend on x. Since the expression

$\left(x + \dfrac{b}{2a}\right)^2$ is a perfect square, its value is either positive or 0. Since a is positive,

the term $a\left(x + \dfrac{b}{2a}\right)^2$ is nonnegative and has a minimum value of 0 at the point

where the value of this term is smallest.

$$a\left(x + \frac{b}{2a}\right)^2 = 0$$

$$\left(x + \frac{b}{2a}\right)^2 = 0 \qquad \text{Divide each side of the equation by } a.$$

$$x + \frac{b}{2a} = 0 \qquad \text{Use the square root property of equality.}$$

$$x = -\frac{b}{2a} \qquad \text{Subtract } \frac{b}{2a} \text{ from each side of the equation.}$$

Since we assumed that a is positive, we conclude that the minimum value of $a\left(x + \dfrac{b}{2a}\right)^2$ occurs when $x = -\dfrac{b}{2a}$ and is therefore the x-value of the vertex. If we had assumed that a is negative, we could then show that the maximum value occurs when $x = -\dfrac{b}{2a}$ and so again is the x-value of the vertex.

Mathematically Speaking

Fill in each blank with the most appropriate term or phrase from the given list.

zero	minimum	vertex
maximum	downward	x-coordinate
axis of symmetry	quadratic	upward
y-coordinate	parabola	two
circle	squared	one

1. Functions of the form $f(x) = ax^2 + bx + c$ are called _____ functions.

2. The graph of a quadratic function is called a(n) _____.

3. The highest or lowest point of a parabola is called a(n) _____.

4. For a parabola given by the function $f(x) = ax^2 + bx + c$, the _____ of the vertex is $-\dfrac{b}{2a}$.

5. When a is positive, the parabola given by the function $f(x) = ax^2 + bx + c$ opens _____.

6. When a is negative, the parabola given by the function $f(x) = ax^2 + bx + c$ has a(n) _____ point.

7. If the discriminant of the quadratic equation given by the function $f(x) = ax^2 + bx + c$ is positive, the parabola has _____ x-intercept(s).

8. For a parabola given by the function $f(x) = ax^2 + bx + c$, the equation of the _____ is $x = -\dfrac{b}{2a}$.

Complete the table. Then plot the points and sketch the graph of the parabola.

9.

x	y = f(x) = 2x²	(x, y)
−2		
−1		
0		
1		
2		

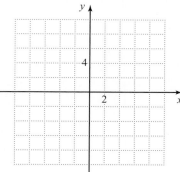

10.

x	y = f(x) = −2x²	(x, y)
−2		
−1		
0		
1		
2		

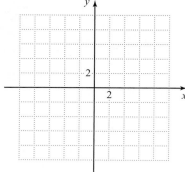

11.

x	$y = f(x) = \frac{1}{2}x^2$	(x, y)
−4		
−2		
0		
2		
4		

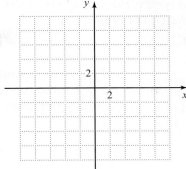

12.

x	$y = f(x) = \frac{1}{3}x^2$	(x, y)
−6		
−3		
0		
3		
6		

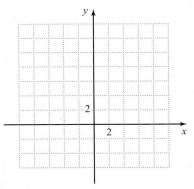

13.

x	$y = f(x) = 2 - x^2$	(x, y)
−3		
−2		
−1		
0		
1		
2		
3		

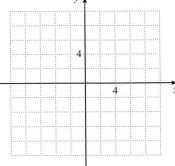

14.

x	$y = f(x) = x^2 - 3$	(x, y)
−3		
−2		
−1		
0		
1		
2		
3		

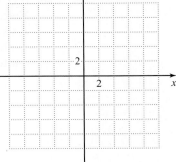

For each of the following functions, find the vertex, the axis of symmetry, and the x- and y-intercepts of the graph. Then sketch the graph.

15. $f(x) = x^2 - 8x$
Vertex:
Axis of symmetry:
x-intercept(s):
y-intercept:

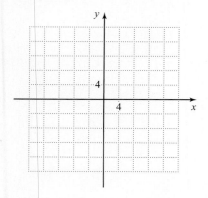

16. $f(x) = x^2 + 6x$
Vertex:
Axis of symmetry:
x-intercept(s):
y-intercept:

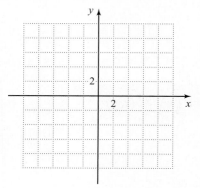

17. $f(x) = x^2 - 2x - 3$
Vertex:
Axis of symmetry:
x-intercept(s):
y-intercept:

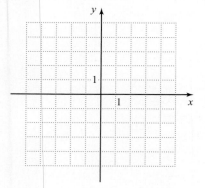

18. $f(x) = x^2 + 4x - 5$
Vertex:
Axis of symmetry:
x-intercept(s):
y-intercept:

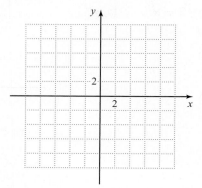

19. $f(x) = -x^2 + 4x + 12$
Vertex:
Axis of symmetry:
x-intercept(s):
y-intercept:

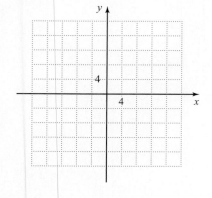

20. $f(x) = -x^2 - 2x + 8$
Vertex:
Axis of symmetry:
x-intercept(s):
y-intercept:

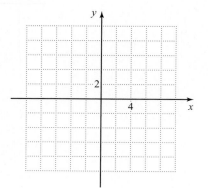

21. $f(x) = x^2 - 1$
Vertex:
Axis of symmetry:
x-intercept(s):
y-intercept:

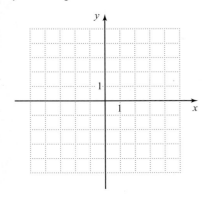

22. $f(x) = 4 - x^2$
Vertex:
Axis of symmetry:
x-intercept(s):
y-intercept:

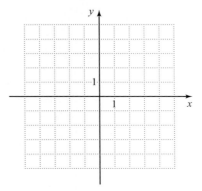

23. $f(x) = (x + 1)^2$
Vertex:
Axis of symmetry:
x-intercept(s):
y-intercept:

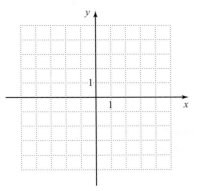

24. $f(x) = (x - 1)^2$
Vertex:
Axis of symmetry:
x-intercept(s):
y-intercept:

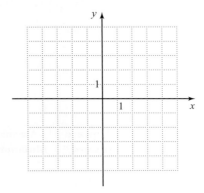

25. $f(x) = -x^2 + 6x - 9$
Vertex:
Axis of symmetry:
x-intercept(s):
y-intercept:

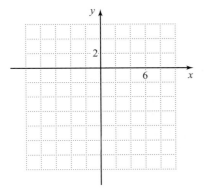

26. $f(x) = x^2 + 6x + 9$
Vertex:
Axis of symmetry:
x-intercept(s):
y-intercept:

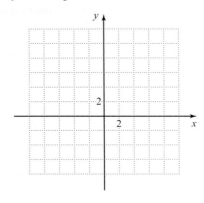

27. $f(x) = x^2 - 3x - 10$

Vertex:

Axis of symmetry:

x-intercept(s):

y-intercept:

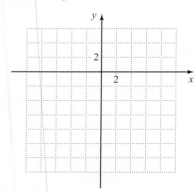

28. $f(x) = x^2 + x - 6$

Vertex:

Axis of symmetry:

x-intercept(s):

y-intercept:

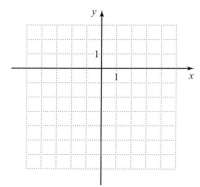

Complete the table.

29.

Function	Opens Upward or Downward?	Maximum or Minimum Point?	Number of x-intercepts	Number of y-intercepts
$f(x) = x^2 + 5$				
$f(x) = 1 - 4x + 4x^2$				
$f(x) = 2 - 3x^2$				
$f(x) = -2x^2 - 5x - 8$				
$f(x) = 4x^2 - 4x - 1$				

30.

Function	Opens Upward or Downward?	Maximum or Minimum Point?	Number of x-intercepts	Number of y-intercepts
$f(x) = -6 - x^2$				
$f(x) = 2x^2$				
$f(x) = 3x^2 + 8x + 7$				
$f(x) = 5 + 2x - 3x^2$				
$f(x) = 9x^2 + 12x + 4$				

Graph each function. Then determine the domain and range.

31. $f(x) = 2x^2 - 1$

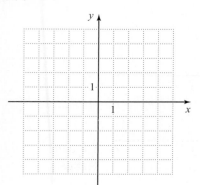

Domain:

Range:

32. $f(x) = 1 - 2x^2$

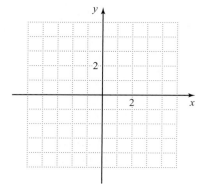

Domain:

Range:

33. $g(x) = -3x^2 + 6x - 2$

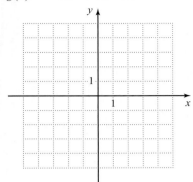

Domain:
Range:

34. $g(x) = 3x^2 - 12x + 6$

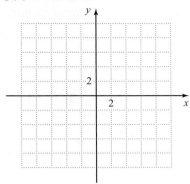

Domain:
Range:

35. $f(x) = 0.5x^2 + 2$

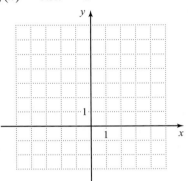

Domain:
Range:

36. $f(x) = -2 - 0.5x^2$

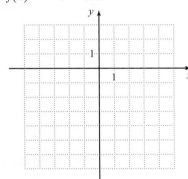

Domain:
Range:

37. $f(x) = x^2 - 3x - 4$

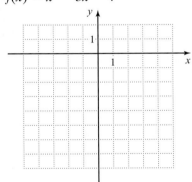

Domain:
Range:

38. $h(x) = x^2 + 5x + 1$

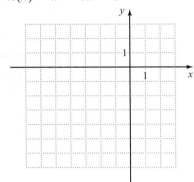

Domain:
Range:

39. $h(x) = -2x^2 + 2x - 3$

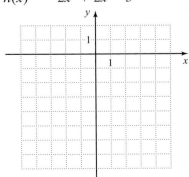

Domain:
Range:

40. $g(x) = 3x^2 - 3x + 1$

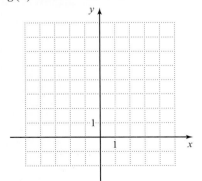

Domain:
Range:

▦ *Graph each function on a grapher. Then, use the grapher to find the vertex and x- and y-intercepts, rounding to the nearest hundredth.*

41. $f(x) = x^2 + 0.2x - 1$ Vertex: x-intercept(s):   y-intercept:

42. $f(x) = x^2 - 1.1x - 3$ Vertex: x-intercept(s):   y-intercept:

43. $f(x) = -0.15x^2 - x + 0.5$ Vertex: x-intercept(s):   y-intercept:

44. $f(x) = -1.4x^2 + 2x + 7.1$ Vertex: x-intercept(s):   y-intercept:

45. $f(x) = 5x^2 + 3x + 7$ Vertex: x-intercept(s):   y-intercept:

46. $f(x) = -3x^2 + 2x - 8$ Vertex: x-intercept(s):   y-intercept:

Mixed Practice

Solve.

47. Complete the table. Then, plot the points and sketch the graph of the parabola.

x	$y = f(x) = -3 + x^2$	(x, y)
−3		
−2		
−1		
0		
1		
2		
3		

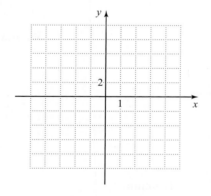

48. Complete the table.

Function	Opens Upward or Downward?	Maximum or Minimum Point?	Number of x-intercepts	Number of y-intercepts
$f(x) = -x^2 - 5x + 5$				
$f(x) = 3x^2 + 8x + 7$				
$f(x) = x^2 - 6x + 9$				
$f(x) = x^2 - 2$				
$f(x) = 6x - 5 - 3x^2$				

Find the vertex, the axis of symmetry, and the x- and y-intercepts of the graph of the function. Then sketch the graph.

49. $f(x) = -x^2 - 4x + 5$

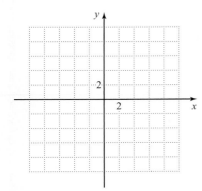

Vertex:

Axis of symmetry:

x-intercept(s):

y-intercept:

50. $f(x) = x^2 - 4x + 4$

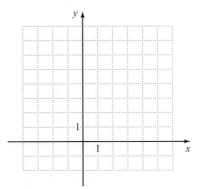

Vertex:

Axis of symmetry:

x-intercept(s):

y-intercept:

Graph each function. Then determine the domain and range.

51. $g(x) = -x^2 + 2x + 1$

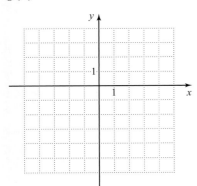

Domain: Range:

52. $h(x) = 2x^2 - 4x + 3$

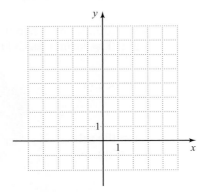

Domain: Range:

Applications

Solve.

53. A stone is thrown straight upward with an initial velocity of 48 feet per second from a bridge 280 ft above a river. The height of the stone above the river t sec after it is thrown is given by the function $s(t) = -16t^2 + 48t + 280$.

   a. Find the vertex of the graph.

   b. Graph the function.

   c. When does the stone reach its maximum height? What is that height?

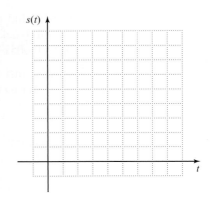

54. A company determines that its daily revenue, R (in dollars), for selling x units of a product is modeled by the function $R(x) = x(120 - x)$.

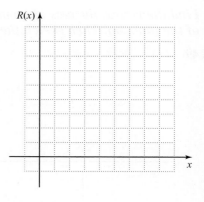

  a. Find the vertex of the graph of this function.

  b. Graph the function.

  c. What number of units must be sold in order to maximize the revenue? What is the maximum revenue?

55. A homeowner has 150 ft of fencing with which to build a rectangular exercise yard for her dog.

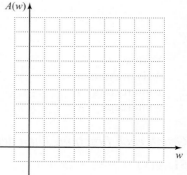

  a. Express the length, l, of the exercise yard in terms of the width, w.

  b. Express in function notation the relationship between the area, $A(w)$, and the width, w, of the exercise yard.

  c. Graph this function.

  d. What dimensions will maximize the area of the exercise yard? What is the maximum area?

56. A farmer plans to build a rectangular animal pen using the barn as one side, as shown in the figure to the right. The farmer has 300 ft of fencing with which to build the pen.

  a. Express the length, l, of the pen in terms of the width, w.

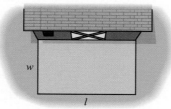

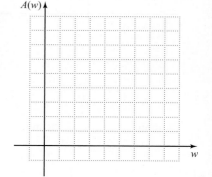

  b. Express in function notation the relationship between the area, $A(w)$, and the width, w, of the pen.

  c. Graph this function.

  d. What dimensions will maximize the area of the animal pen? What is the maximum area?

57. A toy manufacturer determines that the daily cost, C, for producing x units of a dump truck can be approximated by the function $C(x) = 0.005x^2 - x + 100$.

  a. How many toy dump trucks must the manufacturer produce per day in order to minimize the cost?

  b. What is the minimum daily cost?

58. An object is launched straight upward with an initial velocity of 39.2 meters per second from a height of 100 m above the ground. The height, s, of the object above the ground t sec after it is launched is given by $s(t) = -4.9t^2 + 39.2t + 100$.

  a. When will the object reach its maximum height?

  b. What is the maximum height?

● *Check your answers on page A-39.*

MINDSTRETCHERS

Investigation

1. Consider the graph of the equation $f(x) = x^2$.

a. Graph the equation $f(x) = x^2 + c$ for $c = -3, -1, 1,$ and 2. In each case, describe the change in the position of the graph of $f(x) = x^2 + c$ relative to the graph of $f(x) = x^2$.

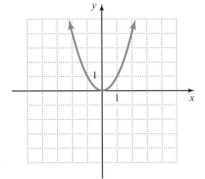

b. Graph the equation $f(x) = (x + c)^2$ for $c = -3, -1, 1,$ and 2. In each case, describe the change in the position of the graph of $f(x) = (x + c)^2$ relative to the graph of $f(x) = x^2$.

c. Use your results from parts (a) and (b) to describe the change in the position of the graph relative to the graph of $f(x) = x^2$ of each of the following:

  i. $f(x) = x^2 + c, c > 0$

  ii. $f(x) = x^2 + c, c < 0$

  iii. $f(x) = (x + c)^2, c > 0$

  iv. $f(x) = (x + c)^2, c < 0$

d. Without graphing, describe the position of the graph of $f(x) = (x - 2)^2 - 3$ relative to the graph of $f(x) = x^2$. Then check your answer by graphing.

Critical Thinking

2. Suppose you are given three arbitrary points (x_1, y_1), (x_2, y_2), and (x_3, y_3) that lie on the graph of a parabola. The equation of the parabola, $y = ax^2 + bx + c$, that contains the points can be found by solving the following system of equations:

$$\textbf{(1)} \quad ax_1^2 + bx_1 + c = y_1$$
$$\textbf{(2)} \quad ax_2^2 + bx_2 + c = y_2$$
$$\textbf{(3)} \quad ax_3^2 + bx_3 + c = y_3$$

Use this system to find the equation of the parabola that passes through the points $(-2, 0)$, $(-1, -4)$, and $(1, 6)$.

Mathematical Reasoning

3. Suppose the x-intercepts of the graph of a parabola are $(x_1, 0)$ and $(x_2, 0)$. What is the equation of the axis of symmetry of this graph? Give examples to justify your reasoning.

9.5   Solving Quadratic and Rational Inequalities

Recall that in Section 2.6 we discussed situations that could be described and solved by a linear inequality in one variable, such as $4x + 5 \leq 3x + 8$. There we saw that the solutions of an inequality are any values of the variable that make the inequality true. We expressed the solutions in one of three ways: as a simplified inequality, such as $x \leq 3$; in interval notation, such as $(-\infty, 3]$; and as a graph on the number line, such as

In this section, we consider situations that can be described and solved by a quadratic or a rational inequality in one variable. For instance, consider the following problem:

The owner of a bicycle shop determines that his monthly revenue for selling n bikes can be represented by $R = 430n - 6n^2$. He wants to find the number of bikes he must sell to generate revenue of at least \$7000 per month. To find the number of bikes he must sell, he can solve the following inequality:

$$430n - 6n^2 \geq 7000$$

In this section, we discuss techniques for solving such inequalities.

Quadratic Inequalities

A **quadratic inequality** in one variable is an inequality that contains a quadratic expression. Here are some examples of quadratic inequalities in one variable.

$$x^2 + 4x < -3 \qquad 2x^2 + 11x + 15 \leq 10$$
$$x^2 + x - 2 > 0 \qquad t^2 + 2t - 3 \geq 0$$

Consider, for instance, a quadratic inequality such as $x^2 + x - 2 > 0$. To solve, we find all the values of x for which the expression $x^2 + x - 2$ is greater than 0, that is, all values for which the expression is positive. Let's take a look at the graph of the function $f(x) = x^2 + x - 2$ to understand how to solve the related inequality.

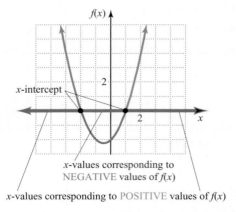

We observe from this graph that the solutions of $x^2 + x - 2 > 0$ are all the real numbers less than -2 or greater than 1. In interval notation, we can express the solutions as $(-\infty, -2) \cup (1, \infty)$. Note that the x-values for which $f(x)$ is positive are separated from those for which $f(x)$ is negative by the x-intercepts, that is, by the points where $f(x)$ is equal to 0.

It is usually easier to solve quadratic inequalities algebraically rather than graphically. The key to solving a quadratic inequality such as $x^2 + x - 2 > 0$ algebraically is to solve the related equation, $x^2 + x - 2 = 0$. Then we draw a number line on which the solutions are plotted. The solutions serve as boundary points, separating the number line into intervals in which the expression $x^2 + x - 2$ is either greater than 0 or less than 0. To determine if the value of the expression is greater than or less than 0 in a particular interval, we use a *test value*, that is, an arbitrary point on the number line that lies within the interval.

Let's solve the inequality $x^2 + x - 2 > 0$ algebraically. To do this, we first solve the related equation $x^2 + x - 2 = 0$.

$$x^2 + x - 2 = 0$$
$$(x + 2)(x - 1) = 0 \qquad \text{Factor.}$$
$$x + 2 = 0 \quad \text{or} \quad x - 1 = 0 \qquad \text{Use the zero-product property.}$$
$$x = -2 \qquad\qquad x = 1$$

Next, we plot the two solutions, -2 and 1, on a number line. The solutions serve as *boundary points*, separating the line into three intervals.

To determine if the quadratic expression is greater than 0 or less than 0 in a particular interval, we choose test values: say, -3 in Interval A, 0 in Interval B, and 2 in Interval C. Then, we evaluate the expression for each test value.

Interval	Test Value	Value of $x^2 + x - 2$	Conclusion
A	-3	$(-3)^2 + (-3) - 2 = 4$	$4 > 0$
B	0	$(0)^2 + (0) - 2 = -2$	$-2 < 0$
C	2	$(2)^2 + (2) - 2 = 4$	$4 > 0$

Value of $x^2 + x - 2$: > 0 $\quad -2 \quad$ < 0 $\quad 1 \quad$ > 0

Because the inequality that we are considering is $x^2 + x - 2 > 0$, we are looking for intervals in which the value of the expression is positive. We conclude that the solutions of the original inequality are all real numbers either less than -2 or greater than 1. We can express these solutions in interval notation as $(-\infty, -2) \cup (1, \infty)$ and graph them as follows:

$-6\ -5\ -4\ -3\ -2\ -1\ \ 0\ \ 1\ \ 2\ \ 3\ \ 4\ \ 5\ \ 6$

Note that the algebraic solution agrees with the solutions we obtained earlier from graphing the corresponding function $f(x) = x^2 + x - 2$ on a coordinate plane.

EXAMPLE 1

Solve $x^2 - 2x - 3 \le 0$. Then graph the solutions.

Solution We begin by solving the related equation $x^2 - 2x - 3 = 0$.

$$x^2 - 2x - 3 = 0$$
$$(x - 3)(x + 1) = 0 \qquad \text{Factor.}$$
$$x - 3 = 0 \quad \text{or} \quad x + 1 = 0 \qquad \text{Use the zero-product property.}$$
$$x = 3 \qquad\qquad x = -1$$

Next, we plot on a number line the two solutions, 3 and −1, which are the boundary points separating the number line into three intervals.

Then, in each interval, we check to determine whether the expression is greater than 0 or less than 0. To do this, let's use the test values −2, 0, and 4.

Interval	Test Value	Value of $x^2 - 2x - 3$	Conclusion
A	−2	$(-2)^2 - 2(-2) - 3 = 5$	$5 > 0$
B	0	$(0)^2 - 2(0) - 3 = -3$	$-3 < 0$
C	4	$(4)^2 - 2(4) - 3 = 5$	$5 > 0$

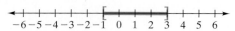

Because $x^2 - 2x - 3 \le 0$, we are looking for values of x for which the expression is either negative or 0. We conclude that the solutions of the inequality are all real numbers between and including −1 and 3, or in interval notation, $[-1, 3]$. The corresponding graph is shown below.

EXAMPLE 2

Solve $x^2 + 6x + 10 > 2$. Graph the solutions.

Solution First, we solve the related equation $x^2 + 6x + 10 = 2$.

$$x^2 + 6x + 10 = 2$$
$$x^2 + 6x + 8 = 0 \qquad \text{Write the equation in standard form.}$$
$$(x + 2)(x + 4) = 0 \qquad \text{Factor.}$$
$$x + 2 = 0 \quad \text{or} \quad x + 4 = 0 \qquad \text{Use the zero-product property.}$$
$$x = -2 \qquad\qquad x = -4$$

PRACTICE 1

Solve $x^2 - 3x - 4 \ge 0$. Then graph the solutions.

PRACTICE 2

Solve $x^2 - 6x - 2 < -7$ and graph the solutions.

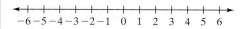

Plotting the solutions -2 and -4 as boundary points on a number line gives us the following intervals:

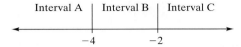

Now, let's use the test values -5, -3, and 0 to determine whether the expression is greater than 2 or less than 2 in each interval.

Interval	Test Value	Value of $x^2 + 6x + 10$	Conclusion
A	-5	$(-5)^2 + 6(-5) + 10 = 5$	$5 > 2$
B	-3	$(-3)^2 + 6(-3) + 10 = 1$	$1 < 2$
C	0	$(0)^2 + 6(0) + 10 = 10$	$10 > 2$

Value of $x^2 + 6x + 10$: > 2 -4 < 2 -2 > 2

Because we are considering the inequality $x^2 + 6x + 10 > 2$, we are looking for values of x for which the expression is greater than 2. So the solutions of the inequality $x^2 + 6x + 10 > 2$ are all real numbers less than -4 or greater than -2, that is, $(-\infty, -4) \cup (-2, \infty)$. We can also represent the solutions graphically as follows:

Rational Inequalities

A **rational inequality** is an inequality that contains a rational expression. Some examples of rational inequalities are

$$\frac{x-1}{x+3} < 0 \qquad \frac{2}{x} > -2 \qquad \frac{x}{2x-5} \leq 1 \qquad \frac{3x-1}{x+4} \geq 0$$

To solve rational inequalities, we use a procedure similar to that used for solving quadratic inequalities. However, when finding the boundary points that separate a number line into intervals, we must consider not only the solutions of the related equation but also the values for which the inequality is undefined.

EXAMPLE 3

Solve $\dfrac{x - 1}{x + 3} < 0$. Graph the solutions.

Solution The inequality is undefined when its denominator, $x + 3$, is 0. So first, we set the denominator equal to 0 and solve for x:

$$x + 3 = 0$$
$$x = -3$$

Next, we solve the related equation $\dfrac{x - 1}{x + 3} = 0$.

$\dfrac{x - 1}{x + 3} \cdot (x + 3) = 0 \cdot (x + 3)$ **Multiply each side of the equation by the LCD.**

$x - 1 = 0$ **Simplify.**

$x = 1$

Then, we plot the boundary points -3 and 1, separating the number line into three intervals.

Interval A | Interval B | Interval C

-3 1

Now, we check a test value in each interval to determine whether the rational expression $\dfrac{x - 1}{x + 3}$ is greater than 0 or less than 0. Let's use -4, 0, and 2.

Interval	Test Value	Value of $\dfrac{x - 1}{x + 3}$	Conclusion
A	-4	$\dfrac{-4 - 1}{-4 + 3} = 5$	$5 > 0$
B	0	$\dfrac{0 - 1}{0 + 3} = -\dfrac{1}{3}$	$-\dfrac{1}{3} < 0$
C	2	$\dfrac{2 - 1}{2 + 3} = \dfrac{1}{5}$	$\dfrac{1}{5} > 0$

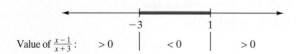

Value of $\frac{x-1}{x+3}$: > 0 < 0 > 0

Since we are solving the inequality $\dfrac{x - 1}{x + 3} < 0$, we are looking for values of x for which the expression is less than 0. The solutions are all real numbers between, but not including, -3 and 1, or in interval notation, $(-3, 1)$. The graph of the solutions is

$-6 -5 -4 -3 -2 -1\ 0\ 1\ 2\ 3\ 4\ 5\ 6$

PRACTICE 3

Solve $\dfrac{x - 5}{x + 5} < 0$. Graph the solutions.

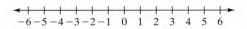

$-6 -5 -4 -3 -2 -1\ 0\ 1\ 2\ 3\ 4\ 5\ 6$

EXAMPLE 4	PRACTICE 4

EXAMPLE 4

Solve $\dfrac{x + 2}{x - 4} \geq -2$. Then graph the solutions.

Solution The inequality is undefined when the expression in the denominator, $x - 4$, is 0. Solving for x, we get

$$x - 4 = 0$$
$$x = 4$$

Now, we solve the related equation $\dfrac{x + 2}{x - 4} = -2$.

$$\frac{x + 2}{x - 4} \cdot (x - 4) = -2 \cdot (x - 4)$$
$$x + 2 = -2x + 8$$
$$3x = 6$$
$$x = 2$$

Plotting 2 and 4 as boundary points separates a number line into three intervals.

Next, we choose a test value in each interval: say, 0, 3, and 5.

Interval	Test Value	Value of $\dfrac{x + 2}{x - 4}$	Conclusion
A	0	$\dfrac{0 + 2}{0 - 4} = -\dfrac{1}{2}$	$-\dfrac{1}{2} > -2$
B	3	$\dfrac{3 + 2}{3 - 4} = -5$	$-5 < -2$
C	5	$\dfrac{5 + 2}{5 - 4} = 7$	$7 > -2$

Value of $\frac{x+2}{x-4}$: > -2 < -2 > -2

Since the inequality under consideration is $\dfrac{x + 2}{x - 4} \geq -2$, we are looking for values of x for which the expression is greater than or equal to -2. So the solutions are all real numbers less than or equal to 2 or greater than 4, or in interval notation, $(-\infty, 2] \cup (4, \infty)$. The graph of the solutions follows:

PRACTICE 4

Solve $\dfrac{x - 2}{x + 3} \geq -1$. Then graph the solutions.

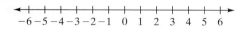

EXAMPLE 5

The monthly profit (in thousands of dollars) that a store makes on the sale of computers is modeled by the function $P(x) = x^2 - 6x - 30$, where x is the number of computers sold. How many computers must the store sell in order to make a profit of more than $10,000?

Solution To find the number of computers the store must sell, we need to solve the inequality $x^2 - 6x - 30 > 10$. To do this, we first solve the related equation.

$$x^2 - 6x - 30 = 10$$
$$x^2 - 6x - 40 = 0$$
$$(x - 10)(x + 4) = 0$$
$$x - 10 = 0 \quad \text{or} \quad x + 4 = 0$$
$$x = 10 \qquad \qquad x = -4$$

Plotting the solutions 10 and −4 as boundary points on a number line gives us the three intervals.

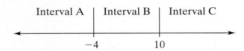

Let's use the test values −5, 0, and 11 for the three intervals.

Interval	Test Value	Value of $P(x) = x^2 - 6x - 30$	Conclusion
A	−5	$(-5)^2 - 6(-5) - 30 = 25$	$P(-5) > 10$
B	0	$(0)^2 - 6(0) - 30 = -30$	$P(0) < 10$
C	11	$(11)^2 - 6(11) - 30 = 25$	$P(11) > 10$

We are looking for those values of x for which $P(x)$ is greater than 10. Since the number of computers sold cannot be negative, we do not include the solutions in Interval A. So for the store to make a profit greater than $10,000, each month it must sell more than 10 computers, that is, 11 or more computers.

PRACTICE 5

The cost of operating a bookstand for x weeks can be approximated by the cost function $C(x) = 5x^2 + 25x + 70$. For how many weeks can the stand run if the cost must not exceed $400?

9.5   Exercises

 MyMathLab

 PRACTICE   WATCH   DOWNLOAD   READ   REVIEW

Mathematically Speaking

Fill in each blank with the most appropriate term or phrase from the given list.

boundary point	quadratic	denominator
numerator	test value	linear
ratio	rational inequality	

1. To determine if the value of an expression is greater than or less than 0 in a particular interval, a(n) _____ is used.

2. A(n) _____ inequality in one variable is an inequality that contains a quadratic expression.

3. A(n) _____ is an inequality that contains a rational expression.

4. A rational inequality is undefined when the value of its _____ is 0.

Solve. Then graph the solutions.

5. $x^2 - x > 0$

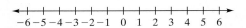

6. $x^2 + 4x > 0$

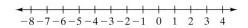

7. $x^2 < 4$

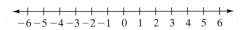

8. $9 > x^2$

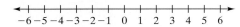

9. $x^2 - x - 2 \leq 10$

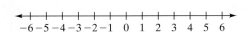

10. $x^2 + 2x - 7 \leq 8$

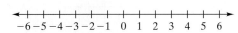

11. $x^2 + 6x + 9 \geq 0$

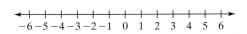

12. $x^2 - 4x + 4 \geq 0$

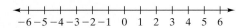

13. $6 + x - x^2 \leq 0$

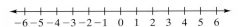

14. $20 - x - x^2 \leq 0$

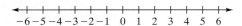

15. $x^2 + 5x + 4 < 0$

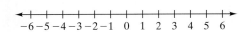

16. $x^2 - 4x + 3 \leq 0$

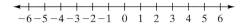

17. $2x^2 - 3x + 1 \geq 0$

18. $2x^2 + 5x + 2 > 0$

19. $1 - 4x^2 \leq 0$

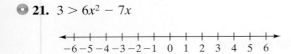

20. $9x^2 - 4 < 0$

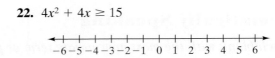

21. $3 > 6x^2 - 7x$

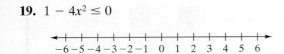

22. $4x^2 + 4x \geq 15$

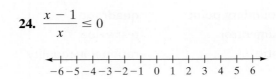

23. $\dfrac{x}{x + 2} \leq 0$

24. $\dfrac{x - 1}{x} \leq 0$

25. $\dfrac{1}{6 - x} < 0$

26. $\dfrac{2}{3 - x} < 0$

27. $\dfrac{x + 3}{x - 3} > 2$

28. $\dfrac{x - 6}{x + 6} > 4$

29. $\dfrac{2x - 1}{x + 4} \geq 0$

30. $\dfrac{x - 2}{3x + 2} \geq 0$

31. $\dfrac{2x - 1}{2x + 5} \leq 2$

32. $\dfrac{3x + 9}{2x - 3} \leq 3$

Mixed Practice

Solve. Then, graph the solutions.

33. $x^2 + x - 5 < 7$

34. $\dfrac{3}{5 - x} \geq 1$

35. $\dfrac{x + 2}{x - 4} < 2$

36. $-x^2 + 3x + 4 \leq 0$

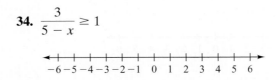

37. $2x^2 + 5x - 3 \geq 0$

```
 ←—+—+—+—+—+—+—+—+—+—+—+—+—+→
  -6 -5 -4 -3 -2 -1  0  1  2  3  4  5  6
```

38. $\dfrac{x-1}{2x+3} \geq 0$

```
 ←—+—+—+—+—+—+—+—+—+—+—+—+—+→
  -6 -5 -4 -3 -2 -1  0  1  2  3  4  5  6
```

Applications

Solve.

39. A penny is tossed upward with an initial velocity of 48 ft per sec. The height, h, of the penny relative to the point of release t sec after it is tossed is modeled by the function $h(t) = -16t^2 + 48t$. For what interval of time is the penny above the point of release?

40. An outerwear manufacturer determines that its weekly revenue, R, for selling a rain parka at a price of p dollars is modeled by the function $R(p) = 150p - p^2$. What range of prices for the parka will generate weekly revenue of at least $5000?

41. A furniture maker determines that the weekly cost, C, for producing x end tables is given by $C(x) = 2x^2 - 60x + 900$. How many end tables can be produced to keep the weekly cost under $500?

42. A ball is thrown straight upward with an initial velocity of 40 ft per sec from a height of 180 ft. The height, h, of the ball above the ground t sec after it is thrown upward is given by $h(t) = -16t^2 + 40t + 180$. For what values of t is the ball at least 196 ft above the ground?

43. A gardener wants to enclose a rectangular flower bed using 90 ft of fencing. For what range of lengths will the area exceed 450 ft²?

44. The base of a box used for packaging has a perimeter of 60 in. For what range of widths will the area of the base be at least 200 in²?

45. A company determines that its average cost C, in dollars, for selling x units of a product is modeled by the function $C(x) = \dfrac{864 + 2x}{x}$. For what number of units will the average cost be less than $8?

46. A publishing company had revenue of $18 million last year. The company's financial analyst uses the formula $P(R) = \dfrac{100R - 1800}{R}$ to determine the percent growth, P, in the company's revenue this year over last year, where R is in millions of dollars. For what revenues R will the company's revenue grow by more than 10%?

● *Check your answers on page A-41.*

MINDSTRETCHERS

Writing

1. The inequalities $(x - 1)(x + 3) < 0$ and $\dfrac{x - 1}{x + 3} < 0$ have the same solutions. Explain why.

Mathematical Reasoning

2. Solve the compound inequality $4 < x^2 - 3x + 6 < 10$.

Groupwork

3. Working with a partner, describe a strategy for solving each of the following inequalities and then solve the inequalities.

   a. $x^3 - 2x^2 - 8x \geq 0$

   b. $x^4 - 5x^2 + 4 < 0$

   c. $\dfrac{x - 2}{x^2 - 16} \leq 0$

   d. $\dfrac{x^2 - 4x}{x^2 - x - 6} > 0$

KEY CONCEPTS AND SKILLS   (CONCEPT)   (SKILL)

Concept/Skill	Description	Example
[9.1] Quadratic Equation	An equation that can be written in the form $ax^2 + bx + c = 0$, where a, b, and c are real numbers and $a \neq 0$.	$x^2 - 2x + 8 = 0$
[9.1] Square Root Property of Equality	If b is a real number and $x^2 = b$, then $x = \pm\sqrt{b}$, that is, $x = \sqrt{b}$ or $x = -\sqrt{b}$.	$x^2 = 63$ $x = \pm\sqrt{63}$ $x = 3\sqrt{7}$ or $x = -3\sqrt{7}$
[9.1] To Complete the Square for the Expression $x^2 + bx$	Take one-half the coefficient of the linear term, x, then square it. Add this value to the expression. $$x^2 + bx + \left(\frac{1}{2}b\right)^2$$	Find the value of c that will make $x^2 + 6x + c$ a perfect square trinomial. $\left(\frac{1}{2} \cdot 6\right)^2 = 3^2 = 9$, so $c = 9$ $x^2 + 6x + 9$
[9.1] To Solve a Quadratic Equation ($ax^2 + bx + c = 0$) by Completing the Square	• If $a = 1$, then proceed to the next step. If $a \neq 1$, then divide each side of the equation by a, the coefficient of the quadratic (or second-degree) term. • Move all terms with variables to one side of the equal sign and all constants to the other side. • Take one-half the coefficient of the linear term. Then square it and add this value to each side of the equation. • Factor the side of the equation containing the variables, writing it as the square of a binomial. • Use the square root property of equality and solve the resulting equations.	$2x^2 + 8x - 6 = 0$ $x^2 + 4x - 3 = 0$ $x^2 + 4x = 3$ Since $\left(\frac{1}{2} \cdot 4\right)^2 = 2^2 = 4$, add 4 to both sides. $x^2 + 4x + 4 = 3 + 4$ $(x + 2)^2 = 7$ $x + 2 = \pm\sqrt{7}$ $x = -2 \pm \sqrt{7}$ Solutions: $-2 + \sqrt{7}$ and $-2 - \sqrt{7}$
[9.2] To Solve a Quadratic Equation ($ax^2 + bx + c = 0$) Using the Quadratic Formula	• Write the equation in standard form, if necessary. • Identify the coefficients a, b, and c. • Substitute the values for a, b, and c in the **quadratic formula** $$x = \frac{-b \pm \sqrt{b^2 - 4ac}}{2a}$$ • Simplify.	$-2x^2 + 6x = 7$ $-2x^2 + 6x - 7 = 0$ $a = -2, b = 6$, and $c = -7$ $x = \frac{-6 \pm \sqrt{6^2 - 4(-2)(-7)}}{2(-2)}$ $= \frac{-6 \pm \sqrt{36 - 56}}{-4}$ $= \frac{-6 \pm \sqrt{-20}}{-4}$ $= \frac{-6 \pm 2i\sqrt{5}}{-4}$ $x = \frac{3 \pm i\sqrt{5}}{2}$ Solutions: $\frac{3 + i\sqrt{5}}{2}$ and $\frac{3 - i\sqrt{5}}{2}$

continued

Concept/Skill	Description	Example
[9.2] Discriminant of a Quadratic Equation $(ax^2 + bx + c = 0)$	The expression in the radicand of the quadratic formula: $$b^2 - 4ac$$	For $-2x^2 + 6x - 7 = 0$, the value of the discriminant is $6^2 - 4(-2)(-7) = -20$.
[9.3] Equation Quadratic in Form	An equation that can be written in the form $au^2 + bu + c = 0$, where $a \neq 0$ and u represents an algebraic expression.	$x^4 - 3x^2 + 2 = 0$ is quadratic in form, since it can be written as $u^2 - 3u + 2 = 0$, where $u = x^2$.
[9.4] Quadratic Function	A function of the form $f(x) = ax^2 + bx + c$.	$f(x) = 5x^2 - 3x + 9$
[9.4] Parabola	The graph of a quadratic function.	The graph of $f(x) = x^2 + 2x - 3$ is the following parabola:
[9.4] Vertex of a Parabola	The highest point (with maximum y-value) or lowest point (with minimum y-value) of a parabola.	

continued

Concept/Skill	Description	Example
[9.4] To Find the Vertex of a Parabola	For the parabola given by the function $f(x) = ax^2 + bx + c$, • the x-coordinate of the vertex is $-\dfrac{b}{2a}$, and • the y-coordinate of the vertex is found by substituting $-\dfrac{b}{2a}$ for x in $f(x)$.	$f(x) = x^2 + 4x - 5$ The x-coordinate of the vertex is $-\dfrac{4}{2(1)} = -2$ The y-coordinate of the vertex is $f(-2) = (-2)^2 + 4(-2) - 5$ $= 4 - 8 - 5$ $= -9$ The vertex is $(-2, -9)$.
[9.4] Equation of the Axis of Symmetry	For the graph of $f(x) = ax^2 + bx + c$, the equation of the axis of symmetry is $$x = -\frac{b}{2a}$$	For $f(x) = x^2 + 4x - 5$, the equation of the axis of symmetry is $$x = -\frac{4}{2(1)} = -2$$
[9.5] Quadratic Inequality	An inequality that contains a quadratic expression.	$x^2 + 2x < -1$ and $t^2 + 3t - 5 \geq 0$
[9.5] To Solve a Quadratic Inequality	• Solve the related equation and plot the solutions on a number line separating it into intervals. • Choose a test value in each interval to check if it satisfies the original inequality. • Identify the intervals that contain the solutions of the inequality.	$x^2 + 3x - 10 < 0$ **Related equation** $x^2 + 3x - 10 = 0$ $(x - 2)(x + 5) = 0$ $x - 2 = 0$ or $x + 5 = 0$ $x = 2$ $x = -5$ Interval A | Interval B | Interval C ←———————————————→ -5 2 Let's use values -6, 0, and 3: $$x^2 + 3x - 10$$ $(-6)^2 + 3(-6) - 10 = 8 > 0$ $(0)^2 + 3(0) - 10 = -10 < 0$ $(3)^2 + 3(3) - 10 = 8 > 0$ Value of $x^2 + 3x - 10$: >0 | <0 | >0 Solutions: $(-5, 2)$ ←+—(—+—+—+—+—+—)—+—+—+—+→ $-6\ -5\ -4\ -3\ -2\ -1\ \ 0\ \ 1\ \ 2\ \ 3\ \ 4\ \ 5\ \ 6$

continued

Concept/Skill	Description	Example
[9.5] Rational Inequality	An inequality that contains a rational expression.	$\dfrac{x + 1}{x - 3} \geq -5$ and $\dfrac{2x + 4}{2x + 3} < 0$
[9.5] To Solve a Rational Inequality	• Find all values for which the inequality is undefined. • Solve the inequality's related equation. • Plot the values found in the previous steps on a number line, separating it into intervals. • Choose a test value in each interval to check if it satisfies the original inequality. • Identify the intervals that contain the solutions of the inequality.	$\dfrac{3x}{x + 1} \geq 2$ The rational expression is undefined when $x + 1 = 0$ or $x = -1$. $\dfrac{3x}{x + 1} = 2$ $\dfrac{3x}{x + 1} \cdot (x + 1) = 2 \cdot (x + 1)$ $3x = 2x + 2$ $x = 2$ Interval A | Interval B | Interval C $\xleftarrow{\hspace{1cm}} \underset{-1}{\|} \underset{2}{\|} \xrightarrow{\hspace{1cm}}$ Let's use test values $-2, 0,$ and 3: $\dfrac{3x}{x + 1}$ $\dfrac{3(-2)}{-2 + 1} = 6 > 2$ $\dfrac{3(0)}{0 + 1} = 0 < 2$ $\dfrac{3(3)}{3 + 1} = \dfrac{9}{4} > 2$ $\xleftarrow{\hspace{1cm}} \underset{-1}{\quad} \underset{2}{\quad} \xrightarrow{\hspace{1cm}}$ Value of $\dfrac{3x}{x+1}$: >2 | <2 | >2 Solutions: $(-\infty, -1) \cup [2, \infty)$ $\xleftarrow{\quad} \ )\ +\ +\ [\ +\ +\ \xrightarrow{\quad}$ $-6\ -5\ -4\ -3\ -2\ -1\ \ 0\ \ 1\ \ 2\ \ 3\ \ 4\ \ 5\ \ 6$

Solving a Quadratic Equation ($ax^2 + bx + c = 0$)

Method	When to Use
Factoring	Use when the polynomial can be factored.
The Square Root Property	Use when $b = 0$.
The Quadratic Formula	Use when the first two methods do not apply.
Completing the Square	Use only when specified. This method is easier to use when $a = 1$ and b is even.

Number and Type of Solutions of a Quadratic Equation ($ax^2 + bx + c = 0$)

Value of the Discriminant ($b^2 - 4ac$)	Number and Type of Solutions
Zero	One real solution
Positive	Two real solutions
Negative	Two complex solutions (containing i)

Chapter 9   Review Exercises

To help you review this chapter, solve these problems.

[9.1]   *Solve by using the square root property.*

1. $x^2 - 81 = 0$

2. $3n^2 - 7 = 8$

3. $4(a - 5)^2 = 1$

4. $(2x + 1)^2 + 10 = 6$

5. $x^2 + 8x + 16 = 2$

6. $(n - 5)(n + 5) = -33$

Fill in the blank to make each trinomial a perfect square.

7. $x^2 + 10x + [\ \ ]$

8. $n^2 - 9n + [\ \ ]$

Solve by completing the square.

9. $x^2 - 6x + 2 = 0$

10. $a^2 + a - 3 = 0$

11. $2n^2 + 2n + 9 = 3 - 4n$

12. $3x^2 - 2x - 9 = 0$

[9.2]   *Solve by using the quadratic formula.*

13. $x^2 + 7x + 6 = 0$

14. $x^2 - 4x + 5 = 0$

15. $3x^2 - 13x = 5 - 7x$

16. $4x^2 + 12x = -9$

17. $\dfrac{1}{3}x^2 + \dfrac{3}{2}x + 1 = 0$

18. $0.01x^2 + 0.1x + 0.34 = 0$

Use the discriminant to determine the number and type of solutions for each of the following equations.

19. $2x^2 - x = 2x - 5$

20. $4x^2 + 9x - 3 = 0$

[9.3]   *Solve.*

21. $\dfrac{2}{n - 3} + \dfrac{1}{n - 1} = -1$

22. $x^4 - 2x^2 - 24 = 0$

23. $x - 7\sqrt{x} + 12 = 0$

24. $(p - 1)^2 + 3(p - 1) + 2 = 0$

Find a quadratic equation that has the given solutions.

25. $x = -3$

26. $n = -7$

[9.4]   *Find the vertex, axis of symmetry, and x- and y-intercepts of the graph of each of the following functions. Then sketch the graph.*

27. $f(x) = x^2 - 6x + 5$
Vertex:
Axis of symmetry:
x-intercept(s):
y-intercept:

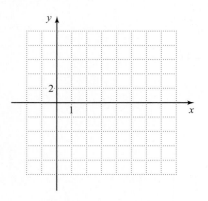

28. $f(x) = 3 + 2x - x^2$
Vertex:
Axis of symmetry:
x-intercept(s):
y-intercept:

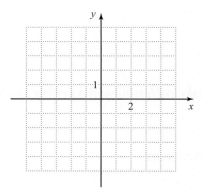

29. $f(x) = x^2 + 8x + 16$
Vertex:
Axis of symmetry:
x-intercept(s):
y-intercept:

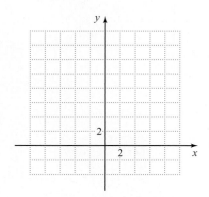

30. $f(x) = x^2 - 5x - 6$
Vertex:
Axis of symmetry:
x-intercept(s):
y-intercept:

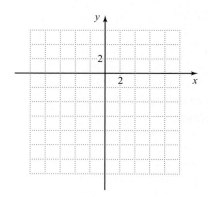

Complete the table.

	Function	Opens Upward or Downward?	Maximum or Minimum Point?	Number of x-intercepts	Number of y-intercepts
31.	$f(x) = x^2 + 9$				
32.	$f(x) = 1 + 3x - 2x^2$				

Graph the function. Then determine the domain and range.

33. $f(x) = 3x - 4x^2$

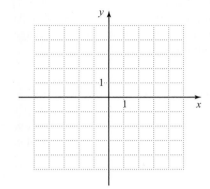

Domain: Range:

34. $f(x) = 2x^2 - x - 1$

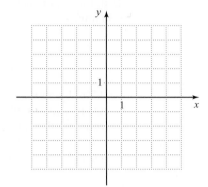

Domain: Range:

[9.5]   *Solve. Then graph the solutions.*

35. $x^2 - 4x < 12$

36. $10 + 3x - x^2 \leq 0$

37. $2x^2 - 9x - 4 \geq -8$

38. $3 > 4x^2 - 4x$

39. $\dfrac{x + 3}{x - 5} > -3$

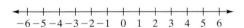

40. $\dfrac{4x - 12}{3x} \leq 0$

Mixed Applications

Solve.

41. The distance s, in feet, that an object falls t sec after it is dropped is given by the equation $s = 16t^2$. How long will it take an object to fall 400 ft?

42. A cable television technician places a 10-ft ladder against a house. The ladder reaches a point on the house that is three times the distance, d, from the base of the house, as shown in the figure. To the nearest tenth of a foot, how far up the side of the house does the ladder reach?

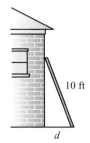

43. Two trains leave the same station at the same time. The train going south travels at an average speed that is 8 mph faster than the train traveling west. Find the average speed, to the nearest mile per hour, of each train if they are 35 mi apart after 30 min.

44. A financial analyst advises a client to invest $5000 in a high-risk fund. To determine the average rate of return, r (in decimal form), on the investment after two years, the broker uses the equation $A = 5000[(1 + r)^2 - 1]$, where A is the amount earned on the investment. To the nearest tenth of a percent, what was the average rate of return if the investment earned $1100?

45. An open box is to be made from a 25-in. by 40-in. rectangular piece of cardboard by cutting squares of equal area from each corner and turning up the sides. If the specifications require the base of the box to have an area of 700 in², what size squares should be cut from each corner?

46. The owner of a small office supply store uses the equation $R = 100x - 2x^2$ to determine the number of cases of paper, x, he needs to sell each week for revenue of R dollars. How many cases does he need to sell each week for revenue of $1250?

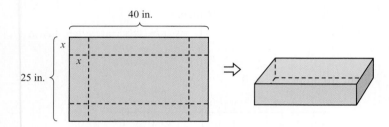

47. The number of Krispy Kreme stores, S, in operation for each year from 1999 to 2003 can be approximated by the equation $S = 7n^2 + 8n + 131$, where n represents the number of years after 1999. In what year were there 218 stores? (*Source:* Based on data from Krispy Kreme 2003 Annual Report)

48. The number of sales associates A, in millions, working in Wal-Mart stores for each year from 1995 to 2002 can be approximated by the equation $A = 0.009x^2 + 0.05x + 0.6$, where x is the number of years after 1995. In what year were there approximately 1.2 million sales associates? (*Source:* Based on data from Wal-Mart Stores, Inc.)

49. A student bicycles to and from school each day. On a particular day, she bicycled to school at a rate that was 2 mph faster than her rate bicycling home. If she lives 8 mi from school and it took her 6 min longer to bicycle home, to the nearest tenth of a mile per hour, at what rate did she bicycle to school that day?

50. It takes an older printing press 5 hr longer to complete a print job than a newer printing press. If the printing presses work together, they can complete the job in 2 hr. To the nearest tenth of an hour, how long does it take each press to complete the job working alone?

51. A city parks department uses 300 ft of fencing to enclose a rectangular playground in one of its parks.

a. Express the width, *w*, of the playground in terms of the length, *l*.

b. Express in function notation the relationship between the area, $A(l)$, and the length of the playground.

c. Graph the function.

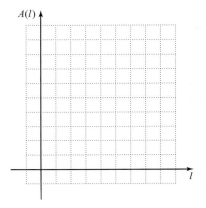

d. What dimensions will maximize the area of the playground? What is the maximum area?

52. The owner of a factory determines that the daily cost in dollars of fabricating *x* units of a machine part is modeled by the function $C(x) = 0.05x^2 - 2x + 100$.

a. Graph the function.

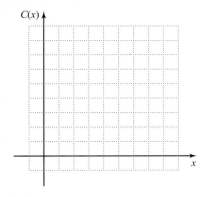

b. What number of units must be fabricated per day in order to minimize the cost?

c. What is the minimum cost?

53. A baseball is hit straight upward with an initial velocity of 80 ft per sec from a height of 4 ft. The height *h*, in feet, of the baseball above the ground *t* sec after it is hit is modeled by the function $h(t) = -16t^2 + 80t + 4$. For what values of *t* is the baseball 40 ft above the ground?

54. A homeowner wants to build a rectangular patio that has a perimeter of 44 ft. For what range of widths will the patio have an area of at least 120 ft²?

● *Check your answers on page A-42.*

Chapter 9   **POSTTEST**

To see if you have mastered the topics in this chapter, take this test.

1. Solve by using the square root property:
$5n^2 - 11 = 29$

2. Solve by completing the square: $3p^2 - 6p = -24$

3. Solve by using the quadratic formula:
$4x^2 + 4x - 3 = 0$

4. Use the discriminant to determine the number and type of solutions of $2x^2 + 7x + 9 = 0$.

Solve.

5. $5(3n + 2)^2 - 90 = 0$

6. $x^2 + 8x = 6$

7. $x^2 - x - 2 = 4x - 13$

8. $4x^2 + 3x = 7 + 6x$

9. $2x^2 - 12x + 20 = 1$

10. $\frac{1}{2}x^2 + x - 2 = 0$

11. $x^{1/2} - 2x^{1/4} = -1$

12. Find a quadratic equation in n that has the solutions $\frac{1}{2}$ and $\frac{2}{3}$.

Identify the vertex, the equation of the axis of symmetry, and the x- and y-intercepts of the graph. Then graph.

13. $f(x) = x^2 - 6x + 8$

14. $f(x) = -x^2 + 3x + 10$

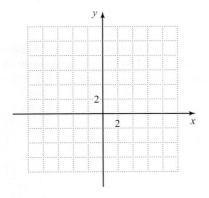

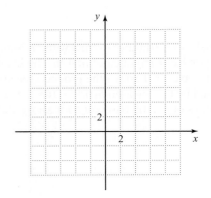

15. Graph the function $f(x) = 2x^2 - 4x - 1$. Then identify the domain and range.

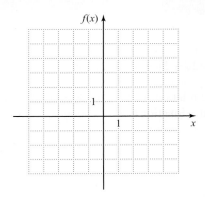

16. Solve. Then graph the solutions: $-x^2 - 3x + 18 < 0$

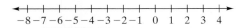

17. Two airplanes leave the same airport at the same time. The airplane flying west travels at an average speed that is 20 mph faster than the airplane flying south. If the two airplanes are 650 mi apart after 1 hr, find the average speed of each airplane to the nearest mile per hour.

18. A junior credit clerk takes 3 hr longer to process the same number of credit applications as a senior credit clerk. If they work together, they can process the applications in 2 hr. To the nearest tenth of an hour, how long does it take each clerk to process the applications working alone?

19. A toy rocket is launched straight upward with an initial velocity of 96 ft per sec from a 3-foot-high platform. The height h, in feet, of the rocket above the ground t sec after it is launched is modeled by the function $h(t) = -16t^2 + 96t + 3$. When will the rocket reach its maximum height? What is that height?

20. The owner of a factory determines that the daily cost C, in dollars, to produce x units of a product is modeled by the function $C(x) = 0.025x^2 - 8x + 800$. How many units can the factory produce each day in order to keep the daily cost below $320?

● *Check your answers on page A-43.*

Cumulative Review Exercises

To help you review, solve the following problems.

1. Simplify: $4n^5(2n^3)^{-1}$

2. Solve: $|2x - 3| + 1 = 7$

3. Graph: $x + 4y - 8 = 0$

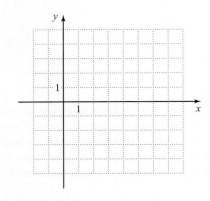

4. Does the relation $\{(-5, 2), (-3, -4), (0, 1), (3, 1), (4, 2)\}$ represent a function?

5. Multiply: $(a - 2)(a^2 + 2a + 4)$

6. Factor: $2x^2y^4 - 12x^3y^3 + 16x^4y^2$

7. Solve: $6x^2 - 2x = -1$

8. Simplify: $\sqrt{27} - 3\sqrt{18} + 7\sqrt{3}$

9. A farmer has less than 1000 ft of fencing to enclose a rectangular grazing area on his farm.

    a. Express this relationship as a linear inequality, letting l and w represent the length and the width of the fence, respectively.

    b. Graph the inequality.

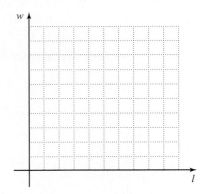

    c. What is one possible set of dimensions for the grazing area?

10. A long-distance phone service provider charges its customers a flat monthly fee and an additional fee for each minute of long-distance calls made during the month. A customer's long-distance phone bill is $13.50 when she makes 120 min of long-distance phone calls and $19.90 when she makes 248 min of long-distance phone calls.

    a. Write an equation in slope-intercept form that relates the monthly phone bill, B, in dollars, and the number of minutes, m, of long-distance calls made that month.

    b. Identify the slope of the line and interpret its meaning in the context of the situation.

    c. Identify the y-intercept and interpret its meaning in the context of the situation.

● *Check your answers on page A-43.*

Exponential and Logarithmic Functions

Ecology and Exponential Functions

Ecologists study the effect on natural resources of growing industrialization. One of their concerns is the dwindling supply of fossil fuels, such as oil, natural gas, and coal. The known world coal reserves are estimated to be about 1×10^{12} tons, with consumption currently running at approximately 5×10^9 tons per year.

Suppose we assume that coal consumption increases by 5% per year, that is, each year it is 1.05 times that of the previous year. In this model, described by $C(t) = (5 \times 10^9)(1.05)^t$, consumption is an *exponential function* of time because consumption changes by a constant *factor* as time increases by one year. (*Sources:* U.S. Energy Information Administration; Robert Ricklefs and Gary Miller, *Ecology*, New York: W. H. Freeman and Company, 2000)

Chapter 10   PRETEST

To see if you have already mastered the topics in this chapter, take this test.
Use a calculator where necessary. Round answers to four decimal places.

1. Consider the functions $f(x) = 2x - 1$ and $g(x) = 2x^2 + 5x - 3$. Find:

   a. $(f + g)(x)$

   b. $(f - g)(x)$

   c. $(f \cdot g)(x)$

   d. $\left(\dfrac{f}{g}\right)(x)$

2. Given $f(x) = \dfrac{5}{x}$, $x \neq 0$, and $g(x) = 3x - 4$, find:

   a. $(f \circ g)(x)$

   b. $(g \circ f)(x)$

   c. $(f \circ g)(3)$

   d. $(g \circ f)(-5)$

3. Determine whether the function whose graph is shown on the following coordinate plane is one-to-one. If it is, sketch the graph of its inverse.

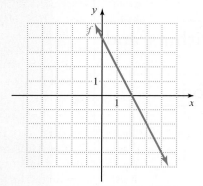

4. Find the inverse of the one-to-one function $f(x) = 3x - 7$.

5. Evaluate each function for the given value. If necessary, round the answer to the nearest thousandth.

   a. $f(x) = 2^{x-5}$; $f(3)$

   b. $f(x) = -e^{-x}$; $f(-2)$

6. Evaluate:

   a. $\log_7 1$

   b. $\log_9 \dfrac{1}{81}$

7. Graph:

   a. $f(x) = 2^x + 1$

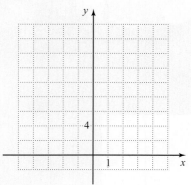

   b. $f(x) = -\log_4 x$

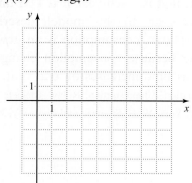

8. Evaluate:

    a. $\log_6 6^5$

    b. $\log_4 1$

9. Evaluate:

    a. $\log \dfrac{1}{10}$

    b. $\ln e^5$

10. Find and round to four decimal places: $\log_3 8$

11. Write each expression as a sum or difference of logarithms.

    a. $\log_5(5x^2)$

    b. $\log_8 \dfrac{3x^3}{y}$

12. Write each expression as a single logarithm.

    a. $3\log_7 2 + \log_7 5$

    b. $4\log_6 x - 2\log_6(x + 2)$

Solve.

13. $5^x = 25$

14. $4^{x+1} = 32^x$

15. $\log_x \dfrac{1}{2} = -1$

16. $\log_2(x - 4) = 3$

17. $\log_2 x + \log_2(x + 6) = 4$

18. Inflation is a sustained increase in the cost of goods and services. Suppose the annual rate of inflation will be 4% during the next 10 yr. Then the cost, C, of goods and services in any year, x, during that decade will be modeled by the function $C(x) = P(1.04)^x$, where P is the present cost. If the present cost of a gallon of milk is $3.89, what will it cost in 5 yr?

19. The effective annual interest rate r on an investment earning interest rate k, compounded continuously, can be determined from the logarithmic equation $\ln(r + 1) = k$, where r and k are expressed in decimal form. To the nearest tenth of a percent, find the effective annual interest rate on an investment earning 4.5% interest compounded continuously.

20. Gallium-67, a radioactive isotope used to detect tumors, has a half-life of about 3.3 days. The amount, C, of the isotope remaining after t days is given by the equation $C = C_0 e^{-0.21t}$, where C_0 is the initial amount present. Approximately how long will it take 100 mg of gallium-67 to decay to 12 mg?

● *Check your answers on page A-43.*

10.1   The Algebra of Functions and Inverse Functions

In this section, we return to the discussion of functions that began in Section 3.6. There we considered how to identify, evaluate, and graph a function and also how to determine the domain and range of a function. Here, we learn about a new function, then deal with operations on functions, with particular attention to the operation of composition. We then move on to one-to-one and inverse functions. This final topic will be of particular importance in making the transition from Section 10.2 to Section 10.3.

The Absolute Value Function

We have already considered various types of functions, including linear, rational, radical, and quadratic functions. Now let's look at the absolute value function.

The *absolute value function* is the function $f(x) = |x|$, shown in the following graph. Note that except for the origin, the graph of the absolute value function lies entirely above the x-axis.

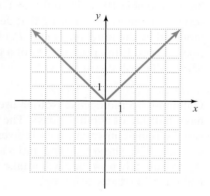

We see from the graph of this function that its domain is $(-\infty, \infty)$. Since the y-values are all nonnegative numbers, the range is $[0, \infty)$, that is, the set including all positive numbers and 0.

EXAMPLE 1	PRACTICE 1

EXAMPLE 1

Graph $f(x) = -|x|$. Identify the function's domain and range.

Solution We find some ordered pairs and then plot them.

x	$f(x) = -\lvert x \rvert$	(x, y)
-2	$f(-2) = -\lvert -2 \rvert = -2$	$(-2, -2)$
-1	$f(-1) = -\lvert -1 \rvert = -1$	$(-1, -1)$
0	$f(0) = -\lvert 0 \rvert = 0$	$(0, 0)$
1	$f(1) = -\lvert 1 \rvert = -1$	$(1, -1)$
2	$f(2) = -\lvert 2 \rvert = -2$	$(2, -2)$

PRACTICE 1

Graph $f(x) = |x + 1|$. Identify the function's domain and range.

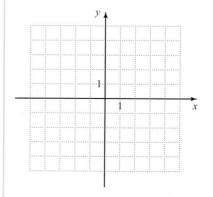

The domain is $(-\infty, \infty)$. We see from the graph that the dependent variable y can take any nonpositive value, so the range is $(-\infty, 0]$.

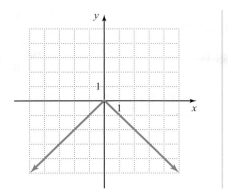

Operations on Functions

Recall that the sum of two polynomials is a third polynomial. Similarly, the sum of two functions is a function. For example, suppose that $f(x) = 3x^2$ and $g(x) = 7x - 1$. We can define a new function (the "sum function") represented by $(f + g)(x)$ as follows:

$$(f + g)(x) = f(x) + g(x) = 3x^2 + 7x - 1$$

Similarly, we can subtract, multiply, or divide given functions to define other new functions.

There are many situations in which we define a new function in terms of old ones. For instance, suppose that a company is producing and selling a certain product. In this case, the profit, P, that the company makes can be computed by subtracting the cost, C, from the revenue, R. If each of these three variables is a function of the number of products, x, produced and sold, then

$$P(x) = R(x) - C(x)$$

and we can write the profit function $P(x)$ as the difference function $(R - C)(x)$.

In general, we define the sum, difference, product, and quotient functions as follows:

Definitions

If f and g are functions, then

$(f + g)(x) = f(x) + g(x)$ **The sum of two functions**

$(f - g)(x) = f(x) - g(x)$ **The difference of two functions**

$(f \cdot g)(x) = f(x) \cdot g(x)$ **The product of two functions**

$\left(\dfrac{f}{g}\right)(x) = \dfrac{f(x)}{g(x)},$ for $g(x) \neq 0$ **The quotient of two functions**

EXAMPLE 2

Given that $f(x) = 3x - 5$ and $g(x) = 2x + 1$, find:

a. $(f + g)(x)$ **b.** $(f - g)(x)$

c. $(f \cdot g)(x)$ **d.** $\left(\dfrac{f}{g}\right)(x)$

PRACTICE 2

Given that $f(x) = x^2 - 1$ and $g(x) = 3x - 2$, find:

a. $(f + g)(x)$

b. $(f - g)(x)$

Solution

a. $(f + g)(x) = f(x) + g(x)$

$\quad\quad\quad\quad = (3x - 5) + (2x + 1)$ **Substitute $(3x - 5)$ for $f(x)$ and $(2x + 1)$ for $g(x)$.**

$\quad\quad\quad\quad = 5x - 4$

b. $(f - g)(x) = f(x) - g(x)$

$\quad\quad\quad\quad = (3x - 5) - (2x + 1)$

$\quad\quad\quad\quad = 3x - 5 - 2x - 1$

$\quad\quad\quad\quad = x - 6$

c. $(f \cdot g)(x) = f(x) \cdot g(x)$

$\quad\quad\quad\quad = (3x - 5)(2x + 1)$

$\quad\quad\quad\quad = 6x^2 + 3x - 10x - 5$

$\quad\quad\quad\quad = 6x^2 - 7x - 5$

d. $\left(\dfrac{f}{g}\right)(x) = \dfrac{f(x)}{g(x)}$

$\quad\quad\quad\quad = \dfrac{3x - 5}{2x + 1}, \quad$ where $x \neq -\dfrac{1}{2}$

c. $(f \cdot g)(x)$

d. $\left(\dfrac{f}{g}\right)(x)$

Can you explain in Example 2(d) why the value $-\dfrac{1}{2}$ is excluded?

Consider the functions $f(x) = 3x - 5$ and $g(x) = 2x + 1$ given in Example 2. Suppose we want to evaluate $(f + g)(x)$ for a particular value of x, for example, 3. Since $(f + g)(x) = f(x) + g(x)$, we get:

$$(f + g)(3) = f(3) + g(3)$$
$$= [3(3) - 5] + [2(3) + 1]$$
$$= (9 - 5) + (6 + 1)$$
$$= 4 + 7$$
$$= 11$$

So $(f + g)(3) = 11$.

Recall that in Example 2, we found the sum function $(f + g)(x) = 5x - 4$. Evaluating this function for $x = 3$, we get:

$$(f + g)(3) = 5(3) - 4$$
$$= 15 - 4$$
$$= 11$$

The preceding examples suggest that to evaluate the sum of two functions for a particular value of x, we can *either evaluate the individual functions and then add the values, or find the sum function and then evaluate that function.* Similar statements can be made for the difference, product, and quotient of two functions.

EXAMPLE 3	PRACTICE 3

EXAMPLE 3

Given that $f(x) = -x + 7$ and $g(x) = x^2 + 3x$, find:

a. $(f + g)(2)$ b. $(f - g)(-3)$

c. $(f \cdot g)(0)$ d. $\left(\dfrac{f}{g}\right)(1)$

Solution

a. $(f + g)(2) = f(2) + g(2)$ — Express as the sum of two functions.

$= (-2 + 7) + [2^2 + 3(2)]$ — Evaluate $f(2)$ and $g(2)$.

$= (5) + (4 + 6)$

$= 5 + 10$

$= 15$

b. $(f - g)(-3) = f(-3) - g(-3)$

$= [-(-3) + 7] - [(-3)^2 + 3(-3)]$

$= (3 + 7) - (9 - 9)$

$= 10 - 0$

$= 10$

c. $(f \cdot g)(0) = f(0) \cdot g(0)$

$= (-0 + 7) \cdot [0^2 + 3(0)]$

$= 7 \cdot 0$

$= 0$

d. $\left(\dfrac{f}{g}\right)(1) = \dfrac{f(1)}{g(1)}$

$= \dfrac{-1 + 7}{1^2 + 3(1)}$

$= \dfrac{6}{4}$

$= \dfrac{3}{2}$

PRACTICE 3

Given that $f(x) = x - x^2$ and $g(x) = 2x + 5$, find:

a. $(f + g)(-1)$

b. $(f - g)(5)$

c. $(f \cdot g)(1)$

d. $\left(\dfrac{f}{g}\right)(2)$

The Composition of Functions

Another operation that can be performed on functions is *composition*, represented by the symbol $\circ$. When we take the composition of two functions, we find the value of one function and then evaluate the other function at that value.

For instance, consider the following example. A size-4 dress in the United States is a size 36 in France. A function that converts dress sizes in the United States to those in France is $g(x) = x + 32$, where x is the U.S. size and $g(x)$ is the French size. Similarly, the function $f(x) = 2x - 40$ relates dress sizes in France to those in Italy, where x is the French size and $f(x)$ is the Italian size. Using these two functions, we can define a function—say, $h(x)$—that relates dress sizes in the United States to those in Italy.

If $g(x) = x + 32$ and $f(x) = 2x - 40$, then

$$h(x) = f(g(x)) = f(x + 32) = 2(x + 32) - 40 = 2x + 64 - 40 = 2x + 24$$

We call the function $h(x)$ the *composition of f and g*.

Definition

If f and g are functions, then the **composition of f and g** is the function, represented by $f \circ g$, where

$$(f \circ g)(x) = f(g(x))$$

Note that the *composite function $f \circ g$* can be read either "f composed with g" or "f circle g." The composite function is defined for those values of x in the domain of g where $g(x)$ is also in the domain of f.

EXAMPLE 4

If $f(x) = 3x - 1$ and $g(x) = 2x + 1$, find each composition:

a. $(f \circ g)(x)$ **b.** $(g \circ f)(x)$

c. $(f \circ g)(2)$ **d.** $(g \circ f)(2)$

Solution

a. $(f \circ g)(x) = f(g(x))$

$\qquad = f(2x + 1)$ Substitute $2x + 1$ for $g(x)$.

$\qquad = 3(2x + 1) - 1$ Substitute $2x + 1$ for x in $f(x)$.

$\qquad = 6x + 3 - 1$ Use the distributive property.

$\qquad = 6x + 2$

b. $(g \circ f)(x) = g(f(x))$

$\qquad = g(3x - 1)$ Substitute $3x - 1$ for $f(x)$.

$\qquad = 2(3x - 1) + 1$ Substitute $3x - 1$ for x in $g(x)$.

$\qquad = 6x - 2 + 1$ Use the distributive property.

$\qquad = 6x - 1$

c. $(f \circ g)(2) = f(g(2))$

$\qquad = f(2(2) + 1)$ Substitute 2 for x in $g(x)$.

$\qquad = f(5)$ Simplify.

$\qquad = 3(5) - 1$ Substitute 5 for x in $f(x)$.

$\qquad = 15 - 1$

$\qquad = 14$

PRACTICE 4

If $f(x) = 4x - 3$ and $g(x) = -x + 5$, find each composition:

a. $(f \circ g)(x)$

b. $(g \circ f)(x)$

c. $(f \circ g)(3)$

d. $(g \circ f)(3)$

d. $(g \circ f)(2) = g(f(2))$

$\qquad\qquad\quad = g(3(2) - 1)$ Substitute 2 for x in $f(x)$.

$\qquad\qquad\quad = g(5)$ Simplify.

$\qquad\qquad\quad = 2(5) + 1$ Substitute 5 for x in $g(x)$.

$\qquad\qquad\quad = 11$

Note that in Example 4, the answers to parts (a) and (b) are different, as are the answers to parts (c) and (d), showing that the operation of composition is not commutative.

> **Tip** Remember that the composite function $(f \circ g)(x) = f(g(x))$, whereas the product function $(f \cdot g)(x) = f(x) \cdot g(x)$.

One-to-One Functions

Recall that a function is a relation in which no two ordered pairs have the same *first* coordinates. Some functions satisfy the additional condition that no two ordered pairs have the same *second* coordinates. These functions are said to be *one-to-one.*

For instance, suppose that f is defined by the relation $\{(2, 5), (-4, 8), (1, 2)\}$. Note that every ordered pair has a different second coordinate, so that the function is one-to-one. By contrast, the function g defined by $\{(-4, \mathbf{1}), (2, 3), (0, \mathbf{1})\}$ is not one-to-one, since the second coordinate, 1, corresponds to two first coordinates, -4 and 0.

> **One-to-One Function**
>
> A function f is said to be **one-to-one** if each y-value in the range corresponds to exactly one x-value in the domain.

In other words, a function is one-to-one if all the ordered pairs that define the function have different second coordinates.

Any linear function $f(x) = mx + b$, where $m \neq 0$, such as $f(x) = 3x - 1$, is one-to-one. One way to see this is to assume that the y-values for two x-values, for example, a and $b,$ are equal. In this case, we get

$$f(a) = f(b)$$
$$3a - 1 = 3b - 1$$
$$a = b$$

So there can be only one x-value with this y-value. Therefore, the function is one-to-one.

By contrast, let's consider the function $g(x) = x^2$. This function has the same value, namely 4, for $x = 2$ and $x = -2$. Since there is more than one x-value for this y-value, g is not one-to-one.

Since the ordered pairs of a one-to-one function must have different second coordinates, it follows that the graph of a one-to-one function cannot have two points with different x-coordinates and the same y-coordinate. This suggests a test for determining if a graph represents a one-to-one function.

The Horizontal-Line Test

If every horizontal line intersects the graph of a function at most once, then the function is one-to-one.

Let's look again at the functions $f(x) = 3x - 1$ and $g(x) = x^2$. We see from the following figure that no horizontal line intersects the graph $f(x) = 3x - 1$ at more than one point. So f passes the horizontal-line test and is a one-to-one function.

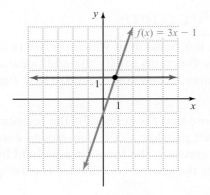

On the other hand, we see that a horizontal line intersects the graph of $g(x) = x^2$ more than once. So g fails the horizontal-line test and is not a one-to-one function.

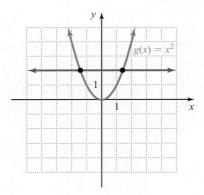

Recall from Section 3.6 that a vertical-line test can be used to see if a graph defines a function.

Tip   To check if a graph defines a function, we use the *vertical-line* test. To check if a function is one-to-one, we use the *horizontal-line* test.

EXAMPLE 5

Consider the following graphs of functions. Determine whether each function is one-to-one.

a.

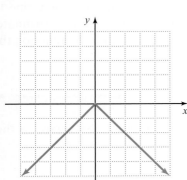

b.

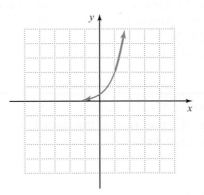

Solution

a. Since any horizontal line in Quadrants III and IV intersects the graph more than once, the graph fails the horizontal-line test. So the function is not one-to-one.

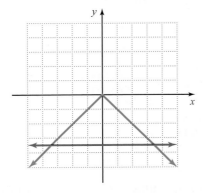

b. Since no horizontal line intersects the graph at more than one point, the graph passes the horizontal-line test. So the function is one-to-one.

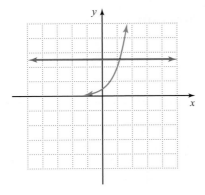

PRACTICE 5

Determine whether the function whose graph is shown is a one-to-one function.

a.

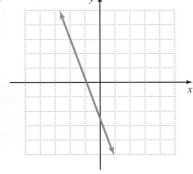

b.

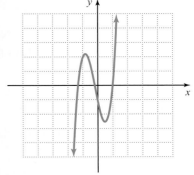

The Inverse of a Function

Every one-to-one function has an *inverse function* that is formed by interchanging the coordinates of the ordered pairs that define the original function.

For instance, let's consider a function defined by the following set of ordered pairs: $\{(2, 5), (-4, 8), (1, 2)\}$. We have already noted that this function is one-to-one. To find the inverse function, we interchange the first and second coordinates, getting $\{(5, 2), (8, -4), (2, 1)\}$. This set of ordered pairs defines the function that is the inverse of the original function.

> **Definition**
>
> The **inverse of a one-to-one function** f, written f^{-1} and read "f inverse," is the set of all ordered pairs (y, x) where f is the set of ordered pairs (x, y).

From this definition, it follows that if $f(x) = y$, then $f^{-1}(y) = x$. Since f^{-1} is found by interchanging x and y, the range of f^{-1} is the domain of f and the domain of f^{-1} is the range of f.

Not every function has an inverse function. For instance, the function $f(x) = x^2$ does not have an inverse function. Can you explain why?

EXAMPLE 6	**PRACTICE 6**
Find the inverse of the function defined by $\{(2, -5), (1, 0), (3, 3), (0, -1)\}$.	If f is defined by $\{(-3, -1), (-1, 0), (0, 2), (4, 3)\}$, find f^{-1}.
Solution The inverse function is found by switching the coordinates of each ordered pair: $\{(-5, 2), (0, 1), (3, 3), (-1, 0)\}$	

When a one-to-one function is defined by an equation rather than as a set of ordered pairs, we can find its inverse function by interchanging the variables. The following rule gives the procedure for finding the inverse of such a function.

> **To Find the Inverse of a Function**
> - Substitute y for $f(x)$.
> - Interchange x and y.
> - Solve the equation for y.
> - Substitute $f^{-1}(x)$ for y.

EXAMPLE 7	PRACTICE 7
Find the inverse function of $f(x) = 2x + 5$.	Find the inverse function of $g(x) = 3x - 1$.

Solution

$$f(x) = 2x + 5$$
$$y = 2x + 5 \quad \text{Substitute } y \text{ for } f(x).$$
$$x = 2y + 5 \quad \text{Interchange } x \text{ and } y.$$
$$x - 5 = 2y \quad \text{Solve for } y.$$
$$\frac{x - 5}{2} = y$$
$$y = \frac{x - 5}{2}$$
$$f^{-1}(x) = \frac{x - 5}{2} \quad \text{Substitute } f^{-1}(x) \text{ for } y.$$

So the inverse function of $f(x)$ is $f^{-1}(x) = \dfrac{x - 5}{2}$.

Recall that if $f(x) = y$, then $f^{-1}(y) = x$. Therefore, $f^{-1}(f(x)) = f^{-1}(y) = x$. Similarly, $f(f^{-1}(y)) = f(x) = y$. These observations suggest the following property.

The Composition of a Function and Its Inverse

For any functions f and g, the following two conditions are equivalent:
- $f(x)$ and $g(x)$ are inverse functions of each other.
- $f(g(x)) = x$ and $g(f(x)) = x$.

We can use this property to check whether two functions are inverses of each other. For instance, consider the function in Example 7, $f(x) = 2x + 5$, where we found that $f^{-1}(x) = \dfrac{x - 5}{2}$. Now let's evaluate the composition $(f^{-1} \circ f)(x)$.

$$(f^{-1} \circ f)(x) = f^{-1}(f(x))$$
$$= f^{-1}(2x + 5)$$
$$= \frac{(2x + 5) - 5}{2}$$
$$= \frac{2x}{2}$$
$$= x$$

So $(f^{-1} \circ f)(x) = x$, as the property predicts. Confirm that $(f \circ f^{-1})(x) = x$.

EXAMPLE 8	PRACTICE 8

Is $g(x) = \dfrac{1}{4}x + 2$ the inverse of $f(x) = 4x - 8$?

Determine whether $p(x) = 5x - 2$ is the inverse of $q(x) = -5x + 2$.

Solution We can decide if g is the inverse of f by determining whether $f(g(x)) = x$ and $g(f(x)) = x$.

$$f(g(x)) = f\left(\dfrac{1}{4}x + 2\right)$$

$$= 4\left(\dfrac{1}{4}x + 2\right) - 8$$

$$= x + 8 - 8$$

$$= x$$

$$g(f(x)) = g(4x - 8)$$

$$= \dfrac{1}{4}(4x - 8) + 2$$

$$= x - 2 + 2$$

$$= x$$

Since $f(g(x)) = x$ and $g(f(x)) = x$, the functions are inverses of each another.

The graph of a one-to-one function and its inverse function have a special relationship. If a point, say (a, b), lies on the graph of a one-to-one function, then the point (b, a) with the coordinates interchanged lies on the graph of its inverse. Therefore, the graph of a function and the graph of its inverse function are mirror images of each other: They are symmetric about the line $y = x$.

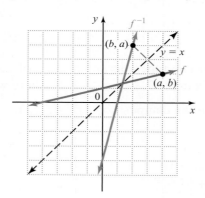 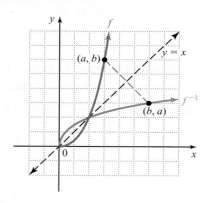

Consider, for instance, the one-to-one function $f(x) = 4x - 8$ and its inverse $f^{-1}(x) = \dfrac{1}{4}x + 2$. We graph both functions on the coordinate plane to the right.

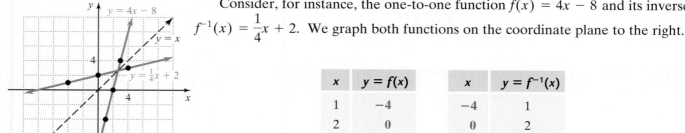

x	$y = f(x)$		x	$y = f^{-1}(x)$
1	−4		−4	1
2	0		0	2
3	4		4	3

Both functions are linear, so their graphs are straight lines. Note that, as predicted, each graph is the mirror image of the other, reflected about the line $y = x$. A point on either graph corresponds to a point on the other graph with the coordinates interchanged. For instance, $(2, 0)$ lies on the graph of $f(x)$, and $(0, 2)$ lies on the graph of $f^{-1}(x)$. We say that the graphs of f and f^{-1} are *symmetric* about the line $y = x$.

EXAMPLE 9	PRACTICE 9

Given the graph of a function, graph its inverse.

Given the graph of a function, graph its inverse.

a.

b.

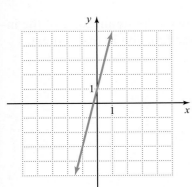

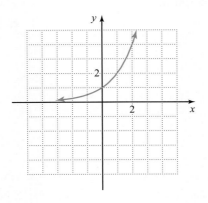

a.

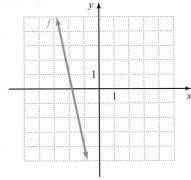

Solution

a.

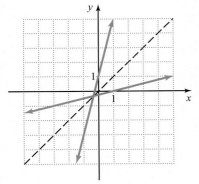

The graph of the inverse function is the mirror image of the original graph about the line $y = x$.

b.

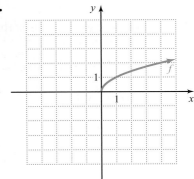

b.

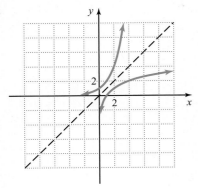

EXAMPLE 10

The function $f(x) = \dfrac{9}{5}x + 32$ can be used to convert a Celsius temperature, x, to the equivalent Fahrenheit temperature, $f(x)$.

a. Is f a one-to-one function? Explain how you know.

b. Identify the inverse function of f.

c. Explain the significance of the inverse function in this context.

Solution

a. To determine if f is one-to-one, let's graph it.

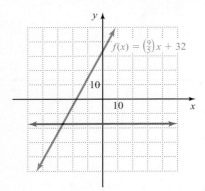

We see that the graph passes the horizontal-line test. So the function is one-to-one.

b. We need to identify the inverse function of f as follows:

$$f(x) = \frac{9}{5}x + 32$$

$$y = \frac{9}{5}x + 32$$

$$x = \frac{9}{5}y + 32$$

$$x - 32 = \frac{9}{5}y$$

$$\frac{5}{9}(x - 32) = y$$

$$y = \frac{5}{9}(x - 32)$$

$$f^{-1}(x) = \frac{5}{9}(x - 32)$$

c. Because f^{-1} is the inverse of f, it can be used to convert a Fahrenheit temperature x to the equivalent Celsius temperature.

PRACTICE 10

A rental agency charges $C(x)$ dollars to rent a six-passenger van for x hr, where $C(x) = 40x$.

a. Is C a one-to-one function? Explain how you know.

b. Identify C^{-1}.

c. Explain the significance of the inverse function in this context.

Mathematically Speaking

Fill in each blank with the most appropriate term or phrase from the given list.

product of $(f + g)$ and x	one-to-one	the y-axis
the line $y = x$	sum of f and g	product of f, g, and x
linear	product of f and g	composition of f and g
horizontal	linear function	vertical
absolute value function	composite functions	inverse functions of each other

1. The function $f(x) = |x|$ is the ———.

2. If f and g are functions, then the ——— is defined as $(f + g)(x) = f(x) + g(x)$.

3. If f and g are functions, then the ——— is defined as $(f \cdot g)(x) = f(x) \cdot g(x)$.

4. If f and g are functions, then the ——— is the function, represented by $f \circ g$, where $(f \circ g)(x) = f(g(x))$.

5. A function f is said to be ——— if each y-value in the range corresponds to exactly one x-value in the domain.

6. If every ——— line intersects the graph of a function at most once, then the function is one-to-one.

7. If f and g are functions and $f(g(x)) = x$ and $g(f(x)) = x$, then $f(x)$ and $g(x)$ are ———.

8. The graphs of a function and its inverse are symmetric about ———.

Evaluate each function for the given values.

9. $f(x) = \left| \dfrac{1}{2}x + 3 \right|$

    a. $f(0)$ **b.** $f(-8)$

    c. $f(-4t)$ **d.** $f(t - 6)$

10. $h(x) = 1 - |2x|$

    a. $h(-5)$ **b.** $h\left(\dfrac{1}{2}\right)$

    c. $h(-a)$ **d.** $h(a + 1)$

Graph each function. Then identify its domain and range.

11. $g(x) = |x| + 2$

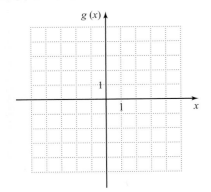

12. $g(x) = -|x| + 3$

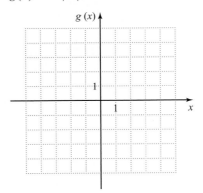

13. $h(x) = |x + 2|$

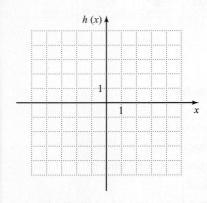

14. $g(x) = |x - 4|$

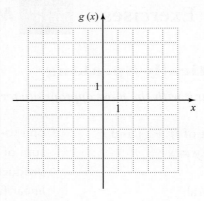

Given f(x) and g(x), find (f + g)(x), (f − g)(x), (f · g)(x), and $\left(\dfrac{f}{g}\right)(x)$.

15. $f(x) = 2x + 3; g(x) = 4x^2 - 9$

16. $f(x) = 3x^2 - 6x; g(x) = x - 2$

17. $f(x) = 3x^2 + 11x + 6; g(x) = 2x^2 + 5x - 3$

18. $f(x) = 1 - 4x - 5x^2; g(x) = x^2 + x$

19. $f(x) = \dfrac{1}{x + 3}; g(x) = x - 1$

20. $f(x) = \dfrac{-1}{x + 2}; g(x) = 2$

21. $f(x) = 2\sqrt{x}; g(x) = 4\sqrt{x} - 1$

22. $f(x) = \sqrt{x} + 3; g(x) = \sqrt{x} - 3$

Given $f(x) = x^2 - 3x - 4$ and $g(x) = 1 - x^2$, find each of the following.

23. $(f + g)(3)$

24. $(f + g)(-2)$

25. $(f - g)(-1)$

26. $(g - f)(0)$

27. $(f \cdot g)(2)$

28. $(f \cdot g)(-1)$

29. $\left(\dfrac{g}{f}\right)(5)$

30. $\left(\dfrac{f}{g}\right)(-4)$

Given $f(x)$ and $g(x)$, find each composition.

31. $f(x) = 2x - 4$ and $g(x) = \dfrac{1}{2}x - 5$

  a. $(f \circ g)(x)$

  b. $(g \circ f)(x)$

  c. $(f \circ g)(3)$

  d. $(g \circ f)(-1)$

32. $f(x) = \dfrac{1}{3}x + 2$ and $g(x) = -3x + 6$

  a. $(f \circ g)(x)$

  b. $(g \circ f)(x)$

  c. $(f \circ g)(-2)$

  d. $(g \circ f)(4)$

33. $f(x) = x - 1$ and $g(x) = x^2 + 4x - 10$

  a. $(f \circ g)(x)$

  b. $(g \circ f)(x)$

  c. $(f \circ g)(1)$

  d. $(g \circ f)(-1)$

34. $f(x) = 4 + 5x - x^2$ and $g(x) = x + 3$

  a. $(f \circ g)(x)$

  b. $(g \circ f)(x)$

  c. $(f \circ g)(-3)$

  d. $(g \circ f)(3)$

35. $f(x) = \dfrac{3}{x}$ and $g(x) = 2x + 5$

  a. $(f \circ g)(x)$

  b. $(g \circ f)(x)$

  c. $(f \circ g)\left(\dfrac{1}{2}\right)$

  d. $(g \circ f)(-2)$

36. $f(x) = x - 6$ and $g(x) = \dfrac{2}{x + 4}$

  a. $(f \circ g)(x)$

  b. $(g \circ f)(x)$

  c. $(f \circ g)(-2)$

  d. $(g \circ f)(0)$

37. $f(x) = -\sqrt{x}$ and $g(x) = 1 - 4x$

  a. $(f \circ g)(x)$

  b. $(g \circ f)(x)$

  c. $(f \circ g)(-6)$

  d. $(g \circ f)(16)$

38. $f(x) = 3x + 5$ and $g(x) = 2\sqrt[3]{x}$

  a. $(f \circ g)(x)$

  b. $(g \circ f)(x)$

  c. $(f \circ g)(-8)$

  d. $(g \circ f)(1)$

Determine whether the function whose graph is shown is a one-to-one function.

39.

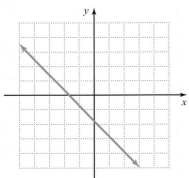

40.

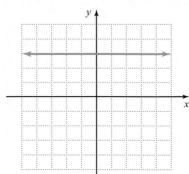

41.

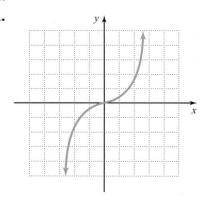

42.

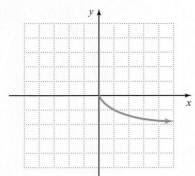

43.

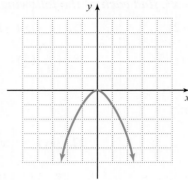

44.

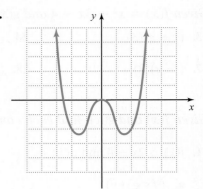

45.

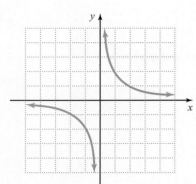

46.

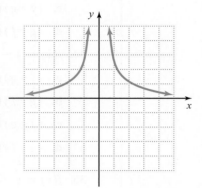

Match the graph of the function on the left with its inverse on the right.

47.

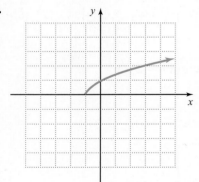

a.

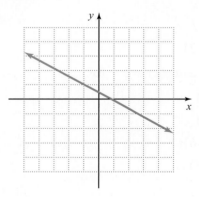

48.

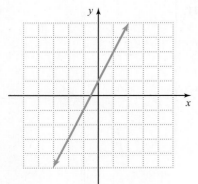

b.

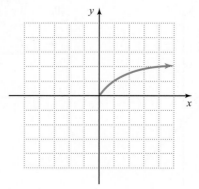

49.

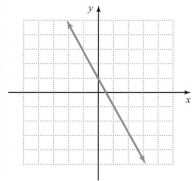

c.

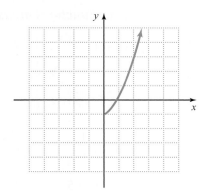

50.

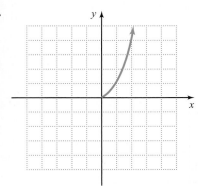

d.

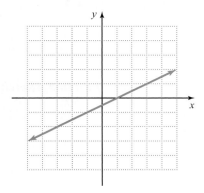

Find the inverse of the function represented by the set of ordered pairs.

51. $\{(-4, 5), (-2, 3), (0, 1), (2, -1), (4, -3)\}$

52. $\{(-9, 2), (-5, 3), (-2, 4), (2, -9), (3, -5), (4, -2)\}$

53. $\{(-27, -3), (-8, -2), (-1, -1), (0, 0), (1, 1), (8, 2), (27, 3)\}$

54. $\{(36, -6), (25, -5), (16, -4), (9, -3), (4, -2), (1, -1), (0, 0)\}$

Find the inverse of each of the following one-to-one functions.

55. $f(x) = 4x$

56. $f(x) = 7x$

57. $g(x) = \dfrac{1}{4}x$

58. $g(x) = \dfrac{1}{7}x$

59. $f(x) = -x - 5$

60. $f(x) = x + 10$

61. $f(x) = 5x + 2$

62. $h(x) = 4x - 9$

◉ 63. $g(x) = \dfrac{1}{2}x - 3$

64. $f(x) = \dfrac{1}{3}x + 1$

65. $f(x) = x^3 - 4$

66. $f(x) = x^3 + 2$

67. $h(x) = \dfrac{3}{x + 4}$

68. $g(x) = \dfrac{1}{x - 3}$

69. $f(x) = \dfrac{x}{2x - 1}$

70. $f(x) = \dfrac{2x}{3x + 2}$

◉ 71. $f(x) = \sqrt[3]{x + 4}$

72. $f(x) = \sqrt[3]{5 - x}$

Determine whether the two given functions are inverses of each other.

73. $f(x) = \dfrac{x - 1}{3}$ and $g(x) = 3x + 1$

74. $f(x) = \dfrac{x + 7}{2}$ and $g(x) = 2x - 7$

75. $p(x) = 5x - 10$ and $q(x) = \dfrac{1}{5}x + 2$

76. $f(x) = 3x + 12$ and $g(x) = \dfrac{1}{3}x + 4$

77. $f(x) = -4x - 1$ and $g(x) = \dfrac{x + 1}{4}$

78. $p(x) = -6x + 7$ and $q(x) = \dfrac{7 - x}{6}$

79. $g(x) = (x + 5)^3$ and $h(x) = \sqrt[3]{x} - 5$

80. $f(x) = x^3 - 10$ and $g(x) = \sqrt[3]{x + 10}$

81. $p(x) = \sqrt[3]{2x - 7}$ and $q(x) = 2x^3 + 7$

82. $p(x) = \sqrt[5]{x} - 1$ and $q(x) = (1 - x)^5$

83. $f(x) = \dfrac{2}{x} - 3$ and $g(x) = \dfrac{2}{x + 3}$

84. $f(x) = \dfrac{-5}{x - 8}$ and $g(x) = -\dfrac{5}{x} + 8$

Given the graph of a one-to-one function, sketch the graph of its inverse.

85.

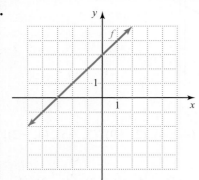

86.

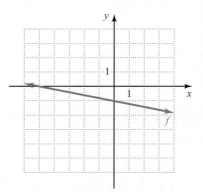

87.

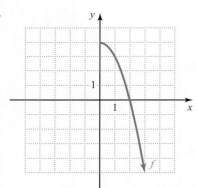

88.

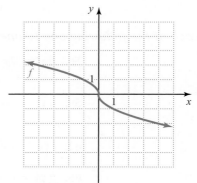

Mixed Practice

Determine whether the given functions are inverses of each other.

89. $g(x) = -4x + 9$ and $h(x) = \dfrac{9 - x}{4}$

90. $p(x) = \sqrt[4]{x} + 5$ and $q(x) = (x - 5)^4$

Solve.

91. Given $f(x) = 3x^2 - 5x + 2$ and $g(x) = x^2 - 8$, find $(g - f)(-3)$.

92. Determine whether the function shown by the graph is a one-to-one function.

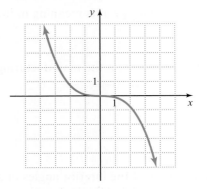

93. Given $f(x) = -2x + 7$ and $g(x) = \dfrac{4}{x + 2}$, find each composition:

    a. $(f \circ g)(x)$

    b. $(g \circ f)(x)$

    c. $(f \circ g)(6)$

    d. $(g \circ f)(0)$

94. Given $f(x) = 2x^2 - 5x + 7$ and $g(x) = -x + 2$, find $(f + g)(x)$, $(f - g)(x)$, $(f \cdot g)(x)$, and $\left(\dfrac{f}{g}\right)(x)$.

95. Given the following graph of a one-to-one function, sketch the graph of its inverse:

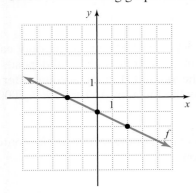

Find the inverse of each of the following one-to-one functions.

96. $f(x) = \dfrac{1}{4}x - 7$

97. $f(x) = \dfrac{3x}{2x + 3}$

98. $f(x) = x^3 - 6$

Applications

Solve.

99. The accounting department of a manufacturing plant determines that the weekly cost, C, in dollars, of producing x units of a product is approximated by the function $C(x) = 4.2x + 1000$. The number of units the plant produces in d days is modeled by the function $x(d) = 500d$.

    a. Find $(C \circ x)(d)$ and interpret its meaning in this situation.

    b. Determine the weekly cost if the manufacturing plant operates 5 days per week.

100. Each day, an employee at a toy factory receives a salary of $70 plus $0.50 for each item of doll clothing he produces. The number of items of doll clothing the employee can produce in h hr is modeled by the function $x(h) = 8h$.

    a. Find the function $S(x)$ that expresses the employee's salary in terms of x, the number of items of doll clothing produced each day.

    b. Find $(S \circ x)(h)$ and interpret its meaning in this situation.

    c. Determine the employee's daily salary if he works 7.5 hours per day.

101. The sum of the measures of the interior angles of a polygon (in degrees) is modeled by the function $S(n) = 180n - 360$, where n is the number of sides of the polygon.

    a. Identify the inverse function of S.

    b. Explain the meaning of the inverse function in this situation.

    c. Determine the number of sides of a polygon if the sum of the interior angles is 540 degrees.

102. A clinical dietitian can estimate the ideal body weight, w (in pounds) for men using the function $w(h) = 6h - 254$, where h is height (in inches).

    a. Identify the inverse function of w.

    b. Explain the meaning of the inverse function in this situation.

    c. Find the height of a man whose ideal body weight is 178 lb.

103. A long-distance telephone service provider charges a monthly service fee of $4.50 plus $0.05 for each minute (or part thereof) of long-distance calling.

    a. Find the function $B(t)$ that expresses the monthly bill in terms of t, the number of minutes of long-distance calling.

    b. Find $B^{-1}(t)$.

    c. Explain the significance of the inverse within the context of the situation.

104. A young couple deposits $5000 into an account earning 3.5% simple interest. If no additional deposits are made, then the amount A in the account after t yr is modeled by the function $A(t) = 175t + 5000$.

    a. Find $A^{-1}(t)$.

    b. Explain the meaning of the inverse in the context of this situation.

● *Check your answers on page A-44.*

MINDSTRETCHERS

Critical Thinking

1. Consider the graphs of the functions f and g shown on the coordinate plane below on the left. Use these graphs to graph $h(x) = (f + g)(x)$ and $k(x) = (f - g)(x)$ on the following coordinate plane.

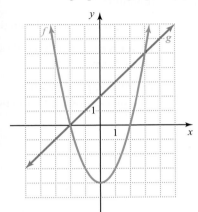 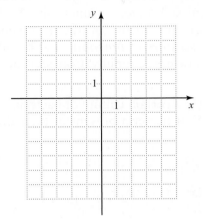

Mathematical Reasoning

2. Consider the function $h(x) = 6x - 1$. If $f(x) = 3x + 2$, find $g(x)$ such that $h(x) = (f \circ g)(x)$.

Groupwork

3. Working with a partner, give three examples of functions where each function is equal to its own inverse and explain why.

10.2 Exponential Functions

What Exponential Functions Are and Why They Are Important

OBJECTIVES

■ To evaluate exponential functions

■ To graph exponential functions

■ To solve exponential equations of the form $b^x = b^n$

■ To solve applied problems involving exponential functions

In the previous section, we discussed inverse functions. In the remaining sections of this chapter, we focus on the *exponential function* and its inverse, the *logarithmic function*. Exponential functions model data that increase or decrease rapidly. Consider the following example.

Suppose a bacteriologist is growing a certain culture of bacteria. The following table shows the number N of bacteria present in the culture after t hr.

Time, t	0	1	2	3	4	5
Number of Bacteria Present, N	1	2	4	8	16	32

From the table, we see that the number of bacteria in the culture doubles every hour. So after 24 hr, the number of bacteria in the culture will have grown from 1 bacterium to 2^{24}, or 16,777,216 bacteria. This type of growth can be modeled by the exponential function $N(t) = 2^t$.

Some other examples of real-life situations that are described by exponential functions include population growth, growth of an investment in which interest is compounded, and decay of radioactive substances.

We devote this section to a discussion of exponential functions (and equations) and how they are used as models in solving applied problems.

Evaluating Exponential Functions

In previous sections, we discussed polynomial functions, such as x^2 and $-5y^3$, in which a variable is raised to a constant power. In this section, we focus on *exponential functions*, such as 2^x and 5^{-n}, where a constant is raised to a power that is a variable expression.

> **Definition**
>
> An **exponential function** is any function that can be written in the form
> $$f(x) = b^x$$
> where x is a real number, $b > 0$, and $b \neq 1$.

Note that we call b the *base* of the exponential function. In more advanced courses, it is shown that b^x exists for all real values of x, both rational and irrational, as long as b is positive. Here are some other examples of exponential functions.

$$f(x) = 4^x \qquad g(x) = \left(\frac{1}{6}\right)^x \qquad h(x) = 10^{x+1}$$

In previous sections, we evaluated linear, polynomial, rational, and quadratic functions. Now, let's evaluate exponential functions.

EXAMPLE 1

Evaluate:

a. $f(x) = 3^x$ for $x = 2$ **b.** $g(x) = \left(\frac{1}{2}\right)^x$ for $x = -4$

Solution

a. $f(x) = 3^x$
$\quad f(2) = 3^2$ Substitute 2 for x.
$\quad\quad\; = 9$

b. $g(x) = \left(\frac{1}{2}\right)^x$

$\quad g(-4) = \left(\frac{1}{2}\right)^{-4}$ Substitute -4 for x.

$\quad\quad\quad = 2^4$
$\quad\quad\quad = 16$

PRACTICE 1

Evaluate:

a. $f(x) = 4^x$ for $x = 3$

b. $g(x) = \left(\frac{1}{3}\right)^x$ for $x = -2$

EXAMPLE 2

Given $f(x) = 3^{2x-1}$, find:

a. $f(1)$ **b.** $f(-1)$

Solution

a. $f(x) = 3^{2x-1}$
$\quad f(1) = 3^{2(1)-1}$ Substitute 1 for x.
$\quad\quad\; = 3^1 = 3$

b. $f(x) = 3^{2x-1}$
$\quad f(-1) = 3^{2(-1)-1}$ Substitute -1 for x.
$\quad\quad\quad = 3^{-3}$
$\quad\quad\quad = \frac{1}{3^3} = \frac{1}{27}$

PRACTICE 2

Given $f(x) = 2^{3x-1}$, find:

a. $f(2)$

b. $f(-2)$

The exponential functions in previous examples have bases that are rational numbers. However, an important and frequently used base in applications of exponential functions is the irrational number e. Like the irrational number π, e has a nonterminating, nonrepeating decimal representation, which is approximately 2.7182818. The number e is studied in later mathematics courses, including calculus and statistics, and has important applications, as we shall see.

Definition
The function defined by $f(x) = e^x$ is called the **natural exponential function.**

To evaluate the natural exponential function, we can use the $\boxed{e^x}$ feature on a calculator.

EXAMPLE 3

Evaluate the function for the given value. Round the answer to the nearest thousandth.

a. $f(x) = e^{3x}$ for $x = 2$ **b.** $g(x) = e^{x+3}$ for $x = -5$

Solution

a. $f(x) = e^{3x}$

$f(2) = e^{3(2)}$

$= e^6 \approx 403.429$

b. $g(x) = e^{x+3}$

$g(-5) = e^{-5+3}$

$= e^{-2} \approx 0.135$

PRACTICE 3

Evaluate and round to the nearest thousandth.

a. $f(x) = e^{5x}$ for $x = -1$

b. $g(x) = e^{x-1}$ for $x = 3$

Graphing Exponential Functions

Now, let's focus on graphing exponential functions.

EXAMPLE 4

Graph the exponential functions $f(x) = 2^x$ and $g(x) = 4^x$ on the same coordinate plane.

Solution For each function, we find several values and list the results in a table. Then, we plot the ordered pairs and connect them with a smooth curve.

x	$f(x) = 2^x$	(x, y)	x	$g(x) = 4^x$	(x, y)
-2	$f(-2) = 2^{-2} = \frac{1}{4}$	$\left(-2, \frac{1}{4}\right)$	-2	$g(-2) = 4^{-2} = \frac{1}{16}$	$\left(-2, \frac{1}{16}\right)$
-1	$f(-1) = 2^{-1} = \frac{1}{2}$	$\left(-1, \frac{1}{2}\right)$	-1	$g(-1) = 4^{-1} = \frac{1}{4}$	$\left(-1, \frac{1}{4}\right)$
0	$f(0) = 2^0 = 1$	$(0, 1)$	0	$g(0) = 4^0 = 1$	$(0, 1)$
1	$f(1) = 2^1 = 2$	$(1, 2)$	1	$g(1) = 4^1 = 4$	$(1, 4)$
2	$f(2) = 2^2 = 4$	$(2, 4)$	2	$g(2) = 4^2 = 16$	$(2, 16)$
3	$f(3) = 2^3 = 8$	$(3, 8)$	3	$g(3) = 4^3 = 64$	$(3, 64)$

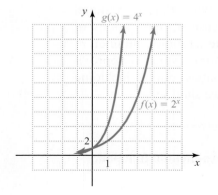

PRACTICE 4

Graph the exponential functions $f(x) = 3^x$ and $g(x) = 5^x$ on the same coordinate plane.

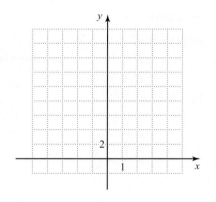

We observe from the graphs in Example 4 that each function is defined for any real number x and takes on any positive value. The y-intercept for both graphs is $(0, 1)$, and neither graph has an x-intercept. Also, each graph is increasing and passes both the vertical- and horizontal-line tests.

EXAMPLE 5	PRACTICE 5

Graph the exponential functions $f(x) = \left(\frac{1}{2}\right)^x$ and $g(x) = \left(\frac{1}{4}\right)^x$ on the same set of axes.

Solution To graph each function, we find several values, listing the results in a table. Then, we plot the ordered pairs and connect them with a smooth curve.

x	$f(x) = \left(\frac{1}{2}\right)^x$	(x, y)
-3	$f(-3) = \left(\frac{1}{2}\right)^{-3} = 8$	$(-3, 8)$
-2	$f(-2) = \left(\frac{1}{2}\right)^{-2} = 4$	$(-2, 4)$
-1	$f(-1) = \left(\frac{1}{2}\right)^{-1} = 2$	$(-1, 2)$
0	$f(0) = \left(\frac{1}{2}\right)^{0} = 1$	$(0, 1)$
1	$f(1) = \left(\frac{1}{2}\right)^{1} = \frac{1}{2}$	$\left(1, \frac{1}{2}\right)$
2	$f(2) = \left(\frac{1}{2}\right)^{2} = \frac{1}{4}$	$\left(2, \frac{1}{4}\right)$

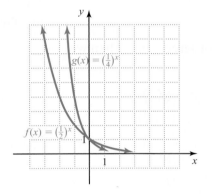

x	$g(x) = \left(\frac{1}{4}\right)^x$	(x, y)
-3	$g(-3) = \left(\frac{1}{4}\right)^{-3} = 64$	$(-3, 64)$
-2	$g(-2) = \left(\frac{1}{4}\right)^{-2} = 16$	$(-2, 16)$
-1	$g(-1) = \left(\frac{1}{4}\right)^{-1} = 4$	$(-1, 4)$
0	$g(0) = \left(\frac{1}{4}\right)^{0} = 1$	$(0, 1)$
1	$g(1) = \left(\frac{1}{4}\right)^{1} = \frac{1}{4}$	$\left(1, \frac{1}{4}\right)$
2	$g(2) = \left(\frac{1}{4}\right)^{2} = \frac{1}{16}$	$\left(2, \frac{1}{16}\right)$

Graph the exponential functions $f(x) = \left(\frac{1}{3}\right)^x$ and $g(x) = \left(\frac{1}{5}\right)^x$ on the same coordinate plane.

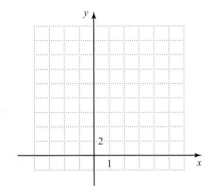

From the graphs in Example 5, we see that each function is defined for any real number x and takes on any positive value. The y-intercept for both graphs is $(0, 1)$, and neither graph has an x-intercept. Also, each graph is decreasing and passes both the vertical- and horizontal-line tests.

Explain why the graphs in Example 4 are *increasing*, whereas the graphs in Example 5 are *decreasing*.

The following table summarizes some characteristics of exponential functions as shown in Examples 4 and 5.

Characteristics of the Exponential Function $f(x) = b^x$

- The domain of the function is the set of all real numbers and the range is the set of all positive numbers.

- The vertical- and horizontal-line tests show that $f(x) = b^x$ is a one-to-one function.

- The function is increasing if the base is greater than 1. The function is decreasing if the base is between 0 and 1.

- The y-intercept of the graph of the function is $(0, 1)$, but the graph has no x-intercept. The graph approaches the x-axis, but never crosses it.

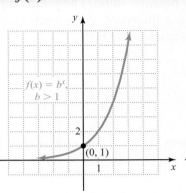

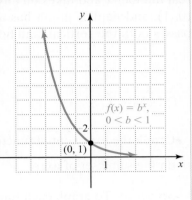

EXAMPLE 6

Graph: $f(x) = 2^{x-1}$

Solution First, we find and plot some ordered pairs. Then we connect them with a smooth curve.

x	$f(x) = 2^{x-1}$	(x, y)
-2	$f(-2) = 2^{-2-1} = 2^{-3} = \frac{1}{8}$	$\left(-2, \frac{1}{8}\right)$
-1	$f(-1) = 2^{-1-1} = 2^{-2} = \frac{1}{4}$	$\left(-1, \frac{1}{4}\right)$
0	$f(0) = 2^{0-1} = 2^{-1} = \frac{1}{2}$	$\left(0, \frac{1}{2}\right)$
1	$f(1) = 2^{1-1} = 2^0 = 1$	$(1, 1)$
2	$f(2) = 2^{2-1} = 2^1 = 2$	$(2, 2)$
3	$f(3) = 2^{3-1} = 2^2 = 4$	$(3, 4)$

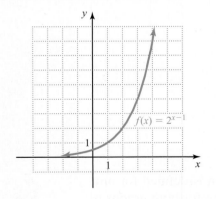

PRACTICE 6

Graph: $g(x) = 3^{x-1}$

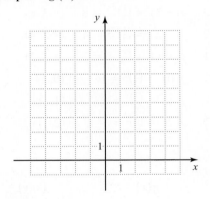

Solving Exponential Equations of the Form $b^x = b^n$

An equation that contains a variable as an exponent is called an *exponential equation*. Some examples of exponential equations are

$$3^{x-1} = 3^{2x-1} \qquad 2^x = 8 \qquad 4^x = 9$$

We can use the fact that exponential functions are one-to-one to solve many exponential equations.

One-to-One Property of Exponential Functions

For $b > 0$ and $b \neq 1$, if $b^x = b^n$, then $x = n$.

We can use this property to solve exponential equations in which each side of the equation can be expressed as a power of the same base. When this is the case, we can equate the exponents.

EXAMPLE 7	PRACTICE 7
Solve:	Solve:

EXAMPLE 7

Solve:

a. $3^x = 27$ **b.** $4^x = 32$

Solution

a. The number 27 is a power of 3, so to solve this equation we first write 27 as a power of 3:

$$3^x = 27$$
$$3^x = 3^3$$

Since the bases on each side of the equation are the same, we can use the one-to-one property of exponential functions and set the exponents equal. So $x = 3$. Substituting 3 for x in the original equation, we see that the solution checks.

b. Because both 4 and 32 are powers of 2, we get

$$4^x = 32$$
$$(2^2)^x = 2^5$$
$$2^{2x} = 2^5$$

Since the bases are the same, we apply the one-to-one property of exponential functions:

$$2x = 5$$
$$x = \frac{5}{2}$$

So $4^x = 32$ when $x = \frac{5}{2}$. We can check the answer in the original equation.

PRACTICE 7

Solve:

a. $2^x = 64$

b. $9^x = 27$

Note that in some exponential equations, such as $4^x = 9$, the two sides cannot easily be expressed in terms of the same base. In a later section, we focus on solving equations like these using *logarithms*.

Exponential functions (and equations) are used to solve many real-life applications, such as those dealing with interest rates on investments or loans. Let's consider the exponential equation

$$A = P\left(1 + \frac{r}{n}\right)^{nt}$$

This equation is used to compute the amount A accumulated (or owed) after P dollars are invested (or borrowed) at an annual interest rate r, compounded n times per year for t years. We call this equation the *compound interest formula*.

EXAMPLE 8

When a computer company downsized 3 yr ago, an employee received a lump-sum severance check for $10,000. At that time, she invested one-half of the money in a growth fund that yielded an annual rate of 9% in dividends, compounded quarterly. To the nearest cent, what is the value of her investment after 3 yr?

Solution

Use the formula $A = P\left(1 + \dfrac{r}{n}\right)^{nt}$, where

$P = \$5000$ (the amount invested that is one-half of $10,000)

$r = 9\% = 0.09$ (the annual interest rate)

$n = 4$ (the number of times interest is compounded each year: quarterly means 4 times)

$t = 3$ (the length of time of the investment, in years)

Substituting these values in the compound interest formula, we get:

$$A = P\left(1 + \frac{r}{n}\right)^{nt}$$
$$= 5000\left(1 + \frac{0.09}{4}\right)^{4 \cdot 3}$$
$$= 5000(1 + 0.0225)^{12}$$
$$= 5000(1.0225)^{12}$$

To find the approximate value of A, we use the $\boxed{\wedge}$ or $\boxed{y^x}$ key on a calculator.

$$A \approx 6530.25$$

So the amount that the fund is worth after 3 yr, to the nearest cent, is $6530.25.

PRACTICE 8

The half-life of a radioactive substance is the time it takes for half of the material to decay. Arsenic-74, with a half-life of about 18 days, is used to locate brain tumors. The amount A of a 90-mg sample remaining after x days is modeled by the function $A(x) = 90\left(\dfrac{1}{2}\right)^{x/18}$.

Approximate, to the nearest milligram, the amount of the sample remaining after 3 days.

Mathematically Speaking

Fill in each blank with the most appropriate term or phrase from the given list.

imaginary	decreasing	exponential function
crosses	approaches	increasing
natural exponential function	real	
	irrational	

1. A(n) _____ is any function that can be written in the form $f(x) = b^x$, where x is a real number, $b > 0$, and $b \neq 1$.

2. The number e is _____.

3. The function defined by $f(x) = e^x$ is called the _____.

4. The domain of an exponential function is the set of all _____ numbers.

5. An exponential function is _____ if the base is between 0 and 1.

6. The graph of an exponential function _____ the x-axis.

Determine whether each of the following functions is a polynomial function, a rational function, a radical function, or an exponential function.

7. $f(x) = 2x^3 - 1$

8. $g(x) = 8^{x+2}$

9. $h(x) = x^{1/2}$

10. $f(x) = 3x^{-2} + 5$

11. $g(x) = \left(\dfrac{1}{2}\right)^{2x}$

12. $f(x) = e^{2x}$

13. $f(x) = -3x^2 + 2$

14. $g(x) = -3^x + 2$

Evaluate each function for the given values.

15. $f(x) = 2^x$
  a. $f(-3)$
  b. $f(0)$
  c. $f(4)$

16. $f(x) = 6^x$
  a. $f(-1)$
  b. $f(2)$
  c. $f(-2)$

17. $g(x) = \left(\dfrac{1}{9}\right)^x$
  a. $g\left(-\dfrac{1}{2}\right)$
  b. $g\left(\dfrac{1}{2}\right)$
  c. $g(2)$

18. $h(x) = \left(\dfrac{1}{8}\right)^x$
  a. $h(-2)$
  b. $h\left(-\dfrac{1}{3}\right)$
  c. $h\left(\dfrac{1}{3}\right)$

19. $f(x) = 3^{x-4}$
  a. $f(1)$
  b. $f(3)$
  c. $f(6)$

20. $g(x) = 4^{x+1}$
  a. $g(-4)$
  b. $g(1)$
  c. $g(3)$

21. $h(x) = -2^x + 3$
  a. $h(-3)$
  b. $h(0)$
  c. $h(4)$

22. $f(x) = 10 - 3^x$
  a. $f(-2)$
  b. $f(0)$
  c. $f(2)$

▦ *Evaluate each function for the given values. Round to the nearest thousandth.*

23. $f(x) = -e^x$

   a. $f(-5)$

   b. $f(1)$

26. $g(x) = -e^{4x}$

   a. $g\left(-\frac{1}{4}\right)$

   b. $g(0)$

24. $f(x) = e^{-x}$

   a. $f(-3)$

   b. $f(1)$

27. $f(x) = e^{3x-2}$

   a. $f\left(-\frac{1}{3}\right)$

   b. $f(0)$

25. $g(x) = e^{-2x}$

   a. $g(2)$

   b. $g\left(-\frac{1}{2}\right)$

28. $f(x) = e^{2x+1}$

   a. $f(2)$

   b. $f\left(\frac{1}{2}\right)$

Graph each function.

29. $f(x) = 3^x$

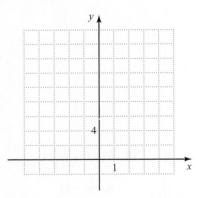

30. $f(x) = 4^x$

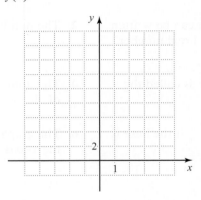

31. $f(x) = \left(\frac{1}{2}\right)^x$

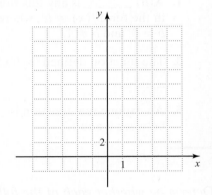

32. $f(x) = \left(\frac{1}{3}\right)^x$

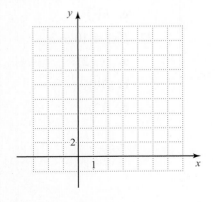

33. $f(x) = -2^x$

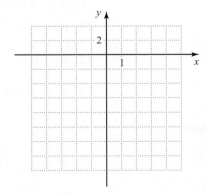

34. $f(x) = -3^x$

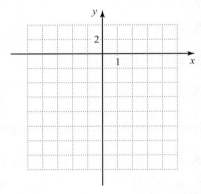

35. $f(x) = 2^{x-2}$

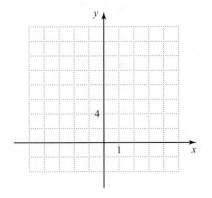

36. $f(x) = 3^{x+1}$

37. $f(x) = \left(\dfrac{1}{3}\right)^x + 2$

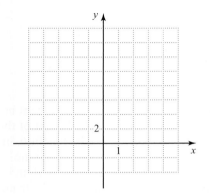

38. $f(x) = \left(\dfrac{1}{4}\right)^x - 3$

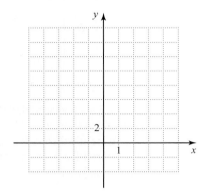

Solve.

39. $5^x = 25$

40. $10^x = 10{,}000$

41. $2^x = \dfrac{1}{32}$

42. $3^x = \dfrac{1}{81}$

43. $36 = 6^{-x}$

44. $125 = 5^{-x}$

45. $8^x = 16$

46. $9^x = 243$

47. $4^{x+1} = 64$

48. $2^{x-3} = 32$

49. $7^{x-4} = 1$

50. $6^{x+2} = 1$

51. $16^{x+1} = 32$

52. $27^{x-1} = 81$

53. $9^{-x+3} = \dfrac{1}{27}$

54. $32^{-x+2} = \dfrac{1}{8}$

55. $100^{x-5} = 100{,}000^x$

56. $256^x = 32^{x+1}$

57. $64^{x-2} = 128^{x-3}$

58. $125^{x+1} = 625^{x-1}$

Mixed Practice

Solve.

59. $4^{-x+3} = \dfrac{1}{32}$

60. Graph the function $f(x) = -\left(\dfrac{1}{2}\right)^x$.

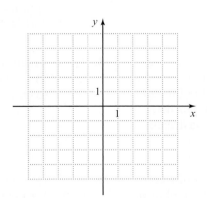

Determine whether each of the following functions is a polynomial or a rational, radical, or exponential function.

61. $g(x) = x^{3/2}$

62. $f(x) = 3^{2x}$

Evaluate each function for the given value.

63. $g(x) = -5^{x-2}$ for $x = 0$

64. $f(x) = -5^x + 1$ for $x = -1$

▦ *Evaluate each function for the given values. Round to the nearest thousandth.*

65. $f(x) = e^{2x}$ for $x = -1$

66. $g(x) = e^{-3x}$ for $x = \dfrac{1}{4}$

Applications

Solve. Use a calculator where appropriate.

67. A new employee is given a starting salary of $28,000 and is guaranteed a 4% annual salary increase. The function $S(n) = 28,000(1.04)^n$ can be used to calculate the employee's salary S after n years of employment. What will the employee's salary be after 2 yr of employment?

68. A botanist notes that the height of a certain plant increases by 10% each week. Suppose the height of the plant is measured at 2 in. today. Then its height h, in inches, x weeks from today can be found using the function $h(x) = 2(1.1)^x$. To the nearest tenth of an inch, determine the height of the plant 8 weeks from today.

69. An environmental group estimates that the amount of pollution in a pond will decrease by 5% each month. To determine the concentration, C, of pollutants (in parts per million) after m mo, the group uses the function $C(m) = 40(0.95)^m$.

  a. To the nearest tenth, what will the concentration of pollutants in the pond be after 1 yr?

  b. What was the initial concentration of pollutants in the pond?

70. The owner of a small business determines that a piece of office equipment purchased new depreciates at a rate of 15% per year. The value, v, of the equipment t yr after it is purchased is modeled by the function $v(t) = 1250(0.85)^t$.

  a. To the nearest dollar, what will be the value of the equipment after 5 yr?

  b. What was the value of the office equipment when it was purchased?

71. An investor puts $8000 into an account with an annual interest rate of 6%, compounded monthly. To the nearest cent, calculate the amount in the account after 5 yr.

72. Strontium-85, a radioactive isotope used in bone imaging, has a half-life of about 65 days. The amount A of a 30-mg sample remaining after t days is modeled by the function $A(t) = 30\left(\dfrac{1}{2}\right)^{t/65}$. Determine, to the nearest tenth of a mg, the amount of the sample remaining after 130 days.

73. If the initial concentration of a certain drug in a patient's bloodstream is 50 milligrams per liter, then the concentration C after t hr is modeled by the function $C(t) = 50e^{-0.125t}$. To the nearest milligram per liter, approximate the concentration in the bloodstream after 8 hr.

74. The population, P, of a town is modeled by the function $P(t) = 300e^{0.3t}$, where t is measured in years. What was the size of the population after 4 yr?

75. A bacteriologist found that the population of a certain bacterium doubles every 30 min. The number of bacteria, N, present after h hr is given by the equation $N = N_0(2)^{2h}$, where N_0 is the number of bacteria present in the initial population. If 10 bacteria are present in the initial population, in how many hours will 1280 bacteria be present?

76. A professional organization determined that its membership tripled every 5 yr since its inception. If the organization had 500 members at its inception, then t yr after its inception the number of members m is given by the equation $m = 500(3)^{t/5}$. How many years after its inception were there 13,500 members?

● *Check your answers on page A-45.*

MINDSTRETCHERS

Patterns

1. Consider the exponential expressions 2^x and 3^x.

a. Complete the table.

x	−4	−3	−2	−1	0	1	2	3	4
2^x									
3^x									

b. What do you notice about the value of 2^x relative to the value of 3^x when x is negative? What do you notice when x is positive?

c. Use your results from parts (a) and (b) to solve the following inequalities:

   i. $2^x > 3^x$ **ii.** $2^x < 3^x$

d. Consider the exponential expressions $\left(\dfrac{1}{2}\right)^x$ and $\left(\dfrac{1}{3}\right)^x$. Without making any calculations, solve the following

inequalities:

   i. $\left(\dfrac{1}{2}\right)^x > \left(\dfrac{1}{3}\right)^x$ **ii.** $\left(\dfrac{1}{2}\right)^x < \left(\dfrac{1}{3}\right)^x$

Writing

2. In the definition of an exponential function, there are two restrictions on b, namely, $b > 0$ and $b \neq 1$. Explain the importance of each.

Mathematical Reasoning

3. The exponential function $f(x) = b^x$ is a one-to-one function. Use the graph of $f(x) = b^x$, where $b > 1$, to sketch a graph of its inverse. Then identify the domain and range of the inverse function.

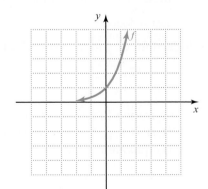

10.3 Logarithmic Functions

OBJECTIVES

- To write equivalent exponential and logarithmic equations
- To evaluate logarithms
- To solve logarithmic equations
- To graph logarithmic functions
- To solve applied problems involving logarithms

What Logarithmic Functions Are and Why They Are Important

As with exponential functions, *logarithmic functions* help us to understand many natural phenomena. Consider, for instance, the magnitude of an earthquake. This magnitude is calculated from the amplitude of the largest seismic wave that the earthquake registers. The Richter scale is a measure of earthquake magnitude. From the Richter scale shown, we see that an earthquake of magnitude 6 has amplitude 10 times as great as an earthquake of magnitude 5.

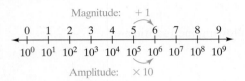

The exponents used on the Richter scale are examples of *logarithms*.

Writing Equivalent Exponential and Logarithmic Equations

From the previous section, we know that the exponential function $f(x) = b^x$ is one-to-one and so has an inverse. Its inverse function, $f^{-1}(x)$, is called a logarithmic function.

Definition

The inverse of $f(x) = b^x$ is represented by the **logarithmic function**

$$f^{-1}(x) = \log_b x$$

where $\log_b x$ is read either "the **logarithm** of x *to the base b*" or "the logarithm, base b, of x."

In general, we know that for a one-to-one function f, $f^{-1}(x) = y$ is equivalent to $f(y) = x$. Since the function $f^{-1}(x) = \log_b x$ is the inverse of the function $f(x) = b^x$, it follows that $y = \log_b x$ is equivalent to $b^y = x$.

Equivalent Logarithmic and Exponential Equations

If $b > 0$, $b \neq 1$, $x > 0$, and y is any real number, then $y = \log_b x$ is equivalent to $b^y = x$.

In words, this property states that the logarithm of x to the base b is the exponent to which the base b must be raised to get x.

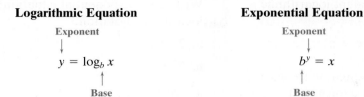

Logarithmic Equation

Exponent

$$y = \log_b x$$

Base

Exponential Equation

Exponent

$$b^y = x$$

Base

Tip   Remember that a logarithm is an exponent.

The key to solving many logarithmic problems is to be able to write equivalent exponential and logarithmic equations. Examples of such equivalent equations are shown in the following table.

Logarithmic Equation	Exponential Equation
$\log_2 16 = 4$	$2^4 = 16$
$\log_3 9 = 2$	$3^2 = 9$
$\log_8 1 = 0$	$8^0 = 1$
$\log_4 \dfrac{1}{64} = -3$	$4^{-3} = \dfrac{1}{64}$
$\log_9 3 = \dfrac{1}{2}$	$9^{1/2} = 3$

Let's consider some other examples of writing exponential equations in logarithmic notation, and vice versa.

EXAMPLE 1

Write each logarithmic equation as an exponential equation.

a. $\log_{10} 1000 = 3$ **b.** $\log_7 7 = 1$ **c.** $\log_5 \sqrt{5} = \dfrac{1}{2}$

Solution We can write each logarithmic equation in its equivalent exponential form using the definition of a logarithm.

a. $\log_{10} 1000 = 3$ means $10^3 = 1000$

Logarithms are exponents.

b. $\log_7 7 = 1$ means $7^1 = 7$

c. $\log_5 \sqrt{5} = \dfrac{1}{2}$ means $5^{1/2} = \sqrt{5}$

PRACTICE 1

Write each logarithmic equation using exponential notation.

a. $\log_8 64 = 2$

b. $\log_5 \dfrac{1}{5} = -1$

c. $\log_2 \sqrt[3]{2} = \dfrac{1}{3}$

EXAMPLE 2

Write each exponential equation in its equivalent logarithmic form.

a. $5^4 = 625$ **b.** $8^{-1} = \dfrac{1}{8}$ **c.** $7^{1/3} = \sqrt[3]{7}$

Solution To find the equivalent logarithmic equation for each exponential equation, we use the definition of a logarithm.

a. $5^4 = 625$ is equivalent to $\log_5 625 = 4$.

b. $8^{-1} = \dfrac{1}{8}$ is equivalent to $\log_8 \dfrac{1}{8} = -1$.

c. $7^{1/3} = \sqrt[3]{7}$ is equivalent to $\log_7 \sqrt[3]{7} = \dfrac{1}{3}$.

PRACTICE 2

Write each exponential equation in logarithmic notation.

a. $2^5 = 32$

b. $7^0 = 1$

c. $10^{1/2} = \sqrt{10}$

Evaluating Logarithms

Because logarithms are exponents, we can evaluate some logarithms by inspection.

EXAMPLE 3

Evaluate:

a. $\log_2 8$ **b.** $\log_4 16$ **c.** $\log_{49} 7$

Solution We know that the logarithm of x with base b, written as $\log_b x$, is the exponent to which b must be raised to get x. We use this relationship to evaluate each of the following expressions.

a. $\log_2 8 = 3$ because $2^3 = 8$.

b. $\log_4 16 = 2$ because $4^2 = 16$.

c. $\log_{49} 7 = \dfrac{1}{2}$ because $49^{1/2} = \sqrt{49} = 7$.

PRACTICE 3

Evaluate:

a. $\log_{10} 10$

b. $\log_3 27$

c. $\log_8 2$

Since logarithms are exponents, we can use properties of exponents to verify the following two special properties of logarithms. Both properties of logarithms are useful in evaluating certain logarithms.

Special Properties of Logarithms

- $\log_b b = 1$ because $b^1 = b$; that is, 1 is the exponent to which b must be raised to get b.
- $\log_b 1 = 0$ because $b^0 = 1$; that is, 0 is the exponent to which b must be raised to get 1.

EXAMPLE 4

Find the value of each logarithmic expression:

a. $\log_8 8$ **b.** $\log_4 1$

Solution

a. Because $\log_b b = 1$, we conclude that $\log_8 8 = 1$.

b. Because $\log_b 1 = 0$, we conclude that $\log_4 1 = 0$.

PRACTICE 4

Find the value of each logarithmic expression.

a. $\log_6 6$

b. $\log_3 1$

Solving Logarithmic Equations

Now, let's look at how to solve equations involving logarithms. The key is to write and then solve the equivalent exponential equations.

EXAMPLE 5

Solve:

a. $\log_3 x = 4$ **b.** $\log_5 x = -2$

Solution

a. $\log_3 x = 4$

$3^4 = x$ Write the equivalent exponential equation.

$x = 81$

So the solution is 81.

b. $\log_5 x = -2$

$5^{-2} = x$ Write the equivalent exponential equation.

$x = \dfrac{1}{25}$

So the solution is $\dfrac{1}{25}$.

PRACTICE 5

Solve:

a. $\log_4 x = 3$

b. $\log_2 x = -1$

EXAMPLE 6

Solve:

a. $\log_x 64 = 2$ **b.** $\log_x 5 = \dfrac{1}{2}$

Solution

a. $\log_x 64 = 2$

$x^2 = 64$ Write the equivalent exponential equation.

$x = 8$

Note that even though $(-8)^2 = 64$, the base of a logarithm must be positive. So the only solution is 8.

PRACTICE 6

Solve:

a. $\log_x 125 = 3$

b. $\log_x 2 = \dfrac{1}{4}$

b. $\log_x 5 = \dfrac{1}{2}$

  $x^{1/2} = 5$ **Write the equivalent exponential equation.**

   $x = 25$

So the solution is 25.

EXAMPLE 7	PRACTICE 7
Solve: $\log_8 32 = x$	Solve: $\log_9 27 = x$.

Solution

  $\log_8 32 = x$

    $8^x = 32$ **Write the equivalent exponential equation.**

  $(2^3)^x = 2^5$ **Write each side of the equation as a power of 2.**

   $2^{3x} = 2^5$

    $3x = 5$ **Use the one-to-one property of exponential functions.**

     $x = \dfrac{5}{3}$

So the solution is $\dfrac{5}{3}$.

Note that in Example 5, we solved the equation given its base and power. In Example 6, we solved for the base. And in Example 7, we found the exponent. Rewriting a logarithmic equation as the equivalent exponential equation and then solving for the unknown works in all three cases.

Graphing Logarithmic Functions

Recall that we defined a logarithmic function as the inverse of the corresponding exponential function. For example, if $f(x) = 3^x$, then its inverse is $f^{-1}(x) = \log_3 x$. Since a logarithmic function is the inverse of an exponential function, its graph is symmetric to the graph of the exponential function about the line $y = x$. So to graph $f^{-1}(x) = \log_3 x$, we can first graph $f(x) = 3^x$ and then use symmetry about the line $y = x$ to graph the inverse function. See the graph below.

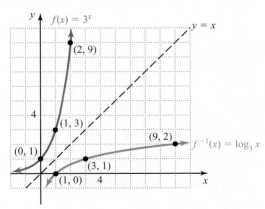

In graphing a logarithmic function, it is unnecessary to graph the corresponding exponential function. Instead, we can write the logarithmic function in exponential form and then graph, as shown in Examples 8 and 9.

EXAMPLE 8

Graph: $f(x) = \log_2 x$

Solution First, we substitute y for $f(x)$ in the given function: $y = \log_2 x$. Next, we write $y = \log_2 x$ in exponential form: $x = 2^y$. Since $x = 2^y$ is already solved for x, let's choose some y-values and find the corresponding x values.

If $y = 0$, then $x = 2^0 = 1$.

If $y = 1$, then $x = 2^1 = 2$.

If $y = 2$, then $x = 2^2 = 4$.

If $y = -1$, then $x = 2^{-1} = \dfrac{1}{2}$.

Now, we plot the ordered pairs $(1, 0)$, $(2, 1)$, $(4, 2)$, and $\left(\frac{1}{2}, -1\right)$, then connect the points with a smooth curve.

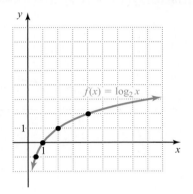

PRACTICE 8

Graph: $f(x) = \log_5 x$

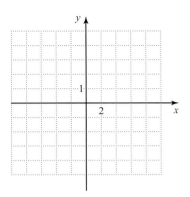

Note that the graph in Example 8 is defined for any positive value of x and takes on any real value of y. It is increasing and has x-intercept $(1, 0)$ but no y-intercept.

EXAMPLE 9

Graph: $f(x) = \log_{1/2} x$

Solution First, we substitute y for $f(x)$ in the given function: $y = \log_{1/2} x$. Next, we write $y = \log_{1/2} x$ in exponential form:

$x = \left(\dfrac{1}{2}\right)^y$. Then, we find some ordered pair solutions of the equation. Finally, we plot the ordered pairs and connect the corresponding points with a smooth curve.

If $y = 0$, then $x = \left(\dfrac{1}{2}\right)^0 = 1$.

If $y = 1$, then $x = \left(\dfrac{1}{2}\right)^1 = \dfrac{1}{2}$.

If $y = 2$, then $x = \left(\dfrac{1}{2}\right)^2 = \dfrac{1}{4}$.

If $y = -1$, then $x = \left(\dfrac{1}{2}\right)^{-1} = 2$.

PRACTICE 9

Graph: $f(x) = \log_{1/5} x$

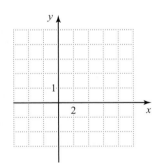

The graph in Example 9 is defined for any positive value x and takes on any real value of y. It is decreasing, has x-intercept $(1, 0)$, and has no y-intercept.

Some characteristics of logarithmic functions can be seen from the graphs in Examples 8 and 9.

Characteristics of the Logarithmic Function $f(x) = \log_b x$

- The domain of the function is the set of positive real numbers and the range is the set of real numbers.

- The function is increasing if the base is greater than 1. The function is decreasing if the base is between 0 and 1.

- The x-intercept of the graph of the function is $(1, 0)$, but the graph has no y-intercept. The graph approaches the y-axis but never intersects it.

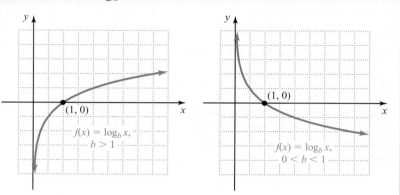

EXAMPLE 10

The loudness, L, of a sound, measured in decibels (dB), can be defined by the logarithmic equation $L = 10 \log_{10}\left(\dfrac{I}{I_0}\right)$, where I is the intensity of the sound (in watts per square meter) and I_0 is a constant equal to 10^{-12}. I_0 is called the zero decibel level and is the intensity of a barely audible sound.

Use the given logarithmic equation to complete the following table. Recall that the logarithm of a number is the power to which we raise the base in order to get that number.

Type of Sound	Intensity	Loudness
Busy street	10^{-5}	
Normal conversation	10^{-6}	
Quiet office	10^{-8}	

(*Source: Webster's New World Book of Facts*, 1999)

PRACTICE 10

In a typical hurricane, the barometric pressure is a function of the distance from the hurricane's eye. Researchers have found that x mi from the eye, the barometric pressure (in inches of mercury) can be approximated by the following function:

$$f(x) = 27 + 1.1 \log_{10}(x + 1)$$

Use this function to find the barometric pressure, to the nearest tenth of an inch, 9 mi from a hurricane's eye.

(*Source:* A. Miller and R. Anthes, *Meteorology*)

Solution Note that $\log_{10} 10^7 = 7$.

Type of Sound	Intensity	Loudness
Busy street	10^{-5}	$L = 10 \log_{10}\left(\dfrac{I}{I_0}\right)$
		$= 10 \log_{10}\left(\dfrac{10^{-5}}{10^{-12}}\right)$
		$= 10 \underbrace{\log_{10} 10^7}$
		$= 10 \cdot 7$
		$= 70$
		So the loudness is 70 dB.
Normal conversation	10^{-6}	$L = 10 \log_{10}\left(\dfrac{I}{I_0}\right)$
		$= 10 \log_{10}\left(\dfrac{10^{-6}}{10^{-12}}\right)$
		$= 10 \underbrace{\log_{10} 10^6}$
		$= 10 \cdot 6$
		$= 60$
		So the loudness is 60 dB.
Quiet office	10^{-8}	$L = 10 \log_{10}\left(\dfrac{I}{I_0}\right)$
		$= 10 \log_{10}\left(\dfrac{10^{-8}}{10^{-12}}\right)$
		$= 10 \underbrace{\log_{10} 10^4}$
		$= 10 \cdot 4$
		$= 40$
		So the loudness is 40 dB.

Note in Example 10 that for each sound, after substituting the values for I and I_0 in the expression $\log_{10}\left(\dfrac{I}{I_0}\right)$ and then simplifying, we get an expression of the form $\log_{10} 10^n$. In each case, regardless of the value of n, we find that $\log_{10} 10^n = n$. That is, the logarithm, base 10, of 10 raised to a power equals that power.

10.3 Exercises FOR EXTRA HELP

Mathematically Speaking

Fill in each blank with the most appropriate term or phrase from the given list.

intersects	exponential function	increasing
decreasing	logarithmic function	approaches
identical		base
exponent	x to the base b	symmetric
b to the base x		

1. The inverse of $f(x) = b^x$ is represented by the _____ $f^{-1}(x) = \log_b x$.

2. The term $\log_b x$ may be read as "the logarithm of _____."

3. A logarithm is a(n) _____.

4. The graph of a logarithmic function is _____ to the graph of the corresponding exponential function.

5. The logarithmic function is _____ if the base is between 0 and 1.

6. The graph of a logarithmic function _____ the x-axis.

Write each logarithmic equation in its equivalent exponential form.

7. $\log_3 81 = 4$

8. $\log_4 16 = 2$

9. $\log_{1/2} \dfrac{1}{32} = 5$

10. $\log_{1/3} \dfrac{1}{27} = 3$

11. $\log_5 \dfrac{1}{25} = -2$

12. $\log_{10} \dfrac{1}{10,000} = -4$

13. $\log_{1/4} 4 = -1$

14. $\log_{1/6} 36 = -2$

15. $\log_{16} 2 = \dfrac{1}{4}$

16. $\log_{64} 4 = \dfrac{1}{3}$

17. $\log_{10} \sqrt{10} = \dfrac{1}{2}$

18. $\log_6 \sqrt[5]{6} = \dfrac{1}{5}$

Write each exponential equation in its equivalent logarithmic form.

19. $3^5 = 243$

20. $10^6 = 1,000,000$

21. $\left(\dfrac{1}{4}\right)^1 = \dfrac{1}{4}$

22. $\left(\dfrac{1}{5}\right)^2 = \dfrac{1}{25}$

23. $2^{-4} = \dfrac{1}{16}$

24. $9^{-2} = \dfrac{1}{81}$

25. $\left(\dfrac{1}{3}\right)^{-4} = 81$

26. $\left(\dfrac{1}{6}\right)^{-1} = 6$

27. $49^{1/2} = 7$

28. $256^{1/4} = 4$

29. $11^{1/5} = \sqrt[5]{11}$

30. $15^{1/2} = \sqrt{15}$

Evaluate.

31. $\log_5 125$

32. $\log_2 64$

33. $\log_3 9$

34. $\log_{10} 100$

35. $\log_6 6$

36. $\log_7 7$

37. $\log_{1/2} \dfrac{1}{16}$

38. $\log_{1/3} \dfrac{1}{9}$

39. $\log_4 \dfrac{1}{64}$

40. $\log_2 \dfrac{1}{2}$

41. $\log_{1/4} 16$

42. $\log_{1/10} 1000$

43. $\log_9 1$

44. $\log_4 1$

45. $\log_{27} 3$

46. $\log_{25} 5$

47. $\log_{1/16} \dfrac{1}{2}$

48. $\log_{1/64} \dfrac{1}{4}$

49. $\log_{36} \dfrac{1}{6}$

50. $\log_8 \dfrac{1}{2}$

Solve.

51. $\log_3 x = 3$

52. $\log_2 x = 6$

53. $\log_6 x = -2$

54. $\log_3 x = -5$

55. $\log_{2/3} x = -1$

56. $\log_{1/6} x = -2$

57. $\log_4 x = \dfrac{1}{2}$

58. $\log_{64} x = \dfrac{1}{3}$

59. $\log_x 216 = 3$

60. $\log_x 81 = 2$

61. $\log_x 2 = \dfrac{1}{3}$

62. $\log_x 3 = \dfrac{1}{4}$

63. $\log_x \dfrac{9}{16} = 2$

64. $\log_x \dfrac{27}{8} = 3$

65. $\log_x 7 = -1$

66. $\log_x 16 = -4$

67. $\log_4 8 = x$

68. $\log_9 3 = x$

69. $\log_{27} 81 = x$

70. $\log_{16} 64 = x$

71. $\log_{1/3} 81 = x$

72. $\log_{1/2} 32 = x$

Graph each function.

73. $f(x) = \log_3 x$

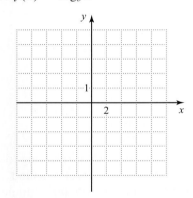

74. $f(x) = \log_4 x$

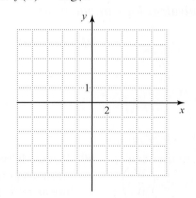

75. $f(x) = \log_{1/3} x$

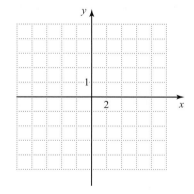

76. $f(x) = \log_{1/4} x$

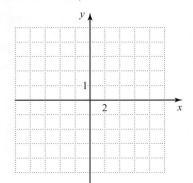

77. $f(x) = -\log_3 x$

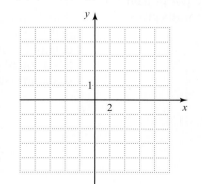

78. $f(x) = -\log_2 x$

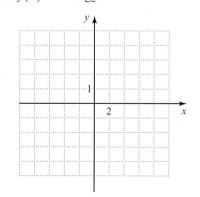

Mixed Practice

Solve.

79. $\log_x 5 = \dfrac{1}{3}$

80. $\log_x \dfrac{25}{64} = 2$

81. $\log_9 27 = x$

82. Graph the function $f(x) = -\log_4 x$.

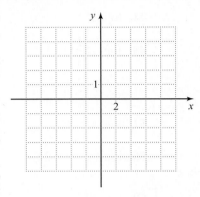

Evaluate.

83. $\log_2 \dfrac{1}{16}$

84. $\log_{81} 3$

Write each logarithmic equation in its equivalent exponential form.

85. $\log_{1/2} 16 = -4$

86. $\log_{81} 3 = \dfrac{1}{4}$

Write each exponential equation in its equivalent logarithmic form.

87. $\left(\dfrac{1}{7}\right)^{-2} = 49$

88. $32^{1/5} = 2$

Applications

Solve.

89. The power gain, P (in decibels), in an amplifier can be determined using the equation $P = 10 \log_{10}\left(\dfrac{W_2}{W_1}\right)$, where W_1 and W_2 represent the input and output power in watts, respectively. What is the power gain in an amplifier if an input power of 10^{-2} watts gives an output power of 100 watts?

90. Scientists speak of *sound pressure* when referring to noise as perceived by a listener. The level of sound pressure S (in decibels) is defined by the logarithmic equation $S = 20 \log_{10}\left(\dfrac{P}{P_0}\right)$, where P is the actual sound pressure in newtons per square meter (N/m^2) and P_0 is the threshold of human hearing, which equals 2×10^{-5} N/m^2. A sound at the pain threshold has a sound pressure of about 20 N/m^2. Calculate the sound pressure level at the pain threshold.

91. The population of a certain bacterium doubles every hour. The equation $h = \log_2\left(\dfrac{N}{N_0}\right)$ can be used to determine the number of hours, h, it takes for the initial population N_0 to increase to a population of N bacteria. If the initial population consists of 8 bacteria, how many hours will it take for the population to reach 512 bacteria?

92. The half-life of radium-226, a radioactive isotope used in the treatment of cervical cancer, is about 1600 yr. The equation $t = -1600\log_2\left(\dfrac{b}{a}\right)$ can be used to determine the number of years it will take a mg of the radium to decay to b mg. How long will it take 40 mg of radium-226 to decay to 5 mg?

93. The formula $\log_2 I = t$ relates amount of electrical current I (in amperes) flowing through a circuit to the length of time t (in seconds) that it takes for the current to pass through the circuit. Find the amount of current in a circuit when $t = 3$ sec.

94. A colony of bacteria triples in size every hour. The number, N, of bacteria after t hr can be found using the logarithmic equation $\log_3 N = t$. After how many hours will there be 27 bacteria in the colony?

95. A picosecond is one trillionth (10^{-12}) of a second. Engineers found that the time t (in picoseconds) that a particular computer requires to carry out N computations is given by the equation $t = N + \log_2 N$. How long does it take this computer to carry out 64 computations?

96. In a certain chemical reaction, the time t (in seconds) required for 1 kg of a substance to be converted to N kg of a second substance is given by the equation $t = -10\log_{10} N$. How long will it take for the reaction to produce $\dfrac{1}{10}$ kg of the second substance?

97. The length of the side of the fence surrounding the square garden shown in the diagram can be found by solving the logarithmic equation $\log_{36} x = \dfrac{1}{2}$.

x ft

a. What is the length of the side of the fence?

b. In this situation, what does the base in the equation represent?

98. The length of the side of the cube-shaped box shown in the diagram can be found by solving the logarithmic equation $\log_{64} x = \dfrac{1}{3}$.

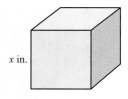

x in.

a. What is the length of the side of the box?

b. What does the base in the equation represent in this context?

● *Check your answers on page A-45.*

MINDSTRETCHERS

Writing

1. Consider the logarithmic equation $\log_b y = x$. Explain why the value of y can never be equal to 0.

Research

2. Either in your college library or on the Web, investigate the work of John Napier in the development of logarithms. Summarize your findings.

Critical Thinking

3. Suppose $f(x) = \log_2 x$ and $g(x) = 2^x$.

    a. Find $(f \circ g)(x)$.

    b. How could you have predicted the result you found in part (a)?

CULTURAL NOTE

Marie Curie, shown meeting with the physicist Albert Einstein, was a pioneer in the field of radioactivity. At the beginning of the twentieth century, she discovered radium and polonium—elements whose radioactive decay can be modeled by *exponential functions*. Madame Curie was the first person to win two Nobel prizes. Her work tremendously influenced basic science and ushered in a new era in medical research and treatment.

10.4   Properties of Logarithms

In the previous section, we introduced the concept of logarithmic functions. In this section, we consider *properties of logarithms,* which we can use to evaluate and to rewrite logarithmic expressions. These properties are also useful in solving other applied problems involving logarithms.

Using the Product Property

Since logarithms are exponents, they follow the laws of exponents. These laws lead to the properties of logarithms. To begin, we use the product rule of exponents, $b^x \cdot b^y = b^{x+y}$, to derive the *product property of logarithms*.

Suppose we let $x = \log_b M$ and $y = \log_b N$. If we write each equation in exponential form, we get $b^x = M$ and $b^y = N$. Forming the product of M and N, we have

$$MN = b^x \cdot b^y = b^{x+y}$$

Writing $MN = b^{x+y}$ in logarithmic form gives us

$$\log_b MN = x + y$$

Since $x = \log_b M$ and $y = \log_b N$, we can write

$$\log_b MN = \log_b M + \log_b N$$

which is called the *product property of logarithms*.

Product Property of Logarithms

For any positive real numbers M, N, and b, $b \neq 1$,

$$\log_b MN = \log_b M + \log_b N$$

This property states that the logarithm of a product is the sum of the logarithms of the factors. Let's use this property to rewrite logarithmic expressions.

EXAMPLE 1

Write each expression using the product property.

a. $\log_4(6 \cdot 5)$ **b.** $\log_8 8x$

Solution We use the product property of logarithms as follows:

a. $\log_4(6 \cdot 5) = \log_4 6 + \log_4 5$

b. $\log_8 8x = \log_8 8 + \log_8 x$
    $= 1 + \log_8 x$ Evaluate $\log_8 8$.

PRACTICE 1

Use the product property to rewrite each expression.

a. $\log_6(4 \cdot 9)$

b. $\log_2 2x$

EXAMPLE 2

Write as a single logarithm.

a. $\log_5 16 + \log_5 4$ **b.** $\log_3 x + \log_3(x + 1)$

Solution Using the product property of logarithms, we get

a. $\log_5 16 + \log_5 4 = \log_5(16 \cdot 4)$
$= \log_5 64$

b. $\log_3 x + \log_3(x + 1) = \log_3[x(x + 1)]$
$= \log_3(x^2 + x)$

PRACTICE 2

Write as a single logarithm.

a. $\log_2 9 + \log_2 15$

b. $\log_7(x + y) + \log_7 x$

Using the Quotient Property

Now, let's use the quotient rule of exponents, $\dfrac{b^x}{b^y} = b^{x-y}$, to derive the *quotient property of logarithms*. Again, we let $x = \log_b M$ and $y = \log_b N$. Next, we write each equation in exponential form: $M = b^x$ and $N = b^y$. Forming the quotient of M and N gives us

$$\frac{M}{N} = \frac{b^x}{b^y} = b^{x-y}$$

We can write $\dfrac{M}{N} = b^{x-y}$ in logarithmic form as

$$\log_b \frac{M}{N} = x - y$$

Substituting $\log_b M$ for x and $\log_b N$ for y, we have

$$\log_b \frac{M}{N} = \log_b M - \log_b N$$

which is called the *quotient property of logarithms*.

> **Quotient Property of Logarithms**
> For any positive real numbers M, N, and b, $b \neq 1$,
> $$\log_b \frac{M}{N} = \log_b M - \log_b N$$

This property states that the logarithm of a quotient is the logarithm of the numerator minus the logarithm of the denominator.

EXAMPLE 3

Write each expression using the quotient property.

a. $\log_7 \dfrac{9}{5}$

b. $\log_a \dfrac{a}{x}$

PRACTICE 3

Write each expression using the quotient property.

a. $\log_4 \dfrac{3}{4}$

b. $\log_b \dfrac{u}{v}$

Solution We use the quotient property of logarithms.

a. $\log_7 \dfrac{9}{5} = \log_7 9 - \log_7 5$

b. $\log_a \dfrac{a}{x} = \log_a a - \log_a x$

$= 1 - \log_a x$ Evaluate $\log_a a$.

EXAMPLE 4

Write as a single logarithm.

a. $\log_{10} 40 - \log_{10} 10$ **b.** $\log_2(x+1) - \log_2(x-3)$

Solution Using the quotient property of logarithms, we get

a. $\log_{10} 40 - \log_{10} 10 = \log_{10} \dfrac{40}{10} = \log_{10} 4$

b. $\log_2(x+1) - \log_2(x-3) = \log_2 \dfrac{x+1}{x-3}$

PRACTICE 4

Write as a single logarithm.

a. $\log_7 49 - \log_7 98$

b. $\log_3(x-y) - \log_3(x+y)$

Using the Power Property

Finally, we consider the *power property of logarithms*. To derive this property, we use the power rule of exponents, $(b^x)^r = b^{xr}$. Again, if we let $x = \log_b M$ and then write this equation in exponential form, we get $b^x = M$. Raising both sides of this equation to the rth power gives:

$$(b^x)^r = M^r \quad \text{or} \quad b^{xr} = M^r$$

Writing $b^{xr} = M^r$ as a logarithmic equation, we have $\log_b M^r = xr$. Since $x = \log_b M$, we can now write

$$\log_b M^r = xr = (\log_b M)r = r \log_b M$$

which is called the *power property of logarithms*.

> **Power Property of Logarithms**
> For any positive real numbers M and b, $b \neq 1$, and any real number r,
> $$\log_b M^r = r\log_b M$$

This property states that the logarithm of a power of a number is that power times the logarithm of that number.

EXAMPLE 5

Write each expression using the power property.

a. $\log_8 n^2$

b. $\log_3 \sqrt{n}$

PRACTICE 5

Use the power property to write each expression.

a. $\log_2 x^3$

b. $\log_5 \sqrt[3]{x}$

Solution Using the power property, we get

a. $\log_8 n^2 = 2 \log_8 n$

b. $\log_3 \sqrt{n} = \log_3 n^{1/2}$ **Write the radical using a rational exponent.**

$\qquad = \dfrac{1}{2} \log_3 n$

In the previous section, we noted that $\log_b b = 1$ and $\log_b 1 = 0$. Now, we consider two other special properties of logarithms. Both properties follow from the inverse relationship between exponential and logarithmic functions.

Special Properties of Logarithms

For $b > 0$ and $b \neq 1$,

- $\log_b b^x = x$
- $b^{\log_b x} = x$, where $x > 0$

In words, the first property states that the logarithm with base b of b raised to a power equals that power. The second property states that b raised to the logarithm with base b of a number equals that number.

EXAMPLE 6	PRACTICE 6

EXAMPLE 6

Find the value of each logarithm.

a. $\log_2 2^5$ **b.** $10^{\log_{10} 9}$

Solution

a. Since $\log_b b^x = x$, we conclude that

$$\log_2 2^5 = 5$$

b. Since $b^{\log_b x} = x$, we conclude that

$$10^{\log_{10} 9} = 9$$

PRACTICE 6

Evaluate:

a. $\log_{10} 10^7$

b. $3^{\log_3 29}$

The following table summarizes the properties of logarithms.

Properties of Logarithms

For any positive real numbers M, N, and b, where $b \neq 1$, and for any real numbers r and x,

Product Property $\log_b MN = \log_b M + \log_b N$

Quotient Property $\log_b \dfrac{M}{N} = \log_b M - \log_b N$

Power Property $\log_b M^r = r \log_b M$

Special Properties $\log_b b = 1$

$\qquad\qquad\qquad\quad \log_b 1 = 0$

$\qquad\qquad\qquad\quad \log_b b^x = x$

$\qquad\qquad\qquad\quad b^{\log_b x} = x$, where $x > 0$

Using the Properties Together

Using the properties of logarithms together, we can write logarithmic expressions in alternative forms. These forms are important for solving equations with logarithms and in more advanced courses, such as calculus.

EXAMPLE 7

Use the properties of logarithms to write as a single logarithm.

a. $3 \log_4 2 + 2 \log_4 5$

b. $2(\log_5 x - \log_5 y)$

Solution

a. $3 \log_4 2 + 2 \log_4 5 = \log_4 2^3 + \log_4 5^2$ Use the power property.

$\qquad = \log_4 8 + \log_4 25$

$\qquad = \log_4(8 \cdot 25)$ Use the product property.

$\qquad = \log_4 200$

b. $2(\log_5 x - \log_5 y) = 2 \log_5 x - 2 \log_5 y$ Use the distributive property.

$\qquad = \log_5 x^2 - \log_5 y^2$ Use the power property.

$\qquad = \log_5 \dfrac{x^2}{y^2}$ Use the quotient property.

PRACTICE 7

Write as a single logarithm using the properties of logarithms.

a. $2 \log_6 4 + 3 \log_6 2$

b. $3(\log_2 u - \log_2 v)$

EXAMPLE 8

Use the properties of logarithms to write each expression as a sum or difference of logarithms.

a. $\log_2 2x^5$ **b.** $\log_5 \sqrt[3]{\dfrac{x}{y}}$ **c.** $\log_n \dfrac{r^3 s}{t^2}$

Solution

a. $\log_2 2x^5 = \log_2 2 + \log_2 x^5$ Use the product property.

$\qquad = 1 + 5 \log_2 x$ Evaluate $\log_2 2$ and use the power property.

b. $\log_5 \sqrt[3]{\dfrac{x}{y}} = \log_5 \left(\dfrac{x}{y}\right)^{1/3}$ Write the radical using a rational exponent.

$\qquad = \dfrac{1}{3} \log_5 \dfrac{x}{y}$ Use the power property.

$\qquad = \dfrac{1}{3}(\log_5 x - \log_5 y)$ Use the quotient property.

$\qquad = \dfrac{1}{3} \log_5 x - \dfrac{1}{3} \log_5 y$ Use the distributive property.

c. $\log_n \dfrac{r^3 s}{t^2} = \log_n (r^3 s) - \log_n t^2$ Use the quotient property.

$\qquad = \log_n r^3 + \log_n s - \log_n t^2$ Use the product property.

$\qquad = 3 \log_n r + \log_n s - 2 \log_n t$ Use the power property.

PRACTICE 8

Use the properties of logarithms to write each expression as a sum or a difference of logarithms.

a. $\log_5 5x^2$

b. $\log_7 \sqrt{\dfrac{u}{v}}$

c. $\log_u \dfrac{b^3}{ac^5}$

The properties of logarithms are used in solving problems in many areas, such as chemistry, banking, and finance.

EXAMPLE 9

Chemists use a logarithmic scale called the pH scale for determining the acidity of a solution by measuring the concentration of hydrogen ions in the solution. They use the formula $pH = -\log_{10}[H^+]$. A pH level of 7 is neutral. A level greater than 7 is basic, and a level less than 7 is acidic.

a. Find the pH of vinegar that has a hydrogen ion concentration, $[H^+]$, of 1×10^{-3}.

b. Determine whether vinegar is basic or acidic.

Solution

a. Using the formula $pH = -10\log_{10}[H^+]$, we get

$$
\begin{aligned}
pH &= -\log_{10}[H^+] \\
&= -\log_{10}(1 \times 10^{-3}) \\
&= -(\log_{10} 1 + \log_{10} 10^{-3}) \\
&= -\log_{10} 1 - \log_{10} 10^{-3} \\
&= 0 - (-3) \\
&= 3
\end{aligned}
$$

So the pH of vinegar is 3.

b. Since $3 < 7$, vinegar is acidic.

PRACTICE 9

The number, N, of bacteria present in a culture after x hr can be determined by using the formula
$$\log_{10} N = x(\log_{10} 6 - \log_{10} 2).$$

a. How many bacteria are present in the culture after 1 hr?

b. By what factor does the number of bacteria present in the culture increase each hour?

FOR EXTRA HELP

Mathematically Speaking

Fill in each blank with the most appropriate term or phrase from the given list.

base	divided by	sum
plus	product	times
number	minus	power

1. The logarithm of a product is the _____ of the logarithms of the factors.

2. The logarithm of a quotient is the logarithm of the numerator _____ the logarithm of the denominator.

3. The logarithm of a power of a number is the power _____ the logarithm of the number.

4. For any _____ , the logarithm of 1 is 0.

5. The logarithm with base b of b raised to a power equals that _____ .

6. The number b raised to the logarithm with base b of a number equals that _____ .

Write each expression as the sum or difference of logarithms.

7. $\log_2(16 \cdot 5)$

8. $\log_3(2 \cdot 27)$

9. $\log_6 \dfrac{7}{36}$

10. $\log_2 \dfrac{4}{9}$

11. $\log_3 8x$

12. $\log_5 15y$

13. $\log_n uv$

14. $\log_a pq$

15. $\log_b[a(b-1)]$

16. $\log_y[x(y-3)]$

17. $\log_4 \dfrac{64}{n}$

18. $\log_{10} \dfrac{p}{1000}$

19. $\log_7 \dfrac{v}{u}$

20. $\log_4 \dfrac{q}{p}$

21. $\log_a \dfrac{b-a}{b+a}$

22. $\log_k \dfrac{k+1}{k+5}$

Write each expression as a single logarithm.

23. $\log_2 5 + \log_2 7$

24. $\log_5 2 + \log_5 6$

25. $\log_7 10 - \log_7 2$

26. $\log_6 3 - \log_6 21$

27. $1 + \log_x 6$

28. $\log_a 18 + 1$

29. $1 - \log_b 16$

30. $\log_a 7 - 1$

31. $\log_5 2 + \log_5(x-5)$

32. $\log_3(y+2) + \log_3 6$

33. $\log_n x + \log_n z$

34. $\log_b c + \log_b a$

◉ 35. $\log_5 k - \log_5 n$ **36.** $\log_6 y - \log_6 x$ **37.** $\log_{10}(a + b) - \log_{10}(a - b)$

38. $\log_4 x - \log_4(x - y)$

Use the power property to rewrite each expression.

39. $\log_6 8^2$ **40.** $\log_2 3^3$ **◉ 41.** $\log_3 x^4$ **42.** $\log_5 y^6$

43. $\log_a \sqrt[5]{b}$ **44.** $\log_x \sqrt[4]{y}$ **◉ 45.** $\log_4 a^{-1}$ **46.** $\log_7 y^{-2}$

Evaluate.

47. $\log_5 5^4$ **48.** $\log_3 3^7$ **◉ 49.** $\log_2 2^{1/2}$ **50.** $\log_6 6^{1/3}$

51. $\log_x x^{-3}$ **52.** $\log_a a^{-1}$ **53.** $8^{\log_8 15}$ **54.** $7^{\log_7 20}$

◉ 55. $9^{\log_9 x}$ **56.** $4^{\log_4 b}$ **57.** $x^{\log_x 10}$ **58.** $n^{\log_n 23}$

Write each expression as a single logarithm.

59. $2 \log_2 6 + \log_2 3$ **60.** $3 \log_{10} 3 + \log_{10} 2$ **◉ 61.** $4 \log_7 2 - 3 \log_7 4$

62. $2 \log_3 8 - 3 \log_3 5$ **63.** $-\log_4 x + 6 \log_4 y$ **64.** $-3 \log_6 a + 5 \log_6 b$

65. $2 \log_b 3 + 3 \log_b 2 - \log_b 9$ **66.** $3 \log_x 4 - 2 \log_x 6 - 4 \log_x 2$ **◉ 67.** $\dfrac{1}{2}(\log_4 x - 2 \log_4 y)$

68. $\dfrac{1}{3}(-6 \log_7 u + \log_7 v)$ **69.** $2 \log_5 x + \log_5(x - 1)$ **70.** $4 \log_2 n + \log_2(n + 3)$

◉ 71. $\dfrac{1}{3}[\log_6(x^2 - y^2) - \log_6(x + y)]$ **72.** $\dfrac{1}{2}[\log_2(a^2 - b^2) - \log_2(a - b)]$ **73.** $-\log_2 z + 4 \log_2 x - 5 \log_2 y$

74. $-4 \log_8 b + 5 \log_8 a - 2 \log_8 c$ **75.** $\dfrac{1}{4} \log_5 a - 8 \log_5 b + \dfrac{3}{4} \log_5 c$ **76.** $\dfrac{2}{3} \log_7 u + \dfrac{1}{3} \log_7 v - 7 \log_7 w$

Write each expression as a sum or difference of logarithms.

77. $\log_6 3y^2$ **78.** $\log_2 5n^4$ **79.** $\log_x x^3 y^2$ **80.** $\log_b a^4 b^8$

◉ 81. $\log_2 4x^2 y^5$ **82.** $\log_3 3uv^3$ **83.** $\log_8 \sqrt{10x}$ **84.** $\log_5 \sqrt[3]{9v^2}$

85. $\log_4 \dfrac{x^2}{y}$ **86.** $\log_5 \dfrac{a^3}{b^4}$ **87.** $\log_7 \sqrt[4]{\dfrac{u^3}{v}}$ **88.** $\log_3 \sqrt[3]{\dfrac{x^2}{y^2}}$

89. $\log_c \dfrac{a^2 c^4}{b^3}$ **90.** $\log_z \dfrac{x^5 y^4}{z^2}$ **◉ 91.** $\log_6 \dfrac{x^3}{(x - y)^2}$

92. $\log_5 \dfrac{a + b}{a^4}$ **93.** $\log_a x^2 \sqrt[5]{y^3 z^2}$ **94.** $\log_n \dfrac{\sqrt[4]{a^3 c^2}}{b^5}$

Mixed Practice

Write each expression as a single logarithm.

95. $\log_b 8 - 1$

96. $\log_2 5 + \log_2(k+3)$

97. $2\log_a 6 - 4\log_a 2 - 3\log_a 3$

98. $\dfrac{1}{4}(-\log_5 a + 8\log_5 b)$

Solve.

99. Use the power property to rewrite the expression $\log_p \sqrt[3]{q}$.

100. Evaluate: $4^{\log_4 25}$

Write each expression as a sum or difference.

101. $\log_2 8a^3 b^2$

102. $\log_5 \sqrt[4]{\dfrac{x^3}{y^2}}$

103. $\log_4 64a$

104. $\log_5 \dfrac{b}{25}$

Applications

Solve.

105. The magnitude, M, of an earthquake on the Richter scale is given by the logarithmic equation $M = \log_{10} I$, where I is the intensity of the earthquake. The largest earthquake in the year 2003, with an intensity of $10^{8.3}$, occurred in Hokkaido, Japan. Calculate the magnitude of that earthquake. (*Source:* U.S. Geological Survey, Earthquake Hazards Program)

106. The formula for the pH of a solution is given by the logarithmic equation pH $= -\log_{10}[H^+]$, where $[H^+]$ is the concentration of hydrogen ions. The concentration of hydrogen ions in pure water is 10^{-7}. Calculate the pH of pure water.

107. The sound pressure level, S, of acoustic sound pressure waves is defined by the logarithmic equation $S = 10\log_{10}\left(\dfrac{P}{P_0}\right)^2$, where P is the actual pressure and P_0 is the pain threshold for human hearing. Use the properties of logarithms to write the expression on the right so that it does not contain the logarithm of a quotient.

108. The loudness of sound, in decibels, can be defined by the logarithmic equation $L = 10\log_{10}\dfrac{I}{10^{-16}}$, where I is the intensity of the sound in watts per square centimeter. Use the properties of logarithms to rewrite the right side of the equation so that it does not contain the quotient of a logarithm.

109. The half-life of cobalt-60, a radioactive isotope used in industry to identify structural defects in metal parts, is approximately 5 yr. The amount, C, of cobalt remaining after t yr and the amount C_0 initially present are related by the equation $\log_{10} C = \log_{10} C_0 + 0.2t\log_{10}\dfrac{1}{2}$.

 a. Use the properties of logarithms to show that the equation can be written as
$$C = C_0\left(\dfrac{1}{2}\right)^{0.2t}.$$

 b. If 50 g of cobalt-60 are initially present, how many grams will remain after 10 yr?

110. The amount, A, in an account earning 5% interest, compounded annually, after t yr can be determined using the logarithmic equation $\log_{10} A = \log_{10} P + t \log_{10}(1.05)$, where P is the initial amount deposited into the account.

    a. Use the properties of logarithms to show that the equation can be written as $A = P(1.05)^t$.

    b. Determine the amount in the account after 1 yr if $1000 is initially deposited.

● *Check your answers on page A-46.*

MINDSTRETCHERS

Mathematical Reasoning

1. Show that the following properties of logarithms hold:

    a. $\log_b b^x = x$, for any positive real number b, where $b \neq 1$, and for any real number x

    b. $b^{\log_b x} = x$, for any positive real numbers b and x, where $b \neq 1$

Critical Thinking

2. Show that $3 + \log_3 2 = 1 + \log_3 18$.

Groupwork

3. Suppose $\log_b x = 3.5$, $\log_b y = 0.2$, and $\log_b z = 1.4$. Find the value of each of the following logarithms:

    a. $\log_b(xy)$ **b.** $\log_b \dfrac{y}{x}$ **c.** $\log_b \dfrac{xy}{z}$

    d. $\log_b z^5$ **e.** $\log_b(yz)^2$

10.5 Common Logarithms, Natural Logarithms, and Change of Base

In the previous sections of this chapter, we discussed logarithms in which the base was any positive real number. In this section, we focus on two special logarithmic bases, 10 and e. A logarithm to the base 10 is called a *common logarithm*, and a logarithm to the base e is called a *natural logarithm*. Common logarithms and natural logarithms occur frequently in real-world situations.

OBJECTIVES

- To evaluate common logarithms
- To evaluate natural logarithms
- To use the change-of-base formula
- To solve applied problems using common or natural logarithms

Evaluating Common Logarithms

Recall the applications that we have considered involving logarithms to the base 10, such as the Richter scale used to measure the magnitude of an earthquake and the decibel scale used to find the loudness of sound. Base 10 is commonly used for logarithms, since this base makes it easy to find the logarithm of both powers of 10 and multiples of 10.

Definition

A **common logarithm** is a logarithm to the base 10, where **log x** means $\log_{10} x$.

Note that when we write a common logarithm it is customary to omit the base. To evaluate common logarithms using a calculator, we use the $\boxed{\log}$ key.

EXAMPLE 1	PRACTICE 1
Use a calculator to approximate each logarithm to four decimal places.	Using a calculator, approximate each logarithm to four decimal places.
a. log 9 $\qquad$ **b.** log 35	**a.** log 6
Solution Using the $\boxed{\log}$ key on the calculator, we get	**b.** log 27
a. log 9 ≈ 0.9542 $\qquad$ **b.** log 35 ≈ 1.5441	

We can evaluate common logarithms of powers of 10 without using a calculator. Recall from the properties of logarithms that

$$\log_b b^x = x$$

So substituting 10 for b gives us

$$\log_{10} 10^x = x \quad \text{or} \quad \log 10^x = x$$

EXAMPLE 2

Evaluate each logarithm.

a. log 10,000   **b.** $\log \dfrac{1}{100}$ **c.** $\log \sqrt{10}$

Solution First we write each logarithm as a power of 10. Then we use the relationship $\log 10^x = x$ to simplify.

a. $\log 10{,}000 = \log 10^4 = 4$

b. $\log \dfrac{1}{100} = \log \dfrac{1}{10^2}$

$\qquad\quad = \log 10^{-2}$

$\qquad\quad = -2$

c. $\log \sqrt{10} = \log 10^{1/2} = \dfrac{1}{2}$

PRACTICE 2

Find the value of each logarithm.

a. log 1000

b. $\log \dfrac{1}{10}$

c. $\log \sqrt[3]{10}$

Evaluating Natural Logarithms

Now, let's focus our attention on another special logarithmic base, namely, e. Recall from Section 10.2 that e is an irrational number that is approximately equal to 2.7182818. For the natural exponential function $f(x) = e^x$, the inverse is represented by

$$f^{-1}(x) = \log_e x$$

Logarithms to the base e, or *natural logarithms*, model a wide range of real-life applications, such as continuous compound interest and population growth. The notation ln x (read "el en x" or "the natural logarithm of x") is used to represent a natural logarithm.

> **Definition**
> A **natural logarithm** is a logarithm to the base e, where **ln x** means $\log_e x$.

Natural logarithms, like common logarithms, can be evaluated using a calculator. To do this, we use the natural logarithm key ▏ **ln** ▏ found on the calculator.

EXAMPLE 3

Use a calculator to approximate each logarithm to four decimal places.

a. ln 2   **b.** ln 30

Solution Using the ▏ **ln** ▏ key, we get the following:

a. ln 2 ≈ 0.6931   **b.** ln 30 ≈ 3.4012

PRACTICE 3

Approximate each logarithm to four decimal places using a calculator.

a. ln 4

b. ln 25

We can evaluate natural logarithms of powers of e without using a calculator. From the property $\log_b b^x = x$, we know that $\log_e e^x = x$. So $\ln e^x = x$.

EXAMPLE 4	PRACTICE 4
Find the value of each logarithm.	Evaluate each logarithm.

EXAMPLE 4

Find the value of each logarithm.

a. $\ln e$ **b.** $\ln \dfrac{1}{e^4}$ **c.** $\ln \sqrt[3]{e}$

Solution Here we use the relationship $\ln e^x = x$ to evaluate each logarithm.

a. $\ln e = \ln e^1 = 1$

b. $\ln \dfrac{1}{e^4} = \ln e^{-4} = -4$

c. $\ln \sqrt[3]{e} = \ln e^{1/3} = \dfrac{1}{3}$

PRACTICE 4

Evaluate each logarithm.

a. $\ln e^2$

b. $\ln \dfrac{1}{e^7}$

c. $\ln \sqrt[4]{e^3}$

Tip Remember that the base of common logarithms is understood to be 10, whereas the base of natural logarithms is understood to be e.

Using the Change-of-Base Formula

As we have just seen, calculators can be used to evaluate common and natural logarithms. However, in general, calculators do not directly evaluate logarithms with bases other than 10 or e. For these logarithms, we use the *change-of-base formula*.

Change-of-Base Formula

For positive real numbers N, b, and c, where $b \neq 1$ and $c \neq 1$,

$$\log_b N = \frac{\log_c N}{\log_c b}$$

According to this formula, if we know the logarithm of a number N to one base (the reference base, c), we can use the formula to find the logarithm of N to some other base, b, simply by dividing by the logarithm of b to the base c. Since we can readily use a calculator to approximate common logarithms and natural logarithms, it is convenient to use either 10 or e as the reference base:

$$\log_b N = \frac{\log N}{\log b} \quad \text{and} \quad \log_b N = \frac{\ln N}{\ln b}$$

EXAMPLE 5

Use a calculator to approximate $\log_3 12$ to four decimal places.

Solution Using the change-of-base formula, we write the logarithm as the quotient of logarithms to base 10.

$$\log_b N = \frac{\log_c N}{\log_c b}$$

$$\log_3 12 = \frac{\log 12}{\log 3} \qquad \text{Substitute 12 for } N \text{ and 3 for } b.$$

$$\approx 2.2618595$$

So $\log_3 12 \approx 2.2619$.

In Example 5, we used 10 as the reference base to evaluate the given logarithm. Do you think that we would get the same answer if we used natural logarithms? Explain.

In previous sections, we saw several applications involving common logarithms. Now, let's consider some applications that use natural logarithms.

PRACTICE 5

Using a calculator, approximate $\log_4 2.5$ to four decimal places.

EXAMPLE 6

Some psychologists who study learning theory use the following formula:

$$t = \frac{1}{c} \ln \frac{A}{A - N}$$

Here, t represents the time in weeks that passes before mastery of a task is achieved, N measures the learning achieved, A measures the maximum learning possible, and c, a constant, measures an individual's learning style. Suppose that in a certain study, a psychologist measures human capability to memorize random words. If a subject takes 4.5 weeks to learn 20 random words and the maximum possible number of words to learn is 32, how many weeks should it take to reach the maximum? Assume that c is 0.22.

Solution Substituting 0.22 for c, 32 for A, and 20 for N, we get

$$t = \frac{1}{c} \ln \frac{A}{A - N}$$

$$= \frac{1}{0.22} \ln \frac{32}{32 - 20}$$

$$= \frac{1}{0.22} \ln \frac{32}{12}$$

$$\approx (4.55)(0.9808)$$

$$\approx 4.5$$

So it will take the subject approximately 4.5 weeks to learn the maximum possible number of words.

PRACTICE 6

The half-life of a radioactive substance is the time it takes for half of the material to decay. The amount A (in pounds) of substance remaining after t yr can be determined using the formula

$$\ln \frac{A}{C} = -\frac{t}{h} \ln 2$$

where C is the initial amount (in pounds) and h is its half-life (in years). If the half-life of lead-210 is approximately 19 yr, how long, to the nearest year, will it take for 10 lb of this substance to decay to $\frac{1}{2}$ lb?

10.5   Exercises

Mathematically Speaking

Fill in each blank with the most appropriate term or phrase from the given list.

change-of-base formula	common logarithm	quotient property
absolute value function	natural exponential function	natural logarithm

1. A(n) ———— is a logarithm to the base 10.

2. The inverse of the ———— $f(x) = e^x$ is $f^{-1}(x) = \log_e x$.

3. A(n) ———— is a logarithm to the base e.

4. According to the ————, $\log_2 5 = \dfrac{\log_6 5}{\log_6 2}$.

Use a calculator to approximate each logarithm to four decimal places.

5. $\log 4$

6. $\log 7$

7. $\log 18$

8. $\log 22$

9. $\log \dfrac{2}{3}$

10. $\log \dfrac{1}{7}$

11. $\log 1.3$

12. $\log 0.9$

13. $\ln 3$

14. $\ln 6$

15. $\ln 17$

16. $\ln 13$

17. $\ln \dfrac{5}{8}$

18. $\ln \dfrac{3}{4}$

19. $\ln 1.2$

20. $\ln 0.24$

Evaluate.

21. $\log 1,000,000$

22. $\log 100,000$

23. $\log \dfrac{1}{1000}$

24. $\log \dfrac{1}{10,000}$

25. $\log 0.01$

26. $\log 0.000001$

27. $\log \sqrt[4]{1000}$

28. $\log \sqrt[3]{100}$

29. $\log 10^x$

30. $\log 10^a$

31. $\ln e^4$

32. $\ln e^{10}$

33. $\ln \dfrac{1}{e}$

34. $\ln \dfrac{1}{e^2}$

35. $\ln \sqrt{e}$

36. $\ln \sqrt[3]{e^2}$

37. $\ln e^b$

38. $\ln e^y$

39. $10^{\log 6}$

40. $10^{\log 2}$

41. $e^{\ln 3}$

42. $e^{\ln 5}$

Use a calculator to approximate each logarithm to four decimal places.

43. $\log_2 7$

44. $\log_5 10$

45. $\log_6 21$

46. $\log_8 14$

47. $\log_{1/2} 11$

48. $\log_{1/3} 4$

49. $\log_3 \dfrac{1}{2}$

50. $\log_4 \dfrac{1}{3}$

51. $\log_5 3.6$

52. $\log_9 6.8$

53. $\log_7 0.023$

54. $\log_6 0.0044$

Mixed Practice

Evaluate.

55. $\log \sqrt[2]{10{,}000}$

56. $\log 10^6$

57. $\ln \dfrac{1}{e^3}$

58. $e^{\ln 7}$

⊞ *Use a calculator to approximate each logarithm to four decimal places.*

59. $\log_{1/2} 7$

60. $\log_6 \dfrac{1}{4}$

61. $\log 2.7$

62. $\ln \dfrac{2}{3}$

Applications

Solve. Use a calculator where appropriate.

63. In chemistry, the pH of a substance can be determined using the formula pH $= -\log[H^+]$, where H^+ represents the hydrogen ion concentration. To the nearest tenth, determine the pH of household bleach, which has a hydrogen ion concentration of 3.2×10^{-13}.

64. Based on a recent census, the population of a town is increasing according to the equation $t = 15.2 \ln\left(\dfrac{P}{P_0}\right)$, where P_0 is the initial population and P is the population after t yr. If the initial population of the town is 24,342 people, then to the nearest year, when will the population reach 50,000 people?

65. The annual revenue, R, in millions of dollars, of a small computer company can be modeled by the logarithmic function $R(x) = 1.45 + 2\log(x + 1)$, where x represents the number of years after 2000. To the nearest million dollars, calculate the company's revenue in the year 2005.

66. Students in a math class were given an exam at the end of the semester. To see how well the students remembered the material, they were given a similar exam each month for 1 yr after the semester ended. The average exam score, s, for the class on a test given x mo after the original exam is modeled by the equation $s = 80 - 8\ln(x + 1)$. To the nearest whole number, what was the average exam score for the class after 6 mo?

67. The Beer-Lambert law relates the decay of the intensity of light to the properties of the material the light is passing through. The relationship is described by the equation $kd = -(\ln I - \ln I_0)$, where I is the intensity of light at depth d, I_0 is the initial intensity of the light, and k is a constant. Show that this equation can be written as $I = I_0 e^{-kd}$.

68. In the theory of rocket flight, the velocity, v, gained by a launch vehicle when its propellant is burned off is given by the equation $v = c \ln W - c \ln W_0$, where c is the exhaust velocity, W is the burnout weight, and W_0 is the takeoff weight. Write the equation so that the right side contains a single logarithm.
(*Source:* NASA)

69. The population of a certain bacterium doubles every hour. If a culture has an initial population of 1 bacterium, then the number of hours it takes for the population to grow to 60 bacteria is given by the expression $\log_2 60$. Evaluate this expression, rounding the answer to the nearest tenth.

70. The half-life of carbon-14, a radioactive isotope used in carbon dating, is about 5730 yr. The time it takes for 35 g of carbon-14 to decay to 5 g is given by the expression $-5730 \log_2 \dfrac{1}{7}$. Evaluate this expression. Round the answer to the nearest year.

● *Check your answers on page A-46.*

MINDSTRETCHERS

Groupwork

1. Working with a partner, choose particular positive values of b, c, and N in the change-of-base formula $\log_b N = \dfrac{\log_c N}{\log_c b}$, where $b \neq 1$ and $c \neq 1$. Then use a calculator to verify that the formula holds for these values. (*Hint:* Choose values of N and b so that N is a power of b.)

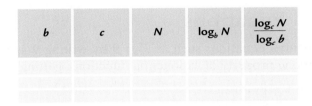

b	c	N	$\log_b N$	$\dfrac{\log_c N}{\log_c b}$

Mathematical Reasoning

2. Suppose $f(x) = x + 3$. Find all values of x such that $5^{2\log_5 f(x)} = f(x)$.

Technology

3. Consider the equation $y = \log_3 x$. Explain how you could graph this equation on a grapher that has the capability of evaluating only common and natural logarithms. Then graph the equation on a grapher.

In Section 10.2, we solved exponential equations of the form $b^x = b^n$. Similarly, in Section 10.3, we solved logarithmic equations of the form $y = \log_b x$. In this section, we solve a wider range of exponential and logarithmic equations by using the properties of logarithms. Then we consider problems modeled by these types of equations.

Solving Exponential Equations

Recall that we can solve exponential equations such as $3^x = 9$ by writing 9 as a power of 3 and using the one-to-one property of exponential functions, getting

$$3^x = 9$$
$$3^x = 3^2$$
$$x = 2$$

By contrast, the exponential equation $3^x = 12$ cannot be solved this way because 12 is not a power of 3. Since $3^x = 12$, we can take the common logarithm of each side of the equation and set the two logarithms equal to each other, as follows:

$$\log 3^x = \log 12$$
$$x \log 3 = \log 12$$
$$x = \frac{\log 12}{\log 3}$$
$$x \approx 2.2619$$

So x is approximately 2.2619.

Consider the following example in which we solve an exponential equation by taking the logarithm of each side.

EXAMPLE 1	PRACTICE 1
Solve $2^{x-1} = 7$. Approximate the answer to four decimal places.	Solve $4^{x+3} = 15$. Approximate the answer to four decimal places.

Solution

$$2^{x-1} = 7$$

$$\log 2^{x-1} = \log 7 \qquad \text{Take the common logarithm of each side of the equation.}$$

$$(x - 1)\log 2 = \log 7 \qquad \text{Use the power property of logarithms.}$$

$$x - 1 = \frac{\log 7}{\log 2} \qquad \text{Divide each side of the equation by log 2.}$$

$$x = \frac{\log 7}{\log 2} + 1 \qquad \text{Add 1 to each side of the equation.}$$

$$x \approx 2.8074 + 1$$

$$x \approx 3.8074$$

So x is approximately 3.8074.

Do you think you would get the same answer if you took the natural logarithm of each side of the equation in Example 1? Explain.

Solving Logarithmic Equations

When solving a logarithmic equation, we use the properties of logarithms to try to write the equation so that it contains a single logarithm on one side. Then we write the equivalent exponential equation and solve it.

EXAMPLE 2	PRACTICE 2
Solve: $\log_2(x - 4) = 4$	Solve: $\log_4(3x + 1) = 2$

Solution

$$\log_2(x - 4) = 4$$

$\quad\quad 2^4 = x - 4 \quad$ **Write the equivalent exponential equation.**

$\quad\quad 16 = x - 4$

$\quad\quad x = 20$

Check

$$\log_2(x - 4) = 4$$

$\log_2(20 - 4) \stackrel{?}{=} 4 \quad$ **Substitute 20 for x.**

$\quad\quad \log_2 16 \stackrel{?}{=} 4$

$\quad\quad\quad 4 = 4 \quad$ **True**

So x is 20.

EXAMPLE 3	PRACTICE 3
Solve: $\log_2 x + \log_2(x + 4) = 5$	Solve: $\log_3 x + \log_3(x - 6) = 3$

Solution

$$\log_2 x + \log_2(x + 4) = 5$$

$\quad\quad \log_2[x(x + 4)] = 5 \quad$ **Use the product property.**

$\quad\quad\quad 2^5 = x(x + 4) \quad$ **Write the equivalent exponential equation.**

$\quad\quad\quad 32 = x^2 + 4x$

$\quad\quad\quad 0 = x^2 + 4x - 32 \quad$ **Subtract 32 from each side of the equation.**

$\quad\quad\quad 0 = (x + 8)(x - 4) \quad$ **Factor.**

$x + 8 = 0 \quad\text{or}\quad x - 4 = 0$

$\quad x = -8 \quad\quad\quad x = 4$

Check

Substitute -8 for x.

$$\log_2 x + \log_2(x + 4) = 5$$
$$\log_2(-8) + \log_2(-8 + 4) \overset{?}{=} 5$$

Logarithms of negative
numbers are undefined.

Substitute 4 for x.

$$\log_2 x + \log_2(x + 4) = 5$$
$$\log_2 4 + \log_2(4 + 4) \overset{?}{=} 5$$
$$\log_2 4 + \log_2 8 \overset{?}{=} 5$$
$$2 + 3 \overset{?}{=} 5$$
$$5 = 5 \qquad \textbf{True}$$

Since -8 is an extraneous solution, the only solution is 4.

EXAMPLE 4

Solve: $\log_5(2x) - \log_5(x - 1) = 1$

Solution

$$\log_5(2x) - \log_5(x - 1) = 1$$

$$\log_5 \frac{2x}{x - 1} = 1 \qquad \text{Use the quotient property.}$$

$$5^1 = \frac{2x}{x - 1} \qquad \begin{array}{l}\text{Write the equivalent exponential}\\\text{equation.}\end{array}$$

$$5 = \frac{2x}{x - 1}$$

$$5(x - 1) = 2x$$

$$5x - 5 = 2x$$

$$3x = 5$$

$$x = \frac{5}{3}$$

So the solution is $\frac{5}{3}$. We can check the solution in the original equation.

PRACTICE 4

Solve: $\log_2(2x + 3) - \log_2 x = 3$

Recall that in Section 10.2 we discussed the compound interest formula,

$$A = P\left(1 + \frac{r}{n}\right)^{nt}$$

which is used to calculate the number of dollars, A, accrued (or owed) after P dollars are invested (or loaned) at an annual interest rate r compounded n times per year for t years.

When the number of times per year, n, that interest is compounded is high (for instance, when the compounding is daily instead of monthly), the amount of money, A, can be approximated using the exponential equation

$$A = Pe^{rt}$$

This equation is called the *continuous compound interest formula*. Many investment firms and banks use this formula to compound interest. In this method, which is known as *continuous compounding,* interest is said to be compounded every instant.

EXAMPLE 5

An account pays 10% interest compounded continuously. To the nearest year, how long does it take a deposit to triple at this rate?

Solution If P dollars are deposited in the account, the amount of money after t yr is given by $A = Pe^{0.10t}$. Since we want to know how long it takes for this amount to grow to $3P$, we solve the exponential equation $3P = Pe^{0.10t}$ for t.

$$3P = Pe^{0.10t}$$
$$3 = e^{0.10t}$$
$$\ln 3 = \ln e^{0.10t}$$
$$\ln 3 = 0.10t \ln e$$
$$\ln 3 = 0.10t$$
$$\frac{\ln 3}{0.10} = t$$
$$t \approx 10.9861$$

So it takes approximately 11 yr for a deposit in this account to triple.

PRACTICE 5

In predicting population growth, scientists use the formula $P = P_0 e^{kt}$. This formula models the size, P, of a population that has an annual growth rate k in decimal form, where t is time in years and P_0 is the initial population at time 0. In 1990, the population of a town was 36,695. By 2000, the population had grown to 61,102. To the nearest tenth of a percent, find the annual growth rate over this period.

In Example 5, note that since the base of the exponential function is e, we take the natural logarithm of each side of the equation.

10.6   **Exercises**

Solve. Round each answer to four decimal places.

1. $3^x = 18$

2. $6^x = 56$

3. $\left(\dfrac{1}{2}\right)^x = 9$

4. $\left(\dfrac{1}{3}\right)^x = 15$

5. $4^x = \dfrac{3}{4}$

6. $7^x = \dfrac{6}{7}$

7. $5.4^x = 0.0034$

8. $8.1^x = 0.097$

9. $2^{3x} = 4.6$

10. $5^{2x} = 7.5$

11. $100^x = 55$

12. $1000^{-x} = 95$

13. $3^{x+4} = 38$

14. $2^{x+1} = 68$

15. $5^{2x-3} = 43$

16. $6^{3x-5} = 324$

17. $e^{0.25x} = 3.1$

18. $e^{-0.012x} = 7.5$

Solve.

19. $\log_3(x + 10) = 4$

20. $\log_2(x + 7) = 5$

21. $\log_5(4x - 3) = 1$

22. $\log_4(3x - 5) = 3$

23. $\log_2(x + 1) = -3$

24. $\log_3(x - 1) = -2$

25. $\log(x^2 - 21) = 2$

26. $\log_3(x^2 - 19) = 4$

27. $\log_4 x + \log_4 6 = 2$

28. $\log_2 x + \log_2 12 = 6$

29. $\log_3 x - \log_3 7 = 2$

30. $\log_5 x - \log_5 2 = 1$

31. $\log_2 x + \log_2(x - 3) = 2$

32. $\log_3(x + 6) + \log_3 x = 3$

33. $\log_2(x + 2) + \log_2(x - 5) = 3$

34. $\log_7(x - 3) + \log_7(x + 3) = 1$

35. $\log_4(2x - 1) + \log_4(6x - 1) = 2$

36. $\log(3x + 4) + \log(5x - 9) = 1$

37. $\log_3(7x) - \log_3(x - 1) = 2$

38. $\log_2(10x + 4) - \log_2 x = 4$

39. $\log_2(3x + 7) - \log_2(x - 1) = 3$

40. $\log_5(2x - 1) - \log_5(x - 8) = 1$

41. $\log_8 x + \log_8(x + 1) = \dfrac{1}{3}$

42. $\log_9(x - 6) + \log_9 x = \dfrac{3}{2}$

43. $\log_2(3x - 8) - \log_2(x + 4) = -1$

44. $\log_3(x - 9) - \log_3(2x + 3) = -2$

Mixed Practice

Solve.

45. $\log_3(x^2 - 16) = 2$

46. $\log_4 x - \log_4 3 = 2$

47. $\log_2 x - \log_2(x - 4) = 2$

48. $\log_4(x - 2) + \log_4 x = \dfrac{3}{2}$

Use a calculator to solve. Round each answer to four decimal places.

49. $\left(\dfrac{1}{4}\right)^x = 11$

50. $4^{2x+5} = 75$

Applications

Solve. Use a calculator where appropriate.

51. Between the years 2000 and 2050, the population, P, in millions, of the United States is projected to grow according to the model $P = 284(1.007)^t$, where t represents the number of years after the year 2000. To the nearest year, when is the population of the United States expected to be 350 million people? (*Source:* U.S. Bureau of the Census)

52. In the United States, total private health care expenditures, E (in billions of dollars), is projected to grow according to the model $E = 1334(1.073)^t$, where t represents the number of years after the year 2000. To the nearest year, when are the private expenditures expected to reach \$2500 billion? (*Source:* U.S. Centers for Medicare and Medicaid Services)

53. The compound interest formula, $A = P(1 + \frac{r}{n})^{nt}$, is used to calculate the amount, A, in an account t yr after P dollars are invested at interest rate r (in decimal form) compounded n times per year. Suppose a student invests \$1000 in an account at 5% interest, compounded quarterly. Approximately how long will it take for the amount in the account to double?

54. The amount, C, of a radioactive substance remaining after t yr is given by the formula $C = C_0(2)^{-t/h}$, where C_0 is the initial amount of the substance present and h is the half-life of the substance in years. The half-life of radium-226, a radioactive isotope, is 1600 yr. Approximately how long will it take an 18-g sample to decay to 1 g?

55. The pH of a substance is given by the formula $pH = -\log[H^+]$, where $[H^+]$ is the concentration of hydrogen ions. Find the approximate concentration of hydrogen ions in blood, which has a pH of 7.4.

56. The loudness, L, of a sound in decibels (dB) is defined by the equation $L = 10\log\left(\frac{I}{I_0}\right)$, where I is the intensity of the sound in watts per square meter and I_0 equals 10^{-12} W/m². An airplane taking off has a loudness of about 140 dB. What is the intensity of the sound?

57. A magazine publisher determines that the circulation, C, in thousands, of its magazine has grown according to the equation $C = 300e^{0.2x}$, where x represents the number of years after the magazine was launched.

   a. Solve the equation for x.

   b. Approximately how many years after it was launched did the magazine circulation reach 1,000,000?

58. The effective annual interest rate, R, on an investment with an annual interest rate r, expressed as a decimal, compounded continuously, is given by the equation $R = e^r - 1$.

   a. Solve the equation for r.

   b. To the nearest tenth of a percent, calculate the annual interest rate on an investment if the effective annual interest rate is 5.65%.

59. The amount, A, in dollars, in an account with an annual interest rate r, compounded continuously, is given by the formula $A = 12{,}000e^{rt}$, where t is time in years.

   a. What was the initial amount invested in the account? Explain how you know.

   b. What is the annual interest rate, to the nearest tenth of a percent, if the amount in the account after 15 yr will be \$31,814?

60. The number, N, of bacteria present in a culture after h hr is given by the equation $N = 8e^{0.139h}$.

   a. How many bacteria were initially present in the culture? Explain how you know.

   b. After how many hours, to the nearest hour, will the population double?

61. A toy company determines that its revenue, R (in millions of dollars), on the sales of a new line of action figures is given by the equation $R = 1.7 + 2.3 \ln(x + 1)$, where x is the number of months the action figures have been on the market. After how many months, to the nearest month, did the company's revenue reach $7 million?

62. The time t (in hours) it takes for the initial concentration, C_0 (in milligrams per liter), of a certain medication in the bloodstream to decrease to a concentration of C milligrams per liter is modeled by the equation $t = \dfrac{1}{k}(\ln C - \ln C_0)$. Suppose that half of the initial concentration of 4 milligrams per liter remains in the bloodstream 3 hr after it is administered to a patient intravenously. Calculate the value of k to the nearest thousandth and interpret its meaning in the situation.

● *Check your answers on page A-46.*

MINDSTRETCHERS

Technology

1. Consider the exponential equation $2^{3x} = 6$.

    a. Explain how you can use a grapher to solve this equation. Then solve with a grapher, rounding the answer to four decimal places.

    b. Verify your solution algebraically.

Mathematical Reasoning

2. Financial analysts use the *Rule of 72* to estimate how long it takes for an investment generating interest to double in value. The rule states that an investment with annual interest rate r (expressed as a percent) will double in approximately $\dfrac{72}{r}$ yr. According to this rule, an investment paying 8% annual interest, for example, will be worth twice its original value in about 9 yr, since $\dfrac{72}{8}$ equals 9. Consider the continuous interest formula $A = Pe^{rt}$, which gives the current value (in dollars) of an investment after t yr on an initial investment of P dollars with interest rate r (in decimal form). Use this formula to show why the Rule of 72 works for this case.

Groupwork

3. Working with a partner, solve the following inequalities:

    a. $3^x \geq 9$ **b.** $\log_2(x - 1) > 1$

    c. $\log(x^2 + 1) \leq 1$ **d.** $\log_2(x - 3) < 2$

KEY CONCEPTS AND SKILLS

(CONCEPT)   (SKILL)

Concept/Skill	Description	Example
[10.1] Operations on Functions	If f and g are functions, then **Sum:** $(f + g)(x) = f(x) + g(x)$ **Difference:** $(f - g)(x) = f(x) - g(x)$ **Product:** $(f \cdot g)(x) = f(x) \cdot g(x)$ **Quotient:** $\left(\dfrac{f}{g}\right)(x) = \dfrac{f(x)}{g(x)}$, for $g(x) \neq 0$	$f(x) = x + 2$ and $g(x) = x^2 - 4$ $\begin{aligned}(f + g)(x) &= f(x) + g(x)\\ &= (x + 2) + (x^2 - 4)\\ &= x^2 + x - 2\end{aligned}$ $\begin{aligned}(f - g)(x) &= f(x) - g(x)\\ &= (x + 2) - (x^2 - 4)\\ &= -x^2 + x + 6\end{aligned}$ $\begin{aligned}(f \cdot g)(x) &= f(x) \cdot g(x)\\ &= (x + 2)(x^2 - 4)\\ &= x^3 + 2x^2 - 4x - 8\end{aligned}$ $\begin{aligned}\left(\dfrac{f}{g}\right)(x) &= \dfrac{f(x)}{g(x)}\\ &= \dfrac{x + 2}{x^2 - 4}\\ &= \dfrac{1}{x - 2}, x \neq -2, 2\end{aligned}$
[10.2] Composition of Two Functions	The function represented by $f \circ g$, where $(f \circ g)(x) = f(g(x))$	$f(x) = 3x + 5$ and $g(x) = 2x - 1$ $\begin{aligned}(f \circ g)(x) &= f(2x - 1)\\ &= 3(2x - 1) + 5\\ &= 6x - 3 + 5\\ &= 6x + 2\end{aligned}$
[10.1] One-to-One Function	A function f in which each y-value in the range corresponds to exactly one x-value in the domain.	$f(x) = 4x + 5$ $g(x) = \sqrt{x}$
[10.1] Horizontal-Line Test	If every horizontal line intersects the graph of a function at most once, then the function is one-to-one.	One-to-one function:

continued

Concept/Skill	Description	Example
		Not a one-to-one function:
[10.1] Inverse of a One-to-One Function	The set of all ordered pairs (y, x), where f is the set of ordered pairs (x, y); written f^{-1} and read "f inverse."	f is defined by $\{(1, 4), (3, -1), (2, 5)\}$, so f^{-1} is defined by $\{(4, 1), (-1, 3), (5, 2)\}$.
[10.1] To Find the Inverse of a Function	• Substitute y for $f(x)$. • Interchange x and y. • Solve the equation for y. • Substitute $f^{-1}(x)$ for y.	$f(x) = 2x - 1$ $y = 2x - 1$ $x = 2y - 1$ $y = \dfrac{x + 1}{2}$ $f^{-1}(x) = \dfrac{x + 1}{2}$
[10.1] Composition of a Function and Its Inverse	For any functions f and g, the following two conditions are equivalent: • $f(x)$ and $g(x)$ are inverse functions of one another. • $f(g(x)) = x$ and $g(f(x)) = x$.	$f(x) = \dfrac{1}{4}x + 3$ and $g(x) = 4x - 12$ $f(g(x)) = f(4x - 12)$ $\qquad = \dfrac{1}{4}(4x - 12) + 3$ $\qquad = x - 3 + 3$ $\qquad = x$ $g(f(x)) = g\left(\dfrac{1}{4}x + 3\right)$ $\qquad = 4\left(\dfrac{1}{4}x + 3\right) - 12$ $\qquad = x + 12 - 12$ $\qquad = x$
[10.2] Exponential Function	Any function that can be written in the form $\qquad f(x) = b^x$ where x is a real number, $b > 0$, and $b \neq 1$.	$f(x) = 4^x$ $g(x) = \left(\dfrac{1}{6}\right)^x$
[10.2] Natural Exponential Function	The function defined by $f(x) = e^x$.	$f(t) = e^t$

continued

Concept/Skill	Description	Example
[10.2] One-to-One Property of Exponential Functions	For $b > 0$ and $b \neq 1$, if $b^x = b^n$, then $x = n$.	$2^x = 64$ $2^x = 2^6$ $x = 6$
[10.3] Logarithmic Function	The inverse of $f(x) = b^x$ is represented by the logarithmic function $f^{-1}(x) = \log_b x$, where $\log_b x$ is read either "the logarithm of x to the base b" or "the logarithm, base b, of x."	If $f(x) = 3^x$, then $f^{-1}(x) = \log_3 x$ is the logarithmic function associated with f. $g(x) = \log_5 x$ is another logarithmic function.
[10.3] Equivalent Logarithmic and Exponential Equations	If $b > 0$, $b \neq 1$, $x > 0$, and y is any real number, then $$y = \log_b x$$ is equivalent to $b^y = x$.	Logarithmic equation: $y = \log_4 x$ Exponential equation: $4^y = x$
[10.3] Special Properties of Logarithms	• $\log_b b = 1$ because $b^1 = b$, that is, 1 is the exponent to which b must be raised to get b. • $\log_b 1 = 0$ because $b^0 = 1$, that is, 0 is the exponent to which b must be raised to get 1.	$\log_5 5 = 1$ because $5^1 = 5$ $\log_5 1 = 0$ because $5^0 = 1$
[10.4] Product Property of Logarithms	For any positive real numbers M, N, and b, $b \neq 1$, $$\log_b MN = \log_b M + \log_b N$$	$\log_n 6x = \log_n 6 + \log_n x$
[10.4] Quotient Property of Logarithms	For any positive real numbers M, N, and b, $b \neq 1$, $$\log_b \frac{M}{N} = \log_b M - \log_b N$$	$\log_5 \frac{a}{2} = \log_5 a - \log_5 2$
[10.4] Power Property of Logarithms	For any positive numbers M and b, $b \neq 1$, and any real number r, $$\log_b M^r = r \log_b M$$	$\log_3 x^5 = 5 \log_3 x$
[10.4] Special Properties of Logarithms	For $b > 0$ and $b \neq 1$, • $\log_b b^x = x$ • $b^{\log_b x} = x$, where $x > 0$	$\log_n n^3 = 3$ $n^{\log_n 7} = 7$
[10.5] Common Logarithm	A logarithm to the base 10, where **log x** means $\log_{10} x$.	$\log 15 = \log_{10} 15 \approx 1.1761$ $\log \frac{1}{2} = \log_{10} \frac{1}{2} \approx -0.3010$
[10.5] Natural Logarithm	A logarithm to the base e, where **ln x** means $\log_e x$.	$\ln 9 = \log_e 9 \approx 2.1972$ $\ln \frac{2}{3} = \log_e \frac{2}{3} \approx -0.4055$
[10.5] Change-of-Base Formula	For positive real numbers N, b, and c, where $b \neq 1$ and $c \neq 1$, $$\log_b N = \frac{\log_c N}{\log_c b}$$	$\log_6 43 = \frac{\log 43}{\log 6} \approx 2.0992$

continued

Characteristics of the Exponential Function $f(x) = b^x$

- The domain of the function is the set of all real numbers and the range is the set of all positive numbers.
- The vertical- and horizontal-line tests show that $f(x) = b^x$ is a one-to-one function.
- The function is increasing if the base is greater than 1. The function is decreasing if the base is between 0 and 1.
- The y-intercept of the graph of the function is $(0, 1)$, but the graph has no x-intercept. The graph approaches the x-axis, but never intersects it.

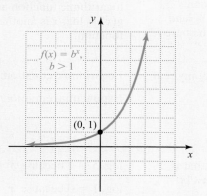
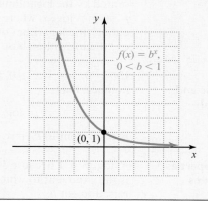

Characteristics of the Logarithmic Function $f(x) = \log_b x$

- The domain of the function is the set of positive real numbers and the range is the set of real numbers.
- The function is increasing if the base is greater than 1. The function is decreasing if the base is between 0 and 1.
- The x-intercept of the graph of the function is $(1, 0)$, but the graph has no y-intercept. The graph approaches the y-axis, but never intersects it.

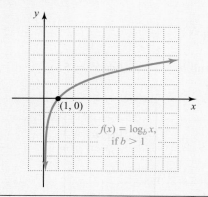
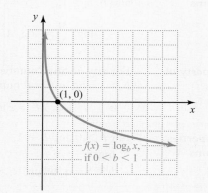

Chapter 10   Review Exercises

[10.1] *Graph each function. Then identify its domain and range.*

1. $g(x) = -|x| + 5$

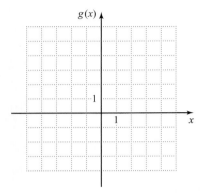

2. $g(x) = |x + 2| - 2$

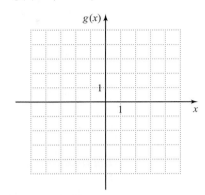

Given f(x) and g(x), find $(f + g)(x)$, $(f - g)(x)$, $(f \cdot g)(x)$, and $\left(\dfrac{f}{g}\right)(x)$.

3. $f(x) = 5 - 6x; g(x) = x - 3$

4. $f(x) = 3x^2 + 1; g(x) = 2x^2 - 4x$

5. $f(x) = \dfrac{2}{x - 3}; g(x) = \dfrac{3}{x^2 - 9}$

6. $f(x) = 7\sqrt{x} + 2; g(x) = \sqrt{x} - 2$

Given $f(x) = x^2 - 6x + 5$ and $g(x) = x^2 - x$, find each of the following.

7. $(f + g)(-2)$

8. $(f - g)(3)$

9. $(f \cdot g)(1)$

10. $\left(\dfrac{f}{g}\right)(-4)$

Given f(x) and g(x), find each composition.

11. $f(x) = x + 2$ and $g(x) = x^2 - 5x + 9$

   a. $(f \circ g)(x)$

   b. $(g \circ f)(x)$

   c. $(f \circ g)(0)$

   d. $(g \circ f)(-2)$

12. $f(x) = \sqrt{x}$ and $g(x) = 2x - 9$

   a. $(f \circ g)(x)$

   b. $(g \circ f)(x)$

   c. $(f \circ g)(9)$

   d. $(g \circ f)(5)$

Determine whether the function whose graph is shown is one-to-one.

13.

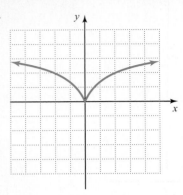

14.

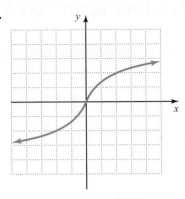

Find the inverse of each of the following functions.

15. $f(x) = 8x + 3$

16. $f(x) = \dfrac{1}{6}x - 1$

17. $f(x) = x^3 - 5$

18. $f(x) = \sqrt[3]{x + 10}$

19. $f(x) = \dfrac{2}{x + 3}$

20. $f(x) = \dfrac{x}{3x + 1}$

Determine whether the two given functions are inverses of each other.

21. $f(x) = \dfrac{2}{3}x + 2$ and $g(x) = \dfrac{3}{2}x - 3$

22. $f(x) = \dfrac{6}{x} - 4$ and $g(x) = \dfrac{6}{x} + 4$

Given the graph of a one-to-one function, sketch the graph of its inverse.

23.

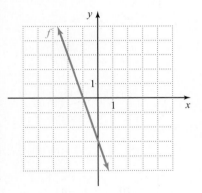

24.

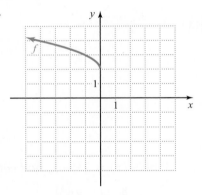

[10.2] *Evaluate each function for the given value. Round each answer to four decimal places, if necessary.*

25. $f(x) = -3^x; f(-3)$

26. $f(x) = \left(\dfrac{1}{4}\right)^{-x}; f(2)$

27. $g(x) = 2^{x+3}; g(-1)$

28. $g(x) = 9^x - 5; g\left(\dfrac{3}{2}\right)$

29. $f(x) = e^{2x}; f\left(\dfrac{1}{2}\right)$

30. $h(x) = e^{-x+1}; h(3)$

Graph each function.

31. $f(x) = 3^{-x}$

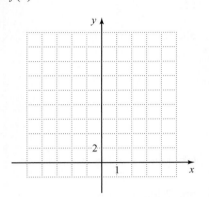

32. $f(x) = 2^x + 1$

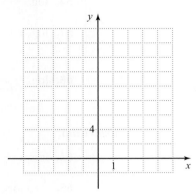

Solve.

33. $4^x = 32$

34. $27^x = \dfrac{1}{9}$

35. $2^{x-3} = 16$

36. $25^{x+2} = 125^{3-x}$

[10.3]   *Write each logarithmic equation as an exponential equation.*

37. $\log_6 216 = 3$

38. $\log_{1/3} 9 = -2$

Write each exponential equation as a logarithmic equation.

39. $5^{-2} = \dfrac{1}{25}$

40. $81^{1/4} = 3$

Evaluate.

41. $\log_8 8$

42. $\log_{1/2} \dfrac{1}{32}$

43. $\log_6 1$

44. $\log_{16} 2$

Solve.

45. $\log_4 x = 3$

46. $\log_2 x = -6$

47. $\log_x 7 = -1$

48. $\log_x \dfrac{4}{9} = 2$

49. $\log_4 64 = x$

50. $\log_{1/2} 4 = x$

Graph each function.

51. $f(x) = -\log_4 x$

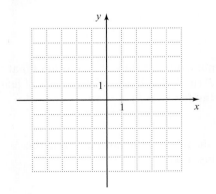

52. $f(x) = \log_2 x$

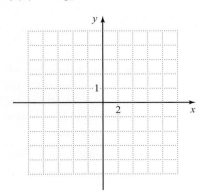

[10.4]   *Use the properties of logarithms to write each expression as the sum or difference of logarithms.*

53. $\log_3 27x$

54. $\log_6 \dfrac{6}{a}$

55. $\log_n \dfrac{x}{x-5}$

56. $\log_9 [u(u+1)]$

57. $\log_5 x^3 y$

58. $\log_2 \sqrt[5]{x^2 y^3}$

59. $\log_3 \dfrac{x^2 z^3}{y^2}$

60. $\log_b \dfrac{b^2}{(a-b)^4}$

Write each expression as a single logarithm.

61. $\log_4 10 + \log_4 n$

62. $\log_9 x - \log_9 5$

63. $\log_3 a + 2 \log_3 b$

64. $5 \log_b a - 4 \log_b c$

65. $\dfrac{1}{4}(3 \log_6 x - 8 \log_6 y)$

66. $2 \log_a x - \log_a y$

Evaluate.

67. $\log_8 8^3$

68. $\log_x x^{-1/3}$

69. $5^{\log_5 y}$

70. $a^{\log_a 14}$

[10.5]   *Use a calculator to approximate each logarithm to four decimal places.*

71. $\log 23$

72. $\log 9.4$

73. $\ln 48$

74. $\ln \dfrac{2}{3}$

Evaluate.

75. $\log 0.1$

76. $\log \sqrt[4]{1000}$

77. $\ln e^{100}$

78. $\ln \dfrac{1}{e^y}$

Use a calculator to approximate each logarithm to four decimal places.

79. $\log_5 121$

80. $\log_9 6.1$

[10.6]   *Solve. Round each answer to four decimal places.*

81. $7^x = 72$

82. $6^{2x} = 0.58$

83. $2^{x-5} = 20$

84. $3^{2x+3} = 63$

Solve.

85. $\log_7(x+8) = 0$

86. $\log_2 12 - \log_2 x = -3$

87. $\log_6 x + \log_6(x-5) = 2$

88. $\log_5(x+2) + \log_5(x+6) = 1$

89. $\log_4(x-3) - \log_4 x = -1$

90. $\log_3(2x-1) - \log_3(x-4) = 2$

Mixed Applications

Solve.

91. A real-estate agent makes a monthly base salary of $2500 plus a commission, c, on his total sales for the month. His total monthly salary can be modeled by the function $f(c) = 2500 + c$. The function $c(s) = 0.008s$ represents his commission on his total sales s.

  a. Find $(f \circ c)(s)$ and interpret its meaning in this situation.

  b. Determine the agent's total monthly salary if his monthly sales totaled $400,000.

92. A clothing manufacturing company determines that its weekly profit, P (in dollars), for selling x pairs of jeans is modeled by the function $P(x) = 10x - 150$.

  a. Find $P^{-1}(x)$.

  b. Explain the meaning of the inverse within the context of this situation.

  c. How many pairs of jeans must the company sell in order to make a weekly profit of $1200?

93. The compound interest formula, $A = P\left(1 + \dfrac{r}{n}\right)^{nt}$, is used to compute amount A (in dollars) in an account after P dollars are invested at an annual interest rate r, compounded n times per year for t yr.

    a. If an investor puts \$6500 in an account earning 5.4% interest, compounded twice a year, how much will be in the account after 5 yr?

    b. To the nearest year, how long will it take for this investment to double?

94. Cesium-137 is a radioactive isotope that is used in radiation therapy. The amount, A, of the isotope remaining after t yr is given by the equation $A = A_0 e^{-0.0231t}$, where A_0 is the initial amount.

    a. To the nearest tenth of a milligram, find the amount of a 36-mg sample of cesium-137 remaining after 60 yr.

    b. From your answer in part (a), determine the half-life of cesium-137.

95. A bacteriologist determines that the population of a certain bacterium in a colony doubles every 40 min. The number of bacteria, N, present after t hr is given by the equation $t = \dfrac{2}{3} \log_2 \dfrac{N}{N_0}$, where N_0 is the number of bacteria present in the initial population. If the initial population consists of 6 bacteria, in how many hours will there be 384 bacteria in the colony?

96. A stamp collector determines that the value of a particular stamp quadruples every 15 yr. The value of the stamp, v (in dollars), in t yr is given by the equation $t = 15 \log_4 \dfrac{v}{v_0}$, where v_0 is the initial value of the stamp. If the initial value of the stamp is \$3, to the nearest tenth of a year, when will it have a value of \$24?

97. Students in a physics class were given a final exam at the end of the course. As part of a study on memory, the students were given an equivalent exam each month for 1 yr. The average exam score for the class on a test given x months after the course ended is modeled by the equation $s = \log \dfrac{10^{90}}{(x + 1)^{16}}$. Use the properties of logarithms to show that the equation can be written as $s = 90 - 16 \log(x + 1)$.

98. To determine the approximate time of death of a person, a medical examiner uses the equation $\ln(T - A) = \ln(98.6 - A) - kt \ln e$, where T is the temperature t hr after death, A is the temperature of the surrounding environment, and k is a constant. Show that this equation can be written as $T - A = (98.6 - A)e^{-kt}$.

99. In chemistry, the pH of a substance can be determined using the formula $\text{pH} = -\log[\text{H}^+]$, where $[\text{H}^+]$ represents the hydrogen ion concentration. To the nearest tenth, determine the pH of milk, which has a hydrogen ion concentration of 2.5×10^{-7}.

100. The time t in years it takes for a continuously-compounded investment of P dollars at an annual interest rate r (in decimal form) to grow to x times the original amount is modeled by the equation $t = \dfrac{\ln x}{r}$. How long, to the nearest year, will it take an investment earning 4.4% interest to triple?

101. A car purchased new for A dollars depreciates according to the equation $V = A(0.85)^t$, where V is the value of the car t yr after it is purchased. To the nearest year, how many years after a car is purchased new for \$28,500 will its value depreciate to \$15,000?

102. The amount, A (in dollars), in an account at time t (in years), with an annual interest rate r, compounded continuously, is given by the formula $A = Pe^{rt}$, where P is the initial amount invested. If \$8000 is invested in an account at 4.8% interest compounded continuously, then to the nearest year, how long will it take the amount in the account to grow to \$12,300?

● *Check your answers on page A-46.*

Chapter 10   **POSTTEST**

To see if you have mastered the topics in this chapter, take this test. Use a calculator where necessary.

1. Consider the functions $f(x) = 3x^2 - 4x - 4$ and $g(x) = 3x^2 + 2x$. Find:

   a. $(f + g)(x)$

   b. $(f - g)(x)$

   c. $(f \cdot g)(x)$

   d. $\left(\dfrac{f}{g}\right)(x)$

2. Given $f(x) = \sqrt{x}$ and $g(x) = 2x + 5$, find:

   a. $(f \circ g)(x)$

   b. $(g \circ f)(x)$

   c. $(f \circ g)(2)$

   d. $(g \circ f)(16)$

3. Determine whether the following function whose graph is shown is one-to-one. If it is, sketch the graph of its inverse.

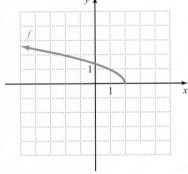

4. Find the inverse of the one-to-one function $f(x) = \dfrac{1}{x - 7}$.

5. Evaluate each function for the given value. Round to four decimal places.

   a. $f(x) = 3^x - 2; f(-2)$ **b.** $f(x) = e^{x+2}; f(5)$

6. Evaluate:

   a. $\log_9 1$ **b.** $\log_4 \dfrac{1}{2}$

7. Graph:

   a. $f(x) = -\left(\dfrac{1}{2}\right)^{-x}$

   b. $f(x) = -\log_3 x$

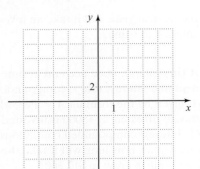

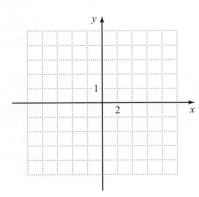

8. Write each expression as a sum or difference of logarithms.

   a. $\log_3(9x^4)$

   b. $\log_6 \dfrac{x^5}{6y^2}$

10. Evaluate:

   a. $\log_9 9^7$ **b.** $6^{\log_6 x}$

12. Evaluate, rounding to four decimal places: $\log_5 42$

Solve.

13. $7^x = 49$

14. $27^{2x-3} = \left(\dfrac{1}{3}\right)^{x-5}$

16. $\log_5(2x + 3) = 2$

18. The population, P (in millions), of India is modeled by the function $P(t) = 1020(1.014)^t$, where t represents the number of years after 2001. Calculate the expected population of India in the year 2009. (*Source:* U.S. Bureau of the Census, International Database)

20. If the initial concentration of a certain drug in a patient's bloodstream is 20 mg/L, then the concentration C after t hr is given by the equation $C = 20e^{-0.125t}$. To the nearest hour, approximate the time it takes for the concentration to decrease to 5 mg/L.

9. Write each expression as a single logarithm.

   a. $2 \log_5 n + 5 \log_5 2$

   b. $3 \log_b(a - 2) - 4 \log_b(a + 2)$

11. Evaluate:

   a. $\log \sqrt{1000}$ **b.** $\ln \dfrac{1}{e^5}$

15. $\log_x 4 = -2$

17. $\log_2(x + 3) + \log_2(x - 4) = 3$

19. The loudness, L (in decibels), of a sound can be determined by the equation $L = 10(\log I + 12)$, where I is the intensity of the sound (in watts per square meter). Normal conversation has a loudness of 60 dB. What is the intensity of the sound?

● *Check your answers on page A-47.*

Cumulative Review Exercises

To help you review, solve the following problems.

1. Solve: $3(2x - 7) - 4x = -\dfrac{1}{3}(9x + 12) - 8$

2. Solve: $|5x - 1| > 9$

3. Find the point-slope form and the slope-intercept form of the equation of the line that contains the points $(4, 1)$ and $(-2, 7)$.

4. Factor: $81x^4y^4 - 16$

5. Solve: $2x^2 + 4x - 5 = 0$

6. Multiply: $(4 - \sqrt{-16})(7 - \sqrt{-1})$

7. Solve: $\log_4(x + 3) - \log_4 x = z$

8. Meteorologists use the formula
$$W = 35.74 + 0.6215T - 35.75V^{0.16} + 0.4275TV^{0.16},$$
where T is the air temperature (in degrees Fahrenheit) and V is the speed of the wind (in miles per hour), to calculate the windchill temperature W (in degrees Fahrenheit). What is the windchill temperature, to the nearest degree Fahrenheit, if the air temperature is 40°F and the speed of the wind is 18 mph?

9. An investor puts a part of his $15,000 in a low-risk fund and the rest in a high-risk fund. After 1 yr, the low-risk fund gained 6% and the high-risk fund lost 2%. If the total return on the investments was $100, how much was invested in each fund?

10. The height of a model rocket t sec after it is launched upward from a platform 6 ft above the ground with an initial velocity of 96 ft per sec is modeled by the function $s(t) = -16t^2 + 96t + 6$. Find the time it takes for the rocket to reach its maximum height.

● *Check your answers on page A-47.*

CHAPTER 11

Conic Sections and Nonlinear Systems

Medicine and Conic Sections

The ellipse, a kind of conic section, has a surprising property that underlies many applications: Any light, sound, or energy that emanates from one of the foci and reflects off the mirrored surface of the ellipse will be reflected to the other focus. This property is used in the medical procedure known as lithotripsy, in which a kidney stone is pulverized by placing the patient in an elliptical tank of water, with the kidney stone at one focus. At the other focus, high-energy sound waves, which are concentrated on the kidney stone, are generated. (*Sources:* http://www.mwstone.com/STONES/index.htm; Davi-Ellen Chabner, *The Language of Medicine*, Philadelphia: WB Saunders, 1996)

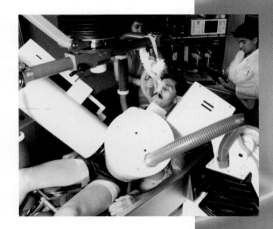

Chapter 11   PRETEST

To see if you have already mastered the topics in this chapter, take this test.

1. Write the equation of the parabola $y = 3x^2 - 12x + 23$ in standard form. Then, identify the vertex.

2. Find the distance between each pair of points:

   a. $(-5, 9)$ and $(-7, 13)$

   b. $(1.2, -1)$ and $(0.8, -0.7)$

3. Find the midpoint of the line segment joining each pair of points:

   a. $(11, -6)$ and $(-3, 10)$

   b. $(1, -4)$ and $(0, 1)$

4. Find the equation of a circle with center $(7, -3)$ and radius $\sqrt{10}$.

5. Find the center and radius of the circle $x^2 + y^2 - 2x - 5 = 0$.

Graph.

6. $x = -2(y + 1)^2 + 3$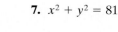

7. $x^2 + y^2 = 81$

8. $(x + 2)^2 + (y - 1)^2 = 16$

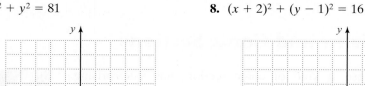

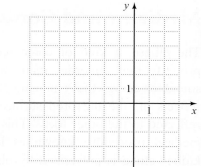

9. $\dfrac{x^2}{9} + y^2 = 1$

10. $12x^2 + 3y^2 = 48$

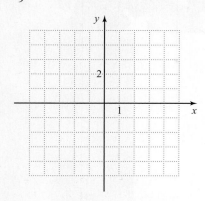

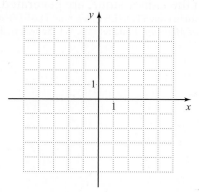

11. $\dfrac{y^2}{25} - \dfrac{x^2}{9} = 1$

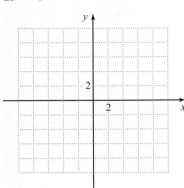

12. $x^2 - 4y^2 = 16$

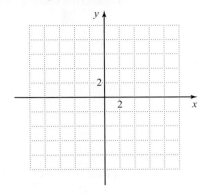

Solve each nonlinear system.

13. $x^2 + y^2 = 9$
$\quad\quad x = y - 3$

14. $y = x^2 - 2$
$\quad\quad y = -x^2 - 6$

15. $4x^2 + 2y^2 = 26$
$\quad\quad 5x^2 - 2y^2 = 28$

Solve.

16. $x^2 - 9y^2 < 36$

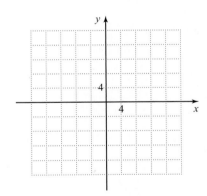

17. $x^2 + (y + 3)^2 \le 4$
$\quad\quad 8x^2 + 32y^2 \ge 128$

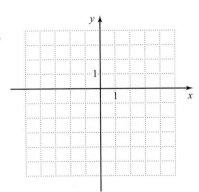

18. The cross section of a television satellite dish is parabolic. Find an equation that represents the parabola formed by the cross section of the satellite dish shown in the diagram at the right.

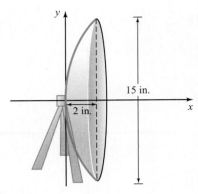

15 in.

2 in.

19. The comet Encke travels in an elliptical orbit with the Sun at one focus. An equation that approximates the comet's orbit is $\dfrac{x^2}{354^2} + \dfrac{y^2}{181^2} = 1$, where the units are in millions of kilometers. The maximum distance between the comet and the Sun is 658 million km. What is the minimum distance between the comet and the Sun, to the nearest 10 million km? (*Source:* Patrick Moore, *Guide to Comets*, Lutterworth Press, 1977)

20. The security officers on a campus communicate using walkie-talkies. The walkie-talkie has a maximum range of 1000 ft.

a. Write an inequality to represent the situation.

b. Graph the inequality.

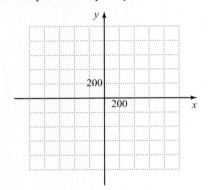

• *Check your answers on page A-48.*

11.1   Introduction to Conics; The Parabola

What Conics Are and Why They Are Important

OBJECTIVES

- To graph a parabola
- To write the equation of a parabola in standard form
- To solve applied problems involving parabolas

The *conic sections* are among the earliest mathematical objects to be systematically studied. Originally viewed in geometric terms, each **conic section** was defined as the intersection of a plane and a (double) cone. Changing the angle at which the plane slices the cone results in a different curve. Depending on the angle, the curve could be a *parabola*, a *circle*, an *ellipse,* or a *hyperbola.*

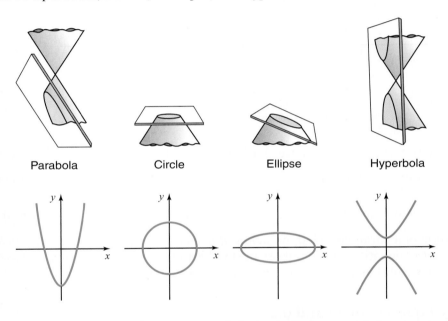

Parabola   Circle   Ellipse   Hyperbola

Conic sections can also be represented algebraically by second-degree equations in two variables. For instance, the following equations represent particular conic sections:

$$y = x^2 \qquad x^2 + y^2 = 1 \qquad \frac{x^2}{4} + \frac{y^2}{9} = 1 \qquad \frac{x^2}{9} - \frac{y^2}{16} = 1$$

     Parabola   Circle   Ellipse   Hyperbola

Much of this chapter is devoted to investigating equations of conic sections and finding their graphs on a coordinate plane.

The first few sections of the chapter focus on particular conic sections: the parabola, the circle, the ellipse, and the hyperbola, in that order. In the rest of the chapter, we consider how to solve *nonlinear* systems of equations—that is, systems containing equations that are not linear—and how to solve nonlinear inequalities and nonlinear systems of inequalities.

The range of applications of conic sections is wide, from the path of moving objects to medical procedures, from the shape of objects in nature to that of the most modern architectural structures.

The Parabola

The first of the conic sections that we will discuss is one that we have considered previously, namely, the *parabola*. Let's review the earlier discussion before extending it.

Recall from Section 9.4 that the graph of a quadratic equation such as $y = x^2 + 3x - 1$ is a U-shaped curve called a parabola. Every parabola has a *vertex* and an *axis of symmetry,* features of the curve that are useful in graphing the corresponding equation. In general, the graph of $y = ax^2 + bx + c$ opens upward if a is positive and downward if a is negative. In either case, the axis of symmetry is parallel to the y-axis.

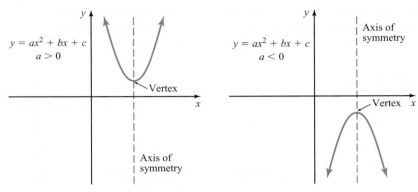

In our earlier discussion, we showed that for the equation $y = ax^2 + bx + c$, the x-coordinate of the graph's vertex is $-\dfrac{b}{2a}$, and the axis of symmetry is the vertical line with equation $x = -\dfrac{b}{2a}$. Now, let's consider an alternative form of the equation of a parabola called the *standard form* of the equation. The advantage of writing the equation of a parabola in standard form is that its vertex and axis of symmetry can then be read directly from the equation.

> ### The Equation of a Parabola
>
> The equation of a parabola that opens upward or downward is in **standard form** if it is written as $y = a(x - h)^2 + k$, where (h, k) is the vertex of the associated parabola and the equation of the axis of symmetry is $x = h$.

EXAMPLE 1

Graph: $y = -2(x + 3)^2 + 8$

Solution The given equation can be written in standard form as $y = -2[x - (-3)]^2 + 8$, where $a = -2$, $h = -3$, and $k = 8$. So the vertex is the point $(-3, 8)$, and the axis of symmetry is the line with equation $x = -3$.

To graph, let's find the intercepts. To identify any y-intercepts, we substitute 0 for x:

$$y = -2(x + 3)^2 + 8$$
$$y = -2(0 + 3)^2 + 8$$
$$y = -10$$

So the y-intercept is the point $(0, -10)$.

PRACTICE 1

Graph: $y = 5(x + 1)^2 - 3$

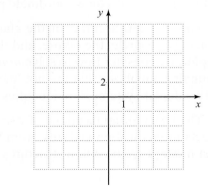

To find any x-intercepts, we substitute 0 for y:

$$y = -2(x + 3)^2 + 8$$
$$0 = -2(x + 3)^2 + 8$$
$$-8 = -2(x + 3)^2$$
$$4 = (x + 3)^2$$
$$\pm 2 = x + 3$$
$$2 = x + 3 \quad \text{or} \quad -2 = x + 3$$
$$-1 = x \qquad\qquad -5 = x$$
$$x = -1 \qquad\qquad x = -5$$

So the x-intercepts are $(-1, 0)$ and $(-5, 0)$.

We know that the graph of our equation is a parabola that opens downward, since a is negative. To sketch the graph, first plot the vertex and the x- and y-intercepts. Then find a few additional points on each side of the axis of symmetry and plot them.

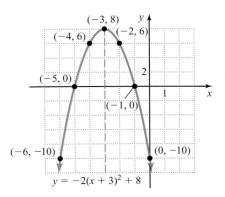

When the equation of a parabola is given in the form $y = ax^2 + bx + c$, we can rewrite the equation in standard form by completing the square. Consider, for instance, the equation $y = 4x^2 - 24x - 13$.

To write this equation in standard form, we first group the x-terms and factor out the coefficient of x^2.

$$y = (4x^2 - 24x) - 13$$
$$y = 4(x^2 - 6x) - 13$$

Next, we complete the square for $x^2 - 6x$ by taking one-half of the coefficient of the x-term and squaring it. That is, $\left[\dfrac{1}{2}(-6)\right]^2 = (-3)^2 = 9$. Now, we add this value to $x^2 - 6x$ in the parentheses, which gives us $x^2 - 6x + 9$. Since each term of the trinomial is multiplied by 4, when we add 9 within the parentheses we are actually *adding* 4(9), or 36. In order to keep the value of the right side of the equation unchanged, we must also *subtract* the same amount, 36, outside the parentheses since the terms are on the same side of the equal sign, as shown below.

Add 4(9) = 36.

$$y = 4(x^2 - 6x + 9) - 13 - 36$$

Subtract 4(9) = 36.

Finally, we write the perfect square trinomial $x^2 - 6x + 9$ as the square of a binomial and simplify, getting

$$y = 4(x - 3)^2 - 49$$

So the equation $y = 4x^2 - 24x - 13$ written in standard form is $y = 4(x - 3)^2 - 49$.

EXAMPLE 2

Consider the equation $y = 3x^2 - 6x + 8$, whose graph is a parabola.

a. Write this equation in standard form.

b. Identify the vertex and the axis of symmetry of the parabola.

c. Graph this parabola.

Solution

a. To write an equation of the form $y = ax^2 + bx + c$ in standard form, we complete the square as follows:

$$y = 3x^2 - 6x + 8$$
$$y = (3x^2 - 6x) + 8 \qquad \text{Group the } x\text{-terms.}$$
$$y = 3(x^2 - 2x) + 8 \qquad \text{Factor out the coefficient of } x^2.$$

Add $3(1) = 3$.

$$y = 3(x^2 - 2x + 1) + 8 - 3 \qquad \text{Complete the square of the polynomial in parentheses, and subtract } 3(1), \text{ or } 3,$$

Subtract $3(1) = 3$. to keep the value of the right side of the equation unchanged.

$$y = 3(x - 1)^2 + 5 \qquad \text{Write } x^2 - 2x + 1 \text{ as the square of a binomial and simplify.}$$

So the equation $y = 3x^2 - 6x + 8$ written in standard form is $y = 3(x - 1)^2 + 5$.

b. From the standard form of the equation, we see that $h = 1$ and $k = 5$. So the vertex of the parabola is the point $(1, 5)$ and the axis of symmetry is $x = 1$.

c. To graph the parabola, let's find the intercepts. To find any y-intercepts, we substitute 0 for x:

$$y = 3x^2 - 6x + 8$$
$$y = 3(0)^2 - 6(0) + 8$$
$$y = 8$$

So the y-intercept is $(0, 8)$.

To find any x-intercepts, we substitute 0 for y:

$$y = 3x^2 - 6x + 8$$
$$0 = 3x^2 - 6x + 8$$

Recall from Section 9.2 that we can use the discriminant of a quadratic equation to determine if the solutions are real numbers or complex numbers (containing i). Evaluating the discriminant $b^2 - 4ac$ of the equation $0 = 3x^2 - 6x + 8$ yields:

$$(-6)^2 - 4(3)(8) = -60$$

Since the discriminant is negative, the equation has no real solution. So the parabola has no x-intercept.

PRACTICE 2

The graph of the equation $y = 4x^2 - 8x + 7$ is a parabola.

a. Write this equation in standard form.

b. Identify the vertex and the axis of symmetry of the parabola.

c. Graph this parabola.

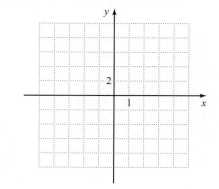

We know that the graph of our equation is a parabola that opens upward, since a is positive. To sketch the graph, let's first plot the vertex and the y-intercept. Then we plot a few additional points on each side of the axis of symmetry.

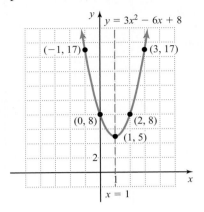

So far in our discussion of parabolas we have considered only parabolas that open either upward or downward. However, some parabolas open to the left or to the right. Equations of the form $x = ay^2 + by + c$ have as their graphs parabolas that open to the right if $a > 0$ and to the left if $a < 0$. In either case, the axis of symmetry is parallel to the x-axis.

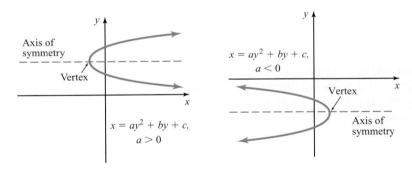

Just as we can write in standard form the equation of a parabola that opens upward or downward, we can also write in standard form the equation of a parabola that opens either to the left or to the right.

The Equation of a Parabola

The equation of a parabola that opens to the left or to the right is in **standard form** if it is written $x = a(y - k)^2 + h$, where (h, k) is the vertex of the associated parabola and the equation of the axis of symmetry is $y = k$.

EXAMPLE 3

Graph: $x = 2(y - 2)^2 - 2$

Solution The given equation is of the form $x = a(y - k)^2 + h$, where $h = -2$ and $k = 2$. So the vertex is the point $(-2, 2)$, and the axis of symmetry is the horizontal line with equation $y = 2$.

To graph, let's find the intercepts. To identify any y-intercepts, we substitute 0 for x:

$$x = 2(y - 2)^2 - 2$$
$$0 = 2(y - 2)^2 - 2$$
$$2 = 2(y - 2)^2$$
$$1 = (y - 2)^2$$
$$\pm 1 = y - 2$$
$$1 = y - 2 \quad \text{or} \quad -1 = y - 2$$
$$y = 3 \qquad\qquad y = 1$$

So there are two y-intercepts, $(0, 3)$ and $(0, 1)$.

To find any x-intercepts, we substitute 0 for y:

$$x = 2(y - 2)^2 - 2$$
$$x = 2(0 - 2)^2 - 2$$
$$x = 2(-2)^2 - 2$$
$$x = 6$$

So there is only one x-intercept, $(6, 0)$.

Since $a = 2$ is positive, the graph opens to the right. We plot the vertex, the x- and y-intercepts, and a few additional points on each side of the axis of symmetry, $y = 2$. Then we sketch the smooth curve passing through them, obtaining the graph of $x = 2(y - 2)^2 - 2$.

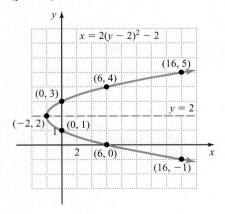

Note that the parabola in Example 3 is not the graph of a function. Can you explain why?

PRACTICE 3

Graph: $x = -(y - 3)^2 + 4$

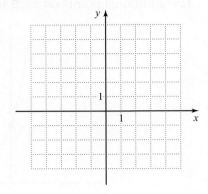

EXAMPLE 4

Write the equation of the parabola $x = -4y^2 + 16y - 3$ in standard form. Then identify its vertex.

Solution To write the given equation in standard form, we complete the square.

$x = -4y^2 + 16y - 3$

$x = (-4y^2 + 16y) - 3$ Group the y-terms.

$x = -4(y^2 - 4y) - 3$ Factor out the coefficient of y^2.

Add $-4(4) = -16$.

$x = -4(y^2 - 4y + 4) - 3 - (-16)$ Complete the square of the polynomial in parentheses and subtract $-4(4)$, or -16, to keep the value of the right side of the equation unchanged.

Subtract $-4(4) = -16$.

$x = -4(y - 2)^2 + 13$ Write $y^2 - 4y + 4$ as the square of a binomial and simplify.

So the equation of the parabola $x = -4y^2 + 16y - 3$ in standard form is $x = -4(y - 2)^2 + 13$. The vertex of the graph of this equation is the point $(13, 2)$.

EXAMPLE 5

The Gateway Arch, built in 1965, has become the symbol of the city of St. Louis. Approximately parabolic in shape, the arch is 630 ft high and 630 ft wide at the base. (See diagram.)

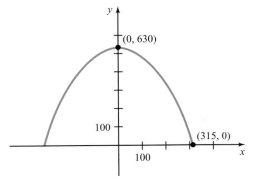

a. Find the equation of the arch.

b. Find the height of the arch 105 ft from either base.

PRACTICE 4

Express in standard form the equation of the parabola $x = -5y^2 - 20y - 9$. Find its vertex.

PRACTICE 5

A rocket is launched straight upward from the ground with an initial velocity of 64 feet per second. Two seconds after it is launched, the rocket reaches its maximum height of 64 ft, as shown.

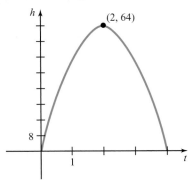

a. Find an equation in standard form for the height, h, of the rocket t sec after it is launched.

b. Find the height of the rocket 1 sec after it is launched.

Solution

a. Since the arch is approximately parabolic, we can write its equation in standard form as $y = a(x - h)^2 + k$. We know that the vertex of the parabola is the point $(0, 630)$, so $h = 0$ and $k = 630$. Therefore, the equation is $y = a(x - 0)^2 + 630$, or $y = ax^2 + 630$. Since the width of the parabola at its base is 630 ft, the point $(315, 0)$ lies on the graph. We can substitute the coordinates of the point $(315, 0)$ into the equation and solve for a.

$$y = ax^2 + 630$$
$$0 = a(315)^2 + 630$$
$$0 = 99,225a + 630$$
$$-630 = 99,225a$$
$$\frac{-630}{99,225} = a$$
$$a = -\frac{2}{315}$$

So the equation of the arch is $y = -\dfrac{2}{315}x^2 + 630$.

b. We need to find the height of the arch 105 ft from either base, in particular, above the point on the x-axis with x-coordinate $315 - 105$, or 210.

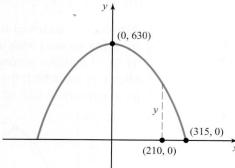

To find y, we substitute 210 for x:

$$y = -\frac{2}{315}x^2 + 630$$

$$y = -\frac{2}{315}(210)^2 + 630$$

$$y = -280 + 630$$

$$y = 350$$

So the height of the arch 105 ft from either base is 350 ft.

Mathematically Speaking

Fill in each blank with the most appropriate term or phrase from the given list.

solving for *x*	function	upward or downward
to the left or to the right	completing the square	equation

1. The equation of a parabola that opens _____ is in standard form if it is written as $y = a(x - h)^2 + k$, where (h, k) is the parabola's vertex and $x = h$ is the equation of the axis of symmetry.

2. When the equation of a parabola is given in the form $y = ax^2 + bx + c$, it can be written in standard form by _____.

3. The equation of a parabola that opens _____ is in standard form if it is written as $x = a(y - k)^2 + h$, where (h, k) is the parabola's vertex and $y = k$ is the equation of the axis of symmetry.

4. A parabola that opens to the left or to the right is not the graph of a(n) _____.

Identify the vertex and axis of symmetry of each parabola. Then graph.

5. $y = (x - 3)^2 - 4$

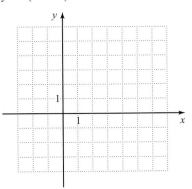

6. $y = (x - 1)^2 + 1$

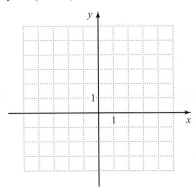

7. $y = -\dfrac{1}{2}(x + 2)^2 - 5$

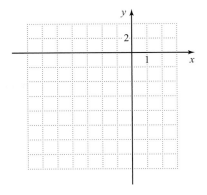

8. $y = -\dfrac{1}{3}(x + 3)^2 - 1$

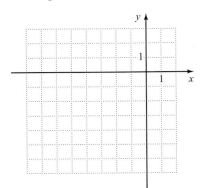

9. $x = 4(y - 1)^2$

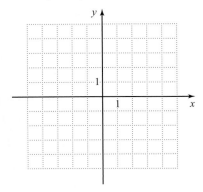

10. $x = 2y^2 - 8$

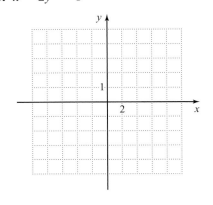

11. $x = -y^2 + 2y + 2$

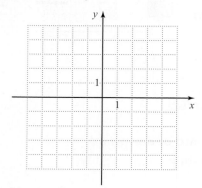

12. $x = -y^2 - 4y + 1$

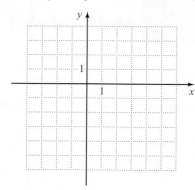

13. $y = -2x^2 + 12x - 10$

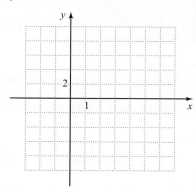

14. $y = 3x^2 - 6x - 9$

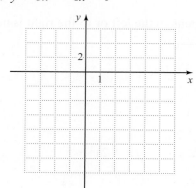

15. $x = 3y^2 + 12y + 10$

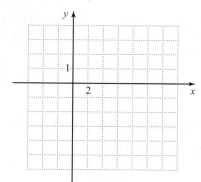

16. $x = -4y^2 + 8y - 1$

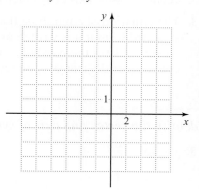

Write each equation in standard form. Then identify the vertex of the parabola.

17. $y = x^2 - x + 1$

18. $y = x^2 + 3x - 5$

19. $y = 5x^2 - 50x + 57$

20. $y = 4x^2 + 56x + 207$

21. $x = -2y^2 - 32y - 95$

22. $x = -3y^2 + 36y - 136$

23. $x = 3y^2 + 9y + 11$

24. $x = 6y^2 - 6y + 5$

Find the equation in standard form of the parabola whose graph is shown.

25.

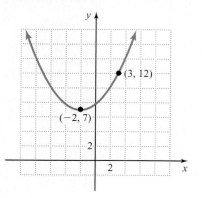

26.

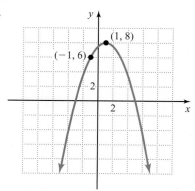

27.

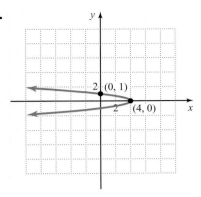

28.

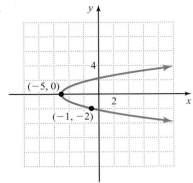

Mixed Practice

Write each equation in standard form. Then identify the vertex of the parabola.

29. $y = x^2 + 5x + 2$

30. $x = -4y^2 + 24y - 40$

31. Find the equation in standard form of the parabola whose graph is shown.

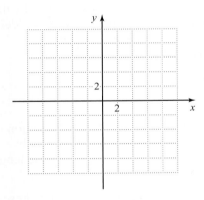

Identify the vertex and axis of symmetry of each parabola. Then graph.

32. $y = \dfrac{1}{2}(x + 3)^2 + 2$

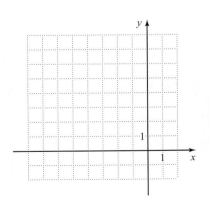

33. $x = -2y^2 - 4y + 1$

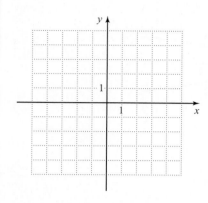

34. $y = -2x^2 + 4x - 6$

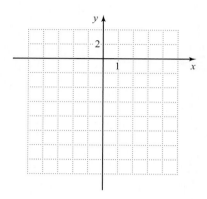

Applications

Solve.

35. The owner of a cosmetics store determines that her weekly revenue R (in dollars) for selling x bottles of perfume is given by the equation of the parabola $R = -(x - 40)^2 + 1600$. Identify the vertex of the parabola and explain its significance in this situation.

36. The height h (in feet) of an object relative to the point of release t sec after it is thrown straight upward with an initial velocity of 32 feet per second is given by the equation of the parabola $h = -16(t - 1)^2 + 16$. Identify the vertex of the parabola and explain its significance in this situation.

37. A farmer has 500 ft of fencing to enclose a rectangular grazing field.

   a. Write an equation in standard form that represents the area, A, of the enclosed field in terms of the width, w, of the field.

   b. What dimensions will produce a field with a maximum area? What is the maximum area?

38. A homeowner decides to change the configuration of her rectangular garden by increasing the width by twice the number of feet that she decreases the length (see figure).

   a. Write an equation in standard form that represents the area, A, of the new garden in terms of x.

   b. What dimensions will maximize the area of the garden? What is the maximum area?

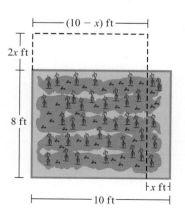

39. The cross section of a radio telescope is parabolic. Find an equation that represents the parabola formed by the cross section of the radio telescope shown in the diagram.

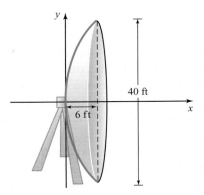

40 ft

6 ft

40. The cables suspended between two towers of a suspension bridge form a curve that can be approximated by a parabola. Suppose that plans for a new suspension bridge require the distance between two towers that rise 116 m above the road surface to be 1120 m. If the center of the cable is 4 m above the road surface, find an equation in standard form that represents the parabola formed by the cable as shown in the diagram.

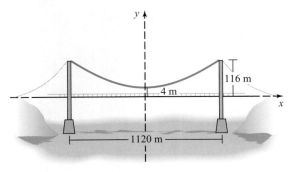

116 m

4 m

1120 m

• *Check your answers on page A-48.*

MINDSTRETCHERS

Research

1. Either in your college library or on the Web, investigate the contribution to the study of conic sections by the Greek mathematician Apollonius of Perga, who lived more than 2000 years ago.

Writing

2. In addition to the four conic sections shown at the beginning of the section, there are three "degenerate" conic sections—a point, two intersecting straight lines, and a single straight line. Explain how a plane could intersect a double cone to produce each of these.

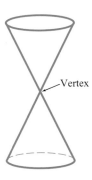

Vertex

Critical Thinking

3. Consider a fixed line *l* and a fixed point *F* not on the line. The set of all points on a coordinate plane that are the same distance from *F* and from *l* form a parabola.

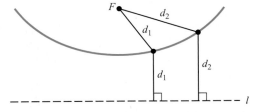

F

d_2

d_1

d_2

d_1

l

Determine the coordinates of the vertex of a parabola formed by the set of all points that are the same distance from the line $y = 4$ and the fixed point $(3, 10)$.

CULTURAL NOTE

Thousands of years ago, the ancient Chinese, Greeks, Incas, Romans—and possibly others—discovered that parabolic mirrors could concentrate the Sun's rays onto a flammable object, causing the object to ignite in seconds. In the ancient Chinese kitchen, these "burning mirrors" were as common as pots. According to legend, the mathematician Archimedes used a parabolic mirror to destroy the Roman fleet invading Syracuse. Today, parabolic mirrors in solar towers generate steam, which in turn drives turbines.

(*Sources: Greek Mathematical Works*, Loeb Classical Library. Cambridge, MA: Harvard University Press, 1941; J. Perlin, "Burning Mirrors: Snagging Pure Fire from the Rays of the Sun," *Whole Earth*, Winter 1999; G. J. Toomer, *Diocles on Burning Mirrors*. New York: Springer Verlag, 1976)

11.2   The Circle

In this section, we discuss the most familiar of all conic sections, the *circle*. But first, we consider two formulas about pairs of points on the coordinate plane—the *distance formula* and the *midpoint formula*.

The Distance Formula and the Midpoint Formula

Recall from Section 8.3 that for any two points on a coordinate plane, (x_1, y_1) and (x_2, y_2), we can determine the distance d between them by finding the length of the line segment connecting them. To do this, we draw a vertical and a horizontal line through the points to form a right triangle, as shown in the diagram below. Then we apply the Pythagorean theorem to find the length of the hypotenuse. Note that the lengths of the two legs of the right triangle are $x_2 - x_1$ and $y_2 - y_1$:

$$d^2 = (x_2 - x_1)^2 + (y_2 - y_1)^2$$
$$d = \sqrt{(x_2 - x_1)^2 + (y_2 - y_1)^2}$$

which is the *distance formula*.

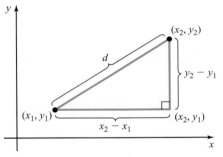

OBJECTIVES

- To use the distance formula and the midpoint formula
- To graph a circle
- To find the equation of a circle
- To write the equation of a circle in standard form
- To solve applied problems involving circles

Distance Formula

The distance between any two points (x_1, y_1) and (x_2, y_2) on a coordinate plane is given by the formula $d = \sqrt{(x_2 - x_1)^2 + (y_2 - y_1)^2}$.

EXAMPLE 1

Find the distance between the points $(3, 0)$ and $(-1, 4)$, rounded to the nearest tenth.

Solution Let $(3, 0)$ stand for (x_1, y_1) and $(-1, 4)$ stand for (x_2, y_2). Substituting into the distance formula, we get

$$d = \sqrt{(x_2 - x_1)^2 + (y_2 - y_1)^2}$$
$$= \sqrt{(-1 - 3)^2 + (4 - 0)^2}$$
$$= \sqrt{(-4)^2 + (4)^2}$$
$$= \sqrt{16 + 16}$$
$$= \sqrt{32}$$
$$= 4\sqrt{2}$$
$$\approx 5.7$$

So the distance between the two points is approximately 5.7 units.

PRACTICE 1

Find the distance between the points $(1, -2)$ and $(0, -5)$, rounded to the nearest tenth.

Our second formula concerns the *midpoint* of the line segment, which connects two given points called the *endpoints* of the segment. By midpoint, we mean the point on the segment located halfway between the two endpoints. The following formula gives the coordinates of the midpoint in terms of the coordinates of the endpoints.

Midpoint Formula

The line segment with endpoints (x_1, y_1) and (x_2, y_2) has as its midpoint

$$\left(\frac{x_1 + x_2}{2}, \frac{y_1 + y_2}{2} \right)$$

The coordinates of the midpoint are the average of the x-coordinates and the average of the y-coordinates of the endpoints.

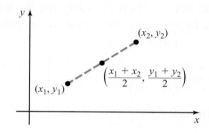

Can you show that the point $\left(\dfrac{x_1 + x_2}{2}, \dfrac{y_1 + y_2}{2} \right)$ lies on this segment?

EXAMPLE 2	PRACTICE 2
Find the midpoint of the line segment joining the points $(-3, 2)$ and $(1, 5)$.	Find the midpoint of the line segment joining the points $(3, 1)$ and $(7, 4)$.

Solution Let $(-3, 2)$ stand for (x_1, y_1) and $(1, 5)$ stand for (x_2, y_2). Substituting into the midpoint formula, we get

$$\frac{x_1 + x_2}{2} = \frac{-3 + 1}{2} = \frac{-2}{2} = -1$$

$$\frac{y_1 + y_2}{2} = \frac{2 + 5}{2} = \frac{7}{2}$$

So the midpoint is $\left(-1, \dfrac{7}{2} \right)$.

The Circle

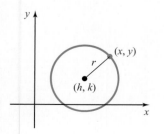

A **circle** is the set of all points on a coordinate plane that are a fixed distance from a given point. The given point is called the **center** of the circle and is represented by (h, k). The fixed distance is called the **radius** of the circle and is represented by r.

Let's consider the simplest case—a circle centered at the origin. Suppose, for instance, we want to find the equation of a circle with radius 2 and with center

$(0, 0)$ and then to graph it. If (x, y) is an arbitrary point on the circle, we can apply the distance formula.

$$d = \sqrt{(x_2 - x_1)^2 + (y_2 - y_1)^2}$$
$$2 = \sqrt{(x - 0)^2 + (y - 0)^2}$$
$$4 = x^2 + y^2$$
$$x^2 + y^2 = 4$$

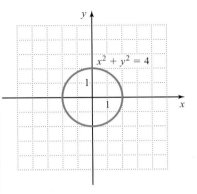

This last equation is the equation of the circle with radius 2 centered at the origin. To graph $x^2 + y^2 = 4$, we start at the center $(0, 0)$ and then plot points 2 units to the left, to the right, up, and down from the center. Finally, we sketch a smooth curve through the four points $(-2, 0)$, $(2, 0)$, $(0, 2)$, and $(0, -2)$, as shown in the figure.

> ### The Equation of a Circle Centered at the Origin
>
> The equation of a circle with center $(0, 0)$ and with radius r is
>
> $$x^2 + y^2 = r^2$$

EXAMPLE 3

Graph: $x^2 + y^2 = 25$

Solution To get the equation in the form $x^2 + y^2 = r^2$, we can write it as $x^2 + y^2 = 5^2$. From the standard form we see that the center is the point $(0, 0)$ and the radius is 5. The graph of the equation is as follows:

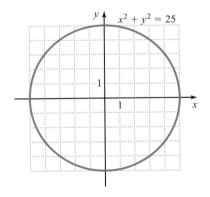

PRACTICE 3

Graph: $x^2 + y^2 = 9$

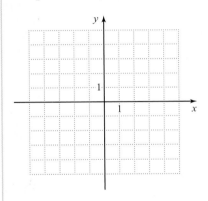

Note from Example 3 that in the equation $x^2 + y^2 = 25$, y is not a function of x. Can you explain why?

More generally, suppose that we wish to find the equation of a circle with given radius r and with arbitrary center (h, k). If (x, y) is any point on the circle, the distance formula gives us:

$$r = \sqrt{(x - h)^2 + (y - k)^2}$$
$$r^2 = (x - h)^2 + (y - k)^2$$
$$(x - h)^2 + (y - k)^2 = r^2$$

which is the equation of the circle in *standard form*.

> ### The Equation of a Circle Centered at (h, k)
>
> The equation of the circle with center (h, k) and with radius r is
>
> $$(x - h)^2 + (y - k)^2 = r^2$$

Note that we can determine the center and radius of a circle just by looking at the equation in standard form.

EXAMPLE 4

Consider the circle whose equation is $(x - 2)^2 + (y + 1)^2 = 16$.

a. Identify the center and radius of the circle.

b. Graph the circle.

Solution

a. To get the equation in the form $(x - h)^2 + (y - k)^2 = r^2$, we write it as $(x - 2)^2 + [y - (-1)]^2 = 4^2$. From the standard form we see that $h = 2$, $k = -1$, and $r = 4$. So the center of the circle is the point $(2, -1)$, and the radius is 4.

b. To graph the circle, we first plot the center $(2, -1)$. From the center point we plot points 4 units to the left, to the right, up, and down and then sketch a smooth curve through these four points.

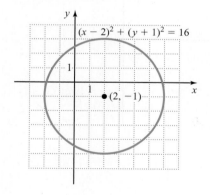

PRACTICE 4

The graph of the equation $(x + 6)^2 + (y - 2)^2 = 100$ is a circle.

a. Identify the center and radius of the circle.

b. Graph the circle.

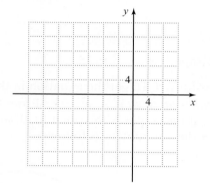

EXAMPLE 5

Find the equation of the circle with center $(-2, 5)$ and with radius 6.

Solution For this circle, $h = -2$ and $k = 5$, and $r = 6$. So the equation of the circle is $[x - (-2)]^2 + (y - 5)^2 = 6^2$, or $(x + 2)^2 + (y - 5)^2 = 36$.

PRACTICE 5

Determine the equation of the circle centered at $(7, -1)$ with radius 8.

Sometimes, the equation of a circle is not given in standard form. We can write the equation in standard form by *completing the square*.

EXAMPLE 6

Find the center and the radius of the circle with equation $x^2 + y^2 + 2x - 8y + 15 = 0$.

Solution To find the center and radius of the circle, first let's write the given equation in standard form by completing the squares for both x-terms and y-terms.

$$x^2 + y^2 + 2x - 8y + 15 = 0$$

$$x^2 + y^2 + 2x - 8y = -15 \quad \text{Subtract 15 from each side of the equation.}$$

$$(x^2 + 2x) + (y^2 - 8y) = -15 \quad \text{Group the } x\text{- and } y\text{-terms.}$$

$$(x^2 + 2x + 1) + (y^2 - 8y + 16) = -15 + 1 + 16 \quad \text{Complete the squares, adding both 1 and 16 to each side of the equation.}$$

$$(x + 1)^2 + (y - 4)^2 = 2 \quad \text{Express both the polynomial in } x \text{ and the polynomial in } y \text{ as the square of a binomial.}$$

Since $(x + 1)^2 + (y - 4)^2 = 2$ is in the form $(x - h)^2 + (y - k)^2 = r^2$, we see that $h = -1$ and $k = 4$. Note that since $r^2 = 2$, $r = \sqrt{2}$. So the center of the circle is the point $(-1, 4)$ and its radius is $\sqrt{2}$.

PRACTICE 6

Find the center and radius of the circle with equation $x^2 + y^2 - 4x - 2y - 19 = 0$.

There are many applications of circles found in nature, from the ripples on the surface of a lake to the cross section of a planet. Some applications of circles involve traveling a fixed distance from a given point.

EXAMPLE 7

A family is on a sailboat that is sinking at sea. A Coast Guard cutter, located 7 mi east and 5 mi north of a harbor, can travel 2 mi in the time that it takes the sailboat to sink.

a. Find the equation of the circle that consists of the farthest points that the cutter can reach before the sailboat sinks.

b. If the sailboat is 8 mi east and 6 mi north of the harbor, can the cutter reach the sailboat before it sinks?

PRACTICE 7

A computer hub in a public park provides wireless access to the Web for a laptop computer up to 10 yd away.

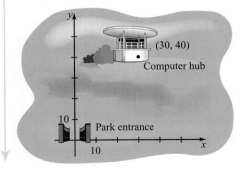

Solution

a. Let's use the harbor as the origin of the coordinate system. The farthest points that the cutter can reach are all 2 mi from the point $(7, 5)$. So the equation of this circle is $(x - 7)^2 + (y - 5)^2 = 2^2$, or $(x - 7)^2 + (y - 5)^2 = 4$, where units are in miles.

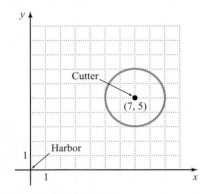

b. One way to see if the cutter can reach the sailboat in time is to calculate the distance between the cutter and the sailboat. The cutter starts at the point $(7, 5)$, and the sailboat is at $(8, 6)$. We can find out how far apart they are by using the distance formula.

$$
\begin{aligned}
d &= \sqrt{(x_2 - x_1)^2 + (y_2 - y_1)^2} \\
&= \sqrt{(8 - 7)^2 + (6 - 5)^2} \\
&= \sqrt{1 + 1} \\
&= \sqrt{2} \\
&\approx 1.4
\end{aligned}
$$

So they are about 1.4 mi apart. Since $1.4 < 2$, the cutter can reach the sailboat before it sinks.

What is another way to solve part (b)?

a. Find the equation of the circle that represents the boundary for surfing the Web in the park. Let units be in yards.

b. Can a visitor sitting 35 yd east and 18 yd north of the entrance in the park surf the Web?

Mathematically Speaking

Fill in each blank with the most appropriate term or phrase from the given list.

solving	the difference between	the sum of
length	completing the squares	radius
centered at	midpoint	at the origin
circle	at any point	passing through

1. The distance between any two points (x_1, y_1) and (x_2, y_2) on a coordinate plane is equal to the square root of _____ $(x_2 - x_1)^2$ and $(y_2 - y_1)^2$.

2. The _____ of a line segment with endpoints (x_1, y_1) and (x_2, y_2) is $\left(\dfrac{x_1 + x_2}{2}, \dfrac{y_1 + y_2}{2}\right)$.

3. A(n) _____ is the set of all points on a coordinate plane that are a fixed distance from a given point.

4. The equation of a circle centered _____ with radius r is $x^2 + y^2 = r^2$.

5. The equation of a circle _____ (h, k) with radius r is $(x - h)^2 + (y - k)^2 = r^2$.

6. The equation $x^2 + y^2 + 3x - 2y = 0$ can be put in standard form by _____ for both x-terms and y-terms.

Find the distance between the two points on the coordinate plane. If necessary, round the answer to the nearest tenth.

7. $(3, 2)$ and $(9, 10)$

8. $(2, 8)$ and $(7, -4)$

9. $(1, 6)$ and $(-1, 2)$

10. $(3, 12)$ and $(4, 5)$

11. $(8, -4)$ and $(5, -6)$

12. $(-7, 10)$ and $(-9, 8)$

13. $(-3.2, 1.7)$ and $(-1.2, -5.3)$

14. $(8.1, -2.6)$ and $(9.1, -14.6)$

15. $\left(\dfrac{1}{4}, \dfrac{1}{3}\right)$ and $\left(\dfrac{1}{2}, \dfrac{2}{3}\right)$

16. $\left(\dfrac{1}{6}, \dfrac{3}{4}\right)$ and $\left(\dfrac{5}{6}, \dfrac{1}{4}\right)$

17. $(-4\sqrt{3}, -2\sqrt{2})$ and $(\sqrt{3}, -3\sqrt{2})$

18. $(\sqrt{5}, -\sqrt{6})$ and $(2\sqrt{5}, 3\sqrt{6})$

Find the coordinates of the midpoint of the line segment joining the points.

19. $(5, 9)$ and $(7, 1)$

20. $(2, 8)$ and $(4, -4)$

21. $(3, 11)$ and $(-12, -1)$

22. $(6, -7)$ and $(-5, 3)$

23. $(-8, -11)$ and $(-13, -2)$

24. $(-12, -9)$ and $(-5, 10)$

25. $(3.4, -1.1)$ and $(0.6, -3.9)$

26. $(-2.7, 0.3)$ and $(-6.3, 5.7)$

27. $\left(\dfrac{3}{4}, -\dfrac{2}{5}\right)$ and $\left(-\dfrac{1}{3}, \dfrac{1}{2}\right)$

28. $\left(\dfrac{1}{9}, \dfrac{1}{6}\right)$ and $\left(-\dfrac{2}{3}, -\dfrac{3}{2}\right)$

29. $(-7\sqrt{3}, 5\sqrt{6})$ and $(7\sqrt{3}, 3\sqrt{6})$

30. $(-9\sqrt{2}, 5\sqrt{5})$ and $(-3\sqrt{2}, -5\sqrt{5})$

Find the center and radius of each circle. Then graph.

31. $x^2 + y^2 = 36$

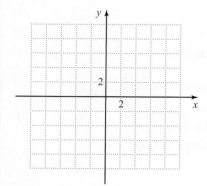

32. $x^2 + y^2 = 16$

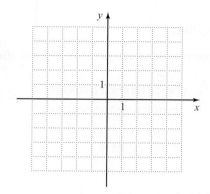

33. $(x - 3)^2 + (y - 2)^2 = 1$

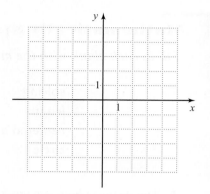

34. $(x + 1)^2 + (y + 4)^2 = 9$

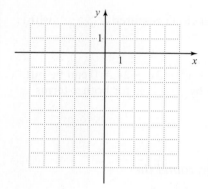

35. $(x + 2)^2 + (y - 4)^2 = 25$

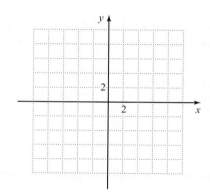

36. $(x - 3)^2 + (y + 3)^2 = 4$

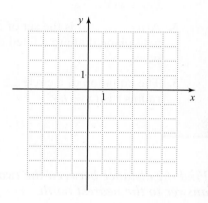

37. $x^2 + y^2 - 2y = 8$

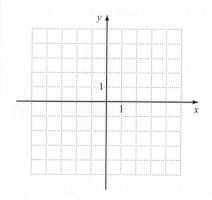

38. $x^2 + y^2 - 4x = 12$

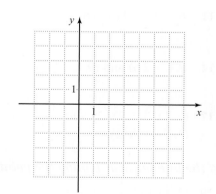

39. $x^2 + y^2 + 2x + 2y - 23 = 0$

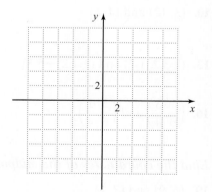

40. $x^2 + y^2 + 4x + 6y - 3 = 0$ **41.** $x^2 + y^2 - 8x - 6y + 16 = 0$ **42.** $x^2 + y^2 + 4x + 2y + 1 = 0$

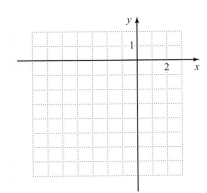   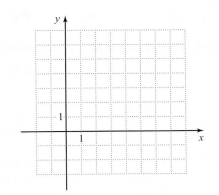   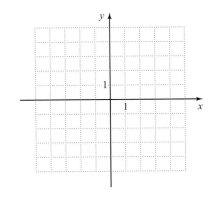

Use the given information to find the equation of each circle.

43. Center: $(-7, 2)$; radius: 9

44. Center: $(6, -10)$; radius: 8

45. Center: $(0, 4)$; radius: $\sqrt{5}$

46. Center: $(-1, 0)$; radius: $\sqrt{2}$

47. Center: $(3, 5)$; passes through the point $(-1, 9)$

48. Center: $(-2, 8)$; passes through the point $(-3, 5)$

49. Endpoints of a diameter: $(-6, 1)$ and $(2, 11)$

50. Endpoints of a diameter: $(7, 7)$ and $(3, -5)$

Find the center and radius of each circle.

51. $x^2 + y^2 + 3x + 4y = 0$

52. $x^2 + y^2 - 8x - y + 4 = 0$

53. $x^2 + y^2 - 10x + 6y - 4 = 0$

54. $x^2 + y^2 + 14x - 2y = 0$

55. $3x^2 + 3y^2 - 12x - 24 = 0$

56. $2x^2 + 2y^2 + 20y + 10 = 0$

Mixed Practice

Find the center and radius of each circle. Then graph.

57. $(x - 4)^2 + (y + 1)^2 = 9$

58. $x^2 + y^2 + 2x - 4y - 11 = 0$

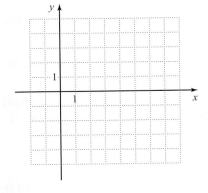   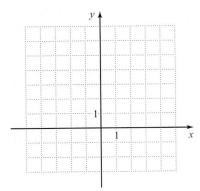

Solve.

59. Find the equation of the circle with center $(2, -4)$ that passes through the point $(-1, -2)$.

60. Find the center and radius of the circle:

$$x^2 + y^2 - 3x + 6y + 9 = 0$$

Find the distance between the two points on the coordinate plane. If necessary, round the answer to the nearest tenth.

61. $(-4, 7)$ and $(-6, 2)$

62. $(-6.8, -2.9)$ and $(-3.8, 3.1)$

Find the coordinates of the midpoint of the line segment joining the points.

63. $(8, 5)$ and $(-3, -11)$

64. $(-4.8, 0.7)$ and $(-9.4, 5.3)$

Applications

Solve.

65. A lawyer drives from her home, located 11 mi east and 8 mi north of the town courthouse, to her office, located 1 mi west and 1 mi south of the courthouse. Find the distance between the lawyer's home and her office to the nearest mile.

66. On a particular sales route, a sales representative first drives to a community college located 20 mi east and 17 mi south from his home. From there, he drives to a state college located 25 mi east and 7 mi south of the community college. Then he drives home. Calculate the distance between the state college and the sales representative's home.

67. In geometry, an *altitude* of a triangle is a line segment through a vertex that is perpendicular to the opposite side of the triangle. Consider an isosceles triangle formed by the points $A(2, 1)$, $B(1, 6)$, and $C(6, 5)$.

   a. Find the coordinates of the midpoint D of the base $\overline{AC}$.

   b. By considering its slope, show that $\overline{BD}$ is an altitude of triangle ABC.

   c. Calculate the area of triangle ABC.

68. The *perpendicular bisector* of a segment is the line perpendicular to the segment that intersects the segment at its midpoint. Consider the line segment with endpoints $(3, 5)$ and $(9, 7)$.

   a. Find the equation of the perpendicular bisector of the line segment in slope-intercept form.

   b. Find the coordinates of the x-intercept of the perpendicular bisector.

   c. Show that the x-intercept of the perpendicular bisector is equidistant from the endpoints of the line segment.

69. An oil barge runs aground, producing a circular oil slick. If the slick is 6 mi across, find an equation that represents the boundary of the slick. Assume the center of the oil slick is at the origin of a coordinate plane.

70. Circus acts are performed in a circular area of an arena called a ring. If a ring has a diameter of 42 ft, find an equation that represents the boundary of the ring. Assume that the center of the ring is at the origin of a coordinate plane.

71. A radio station broadcasts a signal that can be received by a radio within a 36-mi radius of the station.

   a. If the radio station is located 8 mi west and 15 mi north of the center of City A, find an equation of the circle that represents the boundary of the radio signal.

   b. Can a student whose apartment is located 15 mi east and 3 mi south of the center of City A listen to the station's broadcast on his radio?

72. A power outage affected all homes and businesses within a 5-mi radius of the power station.

   a. If the power station is located 2 mi east and 6 mi south of the center of town, find an equation of the circle that represents the boundary of the power outage.

   b. Will a mall located 4 mi east and 7 mi north of the power station be affected by the outage?

• *Check your answers on page A-49.*

MINDSTRETCHERS

Mathematical Reasoning

1. Use the distance formula to show that $\left(\dfrac{x_1 + x_2}{2}, \dfrac{y_1 + y_2}{2}\right)$ is equidistant from (x_1, y_1) and (x_2, y_2).

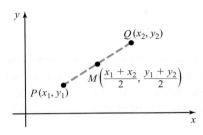

Critical Thinking

2. The equation of the smallest circle shown is $x^2 + y^2 = r^2$. Find the equation of the largest circle.

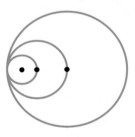

Technology

3. Consider the equation $(x - 5)^2 + (y - 2)^2 = 4$.

   a. Is y a function of x? Explain how you know.

   b. Solve the equation for y.

   c. Explain how you can use a graphing calculator or computer software to graph this equation.

11.3 The Ellipse and the Hyperbola

In this section, we discuss two additional conic sections, namely, the *ellipse* and the *hyperbola*. In considering these conics, we restrict our discussion to ellipses and hyperbolas that are centered at the origin.

OBJECTIVES

- To graph an ellipse with center at the origin

- To graph a hyperbola with center at the origin

- To solve applied problems involving ellipses or hyperbolas

The Ellipse

An **ellipse** is the set of all points on a coordinate plane the *sum* of whose distances from two fixed points is constant. The fixed points are called **foci** (the plural of **focus**). The point halfway between the two foci is called the **center** of the ellipse. In the ellipse shown, the center is the origin.

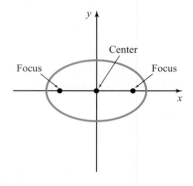

Note that an ellipse is not the graph of a function. Can you explain why?

An ellipse centered at the origin has two *x*-intercepts and two *y*-intercepts. The equation of such an ellipse has the following *standard form:*

> **The Equation of an Ellipse Centered at the Origin**
>
> The equation of an ellipse with center $(0, 0)$, *x*-intercepts $(-a, 0)$ and $(a, 0)$, and
>
> *y*-intercepts $(0, -b)$ and $(0, b)$ is $\dfrac{x^2}{a^2} + \dfrac{y^2}{b^2} = 1$.

To graph an ellipse of the form $\dfrac{x^2}{a^2} + \dfrac{y^2}{b^2} = 1$, we begin by plotting the *x*- and *y*-intercepts, which we can read from its equation. We then sketch an oval-shaped curve passing through these four points, as shown.

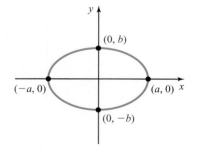

EXAMPLE 1

Graph: $\dfrac{x^2}{25} + \dfrac{y^2}{16} = 1$

Solution We recognize the equation as that of an ellipse centered at the origin in standard form: $\dfrac{x^2}{a^2} + \dfrac{y^2}{b^2} = 1$. Since $a^2 = 25$ and $b^2 = 16$, it follows that $a = 5$ and $b = 4$. So the x-intercepts are $(-5, 0)$ and $(5, 0)$, and the y-intercepts are $(0, -4)$ and $(0, 4)$. Finally, we sketch the smooth curve passing through these intercepts.

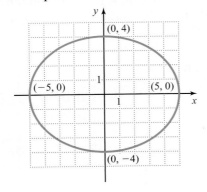

PRACTICE 1

Graph: $\dfrac{x^2}{100} + \dfrac{y^2}{49} = 1$

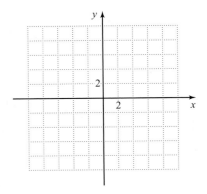

EXAMPLE 2

Graph: $9x^2 + 4y^2 = 36$

Solution Let's divide both sides of the equation by 36, getting 1 on the right side.

$$9x^2 + 4y^2 = 36$$

$$\frac{x^2}{4} + \frac{y^2}{9} = 1$$

$$\frac{x^2}{2^2} + \frac{y^2}{3^2} = 1$$

The result is the equation of an ellipse centered at the origin in standard form, where $a = 2$ and $b = 3$. The x-intercepts of the graph are $(-2, 0)$ and $(2, 0)$, and the y-intercepts are $(0, -3)$ and $(0, 3)$. We can sketch the smooth curve passing through these points.

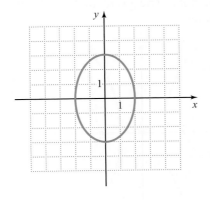

PRACTICE 2

Graph: $25x^2 + 4y^2 = 100$

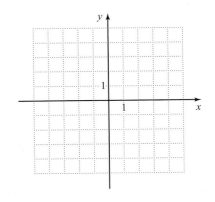

There are many applications of ellipses, including the orbits of planets, supporting arches of bridges, and medical procedures.

EXAMPLE 3	**PRACTICE 3**
The planet Mercury travels in an elliptical orbit with the Sun at one focus. An equation that approximates Mercury's orbit is given by $\frac{x^2}{1296} + \frac{y^2}{1247} = 1$, where the units are millions of miles. The minimum distance between Mercury and the Sun is approximately 29 million mi. What is the maximum distance between Mercury and the Sun? 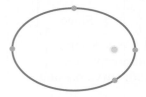	The arch supporting the bridge shown below is in the shape of half an ellipse. The equation of the ellipse is $\frac{x^2}{100} + \frac{y^2}{144} = 1$, where x and y are in feet. Determine the maximum width and maximum height of the arch. 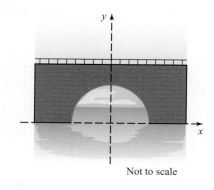 Not to scale

Solution The distance from the Sun, at point S, to the nearest point in Mercury's orbit Q is approximately 29. We are looking for the distance from S to the point P, the most distant point from the Sun.

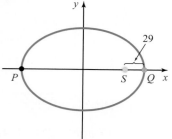

Diagram is not to scale.

From the equation $\frac{x^2}{1296} + \frac{y^2}{1247} = 1$, we see that $a^2 = 1296$, so $a = 36$. The length of $\overline{PQ}$ is therefore 2(36), or 72. So the length of $\overline{PS}$ is $72 - 29$, or 43. So the maximum distance between Mercury and the Sun is approximately 43 million mi.

The Hyperbola

The fourth and last conic section we consider is the **hyperbola.** A hyperbola is the set of all points on a coordinate plane that satisfy the following property: The *difference* of the distances from two fixed points is constant. Each fixed point is said to be a **focus** of the hyperbola. The point halfway between the two foci is called the **center** of the hyperbola.

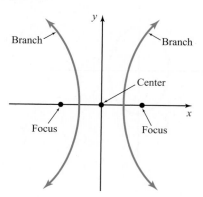

Note that each **branch** of a hyperbola extends indefinitely on both sides. Although a hyperbola looks somewhat like two parabolas, in fact the shape of a parabola and the shape of the branch of a hyperbola differ.

Some hyperbolas centered at the origin have two x-intercepts but no y-intercepts, as shown to the right. These hyperbolas have equations of the form

$$\frac{x^2}{a^2} - \frac{y^2}{b^2} = 1$$

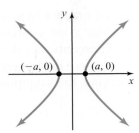

The x-intercepts are $(-a, 0)$ and $(a, 0)$.

Other hyperbolas centered at the origin have two y-intercepts but no x-intercepts, as shown to the right. These hyperbolas have equations of the form

$$\frac{y^2}{b^2} - \frac{x^2}{a^2} = 1$$

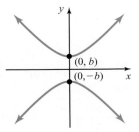

The y-intercepts are $(0, -b)$ and $(0, b)$.

The equation of a hyperbola in the form of either $\frac{x^2}{a^2} - \frac{y^2}{b^2} = 1$ or $\frac{y^2}{b^2} - \frac{x^2}{a^2} = 1$ is said to be in *standard form*.

The Equation of a Hyperbola Centered at the Origin

The equation of a hyperbola with center $(0, 0)$ and x-intercepts $(-a, 0)$ and $(a, 0)$ is

$$\frac{x^2}{a^2} - \frac{y^2}{b^2} = 1.$$

The equation of a hyperbola with center $(0, 0)$ and y-intercepts $(0, -b)$ and $(0, b)$ is

$$\frac{y^2}{b^2} - \frac{x^2}{a^2} = 1.$$

As we move farther from the origin, each branch of a hyperbola gets closer and closer to a straight line called an *asymptote*. Sketching the two asymptotes of a hyperbola allows us to graph the hyperbola more accurately. The hyperbolas that we are considering have as their asymptotes the lines $y = \frac{b}{a}x$ and $y = -\frac{b}{a}x$, as shown on the following coordinate planes.

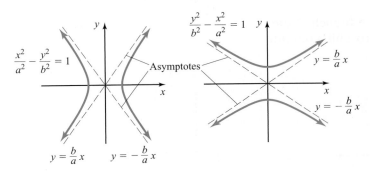

Each branch of a hyperbola approaches, but never intersects, an asymptote. Note that the asymptotes themselves are not part of the hyperbola, as indicated by the broken lines.

To graph a hyperbola centered around the origin, given its equation, we first find the *x*- or *y*-intercepts and the equations of the asymptotes. Then, we plot the intercepts and sketch the asymptotes using broken lines. Finally, we draw the branches of the hyperbola passing through the intercepts and approaching, but not touching, the asymptotes.

EXAMPLE 4	PRACTICE 4

Graph: $\dfrac{y^2}{25} - \dfrac{x^2}{4} = 1$

Solution The equation is of the form $\dfrac{y^2}{b^2} - \dfrac{x^2}{a^2} = 1$. Let's find the intercepts of this hyperbola.

Writing the equation in the form $\dfrac{y^2}{5^2} - \dfrac{x^2}{2^2} = 1$ with $a = 2$ and $b = 5$, we see that the *y*-intercepts are $(0, -5)$ and $(0, 5)$, and that there are no *x*-intercepts. Next, we find the equations of the asymptotes of the hyperbola, getting

$$y = \frac{b}{a}x \quad \text{and} \quad y = -\frac{b}{a}x$$
$$y = \frac{5}{2}x \qquad\qquad y = -\frac{5}{2}x$$

Then, we plot the intercepts and sketch the asymptotes using broken lines.

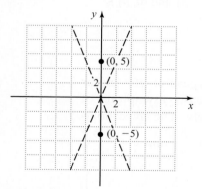

Finally, we draw the graph, sketching each branch of the hyperbola through a *y*-intercept and approaching but not touching its asymptote.

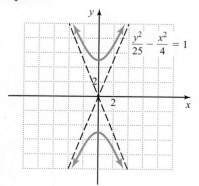

Graph: $\dfrac{y^2}{16} - \dfrac{x^2}{9} = 1$

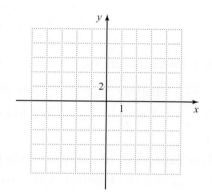

EXAMPLE 5

Graph: $64x^2 - 4y^2 = 64$

Solution To write the equation in standard form, we divide each side of the equation by 64, getting 1 on the right side.

$$64x^2 - 4y^2 = 64$$

$$\frac{x^2}{1} - \frac{y^2}{16} = 1$$

$$\frac{x^2}{1^2} - \frac{y^2}{4^2} = 1$$

This equation is of the form $\dfrac{x^2}{a^2} - \dfrac{y^2}{b^2} = 1$, with $a = 1$ and $b = 4$. So the x-intercepts are $(-1, 0)$ and $(1, 0)$, and there are no y-intercepts. The equations of the asymptotes are $y = \dfrac{4}{1}x$ and $y = -\dfrac{4}{1}x$, or $y = 4x$ and $y = -4x$. To graph, plot the intercepts and sketch the asymptotes. Then, draw a smooth curve through each intercept approaching, but not touching, the asymptotes.

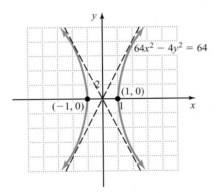

PRACTICE 5

Graph: $4x^2 - 16y^2 = 144$

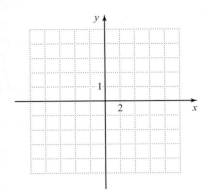

There are many applications of hyperbolas. For example, the paths of certain comets are hyperbolic, as are the mirrors of some telescopes.

EXAMPLE 6

When a jet breaks the sound barrier, the region on the ground bounded by the hyperbola $\frac{x^2}{9} - \frac{y^2}{25} = 1$, where the units are in miles, experiences a sonic boom shock wave. In the following diagram, the coordinate axes are positioned so that the origin is at the center of the hyperbola. Find, to the nearest mile, the width of the region 3 mi from the hyperbola's x-intercept.

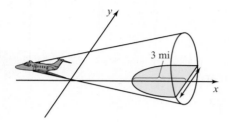

Solution Since the equation is of the form $\frac{x^2}{3^2} - \frac{y^2}{5^2} = 1$ with $a = 3$ and $b = 5$, the x-intercept is $(3, 0)$. The point $(6, 0)$ on the x-axis is 3 mi from the hyperbola's x-intercept. To find the width of the hyperbola at $x = 6$, we calculate the value of y when x is 6, and then double it.

$$\frac{x^2}{9} - \frac{y^2}{25} = 1$$

$$\frac{6^2}{9} - \frac{y^2}{25} = 1$$

$$-\frac{y^2}{25} = 1 - \frac{36}{9}$$

$$-\frac{y^2}{25} = -3$$

$$y^2 = 75$$

$$y = \pm\sqrt{75}, \text{ or } \pm 5\sqrt{3}$$

Since y cannot be negative, we conclude that y is $5\sqrt{3}$. Doubling this value, we find that the width of the region experiencing the sonic boom shock wave is $10\sqrt{3}$ mi, or approximately 17 mi.

PRACTICE 6

Many comets have a hyperbolic path, with the Sun as one of the foci of the hyperbola. A comet has as its path the graph of the equation

$$\frac{x^2}{8100} - \frac{y^2}{2700} = 1, \text{ where the units}$$

are in millions of miles. If the Earth is the center of the hyperbola, how close does the comet come to the Earth?

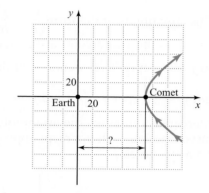

| 11.3 | Exercises |  |

Mathematically Speaking

Fill in each blank with the most appropriate term or phrase from the given list.

circle	ellipse	hyperbola
sum of	difference between	asymptotes

1. A(n) _____ is the set of all points on a coordinate plane the sum of whose distances from two fixed points is constant.

2. If an ellipse has center $(0, 0)$ and passes through $(-a, 0)$, $(a, 0)$, $(0, -b)$, and $(0, b)$, then the _____ $\dfrac{x^2}{a^2}$ and $\dfrac{y^2}{b^2}$ is 1.

3. A(n) _____ is the set of all points on a coordinate plane the difference of whose distances from two fixed points is constant.

4. If a hyperbola has center $(0, 0)$ and x-intercepts $(-a, 0)$ and $(a, 0)$, then the _____ $\dfrac{x^2}{a^2}$ and $\dfrac{y^2}{b^2}$ is 1.

Graph each ellipse.

5. $\dfrac{x^2}{16} + \dfrac{y^2}{4} = 1$

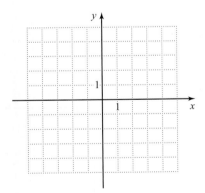

6. $\dfrac{x^2}{25} + \dfrac{y^2}{9} = 1$

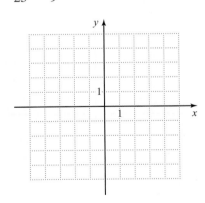

⊙ 7. $\dfrac{x^2}{36} + \dfrac{y^2}{81} = 1$

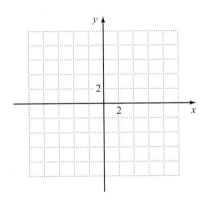

8. $\dfrac{x^2}{16} + \dfrac{y^2}{49} = 1$

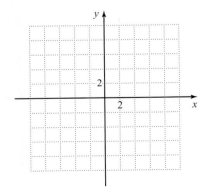

9. $64x^2 + 16y^2 = 64$

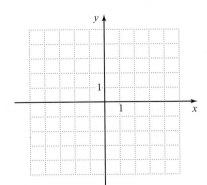

10. $4x^2 + 36y^2 = 36$

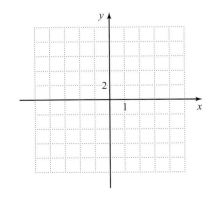

11. $x^2 + 4y^2 = 100$

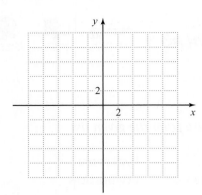

12. $36x^2 + 9y^2 = 144$

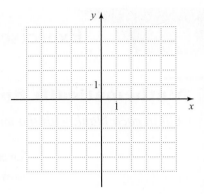

13. $2x^2 + 8y^2 = 128$

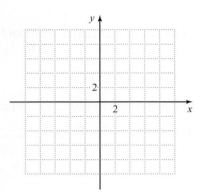

14. $18x^2 + 2y^2 = 162$

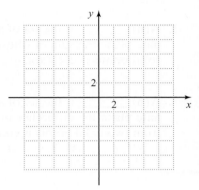

Graph each hyperbola.

15. $\dfrac{x^2}{64} - \dfrac{y^2}{9} = 1$

16. $\dfrac{x^2}{25} - \dfrac{y^2}{4} = 1$

17. $\dfrac{y^2}{81} - \dfrac{x^2}{36} = 1$

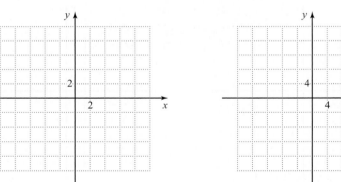

18. $\dfrac{y^2}{49} - \dfrac{x^2}{9} = 1$

19. $x^2 - 9y^2 = 225$

20. $4x^2 - y^2 = 64$

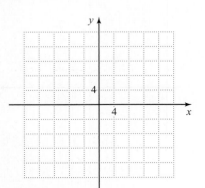

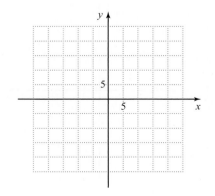

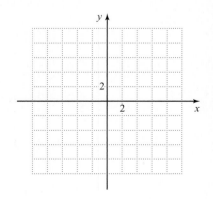

21. $3y^2 - 12x^2 = 108$

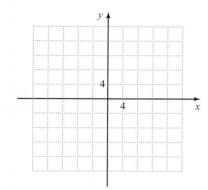

22. $2y^2 - 8x^2 = 200$

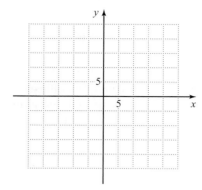

Identify whether the conic section whose equation is given is a parabola, a circle, an ellipse, or a hyperbola.

23. $8x^2 + 8y^2 = 32$ **24.** $3x^2 + 5y^2 = 45$ **25.** $x^2 - 2y = 10$

26. $6x^2 + 6y^2 = 18$ **27.** $10y^2 - 12x^2 = 120$ **28.** $9y^2 + x = 27$

29. Find the equation in standard form of an ellipse centered at the origin that passes through the points $(-11, 0)$, $(11, 0)$, $(0, -3)$, and $(0, 3)$.

30. Find the equation in standard form of an ellipse centered at the origin that passes through the points $(-6, 0)$, $(6, 0)$, $(0, -13)$, and $(0, 13)$.

31. Find the equation in standard form of an ellipse centered at the origin that passes through the points $(-\sqrt{7}, 0)$, $(\sqrt{7}, 0)$, $(0, -2\sqrt{2})$, and $(0, 2\sqrt{2})$.

32. Find the equation in standard form of an ellipse centered at the origin that passes through the points $(-4\sqrt{3}, 0)$, $(4\sqrt{3}, 0)$, $(0, -\sqrt{5})$, and $(0, \sqrt{5})$.

33. Find the equation in standard form of a hyperbola centered at the origin that passes through the points $(-4, 0)$ and $(4, 0)$ and whose graph approaches the asymptotes $y = -\dfrac{1}{4}x$ and $y = \dfrac{1}{4}x$.

34. Find the equation in standard form of a hyperbola centered at the origin that passes through the points $(0, -9)$ and $(0, 9)$ and whose graph approaches the asymptotes $y = -\dfrac{9}{7}x$ and $y = \dfrac{9}{7}x$.

Mixed Practice

Graph each hyperbola.

35. $\dfrac{y^2}{4} - \dfrac{x^2}{16} = 1$

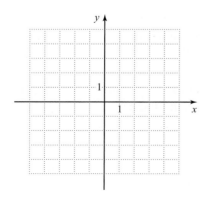

36. $2x^2 - 8y^2 = 72$

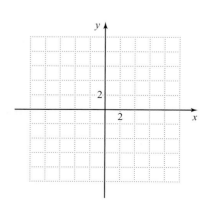

Identify whether the conic section whose equation is given is a parabola, a circle, an ellipse, or a hyperbola.

37. $3x^2 + 6y^2 = 24$

38. $2x^2 - 3y = 9$

Solve.

39. Find the equation in standard form of an ellipse centered at the origin that passes through the points $(-3\sqrt{5}, 0), (3\sqrt{5}, 0), (0, -\sqrt{3}), (0, \sqrt{3})$.

40. Find the equation in standard form of a hyperbola centered at the origin that passes through the points $(0, -3), (0, 3)$ and whose graph approaches the asymptotes $y = -\dfrac{3}{4}x$ and $y = \dfrac{3}{4}x$.

Graph each ellipse.

41. $\dfrac{x^2}{49} + \dfrac{y^2}{25} = 1$

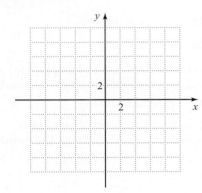

42. $12x^2 + 3y^2 = 108$

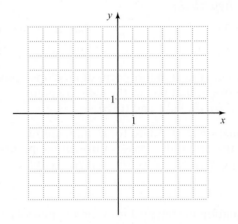

Applications

Solve.

43. According to the laws of the game, the playing field in Australian Rules Football is elliptical and must have a length between 135 m and 185 m and a width between 110 m and 155 m. An equation of the ellipse that represents a team's playing field is $\dfrac{x^2}{8100} + \dfrac{y^2}{5625}$, where units are in meters. (*Source:* Australian Football League)

   a. What are the length and the width of this playing field?

   b. Does the team's field meet the requirements stated?

44. An elliptical bicycle path is constructed in a local park. City planners required the length of the path to be between 20 and 40 yd longer than the width. The equation of the ellipse that represents a proposed path is given by $\dfrac{x^2}{4225} + \dfrac{y^2}{2500} = 1$, where units are in yards.

   a. What are the length and the width of the bicycle path?

   b. Does the proposed path meet the city planners' requirement?

45. The amount of rubber, A, in a 25-foot-long cylindrical rubber hose is given by the equation

$$A = 25\pi(R^2 - r^2)$$

where the units are in feet, R is the radius of the outside of the hose, and r is the inside radius. The amount of rubber in the hose is 100π.

a. Sketch a graph of the relationship between the two radii on a coordinate plane.

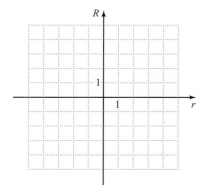

b. Explain why the part of the graph that lies in the first quadrant is of particular interest in this situation.

46. A square parcel of land has sides y ft in length. At each of the four corners of the land, a square house is built with side x ft long. The area of the land on which houses are not built is given by the expression $y^2 - 4x^2$, where units are in feet. This area is measured to be 6400 ft².

a. Sketch a graph of the relationship between x and y on the coordinate plane.

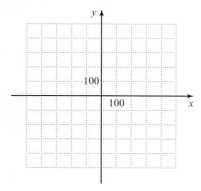

b. Describe the portion of the graph in Quadrant I.

47. The cross section of an ellipsoidal passenger capsule on an observation wheel is an ellipse (see figure). The cross section is 6 m long and 4 m wide. Find an equation of the ellipse that represents the cross section of the capsule.

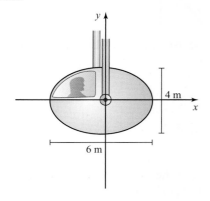

48. A bridge with a semi-elliptical arch extends over a river as shown. Find an equation of the ellipse that represents the arch of the bridge.

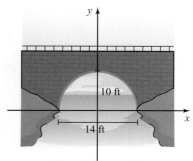

Figure is not to scale.

● *Check your answers on page A-50.*

MINDSTRETCHERS

Investigation

1. An ellipse can be drawn by using a string attached to two fixed tacks and a pencil, as shown in the figure.

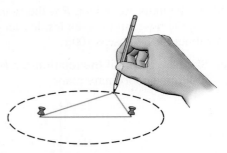

Varying the length of the string and the distance between the tacks, draw several ellipses. What do you notice about the shape of the ellipse as the tacks are moved closer together? What do you notice about the shape as the tacks are moved farther apart?

Mathematical Reasoning

2. Consider a hyperbola with equation either $\dfrac{x^2}{a^2} - \dfrac{y^2}{b^2} = 1$ or $\dfrac{y^2}{b^2} - \dfrac{x^2}{a^2} = 1$. The points $(a, b), (-a, b), (-a, -b)$, and $(a, -b)$ can be used to form the *fundamental rectangle* of the hyperbola.

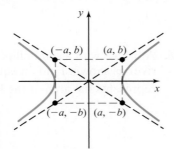
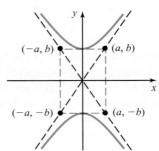

Show that the lines containing the diagonals of this rectangle are the asymptotes of the hyperbola.

Groupwork

3. Working with a partner, use a flashlight to project a cone of light from the bulb. By projecting the light from the flashlight on a wall, create a parabola, a circle, an ellipse, and a hyperbola. Describe your results.

Solving Nonlinear Systems of Equations

In the previous sections of this chapter, we discussed nonlinear equations, such as $y = 2x^2$, $x^2 + y^2 = 25$, and $4x^2 - y^2 = 16$, and their graphs. Here, we focus on solving **nonlinear systems of equations.** These are systems in which at least one equation is nonlinear. For example, consider the following system:

$$x^2 + y^2 = 25$$
$$x - y = 1$$

Recall that in Sections 4.2 and 4.3, we solved systems of linear equations in two variables by using the substitution and elimination methods. We now apply these two methods in solving nonlinear systems of equations in two variables. As with systems of linear equations, a **solution of a nonlinear system of equations** in two variables is an ordered pair of numbers that makes each equation in the system true.

First, we consider the substitution method and then the elimination method.

Solving Nonlinear Systems by Substitution

From the following example, we see that the substitution method particularly lends itself to solving nonlinear systems in which one of the equations contains linear terms.

OBJECTIVES

- To solve a nonlinear system by substitution

- To solve a nonlinear system by elimination

- To solve applied problems involving nonlinear systems

EXAMPLE 1	PRACTICE 1

Solve by substitution.

$$\textbf{(1)} \quad x^2 + y^2 = 25$$
$$\textbf{(2)} \quad y - x = 1$$

Solution Since equation (2) is linear, we first solve it for x or y. Solving for y, we get:

$$\textbf{(2)} \quad y - x = 1$$
$$y = x + 1$$

Next, we substitute the expression $(x + 1)$ for y in equation (1) and solve the resulting equation for x.

$$\textbf{(1)} \qquad x^2 + y^2 = 25$$
$$x^2 + (x + 1)^2 = 25 \qquad \text{Substitute } (x + 1) \text{ for } y.$$
$$x^2 + x^2 + 2x + 1 = 25$$
$$2x^2 + 2x - 24 = 0$$
$$2(x^2 + x - 12) = 0$$
$$x^2 + x - 12 = 0$$
$$(x + 4)(x - 3) = 0$$
$$x + 4 = 0 \quad \text{or} \quad x - 3 = 0$$
$$x = -4 \qquad\qquad x = 3$$

Solve by substitution.

$$\textbf{(1)} \quad x^2 + y^2 = 100$$
$$\textbf{(2)} \quad y - x = 2$$

To find the corresponding values of y, we substitute the values of x in either equation of the original system. Substituting in equation (2), we get:

(2) $y - x = 1$ $y - x = 1$

  $y - (-4) = 1$ Substitute -4 for x. $y - 3 = 1$ Substitute 3 for x.

       $y = -3$ $y = 4$

We see that when $x = -4$, $y = -3$, and when $x = 3$, $y = 4$.

When we check $(-4, -3)$ and $(3, 4)$ in the equations of the system, we find that both ordered pairs make the equations true. So the solutions are $(-4, -3)$ and $(3, 4)$.

The graph of each equation in this system is shown to the right. Note that the line intersects the circle at two points, $(-4, -3)$ and $(3, 4)$, which also confirms the solutions of the system.

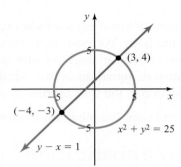

EXAMPLE 2	PRACTICE 2

EXAMPLE 2

Solve by substitution.

(1) $x^2 + y^2 = 4$

(2) $y = x^2 - 2$

Solution Since both equations in the system have an x^2 term, we can solve equation (2) for x^2. Then, we substitute for x^2 in equation (1).

(2) $y = x^2 - 2$

             $y + 2 = x^2$

(1) $x^2 + y^2 = 4$

      $(y + 2) + y^2 = 4$ Substitute $(y + 2)$ for x^2.

       $y + 2 + y^2 = 4$

       $y^2 + y + 2 = 4$

       $y^2 + y - 2 = 0$

      $(y + 2)(y - 1) = 0$

   $y + 2 = 0$ or $y - 1 = 0$

       $y = -2$ $y = 1$

Next, we find the corresponding values of x by substituting into equation (2) the values found for y.

(2) $y = x^2 - 2$ $y = x^2 - 2$

  $-2 = x^2 - 2$ Substitute -2 for y. $1 = x^2 - 2$ Substitute 1 for y.

   $0 = x^2$ $3 = x^2$

   $x = 0$ $x = \pm\sqrt{3}$

PRACTICE 2

Use the substitution method to solve.

(1) $y = x^2 - 4$

(2) $x^2 + y^2 = 4$

We see that when $y = -2$, $x = 0$, and when $y = 1$, $x = \pm\sqrt{3}$. So the solutions to the system are $(0, -2)$, $(\sqrt{3}, 1)$, and $(-\sqrt{3}, 1)$.

The graph of each equation in this system is shown to the right. Note the parabola intersects the circle at three points, namely $(0, -2)$, $(\sqrt{3}, 1)$, and $(-\sqrt{3}, 1)$, which confirms the solutions of the system.

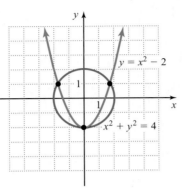

Solving Nonlinear Systems by Elimination

When the variable terms in each equation of a nonlinear system are of second degree, we can use the elimination method to solve the system. In this case, we can eliminate an x^2- or y^2-term using a procedure similar to the one shown in Section 4.2.

EXAMPLE 3	PRACTICE 3

EXAMPLE 3

Solve by elimination.

$$\text{(1)} \quad x^2 + y^2 = 10$$
$$\text{(2)} \quad 9x^2 + y^2 = 18$$

Solution Note that if we multiply equation (1) by -1 and then add the equations, we eliminate the y^2-terms.

(1) $x^2 + y^2 = 10$ $\xrightarrow{\text{Multiply by } -1.}$ $-x^2 - y^2 = -10$
(2) $9x^2 + y^2 = 18$

$$\begin{aligned} -x^2 - y^2 &= -10 \\ \underline{9x^2 + y^2 } &= \underline{18} \\ 8x^2 &= 8 \\ x^2 &= 1 \\ x &= \pm 1 \end{aligned}$$

When $x = 1$ or $x = -1$, $x^2 = 1$. So to find the corresponding values of y, we can substitute 1 for x^2 in equation (2).

$$\begin{aligned} \text{(2)} \quad 9x^2 + y^2 &= 18 \\ 9(1) + y^2 &= 18 \qquad \text{Substitute 1 for } x^2. \\ 9 + y^2 &= 18 \\ y^2 &= 9 \\ y &= \pm 3 \end{aligned}$$

We see that when $x = 1$, $y = 3$ or -3, and when $x = -1$, $y = 3$ or -3. So the solutions are $(1, 3)$, $(1, -3)$, $(-1, 3)$, and $(-1, -3)$.

PRACTICE 3

Use the elimination method to solve.

$$\text{(1)} \quad x^2 + y^2 = 9$$
$$\text{(2)} \quad 2x^2 - y^2 = -6$$

The graph of each equation in this system is shown to the right. Note that the circle intersects the ellipse at $(1, 3)$, $(1, -3)$, $(-1, 3)$, and $(-1, -3)$, confirming the solutions of the system.

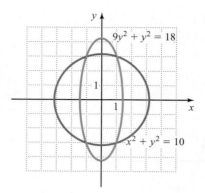

$9y^2 + y^2 = 18$

$x^2 + y^2 = 10$

EXAMPLE 4

Use the elimination method to solve.

$$\textbf{(1)}\quad x^2 - 4y^2 = 16$$
$$\textbf{(2)}\quad x^2 + y^2 = 1$$

Solution Here, the goal is to eliminate the x^2-term and then solve for y.

$\textbf{(1)}\ x^2 - 4y^2 = 16\ \xrightarrow{\text{Multiply by } -1.}\ -x^2 + 4y^2 = -16$

$\textbf{(2)}\ x^2 + y^2 = 1 \qquad\qquad\quad\ \underline{x^2 + y^2 = 1}$

$$5y^2 = -15 \qquad \textbf{Add the equations.}$$
$$y^2 = -3$$
$$y = \pm i\sqrt{3}$$

When $y = \pm i\sqrt{3}$, $y^2 = -3$. So we can substitute -3 for y^2 in equation (2) to find the corresponding x-values.

$$\textbf{(2)}\quad x^2 + y^2 = 1$$
$$x^2 - 3 = 1 \qquad \textbf{Substitute } -3 \textbf{ for } y^2.$$
$$x^2 = 4$$
$$x = \pm 2$$

We see that when $y = \pm i\sqrt{3}$, $x = \pm 2$. So the solutions are $(2, i\sqrt{3})$, $(-2, i\sqrt{3})$, $(2, -i\sqrt{3})$, and $(-2, -i\sqrt{3})$.

The graph of each equation in this system is shown to the right. Note that the hyperbola does not intersect the circle, which confirms that there are no real solutions.

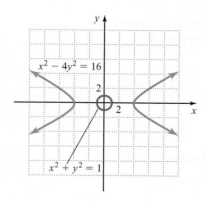

$x^2 - 4y^2 = 16$

$x^2 + y^2 = 1$

PRACTICE 4

Solve by elimination.

$$\textbf{(1)}\quad x^2 - y^2 = -9$$
$$\textbf{(2)}\quad 4x^2 + y^2 = 4$$

Many applications in astronomy, astrophysics, and communications involve nonlinear systems.

EXAMPLE 5

Suppose the equation $\dfrac{x^2}{4} + \dfrac{y^2}{16} = 1$ describes the elliptical orbit of a planet and the equation $y = x^2 - 4$ describes the parabolic path of a comet. Where will the path of the comet intersect the orbit of the planet?

Solution The nonlinear system that we consider is

(1) $\quad \dfrac{x^2}{4} + \dfrac{y^2}{16} = 1$

(2) $\quad\quad y = x^2 - 4$

To eliminate the fractions in equation (1), we multiply each side of the equation by 16, getting:

(1) $\quad 4x^2 + y^2 = 16$

(2) $\quad\quad y = x^2 - 4$

Solving equation (2) for x^2, we have:

(2) $\quad y = x^2 - 4$

$y + 4 = x^2$

Now, we substitute $(y + 4)$ for x^2 in equation (1), which gives us:

(1) $\quad 4x^2 + y^2 = 16$

$4(y + 4) + y^2 = 16$

$4y + 16 + y^2 = 16$

$y^2 + 4y = 0$

$y(y + 4) = 0$

$y = 0 \quad \text{or} \quad y + 4 = 0$

$y = -4$

Finally, substituting the y-values in equation (2), we find the corresponding x-values:

(2) $\quad y = x^2 - 4$ 　　　　 **(2)** $\quad y = x^2 - 4$

$0 = x^2 - 4$ 　　　　　　　　 $-4 = x^2 - 4$

$0 = (x + 2)(x - 2)$ 　　　　　 $0 = x^2$

$x + 2 = 0 \quad \text{or} \quad x - 2 = 0$ 　　 $0 = x$

$x = -2 \quad\quad\quad x = 2$ 　　　　 $x = 0$

The points of intersection are the ordered pairs $(-2, 0)$, $(2, 0)$, and $(0, -4)$. So the path of the comet intersects the orbit of the planet at $(-2, 0)$, $(2, 0)$, and $(0, -4)$.

PRACTICE 5

Suppose a baseball is dropped from the top of a 144-ft building. At any given time t, the height (in feet) of the baseball above the ground is represented by the equation $y = -16t^2 + 144$. At the same time that the baseball is dropped, a soccer ball is thrown upward from the ground with an initial velocity of 64 feet per second. At any given time t, its height above the ground is represented by the equation $y = -16t^2 + 64t$. At what time will the two balls be the same height above the ground?

11.4   Exercises   FOR EXTRA HELP

Solve.

1. $y = x^2 - 2$
$y = 2x + 1$

2. $y = x^2 + 7$
$y = 3x + 5$

3. $x^2 + y^2 = 12$
$x = y^2 - 6$

4. $x^2 + y^2 = 16$
$x = y^2 - 4$

5. $y = x^2 - 5$
$y = -x^2 + 11$

6. $x = y^2 - 8$
$x = -y^2 + 4$

7. $x^2 + y^2 = 16$
$x - y = 4$

8. $x^2 + y^2 = 25$
$x + y = 5$

9. $x = 2y^2 - y - 3$
$y = \dfrac{1}{4}x$

10. $y = 3x^2 - 2x - 4$
$x = \dfrac{1}{2}y$

11. $x^2 + y^2 = 32$
$y = x^2 - 2$

12. $x^2 + y^2 = 84$
$x = y^2 + 6$

13. $-x^2 + 2y^2 = -8$
$x^2 + 3y^2 = 18$

14. $4x^2 + y^2 = 30$
$5x^2 - y^2 = 15$

15. $3x^2 + y^2 = 24$
$x^2 + y^2 = 16$

16. $x^2 + y^2 = 20$
$x^2 + 4y^2 = 32$

17. $5x^2 + 6y^2 = 24$
$5x^2 + 5y^2 = 25$

18. $2x^2 + 8y^2 = 40$
$5x^2 + 8y^2 = 16$

19. $5x^2 - 3y^2 = 35$
$3x^2 - 3y^2 = 3$

20. $4x^2 - 6y^2 = 16$
$4x^2 - 2y^2 = 32$

21. $6x^2 + 6y^2 = 96$
$x^2 + 9y^2 = 144$

22. $3x^2 - 3y^2 = 108$
$2x^2 + y^2 = 72$

23. $\dfrac{x^2}{4} + \dfrac{y^2}{16} = 1$
$\dfrac{x^2}{2} + \dfrac{y^2}{24} = 1$

24. $\dfrac{x^2}{3} + \dfrac{y^2}{9} = 1$
$\dfrac{x^2}{4} + \dfrac{y^2}{6} = 1$

25. $\dfrac{y^2}{8} - \dfrac{x^2}{4} = 1$
$\dfrac{y^2}{11} + \dfrac{x^2}{11} = 1$

26. $-\dfrac{x^2}{9} + y^2 = 1$
$\dfrac{x^2}{16} - \dfrac{y^2}{4} = 1$

27. $(x - 4)^2 + y^2 = 4$
$(x + 2)^2 + y^2 = 16$

28. $x^2 + (y + 1)^2 = 9$
$y = x^2 + 2$

Mixed Practice

Solve.

29. $x^2 + y^2 = 27$
$\quad\quad y = x^2 + 3$

30. $x^2 + y^2 = 13$
$\quad\quad 3x^2 + y^2 = 21$

31. $x^2 + y^2 = 25$
$\quad\quad x = y^2 - 5$

32. $y = x^2 - 7$
$\quad\quad y = -x^2 + 3$

33. $3x^2 + 6y^2 = 24$
$\quad\quad 3x^2 + 5y^2 = 19$

34. $\dfrac{x^2}{18} + \dfrac{y^2}{3} = 1$
$\quad\quad \dfrac{x^2}{9} + \dfrac{y^2}{6} = 1$

Applications

Solve.

35. The revenue, R, of an electronics manufacturer for selling x handheld organizers is given by $R = 0.25x^2 - 100x$. The revenue of a competing manufacturer for selling similar products is $R = 75x$. Under what circumstances will both manufacturers have the same revenue?

36. A company's annual operating cost, C (in millions of dollars), is given by $C = t^2 - t + 4$, where t is the number of years after the year 2000. The corresponding equation for the company's annual revenue, R (in millions of dollars), is $R = 0.5t^2 + 1.5t + 2$. In what year(s) did the company break even?

37. Suppose the equation $x^2 + 2y^2 = 7$ represents the elliptical path traveled by one comet and that equation the $x^2 - 3y^2 = 2$, where $x \geq 0$, represents the hyperbolic path traveled by another comet. At what point(s) will the paths of the two comets intersect?

38. Two elliptical pathways in a park intersect as shown. If the equations of the ellipses that represent the outermost perimeter of the pathways are $x^2 + 4y^2 = 10,000$ and $4x^2 + y^2 = 10,000$, at what points do the pathways intersect?

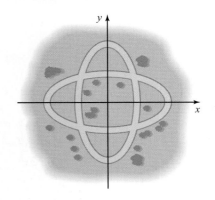

39. One square has a side length 3 in. greater than that of a second square. The difference between the areas of the two squares is 75 in². Find the dimensions of the two squares.

40. Points A and B lie on a coordinate plane, as shown below. If $AC = \sqrt{41}$ and $BC = \sqrt{13}$, find the possible locations of C.

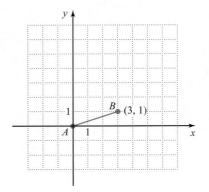

• *Check your answers on page A-51.*

MINDSTRETCHERS

Technology

1. Use a graphing calculator or a computer with graphing software to solve the following nonlinear system. Round to the nearest hundredth.

$$x^2 + y^2 = 9$$
$$y = e^x$$

Mathematical Reasoning

2. Show algebraically that the graph of the hyperbola whose equation is $4x^2 - y^2 = 16$ does not intersect its asymptotes.

Critical Thinking

3. Solve.

$$\begin{array}{ll} \textbf{(1)} & x^2 + y^2 = 15 \\ \textbf{(2)} & 2x^2 + y^2 = 22 \\ \textbf{(3)} & 3x^2 - y^2 = 13 \end{array}$$

11.5   Solving Nonlinear Inequalities and Nonlinear Systems of Inequalities

In this section, we solve nonlinear inequalities and nonlinear systems of inequalities by using the same general methods for graphing linear inequalities and systems of linear inequalities that we discussed in Chapters 3 and 4.

Solving Nonlinear Inequalities by Graphing

Let's look at how to graph a nonlinear inequality in two variables. Recall from Section 3.5 that to graph a linear inequality such as $x + 2y < 2$, first we graph the corresponding linear equation, which is called the boundary line. Then, we choose a test point not on the boundary to determine the region that satisfies the inequality. This method can also be used to graph a nonlinear inequality such as $x^2 + y^2 < 49$.

EXAMPLE 1	**PRACTICE 1**

EXAMPLE 1

Graph the solutions of $x^2 + y^2 < 49$.

Solution The corresponding equation is $x^2 + y^2 = 49$, which is a circle with center $(0, 0)$ and radius 7. Since the original inequality involves the symbol $<$, we draw a *broken curve* to indicate that the boundary $x^2 + y^2 = 49$ is *not* part of the graph of $x^2 + y^2 < 49$. The following graph is a circle that divides the plane into two regions—the "inner" and "outer" regions of the circle:

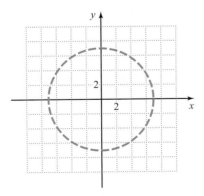

To determine in which region the original inequality is satisfied, we choose a test point in either region and substitute the coordinates of this point in the inequality. Let's choose the test point $(0, 0)$. Substituting in the inequality, we get:

$$x^2 + y^2 < 49$$
$$0^2 + 0^2 < 49 \qquad \text{Substitute 0 for } x \text{ and 0 for } y.$$
$$0 + 0 < 49$$
$$0 < 49 \qquad \text{True}$$

PRACTICE 1

Graph: $4x^2 + y^2 > 4$

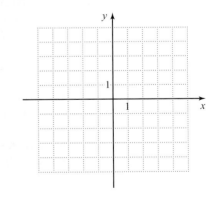

Since the inequality is a true statement, the region containing the test point is the graph $x^2 + y^2 < 49$. So the inner region of the circle, shown here, is the graph of the solutions of the inequality.

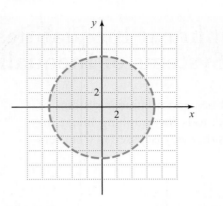

EXAMPLE 2

Graph: $x^2 - 9y^2 \geq 36$

Solution The corresponding equation is $x^2 - 9y^2 = 36$, or $\dfrac{x^2}{36} - \dfrac{y^2}{4} = 1$.

Since the original inequality involves the symbol $\geq$, we draw a *solid curve* to indicate that the boundary $x^2 - 9y^2 = 36$ is part of the graph of $x^2 - 9y^2 \geq 36$. The graph to the right is a hyperbola that divides the plane into three regions.

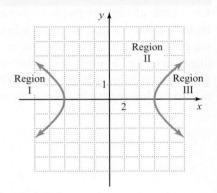

To determine in which region the original inequality is satisfied, we choose a test point in each region, but not on the boundary.

Region I	**Region II**	**Region III**
Test point $(-7, 0)$	Test point $(0, 0)$	Test point $(7, 0)$
$x^2 - 9y^2 \geq 36$	$x^2 - 9y^2 \geq 36$	$x^2 - 9y^2 \geq 36$
$(-7)^2 - 9(0)^2 \geq 36$	$(0)^2 - 9(0)^2 \geq 36$	$(7)^2 - 9(0)^2 \geq 36$
$49 - 0 \geq 36$	$0 - 0 \geq 36$	$49 - 0 \geq 36$
$49 \geq 36$ **True**	$0 \geq 36$ **False**	$49 \geq 36$ **True**

Since the inequality is a true statement for Regions I and III, those regions are part of the graph of $x^2 - 9y^2 \geq 36$. So Region I, Region III, and the boundary hyperbola constitute the graph of the inequality, which is shown on the coordinate plane to the right:

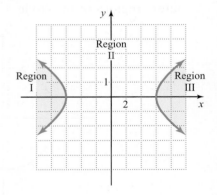

PRACTICE 2

Graph the solutions of $4x^2 - 25y^2 \leq 100$.

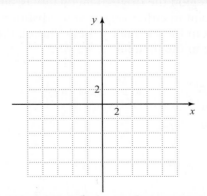

Tip When graphing nonlinear inequalities:

- draw a solid curve for an inequality that involves either the symbol $\leq$ or $\geq$.
- draw a broken curve for an inequality that involves either the symbol $<$ or $>$.

Solving Nonlinear Systems of Inequalities

In Section 4.4, we solved systems of linear inequalities in two variables by graphing each inequality on the same coordinate plane. In such systems, every point in the region of overlap is a solution of the system. Here, we use the same general methods for graphing nonlinear systems of inequalities in two variables. As with systems of linear inequalities, **a solution of a system of nonlinear inequalities** is an ordered pair of real numbers that satisfies each inequality in the system.

EXAMPLE 3	PRACTICE 3

EXAMPLE 3

Graph the solutions of the system:

$$x^2 + y^2 \leq 16$$
$$y \leq x - 4$$

Solution Let's begin by graphing each inequality. First, we graph $x^2 + y^2 \leq 16$. The boundary is a circle, the graph of the equation $x^2 + y^2 = 16$, and is a solid curve because the original inequality involves the symbol $\leq$. The test point $(0, 0)$ gives us a true statement for the inequality $x^2 + y^2 \leq 16$. So the inner region of the circle is our graph and we shade it in as shown.

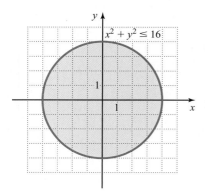

Next, we graph $y \leq x - 4$. The boundary line is the graph of $y = x - 4$, drawn as a solid line since the original inequality involves the symbol $\leq$. The test point $(0, 0)$ gives us a false statement for the inequality $y \leq x - 4$. So we shade the region below the line as shown.

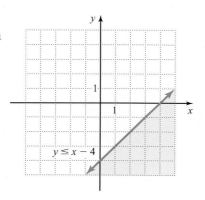

PRACTICE 3

Solve by graphing:

$$x + y \geq 3$$
$$x^2 + y^2 \leq 25$$

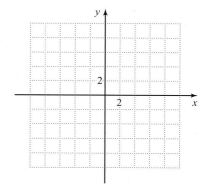

Finally, we draw each inequality on the same coordinate plane. Since a solution of the nonlinear system of inequalities must satisfy each inequality, the solutions are all the points that lie in *both* shaded regions, that is, in the purple overlapping region of the two graphs, including its boundary.

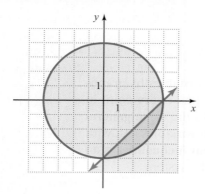

In Example 3, can you describe the overlapping region of the two graphs? Explain.

EXAMPLE 4	PRACTICE 4

Solve by graphing:

$$\frac{x^2}{16} + \frac{y^2}{25} < 1$$

$$\frac{x^2}{25} - \frac{y^2}{4} > 1$$

Solution We graph each inequality separately on the same coordinate plane. First, we draw the ellipse $\frac{x^2}{16} + \frac{y^2}{25} = 1$ as a broken curve. The test point $(0, 0)$ gives us a true statement in the inequality $\frac{x^2}{16} + \frac{y^2}{25} < 1$, so we shade the region containing this point. Next, we graph the hyperbola $\frac{x^2}{25} - \frac{y^2}{4} = 1$ as a broken curve. In the inequality $\frac{x^2}{25} - \frac{y^2}{4} > 1$, the test points $(-6, 0)$ and $(6, 0)$ give us true statements and the test point $(0, 0)$ gives us a false statement. So we shade the regions containing the points $(-6, 0)$ and $(6, 0)$. The graphs have no overlapping regions. So there are no real solutions.

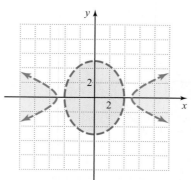

Graph the solutions of the system:

$$x^2 + y^2 \le 1$$

$$\frac{x^2}{9} + \frac{y^2}{4} \ge 1$$

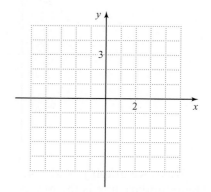

Now, let's look at how nonlinear systems of inequalities are applied in real-world situations.

EXAMPLE 5

In business, a company breaks even when the cost, C, of producing its product equals its revenue, R. When its cost is greater than its revenue, the company has a loss. When its revenue exceeds its cost, the company makes a profit. The profit region of a company that produces novelty items can be found by solving the following system:

$$C \geq 40x + 500$$
$$R \leq 100x - 0.2x^2$$

where x is the number of novelty items produced and sold each month. Find the profit region of this company.

Solution To solve the system, we graph each inequality on the same coordinate plane. First, we graph $C \geq 40x + 500$ as a solid line. Choose a test point, say $(0, 0)$. Since this test point does not satisfy the inequality, we shade the region above the line.

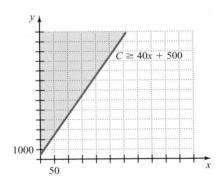

Next, we draw the parabola $R = 100x - 0.2x^2$ as a solid curve. Then we choose a test point, say $(10, 0)$. Since the test point $(10, 0)$ satisfies the inequality, we shade the region below the parabola.

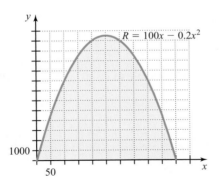

The profit region of the company is the overlapping region.

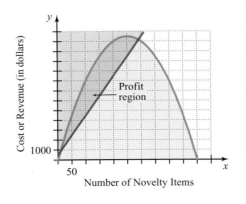

PRACTICE 5

Two underground water sprinklers are placed 2 ft apart. Each sprinkler waters a circular area of the lawn within a 3-ft radius. The region of lawn that the two sprinklers both water can be found by solving the following system:

$$x^2 + y^2 \leq 9$$
$$(x - 2)^2 + y^2 \leq 9$$

Find this region.

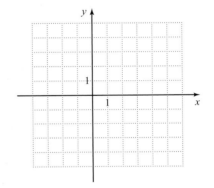

11.5   **Exercises**   FOR EXTRA HELP

Mathematically Speaking

Fill in each blank with the most appropriate term or phrase from the given list.

solid	each	broken
below	at least one	above

1. In graphing a nonlinear inequality that involves either the $<$ or the $>$ symbol, a(n) _____ curve is drawn.

2. In graphing a nonlinear inequality that involves either the $\leq$ or the $\geq$ symbol, a(n) _____ curve is drawn.

3. A solution of a nonlinear system of inequalities must satisfy _____ inequality in the system.

4. The graph of $y < 3x^2 - 7x$ is the region _____ the parabola $y = 3x^2 - 7x$.

Graph the solutions of each nonlinear inequality.

5. $x^2 + y^2 > 16$

6. $x^2 + y^2 \geq 4$

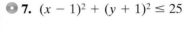 7. $(x - 1)^2 + (y + 1)^2 \leq 25$

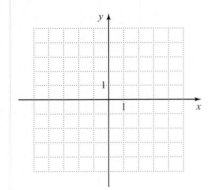   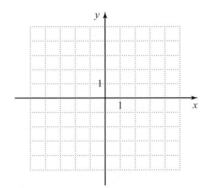   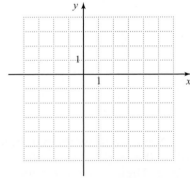

8. $(x + 3)^2 + (y - 2)^2 < 1$

9. $y > -x^2 + 5x - 4$

10. $y < 2x^2 - 8x$

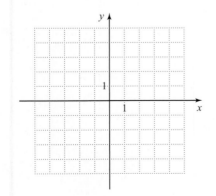   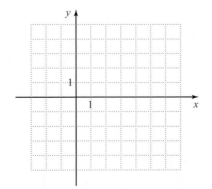   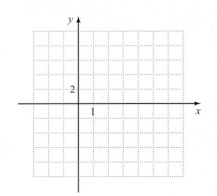

11. $9x^2 + y^2 \geq 36$

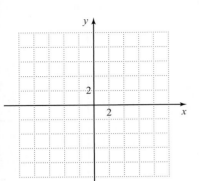

12. $x^2 + 16y^2 > 64$

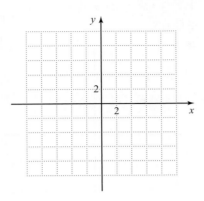

13. $16x^2 + 36y^2 < 144$

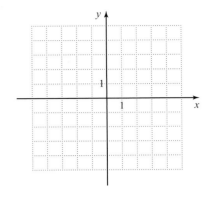

14. $25x^2 + 9y^2 \leq 225$

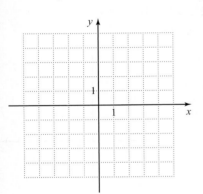

15. $4x^2 - y^2 > 100$

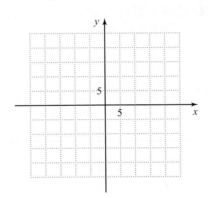

16. $2y^2 - 8x^2 < 32$

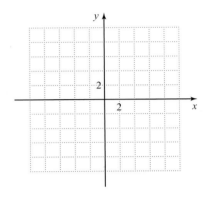

Graph the solutions of each nonlinear system of inequalities.

17. $y > x^2 - 3$
  $y \leq -x + 1$

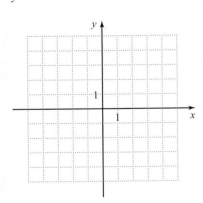

18. $y \leq -x^2 + 5$
  $y \leq \dfrac{1}{2}x - 1$

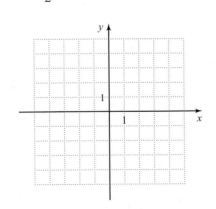

19. $x^2 + y^2 \geq 9$
  $x^2 + y^2 < 25$

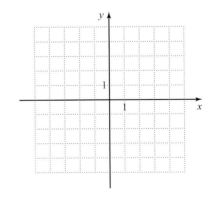

20. $x^2 + y^2 > 16$
$x^2 + y^2 < 4$

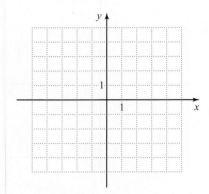

21. $y > x^2 - 3x + 2$
$y \le -x^2 - 2x - 4$

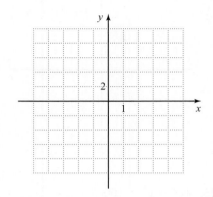

22. $y < x^2 + 2x + 1$
$y < -x^2 - 2x - 4$

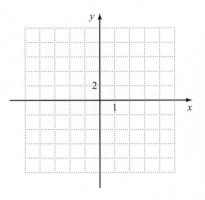

23. $8x^2 + 2y^2 \le 72$
$x^2 + 4y^2 \le 36$

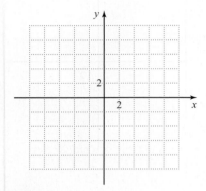

24. $12x^2 + 27y^2 > 108$
$25x^2 + y^2 < 25$

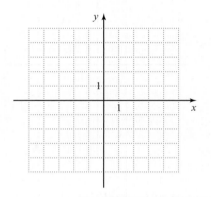

25. $x^2 - 4y^2 < 16$
$x^2 + y^2 \ge 1$

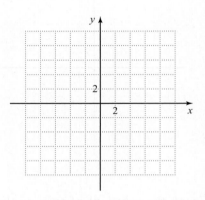

26. $16x^2 + 4y^2 > 64$
$y^2 - 9x^2 \le 36$

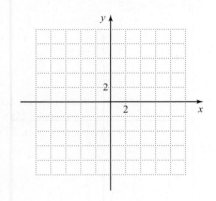

Mixed Practice

Graph the solutions of each nonlinear system of inequalities.

27. $x^2 + y^2 < 4$
$y \geq x^2 + 3$

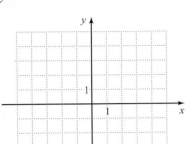

28. $x > y^2 - 2y + 3$
$x \geq -y^2 + 2y - 3$

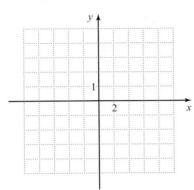

29. $x^2 + 16y^2 > 16$
$4x^2 + 9y^2 < 36$

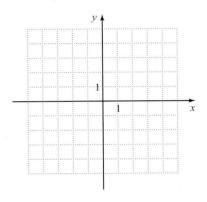

Graph the solutions of each nonlinear inequality.

30. $y \leq -x^2 + 2x + 3$

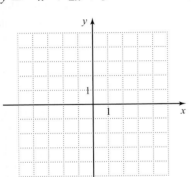

31. $4x^2 + y^2 > 16$

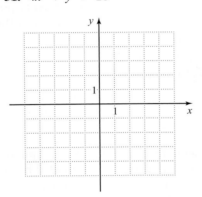

32. $9x^2 - y^2 \leq 36$

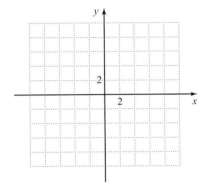

Applications

Solve.

33. In a search-and-rescue mission, a team maps out an elliptical search region from the last known location of a pair of hikers. The search region can be represented by the inequality $4x^2 + 9y^2 \leq 144$. Graph this region.

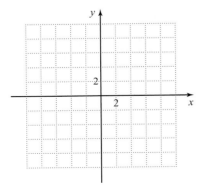

34. As a safety precaution, emergency workers evacuated residents within a 2-mi radius of a chemical spill. The evacuated region can be represented by the inequality $x^2 + y^2 \leq 4$. Graph this region.

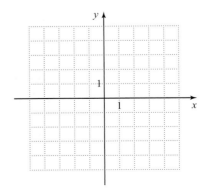

35. A homeowner wants to plant a flower garden and a vegetable garden on two separate square plots of land. She determines that the length of the vegetable garden is to be at least 2 ft longer than the length of the flower garden and that the combined area of the two gardens is not to exceed 2500 ft².

  a. Write a system of inequalities to represent the situation, letting x represent the length of the flower garden and y represent the length of the vegetable garden.

  b. Graph the system.

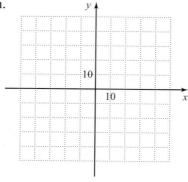

  c. Are all points in the solution region possible solutions within the context of the situation? Explain.

36. A company determines that its cost (in dollars) of producing x machine parts is given by $C = 0.3x^2 + 9$ and that its revenue (in dollars) for selling x machine parts is given by $R = 12x - 0.2x^2$.

  a. Write a system of inequalities that would represent the profit region for producing and selling machine parts.

  b. Graph the system.

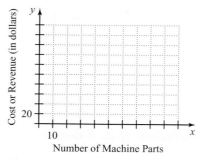

Number of Machine Parts

  c. Is every point in the solution region a possible solution within the context of the situation? Explain.

37. A classic rock radio station broadcasts a signal that can be received within a 30-mi radius of the station's transmitting tower. An R&B radio station located 50 mi away broadcasts a signal that can be received within a 40-mi radius of its tower. The system of inequalities

$$x^2 + y^2 \leq 900$$
$$(x - 50)^2 + y^2 \leq 1600$$

can be used to determine the region in which both radio station signals can be received. Find this region.

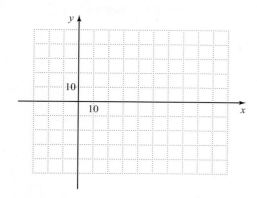

38. A hotel builds an elliptical swimming pool surrounded by an elliptical walkway. The system of inequalities

$$\frac{x^2}{400} + \frac{y^2}{225} \geq 1$$
$$\frac{x^2}{625} + \frac{y^2}{400} \leq 1$$

represents the region used for the walkway. Show this region.

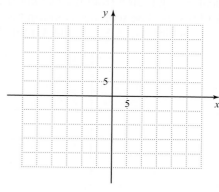

● *Check your answers on page A-51.*

MINDSTRETCHERS

Mathematical Reasoning

1. Is it possible for a single point to be the only solution of a system of nonlinear inequalities? If so, give an example of such a system. If not, explain why.

Critical Thinking

2. Graph the solution of the following system:

$$x^2 + 9y^2 \geq 36$$
$$x^2 - 4y^2 < 16$$
$$x - 2y > 4$$

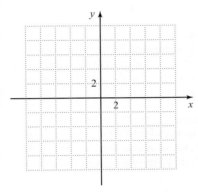

Groupwork

3. Working with a partner, write a system of inequalities with the solutions shown on the following graph.

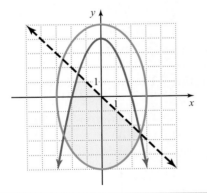

KEY CONCEPTS AND SKILLS

(CONCEPT)   (SKILL)

Concept/Skill	Description	Example
[11.1] Conic Section	A curve formed by the intersection of a plane and a (double) cone.	 Circle Ellipse Parabola Hyperbola
[11.1] Equation of a Parabola That Opens Upward or Downward	The equation $y = a(x - h)^2 + k$ is written in **standard form**, where (h, k) is the vertex of the associated parabola and the equation of the axis of symmetry is $x = h$.	The equation $y = 2(x - 4)^2 + 5$ is in standard form with vertex $(4, 5)$ and axis of symmetry $x = 4$.
[11.1] Equation of a Parabola That Opens to the Left or Right	The equation $x = a(y - k)^2 + h$ is written in **standard form**, where (h, k) is the vertex of the associated parabola and the equation of the axis of symmetry is $y = k$.	The equation $x = -2(y - 5)^2 + 4$ is in standard form with vertex $(4, 5)$ and axis of symmetry $y = 5$.
[11.2] Distance Formula	The distance between any two points (x_1, y_1) and (x_2, y_2) on a coordinate plane is given by the formula $$d = \sqrt{(x_2 - x_1)^2 + (y_2 - y_1)^2}$$	The distance between $(2, 6)$ and $(-3, 1)$ is $$\begin{aligned} d &= \sqrt{(-3 - 2)^2 + (1 - 6)^2} \\ &= \sqrt{(-5)^2 + (-5)^2} \\ &= \sqrt{25 + 25} \\ &= \sqrt{50} \\ &= 5\sqrt{2} \quad \text{units} \end{aligned}$$
[11.2] Midpoint Formula	The line segment with endpoints (x_1, y_1) and (x_2, y_2) has as its midpoint $$\left(\frac{x_1 + x_2}{2}, \frac{y_1 + y_2}{2} \right)$$	The midpoint of the line segment joining $(6, 7)$ and $(-3, 5)$ is $$\left(\frac{6 + (-3)}{2}, \frac{7 + 5}{2} \right) = \left(\frac{3}{2}, 6 \right)$$
[11.2] Circle	The set of all points on a coordinate plane that are a fixed distance from a given point. The given point is called the **center** and is represented by (h, k). The fixed distance is called the **radius** of the circle and is represented by r.	
[11.2] Equation of a Circle Centered at the Origin	The equation of a circle with center $(0, 0)$ and with radius r is $x^2 + y^2 = r^2$.	$x^2 + y^2 = 9^2$ is the equation of a circle with center $(0, 0)$ and radius 9.
[11.2] Equation of a Circle Centered at (h, k)	The equation of the circle with center (h, k) and with radius r is $(x - h)^2 + (y - k)^2 = r^2$.	$(x - 3)^2 + (y - 2)^2 = 7^2$ is the equation of a circle with center $(3, 2)$ and radius 7.

continued

Concept/Skill	Description	Example
[11.3] **Ellipse**	The set of all points on a coordinate plane the *sum* of whose distances from two fixed points is constant. The fixed points of this graph are called **foci** (the plural of **focus**). The point halfway between the two foci is called the **center** of the ellipse.	
[11.3] **Equation of an Ellipse Centered Around the Origin**	The equation of an ellipse with center $(0, 0)$, x-intercepts $(-a, 0)$ and $(a, 0)$, and y-intercepts $(0, -b)$ and $(0, b)$ is $\dfrac{x^2}{a^2} + \dfrac{y^2}{b^2} = 1$.	$\dfrac{x^2}{3^2} + \dfrac{y^2}{4^2} = 1$ is the equation of an ellipse with center $(0, 0)$, x-intercepts $(-3, 0)$ and $(3, 0)$, and y-intercepts $(0, -4)$ and $(0, 4)$.
[11.3] **Hyperbola**	The set of all points on a coordinate plane that satisfy the following property: The *difference* of the distances from two fixed points is constant. Each fixed point is said to be a **focus** of the hyperbola. The point halfway between the two foci is called the **center** of the hyperbola. Each **branch** of a hyperbola extends indefinitely.	
[11.3] **Equation of a Hyperbola Centered at the Origin**	The equation of a hyperbola with center $(0, 0)$ and x-intercepts $(-a, 0)$ and $(a, 0)$ is $\dfrac{x^2}{a^2} - \dfrac{y^2}{b^2} = 1$. The equation of a hyperbola with center $(0, 0)$ and y-intercepts $(0, -b)$ and $(0, b)$ is $\dfrac{y^2}{b^2} - \dfrac{x^2}{a^2} = 1$.	$\dfrac{x^2}{5^2} - \dfrac{y^2}{6^2} = 1$ is the equation of a hyperbola with center $(0, 0)$ and x-intercepts $(-5, 0)$ and $(5, 0)$. $\dfrac{y^2}{6^2} - \dfrac{x^2}{5^2} = 1$ is the equation of a hyperbola with center $(0, 0)$ and y-intercepts $(0, -6)$ and $(0, 6)$.
[11.4] **Nonlinear System of Equations**	A system of equations in which at least one equation is nonlinear.	$x^2 + y^2 = 25$ $x - y = 1$
[11.4] **Solution of a System of Nonlinear Equations**	An ordered pair of numbers that makes each equation in the system true.	The ordered pairs $(4, 3)$ and $(-3, -4)$ are solutions of the system $x^2 + y^2 = 25$ $x - y = 1$
[11.5] **Solution of a System of Nonlinear Inequalities**	An ordered pair of real numbers that makes each inequality in the system true.	The ordered pair $(1, 2)$ is a solution of the system of inequalities $x^2 + y^2 \leq 16$ $y \leq x + 4$

continued

Summary of Conic Sections

Conic Section	Graph	Equation	Comment on Equation
Parabola		$y = ax^2 + bx + c$ $y = a(x - h)^2 + k$ for $a > 0$	Contains an x^2-term, but no y^2-term.
		$y = ax^2 + bx + c$ $y = a(x - h)^2 + k$ for $a < 0$	Contains an x^2-term, but no y^2-term.
		$x = ay^2 + by + c$ $x = a(y - k)^2 + h$ for $a > 0$	Contains a y^2-term, but no x^2-term.
		$x = ay^2 + by + c$ $x = a(y - k)^2 + h$ for $a < 0$	Contains a y^2-term, but no x^2-term.
Circle Centered at the Origin and Circle Centered at (h, k)		$x^2 + y^2 = r^2$ $(x - h)^2 + (y - k)^2 = r^2$	Contains an x^2-term and a y^2-term with the same positive coefficient.
Ellipse Centered at the Origin		$\dfrac{x^2}{a^2} + \dfrac{y^2}{b^2} = 1$	Contains an x^2-term and a y^2-term with different positive coefficients.
Hyperbola Centered at the Origin		$\dfrac{x^2}{a^2} - \dfrac{y^2}{b^2} = 1$	Contains an x^2-term with a positive coefficient and a y^2-term with a negative coefficient.
		$\dfrac{y^2}{b^2} - \dfrac{x^2}{a^2} = 1$	Contains a y^2-term with a positive coefficient and an x^2-term with a negative coefficient.

Chapter 11   Review Exercises

To help you review this chapter, solve these problems.

[11.1]   *Identify the vertex and axis of symmetry of each parabola. Then graph.*

1. $y = -2(x + 3)^2 + 5$

2. $x = 3(y - 2)^2 - 10$

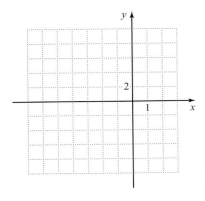

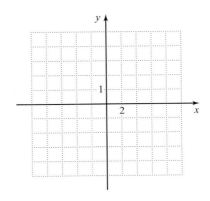

3. $x = 4y^2 + 32y + 64$

4. $y = -x^2 - 6x - 7$

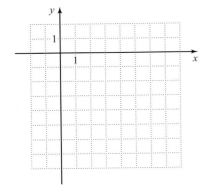

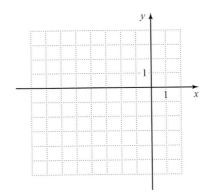

Write each equation in standard form. Then identify the vertex of each parabola.

5. $y = 8x^2 - 56x + 74$

6. $x = 5y^2 + 15y + 7$

[11.2]   *Find the distance between the two points on the coordinate plane. If necessary, round the answer to the nearest tenth.*

7. $(9, -5)$ and $(6, 1)$

8. $(-8, 10)$ and $(-2, 11)$

9. $(-1.3, -0.7)$ and $(-1.8, 0.5)$

10. $(\sqrt{7}, -\sqrt{2})$ and $(-2\sqrt{7}, -5\sqrt{2})$

Find the coordinates of the midpoint of the line segment joining each pair of points.

11. $(-16, 9)$ and $(24, 17)$

12. $(-7, -13)$ and $(10, -5)$

13. $\left(\dfrac{5}{2}, \dfrac{1}{5}\right)$ and $\left(-\dfrac{3}{4}, \dfrac{3}{10}\right)$

14. $(-0.35, -2.8)$ and $(-1.05, -4.5)$

Find the center and radius of each circle. Then graph.

15. $x^2 + y^2 = 64$

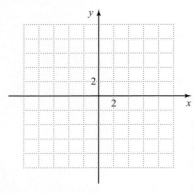

16. $(x + 2)^2 + (y - 4)^2 = 4$

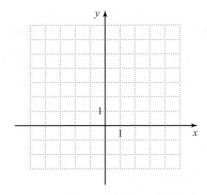

17. $x^2 + y^2 - 4x + 6y - 12 = 0$

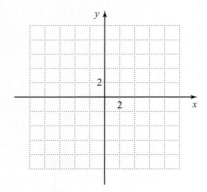

18. $x^2 + y^2 + 8x - 2y + 8 = 0$

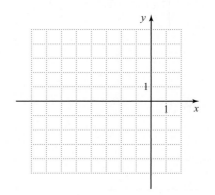

Use the given information to find the equation of each circle.

19. Center: $(9, 0)$; radius: 13

20. Center: $(-6, 10)$; radius: $2\sqrt{5}$

Find the center and radius of each circle.

21. $x^2 + y^2 - 12x - 14y - 35 = 0$

22. $x^2 + y^2 + 16x + 10y + 9 = 0$

[11.3]   *Graph.*

23. $\dfrac{x^2}{64} + \dfrac{y^2}{9} = 1$

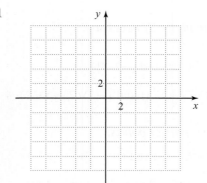

24. $\dfrac{x^2}{16} - \dfrac{y^2}{36} = 1$

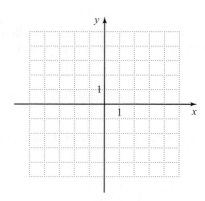

25. $y^2 - 4x^2 = 4$

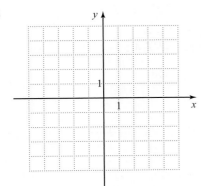

26. $4x^2 + y^2 = 64$

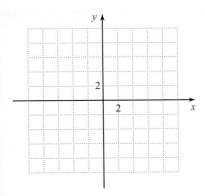

[11.4]   *Solve.*

27. $y = 2x^2 + 1$
  $y = 2x + 5$

28. $x^2 + y^2 = 10$
  $y = 3x$

29. $x^2 + 4y^2 = 16$
  $x = y^2 - 4$

30. $x^2 + y^2 = 8$
  $3x^2 + y^2 = 12$

31. $6x^2 + 2y^2 = 16$
  $2x^2 - y^2 = 17$

32. $4x^2 - 9y^2 = 36$
  $6x^2 + 6y^2 = 54$

[11.5]   *Graph the solutions of each nonlinear inequality.*

33. $y < -x^2 + 1$

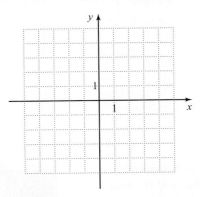

34. $(x + 3)^2 + y^2 \geq 4$

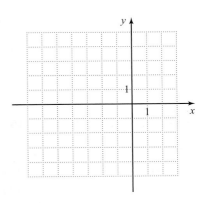

35. $8x^2 + 2y^2 \leq 32$

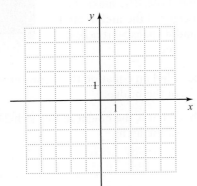

36. $9y^2 - 36x^2 > 144$

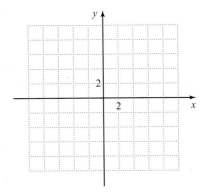

Solve each nonlinear system of inequalities by graphing.

37. $x^2 + y^2 < 64$

$y \geq \dfrac{1}{2}x + 3$

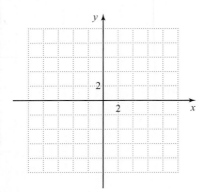

38. $x \geq y^2 - 4$

$x \leq -y^2 + 4$

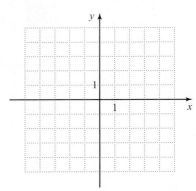

39. $x^2 + y^2 < 16$

$x^2 + y^2 > 49$

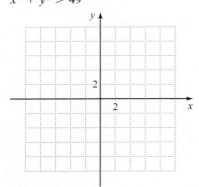

40. $4x^2 + 25y^2 \leq 100$

$9x^2 - y^2 \geq 9$

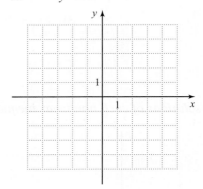

Mixed Applications

Solve.

41. A small manufacturing plant determines that its revenue, R (in dollars), for selling x units of a product is given by $R = 20x - 0.5x^2$.

   a. Write the equation in standard form.

   b. Identify the vertex and explain its significance within the context of the situation.

42. An arch supporting a bridge is in the shape of a parabola. The maximum height of the arch is 50 ft and it spans 80 ft across a river.

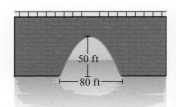

   a. Find an equation that represents the parabolic arch of the bridge.

   b. What is the height of the arch 24 ft from the center of the arch?

43. A student leaving his apartment drives 8 mi east and 15 mi south to get to school. From there he drives 4 mi west and 7 mi north to get to work. Find the distance between the student's apartment and work.

44. An amusement park Ferris wheel has a diameter of 120 ft. Find an equation that represents the wheel. Assume the center of the wheel is at the origin of a coordinate plane.

45. A satellite 500 mi above the Earth's surface has a circular orbit. The radius of the Earth is approximately 4000 mi. Find an equation that represents the circular orbit of the satellite.

46. Pluto travels in an elliptical orbit with the Sun at one focus. An equation that approximates Pluto's orbit is given by $\dfrac{x^2}{16} + \dfrac{y^2}{15} = 1$, where units are in billions of kilometers. The minimum distance between Pluto and the Sun is approximately 3 billion km. What is the maximum distance between Pluto and the Sun?

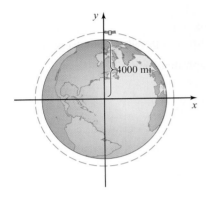

47. Suppose the playing surface of an elliptical pool table is 8 ft long and 4 ft wide. Find an equation of the ellipse that represents the playing surface of the pool table. Assume that the center of the pool table is at the origin of a coordinate plane.

48. As part of a physics experiment, a student throws a tennis ball straight downward with an initial velocity of 32 feet per second from a height of 120 ft. The height, h (in feet), of the ball above the ground t sec after it is thrown is given by $h = -16t^2 - 32t + 120$. At the same time, another student drops a golf ball from a height of 80 ft. The height of the golf ball above the ground t sec after it is dropped is given by $h = -16t^2 + 80$. How long after the balls are released will they be at the same height?

49. In a particular school district, a student living within a 1-mi radius of the school is not eligible for bus service.

   a. Write an inequality that represents the eligibility region. Assume that the school is at the origin of a coordinate plane.

   b. Graph this region.

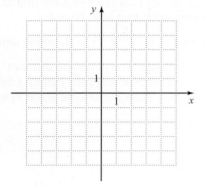

50. The owner of a daycare center wants to build two square sandboxes: one in the toddler playground and one in the preschool playground. The combined areas of the two sandboxes must not exceed 225 ft^2. The length of the sandbox in the preschool playground is to be greater than the length of the sandbox in the toddler playground. The system of inequalities

$$x^2 + y^2 \le 225$$
$$x < y$$

where x represents the length of the sandbox in the toddler playground and y represents the length of the sandbox in the preschool playground, can be used to determine the possible dimensions of the sandboxes.

   a. Graph this system.

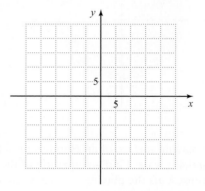

   b. Are all points in the solution region possible solutions within the context of the situation? Explain.

● *Check your answers on page A-52.*

Chapter 11   POSTTEST

    Test solutions are found on the enclosed CD.

To see if you have mastered the topics in this chapter, take this test.

1. Write in standard form the equation of the parabola $x = -2y^2 + 16y - 91$. Then identify the vertex.

2. Find the distance between each pair of points.

   a. $(-2, -1)$ and $(-4, 1)$

   b. $(7.9, -2.4)$ and $(-4.1, 2.6)$

3. Find the midpoint of the line segment joining each pair of points.

   a. $(-8, -15)$ and $(-4, 9)$

   b. $(-3, 7)$ and $(-6, 3)$

4. Find the equation of a circle with center $(-2, 8)$ and radius $3\sqrt{2}$.

5. Find the center and radius of the circle:
$x^2 + y^2 + 10x - 2y + 14 = 0$

Graph.

6. $x = 3(y - 1)^2 - 5$

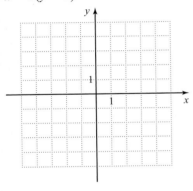

7. $x^2 + y^2 = 144$

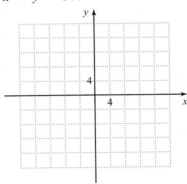

8. $(x - 4)^2 + (y + 5)^2 = 1$

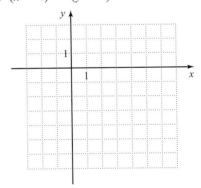

9. $\dfrac{x^2}{100} + \dfrac{y^2}{64} = 1$

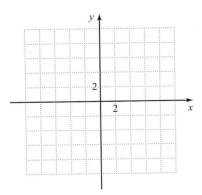

10. $64x^2 + 4y^2 = 256$

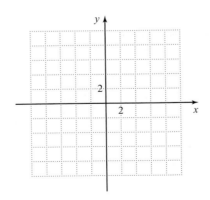

11. $\dfrac{y^2}{49} - \dfrac{x^2}{25} = 1$

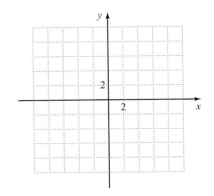

12. $6x^2 - 54y^2 = 216$

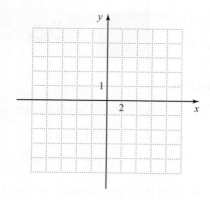

Solve.

13. $x^2 + y^2 = 25$
$y = x + 5$

14. $y = x^2 + 4$
$y = 2x^2 - 6$

15. $3x^2 + y^2 = 9$
$9x^2 + 2y^2 = 15$

Solve by graphing.

16. $x^2 + 9y^2 > 36$

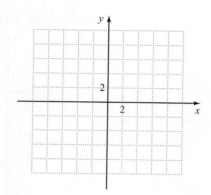

17. $4x^2 + 16y^2 \leq 64$
$4x^2 - y^2 \leq 100$

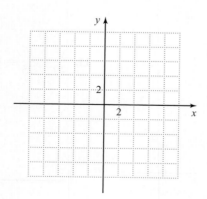

18. The curves of the cables suspended between two towers of a suspension bridge can be approximated by a parabola. Suppose that the distance between two towers that rise 110 m above the road is 900 m. If the center of the cable is 2 m above the road, find an equation in standard form that represents the parabola formed by the cable. (See diagram.)

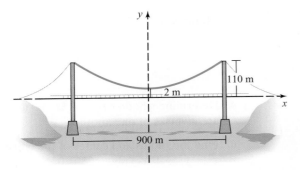

19. Cricket, a team sport similar to baseball, is played on an elliptical field. Find an equation of the ellipse shown.

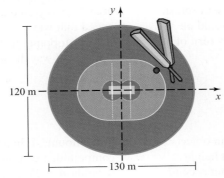

20. The owner of a factory determines that his cost, C (in dollars), for producing x units of machine parts is $C = 0.5x^2 + 500$ and that his revenue, R (in dollars), for selling x units is $R = 50x$.

a. Write a system of inequalities that represents the profit region of the factory.

b. Graph the system.

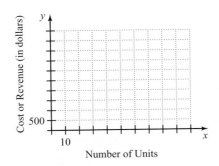

● *Check your answers on page A-54.*

Cumulative Review Exercises

To help you review, solve the following problems.

1. Find the slope-intercept form of the equation of a line perpendicular to the line $y = \frac{3}{4}x - 1$ that contains the point (9, 4).

2. Multiply: $(2x + y)(x - 3y)^2$

3. Solve: $\sqrt{2n + 5} - 2 = \sqrt{4n - 7}$

4. Determine whether $f(x) = 2x - 3$ and $g(x) = \frac{x + 3}{2}$ are inverses of one another.

5. Solve: $32^{x-4} = 16^{2x}$

6. Find the radius and center of the circle: $x^2 + y^2 - 10x - 4y + 25 = 0$

7. An investor wants to invest twice the amount of money in a high-risk fund that has a guaranteed rate of return of $5\frac{1}{2}\%$ as in a fund that has a guaranteed rate of return of 4%. What is the minimum amount that should be invested in each fund if she wants to have a return of at least $1500?

8. On a road trip, a group of students traveled 192 mi before stopping at a rest area. After leaving the rest area, they traveled an additional 224 mi at an average speed that was 8 mph slower than their average speed before stopping. If the entire trip took 7 hr of driving time, what was the average speed during each part of the trip?

9. The height, h (in feet), above the ground of an object t sec after it is thrown straight upward from a height of 144 feet with an initial velocity of 24 feet per second is given by the function $h(t) = -16t^2 + 24t + 144$.

   a. How far above the ground is the object 2 sec after it is thrown?

   b. When does the object reach its maximum height? What is the maximum height?

   c. When is the object 104 ft above the ground?

10. The amount, A, in an account at an annual interest rate r compounded continuously after t yr is given by the formula $A = Pe^{rt}$, where P is the initial amount invested. If $6800 is invested in an account at 3.85% interest compounded continuously, approximately how long will it take the amount in the account to grow to $10,000?

● *Check your answers on page A-54.*

Transition to Intermediate Algebra

A.1 Review of Introductory Algebra

To help you review introductory algebra (covered in Chapters 1–7), try the following selected exercises, folding the page to hide the answer column.

Topic	Exercise	Answer	Section
1. Calculating with real numbers	Simplify: $(-10) \div 2 - 4(-5)$	15	[1.2–1.4]
2. Translating	Translate to an algebraic expression: twice the sum of x and 6	$2(x + 6)$	[1.6]
3. Combining like terms	Simplify: $8n - 3 - 2(n + 1)$	$6n - 5$	[1.7]
4. Evaluating algebraic expressions	Evaluate for $a = -5$ and $b = 10$: $2a^3 + b^2$	-150	[1.8]
5. Solving linear equations	Solve and check: $\frac{3}{8}x = -9$	$x = -24$	[2.2]
6. Solving linear equations by combining properties	Solve and check: $2y + 1 = 6y - 1$	$y = \frac{1}{2}$ or 0.5	[2.3]
7. Formulas	Solve for a in terms of b and c: $7a + 2b = c$	$a = \dfrac{c - 2b}{7}$	[2.4]
8. Percent	20% of what amount is $4.50?	$22.50	[2.5]
9. Solving linear inequalities	Solve and graph: $5x + 12 \le 8x$ (number line: $-2\ -1\ 0\ 1\ 2\ 3\ 4\ 5\ 6$)	$x \ge 4$ (number line with solid dot at 4, arrow right: $-2\ -1\ 0\ 1\ 2\ 3\ 4\ 5\ 6$)	[2.6]

continued

Topic	Exercise	Answer	Section
10. The coordinate plane	Plot the point with coordinates $(2, -1)$.		[3.1]
11. Slope of a line	Is the slope of the line positive, negative, zero, or undefined?	Positive	[3.2]
12. Graphing linear equations	Graph: $2x + 5y = 10$		[3.3–3.4]

continued

Topic	Exercise	Answer	Section
13. Graphing linear inequalities	Graph: $y > 2x$		[3.5]
14. Functions	Identify the domain and range of $f(x) = 3x^2$	Domain: $(-\infty, \infty)$; range: $[0, \infty)$	[3.6]
15. Solving systems of linear equations by graphing	Solve by graphing: $x + y = 3$ $x - y = 5$	$(4, -1)$	[4.1]
16. Solving systems of linear equations algebraically	Solve: $x = 2y + 13$ $x - y = 9$	$(5, -4)$	[4.2–4.3]
17. Laws of exponents	Simplify: $\dfrac{6x^{-1}}{x}$	$\dfrac{6}{x^2}$	[5.1]
18. Scientific notation	Express in scientific notation: 0.00091	9.1×10^{-4}	[5.2]
19. Polynomials	Simplify and write in descending order of powers: $x + x^2 - 1 - 3x^2 + 4$	$-2x^2 + x + 3$	[5.3]
20. Adding and subtracting polynomials	Subtract: $(4n^2 + n) - (3n^2 + 5n - 8)$	$n^2 - 4n + 8$	[5.4]

continued

Topic	Exercise	Answer	Section
21. Multiplying polynomials	Multiply: $(2x - 5)(x + 3)$	$2x^2 + x - 15$	[5.5]
22. Special products	Multiply: $2p(3p + 1)^2$	$18p^3 + 12p^2 + 2p$	[5.6]
23. Common factoring	Factor: $6x^2y - 10xy + 12xy^3$	$2xy(3x - 5 + 6y^2)$	[6.1]
24. Factoring trinomials	Factor: $10x^2 - 11x - 6$	$(2x - 3)(5x + 2)$	[6.2–6.3]
25. Special factoring	Factor: $9n^2 - 24n + 16$	$(3n - 4)^2$	[6.4]
26. Solving quadratic equations by factoring	Solve: $x(3x + 20) = -12$	$-6, -\dfrac{2}{3}$	[6.6]
27. Multiplying and dividing rational expressions	Perform the indicated operation: $\dfrac{3p + 12}{3p - 15} \div \dfrac{p + 4}{2p - 10}$	2	[7.2]
28. Adding and subtracting rational expressions	Simplify: $\dfrac{1}{x + y} - \dfrac{1}{y}$	$-\dfrac{x}{y(x + y)}$	[7.3]
29. Solving rational equations	Solve: $\dfrac{3}{4} - \dfrac{1}{x} = \dfrac{1}{x - 2}$	$\dfrac{2}{3}, 4$	[7.5]
30. Ratio and proportion	Solve: $\dfrac{3}{n} = \dfrac{10}{21}$	6.3 or $\dfrac{63}{10}$	[7.6]

If you had trouble solving any exercises, you can review the appropriate material in a variety of ways. Study the explanations in the related sections and in the Key Concepts and Skills tables found at the end of each chapter. Get additional practice by trying the Exercises at the end of the sections or by trying the Posttests and Review Exercises at the end of the chapters. Take advantage of the student supplements, described in detail in the preface, that go along with this text:

- Student's Solution Manual
- Video lectures on CD or DVD
- Pass the Test CD
- Worksheets
- MathXL Tutorials on CD
- Pearson Tutor Center

A.2   Solving Compound Inequalities

In Section 2.6, we discussed how to solve linear inequalities. Here, we consider more complex inequalities that result from joining linear inequalities. Consider this example.

White blood cells, also called leukocytes, help protect the body against disease and infection. A cubic millimeter of blood contains from 4000 to 10,000 leukocytes. The table shows the condition of white blood cells depending on the number N of leukocytes in 1 mm^3. (*Source:* Priscilla Lemone and Karen M. Burke, *Medical Surgical Nursing: Critical Thinking and Client Care*, Prentice-Hall, 2000)

Leukopenia (more at risk of catching an infection)	Normal Range	Leukocytosis (indication of a possible infection in the body)
$N \leq 4000$	$4000 < N < 10{,}000$	$N \geq 10{,}000$

A normal range of white blood cells in 1 mm^3 of blood can be expressed by two inequalities,

$$4000 < N \quad \text{and} \quad N < 10{,}000,$$

or by joining these inequalities to get

$$4000 < N < 10{,}000.$$

Definition

Two inequalities that are joined by the word *and* or the word *or* form a **compound inequality**.

Let's look at the graphs of $4000 < N$, $N < 10{,}000$, and $4000 < N < 10{,}000$. For the first graph, recall that $4000 < N$ can be written as $N > 4000$.

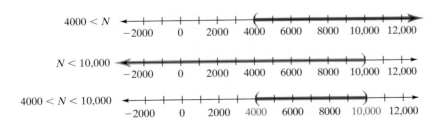

The last graph shows that a solution of $4000 < N < 10{,}000$ is the intersection of the solutions of the inequalities $4000 < N$ and $N < 10{,}000$.

We read $4000 < N < 10{,}000$ as "N is greater than 4000 and less than 10,000." Another way to read this inequality is "N is between 4000 and 10,000, excluding 4000 and 10,000."

We can use interval notation, inequality notation, and graphs to represent the solutions of a compound inequality in one variable, as shown in the following table. Note that when showing the graphs of compound inequalities, it is common practice to use parentheses and brackets rather than circles to indicate the endpoints of intervals on the number line. Parentheses replace open circles, and brackets replace closed circles.

INTERVALS, INEQUALITIES, AND GRAPHS

Interval Notation	Inequality Notation	Graph*	Meaning
(a, b)	$a < x < b$		All real numbers between a and b, excluding a and b
$[a, b]$	$a \le x \le b$		All real numbers between a and b, including a and b
$[a, b)$	$a \le x < b$		All real numbers between a and b, including a and excluding b
$(a, b]$	$a < x \le b$		All real numbers between a and b, excluding a and including b

Solving Inequalities with the Word *And*

A *solution of a compound inequality* joined by *and* is the intersection of the solutions of the individual inequalities. In other words, a solution is any value of the variable that makes *both* inequalities true.

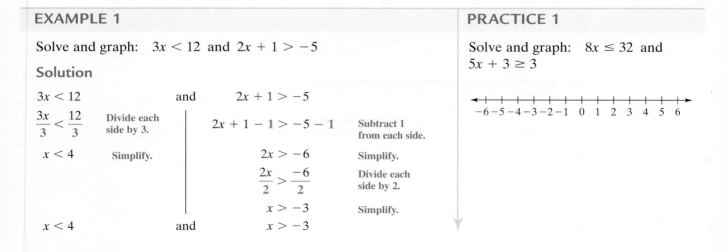

EXAMPLE 1

Solve and graph: $3x < 12$ and $2x + 1 > -5$

Solution

$3x < 12$ and $2x + 1 > -5$

$\dfrac{3x}{3} < \dfrac{12}{3}$	Divide each side by 3.	$2x + 1 - 1 > -5 - 1$ Subtract 1 from each side.
$x < 4$	Simplify.	$2x > -6$ Simplify.
		$\dfrac{2x}{2} > \dfrac{-6}{2}$ Divide each side by 2.
		$x > -3$ Simplify.

$x < 4$ and $x > -3$

PRACTICE 1

Solve and graph: $8x \le 32$ and $5x + 3 \ge 3$

Since $x > -3$ is the same as $-3 < x$, we can write $x < 4$ and $x > -3$ as one inequality, getting

$$-3 < x < 4.$$

The solutions of this inequality are all values of x that satisfy both $x < 4$ and $x > -3$, and its graph is the intersection, or overlap, of the graphs of $x < 4$ and $x > -3$, as shown.

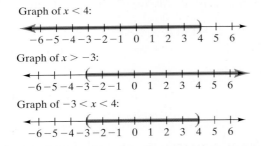

The middle graph can be written as $(-3, \infty)$, where the infinity sign indicates that the interval extends without bound in the positive direction. The top graph can be written as $(-\infty, 4)$, where the negative infinity symbol indicates that the interval extends without bound in the negative direction. The graph of $-3 < x < 4$ is the intersection of the intervals $(-\infty, 4)$ and $(-3, \infty)$. This graph can also be written $(-\infty, 4) \cap (-3, \infty)$, where the symbol $\cap$ denotes the intersection of the solution intervals. Note that the graph of $-3 < x < 4$ includes all points between 23 and 4, excluding 23 and 4.

So the solutions of $3x < 12$ and $2x + 1 > -5$ are all real numbers between -3 and 4, or, in interval notation, $(-3, 4)$.

EXAMPLE 2

Solve and graph: $-3 \le 8x + 5 \le 21$

Solution Write the compound inequality as two inequalities joined by *and*, then solve each inequality.

$$-3 \le 8x + 5 \le 21$$

$-3 \le 8x + 5$	and	$8x + 5 \le 21$
$-3 - 5 \le 8x + 5 - 5$		$8x + 5 - 5 \le 21 - 5$	Subtract 5 from each side.
$-8 \le 8x$		$8x \le 16$	Simplify.
$\dfrac{-8}{8} \le \dfrac{8x}{8}$		$\dfrac{8x}{8} \le \dfrac{16}{8}$	Divide each side by 8.
$-1 \le x$	and	$x \le 2$	

We can combine both inequalities to get

$$-1 \le x \le 2.$$

PRACTICE 2

Solve and graph: $-5 < 6x + 1 < 7$

The graph of $-1 \leq x \leq 2$ is

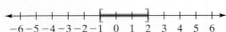

So the solutions of $-3 \leq 8x + 5 \leq 21$ are all real numbers between -1 and 2, including -1 and 2, or, in interval notation, $[-1, 2]$.

Let's consider an alternative way of solving the inequality in Example 2.

$$-3 \leq 8x + 5 \leq 21$$
$$-3 - 5 \leq 8x + 5 - 5 \leq 21 - 5 \qquad \text{Subtract 5 from each part of the compound inequality.}$$
$$-8 \leq 8x \leq 16$$
$$\frac{-8}{8} \leq \frac{8x}{8} \leq \frac{16}{8} \qquad \text{Divide each part of the compound inequality by 8.}$$
$$-1 \leq x \leq 2$$

Note that this method has fewer steps than the method shown above.

EXAMPLE 3	PRACTICE 3

Solve: $1 - 2x \geq 0$ and $2x + 7 \geq 11$

Solution

$$\begin{array}{ccc}
1 - 2x \geq 0 & \text{and} & 2x + 7 \geq 11 \qquad \text{Solve each inequality.}\\
-2x \geq -1 & & 2x \geq 4\\
\dfrac{-2x}{-2} \leq \dfrac{-1}{-2} & & \dfrac{2x}{2} \geq \dfrac{4}{2}\\
x \leq \dfrac{1}{2} & \text{and} & x \geq 2
\end{array}$$

Since there is no real number that is both less than or equal to $\frac{1}{2}$ and greater than or equal to 2, there is no solution. By examining the graphs of $x \leq \frac{1}{2}$ and $x \geq 2$, we can also see that there are no intersecting points.

Solve: $3x - 2 < 1$ and $4x - 1 > 19$

Solving Inequalities with the Word *Or*

Now let's focus on compound inequalities that are joined by *or*. A *solution of a compound inequality* joined by *or* is the union of the solutions of the individual inequalities. That is, a solution is any value of the variable that makes *either* inequality true.

EXAMPLE 4

Solve and graph: $7 - 3x \geq 1$ or $5x + 2 \geq 22$

Solution

$$7 - 3x \geq 1 \qquad\qquad\qquad \text{or} \qquad 5x + 2 \geq 22$$
$$-3x \geq -6 \qquad\qquad\qquad\qquad\qquad 5x \geq 20$$
$$\frac{-3x}{-3} \leq \frac{-6}{-3} \quad \substack{\text{Divide each side by } -3 \\ \text{and reverse the direc-} \\ \text{tion of the inequality.}} \qquad \frac{5x}{5} \geq \frac{20}{5}$$
$$x \leq 2 \qquad\qquad\qquad \text{or} \qquad\qquad x \geq 4$$

The solutions of this inequality are all values of x that satisfy either $x \leq 2$ or $x \geq 4$. Its graph is the union of the graphs of $x \leq 2$ and $x \geq 4$.

Graph of $x \leq 2$:

Graph of $x \geq 4$:

Graph of $x \leq 2$ or $x \geq 4$:

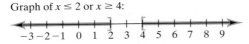

The solutions of $7 - 3x \geq 1$ or $5x + 2 \geq 22$ are all real numbers less than or equal to 2 *or* greater than or equal to 4, or in interval notation, the union of the intervals $(-\infty, 2]$ and $[4, \infty)$. We can also write this as $(-\infty, 2] \cup [4, \infty)$, where the symbol $\cup$ denotes the union of the solution intervals.

Are numbers in the interval $(2, 4)$ solutions of the compound inequality $7 - 3x \geq 1$ or $5x + 2 \geq 22$ in Example 4? Explain.

PRACTICE 4

Solve and graph:
$4x - 9 > 3$ or $8 - x > 10$

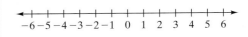

EXAMPLE 5

Solve and graph: $3x - 4 < 11$ or $-2x + 5 \leq 3$

Solution

$$3x - 4 < 11 \qquad \text{or} \qquad -2x + 5 \leq 3$$
$$3x < 15 \qquad\qquad\qquad\qquad -2x \leq -2$$
$$\frac{3x}{3} < \frac{15}{3} \qquad\qquad \frac{-2x}{-2} \geq \frac{-2}{-2} \quad \substack{\text{Divide each side by } -2 \\ \text{and reverse the direction} \\ \text{of the inequality.}}$$
$$x < 5 \qquad \text{or} \qquad x \geq 1$$

PRACTICE 5

Solve and graph: $-3x - 2 < 7$ or $5x + 1 \leq 11$

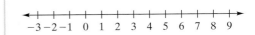

The solutions of this inequality are all values of x that satisfy either $x < 5$ or $x \geq 1$. By looking at the graphs of $x < 5$ and $x \geq 1$, we see that the union of the graphs is the entire real-number line.

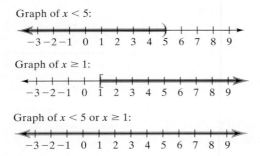

Graph of $x < 5$:

Graph of $x \geq 1$:

Graph of $x < 5$ or $x \geq 1$:

The solutions of $3x - 4 < 11$ or $-2x + 5 \leq 3$ are all real numbers less than 5 *or* greater than or equal to 1. Another way to express the solution is the union of the intervals $(-\infty, 5)$ and $[1, \infty)$. Since every real number is either less than 5 or greater than or equal to 1, the solutions are all real numbers, or $(-\infty, \infty)$ in interval notation.

Rewrite the inequality using interval notation. Then graph the inequality.

1. $2 < x < 5$

2. $1 \le x \le 6$

3. $-3 \le x < 0$

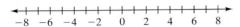

4. $-5 < x \le -0.5$

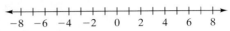

5. $x < -4$ or $x > 3$

6. $x \le 1$ or $x \ge 2.5$

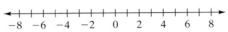

7. $x < -2$ or $x \ge 0.5$

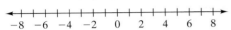

8. $x \le -3$ or $x > 0$

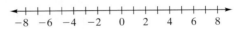

Solve and graph the inequality.

9. $2x > -10$ and $3x - 1 < 8$

10. $5x - 7 > 8$ and $4x < 16$

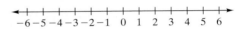

11. $\frac{2}{3}t > -4$ and $9 - 4t \ge -11$

12. $-\frac{n}{4} \le 1$ and $2n - 17 < -17$

13. $7 \le 6x + 1$ and $-4x \ge 28$

14. $6 < -5x - 4$ and $2x > 11$

15. $-41 < 9a - 5 \le 13$

16. $-17 \le 3y + 10 < 1$

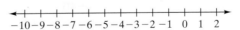

17. $-13 < \frac{4x}{5} - 9 < -7$

18. $-13 \le \frac{2n}{3} - 7 \le -10$

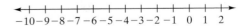

19. $8 \le 12 - h$ and $-\frac{2}{3}h > 1$

20. $-6 > -a - 1$ and $\frac{2}{5}a > 4$

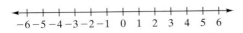

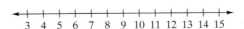

21. $0 \le -8x - 20 \le 24$

22. $-9 < 15 - 6x < 30$

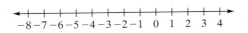

23. $2x - 3 < -5$ or $3x + 4 > 10$

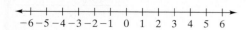

24. $5x + 2 \leq -18$ or $4x - 1 \geq 15$

25. $6 - r > 14$ or $-11 \leq 3r - 8$

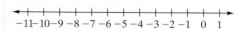

26. $-16 \geq 19 - 7n$ or $11n + 23 < -10$

27. $18 - 10x \geq 23$ or $26 + 9x \geq -28$

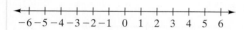

28. $8x + 13 < 25$ or $-9 - 6x < 33$

29. $-17 \leq 16 - 11a$ or $-5 < -7a - 12$

30. $19 > -5t + 24$ or $-27 \leq 13t - 14$

31. $\dfrac{n}{2} + 8 > 11$ or $9 - \dfrac{2n}{3} \geq 6$

32. $\dfrac{x}{7} - 10 \leq -10$ or $\dfrac{3x}{4} + 5 > 8$

33. $4x + 19 > -19$ or $15 - 3x \geq 26$

34. $-7 < 2x + 7$ or $6x + 21 \leq 30$

Solve.

35. $7x - 9 > 4x - 18$ and $6x + 5 < x$

36. $2x + 11 \leq 3x + 17$ and $8x - 9 \leq x + 12$

37. $5x + 9 \leq 6x - 3$ or $10x - 7 < 4x + 5$

38. $12x < 14x - 10$ or $3x + 8 \geq 7x - 8$

39. $3x - 9 \leq -2 - \dfrac{x}{2}$ and $15x - 20x < -40$

40. $13x + 2 > 16x$ and $4 - \dfrac{x}{3} \leq x - 2$

41. $\dfrac{1}{2}x - 21 < \dfrac{1}{4}x - 10$ or $6x + 19 > 4x + 5$

42. $x + 1 > \dfrac{7}{8}x + \dfrac{3}{4}$ or $12x - 23 < 8x - 15$

43. $2(5x + 1) > 9x + 1$ and $11x - 7 \geq 4x$

44. $13 - 3x > -8$ and $12x + 7 < -(1 - 10x)$

45. $-15 \leq 5(x - 7) < 20$

46. $-16 < -4(9 - x) \leq 24$

47. $1 \leq -\dfrac{1}{3}(4x - 27) < 17$

48. $-7 < \dfrac{1}{2}(16 - 5x) < 23$

49. $13 \leq 8x - 3(4x + 1) \leq 25$

50. $-15 < 21 - 2(3x + 9) < 0$

51. $-8.3 < 1.7 - 0.5(7x - 1) < 3.95$

52. $-10.3 < 2(1.4x - 5.1) - 0.1 \leq 6.5$

● Check your answers on page A-54.

A.3 Absolute Value Equations and Inequalities

Recall from Section 1.1 that the *absolute value* of a number is its distance from 0 on a number line. The absolute value of the number x is written $|x|$.

Definition

For any real number x,

$$|x| = x \text{ when } x \geq 0 \quad \text{and} \quad |x| = -x \text{ when } x < 0.$$

Let's look at $|4|$ and $|-4|$ on the number line.

The distance from 0 to 4 or from 0 to -4 is 4 units. So we write:

$$|4| = 4 \qquad |-4| = 4$$

Now let's look at the distance between -2 and 4 on the number line.

Another way to find the distance between -2 and 4 on the number line is to find the absolute value of the difference of the numbers, as shown:

$$|-2 - 4| = |-6| = 6 \quad \text{or} \quad |4 - (-2)| = |6| = 6$$

Explain why the order in which we subtract does not matter when finding the distance between two points on a number line.

Definition

For any real numbers a and b, the *distance* between them on a number line is $|a - b|$ or $|b - a|$.

Solving Absolute Value Equations

We can use a number line to solve an *absolute value equation*.

EXAMPLE 1

Solve $|x| = 2$ using a number line.

Solution Using a number line, we find all the numbers that are 2 units from 0.

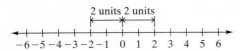

Both -2 and 2 are two units away from 0 on the number line.

Check

$|x| = 2$		
$|-2| \overset{?}{=} 2$	Substitute -2 for x.	
$2 = 2$	True	

$|x| = 2$		
$|2| \overset{?}{=} 2$	Substitute 2 for x.	
$2 = 2$	True	

So the solutions are -2 and 2.

PRACTICE 1

Solve $|x| = 5$ using a number line.

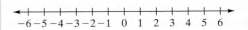

EXAMPLE 2

Solve:

a. $|x| = 0$ **b.** $|x| = -1$

Solution

a. $|x| = 0$
    $x = 0$

Zero is the only number with absolute value 0. So the solution is 0.

b. $|x| = -1$
No solution

Since the absolute value represents distance, it cannot be negative. So there is no solution.

PRACTICE 2

Solve.

a. $0 = |y|$

b. $|y| = -\dfrac{1}{2}$

The previous examples suggest the following rule for solving linear absolute value equations:

> ### To Solve Absolute Value Equations
>
> For any positive number a and any algebraic expression X,
> * if $|X| = a$, then $X = a$ or $X = -a$.
> * if $|X| = 0$, then $X = 0$.
> * if $|X| = -a$, then the equation has no solution.

We can use this rule and the properties of equality to solve absolute value equations.

EXAMPLE 3	PRACTICE 3
Solve: $5\|x\| - 4 = 11$	Solve: $2\|y\| - 9 = 1$

Solution

$$5|x| - 4 = 11$$
$$5|x| - 4 + 4 = 11 + 4 \qquad \text{Add 4 to each side of the equation.}$$
$$5|x| = 15 \qquad \text{Simplify.}$$
$$\frac{5|x|}{5} = \frac{15}{5} \qquad \text{Divide each side of the equation by 5.}$$
$$|x| = 3$$
$$x = 3 \quad \text{or} \quad x = -3$$

So the solutions are 3 and -3. Note that since the absolute value is equal to a positive number, 3, there are two solutions.

EXAMPLE 4	PRACTICE 4
Solve: $\|x - 8\| = 5$	Solve: $\|x + 6\| = 4$

Solution

$$|x - 8| = 5$$
$$|X| \quad = a$$

Since 5 is positive, $x - 8 = 5$ or $x - 8 = -5$. Solving each equation, we get

$$x - 8 = 5 \qquad \text{or} \quad x - 8 = -5$$
$$x = 13 \quad \text{or} \qquad x = 3$$

So the solutions are 13 and 3.

EXAMPLE 5	PRACTICE 5
Solve: $\|2y + 5\| = 9$	Solve: $\|3x - 1\| = 8$

Solution

$$|2y + 5| = 9$$
$$|X| \quad = a$$

Since 9 is positive, $2y + 5 = 9$ or $2y + 5 = -9$.

$$2y + 5 = 9 \qquad \text{or} \quad 2y + 5 = -9$$
$$2y = 4 \qquad \qquad 2y = -14$$
$$y = 2 \quad \text{or} \qquad y = -7$$

So the solutions are 2 and -7.

EXAMPLE 6

Solve: $-3|x - 3| = 9$

Solution

$$-3|x - 3| = 9$$
$$|x - 3| = -3 \qquad \text{Divide each side of the equation by } -3.$$
$$|X| = -a$$

Since the absolute value cannot be negative, the equation has no solution.

PRACTICE 6

Solve: $-\dfrac{1}{2}|x - 1| = 4$

Some equations have two absolute value expressions. Consider, for instance, the equation $|x| = |y|$. We are told that x and y are equal in absolute value. This means that they are the same distance from 0 on the number line. So they must be either equal to each other or opposites of each other. We can express this as follows:

$$|x| = |y| \qquad \text{means} \qquad x = y \qquad \text{or} \qquad x = -y$$

Equal in absolute value   Equal to each other   Opposites of each other

EXAMPLE 7

Solve and check: $|2y + 1| = |y + 2|$

Solution Since the expressions have the same absolute value, either $2y + 1 = y + 2$ or $2y + 1 = -(y + 2)$. Solving each equation, we get

$$|2y + 1| = |y + 2|.$$

$$2y + 1 = y + 2 \quad \text{or} \quad 2y + 1 = -(y + 2)$$
$$y = 1 \qquad\qquad 2y + 1 = -y - 2$$
$$3y = -3$$
$$\text{or} \qquad y = -1$$

Check

Substitute 1 for y.

$$|2y + 1| = |y + 2|$$
$$|2(1) + 1| \stackrel{?}{=} |1 + 2|$$
$$|2 + 1| \stackrel{?}{=} |1 + 2|$$
$$|3| = |3|$$
$$3 = 3 \quad \text{True}$$

Substitute -1 for y.

$$|2y + 1| = |y + 2|$$
$$|2(-1) + 1| \stackrel{?}{=} |-1 + 2|$$
$$|-2 + 1| \stackrel{?}{=} |-1 + 2|$$
$$|-1| = |1|$$
$$1 = 1 \quad \text{True}$$

So the solutions are 1 and -1.

PRACTICE 7

Solve and check: $|5x + 2| = |4x + 1|$

EXAMPLE 8

Solve: $|x - 2| = |x - 1|$

Solution We know that $x - 2 = x - 1$ or $x - 2 = -(x - 1)$. Solving each equation, we get:

$$|x - 2| = |x - 1|$$

$$x - 2 = x - 1 \qquad \text{or} \quad x - 2 = -(x - 1)$$
$$-2 = -1 \quad \text{False} \qquad \qquad x - 2 = -x + 1$$
$$2x = 3$$
$$x = \frac{3}{2}$$

Since $-2 = -1$ is a false statement, there is no solution to the equation $x - 2 = x - 1$. So the only solution to the original equation is $\frac{3}{2}$.

PRACTICE 8

Solve: $|3 - x| = |x + 4|$

Solving Absolute Value Inequalities

As with absolute value equations, we can use a number line to solve an *absolute value inequality*.

EXAMPLE 9

Solve $|x| < 2$ using a number line.

Solution The inequality $|x| < 2$ represents all real numbers whose distance from 0 is less than 2 units. The graph of $|x| < 2$ is

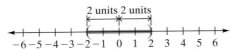

We can check the graph by substituting any number between -2 and 2, say -1, in the original inequality.

$$|x| < 2$$
$$|-1| \overset{?}{<} 2$$
$$1 < 2 \quad \text{True}$$

So the solutions of $|x| < 2$ are all real numbers between -2 and 2, that is, all real numbers that satisfy the compound inequality $-2 < x < 2$. We can also express the solutions in interval notation as $(-2, 2)$.

PRACTICE 9

Solve $|x| < 5$ using a number line.

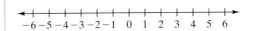

EXAMPLE 10

Solve $|x| > 2$ using a number line.

Solution The inequality $|x| > 2$ represents all real numbers whose distance from 0 is greater than 2 units. The graph of $|x| > 2$ is

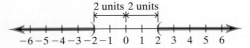

So the solutions of $|x| > 2$ are all real numbers that are less than -2 or greater than 2, which can be written as the compound inequality $x < -2$ or $x > 2$. We can express the solutions in interval notation as the union of $(-\infty, -2)$ and $(2, \infty)$; that is, $(-\infty, -2) \cup (2, \infty)$.

PRACTICE 10

Solve $|x| > 5$ using a number line.

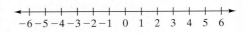

The preceding examples suggest the following rule for solving absolute value inequalities:

> **To Solve Absolute Value Inequalities**
>
> For any positive number a and any algebraic expression X,
> - if $|X| < a$, then $-a < X < a$.
> - if $|X| > a$, then $X < -a$ or $X > a$.
>
> Similar rules hold for $|X| \le a$ and $|X| \ge a$.

We can use this rule to solve absolute value inequalities.

EXAMPLE 11

Solve and graph: $|x - 3| \le 4$

Solution

$|x - 3| \le 4$

$-4 \le x - 3 \le 4$ Write a compound inequality.

$-4 + 3 \le x - 3 + 3 \le 4 + 3$ Add 3 to each part of the compound inequality.

$-1 \le x \le 7$ Simplify.

The graph of $|x - 3| \le 4$ is

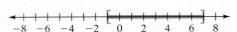

So the solutions of $|x - 3| \le 4$ are all real numbers between -1 and 7, including -1 and 7, written in interval notation as $[-1, 7]$.

PRACTICE 11

Solve and graph: $|x + 2| < 5$

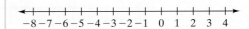

EXAMPLE 12

Solve and graph: $|x + 5| > 3$

Solution

$|x + 5| > 3$

$\quad x + 5 < -3 \quad$ or $\quad x + 5 > 3 \qquad$ Write a compound inequality.

$\qquad x < -8 \qquad\qquad x > -2 \qquad$ Subtract 5 from each side.

The graph of $|x + 5| > 3$ is

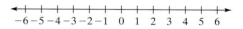

So the solutions of $|x + 5| > 3$ are all real numbers less than -8 or greater than -2, written in interval notation as the union of $(-\infty, -8)$ and $(-2, \infty)$; that is, $(-\infty, -8) \cup (-2, \infty)$.

PRACTICE 12

Solve and graph: $|x - 2| \geq 4$

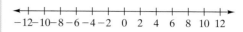

EXAMPLE 13

Solve and graph: $|2x - 5| < 7$

Solution

$|2x - 5| < 7$

$\quad -7 < 2x - 5 < 7 \qquad$ Write a compound inequality.

$\quad -2 < 2x < 12 \qquad\qquad$ Add 5 to each part of the compound inequality.

$\quad -1 < x < 6 \qquad\qquad$ Divide each part of the compound inequality by 2.

The graph of $-1 < x < 6$ is

So the solutions of $|2x - 5| < 7$ are all real numbers between -1 and 6, written in interval notation as $(-1, 6)$.

PRACTICE 13

Solve and graph: $|3x - 1| \leq 4$

EXAMPLE 14

Solve and graph: $|1 - 2x| \geq 2$

Solution

$$|1 - 2x| \geq 2$$

$1 - 2x \leq -2$ or	$1 - 2x \geq 2$	Write a compound inequality.
$-2x \leq -3$	$-2x \geq 1$	Subtract 1 from each side.
$x \geq \dfrac{3}{2}$	$x \leq -\dfrac{1}{2}$	Divide each side by -2 and reverse the inequality.

The graph of $x \geq \dfrac{3}{2}$ or $x \leq -\dfrac{1}{2}$ is

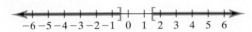

So the solutions of $|1 - 2x| \geq 2$ are all real numbers less than or equal to $-\dfrac{1}{2}$ or greater than or equal to $\dfrac{3}{2}$. We can express the solutions in interval notation as the union of $\left(-\infty, -\dfrac{1}{2}\right]$ and $\left[\dfrac{3}{2}, \infty\right)$, that is, $\left(-\infty, -\dfrac{1}{2}\right] \cup \left[\dfrac{3}{2}, \infty\right)$.

PRACTICE 14

Solve and graph: $|3 - 4x| > 1$

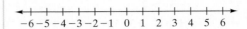

A.3 Exercises

Find the distance between the two real numbers on a number line.

1. -3 and 3
2. -5 and 6
3. -11 and -4
4. -13 and -2
5. 1.5 and 9
6. 0.6 and 8.3

Solve.

7. $|x| = 7$
8. $|x| = 1$
9. $|n| = \dfrac{2}{3}$
10. $|y| = \dfrac{5}{2}$

11. $|6x| = 24$
12. $|5n| = 10$
13. $2|y| = -1.6$
14. $-3|x| = 12.3$

15. $|x| + 1 = 1$
16. $|x| + 7 = 7$
17. $|x| - 4 = -2$
18. $|z| - 5 = 6$

19. $3|n| + 10 = 7$
20. $2|x| - 4 = 8$
21. $12 - 4|y| = -16$
22. $-6|x| - 3 = -33$

23. $|4x| + 9 = 15$
24. $|3x| + 11 = 12$
25. $\left|\dfrac{2}{3}n\right| - 5 = -1$
26. $\left|\dfrac{1}{2}x\right| - 5 = 0$

27. $|y + 7| = 4$
28. $|n - 9| = 12$
29. $|x - 3| = -1$
30. $|x + 6| = 10$

31. $|2z + 13| = 21$
32. $|3n - 8| = -5$
33. $|4x - 11| = 17$
34. $|5x + 3| = 2$

35. $2|3n - 1| = 16$
36. $4|4 - 2x| = 24$
37. $-|10 - 6z| = 7$
38. $3|3n + 2| = -9$

39. $-\dfrac{1}{3}|8 - 7n| = -12$
40. $-\dfrac{1}{4}|5x + 6| = -4$
41. $|5x| = |7x - 24|$
42. $|4y + 1| = |6y|$

43. $\left|\dfrac{1}{2}x + 3\right| = \left|\dfrac{1}{4}x\right|$
44. $\left|\dfrac{1}{3}n - 8\right| = \left|\dfrac{1}{5}n\right|$
45. $|2x - 3| = |x + 9|$
46. $|n - 4| = |3n + 2|$

47. $|5y + 1| = |6y - 1|$
48. $|4z - 10| = |3z + 10|$
49. $|11 - 7x| = |5 - 9x|$
50. $|12 - 3z| = |15 - 6z|$

51. $|x + 7| = |x + 1|$
52. $|x - 8| = |x + 6|$
53. $|4 - n| = |n - 2|$
54. $|y - 9| = |5 - y|$

55. $-|3t + 4| = -|2t - 1|$
56. $-|5n - 3| = -|n - 4|$
57. $|4a + 7| = -|8 - a|$
58. $-|2y + 6| = |3y - 2|$

Solve and then graph.

59. $|x| > 3$
60. $|x| \ge 4$

61. $|x| \le \dfrac{1}{2}$

$-6\ -5\ -4\ -3\ -2\ -1\ \ 0\ \ 1\ \ 2\ \ 3\ \ 4\ \ 5\ \ 6$

62. $|x| < \dfrac{3}{2}$

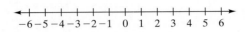

63. $|7x| \ge 21$

$-6\ -5\ -4\ -3\ -2\ -1\ \ 0\ \ 1\ \ 2\ \ 3\ \ 4\ \ 5\ \ 6$

64. $|5x| \ge 10$

$-6\ -5\ -4\ -3\ -2\ -1\ \ 0\ \ 1\ \ 2\ \ 3\ \ 4\ \ 5\ \ 6$

65. $|x + 4| < 4$

$-12\ -11\ -10\ -9\ -8\ -7\ -6\ -5\ -4\ -3\ -2\ -1\ \ 0$

66. $|x - 1| \le 1$

$-6\ -5\ -4\ -3\ -2\ -1\ \ 0\ \ 1\ \ 2\ \ 3\ \ 4\ \ 5\ \ 6$

⊙ 67. $|x - 5| \ge 3$

$-1\ \ 0\ \ 1\ \ 2\ \ 3\ \ 4\ \ 5\ \ 6\ \ 7\ \ 8\ \ 9\ \ 10\ \ 11$

68. $|x + 7| > 1$

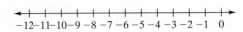

69. $|x| + 5 \ge 6$

$-6\ -5\ -4\ -3\ -2\ -1\ \ 0\ \ 1\ \ 2\ \ 3\ \ 4\ \ 5\ \ 6$

70. $|x| + 2 > 4$

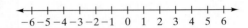

71. $|x| - 3 < -2$

$-6\ -5\ -4\ -3\ -2\ -1\ \ 0\ \ 1\ \ 2\ \ 3\ \ 4\ \ 5\ \ 6$

72. $|x| - 5 < -1$

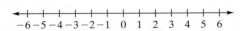

73. $|2x + 3| \ge 11$

$-12\ -10\ -8\ -6\ -4\ -2\ \ 0\ \ 2\ \ 4\ \ 6\ \ 8\ \ 10\ \ 12$

74. $|4x + 5| > 7$

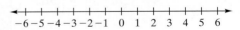

⊙ 75. $|6 - 3x| < 9$

$-6\ -5\ -4\ -3\ -2\ -1\ \ 0\ \ 1\ \ 2\ \ 3\ \ 4\ \ 5\ \ 6$

76. $|7 - 2x| \le 1$

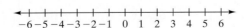

77. $\left|\dfrac{2}{3}x + 4\right| \ge 0$

$-6\ -5\ -4\ -3\ -2\ -1\ \ 0\ \ 1\ \ 2\ \ 3\ \ 4\ \ 5\ \ 6$

78. $\left|\dfrac{1}{2}x - 1\right| \ge 0$

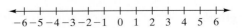

79. $-|2x - 7| < -6$

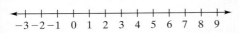

80. $-|5x - 8| \geq -2$

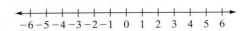

81. $|4x - 3| - 6 \leq 5$

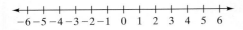

82. $|3x + 1| + 2 \geq 9$

• *Check your answers on page A-55.*

A.4   Systems of Linear Equations in Three Variables

In Chapter 4, we solved systems of linear equations in two variables. In this appendix, we learn how to solve systems of linear equations in three variables.

Introduction to Systems of Linear Equations in Three Variables

A solution of a system of linear equations is the values for the variables that satisfy all three equations of the system.

> **Definition**
>
> A **solution** of a system of linear equations in three variables is an **ordered triple** of numbers that makes all three equations in the system true.

EXAMPLE 1

Consider the following system of equations:

$$2x + y + 3z = 1$$
$$x - y - 2z = 6$$
$$3x + 2y = 11$$

a. Is $(3, 1, -2)$ a solution of the system?

b. Is $(0, 4, -1)$ a solution of the system?

Solution

a. To determine if the ordered triple $(3, 1, -2)$ is a solution of this system, we substitute 3 for x, 1 for y, and -2 for z and check if all three equations are true.

		Simplifies to	
$2x + y + 3z = 1 \rightarrow$	$2(3) + 1 + 3(-2) \stackrel{?}{=} 1$	$\longrightarrow 1 = 1$	True
$x - y - 2z = 6 \rightarrow$	$3 - 1 - 2(-2) \stackrel{?}{=} 6$	$\longrightarrow 6 = 6$	True
$3x + 2y = 11 \rightarrow$	$3(3) + 2(1) \stackrel{?}{=} 11$	$\longrightarrow 11 = 11$	True

The ordered triple $(3, 1, -2)$ is a solution of the system because it satisfies all three equations.

b. To determine if the ordered triple $(0, 4, -1)$ is a solution of this system, we substitute 0 for x, 4 for y, and -1 for z and check if all three equations are true.

		Simplifies to	
$2x + y + 3z = 1 \rightarrow$	$2(0) + 4 + 3(-1) \stackrel{?}{=} 1$	$\longrightarrow 1 = 1$	True
$x - y - 2z = 6 \rightarrow$	$0 - 4 - 2(-1) \stackrel{?}{=} 6$	$\longrightarrow -2 = 6$	False
$3x + 2y = 11 \rightarrow$	$3(0) + 2(4) \stackrel{?}{=} 11$	$\longrightarrow 8 = 11$	False

The ordered triple $(0, 4, -1)$ is not a solution of the system because it does not satisfy all three equations.

PRACTICE 1

Determine whether the ordered triple is a solution of the system.

$$x + 3y - 2z = 9$$
$$-2x + 4z = 0$$
$$3x - 5y + z = -29$$

a. $(6, 1, 3)$

b. $(-4, 3, -2)$

We can graph a linear equation in three variables, such as $2x + y + 3z = 1$. Although we are not considering such graphs in detail, we give a brief discussion of them since they can help us to visualize the possible number of solutions a system of three linear equations can have.

To begin with, an ordered triple of numbers (x, y, z) can be thought of as the coordinates of a point in space. The coordinates are defined with respect to three axes. The point $(2, 4, 3)$, for instance, is shown here.

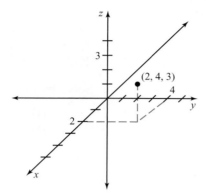

If we were to graph in this coordinate system any linear equation in three variables, we would find that the graph is a plane. So a system of three such equations has as its graph the points where three planes in space intersect. These planes can intersect in a variety of ways. The following graphs illustrate some of these ways.

One solution:
three planes that intersect
at one common point

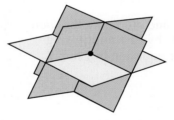

No solution:
three planes that do not
intersect

Infinitely many solutions:
three planes that intersect
at infinitely many points
along a line

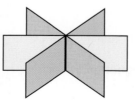

We see from these graphs that a system of linear equations in three variables can have one solution, no solution, or infinitely many solutions.

Solving Systems of Three Linear Equations by the Elimination Method

Just as we used the elimination method to solve systems of two linear equations, so we can use the elimination method to solve systems of three linear equations. To use this method, we eliminate one of the variables in order to get a system of two equations in two variables. Then we solve this system of two equations using the elimination method, as we did in Section 4.3.

EXAMPLE 2

Solve by the elimination method:

$$\begin{array}{rl} \textbf{(1)} & x - y - 2z = 4 \\ \textbf{(2)} & -x + 2y + z = 1 \\ \textbf{(3)} & -x + y - 3z = 11 \end{array}$$

Solution First, we decide which two equations to choose in order to eliminate one of the three variables. Since the coefficients of the x-terms in equations (1) and (2) are opposites, we can eliminate x if we add these equations, getting a linear equation in y and z. Let's call this equation in y and z equation (4).

$$\begin{array}{rl} \textbf{(1)} & x - y - 2z = 4 \\ \textbf{(2)} & \underline{-x + 2y + z = 1} \\ \textbf{(4)} & y - z = 5 \quad \text{Add the equations.} \end{array}$$

Next, we use a different pair of equations and eliminate the *same* variable, x. Let's add equations (2) and (3) by first multiplying equation (2) by -1. We get another linear equation in y and z. Let's call it equation (5).

$$\begin{array}{ll} \textbf{(2)} \;\; -x + 2y + z = 1 & \xrightarrow{\textbf{Multiply by } -1.} \quad x - 2y - z = -1 \\ \textbf{(3)} \;\; \underline{-x + y - 3z = 11} & \hspace{3.2cm} \underline{-x + y - 3z = 11} \quad \text{Add the} \\ & \textbf{(5)} \hspace{1.8cm} -y - 4z = 10 \quad\;\; \text{equations.} \end{array}$$

Now we solve equations (4) and (5) for y and z using the elimination method again. Note that if we add the equations, we can eliminate the y-terms and then solve for z.

$$\begin{array}{rl} \textbf{(4)} & y - z = 5 \\ \textbf{(5)} & \underline{-y - 4z = 10} \\ & \;\;\;-5z = 15 \quad \text{Add the equations.} \\ & \;\;\;\;\;\;\; z = -3 \end{array}$$

To find y, we can substitute -3 for z in either equation (4) or (5). Let's use equation (4).

$$\begin{array}{rl} \textbf{(4)} & y - z = 5 \\ & y - (-3) = 5 \\ & y + 3 = 5 \\ & y = 2 \end{array}$$

PRACTICE 2

Solve by the elimination method:

$$\begin{array}{rl} \textbf{(1)} & x + y + z = 2 \\ \textbf{(2)} & 2x - y + 5z = -5 \\ \textbf{(3)} & -x + 2y + 2z = 1 \end{array}$$

Finally, we substitute 2 for y and -3 for z in one of the equations of the *original* system and then solve for x. Let's use equation (1).

$$\begin{aligned}
\textbf{(1)} \qquad x - y - 2z &= 4 \\
x - 2 - 2(-3) &= 4 \\
x - 2 + 6 &= 4 \\
x + 4 &= 4 \\
x &= 0
\end{aligned}$$

Since $x = 0$, $y = 2$, and $z = -3$, the ordered triple $(0, 2, -3)$ is the solution.

Check To determine if the ordered triple $(0, 2, -3)$ is the solution to the original system, we check to see if this ordered triple satisfies all three equations.

$$\begin{aligned}
\textbf{(1)} \qquad x - y - 2z &= 4 \\
0 - 2 - 2(-3) &\overset{?}{=} 4 \\
-2 + 6 &\overset{?}{=} 4 \\
4 &= 4 \qquad \text{True}
\end{aligned}$$

$$\begin{aligned}
\textbf{(2)} \qquad -x + 2y + z &= 1 \\
-0 + 2(2) + (-3) &\overset{?}{=} 1 \\
4 + (-3) &\overset{?}{=} 1 \\
1 &= 1 \qquad \text{True}
\end{aligned}$$

$$\begin{aligned}
\textbf{(3)} \qquad -x + y - 3z &= 11 \\
-0 + 2 - 3(-3) &\overset{?}{=} 11 \\
2 + 9 &\overset{?}{=} 11 \\
11 &= 11 \qquad \text{True}
\end{aligned}$$

The ordered triple $(0, 2, -3)$ satisfies each of the original equations, so the solution is $(0, 2, -3)$. Throughout the remainder of this section, the check is left as an exercise.

Example 2 suggests the following procedure for solving systems of linear equations in three variables using the elimination method.

To Solve a System of Linear Equations in Three Variables by Elimination

- Write all equations in the general form $Ax + By + Cz = D$.
- Choose a pair of equations and use them to eliminate a variable, getting an equation in two variables.
- Next, use a different pair of equations to eliminate the *same* variable that was eliminated in the previous step, getting another equation in the same two variables.
- Solve the system of equations formed by the equations found in the previous two steps, getting the values of two of the variables.
- Substitute the values found in the previous step in any of the original equations to find the value of the third variable.
- Check by substituting the values found in the previous two steps in all three equations of the original system.

EXAMPLE 3

Solve by elimination:

$$\begin{array}{ll} \textbf{(1)} & 2x + 3y + 12z = 4 \\ \textbf{(2)} & 4x - 6y + 6z = 1 \\ \textbf{(3)} & x + y + z = 1 \end{array}$$

Solution Let's use equations (2) and (3) to eliminate y. First we multiply equation (3) by 6. Then we add the resulting equation to equation (2), getting an equation in x and z, which we call equation (4).

$$\begin{array}{ll} \textbf{(2)} & 4x - 6y + 6z = 1 \\ \textbf{(3)} & x + y + z = 1 \end{array} \xrightarrow{\text{Multiply by 6.}} \begin{array}{l} 4x - 6y + 6z = 1 \\ 6x + 6y + 6z = 6 \\ \hline \textbf{(4)} \quad 10x + 12z = 7 \end{array} \begin{array}{l} \text{Add the} \\ \text{equations.} \end{array}$$

Next, we use a different pair of equations and eliminate the same variable y. Let's multiply equation (3) by -3. Then we add the resulting equation to equation (1) to get another equation in x and z, which we call equation (5).

$$\begin{array}{ll} \textbf{(1)} & 2x + 3y + 12z = 4 \\ \textbf{(3)} & x + y + z = 1 \end{array} \xrightarrow{\text{Multiply by } -3.} \begin{array}{l} 2x + 3y + 12z = 4 \\ -3x - 3y - 3z = -3 \\ \hline \textbf{(5)} \quad -x + 9z = 1 \end{array} \begin{array}{l} \text{Add the} \\ \text{equations.} \end{array}$$

Now we solve equations (4) and (5) for x and z using elimination. If we multiply equation (5) by 10, we can eliminate the x-terms. Then we can solve for z.

$$\begin{array}{ll} \textbf{(4)} & 10x + 12z = 7 \\ \textbf{(5)} & -x + 9z = 1 \end{array} \xrightarrow{\text{Multiply by 10.}} \begin{array}{l} 10x + 12z = 7 \\ -10x + 90z = 10 \\ \hline 102z = 17 \end{array} \begin{array}{l} \text{Add the} \\ \text{equations.} \end{array}$$

$$z = \frac{17}{102} = \frac{1}{6}$$

We can substitute $\frac{1}{6}$ for z in either equation (4) or equation (5) to find x. Using equation (5), we get:

$$\begin{array}{ll} \textbf{(5)} & -x + 9z = 1 \end{array}$$

$$-x + 9\left(\frac{1}{6}\right) = 1$$

$$-x + \frac{9}{6} = 1$$

$$-x = -\frac{3}{6}$$

$$x = \frac{1}{2}$$

Finally, we substitute $\frac{1}{2}$ for x and $\frac{1}{6}$ for z in one of the original equations and solve for y. Let's use equation (3).

PRACTICE 3

Solve by elimination:

$$\begin{array}{ll} \textbf{(1)} & 2x + y + 2z = 1 \\ \textbf{(2)} & x + 2y + z = 2 \\ \textbf{(3)} & x - y - z = 0 \end{array}$$

$$(3) \; x + y + z = 1$$

$$\frac{1}{2} + y + \frac{1}{6} = 1$$

$$y + \frac{4}{6} = 1$$

$$y = \frac{2}{6}$$

$$y = \frac{1}{3}$$

So the solution is $x = \dfrac{1}{2}$, $y = \dfrac{1}{3}$, and $z = \dfrac{1}{6}$, that is, $\left(\dfrac{1}{2}, \dfrac{1}{3}, \dfrac{1}{6}\right)$.

Tip   When solving a system of linear equations in three variables, make sure that you eliminate the *same* variable in two pairs of equations to get a system of equations in two variables.

Some linear systems of equations in three variables have missing terms. When this is the case, we can omit one elimination step. In Example 4, each of the equations has a missing variable.

EXAMPLE 4	PRACTICE 4

EXAMPLE 4

Solve by the elimination method: **(1)** $\quad x - 2z = -5$
(2) $\;-2x + \; z = 4$
(3) $\quad -y + 3z = 3$

Solution The variable y is already eliminated in equations (1) and (2). Let's use these equations to eliminate x.

(1) $\quad x - 2z = -5 \xrightarrow{\text{Multiply by 2.}} 2x - 4z = -10$
(2) $\;-2x + \; z = \;\; 4 \qquad\qquad\quad \underline{-2x + z = 4}$
$\qquad\qquad\qquad\qquad\qquad\qquad\qquad -3z = -6 \quad$ **Add the equations.**
$\qquad\qquad\qquad\qquad\qquad\qquad\qquad\quad z = 2$

Now we can substitute 2 for z in either equation (1) or equation (2) to find x. Using equation (1), we get

(1) $\qquad x - 2z = -5$
$\qquad\qquad x - 2(2) = -5$
$\qquad\qquad\quad x - 4 = -5$
$\qquad\qquad\qquad\quad x = -1$

Finally, we substitute 2 for z in equation (3) to find y.

(3) $\qquad -y + 3z = 3$
$\qquad\qquad -y + 3(2) = 3$
$\qquad\qquad\quad -y + 6 = 3$
$\qquad\qquad\qquad\quad -y = -3$
$\qquad\qquad\qquad\qquad y = 3$

So the solution is $(-1, 3, 2)$.

PRACTICE 4

Solve by the elimination method:

(1) $\quad 3x + 4z = 5$
(2) $\quad 2x - 5y = 8$
(3) $\quad 2y + 3z = 2$

Just as with systems of linear equations in two variables, a linear system with three variables may have no solution or infinitely many solutions.

EXAMPLE 5

Solve by elimination:

(1) $x + 2y - 4z = 7$
(2) $x - y + z = 5$
(3) $2x + y - 3z = 6$

Solution Since the coefficients of y are opposites in equations (2) and (3), we can eliminate y by adding these equations.

(2) $x - y + z = 5$
(3) $\underline{2x + y - 3z = 6}$
(4) $\quad 3x - 2z = 11$ Add the equations.

Next, we eliminate y again by using equations (1) and (3).

(1) $x + 2y - 4z = 7$ $x + 2y - 4z = 7$
(3) $2x + y - 3z = 6$ $\xrightarrow{\text{Multiply by } -2.}$ $\underline{-4x - 2y + 6z = -12}$ Add the
(5) $\quad -3x + 2z = -5$ equations.

Now using equations (4) and (5), we get

(4) $\quad 3x - 2z = 11$
(5) $\underline{-3x + 2z = -5}$
$\quad\quad 0 = 6$ Add the equations.

Since the statement $0 = 6$ is false for all values of the variables, the original system has no solution common to all three equations in the system.

PRACTICE 5

Solve by elimination:

(1) $x - 3y + 2z = 1$
(2) $x - 2y + 3z = 5$
(3) $2x - 6y + 4z = 3$

EXAMPLE 6

Use the elimination method to solve:

(1) $x + 2y - 4z = 7$
(2) $2x + y - 3z = 12$
(3) $x - y + z = 5$

Solution Let's add equations (2) and (3) to eliminate y.

(2) $2x + y - 3z = 12$
(3) $\underline{x - y + z = 5}$
(4) $\quad 3x - 2z = 17$ Add the equations.

PRACTICE 6

Use the elimination method to solve:

(1) $x - y + z = 3$
(2) $2x - 2y - 2z = 6$
(3) $-4x + 4y - 4z = -12$

Next, we use equations (1) and (2) to eliminate y again.

(1) $x + 2y - 4z = 7$

(2) $2x + y - 3z = 12$ $\xrightarrow{\text{Multiply by } -2.}$

$\begin{array}{l} x + 2y - 4z = 7 \\ \underline{-4x - 2y + 6z = -24} \\ \textbf{(5)}\ \ -3x + 2z = -17 \end{array}$ Add the equations.

Now, we add equations (4) and (5) to solve for x and z.

(4) $3x - 2z = 17$

(5) $-3x + 2z = -17$

   $0 = 0$ **Add the equations.**

Since the statement $0 = 0$ is true for all values of the variables, the original system has infinitely many solutions.

How would you describe the graph of the original system in Example 6?

Tip   Using algebra to solve a system of linear equations in three variables:

- If the result is a false statement, then the system has no solution.
- If the result is a true statement, then the system has infinitely many solutions.

Determine if the given ordered triple is a solution of the given system.

1. $x + y - z = -6 \quad (2, -3, 5)$
$2x - y + z = 12$
$-3x - 2y - 4z = -20$

2. $4x - 2y + z = -1 \quad (-1, 0, 3)$
$x - 3y - 5z = -16$
$3x + y + z = 6$

3. $2x - 2y - 4z = 1 \quad \left(-2, -\dfrac{1}{2}, -1\right)$
$5x - 8y + 3z = 9$
$-x + 6y + 5z = -8$

4. $6x + 3y + 2z = 4 \quad \left(\dfrac{1}{3}, -2, 4\right)$
$-12x + 2y + 4z = 8$
$9x - 10y - 5z = 3$

5. $\dfrac{3}{5}x - \dfrac{1}{4}y \qquad = 10 \quad (20, 8, 7)$
$\dfrac{3}{20}x \qquad + 2z = 17$
$\qquad -9y + 10z = -2$

6. $\qquad \dfrac{1}{2}y - \dfrac{2}{3}z = 0 \quad \left(\dfrac{1}{10}, 12, 9\right)$
$30x \qquad + \dfrac{4}{9}z = 7$
$10x + \dfrac{1}{4}y \qquad = 4$

Solve by the elimination method.

7. $2x \qquad = 8$
$x \qquad - 4z = 12$
$3x - 2y + z = 0$

8. $x + 3y + z = 1$
$5y + 8z = 7$
$-3z = 3$

9. $x + y + z = -1$
$2x - y - z = -5$
$x + 2y - z = 6$

10. $x - y - z = 3$
$-x + y + 3z = 3$
$-x - 3y + 2z = 8$

11. $3x - y + 5z = -20$
$2x + 4y - 2z = 15$
$-4x - 2y - z = -1$

12. $2x + 5y - z = 4$
$x + 3y + z = -2$
$-3x - 2y + 3z = 1$

13. $-2x + 6y - 3z = 0$
$x - y + z = 2$
$2x - 5y - z = 1$

14. $6x - 2y + 7z = 10$
$-3x - y + z = 0$
$x + y + z = 4$

15. $8y + 3z = 3$
$5x + 4y = 2$
$10x + 6z = 4$

16. $12x \qquad - 14z = -2$
$6y - 7z = 7$
$6x + 2y \qquad = 4$

17. $\dfrac{2}{3}x - \dfrac{1}{2}y \qquad = 1$
$\dfrac{1}{4}y + \dfrac{1}{3}z = 2$
$\dfrac{1}{3}x \qquad - \dfrac{1}{2}z = -5$

18. $\qquad \dfrac{2}{3}y + \dfrac{3}{5}z = 2$
$\dfrac{3}{4}x - \dfrac{1}{2}y \qquad = 0$
$\dfrac{1}{4}x \qquad + \dfrac{2}{5}z = 1$

19. $x + 4y - 2z = 6$
$2x + 3y + 2z = 7$
$x - y + 4z = 1$

20. $6x + y - 5z = 8$
$3x - 2y + z = 4$
$x + y - 2z = 0$

21. $4x + 3y - 2z = -7$
$5x - 3y + 4z = 24$
$8x + 2y + z = 10$

22. $2x \qquad - 4z = 4$
$3x + 2y + 2z = 3$
$-x + 2y - 6z = -1$

23. $4x - y + 8z = -2$
$x + 2y + 4z = 3$
$2x - 5y \qquad = 1$

24. $x + 7y + 5z = 14$
$3x + y - 15z = 44$
$2x + 4y - 5z = 29$

25. $7x - 6y + 3z = 13$
$2x + 3y + 2z = -2$
$6x - 5y + 7z = -11$

26. $2x + 3y + 5z = 0$
$5x + 2y + 3z = 12$
$3x + 5y + 2z = 18$

• *Check your answers on page A-55.*

A.5 Solving Systems of Linear Equations by Using Matrices

In this appendix, we focus on a method of solving systems of two or three linear equations by using *matrices*. This procedure is based on the elimination method without writing the variables, that is, writing just the coefficients and the constants of the equations. This *matrix method* is often used on a graphing calculator or a computer when solving systems with many equations.

Solving Systems by Using Matrices

A rectangular array of numbers such as

$$
\begin{array}{ccc}
\text{Column} & \text{Column} & \text{Column} \\
1 & 2 & 3 \\
\downarrow & \downarrow & \downarrow
\end{array}
$$

$$
\begin{array}{l}
\text{Row 1} \longrightarrow \\
\text{Row 2} \longrightarrow
\end{array}
\begin{bmatrix}
2 & 3 & 8 \\
1 & 5 & 3
\end{bmatrix}
$$

is called a *matrix*.

Definitions

A **matrix** (plural, **matrices**) is a rectangular array of numbers such as $\begin{bmatrix} 3 & -2 \\ 0 & 8 \end{bmatrix}$. The numbers of the matrix are called **elements** or **entries**. The **rows** of a matrix are horizontal, and the **columns** are vertical. A **square matrix** has the same number of rows as columns.

Here are other examples of matrices.

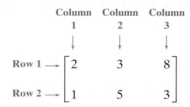

$$
\begin{array}{cc}
\text{Column 1} & \text{Column 2} \\
\downarrow & \downarrow
\end{array}
$$

$$
\begin{array}{l}
\text{Row 1} \longrightarrow \\
\text{Row 2} \longrightarrow
\end{array}
\begin{bmatrix}
-1 & 0 \\
0 & -1
\end{bmatrix}
$$

This square matrix has 2 rows and 2 columns. It is called a 2 × 2 (read "two by two") matrix.

$$
\begin{array}{cccc}
\text{Column} & \text{Column} & \text{Column} & \text{Column} \\
1 & 2 & 3 & 4 \\
\downarrow & \downarrow & \downarrow & \downarrow
\end{array}
$$

$$
\begin{array}{l}
\text{Row 1} \longrightarrow \\
\text{Row 2} \longrightarrow \\
\text{Row 3} \longrightarrow
\end{array}
\begin{bmatrix}
4 & 1 & 1 & 3 \\
-1 & 1 & -2 & -11 \\
1 & 2 & 2 & -1
\end{bmatrix}
$$

This matrix has 3 rows and 4 columns. It is called a 3 × 4 (read "three by four") matrix.

How many elements are there in an $n \times n$ matrix? In an $m \times n$ matrix? Explain.

Let's look at how a system of linear equations relates to a matrix.

System of Equations		Augmented Matrix
Equation (1)	$5x + y = 3$	Row (1)
Equation (2)	$-2x + 2y = 4$	Row (2)

$$\begin{bmatrix} 5 & 1 & | & 3 \\ -2 & 2 & | & 4 \end{bmatrix}$$

Note that coefficients of the variables and the constants in equations (1) and (2) correspond, respectively, to the elements in rows (1) and (2) of the *augmented matrix*. The augmented matrix is sometimes called the *corresponding matrix*. Note that in the augmented matrix there is a vertical bar that separates the coefficients on the left from the constants on the right.

We have already solved systems of equations by using multiples of one or more of the equations to eliminate variables. We can do the same thing to an augmented matrix by using *row operations*. When we perform row operations on a matrix, the result is an **equivalent matrix**, that is, a matrix corresponding to a system of equations equivalent to the original system. Using row operations, we can easily solve the system.

Matrix Row Operations

The following row operations produce an equivalent matrix:

- Interchanging any two rows
- Multiplying (or dividing) the elements of any row by the same nonzero number
- Multiplying (or dividing) the elements of any row by a nonzero number and adding the products to their corresponding elements in any other row

In solving a system of linear equations using matrices, we use these row operations to write a series of equivalent matrices to obtain a simplified matrix of the form

$$\begin{bmatrix} 1 & a & | & b \\ 0 & 1 & | & c \end{bmatrix}$$

for a system of linear equations in two variables, or

$$\begin{bmatrix} 1 & a & b & | & d \\ 0 & 1 & c & | & e \\ 0 & 0 & 1 & | & f \end{bmatrix}$$

for a system of linear equations in three variables.

Note that the first nonzero element of each row is 1 and that the elements below them are 0 in both of the simplified forms.

EXAMPLE 1

Use matrices to solve the system:

$$(1) \quad 2x - y = -4$$
$$(2) \quad x + 3y = 5$$

Solution First, we write the augmented matrix for the system.

System of Equations	**Augmented Matrix**
$2x - y = -4$	$\begin{bmatrix} 2 & -1 & \vert & -4 \\ 1 & 3 & \vert & 5 \end{bmatrix}$
$x + 3y = 5$	

Next, we use matrix row operations to get a 1 in row (1), column (1). To do this, we can interchange row (1) and row (2), getting

$$\begin{bmatrix} 1 & 3 & \vert & 5 \\ 2 & -1 & \vert & -4 \end{bmatrix}$$

Now, we need 0 below the 1 in column (1); that is, we want to replace the 2 in row (2), column (1) with a 0. To do this, we can multiply row (1) by -2 and add the products to row (2).

$$\begin{bmatrix} 1 & 3 & \vert & 5 \\ 2 + (-2)(1) & -1 + (-2)(3) & \vert & -4 + (-2)(5) \end{bmatrix} = \begin{bmatrix} 1 & 3 & \vert & 5 \\ 0 & -7 & \vert & -14 \end{bmatrix}$$

Finally, to get a 1 in row (2), column (2), we can divide row (2) by -7.

$$\begin{bmatrix} 1 & 3 & \vert & 5 \\ 0 \div (-7) & -7 \div (-7) & \vert & -14 \div (-7) \end{bmatrix} = \begin{bmatrix} 1 & 3 & \vert & 5 \\ 0 & 1 & \vert & 2 \end{bmatrix}$$

This matrix is now in the desired form and corresponds to the system

$$x + 3y = 5$$
$$y = 2$$

From this system, we see that $y = 2$. To find x, we substitute 2 for y in the equation $x + 3y = 5$.

$$x + 3y = 5$$
$$x + 3(2) = 5$$
$$x + 6 = 5$$
$$x = -1$$

So the solution is $(-1, 2)$.

PRACTICE 1

Use matrices to solve the system:

$$(1) \quad 2x + y = 3$$
$$(2) \quad x - 3y = 12$$

Example 1 suggests the following rule for solving linear systems by using matrices.

To Solve a System of Linear Equations by Using Matrices

- Write the augmented matrix for the system.
- Use matrix row operations to get a matrix of the form

$$\begin{bmatrix} 1 & a & | & b \\ 0 & 1 & | & c \end{bmatrix} \qquad \text{or} \qquad \begin{bmatrix} 1 & a & b & | & d \\ 0 & 1 & c & | & e \\ 0 & 0 & 1 & | & f \end{bmatrix}.$$

    System of two equations   **System of three equations**

- Write the system of linear equations that corresponds to the matrix in the previous step and find the solution.

Now let's apply the matrix method to solve linear systems with three variables.

EXAMPLE 2

Solve the following system using matrices:

$$\begin{aligned} \textbf{(1)} \quad & x + 3y - 6z = 7 \\ \textbf{(2)} \quad & 2x - y + 2z = 0 \\ \textbf{(3)} \quad & x + y + 2z = -1 \end{aligned}$$

Solution We begin by writing the augmented matrix for the system.

    System of Equations   **Augmented Matrix**

$$\begin{aligned} x + 3y - 6z &= 7 \\ 2x - y + 2z &= 0 \\ x + y + 2z &= -1 \end{aligned} \qquad \begin{bmatrix} 1 & 3 & -6 & | & 7 \\ 2 & -1 & 2 & | & 0 \\ 1 & 1 & 2 & | & -1 \end{bmatrix}$$

Since there is already a 1 in row (1), column (1), we start simplifying the matrix by getting 0's under the 1 in column (1). To do this, we can multiply row (3) by -2 and add it to row (2).

$$\begin{bmatrix} 1 & 3 & -6 & | & 7 \\ 2 + (-2)(1) & -1 + (-2)(1) & 2 + (-2)(2) & | & 0 + (-2)(-1) \\ 1 & 1 & 2 & | & -1 \end{bmatrix}$$

$$= \begin{bmatrix} 1 & 3 & -6 & | & 7 \\ 0 & -3 & -2 & | & 2 \\ 1 & 1 & 2 & | & -1 \end{bmatrix}$$

To get a 0 in row (3), column (1), we can multiply row (1) by -1 and add it to row (3).

$$\begin{bmatrix} 1 & 3 & -6 & | & 7 \\ 0 & -3 & -2 & | & 2 \\ 1 + (-1)(1) & 1 + (-1)(3) & 2 + (-1)(-6) & | & -1 + (-1)(7) \end{bmatrix}$$

$$= \begin{bmatrix} 1 & 3 & -6 & | & 7 \\ 0 & -3 & -2 & | & 2 \\ 0 & -2 & 8 & | & -8 \end{bmatrix}$$

PRACTICE 2

Use matrices to solve the following system:

$$\begin{aligned} \textbf{(1)} \quad & x - y + 2z = -2 \\ \textbf{(2)} \quad & x + y - 4z = 5 \\ \textbf{(3)} \quad & 3x + 2y + 4z = 18 \end{aligned}$$

To get a 1 in row (2), column (2), we first multiply row (2) by -1.

$$\left[\begin{array}{ccc|c} 1 & 3 & -6 & 7 \\ 0 & 3 & 2 & -2 \\ 0 & -2 & 8 & -8 \end{array}\right]$$

Then add row (3) to row (2).

$$\left[\begin{array}{ccc|c} 1 & 3 & -6 & 7 \\ 0 & 3+(-2) & 2+8 & -2+(-8) \\ 0 & -2 & 8 & -8 \end{array}\right] = \left[\begin{array}{ccc|c} 1 & 3 & -6 & 7 \\ 0 & 1 & 10 & -10 \\ 0 & -2 & 8 & -8 \end{array}\right]$$

To get a 0 in row (3), column (2), we can multiply row (2) by 2 and add it to row (3).

$$\left[\begin{array}{ccc|c} 1 & 3 & -6 & 7 \\ 0 & 1 & 10 & -10 \\ 0+2(0) & -2+2(1) & 8+2(10) & -8+2(-10) \end{array}\right]$$

$$= \left[\begin{array}{ccc|c} 1 & 3 & -6 & 7 \\ 0 & 1 & 10 & -10 \\ 0 & 0 & 28 & -28 \end{array}\right]$$

Next we multiply row (3) by $\dfrac{1}{28}$ to get a 1 in row (3), column (3).

$$\left[\begin{array}{ccc|c} 1 & 3 & -6 & 7 \\ 0 & 1 & 10 & -10 \\ \frac{1}{28}(0) & \frac{1}{28}(0) & \frac{1}{28}(28) & \frac{1}{28}(-28) \end{array}\right] = \left[\begin{array}{ccc|c} 1 & 3 & -6 & 7 \\ 0 & 1 & 10 & -10 \\ 0 & 0 & 1 & -1 \end{array}\right]$$

The matrix is now in the desired form and corresponds to the following system:

$$\begin{aligned} x + 3y - 6z &= 7 \\ y + 10z &= -10 \\ z &= -1 \end{aligned}$$

From this system, we see that $z = -1$. To find y, we substitute -1 for z in the equation $y + 10z = -10$.

$$\begin{aligned} y + 10z &= -10 \\ y + 10(-1) &= -10 \\ y - 10 &= -10 \\ y &= 0 \end{aligned}$$

Finally, substituting 0 for y and -1 for z in the equation $x + 3y - 6z = 7$, we can solve for x.

$$\begin{aligned} x + 3y - 6z &= 7 \\ x + 3(0) - 6(-1) &= 7 \\ x + 0 + 6 &= 7 \\ x &= 1 \end{aligned}$$

So the solution is $x = 1$, $y = 0$, and $z = -1$, or $(1, 0, -1)$.

EXAMPLE 3

Solve the system using matrices:

$$
\begin{aligned}
\textbf{(1)} \qquad x - y &= -5 \\
\textbf{(2)} \quad 2x + 3y - z &= 6 \\
\textbf{(3)} \qquad y + z &= 2
\end{aligned}
$$

Solution Note that there is no z-term in equation (1) and no x-term in equation (3). When a term is missing in an equation, we use 0 in the augmented matrix.

System of Equations	Augmented Matrix

$$
\begin{aligned}
x - y \qquad &= -5 \\
2x + 3y - z &= 6 \\
y + z &= 2
\end{aligned}
\qquad
\begin{bmatrix}
1 & -1 & 0 & -5 \\
2 & 3 & -1 & 6 \\
0 & 1 & 1 & 2
\end{bmatrix}
$$

To get a 0 in row (2), column (1), we multiply row (1) by -2 and add it to row (2).

$$
\begin{bmatrix}
1 & -1 & 0 & -5 \\
2 + (-2)(1) & 3 + (-2)(-1) & -1 + (-2)(0) & 6 + (-2)(-5) \\
0 & 1 & 1 & 2
\end{bmatrix}
$$

$$
=
\begin{bmatrix}
1 & -1 & 0 & -5 \\
0 & 5 & -1 & 16 \\
0 & 1 & 1 & 2
\end{bmatrix}
$$

Interchanging row (2) and row (3), we get a 1 in row (2), column (2).

$$
\begin{bmatrix}
1 & -1 & 0 & -5 \\
0 & 1 & 1 & 2 \\
0 & 5 & -1 & 16
\end{bmatrix}
$$

To get 0 in row (3), column (2), we multiply row (2) by -5 and add it to row (3).

$$
\begin{bmatrix}
1 & -1 & 0 & -5 \\
0 & 1 & 1 & 2 \\
0 + (-5)(0) & 5 + (-5)(1) & -1 + (-5)(1) & 16 + (-5)(2)
\end{bmatrix}
$$

$$
=
\begin{bmatrix}
1 & -1 & 0 & -5 \\
0 & 1 & 1 & 2 \\
0 & 0 & -6 & 6
\end{bmatrix}
$$

Finally, we multiply row (3) by $-\dfrac{1}{6}$ to get a 1 in row (3), column (3).

$$
\begin{bmatrix}
1 & -1 & 0 & -5 \\
0 & 1 & 1 & 2 \\
(-\frac{1}{6})(0) & (-\frac{1}{6})(0) & (-\frac{1}{6})(-6) & (-\frac{1}{6})(6)
\end{bmatrix}
=
\begin{bmatrix}
1 & -1 & 0 & -5 \\
0 & 1 & 1 & 2 \\
0 & 0 & 1 & -1
\end{bmatrix}
$$

PRACTICE 3

Use matrices to solve the system:

$$
\begin{aligned}
\textbf{(1)} \quad x + y + z &= -3 \\
\textbf{(2)} \qquad 3y + 2z &= 0 \\
\textbf{(3)} \quad 4x - y \qquad &= -8
\end{aligned}
$$

The matrix is now in the desired form and corresponds to the following system:

$$x - y \quad\;\; = -5$$
$$y + z = 2$$
$$z = -1$$

From this system, we see that $z = -1$. To find y, we substitute -1 for z in the equation $y + z = 2$.

$$y + z = 2$$
$$y + (-1) = 2$$
$$y = 3$$

Substituting 3 for y in the equation $x - y = -5$, we can find x.

$$x - y = -5$$
$$x - 3 = -5$$
$$x = -2$$

So the solution is $(-2, 3, -1)$.

Solve each system using matrices.

1. $x - 7y = 12$
$x - 4y = 6$

2. $6x + y = 7$
$8x + y = 11$

3. $3x + 4y = -1$
$2x + 3y = -1$

4. $5x - 2y = 6$
$-4x + 3y = 5$

5. $9x - 15y = 12$
$-3x + 5y = -4$

6. $7x - 4y = 3$
$14x - 8y = -8$

7. $\frac{1}{2}x - \frac{1}{3}y = -7$
$3x + 5y = 21$

8. $2x + 6y = 0$
$\frac{2}{3}x - y = -3$

9. $x + 3y - 2z = 10$
$3x - 2y - z = 9$
$4x - y + 5z = 1$

10. $5x - 2y + 2z = -2$
$-x + 3y + 6z = -20$
$x + y - z = 8$

11. $6x - 4y - 3z = -4$
$2x + 8y - 7z = -11$
$-x + 2y + z = 2$

12. $2x + 9y - 6z = -4$
$-3x - 15y + 18z = 4$
$-4x - 11y + 8z = 11$

13. $5x + 7y + 3z = -48$
$-8x + 2y + 4z = 24$
$12x - 6y - 9z = -21$

14. $15x - 10y + 5z = -10$
$6x + 4y - 2z = 4$
$2x - 5y - 3z = 28$

15. $7x - y + 8z = -9$
$3x + 3y - 10z = -5$
$3.5x - 0.5y + 4z = -14$

16. $1.5x + 2y + 3z = 3$
$3x + 4y - 2z = 2$
$6x + 8y + 4z = 8$

17. $y + 5z = -15$
$4x - 4y = -20$
$8x - 2z = 8$

18. $7x + 9y = -3$
$-2x + 3z = 3$
$5y - z = 11$

19. $2x + 3y = 7$
$-11y - 4z = -5$
$x + y - 6z = 0$

20. $10x + 8y - 4z = 1$
$3x - 4y = -4$
$-x + 2y + 3z = 1$

• *Check your answers on page A–55.*

Additional Topics

B.1   Determinants and Cramer's Rule

In Chapter 4 and Appendixes A.4 and A.5, we discussed several methods of solving systems of two or three linear equations. Here we consider another method, based on the concept of *determinants*.

To begin, recall that a **matrix** is a rectangular array of numbers. The numbers of the matrix are called **elements** or **entries**. The **rows** of the matrix are horizontal, and the **columns** are vertical.

The following are examples of **square matrices**, that is, matrices that have the same number of rows as columns.

$$\begin{bmatrix} -2 & 0 \\ 0 & 3 \end{bmatrix} \longleftarrow 2 \times 2 \text{ (read "two by two") matrix}$$

$$\begin{bmatrix} 5 & 1 & 2 \\ 0 & 3 & -1 \\ 4 & -2 & 1 \end{bmatrix} \longleftarrow 3 \times 3 \text{ matrix}$$

Determinants of 2×2 Matrices and Cramer's Rule

A determinant is a real number associated with a square matrix. Let's consider determinants of 2×2 matrices.

Definition

The **determinant of the 2×2 matrix** $\begin{bmatrix} a & b \\ c & d \end{bmatrix}$, written $\begin{vmatrix} a & b \\ c & d \end{vmatrix}$, is the real number $ad - bc$.

EXAMPLE 1

Evaluate $\begin{vmatrix} 4 & -1 \\ 3 & 7 \end{vmatrix}$

Solution Here $a = 4$, $b = -1$, $c = 3$, and $d = 7$.

$$\begin{vmatrix} 4 & -1 \\ 3 & 7 \end{vmatrix} = ad - bc$$
$$= (4)(7) - (-1)(3)$$
$$= 28 + 3$$
$$= 31$$

PRACTICE 1

Find the value of the determinant:

$$\begin{vmatrix} 5 & -2 \\ 8 & -3 \end{vmatrix}$$

We can apply the concept of determinant in solving a system of linear equations. Consider the general system of two linear equations in two unknowns, x and y.

$$a_1 x + b_1 y = k_1$$
$$a_2 x + b_2 y = k_2$$

If we solve this system by elimination, the result is:

$$x = \frac{k_1 b_2 - k_2 b_1}{a_1 b_2 - a_2 b_1} \text{ and } y = \frac{a_1 k_2 - a_2 k_1}{a_1 b_2 - a_2 b_1},$$

where $a_1 b_2 - a_2 b_1 \neq 0$.

We can write the values of x and y in terms of the following determinants:

$$x = \frac{\begin{vmatrix} k_1 & b_1 \\ k_2 & b_2 \end{vmatrix}}{\begin{vmatrix} a_1 & b_1 \\ a_2 & b_2 \end{vmatrix}} \text{ and } y = \frac{\begin{vmatrix} a_1 & k_1 \\ a_2 & k_2 \end{vmatrix}}{\begin{vmatrix} a_1 & b_1 \\ a_2 & b_2 \end{vmatrix}}$$

It is common practice to represent the numerator for x as D_x, the numerator for y as D_y, and the denominator of both variables as D. Note the relationship between these determinants and the numbers in the original system:

$$D_x = \begin{vmatrix} k_1 & b_1 \\ k_2 & b_2 \end{vmatrix} \qquad D_y = \begin{vmatrix} a_1 & k_1 \\ a_2 & k_2 \end{vmatrix} \qquad D = \begin{vmatrix} a_1 & b_1 \\ a_2 & b_2 \end{vmatrix}$$

y coefficients

x coefficients

Constants

Substituting the names of these determinants in the equations for x and y, we get:

$$x = \frac{D_x}{D} \text{ and } y = \frac{D_y}{D}$$

We can now state the determinant formula for solving a system of two linear equations in two variables that is known as *Cramer's Rule*.

Cramer's Rule for Solving a System of Two Linear Equations in Two Unknowns

The solution to the system

$$a_1x + b_1y = k_1$$
$$a_2x + b_2y = k_2$$

is $x = \dfrac{D_x}{D}$ and $y = \dfrac{D_y}{D}$, where $D_x = \begin{vmatrix} k_1 & b_1 \\ k_2 & b_2 \end{vmatrix}$, $D_y = \begin{vmatrix} a_1 & k_1 \\ a_2 & k_2 \end{vmatrix}$, and $D = \begin{vmatrix} a_1 & b_1 \\ a_2 & b_2 \end{vmatrix}$,

where $D \neq 0$.

Note that for a given system of linear equations, if the value of D is equal to 0, then Cramer's Rule does not apply. In this case, there is not a unique solution, so either the system has no solution (the lines are parallel) or infinitely many solutions (the lines are identical).

EXAMPLE 2

Use Cramer's Rule to solve the system:

$$x + 3y = 2$$
$$x - 4y = 9$$

Solution

To begin, let's evaluate D_x, D_y, and D.

$$D_x = \begin{vmatrix} k_1 & b_1 \\ k_2 & b_2 \end{vmatrix} = \begin{vmatrix} 2 & 3 \\ 9 & -4 \end{vmatrix} = (2)(-4) - (3)(9) = -35$$

(Constants / y coefficients)

$$D_y = \begin{vmatrix} a_1 & k_1 \\ a_2 & k_2 \end{vmatrix} = \begin{vmatrix} 1 & 2 \\ 1 & 9 \end{vmatrix} = (1)(9) - (2)(1) = 7$$

(Constants / x coefficients)

$$D = \begin{vmatrix} a_1 & b_1 \\ a_2 & b_2 \end{vmatrix} = \begin{vmatrix} 1 & 3 \\ 1 & -4 \end{vmatrix} = (1)(-4) - (3)(1) = -7$$

(x coefficients / y coefficients)

Cramer's Rule gives us

$$x = \frac{D_x}{D} = \frac{-35}{-7} = 5 \text{ and } y = \frac{D_y}{D} = \frac{7}{-7} = -1.$$

So the solution is $(5, -1)$. We leave the check as an exercise.

PRACTICE 2

Use Cramer's Rule to solve:

$$2x - y = 5$$
$$x + 5y = -14$$

Determinants of 3 × 3 Matrices and Cramer's Rule

Determinants of 3 × 3 matrices can be used to solve systems of three linear equations in three variables. Such determinants are more complex to evaluate than determinants of 2 × 2 matrices.

Definition

The **3 × 3 determinant** $\begin{vmatrix} a_1 & b_1 & c_1 \\ a_2 & b_2 & c_2 \\ a_3 & b_3 & c_3 \end{vmatrix}$ is equal to

$$a_1 \cdot \begin{vmatrix} b_2 & c_2 \\ b_3 & c_3 \end{vmatrix} - a_2 \cdot \begin{vmatrix} b_1 & c_1 \\ b_3 & c_3 \end{vmatrix} + a_3 \cdot \begin{vmatrix} b_1 & c_1 \\ b_2 & c_2 \end{vmatrix}.$$

Note that the determinant of a 3 × 3 matrix is defined in terms of three related 2 × 2 matrices. Each of these smaller matrices is called a *minor* and can be found by considering the larger matrix and then crossing out the appropriate row and column in which each value of a appears, as shown below.

$$\begin{array}{ccc} a_1 & b_1 & c_1 \\ a_2 & b_2 & c_2 \\ a_3 & b_3 & c_3 \end{array} \quad \text{The minor of } a_1 \text{ is } \begin{vmatrix} b_2 & c_2 \\ b_3 & c_3 \end{vmatrix}.$$

$$\begin{array}{ccc} a_1 & b_1 & c_1 \\ a_2 & b_2 & c_2 \\ a_3 & b_3 & c_3 \end{array} \quad \text{The minor of } a_2 \text{ is } \begin{vmatrix} b_1 & c_1 \\ b_3 & c_3 \end{vmatrix}.$$

$$\begin{array}{ccc} a_1 & b_1 & c_1 \\ a_2 & b_2 & c_2 \\ a_3 & b_3 & c_3 \end{array} \quad \text{The minor of } a_3 \text{ is } \begin{vmatrix} b_1 & c_1 \\ b_2 & c_2 \end{vmatrix}.$$

So we conclude that:

$$\begin{vmatrix} a_1 & b_1 & c_1 \\ a_2 & b_2 & c_2 \\ a_3 & b_3 & c_3 \end{vmatrix}$$

$$= a_1(\text{the minor of } a_1) - a_2(\text{the minor of } a_2) + a_3(\text{the minor of } a_3)$$

Evaluating the determinant of a 3 × 3 matrix in this way is called *expanding by the minors of the first column*. Such a determinant can, in fact, be evaluated by considering minors of any row or column as long as they are preceded by the appropriate sign, a subject pursued in more advanced texts.

EXAMPLE 3

Evaluate $\begin{vmatrix} 5 & -2 & 1 \\ 1 & -1 & 0 \\ -4 & 3 & 2 \end{vmatrix}$.

Solution We evaluate the determinant of this 3×3 matrix expanding by the minors of the first column.

$$\begin{vmatrix} 5 & -2 & 1 \\ 1 & -1 & 0 \\ -4 & 3 & 2 \end{vmatrix} = 5 \cdot \begin{vmatrix} -1 & 0 \\ 3 & 2 \end{vmatrix} - 1 \cdot \begin{vmatrix} -2 & 1 \\ 3 & 2 \end{vmatrix} + (-4) \cdot \begin{vmatrix} -2 & 1 \\ -1 & 0 \end{vmatrix}$$

Alternate the signs.

$$= 5[(-1)(2) - (0)(3)] - 1[(-2)(2) - (1)(3)] + (-4)[(-2)(0) - (1)(-1)]$$
$$= (5)(-2) - (1)(-7) + (-4)(1)$$
$$= -7$$

Just as with a system of two linear equations in two unknowns, we can also use Cramer's Rule to solve a system of three linear equations in three unknowns.

PRACTICE 3

Evaluate $\begin{vmatrix} 3 & -1 & 5 \\ -2 & 2 & -3 \\ 1 & 0 & 4 \end{vmatrix}$.

Cramer's Rule for Solving a System of Three Linear Equations in Three Unknowns

The solution to the system

$$a_1x + b_1y + c_1z = k_1$$
$$a_2x + b_2y + c_2z = k_2$$
$$a_3x + b_3y + c_3z = k_3$$

is $x = \dfrac{D_x}{D}, y = \dfrac{D_y}{D}$, and $z = \dfrac{D_z}{D}$, where $D_x = \begin{vmatrix} k_1 & b_1 & c_1 \\ k_2 & b_2 & c_2 \\ k_3 & b_3 & c_3 \end{vmatrix}, D_y = \begin{vmatrix} a_1 & k_1 & c_1 \\ a_2 & k_2 & c_2 \\ a_3 & k_3 & c_3 \end{vmatrix},$

$D_z = \begin{vmatrix} a_1 & b_1 & k_1 \\ a_2 & b_2 & k_2 \\ a_3 & b_3 & k_3 \end{vmatrix}$, and $D = \begin{vmatrix} a_1 & b_1 & c_1 \\ a_2 & b_2 & c_2 \\ a_3 & b_3 & c_3 \end{vmatrix}$, where $D \neq 0$.

EXAMPLE 4

Solve using Cramer's Rule:

$$3x - 4y + 2z = 9$$
$$2x - 3y + z = 5$$
$$x + 2y + 3z = 8$$

Solution

$$D_x = \begin{vmatrix} k_1 & b_1 & c_1 \\ k_2 & b_2 & c_2 \\ k_3 & b_3 & c_3 \end{vmatrix} = \begin{vmatrix} 9 & -4 & 2 \\ 5 & -3 & 1 \\ 8 & 2 & 3 \end{vmatrix}$$

$$= 9\begin{vmatrix} -3 & 1 \\ 2 & 3 \end{vmatrix} - 5\begin{vmatrix} -4 & 2 \\ 2 & 3 \end{vmatrix} + 8\begin{vmatrix} -4 & 2 \\ -3 & 1 \end{vmatrix}$$

$$= 9(-11) - 5(-16) + 8(2)$$

$$= -3$$

$$D_y = \begin{vmatrix} a_1 & k_1 & c_1 \\ a_2 & k_2 & c_2 \\ a_3 & k_3 & c_3 \end{vmatrix} = \begin{vmatrix} 3 & 9 & 2 \\ 2 & 5 & 1 \\ 1 & 8 & 3 \end{vmatrix}$$

$$= 3\begin{vmatrix} 5 & 1 \\ 8 & 3 \end{vmatrix} - 2\begin{vmatrix} 9 & 2 \\ 8 & 3 \end{vmatrix} + 1\begin{vmatrix} 9 & 2 \\ 5 & 1 \end{vmatrix}$$

$$= 3(7) - 2(11) + 1(-1)$$

$$= -2$$

$$D_z = \begin{vmatrix} a_1 & b_1 & k_1 \\ a_2 & b_2 & k_2 \\ a_3 & b_3 & k_3 \end{vmatrix} = \begin{vmatrix} 3 & -4 & 9 \\ 2 & -3 & 5 \\ 1 & 2 & 8 \end{vmatrix}$$

$$= 3\begin{vmatrix} -3 & 5 \\ 2 & 8 \end{vmatrix} - 2\begin{vmatrix} -4 & 9 \\ 2 & 8 \end{vmatrix} + 1\begin{vmatrix} -4 & 9 \\ -3 & 5 \end{vmatrix}$$

$$= 3(-34) - 2(-50) + 1(7)$$

$$= 5$$

$$D = \begin{vmatrix} a_1 & b_1 & c_1 \\ a_2 & b_2 & c_2 \\ a_3 & b_3 & c_3 \end{vmatrix} = \begin{vmatrix} 3 & -4 & 2 \\ 2 & -3 & 1 \\ 1 & 2 & 3 \end{vmatrix}$$

$$= 3\begin{vmatrix} -3 & 1 \\ 2 & 3 \end{vmatrix} - 2\begin{vmatrix} -4 & 2 \\ 2 & 3 \end{vmatrix} + 1\begin{vmatrix} -4 & 2 \\ -3 & 1 \end{vmatrix}$$

$$= 3(-11) - 2(-16) + 1(2)$$

$$= 1$$

By Cramer's Rule, we get:

$$x = \frac{D_x}{D} = \frac{-3}{1} = -3 \qquad y = \frac{D_y}{D} = \frac{-2}{1} = -2 \qquad z = \frac{D_z}{D} = \frac{5}{1} = 5$$

So the solution is $(-3, -2, 5)$. We leave the check as an exercise.

PRACTICE 4

Use Cramer's Rule to solve:

$$4x + y - z = 5$$
$$x - y + 2z = 8$$
$$x + y + z = 5$$

B.1 **Exercises** FOR EXTRA HELP *MyMathLab*
PRACTICE WATCH DOWNLOAD READ REVIEW

Evaluate each determinant.

1. $\begin{vmatrix} 2 & 5 \\ 1 & -1 \end{vmatrix}$

2. $\begin{vmatrix} 8 & -5 \\ 1 & 2 \end{vmatrix}$

3. $\begin{vmatrix} 7 & 2 \\ 9 & 4 \end{vmatrix}$

4. $\begin{vmatrix} 11 & 3 \\ 3 & 0 \end{vmatrix}$

5. $\begin{vmatrix} -6 & 4 \\ -12 & 8 \end{vmatrix}$

6. $\begin{vmatrix} 3 & 1 \\ 6 & 2 \end{vmatrix}$

Solve using Cramer's Rule.

7. $\begin{aligned} 4x - 5y &= 13 \\ 3x - y &= 7 \end{aligned}$

8. $\begin{aligned} x + 2y &= 4 \\ -x + y &= 5 \end{aligned}$

9. $\begin{aligned} 3x + 2y &= 11 \\ x - 2y &= 1 \end{aligned}$

10. $\begin{aligned} -2x + y &= -10 \\ 2x + 6y &= 10 \end{aligned}$

11. $\begin{aligned} 3x + 2y &= 1 \\ -7x + 10y &= -6 \end{aligned}$

12. $\begin{aligned} 2x - 6y &= 6 \\ -x + 9y &= 1 \end{aligned}$

Evaluate each determinant.

13. $\begin{vmatrix} 2 & 1 & 5 \\ 0 & 3 & 1 \\ 4 & 2 & 4 \end{vmatrix}$

14. $\begin{vmatrix} 0 & 2 & 1 \\ 5 & 4 & 3 \\ 1 & -1 & 3 \end{vmatrix}$

15. $\begin{vmatrix} 3 & 3 & 1 \\ 2 & 2 & -6 \\ 2 & 3 & -4 \end{vmatrix}$

16. $\begin{vmatrix} 6 & -3 & 5 \\ 1 & -2 & 4 \\ 2 & -4 & 7 \end{vmatrix}$

Solve using Cramer's Rule.

17. $\begin{aligned} 2x - 3y + z &= 8 \\ 5x + 4y + 2z &= -1 \\ 7x + 2y + 3z &= 5 \end{aligned}$

18. $\begin{aligned} 2x - y + z &= 2 \\ 3x - 2y - z &= -1 \\ x - 3y + 6z &= -1 \end{aligned}$

19. $\begin{aligned} x + 3y - 4z &= -12 \\ 3x \quad\quad + z &= 2 \\ 5x - y + z &= -3 \end{aligned}$

20. $\begin{aligned} 2x + 3y \quad\quad &= 4 \\ 3x + 7y - 4z &= -3 \\ x - y + 2z &= 9 \end{aligned}$

• *Check your answers on page A-55.*

B.2   Synthetic Division

In Section 5.7, we discussed the division of polynomials. Here we consider a short-cut for dividing polynomials, known as *synthetic division*, which eliminates the repetition in the division process. This procedure is only used in a particular situation, namely, when we are dividing a polynomial by a binomial of the form $x - k$.

Consider the following division:

$$
\begin{array}{r}
3x \quad + 7 \\
x - 5 \overline{\smash{\big)}\ 3x^2 -\ \ 8x -\ 40} \\
\underline{3x^2 - 15x} \\
7x -\ 40 \\
\underline{7x -\ 35} \\
-5 \quad \leftarrow \text{Remainder}
\end{array}
$$

To see how this shortcut was developed, let's first rewrite the example above by omitting the variables as shown:

$$
\begin{array}{r}
3 \quad\ \ 7 \\
1 - 5 \overline{\smash{\big)}\ 3 -\ \ 8 -\ 40} \\
\underline{3 - 15} \\
7 -\ 40 \\
\underline{7 -\ 35} \\
-5 \quad \leftarrow \text{Remainder}
\end{array}
$$

Note that several of the numbers in this division are exactly the same as numbers written directly above. So we can simplify the division process further by omitting these repeated numbers. We can also omit the 1 in the divisor, since we are only considering divisors with leading coefficient 1:

$$
\begin{array}{r}
3 \quad\ \ 7 \\
1 - 5 \overline{\smash{\big)}\ 3 -\ \ 8 -\ 40} \\
\underline{3 - 15} \\
7 -\ 40 \\
\underline{7 -\ 35} \\
-5 \quad \leftarrow \text{Remainder}
\end{array}
\qquad
\begin{array}{r}
3 \quad\ \ 7 \\
- 5 \overline{\smash{\big)}\ 3 -\ \ 8 -\ 40} \\
\underline{- 15} \\
7 \\
\underline{- 35} \\
-5 \quad \leftarrow \text{Remainder}
\end{array}
$$

Finally, we can simplify the arithmetic by changing the -5 to its additive inverse 5 so that we can add rather than subtract in each column. Using a compressed format, we have:

$$
\begin{array}{r l}
& \text{Coefficients of the dividend} \\
& \overbrace{\qquad\qquad\qquad\qquad} \\
\underline{5}\rvert \quad 3 \qquad -8 \qquad -40 \\
\qquad\qquad\quad 15 \qquad\ \ 35 \\
\hline
\quad 3 \qquad\ \ 7 \qquad\ -5 \\
\underbrace{\qquad\qquad\quad} \qquad \uparrow
\end{array}
$$

Coefficients of   Remainder
the quotient

We can use the division method modeled on the previous page to see how synthetic division works step by step.

EXAMPLE 1	PRACTICE 1

EXAMPLE 1

Use synthetic division to divide $2x^2 - x + 1$ by $x - 4$.

Solution Since the divisor is $x - 4$, k is 4. So we write 4 as $4\rfloor$. Turning to the dividend, we write the coefficients and constant term without the variable: 2, −1, and 1.

$$4\rfloor \quad 2 \quad -1 \quad 1$$
$$\downarrow$$
$$2$$

Draw a line, and bring down the leading coefficient of the dividend, 2.

$$4\rfloor \quad 2 \quad -1 \quad 1$$
$$8$$
$$2 \quad 7$$

Multiply the 2 by 4, getting 8. Write the 8 under the −1, adding to get 7.

$$4\rfloor \quad 2 \quad -1 \quad 1$$
$$8 \quad 28$$
$$2 \quad 7 \quad 29$$

Multiply the 7 by 4, getting 28. Write the 28 under the 1, adding to get 29.

Since the dividend is a second-degree polynomial, the quotient is first degree, that is, one degree less. From the bottom row, we see that the solution is $2x + 7$ with remainder 29. That is,

$$\frac{2x^2 - x + 1}{x - 4} = 2x + 7 + \frac{29}{x - 4}.$$

PRACTICE 1

Use synthetic division to find the quotient: $(5x^2 - 3x - 1) \div (x - 3)$

EXAMPLE 2	PRACTICE 2

EXAMPLE 2

Find the quotient using synthetic division: $\dfrac{x^3 + 6x - 1}{x + 2}$

Solution The divisor is $x + 2$, which in the form $x - k$ is $x - (-2)$. So k is −2, which we write as $-2\rfloor$. Since there is no x^2-term in the dividend, we insert a coefficient of 0 for the missing term. So the dividend coefficients and constant term are: 1, 0, 6, and −1.

$$-2\rfloor \quad 1 \quad 0 \quad 6 \quad -1$$
$$-2 \quad 4 \quad -20$$
$$1 \quad -2 \quad 10 \quad -21$$

Since the dividend is a third-degree polynomial, the quotient is second degree. The quotient is $x^2 - 2x + 10$ with a remainder of −21. So $\dfrac{x^3 + 6x - 1}{x + 2} = x^2 - 2x + 10 - \dfrac{21}{x + 2}.$

PRACTICE 2

Divide using synthetic division:
$$\frac{x^3 + 4x^2 + 7}{x + 5}$$

B.2   Exercises

FOR EXTRA HELP

Find the quotient, using synthetic division.

1. $\dfrac{x^2 + x - 1}{x - 2}$

2. $\dfrac{x^2 - 4x - 2}{x - 4}$

3. $(x^2 - 5x - 7) \div (x + 3)$

4. $(x^2 + 2x - 4) \div (x + 5)$

5. $\dfrac{5x^2 - 3x + 2}{x + 1}$

6. $\dfrac{2x^2 - 4x - 21}{x + 3}$

7. $(3x^2 - 8x - 3) \div (x - 3)$

8. $(9x^2 - 4x - 5) \div (x - 1)$

9. $(3x^2 - 8x) \div (x - 4)$

10. $(2x^2 - 11) \div (x - 2)$

11. $\dfrac{4x^3 + x^2 - 3x + 5}{x - 1}$

12. $\dfrac{2x^3 + x^2 - 6x - 24}{x - 2}$

13. $(-y^3 + 17y - 40) \div (y + 5)$

14. $(7t^3 + 8t^2 - 1) \div (t + 1)$

15. $(x^3 - 5x) \div (x - 2)$

16. $(x^3 + 3) \div (x - 1)$

17. $\dfrac{x^4 + 3x^3 + 5x}{x + 3}$

18. $\dfrac{x^4 - 4x^2 - 8}{x + 2}$

19. $(2x^3 + 7x^2 - 2x + 3) \div (x - 0.5)$

20. $(6x^3 + 5x^2 - 9x + 2) \div (x + 0.5)$

• *Check your answers on page A-55.*

Symbols and Graphing Calculators

C.1 Table of Symbols

$+$	add
$-$	subtract
$\times, \cdot, (a)(b), 2y$	multiply
$\div, \dfrac{a}{b}, x + 1\overline{)x^2 - 1}$	divide
$=$	is equal to
$\approx$	is approximately equal to
$\neq$	is not equal to
$<$	is less than
$\leq$	is less than or equal to
$>$	is greater than
$\geq$	is greater than or equal to
$\cup$	the union of two sets
$\cap$	the intersection of two sets
$(\ )$	parentheses (a grouping symbol)
$[\ ]$	brackets (a grouping symbol)
$\{\ \}$	braces (a grouping symbol)
(a, b)	the set of real numbers between a and b, excluding a and b
$[a, b]$	the set of real numbers between a and b, including a and b

$[a, b)$	the set of real numbers between a and b, including a but excluding b
$(a, b]$	the set of real numbers between a and b, excluding a but including b
$\{a, b, c\}$	the set with elements a, b, and c
∞	infinity
π	pi, an irrational number approximately equal to 3.14
e	an irrational number approximately equal to 2.718
$-a$	the opposite, or additive inverse, of a
$\dfrac{1}{a}$	the reciprocal, or multiplicative inverse, of a
$|n|$	the absolute value of n
x^n	x raised to the power n
$\sqrt{a}$	the principal square root of a
$\sqrt[n]{a}$	the nth root of a
(x, y)	an ordered pair whose first coordinate is x and whose second coordinate is y
$f(x)$	the function f of x
$f^{-1}(x)$	the inverse function of $f(x)$
$f \circ g$	the composition of functions f and g
i	the imaginary number $\sqrt{-1}$
$a + bi$	a complex number
log	the common logarithm
ln	the natural logarithm
$^\circ$	degree (for angles)
$\overleftrightarrow{AB}$	line AB
$\overline{AB}$	line segment AB
$\angle A$	angle A

C.2   Introduction to Graphing Calculators

This appendix covers the basic graphing features of a graphing calculator (or graphing software) used in this text. Note that the keystrokes and screens presented here and in the calculator inserts found throughout the text may be different from those on your graphing calculator. Refer to your user's manual for specific information and instructions on accessing and using the features of your particular model.

Graphing Equations

To graph an equation, it must be entered into the graphing calculator's *equation editor* in "$y =$" form. On many graphing calculators, the equation editor can be displayed by pressing the $\boxed{Y=}$ key. Note that you may enter more than one equation in the equation editor.

Equation Editor

For example, to graph the equation $4x + 2y = 6$, you must first solve the equation for y, getting $y = -2x + 3$. Then enter the expression $-2x + 3$ to the right of **\Y1 =** in the equation editor. Finally pressing the $\boxed{\text{GRAPH}}$ key displays the graph on a coordinate plane.

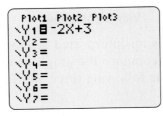
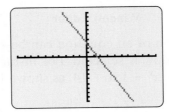

Note that most graphing calculators have two separate keys to distinguish between subtraction and negation. The $\boxed{-}$ key is used for subtraction and the $\boxed{(-)}$ key is used for negation.

   You must be careful when entering equations containing fractions. To ensure that the graphing calculator interprets the input correctly, you may need to enclose all or part of the fraction in parentheses. For instance, to graph the equation $y = \dfrac{1}{2}x$, enter the expression $(1/2)x$ to the right of **\Y1 =** in the equation editor. Then press $\boxed{\text{GRAPH}}$ to display the graph of the equation.

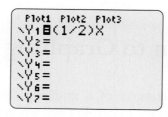
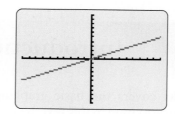

Viewing Windows

The screen on which a graph is displayed is called the *viewing window* and it represents a portion of a coordinate plane. The viewing window is defined by the following values:

Xmin: the minimum x value displayed on the x-axis
Xmax: the maximum x value displayed on the x-axis
Xscl: the distance between adjacent tick marks on the x-axis
Ymin: the minimum y value displayed on the y-axis
Ymax: the maximum y value displayed on the y-axis
Yscl: the distance between adjacent tick marks on the y-axis

You can set the viewing window by entering the minimum and maximum values and the scales for the axes in the *window editor*. The window editor can be displayed by pressing the WINDOW key. The screen on the left shows the window editor and displays the settings for the corresponding *standard viewing window* shown on the right.

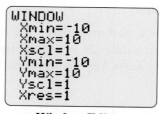

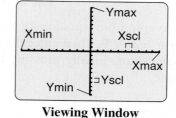

| **Window Editor** | **Viewing Window** |

The graph of an equation can be easily misinterpreted if an inappropriate viewing window is selected. For instance, compare the graph of the quadratic equation $y = x^2 - 15x + 14$, as shown in the following three viewing windows.

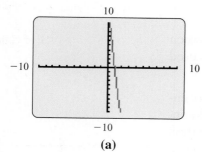
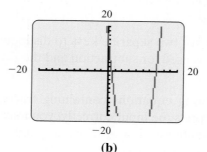
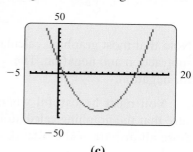

(a)   **(b)**   **(c)**

Although each of these viewing windows displays the graph of the equation, the viewing window in (c) is best because it shows all the key features of the graph, that is, the x- and y-intercepts and the vertex. Selecting an appropriate viewing window may require some practice, but familiarity with key features of the graphs of the equations discussed in the text will facilitate this process.

Other graphing calculator features, such as TRACE, ZOOM and INTERSECT, are discussed in calculator inserts throughout the text.

Answers

Chapter R

Pretest: Chapter R, *p. 2*

1. 7^3 **2.** 400 **3.** 8 **4.** 16 **5.** 1, 2, 3, 4, 6, 12
6. $2^2 \cdot 5$ **7.** $\frac{2}{3}$ **8.** $\frac{2}{5}$ **9.** $7\frac{1}{2}$ **10.** $\frac{4}{27}$ **11.** $2\frac{2}{3}$
12. Nine and thirteen thousandths
13. 3.1 **14.** 20.3103 **15.** 5.66 **16.** 37.843
17. 23.5 **18.** 0.18 **19.** 0.0031 **20.** 0.07
21. $2 million **22.** $\frac{1}{3}$ mi **23.** Two times **24.** 10%
25. 0.95

Practices: Section R.1 pp. *3–5*

1, *p. 3:* 2^5 **2,** *p. 3:* 784 **3,** *p. 4:* 10^9 **4,** *p. 5:* 16
5, *p. 5:* 1 **6,** *p. 5:* No; the average monthly bill this year was $100.

Practices: Section R.2 pp. *6–7*

1, *p. 6:* 50 **2,** *p. 7:* 180 **3,** *p. 7:* 24 hr

Practices: Section R.3 pp. *8–14*

1, *p. 8:* $\frac{29}{9}$ **2,** *p. 9:* $2\frac{2}{3}$ **3,** *p. 9:* $\frac{4}{5}$ **4,** *p. 10:* $\frac{2}{3}$
5, *p. 10:* $\frac{2}{5}$ **6,** *p. 10:* $\frac{5}{3}$, or $1\frac{2}{3}$ **7,** *p. 11:* $\frac{3}{10}$
8, *p. 11:* $8\frac{1}{8}$ **9,** *p. 12:* $5\frac{1}{6}$ **10,** *p. 12:* $\frac{3}{8}$
11, *p. 13:* $\frac{7}{22}$ **12,** *p. 13:* $\frac{63}{8}$, or $7\frac{7}{8}$ **13,** *p. 14:* 6
14, *p. 14:* $\frac{8}{5}$, or $1\frac{3}{5}$ **15,** *p. 14:* 400 students

Practices: Section R.4 pp. *16–20*

1, *p. 16:* **a.** $\frac{5}{10} = \frac{1}{2}$ **b.** $2\frac{73}{1000}$ **2,** *p. 16:* Four and three thousandths **3,** *p. 17:* 748.08 **4,** *p. 17:* 42.092
5, *p. 17:* 1.179 **6,** *p. 18:* 9.835 **7,** *p. 18:* 327,000
8, *p. 19:* 18.04 **9,** *p. 19:* 0.00086 **10,** *p. 20:* 1.5

Practices: Section R.5 pp. *21–22*

1, *p. 21:* $\frac{7}{100}$ **2,** *p. 21:* 0.05 **3,** *p. 22:* 2.5%
4, *p. 22:* 25% **5,** *p. 22:* 0.40, or 0.4

Review Exercises: Chapter R, *p. 23*

1. 6^5 **2.** 7^8 **3.** $2^2 \cdot 10^3$ **4.** $5^4 \cdot 4^3$ **5.** 25,000 **6.** 576
7. 15 **8.** 0 **9.** 14 **10.** 123 **11.** 4 **12.** 2 **13.** 1, 2, 3, 5, 6, 10, 15, 25, 30, 50, 75, 150 **14.** 1, 3, 19, 57
15. Prime **16.** Composite **17.** Composite **18.** Prime
19. $2 \cdot 3 \cdot 7$ **20.** $2 \cdot 3^3$ **21.** $2^4 \cdot 3$ **22.** $2^2 \cdot 5^2$ **23.** 24
24. 60 **25.** 72 **26.** 84 **27.** $\frac{19}{5}$ **28.** $\frac{73}{10}$ **29.** $5\frac{3}{4}$ **30.** $3\frac{4}{9}$

31. $\frac{1}{2}$ **32.** $\frac{2}{3}$ **33.** $5\frac{1}{2}$ **34.** $6\frac{2}{7}$ **35.** $\frac{5}{9}$ **36.** 1
37. $\frac{1}{4}$ **38.** 1 **39.** $\frac{34}{35}$ **40.** $1\frac{7}{18}$ **41.** $\frac{1}{4}$
42. $\frac{7}{20}$ **43.** $6\frac{1}{2}$ **44.** $7\frac{2}{5}$ **45.** $10\frac{3}{5}$ **46.** 8
47. $3\frac{1}{3}$ **48.** $\frac{1}{9}$ **49.** $1\frac{4}{5}$ **50.** $2\frac{1}{2}$ **51.** $6\frac{1}{2}$
52. $1\frac{1}{5}$ **53.** $2\frac{7}{8}$ **54.** $9\frac{7}{12}$ **55.** $\frac{2}{15}$ **56.** $\frac{2}{3}$
57. 14 **58.** $36\frac{2}{3}$ **59.** $15\frac{1}{6}$ **60.** 3 **61.** $5\frac{2}{3}$
62. $1\frac{2}{3}$ **63.** $\frac{1}{9}$ **64.** $\frac{1}{40}$ **65.** $3\frac{3}{7}$ **66.** $1\frac{4}{5}$
67. $\frac{1}{2}$ **68.** $20\frac{2}{5}$ **69.** $\frac{7}{8}$ **70.** $2\frac{3}{500}$
71. Hundredths **72.** Tenths **73.** Seventy-two hundredths
74. Five hundredths **75.** Three and nine thousandths
76. Twelve and two hundred thirty-five thousandths **77.** 7.3
78. 0.039 **79.** 4.39 **80.** $899 **81.** 18.11 **82.** 24.13
83. 1.873 **84.** 0.9 **85.** 2.912 **86.** 0.000010, or 0.00001 **87.** 2710 **88.** 530 **89.** 0.0015 **90.** 3.19
91. 5 **92.** 23.7 **93.** 7.35 **94.** 14.5 **95.** $\frac{3}{4}$
96. $\frac{1}{25}$ **97.** $1\frac{3}{50}$ **98.** $2\frac{1}{2}$ **99.** 0.06 **100.** 0.29 **101.** 1.5
102. 0.002 **103.** 31% **104.** 5% **105.** 1.45%
106. 200% **107.** 10% **108.** 37.5% **109.** 80%
110. 35% **111.** 73°F **112.** 8:10 A.M. **113.** $\frac{3}{4}$ mi
114. 7 lb **115.** $18 **116.** 1.66 m **117.** $1.30
118. 14.42 in. **119.** $\frac{3}{20}$ **120.** 10^6

Posttest: Chapter R, *p. 26*

1. 512 **2.** 37 **3.** 1, 2, 4, 5, 10, 20 **4.** $\frac{13}{4}$ **5.** $\frac{5}{18}$
6. $1\frac{1}{2}$ **7.** $12\frac{1}{24}$ **8.** $\frac{13}{90}$ **9.** $3\frac{19}{20}$ **10.** $\frac{3}{5}$ **11.** 2
12. $2\frac{3}{16}$ **13.** Two and three hundred ninety-six thousandths
14. 16.202 **15.** 6.99 **16.** 44.678 **17.** 2070
18. 0.0005 **19.** 0.125; 12.5% **20.** 70%; $\frac{7}{10}$ **21.** 6 times
22. 1.8 **23.** 625.3 sq m **24.** 15 times **25.** 0.2

Chapter 1

Pretest: Chapter 1, *p. 28*

1. +$2000 **2.** yes **3.** **4.** 5
5. $\frac{2}{3}$ **6.** < **7.** commutative property of addition **8.** -2
9. -26 **10.** $\frac{4}{1}$, or 4 **11.** 9 **12.** Ten less than the product of three and *n* (Answers may vary.) **13.** -6^4 **14.** -8
15. $4n - 7$ **16.** $14x - 10$ **17.** 6051 m **18.** $\frac{L}{3}$, or $\frac{1}{3}L$
19. 30 cm **20.** The third round

Practices: Section 1.1, *pp. 30–37*

1, *p. 30:* −5°F   **2,** *p. 32:*

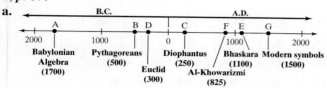

3, *p. 33:* **a.** 41 **b.** $\frac{8}{9}$ **c.** −1.7 **d.** $\frac{2}{5}$ **4,** *p. 34:* **a.** $\frac{1}{2}$ **b.** 0
c. 9 **d.** −3   **5,** *p. 35:* **a.** True **b.** False **c.** True **d.** True
e. False   **6,** *p. 35:* ; 3, $-\frac{1}{2}$, −1.6,
and −2.4   **7,** *p. 35:* Caspian Sea; −92 < −52 < 0
8, *p. 37:*

a.

b. A, B, D, C, F, E, and G

Exercises 1.1, *p. 38*

1. empty   **3.** whole numbers   **5.** integers   **7.** irrational
numbers   **9.** real numbers   **11.** −5 km   **13.** −22.5°C
15. −$160   **17.**
19.
21.
23.

	Whole Numbers	Integers	Rational Numbers	Real Numbers
25. −7		✓	✓	✓
27. −1.9			✓	✓
29. 10	✓	✓	✓	✓

31. 3   **33.** 0   **35.** 3.5   **37.** 4   **39.** 0   **41.** 4.6
43. $-\frac{1}{2}$ **45.** 4 and −4   **47.** Impossible; absolute
value is always positive or zero.   **49.** True   **51.** True
53. True   **55.** >   **57.** >   **59.** =   **61.** =   **63.** <
65. ; $3\frac{1}{2}$, 0, $-\frac{1}{2}$, $-1\frac{1}{2}$
67. ; 3.5, 3, −3, −3.5
69. −$53   **71.** Rational number and real number
73. −1.5   **75.** False   **77.** <   **79.** Today   **81.** −64.8°C
83. a. Sirius **b.** −13
c.
85. a. Socrates **b.** Pythagoras **c.** Euclid **d.** Woods

Practices: Section 1.2, *pp. 45–48*

1, *p. 45:* −1;

2, *p. 45:* 0;

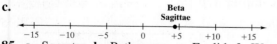

3, *p. 46:* 0.5;

4, *p. 46:* −31   **5,** *p. 47:* 3.5   **6,** *p. 47:* 0   **7,** *p. 47:* $-\frac{5}{18}$
8, *p. 48:* $36.50   **9,** *p. 48:* −3.588

Exercises 1.2, *p. 49*

1. negative   **3.** opposites   **5.** 1   **7.** 0   **9.** $4\frac{1}{2}$ **11.** 18
13. 0   **15.** 23   **17.** −5   **19.** 4   **21.** −80   **23.** −8
25. 0   **27.** 4.3   **29.** 0.6   **31.** −5.7   **33.** −16.3   **35.** −1
37. $-\frac{2}{5}$ **39.** $\frac{2}{5}$ **41.** $\frac{5}{6}$ **43.** −88   **45.** −7   **47.** −12
49. 1.245   **51.** −7   **53.** 15   **55.** −3.7   **57.** −0.3
59. 5° above 0° (+5)   **61.** Lost $16,000 (−$16,000)
63. The San Francisco 49ers by four points   **65.** Yes; he
will have $283.26 in the account.   **67.** 31,300 ft

Practices: Section 1.3, *pp. 52–54*

1, *p. 52:* 5   **2,** *p. 53:* 3   **3,** *p. 53:* −4   **4,** *p. 53:* −15.7
5, *p. 53:* 28   **6,** *p. 54:* 17   **7,** *p. 54:* 870 years older

Exercises 1.3, *p. 55*

1. 17   **3.** −31   **5.** −44   **7.** 71   **9.** −35   **11.** −32
13. 32   **15.** −45   **17.** −62   **19.** 44   **21.** 1000
23. −1.42   **25.** −7.8   **27.** 0   **29.** 10.3   **31.** $-\frac{7}{6}$
33. $-12\frac{1}{4}$ **35.** $6\frac{1}{10}$ **37.** 12   **39.** −7   **41.** −18   **43.** −21
45. −26.487   **47.** 48   **49.** −38   **51.** $1\frac{7}{12}$ **53.** 14
55. 28 centuries   **57.** 9000 ft   **59.** $276,039
61. 10,336 ft   **63. a.** radon: 9.2° neon: 2.7° bromine: 66°
b. bromine **c.** bromine

Practices: Section 1.4, *pp. 58–66*

1, *p. 58:* 100   **2,** *p. 59:* **a.** 8 **b.** $-\frac{5}{27}$
c. 0.12 **d.** −4.75 **e.** 0 **f.** $\frac{2}{3}$ **3,** *p. 59:* 64   **4,** *p. 60:* −28
5, *p. 60:* −30   **6,** *p. 60:* −10   **7,** *p. 61:* About −5.9; the
rock is moving downward at a velocity of 5.9 ft/sec.
8, *p. 62:* **a.** −8 **b.** 7 **c.** $-\frac{1}{2}$ **d.** −0.7 **e.** 60   **9,** *p. 63:* **a.** 0
b. Not possible, because −2 ÷ 0 is undefined **c.** 0
10, *p. 64:* **a.** $\frac{1}{-5}$, or $-\frac{1}{5}$ **b.** $\frac{-8}{1}$, or −8 **c.** $\frac{3}{4}$ **d.** $-\frac{5}{8}$
11, *p. 65:* **a.** $-\frac{4}{3}$, or $-1\frac{1}{3}$ **b.** 25   **12,** *p. 65:* **a.** −12
b. −3   **13,** *p. 66:* −$125 (Down $125)

Exercises 1.4, *p. 67*

1. negative   **3.** Multiplication identity   **5.** absolute values
7. undefined   **9.** one   **11.** Multiplicative inverse property
13. No   **15.** No   **17.** No   **19.** −12   **21.** 21   **23.** −3
25. $-\frac{4}{27}$ **27.** $-\frac{16}{27}$ **29.** 0.9   **31.** −60   **33.** 120   **35.** 0
37. 144   **39.** 120   **41.** $-\frac{1}{27}$ **43.** 0.98   **45.** −23
47. 27   **49.** −15   **51.** −5   **53.** 6   **55. a.** 5 **b.** 1 **c.** −3
d. −7 **e.** −11   **57. a.** $-\frac{2}{1}$, or −2 **b.** $\frac{1}{5}$ **c.** $-\frac{4}{3}$, or $-1\frac{1}{3}$
d. $\frac{5}{16}$ **e.** $\frac{1}{-1}$, or −1   **59.** 8   **61.** −9   **63.** 0   **65.** −25
67. 25   **69.** 4   **71.** 5   **73.** $-\frac{1}{8}$, or −0.125
75. $-\frac{1}{2}$, or −0.5   **77.** $-\frac{6}{5} = -1\frac{1}{5}$ **79.** −32   **81.** $-\frac{1}{8}$
83. −0.5   **85.** −20   **87.** −8   **89.** 10   **91.** 2.54
93. −1.17   **95.** −16   **97.** 1   **99.** 2   **101.** $\frac{1}{2}$ **103.** 4
105. 14   **107.** 3.64   **109.** −120   **111. a.** −6 **b.** −1 **c.** 4
d. 9 **e.** 14   **113.** 4   **115.** $-\frac{3}{2}$ **117.** 0.12   **119.** 5

121. 6 **123.** −$10 (lost $10) **125.** The team scored 2 more points than its opponents (+2). **127.** −15 in. (dropped 15 in.) **129. a.** −135 calories **b.** +106 calories **c.** −852 calories **131.** −26 **133.** A decrease of about 4736 people per year (−4735.5) **135.** An average loss of 4 yd (−4 yd) **137.** Expenses averaged $6000 per month (−$6000). **139.** Yes.

Practices: Section 1.5, *pp. 73–79*

1, *p. 73:* a. $5 + (-3)$ **b.** $3a + b$ **2, *p. 74:* a.** $(2)(-8)$ **b.** $n(-4)$ **3, *p. 74:* a.** $8 + [(-1) + 2]$ **b.** $x + (3y + z)$ **4, *p. 75:* a.** $(-3)[(5)(-2)]$ **b.** $[3(-6)]n$ **5, *p. 75:* −17 6, *p. 76:* a.** -120 **b.** 144 **7, *p. 77:* a.** -5 **b.** $6y$ **c.** -2 **d.** $-5x$ **8, *p. 77:* a.** 2 **b.** $\frac{2}{3}$ **c.** $-y$ **9, *p. 78:* a.** 5 **b.** $\frac{1}{2}$ **c.** $-\frac{1}{5}$ **d.** $-\frac{7}{2}$ **10, *p. 78:* a.** $(-2)(9) + (-2)(4.3)$ **b.** $0.2a + 0.2b$ **c.** $2q - pq$ **11, *p. 79:* (a)** The associative property of multiplication; **(b)** The multiplicative inverse property; **(c)** The multiplicative identity property **12, *p. 79:*** $1n = n$ dollars by the multiplicative identity property.

Exercises 1.5, *p. 80*

1. commutative property **3.** negative **5.** multiplicative identity property **7.** reciprocals **9.** $2 + 3.7$ **11.** $(-1) + [(-6) + 7]$ **13.** -3 **15.** $3 \cdot 1 + 3 \cdot 9$ **17.** $5(2 + 7)$ **19.** $2(a + b)$ **21.** commutative property of multiplication **23.** commutative property of addition **25.** distributive property **27.** multiplicative identity property **29.** additive inverse property **31.** associative property of addition **33.** associative property of multiplication **35.** 0 **37.** 6.33 **39.** 42 **41.** -8 **43.** -700 **45.** -2 **47.** 7 **49.** $\frac{1}{7}$ **51.** -1 **53.** $(-4)(2) + (-4)(5)$ **55.** $3 \cdot x + 3 \cdot 10 = 3x + 30$ **57.** $(-1)(a) + (-1)(6b) = -a + (-6b)$, or $-a - 6b$ **59.** $n \cdot n - n \cdot 2 = n^2 - 2n$ **61. (a)** The commutative property of multiplication **(b)** The associative property of multiplication **(c)** Multiplication of real numbers **63. (a)** The associative property of multiplication **(b)** The multiplicative inverse property **(c)** The multiplicative identity property **65.** 0.2 **67.** associative property of multiplication **69.** multiplicative inverse property **71.** The distance traveled to work is the same as the distance traveled going home. By the commutative property of addition: (distance traveled on bus) + (distance walked) = (distance walked) + (distance traveled on bus). **73.** By the distributive property, $r(p + q) = rp + rq$, so the shopper will pay the same amount regardless of which expression is used. **75.** Using the additive identity property, $p + 0 = p$, so his weight at the end of the week is p lb. **77.** According to the commutative property of multiplication, the calculations give the same result.

Practices: Section 1.6, *pp. 85–91*

1, *p. 85:* a. 3 **b.** 1 **2, *p. 86:*** Answers may vary. **a.** $\frac{1}{3}$ of p **b.** the difference between 9 and x **c.** s divided by -8 **d.** n plus -6 **e.** the product of $\frac{3}{8}$ and m **3, *p. 87:* a.** twice x minus the product of 3 and y **b.** 4 plus

$3m$ **c.** 5 times the difference between a and b **d.** the difference between r and s divided by the sum of r and s **4, *p. 87:* a.** $\frac{1}{6}n$ **b.** $n + (-5)$ **c.** $m - (-4)$ **d.** $\frac{100}{x}$ **e.** $-2y$ **5, *p. 88:* a.** $m + (-n)$ **b.** $5y - 11$ **c.** $\frac{m + n}{mn}$ **d.** $-6(x + y)$ **6, *p. 88:*** $60(m + 1)$ words **7, *p. 90:* a.** -36 **b.** 324 **8, *p. 90:*** $2^4(-5)^2$ **9, *p. 90:* a.** $-x^5$ **b.** $2m^3n^4$ **10, *p. 91:*** The population after 10 hr was $243x$, or $3^5 \cdot x$.

Exercises 1.6, *p. 92*

1. variable **3.** algebraic expression **5.** the quotient of x and 7 **7.** base **9.** 1 **11.** 3 **13.** 2 **15.** 3 plus t **17.** 4 less than x **19.** 7 times r **21.** the quotient of a and 4 **23.** the product of $\frac{4}{5}$ and w **25.** the sum of negative 3 and z **27.** twice n plus one **29.** 4 times the quantity x minus y **31.** one minus 3 times x **33.** the product of a and b divided by the sum of a and b **35.** twice x minus 5 times y **37.** $x + 5$ **39.** $d - 4$ **41.** $-6a$ **43.** $y + (-15)$ **45.** $\frac{1}{8}k$ **47.** $\frac{m}{n}$ **49.** $a - 2b$ **51.** $4z + 5$ **53.** $12(x - y)$ **55.** $\frac{b}{a - b}$ **57.** -9 **59.** -432 **61.** $(-2)^3 \cdot (4)^2$ **63.** $6^2 \cdot (-3)^3$ **65.** $3n^3$ **67.** $-4a^3b^2$ **69.** $-y^3$ **71.** $10a^3b^2c$ **73.** $-x^2y^3$ **75.** 8 times the quantity w minus y **77.** The product of s and r divided by the difference between r and s **79.** $a^2(-b)^2$ **81.** -36 **83.** $x + 2y$ **85.** $90° + x° + y°$ **87.** $\frac{30,000}{p}$ dollars **89.** $(t + x)$ dollars **91.** $(2^3 \cdot 5000)$ dollars **93.** s^2 **95.** $[10,000(\frac{1}{20})(20 - n)]$ dollars, or $500(20 - n)$ dollars **97.** $(ab - cd)$ ft^2

Practices: Section 1.7, *pp. 96–99*

1, *p. 96:* a. Terms: m and $-3m$; Like **b.** Terms: $5x$ and 7; Unlike **c.** Terms: $2x^2y$ and $-3xy^2$; Unlike **d.** Terms: m, $2m$, and $-4m$, Like **2, *p. 97:* a.** $6x$ **b.** $-6y$ **c.** $-2a + b$ **d.** 0 **3, *p. 97:* a.** $-2y^2$ **b.** Cannot be simplified **c.** $3xy^2$ **4, *p. 97:*** $3y - 10$ **5, *p. 98:*** $-2a + 3b$ **6, *p. 98:*** $4y - 1$ **7, *p. 98:*** $-2y - 18$ **8, *p. 99:*** $-10y + 13$ **9, *p. 99:*** $5c + 12(c - 40)$; $(17c - 480)$ dollars

Exercises 1.7, *p. 100*

1. coefficient **3.** distributive property **5.** negative **7.** 7 **9.** -5 **11.** 1 **13.** -1 **15.** -0.1 **17.** $\frac{2}{3}$ **19.** 2; -5 **21.** Terms; $2a$ and $-a$; like **23.** Terms; $5p$ and 3; unlike **25.** Terms; $4x^2$ and $-6x^2$; like **27.** Terms; x^2 and $7x^3$; unlike **29.** $10x$ **31.** $-11n$ **33.** $14a$ **35.** $2y + 2$ **37.** 0 **39.** Cannot be simplified **41.** $4r^2t^2$ **43.** Cannot be simplified **45.** $2x + 2$ **47.** $9x$ **49.** $-3y + 10$ **51.** $4x - 9$ **53.** $-n + 39$ **55.** $5x + 1$ **57.** $-3x + 13$ **59.** $-3a - 14$ **61.** Cannot be simplified **63.** Terms; $3a$ and $3a^2$; unlike **65.** 1 **67.** $3y + 7$ **69.** $x° + x° + 40°$; $2x° + 40°$ **71.** $d + 2(d + 4)$; $(3d + 8)$ dollars **73.** $n + (n + 1) + (n + 2)$; $3n + 3$ **75.** $0.05x + 0.04(1000 - x)$; $(0.01x + 40)$ dollars

Practices: Section 1.8, *pp. 103–106*

1, *p. 103:* a. 15 **b.** 30 **2, *p. 103:* a.** 18 **b.** 16 **c.** -300 **d.** 52 **3, *p. 104:* a.** $\frac{7}{3}$, or $2\frac{1}{3}$ **b.** $\frac{3}{4}$ **c.** 81 **d.** -81 **4, *p. 104:*** 0.1 **5, *p. 105:*** $F = \frac{9}{5}C + 32$ **6, *p. 105:*** The distance d is 80 mi. **7, *p. 106:* a.** $K = C + 273$ **b.** K is 267.

Exercises 1.8, *p. 107*

1. -2 **3.** 16 **5.** -32 **7.** -7 **9.** 0 **11.** 12 **13.** 5
15. 56 **17.** -7 **19.** 15 **21.** -14.5 **23.** -16 **25.** 15

27.

x	0	1	2	−1	−2
2x + 5	5	7	9	3	1

29.

y	0	1	2	3	4
y − 0.5	−0.5	0.5	1.5	2.5	3.5

31.

x	0	2	4	−2	−4
$-\frac{1}{2}x$	0	−1	−2	1	2

33.

n	2	4	6	−2	−4
$\frac{n}{2}$	1	2	3	−1	−2

35.

g	0	1	2	−1	−2
$-g^2$	0	−1	−4	−1	−4

37.

a	0	1	2	−1	−2
$a^2 + 2a - 2$	−2	1	6	−3	−2

39. $-20°$ **41.** $7\frac{1}{2}$ ft **43.** 314 m **45.** 13.5 cm² **47.** 1

49.

x	0	1	2	−1	−2
−2x + 4	4	2	0	6	8

51. -11 **53.** 10.5 in² **55.** $A = \frac{a + b + c}{3}$
57. $P = 2(l + w)$ **59.** $E = mc^2$ **61.** $l = 0.4w + 25$
63. The object falls 64 ft. **65. a.** $m = \frac{100(s - c)}{c}$
b. 40%

Review Exercises: Chapter 1, *p. 115*

1. $+3$ mi **2.** $-\$160$ **3.** [number line: point at 2, range −4 to 4]
4. [number line: point at 3, range −4 to 4]
5. [number line: point at 1, range −4 to 4]
6. [number line: point at −3, range −4 to 4]
7. 4 **8.** -6.5 **9.** $-\frac{2}{3}$ **10.** 0.7 **11.** 4 **12.** 0 **13.** 2.6
14. $\frac{5}{9}$ **15.** True **16.** False **17.** -5 **18.** -4
19. additive inverses **20.** Additive inverse property
21. 0 **22.** 2 **23.** 2 **24.** -5 **25.** -85 **26.** 15.3
27. 9 **28.** -11 **29.** -55 **30.** -3 **31.** -27 **32.** 27
33. 16 **34.** -5 **35.** -1.42 **36.** $-8\frac{5}{8}$ **37.** 5 **38.** 10
39. multiplicative inverse property **40.** Multiplication property of zero **41.** -10 **42.** -21 **43.** -5400
44. 2400 **45.** 27 **46.** $-\frac{1}{4}$ **47.** 6000 **48.** 36
49. -23 **50.** 14 **51.** 26 **52.** 38 **53.** -12
54. 6 **55.** $-\frac{3}{2}$ **56.** $\frac{1}{8}$ **57.** 3 **58.** -6 **59.** $-2\frac{1}{5}$
60. $-\frac{6}{5} = -1\frac{1}{5}$ **61.** 32 **62.** 2 **63.** -1 **64.** -2

65. 13 **66.** -20 **67.** $9 + 3$ **68.** $(-3)(1) + (-3)(9)$
69. associative property of addition **70.** additive inverse property **71.** 7 **72.** -160 **73.** -4 **74.** $\frac{3}{2}$
75. $3a - 12b$ **76.** $-x + 5$ **77.** 3 **78.** 2 **79.** 1
80. 4 For Exs. 81–90, answers may vary. **81.** the sum of negative 6 and w **82.** the product of negative $\frac{1}{3}$ and x
83. 6 more than negative 3 times n **84.** 5 times the quantity p minus q **85.** $x - 10$ **86.** $\frac{1}{2}s$ **87.** $\frac{p}{q}$ **88.** $R - 2V$
89. $6(4n - 2)$ **90.** $\frac{-4a}{5b + c}$ **91.** $(-3)^4$ **92.** $(-5)^3 3^2$
93. $4x^3$ **94.** $-5a^2 b^3 c$ **95.** $\frac{3}{4}$ **96.** $14x - 2y$ **97.** $-2x^2$
98. $r^2 t^2$ **99.** $2a - 9$ **100.** $-3x - 2$ **101.** $-4x - 15$
102. $a - 17$ **103.** 29 **104.** $-\frac{20}{9}$ **105.** -20
106. 60 **107.** 1 **108.** 1 **109.** $+\$700$ **110.** $-\$7000$
111. $-\$2.00$ **112.** Exothermic $(+3°C)$
113. $(I - 0.5h)$ degrees **114.** $14°F$ **115.** Netherlands
116. $6w$ **117.** Tue: $-\$0.15$, Wed: $-\$1.13$, Thu: $+\$0.15$, Fri: $+\$1.59$ **118.** The boiling is $9.2°$ higher. **119.** $3^3 \cdot 10$
120. 71 ft **121.** $3°$ **122.** $[0.05x + 0.07(600 - x)]$ dollars
123. Approximately 406 B.C. **124.** The first loss is 3 times the second loss. **125.** 20 amperes **126.** The account is overdrawn by \$10 (-10). **127.** $(x + 12y)$ dollars
128. $(f + 4s)$ students

Posttest: Chapter 1, *p. 120*

1. $-10,000$ **2.** Yes **3.** [number line: point at −2, range −3 to 3]
4. -7 **5.** 3.5 **6.** True **7.** 7 **8.** 3 **9.** 10
10. $\frac{1}{12}$ **11.** -50 **12.** $x + 2y$ **13.** $(-5)^3$ **14.** -5
15. $7y + 4$ **16.** $2t + 3$ **17.** $1.05d$ dollars
18. An improvement of \$70,000

Chapter 2

Pretest: Chapter 2, *p. 122*

1. No **2.** $n = -8$ **3.** $y = -5$ **4.** $n = -8$ **5.** $x = -9$
6. $x = -\frac{1}{2}$ **7.** $y = 11$ **8.** $x = 3$ **9.** $n = -2$
10. $x = \frac{5}{3}$ **11.** $v = w + 5u$ **12.** 25% **13.** 20
14. [number line: arrow, range −3 to 3] **15.** [number line: open circle, range −3 to 3]
16. 12 min **17.** 20 centerpieces **18.** $m = \frac{2E}{v^2}$
19. \$4000 was invested at 8%, and \$2000 was invested at 5%. **20.** Option A is a better deal if the member uses the gym more than 15 hours per month $(x > 15)$.

Practices: Section 2.1, *pp. 124–128*

1, *p. 124:* No, 4 is not a solution. **2, *p. 124:*** Yes, -8 is a solution. **3, *p. 126:*** $y = 5$ **4, *p. 126:*** $n = -17$
5, *p. 127:* $x = 0.1$ **6, *p. 128:*** 14.88 g

Exercises 2.1, *p. 129*

1. equation **3.** equivalent equations **5.** addition property of equality **7. a.** True **b.** False **c.** True **d.** True
9. Subtract 4 (or add -4). **11.** Add 1. **13.** Subtract 3.5 (or add -3.5). **15.** Add $2\frac{1}{5}$. **17.** $y = -23$ **19.** $t = 0$
21. $a = -12$ **23.** $z = -6$ **25.** $x = 0$ **27.** $t = 6$
29. $r = 6$ **31.** $n = 13$ **33.** $x = -1$ **35.** $y = -3\frac{1}{2}$
37. $m = 2.9$ **39.** $t = -3.6$ **41.** $a = -5$ **43.** $m = -\frac{1}{2}$
45. $y = 1.88$ **47.** $x + 2 = 12; x = 10$ **49.** $n - 4 = 21;$

$n = 25$ **51.** $x + (-3) = -1; x = 2$ **53.** $n + 7 = 11$;
$n = 4$ **55.** d   **57.** a   **59.** Subtract $\frac{2}{3}$ (or add $-\frac{2}{3}$).
61. a. False **b.** True **c.** False **d.** True   **63.** $x - 2.5 = -3.8$;
$x = -1.3$ **65.** $m = -23$ **67.** $n = -4\frac{1}{3}$
69. $x + 10 = 44; x = 34$ mph   **71.** $x - 130 = 220$;
$x = 560$ calories   **73.** $x - 190 = 370; x = 560$ calories
75. $x + 118.5 = 180; x = 61.5°$

Practices: Section 2.2, pp. 134–138

1, p. 134: $y = 63$ **2, p. 135:** $y = 9$ **3, p. 135:** $x = -10$
4, p. 136: $z = 13$ **5, p. 136:** $y = -14$ **6, p. 137:** The
total bill was $758.   **7, p. 138:** It will take about 2.3 hr.

Exercises 2.2, p. 139

1. Multiply by 3.   **3.** Divide by -5.   **5.** Divide by -2.2.
7. Multiply by $\frac{4}{3}$ **9.** Multiply by $-\frac{2}{5}$.   **11.** $x = -5$
13. $n = 18$ **15.** $a = 4.8$ **17.** $x = -0.5$ **19.** $c = -7$
21. $r = -22$ **23.** $x = 12$ **25.** $y = -\frac{5}{2}$ **27.** $n = 8$
29. $c = -3$ **31.** $x = -2.88$ **33.** $a = -2$ **35.** $y = \frac{2}{3}$
37. $x = -2.30$ **39.** $x = -6.82$ **41.** $-4x = 56; x = -14$
43. $\frac{n}{0.2} = 1.1; n = 0.22$ **45.** $\frac{x}{-3} = 20; x = -60$
47. $\frac{1}{6}x = 2\frac{4}{5}; x = \frac{84}{5}$ **49.** c   **51.** a   **53.** $n = 14$
55. $y = 1.14$ **57.** $\frac{x}{5} = 2; x = 10$ **59.** Divide by -5.2.
61. $0.02x = 10.5; x = 525$ yr   **63.** $70r = 3348; r \approx 48$ mph
65. $0.05c = 20; c = 400$ copies   **67.** $\frac{2}{3}x = 800{,}000$;
$x = \$1{,}200{,}000$ **69.** $\frac{1}{5}d = 1000; d = 5000$ m
71. $7.50t = 187.50; t = 25$ hr   **73.** $12x = 10{,}020$;
$x = \$835$ per month

Practices: Section 2.3, pp. 143–150

1, p. 143: $y = 4$ **2, p. 144:** $c = 45$ **3, p. 144:** $b = -3$
4, p. 144: $n = \frac{4}{3}$ **5, p. 145:** $t = 3$ **6, p. 145:** $f = -\frac{3}{4}$
7, p. 146: $z = -5$ **8, p. 146:** $t = -4$ **9, p. 147:**
$y = -2$ **10, p. 147:** The car will have a value of $6500 in
5 yr.   **11, p. 148:** The express train will catch up with the
local train in 2.5 hr, or $2\frac{1}{2}$ hr.   **12, p. 149:** 75 mi
13, p. 150: 20 mph

Exercises 2.3, p. 151

1. $x = 3$ **3.** $t = -2$ **5.** $m = -5$ **7.** $n = 12$
9. $x = -75$ **11.** $t = 2$ **13.** $b = -19$ **15.** $x = 39$
17. $r = -50$ **19.** $y = -2$ **21.** $z = -6$ **23.** $a = -7$
25. $t = 0$ **27.** $y = -1$ **29.** $r = \frac{10}{3}$ **31.** $x = -7$
33. $y = 2$ **35.** $a = 14$ **37.** $t = \frac{3}{2}$ **39.** $y = 0$ **41.** $z = 2$
43. $m = -2$ **45.** $y = 1.32$ **47.** $n = 0.27$ **49.** a   **51.** d
53. $x = -4$ **55.** $z = -2$ **57.** $y = 4.06$
59. $50 + 120x = 1010; x = 8$; The student is carrying
8 credits.   **61.** $x + 2x = 3690; x = 1230$; One candidate
received 1230 votes; the other candidate received 2460
votes.   **63.** $3 + 2(t - 1) = 9; t = 4$; The car was parked in
the garage for 4 hr.   **65.** $0.02x + 0.01(5000 - x) = 85$;
$x = 3500$; 3500 large postcards and 1500 small postcards
can be printed.   **67.** $24\left(t + \frac{1}{3}\right) = 36t; t = \frac{2}{3}$; It took
$\frac{2}{3}$ hr, or 40 min, to catch the bus.   **69.** $27r + 27(r + 2) = 432$;
$r = 7$; One snail is crawling at a rate of 7 cm/min, the other is
crawling at a rate of 9 cm/min.   **71.** $2r + 2(r + 4) = 212$;
$r = 51$; The speed of the slower truck is 51 mph.

Practices: Section 2.4, pp. 155–158

1, p. 155: $p = 1 - q$ **2, p. 155:** $r = \frac{t + s}{3}$
3, p. 156: $x = \frac{5ac}{4}$ **4, p. 156: a.** $x = \frac{y - b}{m}$ **b.** $x = -1$
5, p. 157: a. $r = \frac{A - P}{Pt}$ **b.** $r = 0.025$,   or 2.5%
6, p. 157: a. $h = \frac{2A}{b}$ **b.** $h = 14$ in.
7, p. 158: a. $A = \frac{1}{2}h(b + B)$ **b.** $b = \frac{2A - hB}{h}$ **c.** $b = 5$ cm

Exercises 2.4, p. 159

1. literal equation   **3.** algebraic expression
5. $y = x - 10$ **7.** $d = c + 4$
9. $d = \frac{-3y}{a}$ **11.** $n = 4p$ **13.** $z = \frac{2a}{xy}$ **15.** $x = \frac{7 - y}{3}$
17. $y = \frac{12 - 3x}{4}$ **19.** $y = 4t$ **21.** $b = \frac{p - r}{5}$
23. $l = \frac{2m - h}{4}$ **25.** $r = \frac{I}{pt}$
27. $r = \frac{d}{t}$ **29.** $b = P - a - c$ **31.** $d = \frac{C}{\pi}$
33. $R = \frac{P}{I^2}$ **35.** $a = 3A - b - c$ **37.** $a = S - dn + d$
39. $y = 6 - 3x; y = 42$ **41.** $x = \frac{y + 7}{3}; x = 4$
43. $y = -3x; y = -\frac{3}{2}$
45. $x = \frac{C - by}{a}; x = 14$ **47.** $h = \frac{v}{\pi r^2}$ **49.** $b = \frac{5m}{2ac}$
51. $z = \frac{3 - 4w}{9}$ **53.** $x = \frac{-20 - 2y}{7}; x = -4$ **55. a.** $K = \frac{V}{T}$
b. $V = KT$ **57. a.** $C = \frac{W}{150} \cdot A$ **b.** $A = \frac{150C}{W}$
59. a. $m = \frac{t}{5}$ **b.** $t = 5$ m   **c.** The thunder will be heard in
12.5 sec ($t = 12.5$).   **61. a.** $C = 2\pi r$ **b.** $r = \frac{C}{2\pi}$ **c.** $r \approx 0.8$ ft

Practices: Section 2.5, pp. 163–168

1, p. 163: 20   **2, p. 163:** $25 million   **3, p. 164:** 10.35 m
4, p. 164: 743,600   **5, p. 165:** $87\frac{1}{2}$%
6, p. 165: 31% of the presidents had been vice president.
7, p. 166: The number of nursing homes increased by
20.3%.   **8, p. 166:** The stock index dropped more in 1929.
9, p. 167: 6.5%   ($r = 0.065$)   **10, p. 167:** She invested
$7000 in a mutual fund and $14,000 in bonds.
11, p. 168: 2 g

Exercises 2.5, p. 169

1. base   **3.** times   **5.** 6   **7.** 23   **9.** 2.87 kg   **11.** $40
13. 4   **15.** 32 sq in.   **17.** $120   **19.** $200   **21.** 1.75
23. 4600 m   **25.** $62\frac{1}{2}$%, or 62.5%   **27.** $33\frac{1}{3}$%   **29.** 125%
31. 10%   **33.** $62\frac{1}{2}$%, or 62.5%   **35.** $140   **37.** 40%
39. 20   **41.** 0.035   **43.** $3\frac{1}{3}$%   **45.** 15 oz   **47.** 175%
49. 120   **51.** 9.6   **53.** 105   **55.** There are 80 more
faculty than staff.   **57.** 18.75%   **59.** There are 32
employees.   **61.** 20%   **63.** The workforce is 18.75 million
people.   **65.** 54 tables   **67.** No   **69.** $100   **71.** $250
73. $20,000 was invested at 8% and $14,000 was invested
at 10%.   **75.** $10,000 was invested at 5%.   **77.** $\frac{1}{3}$ cup
additional olive oil   **79.** 6 oz

Practices: Section 2.6, pp. 174–181

1, p. 174: No, 4 is not a solution.
2, p. 174:

3, p. 174: (number line from −5 to 5)

4, p. 175: (number line from −5 to 5)

5, p. 176: $n > -1$; (number line from −5 to 5)

6, p. 177: $x \leq 5\frac{1}{2}$; (number line from −2 to 8)

7, p. 177: $x \geq -7$; (number line from −8 to 2)

8, p. 178: $x \leq 3$; (number line from −5 to 5)

9, p. 178: $x < -5$; (number line from −11 to −1)

10, p. 179: $x > -5$; (number line from −8 to 2)

11, p. 179: $z \leq -5$ **12, p. 179:** $x < 2$

13, p. 180: $(x + 3) + (x + 2) + x \geq 14$; $x \geq 3$. The perimeter will be greater than or equal to 14 in. for any value of x greater than or equal to 3.

14, p. 180: $15(8.50) + 7.5t \geq 300$; $t \geq 23$. She should work at least 23 hr on the second job.

Exercises 2.6, p. 182

1. inequality   **3.** open   **5.** unchanged   **7.** negative
9. a. False   **b.** True   **c.** False   **d.** False

11. (number line from −2 to 8)

13. (number line from −8 to 2)

15. (number line from −5 to 5)

17. (number line from −5 to 5)

19. (number line from −5 to 5)

21. (number line from −5 to 5)

23. (number line from −5 to 5)

25. (number line from −5 to 5)

27. $v < -7$; (number line from −15 to −5)

29. $y > 0$; (number line from −5 to 5)

31. $y \leq 3.5$; (number line from −5 to 5)

33. $v \leq 2$; (number line from −5 to 5)

35. $2 \geq x$, or $x \leq 2$; (number line from −5 to 5)

37. $a < -3$; (number line from −10 to 0)

39. $y < -2$; (number line from −10 to 0)

41. $x \geq 0$; (number line from −5 to 5)

43. $a \leq -4$; (number line from −12 to −2)

45. $9 \geq n$, or $n \leq -9$; (number line from −18 to −8)

47. $n > 3$ **49.** $x \leq 6$ **51.** $y < -7$ **53.** $n \geq 13$
55. $m \geq 7$ **57.** $x > 7$ **59.** $z < 0$ **61.** $x \leq 0.25$
63. $x \leq -3$ **65.** $y > -3$ **67.** $x \leq 0.4$ **69** $x < \frac{1}{2}$
71. $n \geq 4.5$ **73.** $x < -6$ **75.** $y < -625$ **77.** d   **79.** d
81. (number line from −2 to 5)   **83. a.** False   **b.** True
c. True   **d.** True   **85.** (number line from −8 to 8)

87. $a \leq \frac{16}{3}$ **89.** $x < 5$ **91.** $\frac{81 + 85 + 91 + x}{4} > 85$; $x > 83$. The student must score above 83.
93. $\frac{250 + 250 + 150 + 130 + 180 + x}{6} \geq 200$; $x \geq 240$. The store must make at least $240.   **95.** $0.50 + 0.10x \geq 2$; $x \geq 15$. Each call lasts at least 15 min.
97. $1000 + 1500h > 1500 + 1200h$; $h > \frac{5}{3}$, or $1\frac{2}{3}$. He should accept the deal if he sells 2 or more houses each month.   **99.** $200 - 2.5x < 180$; $x > 8$. He will weigh less than 180 lb after 8 mo.

Review Exercises: Chapter 2, p. 190

1. No.   **2.** 0 is a solution.   **3.** $x = -9$ **4.** $t = -2$
5. $a = -14$ **6.** $n = 11$ **7.** $y = 7.9$ **8.** $r = 15.2$
9. $x = -6$ **10.** $z = -10$ **11.** $x = -10$ **12.** $d = -3$
13. $y = 4$ **14.** $x = -3$ **15.** $n = 41$ **16.** $r = -150$
17. $t = -9$ **18.** $y = -12$ **19.** $x = 3$ **20.** $t = -9$
21. $a = -14$ **22.** $r = 54$ **23.** $y = 9$ **24.** $t = 1$
25. $x = 6$ **26.** $y = -3$ **27.** $z = 3$ **28.** $n = 5$
29. $c = -\frac{2}{3}$ **30.** $p = \frac{5}{2}$ **31.** $x = -5$ **32.** $x = -\frac{2}{3}$
33. $n = 0$ **34.** $x = -\frac{17}{6}$ **35.** $x = \frac{8}{5}$ **36.** $x = -1$
37. $a = 2c + 5b$ **38.** $a = \frac{bn}{2}$ **39. a.** $x = \frac{C - By}{A}$
b. $x = \frac{5}{2}$ **40. a.** $h = \frac{2A}{b}$ **b.** $h = 6$ cm   **41.** 40
42. 4   **43.** 160%   **44.** $62\frac{1}{2}$%   **45.** 25.5   **46.** 70
47. (number line from −4 to 4)
48. (number line from −6 to 2)
49. (number line from −2 to 6)
50. (number line from −2 to 6)
51. $y > 5$; (number line from 4 to 10)
52. $t \geq 0$; (number line from −1 to 5)
53. $y \leq 2$; (number line from −3 to 3)
54. $x \geq 2$; (number line from 0 to 6)
55. $n < 5$; (number line from 1 to 7)   **56.** 16,000 Btu
57. 60°   **58.** 140 guests   **59.** 5 sides   **60.** One candidate received 11,925 votes; the other received 27,285 votes.
61. a. $C = 2 + 16y$ **b.** $y = \frac{C - 2}{16}$ **62.** The trucks will meet 4 hr after departure.   **63.** 10:15 P.M.   **64.** 1000 mi
65. Seaver received 99% of the votes cast.   **66.** 35%
67. 28 students   **68.** $30   **69.** Van Buren's electoral vote count dropped 65%.   **70.** 2 L   **71.** 5 pt   **72. a.** $p = 2.2k$
b. $k = \frac{p}{2.2}$ **73.** 4000 books   **74.** Up to 4 songs.

Posttest: Chapter 2, p. 193

1. -2 is not a solution.   **2.** $x = -9$ **3.** $n = -6$
4. $y = 11$ **5.** $y = 8$ **6.** $x = 3$ **7.** $s = 2$ **8.** $x = 2$

9. $a = 1$ **10.** $x = \frac{5}{6}$ **11.** $p = t - 5n$ **12.** 20

13. 200% **14.**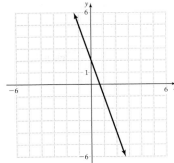

15. $z \geq -3$; **16.** 10 mi

17. $L = \frac{S + 21}{3}$ **18.** 23,000,000 operations **19.** 10 mph and 12 mph **20.** The monthly cost of Plan A exceeds the monthly cost of Plan B if more than 75 min of calls are made outside the network.

Cumulative Review: Chapter 2, *p. 194*

1. −6 yd **2.**   **3.** 2 **4.** True
5. 1 **6.** 5 **7.** $-11x + 36$ **8.** −2 lb **9.** 3^4
10. a. $A = 50 + 25(t - 1)$ **b.** $t = \frac{A - 25}{25}$ **c.** 4 hr

Chapter 3

Pretest: Chapter 3, *p. 196*

1. 1983–1985 and 1995–2001 **2.** 395 thousand

3. a. 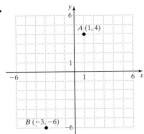   **b.** IV **4.** $m = \frac{1}{2}$

5. The slope of $\overleftrightarrow{PQ}$ is −1 and the slope of $\overleftrightarrow{RS}$ is 1. $\overleftrightarrow{PQ}$ is perpendicular to $\overleftrightarrow{RS}$, since the product of their slopes is −1. **6.** *x*-intercept: (3, 0); *y*-intercept: (0, 4)
7. The slope of the line is positive. As the population of a state increases, the number of representatives in Congress from that state increases. **8.** Variety A grows faster.

9.

x	4	7	$\frac{5}{2}$	2
y	3	9	0	−1

10.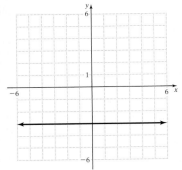

11.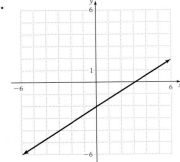

12.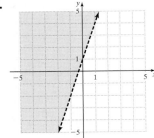

13. a. $y = 5x - 8$ **b.** slope is 5; *y*-intercept is $(0, -8)$.
14. Slope-intercept form: $y = 2x + 8$; point-slope form: $y - 8 = 2(x - 0)$ **15.** Slope-intercept form: $y = x - 3$; point-slope form: $y - 1 = x - 4$

16.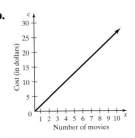

17. a. $c = 2.5x$ **b.**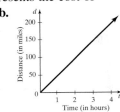

c. The slope of the line is 2.5. It represents the cost of renting a movie. **18. a.** $d = 50t$ **b.**

c. The slope of the graph is 50. It represents the speed the sales representative is driving.

19. 20

20.

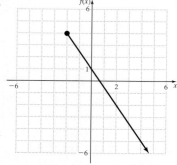

Domain: $[-2, \infty)$; range: $(-\infty, 4]$

Practices: Section 3.1 pp. 200–208

1, p. 200: a. Cattle and calves   **b.** $27 million   **c.** $2 million
2, p. 201: a. July   **b.** 27°F   **c.** In Chicago, mean temperatures increase from January through July and then decrease from July through December.   **3, p. 201:   a.** 47,000 children
b. 2003   **c.** Possible answer: Between 1995 and 2005, the number of children in foster care declined, and the number of children who were adopted increased.

4, p. 205:

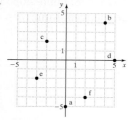

5, p. 205: a. II   **b.** IV   **c.** III   **d.** I
6, p. 206:

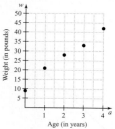

7, p. 207: a.

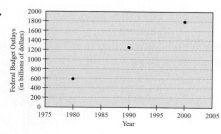

b. The points indicate an upward trend and appear to lie approximately in a straight line. We can use this trend to estimate the federal budget outlays for years in which the exact data are not available.

8, p. 208: From A to B and B to C, the line segments slant upward to the right, indicating that the runner's heartbeats per minute increase over this period of time. From C to D, the line segment slants downward, to the right, indicating that the runner's heartbeats per minute decrease. Possible story: The runner jogs slowly to warm up then runs more quickly, then jogs slowly, and finally rests.

Exercises 3.1, p. 209

1. Bar graph   **3.** Origin   **5.** Ordered pair   **7.** Below

9.

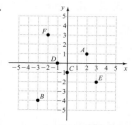

11.

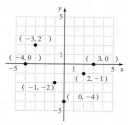

13. III   **15.** IV   **17.** I   **19.** II
21.

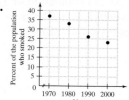

23. II   **25.** I

27. a. Milk, lemon juice, and vinegar   **b.** 8   **c.** Pure water   **29. a.** 90 million subscribers   **b.** 2003   **c.** Every year the number of cell phone subscribers increased.
31. a. $A(20, 40)$, $B(52, 90)$, $C(76, 80)$, and $D(90, 28)$
b. Students A, B, and C scored higher in English than in mathematics.
33.

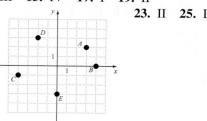

35. a.

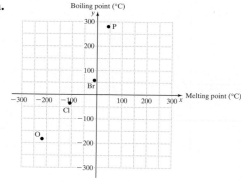

b. The y-coordinate is larger. The pattern shows that for each substance its boiling point is higher than its melting point.   **37.** The number of senators from a state (2) is the same regardless of the size of the state's population.

39. The graph in (a) could describe this motion. As the child moves away from the wall, the distance from the wall increases (line segment slants upward to the right). When the child stands still, the distance from the wall does not change (horizontal line segment). Finally, as the child moves toward the wall, the child's distance from the wall decreases (line segment slants downward to the right).

Practices: Section 3.2 *pp. 218–228*

1, p. 218: $m = \frac{1}{3}$
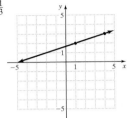

2, p. 219: $m = -\frac{6}{5}$
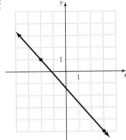

3, p. 220: $m = 0$
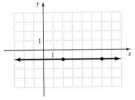

4, p. 220: The slope is undefined.
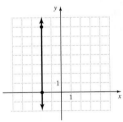

5, p. 222: Slope of $\overleftrightarrow{PQ}$: $-\frac{2}{3}$; slope of $\overleftrightarrow{RS}$: $-\frac{1}{2}$ **6, p. 222:** Scenario A is most desirable. The slope of the line is negative, which indicates a decrease in the number of people ill over time.   **7, p. 223:**
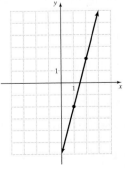

8, p. 224: a.
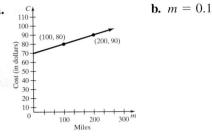
b. $m = 0.1$

c. The slope represents the cost per mile for renting a car.
9, p. 225: The slope of $\overleftrightarrow{EF}$ is $-\frac{5}{4}$ and the slope of $\overleftrightarrow{GH}$ is $-\frac{5}{4}$. Since their slopes are equal, the lines are parallel.
10, p. 226: a. Yes, the lines are parallel since their slopes are both $\frac{15}{4}$. **b.** Yes; the lines on the graph appear to be parallel. **c.** The salaries increased at the same rate.
d. The starting salary of the computer lab technician was about $57,000.   **11, p. 227:** The slope of $\overleftrightarrow{AB}$ is 2 and the slope of $\overleftrightarrow{AC}$ is $\frac{1}{2}$. Since the product of their slopes is not equal to -1, the lines are not perpendicular.   **12, p. 228:** The slope of the diagonal from (0, 0) to (6, 6) is 1. The slope of the diagonal from (0, 6) to (6, 0) is -1. Since the product of the slopes is -1, the diagonals of the square are perpendicular.

Exercises 3.2, *p. 229*

1. Rate of change   **3.** Negative   **5.** Horizontal
7. Parallel
9. $\frac{3}{4}$
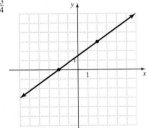

11. undefined
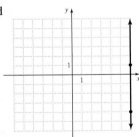

13. $-\frac{2}{5}$
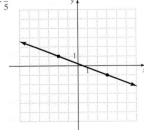

15. 0

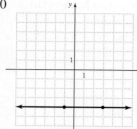

17. −7

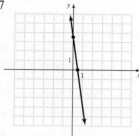

19. The slope of $\overleftrightarrow{AB}$ is −1. The slope of $\overleftrightarrow{CD}$ is 2.

21.

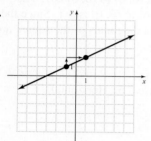

23.

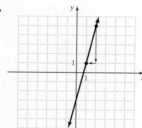

25.

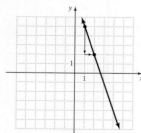

27.

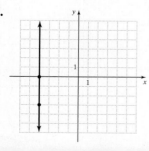

29. Positive slope; neither

31. Negative slope; neither   **33.** Undefined; vertical
35. Zero slope; horizontal   **37. a.** $\overleftrightarrow{PQ}: m = 4$; $\overleftrightarrow{RS}: m = 4$
The lines are parallel. **b.** $\overleftrightarrow{PQ}: m = -\frac{3}{2}$; $\overleftrightarrow{RS}: m = \frac{2}{3}$
The lines are perpendicular.   **39.** Zero slope; horizontal
41.

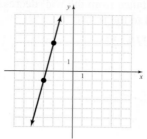

43. a. $\overleftrightarrow{AB}: m = -\frac{1}{2}$; $\overleftrightarrow{CD}: m = 2$; The lines are perpendicular. **b.** $\overleftrightarrow{AB}: m = -\frac{4}{3}$; $\overleftrightarrow{CD}: m = -\frac{4}{3}$; The lines are parallel.

45. Undefined

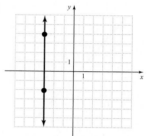

47. a. The slope is positive. **b.** A positive slope indicates that as the temperature of the gas increases, the pressure in the tube increases.   **49. a.** Motorcycle A **b.** Motorcycle B
c. The slope is the change in distance over time, or the average speed of the motorcycles.   **51.** The slope of each line is 1. Since the slopes of the lines are equal, the landfills are growing at the same rate.
53. a.

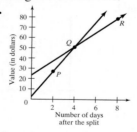

b. The slopes are 12 and 17. Since the slopes are not equal, the rate of increase did change over time.   **55.** The product of the slopes of the two lines is $-5 \cdot \frac{1}{3}$, which is not equal to −1, so $\overleftrightarrow{AD}$ is not the shortest route.   **57. a.** Graph II; As the car travels, its distance increases with time. This implies a positive slope. **b.** Graph I; The car is set for a constant speed. The speed of the car does not change over time. This implies a 0 slope.

Practices: Section 3.3 *pp. 240–250*

1, *p. 240*: a. Not a solution   **b.** A solution
2, *p. 241*:

x	0	5	−3	−$\frac{1}{2}$	−2
y	1	11	−5	0	−3

3, p. 243:

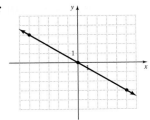

4, p. 244: a.

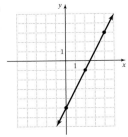

b. $m = 2$ **5, p. 246:** x-intercept: $(4, 0)$; y-intercept: $(0, -2)$

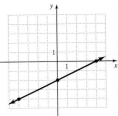

6, p. 247:

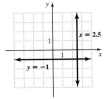

7, p. 248: a. $C = 0.03s + 40$

b.

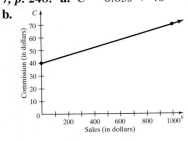

c. $m = 0.03$; for every sale, the commission increases by 0.03 times the value of the sale.

d. $55.   **8, p. 250: a.** $w + 2t = 10$

b.

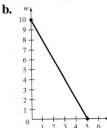

c. For each year the athlete is paid $2 million, the number of years that she could be paid $1 million decreases by 2 years.

d. The t-intercept is the number of years of the contract if she was paid $2 million in each year of the contract. The w-intercept is the number of years of the contract if she was paid $1 million in each year of the contract.

Exercises 3.3, p. 251

1. Solution   **3.** Three points   **5.** y-intercept

7.

x	4	7	$\frac{8}{3}$
y	4	13	0

9.

x	3.5	6	$\frac{1}{10}$	$-\frac{8}{5}$
y	17.5	30	$\frac{1}{2}$	-8

11.

x	0	-4	8	4
y	3	6	-3	0

13.

x	3	6	-3	0
y	0	1	-2	-1

15.

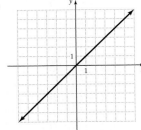

17.

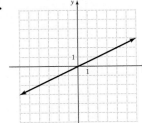

19.

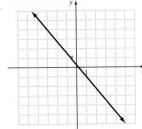

21.

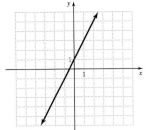

23.

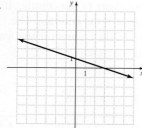

25.

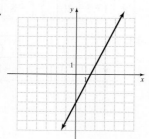

27.

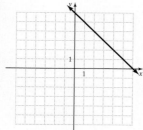

29.

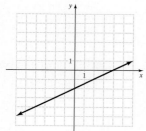

31.   *x*-intercept: (3, 0)
          y-intercept: (0, 5)

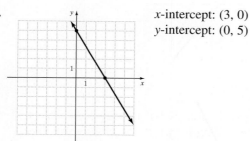

33.   *x*-intercept: (6, 0)
          y-intercept: (0, −3)

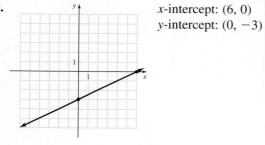

35.   *x*-intercept: (−2, 0)
          y-intercept: $\left(0, \frac{5}{2}\right)$

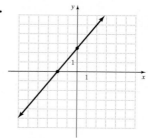

37.   *x*-intercept: $\left(-\frac{3}{2}, 0\right)$
          y-intercept: (0, −1)

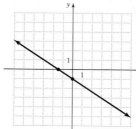

39.   *x*-intercept: (−4, 0)
          y-intercept: (0, 2)

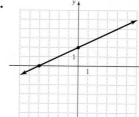

41.

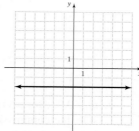

43.

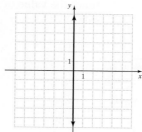

45.

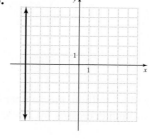

47.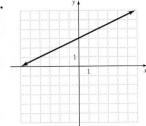

49.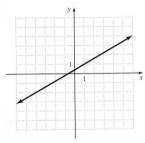

51.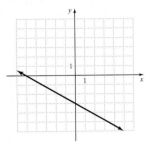

53.

x	−3	$\frac{5}{2}$	8	1
y	12	1	−10	4

55. 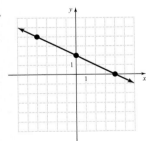   *x*-intercept: (−4, 0)
y-intercept: (0, 2)
possible: (−4, 4)

57. 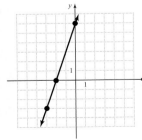   *x*-intercept: (−2, 0)
y-intercept: (0, 6)
possible: (−3, −3)

59. 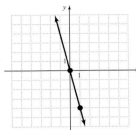   *x*-intercept: (0, 0)
y-intercept: (0, 0)
possible: (1, −4)

61. 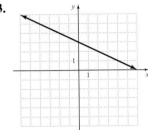   *x*-intercept: $\left(-\frac{1}{2}, 0\right)$
y-intercept: (0, −1)
possible: (−2, 3)

63.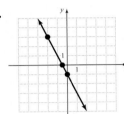

65. a.

t	0	0.5	1	1.5	2
v	0	−6	−22	−38	−54

A positive value of *v* means that the object is moving upward.
A negative value of *v* means that the object is moving
downward. **b.**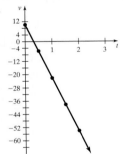

c. The *v*-intercept is the initial velocity of the object.
d. The *t*-intercept represents the time when the object
changes from an upward motion to a downward motion.
67. a. $P = 100m + 500$

b.

m	1	2	3
P	600	700	800

c.

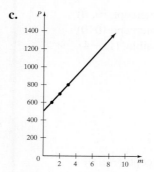

69. a. $0.05n + 0.1d = 2$ **b.**

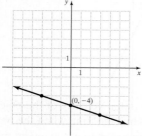

c. Only positive integer values make sense, since you cannot have fractions of nickels or dimes.

71. a. $F = 5d + 40$ **b.**

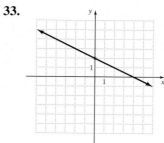

c. The F-intercept represents the fixed cost of renting the computer for 0 days.

Practices: Section 3.4 *pp. 265–271*

1, p. 265: Slope is -2; y-intercept is $(0, 3)$.
2, p. 265: $y = \frac{3}{2}x - 2$ **3, p. 265:** $y = 4x + 6$
4, p. 265: Slope is -1; y-intercept is $(0, 0)$.
5, p. 266:

6, p. 267: $y = 1x + 2$, or $y = x + 2$
7, p. 267: $y = -2x - 1$ **8, p. 268:** $y = -\frac{1}{2}x - 2$
9, p. 268: $w = 45 - 3t$ **10, p. 269:** $y - 0 = 2(x - 7)$
11, p. 269: $y - 7 = 1(x - 7)$ or $y - 0 = 1(x - 0)$
12, p. 269: Point-slope form: $w - 27 = 5(b - 4)$; slope-intercept form: $w = 5b + 7$ **13, p. 271:**

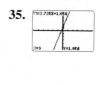

The displayed coordinates of the y-intercept are $x = 0$ and $y = 1.5$

Exercises 3.4, *p. 272*

1. Slope-intercept   **3.** y-intercept   **5.** For $y = 3x - 5$: 3, $(0, -5)$, ╱, $\left(\frac{5}{3}, 0\right)$; for $y = -2x$: -2, $(0, 0)$, ╲, $(0, 0)$; for $y = 0.7x + 3.5$: 0.7, $(0, 3.5)$, ╱, $(-5, 0)$; for $y = \frac{3}{4}x - \frac{1}{2}$: $\frac{3}{4}$, $\left(0, -\frac{1}{2}\right)$, ╱, $\left(\frac{2}{3}, 0\right)$; for $6x + 3y = 12$: -2, $(0, 4)$; ╲, $(2, 0)$; for $y = -5$: 0, $(0, -5)$, —, no x-intercept; for $x = -2$: undefined, no y-intercept, ╎, $(-2, 0)$ **7.** Slope: -1; y-intercept: $(0, 2)$ **9.** Slope: $-\frac{1}{2}$; y-intercept: $(0, 0)$
11. $y = x - 10$ **13.** $y = -\frac{1}{10}x + 1$ **15.** $y = -\frac{3}{2}x + \frac{1}{4}$
17. $y = \frac{2}{5}x - 2$ **19.** $y = 3x + 14$ **21.** b   **23.** a

25.

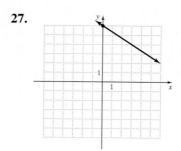

27.

29.

31.

33.

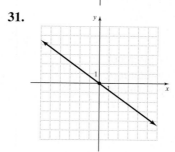

35.

37. $y = 3x + 7$ **39.** $y = 5x - 20$ **41.** $y = -\frac{1}{2}x + 4$
43. $y = -x + 3$ **45.** $y = \frac{1}{6}x - \frac{29}{6}$ **47.** $x = -3$
49. $y = 0$ **51.** $y = \frac{3}{4}x + 3$ **53.** $y = 2$ **55.** $y = -x - 2$
57. For $y = -7x + 2$: -7, $(0, 2)$, \, $(\frac{2}{7}, 0)$; for $y = 4x$: 4,
$(0, 0)$, /, $(0, 0)$; for $y = 2.5x + 10$: 2.5, $(0, 10)$, /, $(-4, 0)$;
for $y = \frac{2}{3}x - \frac{1}{4}$: $\frac{2}{3}$, $(0, -\frac{1}{4})$, /, $(\frac{3}{8}, 0)$; for $5x + 4y = 20$:
$-\frac{5}{4}$, $(0, 5)$, \, $(4, 0)$; for $x = 9$: Undefined, No y-intercept, l,
$(9, 0)$; for $y = -3.2$: 0, $(0, -3.2)$, $-$, No x-intercept
59. $y = 4x - 5$ **61.** d   **63.** $y = \frac{1}{2}x - 1$

65.

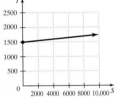

67. $y = -x + 5$

69. a. $\frac{9}{5}$ **b.** $F = \frac{9}{5}C + 32$ **c.** Water boils at 100°C.
71. a. $y - 3500 = 4(x - 500)$ **b.** $y = 4x + 1500$
c. The y-intercept represents the monthly flat fee the utility
company charges its residential customers.
73. a. $t - 1 = \frac{1}{5}(L - 5)$, or $L - 5 = 5(t - 1)$
b. $t = \frac{1}{5}L$, or $L = 5t$ **75. a.** $I = 0.03S + 1500$
b.

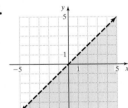

c. \$1686 **77.** $P = \frac{1}{33}d + 1$

79. $L = \frac{5}{6}F + 10$

Practices: Section 3.5 pp. 283–286

1, p. 283: No, $(1, 3)$ is not a solution to the inequality.
2, p. 284:

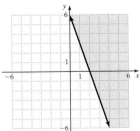

3, p. 285:

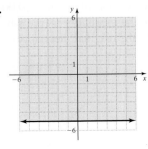

4, p. 285:

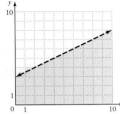

5, p. 286: a. $d + g \leq 3000$ **b.**

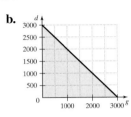

c. The d-intercept represents the maximum number of
gallons of diesel fuel the refinery produces if no gasoline is
produced. The g-intercept represents the maximum number
of gallons of gasoline the refinery produces if no diesel fuel
is produced.

Exercises 3.5, p. 287

1. Half-plane   **3.** Graph   **5.** Broken   **7.** No, not a
solution   **9.** Yes, a solution   **11.** No, not a solution
13.

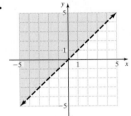

15.

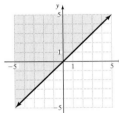

17.

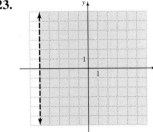

19. d   **21.** b

23.

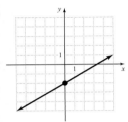

25.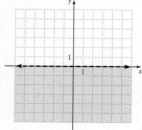

27.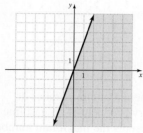

29.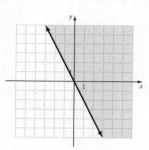

31.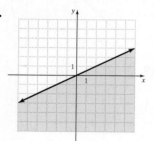

33.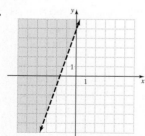

35.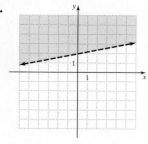

37.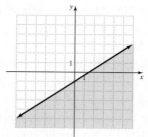

39. 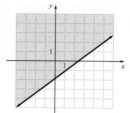   **41.** No, not a solution

43. b   **45.**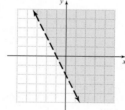

47.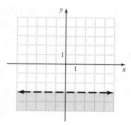

49. a. $h < \frac{1}{4}i$ **b.**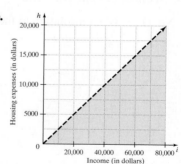

c. Choice of point may vary. Possible point: (20,000, 2500). The guideline holds since the inequality is true when the values are substituted into the original inequality.

51. a. $x + y \geq 200$ **b.**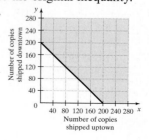

c. Choice of point may vary. Possible point: (200, 60). Check: $200 + 60 \overset{?}{\geq} 200$, $260 \geq 200$, True. At least 200 copies are shipped.

53. a. $30x + 75y \geq 1500$ **b.**

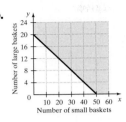

c. Since the point (20, 20) lies in the solution region, selling 20 small and 20 large gift baskets will generate the desired revenue.

55.

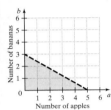

57. a. $10w + 15m \leq 50,000$

b.

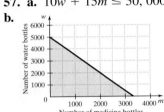

c. Answers may vary. Possible answers: 1500 medicine containers and 1000 bottles of water: (1500, 1000); 300 medicine containers and 500 bottles of water: (300, 500); 1200 medicine containers and 2000 bottles of water: (1200, 2000)

59. a. $8x + 10y \geq 200$

b.

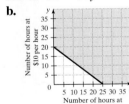

c. Answers may vary. Possible answers: 20 hr at the job paying \$8 per hour and 6 hr at the job paying \$10 per hour; 10 hr at the job paying \$8 per hour and 15 hr at the job paying \$10 per hour

Practices: Section 3.6, *pp. 297–303*

1, *p. 297:* a. Not a function   **b.** A function   **2, *p. 298:***
a. Domain: {2, 5, 8}; range: {−6, 0, 3}   **b.** Domain: {1, 2, 5}; range: {−1, 2, 4};   **c.** Domain: {1, 2, 3, 4, 5}; range: {6.5, 7, 7.5, 8, 9}   **3, *p. 299:* a.** 5   **b.** −4   **c.** $3n − 1$ **d.** $3n + 2$

4, *p. 301:*

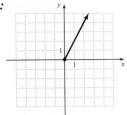

Domain: $[0, \infty)$; range: $[0, \infty)$ **5, *p. 302:* a.** Not a function   **b.** A function   **6, *p. 303:* a.** $C(m) = 0.1m + 40$ **b.** $C(200)$; $C(200) = \$60$

c.

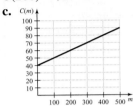

d. Domain: $[0, \infty)$; range: $[40, \infty)$ **e.** The graph in part (c) passes the vertical-line test, showing that the relation is a function.

Exercises 3.6, *p. 304*

1. relation   **3.** dependent   **5.** range   **7.** horizontal line
9. A function   **11.** Not a function   **13.** Not a function
15. A function   **17.** Domain: {−2, −1, 0, 1, 2}; range: {6, 8, 10, 12, 14}   **19.** Domain: {−3, −1, 0, 1, 3}; range: {−27, −1, 0, 1, 27}   **21.** Domain: {−4, −3, −2, −1, 1, 2, 3, 4}; range: {−8, −5, 0, 7}   **23.** Domain: {−4, −2, 0, 1, 2, 3.5, 5}; range: {1, 3, 5}   **25.** Domain: {−7, −3, −1, 2, 4}; range: {−4, −2, 1, 3, 7}   **27.** Domain:{2000, 2001, 2002, 2003, 2004}; range: {20, 23, 32, 34, 38}   **29. a.** −2   **b.** 13   **c.** 5   **d.** −1   **31. a.** 5   **b.** −11.8   **c.** $2.4a − 7$ **d.** $2.4a^2 − 7$

33. a. 3   **b.** 1   **c.** $|−2t + 3|$ **d.** $\left|\dfrac{1}{2}t\right|$ **35. a.** −9   **b.** 0
c. $3n^2 + 6n − 9$ **d.** $12n^2 − 12n − 9$ **37. a.** 10   **b.** 10
c. 10   **d.** 10

39.

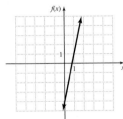

Domain: $(−\infty, \infty)$; range: $(−\infty, \infty)$

41.

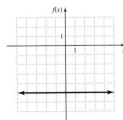

Domain: $(−\infty, \infty)$; range: {−5}

43. 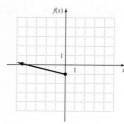   Domain: $(-\infty, 0]$; range: $[-1, \infty)$

45. A function   **47.** Not a function   **49.** A function
51. A function   **53.** A function   **55.** A function
57. Domain: $\{0, 2, 4, 6, 8, 9\}$; range: $\{-3, -1, 0, 1, 5, 6\}$
59. a. 0   **b.** 10   **c.** $3n^2 + n$ **d.** $48n^2 - 4n$
61. a. $d(a) = 0.2a$ **b.** $d(150)$; $d(150) = 30$, so $30 was
saved.   **c.**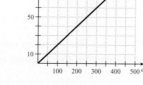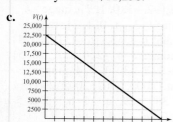

d. Domain: $[0, \infty)$; range: $[0, \infty)$
63. a. $V(t) = 22{,}500 - 1875t$ **b.** $V(6)$ represents the value
of the car six years after it is purchased. The value of the car
after six years is $11,250.

c.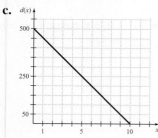

65. a. $d(x) = 500 - 50x$ **b.** $d(2) = 400$; after two weeks the
patient's dosage is 400 mg.

c.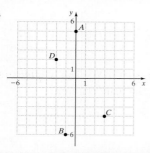

Review Exercises: Chapter 3, *p. 318*

1.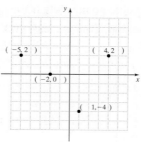

2. 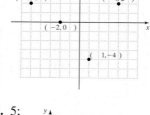   **3.** IV   **4.** III

5. 5; 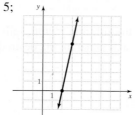   **6.** 0;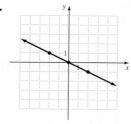

7. 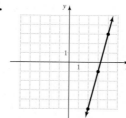   **8.**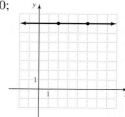

9. Positive slope   **10.** Undefined slope   **11.** Negative
slope   **12.** Zero slope   **13.** Parallel   **14.** Perpendicular
15. $(30, 0)$ **16.** $(0, 50)$

17.

x	0	1	$\frac{5}{2}$	3
y	-5	-3	0	1

18.

x	2	5	-4	8
y	1	-2	7	-5

19. 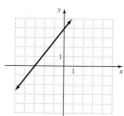   **20.**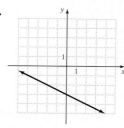

21. 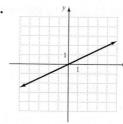   **22.**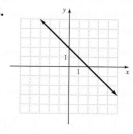

23. $y = x - 10$ **24.** $y = -\frac{1}{2}x - \frac{1}{2}$ **25.** 4, $(0, -16)$,
$\diagup$, $(4, 0)$ **26.** $-\frac{1}{3}$, $(0, 0)$, $\diagdown$, $(0, 0)$ **27.** The slope
of a line perpendicular to this line is -2.   **28.** The slope

of a line parallel to this line is 3.   **29.** Point-slope form: $y - 5 = -(x - 3)$; slope-intercept form: $y = -x + 8$
30. $y = 0$ **31.** Point-slope form: $y - 5 = -5(x - 1)$; slope-intercept form: $y = -5x + 10$ **32.** Point-slope form: $y - 1 = \frac{1}{5}(x - 3)$; slope-intercept form: $y = \frac{1}{5}x + \frac{2}{5}$
33. $y = -\frac{3}{2}x + 3$ **34.** $y = \frac{3}{2}x - 3$ **35.** No, it is not a solution   **36.** Yes, it is a solution
37.

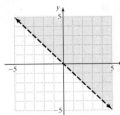

38.

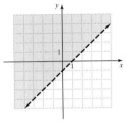

39.

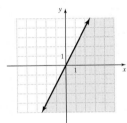

40.

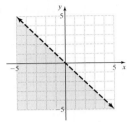

41. A function   **42.** Not a function   **43.** Domain: $\{-7, -5, -3, -1\}$; range: $\{3\}$ **44.** Domain: $\{-6, -4, -3, -2, 0, 2, 3, 4, 6\}$; range: $\{-6, -3, 0, 1, 2, 5, 8\}$ **45.** Domain: $\{-27, -8, -1, 0, 1, 8, 27\}$; range: $\{-3, -2, -1, 0, 1, 2, 3\}$ **46.** Domain: $\{20, 24, 26, 30, 32, 38\}$; range: $\{190, 228, 247, 285, 304, 361\}$ **47. a.** 3 **b.** 7.2 **c.** $a + 6$ **d.** $2a + 2$ **48. a.** 5 **b.** 10 **c.** $|8n - 7|$ **d.** $|n - 3|$

49. Domain: $(-\infty, \infty)$; range: $(-\infty, \infty)$

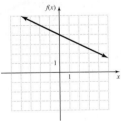

50. Domain: $[1, \infty)$; range: $[-5, \infty)$

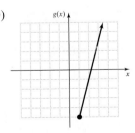

51. Not a function   **52.** A function   **53. a.** 36,000 **b.** 1,500 **c.** The number of movie screens in the United States decreased from 2000 to 2001 and then increased each year from 2001 to 2005   **54. a.** 61 million   **b.** 10%
c. It increases for younger age groups, peaks for 40–44, and decreases for older age groups.   **55. a.** Run number 3
b. Approximately 3 min **c.** With practice, the rat ran through the maze more quickly.   **56. a.** 2005 **b.** $\frac{4}{9}$ **c.** Both out-of-

pocket expenditures (with the exception of the period 2005–2006) and insurance expenditures increased and are expected to continue to increase each year from 2000 through 2010.

57. a.

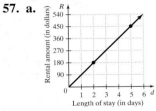

b. The R-intercept is (0, 0). The R-intercept means that the cost for renting a room for 0 days is $0.

58. a.

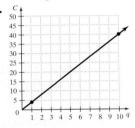

b. The slope of the line is 4. The slope represents the rate the print shop charges for each flyer, which is 4 cents.
59. The graph in part (a) could describe this motion. As the child walks toward the wall, the distance between the child and the wall decreases, implying a negative slope. When the child stands still, the distance between the child and the wall remains the same, as indicated by the horizontal line segment. When the child moves toward the wall again, the distance again decreases, implying a negative slope.
60. In the first part of the flight, the airplane takes off and ascends to a particular altitude (line segment slanting up to the right), then it flies at that same altitude during the second and longest part of the flight (horizontal line segment), and finally in the last part of the flight, it descends and lands (line segment slanting down to the right).
61. a. $i = 20,000 + 0.09s$ **b.**

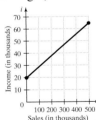

62. a.

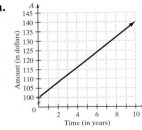

b. The A-intercept of the graph is (0, 100). The A-intercept represents the initial balance in the bank account.

63. a.

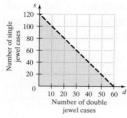

b. Answers may vary. Possible answer: 20 double jewel cases and 30 single jewel cases.

64. a.

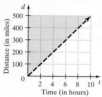

b. Choice of point may vary. Possible answer: (2, 110); the coordinates mean that you caught up and passed your friend if you covered a distance of 110 mi in 2 hr.

65. a. $P(x) = 1.2x + 300$ **b.** $P(200) = 540$; the company makes a monthly profit of $540 for selling 200 bottles of nail polish.

c.

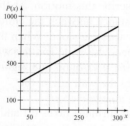

d. Domain: $[0, \infty)$; range: $[300, \infty)$

66. a. $A(m) = 4.95m + 10$ **b.**

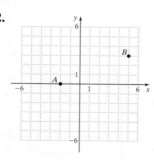

c. A customer pays about $70 for one year of service.

Posttest: Chapter 3, *p. 328*

1. 13,500,000 students   **2.**

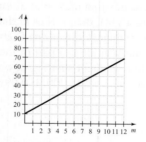

3. II   **4.** $m = 1$
5. The graphs are parallel. The slope of $y = 3x + 1$ is 3 and the slope of $y = 3x - 2$ is also 3. Since the slopes of the two lines are equal, their graphs are parallel.   **6.** The slope

of $\overleftrightarrow{AB}$ is $\frac{7}{2}$. The slope of $\overleftrightarrow{CD}$ is $-\frac{2}{7}$. $\overleftrightarrow{AB}$ is perpendicular to $\overleftrightarrow{CD}$, since the product of their slopes is -1.   **7.** x-intercept: $(-5, 0)$, y-intercept: $(0, 2)$ **8.** The slope is positive. As the number of miles driven increases, the rental cost increases.

9.

x	-3	5	$\frac{1}{3}$	1
y	10	-14	0	-2

10. 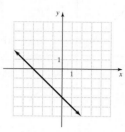   **11.**

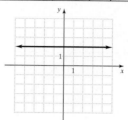

12.

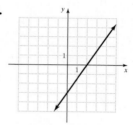

13. Slope: 3; y-intercept: $(0, 1)$ **14.** $y = 2x - 5$
15. $y = -x - 3$ **16.** Point-slope form:
$y - 5 = \frac{3}{7}(x - 3)$; slope-intercept form: $y = \frac{3}{7}x + \frac{26}{7}$
17.

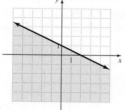

18.

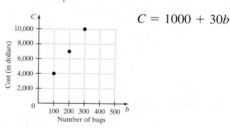

$C = 1000 + 30b$

19. $0.04x + 0.08y \geq 500$

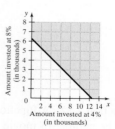

20. $-2a + 5$

Cumulative Review: Chapter 3, *p. 331*

1. -24 **2.** 22   **3.** Distributive property   **4.** 17
5. $x > 2$

6.

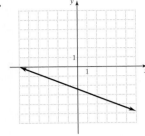

7. The assets of NY Bank are 40% of those of VA Bank ($x = 0.4$).   **8.** There are 20 years in a score.

9. $l = \dfrac{P - 2w}{2}$ **10. a.** $y = 6x$

b.

c. x-intercept: $(0, 0)$, y-intercept: $(0, 0)$

Chapter 4

Pretest: Chapter 4, *p. 334*

1. a. Not a solution **b.** A solution **c.** Not a solution
2. One solution
3. $(-3, 1)$

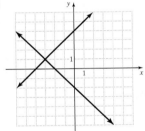

4. $(5, 2)$

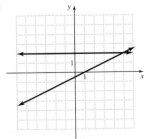

5. $(0, -2)$

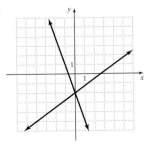

6. $(-5, -6)$ **7.** $(2, 1)$ **8.** $a = -5, b = 1$ **9.** $(1, -3)$
10. Infinitely many solutions   **11.** $(-4, 1)$ **12.** $n = -15$, $m = -7$ **13.** No solution   **14.** $(2, 0)$ **15.** 600 tickets were sold before 5:00 P.M. and 1375 tickets were sold after 5:00 P.M.
16. Fifty $5 tickets were printed.   **17.** $80,000 was invested in the fund at 5% interest, and $120,000 was invested in the fund at 6%.   **18.** The speed of the boat was 6 mph, and the speed of the current was 0.5 mph.
19.

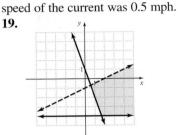

20. a. $x + \quad y \le 10{,}000$
          $0.05x + 0.06y \ge 300$

b.

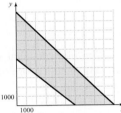

c. Answers may vary. Possible answer: $2500 in the account earning 5% simple interest and $5000 in the account earning 6% simple interest.

Practices: Section 4.1 *pp. 337–345*

1, *p. 337:* a. Yes, it is a solution of the system. **b.** No, it is not a solution of the system.   **2, *p. 339:* a.** One solution **b.** Infinitely many solutions **c.** No solution
3, *p. 340:* $(3, -1)$

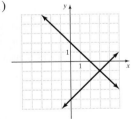

4, *p. 342:* The system has no solution.

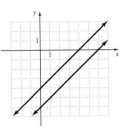

5, *p. 342:* The system has infinitely many solutions.

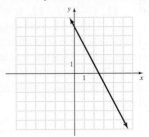

6, *p. 343:* **a.** $\begin{cases} m + v = 1150 \\ v = m - 100 \end{cases}$

b.

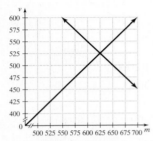

c. (625, 525) **d.** The point of intersection indicates that she got a score of 625 on her test of math skills and a score of 525 on her test of verbal skills.   **7, *p. 344:*** **a.** $y = 1.50x + 450$ **b.** $y = 3x$

c.

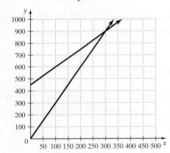

d. The break-even point is (300, 900). So when 300 newsletters are printed, the cost of printing the newsletter and the income from sales will be the same, $900.

8, *p. 345:*

The approximate solution is (0.857, 5.857).

Exercises 4.1, *p. 347*

1. system of equations   **3.** are parallel
5. **a.** Not a solution; **b.** Not a solution; **c.** A solution
7. **a.** Not a solution; **b.** Not a solution; **c.** A solution
9. **a.** III **b.** IV **c.** II **d.** II
11. (3, 1)

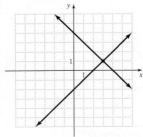

13. (0, 4)

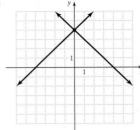

15. (1, 5)

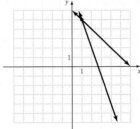

17. (0, 1)

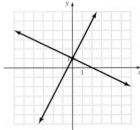

19. $(-2, -1)$

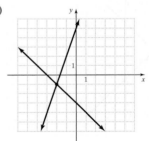

21. Infinitely many solutions

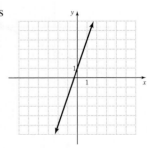

23. No solution

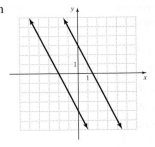

25. Infinitely many solutions

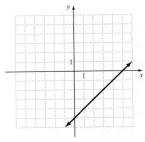

27. No solution

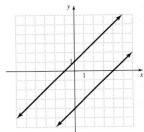

29. $(-2, -2)$

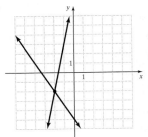

31. No solution

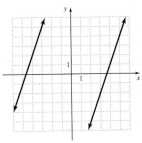

33. $(-1, 2)$

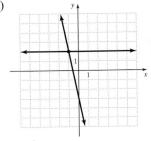

35. $(0, -2)$

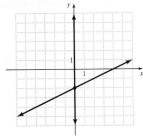

37. $(-6, -4)$

39. Infinitely many solutions

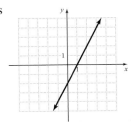

41. $(3, -2)$

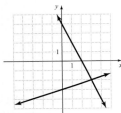

43. $(2, 2)$

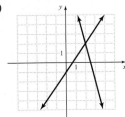

45. b **47. a.** $x + y = 57{,}000$
$$y = x + 3000$$

b.

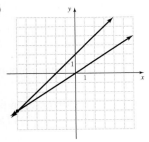

c. The husband made $27,000 and the wife made $30,000.
49. a. $y = 40x + 75$ (Mike)
$y = 30x + 100$ (Sally)

b.

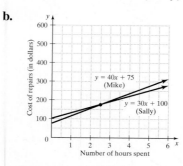

c. The plumbers would charge the same amount for 2.5 hr of work.   **d.** Sally charges less.

51.

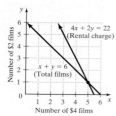

The break-even point for duplicating DVDs is (24, 36).

53.

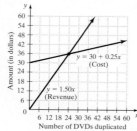

Five $4 films were rented.

55.

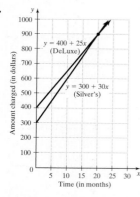

The health clubs charge the same amount ($900) for 20 months.

Practices: Section 4.2 pp. 359–365

1, p. 359: $(4, -3)$ **2, p. 360:** $m = \frac{32}{7}, n = -\frac{5}{7}$
3, p. 361: $(0, -1)$ **4, p. 361:** No solution
5, p. 362: Infinitely many solutions   **6, p. 362:**
a. $c = 35n + 20$ (TV Deal), $c = 25n + 30$ (Movie Deal)
b. $n = 1$ and $c = 55$ **c.** The cost is the same ($55) for both cable deals if you sign up for one month.   **7, p. 363:** 12.5 ounces of the 20% copper alloy and 2.5 ounces of the 50% copper alloy are required.   **8, p. 365:** The manager put

$110,000 in the investment that pays 4% and $88,000 in the investment that pays 5%.

Exercises 4.2, p. 366

1. $(3, 7)$ **3.** $(-4, -3)$ **5.** $(-12, 4)$ **7.** $(2, 1)$ **9.** $(0, 0)$
11. $(-1, 2)$ **13.** $(6, -6)$ **15.** Infinitely many solutions
17. $(\frac{7}{2}, -\frac{1}{2})$ **19.** No solution   **21.** Infinitely many solutions   **23.** No solution   **25.** $p = 3, q = 5$ **27.** $s = -2,$
$t = 1$ **29.** $(-3, 6)$ **31.** No solution   **33.** $(-\frac{3}{2}, \frac{1}{2})$
35. a. $c = 1.25m + 3, c = 1.50m + 2$ **b.** $m = 4, c = 8.$
The solution indicates that both companies charge the same amount ($8) for a 4-mi taxi ride.   **37.** 80 full-price tickets were sold.   **39.** She can combine 2.5 liters of the antiseptic that is 30% alcohol with 7.5 liters of the antiseptic that is 70% alcohol to get the desired concentration.   **41.** There were 3 women in one department and 72 women in the other department.   **43.** The loan at 6% was $4000 and the loan at 7% was $1000.   **45.** $23,000 was invested at 7% and $17,000 was invested at 9%.

Practices: Section 4.3 pp. 369–374

1, p. 369: $(-2, 8)$ **2, p. 370:** $(2, -5)$ **3, p. 370:** $(0, 6)$
4, p. 371: $(2, -2)$ **5, p. 372:** $(10, 8)$ **6, p. 372:** Infinitely many solutions   **7, p. 373:** The whale's speed in calm water is 30 mph and the speed of the current is 10 mph.
8, p. 374: 100 adults and 75 children attended the game.

Exercises 4.3, p. 376

1. $(5, -2)$ **3.** $(-\frac{3}{2}, -\frac{5}{2})$ **5.** $p = -3, q = -16$
7. $(-1, 0)$ **9.** No solution   **11.** Infinitely many solutions
13. $(-2, \frac{1}{2})$ **15.** $s = 0, d = -2$ **17.** $(7, 4)$ **19.** $(-1, 2)$
21. $p = \frac{9}{2}, q = -\frac{11}{2}$ **23.** $(-3.5, 2.5)$ **25.** $(2, -2)$
27. $(\frac{8}{3}, -\frac{4}{3})$ **29.** No solution   **31.** $a = -3, b = -2$
33. Infinitely many solutions   **35.** $(-\frac{3}{4}, \frac{1}{4})$ **37.** The speed of the pass if there were no wind would be 13 yd per sec.
39. The zoo collected 83 full-price admissions and 140 half-price admissions.   **41.** The salary of a senator is $125,100 and the salary of a congressman is $101,900.
43. It takes the computer 3 nanoseconds to carry out one sum and 4 nanoseconds to carry out one product.   **45.** The rate for full-page ads is $950 and for half-page ads is $645.

Practices: Section 4.4, pp. 380–383

1, p. 380:

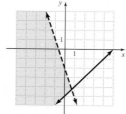

2, p. 381:

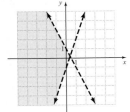

3, p. 382:

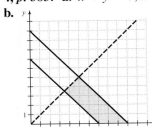

4, p. 383: a. $x + y \geq 7$; $x + y \leq 10$; $x > y$

b.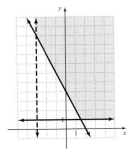

c. (4, 3): 4 freshmen and 3 returning students; (5, 2):
5 freshmen and 2 returning students; (5, 3): 5 freshmen and
3 returning students; (5, 4): 5 freshmen and 4 returning
students; (6, 1): 6 freshmen and 1 returning student; (6, 2):
6 freshmen and 2 returning students; (6, 3): 6 freshmen and
3 returning students; (6, 4): 6 freshmen and 4 returning
students; (7, 0): 7 freshmen and 0 returning students; (7, 1):
7 freshmen and 1 returning student; (7, 2): 7 freshmen and
2 returning students; (7, 3): 7 freshmen and 3 returning
students; (8, 0): 8 freshmen and 0 returning students; (8, 1):
8 freshmen and 1 returning student; (8, 2): 8 freshmen and
2 returning students; (9, 0): 9 freshmen and 0 returning
students; (9, 1): 9 freshmen and 1 returning student; (10, 0):
10 freshmen and 0 returning students. In this situation, only
nonnegative integer solutions can be considered. The number of solutions in the overlapping region is limited to
those listed here.

Exercises 4.4, *pp. 384*

1. a system of   **3.** coordinate plane

5. 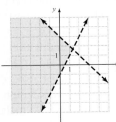   **7.**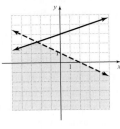

9. 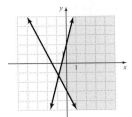   **11.**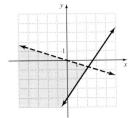

13. 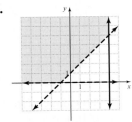   **15.**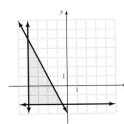

17. 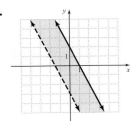   **19.**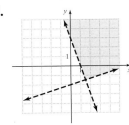

21. 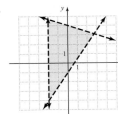   **23.**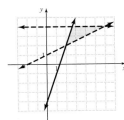

25. 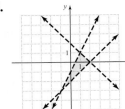   **27.**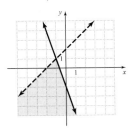

29. 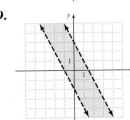   **31.**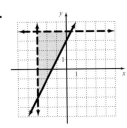

33. a. $x + y \geq 35$; $x \leq 20$; $y \leq 30$
b.

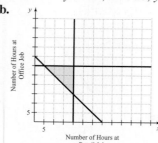

c. She must work between 20 and 30 hr at her office job.
35. a. $2l + 2w \leq 400$; $l \geq w + 25$; $w \geq 25$ **b.**

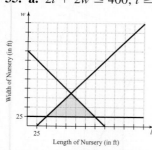

c. The solution region represents all possible dimensions for the nursery. **37. a.** $150x + 225y < 15,000$; $x > y$; $x > 20$
b.

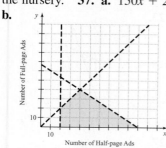

39. a. $2x + 1.5y \leq 360$; $3x + y \leq 400$
b.

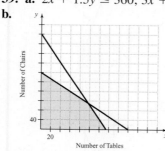

Review Exercises: Chapter 4, *p. 394*

1. No, it is not a solution of the system. **2. a.** No solution
b. One solution **c.** Infinitely many solutions **3. a.** III
b. IV **c.** II **d.** I
4. (1, 5)

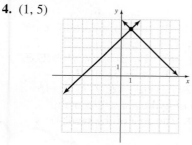

5. No solution

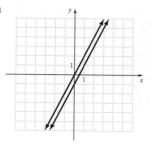

6. Infinitely many solutions

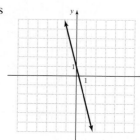

7. (0, 5)

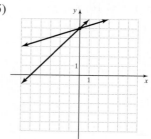

8. $(-1, 4)$ **9.** $a = 2, b = 2$ **10.** No solution
11. Infinitely many solutions **12.** $(4, -3)$ **13.** Infinitely
many solutions **14.** $(6, -3)$ **15.** $(2, -5)$
16.

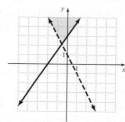

17.

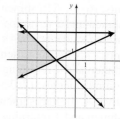

18. a. $y = 0.50x + 1750$, $y = 5.50x$ **b.** The student must
type 350 pages in order to break even. **19. a.** $s = 10h, s = 8h + 50$ **b.** 25 hr **20.** The area of the screen is 6720 sq ft.
21. The tennis court is 31.5 ft wide and 82.5 ft long.
22. The coin box contained 200 nickels and 150 dimes.
23. One train travels at a rate of 60 mph and the other travels
at a rate of 65 mph. **24.** The team made 818 two-point
baskets and 267 three-point baskets. **25.** The pharmacist
should mix 150 mL of the 30% solution and 50 ml of the
10% solution. **26.** 1400 L of the 50% solution and 600 L of
water are needed to fill the tank. **27.** The client put $40,000
in municipal bonds and $10,000 in corporate stocks.
28. $7000 was invested in the high-risk fund and $3000
was invested in the low-risk fund. **29.** The speed of the
slower plane was 400 mph. **30.** The bird flies at a speed
of 21 mph in calm air and the speed of the wind is 5 mph.
31. a. 57 senators **b.** 43 senators **32. a.** The two metals
will be the same temperature after 12 min. **b.** The iron will
be 14° colder than the copper after 14 min.

33. a. $2t + 3a \leq 500; 1.5t + 2a \leq 300$
b.

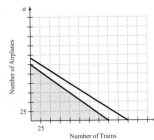

c. The solution region represents all the possible numbers of trains and corresponding number of airplanes the company can assemble and paint each month.
34. a. $x + y \leq 12{,}000; 0.08y > 0.06x; y \geq 4500$
b.

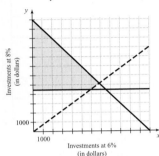

Posttest: Chapter 4, *p. 399*

1. a. A solution **b.** Not a solution **c.** Not a solution
2. The system has no solution.
3. $(3, 0)$

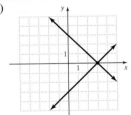

4. No solution

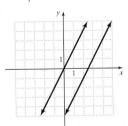

5. $(-1, 2)$

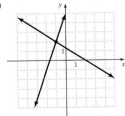

6. $(1, 3)$ **7.** $u = \frac{3}{4}, v = \frac{17}{4}$ **8.** $(2, -5)$ **9.** No solution
10. $p = 0, q = 0.5$ **11.** $(-2, -7)$ **12.** Infinitely many solutions **13.** $(\frac{1}{7}, -\frac{6}{7})$ **14.** The winning candidate got 4204 votes. **15.** One serving of turkey and two servings of salmon **16.** $8000 was invested at 7.5% and $4000 at 6%.

17. 1 gal of the 20% iodine solution and 3 gal of the 60% iodine solution **18.** The speed of the wind was 20 mph and the speed of the plane in still air was 150 mph.
19.

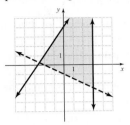

20. a. $64x + 60y \geq 300; x + y \leq 12; x > y$
b.

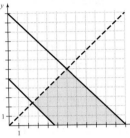

c. Answers may vary. Possible answers: (6, 4): The faster student drives 6 hr and the slower student drives 4 hr. (5, 3): The faster student drives 5 hr and the slower student drives 3 hr.

Cumulative Review: Chapter 4, *p. 402*

1. -26 **2.** No, 5 is not a solution. **3.** $x = -3$
4. ←——+——+——+——⊕——+——+——+——+——→ ; $x > 1$
       $-2\,-1\ \ 0\ \ 1\ \ 2\ \ 3\ \ 4\ \ 5\ \ 6\ \ 7\ \ 8$
5. The plane can fly 1554 mph ($S = 1554$).
6. Slope: $m = -\frac{1}{2}$; y-intercept: $(0, 2)$ **7.** The slope is $\frac{3}{4}$.
8.   **9.** No

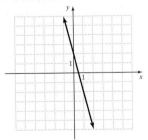

10. The two companies charge the same amount ($95) for a one-day rental if the car is driven 300 miles.

Chapter 5

Pretest: Chapter 5, *p. 404*

1. x^9 **2.** y^4 **3.** -3 **4.** $16x^8y^6$ **5.** $\frac{a^3}{b^{15}}$ **6.** $\frac{x^2}{25y^8}$
7. a. $6x^4, 5x^3, x^2, -7x$, and 8 **b.** 6, 5, 1, -7, and 8 **c.** 4
d. 8 **8.** $3n^2 + n + 2$ **9.** $x^2 + x - 4$
10. $-6a^2 + 9a^2b + 3ab - 4b^2$ **11.** $3x^4 - 12x^3 + 27x^2$
12. $2n^3 + 7n^2 - 3n - 18$ **13.** $4x^2 - 3x - 27$
14. $9y^2 - 49$ **15.** $25 - 20n + 4n^2$ **16.** $t^2 - 2t - 5$
17. $4x + 5$ **18.** 200 mol of hydrogen will contain 1.2×10^{26} molecules. **19.** The area of the sandbox is

$(4x^2 - 25)$ ft^2 **20.** The average monthly cell telephone bill was $48.44 in 2006.

Practices: Section 5.1 pp. 405–411

1, p. 405: a. 10,000 **b.** $-\frac{1}{32}$ **c.** y^6 **d.** $-y$ **2, p. 406: a.** 1 **b.** 1 **c.** 1 **d.** -1 **3, p. 407: a.** 10^{12} **b.** $(-4)^6$ **c.** n^{10} **d.** y^5 **e.** Cannot apply the product rule because the bases are not the same. **4, p. 408: a.** 7^5 **b.** $(-9)^1$, or -9 **c.** s^0, or 1 **d.** r^7 **e.** Cannot apply the quotient rule because the bases are not the same. **5, p. 408: a.** y^9 **b.** x^5y^6 **c.** a^2 **6, p. 409: a.** $\frac{1}{9^2} = \frac{1}{81}$ **b.** $\frac{1}{n^5}$ **c.** $-\frac{1}{3y}$ **d.** $\frac{1}{5^3} = \frac{1}{125}$ **7, p. 410: a.** $\frac{s}{8}$ **b.** $\frac{3}{x}$ **c.** $\frac{1}{r^6}$ **d.** $\frac{3^2}{g^3} = \frac{9}{g^3}$ **e.** x^3 **8, p. 410: a.** a^3 **b.** $\frac{2x^2}{5}$ **c.** $\frac{r^3s}{2}$ **9, p. 411: a.** 2^{23} **b.** 2^{30}

Exercises 5.1, p. 412

1. added **3.** zero power **5.** 125 **7.** $\frac{9}{16}$ **9.** 0.16 **11.** -0.25 **13.** -8 **15.** 81 **17.** $-\frac{1}{8}$ **19.** x^4 **21.** pq **23.** 1 **25.** -1 **27.** 10^{11} **29.** 4^8 **31.** a^6 **33.** Cannot be simplified **35.** n^7 **37.** Cannot be simplified **39.** 8^2 **41.** 5^3 **43.** y^1, or y **45.** a^6 **47.** Cannot be simplified **49.** $x^0 = 1$ **51.** y^6 **53.** p^7q^5 **55.** $x^3y^2z^3$ **57.** a^1, or a **59.** x^2 **61.** $\frac{1}{5}$ **63.** $\frac{1}{x}$ **65.** $-\frac{1}{3a}$ **67.** $\frac{1}{2^4}$ **69.** $-\frac{1}{3^4}$ **71.** $\frac{8}{n^3}$ **73.** $\frac{1}{(-x)^2}$, or $\frac{1}{x^2}$ **75.** $\frac{-x}{3^2}$ **77.** $\frac{y^3}{x^2}$ **79.** $\frac{q}{r}$ **81.** $\frac{4y^2}{x}$ **83.** $\frac{1}{p^5}$ **85.** p^3 **87.** $\frac{1}{a}$ **89.** $2n^4$ **91.** p^4q **93.** $\frac{1}{t^5}$ **95.** x^7 **97.** a^1, or a **99.** $\frac{b^3}{a^3}$ **101.** s^3t^6 **103.** $-\frac{8}{27}$ **105.** Cannot be simplified **107.** $\frac{1}{y^6}$ **109.** $\frac{1}{q^3}$ **111.** x^4y^3

113. a. $35 \cdot 2^5 = 1120$ people were ill on the sixth day of the epidemic; $35 \cdot 2^9 = 17,920$ people were ill on the tenth day. **b.** The number of people ill on the tenth day was 2^4, or 16, times as great as the number ill on the sixth day. **115.** $60 \times (0.95)^{11}$ ppm **117. a.** Volume of small box: $8x^3$, Volume of large box: $125x^3$ **b.** The volume of the larger box is $15\frac{5}{8}$ times the volume of the small box.

Practices: Section 5.2 pp. 416–425

1, p. 416: a. $2^6 = 64$ **b.** $\frac{1}{7^3} = \frac{1}{343}$ **c.** q^8 **d.** $\frac{-1}{p^{15}} = -\frac{1}{p^{15}}$ **2, p. 417: a.** $49a^2$ **b.** $-64x^3$ **c.** $-64x^3$ **3, p. 417: a.** $36a^{18}$ **b.** $q^{16}r^{20}$ **c.** $-2a^3b^{21}$ **d.** $\frac{49}{a^2c^{10}}$ **4, p. 418: a.** $\frac{y^2}{9}$ **b.** $\frac{u^{10}}{v^{10}}$ **c.** $\frac{y^2}{9}$ **d.** $\frac{100a^{10}}{9b^4c^2}$ **e.** $125x^3y^6$ **5, p. 419: a.** $\frac{a^2}{25}$ **b.** $\frac{v}{4u}$ **c.** $\frac{b^6}{a^{10}}$ **6, p. 421:** 253.9 **7, p. 421:** 0.0000000043 **8, p. 422:** 8×10^{12} **9, p. 422:** 7.1×10^{-11} **10, p. 422: a.** 2.464×10^2 **b.** 4×10^{11} **11, p. 423:** 9×10^{14} **12, p. 424:** $1 \times 10^{-3} = 0.001$ **13, p. 424:** 2.5E-17 (Answers may vary.) **14, p. 425:** 7.3E-10 (Answers may vary.) **15, p. 425:** 4.6×10^8, or 460,000,000

Exercises 5.2, p. 426

1. factors **3.** raise each factor to that power **5.** raise the reciprocal of the quotient **7.** left **9.** $2^8 = 256$ **11.** $5^4 = 625$ **13.** $10^{10} = 10,000,000,000$ **15.** $\frac{1}{4^4} = \frac{1}{256}$ **17.** x^{24} **19.** y^8 **21.** $\frac{1}{x^6}$ **23.** n^4 **25.** $64x^3$ **27.** $64y^2$

29. $-64n^{15}$ **31.** $64y^8$ **33.** $\frac{1}{9a^2}$ **35.** $\frac{1}{p^9q^7}$ **37.** $r^{12}t^6$ **39.** $4p^{10}q^2$ **41.** $-2m^{12}n^{24}$ **43.** $-\frac{64m^{15}}{n^{30}}$ **45.** $\frac{1}{a^{12}b^8}$ **47.** $\frac{16y^6}{x^4}$ **49.** $\frac{125}{b^3}$ **51.** $\frac{c^2}{b^2}$ **53.** $-\frac{a^7}{b^7}$ **55.** $\frac{a^6}{27}$ **57.** $-\frac{p^{15}}{q^{10}}$ **59.** $\frac{4}{a}$ **61.** $\frac{8x^{15}}{y^6}$ **63.** $\frac{1}{p^5q^5}$ **65.** $81x^4y^{12}$ **67.** $\frac{v^4}{16u^4}$ **69.** $\frac{y^4z^{16}}{16x^8}$ **71.** $\frac{t^{12}}{r^{10}}$ **73.** $-\frac{b^6}{8a^{12}}$ **75.** 317,000,000 **77.** 0.000001 **79.** 6,200,000 **81.** 0.00004025 **83.** 4.2×10^8 **85.** 3.5×10^{-6} **87.** 2.17×10^{11} **89.** 7.31×10^{-9}

91.

Standard Notation	Scientific Notation (written)	Scientific Notation (displayed on a calculator)
975,000,000	9.75×10^8	9.75E8
487,000,000	4.87×10^8	4.87E8
0.0000000001652	1.652×10^{-10}	1.652E−10
0.000000067	6.7×10^{-8}	6.7E−8
0.0000000000001	1×10^{-13}	1E−13
3,281,000,000	3.281×10^9	3.281E9

93. 9×10^7 **95.** 2.075×10^{-4} **97.** 3.784×10^{-2} **99.** 1.25×10^{10} **101.** 3×10^2 **103.** 3×10^8 **105.** 0.0003067

107.

Standard Notation	Scientific Notation (written)	Scientific Notation (displayed on a calculator)
428,000,000,000	4.28×10^{11}	4.28E11
3,240,000	3.24×10^6	3.24E6
0.000005224	5.224×10^{-6}	5.224E−6
0.000000057	5.7×10^{-8}	5.7E−8
0.000000682	6.82×10^{-7}	6.82E−7
48,360,000	4.836×10^7	4.836E7

109. $\frac{9b^{10}}{a}$ **111.** $-\frac{1}{64y^3}$ **113.** $-\frac{16y^{12}}{x^8z^4}$ **115.** 7×10^{-8} **117.** The larger volume is 8 times the smaller volume. **119.** 4,000,000,000 bytes and 17,000,000,000 bytes **121.** 7×10^{-7} m **123.** 2×10^{11} cells **125.** 0.00000000000000000000000017 g **127.** 780,000,000 **129.** 1.6×10^{13} red blood cells **131. a.** 1.86×10^5 mi per sec **b.** About 8.495×10^8 sec, or about 27 years.

Practices: Section 5.3 pp. 432–436

1, p. 432: (a) Terms: $-10x^2$, $4x$, and 20 (b) Coefficients: -10, 4, and 20 **2, p. 432: a.** Binomial **b.** Monomial **c.** Binomial **3, p. 433: a.** Degree 1 **b.** Degree 2 **c.** Degree 3 **d.** Degree 0

4, p. 433:

Polynomial	Constant Term	Leading Term	Leading Coefficient
$-3x^7 + 9$	9	$-3x^7$	-3
x^5	0	x^5	1
$x^4 - 7x - 1$	-1	x^4	1
$3x + 5x^3 + 20$	20	$5x^3$	5

5, p. 434: a. $9x^5 - 7x^4 + 9x^2 - 8x - 6$ **b.** $7x^5 + x^3 - 3x^2 + 8$ **6, p. 435:** $x^2 + 6x + 20$ **7, p. 435: a.** -1 **b.** 19 **8, p. 436:** 356 ft

Exercises 5.3, p. 437

1. monomial **3.** power **5.** leading term **7.** constant term **9.** Polynomial **11.** Not a polynomial **13.** Polynomial **15.** Not a polynomial **17. a.** Binomial **b.** Monomial **c.** Binomial **d.** Trinomial **19.** $-4x^3 + 3x^2 - 2x + 8$; degree 3 **21.** $-3y + 2$; degree 1 **23.** $-5x^2 + 7x$; degree 2 **25.** $-4y^5 - y^3 - 2y + 2$; degree 5 **27.** $5a^2 - a$; degree 2 **29.** $3p^3 + 9p$; degree 3

31.

Polynomial	Constant Term	Leading Term	Leading Coefficient
$-x^7 + 2$	2	$-x^7$	-1
$2x - 30$	-30	$2x$	2
$-5x + 1 + x^2$	1	x^2	1
$7x^3 - 2x - 3$	-3	$7x^3$	7

33. $10x^3 - 7x^2 + 10x + 6$ **35.** $r^3 + 3r^2 - 8r + 14$ **37.** $0x^2$ **39.** $0x$ **41.** 11; -17 **43.** 37; 79 **45.** 13.73849; -4.74751 **47.** $3x^2 - 5x - 6$ **49.** $2n^3 + 14n^2 + 20n + 10$ **51.** $5a^4 - a^3 + 2a - 4$; degree 4 **53.** For $4x - 20$: 20, $4x$, 4; for $-7x^6 + 9$: 9, $-7x^6$, -7; for $3x + 2 + x^2$: 2, x^2, 1; for $6x - x^5 + 11$: 11, $-x^5$, -1 **55.** Not a polynomial **57.** -1.89375; 83.02657 **59.** $0x^3$ and $0x^0$, or 0 **61.** A polynomial in x; degree 16 **63.** 120 ft **65.** 5,942,000,000 people, or 5942 million people. **67.** About 950 U.S. radio stations.

Practices: Section 5.4 pp. 443–447

1, p. 443: $9x^2 + 3x - 43$ **2, p. 443:** $10p^2 + pq - 10q^2$ **3, p. 444:** $11n^2 + 2n - 3$ **4, p. 444:** $8p^3 + 2p^2q - 2pq^2 - 2q^3 + 25$ **5, p. 445: a.** $3r + 3s$ **b.** $-3q$ **6, p. 446:** $-3x^2 - 13x$ **7, p. 446:** $-5x^2 + 32x - 26$ **8, p. 446:** $-p^2 - 11pq + 17q^2$ **9, p. 447:** The polynomial 7.6 approximates how much greater the life expectancy is for females than for males.

Exercises 5.4, p. 448

1. $2x^2 + 8x + 2$ **3.** $5n^3 + 9n$ **5.** $2p^2 + 3p - 1$ **7.** $11x^2 - 3xy + 2y^2$ **9.** $3p^3 + 2p^2q - 3pq^2 - 3q^3 + 5$ **11.** $10x^2 + 17x - 5$ **13.** $5x^3 + x^2 + 9x + 2$ **15.** $-3a^3 + 6a^2 + 7ab^2$ **17.** $-x^2 - 2x + 11$ **19.** $-2x^3 + 9x^2 - 13x + 11$ **21.** $x^2 - 2x - 9$ **23.** $5x^2 - 5y^2 + 4$

25. $-5p^2 + 7p + 6$ **27.** $t^3 - 12t^2 + 8$ **29.** $3r^3 - 17r^2s - 2$ **31.** $-x - r$ **33.** $2p - 3q - r$ **35.** $3y^2 + 3y + 4$ **37.** $m^3 - 12m + 16$ **39.** $2x^3 - 5x^2 - 10x + 9$ **41.** $9x^2 + 4x + 7$ **43.** $-2x + 12$ **45.** $8x^2y^2 - 12xy - 14$ **47.** $2m^2 - 5m - 9$ **49.** $5x^3 + 3x^2 + 2x - 12$ **51.** $5m - 8$ **53.** $-2x^2 + 6x + 3$ **55.** $5p^2 + 11p - 7$ **57. a.** $6.28r^2 + 6.28rh$ **b.** $12.56r^2$ or $12.56h^2$ **59.** $(32x^3 - 538x^2 + 2061x + 5862)$ million **61.** $(-5x^2 + 154x + 31)$ million

Practices: Section 5.5 pp. 452–457

1, p. 452: $40x^5$ **2, p. 452:** $-350a^4b^5$ **3, p. 452:** $25x^2y^4$ **4, p. 453:** $70s^3 - 21s$ **5, p. 453:** $12m^6n^7 - 4m^4n^4 - 2m^3n^3$ **6, p. 453:** $-14s^5 + 34s^4 + 22s^3 + s^2$ **7, p. 454:** $2a^2 + a - 3$ **8, p. 455:** $16x^2 - 2x - 3$ **9, p. 456:** $14m^2 + 5mn - n^2$ **10, p. 456:** $n^3 - 3n^2 - 6n + 8$ **11, p. 456:** $8n^3 + 15n^2 + n + 6$ **12, p. 457:** $24x^4 + 56x^3 + 27x^2 + 60x - 7$ **13, p. 457:** $p^3 - 3p^2q + 3pq^2 - q^3$ **14, p. 457:** $P + 2Pr + Pr^2$

Exercises 5.5, p. 458

1. $-24x^2$ **3.** $-9t^5$ **5.** $20x^6$ **7.** $-70x^8$ **9.** $16p^3q^3r^2$ **11.** $64x^2$ **13.** $\frac{1}{8}t^{12}$ **15.** $-350a^4$ **17.** $-24a^4b^3c$ **19.** $7x^2 - 5x$ **21.** $45t^2 + 5t^3$ **23.** $24a^5 - 42a^4$ **25.** $12x^3 - 8x^2$ **27.** $x^5 - 2x^4 + 4x^3$ **29.** $15x^3 + 25x^2 + 30x$ **31.** $-45x^3 + 27x^2 + 63x$ **33.** $6x^5 + 24x^4 - 6x^3 - 6x^2$ **35.** $28pq - 4p^3$ **37.** $-7v^2 - 21vw^2$ **39.** $6a^6b^5 + 20a^3b^8$ **41.** $-6x^2 + 26x$ **43.** $8x^3 - 16x^2 + 7x$ **45.** $-23x^3 + 35x^2 - 45x$ **47.** $7x^4y - 8x^3y - 18x^2y^2$ **49.** $-45a^5b^4 + 51a^3b^6$ **51.** $y^2 + 5y + 6$ **53.** $x^2 - 8x + 15$ **55.** $a^2 - 4$ **57.** $2w^2 - w - 21$ **59.** $-10y^2 + 17y - 3$ **61.** $20p^2 - 18p + 4$ **63.** $u^2 - v^2$ **65.** $-2p^2 + 3pq - q^2$ **67.** $3a^2 - 7ab + 2b^2$ **69.** $4pq + 3p - 32q - 24$ **71.** $x^3 - 6x^2 + 10x - 3$ **73.** $2x^3 + 5x^2 - 13x + 5$ **75.** $a^3 - b^3$ **77.** $3x^3 - 6x^2 - 105x$ **79.** $12n^3 - 3n$ **81.** $8m^4n - 6m^3n + 8m^2n^2$ **83.** $40s^7$ **85.** $-36y^3 - 28y^2 + 32y$ **87.** $-24x^{11}$ **89.** $-4a^3b + 2a^2b^4$ **91.** $x^3 - 4x^2 + 5x - 2$ **93.** $x^2 + 30x$ mm^2 **95.** $(-3000x^2 + 55,000x + 1,500,000)$ dollars **97. a.** $(5000 + 15,000r + 15,000r^2 + 5000r^3)$ dollars **b.** $(5000r + 10,000r^2 + 5000r^3)$ dollars **c.** $605

Practices: Section 5.6 pp. 463–467

1, p. 463: $p^2 + 20p + 100$ **2, p. 464:** $s^2 + 2st + t^2$ **3, p. 464:** $16p^2 + 40pq + 25q^2$ **4, p. 465:** $25x^2 - 20x + 4$ **5, p. 465:** $u^2 - 2uv + v^2$ **6, p. 465:** $4x^2 - 36xy + 81y^2$ **7, p. 466:** $t^2 - 100$ **8, p. 466: a.** $r^2 - s^2$ **b.** $64s^2 - 9t^2$ **9, p. 466:** $100 - 49k^4$ **10, p. 467:** $S^2 - s^2$

Exercises 5.6, p. 468

1. plus **3.** negative **5.** $y^2 + 4y + 4$ **7.** $x^2 + 8x + 16$ **9.** $x^2 - 22x + 121$ **11.** $36 - 12n + n^2$ **13.** $x^2 + 2xy + y^2$ **15.** $9x^2 + 6x + 1$ **17.** $16n^2 - 40n + 25$ **19.** $81x^2 + 36x + 4$ **21.** $a^2 + a + \frac{1}{4}$ **23.** $64b^2 + 16bc + c^2$ **25.** $25x^2 - 20xy + 4y^2$ **27.** $x^2 - 6xy + 9y^2$ **29.** $16x^6 + 8x^3y^4 + y^8$ **31.** $a^2 - 1$ **33.** $16x^2 - 9$ **35.** $9y^2 - 100$ **37.** $m^2 - \frac{1}{4}$ **39.** $16a^2 - b^2$ **41.** $9x^2 - 4y^2$ **43.** $1 - 25n^2$ **45.** $x^3 - 25x$ **47.** $5n^4 + 70n^3 + 245n^2$ **49.** $n^4 - m^8$

51. $a^4 - b^4$ **53.** $36n^2 + 48n + 16$ **55.** $16p^2 - 81$
57. $64 - 16a + a^2$ **59.** $-32x^4 + 48x^3 - 18x^2$ **61.** $x^2 + y^2 - 10x - 2y + 26$ **63.** $A + \frac{AP}{50} + \frac{AP^2}{10,000}$
65. $\frac{3m^2 - 2am - 2bm - 2cm + a^2 + b^2 + c^2}{2}$

Practices: Section 5.7 pp. 471–476

1, p. 471: $-4n^5$ **2, p. 471:** $-4pr^3$ **3, p. 472:** $3x^2 - 2x$
4, p. 472: $-7x^5 - 5x^2 + 4$ **5, p. 472:** $-a^6b^3 + \frac{ab}{5} - 3$
6, p. 474: $2x + 3$ **7, p. 475:** $x^2 + 2x + 3 + \frac{2}{3x + 1}$
8, p. 475: $3s - 5$ **9, p. 476:** $n^2 - 4n - 3 + \frac{-13}{4n - 3}$
10, p. 476: a. The future value of the investment after 1 year is $10(1 + r)^1$, or $10(r + 1)$. The future value of the investment after 2 years is $10(1 + r)^2$. **b.** $10r + 10$ and $10r^2 + 20r + 10$ **c.** The future value of the investment after 2 years is $(r + 1)$ times as great as the future value of the investment after 1 year.

Exercises 5.7, p. 477

1. coefficients **3.** remainder **5.** $2x^2$ **7.** $-4a^7$ **9.** $\frac{4x}{3}$
11. $4q^2$ **13.** $3u^2v^2$ **15.** $-\frac{15ab^2}{7}$ **17.** $-\frac{3u^3v^2}{2}$ **19.** $3n + 5$
21. $2b^3 - 1$ **23.** $-6a - 4$ **25.** $3 - 2x^2$
27. $2a^2 - 3a + 5$ **29.** $-\frac{n^2}{5} + 2n + 1$ **31.** $\frac{5a}{2} + \frac{b^2}{2}$
33. $-4xy^2 + 3 + y$ **35.** $2q^2 - pq^2 + \frac{3p^2}{2}$ **37.** $x - 7$
39. $7x - 2$ **41.** $3x - 1$ **43.** $5x + 3$ **45.** $7x + 2$
47. $x + \frac{5}{x + 2}$ **49.** $2x + 1$ **51.** $2x - 3 + \frac{-2}{4x + 3}$
53. $x^2 - 6x + 5$ **55.** $2x^2 - x - 3 + \frac{7}{3x - 4}$
57. $5x + 20 + \frac{78}{x - 4}$ **59.** $2x^2 + 3x + 4 + \frac{15}{2x - 3}$
61. $x^2 - 3x + 9$ **63.** $-7r$ **65.** $-\frac{2}{3}ab$
67. $-3n^3 + 8n + \frac{1}{3}$ **69.** $2x + 1$
71. $x^2 - 1 + \frac{3}{4x + 1}$ **73.** $-4m^2 + 9$ **75. a.** $(x + 6)$ ft
b. 7 ft by 10 ft **77. a.** $t = \frac{d}{r}$ **b.** It takes $(t^2 - 7t + 14)$ hr.
79. There are $(3x - 14)$ thousand subscribers per cell system.

Review Exercises: Chapter 5, p. 484

1. $-x^3$ **2.** -1 **3.** n^{11} **4.** x^7 **5.** n^3 **6.** p^3 **7.** y^7
8. a^3b^3 **9.** y **10.** n^2 **11.** $\frac{1}{(5x)^1}$ **12.** $-\frac{3}{n^2}$ **13.** $\frac{v^4}{8^2}$
14. y^4 **15.** $\frac{1}{x^1}$ **16.** $\frac{y^3}{5^1}$ **17.** a^{10} **18.** $\frac{1}{t^6}$ **19.** $\frac{1}{x^2y}$ **20.** x^2y^1
21. $10^8 = 100,000,000$ **22.** $-x^9$ **23.** $4x^6$
24. $-64m^{15}n^3$ **25.** $\frac{3}{x^{12}}$ **26.** $\frac{b^8}{a^6}$ **27.** $\frac{x^4}{81}$ **28.** $\frac{a^2}{b^6}$ **29.** $\frac{y^6}{x^6}$
30. $x^{10}y^5$ **31.** $\frac{16a^6}{b^8c^2}$ **32.** $\frac{v^4}{49u^{10}w^2}$ **33.** 37,000,000,000
34. 1,630,000,000 **35.** 0.00005022 **36.** 0.00000000006
37. 1.2×10^{12} **38.** 4.27×10^8 **39.** 4×10^{-14}
40. 5.6×10^{-7} **41.** 5.88×10^9 **42.** 6.3×10^3
43. 6×10^6 **44.** 6×10^{-10} **45.** Polynomial **46.** Not a polynomial **47.** Trinomial **48.** Binomial **49.** $-3y^3 + y^2 + 8y - 1$; degree 3, leading term: $-3y^3$, leading coefficient: -3 **50.** $n^4 - 7n^3 - 6n^2 + n$; degree 4, leading term: n^4, leading coefficient: 1 **51.** $-x^3 + x^2 + 2x + 13$
52. $3n^3 + 4n^2 - 6n + 4$ **53.** 12; 0 **54.** 0; -16
55. $x^2 - x + 13$ **56.** $-2y^3 - y^2 - y - 5$ **57.** $a^2 - 2ab - 3b^2$ **58.** $5s^3t + s^2t + 9s^2 - 6st - 3t^2$ **59.** $2x^2 - 8x - 8$ **60.** $-n^3 + 3n^2 + n$ **61.** $5y^4 - 5y^3 + 2y^2 - 6y - 3$ **62.** $2x^3 + 8x^2 - 12x + 3$ **63.** $4t^2 + 4t$
64. $-2x - y$ **65.** $2y^2 - 5y + 2$ **66.** $-5x^2 + 9$

67. $-6x^5$ **68.** $-144a^3b^5$ **69.** $8x^2y^2 - 10xy^3$
70. $-5x^4 + 15x^3 - 5x^2$ **71.** $n^2 + 10n + 21$
72. $3x^2 + 9x - 54$ **73.** $8x^2 - 6x + 1$ **74.** $9a^2 + 3ab - 2b^2$ **75.** $2x^4 + 6x^3 - 5x^2 - 13x + 6$ **76.** $y^3 - 9y^2 + 15y - 2$ **77.** $-6y^2 + 13y$ **78.** $-x^3 + 6x^2 - 6x$
79. $a^2 - 2a + 1$ **80.** $s^2 + 8s + 16$ **81.** $4x^2 + 20x + 25$
82. $9 - 24t + 16t^2$ **83.** $25a^2 - 20ab + 4b^2$
84. $u^4 + 2u^2v^2 + v^4$ **85.** $m^2 - 16$ **86.** $36 - n^2$
87. $49n^2 - 1$ **88.** $4x^2 - y^2$ **89.** $16a^2 - 9b^2$
90. $x^3 - 100x$ **91.** $-48t^4 + 120t^3 - 75t^2$ **92.** $p^4 - 2p^2q^2 + q^4$ **93.** $3x^2$ **94.** $-2a^2b^3c$ **95.** $6x^2 - 2$
96. $5x^3 + 3x^2 - 2x - 1$ **97.** $3x - 7$
98. $x^2 - 2x - 1 + \frac{12}{2x - 1}$ **99.** 1.39×10^{10} yr
100. 6,240,000,000,000,000,000 eV **101.** 3×10^{-5} m
102. 0.00000000011 m **103.** There will be 36 handshakes.
104. The object is 480.4 m above the ground.
105. There were about 249,000 divorces in 1951.
106. There were about 899 two-year colleges in 1970.
107. a. $(3w^2 - 10w)$ ft^2 **b.** $(48w + 24)$ ft^2
c. The area of the concrete walk is 600 ft^2
108. a. $2x + 436$ **b.** 440 acres

Posttest: Chapter 5, p. 488

1. x^7 **2.** n^6 **3.** $\frac{7}{a^1} = \frac{7}{a}$ **4.** $-27x^6y^3$ **5.** $\frac{x^8}{y^{12}}$ **6.** $\frac{y^3}{27x^6}$
7. a. $-x^3, 2x^2, 9x, -1$ **b.** $-1, 2, 9,$ and -1 **c.** 3
d. -1 **8.** $2y^2 - y + 5$ **9.** $-x^2 + x - 9$ **10.** $3x^2y^2 - 4x^2$ **11.** $10m^3n^3 - 20m^2n^3 + 2m^2n^4$ **12.** $y^4 - 3y^3 + 2y^2 + 4y - 4$ **13.** $6x^2 + 19x - 7$ **14.** $49 - 4n^2$
15. $4m^2 - 12m + 9$ **16.** $-4s^2 - 5s + 9$
17. $t^2 - t - 1 + \frac{4}{3t - 2}$ **18.** 1×10^{-7} m
19. a. First house: $(1500x + 140,000)$ dollars; second house: $(800x + 90,000)$ dollars **b.** $(2300x + 230,000)$ dollars **20.** The account balance is \$1060.90.

Cumulative Review: Chapter 5, p. 489

1. $m = \frac{y - b}{x}$ **2.** 61 **3.** 4 **4.** Slope: $\frac{2}{3}$; y-intercept: $(0, -2)$ **5.** 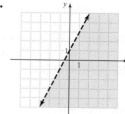 **6.** $(2, -2)$

7. $m^2 - 16m + 16$ **8.** 9×10^{13} kg $\cdot$ m^2/sec^2 **9. a.** $x + 4, x + 8,$ and $x + 12$ **b.** Yes; $1980 - 1972 = 8$, which is a multiple of 4. **10. a.** $0.20b$ **b.** $c = 1.20b$

Chapter 6

Pretest: Chapter 6, p. 492

1. $18a$ **2.** $4p(q + 4)$ **3.** $5xy(2x - x^2y^2 + y)$
4. $(x + 2)(3x + 2)$ **5.** $(n - 3)(n - 8)$ **6.** $(a - 3)(a + 7)$
7. $3y(y - 1)(y - 3)$, or $3y(1 - y)(3 - y)$
8. $(5a - 4b)(a + 2b)$ **9.** $-2(3n + 1)(2n - 7)$
10. $(2x - 7)^2$ **11.** $(5n + 3)(5n - 3)$
12. $y(x + 2y)(x - 2y)$ **13.** $(y^3 - 4)(y^3 - 5)$

14. $0, 6$ **15.** $\frac{2}{3}, -1$ **16.** $3, -5$ **17.** $h = \frac{A - 2lw}{2l + 2w}$

18. $-(16t + 1)(t - 4)$ ft   **19.** $(S + 15)(S - 15)$ ft^2

20. The length of the screen is 32 in. and the width is 24 in.

Practices: Section 6.1 pp. 494–498

1, p. 494: 24 **2, p. 494:** a **3, p. 495:** $6xy^2$ **4, p. 495:** $2y^2(5 + 4y^3)$ **5, p. 496:** $7a(3ab - 2)$ **6, p. 496:** $2ab^2(4a - 3b)$ **7, p. 496:** $12(2a^2 - 4a + 1)$

8, p. 496: $a = \frac{s^2}{b + c}$ **9, p. 497:** $(y - 3)(4 + y)$, or $(y - 3)(y + 4)$ **10, p. 497:** $(x - 1)(3y - 2)$

11, p. 497: $(4 - 3x)(1 + 2x)$ **12, p. 498:** $(b - 5)(a + 4)$

13, p. 498: $(y - z)(5 - y)$ **14, p. 498:** $t(v_0 + \frac{1}{2}at)$

Exercises 6.1, p. 499

1. factoring   **3.** greatest common factor (GCF)   **5.** 27

7. x^3 **9.** $4b$ **11.** $4y^3$ **13.** $3a^2b^2$ **15.** $3x - 1$

17. $x(x + 7)$ **19.** $3(x + 2)$ **21.** $8(3x^2 + 1)$

23. $9(3m - n)$ **25.** $x(2 - 7x)$ **27.** $b^2(5 - 6b)$

29. $5x(2x^2 - 3)$ **31.** $ab(ab - 1)$ **33.** $xy(6y + 7x)$

35. $9pq(3q + 2p)$ **37.** $2x^3y(1 - 6y^3)$ **39.** $3(c^3 + 2c^2 + 4)$

41. $b^2(9b^2 - 3b + 1)$ **43.** $2m^2(m^2 + 5m - 3)$

45. $b^2(5b^3 - 3b + 2)$ **47.** $5x(3x^3 - 2x^2 - 5)$

49. $4ab(a + 2ab - 3)$ **51.** $3cd(3cd + 4c^2 + d^2)$

53. $(x - 1)(x + 3)$ **55.** $(a - 1)(5a - 3)$

57. $(s + 7)(r - 2)$ **59.** $(x - y)(a - b)$

61. $(y + 2)(3x - 1)$ **63.** $(b - 1)(b - 5)$

65. $(y - 1)(y + 5)$ **67.** $(t - 3)(1 + t)$

69. $(b - 7)(9a - 2)$ **71.** $(r + 3)(s + t)$

73. $(x + 6)(y - 4)$ **75.** $(5x - 3z)(3y + 4z)$

77. $(z + 4)(2x + 5y)$ **79.** $P = \frac{TM}{C + L}$ **81.** $l = \frac{S - 2wh}{2w + 2h}$

83. $8p(2p^2 + 3)$ **85.** $12rs(4s - 5r)$ **87.** $6(7j^2 - 1)$

89. $(s - 3)(t - 7)$ **91.** $(y - 4)(3x - 5)$ **93.** $4a^2b$

95. $m(v_2 - v_1)$ **97.** $0.5n(n - 1)$ **99.** $\frac{1}{2}n(n - 3)$

101. $n = \frac{P - D}{C + T}$

Practices: Section 6.2 pp. 504–510

1, p. 504: $(x + 1)(x + 4)$, or $(x + 4)(x + 1)$

2, p. 505: $(y - 4)(y - 5)$, or $(y - 5)(y - 4)$

3, p. 505: Prime polynomial; cannot be factored

4, p. 506: $(y - 4)(y - 8)$ **5, p. 506:** $(p - q)(p - 3q)$, or $(p - 3q)(p - q)$ **6, p. 507:** $(x - 2)(x + 3)$

7, p. 507: $(x + 2)(x - 23)$ **8, p. 508:** $(y - 4)(y + 6)$

9, p. 508: $(a + 3b)(a - 8b)$ **10, p. 509:** $y(y + 1)(y - 10)$

11, p. 509: $8x(x - 1)(x - 2)$ **12, p. 510:** $-(x - 1)(x + 11)$, or $(-x + 1)(x + 11)$, or $(x - 1)(-x - 11)$ **13, p. 510:** $-16(t + 1)(t - 3)$

Exercises 6.2, p. 511

1. not factorable   **3.** have opposite signs   **5.** f   **7.** e

9. b   **11.** $(x - 5)$ **13.** $(x + 4)$ **15.** $(x - 1)$

17. $(x + 2)(x + 4)$ **19.** $(x - 1)(x + 6)$

21. Prime polynomial   **23.** $(x + 1)(x + 4)$

25. $(x - 1)(x - 3)$ **27.** $(y - 4)(y - 8)$

29. $(t + 1)(t - 5)$ **31.** Prime polynomial

33. $(x - 5)(x + 9)$ **35.** $(y - 4)(y - 5)$

37. $(b + 4)(b + 7)$ **39.** Prime polynomial

41. $-(y + 5)(y - 10)$ **43.** $(x - 8)(x - 8)$

45. $(x - 2)(x - 8)$ **47.** $(w - 3)(w - 27)$

49. $(p - q)(p - 7q)$ **51.** $(p + q)(p - 5q)$

53. $(m - 5n)(m - 7n)$ **55.** $(x + y)(x + 8y)$

57. $5(x + 2)(x - 3)$ **59.** $2(x - 2)(x + 7)$

61. $6(t - 1)(t - 2)$ **63.** $3(x + 2)(x + 4)$

65. $y(y - 2)(y + 5)$ **67.** $a(a + 3)(a + 5)$

69. $t^2(t - 2)(t - 12)$ **71.** $4a(a - 1)(a - 2)$

73. $2x(x + 3)(x + 5)$ **75.** $4x(x - 3)(x - 4)$

77. $2s(s - 4)(s + 7)$ **79.** $2c^2(c - 5)(c + 7)$

81. $ax(x - 2)(x - 16)$ **83.** Prime polynomial

85. $5x^2(x - 5)(x + 2)$ **87.** $-(w + 4)(w - 10)$

89. $6(m + 2)(m - 3)$ **91.** $(t - 5)(t - 12)$

93. b   **95.** $(p + q)(p + q) = 1$, or $(p + q)^2 = 1$

97. $16(t - 2)(t + 5)$ **99. a.** $(6x^2 + 24x + 18)$ in^2

b. $6(x + 1)(x + 3)$

Practices: Section 6.3 pp. 516–520

1, p. 516: $(5x + 4)(x + 2)$ **2, p. 516:** $(6x - 7)(x - 3)$

3, p. 517: $(7y - 2)(y + 7)$ **4, p. 518:** $(2x - 5)(x + 2)$

5, p. 518: $3x(3x + 1)(2x - 3)$ **6, p. 519:** $3(6c - 5d)(2c + d)$ **7, p. 519:** $(2x + 1)(x - 4)$

8, p. 520: $x(2x - 5)(2x - 7)$

Exercises 6.3, p. 521

1. e   **3.** b   **5.** c   **7.** $(3x + 1)$ **9.** $(x - 3)$ **11.** $(x - 3)$

13. $(3x + 5)(x + 1)$ **15.** $(2y - 1)(y - 5)$

17. $(3x + 2)(x + 4)$ **19.** Prime polynomial

21. $(6y + 5)(y - 1)$ **23.** $(2y - 7)(y - 2)$

25. $(3a + 2)(3a - 8)$ **27.** $(4x - 1)(x - 3)$

29. $(3y + 2)(4y + 3)$ **31.** $(2m - 3)(m - 7)$

33. $-(3a - 1)(2a + 3)$ **35.** $(8y - 11)(y + 2)$

37. Prime polynomial   **39.** $(8a + 1)(a + 8)$

41. $(3x - 1)(2x + 9)$ **43.** $(4y - 3)(2y - 5)$

45. $2(7y - 5)(y - 2)$ **47.** $4(7a - 1)(a + 1)$

49. $-2(3b + 1)(b - 7)$ **51.** $2y(3y + 2)(2y + 7)$

53. $2a^2(7a - 5)(a - 2)$ **55.** $xy(2x + 3)(x + 5)$

57. $2ab(3b - 1)(b - 7)$ **59.** $(5c - d)(4c - d)$

61. $(2x + y)(x - 3y)$ **63.** $(4a - b)(2a - b)$

65. $3(3x + 2y)(2x - y)$ **67.** $2(4c - 5d)(2c - 3d)$

69. $3(3u + v)(3u + v)$ **71.** $3x(7x - 3y)(2x + 3y)$

73. $-5x^2y(3x + y)(2x - 3y)$ **75.** $a(5x + 2y)(x - 6y)$

77. $(m - 1)(3m - 2)$ **79.** $(4x - 3)(2x + 1)$

81. $2x(x + 3)(7x + 1)$ **83.** $(2m - 3n)(4m - 3n)$

85. $(r - 1)(7r - 2)$ **87.** d   **89.** $-(5t - 4)(t + 5)$

91. a. $(2x^2 + 25x + 72)$ ft^2 **b.** $(2x + 9)(x + 8)$ ft^2

93. $(2n - 5)(2n - 1)$; since the difference of the factors is $(2n - 1) - (2n - 5) = 2n - 1 - 2n + 5 = 4$, the factors represent two integers that differ by 4 no matter what integer n represents.

Practices: Section 6.4 pp. 526–529

1, p. 526: a. The trinomial is a perfect square. **b.** The trinomial is not a perfect square. **c.** The trinomial is a perfect square. **d.** The trinomial is not a perfect square. **e.** The trinomial is a perfect square.   **2, p. 526:** $(n + 10)^2$

3, p. 527: $(t - 2)^2$ **4, p. 527:** $(5c - 4d)^2$ **5, p. 527:** $(x^2 + 4)^2$ **6, p. 528:** **a.** The binomial is a difference of squares. **b.** The binomial is not a difference of squares. **c.** The binomial is not a difference of squares. **d.** The binomial is a difference of squares.   **7, p. 528:** $(y + 11)$ $(y - 11)$ **8, p. 529:** $(3x + 5y)(3x - 5y)$ **9, p. 529:** $(8x^4 + 9y)(8x^4 - 9y)$ **10, p. 529:** $16(4 + t)(4 - t)$

Exercises 6.4, p. 530

1. perfect square trinomial   **3.** difference of squares   **5.** Perfect square trinomial **7.** Neither   **9.** Perfect square trinomial   **11.** Difference of squares   **13.** Difference of squares   **15.** Perfect square trinomial   **17.** Neither **19.** Neither   **21.** Neither   **23.** $(x - 6)^2$ **25.** $(y + 10)^2$ **27.** $(a - 2)^2$ **29.** Prime polynomial   **31.** $(m + 8)(m - 8)$ **33.** $(y + 9)(y - 9)$ **35.** $(12 + x)(12 - x)$ **37.** $(2a - 9)^2$ **39.** $(7x + 2)^2$ **41.** $(6 - 5x)^2$ **43.** $(10m + 9)(10m - 9)$ **45.** Prime polynomial   **47.** $(1 + 3x)(1 - 3x)$ **49.** $(m + 13n)^2$ **51.** $(2a + 9b)^2$ **53.** $(x + 2y)(x - 2y)$ **55.** $(10x + 3y)(10x - 3y)$ **57.** $(y^2 + 1)^2$ **59.** $6(x + 1)^2$ **61.** $3m(3m - 2)^2$ **63.** $4s^2t(t + 10)^2$ **65.** $3k(k + 7)(k - 7)$ **67.** $4y^2(y + 3)(y - 3)$ **69.** $3x^2y(3 + y)(3 - y)$ **71.** $2(ab + 7)(ab - 7)$ **73.** $(16 + r^2)(4 + r)(4 - r)$ **75.** $5(x^2 + 4y^2)(x + 2y)(x - 2y)$ **77.** $(c - d)(x + 2)(x - 2)$ **79.** $(x - y)(4 + a)(4 - a)$ **81.** $(3c + 8d)^2$ **83.** $(a + 15b^2)(a - 15b^2)$ **85.** $6(3a^2b - 1)^2$ **87.** $(9 + w^2)(3 - w)(3 + w)$ **89.** $(6u + 5)^2$ **91.** Difference of squares   **93.** $4\pi(r_1 + r_2)(r_1 - r_2)$ **95.** $16{,}000(1 + r)^2$, or $16{,}000(r + 1)^2$ **97.** $k(v_2 + v_1)(v_2 - v_1)$

Practices: Section 6.5 pp. 534–536

1, p. 534: a. $(4 + 2x - y)(4 - 2x + y)$ **b.** $(8x^4 + 5y)(8x^4 - 5y)$ **2. p. 535: a.** Difference of cubes **b.** Sum of cubes **c.** Neither **3. p. 535: a.** $(5 - y)(25 + 5y + y^2)$ **b.** $(3mn^2 + 1)(9m^2n^4 - 3mn^2 + 1)$ **c.** $2(3x^2 + 1)$ **4. p. 536: a.** $4(5 + 2t)(5 - 2t)$, or $-4(2t + 5)(2t - 5)$

Exercises 6.5, p. 537

1. $(2u - v + 8)(2u - v - 8)$ **3.** $(7 + 2x + 2y)(7 - 2x - 2y)$ **5.** $(p^3 - 11)^2$ **7.** $(3a^4 + 8b)^2$ **9.** $(2a^2 + 15)(2a^2 - 15)$ **11.** $(7x^3 + 12y^2)(7x^3 - 12y^2)$ **13.** $(10p^2q + 3r)(10p^2q - 3r)$ **15.** $5p(p + 2q)$ **17.** Difference of cubes   **19.** Neither **21.** Sum of cubes   **23.** $(x + 1)(x^2 - x + 1)$ **25.** $(p - 2)(p^2 + 2p + 4)$ **27.** Prime polynomial **29.** $(\frac{1}{2} - a)(\frac{1}{4} + \frac{1}{2}a + a^2)$ **31.** $(5x + y)(25x^2 - 5xy + y^2)$ **33.** $(0.4b - 0.3a)(0.16b^2 + 0.12ba + 0.09a^2)$ **35.** $(a^2 - 2)(a^4 + 2a^2 + 4)$ **37.** $(4x^3 + 3y)(16x^6 - 12x^3y + 9y^2)$ **39.** $(2 - a)(a^2 + 5a + 13)$ **41.** $2x(x^2 + 12)$ **43.** Difference of cubes   **45.** $(10a - 5b + 3)(10a - 5b - 3)$ **47.** $2xy(4x + 3y)(4x - 3y)$ **49.** $(x + y)(x^2 - xy + y^2)$

Practices: Section 6.6 pp. 540–544

1, p. 540: $\frac{1}{3}, -5$ **2, p. 541:** $0, -6$ **3, p. 541:** $-\frac{1}{4}, 3$ **4, p. 542:** $1, -5$ **5, p. 542:** The dimensions of the frame should be 8 in. by 10 in.   **6, p. 544:** The scooter going north has traveled 12 mi.

Exercises 6.6, p. 545

1. quadratic equation   **3.** in standard form   **5.** Quadratic **7.** Linear   **9.** Quadratic   **11.** $-3, 4$ **13.** 1 **15.** $0, -\frac{5}{3}$ **17.** $-\frac{1}{2}, 5$ **19.** $-\frac{3}{2}, \frac{3}{2}$ **21.** $0, \frac{2}{3}$ **23.** $0, 2$ **25.** $0, \frac{1}{5}$ **27.** $-2, -3$ **29.** $7, -8$ **31.** $-\frac{1}{2}, 3$ **33.** $\frac{2}{3}, -\frac{1}{2}$ **35.** $\frac{1}{6}$ **37.** $-11, 11$ **39.** $\frac{3}{2}$ **41.** $\frac{1}{2}$ **43.** $\frac{1}{3}, -2$ **45.** $-2, 3$ **47.** $3, 4$ **49.** $2, -4$ **51.** $-\frac{1}{3}, -1$ **53.** $-\frac{1}{2}, -1$ **55.** $0, -5$ **57.** $-\frac{1}{2}, \frac{1}{2}$ **59.** $-\frac{1}{2}, \frac{1}{2}$ **61.** -3 **63.** $-3, 4$ **65.** $-2, 3$ **67.** $4, -5$ **69.** $-\frac{1}{3}, -4$ **71.** $2, 4$ **73.** $1, 2$ **75.** $\frac{1}{4}, -3$ **77.** $\frac{1}{2}, -3$ **79.** $\frac{1}{5}, -4$ **81.** 8 **83.** $0, \frac{5}{3}$ **85.** $0, -\frac{1}{2}$ **87.** There were 15 teams in the league. **89.** One car traveled 6 mi and the other traveled 8 mi. **91.** 200 ft   **93.** The diver will hit the water in $\frac{3}{2}$, or 1.5 sec.

Review Exercises: Chapter 6, p. 550

1. 12 **2.** $3m^2$ **3.** $3(x - 2y)$ **4.** $2pq(8p^2q + 9p - 2q)$ **5.** $(n - 1)(1 + n)$ **6.** $(x - 5)(b - 2)$ **7.** $r = \frac{d}{t_1 + t_2}$ **8.** $x = \frac{c - y}{a - b}$ **9.** Prime polynomial   **10.** Prime polynomial **11.** $(y + 6)(y + 7)$ **12.** $(m - 2n)(m - 5n)$ **13.** $-2(x - 2)(x + 6)$ **14.** $3x(x + y)(x - 5y)$ **15.** $(3x - 1)(x + 2)$ **16.** $(5n + 3)(n + 2)$ **17.** Prime polynomial   **18.** $(3x + 4)(2x - 3)$ **19.** $(2a - 7b)(a + 5b)$ **20.** $-(2a - 3)(2a - 5)$ **21.** $3y(3y - 1)(y + 7)$ **22.** $q(2p + q)(p - 2q)$ **23.** $(b - 3)^2$ **24.** $(8 + x)(8 - x)$ **25.** $(5y - 2)^2$ **26.** $(3a + 4b)^2$ **27.** $(9p + 10q)(9p - 10q)$ **28.** $(2x^4 - 7)^2$ **29.** $3(4x^2 + y^2)(2x + y)(2x - y)$ **30.** $(x - 1)(x + 3)(x - 3)$ **31.** $3(u + 3)(u^2 - 3u + 9)$ **32.** $(4c - 3d)(16c^2 + 12cd + 9d^2)$ **33.** $(x + y + z)(x + y - z)$ **34.** $(3a + 2)(9a^2 + 3a + 1)$ **35.** $2(4u + v)(4u - v)$ **36.** $(x - 1)(x + 3)(x - 3)$ **37.** $-2, 1$ **38.** $0, 4$ **39.** $0, -6$ **40.** $-\frac{1}{2}$ **41.** $2, 8$ **42.** $-\frac{2}{3}, 1$ **43.** $-\frac{5}{2}, 1$ **44.** $3, -4$ **45.** $aL(t_2 - t_1)$ **46.** $(t_2 - t_1)[a - 16(t_2 + t_1)]$ **47.** The distance between the two intersections is 2500 ft.   **48.** The length of the horizontal diagonal of the kite is 32 in.   **49.** The rocket will reach a height of 18 ft above the launch in $\frac{1}{4}$ sec and $\frac{9}{2}$ sec, or in 0.25 sec and 4.5 sec.   **50.** $h = \frac{2A}{b + B}$

Posttest: Chapter 6, p. 552

1. $3x^2$ **2.** $2y(x - 7)$ **3.** $2p(2p - 3q)(2p - q)$ **4.** $(a - b)(x - y)$ **5.** $(n + 3)(n - 16)$ **6.** $(x + 2)(x - 4)$ **7.** $-5x(x + 1)(x - 4)$ **8.** $(4x - 3y)(x + 4y)$ **9.** $-3(2x - 3)^2$ **10.** $(3x + 5y)^2$ **11.** $(11 + 2x)(11 - 2x)$ **12.** $(pq + 1)(pq - 1)$ **13.** $(y + 2)^2(y - 2)^2$ **14.** $(4 - n)(16 + 4n + n^2)$ **15.** $-8, 1$ **16.** $\frac{1}{3}, -2$ **17.** $-\frac{3}{2}, 2$ **18.** 8 m   **19.** $mg(y_2 - y_1)$ **20.** $2x(2x + 55)$ ft^2

Cumulative Review: Chapter 6, *p. 553*

1. -3 **2.** $8a + 7b + 2$ **3.** 8 **4.** $(-1, 7)$

5.

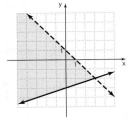

6. n **7.** $21x^2 + x - 2$ **8.** The charge is $(f + 8c)$ dollars.   **9.** The company must sell 50 shirts per day in order to break even.   **10.** $2\pi r(r + h)$

Chapter 7

Pretest: Chapter 7, *p. 556*

1. The expression is undefined when $x = -6$.

2. $\dfrac{n - 2}{3} = \dfrac{(n - 2) \cdot n}{3 \cdot n} = \dfrac{n^2 - 2n}{3n}$ **3.** $\dfrac{4x}{y^2}$ **4.** $4a$

5. $-\dfrac{w}{w + 6}$ **6.** $\dfrac{10}{n^2}$ **7.** $\dfrac{5y - 2}{y + 1}$ **8.** $\dfrac{15x + 2}{12x^2}$ **9.** $\dfrac{2c + 12}{(c - 3)(c + 3)}$

10. $\dfrac{3x - 1}{(x - 1)^2(x + 1)}$ **11.** $\dfrac{4a^2}{3n^2}$ **12.** $\dfrac{y - 4}{5y(y + 4)}$ **13.** $\dfrac{1}{x + 4}$ **14.** 5

15. $-\dfrac{2}{3}$ **16.** $\dfrac{3}{2}$ and 3 **17.** $\dfrac{2x + 1500}{x}$ dollars

18. The average speed during the first part of the trip was 65 mph and the average speed during the second part was 55 mph.   **19.** It takes the photocopier $2\frac{1}{2}$ (or 2.5) minutes to make 30 copies.   **20.** The spring will stretch $7\frac{1}{2}$ in.

Practices: Section 7.1 *pp. 558–563*

1, *p. 558:* **a.** The expression is undefined when n is equal to 3.   **b.** The expression is undefined when n is equal to -3 or 3.   **2,** *p. 559:* **a.** The expressions are equivalent.   **b.** The expressions are equivalent.   **3,** *p. 560:* **a.** $\dfrac{2n^2}{m}$ **b.** $-\dfrac{3x^2}{2}$ **4,** *p. 561:* **a.** $\dfrac{y - 4}{2y + 3}$ **b.** The expression cannot be simplified.   **c.** $\dfrac{1}{3}$ **5,** *p. 561:* **a.** $\dfrac{t - x}{z - 3g}$ **b.** $\dfrac{4(n - 1)}{n + 2}$ **c.** $\dfrac{y - 1}{y + 3}$ **6,** *p. 562:* **a.** -1 **b.** -3 **c.** $-\dfrac{3}{n + 5}$ **d.** $-\dfrac{3s + 2}{s + 1}$ **7,** *p. 563:* **a.** $\dfrac{r}{2}$ **b.** The expression is undefined when $r = 0$.

Exercises 7.1, *p. 564*

1. rational expression   **3.** has an asymptote at   **5.** equivalent   **7.** $x = 0$ **9.** $y = 2$ **11.** $x = -5$ **13.** $n = \frac{1}{2}$ **15.** $x = -1$ or $x = 1$ **17.** $x = -4$ or $x = 5$ **19.** $x = 0$ **21.** The expressions are equivalent.   **23.** The expressions are equivalent.   **25.** The expressions are not equivalent.   **27.** The expressions are equivalent.   **29.** The expressions are not equivalent.   **31.** $\dfrac{5a^3}{6}$ **33.** $\dfrac{1}{4x^3}$ **35.** $\dfrac{3t}{2s^2}$ **37.** $8a^3b^3$ **39.** $\dfrac{p}{5q}$ **41.** $\dfrac{5}{4}$ **43.** $-\dfrac{8x}{3}$ **45.** $x - 2$ **47.** $\dfrac{x + 1}{2x + 3}$ **49.** $\dfrac{a}{b}$ **51.** $\dfrac{2}{3}$ **53.** $t - 1$ **55.** -1 **57.** $-\dfrac{1}{2}$ **59.** $\dfrac{b - 4}{b + 4}$ **61.** $\dfrac{x + 3}{5}$ **63.** $\dfrac{a(a + 7)}{a - 12}$ **65.** $\dfrac{t + 1}{t + 2}$ **67.** $-\dfrac{4d + 3}{4d - 3}$ **69.** $-\dfrac{2s}{2s - 3}$ **71.** $\dfrac{2x + 1}{2x - 1}$ **73.** $\dfrac{2y - 1}{y(2y + 3)}$ **75.** $\dfrac{2ab}{2b + a}$ **77.** $\dfrac{m + 7n}{2(m + 6n)}$ **79.** $\dfrac{2y^2}{3x}$ **81.** $\dfrac{3(x - 1)}{x + 3}$ **83.** -1 **85.** $\dfrac{2x + 3}{3x}$

87. $P = \dfrac{3}{2}$ **89.** The expressions are equivalent.   **91.** $u + v$
93. **a.** $x^2 + 7x + 10$ **b.** $x^2 + 2x$ **c.** $\dfrac{x^2 + 7x + 10}{x^2 + 2x}; \dfrac{x + 5}{x}$
d. $\dfrac{13}{8}$ **95.** **a.** $\dfrac{\pi r_1^2}{\pi r_2^2 - \pi r_1^2}; \dfrac{r_1^2}{r_2^2 - r_1^2}$ **b.** $\dfrac{\pi r_2^2 - \pi r_1^2}{\pi r_3^2 - \pi r_2^2}; \dfrac{r_2^2 - r_1^2}{r_3^2 - r_2^2}$

Practices: Section 7.2 *pp. 568–572*

1, *p. 568:* **a.** $\dfrac{2n}{7m}$ **b.** $\dfrac{p}{15q}$ **2,** *p. 569:* **a.** $\dfrac{3}{2t}$ **b.** $\dfrac{2}{6g - 1}$
3, *p. 569:* **a.** $2(y + 7)(y - 2)$ **b.** $-\dfrac{x + 5}{x + 1}$ **4,** *p. 571:* **a.** $\dfrac{3a}{2b}$
b. $\dfrac{15}{7p^4q}$ **5,** *p. 571:* **a.** $\dfrac{(x + 3)(x + 3)}{(x - 10)(x + 1)}$, or $\dfrac{(x + 3)^2}{(x - 10)(x + 1)}$
b. $\dfrac{1}{3}$ **c.** $\dfrac{y(y + 3)}{5(y + 2)(y - 1)}$ **6,** *p. 572:* $\dfrac{Wv^2}{gr}$

Exercises 7.2, *p. 573*

1. $\dfrac{1}{4t}$ **3.** $\dfrac{6}{ab}$ **5.** $\dfrac{10}{3x^9}$ **7.** $-\dfrac{14}{x}$ **9.** $\dfrac{5}{x^3}$ **11.** $\dfrac{8n - 3}{n}$ **13.** $\dfrac{4}{5}$ **15.** $-\dfrac{1}{2}$ **17.** $\dfrac{3(x + 2y)}{4}$ **19.** $\dfrac{p^2 - 1}{p^2 - 4}$ **21.** $\dfrac{n + 3}{n + 1}$ **23.** $\dfrac{(2y - 1)}{(2y - 5)}$ **25.** $\dfrac{16}{x^4}$ **27.** $3x$ **29.** $\dfrac{1}{2}$ **31.** 2 **33.** $\dfrac{10}{x}$ **35.** $-\dfrac{3}{t^2}$ **37.** $\dfrac{6y^2}{x^4}$ **39.** $\dfrac{(c + 3)(c - 7)}{(c - 5)(c + 9)}$ **41.** $\dfrac{5}{12}$ **43.** $\dfrac{1}{2(1 - x)}$ **45.** $-p$ **47.** $\dfrac{x + y}{5x}$ **49.** $-\dfrac{1}{t + 2}$ **51.** $\dfrac{x + 5}{x + 6}$ **53.** $\dfrac{3(p + 2)(p - 4)}{(p + 5)^2}$ **55.** $-\dfrac{1}{r^2}$ **57.** $\dfrac{2}{3 - x}$ **59.** $-\dfrac{1}{6p^2}$ **61.** $\dfrac{3(y + 3)}{4}$ **63.** $\dfrac{3x(x - 1)}{3x + 1}$ **65.** $\dfrac{A - p}{pr}$ **67.** $\dfrac{pqB}{10,000}$ **69.** $\dfrac{V^2r}{(R + r)^2}$ **71.** **a.** $\dfrac{4r}{3h}$ **b.** $\dfrac{2r}{h + r}$ **c.** $\dfrac{2(h + r)}{3h}$
d. 1

Practices: Section 7.3 *pp. 577–586*

1, *p. 577:* **a.** $\dfrac{10}{y + 2}$ **b.** $\dfrac{5r}{s}$ **c.** 6 **d.** $3n - 1$ **2,** *p. 578:*
a. $\dfrac{5}{v}$ **b.** $\dfrac{3t}{5}$ **c.** $-\dfrac{p}{3q}$ **3,** *p. 578:* **a.** 2 **b.** $2x$ **c.** $-\dfrac{3}{x - 5}$
4, *p. 580:* **a.** LCD $= 20y$ **b.** LCD $= 6t^2$
c. LCD $= 30x^2y^3$ **5,** *p. 581:* **a.** LCD $= 2n(n + 1)^2$
b. LCD $= (p + 2)(p - 1)(p + 5)$ **c.** LCD $=$
$(s + 2t)^2(s - 2t)^2$ **6,** *p. 582:* **a.** $\dfrac{4}{14p^3}$ and $\dfrac{7p^2(p + 3)}{14p^3}$
b. $\dfrac{(3y - 2)(y - 3)}{(y + 3)(y - 3)^2}$ and $\dfrac{y(y + 3)}{(y + 3)(y - 3)^2}$ **7,** *p. 583:* **a.** $\dfrac{11}{12p}$
b. $\dfrac{3y - 2}{15y^2}$ **8,** *p. 584:* **a.** $\dfrac{9x + 6}{x(x + 3)} = \dfrac{3(3x + 2)}{x(x + 3)}$ **b.** $\dfrac{2x - 5}{x - 1}$
9, *p. 584:* **a.** $\dfrac{9x + 4}{4(x - 4)(x + 4)}$ **b.** $\dfrac{x - 9}{(x + 1)(x - 1)}$
10, *p. 585:* **a.** $\dfrac{-3y^2 - 12y - 3}{(y + 3)(y + 2)(y + 1)} = \dfrac{-3(y^2 + 4y + 1)}{(y + 3)(y + 2)(y + 1)}$
b. $\dfrac{8x^3 + 3x^2 - 20x + 5}{20x(x + 1)}$ **11,** *p. 586:* $\dfrac{100(C_1 - C_0)}{C_0}$

Exercises 7.3, *p. 587*

1. $\dfrac{4a}{3}$ **3.** t **5.** $\dfrac{11}{5x}$ **7.** $\dfrac{5}{7y}$ **9.** $\dfrac{3x}{y}$ **11.** $-\dfrac{p}{5q}$ **13.** $\dfrac{9}{x + 1}$ **15.** $-\dfrac{4}{x + 2}$ **17.** $\dfrac{a + 1}{a + 3}$ **19.** $\dfrac{5x + 1}{x - 8}$ **21.** $\dfrac{4x + 1}{5x + 2}$ **23.** 3 **25.** 4 **27.** $\dfrac{x - 4}{x^2 - 4x - 2}$ **29.** 1 **31.** $\dfrac{-x + 7}{3x^2 - x + 2}$ **33.** LCD $= 15(x + 2)$ **35.** LCD $= (p - 3)(p + 8)(p - 8)$ **37.** LCD $= t(t + 3)(t - 3)$ **39.** LCD $= (t + 2)(t + 5)$ $(t - 5)$ **41.** LCD $= (3s - 2)(s - 3)(s + 2)$ **43.** $\dfrac{4x}{12x^2}$ and $\dfrac{15}{12x^2}$ **45.** $\dfrac{35b}{14a^2b}$ and $\dfrac{2a(a - 3)}{14a^2b}$ **47.** $\dfrac{8(n + 1)}{n(n + 1)^2}$ and $\dfrac{5n}{n(n + 1)^2}$ **49.** $\dfrac{3n(n - 1)}{4(n + 1)(n - 1)}$ and $\dfrac{8n}{4(n + 1)(n - 1)}$ **51.** $\dfrac{2n(n - 3)}{(n + 1)(n + 5)(n - 3)}$ and $\dfrac{3n(n + 1)}{(n + 1)(n + 5)(n - 3)}$ **53.** $\dfrac{13}{6x}$ **55.** $\dfrac{4 - 5x}{6x^2}$ **57.** $\dfrac{-2y + 4x}{3x^2y^2} = -\dfrac{2(y - 2x)}{3x^2y^2}$ **59.** $\dfrac{2x}{(x + 1)(x - 1)}$ **61.** $\dfrac{p + 39}{21}$ **63.** $3x - 5$ **65.** $\dfrac{a + 6}{6a^2}$

67. $\frac{a^2 + 1}{a - 1}$ **69.** -1 **71.** $\frac{-8x - 2}{x(x + 1)} = -\frac{2(4x + 1)}{x(x + 1)}$ **73.** $\frac{7x - 6}{x - 4}$

75. $\frac{11x - 6}{(x - 1)(x + 2)}$ **77.** $\frac{n + 20}{3(n - 3)(n + 5)}$ **79.** $\frac{3t - 4}{(t + 5)(t - 5)}$

81. $\frac{2x^2 + 5x - 5}{(x + 1)^2(x + 3)}$ **83.** $\frac{6t + 5}{(2t + 3)(t - 1)}$ **85.** $\frac{12x^3 + 17x^2 - 2}{3x(x - 1)(x + 1)}$

87. $\frac{5y^2 - 28y + 4}{(3y - 1)(y - 4)}$ **89.** $\frac{-9a^2 - 11a + 32}{4(a + 3)^2}$ **91.** $\frac{4q(p + 3)}{24p^2q}, \frac{15p}{24p^2q}$

93. $\frac{3(3x + 2y)}{x^2y^2}$ **95.** $-\frac{3}{c(c + 1)}$ **97.** $\frac{b^2 - 4b - 13}{(b - 3)^2(b + 1)}$

99. $-\frac{5}{m + n}$ **101.** $\frac{r - 1}{r + 2}$ **103.** $\frac{2vt + at^2}{2}$ **105.** $\frac{1000r}{(1 + r)^2}$ dollars

107. The trip took $\frac{30}{r}$ hr. **109.** $\frac{3 + 0.1x}{x}$ dollars

Practices: Section 7.4 *pp. 593–597*

1, p. 593: a. $\frac{3}{5x^3}$ b. $\frac{8}{x + 2}$ **2, p. 594:** $\frac{2n - 1}{2n + 1}$ **3, p. 595:**
a. $2x^2$ b. $\frac{y + 3}{2y^3}$ **4, p. 596:** a. $\frac{4y^2 + y}{4y^2 - 1}$ b. $\frac{b - a}{10ab}$
5, p. 597: $\frac{3abc}{bc + ac + ab}$

Exercises 7.4, *p. 598*

1. $\frac{2}{x}$ **3.** $\frac{a + 1}{a - 1}$ **5.** $\frac{x(3x + 1)}{3x^2 - 1}$ **7.** $\frac{1}{3d(d + 3)}$ **9.** $\frac{x - 2y}{3x}$

11. $\frac{10 - y}{5(5 - y)}$ **13.** $\frac{x + 2}{x + 3}$ **15.** $\frac{3y + 4}{5y + 4}$ **17.** $\frac{xy}{16(y + 1)}$

19. $\frac{3(x^2 - 2x - 2)}{x^2(x + 1)}$ **21.** $\frac{m + 2}{3(m - 1)}$ **23.** $\frac{4(u + 2)}{u^2(u + 1)}$ **25.** $\frac{4y + 3}{3y(2y + 1)}$

27. $\frac{2VR(R + 1)}{2R + 1}$ **29.** $\frac{9E}{I}$ **31.** $\frac{2ab}{a + b}$ mph

33. $\frac{w}{\left(1 + \frac{h}{6400}\right)^2} = \frac{w}{\left(\frac{6400 + h}{6400}\right)^2} = \frac{w}{\frac{(6400 + h)^2}{6400^2}} = \frac{6400^2 w}{(6400 + h)^2}$

Practices: Section 7.5 *pp. 601–607*

1, p. 601: $\frac{1}{2}$ **2, p. 602:** 2, 5 **3, p. 604:** 2 **4, p. 604:** 3
5, p. 605: Working together, it will take both pumps $3\frac{3}{4}$ hr (or 3 hr 45 min) to fill the tank. **6, p. 606:** The speed of the propeller plane was 150 mph. **7, p. 607:**
a. $x = \frac{500}{D - 50}$ b. $1.25 per unit

Exercises 7.5, *p. 608*

1. -3 **3.** $\frac{1}{2}$ **5.** 1 **7.** 4 **9.** 14 **11.** no solution
13. 1, 2 **15.** 6 **17.** 2 **19.** -3 **21.** $-\frac{7}{2}$ **23.** 2, -8
25. 4, -1 **27.** $-\frac{5}{2}$ **29.** $-\frac{1}{3}$, 3 **31.** No solution
33. It will take them 18 min to clean the attic. **35.** It would take the clerical worker 15 hr to finish the job working alone.
37. The speed on the dry road was 60 mph. **39.** $D = \frac{P}{Lp}$

Practices: Section 7.6 *pp. 611–623*

1, p. 611: 8 **2, p. 612:** 40 lb of sodium hydroxide are needed to neutralize 49 lb of sulfuric acid. **3, p. 613:** She would be paying $225 less if she had a $75,000 mortgage at the same rate. **4, p. 614:** The length of DE is 4 in.
5, p. 615: The speed of the plane in still air is 500 mph.

6, p. 617: $k = \frac{4}{5}; y = \frac{4}{5}x$ **7, p. 617:** 220 mg of the drug should be administered. **8, p. 619:** $k = 48; y = \frac{48}{x}$
9, p. 619: The f-stop of the lens is 4. **10, p. 620:** $k = \frac{1}{2}$;
$y = \frac{1}{2}xz$ **11, p. 621:** The other employee must invest $2250. **12, p. 622:** a. $w = \frac{kxy}{z^2}$ b. $k = 24$ c. 9
13, p. 623: The BMI is approximately 19.

Exercises 7.6, *p. 624*

1. ratio **3.** proportion **5.** constant of variation
7. inverse variation **9.** 8 **11.** 80 **13.** 24 **15.** $-\frac{2}{3}$
17. 3 **19.** 5 **21.** $-4, 4$ **23.** 8, -4 **25.** $-\frac{6}{5}, 2$
27. No solution **29.** Decreases; inverse variation
31. Increases; direct variation **33.** Increases; direct variation
35. $k = 3; y = 3x$ **37.** $k = \frac{1}{6}; y = \frac{1}{6}x$
39. $k = 9; y = 9x$ **41.** $k = \frac{3}{2}; y = \frac{3}{2}x$
43. $k = 39; y = \frac{39}{x}$ **45.** $k = 27; y = \frac{27}{x}$
47. $k = \frac{7}{25}; y = \frac{7}{25x}$ **49.** $k = 18; y = \frac{18}{x}$ **51.** $k = 4$;
$y = 4xz$ **53.** $k = \frac{6}{5}; y = \frac{6}{5}xz$ **55.** $k = 3; y = 3xz$
57. $k = 25; y = 25xz$ **59.** $k = 125; y = \frac{125x}{z^2}$
61. $k = 500; y = \frac{500}{xz^2}$ **63.** $k = \frac{1}{10}; y = \frac{xw}{10z^2}$ **65.** 7, -3
67. 11 **69.** $\frac{2}{3}$ **71.** $k = 30; y = \frac{30xw}{z^3}$ **73.** $k = \frac{2}{3}; y = \frac{2}{3}x$
75. $k = 24; y = \frac{24}{x}$ **77.** Decreases; inverse variation
79. It would take $8\frac{1}{3}$ min (or 8 min 20 sec) to print a 25-page report. **81.** It will take 16 gal of gas to drive 120 mi.
83. The cyclist's speed was 20 mph. **85.** The speed of the bus is 50 mph and the speed of the train is 80 mph.
87. $AB = 8$ ft **89.** The height of the tree is 18 feet.
91. There are 8 women at the party. **93.** a. $A = \frac{3}{100}i$, or $A = 0.03i$, the constant of variation represents the income tax rate, which is 3%. b. A person will pay $795 in income tax.
95. 3.06 m **97.** 28.4 J **99.** 5.4 lumens per square meter

Review Exercises: Chapter 7, *p. 635*

1. a. $x = -1$ b. $x = 3$ and $x = -2$ **2.** a. Equivalent
b. Equivalent **3.** $\frac{3}{5m}$ **4.** $\frac{5n - 6}{3n + 2}$ **5.** $-\frac{x + 4}{x + 2}$ **6.** $\frac{2x + 5}{3x - 1}$
7. $\frac{6n^2}{pm}$ **8.** $\frac{1}{2}$ **9.** $\frac{x - 5}{2x + 5}$ **10.** -1 **11.** $\frac{1}{2y}$ **12.** $\frac{5(7m - 10)}{7(m - 10)}$
13. $\frac{y}{5(x + 6)}$ **14.** $\frac{x + 7}{3x + 2}$ **15.** $\frac{4x}{20x^2}$ and $\frac{3}{20x^2}$
16. $\frac{4(n + 4)}{(n - 1)(n + 4)}$ and $\frac{n(n - 1)}{(n - 1)(n + 4)}$
17. $\frac{x + 1}{3(x + 3)(x + 1)}$ and $\frac{3x}{3(x + 3)(x + 1)}$
18. $\frac{2(x + 2)}{(3x + 1)(x - 2)(x + 2)}$ and $-\frac{3x + 1}{(3x + 1)(x - 2)(x + 2)}$ **19.** 2
20. 4 **21.** $\frac{2y + 4}{y(2y - 1)}$ **22.** $\frac{n^2 + 3n - 6}{3n(n + 5)}$ **23.** $\frac{8x + 13}{(x - 3)(x + 3)}$
24. $\frac{2y + 5}{(2 + y)(2 - y)}$ **25.** $\frac{8m - 8}{(m + 1)(m - 3)}$ **26.** $\frac{-x^2 - 3x - 11}{(x + 3)(x - 4)}$
27. $\frac{x^2 - 9x + 2}{(x + 2)^2(x - 4)}$ **28.** $\frac{2n^2 + 7n + 11}{(2n - 1)(n - 1)(n + 3)}$ **29.** $\frac{7}{6x}$
30. $\frac{y}{y + 9}$ **31.** 2 **32.** $\frac{4x + 1}{2x - 3}$ **33.** 7 **34.** $\frac{1}{2}$ **35.** -3
36. 3, -1 **37.** -1 **38.** 2 **39.** 45 **40.** 33 **41.** $\frac{1}{8}$
42. 4, -9 **43.** $k = 0.4; y = 0.4x$ **44.** $k = \frac{3}{2}; y = \frac{3}{2x}$
45. $k = 6; y = 6xz$ **46.** $k = 8; y = \frac{8x}{z^2}$ **47.** $\frac{0.72x + 200}{x}$ dollars
48. The total cost of the car rental is $200.
49. $\frac{2rs}{s + r}$ **50.** It will take 30 min to fill the tub.
51. The family drove 100 miles at 50 mph.
52. $\frac{2x + 1}{x(x + 1)}$ of the job will be done in an hour.
53. She should expect to spend about $41,333.
54. $\frac{3n^2 + 6n + 2}{n(n + 1)(n + 2)}$ **55.** Its velocity after 5 sec is 49 meters per second. **56.** The accommodation is 10 diopters.

Posttest: Chapter 7, *p. 638*

1. The expression is undefined when $x = 8$.

2. $-\dfrac{3y - y^2}{y^2} = -\dfrac{y(3 - y)}{y^2} = -\dfrac{\overset{1}{y}(3 - y)}{\underset{y}{y^2}} = \dfrac{-(3 - y)}{y} = \dfrac{y - 3}{y}$

3. $\dfrac{5a^2}{4b}$ **4.** $\dfrac{x}{y}$ **5.** $\dfrac{3(b - 3)}{b - 7}$ **6.** $\dfrac{1}{x + 3}$

7. $\dfrac{4(4n - 1)}{4(n + 8)(n - 2)}$, $\dfrac{8(n - 2)}{4(n + 8)(n - 2)}$, and $\dfrac{n(n + 8)}{4(n + 8)(n - 2)}$ **8.** 2

9. $\dfrac{3y + 1}{2y(y - 4)}$ **10.** $\dfrac{4d + 14}{(d - 3)(d + 2)} = \dfrac{2(2d + 7)}{(d - 3)(d + 2)}$

11. $\dfrac{2x^2 + 6x + 10}{(2x + 1)(x - 2)(x + 2)} = \dfrac{2(x^2 + 3x + 5)}{(2x + 1)(x - 2)(x + 2)}$

12. $\dfrac{1}{18n(n - 1)}$ **13.** $\dfrac{a + 5}{a + 4}$ **14.** $\dfrac{2(x + 4)(x + 2)}{x - 1}$ **15.** 6

16. -14 **17.** $2, -3$ **18.** $R_1 = \dfrac{RR_2}{R_2 - R}$ **19.** Working alone, the newer machine can process 1000 pieces of mail in 30 min and the older machine can process 1000 pieces of mail in 60 min. **20.** The height of the tree is 60 m.

Cumulative Review: Chapter 7, *p. 640*

1. $2y - 3$ **2.** $n > -3$;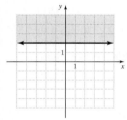

3.

4. $(3, 5)$ **5.** $-5, 2$ **6.** $(10y + 9)(10y - 9)$ **7.** $-3, 2$
8. She should invest $7500 in the fund at 4% and $15,000 in the fund at 6% **9.** It will take the Reston bus 4 hr to overtake the Arlington bus. **10.** It would take about 9 min to print a 20-page report.

Chapter 8

Pretest: Chapter 8, *p. 642*

1. a. 12 **b.** -4 **2.** $10uv^2$ **3.** $\left(\sqrt[4]{81x^4}\right)^3$; $27x^3$ **4. a.** $\dfrac{4}{3}$
b. $\sqrt[3]{xy^2}$ **5.** $\sqrt{42ab}$ **6.** $\sqrt[3]{6r}$ **7.** $7x^2y^2\sqrt{3x}$ **8.** $\dfrac{\sqrt{2p}}{5q^4}$
9. $7\sqrt{3}$ **10.** $2x - 15\sqrt{x} - 8$ **11.** $\dfrac{\sqrt{6xy}}{9y}$
12. $\dfrac{n - \sqrt{3n}}{n - 3}$ **13.** 32 **14.** 8 **15.** 2 **16.** $5 - 37i$
17. $-\dfrac{3}{5} + \dfrac{4}{5}i$ **18. a.** $\dfrac{\sqrt{h - d}}{4}$ **b.** It takes the stone $\dfrac{5\sqrt{3}}{4}$ sec, or approximately 2.2 sec, to be 125 ft above the ground. **19.** The diagonal of the court is $2\sqrt{2834}$ ft, or about 106.5 ft. **20.** $p = \dfrac{f^2}{14,400}$

Practices: Section 8.1, *pp. 644–650*

1, *p. 644:* **a.** 5 **b.** $\dfrac{3}{10}$ **c.** -36 **d.** Not a real number
2, *p. 644:* 2.449 **3,** *p. 645:* **a.** $-2y$ **b.** $6x^3y^3$
4, *p. 646:* **a.** 6 **b.** -3 **c.** $\dfrac{1}{5}$ **d.** $-4x^2$ **5,** *p. 648:* **a.** 2
b. -2 **c.** $4y^2$ **6,** *p. 648* **a.** 9 **b.** 14 **c.** $3\sqrt{5} \approx 6.71$

d. $3 + 4\sqrt{3} \approx 9.93$ **7,** *p. 650*

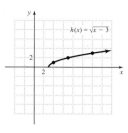

Exercises 8.1, *p. 651*

1. square root **3.** radicand **5.** irrational **7.** index
9. square root function **11.** 8 **13.** -10 **15.** Not a real number **17.** 8 **19.** 3 **21.** -10 **23.** 4 **25.** -8
27. $\dfrac{3}{4}$ **29.** $-\dfrac{2}{5}$ **31.** 0.2 **33.** 4.583 **35.** 6.782
37. 3.775 **39.** 4.820 **41.** 2.724 **43.** x^4 **45.** $4a^3$
47. $9p^4q^2$ **49.** $2x^5y$ **51.** $-5u^3$ **53.** $12uv^4$ **55.** $2t^3$
57. pq^3 **59.** 5 **61.** -2 **63.** $\sqrt{5}$ **65.** $2\sqrt[3]{5}$
67. **69.**

71. $\dfrac{5}{11}$ **73.** -10 **75.** 0.095 **77.** $2a^3b^2$ **79.** -2
81.

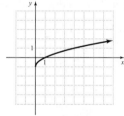

83. It takes the stone $\dfrac{5}{2}$, or 2.5, sec to reach the ground.
85. a. The length of the side of the frame is 5 in.
b. A 3-in. by 3-in. photograph fits in the frame.

Practices: Section 8.2, *pp. 655–659*

1, *p. 655:* **a.** $\sqrt{36}$; 6 **b.** $\sqrt[3]{27}$; 3
c. $-\sqrt[4]{n}$ **d.** $\sqrt[3]{-125y^9}$; $-5y^3$ **2,** *p. 656:* **a.** $\left(\sqrt[4]{81}\right)^3$; 27
b. $\left(\sqrt[3]{-64}\right)^2$; 16 **c.** $\left(\sqrt[4]{\dfrac{4}{9}y^4}\right)^5$; $\dfrac{32}{243}y^{10}$ **3,** *p. 657:* **a.** $-\dfrac{1}{3}$
b. $\dfrac{1}{81a^8}$ **4,** *p. 658:* **a.** 27 **b.** $\sqrt[5]{n}$ **c.** $\sqrt[4]{r}$ **d.** $\dfrac{x^2}{2y}$
5, *p. 659:* $\dfrac{2\pi d^{3/2}}{(Gm)^{1/2}}$, or $\dfrac{2\pi d^{3/2}}{G^{1/2}m^{1/2}}$

Exercises 8.2, *p. 660*

1. raised to the power **3.** index **5.** $\sqrt{16}$; 4
7. $-\sqrt{16}$; -4 **9.** $\sqrt[3]{-64}$; -4 **11.** $6\sqrt[4]{x}$
13. $\sqrt{36a^2}$; $6a$ **15.** $\sqrt[3]{-216u^6}$; $-6u^2$ **17.** $\left(\sqrt[3]{27}\right)^4$; 81
19. $-\left(\sqrt{16}\right)^3$; -64 **21.** $\left(\sqrt[3]{-27y^3}\right)^2$; $9y^2$
23. $-\dfrac{1}{\left(\sqrt[4]{81}\right)^3}$; $-\dfrac{1}{27}$ **25.** $\sqrt{\dfrac{4}{x^{10}}}$; $\dfrac{2}{x^5}$ **27.** 64 **29.** $\sqrt[5]{6}$

31. $\frac{1}{2}$ **33.** 3   **35.** $4\sqrt[5]{n^3}$ **37.** $\sqrt{y}$ **39.** $\frac{1}{2x}$ **41.** $5\sqrt[4]{r}$

43. $\frac{\sqrt[3]{a^2}}{b^2}$ **45.** $12x^5$ **47.** $\frac{4}{p^2}$ **49.** $(\sqrt{9})^3$; 27

51. $(\sqrt[3]{-125u^3})^2$; $25u^2$ **53.** $3\sqrt{n}$ **55.** $8x^3$

57. a. $8000\sqrt[3]{0.5^t}$ **b.** The value of the equipment 6 yr after it was purchased is $2000.

Practices: Section 8.3, pp. 662–667

1, p. 662: a. $\sqrt[4]{y}$ **b.** $\sqrt[3]{5}$ **c.** $\sqrt[10]{n^7}$ **d.** $\sqrt[4]{t}$

2, p. 662: a. $\sqrt[12]{n}$ **b.** $\sqrt[4]{ab^3}$ **c.** Cannot be simplified

3, p. 663: a. $\sqrt{21}$ **b.** $\sqrt[5]{2y^3}$ **c.** $\sqrt[3]{28p^2}$ **d.** $\sqrt{\frac{15}{uv}}$

4, p. 663: a. $6\sqrt{2}$ **b.** $-21\sqrt{2}$ **c.** $2\sqrt[3]{7}$ **d.** $2\sqrt[4]{3}$

5, p. 664: a. $y^3\sqrt{y}$ **b.** $3x^2\sqrt{2x}$ **c.** $3ab^2\sqrt[3]{3a}$

6, p. 665: a. $\sqrt{7n}$ **b.** 6   **c.** $\sqrt[3]{2y}$ **7, p. 665: a.** $\frac{4}{5}$ **b.** $\frac{\sqrt{n}}{8}$

c. $\frac{4\sqrt[3]{x}}{5y^3}$ **d.** $\frac{p^3q\sqrt{7}}{6r^2}$ **8, p. 666:** $2\sqrt{10}$ units

9, p. 667: a. $(-20, 50)$ and $(10, 60)$ **b.** The drivers were $10\sqrt{10}$ mi (or about 31.6 mi) apart.

Exercises 8.3, p. 668

1. the same index   **3.** radicands   **5.** $\sqrt[3]{n}$ **7.** $\sqrt{2}$

9. $\sqrt{x}$ **11.** Cannot be simplified   **13.** $\sqrt[12]{x}$ **15.** $\sqrt[4]{y}$

17. $x^2\sqrt{y}$ **19.** x **21.** $\sqrt[3]{y}$ **23.** $\sqrt{pq}$ **25.** $\sqrt{30}$

27. $\sqrt{6xy}$ **29.** $\sqrt[3]{36a^2b}$ **31.** $\sqrt[4]{21n^3}$ **33.** $\sqrt{\frac{3x}{y}}$

35. $2\sqrt{6}$ **37.** $-12\sqrt{5}$ **39.** $3\sqrt[3]{3}$ **41.** $2\sqrt[4]{6}$ **43.** $2\sqrt[5]{2}$

45. $6x^3\sqrt{x}$ **47.** $2\sqrt{5y}$ **49.** $10r^2\sqrt{2}$ **51.** $3x^2y^3\sqrt{6xy}$

53. $2n^3\sqrt{4n^2}$ **55.** $2n\sqrt[5]{2n^3}$ **57.** $2x^2y^3\sqrt[3]{9x}$

59. $2xy^2\sqrt[4]{4xy^2}$ **61.** 3   **63.** $\sqrt{5n}$ **65.** $2x\sqrt{y}$

67. $\sqrt[3]{4u}$ **69.** $\sqrt[4]{6ab}$ **71.** $\frac{5}{4}$ **73.** $\frac{\sqrt{7}}{9}$ **75.** $\frac{\sqrt{2}}{n^3}$

77. $\frac{\sqrt{7a}}{11}$ **79.** $\frac{\sqrt{a}}{3b^2}$ **81.** $\frac{3\sqrt{u}}{5v}$ **83.** $\frac{3\sqrt[3]{a^2}}{4}$ **85.** $\frac{\sqrt[3]{9a}}{2b^2c^3}$

87. $\frac{a\sqrt[4]{2b^3}}{3c^2}$ **89.** $\frac{a^2\sqrt{13b}}{3c^3d}$ **91.** $6\sqrt{2}$ units

93. $3\sqrt{5}$ units   **95.** 5 units   **97.** $\sqrt[4]{30x^3y^3}$ **99.** x^2

101. $5r^3s^3\sqrt{2s}$ **103.** $\frac{\sqrt[3]{25x}}{4yz^3}$ **105.** $4uv^2\sqrt[3]{2u}$ **107.** $8\sqrt{d}$

109. The width of the screen is $12\sqrt{6}$ in., or approximately 29 in.   **111.** The length of string let out is $20\sqrt{13}$ ft, or about 72.1 ft.   **113. a.** The second college is located at $(48, 20)$.   **b.** The second college is 52 mi from her home.

Practices: Section 8.4, pp. 672–675

1, p. 672: a. $2\sqrt[4]{9}$ **b.** $5\sqrt{p}$ **c.** $-4\sqrt{t^2 - 4}$

d. Cannot be combined   **2, p. 673: a.** $8\sqrt{3}$ **b.** $-19\sqrt{6}$

c. $-5\sqrt[4]{2}$ **d.** $8 + \sqrt{7}$ **3, p. 674: a.** $21\sqrt{n}$

b. $(3b - 11)\sqrt[3]{a}$ **c.** $(5 - 4x)\sqrt{2x}$

4, p. 674: a. $p(x) + q(x) = 5x\sqrt[3]{3x}$

b. $p(x) - q(x) = x\sqrt[3]{3x}$ **5, p. 675: a.** $\sqrt{337}$ in. and 20 in.   **b.** $(20 - \sqrt{337})$ in., or approximately 1.6 in.

Exercises 8.4, p. 676

1. like   **3.** simplified   **5.** $5\sqrt{3}$ **7.** $-3\sqrt[3]{2}$ **9.** Cannot be combined   **11.** $-20\sqrt{y}$ **13.** Cannot be combined

15. 0   **17.** $8\sqrt[4]{p}$ **19.** $13y\sqrt{3x}$ **21.** $-4\sqrt{2r - 1}$

23. $2\sqrt{x^2 - 9}$ **25.** $3\sqrt{6}$ **27.** $4\sqrt{2}$ **29.** $10\sqrt{2}$

31. $-4\sqrt[3]{3}$ **33.** $-2x\sqrt{6x}$ **35.** $(2n^2 + 11)\sqrt{2n}$

37. $(-4y^2 + 7x^2)\sqrt{xy}$ **39.** $-7p^3$

41. $(10 - 2a)\sqrt[4]{3ab^3}$ **43.** $7\sqrt{5} + 2$ **45.** $\sqrt[3]{6} - 3$

47. $f(x) + g(x) = 13x\sqrt{5x}$; $f(x) - g(x) = 7x\sqrt{5x}$

49. $f(x) + g(x) = 3x\sqrt[4]{4x}$; $f(x) - g(x) = 5x\sqrt[4]{4x}$

51. $-9\sqrt{6}$ **53.** $-5\sqrt{3b}$ **55.** $3(k - 2)\sqrt{5k}$

57. The difference in velocity is $16\sqrt{5}$ feet per second, or 35.8 feet per second.   **59.** The side of the larger tile is $2\sqrt{2}$ in., or about 2.8 in. longer than the smaller tile.

61. a. The length of the ramp is $2\sqrt{5}$ ft.   **b.** The length of the ramp increases by approximately 1.9 ft.

Practices: Section 8.5, pp. 679–686

1, p. 679: a. $6y^4\sqrt{2}$ **b.** $-14\sqrt[3]{15}$

2, p. 679: a. $\sqrt[4]{35} + 6\sqrt[4]{10}$ **b.** $6x + \sqrt{x} - 1$

3, p. 680: a. -3 **b.** $2x - 9$ **c.** $p - 16q$

4, p. 680: a. $3 + 4\sqrt{3b} + 4b^2$, or $4b^2 + 4\sqrt{3b} + 3$

b. $n - 6\sqrt{n + 1} + 10$ **5, p. 681:** $2\sqrt{30}$ cm²

6, p. 681: a. 10   **b.** $-\frac{\sqrt[3]{9p^2}}{5}$ **7, p. 682: a.** $\frac{y}{3x^4}$

b. $\frac{n\sqrt[3]{2n^2}}{3}$ **8, p. 683: a.** $\frac{4\sqrt{7}}{7}$ **b.** $\frac{\sqrt{2y}}{8}$ **c.** $\frac{p^2\sqrt{2}}{3}$

d. $\frac{\sqrt[3]{20pqr}}{2r}$ **9, p. 683: a.** $\frac{\sqrt{3xy}}{3y}$ **b.** $\frac{\sqrt[5]{5a^2}}{a}$

10, p. 684: a. $\sqrt{2} - 3\sqrt{3}$ **b.** $\frac{5\sqrt{b} + b}{b}$

11, p. 685: a. $4 - 2\sqrt{2}$ **b.** $\frac{a\sqrt{b} + a\sqrt{c}}{b - c}$

12, p. 686: a. $6x^2\sqrt[3]{9x^2}$ **b.** $\frac{3}{2}$ **13, p. 686: a.** $\frac{\sqrt[3]{6\pi^2 V}}{2\pi}$

b. No; $\frac{\sqrt[3]{6\pi^2(2V)}}{2\pi} = \frac{\sqrt[3]{12\pi^2 V}}{2\pi} \neq 2\left(\frac{\sqrt[3]{6\pi^2 V}}{2\pi}\right)$

Exercises 8.5, p. 687

1. distributive   **3.** quotient   **5.** perfect power   **7.** $4\sqrt{6}$

9. $-4\sqrt{21}$ **11.** $45\sqrt[3]{2}$ **13.** $70x^3\sqrt{2}$ **15.** $-16a^2b^2$

17. $-4xy\sqrt[3]{3y^2}$ **19.** $4 - 4\sqrt{2}$ **21.** $12\sqrt{2}$

23. $-4\sqrt{15} + 36$ **25.** $10\sqrt[3]{6} + 2\sqrt[3]{12}$ **27.** $x^2 + x\sqrt{2}$

29. $3a\sqrt[4]{2a} - a\sqrt[4]{10}$ **31.** $-10 + \sqrt{2}$

33. $-28 - 10\sqrt{3}$ **35.** $10 + 5\sqrt{6}$

37. $16r - 20\sqrt{r} - 24$ **39.** 4   **41.** $x - 64$ **43.** $5x - y$

45. $x - 26$ **47.** $9x - 12\sqrt{x} + 4$ **49.** $6a + 8\sqrt{6a} + 16$

51. $n - 2\sqrt{n + 7} + 8$ **53.** -1 **55.** $2x\sqrt{3}$

57. $-n^3\sqrt{3}$ **59.** $\frac{\sqrt[4]{2r^3}}{3}$ **61.** $\frac{\sqrt{5a}}{4}$ **63.** $\frac{2\sqrt{3x}}{y^3}$

65. $\frac{3\sqrt[3]{2}}{n^3}$ **67.** $\frac{4a^2\sqrt{2a}}{3b^4}$ **69.** $\frac{4\sqrt{5}}{5}$ **71.** $\frac{\sqrt{10y}}{20}$

73. $\frac{\sqrt{6a}}{6}$ **75.** $\frac{\sqrt{uv}}{2v}$ **77.** $\frac{5x^2\sqrt{3y}}{3y}$ **79.** $\frac{\sqrt{10xyz}}{3z}$

81. $\frac{\sqrt[3]{36xy}}{2y}$ **83.** $\frac{\sqrt{11x}}{x}$ **85.** $\frac{\sqrt{21xy}}{3y}$ **87.** $\frac{x\sqrt{15xy}}{12y^2}$

89. $\frac{\sqrt[3]{2a^2bc^2}}{4c^2}$ **91.** $\frac{2\sqrt{6} - 3\sqrt{2}}{6}$ **93.** $\frac{\sqrt{ab} - b}{b}$

95. $\frac{\sqrt{3t} + 2t\sqrt{15}}{3t}$ **97.** $\frac{\sqrt[3]{x^2} - 4\sqrt[3]{x}}{x}$ **99.** $\frac{2 - \sqrt{2}}{2}$

101. $-2\sqrt{2} - 2\sqrt{5}$ **103.** $\frac{8 - 4\sqrt{2x}}{2 - x}$ **105.** $\frac{x - y\sqrt{x}}{x - y^2}$

107. $\frac{a + \sqrt{2a} + 3\sqrt{a} + 3\sqrt{2}}{a - 2}$ **109.** $\frac{x - 2\sqrt{xy} + y}{x - y}$

111. $\dfrac{6a + 13\sqrt{ab} + 6b}{9a - 4b}$

113. $f(x) \cdot g(x) = 2x^2\sqrt{3}; \dfrac{f(x)}{g(x)} = 3x\sqrt{3}$

115. $f(x) \cdot g(x) = x - 1; \dfrac{f(x)}{g(x)} = \dfrac{x + 2\sqrt{x} + 1}{x - 1}$

117. $9 + 9\sqrt{2}$ **119.** $-56a^3b^2$ **121.** $\dfrac{5\sqrt{3y}}{z^2}$

123. $\dfrac{\sqrt{35xy}}{7y}$ **125.** $\dfrac{p + \sqrt{3p} - 4\sqrt{p} - 4\sqrt{3}}{p - 3}$

127. a. $A = \dfrac{\sqrt{3}}{4}a^2$ **b.** $16\sqrt{3}$ in.²   **129.** $\dfrac{\pi\sqrt{2L}}{4}$

131. $\dfrac{\sqrt{\pi A}}{2\pi}$

Practices: Section 8.6, pp. 692–698

1, p. 692: 64   **2, p. 693:** 10   **3, p. 693:** No solution
4, p. 694: 3   **5, p. 695:** 3, 11   **6, p. 696:** 3, 4
7, p. 696: 0   **8, p. 697:** $P = I^2R$
9, p. 698: a. $A = P(r + 1)^3$ **b.** The value after 3 yr is about \$10,927.27.

Exercises 8.6, p. 699

1. 12   **3.** 3   **5.** 2   **7.** -6 **9.** No solution   **11.** 121
13. 27   **15.** 24   **17.** 85   **19.** -2 **21.** 6   **23.** No solution
25. -15 **27.** 2, 3   **29.** 5, -1 **31.** -2 **33.** $\dfrac{3}{4}$ **35.** 3, 5
37. $-\dfrac{1}{4}, 0$ **39.** 3   **41.** No solution   **43.** $\dfrac{5}{4}$ **45.** -1
47. $-2, 0$ **49.** No solution   **51.** 8   **53.** $-6, -4$
55. The satellite is 100 mi above the Earth's surface.
57. a. $L = \dfrac{S^2}{30f}$ **b.** 60 ft   **59.** The painter must place the ladder 9 ft from the side of the house.

Practices: Section 8.7, pp. 703–708

1, p. 703: a. $6i$ **b.** $i\sqrt{2}$ **c.** $-5i$ **2, p. 704: a.** $4 - i$
b. $-12 + 4i$ **c.** $5 + 19i$ **3, p. 704: a.** -12 **b.** $-\sqrt{15}$
c. 40   **d.** $12 - 10i$ **4, p. 705: a.** $1 + 18i$ **b.** $-3 - 11i$
5, p. 705: a. $-1 + 7i$ **b.** $8 - 9i$ **c.** $3i$ **6, p. 706:** 29
7, p. 706: a. $\dfrac{7}{5} + \dfrac{14}{5}i$ **b.** $\dfrac{2}{5}i$ **8, p. 707: a.** $-\dfrac{1}{10} + \dfrac{7}{10}i$
b. $\dfrac{1}{4}$ **9, p. 708:** -1 **10, p. 708: a.** $I = \dfrac{V}{Z}$
b. $(-1 + 4i)$ amps   **c.** $V = (-1 + 4i)(1 - i) = 3 + 5i$

Exercises 8.7, p. 709

1. imaginary number   **3.** real part   **5.** complex conjugates
7. $2i$ **9.** $\dfrac{1}{4}i$ **11.** $i\sqrt{3}$ **13.** $3i\sqrt{2}$ **15.** $10i\sqrt{5}$ **17.** $-3i$
19. $\dfrac{3i\sqrt{5}}{2}$ **21.** $-\dfrac{i\sqrt{3}}{2}$ **23.** $1 + 20i$ **25.** $5 - 6i$
27. $-6i$ **29.** $-7 - 3i$ **31.** $-2 - 2i$ **33.** $12 - 17i$
35. -10 **37.** 9   **39.** -63 **41.** 28   **43.** $3 + 3i$
45. $7 - 12i$ **47.** $-12 + 21i$ **49.** $-4 + 6i$ **51.** $2 + 16i$
53. $46 + 56i$ **55.** $14 + 56i$ **57.** $-16 + 12i$ **59.** 61
61. $5 + 12i$ **63.** $-5 - 12i$ **65.** $-7 + i\sqrt{2}$
67. $1 - 10i$; 101   **69.** $4 + 3i$; 25   **71.** $-9 - 6i$; 117
73. $-8i$; 64   **75.** $11i$; 121   **77.** $\dfrac{28}{17} - \dfrac{7}{17}i$
79. $-\dfrac{3}{26} - \dfrac{15}{26}i$ **81.** $-\dfrac{5}{4}i$ **83.** $\dfrac{2}{7}i$ **85.** $-3 - 4i$
87. $4 + i$ **89.** $\dfrac{3}{4} + \dfrac{9}{4}i$ **91.** $\dfrac{2}{5}$ **93.** $-i$ **95.** -1
97. $-i$ **99.** i **101.** $-i$ **103.** $-8 - 5i$; 89
105. $\dfrac{4i}{5}\sqrt{3}$ **107.** $\dfrac{9}{2} - \dfrac{5i}{2}$ **109.** $-8 + 10i$
111. $-2 - 4i$ **113.** $(8 + i)$ ohms   **115.** $(57 + 69i)$ volts

Review Exercises: Chapter 8, p. 716

1. -66 **2.** -10 **3.** $\dfrac{1}{3}$ **4.** 0.6   **5.** $9y^4$ **6.** $-7a^3b$
7. $-6x^3$ **8.** $3p^3$ **9.** 6
10. $2\sqrt{10}$

11. 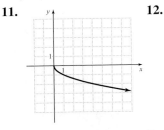   **12.**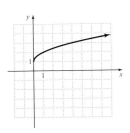

13. $-\sqrt{64}$; -8 **14.** $7\sqrt[3]{x}$
15. $-\left(\sqrt[4]{16n^4}\right)^3$; $-8n^3$ **16.** $\dfrac{1}{(\sqrt[3]{8})^2}$; $\dfrac{1}{4}$ **17.** $\sqrt[4]{x^3}$
18. $\dfrac{\sqrt{r}}{6}$ **19.** $\dfrac{1}{5y}$ **20.** $\dfrac{\sqrt[3]{a}}{12}$ **21.** $\sqrt[4]{x}$ **22.** n **23.** $\sqrt[4]{y}$
24. $q\sqrt{p}$ **25.** $\sqrt[6]{a}$ **26.** $\sqrt{xy}$ **27.** $\sqrt{30rs}$
28. $\sqrt[3]{28p^2q^2}$ **29.** $10n\sqrt{3n}$ **30.** $3x^2y^2\sqrt{5x}$
31. $4t^2\sqrt[3]{2t}$ **32.** $2ab^2\sqrt[4]{6ab^2}$ **33.** $\sqrt{5a}$ **34.** $\sqrt[3]{2p}$
35. $\dfrac{\sqrt{n}}{5}$ **36.** $\dfrac{\sqrt{6}}{7y^2}$ **37.** $\dfrac{4\sqrt[3]{u^2}}{5v^3}$ **38.** $\dfrac{p\sqrt[4]{4q^3}}{3rs^2}$ **39.** $4\sqrt{x}$
40. $11\sqrt[3]{q^2}$ **41.** $7\sqrt{3}$ **42.** 0   **43.** $(12a - 3)\sqrt[3]{7a}$
44. $3p\sqrt[3]{p^2q}$ **45.** $-18a\sqrt{2}$ **46.** $20\sqrt{2} - 10$
47. $n\sqrt[3]{2} - 2\sqrt[3]{n}$ **48.** $4t - 17\sqrt{t} + 15$ **49.** $6 - x$
50. $2y - 2\sqrt{2y} + 1$ **51.** $\dfrac{n\sqrt{3n}}{2}$ **52.** $\dfrac{4\sqrt{2a}}{3b^2}$ **53.** $\dfrac{\sqrt{2}}{4}$
54. $\dfrac{2\sqrt{xy}}{y}$ **55.** $\dfrac{p\sqrt{42q}}{3q}$ **56.** $\dfrac{\sqrt[3]{20uv}}{6u^2}$ **57.** $\dfrac{5\sqrt{2} - 3\sqrt{5}}{5}$
58. $\dfrac{2a\sqrt{3} + \sqrt{a}}{a}$ **59.** $2\sqrt{3} + 2$ **60.** $\dfrac{x + \sqrt{5x} + 2\sqrt{x} + 2\sqrt{5}}{x - 5}$
61. 8   **62.** 25   **63.** No solution   **64.** 7   **65.** $-\dfrac{5}{2}$
66. 4   **67.** $6i$ **68.** $5i\sqrt{5}$ **69.** $8 + 5i$ **70.** $4 - 2i$
71. -9 **72.** $10 - 2i$ **73.** $33 - 15i$ **74.** $24 - 10i$
75. $-\dfrac{1}{8} - \dfrac{1}{8}i$ **76.** $\dfrac{13}{20} - \dfrac{9}{20}i$ **77.** -1 **78.** i
79. a. $N\sqrt[20]{2^t}$ **b.** 80 bacteria are present after 1 hr.
80. $8\sqrt{R}$ **81. a.** The velocity of the car is $10\sqrt{2}$ meters per second (or about 14.1 meters per second).   **b.** The velocity is $(20 - 10\sqrt{2})$ meters per second (or about 5.9 meters per second) greater.   **82.** $\dfrac{\sqrt{kI}}{I}$ **83.** $\dfrac{2\pi\sqrt{mk}}{k}$ **84.** The office building is $4\sqrt{5}$ mi (or approximately 8.9 mi) from her home.
85. The wire needs to be anchored to the ground $4\sqrt{11}$ ft (or approximately 13.3 ft) from the pole.
86. 85 bottles per week are demanded.

Posttest: Chapter 8, p. 719

1. a. -27 **b.** -6 **2.** $12a^3b$ **3.** $\sqrt[5]{(32x^{10})^2}$; $4x^4$
4. a. $8\sqrt[6]{p}$ **b.** $\sqrt[4]{x^3y}$ **5.** $\sqrt[3]{20p^2q}$ **6.** $2\sqrt{2}$
7. $3xy^3\sqrt{13xy}$ **8.** $\dfrac{\sqrt{6u}}{7v^3}$ **9.** $-9\sqrt{6}$ **10.** $1 - 14\sqrt{2}$
11. $\dfrac{\sqrt{6ab}}{10b}$ **12.** $\dfrac{x + \sqrt{xy}}{x - y}$ **13.** No solution   **14.** 1, 2
15. -2 **16.** $31 - 5i$ **17.** $-\dfrac{9}{13} - \dfrac{19}{13}i$ **18. a.** $\dfrac{\sqrt{\pi S}}{2\pi}$
b. The radius of the beach ball is $\dfrac{8\sqrt{2\pi}}{\pi}$ in.

19. The distance to the horizon is $200\sqrt{41}$ mi (or approximately 1280.6 mi).   **20.** The daily demand is 149 units.

Cumulative Review: Chapter 8, *p. 720*

1. $x < -5$

2. $2x^2 - 3x - 3$

3.

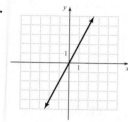

4.

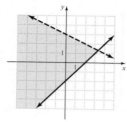

5. $(x + 2)(x - 2)^2$ **6.** $-1, 3$ **7.** $\frac{x + 14}{(x - 1)^2(x + 4)}$

8. a. $p = 5.5h$ **b.** The slope represents the employee's hourly rate, which is $5.50.   **9.** The object will be 115 ft above the ground $\frac{15}{4}$ sec, or 3.75 sec, after it is thrown upward.
10. It will take 8 painters 12.5 hr to paint the house.

Chapter 9

Pretest: Chapter 9, *p. 722*

1. $-2i, 2i$ **2.** $-\frac{7}{2}, \frac{1}{2}$ **3.** $\frac{3 + \sqrt{3}}{2}, \frac{3 - \sqrt{3}}{2}$
4. Two real solutions   **5.** $\frac{3 + 2\sqrt{2}}{2}, \frac{3 - 2\sqrt{2}}{2}$
6. $2 + 2i\sqrt{2}, 2 - 2i\sqrt{2}$ **7.** $-2, -\frac{8}{3}$
8. $\frac{-4 + 3\sqrt{2}}{2}, \frac{-4 - 3\sqrt{2}}{2}$ **9.** $\frac{5 + i\sqrt{5}}{5}, \frac{5 - i\sqrt{5}}{5}$ **10.** $\frac{3}{2}$
11. $\pm 3, \pm 2i\sqrt{2}$ **12.** $2x^2 - 5x - 12 = 0$
13.

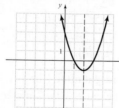

$(2, -1); x = 2; (1, 0)$ and $(3, 0); (0, 3)$

14.

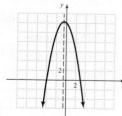

$\left(-\frac{1}{2}, \frac{49}{4}\right); x = -\frac{1}{2}; (-4, 0)$ and $(3, 0); (0, 12)$

15.

Domain: $(-\infty, \infty)$; range: $\left[-\frac{9}{2}, \infty\right)$

16.

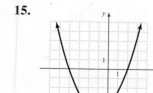

$(-\infty, 2) \cup (4, \infty)$

17. The object is 50 ft below the point of release $\frac{5}{2}$ sec, or 2.5 sec, after it is tossed upward.   **18.** The speed of the current is 3 mph.   **19.** The radius is approximately 6.1 ft. The maximum planting area is approximately 115.7 ft^2.
20. The company must sell between and including 30 and 80 bottles for revenue of at least $1200.

Practices: Section 9.1, *pp. 724–731*

1, p. 724: a. $-7, 7$ **b.** $-3i, 3i$ **2, p. 725: a.** $\pm 2\sqrt{2}$
b. $\pm 3i\sqrt{2}$ **3, p. 726: a.** $\frac{-1 + 2\sqrt{3}}{2}, \frac{-1 - 2\sqrt{3}}{2}$
b. $-3 + i\sqrt{2}, -3 - i\sqrt{2}$ **c.** $-\frac{4}{3}, \frac{10}{3}$
4, p. 727: $3 + 2\sqrt{6}, 3 - 2\sqrt{6}$ **5, p. 728: a.** 9 **b.** $\frac{1}{4}$
6, p. 729: $\frac{-3 + \sqrt{13}}{2}, \frac{-3 - \sqrt{13}}{2}$ **7, p. 730:** $\frac{1 + i\sqrt{7}}{4}, \frac{1 - i\sqrt{7}}{4}$
8, p. 730: $-3 + \sqrt{11}, -3 - \sqrt{11}$ **9, p. 731:** 100

Exercises 9.1, *p. 732*

1. $-4, 4$ **3.** $-2\sqrt{6}, 2\sqrt{6}$ **5.** $-5i, 5i$ **7.** $-\sqrt{2}, \sqrt{2}$
9. $-6\sqrt{2}, 6\sqrt{2}$ **11.** $-\sqrt{5}, \sqrt{5}$ **13.** $-\frac{i\sqrt{6}}{3}, \frac{i\sqrt{6}}{3}$
15. $1 + 4\sqrt{3}, 1 - 4\sqrt{3}$ **17.** $\frac{-5 + 5\sqrt{3}}{2}, \frac{-5 - 5\sqrt{3}}{2}$
19. $\frac{3 + 2i\sqrt{2}}{4}, \frac{3 - 2i\sqrt{2}}{4}$ **21.** $-\frac{25}{4}, -\frac{7}{4}$
23. $3 + 4\sqrt{5}, 3 - 4\sqrt{5}$ **25.** $\frac{-3 + 4\sqrt{2}}{2}, \frac{-3 - 4\sqrt{2}}{2}$
27. $-\frac{4i\sqrt{3}}{3}, \frac{4i\sqrt{3}}{3}$ **29.** $\frac{1 + 3\sqrt{2}}{2}, \frac{1 - 3\sqrt{2}}{2}$ **31.** $r = \frac{\sqrt{\pi V h}}{\pi h}$
33. $v = \pm\frac{\sqrt{Frm}}{m}$ **35.** 36 **37.** $\frac{49}{4}$ **39.** $\frac{4}{9}$ **41.** $0, 8$
43. $-1, 4$ **45.** $-2 + \sqrt{6}, -2 - \sqrt{6}$ **47.** -2
49. $5 + 2i, 5 - 2i$ **51.** $2 + 2i\sqrt{2}, 2 - 2i\sqrt{2}$
53. $2 + 4\sqrt{2}, 2 - 4\sqrt{2}$ **55.** $\frac{-5 + \sqrt{37}}{2}, \frac{-5 - \sqrt{37}}{2}$
57. $\frac{3 + i\sqrt{11}}{2}, \frac{3 - i\sqrt{11}}{2}$ **59.** $\frac{2 + 2\sqrt{10}}{3}, \frac{2 - 2\sqrt{10}}{3}$ **61.** $-2, -\frac{3}{4}$
63. $\frac{-1 + 3i\sqrt{7}}{4}, \frac{-1 - 3i\sqrt{7}}{4}$ **65.** $2 + \sqrt{7}, 2 - \sqrt{7}$
67. $-1 + \sqrt{3}, -1 - \sqrt{3}$ **69.** $r = \frac{\sqrt{3V\pi h}}{\pi h}$
71. $3 + \sqrt{7}, 3 - \sqrt{7}$ **73.** $-4 + 3\sqrt{2}, -4 - 3\sqrt{2}$
75. $5 + 3\sqrt{5}, 5 - 3\sqrt{5}$ **77.** The team is searching 5 mi from the last known location of the hikers.
79. The interest rate is 5%.   **81. a.** The sandbag will be 1000 ft above the ground approximately 7.9 sec after it is dropped. **b.** No; the sandbag will reach the ground about 11.2 sec after it is dropped, which is not equal to 2(7.9) sec, or 15.8 sec.   **83. a.** The patio is 18 ft by 18 ft.
b. It will cost $508.30 to enclose the patio.   **85.** The average speed of the truck driving north is 48 mph, and the average speed of the truck driving east is 64 mph.

Practices: Section 9.2, *pp. 739–743*

1, p. 739: $\frac{-3 + \sqrt{29}}{2}, \frac{-3 - \sqrt{29}}{2}$ **2, p. 739:** $\frac{1 + i\sqrt{5}}{6}, \frac{1 - i\sqrt{5}}{6}$
3, p. 740: $-2.820, 5.320$ **4, p. 740:** -9
5, p. 743: a. $1, 6, 9, 0, 1,$ Real number **b.** $1, 6, -9, 72, 2,$ Real numbers   **c.** $4, 2, 3, -44, 2,$ Complex numbers containing i **6, p. 743:** The value of the index was approximately 1000 in 1998.

Exercises 9.2, p. 744

1. quadratic formula   **3.** one   **5.** $-2, -1$
7. $3 + \sqrt{10}, 3 - \sqrt{10}$ **9.** $\frac{1 + 3\sqrt{5}}{2}, \frac{1 - 3\sqrt{5}}{2}$
11. $2 + i, 2 - i$ **13.** $\frac{-3 + \sqrt{30}}{3}, \frac{-3 - \sqrt{30}}{3}$ **15.** $\frac{1}{2}, \frac{4}{3}$
17. $\frac{-4 + i\sqrt{2}}{2}, \frac{-4 - i\sqrt{2}}{2}$ **19.** $\frac{-1 + \sqrt{2}}{3}, \frac{-1 - \sqrt{2}}{3}$ **21.** $\frac{3}{2}$
23. $1 + i\sqrt{11}, 1 - i\sqrt{11}$ **25.** $\frac{-2 + \sqrt{19}}{3}, \frac{-2 - \sqrt{19}}{3}$
27. $-4, -1$ **29.** $2 + 2i, 2 - 2i$
31. $-4 + 2\sqrt{3}, -4 - 2\sqrt{3}$ **33.** $\frac{5 + 2\sqrt{6}}{2}, \frac{5 - 2\sqrt{6}}{2}$
35. $-0.036, 1.965$ **37.** $-71.421, -5.246$
39. Two complex solutions (containing i)
41. One real solution   **43.** Two real solutions   **45.** Two
complex solutions (containing i)   **47.** two complex solutions
(containing i)   **49.** Two complex solutions (containing i)
51. $\frac{2 + i\sqrt{6}}{2}, \frac{2 - i\sqrt{6}}{2}$ **53.** $\frac{4 + \sqrt{22}}{3}, \frac{4 - \sqrt{22}}{3}$ **55.** The stone is
300 ft above the ground 5 sec after it is thrown downward.
57. Squares measuring 2.5 in. by 2.5 in. should be cut from
each corner.   **59.** The company must sell 16 tickets to generate revenue of $3520.   **61.** There were approximately
215,000 reports of identity theft in the year 2003.

Practices: Section 9.3, pp. 748–753

1, p. 748: $-1, 0$ **2, p. 749:** $\pm 1, \pm 2\sqrt{2}$ **3, p. 750:** 25
4, p. 751: $-64, 125$ **5, p. 751:** Working alone, it would
take the slower painter about 10.5 hr and his partner about
9.5 hr to paint the apartment.   **6, p. 753:** $y^2 + 3y = 0$
7, p. 753: $24n^2 - 22n + 3 = 0$

Exercises 9.3, p. 754

1. quadratic   **3.** in terms of a new variable   **5.** $\frac{1}{2}, 1$
7. 2, 10   **9.** $\pm\sqrt{2}, \pm\sqrt{6}$ **11.** $\pm\frac{1}{2}, \pm i\sqrt{3}$
13. 4   **15.** 256   **17.** $-8, -1$ **19.** 5, 6   **21.** 6
23. $x^2 - 9x + 14 = 0$ **25.** $y^2 - 5y - 36 = 0$
27. $3t^2 + 4t - 4 = 0$ **29.** $6n^2 + 17n + 12 = 0$
31. $x^2 - 8x + 16 = 0$ **33.** $t^2 - 2 = 0$ **35.** $x^2 + 9 = 0$
37. $-1, 8$ **39.** $\pm 3i, \pm\sqrt{2}$ **41.** $5a^2 - 13a - 6 = 0$
43. The person's height is approximately 1.6 m.
45. She drives to work at an average speed of about 35 mph
and she drives home at an average speed of about 27 mph.
47. If only one drain is open, it takes the large drain 3.4 hr
and the small drain 4.9 hr to empty the pool.

Practices: Section 9.4, pp. 758–767

1, p. 758:

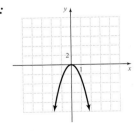

2, p. 759:

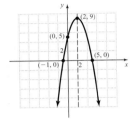

3, p. 760: **a.** $(2, 9); x = 2$ **b.** Opens downward **c.** $(-1, 0)$
and $(5, 0); (0, 5)$ **d.**

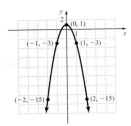

4, p. 762: **a.** Opens upward and has a minimum point; two
x-intercepts and one y-intercept **b.** Opens downward and
has a maximum point; no x-intercepts and one y-intercept
c. Opens downward and has a maximum point;
one x-intercept and one y-intercept
5, p. 763: **a.** $\left(-\frac{3}{2}, -\frac{7}{4}\right)$ **b.** Opens downward
c.

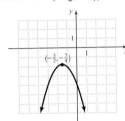

d. Domain: $(-\infty, \infty)$;
range: $\left(-\infty, -\frac{7}{4}\right]$

6, p. 764: **a.** $l = 100 - w$ **b.** $A(w) = 100w - w^2$
c.

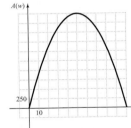

d. A garden measuring 50 ft
by 50 ft will maximize the area
of the garden. The maximum
area is 2500 ft².

7, p. 767: **a.**

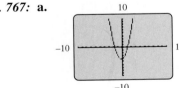

b. $(-0.25, -4.13)$
c. x-intercepts:
$(-1.69, 0)$ and
$(1.19, 0)$;
y-intercept: $(0, -4)$

Exercises 9.4, p. 770

1. quadratic   **3.** vertex   **5.** upward   **7.** two

9. 8, (−2, 8); 2, (−1, 2); 0, (0, 0);
2, (1, 2); 8, (2, 8)

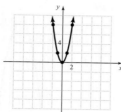

11. 8, (−4, 8); 2, (−2, 2); 0, (0, 0);
2, (2, 2); 8, (4, 8)

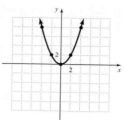

13. −7, (−3, −7); −2, (−2, −2);
1, (−1, 1); 2, (0, 2); 1, (1, 1);
−2, (2, −2); −7, (3, −7)

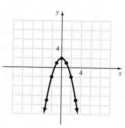

15. (4, −16); x = 4; (0, 0) and
(8, 0); (0, 0)

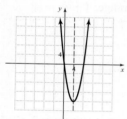

17. (1, −4); x = 1; (−1, 0) and
(3, 0); (0, −3)

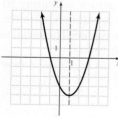

19. (2, 16); x = 2; (−2, 0) and
(6, 0); (0, 12)

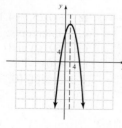

21. (0, −1); x = 0; (−1, 0) and
(1, 0); (0, −1)

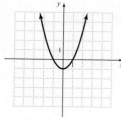

23. (−1, 0); x = −1; (−1, 0); (0, 1)

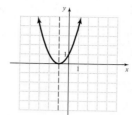

25. (3, 0); x = 3; (3, 0); (0, −9)

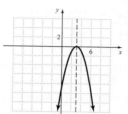

27. $\left(\frac{3}{2}, -\frac{49}{4}\right)$; $x = \frac{3}{2}$; (−2, 0) and
(5, 0); (0, −10)

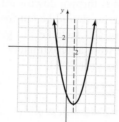

29. Upward, Minimum, 0, 1; Upward, Minimum, 1, 1;
Downward, Maximum, 2, 1; Downward, Maximum, 0, 1;
Upward, Minimum, 2, 1

31. (−∞, ∞), [−1, ∞)

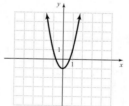

33. (−∞, ∞), (−∞, 1]

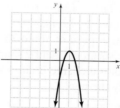

35. (−∞, ∞), [2, ∞)

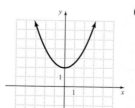

37. 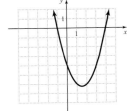 $(-\infty, \infty), \left[-\frac{25}{4}, \infty\right)$

39. 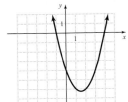 $(-\infty, \infty), \left(-\infty, -\frac{5}{2}\right]$

41. $(-0.10, -1.01); (-1.10, 0), (0.90, 0); (0, -1)$
43. $(-3.33, 2.17); (-7.13, 0), (0.47, 0); (0, 0.5)$
45. $(-0.30, 6.55)$; None; $(0, 7)$ **47.** $6, (-3, 6); 1, (-2, 1);$
$-2, (-1, -2); -3, (0, -3); -2, (1, -2); 1, (2, 1); 6, (3, 6)$

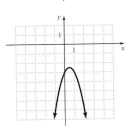

49. 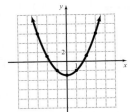 $(-2, 9); x = -2; (-5, 0)$ and $(1, 0); (0, 5)$

51. 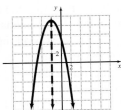 $(-\infty, \infty), (-\infty, 2]$

53. a. $\left(\frac{3}{2}, 316\right)$ **b.**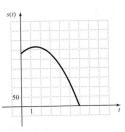

c. At $\frac{3}{2}$ sec (or 1.5 sec) after it is thrown, the stone reaches its maximum height of 316 ft.
55. a. $l = 75 - w$ **b.** $A(w) = 75w - w^2$

c. 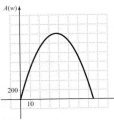 **d.** An exercise yard measuring 37.5 ft by 37.5 ft will produce a maximum area of 1406.25 ft².

57. a. 100 units must be produced in order to minimize the cost. **b.** The minimum daily cost is $50.

Practices: Section 9.5, *pp. 782–786*

1, *p. 782:* $(-\infty, -1] \cup [4, \infty)$

2, *p. 782:* $(1, 5)$

3, *p. 784:* $(-5, 5)$

4, *p. 785:* $(-\infty, -3) \cup [-\frac{1}{2}, \infty)$

5, *p. 786:* The stand can run for six weeks or less.

Exercises 9.5, *p. 787*

1. test value **3.** rational inequality
5. $(-\infty, 0) \cup (1, \infty)$

7. $(-2, 2)$

9. $[-3, 4]$

11. $(-\infty, \infty)$

13. $(-\infty, -2] \cup [3, \infty)$

15. $(-4, -1)$

17. $(-\infty, \frac{1}{2}] \cup [1, \infty)$

19. $(-\infty, -\frac{1}{2}] \cup [\frac{1}{2}, \infty)$

21. $(-\frac{1}{3}, \frac{3}{2})$

23. $(-2, 0]$

25. $(6, \infty)$

27. $(3, 9)$

29. $(-\infty, -4) \cup [\frac{1}{2}, \infty)$

31. $(-\infty, -\frac{11}{2}] \cup (-\frac{5}{2}, \infty)$

33. $(-4, 3)$

35. $(-\infty, 4) \cup (10, \infty)$

37. $(-\infty, -3] \cup [\frac{1}{2}, \infty)$

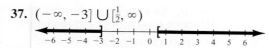

39. The penny is above the point of release between and excluding 0 sec and 3 sec.   **41.** Producing between and excluding 10 and 20 end tables per week will keep the cost under $500.   **43.** The area will exceed 450 ft² for any length between and excluding 15 ft and 30 ft.   **45.** The average cost will be less than $8 if more than 144 units are sold.

Review Exercises: Chapter 9, *p. 795*

1. $-9, 9$ **2.** $-\sqrt{5}, \sqrt{5}$ **3.** $\frac{9}{2}, \frac{11}{2}$ **4.** $\frac{-1 + 2i}{2}, \frac{-1 - 2i}{2}$
5. $-4 + \sqrt{2}, -4 - \sqrt{2}$ **6.** $2i\sqrt{2}, -2i\sqrt{2}$
7. 25 **8.** $\frac{81}{4}$ **9.** $3 + \sqrt{7}, 3 - \sqrt{7}$
10. $\frac{-1 + \sqrt{13}}{2}, \frac{-1 - \sqrt{13}}{2}$ **11.** $\frac{-3 + i\sqrt{3}}{2}, \frac{-3 - i\sqrt{3}}{2}$
12. $\frac{1 + 2\sqrt{7}}{3}, \frac{1 - 2\sqrt{7}}{3}$ **13.** $-1, -6$ **14.** $2 + i, 2 - i$
15. $\frac{3 + 2\sqrt{6}}{3}, \frac{3 - 2\sqrt{6}}{3}$ **16.** $-\frac{3}{2}$ **17.** $\frac{-9 + \sqrt{33}}{4}, \frac{-9 - \sqrt{33}}{4}$
18. $-5 + 3i, -5 - 3i$ **19.** Two complex solutions (containing i)   **20.** Two real solutions   **21.** $-1, 2$
22. $\pm 2i, \pm\sqrt{6}$ **23.** $9, 16$ **24.** $-1, 0$
25. $2x^2 + x - 15 = 0$ **26.** $n^2 + 14n + 49 = 0$
27.

(3, −4); x = 3; (1, 0) and (5, 0); (0, 5)

28.

(1, 4); x = 1; (−1, 0) and (3, 0); (0, 3)

29.

(−4, 0); x = −4; (−4, 0); (0, 16)

30.

$(\frac{5}{2}, -\frac{49}{4})$; $x = \frac{5}{2}$; (−1, 0) and (6, 0); (0, −6)

31. Upward, Minimum, None, One

32. Downward, Maximum, Two, One
33.

$(-\infty, \infty), (-\infty, \frac{9}{16}]$

34.

$(-\infty, \infty), [-\frac{9}{8}, \infty)$

35. $(-2, 6)$
36. $(-\infty, -2] \cup [5, \infty)$
37. $(-\infty, \frac{1}{2}] \cup [4, \infty)$
38. $(-\frac{1}{2}, \frac{3}{2})$
39. $(-\infty, 3) \cup (5, \infty)$
40. $(0, 3]$

41. It will take 5 sec for the object to fall 400 ft.
42. The ladder reaches approximately 9.5 ft up the side of the house.   **43.** The speed of the train traveling south is 53 mph and the speed of the train traveling west is 45 mph.
44. The average rate of return was about 10.5%.
45. $\frac{5}{2}$-in. by $\frac{5}{2}$-in. squares should be cut from each corner.   **46.** He needs to sell 25 cases per week.
47. There were 218 stores in 2002.   **48.** There were approximately 1.2 million sales associates in 2001.
49. She bicycled to school at a rate of 13.7 mph that day.
50. Working alone, it takes the newer printing press about 2.7 hr and the older printing press about 7.7 hr to complete the job.   **51. a.** $w = 150 - l$ **b.** $A(l) = 150l - l^2$
c.

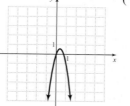

d. A park measuring 75 ft by 75 ft will produce a maximum area of 5625 ft².

52. a.

b. 20 units must be fabricated per day in order to minimize the cost. **c.** The minimum cost is $80.

53. The baseball is 40 ft above the ground at $\frac{1}{2}$ sec and $\frac{9}{2}$ sec.   **54.** Any width from 10 ft and 12 ft will produce an area of at least 120 ft².

Posttest: Chapter 9, *p. 800*

1. $-2\sqrt{2}, 2\sqrt{2}$ **2.** $1 + i\sqrt{7}, 1 - i\sqrt{7}$ **3.** $-\frac{3}{2}, \frac{1}{2}$
4. Two complex solutions (containing i)
5. $\frac{-2 + 3\sqrt{2}}{3}, \frac{-2 - 3\sqrt{2}}{3}$ **6.** $-4 + \sqrt{22}, -4 - \sqrt{22}$
7. $\frac{5 + i\sqrt{19}}{2}, \frac{5 - i\sqrt{19}}{2}$ **8.** $-1, \frac{7}{4}$ **9.** $\frac{6 + i\sqrt{2}}{2}, \frac{6 - i\sqrt{2}}{2}$
10. $-1 + \sqrt{5}, -1 - \sqrt{5}$ **11.** 1   **12.** $6n^2 - 7n + 2 = 0$
13. $(3, -1)$; $x = 3$; $(2, 0)$
and $(4, 0)$; $(0, 8)$

14. $(\frac{3}{2}, \frac{49}{4})$; $x = \frac{3}{2}$; $(-2, 0)$
and $(5, 0)$; $(0, 10)$

15. 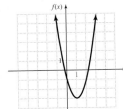   Domain: $(-\infty, \infty)$;
Range: $[-3, \infty)$

16. $(-\infty, -6) \cup (3, \infty)$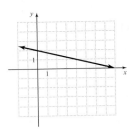

17. The average speed of the airplane flying west is approximately 470 mph and the average speed of the airplane flying south is approximately 450 mph.   **18.** Working alone, the senior clerk can process the applications in 3 hr and the junior clerk can process the applications in 6 hr.
19. The rocket will reach its maximum height of 147 ft in 3 sec.   **20.** The factory can produce more than 80 but fewer than 240 units each day.

Cumulative Review: Chapter 9, *p. 802*

1. $2n^2$ **2.** $-\frac{3}{2}, \frac{9}{2}$ **3.**

4. Yes, the relation represents a function.   **5.** $a^3 - 8$
6. $2x^2y^2(y - 2x)(y - 4x)$ **7.** $\frac{1 + i\sqrt{5}}{6}, \frac{1 - i\sqrt{5}}{6}$
8. $10\sqrt{3} - 9\sqrt{2}$
9. a. $2l + 2w < 1000$ **b.**

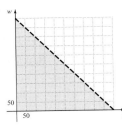

c. 300 ft by 150 ft   **10. a.** $B = 0.05m + 7.5$ **b.** The slope is 0.05. It represents the per-minute fee for long-distance calls.   **c.** The y-intercept is $(0, 7.5)$. It represents the flat monthly fee that the phone service charges.

Chapter 10

Pretest: Chapter 10, *p. 804*

1. a. $2x^2 + 7x - 4$ **b.** $-2x^2 - 3x + 2$ **c.** $4x^3 + 8x^2 - 11x + 3$ **d.** $\frac{1}{x + 3}, x \neq -3, \frac{1}{2}$ **2. a.** $\frac{5}{3x - 4}, x \neq \frac{4}{3}$
b. $\frac{15}{x} - 4, x \neq 0$ **c.** 1 **d.** -7
3. One-to-one

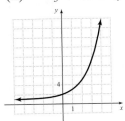

4. $f^{-1}(x) = \frac{x + 7}{3}$ **5. a.** $\frac{1}{4}$ **b.** -7.389 **6. a.** 0 **b.** -2
7. a.   **b.**

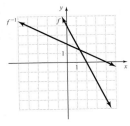

8. a. 5 **b.** 0   **9. a.** -1 **b.** 5   **10.** 1.8928
11. a. $1 + 2\log_5 x$ **b.** $\log_8 3 + 3\log_8 x - \log_8 y$
12. a. $\log_7 40$ **b.** $\log_6 \frac{x^4}{(x + 2)^2}$ **13.** 2   **14.** $\frac{2}{3}$ **15.** 2
16. 12   **17.** 2   **18.** A gallon of milk will cost about $4.73 in 5 yr.   **19.** The effective annual interest rate is approximately 4.6%.   **20.** It will take about 10.1 days for 100 mg of gallium-67 to decay to 12 mg.

Practices: Section 10.1, *pp. 806–818*

1, *p. 806*   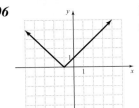   Domain: $(-\infty, \infty)$;
range: $[0, \infty)$

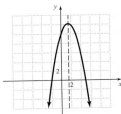
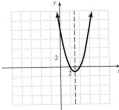

2, p. 807: **a.** $(f + g)(x) = x^2 + 3x - 3$ **b.** $(f - g)(x) = x^2 - 3x + 1$ **c.** $(f \cdot g)(x) = 3x^3 - 2x^2 - 3x + 2$
d. $(\frac{f}{g})(x) = \frac{x^2 - 1}{3x - 2}, x \neq \frac{2}{3}$ **3, p. 809:** **a.** 1 **b.** -35 **c.** 0
d. $-\frac{2}{9}$ **4, p. 810:** **a.** $-4x + 17$ **b.** $-4x + 8$ **c.** 5 **d.** -4
5, p. 813: **a.** One-to-one function **b.** Not a one-to-one function **6, p. 814:** $\{(-1, -3), (0, -1), (2, 0), (3, 4)\}$
7, p. 815: $g^{-1}(x) = \frac{x + 1}{3}$ **8, p. 816:** p is not the inverse of q.
9, p. 817:

a. 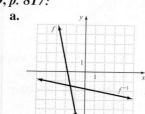 **b.**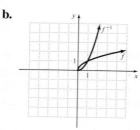

10, p. 818: **a.** C is a one-to-one function since its graph is linear and passes the horizontal-line test. **b.** $C^{-1}(x) = \frac{1}{40}x$ **c.** The inverse can be used to determine the number of hours that the van was rented based on the amount of money charged.

Exercises 10.1, p. 819

1. absolute value function **3.** product of f and g **5.** one-to-one **7.** inverse functions of each other
9. a. 3 **b.** 1 **c.** $|-2t + 3|$ **d.** $|\frac{1}{2}t|$
11. 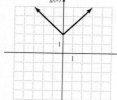 Domain: $(-\infty, \infty)$; range: $[2, \infty)$

13. 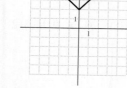 Domain: $(-\infty, \infty)$; range: $[0, \infty)$

15. $(f + g)(x) = 4x^2 + 2x - 6;$
$(f - g)(x) = -4x^2 + 2x + 12;$
$(f \cdot g)(x) = 8x^3 + 12x^2 - 18x - 27;$
$(\frac{f}{g})(x) = \frac{1}{2x - 3}, x \neq -\frac{3}{2}, \frac{3}{2}$
17. $(f + g)(x) = 5x^2 + 16x + 3;$
$(f - g)(x) = x^2 + 6x + 9;$
$(f \cdot g)(x) = 6x^4 + 37x^3 + 58x^2 - 3x - 18;$
$(\frac{f}{g})(x) = \frac{3x + 2}{2x - 1}, x \neq -3, \frac{1}{2}$

19. $(f + g)(x) = \frac{x^2 + 2x - 2}{x + 3}, x \neq -3;$
$(f - g)(x) = \frac{-x^2 - 2x + 4}{x + 3}, x \neq -3;$
$(f \cdot g)(x) = \frac{x - 1}{x + 3}, x \neq -3;$
$(\frac{f}{g})(x) = \frac{1}{(x + 3)(x - 1)}, x \neq -3, 1$
21. $(f + g)(x) = 6\sqrt{x} - 1; (f - g)(x) = -2\sqrt{x} + 1;$
$(f \cdot g)(x) = 8x - 2\sqrt{x}; (\frac{f}{g})(x) = \frac{8x + 2\sqrt{x}}{16x - 1}, x \neq \frac{1}{16}$
23. -12 **25.** 0 **27.** 18 **29.** -4 **31. a.** $x - 14$ **b.** $x - 7$ **c.** -11 **d.** -8 **33. a.** $x^2 + 4x - 11$ **b.** $x^2 + 2x - 13$ **c.** -6 **d.** -14 **35. a.** $\frac{3}{2x + 5}, x \neq -\frac{5}{2}$ **b.** $\frac{6}{x} + 5, x \neq 0$ **c.** $\frac{1}{2}$ **d.** 2 **37. a.** $-\sqrt{1 - 4x}, x \leq \frac{1}{4}$ **b.** $1 + 4\sqrt{x}, x \geq 0$ **c.** -5 **d.** 17 **39.** One-to-one **41.** One-to-one **43.** Not one-to-one **45.** One-to-one **47.** c **49.** a **51.** $\{(5, -4), (3, -2), (1, 0), (-1, 2), (-3, 4)\}$ **53.** $\{(-3, -27), (-2, -8), (-1, -1), (0, 0), (1, 1), (2, 8), (3, 27)\}$
55. $f^{-1}(x) = \frac{1}{4}x$ **57.** $g^{-1}(x) = 4x$ **59.** $f^{-1}(x) = -x - 5$
61. $f^{-1}(x) = \frac{x - 2}{5}$ **63.** $g^{-1}(x) = 2x + 6$
65. $f^{-1}(x) = \sqrt[3]{x} + 4$ **67.** $h^{-1}(x) = \frac{3 - 4x}{x}$
69. $f^{-1}(x) = -\frac{x}{1 - 2x}$, or $\frac{x}{2x - 1}$ **71.** $f^{-1}(x) = x^3 - 4$
73. Inverses **75.** Inverses **77.** Not inverses
79. Inverses **81.** Not inverses **83.** Inverses
85.  **87.**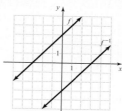

89. Inverses **91.** -43 **93. a.** $-\frac{8}{x + 2} + 7, x \neq -2$ **b.** $\frac{4}{-2x + 9}, x \neq \frac{9}{2}$ **c.** 6 **d.** $\frac{4}{9}$ **95.**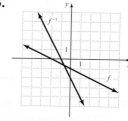

97. $f^{-1}(x) = \frac{3x}{3 - 2x}$ **99. a.** $(C \circ x)(d) = 2100d + 1000$; it represents the plant's weekly cost for d days of operation. **b.** The weekly cost is $11,500. **101. a.** $S^{-1}(n) = \frac{n + 360}{180}$ **b.** The inverse can be used to calculate the number of sides of a polygon if the sum of the interior angles is known. **c.** The polygon has five sides. **103. a.** $B(t) = 0.05t + 4.5$ **b.** $B^{-1}(t) = 20t - 90$ **c.** The inverse can be used to determine the number of minutes of long-distance calling time a customer was billed for.

Practices: Section 10.2, pp. 829–834

1, p. 829: **a.** 64 **b.** 9 **2, p. 829:** **a.** 32 **b.** $\frac{1}{128}$
3, p. 830: **a.** 0.007 **b.** 7.389

4, *p. 830:*

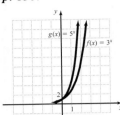

5, *p. 831:*

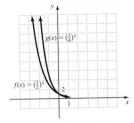

6, *p. 832:*

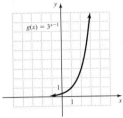

7, *p. 833:* **a.** 6 **b.** $\frac{3}{2}$ **8,** *p. 834:* Approximately 80 mg will remain after 3 days.

Exercises 10.2, *p. 835*

1. exponential function   **3.** natural exponential function
5. decreasing   **7.** Polynomial function   **9.** Radical function   **11.** Exponential function   **13.** Polynomial function
15. a. $\frac{1}{8}$ **b.** 1 **c.** 16   **17. a.** 3 **b.** $\frac{1}{3}$ **c.** $\frac{1}{81}$ **19. a.** $\frac{1}{27}$ **b.** $\frac{1}{3}$
c. 9 **21. a.** $\frac{23}{8}$ **b.** 2 **c.** -13 **23. a.** -0.007 **b.** -2.718
25. a. 0.018 **b.** 2.718   **27. a.** 0.050 **b.** 0.135

29.

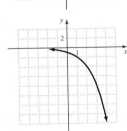

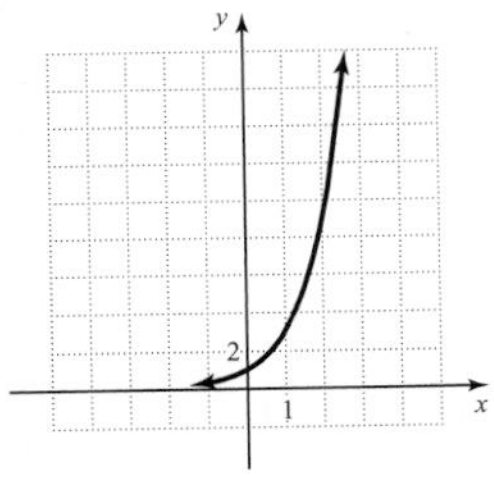

31.

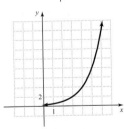

33.

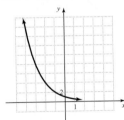

35.

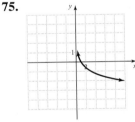

37.

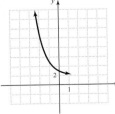

39. 2   **41.** -5 **43.** -2 **45.** $\frac{4}{3}$ **47.** 2   **49.** 4   **51.** $\frac{1}{4}$
53. $\frac{9}{2}$ **55.** $-\frac{10}{3}$ **57.** 9   **59.** $x = \frac{11}{2}$ **61.** radical function
63. $-\frac{1}{25}$ **65.** 0.135   **67.** $30,284.80   **69. a.** The concentration will be approximately 21.6 parts per million after 1 yr. **b.** The initial concentration of pollutants was 40 parts per million.   **71.** The amount in the account after 5 yr

will be $10,790.80.   **73.** The concentration after 8 hr is approximately 18 milligram per liter.   **75.** 1280 bacteria will be present in 3.5 hr.

Practices: Section 10.3, *pp. 841–846*

1, *p. 841:* **a.** $8^2 = 64$ **b.** $5^{-1} = \frac{1}{5}$ **c.** $2^{1/3} = \sqrt[3]{2}$
2, *p. 842:* **a.** $\log_2 32 = 5$ **b.** $\log_7 1 = 0$ **c.** $\log_{10}\sqrt{10} = \frac{1}{2}$
3, *p. 842:* **a.** 1 **b.** 3 **c.** $\frac{1}{3}$ **4,** *p. 843:* **a.** 1 **b.** 0
5, *p. 843:* **a.** 64 **b.** $\frac{1}{2}$ **6,** *p. 843:* **a.** 5 **b.** 16   **7,** *p. 844:* $\frac{3}{2}$
8, *p. 845:*   **9,** *p. 845:*

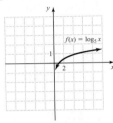   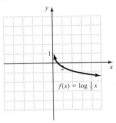

10, *p. 846:* The barometric pressure 9 mi from a hurricane's eye is about 28.1 in. of mercury.

Exercises 10.3, *p. 848*

1. logarithmic function   **3.** exponent   **5.** decreasing
7. $3^4 = 81$ **9.** $(\frac{1}{2})^5 = \frac{1}{32}$ **11.** $5^{-2} = \frac{1}{25}$ **13.** $(\frac{1}{4})^{-1} = 4$
15. $16^{1/4} = 2$ **17.** $10^{1/2} = \sqrt{10}$ **19.** $\log_3 243 = 5$
21. $\log_{1/4}\frac{1}{4} = 1$ **23.** $\log_2\frac{1}{16} = -4$ **25.** $\log_{1/3} 81 = -4$
27. $\log_{49} 7 = \frac{1}{2}$ **29.** $\log_{11}\sqrt[5]{11} = \frac{1}{5}$ **31.** 3   **33.** 2
35. 1   **37.** 4   **39.** -3 **41.** -2 **43.** 0   **45.** $\frac{1}{3}$ **47.** $\frac{1}{4}$
49. $-\frac{1}{2}$ **51.** 27   **53.** $\frac{1}{36}$ **55.** $\frac{3}{2}$ **57.** 2   **59.** 6   **61.** 8
63. $\frac{3}{4}$ **65.** $\frac{1}{7}$ **67.** $\frac{3}{2}$ **69.** $\frac{4}{3}$ **71.** -4

73.   **75.**

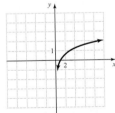   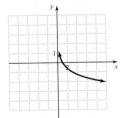

77.

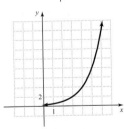

79. 125   **81.** $\frac{3}{2}$ **83.** -4 **85.** $(\frac{1}{2})^{-4} = 16$
87. $\log_{1/7} 49 = -2$ **89.** The power gain is 40 dB.
91. The population reaches 512 bacteria in 6 hr.   **93.** The amount of current is 8 amp.   **95.** It takes the computer 70 picoseconds to carry out 64 computations.   **97. a.** 6 ft
b. The base represents the area enclosed by the fence.

Practices: Section 10.4, *pp. 853–858*

1, *p. 853:* **a.** $\log_6 4 + \log_6 9$ **b.** $1 + \log_2 x$ **2,** *p. 854:*
a. $\log_2 135$ **b.** $\log_7(x^2 + xy)$ **3,** *p. 854:* **a.** $\log_4 3 - 1$

b. $\log_b u - \log_b v$ **4, p. 855: a.** $\log_7 \frac{1}{2}$ **b.** $\log_3 \frac{x-y}{x+y}$

5, p. 855: a. $3\log_2 x$ **b.** $\frac{1}{3}\log_5 x$ **6, p. 856: a.** 7   **b.** 29

7, p. 857: a. $\log_6 128$ **b.** $\log_2 \frac{u^3}{v^3}$ **8, p. 857: a.** $1 + 2\log_5 x$

b. $\frac{1}{2}\log_7 u - \frac{1}{2}\log_7 v$ **c.** $3\log_u b - \log_u a - 5\log_u c$

9, p. 858: a. 3 bacteria are present after 1 hr.   **b.** The number of bacteria in the culture increases by a factor of 3 each hour.

Exercises 10.4, p. 859

1. sum   **3.** times   **5.** power   **7.** $4 + \log_2 5$

9. $\log_6 7 - 2$ **11.** $\log_3 8 + \log_3 x$ **13.** $\log_n u + \log_n v$

15. $\log_b a + \log_b(b-1)$ **17.** $3 - \log_4 n$

19. $\log_7 v - \log_7 u$ **21.** $\log_a(b-a) - \log_a(b+a)$

23. $\log_2 35$ **25.** $\log_7 5$ **27.** $\log_x 6x$ **29.** $\log_b \frac{b}{16}$

31. $\log_5(2x-10)$ **33.** $\log_n xz$ **35.** $\log_5 \frac{k}{n}$ **37.** $\log_{10} \frac{a+b}{a-b}$

39. $2\log_6 8$ **41.** $4\log_3 x$ **43.** $\frac{1}{5}\log_a b$ **45.** $-\log_4 a$

47. 4   **49.** $\frac{1}{2}$ **51.** -3 **53.** 15   **55.** x **57.** 10

59. $\log_2 108$ **61.** $\log_7 \frac{1}{4}$ **63.** $\log_4 \frac{y^6}{x}$ **65.** $\log_b 8$

67. $\log_4 \frac{\sqrt{x}}{y}$ **69.** $\log_5(x^3 - x^2)$ **71.** $\log_6 \sqrt[3]{x-y}$

73. $\log_2 \frac{x^4}{y^5 z}$ **75.** $\log_5 \frac{\sqrt[4]{ac^3}}{b^8}$ **77.** $\log_6 3 + 2\log_6 y$

79. $3 + 2\log_x y$ **81.** $2 + 2\log_2 x + 5\log_2 y$

83. $\frac{1}{2}\log_8 10 + \frac{1}{2}\log_8 x$ **85.** $2\log_4 x - \log_4 y$

87. $\frac{3}{4}\log_7 u - \frac{1}{4}\log_7 v$ **89.** $2\log_c a + 4 - 3\log_c b$

91. $3\log_6 x - 2\log_6(x-y)$ **93.** $2\log_a x + \frac{3}{5}\log_a y + \frac{2}{5}\log_a z$

95. $\log_b \frac{8}{b}$ **97.** $\log_a \frac{1}{12}$ **99.** $\frac{1}{3}\log_p q$

101. $3 + 3\log_2 a + 2\log_2 b$ **103.** $3 + \log_4 a$

105. The magnitude of the earthquake was 8.3.

107. $S = 20\log_{10} P - 20\log_{10} P_0$

109. a. $\log_{10} C = \log_{10} C_0 + \log_{10}\left(\frac{1}{2}\right)^{0.2t}$

$\log_{10} C = \log_{10}\left[C_0\left(\frac{1}{2}\right)^{0.2t}\right]$

$C = 10^{\log_{10}[C_0(1/2)^{0.2t}]}$

$C = C_0\left(\frac{1}{2}\right)^{0.2t}$

b. 12.5 g will remain after 10 yr.

Practices: Section 10.5, pp. 863–866

1, p. 863: a. 0.7782   **b.** 1.4314   **2, p. 864: a.** 3   **b.** -1

c. $\frac{1}{3}$ **3, p. 864: a.** 1.3863   **b.** 3.2189   **4, p. 865: a.** 2

b. -7 **c.** $\frac{3}{4}$ **5, p. 866:** 0.6610   **6, p. 866:** It takes approximately 82 yr.

Exercises 10.5, p. 867

1. common logarithm   **3.** natural logarithm   **5.** 0.6021

7. 1.2553   **9.** -0.1761 **11.** 0.1139   **13.** 1.0986

15. 2.8332   **17.** -0.4700 **19.** 0.1823   **21.** 6   **23.** -3

25. -2 **27.** $\frac{3}{4}$ **29.** x **31.** 4   **33.** -1 **35.** $\frac{1}{2}$ **37.** b

39. 6   **41.** 3   **43.** 2.8074   **45.** 1.6992   **47.** -3.4594

49. -0.6309 **51.** 0.7959   **53.** -1.9386 **55.** 2   **57.** -3

59. -2.8074 **61.** 0.4314   **63.** The pH of household bleach is approximately 12.5.   **65.** The company's revenue was approximately \$3 million in the year 2005.

67. $kd = -\ln\frac{I}{I_0}$ $\Rightarrow$ $kd = \ln\frac{I_0}{I}$ $\Rightarrow$ $e^{kd} = \frac{I_0}{I}$ $\Rightarrow$

$Ie^{kd} = I_0$ $\Rightarrow$ $I = \frac{I_0}{e^{kd}}$ $\Rightarrow$ $I = I_0 e^{-kd}$

69. The population will grow to 60 bacteria in approximately 5.9 hr.

Practices: Section 10.6, pp. 870–873

1, p. 870: -1.0466 **2, p. 871:** 5   **3, p. 871:** 9

4, p. 872: $\frac{1}{2}$ **5, p. 873:** The annual rate of population growth was approximately 5.1%.

Exercises 10.6, p. 874

1. 2.6309   **3.** -3.1699 **5.** -0.2075 **7.** -3.3705

9. 0.7339   **11.** 0.8702   **13.** -0.6889 **15.** 2.6685

17. 4.5256   **19.** 71   **21.** 2   **23.** $-\frac{7}{8}$ **25.** $-11, 11$

27. $\frac{8}{3}$ **29.** 63   **31.** 4   **33.** 6   **35.** $\frac{3}{2}$ **37.** $\frac{9}{2}$ **39.** 3

41. 1   **43.** 4   **45.** $-5, 5$ **47.** $\frac{16}{3}$ **49.** -1.7297 **51.** The population will be 350 million people in the year 2030.

53. It will take approximately 14 yr for the amount in the account to double.   **55.** The concentration of hydrogen ions in blood is about 4.0×10^{-8}.   **57. a.** $x = 5\ln\frac{C}{300}$ **b.** The circulation reached 1,000,000 about 6 yr after it was launched.

59. a. The initial investment was \$12,000; the initial investment was the amount in the account when $t = 0$.   **b.** The annual interest rate is about 6.5%.   **61.** The company's revenue reached \$7 million 9 months after the action figures were on the market.

Review Exercises: Chapter 10, p. 881

1.

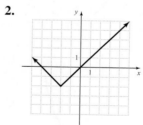

Domain: $(-\infty, \infty)$; range: $(-\infty, 5]$

2.

Domain: $(-\infty, \infty)$; range: $[-2, \infty)$

3. $(f + g)(x) = -5x + 2$, $(f - g)(x) = -7x + 8$, $(f \cdot g)(x) = -6x^2 + 23x - 15$, $\frac{f}{g}(x) = \frac{5-6x}{x-3}, x \neq 3$

4. $(f + g)(x) = 5x^2 - 4x + 1$, $(f - g)(x) = x^2 + 4x + 1$, $(f \cdot g)(x) = 6x^4 - 12x^3 + 2x^2 - 4x$, $\frac{f}{g}(x) = \frac{3x^2 + 1}{2x^2 - 4x}; x \neq 0, 2$

5. $(f + g)(x) = \frac{2x + 9}{(x+3)(x-3)}, x \neq -3, 3$;

$(f - g)(x) = \frac{2x + 3}{(x+3)(x-3)}, x \neq -3, 3$;

$(f \cdot g)(x) = \frac{6}{(x-3)(x^2-9)}, x \neq -3, 3$;

$\frac{f}{g}(x) = \frac{2(x+3)}{3}, x \neq -3, 3$

6. $(f + g)(x) = 8\sqrt{x}$, $(f - g)(x) = 6\sqrt{x} + 4$, $(f \cdot g)(x) = 7x - 12\sqrt{x} - 4$, $\frac{f}{g}(x) = \frac{7x + 16\sqrt{x} + 4}{x - 4}, x \neq 4$

7. 27 **8.** −10 **9.** 0 **10.** $\frac{9}{4}$ **11. a.** $x^2 − 5x + 11$
b. $x^2 − x + 3$ **c.** 11 **d.** 9 **12. a.** $\sqrt{2x − 9}$ **b.** $2\sqrt{x} − 9$
c. 3 **d.** $2\sqrt{5} − 9$ **13.** Not one-to-one **14.** One-to-one
15. $f^{-1}(x) = \frac{x − 3}{8}$ **16.** $f^{-1}(x) = 6x + 6$
17. $f^{-1}(x) = \sqrt[3]{x + 5}$ **18.** $f^{-1}(x) = x^3 − 10$
19. $f^{-1}(x) = \frac{2 − 3x}{x}$ **20.** $f^{-1}(x) = \frac{x}{1 − 3x}$ **21.** Inverses
22. Not inverses **23.**

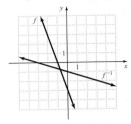

24.

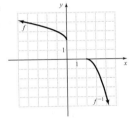

25. $-\frac{1}{27}$ **26.** 16 **27.** 4 **28.** 22 **29.** 2.7183 **30.** 0.1353
31.   **32.**

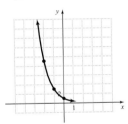 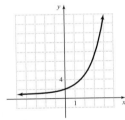

33. $\frac{5}{2}$ **34.** $-\frac{2}{3}$ **35.** 7 **36.** 1 **37.** $6^3 = 216$
38. $\left(\frac{1}{3}\right)^{-2} = 9$ **39.** $\log_5 \frac{1}{25} = −2$ **40.** $\log_{81} 3 = \frac{1}{4}$ **41.** 1
42. 5 **43.** 0 **44.** $\frac{1}{4}$ **45.** 64 **46.** $\frac{1}{64}$ **47.** $\frac{1}{7}$ **48.** $\frac{2}{3}$
49. 3 **50.** −2
51.   **52.**

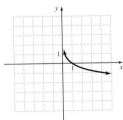 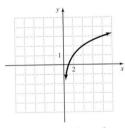

53. $3 + \log_3 x$ **54.** $1 − \log_6 a$ **55.** $\log_n x − \log_n (x − 5)$
56. $\log_9 u + \log_9 (u + 1)$ **57.** $3 \log_5 x + \log_5 y$
58. $\frac{2}{5} \log_2 x + \frac{3}{5} \log_2 y$ **59.** $2 \log_3 x + 3 \log_3 z − 2 \log_3 y$
60. $2 − 4 \log_b (a − b)$ **61.** $\log_4 (10n)$ **62.** $\log_9 \frac{x}{5}$
63. $\log_3 (ab^2)$ **64.** $\log_b \frac{a^5}{c^4}$ **65.** $\log_6 \frac{\sqrt[4]{x^3}}{y^2}$ **66.** $\log_a \frac{x^2}{y}$
67. 3 **68.** $-\frac{1}{3}$ **69.** y **70.** 14 **71.** 1.3617 **72.** 0.9731
73. 3.8712 **74.** −0.4055 **75.** −1 **76.** $\frac{3}{4}$ **77.** 100
78. $−y$ **79.** 2.9798 **80.** 0.8230 **81.** 2.1978
82. −0.1520 **83.** 9.3219 **84.** 0.3856 **85.** −7
86. 96 **87.** 9 **88.** −1 **89.** 4 **90.** 5
91. a. $(f \circ c)(s) = 0.008s + 2500$; it represents the agent's
total monthly salary if his monthly sales totaled s dollars.

b. $5700 **92. a.** $P^{-1}(x) = \frac{1}{10}x + 15$ **b.** The inverse can
be used to determine the number of jeans the company must
sell in order to make a certain profit. **c.** 135 pairs of jeans
93. a. $8484.33 **b.** It will take about 13 yr for this invest-
ment to double. **94. a.** 9 mg remain after 60 yr. **b.** The
half-life of cesium-137 is about 30 yr. **95.** 384 bacteria will
be in the colony in 4 hr. **96.** The stamp will have a value of
$24 in 22.5 yr.
97. $s = \log 10^{90} − \log(x + 1)^{16}$
$\quad\quad s = 90 \log 10 − 16 \log(x + 1)$
$\quad\quad s = 90 − 16 \log(x + 1)$
98. $\ln(T − A) − \ln(98.6 − A) = −kt \ln e$
$\quad\quad \ln \frac{T − A}{98.6 − A} = −kt$
$\quad\quad e^{-kt} = \frac{T − A}{98.6 − A}$
$\quad\quad (98.6 − A)e^{-kt} = T − A$
$\quad\quad T − A = (98.6 − A)e^{-kt}$
99. The pH of milk is about 6.6. **100.** It will take approximately 25 yr for the investment to triple. **101.** Its value
will depreciate to $15,000 about 4 yr after it is purchased.
102. It will take approximately 9 yr for the amount in the
account to grow to $12,300.

Posttest: Chapter 10, *p. 886*

1. a. $6x^2 − 2x − 4$ **b.** $−6x − 4$ **c.** $9x^4 − 6x^3 − 20x^2 − 8x$
d. $\frac{x − 2}{x}, x \neq −\frac{2}{3}, 0$ **2. a.** $\sqrt{2x + 5}$ **b.** $2\sqrt{x} + 5$ **c.** 3
d. 13 **3.** One-to-one

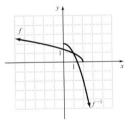

4. $f^{-1}(x) = \frac{7x + 1}{x}$ **5. a.** $-\frac{17}{9}$, or −1.8889 **b.** 1096.6332
6. a. 0 **b.** $-\frac{1}{2}$
7. a.   **b.**

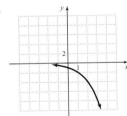 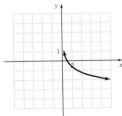

8. a. $2 + 4 \log_3 x$ **b.** $5 \log_6 x − 2 \log_6 y − 1$
9. a. $\log_5 32n^2$ **b.** $\log_b \frac{(a − 2)^3}{(a + 2)^4}$ **10. a.** 7 **b.** x **11. a.** $\frac{3}{2}$
b. −5 **12.** 2.3223 **13.** 2 **14.** 2 **15.** $\frac{1}{2}$ **16.** 11
17. 5 **18.** The population of India will be approximately
1140 million (or 1.14 billion) people. **19.** The intensity is
10^{-6} watts per square meter. **20.** It takes about 11 hr for the
concentration to decrease to 5 mg/L.

Cumulative Review: Chapter 10, *p. 888*

1. $\frac{9}{5}$ **2.** $\left(−\infty, −\frac{8}{5}\right) \cup (2, \infty)$ **3.** Point–slope form: $y − 1 =$
$−(x − 4)$; slope-intercept form: $y = −x + 5$
4. $(3xy + 2)(3xy − 2)(9x^2y^2 + 4)$ **5.** $\frac{−2 + \sqrt{14}}{2}, \frac{−2 − \sqrt{14}}{2}$

6. $24 - 32i$ **7.** $\frac{1}{5}$ **8.** The windchill temperature is approximately 31°F.   **9.** $5000 was invested in the low-risk fund and $10,000 was invested in the high-risk fund.
10. It takes the rocket 3 sec to reach its maximum height.

Chapter 11

Pretest: Chapter 11, *p. 890*

1. $y = 3(x - 2)^2 + 11$; $(2, 11)$ **2. a.** $2\sqrt{5}$ units
b. 0.5 units   **3. a.** $(4, 2)$ **b.** $\left(\frac{1}{2}, -\frac{3}{2}\right)$
4. $(x - 7)^2 + (y + 3)^2 = 10$ **5.** $(1, 0)$; $\sqrt{6}$
6.

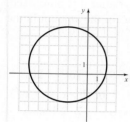

7.

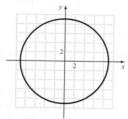

8.

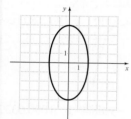

9.

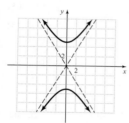

10.

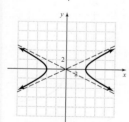

11.

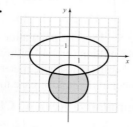

12.

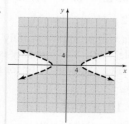

13. $(-3, 0), (0, 3)$

14. $(-i\sqrt{2}, -4), (i\sqrt{2}, -4)$
15. $(-\sqrt{6}, -1), (-\sqrt{6}, 1), (\sqrt{6}, -1), (\sqrt{6}, 1)$

16.

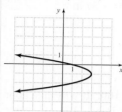

17.

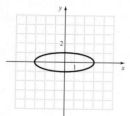

18. $x = \frac{8}{225}y^2$ **19.** The minimum distance is about 50 million km.

20. a. $x^2 + y^2 \le 1000^2$, or $x^2 + y^2 \le 1{,}000{,}000$
b.

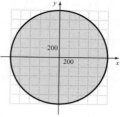

Practices: Section 11.1, *pp. 894–899*

1, *p. 894:*

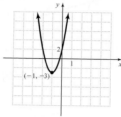

2, *p. 896:* a. $y = 4(x - 1)^2 + 3$
b. Vertex: $(1, 3)$;
axis of symmetry: $x = 1$ **c.**

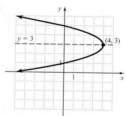

3, *p. 898:*

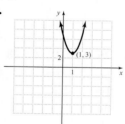

4, *p. 899:* $x = -5(y + 2)^2 + 11$; $(11, -2)$
5, *p. 899:* a. $h = -16(t - 2)^2 + 64$ **b.** 48 ft

Exercises 11.1, *p. 901*

1. Upward or downward   **3.** To the left or to the right
5. Vertex: $(3, -4)$;
axis of symmetry: $x = 3$

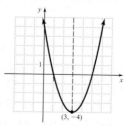

7. Vertex: $(-2, -5)$;
axis of symmetry: $x = -2$

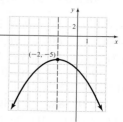

9. Vertex: $(0, 1)$;
axis of symmetry: $y = 1$

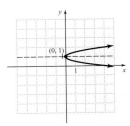

11. Vertex: $(3, 1)$;
axis of symmetry: $y = 1$

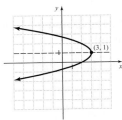

13. Vertex: $(3, 8)$;
axis of symmetry: $x = 3$

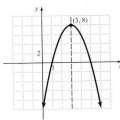

15. Vertex: $(-2, -2)$;
axis of symmetry: $y = -2$

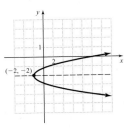

17. $y = \left(x - \frac{1}{2}\right)^2 + \frac{3}{4}; \left(\frac{1}{2}, \frac{3}{4}\right)$
19. $y = 5(x - 5)^2 - 68; (5, -68)$
21. $x = -2(y + 8)^2 + 33; (33, -8)$
23. $x = 3\left(y + \frac{3}{2}\right)^2 + \frac{17}{4}; \left(\frac{17}{4}, -\frac{3}{2}\right)$ **25.** $y = \frac{1}{5}(x + 2)^2 + 7$
27. $x = -4y^2 + 4$ **29.** $y = \left(x + \frac{5}{2}\right)^2 - \frac{17}{4}; \left(-\frac{5}{2}, -\frac{17}{4}\right)$

31. $y = \frac{1}{3}(x - 2)^2 - 5$ **33.** Vertex: $(3, -1)$ Axis of
symmetry: $y = -1$

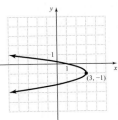

35. $(40, 1600)$; it shows that the company will have maximum revenue of $1600 when 40 bottles of perfume are sold.
37. a. $A = -(w - 125)^2 + 15{,}625$ **b.** Dimensions of 125 ft by 125 ft will produce a field with a maximum area of 15,625 ft². **39.** $x = \frac{3}{200}y^2$

Practices: Section 11.2, *pp. 907–911*

1, *p. 907:* $\sqrt{10} \approx 3.2$ units   **2,** *p. 908:* $\left(5, \frac{5}{2}\right)$

3, *p. 909:*

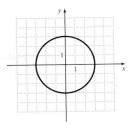

4, *p. 910:* **a.** Center: $(-6, 2)$; radius: 10
b.

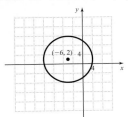

5, *p. 910:* $(x - 7)^2 + (y + 1)^2 = 8^2$, or
$(x - 7)^2 + (y + 1)^2 = 64$ **6,** *p. 911:* Center: $(2, 1)$;
radius: $2\sqrt{6}$ **7,** *p. 911:* **a.** $(x - 30)^2 + (y - 40)^2 = 10^2$, or
$(x - 30)^2 + (y - 40)^2 = 100$
b. No, since $\sqrt{509} \approx 22.6 > 10$.

Exercises 11.2, *p. 913*

1. the sum of   **3.** circle   **5.** centered at   **7.** 10 units
9. $2\sqrt{5} \approx 4.5$ units   **11.** $\sqrt{13} \approx 3.6$ units
13. $\sqrt{53} \approx 7.3$ units   **15.** $\frac{5}{12}$ units   **17.** $\sqrt{77} \approx 8.8$ units
19. $(6, 5)$ **21.** $\left(-\frac{9}{2}, 5\right)$ **23.** $\left(-\frac{21}{2}, -\frac{13}{2}\right)$ **25.** $(2, -2.5)$
27. $\left(\frac{5}{24}, \frac{1}{20}\right)$ **29.** $(0, 4\sqrt{6})$

31. Center: $(0, 0)$; radius: 6

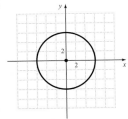

33. Center: $(3, 2)$;
radius: 1

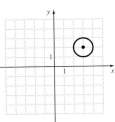

35. Center: $(-2, 4)$;
radius: 5

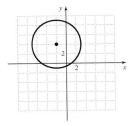

37. Center: (0, 1);
radius: 3

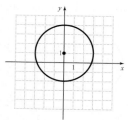

39. Center: (−1, −1);
radius: 5

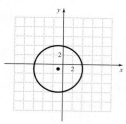

41. Center: (4, 3);
radius: 3

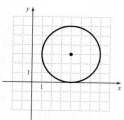

43. $(x + 7)^2 + (y - 2)^2 = 81$ **45.** $x^2 + (y - 4)^2 = 5$
47. $(x - 3)^2 + (y - 5)^2 = 32$ **49.** $(x + 2)^2 + (y - 6)^2 = 41$
51. $\left(-\frac{3}{2}, -2\right); \frac{5}{2}$ **53.** $(5, -3); \sqrt{38}$ **55.** $(2, 0); 2\sqrt{3}$
57. center: (4, −1);
radius: 3

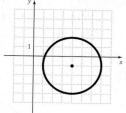

59. $(x - 2)^2 + (y + 4)^2 = 13$ **61.** $\sqrt{29} \approx 5.4$ units
63. $\left(\frac{5}{2}, -3\right)$ **65.** The distance between her home and her
office is 15 mi.   **67. a.** $D(4, 3)$ **b.** The slope of $\overline{BD}$ is −1
and the slope of $\overline{AC}$ is 1. Since the slopes are negative
reciprocals, $\overline{BD}$ is perpendicular to the base $\overline{AC}$ and is an
altitude of the triangle. **c.** 12 sq units   **69.** $x^2 + y^2 = 9$
71. a. $(x + 8)^2 + (y - 15)^2 = 36^2$, or
$(x + 8)^2 + (y - 15)^2 = 1296$
b. Yes, since $\sqrt{853} \approx 29.2 < 36$.

Practices: Section 11.3, *pp. 919–924*
1, *p. 919:*   **2,** *p. 919:*

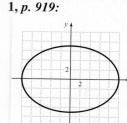   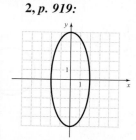

3, *p. 920:* The maximum width is 20 ft and the maximum
height of the arch is 12 ft.

4, *p. 922:*   **5,** *p. 923:*

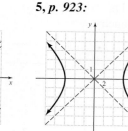

6, *p. 924:* The closest the comet comes to Earth is
90 million mi.

Exercises 11.3, *p. 925*

1. ellipse   **3.** hyperbola
5.   **7.**

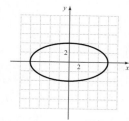

9.   **11.**

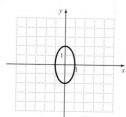

13.   **15.**

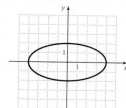 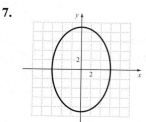

17.   **19.**

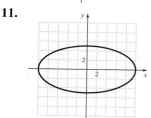 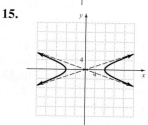

21.

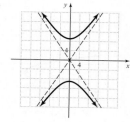 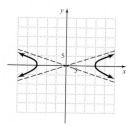

23. Circle   **25.** Parabola   **27.** Hyperbola   **29.** $\frac{x^2}{121} + \frac{y^2}{9} = 1$
31. $\frac{x^2}{7} + \frac{y^2}{8} = 1$ **33.** $\frac{x^2}{16} - \frac{y^2}{1} = 1$, or $\frac{x^2}{16} - y^2 = 1$

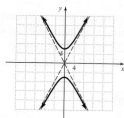

35.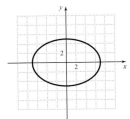

37. Ellipse   **39.** $\frac{x^2}{45} + \frac{y^2}{3} = 1$

41.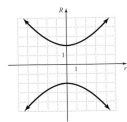

43. a. The length is 180 m and the width is 150 m.
b. Yes

45. a.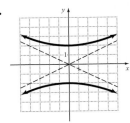

b. In Quadrant I, both the r- and R-coordinate of each point on the graph are non-negative. Neither radius can be negative.

47. $\frac{x^2}{9} + \frac{y^2}{4} = 1$

Practices: Section 11.4, pp. 931–935

1, p. 931: $(-8, -6), (6, 8)$
2, p. 932: $(-2, 0), (2, 0), (-\sqrt{3}, -1), (\sqrt{3}, -1)$
3, p. 933: $(-1, -2\sqrt{2}), (-1, 2\sqrt{2}), (1, -2\sqrt{2}),$ $(1, 2\sqrt{2})$ **4, p. 934:** $(-i, -2\sqrt{2}), (-i, 2\sqrt{2}),$ $(i, -2\sqrt{2}), (i, 2\sqrt{2})$ **5, p. 935:** The two balls will be the same height above the ground $\frac{9}{4}$ sec, or 2.25 sec, after they are released.

Exercises 11.4, p. 936

1. $(-1, -1), (3, 7)$ **3.** $(-3, -\sqrt{3}), (-3, \sqrt{3}),$ $(2, -2\sqrt{2}), (2, 2\sqrt{2})$ **5.** $(-2\sqrt{2}, 3), (2\sqrt{2}, 3)$
7. $(0, -4), (4, 0)$ **9.** $(-2, -\frac{1}{2}), (12, 3)$ **11.** $(-2i, -6),$ $(2i, -6), (-\sqrt{7}, 5), (\sqrt{7}, 5)$ **13.** $(-2\sqrt{3}, -\sqrt{2}),$ $(-2\sqrt{3}, \sqrt{2}), (2\sqrt{3}, -\sqrt{2}), (2\sqrt{3}, \sqrt{2})$
15. $(-2, -2\sqrt{3}), (-2, 2\sqrt{3}), (2, -2\sqrt{3}), (2, 2\sqrt{3})$
17. $(-\sqrt{6}, -i), (-\sqrt{6}, i), (\sqrt{6}, -i), (\sqrt{6}, i)$
19. $(-4, -\sqrt{15}), (-4, \sqrt{15}), (4, -\sqrt{15}), (4, \sqrt{15})$
21. $(0, -4), (0, 4)$ **23.** $(-1, -2\sqrt{3}), (-1, 2\sqrt{3}),$ $(1, -2\sqrt{3}), (1, 2\sqrt{3})$ **25.** $(-1, -\sqrt{10}), (-1, \sqrt{10}),$ $(1, -\sqrt{10}), (1, \sqrt{10})$ **27.** $(2, 0)$ **29.** $(-3i, -6),$ $(3i, -6), (-\sqrt{2}, 5), (\sqrt{2}, 5)$ **31.** $(-5, 0), (4, -3),$ $(4, 3)$ **33.** $(-i\sqrt{2}, -\sqrt{5}), (i\sqrt{2}, -\sqrt{5}),$ $(-i\sqrt{2}, \sqrt{5}), (i\sqrt{2}, \sqrt{5})$ **35.** They will have the same revenue if they sell 0 or 700 organizers. **37.** The paths will intersect at $(\sqrt{5}, 1)$ and $(\sqrt{5}, -1)$. **39.** The smaller

square has dimensions 11 in. by 11 in., in contrast to 14 in. by 14 in. for the larger square.

Practices: Section 11.5, pp. 939–943

1, p. 939: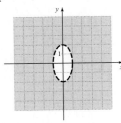

2, p. 940: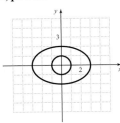

3, p. 941: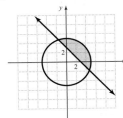

4, p. 942: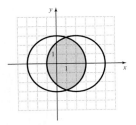

No real solutions

5, p. 943: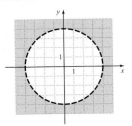

Exercises 11.5, p. 944

1. broken **3.** each
5.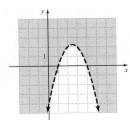

7.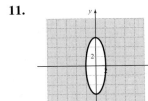

9.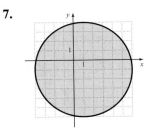

11.

13.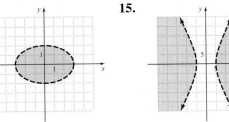

15.

17.

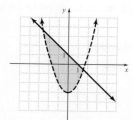

19.

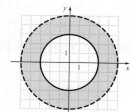

21.

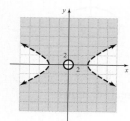

23.

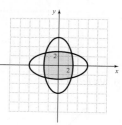

25.

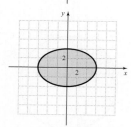

27. No real solutions

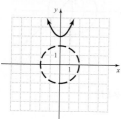

29.

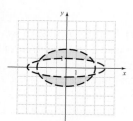

31.

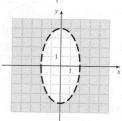

33.

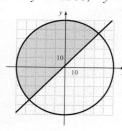

35. a. $x^2 + y^2 \leq 2500$; $y \geq x + 2$

b.

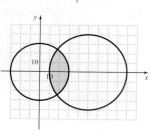

c. No, only points in Quadrant I are possible solutions since the lengths of the gardens cannot be negative.

37.

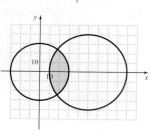

Review Exercises: Chapter 11, *p. 953*

1. Vertex: $(-3, 5)$; axis of symmetry: $x = -3$

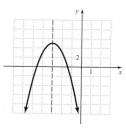

2. Vertex: $(-10, 2)$; axis of symmetry: $y = 2$

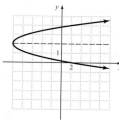

3. Vertex: $(0, -4)$; axis of symmetry: $y = -4$

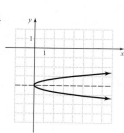

4. Vertex: $(-3, 2)$; axis of symmetry: $x = -3$

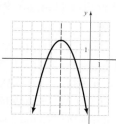

5. $y = 8\left(x - \frac{7}{2}\right)^2 - 24$; $\left(\frac{7}{2}, -24\right)$

6. $x = 5\left(y + \frac{3}{2}\right)^2 - \frac{17}{4}$; $\left(-\frac{17}{4}, -\frac{3}{2}\right)$ **7.** $3\sqrt{5} \approx 6.7$ units

8. $\sqrt{37} \approx 6.1$ units   **9.** 1.3 units   **10.** $\sqrt{95} \approx 9.7$ units

11. $(4, 13)$ **12.** $\left(\frac{3}{2}, -9\right)$ **13.** $\left(\frac{7}{8}, \frac{1}{4}\right)$ **14.** $(-0.7, -3.65)$

15. Center: $(0, 0)$; radius: 8

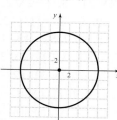

16. Center: $(-2, 4)$; radius: 2

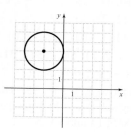

17. Center: $(2, -3)$;
radius: 5
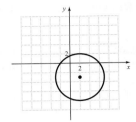

18. Center: $(-4, 1)$;
radius: 3
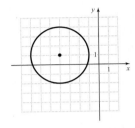

19. $(x - 9)^2 + y^2 = 169$ **20.** $(x + 6)^2 + (y - 10)^2 = 20$
21. Center: $(6, 7)$; radius: $2\sqrt{30}$ **22.** Center: $(-8, -5)$;
radius: $4\sqrt{5}$
23.
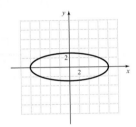
24.
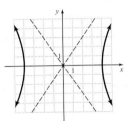

25.
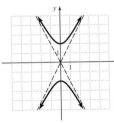
26.
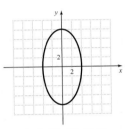

27. $(-1, 3), (2, 9)$ **28.** $(-1, -3), (1, 3)$ **29.** $(-4, 0),$
$(0, -2), (0, 2)$ **30.** $(-\sqrt{2}, -\sqrt{6}), (-\sqrt{2}, \sqrt{6}),$
$(\sqrt{2}, -\sqrt{6}), (\sqrt{2}, \sqrt{6})$
31. $(-\sqrt{5}, -i\sqrt{7}), (-\sqrt{5}, i\sqrt{7}),$
$(\sqrt{5}, -i\sqrt{7}), (\sqrt{5}, i\sqrt{7})$ **32.** $(-3, 0), (3, 0)$
33.
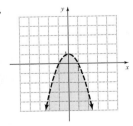
34.
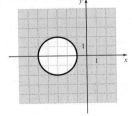

35.
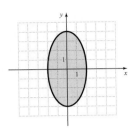
36.
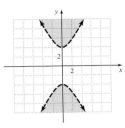

37.
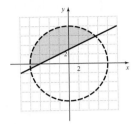
38.
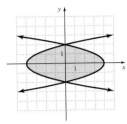

39.
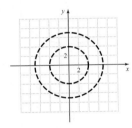
40.
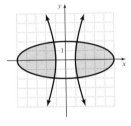

No real solutions
41. a. $R = -0.5(x - 20)^2 + 200$ **b.** $(20, 200)$; it shows
that the manufacturing plant will make a maximum revenue
of $200 when it sells 20 units of the product.
42. a. $y = -\frac{1}{32}x^2 + 50$ **b.** The height of the arch 24 ft
from the center is 32 ft.   **43.** The distance between the
student's apartment and work is $4\sqrt{5}$ mi, or approximately
8.9 mi.   **44.** $x^2 + y^2 = 60^2$, or $x^2 + y^2 = 3600$
45. $x^2 + y^2 = 4500^2$, or $x^2 + y^2 = 20,250,000$
46. The maximum distance is approximately 5 billion km.
47. $\frac{x^2}{4^2} + \frac{y^2}{2^2} = 1$, or $\frac{x^2}{16} + \frac{y^2}{4} = 1$ **48.** They will be at the
same height $\frac{5}{4}$ sec, or 1.25 sec, after they are released.
49. a. $x^2 + y^2 > 1$ **b.**
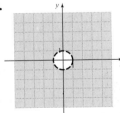

50. a.
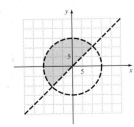
b. No, only points in
Quadrant I are possible
solutions since the
lengths of the sandboxes
must be nonnegative
quantities.

Posttest: Chapter 11, *p. 959*

1. $x = -2(y-4)^2 - 59$; $(-59, 4)$ **2. a.** $2\sqrt{2}$ units
b. 13 units   **3. a.** $(-6, -3)$ **b.** $(-\frac{9}{2}, 5)$
4. $(x+2)^2 + (y-8)^2 = 18$ **5.** $(-5, 1)$; $2\sqrt{3}$
6.   **7.**

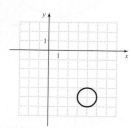
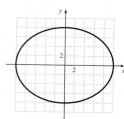

8.   **9.**

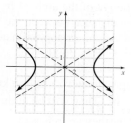
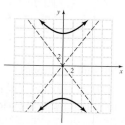

10.   **11.**

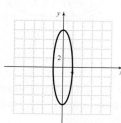
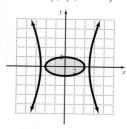

12.   **13.** $(-5, 0)$, $(0, 5)$

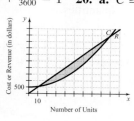

14. $(-\sqrt{10}, 14)$, $(\sqrt{10}, 14)$
15. $(-i, -2\sqrt{3})$, $(-i, 2\sqrt{3})$, $(i, -2\sqrt{3})$, $(i, 2\sqrt{3})$
16.   **17.**

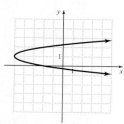
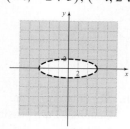

18. $y = \frac{1}{1875}x^2 + 2$ **19.** $\frac{x^2}{65^2} + \frac{y^2}{60^2} = 1$, or
$\frac{x^2}{4225} + \frac{y^2}{3600} = 1$ **20. a.** $C \geq 0.5x^2 + 500$ $R \leq 50x$
b.

Cumulative Review: Chapter 11, *p. 962*

1. $y = -\frac{4}{3}x + 16$ **2.** $2x^3 - 11x^2y + 12xy^2 + 9y^3$
3. 2   **4.** Inverses   **5.** $-\frac{20}{3}$ **6.** Center: (5, 2); radius: 2
7. She must invest at least $10,000 in the fund with a
guaranteed rate of return of 4% and $20,000 in the fund that
has a guaranteed rate of return of $5\frac{1}{2}$%.   **8.** The average speed
during the first part of the trip was 64 mph, and the average
speed during the second part of the trip was 56 mph.
9. a. 128 ft   **b.** The object reaches its maximum height of
153 ft $\frac{3}{4}$ sec, or 0.75 sec, after it is thrown.   **c.** The object is
104 ft above the ground $\frac{5}{2}$ sec, or 2.5 sec, after it is thrown
upward.   **10.** It would take approximately 10 yr for the
amount in the account to grow to $10,000.

Appendix A

Practices: Section A.2, *pp. 968−971*

1, *p. 968:* $0 \leq x \leq 4$, interval notation: [0, 4]

2, *p. 969:* $-1 < x < 1$, interval notation: (−1, 1)

3, *p. 970:* No solution   **4, *p. 971:*** $x < -2$ or $x > 3$; interval notation: $(-\infty, -2) \cup (3, \infty)$

5, *p. 971:* All real numbers; interval notation: $(-\infty, \infty)$

Exercises A.2, *p. 973*

1. (2, 5)

3. [−3, 0)

5. $(-\infty, -4) \cup (3, \infty)$

7. $(-\infty, -2) \cup [0.5, \infty)$

9. $-5 < x < 3$; interval notation: (−5, 3)

11. $-6 < t \leq 5$; interval notation: (−6, 5]

13. No solution

15. $-4 < a \leq 2$; interval notation: (−4, 2]

17. $-5 < x < \frac{5}{2}$; interval notation: $(-5, \frac{5}{2})$

19. $h < -\frac{3}{2}$; interval notation: $(-\infty, -\frac{3}{2})$

21. $-\frac{11}{2} \leq x \leq -\frac{5}{2}$; interval notation: $[-\frac{11}{2}, -\frac{5}{2}]$

23. $x < -1$ or $x > 2$; interval notation: $(-\infty, -1) \cup (2, \infty)$

25. $r < -8$ or $r \geq -1$; interval notation: $(-\infty, -8) \cup [-1, \infty)$

27. All real numbers; interval notation: $(-\infty, \infty)$

29. $a \le 3$; interval notation: $(-\infty, 3]$

31. $n \le \frac{9}{2}$ or $n > 6$; interval notation: $(-\infty, \frac{9}{2}] \cup (6, \infty)$

33. All real numbers; interval notation: $(-\infty, \infty)$

35. $-3 < x < -1$ **37.** $x < 2$ or $x \ge 12$
39. No solution   **41.** All real numbers   **43.** $x \ge 1$
45. $4 \le x < 11$ **47.** $-6 < x \le 6$ **49.** $-7 \le x \le -4$
51. $-0.5 < x < 3$

Practices: Section A.3, *pp. 976–982*

1, *p. 976:* $-5, 5$

2, *p. 976:* **a.** 0 **b.** No solution   **3, *p. 977:*** $5, -5$
4, *p. 977:* $-2, -10$ **5, *p. 977:*** $3, -\frac{7}{3}$ **6, *p. 978:***
No solution   **7, *p. 978:*** $-1, -\frac{1}{3}$ **8, *p. 979:*** $-\frac{1}{2}$
9, *p. 979:* $-5 < x < 5$; interval notation: $(-5, 5)$

10, *p. 980:* $x < -5$ or $x > 5$; interval notation:
$(-\infty, -5) \cup (5, \infty)$

11, *p. 980:* $-7 < x < 3$; interval notation: $(-7, 3)$

12, *p. 981:* $x \le -2$ or $x \ge 6$; interval notation:
$(-\infty, -2] \cup [6, \infty)$

13, *p. 981:* $-1 \le x \le \frac{5}{3}$; interval notation: $[-1, \frac{5}{3}]$

14, *p. 982:* $x < \frac{1}{2}$ or $x > 1$;
interval notation: $(-\infty, \frac{1}{2}) \cup (1, \infty)$

Exercises A.3, *p. 983*

1. 6   **3.** 7   **5.** 7.5   **7.** $-7, 7$ **9.** $-\frac{2}{3}, \frac{2}{3}$ **11.** $-4, 4$
13. No solution   **15.** 0   **17.** $-2, 2$ **19.** No solution
21. $-7, 7$ **23.** $-\frac{3}{2}, \frac{3}{2}$ **25.** $-6, 6$ **27.** $-3, -11$
29. No solution   **31.** $-17, 4$ **33.** $-\frac{3}{2}, 7$ **35.** $-\frac{7}{3}, 3$
37. No solution   **39.** $-4, \frac{44}{7}$ **41.** 2, 12   **43.** $-12, -4$
45. $-2, 12$ **47.** 0, 2   **49.** $-3, 1$ **51.** -4 **53.** 3
55. $-5, -\frac{3}{5}$ **57.** No solution   **59.** $x < -3$ or $x > 3$;
interval notation: $(-\infty, -3) \cup (3, \infty)$

61. $-\frac{1}{2} \le x \le \frac{1}{2}$; interval notation: $[-\frac{1}{2}, \frac{1}{2}]$

63. $x \le -3$ or $x \ge 3$; interval notation: $(-\infty, -3] \cup$
$[3, \infty)$

65. $-8 < x < 0$; interval notation: $(-8, 0)$

67. $x \le 2$ or $x \ge 8$; interval notation: $(-\infty, 2] \cup [8, \infty)$

69. $x \le -1$ or $x \ge 1$; interval notation: $(-\infty, -1] \cup [1, \infty)$

71. $-1 < x < 1$; interval notation: $(-1, 1)$

73. $x \le -7$ or $x \ge 4$; interval notation: $(-\infty, -7] \cup [4, \infty)$

75. $-1 < x < 5$; interval notation: $(-1, 5)$

77. All real numbers; interval notation: $(-\infty, \infty)$

79. $x < \frac{1}{2}$ or $x > \frac{13}{2}$; interval notation: $(-\infty, \frac{1}{2}) \cup (\frac{13}{2}, \infty)$

81. $-2 \le x \le \frac{7}{2}$; interval notation: $[-2, \frac{7}{2}]$

Practices: Section A.4, *pp. 986–992*

1, *p. 986:* **a.** Not a solution **b.** A solution   **2, *p. 988:***
$(1, 2, -1)$ **3, *p. 990:*** $(\frac{1}{2}, 1, -\frac{1}{2})$ **4, *p. 991:*** $(-1, -2, 2)$
5, *p. 992:* No solution   **6, *p. 992:*** Infinitely many solutions

Exercises A.4, *pp. 994*

1. A solution   **3.** Not a solution   **5.** A solution
7. $(4, 5, -2)$ **9.** $(-2, 3, -2)$ **11.** $(\frac{1}{2}, \frac{3}{2}, -4)$
13. $(3, 1, 0)$ **15.** $(\frac{1}{5}, \frac{1}{4}, \frac{1}{3})$ **17.** $(-\frac{3}{2}, -4, 9)$
19. Infinitely many solutions   **21.** $(1, -1, 4)$
23. No solution   **25.** $(4, 0, -5)$

Practices: Section A.5, *pp. 997–1000*

1, *p. 997:* $(3, -3)$ **2, *p. 998:*** $(2, 5, \frac{1}{2})$ **3, *p. 1000:***
$(-1, 4, -6)$

Exercises A.5, *pp. 1002*

1. $(-2, -2)$ **3.** $(1, -1)$ **5.** Infinitely many solutions
7. $(-8, 9)$ **9.** $(3, 1, -2)$ **11.** $(\frac{1}{2}, \frac{1}{4}, 2)$ **13.** $(-5, -2, -3)$
15. No solution   **17.** $(0, 5, -4)$ **19.** $(\frac{25}{8}, \frac{1}{4}, \frac{9}{16})$

Appendix B

Practices: Section B.1, *pp. 1004*

1, *p. 1004:* 1   **2, *p. 1005:*** $(1, -3)$ **3, *p. 1007:*** 9
4, *p. 1008:* $(2, 0, 3)$

Exercises B.1, *p. 1009*

1. -7 **3.** 10   **5.** 0   **7.** $(2, -1)$ **9.** $(3, 1)$ **11.** $(\frac{1}{2}, -\frac{1}{4})$
13. -36 **15.** 20   **17.** $(3, -2, -4)$ **19.** $(-1, 3, 5)$

Practices: Section B.2, *p. 1011*

1, *p. 1011:* $5x + 12 + \frac{35}{x-3}$ **2, *p. 1011:***
$x^2 - x + 5 - \frac{18}{x+5}$

Exercises B.2, *p. 1012*

1. $x + 3 + \frac{5}{x-2}$ **3.** $x - 8 + \frac{17}{x+3}$ **5.** $5x - 8 + \frac{10}{x+1}$
7. $3x + 1$ **9.** $3x + 4 + \frac{16}{x-4}$ **11.** $4x^2 + 5x + 2 + \frac{7}{x-1}$
13. $-y^2 + 5y - 8$ **15.** $x^2 + 2x - 1 - \frac{2}{x-2}$
17. $x^3 + 5 - \frac{15}{x+3}$ **19.** $2x^2 + 8x + 2 + \frac{4}{x-0.5}$

Glossary

The numbers in brackets following each glossary term represent the section that term is discussed in.

absolute value [1.1] The absolute value of a number is its distance from zero on the number line.

addition property of equality [2.1] A property that allows us to add any real number to each side of an equation, resulting in an equivalent equation.

additive identity property [1.5] For any real number a, $a + 0 = a$ and $0 + a = a$.

additive inverse property [1.2] For any real number a, there is exactly one real number $-a$ such that $a + (-a) = 0$ and $(-a) + a = 0$.

additive inverses [1.2] Two numbers that have a sum of 0.

algebraic expression [1.6] An expression in which constants and variables are combined using standard arithmetic operations.

altitude of a triangle [11.2] An altitude of a triangle is a line segment through a vertex that is perpendicular to the opposite side of the triangle.

associative property of addition [1.5] The associative property of addition states that when adding three numbers, regrouping the addends gives the same sum.

associative property of multiplication [1.5] The associative property of multiplication states that when multiplying three numbers, regrouping the factors gives the same product.

axis of symmetry [9.4] The vertical line that passes through the vertex of a parabola.

bar graph [3.1] A graph that represents quantities by thin, parallel rectangles called bars. Each bar's length is proportional to the quantity it represents.

base (exponent) [1.6] The base is the number that is a repeated factor when written with an exponent.

binomial [5.3] A binomial is a polynomial with two terms.

break-even point [4.1] The point at which the income for a business equals its expenses.

center of a circle [11.2] The given point that is a fixed distance from all the points on a circle.

center of a hyperbola [11.3] The point halfway between the two foci is called the center of the hyperbola.

center of an ellipse [11.3] The point halfway between the two foci is called the center of the ellipse.

change-of-base formula [10.5] The change-of-base formula states that for positive real numbers N, b, and c, where $b \neq 1$ and $c \neq 1$, $\log_b N = (\log_c N)/(\log_c b)$.

circle [11.2] A circle is the set of all points on a coordinate plane that are a fixed distance from a given point.

coefficient [1.7, 5.3] The numerical constant of a term; for example in the expression $5x$, 5 is called the coefficient.

columns of a matrix [A.5] Elements in a matrix that are vertical.

common factor [6.1] A common factor of two or more integers is an integer that is a factor of each integer.

common logarithm [10.5] A common logarithm is a logarithm to the base 10, where $\log x$ means $\log_{10} x$.

common multiple [R.2] A number that is a multiple of two or more numbers is called a common multiple.

commutative property of addition [1.5] The commutative property of addition states that changing the order in which two numbers are added does not affect the sum.

commutative property of multiplication [1.5] The commutative property of multiplication states that changing the order in which two numbers are multiplied does not affect the product.

completing the square [9.1] A method for solving quadratic equations in which we transform one side of an equation to a perfect square trinomial.

complex conjugates [8.7] The complex numbers $a + bi$ and $a - bi$ are called complex conjugates.

complex number [8.7] A complex number is any number that can be written in the form $a + bi$ where a and b are real numbers and $i = \sqrt{-1}$.

complex rational expression (complex algebraic fraction) [7.4] An expression that contains a rational expression in its numerator, denominator, or both.

composite number [R.2] A composite number is a whole number that has more than two factors.

composition of f and g [10.1] If f and g are functions, then the composition of f and g is the function, represented by $f \circ g$, where $(f \circ g)(x) = f(g(x))$.

compound inequality [9.5, A.2] Two inequalities that are joined by the word *and* or the word *or*.

conic section [11.1] A curve formed by the intersection of a plane and a double cone.

conjugate [8.5] The conjugate of $a + b$ is $a - b$.

constant of variation (constant of proportionality) [7.6] In the equations, $y = kx$, $y = k/x$ and $y = kxz$, k is called the constant of variation, or the constant of proportionality.

constant term [5.3] In a polynomial, the constant term is the term of degree 0.

coordinate plane [3.1] The flat surface on which we draw graphs.

coordinates [3.1] A pair of numbers in a given order that corresponds to a location on the coordinate plane.

cube root [8.1] The number b is the cube root of a if $b^3 = a$ for any real numbers a and b.

decimal [R.4] A decimal is a number written with three parts: a whole number, the decimal point, and a fraction whose denominator is a power of 10.

decimal places [R.4] The decimal places are the places to the right of the decimal point.

degree of a monomial [5.3] The degree of a monomial is the power of the variable in the monomial.

degree of a polynomial [5.3] The degree of a polynomial is the highest degree of any of its terms.

demand curve [4.1] The demand curve illustrates that when the price of an item increases, the quantity of items sold declines. It is commonly approximated by a straight line with a negative slope.

denominator [R.3] The number below the fraction line in a fraction is called the denominator. It stands for the number of parts into which the whole is divided.

difference of squares [6.4] An expression in the form $a^2 - b^2$, which can be factored as $(a + b)(a - b)$.

difference of two functions [10.1] If f and g are functions, then the difference of two functions is $(f - g)(x) = f(x) - g(x)$.

difference of two perfect cubes [6.5] An expression in the form $a^3 - b^3$, which can be factored as $(a - b)(a^2 + ab + b^2)$.

direct variation [7.6] Direct variation occurs if a relationship between two variables is described by an equation of the form $y = kx$, where k is a positive constant.

direct variation equation [7.6] The equation $y = kx$.

discriminant [9.2] The expression $b^2 - 4ac$ in the quadratic formula that allows us to determine the number and type of solutions to a quadratic equation without actually solving the equation.

distance formula [8.3, 11.2] The formula, $d = \sqrt{(x_2 - x_1)^2 + (y_2 - y_1)^2}$. In words, the distance between two points on the coordinate plane is the square root of the sum of the squares of the difference of the x-values and the difference of the y-values.

distributive property [1.5] The distributive property states that multiplying a factor by the sum of two numbers gives us the same result as multiplying the factor by each of the two numbers and then adding.

domain [3.6] The domain of a function is the set of all values of the independent variable.

elements (entries) of a matrix [A.5] The numbers of a matrix.

elimination (or addition) method [4.3] A method used to solve a system of equations that is based on the property of equality stating: If $a = b$ and $c = d$, then $a + c = b + d$.

ellipse [11.3] An ellipse is the set of all points on a coordinate plane, the *sum* of whose distances from two fixed points is constant.

equation [2.1] An equation is a mathematical statement that two expressions are equal.

equation of a circle centered at (h, k) [11.2] The equation of a circle with center (h, k) and with radius r is $(x - h)^2 + (y - k)^2 = r^2$.

equation of a circle centered at the origin [11.2] The equation of a circle with center $(0, 0)$ and with radius r is $x^2 + y^2 = r^2$.

equation of a hyperbola centered at the origin [11.3] The equation of a hyperbola with center $(0, 0)$ and x-intercepts $(-a, 0)$ and $(a, 0)$, is $\frac{x^2}{a^2} - \frac{y^2}{b^2} = 1$. The equation of a hyperbola with center $(0, 0)$ and y-intercepts $(0, -b)$ and $(0, b)$, is $\frac{y^2}{b^2} - \frac{x^2}{a^2} = 1$.

equation of an ellipse centered at the origin [11.3] The equation of an ellipse with center $(0, 0)$, x-intercepts $(-a, 0)$ and $(a, 0)$, and y-intercepts $(0, -b)$ and $(0, b)$ is $\frac{x^2}{a^2} + \frac{y^2}{b^2} = 1$.

equilibrium [4.1] The point of intersection of the supply curve and the demand curve where all items produced are sold and all customers are satisfied.

equivalent equations [2.1] Equivalent equations are equations that have the same solution.

equivalent fractions [R.3] Equivalent fractions are fractions that represent the same value.

equivalent matrix [A.5] A matrix corresponding to a system of equations equivalent to the original system.

exponent (or power) [R.1, 1.6] An exponent (or power) is a number that indicates how many times another number is multiplied by itself.

exponential form [1.6] Exponential form is a shorthand way of representing a repeated multiplication of the same factor.

exponential function [10.2] An exponential function is any function that can be written in the form $f(x) = b^x$, where x is a real number, $b > 0$ and $b \neq 1$.

extraneous solutions [7.5] Extraneous solutions are *not* solutions of the original equation.

factored completely [6.2] A polynomial is factored completely when it is expressed as the product of a monomial and one or more prime polynomials.

factoring by grouping [6.1] When trying to factor a polynomial that has four terms, it may be possible to group pairs of terms in such a way that a common binomial factor can be found. This method is called factoring by grouping.

factors [R.2] In a multiplication problem, the numbers being multiplied are called the factors.

focus (plural, foci) [11.3] The fixed points that are used to generate an ellipse or a hyperbola.

FOIL method [5.5] A method for multiplying two binomials. Multiply **F**irst terms, **O**utside terms, **I**nside terms, and **L**ast terms. Then combine like terms.

formula [1.8] A formula is an equation that indicates how a number of variables are related to one another.

fraction [R.3] A fraction can mean a part of a whole or the quotient of two whole numbers.

fraction line (or fraction bar) [R.3] The fraction line separates the numerator from the denominator and stands for the phrase *out of* or *divided by*.

function [3.6] A function is a relation in which no two distinct ordered pairs have the same first coordinates.

graph [3.3] The graph of a linear equation in two variables consists of all the points whose coordinates satisfy the equation.

graphing method [4.1] A method of solving a linear system of equations in which we graph the equations that make up the system and any point of intersection is a solution to the system.

greatest common factor (GCF) [6.1] The greatest common factor (GCF) of two or more integers is the greatest integer that is a factor of each integer.

greatest common factor (GCF) of two or more monomials [6.1] The greatest common factor (GCF) of two or more monomials is the product of the greatest common factor of the coefficients and the highest powers of the variable factors common to each monomial.

horizontal-line test [10.1] The horizontal-line test states that if every horizontal line intersects the graph of a function at most once, then the function is one-to-one.

hyperbola [11.3] A hyperbola is the set of all points on a coordinate plane that satisfy the following property: The *difference* of the distances from two fixed points is constant.

imaginary number [8.7] The imaginary number i is $\sqrt{-1}$; in other words, $i^2 = -1$.

imaginary part [8.7] In the complex number $a + bi$, b is called the imaginary part.

improper fraction [R.3] An improper fraction is a fraction greater than or equal to 1, that is, a fraction whose numerator is larger than or equal to its denominator.

index [8.1] In the radical $\sqrt[n]{a}$, n is called the index of the radical.

inequality [1.1, 2.6] An inequality is any mathematical statement containing $<, \leq, >, \geq,$ or $\neq$.

integers [1.1] The integers are the numbers $\ldots, -4, -3, -2, -1, 0, +1, +2, +3, +4, \ldots$ continuing indefinitely in both directions.

inverse of a one-to-one function f [10.1] The inverse of a one-to-one function f, written f^{-1} and read "f inverse," is the set of all ordered pairs (y, x) such that f is the set of ordered pairs (x, y).

inverse variation [7.6] Inverse variation occurs if a relationship between two variables is described by an equation of the form $y = k/x$, where k is a positive constant.

inverse variation equation [7.6] The equation $y = k/x$.

irrational numbers [1.1] Real numbers that cannot be written as the quotient of two integers.

joint variation [7.6] Joint variation occurs if a relationship among three variables is described by an equation of the form $y = kxz$ where k is a positive constant.

joint variation equation [7.6] The equation $y = kxz$.

leading coefficient [5.3] The leading coefficient is the coefficient of the leading term in a polynomial.

leading term [5.3] The leading term of a polynomial is the term in the polynomial with the highest degree.

least common denominator (LCD) [R.3] The least common denominator (LCD) for any set of fractions is the least common multiple of the denominators.

least common multiple (LCM) [R.2] The least common multiple (LCM) of two or more numbers is the smallest nonzero number that is a multiple of each number.

like fractions [R.3] Like fractions are fractions that have the same denominator.

like radicals [8.4] Like radicals are radicals that have the same index and the same radicand.

like terms [1.7] Like terms are terms that have the same variables with the same exponents.

linear equation in one variable [2.1] An equation that can be written in the form $ax + b = c$, where a, b, and c are real numbers and $a \neq 0$.

linear equation in two variables [3.3] An equation that can be written in the *general form* $Ax + By = C$, where A, B, and C are real numbers and A and B are not both 0.

linear inequality in two variables [3.5] A linear inequality in two variables is an inequality that can be written in the form $Ax + By < C$, where A, B, and C are real numbers and A and B are not both 0. The inequality symbol can be $<, >, \leq,$ or $\geq$.

line graph [3.1] A graph that represents quantities as points connected by straight line segments. Read the height of any point on a line graph against the vertical axis.

literal equation [2.4] A literal equation is an equation involving two or more variables.

logarithmic function [10.3] The inverse of $f(x) = b^x$ is represented by the logarithmic function $f^{-1}(x) = \log_b x$.

matrix [A.5] A rectangular array of numbers.

maximum [9.4] A parabola that opens downward has a highest point with a maximum y-value occurring at the vertex.

midpoint formula [11.2] The line segment with endpoints (x_1, y_1) and (x_2, y_2) has as its midpoint $\left(\dfrac{x_1 + x_2}{2}, \dfrac{y_1 + y_2}{2}\right)$.

minimum [9.4] A parabola that opens upward has a lowest point with a minimum y-value occurring at the vertex.

mixed number [R.3] A mixed number consists of a whole number and a proper fraction.

monomial [5.3] A monomial is an expression that is the product of a real number and variables raised to nonnegative integer powers.

multiples [R.2] The multiples of a number are the products of that number and the whole numbers.

multiplication property of equality [2.2] A property that allows us to multiply each side of an equation by any real number to get an equivalent equation.

multiplication property of zero [1.4] The multiplication property of 0 states that the product of any number and 0 is 0.

multiplicative identity property [1.5] The multiplicative identity property states that for any real number a, $a \cdot 1 = a$ and $1 \cdot a = a$.

multiplicative inverse property [1.4] The multiplicative inverse property states that for any nonzero real number a, $a \cdot \dfrac{1}{a} = 1$ and $\dfrac{1}{a} \cdot a = 1$.

multiplicative inverses [1.4] Two nonzero numbers that have a product of 1.

natural exponential function [10.2] The function defined by $f(x) = e^x$ is called the natural exponential function.

natural logarithm [10.5] A natural logarithm is a logarithm to the base e, where $\ln x$ means $\log_e x$.

natural numbers [1.1] The natural numbers are $1, 2, 3, 4, 5, 6, \ldots$.

negative number [1.1] A negative number is a number less than 0.

negative slope [3.2] On a graph, the slope of a line that slants downward from left to right.

nonlinear system of equations [11.4] A system of equations in which at least one equation is nonlinear.

nth root [8.1] The number b is the nth root of a if $b^n = a$ for any real number a and for any positive integer n greater than 1.

numerator [R.3] The number above the fraction line in a fraction is called the numerator. It tells us how many parts of the whole the fraction contains.

one-to-one function [10.1] A function f is said to be one-to-one if each y-value in the range corresponds to exactly one x-value in the domain.

one-to-one property of exponential functions [10.2] The one-to-one property of exponential functions states that for $b > 0$ and $b \neq 1$, if $b^x = b^n$, then $x = n$.

opposites [1.1] Two real numbers that are the same distance from 0 on the number line but on opposite sides of 0 are called opposites.

order of operations rule [R.1, 1.5] The order of operations rule tells us in which order to carry out the operations in an expression so that the value is unambiguous.

ordered pair [3.1] A pair of numbers that represents a point in the coordinate plane.

ordered triple [A.4] A set of three numbers that represent a solution of a system of linear equations in three variables.

origin [1.1, 3.1] On the number line, the point at 0; in the coordinate plane, the point where the axes intersect, $(0, 0)$.

parabola [9.4] A parabola is the graph of a quadratic function.

parallel lines [3.2] Two nonvertical lines are parallel if and only if their slopes are equal.

percent [R.5] A percent is a ratio or fraction with denominator 100. A number written with the % sign means "divided by 100."

percent decrease [2.5] In a percent problem, if the quantity is decreasing, the percent of change is called a percent decrease.

percent increase [2.5] In a percent problem, if the quantity is increasing, the percent of change is called a percent increase.

perfect cube [8.1] A rational number is said to be a perfect cube if it is the cube of another rational number.

perfect square [8.1] A whole number is said to be a perfect square if it is the square of another whole number. A nonnegative rational number is said to be a perfect square if it is the square of another rational number.

perfect square trinomial [6.4] A trinomial that can be factored as the square of a binomial, for example, $a^2 + 2ab + b^2 = (a + b)^2$ and $a^2 - 2ab + b^2 = (a - b)^2$.

perpendicular bisector [11.2] The perpendicular bisector of a segment is the line perpendicular to the segment that intersects the segment at its midpoint.

perpendicular lines [3.2] Two nonvertical lines are perpendicular if and only if the product of their slopes is -1.

point-slope form [3.4] The point-slope form of a linear equation is written as $y - y_1 = m(x - x_1)$, where x_1, y_1 and m are constants and m is the slope and (x_1, y_1) is a point that lies on the graph of the equation.

polynomial [5.3] A polynomial is an algebraic expression with one or more monomials added or subtracted.

positive number [1.1] A positive number is a number greater than 0.

positive slope [3.2] On a graph, the slope of a line that slants upward from left to right.

power (or exponent) [R.1, 1.6] A power (or exponent) is a number that indicates how many times another number is multiplied by itself.

power property of logarithms [10.4] The power property of logarithms states that for any positive real numbers M and b, $b \neq 1$ and any real number r, $\log_b M^r = r \log_b M$.

prime factorization [R.2] Prime factorization is the process of writing a whole number as a product of its prime factors.

prime number [R.2] A prime number is a whole number that has exactly two factors, itself and 1.

prime polynomials [6.2] Prime polynomials are polynomials that are not factorable.

principal square root [8.1] The nonnegative square root of a number.

product of two functions [10.1] If f and g are functions, then the product of two functions is $(f \cdot g)(x) = f(x) \cdot g(x)$.

product property of logarithms [10.4] The product property of logarithms states that for any positive real numbers M, N, and b, $b \neq 1$, $\log_b MN = \log_b M + \log_b N$.

proper fraction [R.3] A proper fraction is a fraction less than 1, that is, a fraction whose numerator is smaller than its denominator.

proportion [7.6] A proportion is a statement that two ratios a/b and c/d are equal, written $\dfrac{a}{b} = \dfrac{c}{d}$, where $b \neq 0$ and $d \neq 0$.

Pythagorean theorem [6.6] The Pythagorean theorem states that for every right triangle, the sum of the squares of the legs equals the square of the hypotenuse: $a^2 + b^2 = c^2$.

quadrant [3.1] One of four regions of a coordinate plane separated by the x- and y-axis.

quadratic equation (second-degree equation) [6.6, 9.1] An equation that can be written in the form $ax^2 + bx + c = 0$, where a, b, and c are real numbers and $a \neq 0$.

quadratic formula [9.2] The quadratic formula states that if $ax^2 + bx + c = 0$, where a, b, and c are real numbers and $a \neq 0$, then $x = -b \pm \dfrac{\sqrt{b^2 - 4ac}}{2a}$.

quadratic function [9.4] A function of the form $f(x) = ax^2 + bx + c$.

quadratic in form [9.3] An equation is quadratic in form if it can be rewritten in the form $au^2 + bu + c = 0$, where $a \neq 0$ and u represents an algebraic expression.

quadratic inequality [9.5] A quadratic inequality in one variable is an inequality that contains a quadratic expression.

quotient of two functions [10.1] If f and g are functions, then the quotient of two functions is $(f/g)(x) = f(x)/g(x)$, for $g(x) \neq 0$.

quotient property of logarithms [10.4] The quotient property of logarithms states that for any positive real numbers M, N, and b, $b \neq 1$, $\log_b (M/N) = \log_b M - \log_b N$.

radical equation [8.6] A radical equation is an equation in which a variable appears in one or more radicands.

radical expressions [8.1] Radical expressions are algebraic expressions that contain *radicals*.

radical sign [8.1] The symbol $\sqrt{}$ is called the radical sign.

radicand [8.1] The radicand is the number under the radical sign.

radius [11.2] The fixed distance from the center of a circle to any point on the circle.

range [3.6] The range of a function is the set of all values of the dependent variable.

rate of change [3.2] Slope can be interpreted as a rate of change. It indicates how fast the graph of a line is changing and if it increases or decreases.

ratio [7.6] A ratio is a comparison of two numbers, expressed as a quotient.

rational equation (fractional equation) [7.5] An equation that contains one or more rational expressions.

rational expression [7.1] A rational expression, $\frac{P}{Q}$, is an algebraic expression that can be written as the quotient of two polynomials, P and Q, where $Q \neq 0$.

rational inequality [9.5] A rational inequality is an inequality that contains a rational expression.

rational numbers [1.1] Numbers that can be written in the form $\frac{a}{b}$, where a and b are integers and $b \neq 0$.

rationalize the denominator [8.5] To rewrite an expression in an equivalent form that contains no radical in its denominator.

real numbers [1.1] Numbers that can be represented as points on a number line.

real part [8.7] In the complex number $a + bi$, a is called the real part.

reciprocals [R.3, 1.4] Two nonzero numbers that have a product of 1. For example, the reciprocal of the fraction $\frac{2}{3}$ is $\frac{3}{2}$.

reduced to lowest terms [R.3] A fraction is said to be reduced to lowest terms when the only common factor of its numerator and its denominator is 1.

relation [3.6] A relation is a set of ordered pairs.

rows of a matrix [A.5] Elements in a matrix that are horizontal.

scientific notation [5.2] A number is in scientific notation if it is written in the form $a \times 10^n$, where n is an integer and a is greater than or equal to 1 but less than 10 ($1 \leq a < 10$).

second-degree equation (quadratic equation) [6.6, 9.1] An equation that can be written in the form $ax^2 + bx + c = 0$, where a, b, and c are real numbers and $a \neq 0$.

set [1.1] A collection of objects called elements or members.

similar triangles [7.6] Similar triangles are triangles that have the same shape but not necessarily the same size.

simplest form [R.3] A fraction is said to be in simplest form when the only common factor of its numerator and its denominator is 1.

slope [3.2] The ratio of the change in y-values to the change in x-values along a line. The slope m of a line passing through the points (x_1, y_1) and (x_2, y_2) is defined to be $m = (y_2 - y_1)/(x_2 - x_1)$, where $x_1 \neq x_2$.

slope-intercept form [3.4] A linear equation is in slope-intercept form if it is written as $y = mx + b$, where m and b are constants, m is the slope, and $(0, b)$ is the y-intercept of the graph of the equation.

solution of a nonlinear system of equations [11.4] A solution of a nonlinear system of equations in two variables is an ordered pair of numbers that makes each equation in the system true.

solution of a system of equations [4.1, A.4] A solution of a system of linear equations in two variables is an ordered pair of numbers that makes both equations in the system true. A solution of a system of linear equations in three variables is an ordered triple of numbers that makes all three equations in the system true.

solution of a system of linear inequalities [4.4] A solution of a system of linear inequalities in two variables is an ordered pair of real numbers that makes both inequalities in the system true.

solution of a system of nonlinear inequalities [11.5] A solution of a system of nonlinear inequalities is an ordered pair of real numbers that satisfies each inequality in the system.

solution of an equation [2.1] A solution of an equation is a value of the variable that makes the equation a true statement.

solution of an equation in two variables [3.2] A solution of an equation in two variables is an ordered pair of numbers that when substituted for the variables makes the equation true.

solution of an inequality [2.6] A solution of an inequality is any value of the variable that makes the inequality true.

solution of an inequality in two variables [3.5] An ordered pair of numbers that, when substituted for the variables, makes the inequality true.

solve an inequality [2.6] To solve an inequality is to find all of its solutions.

square matrix [A.5] A matrix with the same number of rows as columns.

square root [8.1] The number b is a square root of a if $b^2 = a$ for any real numbers a and b and for a nonnegative.

square root property of equality [9.1] If b is a real number and $x^2 = b$, then $x = \pm\sqrt{b}$, that is, $x = \sqrt{b}$ or $x = -\sqrt{b}$.

standard form of a parabola [11.1] The equation of a parabola that opens upward or downward is in standard form if it is written $y = a(x - h)^2 + k$, where (h, k) is the vertex of the associated parabola and the equation of the axis of symmetry is $x = h$. The equation of a parabola that opens to the left or right is in standard form if it is written $x = a(y - k)^2 + h$, where (h, k) is the vertex of the associated parabola and the equation of the axis of symmetry is $y = k$.

substitution method [4.2] A method for solving a system of equations in which one linear equation is solved for one of the variables and then the result is substituted into the other equation.

sum of two functions [10.1] If f and g are functions, then the sum of two functions is $(f + g)(x) = f(x) + g(x)$.

sum of two perfect cubes [6.5] An expression in the form $a^3 + b^3$, which can be factored as $(a + b)(a^2 - ab + b^2)$.

supply curve [4.1] The supply curve illustrates that when the selling price increases, wholesalers are inclined to make more goods available to retailers. It is approximated by a straight line with positive slope.

system of equations [4.1] A system of equations is a group of two or more equations solved simultaneously.

system of linear inequalities [4.4] A system of linear inequalities is two or more linear inequalities considered simultaneously.

term [1.6] A term is a number, a variable, or the product or quotient of numbers and variables.

trinomial [5.3] A trinomial is a polynomial with three terms.

unit fraction [7.3] A fraction with 1 as the numerator is called a unit fraction.

unlike fractions [R.3] Unlike fractions are fractions that have different denominators.

unlike terms [1.7] Unlike terms are terms that do not have the same variable or the variables are not raised to the same powers.

variable [1.6] A quantity that is unknown and can change or vary in value.

varies directly (directly proportional) [7.6] In a relationship between two variables, y varies directly as x if there is a positive constant k such that $y = kx$.

varies inversely (inversely proportional) [7.6] In a relationship between two variables, y varies inversely as x if there is a positive constant k such that $y = k/x$.

varies jointly (jointly proportional) [7.6] In a relationship among three variables, y varies jointly as x and z, if there is a positive constant k such that $y = kxz$.

vertex [9.4] The highest or lowest point of a parabola is called the vertex.

vertical-line test [3.6] If any vertical line intersects a graph at more than one point, the graph does not represent a function. If no such line exists, then the graph represents a function.

whole numbers [1.1] The whole numbers consist of 0 and the natural numbers 0, 1, 2, 3, 4, 5, ...

x-axis [3.1] The horizontal number line in the coordinate plane.

x-coordinate [3.1] The first number in an ordered pair, which represents a horizontal distance in the coordinate plane.

x-intercept [3.3] The x-intercept of a line is the point where the graph crosses the x-axis.

y-axis [3.1] The vertical number line in the coordinate plane.

y-coordinate [3.1] The second number in an ordered pair, which represents a vertical distance in the coordinate plane.

y-intercept [3.3] The y-intercept of a line is the point where the graph crosses the y-axis.

zero-product property [6.6] The zero-product property states that if the product of two factors is 0, then either one or both of the factors must be 0.

Index

Tab Your Way to Success

Use these tabs to mark important pages of your textbook for quick reference and review.

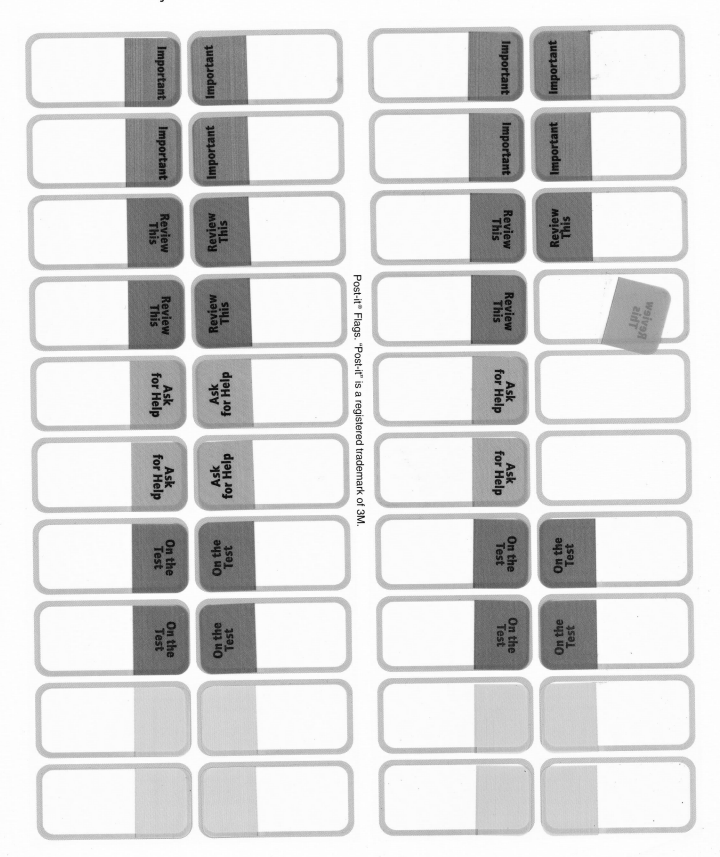

Post-it® Flags. "Post-it" is a registered trademark of 3M.

Using the Tabs

Customize Your Textbook and Make It Work for You!

These removable and reusable tabs offer you five ways to be successful in your math course by letting you bookmark pages with helpful reminders.

 Use these tabs to flag anything your instructor indicates is important.

 Mark important definitions, procedures, or key terms to review later.

 Not sure of something? Need more instruction? Place these tabs in your textbook to address any questions with your instructor during your next class meeting or with your tutor during your next tutoring session.

 If your instructor alerts you that something will be covered on a test, use these tabs to bookmark it.

 Write your own notes or create more of the preceding tabs to help you succeed in your math course.

ISBN-13: 978-0-321-53634-1
ISBN-10: 0-321-53634-7